The Exoplanet Handbook

Exoplanet research is one of the most rapidly developing subjects in astronomy. More than 500 exoplanets are now known, and groups world-wide are actively involved in a broad range of observational and theoretical efforts. This book ties together these many avenues of investigation – from the perspectives of observation, technology and theory – to give a comprehensive, up-to-date review of the field.

All areas of exoplanet investigation are covered, making it a unique and valuable guide for researchers in astronomy and planetary science, including those new to the subject. It treats the many different techniques now available for exoplanet detection and characterisation, the broad range of underlying physics, the overlap with related topics in solar system and Earth sciences, and the concepts underpinning future developments. It emphasises the interconnection between the various topics, and provides extensive references to more in-depth treatments and reviews.

Michael Perryman spent 2010 as a Distinguished Visitor at the University of Heidelberg and at the Max Planck Institute for Astronomy, Heidelberg. During an extensive career with the European Space Agency, he was the scientific leader of the Hipparcos space astrometry mission, a project which provided a fundamental observational basis for many aspects of exoplanet studies. He was professor of astronomy at Leiden University, The Netherlands, between 1993–2009. He chaired the influential European Space Agency–European Southern Observatory (ESA–ESO) working group on extra-solar planets in 2005, and has served on various national working groups and strategy panels for the future development of exoplanet research.

The Exoplanet Handbook

Michael Perryman

Zentrum für Astronomie der Universität Heidelberg
and
Max–Planck–Institut für Astronomie, Heidelberg

CAMBRIDGE UNIVERSITY PRESS
Cambridge, New York, Melbourne, Madrid, Cape Town, Singapore,
São Paulo, Delhi, Dubai, Tokyo, Mexico City

Cambridge University Press
The Edinburgh Building, Cambridge CB2 8RU, UK

Published in the United States of America by Cambridge University Press, New York

www.cambridge.org
Information on this title: www.cambridge.org/9780521765596

© M. Perryman 2011

This publication is copyright. Subject to statutory exception
and to the provisions of relevant collective licensing agreements,
no reproduction of any part may take place without
the written permission of Cambridge University Press.

First published 2011

Printed in the United Kingdom at the University Press, Cambridge

A catalogue record for this publication is available from the British Library

ISBN 978-0-521-76559-6 hardback

Cambridge University Press has no responsibility for
the persistence or accuracy of URLs for external or
third-party internet websites referred to in this publication,
and does not guarantee that any content on such
websites is, or will remain, accurate or appropriate.

Contents

Preface		*page*	xi
1	**Introduction**		**1**
	1.1 The challenge		1
	1.2 Discovery status		1
	1.3 Outline of the treatment		3
	1.3.1 Observational techniques		3
	1.3.2 Host star properties and brown dwarfs		5
	1.3.3 Theoretical considerations		5
	1.4 Astronomical terms and units		5
	1.5 Definition of a planet		7
	1.6 On-line reference compilations		7
2	**Radial velocities**		**9**
	2.1 Description of orbits		9
	2.1.1 Orbits from radial velocity measurements		12
	2.2 Measurement principles and accuracies		16
	2.2.1 Introduction		16
	2.2.2 Cross-correlation spectroscopy		17
	2.2.3 Deriving radial velocities from Doppler shifts		17
	2.2.4 Wavelength calibration		18
	2.2.5 Exposure metering		21
	2.2.6 Accuracy limits		21
	2.2.7 Excluding other sources of periodicity		22
	2.3 Instrument programmes		23
	2.3.1 State-of-the-art in échelle spectroscopy		23
	2.3.2 Externally dispersed interferometry		25
	2.3.3 Future developments		27
	2.4 Results to date		28
	2.4.1 The first radial velocity exoplanets		28
	2.4.2 The present radial velocity census		28
	2.4.3 On-line compilations		28
	2.4.4 Main sequence stars		29
	2.4.5 Evolved stars		32
	2.4.6 Other star categories		34
	2.5 Properties of the radial velocity planets		34
	2.5.1 Frequency of massive planets		35
	2.5.2 Mass distribution		35
	2.5.3 Orbits		36
	2.5.4 Host star dependencies		37

	2.6	Multiple planet systems	37
		2.6.1 General considerations	38
		2.6.2 Resonances	40
		2.6.3 Long-term integration and system stability	43
		2.6.4 Systems with three or more giant planets	47
		2.6.5 Systems in mean motion resonance	49
		2.6.6 Interacting doubles	55
		2.6.7 Non-interacting doubles	55
		2.6.8 Super-Earth systems	55
		2.6.9 Stability of habitable zone systems	55
	2.7	Planets around binary and multiple stars	55
		2.7.1 Configurations and stability	56
		2.7.2 Present inventory	59
		2.7.3 Specific examples	60
3	**Astrometry**		**61**
	3.1	Introduction	61
	3.2	Astrometric accuracy from ground	62
	3.3	Microarcsec astrometry	64
	3.4	Astrophysical limits	65
		3.4.1 Surface structure jitter	65
	3.5	Multiple planets and mandalas	66
	3.6	Modeling planetary systems	67
	3.7	Astrometric measurements from ground	68
	3.8	Astrometric measurements from space	69
		3.8.1 Hipparcos	69
		3.8.2 HST–Fine Guidance Sensor	70
	3.9	Future observations from space	71
4	**Timing**		**75**
	4.1	Pulsars	75
		4.1.1 PSR B1257+12	75
		4.1.2 PSR B1620−26	77
		4.1.3 Other considerations	78
	4.2	Pulsating stars	79
		4.2.1 White dwarfs	79
		4.2.2 Hot subdwarfs	81
	4.3	Eclipsing binaries	81
		4.3.1 Confirmed planets	81
		4.3.2 Unconfirmed planets	82
5	**Microlensing**		**83**
	5.1	Introduction	83
	5.2	Description	84
	5.3	Caustics and critical curves	87
	5.4	Other light curve effects	90
	5.5	Microlens parallax and lens mass	92
	5.6	Astrometric microlensing	94
	5.7	Other configurations	95
	5.8	Microlensing observations in practice	96
	5.9	Exoplanet results	98
		5.9.1 Individual objects	98
		5.9.2 Statistical results	100
	5.10	Summary of limitations and strengths	100
	5.11	Future developments	102

6 Transits — 103

- 6.1 Introduction — 103
- 6.2 Transit searches — 104
  - 6.2.1 Large-field searches from the ground — 105
  - 6.2.2 Other searches from the ground — 108
  - 6.2.3 Searches in open and globular clusters — 109
  - 6.2.4 Future searches from the ground — 110
  - 6.2.5 Searches from space — 110
  - 6.2.6 Follow-up observations from space — 112
  - 6.2.7 Future observations from space — 114
  - 6.2.8 Searches around specific stellar types — 114
- 6.3 Noise limits — 115
- 6.4 Transit light curves — 117
  - 6.4.1 Observables — 117
  - 6.4.2 Theoretical light curves — 117
  - 6.4.3 Circular orbits — 119
  - 6.4.4 Eccentric orbits — 121
  - 6.4.5 Physical quantities — 123
  - 6.4.6 Interferometric observations — 123
  - 6.4.7 Reflected light — 124
  - 6.4.8 Doppler variability — 126
  - 6.4.9 Polarisation — 126
  - 6.4.10 Secondary eclipse — 126
  - 6.4.11 Rossiter–McLaughlin effect — 127
  - 6.4.12 Higher-order photometric effects — 130
  - 6.4.13 Higher-order timing effects — 132
  - 6.4.14 Higher-order spectroscopic effects — 136
- 6.5 Transmission and emission spectroscopy — 137
  - 6.5.1 Background — 137
  - 6.5.2 Observations — 139
- 6.6 Properties of transiting planets — 143
  - 6.6.1 Mass–radius relation — 143
  - 6.6.2 Observed correlations — 146

7 Imaging — 149

- 7.1 Introduction — 149
- 7.2 Techniques — 150
  - 7.2.1 Active optics — 150
  - 7.2.2 Adaptive optics — 150
  - 7.2.3 Coronagraphic masks — 152
  - 7.2.4 Speckle noise — 157
- 7.3 Ground-based imaging instruments — 158
  - 7.3.1 Extreme adaptive optics instruments — 158
  - 7.3.2 Extremely large telescopes — 159
  - 7.3.3 Imaging from the Antarctic — 161
  - 7.3.4 Ground-based interferometry — 162
- 7.4 Space-based imaging — 162
  - 7.4.1 Existing telescopes — 162
  - 7.4.2 Space interferometry — 163
  - 7.4.3 The future: resolved imaging — 167
- 7.5 Imaging results — 169
  - 7.5.1 Searches around nearby stars — 170
  - 7.5.2 Searches around exoplanet host stars — 170
  - 7.5.3 Searches in systems with debris disks — 171
  - 7.5.4 Searches around white dwarfs — 172

7.6		Observations at radio wavelengths	173
	7.6.1	Astrometry	173
	7.6.2	Direct imaging	173
7.7		Observations at mm/sub-mm wavelength	177
7.8		Miscellaneous signatures	177
	7.8.1	Planetary and proto-planet collisions	177
	7.8.2	Collisional debris	178
	7.8.3	Accretion onto the central star	178
	7.8.4	Gravitational wave modulation	179

8 Host stars — 181

8.1		Knowledge from astrometry	181
	8.1.1	Hipparcos distances and proper motions	181
	8.1.2	Nearby star census	181
	8.1.3	Galactic coordinates	183
8.2		Photometry and spectroscopy	184
8.3		Evolutionary models	186
8.4		Element abundances	188
	8.4.1	Metallicity	188
	8.4.2	Possible biases	191
	8.4.3	Origin of the metallicity difference	191
	8.4.4	Refractory and volatile elements	194
	8.4.5	The r- and s-process elements	198
	8.4.6	The alpha elements	198
	8.4.7	Lithium	199
	8.4.8	Beryllium	201
8.5		Asteroseismology	201
	8.5.1	Principles	201
	8.5.2	Application to exoplanet host stars	203
8.6		Activity and X-ray emission	205
	8.6.1	Magnetic and chromospheric activity	205
	8.6.2	X-ray emission	206
8.7		Stellar multiplicity	206

9 Brown dwarfs and free-floating planets — 209

9.1		Brown dwarfs	209
	9.1.1	The role of fusion	209
	9.1.2	Detection	209
	9.1.3	Luminosity and age	211
	9.1.4	Classification	212
	9.1.5	Recognising brown dwarfs	212
	9.1.6	Other properties	213
	9.1.7	Formation	214
9.2		Free-floating objects of planetary mass	215

10 Formation and evolution — 217

10.1		Overview	217
10.2		Star formation	217
10.3		Disk formation	218
	10.3.1	Initial collapse	218
	10.3.2	Young stellar objects	219
	10.3.3	Protoplanetary disks	220
	10.3.4	Debris disks	222
10.4		Terrestrial planet formation	224
	10.4.1	The context	224

	10.4.2	Stages in the formation of terrestrial planets	225
10.5	Size, shape, and internal structure		230
10.6	Giant planet formation		231
	10.6.1	Formation by core accretion	231
	10.6.2	Formation by gravitational disk instability	235
10.7	Formation of planetary satellites		237
10.8	Orbital migration		237
	10.8.1	Evidence for migration	237
	10.8.2	Gas disk migration	238
	10.8.3	Planetesimal disk migration	243
	10.8.4	Planet–planet scattering	243
10.9	Tidal effects		244
	10.9.1	Tidal evolution of close-in planets	244
	10.9.2	Orbital evolution	245
	10.9.3	Spin-up of host stars	248
	10.9.4	Tidal heating	249
	10.9.5	Tidal heating and habitability	250
10.10	Population synthesis		251

11 Interiors and atmospheres — 255

11.1	Introduction		255
11.2	Planetary constituents		255
	11.2.1	Gas, rock, and ice	255
	11.2.2	Chemical composition and condensation	257
11.3	Models of giant planet interiors		260
	11.3.1	Equations of state	261
	11.3.2	Hydrogen and water	261
	11.3.3	Structural models	263
11.4	Predictions of interior models		265
	11.4.1	Dependence on composition	265
	11.4.2	H/He dominated gas giants	266
11.5	Super-Earths		267
	11.5.1	General models	267
	11.5.2	Ocean planets	269
11.6	Diagnostics from rotation		270
11.7	Atmospheres of gas giants		271
11.8	Atmospheres of terrestrial planets		278
	11.8.1	Atmospheric formation	278
	11.8.2	Atmospheric erosion	279
	11.8.3	Atmospheres of ejected planets	281
11.9	Habitability		282
	11.9.1	The habitable zone	283
	11.9.2	Exoplanets in the habitable zone	286
	11.9.3	Spectroscopic indicators of life	287
	11.9.4	SETI	290

12 The solar system — 293

12.1	Birth in clusters	293
12.2	The solar system giants	293
12.3	Minor bodies in the solar system	295
12.4	Solar nebula abundances	295
12.5	Constraints on formation	296
12.6	Orbit considerations	299
12.7	Planetesimal migration in the solar system	302
12.8	Atmosphere of the Earth	305

Appendix A	*Numerical quantities*	**309**
Appendix B	*Notation*	**313**
Appendix C	*Radial velocity planets*	**317**
Appendix D	*Transiting planets*	**325**
References		**329**
Subject Index		**403**

Preface

After centuries of philosophical speculation about the existence of worlds beyond our solar system, the first hints of planetary mass objects orbiting other stars were reported in the late 1980s. A planetary system was discovered around a millisecond pulsar in 1992. Then, in 1995, based on precise radial velocity measurements of the host star, convincing evidence for the first exoplanet surrounding a main-sequence star was announced.

Two further exoplanets were known at the end of that year, and 34 at the end of the millennium. Since then, in just ten years, around 500 have been discovered through various methods, the 500th announced in December 2010 as this volume was going to press. A remarkable advance in understanding their physical, chemical and dynamical properties, and their formation and evolution, has kept pace with this discovery.

As the field has expanded, partly inspired by the vision of finding other Earths and perhaps other life, the stimulus provided to astronomical instrumentation has been equally profound. Amongst the technological advances are radial velocity accuracies unimagined twenty years ago, the accurate photometric monitoring of tens of millions of stars, alert systems which flag and focus attention of the world's planet hunters on gravitational microlensing events for a fleeting insight into invisible exoplanet systems, new coronagraphic techniques, a drive for extreme adaptive optics in unprecedented attempts to image alien worlds, and space missions devoted to their discovery and characterisation.

The advance of exoplanet research has been accompanied by more than 6000 papers in the last fifteen years. Authoritative reviews covers many aspects of the field. But the many different techniques now available for their detection and characterisation, the breadth of the underlying physics, and the extensive overlap with solar system and planetary science investigations, can leave the overall panorama somewhat less clear.

My goal has been to tie these many avenues of investigation – from the perspectives of observation, technology, and theory – into a coherent and objective framework. I have aimed to present an overall review of exoplanet research as it stands at the end of 2010, providing a guide for those new to the field, and with extensive references to more detailed and specialised treatments.

In preparing this overview, on-line journals, combined with the powerful NASA Astrophysics Data System (ADS), have been indispensable. LaTeX was key to its practical development. I have also made regular use of the Exoplanets Encyclopaedia maintained by Jean Schneider, and the Exoplanet Orbit Database maintained by Jason Wright and Geoff Marcy.

I am most grateful to all colleagues who authorised the use of their figures for this work, so indispensable for illustrating the various results. Their names are acknowledged in the corresponding figure captions.

For help in clarifying various points, I thank David Catling, Martin Chaplin, Andrew Collier Cameron, Vik Dhillon, Dainis Dravins, David Erskine, Artie Hatzes, John Heathcote, Vinay Kashyap, Neill Reid, Zsolt Sándor, Birger Schmitz, Glenn Schneider, Tim Schulze-Hartung, Steven Soter, Motohide Tamura, Gerard van Belle, Joachim Wambsganss, and Jason Wright.

Simon Mitton opened the door to having this work published, and I am grateful to him, and to Vince Higgs and Claire Poole at Cambridge University Press for their professional support during its preparation.

Finally, I express my particular gratitude to Joachim Wambsganss, director of the Zentrum für Astronomie der Universität Heidelberg (ZAH/ARI), and Thomas Henning and Hans-Walter Rix, directors of the Max–Planck–Institut für Astronomie, Heidelberg, for their invitation to spend a period in Heidelberg to prepare this work. I am grateful to all three for their practical support and hospitality, and for their enthusiastic encouragement during its preparation.

Our generation is the first to know that other worlds, both large and small, are common. In the long-ago words of William Blake *'What is now proved was once only imagined.'* And I see no need to modify my own view in a broad review of the emerging field ten years ago: *'Developments have been so rapid over the last few years that many significant developments, and many new surprises, can be predicted with confidence.'*

Notification of factual errors or misrepresentations (mac.perryman@gmail.com) would be appreciated.

Michael Perryman
Heidelberg, December 2010

1
Introduction

1.1 The challenge

There are hundreds of billions of galaxies in the observable Universe, with each galaxy such as our own containing some hundred billion stars. Surrounded by this seemingly limitless ocean of stars, mankind has long speculated about the existence of planetary systems other than our own, and the possibility of life existing elsewhere in the Universe.

Only recently has evidence become available to begin to distinguish the extremes of thinking that has pervaded for more than 2000 years, with opinions ranging from *'There are infinite worlds both like and unlike this world of ours'* (Epicurus, 341–270 BCE) to *'There cannot be more worlds than one'* (Aristotle, 384–322 BCE).

Shining by reflected starlight, exoplanets comparable to solar system planets will be billions of times fainter than their host stars and, depending on their distance, at angular separations from their accompanying star of, at most, a few seconds of arc. This combination makes direct detection extraordinarily demanding, particularly at optical wavelengths where the star/planet intensity ratio is large, and especially from the ground given the perturbing effect of the Earth's atmosphere.

Alternative detection methods, based on the dynamical perturbation of the star by the orbiting planet, delivered the first tangible results in the early 1990s. Radio pulsar timing achieved the first convincing detection of planetary mass bodies beyond the solar system in 1992. High-accuracy radial velocity (Doppler) measurements yielded the first suggestions of planetary-mass objects surrounding main sequence stars in 1988, with the first essentially unambiguous detection reported in 1995.

Progress since 1995 This discovery precipitated a changing mindset. Amongst the astronomical community at large, the search for exoplanets, and their characterisation, rapidly became a respectable domain for scientific research, and one equally quickly supported by funding authorities. More planets were discovered by radial velocity search teams in the following years. In 1998, the technique of gravitational microlensing provided evidence for a low-mass planet orbiting a star near the centre of the Galaxy nearly 30 000 light-years away, with the first confirmed microlensing planet reported in 2004. In the photometric search for transiting exoplanets, the first transit of a previously-detected exoplanet was reported in 1999, the first discovery by transit photometry was reported in 2003, the first of the wide-field bright star survey discoveries was reported in 2004, and the first discovery from space observations in 2008. A more complete observational chronology, of necessity both selective and subjective, is given in Table 1.1.

While these manifestations of the presence of exoplanets are also extremely subtle, instrumental advances in Doppler measurements, photometry, microlensing, and others, have since provided the tools for their detection in relatively large numbers. Now, fifteen years after the first observational confirmation of their existence, exoplanet detection and characterisation, and advances in the theoretical understanding of planetary formation and evolution, are moving rapidly on many fronts.

1.2 Discovery status

As of the cut-off date for this review, 2010 November 1, almost 500 exoplanets were known, with more than 50 multiple systems. Some statistics, according to discovery method, are listed in Table 1.2.

Diversity Continuing the trend established by the earliest discoveries, exoplanets do not adhere to the individual or system properties extrapolated from the known architecture of the solar system.

Orbital properties vary widely. Around one third have very elliptical orbits, with $e \gtrsim 0.3$, compared with the largest eccentricities in the solar system, of about 0.2 for Mercury and Pluto (and just 0.05 for Jupiter). More than half are around the mass of Jupiter ($0.3 - 3M_J$), and many of these orbit their host star much closer than Mercury orbits the Sun (0.39 AU): hot highly-irradiated giants piled up towards 0.03 AU that cannot have formed *in situ*. Others are located far out, at distances of 100 AU or more from their host star. Planets with orbits highly inclined to the star's equatorial plane occur frequently, some even with retrograde orbits.

Table 1.1: A selective chronology of exoplanet discoveries. Theoretical contributions are not included, and many other equally important discoveries could have been added. The specified date is the 'received date' of the published journal article. Discoveries are not listed if subsequently contested, and possible discoveries may be listed if subsequently confirmed. The first discoveries of some of the major survey instruments are also included.

Date	Subject	Reference
14-Dec-1987	Possible $1.7M_J$ radial velocity planet (later confirmed): γ Cep	Campbell et al. (1988)
18-Jan-1989	Possible $11M_J$ radial velocity planet (later confirmed): HD 114762	Latham et al. (1989)
21-Nov-1991	Multiple planet system from radio timing of millisec pulsar: PSR B1257+12	Wolszczan & Frail (1992)
10-Dec-1992	Possible $2.9M_J$ radial velocity planet (later confirmed): HD 62509	Hatzes & Cochran (1993)
29-Aug-1995	Radial velocity planet #1 (OHP–ELODIE: $0.47M_J$, $P = 4.2$ d): 51 Peg	Mayor & Queloz (1995)
22-Jan-1996	Radial velocity planet #2 (Lick: $6.6M_J$, $P = 117$ d): 70 Vir	Marcy & Butler (1996)
12-Nov-1999	Photometric transit of a known planet (0.8-m APT): HD 209458	Henry et al. (1999)
19-Nov-1999	" (0.1-m STARE): HD 209458	Charbonneau et al. (2000)
15-Nov-1999	First (six) planets detected with Keck–HIRES	Vogt et al. (2000)
3-Jan-2000	Measurement of Rossiter–McLaughlin effect (ELODIE): HD 209458	Queloz et al. (2000a)
27-Dec-2000	System in (2:1) mean motion resonance (Lick/Keck): GJ 876 b and c	Marcy et al. (2001a)
3-May-2002	'Free-floating' cluster object of planet mass (sub-brown dwarf): S Ori 70	Zapatero Osorio et al. (2002)
27-Nov-2002	First planet discovered by transit photometry surveys (OGLE): OGLE–TR–56	Konacki et al. (2003a)
23-Dec-2003	Atmospheric (escaping) H I, O I, C II detected (HST–STIS): HD 209458 b	Vidal-Madjar et al. (2004)
12-Feb-2004	Microlensing planet #1 (confirmed, $2.6M_J$): OGLE–2003–BLG–235L b	Bond et al. (2004)
4-Mar-2004	First planet detected with HARPS (radial velocity): HD 330075 b	Pepe et al. (2004)
6-Aug-2004	First planet discovered by bright star transit photometry surveys: TrES–1	Alonso et al. (2004)
6-Oct-2004	Imaging of borderline planet/brown dwarf companion (VLT–NACO): GQ Lup	Neuhäuser et al. (2005)
19-Nov-2004	Probable planet detected by imaging (later confirmed): Fomalhaut	Kalas et al. (2005)
3-Feb-2005	Thermal emission (secondary eclipse) detected by Spitzer: TrES–1 b	Charbonneau et al. (2005)
3-Feb-2005	" HD 209458 b	Deming et al. (2005b)
5-Apr-2005	Imaging of planet ($4M_J$) around brown dwarf (VLT–NACO): 2M J1207 b	Chauvin et al. (2005a)
20-May-2005	Microlensing planet #2 ($3.8M_J$): OGLE–2005–BLG–071	Udalski et al. (2005)
24-May-2005	Low-mass planet $< 10M_\oplus$ ($6-8M_\oplus$): GJ 876 d	Rivera et al. (2005)
28-Sep-2005	Low-mass microlensing planet ($5.5M_\oplus$): OGLE–2005–BLG–390L b	Beaulieu et al. (2006)
13-Feb-2006	Astrometric confirmation of radial velocity detection (HST–FGS): ϵ Eri b	Benedict et al. (2006)
10-Mar-2006	System with three Neptune-mass planets ($5-20M_\oplus$, HARPS): HD 69830	Lovis et al. (2006)
12-Aug-2006	First transiting planet from HATNet: HAT–P–1 b	Bakos et al. (2007b)
15-Aug-2006	Detection of day/night variation in thermal emission (Spitzer): υ And b	Harrington et al. (2006)
22-Sep-2006	First transiting planets from SuperWASP: WASP–1 b and WASP–2 b	Collier Cameron et al. (2007a)
5-Oct-2006	Planet in an open cluster (Hyades giant, Okayama): ϵ Tau b	Sato et al. (2007)
20-Dec-2006	Planet around a K giant (Tautenburg): 4 UMa	Döllinger et al. (2007)
19-Jan-2007	Infrared spectrum (Spitzer–IRS): HD 189733 b	Grillmair et al. (2007)
4-Apr-2007	Super-Earth planet ($7.7M_\oplus$) in the habitable zone (HARPS): GJ 581 c	Udry et al. (2007)
6-Apr-2007	Planet detected in timing of p-mode pulsator: V391 Peg b	Silvotti et al. (2007)
8-Apr-2007	Atmospheric H_2O detected (Spitzer–IRAC): HD 189733 b	Tinetti et al. (2007b)
8-May-2007	System with five planets (from 18-yr radial velocity): 55 Cnc	Fischer et al. (2008)
4-Oct-2007	Long-period transiting planet (21.2 d): HD 17156 b	Barbieri et al. (2007)
17-Oct-2007	Planet candidate detected in timing of white dwarf: GD 66 b	Mullally et al. (2008)
19-Oct-2007	Microlensing planets with orbital rotation: OGLE–2006–BLG–109L b,c	Gaudi et al. (2008)
4-Jan-2008	First planet detected by CoRoT (space photometry): CoRoT–1 b	Barge et al. (2008)
7-Aug-2008	Planet detected in timing of eclipsing binary (previously suspected): HW Vir	Lee et al. (2009)
7-Sep-2008	Atmospheric CO_2, CO, H_2O detected (HST–NICMOS): HD 189733 b	Swain et al. (2009c)
30-Sep-2008	Planet detected by imaging (HST–ACS): Fomalhaut	Kalas et al. (2008)
30-Sep-2008	Three planets detected by imaging (Keck/Gemini): HR 8799	Marois et al. (2008b)
10-Nov-2008	Probable planet detected by imaging, later confirmed (VLT–NACO): β Pic	Lagrange et al. (2009b)
28-Jan-2009	Secondary eclipse detection by CoRoT: CoRoT–7 b	Snellen et al. (2009a)
23-Feb-2009	Transiting super-Earth ($3-10M_\oplus$): CoRoT–7 b	Léger et al. (2009)
24-Mar-2009	Low-mass planet $< 2M_\oplus$ ($1.9M_\oplus$, HARPS): GJ 581 e	Mayor et al. (2009a)
20-Jul-2009	Multiple system with one transiting planet: HAT–P–13 b,c	Bakos et al. (2009b)
12-Aug-2009	Possible retrograde orbit (later confirmed): HAT–P–7 b	Winn et al. (2009)
14-Oct-2009	Relative orbit inclinations from astrometry (HST–FGS): υ And c, d	McArthur et al. (2010)
20-Oct-2009	Super-Earth planet ($6.5M_\oplus$) transiting an M star (MEarth): GJ 1214 b	Charbonneau et al. (2009)
2-Nov-2009	Spectrum of an imaged planet (VLT–NACO): HR 8799 c	Janson et al. (2010)
16-Nov-2009	First transiting planet from Kepler (space photometry): Kepler–4 b	Borucki et al. (2010b)
17-Feb-2010	Triple (Laplace) resonance (Keck–HIRES): GJ 876 b, c, e	Rivera et al. (2010)
28-Jul-2010	System with two transiting planets, with timing variations: Kepler–9 b, c	Holman et al. (2010)
12-Aug-2010	System with (possibly) seven planets (HARPS): HD 10180	Lovis et al. (2011)

1.3 Outline of the treatment

Table 1.2: Exoplanet detection statistics, from exoplanet.eu, 2010 November 1. Notes: (1) 358 of these planets are radial velocity discoveries (see Chapter 2 and Appendix C), others are measurements of transiting systems; a few have also been measured by astrometry, although none have been discovered by astrometry; (2) most are transit discoveries, although a few were discovered by radial velocity measurements (see Chapter 6 and Appendix D); some are known to be in multiple systems from subsequent radial velocity measurements; (3) see Chapter 5; (4) see Chapter 7; (5) see Chapter 4; (6) objects of sub-brown dwarf mass, existing in apparent isolation in young clusters (see Chapter 9); (7) is not the sum of the preceding, since some planets have been detected by more than one technique (principally, transit discoveries with radial velocity follow-up).

Category	Systems	Multiple	Planets
Detections			
Radial velocity[1]	390	45	461
Transits[2]	105	7	106
Microlensing[3]	10	1	11
Imaging[4]	10	1	12
Timing[5]	6	3	10
Cluster objects[6]	6	–	6
Total number of planets[7]			494
Unconfirmed or retracted	94	3	97

Exoplanets are being discovered around a wide variety of stellar types. Host stars are not only main sequence stars like the Sun, but they include very low-mass stars, low metallicity stars, giant stars, and other advanced evolutionary stages such as white dwarfs and pulsars. Their internal structure and composition vary widely too. Gas giants with stripped outer envelopes, water worlds formed beyond the snow line, and carbon-dominated terrestrial planets may all exist. The first exoplanet atmospheres have been probed through secondary eclipse photometry and spectroscopy.

Of the multiple exoplanet systems, massive planets orbiting in mean motion resonance are common, presenting a certain challenge to explain their ubiquity. The first triple-planet Laplace resonance has been discovered, as have prominent transit time variations in a two-planet transiting system. Systems with multiple lower-mass planets are being found in increasing numbers as the radial velocity surveys improve their detection threshold and increase their temporal baseline. The five-planet system 55 Cnc has been overtaken by the (possible) six-planet system GJ 581 with a $3.1 M_\oplus$ planet in the habitable zone, and up to seven planets with five of Neptune-mass in the case of HD 10180.

Frequency Based on present knowledge from the radial velocity surveys, at least 5–10% of solar-type stars in the solar neighbourhood harbour massive planets. A much higher fraction, perhaps 30% or more, may have planets of lower mass or with larger orbital radii. If these numbers can be extrapolated, the planets in our Galaxy alone would number many billions.

1.3 Outline of the treatment

The present volume summarises the main areas of exoplanet research, combining a description of techniques, concepts, and underlying physics, with a review of the associated literature through to the end of 2010. It is formulated as an overall introduction to exoplanet research for those new to the field, intended to be accessible to both astronomers and planetary scientists, emphasising the interconnection between the various fields of investigation, and providing extensive pointers to more in-depth treatments and reviews.

1.3.1 Observational techniques

Chapters 2–7 divide the search for and characterisation of exoplanets according to detection technique. In each case, the underlying principles are summarised, along with the principal instruments in use, the status of experimental results, and the instrumentation planned for the future. Figure 1.1 summarises the various detection techniques that are the subject of these chapters.

Radial velocity Chapter 2 covers the many aspects of radial velocity (Doppler) measurements, including the different instrumental approaches being used and under development. It includes a basic treatment of planetary orbits, indicating how radial velocity measurements (as well as astrometric measurements, independently and together) provide access to the planet's orbital parameters. As the most successful of the discovery techniques to date, at least in numerical terms, the text covers the basics of wavelength calibration, the contributory error sources, and an overview of the latest results from Doppler searches, including those around double and multiple stars. For multiple planetary systems, this chapter also provides an introduction to the concepts of resonances, dynamical stability, and dynamical packing. The development of sub-1 m s^{-1}-class accuracies is resulting in the detection of low-mass planets down to just a few Earth masses, which are beginning to appear in large numbers, in multiple systems, and at separations corresponding to the 'habitable zone'.

Astrometry Chapter 3 covers the principles of the detection and characterisation of planetary orbits by astrometric measurement. Since the largest astrometric displacements expected for the most massive nearby planets amount to of order 1 milliarcsec, comparable to the current state-of-the art in astrometric measurements from space, few planets can yet be confirmed through their astrometric displacements, and none have been discovered by astrometry alone. The limiting factors for ground-based and space-based instruments are summarised. The panorama of astrometric discovery and characterisation is expected to change substantially with the advent of microarcsec astrometry, both from the ground and more particularly from space.

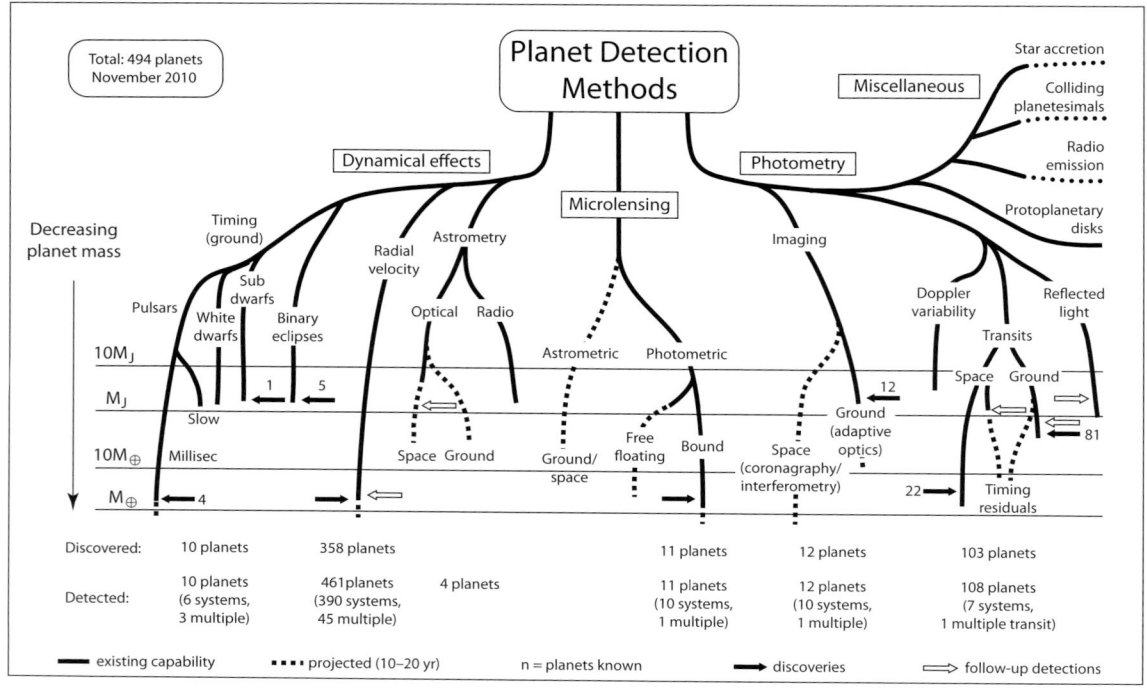

Figure 1.1: Detection methods for exoplanets. The lower limits of the lines indicate the detectable masses that are in principle within reach of present measurements (solid lines), and those that might be expected within the next 10–20 years (dashed). The (logarithmic) mass scale is shown at left. The miscellaneous signatures to the upper right are less well quantified in mass terms. Solid arrows indicate detections according to approximate mass. Open arrows indicate that relevant measurements of previously-detected systems have been made. The figure takes no account of the numbers of planets that may ultimately be detectable by each method. Adapted from Perryman (2000, Figure 1), with permission from the Institute of Physics Publishing Ltd.

Timing Chapter 4 covers exoplanet detection by the measurement of orbit timing residuals, the third discovery technique which makes use of the reflex dynamical motion of the host star. The first non-solar system objects of planetary mass were detected by this technique using radio pulsar timing measurements in 1991. Although pulsars with planets remain the exception, the same technique is being applied to stars which have an underlying periodic photometric signature which is then modulated by an orbiting planet. The technique has recently been applied to detect planets around pulsating white dwarfs, pulsating subdwarfs, and eclipsing binaries. Its success has underlined the diversity of stellar types around which planets remain in orbit.

Microlensing Chapter 5 covers detection by gravitational microlensing. Alone amongst the techniques in sampling primarily very distant exoplanetary systems, its main disadvantage is that it can only provide a single measurement epoch spanning hours or days. A particularly noteworthy result was the measurement, reported in 2008, of a system of two planets in which orbital motion could be measured during the 10-day event duration. The ability to detect true free-floating planets, a sensitivity to Earth-mass planets in the habitable zone, and a measurement technique independent of the host star spectral type or luminosity class, make the prospects of a space-based microlensing survey of particular importance to a broad exoplanet survey census.

Photometry and transits Chapter 6 covers photometric measurements, most importantly the search for exoplanets transiting the disk of their host star, as well as searches for reflected and polarised light. Whilst transits are of relevance only for planets whose orbits happen to lie essentially orthogonal to the plane of the sky, this constrained geometry allows both the mass of the planet to be determined unambiguously (without the $\sin i$ indeterminacy inherent in radial velocity observations) and, of major importance for exoplanet characterisation, its radius. Together yielding the exoplanet density, this offers the first insights into the internal structure and chemical composition of the transiting planet. The search for transit time and transit duration variations offers prospects for detecting accompanying members of the planetary system. Further insights into the atmospheres of the gas giants are being obtained from transit and secondary eclipse spectroscopy.

1.4 Astronomical terms and units

Direct imaging Chapter 7 covers the techniques in use and under development for the direct imaging of an exoplanet in orbit around its host star. The technical challenges, and technological solutions (adaptive optics, coronagraphy, and space-based imaging and interferometry) are described, along with the results obtained to date. This chapter also covers prospects for direct detection based on magnetospheric radio emission, as well as observations at mm/sub-mm wavelength.

1.3.2 Host star properties and brown dwarfs

Host stars Chapter 8 reviews the properties of exoplanet host stars. It includes discussion of their Galactic orbits, their metallicities, and the theories put forward to explain the observed correlation between the occurrence of exoplanets and host-star metallicity. It also reviews the asteroseismology investigations that have been carried out on a few exoplanet host stars, and the characteristics of their X-ray emission.

Brown dwarfs Chapter 9 provides an introduction to the properties of brown dwarfs. The subject overlaps with that of exoplanets both in the definition of a planet, and in the context of so-called free-floating objects of planetary mass which have been discovered in nearby young star-forming regions.

1.3.3 Theoretical considerations

Chapters 10–12 deal with the theories of formation and evolution, and of their interiors and atmospheres.

Formation and evolution Chapter 10 is a summary of the present understanding of planet formation and evolution. Very broadly, the current picture is that formation started with a collapsing protostellar disk, with planets assembled from dust and gas by the progressive agglomeration of material over some 14 orders of magnitude in size. The gas giants, of masses $\gtrsim 10 M_\oplus$, formed by either, perhaps both, core accretion or gravitational disk instability. Close-in planets, high eccentricities, and orbital resonances provide evidence for planetary migration subsequent to formation. Inward, and sometimes, outward migration as a result of interactions between the planet and the gas and residual planetesimal disk, along with planet–planet scattering, provides a compelling picture of the diversity of planetary system architectures observed. For planets that arrive to within ~0.2 AU of the host star, whether by migration or scattering, tidal effects become significant, circularising orbits, synchronising their rotation and orbital periods, and providing an additional source of internal heating.

Interiors and atmospheres Chapter 11 reviews the current knowledge of interiors and atmospheres, deduced primarily from the masses and densities measured for transiting planets, combined with theoretical models based on the equations of hydrostatic and thermodynamic equilibrium. Thermal equilibrium and condensation calculations can predict which chemical species will be present for a given initial elemental composition and, from these, insight is being gained into their internal structures and atmospheric compositions. For terrestrial-mass planets, estimates of the habitable zone, where liquid water could be present, are providing pointers to the first planets which may be habitable.

The solar system Chapter 12 provides a selective summary of solar system properties which are closely linked to developments in exoplanet studies. Solar system observations provide important constraints on theories and properties of exoplanet formation and evolution, while developments in exoplanet formation and evolution, notably planetary migration, are offering insight into the present structure and past evolution of the solar system. Topics covered include the origin of water on Earth, orbital stability, planet obliquities, the origin of the Moon, planet migration, and the origin and evolution of the Earth's atmosphere. Taken together, combined knowledge of exoplanets and the solar system is providing an increasingly detailed picture of planet formation and evolution, further suggesting that the basic models of exoplanet formation, and that of the solar system, are broadly coherent.

1.4 Astronomical terms and units

For those less familiar with astronomical terms and nomenclature, a brief summary of some relevant concepts is as follows.

Astronomical terms Various relevant terms used in astronomy and planetary science may cause some confusion on first encounter. More detailed explanations are given in appropriate places in the text, but advanced warning of some of these may assist orientation.

Metallicity: in astronomy usage, the term 'metal' is divorced from its usual chemical definition related to electrical conductivity and chemical bonding, and instead refers collectively to all elements other than H or He (and essentially therefore to the elements produced by nucleosynthesis in stars or supernovae).

Ice, gas, and rock: in planetary science, 'ices' refer to volatile materials with a melting point between ~100–200 K. In consequence, 'ices' (for example in Uranus and Neptune) are not necessarily H_2O, not necessarily 'cold', and not necessarily solid. Similarly, a 'gas' in planetary science is not defined by phase, but rather as a highly volatile material with a melting point (if at all) below ~100 K. 'Rock' may be defined by its solid phase or present mineralogical composition, but generally also by its presumed chemical composition and highly refractory nature during the epoch of planetary formation.

Notation for star and planet parameters Stars and planets are characterised, amongst other parameters, by their mass M and radius R, with subscripts $\star$ and p referring to star and planet respectively, and the distance to the system d. Orbits are characterised by their period P, semi-major axis a, eccentricity e, inclination with respect to the plane of the sky i ($i = 0°$ face-on, $i = 90°$ edge-on). Further details are given in §2.1.

Masses and radii of stars are usually expressed in solar units ($M_\odot, R_\odot$), while those of planets are typically expressed in either Jupiter units (M_J, R_J) for the gas giants, or Earth units ($M_\oplus, R_\oplus$) for planets closer to terrestrial mass. Numerical values for these and other quantities are given in Appendix A.

It may be noted that the *de facto* definition of R_J in terms of Jupiter's equatorial radius at 10^5 Pa means that, due to its oblateness, Jupiter's own mean radius is actually $0.978 R_J$.

Star distances and masses Stellar distances are given in parsec (pc). As the basic unit of astronomical distance based on the measurement of trigonometric parallax, this is the distance at which the mean Sun–Earth distance (the astronomical unit, or AU) subtends an angle of 1 arcsec (1 pc $\simeq 3.1 \times 10^{16}$ m $\simeq 3.26$ light-years).

For reference, distances to the nearest stars are of order 1 pc; there are about 2000 known stars within 25 pc of the Sun. With the exception of microlensing events, most exoplanet discoveries and detections are restricted to a distance horizon of about 50–100 pc.

In general, stellar masses range from around $0.1 - 30 M_\odot$, with spectral types providing a conventional classification related to the primary stellar properties of temperature and luminosity. The Sun is of spectral type G2V: cooler stars (types K, M) are of lower mass and have longer lifetimes; hotter stars (types F, A, etc.) are of higher mass and have shorter lifetimes. Stellar masses of interest to exoplanet studies are typically in the range $0.1 - 5 M_\odot$, with the majority of targets and detections focused on masses rather close to $1 M_\odot$.

Star names Object names such as 70 Vir (for 70 Virginis) and β Pic (for β Pictoris) reflect constellation-based nomenclature, while other designations reflect discovery catalogues or techniques variously labeled with catalogue running numbers (e.g. HD 114762) or according to celestial coordinates (e.g. PSR B1257+12).

Some of the most commonly referenced star catalogues of relevance are:

HD (Henry Draper): surveyed by Cannon & Pickering (Ann. Astr. Obs. Harvard, Vol. 91–99, 1918–1924).

HIP (Hipparcos): the space-based astrometric catalogue extends to $\simeq 12$ mag, but with a completeness between 7.3–9.0 mag depending on Galactic latitude and spectral type (Perryman et al., 1997), and detailed further in §8.1.1.

BD (Bonner Durchmusterung): the BD was the first of the three-part Durchmusterung (German for survey) covering the entire sky. The northern sky was surveyed from Bonn by Argelander & Schönfeld and published between 1852–1859. The extension southwards was surveyed from Córdoba in Argentina (the Córdoba Durchmusterung, or CD) by Thomme starting in 1892. The southern skies were surveyed from the Cape of Good Hope (the Cape Durchmusterung, or CPD) by Gill & Kapteyn around 1900. Stars tend to be assigned their DM (Durchmustering) number if they are not part of the HD or HIP catalogues.

Nearby stars: these are often designated according to their inclusion in the Catalogue of Nearby Stars (CNS), described further in §8.1.2. Such stars appear in the CDS SIMBAD data base as GJ nnn, although the alternatives Gliese nnn or Gl nnn are often used.

Exoplanet names The *de facto* custom denotes planets around star X as X b, c,... lexically according to discovery sequence (rather than, for example, according to mass or semi-major axis, which would demand constant revision as additional planets are discovered).

The naming convention adopted for microlensing planets, where the host star is generally invisible, is described on page 98.

Other units In aiming for a consistent usage of terms and nomenclature, units referred to in the published literature, or in some of the figures, have occasionally been unified. Usage here follows, as far as is considered reasonable, the International System of Units (SI).

Astronomical measures of density generally use the non-SI unit of $g\,cm^{-3}$. Densities here are expressed in units of $Mg\,m^{-3}$, conforming to SI, and with the same numerical values (and number of keystrokes).

Characterisation of pressure, notably in the description of planetary atmospheres and interiors, is divided in the literature between the SI pascal, $1\,Pa \equiv 1\,N\,m^{-2}$, and the bar ($1\,bar \equiv 10^5\,Pa$). The latter, some 1% smaller than 'standard' atmospheric pressure, is not an SI unit, although it is accepted for use within SI. For uniformity, Pa is used systematically here.

Various units outside of SI are accepted for use, or are consistent with the recommendations of the International Committee for Weights and Measures (CIPM, *Comité International des Poids et Mesures*). These include the measurements of time as minute (min), hour (h), and day (d), and the measurement of angles in seconds of arc: units of arcsec, mas (milli-arcsec) and μas (micro-arcsec) are used accordingly.

Certain units central to astronomy, notably the astronomical units of mass and radius noted above, deviate from SI, but are retained in this treatment. While the astronomical unit is accepted within the SI, consistent with common astronomical usage it is indicated here as AU (although the CIPM designates the unit 'ua').

1.6 On-line reference compilations

Table 1.3: Some on-line information and catalogues of particular relevance to exoplanet research.

URL	Content	Comment
Exoplanet catalogues:		
exoplanet.eu	Extrasolar Planets Encyclopedia	Reference catalogue and bibliography
exoplanets.org	Exoplanet Data Explorer	Orbit database (Jones et al., 2008a)
nsted.ipac.caltech.edu	Star and Exoplanet Database	Exoplanet database (Ali et al., 2007)
Transit data:		
www.inscience.ch/transits	Exoplanet Transit Parameters	Compilation of transiting exoplanets
var.astro.cz/ETD	Exoplanet Transit Database	Compilation of exoplanet transits
idoc-corot.ias.u-psud.fr/	CoRoT Mission	Public access
archive.stsci.edu/kepler/planet_candidates.html	Kepler Mission	Released planetary candidates
www.wasp.le.ac.uk/public/	SuperWASP Project	Public archive
Related catalogues:		
www.dwarfarchives.org	Compendium of M, L, T Dwarfs	Photometry, spectroscopy, astrometry
circumstellardisks.org	Catalogue of Circumstellar Disks	Pre-main sequence and debris disks
General resources:		
adsabs.harvard.edu/abstract_service.html	NASA Astrophysics Data System	Digital bibliographic data, with arXiv
simbad.u-strasbg.fr/simbad	CDS SIMBAD	Astronomical object data base

1.5 Definition of a planet

IAU 2006 resolution In its resolution B5 (IAU, 2006), the IAU classified the solar system bodies into three distinct categories (verbatim):

(1) a *planet* is a celestial body that: (a) is in orbit around the Sun, (b) has sufficient mass for its self-gravity to overcome rigid body forces so that it assumes a hydrostatic equilibrium (nearly round) shape, and (c) has cleared the neighbourhood around its orbit;

(2) a *dwarf planet* is a celestial body that: (a) is in orbit around the Sun, (b) has sufficient mass for its self-gravity to overcome rigid body forces so that it assumes a hydrostatic equilibrium (nearly round) shape, (c) has not cleared the neighbourhood around its orbit, and (d) is not a satellite;

(3) other objects except satellites orbiting the Sun are referred to collectively as *small solar system bodies*.

This classification, which excludes Pluto as a planet under criterion 1c, is nevertheless recognised as being somewhat ambiguous due to the difficulties in formulating precise definitions of shape and orbital 'clearing'.

IAU 2003 recommendation Exoplanet classification is facilitated by the fact that a distinction in mass between planets and smaller bodies is not yet relevant. In contrast, nomenclature is complicated by the problems of distinguishing planets from brown dwarfs.

The IAU 2003 recommendation, by the working group on extrasolar planets (IAU, 2003) is (verbatim):

(1) objects with true masses below the limiting mass for thermonuclear fusion of deuterium (currently calculated to be $13 M_J$ for objects of solar metallicity) that orbit stars or stellar remnants are *planets* (no matter how they formed). The minimum mass required for an extrasolar object to be considered a planet should be the same as that used in the solar system;

(2) substellar objects with true masses above the limiting mass for thermonuclear fusion of deuterium are *brown dwarfs*, no matter how they formed nor where they are located;

(3) free-floating objects in young star clusters with masses below the limiting mass for thermonuclear fusion of deuterium are not planets, but are *sub-brown dwarfs* (or whatever name is most appropriate).

Adopted convention In this treatment, orbiting substellar objects are referred to as planets below $\sim 13 M_J$, and as brown-dwarf planets above that threshold (if appropriate to the context), whatever their likely formation mechanism (see box on page 210).

Isolated objects are referred to as sub-brown dwarfs or brown dwarfs according to their estimated mass lying below or above the deuterium-burning threshold, and only as free-floating planets if evidence points to their having been formed as a planet and subsequently dynamically ejected from the original host system. No such objects have yet been confirmed.

Attempts to formulate a precise definition of a planet are confronted by a number of difficulties, summarised by Basri & Brown (2006). A definition dispensing with upper and lower mass limits is offered, and further quantified, by Soter (2006): *A planet is an end product of disk accretion around a primary star or substar.*

1.6 On-line reference compilations

In general, references to on-line compilations or resources have been avoided because of changing URLs, or the frequent absence of long-term maintenance. Many www sites, nevertheless, are repositories of up-to-date information, and a selection of some of the most relevant are listed in Table 1.3. Many other groups host their own information pages.

2
Radial velocities

The motion of a single planet in orbit around a star causes the star to undergo a reflex motion about the star–planet barycentre (centre of mass). This results in the periodic perturbation of three observable properties of the star, all of which have been detected (albeit typically in different systems): in radial velocity, in angular (or astrometric) position on the sky, and in the time of arrival of some periodic reference signal.

2.1 Description of orbits

As in all orbiting systems, both star and planet orbit the common system barycentre. Under the inverse square law of gravity[1], each moves in a closed elliptical orbit in inertial space, with the centre of mass at one focus (Figure 2.1). Such an ellipse is described in polar coordinates (with respect to a focus) by

$$r = \frac{a(1-e^2)}{1+e\cos v}, \qquad (2.1)$$

or in Cartesian coordinates (with respect to the centre) by

$$\frac{x^2}{a^2} + \frac{y^2}{b^2} = 1, \qquad (2.2)$$

with the semi-major axis a and the semi-minor axis b related to the *eccentricity* e by

$$b^2 = a^2(1-e^2). \qquad (2.3)$$

[The ellipticity, $\eta = (a-b)/a = 1 - \sqrt{(1-e^2)}$, is an alternative measure of non-circularity, not used further.]

The pericentre[2] distance q and apocentre distance Q are given by

$$q = a(1-e),$$

[1] The relations given in this section, their derivations, and more extensive dynamical considerations, can be found in various recent texts (e.g., Murray & Dermott 2000, Chapter 2; Hilditch 2001; Cole & Woolfson 2002, Section M) as well as in many earlier treatments of orbits (e.g. Heintz, 1978a; Roy, 1978).

[2] An *apsis* (plural apsides) is the point on the orbit of minimum distance (*pericentre*, or periapsis) or maximum distance (*apocentre*, or apoapsis) from the barycentre. The line connecting the two is the *line of apsides*, which defines the orbit's

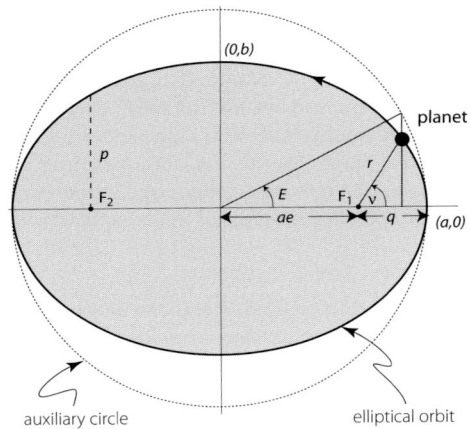

Figure 2.1: Characteristics of an elliptical orbit. The auxiliary circle is of radius equal to the semi-major axis a. Points on the orbit can be described either in terms of the true anomaly v (with respect to the ellipse) or the eccentric anomaly E (with respect to the auxiliary circle). The focus F_1 is the system barycentre. F_2 is the 'empty' focus.

$$Q = a(1+e). \qquad (2.4)$$

A line through a focus and parallel to the minor axis defines the latus rectum, with the *semi-latus rectum* (related to the planet's angular momentum) of length

$$p = a(1-e^2). \qquad (2.5)$$

True, eccentric, and mean anomaly Various angles in the orbit plane, referred to as 'anomalies' (the word dating from the time that planetary motions were considered anomalous), are used to describe the position of a planet along its orbit at a particular time (Figure 2.1).

major axis. Derivative terms refer to the body being orbited: perigee/apogee around the Earth, perihelion/aphelion around the Sun, periastron/apastron around a star, peribac/apobac around a barycentre, and perigalacticon/apogalacticon around a galaxy. The general terms pericentre/apocentre are used preferentially here.

The *true anomaly*, $v(t)$, also frequently denoted $f(t)$, is the angle between the direction of pericentre and the current position of the body measured from the barycentric focus of the ellipse. It is the angle normally used to characterise an observational orbit.

The *eccentric anomaly*, $E(t)$, is a corresponding angle which is referred to the *auxiliary circle* of the ellipse. The true and eccentric anomalies are geometrically related by

$$\cos v(t) = \frac{\cos E(t) - e}{1 - e \cos E(t)}, \qquad (2.6)$$

or, equivalently,

$$\tan \frac{v(t)}{2} = \left(\frac{1+e}{1-e}\right)^{1/2} \tan \frac{E(t)}{2}. \qquad (2.7)$$

The *mean anomaly*, $M(t)$, is an angle related to a fictitious mean motion around the orbit, used in calculating the true anomaly. Over a complete orbit, during which the real planet (or the real star) does not move at a constant angular rate, an average angular rate can nevertheless be specified in terms of the *mean motion*

$$n \equiv 2\pi/P, \qquad (2.8)$$

where P is the orbital period. The mean anomaly at time $t - t_\mathrm{p}$ after pericentre passage is then defined as

$$M(t) = \frac{2\pi}{P}(t - t_\mathrm{p}) \equiv n(t - t_\mathrm{p}). \qquad (2.9)$$

The relation between the mean anomaly, $M(t)$, and the eccentric anomaly, $E(t)$, can be derived from orbital dynamics. This relation, Kepler's equation, is given by

$$M(t) = E(t) - e \sin E(t). \qquad (2.10)$$

The position of an object along its orbit at any chosen time can then be obtained by calculating the mean anomaly M at that time from Equation 2.9, (iteratively) solving the transcendental Equation 2.10 for E, and then using the geometrical identity Equation 2.6 to obtain v.

Orbit specification A Keplerian orbit in three dimensions (Figure 2.2) is described by seven parameters: $a, e, P, t_\mathrm{p}, i, \Omega, \omega$. The first two, a and e, specify the size and shape of the elliptical orbit. P is related to a and the component masses through Kepler's third law (see below), while t_p corresponds to the position of the object along its orbit at a particular reference time, generally with respect to a specified pericentre passage.[3]

The three angles (i, Ω, ω) represent the projection of the true orbit into the observed (apparent) orbit; they

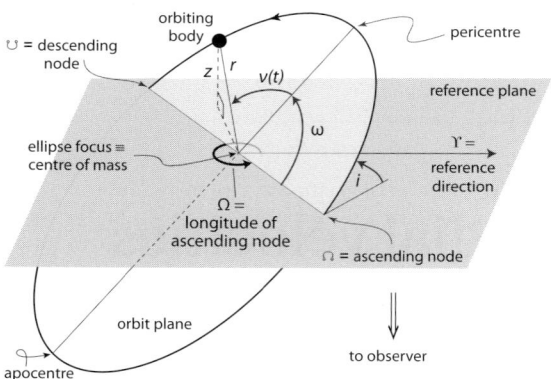

Figure 2.2: *An elliptical orbit in three dimensions. The reference plane is tangent to the celestial sphere. i is the inclination of the orbit plane. The nodes are the points where the orbit and reference planes intersect. Ω defines the longitude of the ascending node, measured in the reference plane. ω is the fixed angle defining the object's argument of pericentre relative to the ascending node. The true anomaly, $v(t)$, is the time-dependent angle characterising the object's position along the orbit.*

depend solely on the orientation of the observer with respect to the orbit. In general, the semi-major axis of the true orbit does not project into the semi-major axis of the apparent orbit.

i specifies the *orbit inclination* with respect to the reference plane, in the range $0 \leq i < 180°$. $i = 0°$ corresponds to a face-on orbit. In the discussion of binary orbits, motion is referred to as prograde (in the direction of increasing position angle on the sky, irrespective of the relation between the rotation and orbit vectors) if $i < 90°$, retrograde if $i > 90°$, and projected onto the line of nodes if $i = 90°$.

Ω specifies the *longitude of the ascending node*, measured in the reference plane. It is the node where the measured object moves away from the observer through the plane of reference. [For solar system objects, it is the node where an orbiting object moves north through the plane of reference.]

ω specifies the *argument of pericentre*, being the angular coordinate of the object's pericentre relative to its ascending node, measured in the orbital plane and in the direction of motion. [For $e = 0$, where pericentre is undefined, $\omega = 0$ can be chosen such that t_p gives the time of nodal passage.]

---

[3]A few remarks are in order: (i) some texts state that just six parameters are required, and omit P, implicitly invoking the relation between a and P (and the component masses) as given by Kepler's third law; (ii) a is the semi-major axis of the orbiting body with respect to the system barycentre, assumed here to be in linear measure unless otherwise noted. If a is determined in angular measure, as in the relative astrometry of binary stars, the system distance d (equivalently the parallax ϖ) is required to establish the linear scale; (iii) the parameters of the two co-orbiting bodies (e.g. a star and planet) with respect to the barycentre are identical, with the exception of their values of a which differ by a factor $M_\mathrm{p}/M_\star$, and their values of ω which differ by $180°$.

2.1 Description of orbits

Three other angles with respect to the adopted reference direction are used in the specification of practical orbits:

$\tilde{\omega} = \Omega + \omega$ the *longitude of pericentre*   (2.11)
$\theta = \tilde{\omega} + \nu$ the *true longitude*   (2.12)
$\lambda = \tilde{\omega} + M$ the *mean longitude*   (2.13)

Since Ω and ω are measured in different planes, the longitude of pericentre, $\tilde{\omega}$, is a 'dog-leg' angle. The true longitude and mean longitude are correspondingly offset with respect to the true anomaly and mean anomaly, respectively. Despite its name, the mean longitude is again a linear function of time and, as for the mean anomaly, has only an auxiliary geometrical interpretation.

Kepler's laws   Kepler's three laws of planetary motion are: (1) the orbit of a planet is an ellipse with the Sun at one focus; (2) the line joining a planet and the Sun sweeps out equal areas in equal intervals of time; (3) the squares of the orbital periods of the planets are proportional to the cubes of their semi-major axes.

The first and third laws are consequences of the inverse square law of gravity, while the second follows from conservation of angular momentum (and is true for any radial law of attraction). Kepler's laws originally referred to relative orbits with respect to the Sun, but corresponding formulations apply also to 'absolute orbits' defined with respect to the barycentre.

For the general two-body problem where the mass of the secondary is not neglected, both orbits are ellipses with their foci at their common barycentre (Figure 2.3). Kepler's third law takes the general form

$$P^2 = \frac{4\pi^2}{GM} a^3, \qquad (2.14)$$

with M and a taking different values according to the type of orbit being measured:

(a) relative orbits: the motion of the planet, now relative to the star rather than the barycentre, can be found by applying an acceleration to the system which cancels that of the star, viz. GM_p/r^2 where r is their instantaneous separation (Figure 2.3). Then

$$P^2 = \frac{4\pi^2}{G(M_\star + M_p)} a_{\rm rel}^3, \qquad (2.15)$$

where the coordinate origin is now the star, not the barycentre, and $a_{\rm rel}$ is the semi-major axis of the relative orbit, i.e. the planet around the star.[4]

---

[4]This measurement of *relative* separation does not arise for exoplanet orbits when the planet is unseen. It is, in contrast, a situation relevant for the relative astrometry of binary stars, where an orbit is measured as a separation and position angle of one star with respect to another; then, the combined system mass can be determined if P and $a_{\rm rel}$ are measurable, while individual masses can only be determined if the mass ratio can be established, either from the ratio of the distances from the barycentre, or the ratio of their speeds around it.

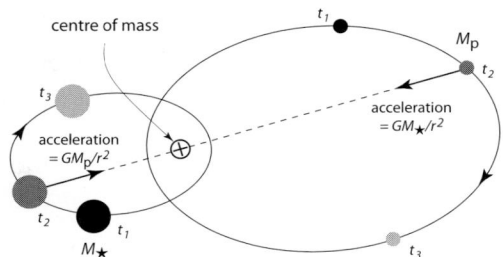

Figure 2.3: Two orbiting bodies, shown at times t_1, t_2, t_3, move about their common barycentre. Both bodies follow orbits having the same shape and period, but of different sizes and with ω differing by $180°$.

For $M_p \ll M_\star$ and in units of Earth's orbit of 1 AU

$$P \simeq 1\,{\rm yr}\left(\frac{a_{\rm rel}}{\rm AU}\right)^{3/2}\left(\frac{M_\star}{M_\odot}\right)^{-1/2}. \qquad (2.16)$$

(b) absolute orbits: the orbit of the star around the system (star–planet) barycentre is given by

$$P^2 = \frac{4\pi^2}{GM'} a_\star^3, \qquad (2.17)$$

where

$$M' \equiv \frac{M_p^3}{(M_\star + M_p)^2}, \qquad (2.18)$$

and $a_\star$ is the semi-major axis of the stellar orbit around the system barycentre. An equivalent expression gives the orbit of the planet around the system barycentre in terms of the semi-major axis of the planet around the system barycentre a_p.

It follows that the sizes of the three orbits are in proportion $a_\star : a_p : a_{\rm rel} = M_p : M_\star : (M_\star + M_p)$, with $a_{\rm rel} = a_\star + a_p$. Furthermore, $e_{\rm rel} = e_\star = e_p$, $P_{\rm rel} = P_\star = P_p$, the three orbits are coplanar, and the orientations of the two barycentric orbits (ω) differ by $180°$.

Since the planet is assumed to be invisible (which is the case for essentially all observations at the present time), the orbital motion of the star around the system barycentre is only correctly determined by astrometry if its position is measured with respect to an 'absolute' (quasi-inertial) reference frame.

That all seven orbital elements are accessible to astrometric measurements for an arbitrary projection geometry essentially follows from two principal considerations: the star position versus time allows the maximum and minimum angular rates to be determined, and hence the position of the line of apsides. With the orientation of the major axis so established, appeal to Kepler's second law fixes the orbit inclination.

Radial velocity measurements of the host star also give information on its barycentric orbital motion, although not all seven Keplerian orbital elements are accessible from the line-of-sight velocity variations alone.

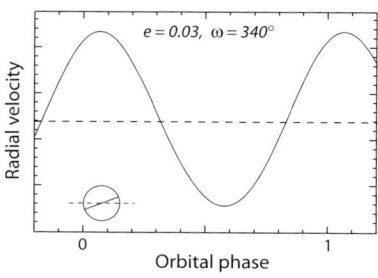 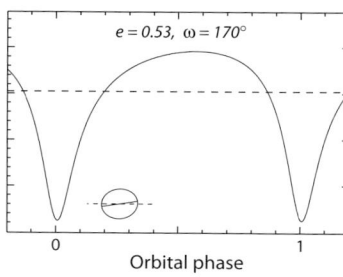 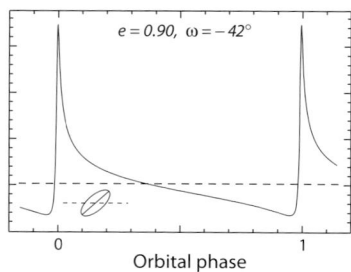

Figure 2.4: Example stellar radial velocity curves, illustrating their dependency on e and ω, for HD 73256 (Udry et al., 2003a, Figure 2), HD 142022 (Eggenberger et al., 2006, Figure 4), and HD 4113 (Tamuz et al., 2008, Figure 1) respectively. Horizontal dashed lines show the systemic velocity (radial velocity of the barycentre). The ellipses at lower left show the viewing geometries.

2.1.1 Orbits from radial velocity measurements

Radial velocity (Doppler) measurements describe the projected motion, along the line-of-sight, of the primary star as it orbits the system barycentre.

Radial velocity semi-amplitude With Figure 2.2 now considered as representing the orbit of the star around the barycentre, the star's z-coordinate along the line-of-sight can be derived from trigonometry

$$z = r(t) \sin i \, \sin(\omega + \nu) \,, \tag{2.19}$$

where $r(t)$ is the distance from the barycentre. Then

$$v_\mathrm{r} \equiv \dot{z} = \sin i \, [\dot{r} \sin(\omega + \nu) + r\dot{\nu} \cos(\omega + \nu)] \,. \tag{2.20}$$

Some algebraic substitutions for r and $\dot{r}$ lead to

$$v_\mathrm{r} = K [\cos(\omega + \nu) + e \cos \omega] \,, \tag{2.21}$$

where the *radial velocity semi-amplitude*[5] is given by

$$K \equiv \frac{2\pi}{P} \frac{a_\star \sin i}{(1 - e^2)^{1/2}} \,. \tag{2.22}$$

From Equation 2.21, it can be seen that v_r varies around the orbit between limits of $K(1 + e \cos \omega)$ and $K(-1 + e \cos \omega)$. This expression for v_r as a function of the true anomaly $\nu(t)$ can be transformed into an expression for v_r as a function of time through Equations 2.6–2.10.

The shape of the radial velocity curve is determined by e and ω (Figure 2.4). Together with P, their combination constrains the value of $a_\star \sin i$ (Equation 2.22). But neither $a_\star$ nor $\sin i$ can be determined separately.

Two alternative expressions for K are instructive. Substituting Equations 2.17 and 2.18 into 2.22 gives

$$K^2 = \frac{G}{(1 - e^2)} \frac{1}{a_\star \sin i} \frac{M_\mathrm{p}^3 \sin^3 i}{(M_\star + M_\mathrm{p})^2} \,. \tag{2.23}$$

With the product $a_\star \sin i$ determined as above, it follows that radial velocity measurements provide a value for the *mass function* (i.e. Equation 2.18 modified by a dependency on the orbit inclination)

$$\mathcal{M} \equiv \frac{M_\mathrm{p}^3 \sin^3 i}{(M_\star + M_\mathrm{p})^2} \,. \tag{2.24}$$

If $M_\mathrm{p} \ll M_\star$, then Equation 2.24 reduces to

$$\mathcal{M} \simeq \frac{M_\mathrm{p}^3 \sin^3 i}{M_\star^2} \,. \tag{2.25}$$

Further, if $M_\star$ can be estimated from its spectral type and luminosity class (or otherwise), then $M_\mathrm{p} \sin i$ can be determined. The mass of the planet nevertheless remains uncertain by the unknown factor $\sin i$.

Equations 2.17, 2.18 and 2.22 can be combined to give an alternative expression for K without the explicit appearance of $a_\star$ (e.g. Cumming et al., 1999, eqn 1)

$$K = \left(\frac{2\pi G}{P} \right)^{1/3} \frac{M_\mathrm{p} \sin i}{(M_\star + M_\mathrm{p})^{2/3}} \frac{1}{(1 - e^2)^{1/2}} \,, \tag{2.26}$$

which can also be written (Torres et al., 2008, eqn 1)

$$\frac{M_\mathrm{p} \sin i}{M_\mathrm{J}} = 4.919 \times 10^{-3} \left(\frac{K}{\mathrm{km\,s^{-1}}} \right) (1 - e^2)^{1/2}$$
$$\left(\frac{P}{\mathrm{days}} \right)^{1/3} \left(\frac{M_\star + M_\mathrm{p}}{M_\odot} \right)^{2/3} \tag{2.27}$$

For a circular orbit with $M_\mathrm{p} \ll M_\star$, for example, the stellar velocity variations are sinusoidal with amplitude

$$K = 28.4 \,\mathrm{m\,s^{-1}} \left(\frac{P}{1\,\mathrm{yr}} \right)^{-1/3} \left(\frac{M_\mathrm{p} \sin i}{M_\mathrm{J}} \right) \left(\frac{M_\star}{M_\odot} \right)^{-2/3} . \tag{2.28}$$

For Jupiter around the Sun ($a = 5.2$ AU, $P = 11.9$ yr) $K_\mathrm{J} = 12.5\,\mathrm{m\,s^{-1}}$. For Earth, $K_\oplus = 0.09\,\mathrm{m\,s^{-1}}$.

Keplerian observables Of the seven elements describing an orbit in three-dimensions ($a, e, P, t_\mathrm{p}, i, \Omega, \omega$), Ω cannot be determined from radial velocity measurements. Furthermore, only the combination $a_\star \sin i$ is determined, with neither $a_\star$ nor $\sin i$ individually.

---

[5]Often denoted as K_1 in binary star work to emphasise that it describes the motion of the primary, this amplitude of radial motion is analogous to the projected semi-major axis measured astrometrically.

2.1 Description of orbits

Radial velocity measurements alone do not provide individual planetary masses, their sum, or their ratio.[6] Taken alone, small periodic radial velocity modulation is insufficient to infer, unambiguously, the presence of a companion of planetary mass. Without additional information, such as an astrometric orbit, a constraint on the orbit inclination i from photometric transits, spectroscopic line profiles, or statistical deconvolution (e.g. Jorissen et al., 2001), small values of K could indicate either a low-mass planet, or an object of significantly higher mass with small orbital inclination, i.e. with the plane of the orbit almost face-on to the line of sight.

Fitting a single planet There are, in consequence, five observables related to the star's Keplerian orbit which can be fit for each orbiting planet on the basis of radial velocity measurements alone: e, P, t_p, and ω, and the combination $K = f(a, e, P, i)$.

Two further terms are usually taken into account: a *systemic velocity* γ describing the constant component of the radial velocity of the system's centre of mass relative to the solar system barycentre, which may also include an instrument-dependent radial velocity offset; and a linear trend parameter, d, which may accommodate instrumental drifts as well as unidentified contributions from massive, long-period companions.

From Equation 2.21, the radial velocity signal of a star with an orbiting planet can then be expressed as a function of the true anomaly as (e.g. Wright & Howard, 2009, eqn 1)

$$v_r(t) = K[\cos(\omega + v(t)) + e\cos\omega] + \gamma + d(t - t_0). \quad (2.29)$$

Establishing the astrometric or radial velocity parameters for a single orbiting planet is typically based on the search of parameter space using χ^2 minimisation.

Good period estimates are invariably used to simplify the search, for example using the well-established Lomb–Scargle algorithm (Figure 2.5). This is a modified periodogram analysis, equivalent to least-squares fitting of sine waves, which targets the efficient and reliable detection of a periodic signal in the case of unevenly-spaced observation times and in the presence of noise (Scargle, 1982; Horne & Baliunas, 1986; Cumming, 2004; Baluev, 2008a). Zechmeister & Kürster (2009) give an analytic solution for the generalisation to a full sine wave fit, taking account of measurement errors and an offset. They include a specific algorithm to search for the period of the best-fit Keplerian orbit to radial velocity data.

---

[6]For double-lined spectroscopic binaries in which two distinct spectra can be measured (not the case for exoplanet orbits), the mass function for each component can be established separately. Then the component mass ratio can also be estimated, but still neither the separate masses, nor the orbit inclination. Further inferences can be made if the system is eclipsing (such that $i \simeq 0$), or if the system can also be resolved astrometrically.

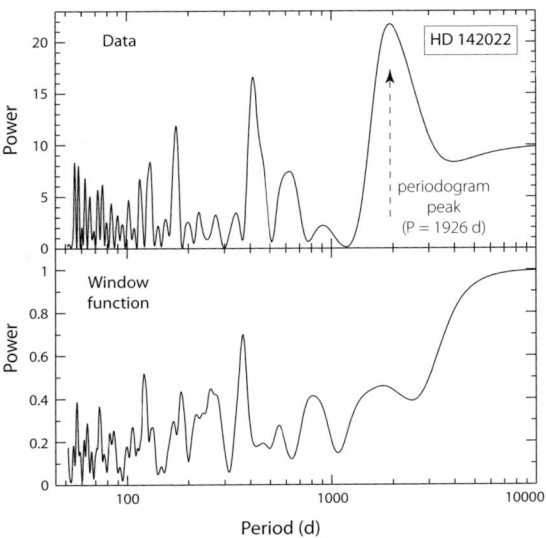

Figure 2.5: The Lomb–Scargle periodogram analysis applied to the radial velocity observations for HD 142022 obtained with the CORALIE échelle spectrograph, La Silla. The power is a measure of the statistical significance of the signal, not of its true amplitude. The window function, determined by the observation epochs, is shown below. The highest peak corresponds to $P = 1926$ d, while the final best-fit solution gives $P = 1928$ d. From Eggenberger et al. (2006, Figure 2), reproduced with permission © ESO.

Multiple planets: kinematic fitting For a system of n_p planets, the total radial velocity signal can be approximated as a linear sum over n_p terms of the form contained in Equation 2.29, giving a total of $5n_p + 1$ Keplerian parameters to be fit, including γ (and optionally d).

This first-order approach to fitting multiple systems considers that the reflex motions of the star caused by each planet are independent, i.e. ignoring the effects of planet–planet gravitational interactions. In this kinematic (or Keplerian) fitting, the dominant planet signal is identified, its Keplerian contribution subtracted from the observational data, and the process repeated until all significant planet signals have been accounted for. A more rigorous multi-planet χ^2 fit to the original data can then be made using these results as starting values. Such an approach was adopted for the five-planet fit to the 18 years of Doppler data for 55 Cnc, with weights assigned to account for signal-to-noise, photospheric motions, and instrument errors (Fischer et al., 2008).

Algorithmic implementation Because the equations describing an astrometric or radial velocity orbit are nonlinear, an unstructured search of parameter space for multiple planets may be computationally prohibitive, and with many false local χ^2 minima. More sophisticated search schemes are therefore desirable.

The Levenberg–Marquardt method is an efficient algorithm for finding a local χ^2 minimum for non-linear models. It varies smoothly between the steepest descent method far from the minimum, and the inverse-Hessian method as the minimum is approached. It has been widely adopted for both radial velocity and astrometric orbit fitting. Implementations are given by Press et al. (2007, mrqmin), while an IDL implementation MPFIT (Markwardt, 2009) applied to radial velocity orbit fitting has been described by Wright & Howard (2009).

Due to sparse sampling, measurement errors, parameter degeneracy and model limitations, there are frequently no unique values of the basic model parameters such as period and eccentricity, and a Bayesian approach can provide more robust estimates of parameter uncertainties (e.g. Brown, 2004; Cumming & Dragomir, 2010). The Markov Chain Monte Carlo method (MCMC) has been used for orbit fitting (e.g. Ford, 2004b; Gregory, 2005; Ford, 2006b; Gregory, 2007a,b), and an implementation specifically tailored to exoplanet radial velocity fitting for a one- or two-planet system, EXOFIT, has been made available by Balan & Lahav (2009).

Various searches have employed genetic algorithms (e.g. Goldberg, 1989; Charbonneau, 1995) to explore the global parameter space, for example in the case of v And (Butler et al., 1999; Stepinski et al., 2000), GJ 876 (Laughlin & Chambers, 2001), 55 Cnc (Marcy et al., 2002), HD 12661 (Goździewski & Maciejewski, 2003, PIKAIA), and μ Ara (Pepe et al., 2007, Stakanof). According to Stepinski et al. (2000) and Goździewski & Maciejewski (2003), the method is inefficient in pin-pointing very accurate best-fit solutions, but provides good starting points for more precise gradient methods such as Levenberg–Marquardt.

Linearisation An alternative to least-squares fitting of Keplerian motions to the radial velocity measurements is to seek linear parameters which can be constructed from certain of the orbital elements.

Konacki & Maciejewski (1999) described a harmonic component analysis which they applied to 16 Cyg. This is based on a Fourier expansion of the Keplerian motion in which the harmonic coefficients are functions of all orbital elements. The coefficients, obtained by linear least-squares, are then used as starting estimates for a local minimisation of the non-linear problem.

Wright & Howard (2009) formulated a linearisation which reduces the search space for n_p planets from the $5n_p+1$ Keplerian parameters to $3n_p$ non-linear variables corresponding to (e, P, t_p) for each planet. In this approach Equation 2.29, generalised to the case of n_p planets, is re-cast as (Wright & Howard, 2009, eqn 4–7)

$$v_r(t) = \sum_{j=1}^{n_p} \left[h_j \cos v_j(t) + c_j \sin v_j(t) \right] + v_0 + d(t-t_0), \quad (2.30)$$

where

$$h_j = K_j \cos \omega_j,$$
$$c_j = -K_j \sin \omega_j,$$
$$v_0 = \gamma + \sum_{j=1}^{n} K_j e_j \cos \omega_j, \quad (2.31)$$

with K_j denoting the semi-amplitudes for each planet.

The non-linear terms (e, P, t_p) are then searched for algorithmically (e.g. using the Levenberg–Marquardt or Markov Chain Monte Carlo methods), with the linear parameters (h, c, v_0), which are transformable back to (ω, K, γ), solved for analytically at each search step.

Multiple planets: dynamical fitting In numerous multiplanet systems, gravitational interaction between two or more planets further modifies the total radial velocity or astrometric signature as a function of time. Such interaction can result in detectable variations in the planetary orbits even over short intervals measured in years. A more complete dynamical (or Newtonian) fit involving N-body integrations must then be used.

Various self-consistent algorithms, which incorporate mutual perturbations in fitting the radial velocity data, were developed for the first resonantly interacting system GJ 876, resulting in substantially improved fits to the radial velocity data (Laughlin & Chambers, 2001; Rivera & Lissauer, 2001; Nauenberg, 2002a; Lee & Peale, 2002). The latter was verified against full hydrodynamic evolution of embedded planets by Kley et al. (2004).

Even good dynamical fits can result in orbital parameter solutions which might be stable over years or decades, but unstable on time scales comparable to the age of the planetary system. Since such formal solutions may be considered as implausible, short-term and long-term dynamical stability is a further constraint that must be satisfied by multiple planet orbit fitting. The issue is considered further in §2.6.

Detectability and selection effects Various studies have been made of planet detectability from radial velocity data as a function of period and signal amplitude (e.g. Nelson & Angel, 1998; Eisner & Kulkarni, 2001b; Cumming, 2004; Narayan et al., 2005).

Cumming (2004) provides analytic expressions for planet detectability as a function of period and eccentricity, and some insight into the detection limits. For short orbital periods, $P \lesssim T$ where T is the duration of the observations, the radial velocity semi-amplitude threshold alone characterises detectability. At long periods, $P \gtrsim T$, the observations cover only part of the orbit by definition, and detectability depends on which part of the orbit is being sampled. If the orbit is close to a velocity maximum/minimum, or to a zero crossing, velocity variations are 'sine-like' or 'cosine-like' respectively, and the velocity variations are

$$\Delta v = K \sin\left(\frac{2\pi T}{P}\right) \approx K\left(\frac{2\pi T}{P}\right), \quad \text{or} \quad (2.32)$$

2.1 Description of orbits

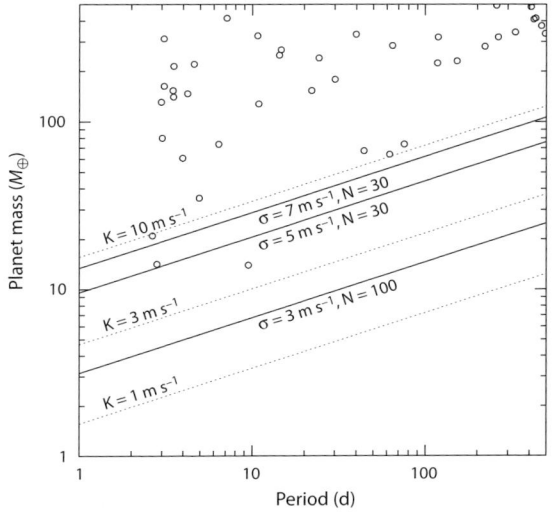

Figure 2.6: Minimum mass limits for a 50% detection threshold as a function of the number of observations, N, and the combined error (measurement and stellar jitter) σ, for $M_\star = 1 M_\odot$ (Equation 2.34). The observation duration is assumed longer than the orbital period. Circles show those planets known in 2005. Dotted lines show various radial velocity semi-amplitudes, K. From Narayan et al. (2005, Figure 7), reproduced by permission of the AAS.

$$\Delta v = K \cos\left(\frac{2\pi T}{P}\right) \approx \frac{K}{2}\left(\frac{2\pi T}{P}\right)^2 \qquad (2.33)$$

respectively. Averaging over phase introduces a dependency on the adopted detection efficiency, ϵ_D: for $\epsilon_D = 0.5$, the amplitude must be large enough that sine-like phases are detected, but cosine-like phases do not have to be, and the velocity threshold scales as $K \propto P$. For $\epsilon_D = 0.99$, almost all phases must be detected, requiring a large amplitude, which then scales as $K \propto P^2$.

While the dependence of K on e given by Equation 2.26 results in large K for highly eccentric orbits, in practice eccentricity acts to make detection more difficult at short periods, where an uneven sampling often results in poor phase coverage during rapid pericentre passages. At longer periods, the increased velocity amplitude and acceleration near the pericentre increase detectability. The transition to the long-period regime occurs for orbital periods $P \approx T/(1-e)^2$. The analysis also allows the completeness of existing surveys to be assessed (Cumming et al., 2008). The results emphasise that there remains a significant selection effect against detecting eccentric orbits for $e \gtrsim 0.6$.

Shen & Turner (2008) found that, once a planet is detected, the eccentricities derived from Keplerian fitting are biased upwards for low signal-to-noise and moderate numbers of observations. They suggest that the numbers of exoplanets with low-eccentricity may be underestimated in current samples. The effect may be evident in results from Keck, for example, where for $P > 10$ d the observed eccentricity distribution is nearly flat for large-amplitude systems, $K > 80$ m s^{-1}, but rises linearly towards low eccentricity at lower amplitudes, $K > 20$ m s^{-1} (Valenti et al., 2009).

Rodigas & Hinz (2009) showed that there is an additional bias due to the presence of an undetected outer companion. For moderate eccentricity $0.1 < e < 0.3$, for example, there is a 13% probability that a modeled eccentric orbit is in fact circular, with the model fit confused by the undetected outer companion.

Narayan et al. (2005) evaluated the detection probability for low mass planets, either in isolation or near mean motion resonance with a hot Jupiter. For a 50% detection rate their expression for the minimum detectable mass is

$$M_{p,\min} \sim 4 M_\oplus \left(\frac{N}{20}\right)^{-\frac{1}{2}} \left(\frac{\sigma}{1\,\mathrm{m\,s^{-1}}}\right) \left(\frac{P}{1\,\mathrm{d}}\right)^{\frac{1}{3}} \left(\frac{M_\star}{M_\odot}\right)^{\frac{2}{3}}, \quad (2.34)$$

for $N \gtrsim 20$, where N is the number of observations, and σ is the quadratic sum of the Doppler velocity measurement error and stellar jitter (Figure 2.6).

Scheduling In a study for the Space Interferometric Mission SIM, Ford (2004a) considered several non-adaptive observing schedules (i.e. schedules fully defined *a priori*) for a targeted astrometric planet search. These included time intervals which are regular periodic (at constant spacing); Golomb ruler[7]; regular power law; regular logarithmic; regular geometric; and random uniform. The efficiency for planet searches was found to be relatively insensitive to the actual observing schedule.

Adaptive scheduling, in which information from previous observations is used to plan future ones most efficiently, can optimise their information content, and can perform significantly better. The objective is to predict future epochs which yield, for example, maximum improvement in orbit parameters and planet masses, the most favourable epochs for distinguishing between alternative models, or an optimum series of future observations. For multiple planet systems, and in particular for resonant orbits, such optimal timings may be concentrated in rather narrow time intervals.

Loredo (2004) outlined a maximum entropy sampling strategy for determining an exoplanet orbit based on Bayesian inference, and Ford (2008) developed a more rigorous adaptive scheduling algorithm exploiting information theory. Based on a small number of initial observations, this proceeds by assuming a prior

---

[7] A Golomb ruler has unique integer intervals between each pair of marks, thus providing the maximum number of unique baselines. The number of marks defines its order; it is 'perfect' if it includes all distances up to its length, and 'optimal' if no shorter ruler of that exists. Thus the (optimal and perfect) Golomb ruler of order 4 and length 6 has marks at 0, 1, 4, 6.

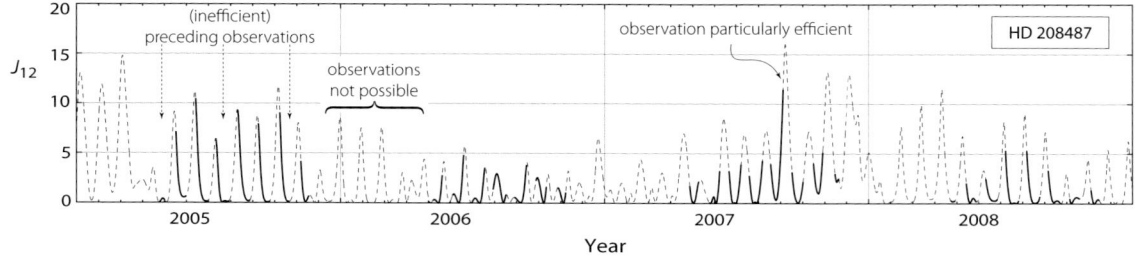

Figure 2.7: Scheduling future radial velocity observations to discriminate between two orbit models for HD 208487. The parts of the curve shown dashed correspond to times when observations are impossible. Observations are optimal when scheduled to coincide with the information peaks J_{12}. Arrows mark the three preceding radial velocity measurements, which in retrospect are seen to have been of little additional information value. From Baluev (2008b, Figure 2), © John Wiley & Sons, Inc.

for the distribution of orbital periods and masses, using Bayesian inference to calculate the posterior probability distribution for all model parameters, using this to calculate the predictive distribution for the radial velocities at some future time, and finally choosing observing times at which additional observations would be most valuable. Monte Carlo simulations demonstrated that such searches are more efficient in terms of detection, and can measure orbital parameters more accurately, than non-adaptive algorithms.

Baluev (2008b, 2010) used criteria from optimal design theory (based on Fisher information, Shannon information, and Kullback–Leibler divergence) to design similar general scheduling formalisms. As an example, Figure 2.7 illustrates the problem of deciding between two orbital solutions based on the 35 measurements of HD 208487 from Butler et al. (2006). Wright et al. (2007) identified periodicity attributable to an additional planet at either 28.6 or 900 d, but their data was inadequate to rule between them due to aliasing.

If the probability densities $p_{1,2}(v)$ describe the distribution of the radial velocity predictions for two possible orbit fits, the expectation of the likelihood ratio statistic considering the first or the second model as true are given by (Baluev, 2008b, eqn 3)

$$I_{2|1} = \int p_1(v) \ln \frac{p_1(v)}{p_2(v)} dv, \quad I_{1|2} = \int p_2(v) \ln \frac{p_2(v)}{p_1(v)} dv \quad (2.35)$$

The maximum information is given by (Baluev, 2008b, eqn 11)

$$J_{12} = I_{2|1} + I_{1|2}$$
$$= -1 + \frac{1}{2}\left(\frac{\sigma_1^2}{\sigma_2^2} + \frac{\sigma_2^2}{\sigma_1^2}\right) + \left(\frac{1}{\sigma_1^2} + \frac{1}{\sigma_2^2}\right)\frac{(v_1 - v_2)^2}{2}, \quad (2.36)$$

under the assumption that the distributions of v_i are close to Gaussian. The largest values of J_{12} then correspond to the most promising times for future observations to rule out one of the alternative models. Although Equation 2.36 is derived by Baluev (2008b) from a formal consideration of information content, its interpretation is straightforward: observations should be made when the two models imply the largest predicted differences in the measured radial velocity, while the uncertainties of these predictions, σ, should be small enough to avoid statistically insignificant differences.

In the case of HD 208487 (Figure 2.7) the radial velocity predictions for the two different orbit models differ by up to $20\,\mathrm{m\,s}^{-1}$. The function J_{12} identifies the epochs which maximise the discrimination between them. It turns out that a single observation at one of the peaks during the 2005 observing season could have decided between the competing models at the $\gtrsim 2\sigma$ level, while the actual observations (indicated by arrows) fell at epochs of low information content, and added little new knowledge to the contested two-planet solution. The subsequent season (2006) offered uniformally limited opportunity for further discrimination.

2.2 Measurement principles and accuracies

2.2.1 Introduction

An instantaneous measurement of the stellar radial velocity about the star–planet barycentre is given by the small, systematic Doppler shift in wavelength of the many absorption lines that make up the star's spectrum.

If, in the observer's reference frame, the source is receding with velocity v at an angle θ relative to the direction from observer to source, the change in wavelength

$$\Delta\lambda = \lambda_{\mathrm{obs}} - \lambda_{\mathrm{em}}, \quad (2.37)$$

is related to the velocity by the expression for the relativistic Doppler shift (e.g. Lang, 1980, eqn 2–226)

$$\lambda_{\mathrm{obs}} = \lambda_{\mathrm{em}} \frac{(1 + \beta\cos\theta)}{(1 - \beta^2)^{1/2}}, \quad (2.38)$$

where $\lambda_{\mathrm{obs}}, \lambda_{\mathrm{em}}$ are observed and emitted wavelengths, and $\beta = (v/c)$. For $v \ll c$ and $\theta \ll \pi/2$, the expression reduces to the classical form

$$v_{\mathrm{r}} = v\cos\theta \approx \left(\frac{\Delta\lambda}{\lambda_{\mathrm{em}}}\right)c, \quad (2.39)$$

2.2 Measurement principles and accuracies

where, conventionally, positive values indicate recession. Special relativistic terms correspond to changes in v_r of several m s^{-1}, and are therefore significant. Equation 2.38 omits the effect of the refractive index of air at the spectrograph, n_{air} (1.000 277 at standard temperature and pressure, STP), which introduces errors of $\lesssim$ 1 m s^{-1} (Marcy & Butler, 1992, eqn 3).

With the radial velocity semi-amplitude values of $K_J \simeq 12.5$ m s^{-1} and $K_\oplus \simeq 0.09$ m s^{-1} as indicative goals, the detection of planets around solar-type stars has demanded long-term radial velocity accuracies of some 15 m s^{-1} or preferably significantly better, corresponding to an accuracy of a few parts in 10^8 in wavelength. This must be maintained over months or years.

High-accuracy radial velocities for exoplanet detection are typically acquired using échelle spectrographs with high spectral resolving power (typically $R \equiv \lambda/\Delta\lambda \sim 50\,000 - 100\,000$), and operated in the optical region (450–700 nm). Many diffraction orders are cross-dispersed, and recorded simultaneously on rectangular format CCDs providing large numbers of resolved absorption lines (Figure 2.8). The principles of an échelle spectrograph are described by, e.g., Vogt (1987) in the case of the Lick Observatory Hamilton spectrometer, by Vogt et al. (1994) in the case of Keck–HIRES, and by Baranne et al. (1996) in the case of ELODIE.

High instrumental stability and accurate wavelength calibration is demanded to minimise effects of gravitational and thermal telescope flexure, and other instrument drifts. Large telescopes and long integration times are still required to achieve the necessary high signal-to-noise, and corresponding sub-pixel accuracies.

Ultra-high spectral resolution Throughput is an important requirement, and ultra-high resolution spectrographs (e.g. the $R = 10^6$ AAT–UHRF, Diego et al. 1995; or the $R = 600\,000$ échelle at Steward Observatory, Ge et al. 2002a) do not provide the efficiencies of $\gtrsim 10\%$ needed for competitive planet searches.

2.2.2 Cross-correlation spectroscopy

Information about the instantaneous Doppler shift is contained in the many thousands of absorption lines present in the high-resolution optical spectrum of solar-type stars. This information can be concentrated into a few parameters by cross-correlation, even at low signal-to-noise levels (Figure 2.9). The technique, using a physical mask, was proposed by Fellgett (1955), first demonstrated by Griffin (1967), and extended to échelle spectroscopy using CORAVEL by Baranne et al. (1979).[8]

---

[8] As Fellgett phrased it, *'It is uneconomical, both in telescope time and in labour of reduction, to observe the details of a stellar spectrum if the sole object of the observation is to measure a radial velocity.'*

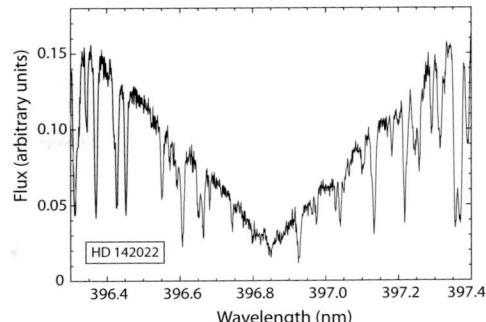

Figure 2.8: The Ca II H absorption line region of the summed HARPS spectra for HD 142022. The figure illustrates the density of absorption lines in 1 nm of the entire 378–691 nm échelle spectrum. From Eggenberger et al. (2006, Figure 1), reproduced with permission © ESO.

Limitations of a fixed physical template, in accuracy and adaptability to spectral type, led to the cross-correlation subsequently implemented numerically as a box-shaped template (Queloz, 1995; Baranne et al., 1996). Finer details of the method have been developed progressively (e.g. Simkin, 1974; Sargent et al., 1977; Tonry & Davis, 1979; Bender, 1990). The essentials are to determine the value of ϵ minimising (Queloz, 1995)

$$C(\epsilon) \propto \int_{-\infty}^{+\infty} S(v) M(v - \epsilon) \, dv \,, \qquad (2.40)$$

where S is the spectrum and M is the mask, both expressed in velocity space v. Associated errors are established from Monte Carlo modeling. Weighting according to the relative line depths further optimises the signal-to-noise, and can also reduce the perturbing effects of telluric lines (Pepe et al., 2002).

The precise shape of the resulting *cross-correlation function* depends on the intrinsic spectral line shapes and on the template line widths, and overall represents a mean profile of all lines in the template. Accordingly, in addition to the radial velocity, the width yields the stellar rotational velocity $v \sin i$, while the equivalent width provides a metallicity estimate if T_{eff} is known approximately (Mayor, 1980; Benz & Mayor, 1981; Queloz, 1995). In the absence of systematic line asymmetry, the underlying shape of the cross-correlation function is well approximated by a Gaussian, with asymmetry reflecting systematic structure in the individual spectral lines.

2.2.3 Deriving radial velocities from Doppler shifts

The measured Doppler shift, Equation 2.37, in practice includes effects other than the line-of-sight velocity of the target star's centre of mass. Contributions from the motion of the observer around the solar system barycentre, due to Earth rotation and orbital motion, are time varying, significant at levels of up to 0.5 and 30 km s^{-1} respectively, and must be accounted for.

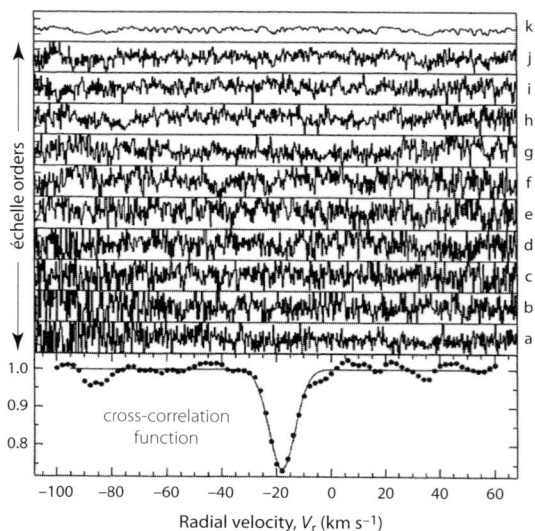

Figure 2.9: Example cross-correlation function for a K0 III star with $S/N \simeq 1$. Observations at $R = 40\,000$ span 411–444 nm in ten orders, each covering 4 nm. Some 1000 lines match the template. The top trace is the same order as that immediately below, but at $S/N = 40$. The cross-correlation function is shown at bottom. From Queloz (1995, Figure 2).

Higher-order relativistic and secular projection effects, and additional spectroscopic line shifts introduced by gravitational redshift and stellar surface effects such as convective flow, pulsations, and star spots (Dravins, 1975; Nidever et al., 2002; Lindegren & Dravins, 2003) may be less critical in determining the relative radial velocities central to exoplanet detection. But there are situations where their contributions must be considered, and all must be accounted for in any *absolute* determination of radial space motion at the $m\,s^{-1}$ level. The magnitude of these effects is as follows.

Effects of Earth motion To detect changes in a star's radial velocity due to an orbiting planet, measured velocities must be referred to a non-moving frame, or at least one with a constant rectilinear space motion, and the solar system barycentre (as opposed to the heliocentre) is consequently adopted as reference.

The time-varying motion of the Earth around the barycentre, which includes effects due to gravitational perturbations from the other planets, is described by the solar system ephemerides provided by JPL (e.g. Konopliv et al., 2006) and IMCCE (Fienga et al., 2008, 2009). By adjusting for known effects, residual velocity terms can be brought below the $1\,m\,s^{-1}$ level (cf. Stumpff, 1980).

Line shifts The contribution from the star's *gravitational redshift* is given by (e.g. Misner et al. 1973, eqn 25.26N; Lang 1980, eqn 2–234)

$$v_r \simeq \frac{GM_\star}{R_\star c}\,. \tag{2.41}$$

This is valid in the Newtonian limit, $R_\star \gg R_S$, where the Schwarzschild radius $R_S \equiv 2GM_\star/c^2$. The contribution amounts to $636\,m\,s^{-1}$ for the Sun, and ranges from $\sim 680\,m\,s^{-1}$ at F5V to $\sim 500\,m\,s^{-1}$ at M5V, and several tens of $km\,s^{-1}$ for white dwarfs.

Convective motions in the photospheres of cool stars lead to spectral line asymmetries (§2.2.7). A net blueshift results from the contribution of photons from the larger and brighter photospheric granules compared with the downward motion in the darker and cooler intergranular lanes. This effect is of order $-0.5\,km\,s^{-1}$ for the Sun, and ranges from about $-1000\,m\,s^{-1}$ at F5V to about $-200\,m\,s^{-1}$ at K0V (Dravins et al., 1981). Effects due to pressure broadening are below about $100\,m\,s^{-1}$ for main sequence stars (Dravins, 1999).

Stellar rotation imposes small radial velocity effects (Gray, 1999), while the contribution of variable meridional flows may be comparable (Beckers, 2007). More critical for high-accuracy relative velocities are effects of surface features such as star spots, which can induce a radial velocity amplitude of a few $m\,s^{-1}$ (e.g. Saar & Donahue, 1997; Hatzes, 2002), resulting in periodic modulation at the stellar rotation period (see §2.2.6).

Stellar space motion Effects of a star's radial motion on the secular evolution of its measured parallax and proper motion are treated by Dravins et al. (1999), and are encapsulated in Equations 3.13. Equivalently, a constant space motion results in a changing radial velocity with time due to the changing projection geometry. For Barnard's star, observed with VLT–UVES over five years (Figure 2.10), Kürster et al. (2006) measured a secular radial velocity acceleration consistent with the predicted value of $4.50\,m\,s^{-1}\,yr^{-1}$ based on the Hipparcos proper motion and parallax combined with the absolute radial velocity of $-110.5\,km\,s^{-1}$ from Nidever et al. (2002).

With knowledge of the star's parallax ϖ, proper motion μ, and systemic radial velocity γ, it is possible to predict this contribution to any secular evolution of the γ velocity which might otherwise be attributed to long-period planetary orbits.

Zero point As a result of these astrophysical as well as instrumental effects, establishing the zero point for absolute radial motions at $\lesssim 50\,m\,s^{-1}$ remains notoriously difficult. Nidever et al. (2002) determined barycentric radial velocities for 889 late-type stars observed at Keck and Lick with typical accuracies of $0.3\,km\,s^{-1}$, and found a difference in zero point of $53\,m\,s^{-1}$ compared with the system of the 38 stable FGK stars adopted as standards by the Geneva group (Udry et al., 1999b,a).

2.2.4 Wavelength calibration

Accurate wavelength calibration is a prerequisite for reaching high radial velocity accuracy. In the early 1970s,

2.2 Measurement principles and accuracies

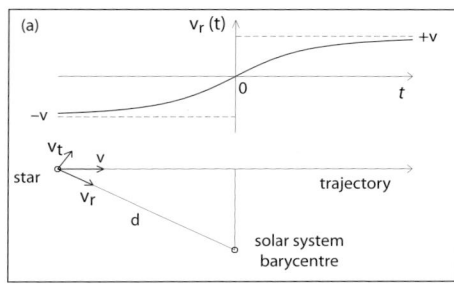
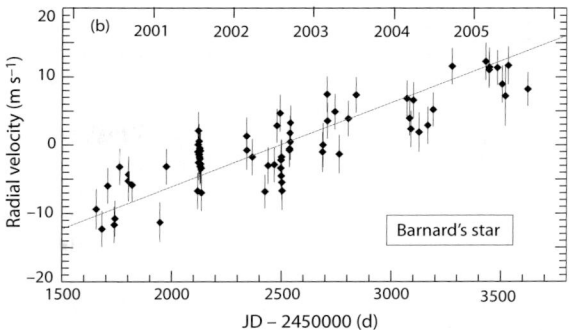

Figure 2.10: (a, bottom) geometry of the space motion of a nearby star: v_t and v_r are, respectively, the transverse and radial components of the space velocity v, and d is the current distance; (a, top) absolute radial velocity as a function of time and its asymptotic limits $\pm v$. (b) barycentric radial velocity measurements of Barnard's star over five years, compared with the predicted secular acceleration (solid line). Adapted from Kürster et al. (2003, Figure 1) and Kürster et al. (2006, Figure 3).

accuracies of around $1\,\mathrm{km\,s^{-1}}$ were limited by photographic plate technology, and by guiding errors at the spectrograph slit which introduced shifts in the stellar spectrum relative to comparison arc lines.

Wavelength calibration was significantly improved by the use of telluric (atmospheric) water vapour lines (Griffin, 1973; Griffin & Griffin, 1973; Walker et al., 1973; Gray & Brown, 2006), largely eliminating errors caused by the different optical paths of the stellar beam and the calibration lamp. Disadvantages include limited spectral ranges where telluric lines of suitable strength are found, in addition to various systematic errors arising from variable path length as a function of zenith distance, producing variations of $20\,\mathrm{m\,s^{-1}}$ per air mass, and from significant winds, producing systematic shifts of $20\,\mathrm{m\,s^{-1}}$ (Gray & Brown, 2006).

Gas cells The use of captive gases to provide a dense and accurate wavelength reference, superimposed on the stellar spectral lines, started with the use of hydrogen fluoride (HF). Although toxic and corrosive, its 3–0 vibration band gave a well-spaced line distribution, with no isotopic confusion, and of similar natural width to those in typical stellar spectra (Campbell & Walker, 1979).

An alternative, iodine (I_2), also mononuclidic, was used by Beckers (1976) and Koch & Woehl (1984) for solar observations, and later by Marcy & Butler (1992) for their precision radial velocity programme at the Lick Observatory 3-m telescope (a retrospective is given by Beckers, 2005). It has a strong line absorption coefficient, and requires a path length of only a few cm. Accuracies improved accordingly, to around $25\,\mathrm{m\,s^{-1}}$ by the early 1990s, and to some $3\,\mathrm{m\,s^{-1}}$ just a few years later (Butler et al., 1996).

The gas cell is placed in the telescope light path, just before the spectrograph slit. Sharp absorption lines of known wavelength are superimposed on the stellar spectrum, thereby providing calibration of both wavelength and the spectrograph point-spread function. For an échelle, the latter is complex with many degrees of freedom (Valenti et al., 1995).

The spectrum through the absorption cell is then modeled as (Marcy & Butler, 1992, and their figure 1)

$$I_{\mathrm{obs}}(\lambda) \propto [I_\star(\lambda + \Delta\lambda_\star)\,T_{I_2}(\lambda + \Delta\lambda_{I_2})] \otimes \mathrm{PSF}\,, \qquad (2.42)$$

where $\Delta\lambda_\star$ and $\Delta\lambda_{I_2}$ are the shifts of the star spectrum and iodine transmission function, determined by least-squares fitting to the composite spectrum, I_{obs}; and $\otimes\mathrm{PSF}$ represents convolution with the spectrograph point-spread function. $I_\star$, via $I_\star \otimes \mathrm{PSF}$, is obtained *a priori* by observing each star without the iodine cell in place. The iodine cell transmission function, T_{I_2}, is obtained from external measurements using a high-resolution Fourier transform spectrometer. The final corrected Doppler shift is then given by $\Delta\lambda = \Delta\lambda_\star - \Delta\lambda_{I_2}$.

Modified calibration sequences have been developed. Iterating from an existing spectral template of a similar star can eliminate the need for an observed stellar template spectrum (Johnson et al., 2006c). Modified calibration sequences to measure binary stars, and in particular double-lined spectroscopic binaries, are described by Konacki (2005b).

The iodine gas cell is now standard in many planet search instruments (including AAT–UCLES, HET–HRS, Keck–HIRES, Lick 3 m, Magellan–MIKE, OAO–HIDES, Subaru–HDS, Tautenburg, TNG–SARG, and VLT–UVES), being of particular relevance for spectrographs which are not intrinsically highly stabilised.

The advantages of the iodine absorption cell are the large number (several thousand) of absorption lines, and the common path of the starlight and the iodine absorption. The measured iodine lines therefore simultaneously track changes in the instrument point-spread function. The disadvantages are the 20–30% loss of light, and the clustering of the absorption line bands in the visible (500–620 nm), making wavelength calibration for redder M dwarfs more problematic.

Thorium–argon calibration HARPS (Mayor et al., 2003), and its predecessor ELODIE (Baranne et al., 1996) employ a thorium–argon emission lamp as the reference wavelength spectrum. In its practical implementation, two optical fibres are used to transfer light to the spectrograph, one collecting the stellar light, the other simultaneously recording either a thorium–argon reference spectrum, or the background sky.

The advantages of the thorium–argon lamp for wavelength calibration are the large numbers of strong emission lines over a wide optical to infrared range (Palmer & Engleman, 1983; Hinkle et al., 2001; Lovis & Pepe, 2007), and the improvement in throughput due to the absence of the iodine absorption cell.

Fibre-optic image scrambling When implemented, a fibre feed from telescope focus to spectrograph offers two further advantages. Firstly, it addresses the variable light illumination of the spectrograph slit, and the associated complications of wavelength calibration. This is because multiple internal reflections within the fibre cause scrambling of the image structure between input and output, decoupling the illumination of the spectrograph optics and detector from guiding errors and calibration source misalignments (Heacox, 1986; Hunter & Ramsey, 1992; Walker et al., 2003b). In addition, the spectrograph can be placed away from the telescope, in a mechanically and thermally stable environment.

The fibre-feed approach, combined with the use of the thorium-argon calibration lamp, was adopted for ELODIE at the OHP 1.9-m telescope (Baranne et al., 1996), and used for the detection of the first exoplanet discovered around a main sequence star, 51 Peg (Mayor & Queloz, 1995). The same concept was subsequently incorporated into HARPS.

Infrared wavelength calibration One issue that has limited high-accuracy radial velocities in the near-infrared is the lack of a suitable wavelength calibrator. Iodine and thorium–argon lines used in the visible, for example, are sparse in the near infrared.

Calibration using imprinted atmospheric lines reaches 10–$20\,\mathrm{m\,s^{-1}}$. For the McMath Fourier transform spectrometer at Kitt Peak, Deming et al. (1987) found variations of order $20\,\mathrm{m\,s^{-1}}$ based on telluric methane. For the $R = 100\,000$ Nasmyth-mounted cryogenic échelle spectrograph CRIRES at the VLT, designed for high-resolution spectroscopy between 0.96–$5.2\,\mu$m, Seifahrt & Käufl (2008) reported telluric N_2O lines near $4.1\,\mu$m stable to around 10–$20\,\mathrm{m\,s^{-1}}$. Both studies used external N_2O gas cells for wavelength calibration.

The use of near-infrared gas cells is in its infancy. Mahadevan & Ge (2009) considered a number of options, and concluded that $H^{13}C^{14}N$, $^{12}C_2H_2$, ^{12}CO, and ^{13}CO together could provide useful calibration in the $1.65\,\mu$m H band. Experiments using an NH_3 cell for VLT-

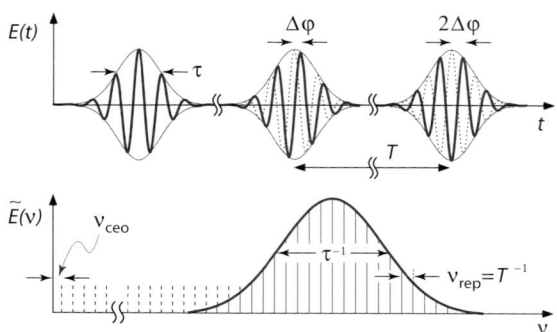

Figure 2.11: Principle of the laser frequency comb. A pulse train (top) produces the optical frequency 'comb' in Fourier space (bottom), with shorter pulse envelopes τ producing broader frequency combs. Within the laser cavity, the round-trip pulse time, T, determines the repetition frequency, $\nu_{\mathrm{rep}} = 1/T$. Dispersive elements cause a difference in the group and phase velocities, shifting the carrier wave with respect to the envelope (top) by $\Delta\varphi$ per pulse. In the frequency domain, the comb shifts by $\nu_{\mathrm{ceo}} = \Delta\varphi/2\pi T$. From Murphy et al. (2007, Figure 1), © John Wiley & Sons, Inc.

CRIRES in the $2.2\,\mu$m K band reach a precision of 3–$5\,\mathrm{m\,s^{-1}}$ over weeks or months (Bean et al., 2010a).

Laser frequency combs The I_2 and Th–Ar wavelength calibration sources depart somewhat from an ideal radial velocity standard. A more robust calibrator would cover the entire optical/infrared range with individually unresolved lines, of uniform spacing and intensity, and with accurately known wavelengths determined by fundamental physics. Laser frequency 'combs', generated from mode-locked femtosec-pulsed lasers (Reichert et al., 1999; Jones et al., 2000a; Udem et al., 2002), seem to offer this possibility (Figure 2.11).

Their application to the calibration of spectrographs for exoplanet detection and other applications, at around the $0.01\,\mathrm{m\,s^{-1}}$ level, has been evaluated (Murphy et al., 2007), with laboratory prototypes demonstrated (Steinmetz et al., 2008; Li et al., 2008a).

The principle is based on the storage of a single laser pulse, maintained on a repetitive path, and circulating in a cavity as a carrier wave. After each circuit, a copy of the pulse is emitted through an output mirror, resulting in an indefinite pulse train, while the energy lost is replenished by stimulated emission in the lasing medium.

The absolute frequencies of the comb lines are set by

$$\nu = \nu_{\mathrm{ceo}} + n\,\nu_{\mathrm{rep}}\,, \qquad (2.43)$$

where $\nu_{\mathrm{rep}} = 1/T$ is the repetition rate of the laser (with T being the round-trip travel time), ν_{ceo} is a carrier-envelope offset frequency which must be calibrated (resulting from non-commensurability of the carrier-wave and repetition frequencies), and n is an integer, with modes given by $n \sim 10^5 - 10^6$. Both ν_{rep} and ν_{ceo} can be synchronised by reference to atomic clocks.

2.2 Measurement principles and accuracies

Table 2.1: Estimated radial velocity jitter, σ'_{rv}, for various spectral types and luminosity classes.

Type	Jitter ($m\,s^{-1}$)	Comment	Reference
F V	7–30	$v_{rot} = 8 - 10\,km\,s^{-1}$	Saar et al. (1998)
"	2–7	inactive stars	Wright (2005)
G V	20–45	$v_{rot} = 8 - 10\,km\,s^{-1}$	Saar et al. (1998)
GK V	4.6 ± 1.8	$v_{rot} \leq 2\,km\,s^{-1}$	Saar et al. (1998)
"	2–5	inactive stars	Wright (2005)
M V	2–7	inactive stars	Wright (2005)
GK IV	5–10	subgiants	Wright (2005)
K III	20	giants	Hekker et al. (2006)

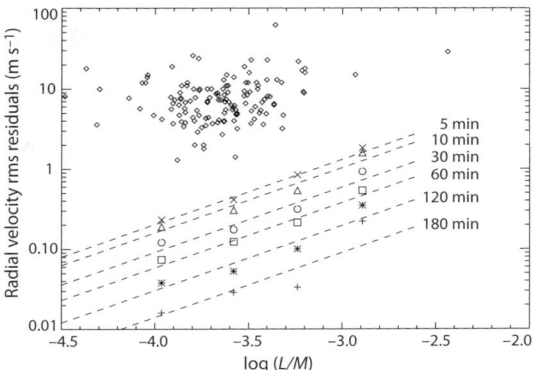

Figure 2.12: Oscillation jitter of AAT–UCLES asteroseismology targets as a function of $\log_{10}(L/M)$ (luminosity–mass ratio) for various simulated integration times. Power laws derived by O'Toole et al. (2008, their eqn 2) are shown as dashed lines. Residual rms values for the sample of known planets from Butler et al. (2006) are shown as small diamonds. From O'Toole et al. (2008, Figure 4), © John Wiley & Sons, Inc.

A laser frequency comb under development at CfA Harvard has been used to calibrate the TRES spectrograph at the Whipple Observatory, and is being further developed for use with HARPS-north (Li et al., 2008a; Latham, 2008b). The prototype uses a mode-locked titanium–sapphire laser operating at ∼ 800 nm, providing $v_{rep} \sim 1\,GHz$. A Fabry–Pérot cavity is introduced to increase the intrinsic line spacing to a more usable 40 GHz. Using GPS time referencing, the comb frequencies are expected to have long-term fractional stability and accuracy better than 10^{-12}, corresponding to a velocity variation below $0.01\,m\,s^{-1}$. By providing a common absolute frequency standard, external time referencing should also allow the direct comparison of measurements made at different observatories.

2.2.5 Exposure metering

Starting with ELODIE (Baranne et al., 1996), many instruments including Keck–HIRES, the Lick Observatory Hamilton Spectrometer, and the Lick Automated Planet Finder (APF) 2.4-m telescope, employ an integral exposure meter to monitor ongoing exposures. This is used to optimise exposure times for a required signal-to-noise, and to calculate the photon-weighted midpoint of each exposure for barycentric correction (Kibrick et al., 2006). HIRES, for example, uses a propellor mirror behind the spectrometer slit to extract a few percent of the light directed to a dedicated photomultiplier tube. APF employs a semi-transparent pellicle mirror, operating at 1 Hz, for autoguiding, exposure metering, and barycentric photon weighting.

2.2.6 Accuracy limits

Improvements in detector technology, wavelength calibration, and various other aspects of instrument design brought the state-of-the-art radial velocity accuracy to some $3-5\,m\,s^{-1}$ by the mid-1990s (e.g. Butler et al., 1996), and to around $0.3-0.5\,m\,s^{-1}$ today (Pepe & Lovis, 2008). Such accuracy represents a displacement of a few nm at the CCD detector, demanding a combination of optical and algorithmic techniques to maintain the required metrology over several years.

Although the radial velocity amplitude is independent of the distance to the star, signal-to-noise considerations limit observations to the brighter stars, typically $V \lesssim 8 - 10$ mag. Equation 2.26 indicates that radial velocity measurements implicitly favour the detection of massive planets, those with small a (i.e. small P) and, with adequate temporal sampling, large e (but see §2.1.1 for detectability in practice).

Error sources presently imposing practical limits on achievable accuracy are of three main types: instrumental terms (currently dominated by a combination of mechanical and thermal stability, and wavelength calibration), stellar noise, and photon noise.

Stellar noise Activity in the stellar atmosphere, stellar oscillations, surface granulation, unrecognised planetary companions, and systematic errors may all contribute 'jitter' to the radial velocity measurements.

Jitter due to inhomogeneities of the stellar atmosphere (spots, plages) is often significant, frequently imposes a limit to the accuracy of radial velocity measurements, and is expected to vary on time scales comparable to the stellar rotation period. It is correlated with stellar chromospheric activity (e.g. Saar & Donahue, 1997; Saar et al., 1998; Butler & Marcy, 1998; Saar & Cuntz, 2001; Tinney et al., 2002; Shkolnik et al., 2005; Jenkins et al., 2006; Desort et al., 2007; Saar, 2009). Emission in the core of the Ca H/K lines is a useful proxy, although active stars seen pole-on can have a high activity index but low radial velocity jitter.

The effects of stellar oscillations on Doppler measurements are typically smaller than those produced by activity, and are most significant for giants and subgiants. O'Toole et al. (2008) made an investigation using data from asteroseismological investigations using AAT–

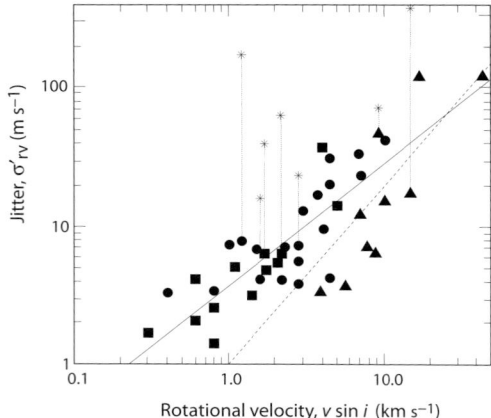

Figure 2.13: Early results from the Lick survey, showing the dependency of astrophysical jitter, σ'_{rv} (Equation 2.44) on stellar rotational velocity, $v \sin i$ for main sequence stars, indicated as F (triangles), G (circles) and K (squares). Stars with planets are plotted before (asterisks) and after (connected by vertical dashed lines) accounting for the planet's contribution. The power-law fits have exponent 0.9 for G and K stars (solid line), and 1.3 for F stars (dashed). From Saar et al. (1998, Figure 2a), reproduced by permission of the AAS.

UCLES. They found a power-law relation between measured noise and both the luminosity-to-mass ratio of the target stars and integration times (Figure 2.12), showing that it is advantageous to average over several p-mode oscillation frequencies (O'Toole et al., 2009b). HARPS results have also shown that 15-min integrations are sufficient to damp resulting radial velocity variations to below 0.2 m s^{-1} (Udry & Mayor, 2008).

Surface granulation can induce variability of order 1 m s^{-1} for solar type stars (Kjeldsen et al., 2005), which may demand measurements over several hours to damp.

The overall effects of jitter can be characterised as an excess in the radial velocity standard error as (Saar et al., 1998; Wright, 2005)

$$\sigma'_{rv} = (\sigma^2_{rv} - \sigma^2_{meas})^{1/2}, \quad (2.44)$$

where σ_{meas} is the contribution of the measurement error to the radial velocity standard error, σ_{rv}, for a given star. σ'_{rv} includes contributions from all of the effects noted above. Astrophysical contributions are dependent on a number of variables, including stellar rotational velocity, age, activity; some indicative estimates are given in Table 2.1 and Figure 2.13.

Photon noise Achievable radial velocity accuracies scale with signal-to-noise. For the HARPS instrument at the ESO 3.6-m telescope, a photon-limited precision of ~ 1 m s^{-1} is reached for a $V = 7.5$ mag G dwarf in around 60 s. In the absence of other noise sources, a single measurement accuracy of 0.1 m s^{-1} requires exposure times ≳ 100 min. Photon-noise limits for CORALIE and HARPS were given by Bouchy et al. (2001).

Earth-mass planets in the habitable zone The sun spot and plage properties of the Sun over one solar cycle, between 1993–2003, have been used to infer the radial velocity curve of a solar-type star exhibiting such features (Lagrange et al., 2010; Meunier et al., 2010). The radial velocity amplitude varies, in a complex way, from a few tenths up to ~ 5 m s^{-1}. Assuming radial velocity accuracies in the range 0.01–0.1 m s^{-1} for a Sun-like star at 10 pc, detection of a $1 M_\oplus$ planet in the habitable zone, between 0.8–1.2 AU, would require weekly sampling over several years, with frequent temporal sampling being crucial. Variations in the Sun's total solar irradiance during cycle 23, as measured by SOHO–VIRGO, is reported by Lanza et al. (2003).

2.2.7 Excluding other sources of periodicity

Effects other than an orbiting exoplanet can result in periodic stellar radial velocity variations, including most of the noise terms considered in the previous section.

Star effects Surface activity such as spots and plages, as well as inhomogeneous convection, can lead to periodic radial velocity variations: a spot with a filling factor of a few percent can induce a radial velocity amplitude of a few m s^{-1} (Saar & Donahue, 1997). Activity may be discounted if the radial velocity period is distinct from the stellar rotation period, or if the star shows only very low level photometric variability at the radial velocity period (although spots are a time-dependent phenomenon). Spectroscopic measurements of the Ca II H and K lines provide a proxy for surface (magnetic) activity (Baliunas et al., 1995), and may reveal rotational modulation attributable to stellar rotation. For close-in planets, magnetic interactions may cause additional stellar activity, as in CoRoT–7 (Lanza et al., 2010).

Stellar pulsations may also cause radial velocity variations, but tend to be discounted if their period and amplitude do not correspond to known excitation mechanisms for the relevant spectral type.

Binary companions Stellar companions can mimic exoplanet signatures, and can be difficult to rule out (see, e.g., HD 41004: Santos et al., 2002b; Zucker et al., 2003, 2004). Hipparcos catalogue astrometry, both the double star annex and the goodness-of-fit statistics for the single-star model (catalogue fields H29–H30), provides a first level of discrimination.

The presence of a lower-level stellar spectrum can also be investigated using TODCOR, a cross-correlation algorithm developed to detect and measure radial velocities of two components of a spectroscopic binary (Mazeh & Zucker, 1994). Its application to HD 41004 is described by Zucker et al. (2003). An extension to triple stellar systems TRICOR/TRIMOR is described, and applied to HD 188753, by Mazeh et al. (2009b).

2.3 Instrument programmes

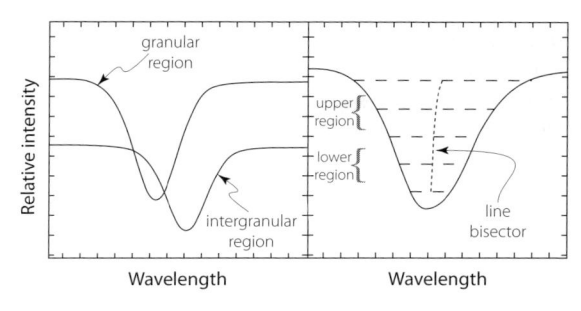

Figure 2.14: Schematic of line profile asymmetry caused by convection, and the resulting line bisector. Left: line profiles resolving individual convection cells would show blueshifted lines from the upflow regions, with redshifted lines from the intergranular downflow lanes. Right: the resulting line profile at low spatial resolution, with a consequently asymmetric line bisector. Adapted from Dravins et al. (1981, Figure 1), reproduced with permission © ESO.

A particularly confusing configuration may arise when a tight binary system orbits the target star. Schneider & Cabrera (2006) suggested that the binary wobble around its own centre of mass can result in the same modulation of the three basic observables (radial velocity, astrometric position, and time of arrival of a periodic signal) as for an orbiting planet. A more detailed analysis by Morais & Correia (2008) showed that the binary actually mimics two planets, rather than a single orbiting planet, but with very similar orbital periods.

Bisector analysis The *spectral line bisector* is the locus of median points midway between equal intensities on either side of a spectral line, thereby dividing it into two halves of equal equivalent width (Figure 2.14). Line bisectors are used to quantify stellar line profile asymmetries, and are important in identifying the underlying cause of certain types of radial velocity variation.

Since the work of Voigt (1956), line bisectors have been used to describe the nature and strength of convection in the Sun, and in star photospheres on the cool side of the granulation boundary where deep convective motion is significant (Dravins, 1975; Dravins et al., 1981; Gray, 1982, 1983; Nowak & Niedzielski, 2008). In such photospheres, local upward motion in the brighter (hotter) convective cells is balanced by a downflow in the darker (cooler) intergranular lanes. In the absence of spatial resolution, averaging over many granules, an overall convective blueshift, of order -0.5 km s^{-1}, results from the dominant contribution of blueshifted photons originating from the larger and brighter granules.

Bisectors due to solar-like granulation are shaped like a distorted 'C', and like the top-half of the letter 'C' for stars somewhat hotter or more luminous than the Sun. They show reversed curvature on the hot side of the granulation boundary, due to the structure of the surface convection zone (Gray, 1989; Gray et al., 2008).

The shape of the spectral line bisector can be quantified in various related ways: (a) the *bisector velocity span* (or 7% span) is constructed as the difference in bisector velocity between upper and lower regions of the line (Figure 2.14), avoiding the wings and cores (Toner & Gray, 1988; Hatzes, 1996; Queloz et al., 2001); (b) the *bisector inverse slope* (Queloz et al., 2001; Santos et al., 2002b) is $v_t - v_b$, where v_t is the mean bisector velocity between 10–40% of the line depth (top), and v_b is that between 55–90% (bottom); (c) the *bisector curvature* is the difference in velocity span between the upper and lower halves of the bisector (e.g. Hatzes, 1996; Nowak & Niedzielski, 2008), for example $(v_3 - v_2) - (v_2 - v_1)$, where v_1, v_2, v_3 are mean velocities at 20–30%, 40–55%, and 75–100% of the line depth (Dall et al., 2006). Various other forms are used (e.g. Dall et al., 2006).

Bisector analysis is usually made directly on the mean line profile given by the cross-correlation function, and is frequently used as an aid in distinguishing planetary signatures from other types of radial velocity modulation (Povich et al., 2001; Martínez Fiorenzano et al., 2005). For a signal of planetary origin the bisector span is expected to be independent of radial velocity, whereas a distinct correlation arises for blended systems (e.g. for HD 41004: Santos et al., 2002b, Figure 3), or for periodic variations resulting from star spots (e.g. for HD 166435: Queloz et al., 2001, Figure 7).

Dall et al. (2006) used HARPS observations of bright stars, both active and inactive, to derive the mean bisectors from individual lines, as well as from the cross-correlation function (Figure 2.15). They showed that different spectral lines can show different bisector shapes, even between lines of the same element, calling for caution in deriving global stellar properties from the bisector derived from the cross-correlation function alone.

2.3 Instrument programmes

A list of radial velocity instruments applied to exoplanet detection and characterisation is given in Table 2.2. It includes both large successful survey instruments, which now typically make use of many tens of nights on each of many telescopes throughout the world, as well as an incomplete list of more modest and less certain plans; it is intended to give a flavour of the activity in this field.

Most instruments are échelle spectrographs; specific 'names' are given when commonly used. The division between 'early' and 'ongoing' is a little arbitrary, but intended to recognise the radial velocity surveys that were being undertaken in advance of the first more secure exoplanet detection announced in 1995.

2.3.1 State-of-the-art in échelle spectroscopy

Of the numerous high-performance radial velocity instruments now operational, this section gives a short description of just two exemplars: HARPS at the ESO 3.6-m

Table 2.2: Overview of spectroscopic instruments applied to radial velocity exoplanet searches. Resolving power, $R = \lambda/\Delta\lambda$, and accuracy σ (in m s^{-1}), are indicative, the latter particularly dependent on magnitude (... → ... indicates that accuracies have improved accordingly over the project's development life). References cover a mix of early and more recent results, and instrumental descriptions, intended to provide an entry to the more detailed literature. The East–Asia network comprises OAO/BOAO/Xinglong (Izumiura, 2005). * instruments use Michelson interferometry and external dispersion (§2.3.2).

Telescope–instrument		Lead institute	R	σ	Start	Reference
Early:						
CFHT 3.6-m	échelle	DAO/UBC	100 000	15	1980	Campbell et al. (1988), Walker et al. (1995)
KPNO 0.9-m	échelle	Kitt Peak	74 000	4	1987	McMillan et al. (1990), McMillan et al. (1994)
HJS 2.7-m	échelle	U. Texas/McDonald	60 000	4–7	1988	Cochran & Hatzes (1994), Endl et al. (2008)
Lick 3-m	échelle	Lick Observatory	40 000	10 → 3	1992	Marcy & Butler (1992), Cumming et al. (1999)
ESO CAT 1.4-m	CES	ESO planet search	100 000	20	1992	Hatzes et al. (1996), Kürster et al. (1999)
OHP 1.9-m	ELODIE	France–Switzerland	42 000	15	1993	Mayor & Queloz (1995), Baranne et al. (1996)
Whipple 1.5-m	AFOE	SAO–HAO	70 000	5	1994	Brown et al. (1994), Noyes et al. (1997)
Ongoing:						
Keck 10-m	HIRES	Keck	85 000	3 → 1	1996	Vogt et al. (1994), Vogt et al. (2000)
AAT 3.9-m	UCLES	Anglo–Australian	80 000	3	1998	Tinney et al. (2001)
Euler 1.2-m	CORALIE	Geneva	50 000	7 → 3	1998	Queloz et al. (2000b), Tamuz et al. (2008)
HET 9.2-m	HRS	U. Texas/McDonald	60 000	3	1998	Tull (1998), Cochran et al. (2007)
ESO VLT 8-m	UVES	ESO	47 000	30	2000	Dekker et al. (2000), Joergens (2008)
TNG 3.6-m	SARG	INAF, Italy	46 000	3	2001	Gratton et al. (2004)
Tautenburg 2-m	échelle	Thüringer LS	67 000	3–10	2001	Hatzes et al. (2003b) Esposito et al. (2006)
OAO 1.9-m	HIDES	Okayama	70 000	6	2001	Sato et al. (2005b), Sato et al. (2008a)
ESO 3.6-m	HARPS	Geneva	115 000	0.3–1	2003	Mayor et al. (2003), Mayor et al. (2009a)
MPG 2.2-m	FEROS	Max Planck–ESO	48 000	10	2003	Kaufer et al. (1999), Setiawan et al. (2008b)
Magellan 6.5-m	MIKE	Magellan	50 000	5	2004	López-Morales et al. (2008), Minniti et al. (2009)
Subaru 8.2-m	HDS	NAOJ	60 000	4–5	2004	Noguchi et al. (2002), Sato et al. (2009)
BOAO 1.8-m	BOES	Bohyunsan, Korea	90 000	6	2004	Izumiura (2005), Kim et al. (2006)
Xinglong 2.2-m	CES	China	44 000	30	2005	Izumiura (2005)
OHP 1.9-m	SOPHIE	OHP, replaces ELODIE	75 000	3	2006	Perruchot et al. (2008)
Sloan 2.5-m	MARVELS*	SDSS/Keck	11 000	3–20	2008	van Eyken et al. (2007), Ge et al. (2009)
Magellan 6.5-m	PFS	Carnegie	38 000	1	2009	Crane et al. (2008)
Mt Abu 1.2-m	échelle	PARAS, India	70 000	3–5	2009	Chakraborty (2008)
APF 2.4-m	APF	Lick	80 000	1	2009	Kibrick et al. (2006)
WHT 4.2-m	HARPS–N	SAO–Geneva	115 000	0.3–1	2010	Latham (2008b)
Near infrared:						
ESO VLT 8-m	CRIRES	ESO (1–5 μm)	100 000	10–20	2007	Käufl et al. (2004), Bean et al. (2010a)
Palomar 5-m	TEDI*	Berkeley (0.9–2.4 μm)	20 000	5	2007	Edelstein et al. (2007)
Gemini 8.1-m	PRVS	Gemini (0.9–1.7 μm)	50 000	10	study	Ramsey et al. (2008), Jones et al. (2008c)
Future:						
ESO VLT 4×8-m	ESPRESSO	ESO	140 000	0.1	2014?	Liske et al. (2009), Pasquini et al. (2009)
E–ELT 42-m	CODEX	ESO	150 000	0.01	2016?	Pepe & Lovis (2008), Pasquini et al. (2008a)
TMT 30-m	MIRES	TMT (2.5–10 μm)	120 000	–	2018?	Tokunaga et al. (2006)
GMT 7×8.4-m	SHARPS	SAO	150 000	–	2018?	Latham (2008b), Johns (2008a)

telescope at La Silla, and HIRES at the Keck I 10-m telescope in Hawaii.

HARPS HARPS (High Accuracy Radial Velocity Planet Searcher) built on experience gained with ELODIE and CORALIE, and was designed to reach accuracies of around 1 m s^{-1} (Mayor et al., 2003). It has been operating at the ESO 3.6-m telescope, La Silla, since 2003 (Rupprecht et al., 2004). It is an échelle spectrograph with $R = 115\,000$, and operates over a spectral range 378–691 nm distributed over échelle orders $N = 89-161$. The detector consists of two CCDs (totaling 4k×4k, 15 μm pixels), with one spectral order ($N = 115$, 530–533 nm) lost in the gap between the two.

The spectrograph is housed in a vacuum vessel, thermally controlled to a few mK to minimise spectral drifts due to temperature and air pressure variations. It is fed by two fibres, one collecting the stellar light, the other simultaneously recording either a thorium–argon reference spectrum, or the background sky. The fibres have an aperture on the sky of 1 arcsec, with image scrambling providing a uniform spectrograph pupil illumination independent of decentering.

In 60 s integration it produces a signal-to-noise ratio of 110 per pixel at 550 nm for a $V = 6$ mag G2V star, and a photon noise error of about 0.9 m s^{-1}. Errors introduced by guiding, focus, and instrumental uncertain-

2.3 Instrument programmes

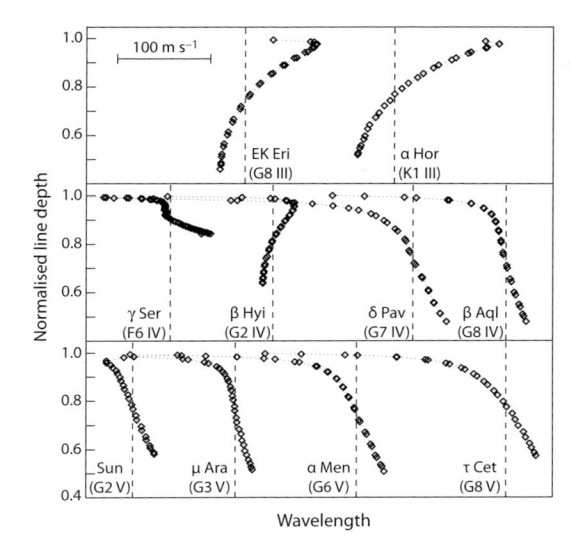

Figure 2.15: *Mean bisectors, constructed from the cross-correlation function, of various bright stars according to luminosity class, III (top row) to V (bottom). Vertical dashed lines indicate the radial velocity. Error bars are the size of the plot symbols or smaller. From Dall et al. (2006, Figure 2), reproduced with permission © ESO.*

ties lead to typical long-term radial velocity accuracies of about $1\,\mathrm{m\,s^{-1}}$ rms for spectral types later than G, and for non-rotating stars with $v \sin i < 2\,\mathrm{km\,s^{-1}}$. Short-term precision of $0.2\,\mathrm{m\,s^{-1}}$ and long-term precision of 0.3–$0.6\,\mathrm{m\,s^{-1}}$ have been achieved (Pepe & Lovis, 2008).

Lovis et al. (2008) summarise the results from the first three years of operations, and detail the global error budget taking account of guiding accuracy and photon noise. The HARPS guaranteed time high-precision sample is presented in Sousa et al. (2008).

A copy of HARPS (HARPS–north) is now installed at the WHT 4.2-m telescope at La Palma (Latham, 2008b).

Keck–HIRES Another example of high-accuracy radial velocity instrumentation is the Keck I 10-m telescope on Mauna Kea, Hawaii, equipped with the HIRES échelle spectrometer. HIRES was one of the five first-light instruments for the Keck I telescope, designed and built at UCO/Lick Observatory (Vogt et al., 1994). The project started in 1988, and achieved first light in 1993. While described as a conventional échelle spectrometer, it was designed to have a relatively large order separation, at the expense of overall wavelength coverage, to allow both for accurate sky subtraction (for faint objects), and for image slicing (for bright objects).

In the configuration described by Vogt et al. (2000), the resolving power is $R = 80\,000$, and the spectra span a wavelength range from 390–620 nm. Wavelength calibration uses an iodine absorption cell.

Given the number and quality of the exoplanet discoveries made by Keck–HIRES (Table 2.3), it is interesting to note that amongst the first-light science originally foreseen (including quasar absorption lines, beryllium in the early universe, lithium abundances, and asteroseismology), exoplanet detection and characterisation did not figure (Vogt et al., 1994).

2.3.2 Externally dispersed interferometry

Background Various interferometric-based techniques have been applied to spectroscopy and radial velocity measurements in the past, although not for exoplanets. In *Fourier transform spectroscopy*, for example, light enters a Michelson interferometer through a beam-splitter. The resulting interference pattern, in frequency space, depends on the path delay within the interferometer, and the recorded intensity is a measure of the light's temporal coherence. One mirror is moved to introduce a variable path length, the temporal coherence is measured as a function of this path delay, $I(\delta)$, and the spectrum is recovered by Fourier transform of $I(\delta)$. Fourier transform spectroscopy provides for very high spectral resolution, but only for very bright sources due to the inefficient use of photons which are sampled only sequentially at each path delay setting.

Fixed delay interferometers with (very) narrow band passes isolating a single spectral line have used this principle for solar measurements since the 1980s, reaching radial velocity accuracies of around $3\,\mathrm{m\,s^{-1}}$ (e.g. Kozhevatov et al., 1995, and references). Other approaches have employed a more extended optical delay path range within the interferometer, without resorting to physical scanning, for example using grisms or holographic gratings (*holographic heterodyne spectroscopy*) in place of the mirrors (e.g. Connes, 1985; Frandsen et al., 1993; Douglas, 1997).

The combination of a Michelson interferometer with post-disperser was suggested for spectroscopic applications already by Edser & Butler (1898). It entered the exoplanet literature with the work of Erskine & Ge (2000), where it was described as a *fringing spectrometer*. Subsequent developments have been described under the epithets of a *dispersed fixed-delay interferometer* or an *externally dispersed interferometer* (Ge, 2002; Ge et al., 2002b; Erskine, 2003; Erskine et al., 2003; van Eyken et al., 2004; Mahadevan et al., 2008b).

Principles In this approach, the high-resolution échelle spectrograph central to other radial velocity instruments (§2.2) is replaced by a fixed-delay 'field widened' (or field compensated) Michelson interferometer, allowing the useful solid angle accepted by the interferometer to be greatly enlarged (see, e.g., Hilliard & Shepherd, 1966). Spatial fringes at a given interference order result from the different path lengths of the input beam through the interferometer. The interferometer output enters a low- to medium-resolution spectrograph with $R \sim 10$–$30\,000$. Doppler measurements are

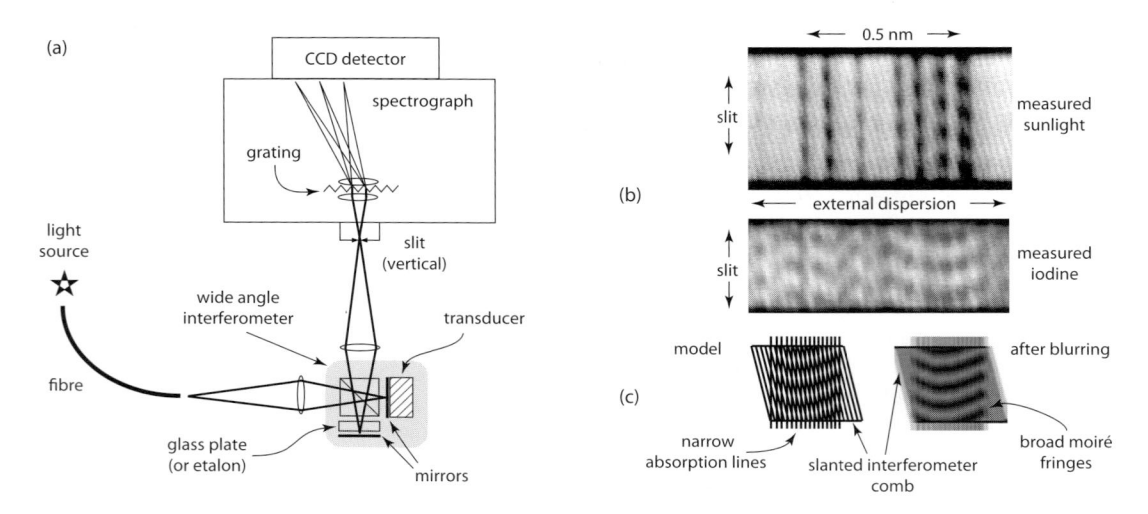

Figure 2.16: Principle of an externally-dispersed interferometer: (a) a wide-angle Michelson interferometer with a fixed internal delay generates spatial fringes which are input to a low-resolution spectrograph which disperses the light in the orthogonal direction, creating a fringing spectrum at the CCD detector. The glass plate, or etalon, serves as the field widening element (e.g., Hilliard & Shepherd, 1966). The transducer steps the interferometer delay in four quarter-wave increments to remove non-fringing artefacts in multiple exposures; (b) fringe pattern resulting from illumination with sunlight (above) and iodine (below), in which the underlying phase-slanted interferometer comb is partially resolved; (c) schematics illustrating the formation of moiré patterns, in which the high detail spectral features are heterodyned to lower resolution where they survive the blurring of the external spectrograph. From Erskine (2003, Figures 1 and 5), reproduced by permission of University of Chicago Press.

then made by monitoring the phase shifts of fringes created spatially at the output of the Michelson interferometer, rather than the tracking of absorption line centroids as in échelle spectroscopy (Figure 2.16).

High Doppler sensitivity is achieved by optimising the fixed optical delay within the interferometer to maximise the combination of fringe contrast and the number of photons per fringe. Ge (2002) selected a total internal path delay of $d \sim 11$ mm, meaning that the interferometer is working (in a number of orders) around $m = d/\lambda \approx 20\,000$. Accordingly, a Doppler shift of 1 part in 20 000 results in one fringe (2π rad) of phase shift.

Sensitivity is further and substantially enhanced by the simultaneous measurement of numerous fringes over a broad wavelength range, which is achieved by coupling the interferometer output to a low- to medium-resolution spectrometer, which disperses the light orthogonal to the fringes. The wavelength range of 34 nm in the system used at Lick by Ge et al. (2002b) is dictated by the post-disperser resolution and CCD dimensions. The function of the disperser being simply to separate the otherwise superimposed fringes, the Doppler error is essentially independent of its resolution.

The output is recorded on a CCD, with fringes sampled across a restricted number of pixels in the direction of the spectrometer slit, and the spectrum sampled in the dispersion direction. In the design described by Ge et al. (2002b), for example, one fringe is sampled by 20 pixels in the spatial direction, and four pixels in the dispersion direction. The interferometric orders themselves provide a frequency comb which extends over the entire bandwidth, thus generating frequency markers analogous to the fiducial lines of an absorption cell, but uniform in spacing, shape, and amplitude.

Operation in 'phase uniform' mode involves stepping the selected interferometer delay in four quarter-wave increments to remove non-fringing artefacts in multiple exposures. In 'phase slanted' mode, the fringes are slanted with respect to the dispersion direction by tilting one of the interferometer mirrors so that the delay has a linear component transverse to the dispersion direction (Erskine, 2003). Moiré fringes then result from the underlying fringe comb generated by the interferometer, and the emission/absorption lines of the source. The combination effectively heterodynes (down converts) the high spatial frequencies of the spectrum to low spatial frequencies resolved by the disperser.

The accuracy of measured Doppler shifts appears competitive with existing high-resolution spectrometers, but with the potential of reducing the manufacturing complexity of the échelle spectrograph, improving mechanical stability, simplifying the point-spread function, improving throughput, and extending the useful wavelength range. Furthermore, by replacing the multi-order cross-dispersed échelle spectrum with a single lower resolution spectrum spread along just one dimension of the CCD (cf. Figure 2.9), the orthogonal dimension can be used for spectra of other objects. These can be fibre-fed to the (wide-angle) interferometer entrance, resulting in a multiplexing multi-object capability.

Instrument developments A prototype instrument (exoplanet tracker) was used to confirm the known exoplanet, 51 Peg, at the KPNO 2.1-m by van Eyken et al. (2004), and as the basis of the W.M. Keck exoplanet tracker instrument (Mahadevan et al., 2008a,b). The first new planet discovered using this technique was HD 102195 (ET1), made at the KPNO 0.9-m, and confirmed with the KPNO 2.1-m and the HET–HRS (Ge et al., 2006). The planet has $P = 4.1$ d and $M_p \sin i = 0.49 M_J$.

The concept also underlies the MARVELS survey at Apache Point Observatory (Multi-object APO Radial-Velocity Exoplanet Large-area Survey). MARVELS forms part of the six-year Sloan Digital Sky Survey SDSS III extension, operating between 2008–14. Its scientific goal is to monitor 11 000 $V = 8 - 12$ stars over 800 sq. deg. for six years. The survey targets 90% F8 and later main sequence stars and subgiants, and 10% G and K giants with $V = 7.6 - 12$ (van Eyken et al., 2007; Ge & Eisenstein, 2009; Ge et al., 2009). The instrument acquires radial velocities, at 3–20 m s^{-1} depending on V magnitude, for 60 objects simultaneously in the 3° field, by lining up the distinct low-resolution spectra across the CCD (Wan et al., 2006). A trial survey of 420 $V = 8 - 12$ mag solar-type stars in 2006 was reported by Ge et al. (2007).

An All-Sky Extrasolar Planet Survey (ASEPS), using wide-field telescopes to monitor stars to $V = 12 - 13$ mag, might increase the number of exoplanets by a factor of 10–100 (Ge, 2007; Ge et al., 2007). Observing strategies for large-scale multi-object surveys have been simulated by Kane et al. (2007).

Extension to the infrared The same technique can be used at longer wavelengths. The TEDI instrument (TripleSpec Exoplanet Discovery Instrument) extends the concept to the near infrared (0.9–2.4 μm) by coupling the wide-angle Michelson interferometer to the existing $R = 3000$ TripleSpec échelle spectrometer at the Palomar 5-m telescope, whose spectral resolution alone is insufficient for exoplanet radial velocity measurement (Edelstein et al., 2007, 2008). The instrument is used in phase uniform mode, and both interferometer outputs are relayed to the spectrograph.

A similar development is underway at APO.

Dispersed FTS Broadly similar principles are used in the *dispersed Fourier transform spectrometer* under development at the USNO (Hajian et al., 2007). But rather than operating at a fixed delay position, the interferogram of the Fourier transform spectrometer is coarsely sampled over a wide range of delay positions, so that a complete high-resolution broad-band spectrum can be reconstructed. The addition of a post-disperser essentially creates several thousand separate narrow-band Fourier transform spectrometers operating in parallel.

2.3.3 Future developments

Planned instruments CODEX is an instrument concept for the European–ELT (Extremely Large Telescope). It aims to detect the expansion of the Universe directly, by measuring the Doppler shift of high-redshift quasar Ly-α absorption lines as a function of time. The experiment targets a Doppler accuracy of around 0.02 m s^{-1} maintained over several decades, which would also have direct application to exoplanet studies (Pepe & Lovis, 2008; Pasquini et al., 2008a,b).

The design has developed from HARPS, and incorporates high stability, slanted volume phase holographic gratings with $R \gtrsim 120\,000$, a wavelength range 370–720 nm, fibre-optic scrambling to reduce the impact of guiding errors (where 0.01 m s^{-1} corresponds to 10μas on the sky), and wavelength calibration using laser frequency combs. The schedule is tied to that of E–ELT, with first light currently envisaged for 2016.

SHARPS is a spectrograph designed for the proposed Giant Magellan Telescope, planned for completion in 2018, with similar scientific objectives.

SAO are developing new instruments for follow-up of the Kepler mission, including the multi-object Hectochelle (Szentgyorgyi & Furész, 2007).

Alternative approaches A different approach to determining radial velocity variations is to measure accelerations directly. Connes (1985, 1994) proposed a system in which the stellar light, and light from a white light source, are passed alternately through a Fabry–Pérot etalon. A feedback loop adjusts the etalon spacing to achieve coincidence. A tunable laser tracks the Fabry–Pérot spacing, and a measurable beat signal is generated from a stabilised laser. The objective is to lock onto the stellar lines, and to monitor fluctuation using terrestrial frequency standards, providing the absolute acceleration of the source with respect to the observer.

Connes (1985) estimated that for a $P = 10$ yr orbit, a 1 m telescope observing a $V = 10$ mag star for 10 yr and an observing time of 10×1 hr per year, would reach an acceleration amplitude of 2×10^{-5} m s^{-2}, corresponding to a radial velocity error of 0.1 m s^{-1}. A laboratory system was reportedly built and tested (Schmitt, 1997; Bouchy et al., 1999).

Riaud & Schneider (2007) have proposed extracting the planet's radial velocity signal from the residual stellar flux halo remaining in coronagraphic adaptive optics imaging in combination with the planned extremely large telescopes. The radial velocity signal is estimated to be detectable within the residual contrast ratio of $10^3 - 10^4$, compared to the contrast ratio of 10^{10} in the absence of a coronagraph.

2.4 Results to date

2.4.1 The first radial velocity exoplanets

Early radial velocity surveys, on a relatively small number of stars, were primarily aimed at characterising the sub-stellar/brown dwarf mass function by searching for binary companions of main-sequence stars with masses below 1 $M_\odot$ (Campbell et al., 1988; Marcy & Moore, 1989; Marcy & Benitz, 1989; McMillan et al., 1990; Duquennoy & Mayor, 1991; Tokovinin, 1992). Some were part of programmes to establish improved radial velocity standards for the IAU (Latham et al., 1989).

As accuracies improved towards plausible planetary signals of around 10–20 m s^{-1}, existing groups intensified their efforts, and others started new observing programmes, leading to the monitoring of many more stars over a number of years.[9]

The first radial velocity detections were announced cautiously, and only substantively confirmed some years later. Thus Campbell et al. (1988) identified a possible $P = 2.7$ yr, $1.7 M_J$ object around γ Cep, parameters which were subsequently questioned (Walker et al., 1992), but which were eventually confirmed by the 1981–2002 study of Hatzes et al. (2003a). Latham et al. (1989) reported a $P = 84$ d, $11 M_J$ companion to HD 114762, which they suggested was a probable brown dwarf. These values were confirmed by Cochran et al. (1991), and further refined by Butler et al. (2006). Hatzes & Cochran (1993) reported a possible $P = 558$ d, $2.9 M_J$ companion to the K giant HD 62509 (β Gem), parameters again substantially confirmed in the 25-year baseline study by Hatzes et al. (2006).

The discovery of 51 Peg b The discovery of a very short-period $P = 4.2$ d ($a = 0.05$ AU) $0.47 M_J$ planet surrounding the star 51 Peg, was announced by Mayor & Queloz (1995). The discovery was promptly confirmed by the Lick Observatory group, who were also quickly able to report two new planets around stars that they had been monitoring: 70 Vir (Marcy & Butler, 1996) and 47 UMa (Butler & Marcy, 1996). The compelling realisation that planetary mass objects existed around main sequence stars marked the start of a substantive and world-wide acceleration in exoplanet research.

The reality of 51 Peg b was the subject of some early and intense controversy. In part this was motivated by its unexpectedly short orbital period and close proximity to the parent star. But an alternative explanation – that the radial velocity shifts arose from non-radial oscillations – was also put forward to explain possible distortions in the absorption line bisector. Studies that followed (Gray, 1997; Hatzes et al., 1997; Marcy et al., 1997; Gray & Hatzes, 1997; Willems et al., 1997; Brown et al., 1998a,b; Gray, 1998; Hatzes et al., 1998a,b) finally resulted in a consensus that the planet hypothesis was the most reasonable.

2.4.2 The present radial velocity census

Apart from the extreme planetary system around the pulsar PSR B1257+12, and until the first transiting planet discovered by photometry in 2003 (OGLE–TR–56: Konacki et al., 2003a), almost the first hundred planets around normal main-sequence stars were discovered using radial velocity techniques. The number of radial velocity discoveries is still rising strongly, as enlarged surveys, higher measurement accuracies, and longer temporal baselines take effect (Table 2.3).

The various measurement groups have now surveyed ~ 2500 stars with masses in the range $0.3 - 2.5 M_\odot$, and brighter than $V = 12$ mag (Marcy et al., 2008), while many more are being added to new survey programmes. As described in subsequent sections, and summarised in Table 2.4, the initial focus on main sequence stars has progressively expanded to encompass surveys which include various other stellar types and environments.

Overall results As of 2010 November 1, there were 494 known exoplanets, of which 461 in 390 systems (45 multiple) had measured radial velocity orbits, and 358 had been discovered by radial velocity measurements (Appendix C). Figure 2.17 shows the planets discovered from radial velocity measurements by year, as a function both of M_p (ordinate) and semi-major axis a (circle size). The decreasing lower mass bound with time reflects the improving instrumental accuracy leading to smaller detectable values of K. For a given planet mass, $K \propto P^{-1/3}$ (Equation 2.26), which explains their preferentially smaller a.

Figure 2.18 illustrates the region probed by the radial velocity discoveries, compared with transit and astrometric detections at current and projected accuracies.

Figure 2.19 shows the number of exoplanets discovered to date by radial velocity monitoring as a function of spectral type.

2.4.3 On-line compilations

There are numerous on-line resources and compilations related to exoplanet surveys and results (cf. Table 1.3). Two are particularly useful in the context of radial velocity exoplanet detections, since they maintain up-to-date records of new discoveries, compilations of related parameters, and a variety of tools for their inspection and analysis, viz:

---

[9] Walker (2008) provides a recent perspective on the radial velocity exoplanet searches carried out in the 1970s–1980s, and states *'It is quite hard nowadays to realise the atmosphere of scepticism and indifference in the 1980s to proposed searches for extra-solar planets. Some people felt that such an undertaking was not even a legitimate part of astronomy'*.

2.4 Results to date

Table 2.3: *Number of planets discovered by specific radial velocity instruments, based exclusively on the compilation at* exoplanets.org *as of 2010–11–01. This listed 317 planets actually discovered by radial velocity measurements (i.e. an incomplete compilation), compared to 461 with measured radial velocities and 494 discovered by all methods given by* exoplanet.eu *at the same date. A planet discovered by observations at N observatories, is counted for 1/N discoveries by each, and final numbers are rounded.*

Observatory/instrument	Number
Keck–HIRES	101
ESO 3.6–HARPS	50
Euler–CORALIE	34
Lick (Shane/APF)	32
AAT–UCLES	28
OHP–ELODIE	15
HET–HRS	13
Okayama–HIDES	10
Magellan–MIKE	7
Tautenburg	6
Subaru–HDS	5
OHP–SOPHIE	3
McDonald HJS	3
Whipple–AFOE	2
Bohyunsan–BOES	2
CFHT (from 1988)	1
ESO–CES	1
ESO–FEROS	1
KPNO–ET	1
Oak Ridge+Coravel (from 1989)	1
Xinglong–CES	1

http://exoplanet.eu, the Exoplanet Encyclopedia, is a compilation of all exoplanet announcements, data, and bibliography, maintained by Jean Schneider at the Observatoire de Paris–Meudon. The data base can be interrogated, for example, according to radial velocity measurements alone.

http://exoplanets.org, the Exoplanet Orbit Database, includes results from radial velocity and transit surveys for stars within 200 pc, aiming to incorporate the most secure spectroscopically-measured orbital parameters. It updates the Catalogue of Nearby Exoplanets (Butler et al., 2006; Jones et al., 2008a), and is maintained by Jason Wright and Geoff Marcy.

2.4.4 Main sequence stars

G and K dwarfs Most of the early Doppler planet searches concentrated on G and K main sequence stars (dwarfs), in the mass range $\sim 0.7 - 1.3 M_\odot$. They have numerous absorption lines, are relatively bright, are relatively slow rotators with low rotational broadening, and have relatively stable atmospheres, with photometric jitter for inactive G dwarfs extending down to 2–3 m s^{-1} (Table 2.1). Early reviews of these radial velocity searches were given by Latham (1997); Butler & Marcy (1998); Latham et al. (1998); Marcy & Butler (1998b); Nelson & Angel (1998); and Marcy & Butler (2000).

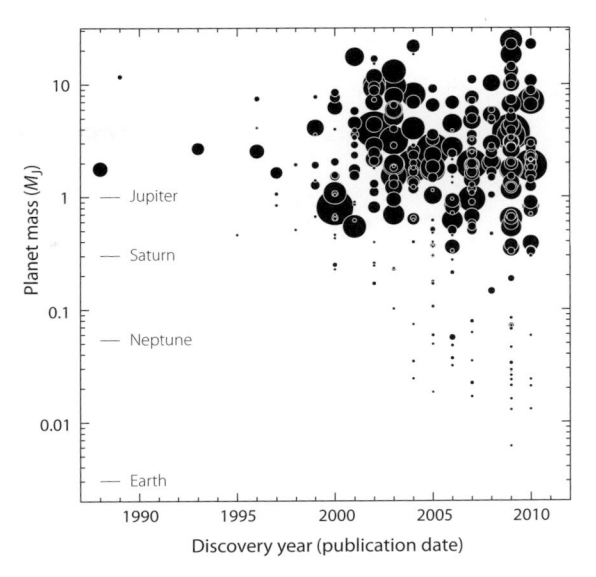

Figure 2.17: *Planets discovered by radial velocity measurements, according to mass and year of discovery. Circle sizes are proportional to the semi-major axis a. Data are for 383 planets from* exoplanets.org, *2010–11–01.*

Programmes have collectively expanded to cover most late-type main sequence stars brighter than $V \sim 7.5 - 8.5$ mag in a systematic manner. Of the larger programmes (Table 2.4), CORALIE is surveying about 1600 stars in the southern hemisphere (Tamuz et al., 2008). In the north, the ELODIE survey of more than 1000 targets focused on metal-rich stars (e.g. da Silva et al., 2006, 2007). ELODIE was replaced by SOPHIE (following the HARPS design) in 2006, and the same programme has continued under the SOPHIE exoplanet consortium (e.g. Santos et al., 2008). In parallel, the California & Carnegie search program uses the Keck, Lick, and AAT telescopes to survey about 1000 stars in the north and south (Valenti & Fischer, 2005).

The N2K consortium is using the Keck, Magellan, and Subaru telescopes to survey the 'next 2000' stars (Fischer et al., 2005; López-Morales et al., 2008). Estimates of $T_{\rm eff}$, [Fe/H], and binarity were constructed from a starting list of more than 100 000 FGK dwarfs (Ammons et al., 2006; Robinson et al., 2006b). Their primary target list was then constructed from the more than 14 000 main sequence and sub-giant stars with $d <$ 110 pc, $V < 10.5$ and $0.4 < B - V < 1.2$. Their final selection of 2000 stars was also biased towards higher metallicity targets, using broad-band photometry to define a subset with [Fe/H] > 0.1.

A large survey, of $\sim 10\,000$ F8 and later dwarfs, started with the Sloan SDSS III MARVELS programme in 2008 (Ge et al., 2009). The first sub-stellar mass companions, including the brown dwarf MARVELS–1 b, have been reported (Fleming et al., 2010; Lee et al., 2011).

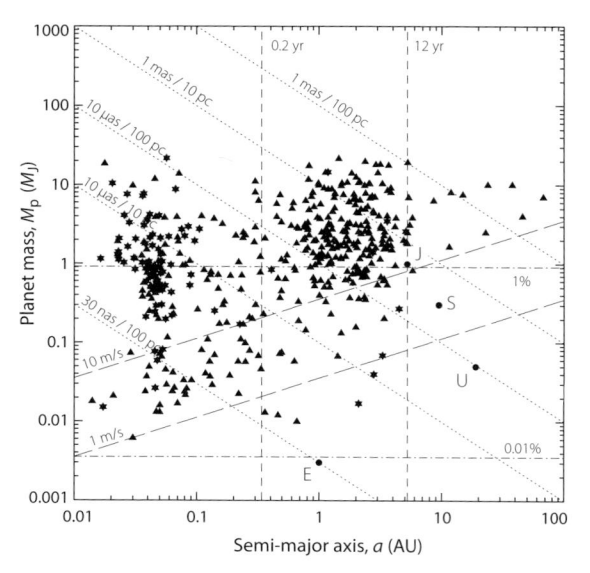

Figure 2.18: Detection domains based on orbital motion. Lines from top left to bottom right show astrometric signatures of 1 mas and 10 µas at distances of 10 and 100 pc (Equation 3.1, assuming $M_\star = M_\odot$). Short- and long-periods cannot be detected by planned space missions: vertical lines show limits at $P = 0.2$ and 12 yr. Lines from top right to bottom left show radial velocity semi-amplitudes of $K = 10$ and $K = 1$ m s^{-1}. Horizontal lines are transit thresholds of 1% and 0.01%, corresponding to ~1R_J and ~1$R_\oplus$ respectively. Positions of E(arth), (J)upiter, (S)aturn and (U)ranus are shown, as are planets discovered from radial velocity ($M_p \sin i$, ▲) and transit photometry (★). Data are from exoplanet.eu, *2010–11–01.*

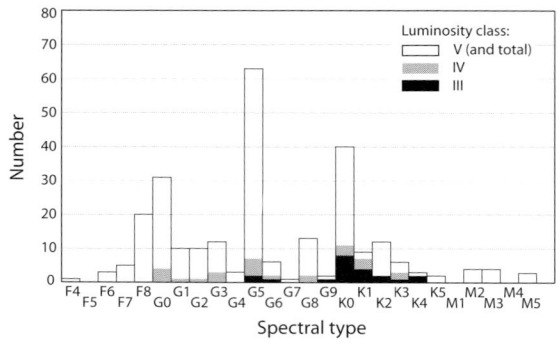

Figure 2.19: The number of exoplanets discovered from radial velocity monitoring versus spectral type (other planets discovered from transit photometry, and confirmed by radial velocity, are not included). Data, from exoplanets.org *on 2010–11–01, includes 261 systems (stars host to a multiple planetary system are counted only once). The majority of exoplanets found from radial velocity measurements orbit G and K dwarfs.*

Examples Out of several hundred planets now detected in this category, a small selection of graphical results is shown in Figures 2.20–2.23 as examples of the variety, measurement quality, and orbit reconstruction possibilities. These figures relate to the following systems (and where all radial velocity derived mass estimates are implicitly uncertain by the factor $\sin i$):

HD 4113 (G5V): Tamuz et al. (2008) used CORALIE to discover a 1.63M_J planet in a highly eccentric ($e = 0.90$) $a = 1.28$ AU, $P = 526$-d orbit. The radial velocity curve shows a sharp, strongly asymmetric form arising from the eccentricity (Figure 2.20), and an underlying linear trend attributable to an outer, longer-period planet.

55 Cnc (G8V): Fischer et al. (2008) used Lick and Keck observations (1989.1–2007.4), to characterise five orbiting planets. Planets were identified by periodogram analysis of residuals to successive Keplerian fits (Figure 2.21). $\{M_p \sin i, P\}$ (M_J, d) are {0.024, 2.8}, {0.83, 14.6}, {0.17, 44.4}, {0.15, 261}, {3.9, 5400}.

HD 40307 (K2.5V): Mayor et al. (2009b) used HARPS to characterise the orbits of three 'super-Earth' planets (Figure 2.22; the total radial velocity semi-amplitude of the combined three-planet signal is only $\sim 6-7$ m s^{-1}). $\{M_p \sin i, P\}$ ($M_\oplus$, d) are {4.2, 4.3}, {6.8, 9.6}, {9.2, 20.5}.

61 Vir (G5V): Vogt et al. (2010b) used 4.6 years of Keck–HIRES and AAT data to characterise this 'super-Earth and two Neptune' planetary system (Figure 2.23). The planets are all in low-eccentricity orbits; their $\{M_p \sin i, P\}$ ($M_\oplus$, d) are {5.1, 4.2}, {18.2, 38}, {24.0, 124}.

M dwarfs, and infrared surveys Towards lower stellar masses, late K and M dwarfs are faint, and high signal-to-noise spectra tend to be limited by telescope size. But special attention has been given to M dwarfs by numerous groups (Table 2.4), since their lower mass implies lower detectable planetary masses for the same radial velocity semi-amplitude (Equation 2.26). At the same time, being of lower luminosity, the corresponding habitable zone defined by the possible presence of liquid water is closer to the host star, with semi-major axes more accessible to radial velocity surveys.

Programmes range from the pioneering survey of 65 M dwarfs started at the Mount Wilson 2.5-m telescope in 1984 (Marcy et al., 1986) to the survey of 100 M dwarfs started in 2007 at the Palomar 5-m using an 'externally-dispersed interferometer' (§2.3.2) in the form of TEDI, a Michelson interferometer coupled to the TripleSpec near-infrared spectrometer (Edelstein et al., 2007).

Results to date suggest that giant gaseous planets occur less frequently around M dwarfs (Endl et al., 2006; Bonfils et al., 2007), a conclusion already reached by Marcy et al. (1986), although a number of lower mass planets, in the super-Earth to Neptune-mass range, have been found. Discoveries include GJ 581 (Bonfils et al., 2005b; Udry et al., 2007; Mayor et al., 2009a), GJ 674 (Bonfils et al., 2007), and GJ 176 (Forveille et al., 2009). More than half are below 25$M_\oplus$, with Endl et al. (2008) quoting a 1.3% detection rate for Jovian-type planets at $a < 1$ AU earlier than M5V. Despite a low overall detection rate, multiplicity rates nevertheless appear high.

2.4 Results to date

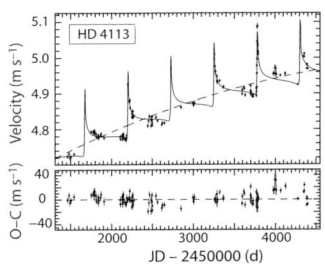

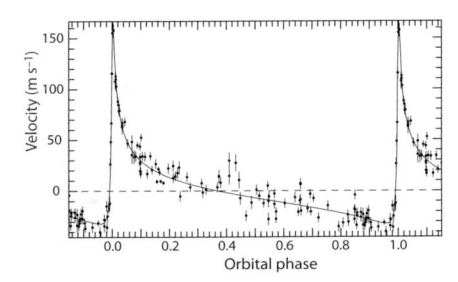

Figure 2.20: Left: radial velocity measurements for HD 4113, observed with CORALIE, with the best-fit Keplerian solution including a linear drift (residuals below). Right: the same measurements folded at $P = 526.58$ d, yielding $K = 97.7$ m s^{-1}, $a = 1.28$ AU, $e = 0.90$, and $M_p \sin i = 1.63 M_J$. The planet–star separation varies between 0.12–2.4 AU. The long-term drift, and its curvature, may imply a brown dwarf companion at 8–20 AU, with $P \sim 20 - 90$ yr. From Tamuz et al. (2008, Figure 1), reproduced with permission © ESO.

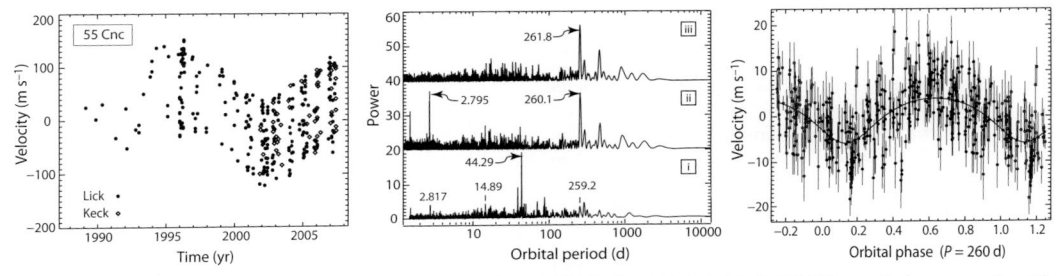

Figure 2.21: The five-planet system 55 Cnc. Left: velocities from Lick/Keck (1989.1–2007.4). Middle: periodograms of residuals to successive Keplerian models (peaks in days): (i) with respect to the 2-planet (14.65 and 5200 d) model: the peak at 44.3 d is due to the third planet; (ii) with respect to the resulting 3-planet model (other major peaks are aliases); (iii) with respect to the 4-planet model (2.8, 14.7, 44.3, and 5200 d); the peak marked is due to the fifth planet (the other major peak is an alias). Right: periodicity of the fifth planet in the Keck data, after subtracting the effect of the other planets. From Fischer et al. (2008, Figures 1–4, 8), reproduced by permission of the AAS.

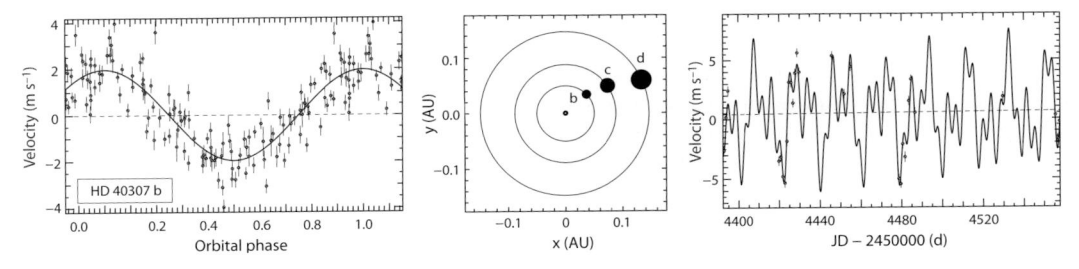

Figure 2.22: The 'three super-Earth' system HD 40307, observed with HARPS. Left: phase-folded velocities and Keplerian curve for the lowest mass planet b, after correction for the other two planets. Middle: pole-on view of the orbits; planet size is proportional to mass. Right: measured velocities with the three Keplerian model superimposed. From Mayor et al. (2009b, Figures 2–4), reproduced with permission © ESO.

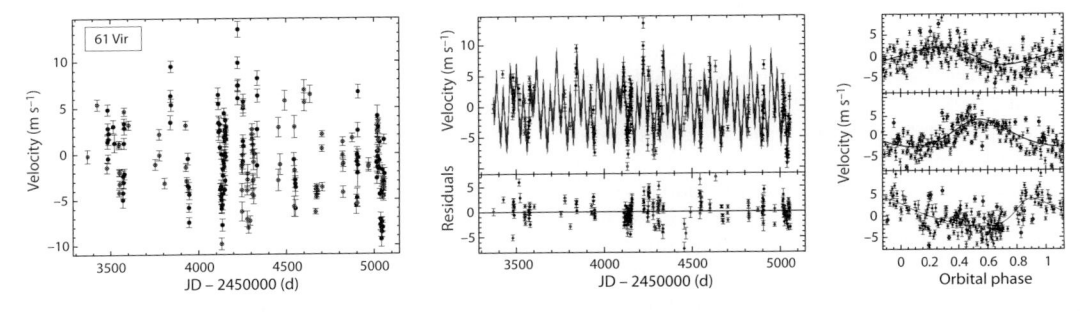

Figure 2.23: The 'super-Earth and two Neptune' system 61 Vir, observed with Keck and AAT. Left: relative radial velocities. Middle: observed velocities, and the three-planet model fit (residuals below). Right: modeled contributions from the three planets b–d (top to bottom, with $e = 0.1, 0.14, 0.35$ respectively). From Vogt et al. (2010b, Figures 2, 6–7), reproduced by permission of the AAS.

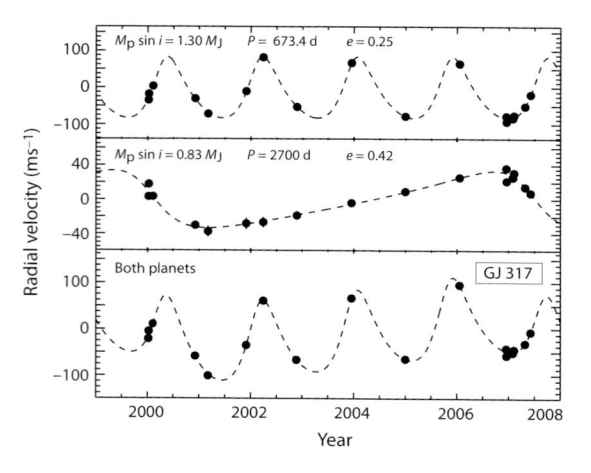

Figure 2.24: *The M dwarf GJ 317, and the best-fitting two-planet model. The top two panels show the single-planet fits with the other component removed. Adding an outer planet with $P \simeq 2700$ d decreases the scatter of the residuals from 12.5 to 6.32 m s^{-1}, and the reduced χ^2 from 2.02 to 1.23. From Johnson et al. (2007a, Figure 4), reproduced by permission of the AAS.*

By late 2010, only seven Doppler-detected giant planets ($M > 0.3 M_J$) were known around six M dwarfs (Johnson et al., 2010c). These include GJ 317 of $1.2 M_J$ (Johnson et al., 2007a, Figure 2.24), and one of the most massive at $2.1 M_J$ around HIP 79431 (Apps et al., 2010). Of particular interest is GJ 876, with its two resonant Jovian-mass planets and an inner super-Earth (Rivera et al., 2005), and a fourth outer Uranus-mass planet in a Laplace resonance configuration, with periods 30.4, 61.1, and 126 d (Rivera et al., 2010).

Johnson et al. (2010c) estimate that $3.4^{+2.2}_{-0.9}\%$ of stars with $M_\star < 0.6 M_\odot$ host planets with $M \sin i > 0.3 M_J$ and $a < 2.5$ AU. Restricted to metal-rich stars with [Fe/H] > +0.2, the occurrence rate rises to $10.7^{+5.9}_{-4.2}\%$.

Early-type dwarfs Observationally, radial velocity measures for higher mass main sequence stars (A–F spectral type) are complicated by three effects: a smaller number of spectral lines due to higher surface temperatures; line broadening due to higher rotational velocities, of order $v \sin i \simeq 100 - 200$ km s^{-1} for A-type stars (Galland et al., 2005a); and large atmospheric 'jitter' due to higher surface inhomogeneities and pulsation, of as much as 50 m s^{-1} or more for mid-F stars (Saar et al., 1998, see also Table 2.1). Together, these limit the applicability of the basic cross-correlation method (Griffin et al., 2000).

Chelli (2000) considered cross-correlation in Fourier space, which allows enhanced suppression of high-frequency noise and low-frequency continuum variations. The correlation is made between each spectrum of the target star and a reference spectrum specific to that star constructed from the sum of all the available spectra. When applied to early-type stars (Galland et al., 2005a), it suggested improved detection possibilities: for example, reaching the planetary domain with ELODIE for A-type main sequence stars with $v \sin i$ up to 100 km s^{-1} for $P < 10$ d, and for all A and F-type stars with HARPS, even for large $v \sin i$ (Figure 2.25).

The resulting search for a sample of A–F main sequence stars, with ELODIE/SOPHIE in the north, and HARPS in the south, led to the first detection of a $9.1 M_J$ companion around the F6V star HD 33564 (Galland et al., 2005b), a $25 M_J$ brown dwarf in orbit around HD 180777, a pulsating A9 dwarf with a high rotation of $v \sin i = 50$ km s^{-1} (Galland et al., 2006b) and a series of subsequent discoveries from the southern survey (Desort et al., 2009, 2010, and references).

2.4.5 Evolved stars

Observational complications inherent to high-mass dwarfs can be somewhat circumvented by observing their counterparts which have evolved away from the main sequence. After main sequence hydrogen burning, radii expand and atmospheres cool, leading to an increase in metal lines. Angular momentum loss as stars cross the subgiant branch results from a coupling of stellar winds to rotationally-generated magnetic fields (Gray & Nagar, 1985; Schrijver & Pols, 1993; do Nascimento et al., 2000). The combination of cooler atmospheres and smaller rotational broadening makes evolved stars well suited for precision radial velocity surveys.

Subgiants Subgiants offer a reasonably favourable region of the HR diagram for radial velocity searches, with low rotational velocities of order $v \sin i \lesssim 5$ km s^{-1}, and reasonably low photometric jitter of around 5–10 m s^{-1} (Wright, 2005, see also Table 2.1). Surveys at Keck and Lick are being made around 150 subgiants, selected from Hipparcos according to $V < 7.6$ mag, $2 < M_V < 3.5$, and $0.55 < B - V < 1.0$ (Johnson et al., 2006b). Numerous planets are being reported (Johnson et al., 2007b; Johnson, 2008; Johnson et al., 2010a,b, 2011; Harakawa et al., 2010), with 30 luminosity class IV listed at exoplanet.eu in late 2010.

G and K giants Radial velocity surveys around G and/or K giants have been made by several groups (Table 2.4): at Lick (Frink et al., 2002; Nidever et al., 2002; Hekker et al., 2006, 2008); Okayama–HIDES and Subaru–HDS (Sato et al., 2003, 2005b, 2008a,b, 2010); La Silla–FEROS (Setiawan et al., 2003a,b, 2004); CORALIE and HARPS (Lovis & Mayor, 2007); Tautenburg (Hatzes et al., 2005; Döllinger et al., 2007, 2009); HET–HRS (Niedzielski et al., 2007; Niedzielski & Wolszczan, 2008; Nowak & Niedzielski, 2008; Niedzielski et al., 2009a,b), and BOAO–BOES (Han et al., 2010). A large survey, of some 1000 GK giants, started with the Sloan SDSS III MARVELS programme in 2008 (Ge et al., 2009).

2.4 Results to date

Table 2.4: *Radial velocity surveys focusing on particular stellar types (excluding brown dwarfs). Surveys, instruments, and descriptions evolve with time, so the table is indicative rather than definitive. References describe the sample or more recent global results. Listings are chronological in survey start date within each category, and there is often overlap between some categories.*

Selection	Telescope/instrument	N(stars)	Start	Reference
Main sequence:				
F8–M5	ESO–CES [ended 1998]	37	1992	Endl et al. (2002)
GK, $V < 7.65$, $v \sin i < 5\,\mathrm{km\,s^{-1}}$	OHP–ELODIE	324	1994	Queloz et al. (2001)
FGKM (California–Carnegie)	Keck+Lick+AAT	1040	1995	Valenti & Fischer (2005)
FGK	Lick–Hamilton	107	1995	Marcy & Butler (1998a)
F7–M5, < 50 pc	Keck–HIRES	530	1996	Vogt et al. (2000)
FGKM	Lick+Keck	844	1998	Nidever et al. (2002)
FGK IV–V, $V < 8$, $\delta < -20°$	AAT–UCLES	300	1998	Jones et al. (2002)
Hipparcos parallax-limited	La Silla–CORALIE	1650	1998	Tamuz et al. (2008)
F–M, metal-rich	AAT–UCLES	20	1999	Tinney et al. (2003)
F–M (N2K), metal-rich	Keck+Magellan+Subaru	2000	2004	Fischer et al. (2005, 2007)
metal-rich, Hipparcos-based	OHP–ELODIE/SOPHIE	1061	2004	da Silva et al. (2006)
low activity (Rocky Planet)	AAT–AAPS	55	2007	O'Toole et al. (2009b,a)
metal-rich (Calan–Hertfordshire)	ESO–HARPS	350	2008	Jenkins et al. (2009a)
GKM (Eta-Earth)	Keck–HIRES	230	2008	Howard et al. (2009, 2011a,b)
F8 and later	SDSS III MARVELS	10 000	2008	Ge et al. (2009)
M dwarfs:				
M2–M5, $V < 11.5$	Mt Wilson 2.5-m	65	1984	Marcy et al. (1986)
Lick/Keck	Lick+Keck	24	1994	Marcy et al. (1998)
$V < 7.5$	AAT–UCLES	7	1998	Jones et al. (2002)
CNS3, $d < 9$ pc, $V < 15$	OHP–ELODIE/SOPHIE (N)	127	1995	Delfosse et al. (1999a)
" $d < 12$ pc	La Silla–FEROS (S)	200	2000	Bonfils et al. (2004)
ESO survey	VLT–UVES	40	2000	Zechmeister et al. (2009)
"	VLT–UVES	26	2002	Guenther & Wuchterl (2003)
McDonald survey	HET–HRS+HJS	100	2002	Endl et al. (2003)
$d < 11$ pc	HARPS	200	2003	Mayor et al. (2009a)
Palomar	Palomar–TEDI	100	2007	Edelstein et al. (2007)
K–M (M2K)	Keck	1600	2009	Apps et al. (2010)
Early-type dwarfs:				
early-type, B8–F7	ELODIE/SOPHIE+HARPS	185	2004	Desort et al. (2009)
Subgiants:				
subgiants	Lick+Keck	159	2006	Johnson et al. (2006b, 2010b)
Giants:				
G/K/clump giants	Lick+Keck	45	1999	Nidever et al. (2002)
"	Lick–CAT	179	1999	Hekker et al. (2006, 2008)
"	Okayama–HIDES/Subaru	300	2001	Sato et al. (2008a,b, 2010)
"	La Silla–FEROS	83	2001	Setiawan et al. (2003a,b)
" (in 13 open clusters)	HARPS	115	2003	Lovis & Mayor (2007)
" (Penn State/Torun)	HET–HRS	1000	2004	Niedzielski & Wolszczan (2008)
"	Tautenburg 2-m	62	2004	Döllinger et al. (2007, 2009)
"	SDSS III–MARVELS	1000	2008	Ge et al. (2009)
"	BOA–BOES	55	2010	Han et al. (2010)
Open clusters:				
Hyades giants	Okayama–HIDES	4	2001	Sato et al. (2007)
Hyades dwarfs	Keck+others	94	2002	Cochran et al. (2002)
13 clusters	CORALIE+HARPS	115	2003	Lovis & Mayor (2007)
Abundance-based:				
metal-poor	FEROS	70	2008	Setiawan et al. (2010)
metal-poor	Keck–HIRES	200	2003	Sozzetti et al. (2006, 2009)
metal-poor	HARPS	105	2003	Santos et al. (2007)
Young stars:				
young stars, Cha I cloud	VLT–UVES	12	2000	Joergens (2006)
young stars, 1–100 Myr	FEROS+HARPS	200	2003	Setiawan et al. (2008a,b)
Binary and multiple stars:				
CORAVEL single-lined (SB1)	ELODIE+CORALIE	101	2001	Eggenberger (2010)
wide binaries, $a = 100 - 1000$ AU	TNG–SARG	100	2003	Gratton et al. (2003)
binary/multiple	Keck–HIRES/HET–HRS	450	2003	Konacki (2005a,b)
visual binaries	Okayama–HIDES	9	2004	Toyota et al. (2009)

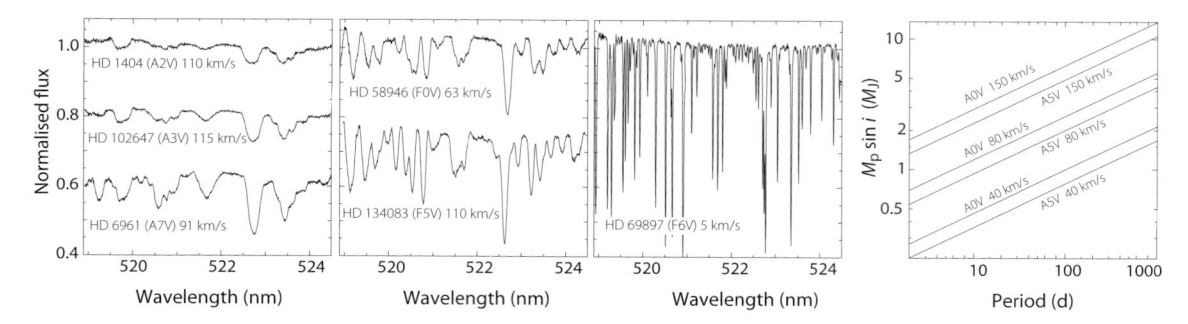

Figure 2.25: Left three panels: spectra of early-type stars acquired with ELODIE, showing the progression in both the number and the broadening of spectral lines for representative examples of A2V, A3V, A7V, F0V, F5V and F6V spectral types. Right: simulated mass detection limits for A-type stars observed with HARPS. From Galland et al. (2005a, Figures 1 and 10), reproduced with permission © ESO.

By 2008, nine planets around stars with $M \gtrsim 1.5 M_\odot$ had been reported by Johnson (2008), with 20 luminosity class III listed at exoplanet.eu in late 2010.

2.4.6 Other star categories

Open clusters The determination of stellar masses for giant stars is complicated by the fact that evolutionary tracks for stars covering a wide range of masses converge to the same region of the Hertzsprung–Russell diagram. This consideration motivated Sato et al. (2007) and Lovis & Mayor (2007, 115 stars in 13 clusters) to target giants in intermediate-age (0.2–2 Gyr) open clusters, where cluster membership combined with the estimated cluster age provides a more secure estimate of the host star masses compared to that of field giants. In contrast to the absence of transiting planets from searches in open clusters (§6.2.3), a few cluster planets have already been discovered from these radial velocity surveys. A 7.6M_J planet orbiting the Hyades giant ϵ Tau was reported by Sato et al. (2007), while two other massive planets (10.6 and 19.8M_J respectively) have been reported around giants in the open clusters NGC 2423 and NGC 4349 by Lovis & Mayor (2007).

Observations of 94 main sequence Hyades stars yielded no close-in giant planets (Cochran et al., 2002; Paulson et al., 2002, 2003, 2004a,b).

Metal-poor stars The observed correlation between exoplanet occurrence and host star metallicity suggests that radial velocity searches around metal-poor stars should have relatively low success rates. Until 2008, the low-metallicity record was held jointly by HD 155358 (Cochran et al., 2007), and HD 47536 (Setiawan et al., 2003a), both with [Fe/H] = –0.68.

Specific searches around low-metallicity stars have been targeted since 2003, for example with HARPS (Santos et al., 2007), FEROS (Setiawan et al., 2010), and with Keck–HIRES (Sozzetti et al., 2006, 2009). The results confirm that giant planets around low-metallicity stars are less common, with only one ([Fe/H] = –0.49) having been detected in five years from the 105 stars surveyed by Santos et al. (2007).

Nevertheless, planets around stars of still lower metallicity have been discovered. These include HIP 13044, a member of a Galactic halo stream (Helmi et al., 1999) with [Fe/H] = –2.0 (Setiawan et al., 2010).

Young stars While there is a general consensus that planets form from the disks of dust and gas around newly-formed stars, details of their early formation, including the time scale of planet formation, remain unclear. Searches for planets around young stars with protoplanetary disks, of age 1–100 Myr, are being targeted by FEROS/HARPS (Setiawan et al., 2008a). Setiawan et al. (2007) reported a 6.1M_J planet around HD 70573, a GV star with an age of about 100 Myr. Setiawan et al. (2008a) reported a planet of mass 9.8M_J around TW Hya, a nearby young star with an age of 8–10 Myr that is surrounded by a well-studied circumstellar disk. The planet orbits the star with $P = 3.56$ d ($a = 0.04$ AU), inside the inner rim of the disk, perhaps demonstrating that planets can form before the disk has been dissipated by stellar winds and radiation.

Binary and multiple stars Results and specific programmes aimed at characterising planets around binary and multiple stars are considered in §2.7.

2.5 Properties of the radial velocity planets

As discoveries proceed, more significant statistical analyses of the planetary population are becoming possible, with statistically distinct properties and correlations becoming evident through principal component or hierarchical clustering analyses (e.g. Marchi, 2007). Various reviews of exoplanet properties have been given as discoveries have advanced (e.g. Marcy et al., 2006; Udry & Santos, 2007; Santos, 2008; Johnson, 2009).

2.5 Properties of the radial velocity planets

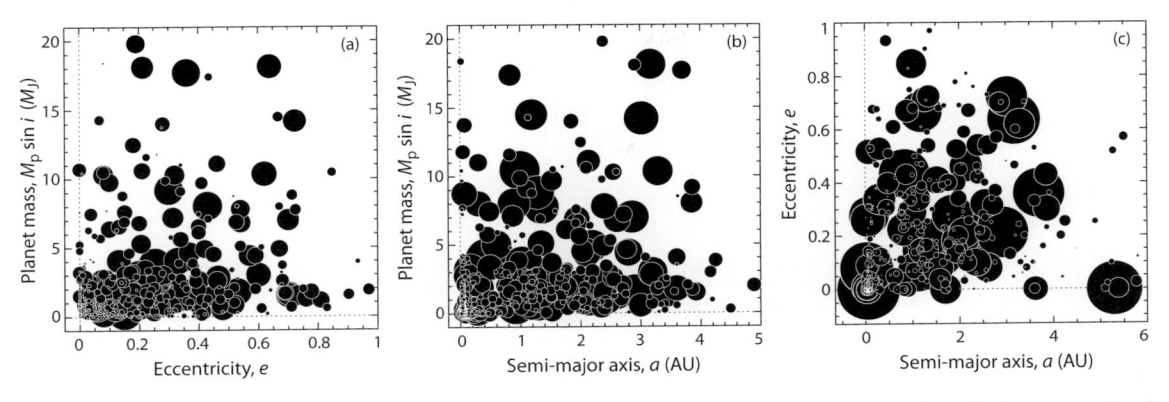

Figure 2.26: Properties of the radial velocity planets: (a) planet mass $M_p \sin i$ versus eccentricity e, with circle sizes proportional to a; (b) planet mass $M_p \sin i$ versus semi-major axis a, with circle sizes proportional to e; (c) eccentricity e versus semi-major axis a, with circle sizes proportional to $M_p \sin i$. Data are from exoplanet.eu, 2010–11–01.

2.5.1 Frequency of massive planets

For the target list of 1200 FGKM dwarfs in the solar neighbourhood monitored by the California and Carnegie Planet Search programme (Wright et al., 2004), Marcy et al. (2008) report that 87% of stars observed for more than a decade show no Doppler variations at a 3σ limit of $10\,\mathrm{m\,s^{-1}}$. These limits largely exclude the existence of a significant number of undiscovered Jupiter-mass planets within 3 AU, and Saturn-mass planets within 1 AU.

At least 6–7% have giant planets, with $M_p > 0.5 M_J$ and $a < 5$ AU, a similar fraction also being reported by the Geneva group (Udry & Mayor, 2008). Some 15% of these fall into the category of 'hot Jupiters', with $P \lesssim 10$ d, and orbiting very close to their host stars.

The remaining 6% of stars show long-term radial velocity trends, often with significant curvature, indicating sub-stellar, brown dwarf, or planetary companions with orbital periods of a decade or more (Patel et al., 2007).

2.5.2 Mass distribution

Various distributions of the radial velocity planet population are shown in Figures 2.26–2.27. In the following discussions, it is recalled that for any individual object the measured mass is uncertain by the factor $\sin i$. Similarly, the underlying distribution of M_p (rather than that of the minimum mass $M_p \sin i$) cannot be derived unambiguously (cf. Jorissen et al., 2001). For a random orientation of orbital planes, $\langle M_p \rangle = (\pi/2)\, M_p \sin i$.

Figure 2.27 shows that the distribution of minimum masses continues to rise at least as far as Saturnian masses. A power-law fit gives $dN/dM \propto M^{-1.15}$ (Marcy et al., 2008), with a cut-off above $\sim 12 M_J$. Incompleteness at smaller masses, due to the increased difficulty of planet detection for decreasing mass, indicates that the mass distribution probably rises even more steeply towards lower masses.

Brown dwarf desert A prominent feature of the early radial velocity exoplanet discoveries was the general absence of substellar objects with masses in the range $10 - 80 M_J$, a phenomenon which was referred to as the 'brown dwarf desert'. This clear mass separation between stellar and planetary companions to solar-type stars has continued, and is taken to signify evidence of a different formation mechanism for these two populations. Doppler surveys of several thousand stars have yielded only a small number of brown dwarf candidates in this mass range. Similar statistics result from comprehensive infrared coronagraphic searches (McCarthy & Zuckerman, 2004). The majority of these may be hydrogen-burning stars with low orbital inclination (Halbwachs et al., 2000; Udry et al., 2000), and this absence of brown dwarfs renders exoplanets distinguishable by their high occurrence at low masses, irrespective of the $\sin i$ uncertainty.

Low-mass planets Below $\sim 0.1 M_J$, and despite the progressively smaller radial velocity variations and observational incompleteness, which together render the detection of lower mass planets more difficult, the distribution of masses appears to rise (Udry & Mayor, 2008).

As of late 2010, eight planets are known with $M \sin i < 0.02 M_J$ ($\sim 5.4 M_\oplus$), of which two were transit discoveries (CoRoT–7 b and GJ 1214 b). The six others are close-in planets ($a < 0.07$ AU, $P < 5$ d except for GJ 581 c), with two of the lowest orbiting the low-mass M dwarf GJ 581. The are, in order of mass, GJ 581 e ($1.9 M_\oplus$, HARPS; Mayor et al., 2009a), HD 156668 b ($4.1 M_\oplus$, Keck; Howard et al., 2011a), HD 40307 b ($4.2 M_\oplus$, HARPS; Mayor et al., 2009b), 61 Vir b ($5.1 M_\oplus$, Keck/AAT; Vogt et al., 2010b), GJ 581 c ($5.4 M_\oplus$, HARPS; Udry et al., 2007), and HD 215497 b ($5.4 M_\oplus$, HARPS; Lo Curto et al., 2010).

Low-mass planets are being frequently found as members of multiple planet systems, and may indeed be found preferentially in multiple systems (Lo Curto

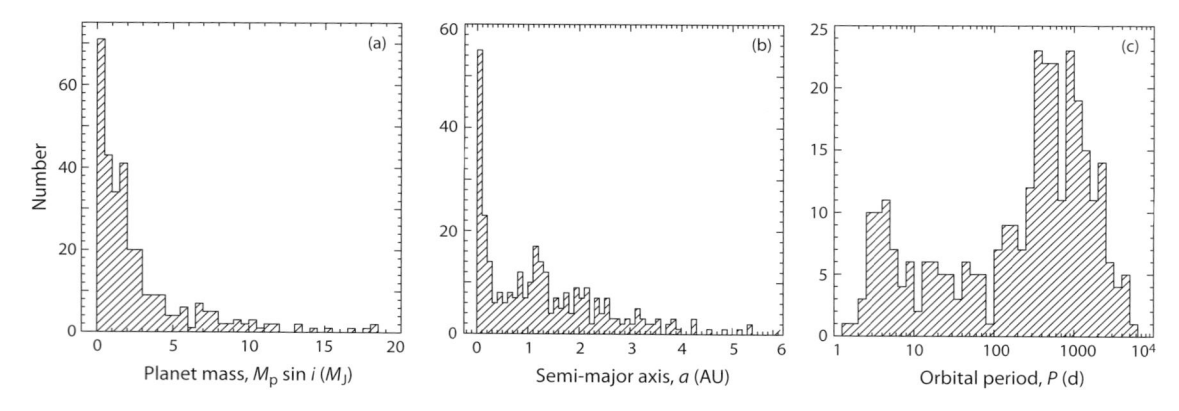

Figure 2.27: Histograms for the planets discovered from radial velocity surveys: (a) planet masses, $M_\mathrm{p}\sin i$ (M_J); (b) semi-major axes, a (AU); (c) orbital periods (log scale), P (d). Data are for 383 planets from exoplanets.org, 2010–11–01.

et al., 2010). Amongst these are the $7.5 M_\oplus$ GJ 876 d in the four-planet system (Rivera et al., 2005), the Neptune-mass three-planet system around HD 69830 (Lovis et al., 2006; Ji et al., 2007), the three super-Earth system HD 40307 (Mayor et al., 2009b), and the two-planet system HD 215497 (Lo Curto et al., 2010). The M dwarf GJ 581 hosts as many as six planets $\lesssim 7 M_\oplus$ (Udry et al., 2007; Vogt et al., 2010a), with two or three within or at the edges of its habitable zone (§11.9.2).

The discovery of so many low-mass planets, close to the detection threshold and over a relatively short period of the high-precision Doppler surveys, suggests the existence of a large population of low-mass planets, perhaps reaching 30% for G and K dwarfs (Mayor et al., 2009b). Some of the scatter in radial velocity measurements for individual objects is probably attributable to undetected, low-amplitude, multi-planet systems.

Bimodality in the mass distribution is seen in the homogeneous HARPS results (Udry & Mayor, 2008), and is predicted by Monte Carlo simulations of accretion-based formation models (Mordasini et al., 2008).

2.5.3 Orbits

Period distribution The distribution of orbital periods is shown in Figure 2.27c. There is a 'pile-up' of planets at $P \sim 3$ d ($a \sim 0.05$ AU), together with a possible minimum in the interval $P \sim 10 - 100$ d ($a \sim 0.1 - 0.4$ AU).

The significant numbers of giant planets with $P \sim 3 - 10$ d ($a \sim 0.05 - 0.1$ AU), orbiting close to their host stars, was essentially unexpected before the early exoplanet discoveries.[10] The early trend has continued: 50 planets have $P < 10$ d. As discussed elsewhere, these 'hot Jupiters' are believed to have formed much further out, before being scattered or migrating inwards to their present location, with some mechanism halting the migration before the planets fall onto their host stars.

The observed masses and orbits of giant planets within 5 AU are reproduced rather well by current theories of their formation and migration (Kley et al., 2005), including their dependence on stellar metallicity and mass (Ida & Lin, 2004b, 2005b; Armitage, 2007b).

The period distribution for $P \gtrsim 1000$ d ($a \gtrsim 2$ AU) is less certain, with the apparent decline beyond 3 AU probably largely attributable to the limited duration, of around 10 years, of the high-precision Doppler surveys (Kholshevnikov & Kuznetsov, 2002). Some of the best constraints, out to some 8–10 AU, come from the almost 30-yr baseline for 17 objects included in the early CFHT and McDonald Observatory observations (Wittenmyer et al., 2006, 2007). Extrapolation out to 20 AU, comparable to the orbit of Uranus, is correspondingly uncertain, but even a flat extrapolation would approximately double the known rate (Marcy et al., 2005a). This suggests that a large population of still undetected Jupiter-mass planets may exist between 3–20 AU, perhaps implying that as many as 20% of Sun-like stars have a giant planet within 20 AU (Cumming et al., 2008).

Attempted adaptive optics imaging of giant planets around young stars from their thermal infrared emission using Gemini north–Altair (Lafrenière et al., 2007a) and VLT–NACO (Apai et al., 2008) has yielded independent limits on their occurrence, also suggesting an absence of giant planets beyond some 20–30 AU.

Current theories support the idea that gas giants may be rare beyond 20 AU. The declining densities of protoplanetary disks with increasing radius, combined with the longer dynamical time scales, together imply a longer time scale for planetary growth compared with protoplanetary disk lifetimes of order 3 Myr (Alibert et al., 2005a; Hubickyj et al., 2004).

---

[10] Although Struve (1952), in considering the timeliness of radial velocity searches, commented that *'It is not unreasonable that a planet might exist at a distance of 1/50 AU... Its period around a star of solar mass would then be about 1 day.'*

2.6 Multiple planet systems

Mass–period relation A correlation is seen between M_p and a (Figure 2.26b). There is a paucity of massive planets on short-period orbits (Zucker & Mazeh, 2002; Udry et al., 2002; Jiang et al., 2007). There is also a rise in the maximum planet mass with increasing a (Udry et al., 2003b). Since the more massive planets are presumably formed further out in the protoplanetary disk, where accretion material is abundant and orbital paths longer, the resulting larger masses are perhaps less easily displaced by whatever migration mechanism forces the inward movement.

Eccentricities Most pre-discovery theories of planetary formation suggested that exoplanets would be in circular orbits similar to those in the solar system (Boss, 1995; Lissauer, 1995). In practice (Figure 2.26), there is a significant correlation between a (or P) and e (Stepinski & Black, 2000). Close-in planets are in preferentially low eccentricity orbits, while exoplanets with $P \geq 6$ d have e spanning the range 0–0.93, with a median $e \sim 0.3$. A few, long-period, low-eccentricity orbits are found, representing a small sample of solar system analogues.

As discussed elsewhere, the origin of the eccentric orbits probably arises from several mechanisms: gravitational interaction between multiple giant planets; interaction between the giant planets and planetesimals in the early stages of planet formation; or the secular influence of an additional passing mass, either unbound or bound (with the four most extreme eccentricities being found orbiting components of binary systems). Furthermore, for small pericentre distances, tidal circularisation appears to be an important damping mechanism.

Solar system analogues The high frequency of eccentric orbits amongst giant planets with semi-major axis $a \gtrsim 1-2$ AU suggests that common perturbation mechanisms may cause non-circular orbits both for the giant planets themselves, and for any lower mass rocky planets closer in. The eccentric giant planets tend to eject lower-mass objects, or enhance orbital eccentricities.

Around solar-mass stars, analogues to Jupiter, both in terms of low eccentricity and comparable orbital semi-major axis, still seem to offer the best prospects for harbouring Earth-mass planets moving in circular orbits within the habitable zone. The $0.94 M_\odot$ star 55 Cnc has a low-eccentricity ($e = 0.06$) $3.8 M_J$ planet with $a = 5.8$ AU, not too dissimilar from Jupiter, although with four large planets orbiting inward of 1 AU.

A better Jupiter analogue orbits the $0.88 M_\odot$ star HD 154345 (Wright et al., 2007). With $M_p \sin i = 0.92 M_J$, $P = 9$ yr, $a = 4.2$ AU, and $e = 0.16$, the planet moves in a nearly circular orbit with no other giant planets lying inward of it. There may be additional planets with $P = 45 - 60$ d, while the star itself has a heavy element abundance close to solar, with [Fe/H] = -0.10 ± 0.04 (Valenti & Fischer, 2005).

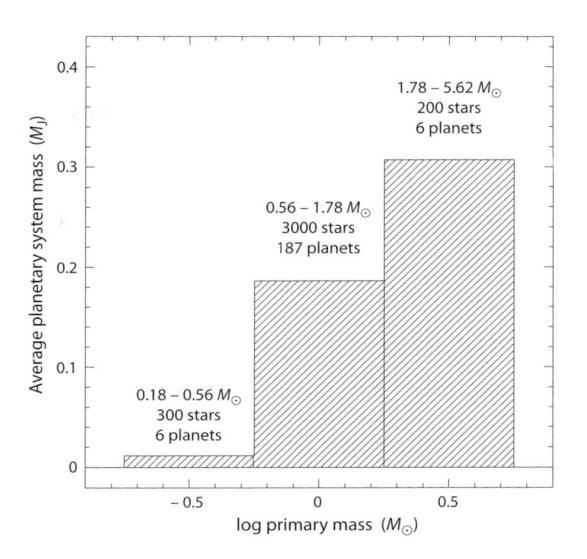

Figure 2.28: Average planetary system mass as a function of stellar mass, for all planets known as of February 2007. More massive stars seem to harbour more massive planetary systems. From Lovis & Mayor (2007, Figure 11), reproduced with permission © ESO.

2.5.4 Host star dependencies

Metallicity of host star A correlation between the presence of gas giant planets and high metallicity of the host star was noted in the early years of exoplanet discoveries (Gonzalez, 1997). More homogeneous studies have confirmed the trend, with investigations extending to the abundances of numerous elements in addition to iron. The consensus is that the effect is attributable to primordial effects rather than to self-enrichment by accretion, and is considered in further detail in §8.4. The correlation between occurrence and metallicity may not extend to the lowest mass planets (Udry et al., 2006), nor to those orbiting giants stars (Pasquini et al., 2007).

Mass of host star The results of the surveys of FGK main sequence stars in the mass range $0.3 - 1.4 M_\odot$, along with the few, typically low-mass planets detected around M dwarfs, and the more massive planets found from the evolved star surveys, suggest that the material trapped in the form of planets is strongly correlated with the mass of the primary star (Figure 2.28).

If the surface density of disk material furthermore scales with the central star mass, there should be an additional correlation between planet occurrence and stellar mass (Laughlin et al., 2004a; Ida & Lin, 2005b; Kennedy & Kenyon, 2008b).

2.6 Multiple planet systems

This section opens with some general considerations relevant for multiplanet systems, including an introduction to resonances, stability, and chaos, before considering a number of specific systems within five broad and non-exclusive categories (following Wright, 2010b): systems with three or more giants (§2.6.4), systems with two giants in mean motion resonance (§2.6.5), other systems in which planet–planet interactions are non-negligible (§2.6.6), non-interacting systems (§2.6.7), and systems containing only lower mass planets with $M_\mathrm{p} \sin i < 20 M_\oplus$ (§2.6.8).

2.6.1 General considerations

Frequency of multiple systems The first radial velocity target known to comprise multiple planets was the triple planetary system υ And (Butler et al., 1999). This was followed by the discovery of a second planet orbiting 47 UMa (Fischer et al., 2002), and a resonant pair orbiting GJ 876 (Marcy et al., 2001a).

Some 10–15% of all planetary systems are presently known to be multiple (Figure 2.29), with a further comparable fraction showing evidence for multiplicity in the form of long-term radial velocity trends.

An increasing number of multiple systems are being discovered, partly as a result of improved accuracies, and partly as a result of longer measurement baselines. The tenth triple system, HIP 14810, was announced by Wright et al. (2009a). μ Ara and GJ 876 each comprise four planets, 55 Cnc counts five, GJ 581 possibly has six, and HD 10180 up to seven.

Longer temporal coverage of known systems is leading to the discovery of outer companions with long orbital periods. Nine radial velocity planets have $P \gtrsim 10$ yr (including μ Ara e, with 55 Cnc d and HD 134987 c at $P \sim 15$ yr, and $P \sim 40$ yr suggested for 47 UMa d). Improved radial velocity accuracy is leading to the discovery of numerous low-mass planets in multiple systems, with around 20 now known (in total) below $0.03 M_\mathrm{J}$ ($10 M_\oplus$).

Multiple systems and theories of formation As developed in detail in Chapter 10, the core accretion model provides a compelling scenario for giant planet formation. In brief, from an embryonic disk of dust and gas, dust particles collide and grow to form progressively larger planetary cores. If a significant amount of gas remains in the disk, a sufficiently massive core can gravitationally accrete more gas, rapidly growing in mass. Such giant planets are likely to form beyond the snow line, at around 3 AU for solar-type stars, where ices can participate in the initial planetary cores. The fact that some 20% of known exoplanets orbit within 0.1 AU, where little ice is available, leads to the hypothesis that short-period planets formed far out and migrated inwards to their present locations.

The discovery of numerous systems in or near mean motion resonance lends support to the migration hypothesis, with hydrodynamical and N-body simulations

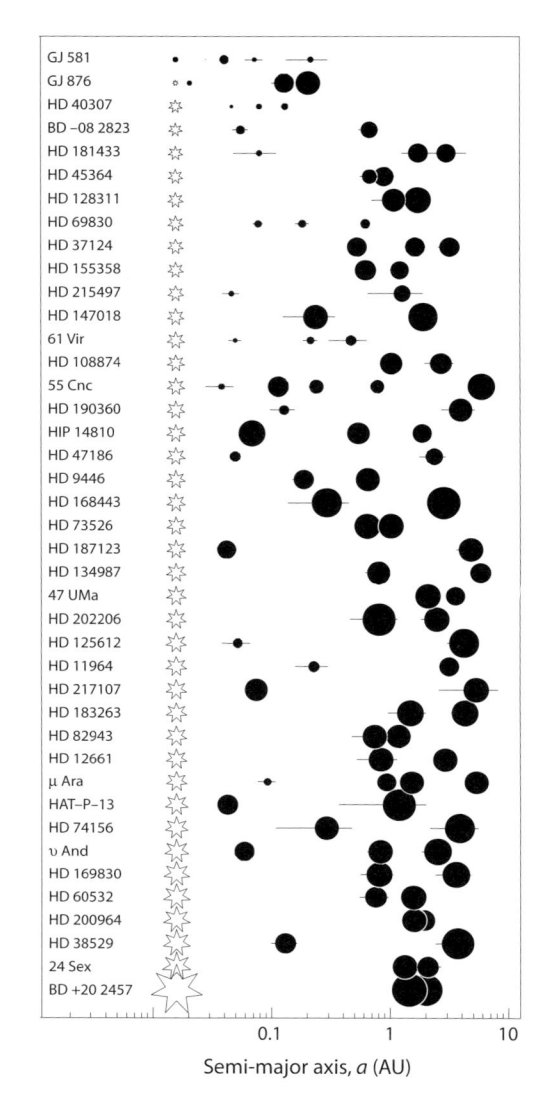

Figure 2.29: Multiple planet systems, ordered by host star mass (indicated at left with size proportional to $M_\star$, ranging from $0.31 M_\odot$ for GJ 581 to $2.8 M_\odot$ for BD +20° 2457). Each planet in the system is shown to the right, with sizes proportional to $\log M_\mathrm{p}$ (ranging from about $0.01 - 20 M_\mathrm{J}$). Horizontal bars through the planets indicate maximum and minimum star–planet distance based on their eccentricities. Data are for 97 planets in 41 systems from exoplanets.org, 2010–11–01. From a concept by Marcy et al. (2008, Figure 13).

with externally applied damping suggesting that the 2:1 mean motion resonance (and others) appear naturally as the planets migrate inwards at different rates.

Various mechanisms may be responsible for the wide distribution of observed eccentricities, some driving eccentricities to larger values while damping others to low values. An early era of strong planet–planet scattering producing large values may be followed by a damping phase as a result of subsequent interactions

2.6 Multiple planet systems

with remaining planetesimals. Evidence for these various mechanisms are detailed in §2.6.4 and §2.6.5.

Coplanarity The $\sin i$ uncertainty for each planet means that the extent to which multiple systems have coplanar orbits is not well constrained by present observations. Numerical simulations, such as those by Thommes & Lissauer (2003), Adams & Laughlin (2003), and others, suggest that a significant fraction of planetary systems involving giant planets may be substantially non-coplanar. Dynamical mechanisms that lead to fast amplification of the relative inclination are especially effective in the first-order resonance configurations (Thommes & Lissauer, 2003). Also, dynamical relaxation and collisional scattering of protoplanets may favour large relative inclinations, even if they initially emerge in a flat protoplanetary disk.

Constraints on the relative orbital inclinations of individual systems are frequently presented on the basis of long-term numerical orbit integrations, which may reveal islands of stability or instability for certain hypothesised relative inclinations. Direct observations, from astrometric measurements, are restricted to systems with large astrometric signatures for each individual planet (§3.1). υ And, observed with HST–FGS, currently provides the most convincing example (§3.8.2), with the large relative inclination between planets c and d ($\Delta i_{cd} = 29°.9 \pm 1°$) lending preliminary support to the simulation results.

Statistics of multiple planet systems The number of multiple systems is presently still relatively small, while initially single systems are progressively becoming multiple systems with the discovery of new planets. Nevertheless some differences are evident between single and multiple systems.

Most striking is that the pile-up of hot Jupiters between 0.03–0.07 AU, and the discontinuity at ~1 AU observed in the distribution of single systems, are both absent from the more uniform distribution seen in multiplanet systems (Figure 2.30). Systems with multiple giant planets appear to lead to a suppressed occurrence of close-in planets. Both characteristics presumably reflect details of migration in the two types of system. Various other differences in the properties of single and multiple systems, in $M_p \sin i$, e, and $M_\star$, the implications for migration models, and the possibilities of selection effects in the present samples, are considered further by Wright et al. (2009b).

Dynamical modeling In the basic kinematic (or 'Keplerian') fitting approach (§2.1.1), the total radial velocity signal due to an orbiting system of n_p planets is assumed to result from the independent reflex motions due to each planet separately.

Many multiple planetary systems may, however, show effects of gravitational planet–planet interactions,

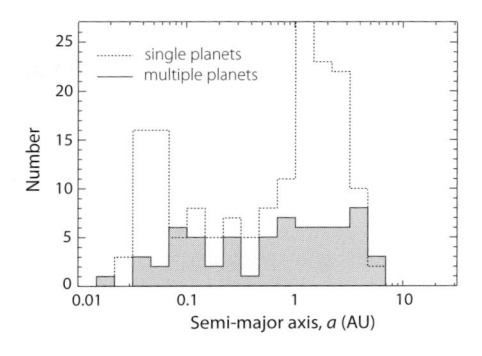

Figure 2.30: Distribution of semi-major axes for multiple planet systems (solid) and apparently single systems (dashed). The pile-up of hot Jupiters, and the jump in abundance beyond 1 AU seen in the single planet systems are not evident in the multiple planet systems. From Wright et al. (2009b, Figure 9), reproduced by permission of the AAS.

even over relatively short time scales. These can lead to observable evolution of the orbital parameters over periods of years, and to radial velocity variations of the central star that differ substantially from velocity variations derived assuming the planets are executing independent Keplerian motions. Dynamical analyses of these more complex systems are made using either of two approaches.

N-body numerical integration methods are used both for dynamical (or 'Newtonian') orbit fitting (see also §2.1.1), and for studying the detailed planetary motions over relatively short time intervals in the future, including testing the validity of analytical results. N-body integration rests on the basic laws of gravity and motion. Results may yield 'deprojected' planetary masses (without the $\sin i$ ambiguity) and, in favourable cases based on stability arguments, relative orbital inclinations.

Analytical methods make use of the *disturbing function*, the difference in gravitational potential due to a star alone, and that due to a star and other perturbing planets. Analytical theory successfully describes two principal phenomena seen in multiple systems: secular (non-periodic) evolution, and resonances.

In *secular theory*, terms that depend on the planet's mean motion, n, as well as other higher-order orbit terms, are ignored, and the theory describes the system's *secular evolution*, essentially predicting how the shape of an average orbit evolves with time. In most two planet systems, secular theory predicts that their eccentricities oscillate, with an increasing eccentricity of one planet (corresponding to an increase of orbital angular momentum) being accompanied by decreasing eccentricity of the other through conservation of angular momentum. Secular theory also predicts the general behaviour of the difference in the two longitudes of pericentre, $\Delta\varpi$: depending on initial conditions, $\Delta\varpi$ may oscillate around 0 (aligned libration), 2π (anti-aligned libration), or circulate through 2π.

In *resonant theory*, terms that depend on the mean motions are included, but only those related to the resonance under study are considered.

2.6.2 Resonances

Orbital resonances arise when two orbiting bodies exert a regular, periodic gravitational influence on each other as a result of simple numerical relationship between frequencies or periods. The repetitive force between them does not average to zero over long time scales, and secular theory is therefore not applicable. The periods involved can be the rotation and orbit periods of a single body (spin–orbit coupling), the orbital periods of two or more bodies (orbit–orbit coupling), or other more complicated combinations of orbit parameters, such as eccentricity versus semi-major axis or orbit inclination (Kozai resonance).

Resonances can result in unstable interactions, in which the bodies exchange momentum and their mutual orbits evolve. In other circumstances a resonant system can be stable and self correcting, so that the bodies remain in resonance indefinitely. Some of the many manifestations of orbital resonances observed in the solar system (§12.6) have already been observed in the orbits of exoplanets (e.g. Kley, 2010).

Mean motion resonances A significant number of multiple exoplanet systems appear to have orbital periods which are related by commensurabilities of the form $P_1/P_2 \simeq i/j$, where subscripts 1, 2 refer to the inner and outer planets, and i and j are small integers. These commensurabilities arise from dynamical considerations; as noted below, exact commensurability is not a resonance condition *per se*.

The concepts and analytical tools used to understand resonance phenomena in the solar system have been developed over more than a century (Peale, 1976). Murray & Dermott (2000) provide an exposition of secular and resonant theory, including a detailed treatment of mean motion resonances. The following summarises the concepts and terminologies most widely encountered in exoplanet studies to date.

Consider two planets moving in circular coplanar orbits which satisfy

$$\frac{n_2}{n_1} \simeq \frac{p}{p+q}, \qquad (2.45)$$

where $n_1 = 2\pi/P_1$ and $n_2 = 2\pi/P_2$ are the mean motions of the inner and outer planets, and p and q are integers. If the two planets are at *conjunction* at time $t = 0$ (in alignment, independently of where along the orbits this might occur), then the next conjunction occurs when

$$n_1 t - n_2 t = 2\pi, \qquad (2.46)$$

and the time interval between successive conjunctions is therefore given by

$$\Delta T = \frac{2\pi}{n_1 - n_2} = \frac{p}{q}\frac{2\pi}{n_2} = \frac{p}{q}P_2 = \frac{p+q}{q}P_1, \qquad (2.47)$$

from which

$$q\Delta T = pP_2 = (p+q)P_1. \qquad (2.48)$$

For $q = 1$, each planet completes an integer number of orbits between successive conjunctions, with every conjunction occurring at the same longitude in inertial space. More generally, every q-th conjunction occurs at the same longitude, with q defining the *resonance order* of the *mean motion resonance*.

If the outer planet moves in an eccentric orbit with $e_2 \neq 0$ and $\dot{\tilde{\omega}}_2 \neq 0$, i.e. if the longitude of the pericentre precesses, then resonances can still occur if

$$\frac{n_2 - \dot{\tilde{\omega}}_2}{n_1 - \dot{\tilde{\omega}}_2} = \frac{p}{p+q}, \qquad (2.49)$$

in which case

$$(p+q)n_2 - p n_1 - q\dot{\tilde{\omega}}_2 = 0. \qquad (2.50)$$

Now, every q-th conjunction takes place at the same point in the outer planet's orbit, but no longer at the same longitude in inertial space. This relation encapsulates the dynamical significance of mean motion resonance. It also shows that the near (but not necessarily exact) commensurabilities in orbital periods are a consequence of (typically) small orbital precession.

Expressed in terms of the mean longitude, λ, and generalised to both orbits, corresponding *resonant arguments* (or resonant angles), ϕ, can be identified as

$$\phi = (p+q)\lambda_2 - p\lambda_1 - q\tilde{\omega}_{1,2}. \qquad (2.51)$$

For the 2:1 and 3:1 resonances, for example, the resonant arguments take the form

$$\phi(2:1) = 2\lambda_2 - \lambda_1 - \tilde{\omega}_{1,2} \quad (p=1, q=1), \qquad (2.52)$$
$$\phi(3:1) = 3\lambda_2 - \lambda_1 - 2\tilde{\omega}_{1,2} \quad (p=1, q=2). \qquad (2.53)$$

The resonant argument measures the angular displacement of the two planets at their point of conjunction, with reference to one or other of the pericentres.

Resonant dynamics are important if ϕ varies slowly relative to the orbital motion. If, rather than circulating, the resonant argument is stationary, or oscillates (librates), then the planets are in resonance. *Exact resonance* describes the particular combination of mean motions and precession rates for which the time variation of a particular resonant argument is zero ($\dot{\phi} = 0$), while the term *deep resonance* is used to describe systems with small libration amplitudes. Outside of exact resonance, the orbital elements $a, e, \tilde{\omega}$ and the resonant argument ϕ all evolve with time.

2.6 Multiple planet systems

> **Physics of resonance:** Peale (1976) gives a heuristic physical description of resonance, reproduced here in outline.
>
> Consider a test particle of negligible mass moving in an eccentric outer orbit around a more massive object. Assume a fixed line of apsides, precise commensurability of the mean motions, and conjunctions occurring close to, but not at, apocentre.
>
> Net tangential forces increase the orbital angular velocity of the test particle, moving the successive conjunctions closer to apocentre. Similarly, conjunctions occurring after apocentre move successive conjunctions back towards apocentre. Conjunctions thus librate stably about the apocentre, preserving commensurability. A secular variation in the pericentre longitude $\tilde{\omega}$ does not alter this fundamental libration. Similarly, pericentre conjunctions correspond to an unstable equilibrium configuration (Figure 2.31).
>
> For conjunctions occurring precisely at the apocentre of the outer test particle, radial forces accelerate it towards the primary, resulting in a trajectory slightly inside its nominal orbit. The test particle accordingly reaches its closest approach to the primary slightly earlier, and the line of apsides has rotated in a retrograde sense as a result. This regressional motion of the line of apsides can reduce, or even dominate, the normal prograde motion resulting from the oblateness of the primary. This type of orbital resonance also leads to a secular increase of the eccentricity.

The more general case In the more general analysis of orbital resonances, the time derivatives of the orbital elements are expressed in terms of partial derivatives of the disturbing function. The general form of the argument which then appears in the expansion of this disturbing function is (Peale 1976, eqn 7; Murray & Dermott 2000, eqn 8.18)

$$\phi = j_1\lambda_1 + j_2\lambda_2 + k_1\tilde{\omega}_1 + k_2\tilde{\omega}_2 + i_1\Omega_1 + i_2\Omega_2 , \quad (2.54)$$

where the variables λ, ω, Ω are confined to the arguments (angular dependencies) of the two orbits. Equation 2.51 is a special case of this more general form. More complex resonances have different numbers of resonant arguments involving various permutations of these terms, classified according to the relevance of the coefficients i, j, k (e.g. Peale, 1976, Table 2). The majority of resonance configurations are characterised by only one librating resonant argument (Michtchenko et al., 2008a,b).

Motion near the 2:1 (and other first-order) resonances depends on initial conditions, and includes libration (in a, e, and ϕ), apocentric libration, and inner and outer circulation (e.g. Murray & Dermott, 2000, Figure 8.16). The phase space boundary which separates librating (or oscillating) from circulating regimes (as in a pendulum) is determined by the total energy, and is referred to as the *separatrix*.

Examples of specific resonant exoplanet systems, and some inferences that can be made from them, are considered in §2.6.5. From the complexity of the mean motion resonance phenomenon it should be apparent that very diverse systems, and complex behaviour in the long-term orbit integrations, can be expected.

Resonance capture and migration The stability of a system once in resonance is understood. The reason for the existence of so many multiple exoplanets in mean motion resonance was, initially, somewhat less evident.

The origin of orbital commensurabilities in the solar system is attributed either to dissipative processes early on in its formation, or to the slow differential increase in the semi-major axes of satellite orbits as a result of tidal transfer of angular momentum from the primary (Goldreich, 1965; Peale, 1976). Goldreich (1965) showed that a pair of satellites under the influence of their planet's tidal forces, initially with non-commensurable mean motions, will change their semi-major axes, the inner spiraling outwards until their periods become commensurable, thereafter maintaining this state as the orbits continue to evolve.

The present consensus is that the observed exoplanet orbital resonances could not have simply formed *in situ*. Rather, the evidence from both hydrodynamic and N-body simulations is that some planets have become trapped in specific resonances as a result of differential convergent migration in which dissipative processes steadily altered their semi-major axes and eccentricities (e.g. Bryden et al., 2000; Kley, 2000; Snellgrove et al., 2001; Lee & Peale, 2002; Nelson & Papaloizou, 2002; Papaloizou, 2003; Thommes & Lissauer, 2003; Kley et al., 2004, 2005; Lee, 2004). The occurrence of orbital resonances constitutes strong evidence for such migration having occurred.

As shown for example by Kley et al. (2004), a sufficiently slow migration of two giant planets embedded in a protoplanetary disk can end with either a 3:1 or 2:1 resonant system, depending on the rate of migration. For masses in the range $1-4M_\oplus$, Papaloizou & Szuszkiewicz (2005) found somewhat similar behaviour using both hydrodynamic and N-body simulations. Slow convergent migration results in period ratios given by first-order commensurabilities $p+1{:}p$ with p as small as 1–2, while more rapid migration (typically characterised by larger mass ratios) results in higher p commensurabilities such as 4:3 or 5:4.

Various studies of the migration of individual planets have characterised the motion by a migration rate, $\dot{a}/a$, and an eccentricity damping rate, $\dot{e}/e$. Corresponding e-folding times are given by

$$\tau_a = -\left(\frac{\dot{a}}{a}\right)^{-1} \quad \text{and} \quad \tau_e = -\left(\frac{\dot{e}}{e}\right)^{-1} . \quad (2.55)$$

For GJ 876, for example, Lee & Peale (2002) found that for a sufficiently slow migration, the final state depends only on the ratio of the e-folding times $K = \tau_a/\tau_e$. Resonant capture is suppressed for high drift or migration rates, but depends on the detailed form of the resonant argument (Quillen, 2006b; Quillen & Faber, 2006).

From hydrodynamical calculations it has been proposed that K assumes values close to unity, reflecting the mass and viscosity of the protoplanetary disk (Kley et al., 2004). If such a system is subject to adiabatic migration, a given value of K results in specific values of the final eccentricities (Lee & Peale, 2002).

More on the observational evidence for resonance capture is given under the description of specific resonance systems (§2.6.5).

Apsidal motion Within the class of 2:1 resonances, for example, configurations of stability or instability are determined by the eccentricities of the inner and outer planets, while their apsidal alignment may also display different behaviour in different systems (Figure 2.31).

As first observed for the two planets GJ 876 b and c, both of the lowest-order eccentricity-type mean motion resonance variables (Equation 2.52)

$$\phi_1 = 2\lambda_2 - \lambda_1 - \tilde{\omega}_1 \quad \text{and}$$
$$\phi_2 = 2\lambda_2 - \lambda_1 - \tilde{\omega}_2 \quad (2.56)$$

librate around $0°$. So therefore does the (apsidal resonance variable) quantity

$$\phi' = \tilde{\omega}_1 - \tilde{\omega}_2 \equiv \phi_1 - \phi_2 , \quad (2.57)$$

meaning that the pericentres remain nearly aligned while librating, with conjunctions occurring (in this case) when both planets are near to their pericentres. The deep resonance in GJ 876 is not only an inference from dynamical modeling, but the associated retrograde pericentre precession has been observed for many orbital periods as a consequence of their relatively close-in orbits ($P_b \sim 61$ d, $P_c \sim 30$ d).

A number of the 2:1 resonant exoplanet systems are observed to be in such a state of apsidal alignment, with their apocentres either aligned (e.g. GJ 876 and HD 37124) or anti-aligned (e.g. HD 60532, HD 73526, HD 155358, HIP 14810). The limited statistics suggest that alignment and anti-alignment might occur with approximately equal frequency (Zhou & Sun, 2003; Barnes & Greenberg, 2006a, 2007a; Barnes, 2008).

When it does occur, small-amplitude *apsidal libration* may manifest itself around equilibrium angles of $0°$ (aligned, or symmetric) or $180°$ (anti-aligned, or antisymmetric), but any other (asymmetric) angle can in principle serve as a stable equilibrium (Beaugé et al., 2003; Ji et al., 2003c,d; Lee, 2004; Voyatzis & Hadjidemetriou, 2005; Beaugé et al., 2006; Libert & Henrard, 2006; Marzari et al., 2006).

Not all theoretically-possible 2:1 resonance configurations can be achieved through migration alone, and their existence would provide further constraints on the system's evolution (Lee, 2004).

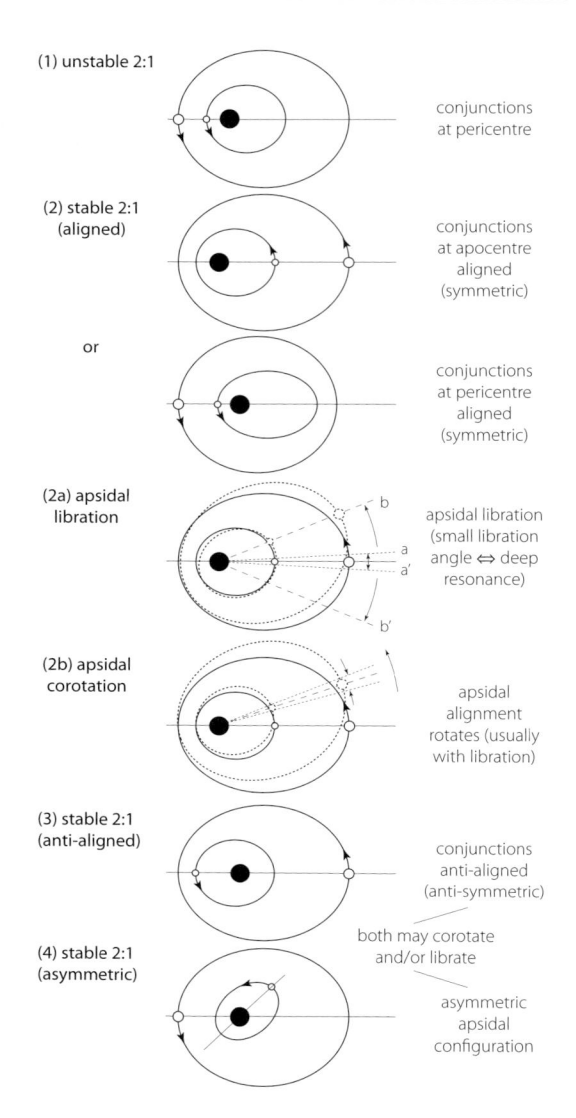

Figure 2.31: Schematic of the 2:1 mean motion resonance, and associated apsidal motion: (1) conjunctions at close approach (here at periapsis) are unstable; (2) for stable resonances, the apsidal alignment may in practice librate over many orbits (2a), and orbital periods will not be precisely commensurate. Small libration angles (of a few degrees) correspond to deep resonance. Libration angles may be many tens of degrees, or the alignment angle may continuously rotate (2b), with additional libration superimposed; (3) conjunctions may also be anti-aligned and, again, either circulating and/or librating or, in theory, they may be asymmetric (4).

One mechanism proposed for establishing a significant apsidal libration in a pair of planets initially on nearly circular orbits invokes an impulsive generation of eccentricity, perhaps caused by the ejection of a planet (Malhotra, 2002). The model can match the aligned orbits observed in υ And (Chiang et al., 2001). An alternative mechanism, invoking adiabatic perturbations,

2.6 Multiple planet systems

> **Hill radius and Hill stability:** The *Hill radius* is the radius within which the gravity of one object dominates that of other bodies within the system. For a planet in a circular orbit of radius r, its Hill radius depends on the planet and central star mass as (e.g. Hamilton & Burns, 1992)
>
> $$R_H = r \left(\frac{M_p}{3 M_\star} \right)^{1/3}. \qquad (2.58)$$
>
> The concept of *Hill stability* is based on the more precise construct of zero-velocity surfaces introduced by Hill (1878) in his treatment of the lunar motion, and has since been used extensively in analytical considerations of both the restricted three-body problem, and the problem of general three-body motions. Although, in general, it is only possible to follow such close encounters in detail by numerical orbit integration of all three masses, the concept of Hill stability provides analytic limits on the range of the system's orbital elements if it is to remain stable, i.e. avoiding either disruption, collision, or positional exchange.
>
> The parameter $p^2 E$, where E is the energy and p is the angular momentum of the system, controls the topology of the zero-velocity surfaces, and determines the regions of allowed motions since the bodies may not cross the zero-velocity surfaces. The binary system is considered to be *Hill stable* against disruption or exchange if the value of $p^2 |E|$ is greater or equal to the critical value derived for the corresponding three-body Hill surface defined by the position of the colinear Lagrange points. The condition is sufficient for stability, but is not necessary, such that exchange cannot occur if the condition is satisfied, but might not occur even if the condition is violated.
>
> The theory was originally applied to coplanar systems in bound orbits. It was extended by Szebehely & Zare (1977) and Marchal & Bozis (1982) to bound stellar systems with mutually inclined orbits, by Gladman (1993) to the stability of planets on circular orbits, and by Veras & Armitage (2004a) and Donnison (2009) to the case of two massive planets with mutually inclined orbits. For further details, see, e.g. Murray & Dermott (2000); Veras & Armitage (2004a); Donnison (2006).
>
> Conservative stability criteria based on multiples of their combined Hill radii (e.g. Hayes & Tremaine, 1998; Cuntz & Yeager, 2009) are also encountered.

& Greenberg, 2006c; Barnes, 2008): *Lagrange stability* in which no planet is ever ejected and the semi-major axes remain bounded for all time; and *Hill stability* or *hierarchical stability* in which the outermost planet might escape but not the inner, implying that the ordering of the bodies remains constant (see box).[11]

The derivation of an orbit solution which matches the radial velocity observables for a multiple planetary system, as derived from either kinematic or dynamical fitting (§2.1.1), does not in itself guarantee that the resulting orbits are stable over long periods of time. The formal best-fit solution might, for example, describe an unstable system which would swiftly disintegrate through ejection of one of the planets, or by causing two of the planets to collide. If such unstable behaviour is predicted to occur on short time scales, the solution is typically rejected as implausible.[12] Conversely stability, even verified over long time periods, cannot prove the correctness of the solution, although it may be considered as a necessary condition for its validity.

The development of efficient numerical methods for exploring parameter space, or for studying long-term orbit evolution, has been important in understanding the large-scale and long-term dynamics of the solar system. These tools are now routinely applied to studies of the stability and evolution of multiple planets (e.g. Barnes & Quinn, 2004). Of particular importance has been the use of *symplectic integrators*, developed for numerical integration of systems governed by Hamilton's equations within a symplectic geometry (see box on page 44).

Wright et al. (2009b) describe their algorithmic approach to assessing system stability which proceeds as follows: orbits are assumed to be coplanar and edge-on, with stellar masses established independently (Takeda et al., 2007a). The radial velocity parameters are converted into initial conditions in a Jacobi coordinate system (Lissauer & Rivera, 2001; Lee & Peale, 2003). Systems are then integrated for $\gtrsim 10^8$ yr using the hybrid integrator Mercury (Chambers, 1999). Solutions are classified as unstable if two planets are predicted to collide, or if a planet falls onto the star ($d < 0.005$ AU), or if it is ejected from the system ($d > 100$ AU).

could follow from tidal torques in a massive primordial disk providing the eccentricity excitation for the outer planet (Chiang & Murray, 2002). Malhotra (2002) suggested that a large apsidal libration amplitude would favour the impulsive mechanism, whereas small amplitudes (as for GJ 876) would favour the adiabatic.

In systems initially in a 3:1 mean motion resonance, dynamical studies show that apsidal librations are also common. Maps of dynamical stability nevertheless show a complicated structure in phase space, with chaos and order coexisting and alternating as the initial eccentricities or apsidal phases evolve (Voyatzis, 2008).

2.6.3 Long-term integration and system stability

Dynamical stability as applied to exoplanet systems can carry at least two somewhat different meanings (Barnes

---

[11] *Hierarchical* has also been used to describe exoplanet systems in which there is a significant ratio of their semi-major axes, $a_1/a_2 \lesssim 0.3$ (Lee & Peale, 2002; Goździewski & Konacki, 2004; Wright, 2010b). With these two distinct definitions, a hierarchical planetary system is not necessarily hierarchically stable. In multiple star terminology, hierarchical structuring describes orbits nested within significantly larger orbits.

[12] An example is the two-planet system HD 82943, with an age of 2.9 Gyr. Despite the genetic search algorithm leading from almost all initial values to the solution given by Mayor et al. (2004), Ferraz-Mello et al. (2005) have shown that the published orbits correspond to a fate in which one of the planets is expelled, or collides with the star, in less than 100 000 yr.

Symplectic integrators: Central to many long-term orbit studies is the use of *symplectic integrators*. In describing the motion of dynamical systems, a sequence of deterministic algebraic steps are executed to follow its evolution from an initial to some final state. Such *algebraic mappings* include the classical (non-symplectic) fourth-order Runge–Kutta iteration for the approximate solution of ordinary differential equations (e.g. Press et al., 2007). These conventional integrators carry the penalty of 'numerical dissipation', which results in a secular change of energy even in a conservative system. The consequence is that positional errors then grow quadratically with time.

Symplectic integrators were developed as a specific numerical integration for the solution of Hamilton's equations, i.e. for particular problems occurring in classical mechanics within the appropriate (symplectic) geometry. Wisdom & Holman (1991), who called their method an 'N-body map', and Kinoshita et al. (1991), who called their method a 'modified symplectic integrator', developed such a mapping for the N-body problem, inspired by efforts to understand the very long-term stability of solar system orbits. Essentially, the Hamiltonian is split into an unperturbed Keplerian part, and an additional contribution describing smaller mutual gravitational interactions; each part can then be integrated separately. Separation involves the introduction of a new system of variables transformed from standard Cartesian to Jacobi coordinates, in which (hierarchically) the masses of bodies m_i and m_j are replaced by a virtual body of mass $m_i + m_j$, and their individual positions replaced by their relative positions and that of their centre of mass. For a step-by-step description see, e.g., Murray & Dermott (2000, Section 9.5.4).

The defining feature of symplectic integrators is that they preserve symplecticity; practically this means that the total energy and angular momentum do not exhibit secular evolution, with the consequent advantage that positional errors grow only linearly with time.

Use of symplectic integrators for the N-body problem is now widespread. Higher-order gravitational effects (e.g. quadrupole moments, general relativity, and stellar mass loss) can be included. Other developments include reducing long-term errors (Saha & Tremaine, 1992), introducing separate time steps for each body (Saha & Tremaine, 1994), improved treatment of close approaches (Levison & Duncan, 1994), allowing for dissipative terms such as drag, inelastic collisions, radiation forces, and tidal distortions (Malhotra, 1994), and special considerations for binary star systems (Chambers et al., 2002).

Symplectic integrators used for exoplanet studies include SWIFT (Levison & Duncan, 1994), the multi-time step SyMBA (Duncan et al., 1998), the hybrid integrator Mercury (Chambers, 1999), SABA4 (Laskar & Robutel, 2001), and HNBody (Rauch & Hamilton, 2002). A hybrid N-body technique allowing for radial zones with distinct time steps has been developed by McNeil & Nelson (2009).

The effects of resonances and chaos can mean that stable and unstable solutions are only marginally separated in parameter space. The evolution of the system must therefore be explored for a range of parameters in the vicinity of the χ^2 minimum to ensure the solution's stability. HD 183263 is an example of a two-planet system where the χ^2 minimum falls close to a stability boundary, such that many solutions consistent with the data are unstable (Figure 2.32).

Dynamical classification Barnes (2008) has classified multiple planet systems according to the type of dynamical interaction that dominates. A system is considered tidally dominated if the semi-major axis $a \lesssim 0.1$ AU (Rasio et al., 1996), resonantly dominated if one or more resonant arguments librates, and secularly dominated otherwise (as is the case for the solar system gas giants).

The general dynamical properties of multiple exoplanet systems, and a summary of their dynamical state, is given by Barnes (2008), where a tabulation provides, for each pair: the resonance, if applicable; the apsidal motion, viz. circulation or (anti-)aligned libration; proximity to the apsidal separatrix (Barnes & Greenberg, 2006b); proximity to the Hill stability boundary (Barnes & Greenberg, 2007b); and whether the orbit evolution is dominated by tidal, resonant, or secular interactions.

Of the 34 planet pairs classified by Barnes (2008), the numbers in each category were 10, 8 and 16 respectively.

Dynamical packing Many planetary systems are found to be dynamically 'full', in the sense that no additional companions can survive in between the observed planets, with most pairs of planets lying close to dynamical instability. This finding has led to the *packed planetary systems* hypothesis (Barnes & Quinn, 2004; Barnes & Raymond, 2004; Raymond & Barnes, 2005; Raymond et al., 2006a), and to general studies of orbital stability for system that are closely spaced (Smith & Lissauer, 2009).

Various searches have been made for regions of Lagrange stability between pairs of planets that are more widely separated (Menou & Tabachnik, 2003; Dvorak et al., 2003b; Barnes & Raymond, 2004; Funk et al., 2004; Raymond & Barnes, 2005; Raymond et al., 2006a; Rivera & Haghighipour, 2007; Barnes & Greenberg, 2007b). In a few systems (for example, in HD 74156 b–c, HD 38529 b–c, and 55 Cnc c–d) there are gaps which are large enough to support additional Saturn-mass planets.

A new Saturn-mass planet in the HD 74156 system was reported by Bean et al. (2008b). It lies between planets b and c, close to the stable orbit identified in the numerical simulations of Raymond & Barnes (2005). The discovery provides some support for the understanding of dynamical stability of multiple systems gained so far.

Long-term dynamical studies of the solar system (§12.6) demonstrate that, in its outer parts, it is not only very stable but also dynamically full (Gladman & Duncan, 1990; Holman & Wisdom, 1993; Levison & Duncan, 1993). The inner solar system is significantly less stable, but it too is dynamically full (Sussman & Wisdom, 1992) and Laskar (1994).

2.6 Multiple planet systems

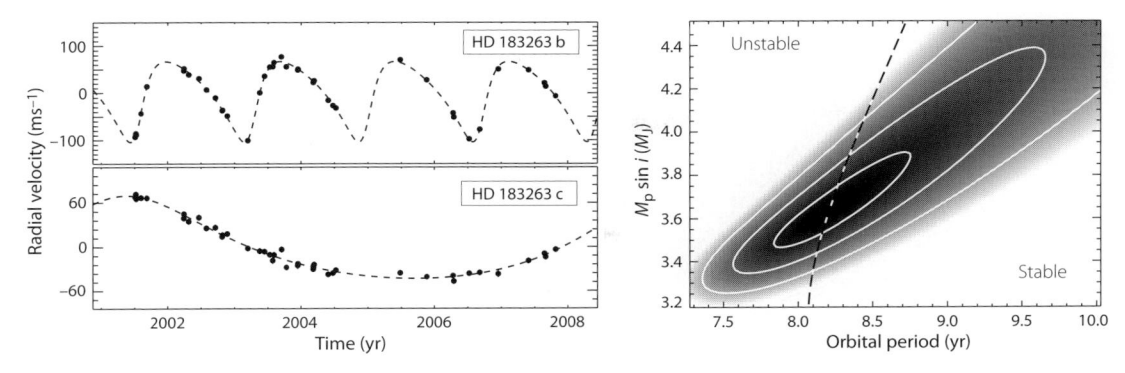

Figure 2.32: HD 183263 observed at the Keck Observatory. Left: the inner planet (b) with $P = 1.7$ yr, $M_p \sin i = 3.7 M_J$ (top) and the outer planet (c) with $P \sim 8$ yr, $M_p \sin i = 3.6 M_J$ (bottom); in each case, the effect of the other planet has been removed. Right: contours of χ^2, with solid contour lines increasing by factors 1, 4, 9 from the minimum. The dashed curve divides stable from unstable orbital solutions, as determined from an ensemble of N-body integrations. From Wright et al. (2009b, Figures 2–3), reproduced by permission of the AAS.

Chaotic orbits An appreciation of the role of chaos has been an important development in the understanding of the solar system's dynamical evolution (Wisdom, 1987a,b; Murray, 1998; Murray & Holman, 2001), and the same principles are directly applicable to the study of multiple exoplanet orbits. In the solar system, chaos is apparent in phenomena such as the 3:1 Kirkwood gap (Wisdom, 1983), the tumbling motion of Saturn's moon Hyperion (Wisdom et al., 1984), and in elements of Pluto's orbit (Sussman & Wisdom, 1988), and has been attributed to result from overlapping resonances (Lecar et al., 2001).

Given a knowledge of initial conditions and forces acting upon it, the evolution of a conservative N-body system follows deterministic laws, in principle allowing the calculation of its past and future state with arbitrary accuracy. An orbit is loosely described as *chaotic* (as opposed to *regular*) if its dynamical state at some future time is sensitively dependent on the initial conditions.

Chaotic orbital motion may only become apparent after long times, for example only after 10^8 yr in the case of Pluto (Sussman & Wisdom, 1988), so that efficient methods of orbit integration, and procedures to quantify chaotic motion, are pivotal to its consideration.

The tendency to chaos can be quantified. The divergence of two trajectories in *phase space* (the six-dimensional mathematical space defined by its position and velocity vectors) with initial infinitesimal vectorial separation δZ_0 can be written

$$|\delta Z(t)| \approx e^{\Lambda t}|\delta Z_0|, \qquad (2.59)$$

where Λ is the *Lyapunov exponent*, or Lyapunov characteristic exponent (e.g. Lyapunov, 1892; Hénon & Heiles, 1964; Oseledec, 1968; Lichtenberg & Lieberman, 1983; Froeschlé, 1984; Murray & Dermott, 2000). The rate of separation can be different for different orientations of the initial separation vector, and the number of Lyapunov exponents is therefore given by the dimensionality of the phase space. Reference is normally made only to the *maximal Lyapunov exponent* constructed from Equation 2.59

$$\Lambda = \lim_{t \to \infty} \frac{1}{t} \ln \frac{|\delta Z(t)|}{|\delta Z_0|}, \qquad (2.60)$$

since that dictates the predictability of the system.

The evolution of Λ, determined from numerical integration of the orbits, typically shows a marked difference between regular and chaotic behaviour: regular orbits tend to a slope of -1 in $\log \Lambda$ versus $\log t$, and chaotic orbits tend to some positive value of Λ. The corresponding *Lyapunov time*, the time taken for the displacement to increase by a factor e, is given by $1/\Lambda$.

As specific examples, both Sussman & Wisdom (1992) and Laskar (1996) found that the inner solar system is chaotic, with a Lyapunov exponent of $1/(5\,\text{Myr})$. The chaotic behaviour is traced to two secular resonances, one related to the perihelia and nodes of Mars and Earth, the other related to those of Mercury, Venus and Jupiter. As a consequence, an error of only 15 m in the Earth's initial position leads to an error of 150 m after 100 Myr, growing to 150×10^6 km after 100 Myr. It is thus conceivable to construct solar system ephemerides over 10 Myr, but practically impossible to predict planetary motions beyond 100 Myr. Similarly, the orbit of Pluto is found to be chaotic, with a Lyapunov exponent of $1/(20\,\text{Myr})$ (Sussman & Wisdom, 1988), although its small mass means that its chaotic orbit has negligible consequences for the other planets.

A system can be stable, at the same time as being chaotic, so long as its future evolution is restricted to within certain bounds. The absence of a clear pattern in the evolution of the orbital elements with time does not in itself imply that the orbit is chaotic, but nei-

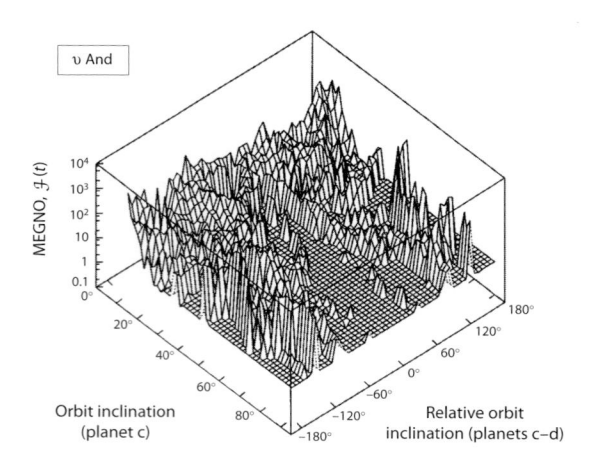

Figure 2.33: Stability analysis for υ And carried out with MEGNO *using the Lick Observatory orbit data from Butler et al. (1999). The parameter $\bar{\mathcal{Y}}$ (Equation 2.64) develops along the vertical axis, as a function of both the orbit inclination of planet c, and the relative orbit inclinations of planets c and d. The flat areas indicate quasi-periodic zones, while the peaks indicate chaotic regions. From Goździewski et al. (2001, Figure 12), reproduced with permission © ESO.*

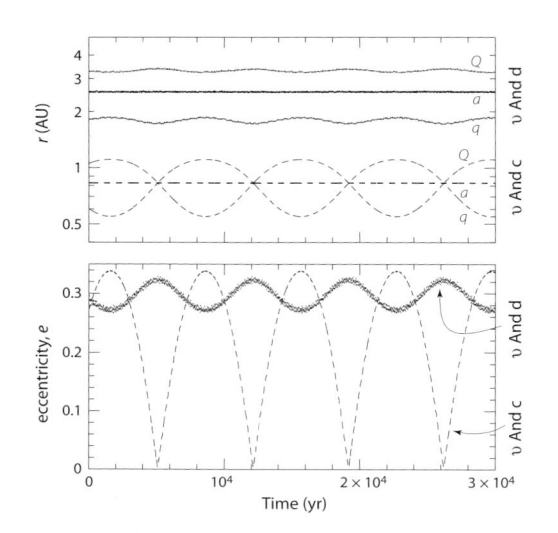

Figure 2.34: Secular evolution of the planetary system around υ And. Top: the semi-major axes, a (central lines), as well as the pericentre and apocentre distances (q and Q respectively), for the outer two planets, c and d. Bottom: evolution of the orbital eccentricities. Both planet c (lower, dashed) and planet d (upper, dotted) have a significant eccentricity at the present time (t = 0), although the eccentricity of c periodically returns to very small values near zero. From Ford et al. (2005, Figure 1), by permission from Macmillan Publishers Ltd, Nature, ©2005.

ther does a chaotic orbit necessarily imply that the system will be catastrophically disrupted. There is no general relation, for example, between the Lyapunov time and the event time when a macroscopic instability of a planetary system occurs (Soper et al., 1990; Lecar et al., 2001; Michtchenko & Ferraz-Mello, 2001b). Neither does the Lyapunov time provide a measure of the 'degree' of chaos, since a system either is, or is not, chaotic.

Indicators of chaos Practical methods for rapidly distinguishing between regular and chaotic orbits include various types of frequency analysis (e.g. Laskar, 1990, 1993) and various types of fast Lyapunov indicators (e.g. Froeschlé et al., 1997; Guzzo et al., 2002).

For the four-planet system μ Ara, for example, Pepe et al. (2007) used the SABA4 symplectic integrator (Laskar & Robutel, 2001) to integrate over 2000 years with a step size of 0.02 yr. Orbit stability was measured using a simplified frequency analysis based on the mean motion of a given planet over two consecutive intervals of $t = 1000$ yr, n_p and n'_p. They used the parameter

$$D = \frac{|n_p - n'_p|}{t} \quad \text{deg yr}^{-1} \qquad (2.61)$$

to provide a simplified measure of the orbit's evolution, with values close to zero implying a regular solution, and with larger values indicating more chaotic motion.

Within the class of fast indicators is MEGNO (Mean Exponential Growth of Nearby Orbits; Cincotta & Simó, 2000), originally developed to study the global dynamics of non-axisymmetric galactic potentials. It has been used to investigate stability and resonant structure in a number of multiple exoplanet systems, and typically performs over a relatively small number ($\sim 10^4$) of orbital periods of the outermost planet. Its essence is to amplify the effects of instability, such that signs of chaotic behaviour can be detected much earlier. This is seen by rewriting Equation 2.60 in integral form as

$$\Lambda = \lim_{T \to \infty} \frac{1}{T} \int_0^T \frac{\delta \dot{\mathbf{Z}}(t)}{\delta \mathbf{Z}(t)} \, dt \,, \qquad (2.62)$$

then introducing the time-amplified *mean exponential growth of nearby orbits* $\mathcal{Y}(T)$ (Cincotta & Simó, 2000, eqn 22) as

$$\mathcal{Y}(T) \equiv \frac{2}{T} \int_0^T \frac{\delta \dot{\mathbf{Z}}(t)}{\delta \mathbf{Z}(t)} t \, dt \,, \qquad (2.63)$$

with its evolving mean value

$$\bar{\mathcal{Y}}(T) \equiv \frac{1}{T} \int_0^T \mathcal{Y}(t) \, dt \,. \qquad (2.64)$$

For regular orbits $\bar{\mathcal{Y}}(T) \to 2$, while for chaotic orbits it grows, proportionally to the Lyapunov exponent, as $\bar{\mathcal{Y}} \to (\Lambda/2)T$. Practical implementation requires the integration of two (straightforward) differential equations in addition to the underlying equations of motion.

Various multiple planet systems have been studied with this indicator, including the triple system υ And (Goździewski et al., 2001, and considered further below), HD 12661 (Goździewski, 2003b; Goździewski &

2.6 Multiple planet systems

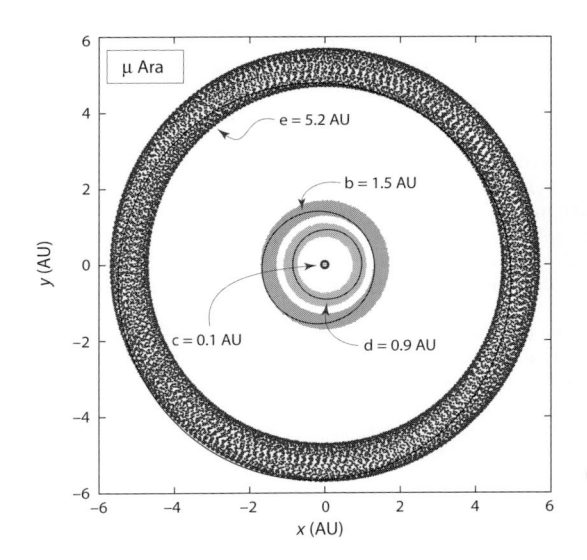

Figure 2.35: Face-on view of μ Ara (HD 160691), illustrating orbital evolution of the four planets over 1 Myr. Present orbits are shown as solid lines, while dots (indistinguishable here for the inner three planets) correspond to positions every 50 years in the dynamical integration. The semi-major axes are constant, while the eccentricities undergo small variations in the range 0.09–0.13, 0.16–0.21, 0–0.19, and 0.08–0.11 for planets b–e respectively. From Pepe et al. (2007, Figure 6), reproduced with permission © ESO.

Maciejewski, 2003), HD 37124 (Goździewski, 2003a; Goździewski et al., 2008a), HD 160691≡μ Ara (Bois et al., 2003; Goździewski et al., 2003, 2005a), HD 169830 (Goździewski & Konacki, 2004), the quintuple 55 Cnc (Gayon et al., 2008), and the possibly retrograde resonance system HD 73526 (Gayon & Bois, 2008a).

A related chaos-detection method is through the use of *relative Lyapunov indicators*, which measure the difference between the convergence of the finite time Lyapunov indicators to the maximal Lyapunov characteristic exponent of two initially very close orbits (Sándor et al., 2000, 2004). The method is considered to be very fast in determining the ordered or chaotic nature of individual orbits, even those with very small Lyapunov characteristic exponents, with results over a few hundred times the period of the longest orbit. Examination of the stability of various exoplanet systems using this indicator has been variously described (e.g. Érdi et al., 2004; Sándor et al., 2004, 2007b).

2.6.4 Systems with three or more giant planets

υ And The first exoplanet system discovered to be multiple was υ And. In addition to the 0.6 M_J object in a low-eccentricity 4.6-d orbit originally detected by Butler et al. (1997), two more distant planets were identified from subsequent radial velocity observations, with $M_p \sin i$ of 2.0 and 4.1 M_J, a of 0.82 and 2.5 AU, and large e of 0.23 and 0.36 respectively (Butler et al., 1999).

The innermost planet, with $e \sim 0$, significantly exceeds the minimum stability requirement given by the Hill radius criterion, suggesting little interaction with the outer two companions. For the two outer planets (c and d) in contrast, the system stability depends strongly on the planet masses, and hence their relative orbital inclinations, with certain combinations implying chaotic or unstable orbits.

A number of numerical studies have been carried out to assess the system stability over long time scales (Holman et al., 1997; Krymolowski & Mazeh, 1999; Laughlin & Adams, 1999; Lissauer, 1999; Rivera & Lissauer, 2000; Stepinski et al., 2000; Barnes & Quinn, 2001; Chiang et al., 2001; Jiang & Ip, 2001; Lissauer & Rivera, 2001; Chiang & Murray, 2002; Michtchenko & Malhotra, 2004; Michtchenko et al., 2006). Mazeh et al. (1999), for example, derived a mass for the outer planet of 10.1 ± 4.7 M_J using Hipparcos astrometry, implying an orbital inclination of 156°, and masses of the inner two planets of 1.8 ± 0.8 and 4.9 ± 2.3 M_J if the orbits are coplanar.

Goździewski et al. (2001) used the fast orbit indicator MEGNO to derive $\langle \mathcal{Y} \rangle (t)$ as a function of trial values of the inclination of planet c, and the relative inclinations of planets c and d (Equation 3.17). Quasi-periodic and chaotic zones show up prominently in various regions of the i_c, i_{c-d} parameter space (Figure 2.33).

From direct numerical integration of the orbits, using Mercury, Ford et al. (2005) found that for a coplanar configuration the system becomes dynamically unstable when $\sin i < 0.5$ while, if the coplanarity condition is relaxed, dynamical instability resulted from relative inclinations exceeding $\sim 40°$. They found that the long-term secular evolution of the orbits (Figure 2.34) was best modeled as resulting from an impulsive perturbation to the outer planet, υ And d, in which a sudden change in its eccentricity was produced by a close encounter with another planet. After a brief period of chaotic evolution lasting $\sim 10^3$ yr, the perturbing planet would have been ejected, leaving the remaining two outer planets in a stable configuration resembling that observed today.

HD 37124 HD 37124 is an example of dynamical stability modeling leading to the discovery of additional planets. Vogt et al. (2000) detected a 150-d Jupiter-mass planet from Keck–HIRES observations. Butler et al. (2003) identified a second planet with $P \sim 6$-yr. The two-planet solution was shown to be unstable by Goździewski (2003a), with further Keck data identifying a third planet (Vogt et al., 2005). $\{M_p \sin i\ (M_J), P\ (d), e\}$ for planets b, c, d are respectively {0.638, 154, 0.055}, {0.697, 2300, 0.2}, {0.624, 844, 0.14}. Goździewski et al. (2006) showed that the orbit of the outer planet is still poorly constrained, while planet–planet interactions are strong enough that kinematic modeling is insufficient to describe the system adequately.

> **Titius–Bode law:** As noted by Johann Titius in 1766, and as reformulated and popularised by Johann Bode (1747–1826), the mean orbital distances of the six planets known at the time (Mercury to Saturn) are well approximated by
>
> $$a = 0.4 + 0.3\,(2)^i\,, \quad \text{where } i = -\infty, 0, 1, 2, 4, 5\,, \tag{2.65}$$
>
> and a is the Sun–planet distance in AU. The discovery of Uranus in 1781 at 19.2 AU ($i = 6$), and the first asteroid Ceres in 1801 at 2.8 AU ($i = 3$) were considered as predictions confirming an underlying structural law of the solar system. As a 'law', however, it has two shortcomings: it has, currently, no physical basis, and it has too many exceptions (notably the presence of Neptune, and the absence of planets between Mercury and Venus). The topic has nevertheless generated considerable debate (Nieto, 1972), which continues to the present, both in numerical and dynamical form (e.g. Lynch, 2003; Neslušan, 2004; Bass & del Popolo, 2005; Ortiz et al., 2007; Bohr & Olsen, 2010), or perhaps as the *'psychological tendency to find patterns where none exist'* (Newman et al., 1994).
>
> Murray & Dermott (2000, Section 1.5) assessed its significance by applying a two-parameter geometrical progression to the major satellites of Uranus (Miranda to Oberon), as a distinct instance of orbital periods also displaying a number of simple numerical relationships. Although generating excellent numerical fits, their Monte Carlo analysis demonstrated that almost any distribution of periods would fit a Titius–Bode-type law equally well. They concluded that *'There is no compelling evidence that the Uranian satellite system is obeying any relation similar to the Titius–Bode law beyond what would be expected by chance. This leads us to suggest that the 'law' as applied to other systems, including the planets themselves, is also without significance.'*
>
> Hayes & Tremaine (1998) made a similar statistical analysis for the solar system, and concluded that the law's significance is simply that stable planetary systems tend to be regularly spaced; they suggested that the conclusion could be strengthened by more rigorous rejection of unstable planetary systems based on long-term orbit integration.
>
> Multiple exoplanet systems have opened an extended panorama for further conjecture (Poveda & Lara, 2008; Kotliarov, 2008, 2009; Panov, 2009; Chang, 2010). Poveda & Lara (2008) made a Titius–Bode type fit to the five planets orbiting 55 Cnc. They predict a further planet with $a \simeq 2.08$ AU ($P \simeq 1130$ d) in the gap between $a = 0.781 - 5.77$ AU (evident in Figure 2.29), and possibly another at $a \simeq 15.3$. A critique is given by Kotliarov (2009).
>
> An alternative structural law discussed by Kotliarov (2008), which he attributes to Butusov (1973), is related to symmetry. It is similarly phenomenological but claims fewer exceptions. It can be expressed as
>
> $$a_Z^2 = a_{Z+k} \times a_{Z-k}\,, \tag{2.66}$$
>
> where a is the star–planet distance, and Z is the planet defining the centre of symmetry, with $k \in \{1, 2, \ldots, Z-1\}$. Applied to the solar system, it implies that there is a Jovian symmetry to planetary distances, counting Jupiter as $k = 6$ with Ceres as $k = 5$. Applied to 55 Cnc, Kotliarov (2008) predicts planets at different locations to Poveda & Lara (2008), viz. at $a \simeq 2.5$ and $a \simeq 16.1$.

HD 74156 The two planet system was originally reported by Naef et al. (2004) from ELODIE data. A third planet was detected from a combination of CORALIE, ELODIE and HET–HRS data by Bean et al. (2008b), with the properties $M_p \sin i = 0.40 \pm 0.02 M_J$, $P = 347$ d, and $e = 0.25 \pm 0.11$.

Raymond & Barnes (2005) had predicted the parameter space that such a third planet could have occupied in this and three other systems, by identifying stability regions for massless test particles defined by the previously known planets. For HD 74156 they had shown that a Saturn mass planet could exist within a broad region of $a = 0.9 - 1.4$ AU and $e \leq 0.15$. Barnes et al. (2008a) gave an alternative slightly improved orbit, and suggested that the confirmed prediction of the third planet points to planet formation being an efficient process, with planetary systems typically containing many planets.

μ Ara (HD 160691) The properties of this four-planet system were progressively revealed by AAT–UCLES (Butler et al., 2001; McCarthy et al., 2004) and HARPS (Santos et al., 2004a; Pepe et al., 2007) data. The fourth planet was also tentatively announced by Goździewski et al. (2007). Simulations demonstrating the system stability are shown in Figure 2.35. The semi-major axes remain almost constant, while the eccentricities of all four planets undergo small variations. Various other dynamical analyses have also been made (Bois et al., 2003; Goździewski et al., 2003, 2005a).

55 CnC The 55 Cnc system was the first known planetary system comprising five planets (Table 2.5). The first planet, 55 Cnc b, was reported by Butler et al. (1997), and the second (and possibly a third) also from Lick data by Marcy et al. (2002). 55 CnC c was confirmed by McArthur et al. (2004) based on Lick, ELODIE and HET Doppler measurements and HST–FGS astrometry. They also announced 55 Cnc e with $P = 2.8$-d.

Fischer et al. (2008) used 18 years of combined Lick and Keck data to confirm the four proposed planets, and to identify a fifth, 55 Cnc f, moving in the large empty zone between two other planets. All five planets reside in low-eccentricity orbits, four having $e < 0.1$. They adopted a multi-planet Keplerian fitting procedure (§2.1.1), using a numerical N-body simulation to show that the system is dynamically stable.

55 Cnc has some basic structural attributes found in our solar system: the orbits are rather circular and nearly coplanar, and a dominant gas giant lies at a distance of about 6 AU. Although planets b and c have a period ratio

2.6 Multiple planet systems

Table 2.5: *The planets of 55 Cnc, ordered by increasing semi-major axis. Data are from Fischer et al. (2008).*

Planet	$M_p \sin i$ (M_J)	a (AU)	P (d)	e
e	0.02	0.038	2.80	0.26
b	0.83	0.115	14.65	0.02
c	0.17	0.240	44.38	0.05
f	0.15	0.789	260.67	0.00
d	3.90	5.888	5371.82	0.06

of 3.027:1.000, hinting at a possible mean motion resonance (Marcy et al., 2002; Ji et al., 2003b; Voyatzis & Hadjidemetriou, 2006), the 3:1 mean motion resonance was excluded by the N-body model of Fischer et al. (2008), as none of the relevant resonant arguments librate.

Other dynamical simulations were made by Gayon et al. (2008) using MEGNO, who found that about 15% of the systems resulting from the nominal orbital elements of the system are highly chaotic. Raymond et al. (2008b) evaluated the stability of the large region between planets f and d using N-body integrations that included an additional, yet-to-be-discovered planet g with a radial velocity amplitude of ~5 m s^{-1}, i.e. $M_p \simeq 0.5-1.2 M_{Saturn}$. They found a large stable zone extending from 0.9–3.8 AU with $e < 0.4$, which could contain 2–3 additional planets each of $M_p \sim 50 M_\oplus$. Any planets exterior to planet d must reside beyond 10 AU.

2.6.5 Systems in mean motion resonance

The 2:1 resonance system GJ 876 GJ 876 was the first known M-dwarf host and, from Hipparcos, the fortieth nearest stellar system to the Sun. The orbit elements are particularly well determined due to their short periods, and orbit modeling shows that the two most massive planets b and c are locked in a 2:1 resonance, the first to be discovered amongst exoplanet systems.

Marcy et al. (1998) using Keck, and Delfosse et al. (1998) using ELODIE/CORALIE, both reported a 61-d, $2M_J$ planet. From further Keck observations, Marcy et al. (2001a) showed that the radial velocity signal was actually the combination of two planets in a 2:1 mean motion resonance, with the inner planet having $P = 30.1$ d and $0.6M_J$. In the original data, the inner companion was indistinguishable from an additional orbital eccentricity of the outer planet. The axes of the two orbits are nearly aligned, and the orbital elements evolve significantly with time. A third, non-interacting, $7.5M_\oplus$ mass planet in a 1.9 d orbit was detected by Rivera et al. (2005).

Laughlin & Chambers (2001), Rivera & Lissauer (2001), and Nauenberg (2002a) independently developed fitting procedures to account for the gravitational interactions between the two outer planets. Rivera & Lissauer (2001) found that most of their solutions are stable for at least 10^8 yr, while test particles orbiting between the two planets are lost in $\lesssim 300$ yr.

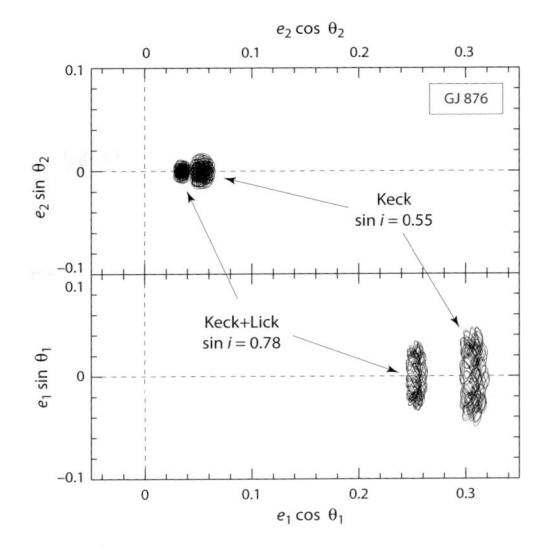

Figure 2.36: *Small-amplitude librations, about 0°, of the two 2:1 mean-motion resonance variables for the GJ 876 planets. Trajectories for 3100 days (the average pericentre precession period for both planets) are shown as plots of $e_j \sin j$ versus $e_j \cos j$ ($j = 1, 2$) for the best-fit solutions to the Keck data alone, and for the combined Keck and Lick data, both from Laughlin & Chambers (2001). The values of $\sin i$ are obtained in the best-fit solution. From Lee & Peale (2002, Figure 1), reproduced by permission of the AAS.*

Lee & Peale (2002) showed that both mean motion resonance variables, $\phi_1 = 2\lambda_2 - \lambda_1 - \tilde{\omega}_1$ and $\phi_2 = 2\lambda_2 - \lambda_1 - \tilde{\omega}_2$ (cf. Equation 2.52), perform very small amplitude librations around 0° (Figure 2.36). Their simultaneous libration about 0° implies that the variable $\phi_3 = \phi_1 - \phi_2 = \tilde{\omega}_1 - \tilde{\omega}_2$ also librates around 0°, and the planets are consequently in three resonances at the 2:1 mean motion commensurability. The small libration of ϕ_3 means that the lines of apsides of the two orbits are therefore nearly aligned, and conjunctions of the two planets occur very close to their longitudes of pericentre (Beaugé & Michtchenko, 2003).[13] The existence of the mean motion resonance is taken as confirmation that the two orbits are essentially coplanar, with the deep resonance (i.e. with small libration amplitudes) implying that the system is stable indefinitely.

Laughlin et al. (2005a) incorporated additional velocity data obtained over 16 years, and confirmed the small libration amplitudes (Figure 2.37). Such a configuration can be explained by a slow differential migration resulting from an interaction of the planets with a protoplanetary disk (Snellgrove et al., 2001; Lee & Peale, 2002; Kley et al., 2005; Beaugé et al., 2006; Crida et al., 2010).

---

[13] The nature of the resonance contrasts with the case of the Io–Europa system, where the lines of apsides are anti-aligned, and where conjunctions therefore occur when Io is near its pericentre, and Europa is near its apocentre.

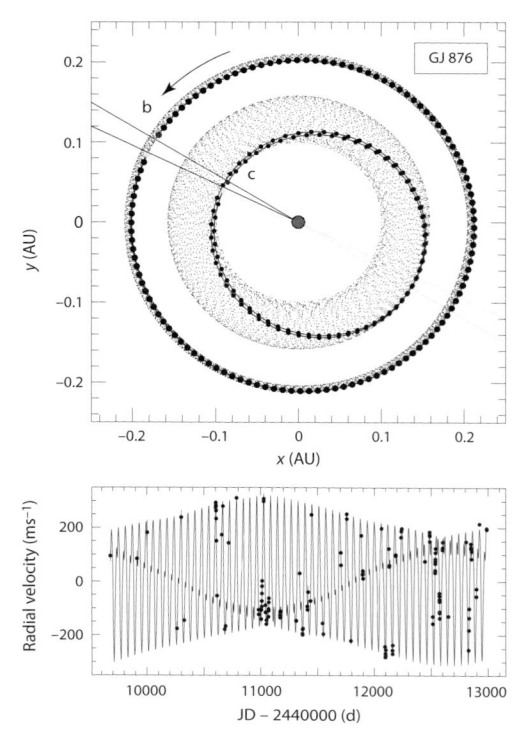

Figure 2.37: *The 2:1 mean motion resonance in GJ 876. Top: the two clouds of small dots show the planet positions every 0.5 d over almost 10 yr, illustrating the precession in the line of apsides of $-41°$ yr^{-1}. Connected circles show the planet positions every 0.5 d for 60 d. The two lines radiating from the central star mark the longitudes of pericentre, $\tilde{\omega}_b$ and $\tilde{\omega}_c$, at JD 2 449 710. They oscillate about alignment with a libration amplitude of $34°$. Bottom: the measured stellar reflex velocities (circles), and the solution from the self-consistent, coplanar, three-body integration. From Laughlin et al. (2005a, Figures 1–2), reproduced by permission of the AAS.*

Laughlin et al. (2005a) also showed that configurations with modest mutual inclinations are possible, which has an interesting corollary: for non-coplanar configurations the line of nodes of the inner planet precesses at about $-4°$ yr^{-1}. The inner planet may therefore be observed to transit the host star at some time in the relatively near future, i.e. when either the ascending or descending node precesses through the line of sight, even though it is not transiting at the present time.

Laplace resonance in GJ 876 A fourth planet, GJ 876 e, was discovered from continued radial velocity monitoring by Rivera et al. (2010). N-body fits show that the four-planet system has an invariable plane with an inclination of $59°.5$, and is stable for more than 1 Gyr. Their model places the fourth planet in a three-body (Laplace) resonance with the two giant planets ($P_c = 30.4$ d, $P_b = 61.1$ d, $P_e = 126.6$ d). Unlike the case of the Galilean satellites of Jupiter (Figure 2.38), the three planets come close to a triple conjunction once per orbit of planet e.

The critical argument for the Laplace resonance, $\phi = \lambda_c - 3\lambda_b + 2\lambda_e$, librates with an amplitude of $40° \pm 13°$.

The 2:1 resonance system HD 128311 The two-planet system HD 128311 offers an interesting insight into planetary migration along similar lines. The first planet was reported by Butler et al. (2003), and the second by Vogt et al. (2005). A Trojan 1:1 solution was suggested by (Goździewski & Konacki, 2006), while the most recent orbit solution is given by Wittenmyer et al. (2009).

Sándor & Kley (2006) performed a three-body numerical integration of the (Newtonian) orbital solution of Vogt et al. (2005), and examined the evolution of the difference in the longitudes of pericentre $\Delta\tilde{\omega} = \tilde{\omega}_2 - \tilde{\omega}_1$, and the two appropriate resonant angles ϕ_1, ϕ_2 (Equation 2.52 corresponding to $\tilde{\omega}_1, \tilde{\omega}_2$ respectively). Of these, only ϕ_1 librates around $0°$ with an amplitude of $\sim 60°$, while ϕ_2 and $\Delta\tilde{\omega}$ rotate. At the same time, the eccentricities show large oscillations (Figure 2.39).

This behaviour is in contrast with the apsidal corotation predicted by adiabatic migration (Figure 2.39, top right). Instead, Sándor & Kley (2006) considered that the system's present state can be attributed to a strong scattering event in the past. Assuming that the two planets were once locked in a 2:1 resonance with apsidal corotation as a result of inward migration, they then studied two distinct perturbation scenarios which might have disrupted the corotation (Figure 2.39, bottom).

In the first, a sudden halting of migration was invoked. This might arise, for example, when the outer planet reaches some disk discontinuity with an empty region inside; such a model would be consistent with Spitzer observations of young stars which have shown that the inner part of protoplanetary disks may indeed contain only very little mass, possibly due to photo-evaporation by the central star (D'Alessio et al., 2005; Calvet et al., 2005). In the second scenario, an encounter with a low-mass $10M_\oplus$ planet approaching from inside or outside (similar to the perturbation proposed by Ford et al. 2005 for υ And) could also break the apsidal libration. The smaller planet would be ejected from the system, or thrown into a larger orbit, in the process.

Particularly in the case of a small planet approaching from the inside, the apsidal corotation of the giant planets is indeed broken. Consistent with the behaviour predicted from long-term orbit integration, ϕ_2 and $\Delta\tilde{\omega}$ then circulate, while the giant planets remain in the 2:1 resonance, with ϕ_1 still librating around $0°$.

The 3:1 antisymmetric resonance system HD 60532
Amongst systems in which a 3:1 mean motion resonance has been suggested, that of the two giant planets orbiting HD 60532 appears reasonably secure (Desort et al., 2008; Laskar & Correia, 2009; Sándor & Kley, 2010). The two orbital solutions found by (Laskar & Correia, 2009)

2.6 Multiple planet systems

(a) Galilean satellites of Jupiter

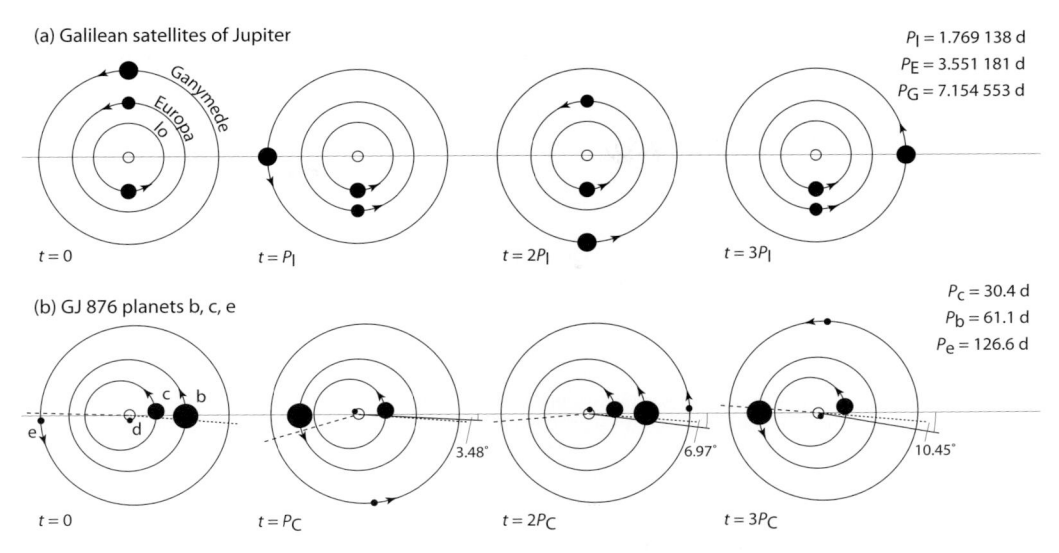

$P_I = 1.769\,138$ d
$P_E = 3.551\,181$ d
$P_G = 7.154\,553$ d

$t = 0$ $t = P_I$ $t = 2P_I$ $t = 3P_I$

(b) GJ 876 planets b, c, e

$P_c = 30.4$ d
$P_b = 61.1$ d
$P_e = 126.6$ d

$t = 0$ $t = P_c$ $t = 2P_c$ $t = 3P_c$

3.48°   6.97°   10.45°

Figure 2.38: The two known examples of Laplace resonance, in which three (or more) orbiting bodies have a simple integer ratio between their orbital periods: (a) Jupiter's inner satellites Ganymede, Europa, and Io are in a 1:2:4 Laplace resonance, with Ganymede completing 1 orbit in the time that Europa makes 2, and Io makes 4. They never experience triple conjunctions. Their mean motions (Equation 2.8) are related by $n_I - 3n_E + 2n_G = 0$, satisfied to nine significant figures (Peale, 1976); (b) planets c, b, e of the M dwarf host star GJ 876 are an exoplanet triplet in a Laplace resonance. The reference frame rotates at the mean orbital precession of planet b, $-10°.45$ over 90 d (solid lines). The apsidal line of b coincides with the x-axis, while those for planets c and e are shown with short-dash and long-dash lines respectively. In each system, the orbits and object radii (assuming $R \propto M^{1/3}$) are shown to scale. Lower figure is adapted from Rivera et al. (2010, Figure 7).

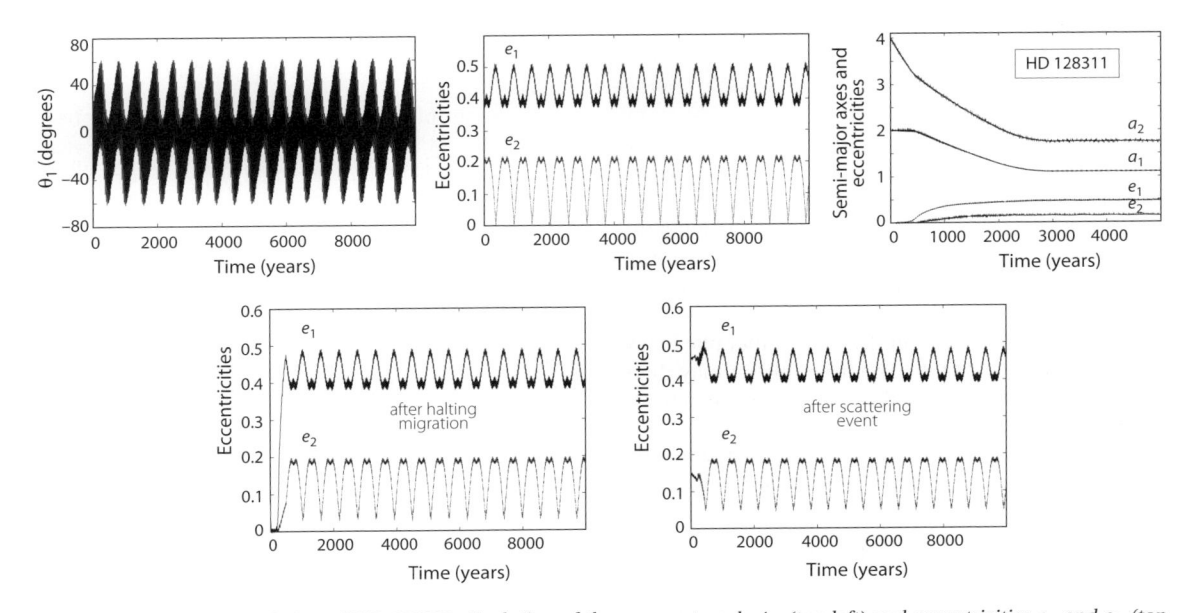

Figure 2.39: Resonant evolution of HD 128311. Evolution of the resonant angle ϕ_1 (top left) and eccentricities e_1 and e_2 (top middle) obtained by numerical integration of the orbital parameters of Vogt et al. (2005). Top right: predicted behaviour of the semi-major axes and eccentricities during adiabatic migration with an e-folding time of 2×10^3 yr; for the simulations, the ongoing migration is progressively stopped between $2 - 3 \times 10^3$ yr. Bottom left: evolution of the planetary eccentricities, both originally in circular orbits, following a sudden halting in the migration of the outer planet. Bottom right: evolution of the eccentricities following a scattering event with a low-mass planet migrating outwards. From Sándor & Kley (2006, Figures 1, 2, 3, 5), reproduced with permission ©ESO.

> **Solar system 1:1 resonances and the Trojans:** In the circular restricted three-body problem, three bodies move in circular coplanar orbits, with the mass of the third being negligible. In 1772, Lagrange proved the existence of five equilibrium points where the third particle has zero velocity and zero acceleration in the rotating frame. These are the three colinear Lagrangian equilibrium points L_1, L_2, L_3, and two (leading and trailing) triangular equilibrium points L_4, L_5. Although L_1-L_3 are unstable, special starting conditions of the third particle, in position and velocity, can nevertheless result in semi-stable periodic orbits in their vicinity. This is exploited by artificial satellites such as SOHO which observes the Sun from L_1 (Domingo et al., 1995), and others at (WMAP, Herschel, Planck) or destined for (e.g. Gaia) L_2.
>
> Stable orbits in the vicinity of L_4 or L_5 were first demonstrated by Routh (1875) for the special case of circular orbits, and by Danby (1964a,b) for elliptical orbits. They comprise a short-term component of motion with a period close to the orbital period, and a longer term periodic motion about the equilibrium point referred to as libration. Resulting orbits can be described as a short-period epicyclic motion around a long-period motion of the epicentre, the relative contributions of the two components determined by the starting conditions. The resulting elongated orbits around L_4 or L_5 are referred to as *tadpole orbits*. For an increased radial separation from L_4 or L_5, the resulting *horseshoe orbits* can encompass both L_4 and L_5.
>
> Examples of solar system bodies in such stable 1:1 resonances are referred to as *Trojans*, and they include both asteroids and satellites. Librating mostly in tadpole orbits around the Sun–Jupiter triangular equilibrium points is the leading group around L_4 (the 'Greeks'), including the first known (588) Achilles discovered by Max Wolf in 1906, and the trailing group around L_5 (the 'Trojans'). Typical libration amplitudes are $15-20°$ (Shoemaker et al., 1989).
>
> Other Trojan asteroids orbit the Sun–Mars system, with the first (5261) Eureka, discovered in 1990, librating about L_5 (Mikkola et al., 1994). Asteroid (3753) Cruithne is in a horseshoe orbit around the Sun–Earth system (Wiegert et al., 1997).
>
> The dynamics of planet–satellite systems around their triangular equilibrium points are identical, and result in the *coorbital* or *Trojan satellites*. Telesto and Calypso librate around the Saturn–Tethys system, and Helene and Polydeuces around the Saturn–Dione system. In the curious Janus–Epimetheus Saturnian system, Epimetheus moves in a horseshoe orbit, exchanging altitude with Janus as they approach in a four-year repeating cycle (Dermott & Murray, 1981a,b), and resulting in forced rotational libration (Tiscareno et al., 2009).

have very different inclinations, $i = 20°$ or $i = 90°$, with planetary masses differing by a factor $1/\sin i \sim 3$.

Sándor & Kley (2010) carried out detailed two-dimensional hydrodynamical simulations of appropriate thin disks with an embedded pair of massive planets, studying the effects of migration and resonant capture using N-body simulations.

For resonant capture, planet pairs must undergo convergent migration, such that if the inner planet migrates inwards at the same speed, or faster than the outer one, then no resonant capture occurs. Resonant capture through migration therefore depends on the inner planet opening a suitably large gap in the inner disk to slow down its inward migration rate. In their simulations, Sándor & Kley (2010) found that capture into the observed 3:1 resonance takes place only for higher planetary masses, thus favouring orbital solutions having the smaller inclination of $i = 20°$ (Figure 2.40).

They also found that the inner disk, between the inner planet and the star, plays a key role in determining the final configuration of this, and by implication other, resonant systems. Specifically, fast inward migration of the outer planet may result in it crossing the 3:1 mean motion resonance without capture, ending the migration in the more robust 2:1 mean motion resonance. In the case of HD 60532, the damping effect of the disk on the inner planet's eccentricity is responsible for an antisymmetric (antialigned) resonance configuration in which $\tilde{\omega}_2 - \tilde{\omega}_1$ oscillates around $180°$, rather than the $0°$ in the case of aligned pericentres.

Other mean motion resonances Other systems believed to be in a 2:1 mean motion resonance include HD 82943 (Goździewski & Maciejewski, 2001; Ji et al., 2003a; Mayor et al., 2004; Ferraz-Mello et al., 2005; Lee et al., 2006); HD 73526 (Tinney et al., 2003, 2006; Sándor et al., 2007a); and μ Ara d–b (Pepe et al., 2007).

Other possible 3:1 resonances include 14 Her (Goździewski et al., 2006, 2008b), and the contested case of 55 Cnc b–c (Marcy et al., 2002; Ji et al., 2003b; Fischer et al., 2008).

A number of other possible resonances have been identified: the 4:1 resonance of HD 108874 (Butler et al., 2003; Goździewski et al., 2006) and perhaps 14 Her (Wittenmyer et al., 2007); the 5:1 resonance of HD 202206 (Udry et al., 2002; Correia et al., 2005; Goździewski et al., 2006); a possible 3:2 resonance of HD 45364 (Correia et al., 2009; Rein et al., 2010); a possible 5:2 resonance of HD 37124 (Goździewski et al., 2006); a possible 6:1 or 11:2 resonance of HD 12661 (Goździewski, 2003b; Lee & Peale, 2003; Rodríguez & Gallardo, 2005; Zhang & Zhou, 2006; Veras & Ford, 2009); and a possible 5:2 or 7:3 resonance in 47 UMa (Laughlin et al., 2002).

Proximity to resonance Proximity to a resonance may still affect the secular motion, as in the case of the near 5:2 resonance between Jupiter and Saturn, Laplace's 'great inequality' (e.g. Varadi et al., 1999; Michtchenko & Ferraz-Mello, 2001a). Such proximity has been suggested for HD 12661 and υ And (Libert & Henrard, 2007).

Inclination resonance In the absence of observational data (as opposed to theoretical dynamical constraints) on the mutual inclination of exoplanet orbits, most studies of mean motion resonances have focused on coplanar configurations. The extension to non-

2.6 Multiple planet systems

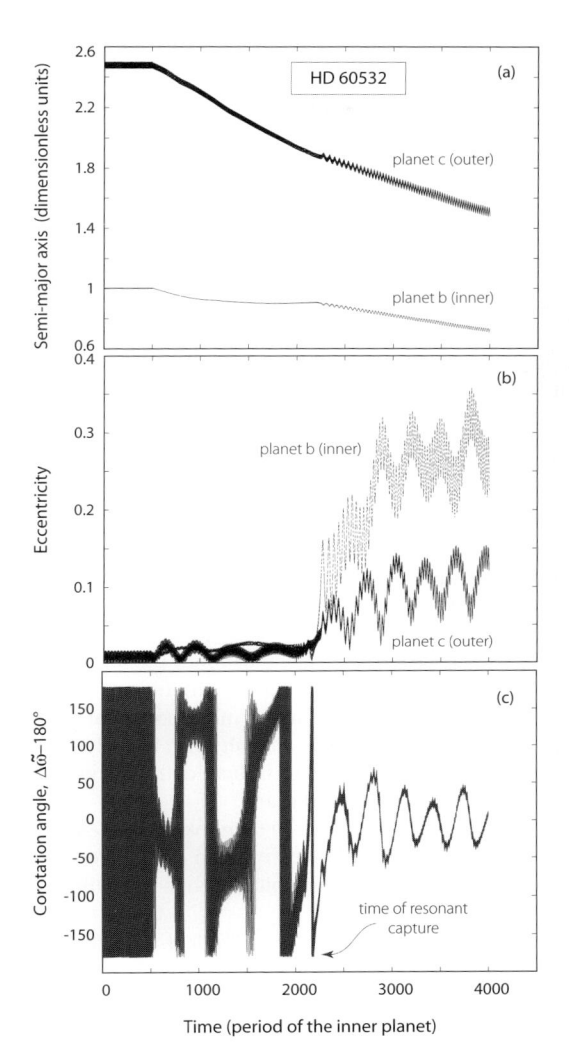

Figure 2.40: Dynamical simulation of the two giant planets orbiting HD 60532, embedded in a protoplanetary disk, showing (a) semi-major axes, (b) eccentricities, and (c) the evolution of the resonant angle $\Delta \tilde{\omega} = \tilde{\omega}_2 - \tilde{\omega}_1$. During the first 500 orbits of the inner planet, the planets are 'fixed' to obtain a steady state in the disk. Capture into the 3:1 mean motion resonance occurs after ~2300 periods of the inner planet, subsequent to which the orbits exhibit (antialigned) apsidal corotation. The planetary masses (3.15M_J and 7.46M_J) correspond to a (coplanar) system inclination of $i = 20°$. From Sándor & Kley (2010, Figures 4 and 5b), reproduced with permission © ESO.

coplanar orbits is relevant for an understanding of eccentricity excitation and migration.

Thus Thommes & Lissauer (2003) found that, subsequent to a 2:1 eccentric resonance capture, a subsequent capture into a 4:2 *inclination resonance* is possible. This is the lowest-order inclination resonance at the 2:1 commensurability (since the mutual inclination appears first as $(\Delta i)^2$ in the series expansion of the disturbing function), and is characterised by the mean motion resonance variables (cf. Equation 2.54)

$$\phi_{11} = 2\lambda_1 - 4\lambda_2 + 2\Omega_1 \quad \text{and}$$
$$\phi_{22} = 2\lambda_1 - 4\lambda_2 + 2\Omega_2 \,, \qquad (2.67)$$

where Ω_j are the longitudes of the ascending nodes. The properties and subsequent evolution of this resonance are detailed by Lee & Thommes (2009).

Capture into higher-order resonances (such as the 3:1, 4:1, and 5:1) can also result in the excitation of inclinations, with mutual inclinations reaching $\Delta i \sim 20-70°$ in the simulations of Libert & Tsiganis (2009b), at least for low-mass inner planets and for one or both planets developing eccentricities $e \gtrsim 0.4$. Amongst the current systems believed to be in higher-order resonance, HD 60532 (3:1), HD 108874 (4:1) and HD 102272 (4:1) do not satisfy these requirements, and are therefore likely to reside in coplanar resonances, an inference verified by numerical simulations.

Kozai resonance Large mutual inclinations between the orbits of multiple planets could be generated by (Kozai) resonant interactions, as well as more directly by planet–planet scattering and planetesimal-driven migration. Whether any of the known systems are in a stable Kozai resonant state, with mutual orbital inclinations of order $\Delta i \sim 40-60°$, was investigated parametrically by Libert & Tsiganis (2009a). They found that four of the systems studied (υ And, HD 12661, HD 74156, and HD 169830) could in principle be in Kozai resonance, provided that their mutual inclination is at least 45°. Direct astrometric determination of the mutual inclinations in υ And by HST–FGS (§3.8.2, McArthur et al., 2010) gave $\Delta i_{cd} = 29°\!.9 \pm 1°$.

The 1:1 resonance A number of objects in the solar system move in 1:1 resonance orbits. They are examples of Lagrange's celebrated solution to the restricted three-body problem, and generally satisfy the condition that the mass of the smallest is negligible compared with that of the other two (see box).

For the more general three-body problem, and in particular for two equal mass planets which share a time-averaged orbital period, Laughlin & Chambers (2002) identified a wide parameter space in which stable coorbital configurations exist (Figure 2.41). These fall into rather distinct regimes. In the first, the star and both planets participate in tadpole-like librations around the vertices of an equilateral triangle. The dynamics resembles that of Jupiter's Trojan asteroids, and the configuration is stable for mass ratios $2M_p/(2M_p+M_\star) < 0.03812$. In the second regime, for larger perturbations, the planets execute symmetrical horseshoe-type orbits similar to those of Janus and Epimetheus. Stability analysis indicates that a pair of Saturn-mass planets could survive in this resonance over long periods (Figure 2.41a).

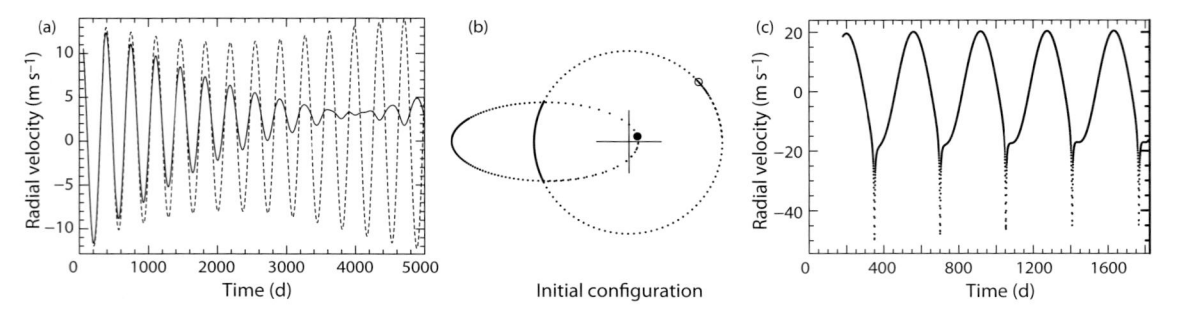

Figure 2.41: Theoretical 1:1 resonance orbits for equal mass planets: (a) synthetic radial velocity variations predicted for the tadpole-type (dotted) and horseshoe-type (solid) orbits; (b) initial configuration for the 1:1 eccentric resonance; solid lines indicate parts of the respective orbits swept out over 15% of the orbital period; (c) associated radial velocity variations for the case of eccentric resonance. From Laughlin & Chambers (2002, Figures 3, 4, 6), reproduced by permission of the AAS.

In practice, recognising such systems from the radial velocity data might be problematic, since the periodogram signal can be indistinguishable from that of a single planet in an eccentric orbit (Laughlin & Chambers, 2002), or from that of a system in 2:1 mean motion resonance (Goździewski & Konacki, 2006).

Their third configuration, described as *eccentric resonance*, is qualitatively distinct from Lagrange's solution (Figure 2.41b). One orbit starts as highly eccentric, the other is more nearly circular, the pericentres are aligned, and conjunctions occur near pericentre; the pair of planets then exchange angular momentum and eccentricity (in the case studied, with a periodicity of ~800 yr). With appropriate initial conditions, they can avoid close encounters indefinitely. In this regime, the radial velocity variations are distinctive, perhaps recognisable by a single deviating measurement superposed on an otherwise sinusoidal pattern (Figure 2.41c).

A similar configuration was found independently by Nauenberg (2002b). These eccentric resonances might arise from the same type of planet–planet interactions responsible for non-resonant high eccentricity systems.

The theoretical existence and properties of the various 1:1 resonance configurations (Hadjidemetriou et al., 2009) has led some investigations to consider them as alternative models for certain specific systems (e.g. Érdi et al., 2007), in particular in cases where long-term orbit integration has identified dominant instabilities.

Goździewski & Konacki (2006) examined the long-term stability of HD 128311 (§2.6.5), searching for alternative 2:1 resonances using a genetic algorithm, and examining their long-term stability using MEGNO. They also searched for 1:1 resonances consistent with the observed radial velocity variations of the host star, and identified extended zones of stability in which such a system would survive (Figure 2.42). They used the term 'Trojan planets' for all 1:1 mean motion resonances, including the eccentric resonance configurations identified by Laughlin & Chambers (2002), i.e. not only for the Lagrange-type solutions, but also for those with similar semi-major axes, but possibly with large relative inclinations and time-varying eccentricities.

In the multiple planet system μ Ara studied by Pepe et al. (2007) one of their solutions, located using a genetic algorithm search, gave an inner planet at 9.64 d, an outer planet at 2741 d, and with an intermediate pair of Trojan planets in a 1:1 mean motion resonance. The solution was discarded as a result of its high dynamical instability; orbit integration indicating that it would be disrupted in less than 100 years.

Schwarz et al. (2007a,c,b) demonstrated the stability of Trojan terrestrial planets in mean motion resonance with a giant planet in the habitable zone for a number of systems.

The formation of a hypothetical terrestrial-type body at the Lagrange points of a giant exoplanet has been studied using an N-body code by Beaugé et al. (2007). Although such planets may form *in situ*, the accretion process is inefficient, and the mass of the final planet never exceeds ~ $0.6 M_\oplus$.

Retrograde resonances All known exoplanets are generally considered to revolve in the same direction around their respective system barycentres, and most fitted radial velocity orbit elements are derived by assuming *prograde* or *regular* orbits, i.e. all orbiting in the same direction. Such behaviour is expected, at least for unperturbed states, according to current theories of planet formation.

The theoretical existence and dynamical stability of *retrograde* mean motion resonances, i.e. with one planet counter-revolving with respect to the other in a two-planet system, or to all other planets in a multiple planet system, has been investigated and demonstrated (Gayon & Bois, 2008a,b; Gayon et al., 2009).

A planet might conceivably be found in a retrograde mean motion resonance either as a result of capture of a pre-existing free-floating planet directly into a retrograde orbit, or through (violent) dynamical evolution as

2.7 Planets around binary and multiple stars

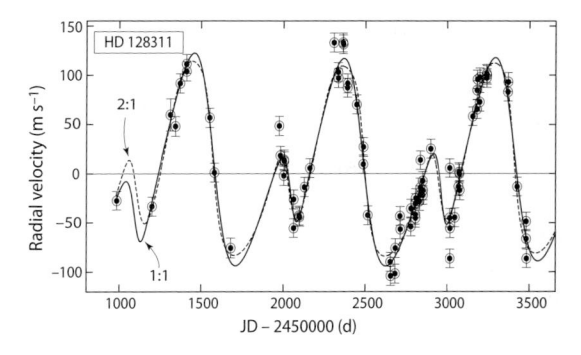

Figure 2.42: The two-planet system HD 128311, showing radial velocity curves for the best-fit solutions corresponding to the 2:1 (dashed) and 1:1 (solid) mean motion resonances. Both curves give an rms of $15\,\mathrm{m\,s^{-1}}$. Error bars include the stellar jitter of $9\,\mathrm{m\,s^{-1}}$. From Goździewski & Konacki (2006, Figure 4), reproduced by permission of the AAS.

a result of planet–planet interactions. The slingshot effects seen in the simulations of Nagasawa et al. (2008) show that close-in planets, some in retrograde motion as a result of Kozai resonances, may indeed result.

In the solar system, a small number of asteroids, a number of asteroid-size planetary moons, and Triton (the largest of Neptune's moons) have retrograde orbits.

Gayon & Bois (2008a) used the genetic algorithm PIKAIA and the fast-chaos indicator MEGNO to search for stable 2:1 retrograde resonant configurations for HD 73526, consistent with the published initial conditions. They found prominent stability islands, which occur when the apsidal longitudes of the two orbits precess at the same average rate, albeit with complex structure, and even though they do not precess in the same direction. The resulting radial velocity curve is very similar to that given by the prograde solutions of Tinney et al. (2006) and Sándor et al. (2007a).

Similar counter-revolving solutions have been investigated by Gayon-Markt & Bois (2009) for eight further compact multiplanetary systems. Their results suggest that six of them (HD 69830, HD 73526, HD 108874, HD 128311, HD 155358 and HD 202206) could be dynamically regulated by such a configuration. Systems with retrograde planets can be packed substantially more closely than prograde systems with an equal number of planets (Smith & Lissauer, 2009).

2.6.6 Interacting doubles

A number of systems contain two giant planets in which planet–planet interactions are significant. These include: HD 12661 (Goździewski & Maciejewski, 2003); HD 14810 (Wright et al., 2009b); HD 155385 (Cochran et al., 2007); HD 169830 (Goździewski & Konacki, 2004); and HD 183263 (Wright et al., 2009b).

2.6.7 Non-interacting doubles

A number of systems contain planet pairs with very large ratios of their orbital periods (sometimes referred to as 'hierarchical', but see the footnote on page 43), which are therefore unlikely to be interacting.

These include: HD 11964 with $P = 38$ d and 5.3 yr (Wright et al., 2009b); HD 68988 with $P = 6.3$ d and 11–60 yr (Wright et al., 2007); HD 168443 with $P = 58$ d and 4.8 yr (Marcy et al., 2001b); HD 187123 with $P = 3$ d and 10.5 yr (Wright et al., 2009b); HD 190360 with $P = 17.1$ d and 8.0 yr (Vogt et al., 2005); HD 217107 with $P = 7.1$ d and 11.7 yr (Wright et al., 2009b); and HD 38529 with $P = 14.3$ d and 5.9 yr (Fischer et al., 2003) with an infrared excess detected by Spitzer (Moro-Martín et al., 2007).

2.6.8 Super-Earth systems

Planetary systems comprising only low-mass planets are starting to be discovered with improved radial velocity accuracies. The first three such systems discovered are:

HD 69830: the triple Neptune system with $P \sim 9$, 32, and 200 d (Lovis et al., 2006). The outer planet is located near the inner edge of the habitable zone. Beichman et al. (2005a) reported Spitzer photometry and spectroscopy which shows an infrared excess characteristic of a large cloud of fine silicate dust within a few AU of the star, suggestive of a large asteroid belt.

GJ 581: a five- or six-planet system (Bonfils et al., 2005b; Udry et al., 2007; Vogt et al., 2010a), with some lying within, or close to the inner and outer edges, of the star's habitable zone (§11.9.2).

HD 40307: the third triple super-Earth system, with masses 4.2, 6.9, and $9.2 M_\oplus$ (Mayor et al., 2009b).

2.6.9 Stability of habitable zone systems

Numerous investigations have been carried out to examine the stability of planets which might exist in the habitable zones of known exoplanet systems (§11.9.1). Using fast indicators of stability or chaos, stability maps can help to establish, for example, whether and where Earth-like planets could exist in systems having one giant planet (Figure 2.43).

Hypothetical orbits of such terrestrial-type planets within the estimated habitable zone have been classified as fully stable if no mean motion resonance exists within that zone, partially stable if only a few resonances exist, marginally stable if many resonances exist, and very unstable if the region is found to be strongly chaotic (Sándor et al., 2007b).

2.7 Planets around binary and multiple stars

Many stars, perhaps most, occur in binary or multiple star systems. Nearby G–K dwarfs, for example, which have been high-priority targets for many Doppler search

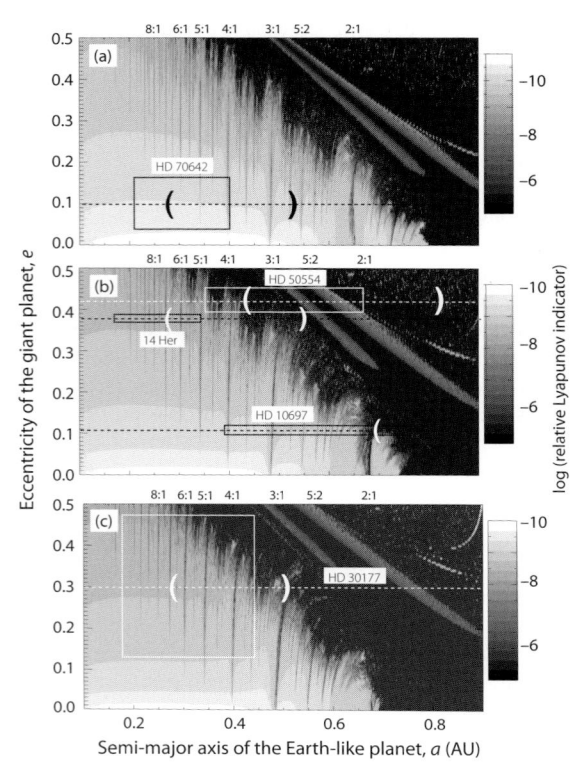

Figure 2.43: Stability maps, computed by the method of relative Lyapunov indicators, for orbits of possible habitable-zone planets interior to the orbit of a known giant planet, for various values of mass ratios $\mu = M_p/(M_\star + M_p)$, where M_p is the mass of the giant planet, and the mass of the hypothetical terrestrial-type planet is negligible: (a) $\mu = 0.002$, (b) $\mu = 0.005$, (c) $\mu = 0.009$. The habitable zones corresponding to the star's zero-age main sequence luminosity are indicated by elongated rectangles, whose heights correspond to the uncertainty on the giant planet's eccentricity, while the current habitable zones are indicated by parentheses. Positions of mean motion resonances are indicated. In (c), for example, the eccentricity of the giant planet in the HD 30177 system is highly uncertain. The present habitable zone is strongly chaotic for large values of e, while its inner part is marginally stable for small e. From Sándor et al. (2007b, Figures 7, 9 and 10), © John Wiley & Sons, Inc.

programmes, are known to occur more often in binary or multiple systems than they do in isolation (Abt, 1979; Duquennoy & Mayor, 1991; Mathieu, 1994; Eggenberger et al., 2004a).

At the same time, many young binaries are known to possess disks—either *circumstellar* (around one of the stars) or *circumbinary* (surrounding both stars)—which, in analogy with single stars, may provide the accretion material necessary for planet formation (e.g. Mathieu, 1994; Akeson et al., 1998; Mathieu et al., 2000; White & Ghez, 2001; Monin et al., 2007).

Understanding the occurrence of planets around binary stars, and how their frequency and properties compare with planets around single stars poses, however, numerous problems: of observability, of quantifying their dynamical stability, and of understanding their detailed origin and formation.

2.7.1 Configurations and stability

The first studies of dynamical configurations for planets in binary systems were carried out well in advance of the first exoplanet detections. Szebehely (1980) classified the possibilities as 'inner orbits' revolving around the primary star, 'satellite orbits' revolving around the secondary, or 'outer orbits' revolving around both.

From studies of the stability of resonant periodic orbits in the circular restricted three-body problem, and as illustrated schematically in Figure 2.44, Dvorak (1982) suggested that a planet may exist in either a circumstellar orbit, i.e. orbiting either the primary or secondary star (also referred to as an *S-type orbit*), or in a circumbinary orbit, i.e. orbiting both stars (*P-type*). Existence in a 1:1 'Trojan' resonance around the L4 or L5 Lagrange points (*L-type*) is also possible dynamically.

Dynamical stability The definition of dynamical stability of a planetary orbit in a binary system has been considered by various authors (e.g. Harrington, 1977; Szebehely, 1980, 1984), following similar stability considerations for multiple planetary systems (§2.6.1).

Harrington (1977), for example, considered a planet orbit as stable if its semi-major axis and orbital eccentricity do not undergo secular changes. More pragmatically, Haghighipour (2008) considered a planet in a binary system stable if, for the duration of study, it does not collide with other bodies, and it does not leave the system's gravitational field.

With the more limited numerical tools then available, the first stability studies were restricted to special cases, or conducted only over a few orbital periods (Graziani & Black, 1981; Black, 1982; Pendleton & Black, 1983). With the development of symplectic integrators (page 44), particular routines have been developed for following the orbits of low-mass bodies in binary systems (Chambers et al., 2002). The long-term stability of planets in various S-type and P-type orbits, at least within certain domains, has been confirmed by such dynamical studies.

Stability of S-type orbits For a planet in an S-type orbit, the gravitational force of the secondary is the principal source of orbital perturbations. Qualitatively, the perturbative effect varies with the companion star mass, and according to the eccentricity and semi-major axis of the binary which together determine the closest approach of the secondary to the planet.

The maximum value of the semi-major axis of a stable planetary orbit is given by (Rabl & Dvorak, 1988; Hol-

2.7 Planets around binary and multiple stars

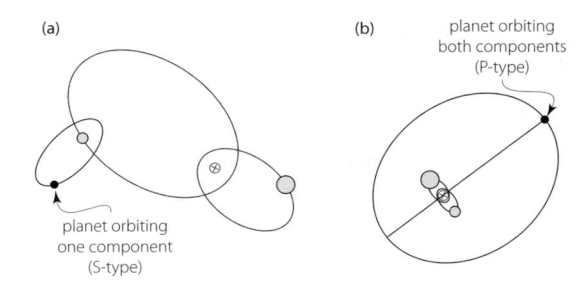

Figure 2.44: Schematic of a planet in a binary star system (⊗ indicates the stellar centre of mass): (a) the planet orbits one component of the stellar binary; (b) the planet is in a circumbinary orbit about both stellar components.

man & Wiegert, 1999; Haghighipour, 2008)

$$\begin{aligned}a_c/a_b =\ &(0.464\pm 0.006)+(-0.380\pm 0.010)\mu\\&+(-0.631\pm 0.034)e_b+(0.586\pm 0.061)\mu e_b\\&+(0.150\pm 0.041)e_b^2+(-0.198\pm 0.047)\mu e_b^2\,,\end{aligned} \quad (2.68)$$

where a_c is the critical semi-major axis, a_b and e_b are the semi-major axis and eccentricity of the binary, $\mu = M_1/(M_1+M_2)$, and M_1 and M_2 are the masses of the primary and secondary stars. The ± signs together define a transitional region, comprising a mixture of stable and unstable states, where stability depends on orbital parameters, mass ratio, and initial conditions. Slightly different limits have been found in later studies (Moriwaki & Nakagawa, 2004; Fatuzzo et al., 2006).

Such results are applicable to planets of Jupiter-mass, as well as of terrestrial mass (Quintana et al., 2002; Quintana & Lissauer, 2006; Quintana et al., 2007) and, discounting the effects of radiation pressure, dust particles and debris disks (Trilling et al., 2007).

Numerous other stability studies for S-type systems have been reported, both for general cases (e.g. Benest, 1993; Pilat-Lohinger & Dvorak, 2002; Pilat-Lohinger et al., 2004; Musielak et al., 2005; Mudryk & Wu, 2006; Cuntz et al., 2007; Marzari & Barbieri, 2007b,a; Sepinsky et al., 2007), as well as for specific systems, notably for α Cen (Benest, 1988; Wiegert & Holman, 1997; Sirius (Benest, 1989), η CrB (Benest, 1996), and γ Cep (Dvorak et al., 2003a, 2004; Haghighipour, 2006; Verrier & Evans, 2006), and others (Szenkovits & Makó, 2008).

A dynamical classification, distinguishing dynamically rigid systems in which the orbital planes of planets precess together as if they were embedded in a rigid disk, and weakly coupled systems in which the mutual inclination angle between initially coplanar planets grows to large values on secular timescales, has been proposed by Takeda et al. (2008a).

Kozai resonance A specific dynamical effect observed in numerical simulations of S-type binaries is the phenomenon of *Kozai resonance*. In three-body systems with two massive objects and a smaller body in a highly inclined orbit, such as in the case of a binary–planet system, angular momentum exchange between the planet and the secondary star results in the planet's inclination and eccentricity oscillating synchronously, i.e. increasing one at the expense of the other, whilst conserving the integral of motion (the Delaunay quantity)

$$H_K = (1-e^2)^{1/2}\cos i\,. \quad (2.69)$$

The Kozai instability was originally studied in the context of high inclination comets and asteroids in the solar system, and sets in at inclinations greater than a critical value of $i_c = \arcsin\sqrt{0.4} \approx 39.°2$ (Kozai, 1962; Innanen et al., 1997).

This behaviour, in which e and i oscillate with the same periodicity but 180° out of phase, is seen in general numerical simulations of S-type systems (Takeda & Rasio, 2006; Takeda et al., 2008a), as well as for specific configurations of individual systems such as γ Cep (Haghighipour, 2004, 2006; Verrier & Evans, 2006, Figure 2.45).

In multiple planetary systems with highly inclined orbits, the Kozai resonance mechanism can result in highly eccentric outer planets. If their orbits cross those of the inner planets, strong planet–planet interactions can result in the ejection of one or more of the planets (Malmberg et al., 2007a; Parker & Goodwin, 2009; Verrier & Evans, 2009).

Stability of P-type orbits Similar numerical simulations have been carried out for P-type circumbinary orbits (Ziglin, 1975; Szebehely & McKenzie, 1981; Dvorak, 1984, 1986; Dvorak et al., 1989; Kubala et al., 1993; Holman & Wiegert, 1999; Broucke, 2001; Pilat-Lohinger et al., 2003; Musielak et al., 2005; Pierens & Nelson, 2008b), and for more general hierarchical triple systems (Verrier & Evans, 2007). Again, stability criteria are dictated by the perturbative effects of the various components.

The P-type planetary orbit is found to be stable for semi-major axes exceeding a critical value given by (Dvorak et al., 1989; Holman & Wiegert, 1999; Haghighipour, 2008)

$$\begin{aligned}a_c/a_b =\ &(1.60\pm 0.04)+(4.12\pm 0.09)\mu+(5.10\pm 0.05)e_b\\&+(-4.27\pm 0.17)\mu e_b+(-2.22\pm 0.11)e_b^2\\&+(-5.09\pm 0.11)\mu^2+(4.61\pm 0.36)e_b^2\mu^2\,,\end{aligned} \quad (2.70)$$

with parameters defined as for Equation 2.68.

Outside of these specific regions, numerical simulations show islands of (in)stability that appear in specific configurations (Hénon & Guyot, 1970; Dvorak, 1984; Rabl & Dvorak, 1988; Dvorak et al., 1989), which are now understood to arise from the existence of $n:1$ mean motion resonances (Holman & Wiegert, 1999; Verrier & Evans, 2006).

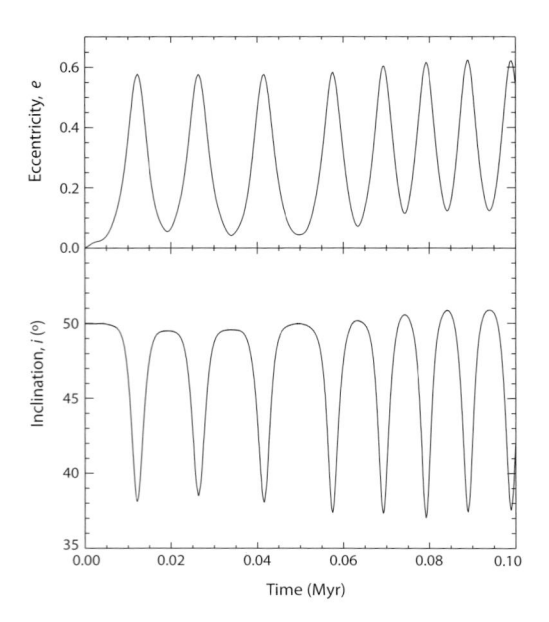

Figure 2.45: Variation of eccentricity and inclination with time for a test particle undergoing Kozai resonance. The configuration is for the system γ Cep, with starting elements $a = 0.5$ AU and $i = 50°$: e and i oscillate strongly, 180° out of phase. From Verrier & Evans (2006, Figure 3), © John Wiley & Sons, Inc.

Planetary formation in binaries In the basic picture of planet formation around single stars, there are generally held to be four key stages of growth: coagulation of dust particles and their growth to cm-size objects; collisional growth to reach km-size planetesimals; the coalescence of planetesimals to form protoplanets or planetary embryos; and the collisional growth of planetary embryos to form terrestrial-size objects.

In a binary star system with moderate to small separation, the secondary star may have a significant effect on each of these processes. Various studies and simulations of the complex processes of planet formation around binary stars have been made (Marzari & Scholl, 2000; Barbieri et al., 2002; Quintana et al., 2002; Lissauer et al., 2004; Pfahl & Muterspaugh, 2006; Haghighipour & Raymond, 2007; Tsukamoto & Makino, 2007; Paardekooper et al., 2008; Pascucci et al., 2008; Xie & Zhou, 2008; Marzari et al., 2009; Payne et al., 2009).

For α Cen A, for example, Thébault et al. (2008) concluded that the planetesimal accretion phase could not have occurred *in situ*, and that outward migration after formation is most likely. A small binary inclination may, however, facilitate accretion (Xie & Zhou, 2009).

Modifying processes Various processes have been identified which are likely to modify planet formation around binary stars. The existence of additional stellar companions may inhibit the formation of circumstellar giant planets with particular orbital separations. Thus Nelson (2000) found that planet formation is unlikely in equal-mass binary systems with $a \sim 50$ AU; Mayer et al. (2005) considered the effects of modified gravitational instabilities; and Thébault et al. (2006) considered regions in $a \lesssim 50$ AU systems in which runaway accretion is strongly affected.

A stellar component on an eccentric orbit can truncate the circumprimary disk to smaller radii, and remove material that would otherwise be available for the formation of terrestrial planets (Artymowicz & Lubow, 1994; Pichardo et al., 2005). More distant but highly inclined companions can modify their orbits over long time scales (Innanen et al., 1997; Takeda & Rasio, 2005). During the protoplanet formation stage, the perturbation of the secondary may increase the planetesimals' relative velocities, resulting in greater fragmentation, or creating regions where the orbits are unstable and therefore emptied of accretion material (Heppenheimer, 1978; Whitmire et al., 1998; Thébault et al., 2004).

Numerical simulations nevertheless show that giant or terrestrial planet formation should still be possible in binary systems. A binary companion can modify the structure of a planet-forming nebula, and create regions where gas and dust densities are locally enhanced. In turn, this may augment planet formation around the primary by gravitational instability (Mayer et al., 2005; Boss, 2006a), or by core accretion (Thébault et al., 2004).

Observational results from adaptive optic searches also support the idea that a stellar companion within 100 AU tends to inhibit, but not entirely suppress, planet formation (Eggenberger et al., 2007a).

Radial velocity surveys of binary stars Early radial velocity surveys tended to exclude spectroscopic and visual binary stars with separations below 2–6 arcsec because of problems of light contamination from the other component at the spectrograph entrance (Udry et al., 2000; Perrier et al., 2003; Marcy et al., 2005b; Jones et al., 2006b). This resulted in limited or biased surveys for binary star separations $\lesssim 100-200$ AU.

At larger separations, viz. for visual binaries with separations $\gtrsim 100$ AU, circumstellar planet searches can treat the targets as distinct single stars, and such surveys have faced limited technical difficulty (Gratton et al., 2003; Desidera & Barbieri, 2007; Toyota et al., 2009).

Observational confirmation that giant planets existed orbiting one component of a spectroscopic binary star came with the discoveries of GJ 86 Ab (Queloz et al., 2000b) with its white dwarf companion, and γ Cep Ab (Hatzes et al., 2003a), a binary with $P = 57$ yr, with the planet orbiting the primary with $P = 2.48$ yr.

The growing interest in exploring broader planetary domains as a way of constraining formation models has led to some more recent surveys preferentially including, rather than excluding, binary systems (Table 2.4).

Amongst specific radial velocity surveys focusing on binaries are a sample of 101 single-lined spectroscopic

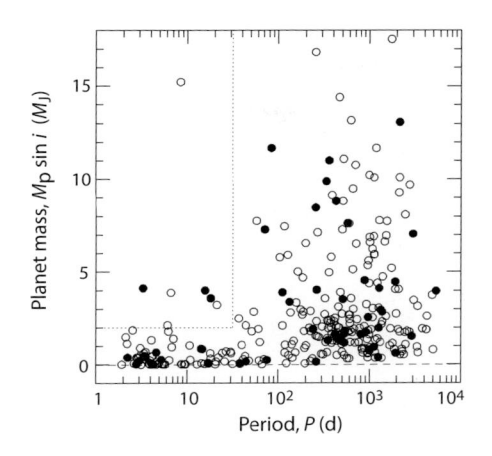

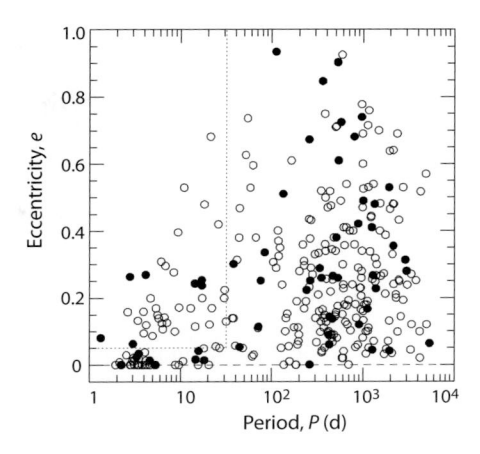

Figure 2.46: Planets detected from radial velocity observations (317 planets from exoplanets.org, 2010–11–01), with those orbiting single stars shown as open circles, and those in binary or multiple systems as filled circles. Left: mass ($M_p \sin i$) versus orbital period. Right: eccentricity versus period. Dashed regions correspond to those delineated by Eggenberger (2010, Figure 1).

binaries (SB1) with $P > 1.5$ yr, selected on the basis of the earlier CORAVEL surveys (Duquennoy & Mayor, 1991; Halbwachs et al., 2003). A programme to supplement the original CORAVEL data with 10–15 additional high-precision measurements of each system, using ELODIE in the north and CORALIE in the south, was initiated in 2001 (Eggenberger, 2010). Analysis employs the spectral deconvolution programme TODCOR. No promising planetary candidates have been reported to date, and present constraints indicate that less than 20% of SB1 systems host a short-period, $P < 40$ d circumprimary giant ($M_p \sin i \gtrsim 0.5 M_J$) planet.

No extensive surveys have yet been undertaken for double-lined spectroscopic binaries (SB2) (Eggenberger, 2010). The principal reason for this is related to the problem of wavelength calibration: the standard technique using iodine cells (§2.2.4) is only applicable to single stars, and the time-varying spectra of SB2s cannot provide the template spectrum required by Equation 2.42. A modified procedure applicable to SB2s, using interleaved observations with and without the iodine cell, has been described by Konacki (2005b).

Other surveys for binary identification In parallel, the fraction of planets known to reside in binary systems has risen as a result of searches for common proper motion companions to known planet-hosting stars.

Other search techniques for planets around binary systems include photometric transit studies around eclipsing binaries (§6.2.8) which may allow the detection of transiting circumstellar or cirmcumbinary planets (Deeg et al., 1998; Doyle et al., 2000; Ofir, 2008), or Trojan planets (Caton et al., 2000), and of non-transiting giant planets in circumbinary orbits through eclipse timing, which has resulted in the discovery of a few circumbinary systems (§4.3).

2.7.2 Present inventory

By late 2010, more than 50 planets have been found to be associated with binary or multiple stars (Eggenberger et al., 2004b; Eggenberger, 2010). Although most are gas giants orbiting the primary component, with projected separations in the range 20–12 000 AU, one possibly circumbinary planet, HD 202206, has been reported (Correia et al., 2005).

At small binary separations, $\lesssim 100$ AU, only a small number of planets have been found, and giant planets may be rarer in these systems than around wider separation binaries or single stars (e.g. Bonavita & Desidera, 2007). This may be consistent with studies which suggest that giant planet formation is inhibited under the influence of a nearby orbiting stellar companion (Nelson, 2000; Mayer et al., 2005; Thébault et al., 2006).

At the smallest binary separations, $\lesssim 20$ AU, the absence of circumstellar planets is consistent with the idea that giant planet formation in these systems is suppressed (Nelson, 2000; Thébault et al., 2004; Mayer et al., 2005; Thébault et al., 2006; Boss, 2006a), although Doppler surveys to confirm this remain incomplete.

Properties Zucker & Mazeh (2002) suggested that planets in binaries follow a different mass–period relation than those orbiting single stars, with the most massive short-period planets being found in binary or multiple systems.

More recent studies have decreased the significance of the earliest results (Eggenberger et al., 2004b; Desidera & Barbieri, 2007; Mugrauer et al., 2007a; Eggenberger, 2010), and the most recent distributions (Figure 2.46a) reveal mixed evidence for such an effect.

Meanwhile, short period ($P < 50$ d) planets around binary stars do appear to have smaller eccentricites, $e \lesssim 0.05$, than around single stars (Figure 2.46b).

Still based on somewhat limited statistics, the few planets with the highest eccentricites, $e > 0.8$, tend to be accompanied by a stellar or brown dwarf companion (Tamuz et al., 2008), a correlation that may extend to lower masses (Ribas & Miralda-Escudé, 2007). This has been taken to support the idea of eccentricities being excited by the Kozai mechanism, in which hierarchical triple systems with high relative inclinations cause large-amplitude periodic oscillations of the eccentricity of the inner pair (Holman et al., 1997; Innanen et al., 1997; Mazeh et al., 1997a; Ford et al., 2000b; Wu & Murray, 2003; Takeda & Rasio, 2005; Moutou et al., 2009).

The coupling of Kozai oscillation with tidal friction, also referred to as Kozai migration, might also lead to short-period planets preferentially following nearly circular orbits (Figure 2.46b). In multiple systems this mechanism may be more effective than type II migration in bringing massive planets close to their host stars. This may also explain why the most massive short-period planets are found in binary or multiple systems (Takeda & Rasio, 2006; Fabrycky & Tremaine, 2007).

2.7.3 Specific examples

γ Cep The binary star system γ Cep harbours a stable giant planet orbiting the primary star at a distance some one tenth of the stellar separation (Hatzes et al., 2003a). The primary and secondary stars, of masses 1.6 and $0.4 M_\odot$, are separated by 18.5 AU ($P \sim 57$ yr, $e = 0.36$), with the planet of mass $1.7 M_J$ orbiting the primary with $a = 2.13$ AU, $P = 2.5$ yr, $e = 0.12$.

Numerical integrations as a function of a_b, e_b, and i_b (the inclination of the planet with respect to the stellar binary) by Haghighipour (2006), show that the planet orbit is stable for values of the binary eccentricity $0.2 \leq e_b \leq 0.45$, with the system becoming unstable within a few thousand years for $e_b > 0.5$. Within the stable range of orbital eccentricity, the planet remains stable for $i_b < 40°$, becoming unstable within a few thousand years for $i_b > 40°$. For large values of the inclination, the system may be locked in a Kozai resonance (Haghighipour, 2004). The stability map for test particles interior to the planet orbit show complex stability patterns influenced by mean motion resonances (Figure 2.47).

16 Cyg B The planet around 16 Cyg B has a highly eccentric orbit, $e \sim 0.7$. 16 Cyg B is one component of a widely separated binary system whose eccentricity is itself very large, $e > 0.54$. The orbital period of the binary star is long and difficult to measure accurately, but from astrometric observations made over some 170 years is estimated to be >18 000 yr (Hauser & Marcy, 1999).

The system is an interesting case for planetary formation theories which try to explain the large eccentricities in binary systems as a result of gravitational perturbations by the other stellar component, here 16 Cyg A.

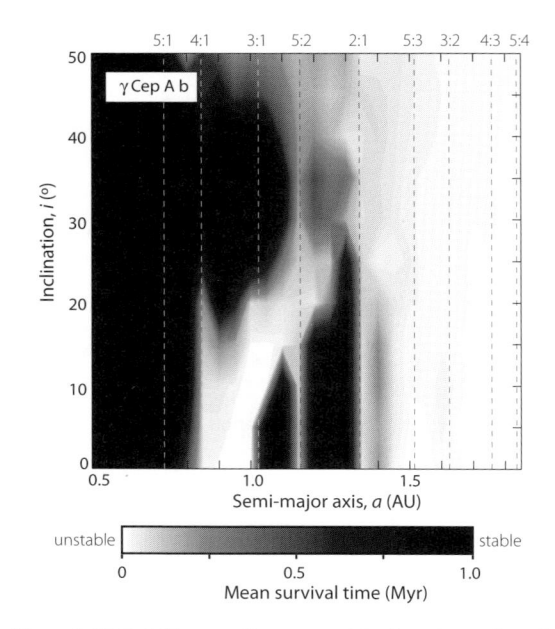

Figure 2.47: Stability map for test particles interior to the orbit of the planet γ Cep Ab, for the coplanar case $i_b = 0$. Stability is indicated by the test particle's survival time averaged over longitudes and normalised to 1 Myr. Darker regions indicate higher stability. Dashed vertical lines are the nominal locations of the mean motion resonances with the planet γ Cep Ab. From Verrier & Evans (2006, Figure 2), © John Wiley & Sons, Inc.

If the planet was originally formed in a circular orbit, with an orbital plane inclined to that of the stellar binary by >45°, then the planet orbit is predicted to oscillate chaotically between high- and low-eccentricity states, on time scales of $10^7 - 10^{10}$ yr, (Holman et al., 1997; Mazeh et al., 1997a; Hauser & Marcy, 1999). The discovery of further planets with $M_p \sim M_J$ within 30 AU would require this conclusion to be revisited.

HD 188753 Some controversy surrounds the triple system HD 188753, a 155 d spectroscopy binary (HD 188753B) in a 25.7-yr orbit ($a = 12.3$ AU) about the primary (HD 188753A). Konacki (2005a) reported a 3.35 d $1.1 M_J$ planet around the primary component based on 10 m s^{-1} accuracy Keck–HIRES data, unconfirmed by Eggenberger et al. (2007b).

According to theoretical studies, the pericentre distance of the AB pair is too small to have allowed giant planet formation. Modeling by Jang-Condell (2007) has shown that the circumstellar disk around the primary would have been truncated at around 1.5–2.7 AU, leaving insufficient material, and in any case too hot, for a Jovian planet to have formed *in situ*. Alternative formation mechanisms were forthcoming, in particular proposing that the primary could have acquired the planet through interactions within an open star cluster that has now dissolved (Pfahl, 2005; Portegies Zwart & McMillan, 2005).

3

Astrometry

Astrometry concerns the measurement of positions and motions of solar system bodies, stars within the Galaxy and, in principle, galaxies and clusters within the Universe. Traditionally, an important goal has been to determine stellar parallaxes and proper motions.

In the present context, repeated high-accuracy astrometry aims to determine the transverse component of the displacement of the host star due the gravitational perturbation of an orbiting planet. This dynamical manifestation of their gravitational influence in the plane of the sky is closely related to radial velocity measurements, which are sensitive to the corresponding photocentre displacement along the line-of-sight.

Current best accuracies, of around 1 mas achieved with Hipparcos and the HST–Fine Guidance Sensors, only just touch the regime where displacements of star positions due to orbiting planets can be detected.

The panorama will change with Gaia, due for launch in 2012, which will measure all billion stars to $V = 20$ mag, at accuracies of $\sim 20-25\,\mu$as at 15 mag. This should lead to several thousand star–planet systems discovered astrometrically. Absolute orbits, planet masses without the $\sin i$ ambiguity, and information on coplanarity will become available. Developments in interferometry, notably with VLTI–PRIMA, or possibly from space as proposed for SIM, also hold promise.

A related manifestation of the dynamical displacement of the primary is the modulation of timing signals of rotating, pulsating, or eclipsing systems. Planets around radio pulsars, pulsating sdB stars, and eclipsing binaries have been discovered using this approach, as described in Chapter 4.

3.1 Introduction

The path of a star orbiting the star–planet barycentre appears projected on the plane of the sky as an ellipse with angular semi-major axis α given by

$$\alpha = \frac{M_{\rm p}}{M_\star + M_{\rm p}} a \simeq \frac{M_{\rm p}}{M_\star} a \quad (3.1)$$

$$\equiv \left(\frac{M_{\rm p}}{M_\star}\right)\left(\frac{a}{1\,{\rm AU}}\right)\left(\frac{d}{1\,{\rm pc}}\right)^{-1} \text{arcsec}, \quad (3.2)$$

where a is the semi-major axis of the planet orbit (here assumed circular), d is the distance, and $M_{\rm p}$ (planet) and $M_\star$ (star) are in common units. The former expresses α in linear measure (e.g. in AU if a is in AU, and the approximation holding for $M_{\rm p} \ll M_\star$), while the latter equivalently expresses it in angular measure through the definition of the parsec.

This *astrometric signature* α is the observable for astrometric planet detection, and is proportional to both $M_{\rm p}$ and a, and inversely to d. Astrometry is therefore particularly sensitive to long orbital periods ($P \gtrsim 1$ yr), applying equally to hot or rapidly rotating stars. The technique aims to discern an orbiting planet on top of the two other classical astrometric effects: the linear path of the system's barycentre (its proper motion) combined with the reflex motion (its parallax) resulting from the Earth's orbital motion around the Sun (Figure 3.1).

Size of the effect The size of the effect calculated for the known planets is shown in Figure 3.2, and some representative values are given in Table 3.1. The accuracy required to detect planets astrometrically is typically sub-mas, although it would reach a few mas for planets of $M_{\rm p} \sim 1\,M_{\rm J}$ orbiting nearby solar-mass stars. For the 424 planets with available data in late 2010 (out of 494 known) the median value is $\sim 16\,\mu$as, or $\sim 10^{-3}$ AU in linear displacement.

The accuracy required for the detection of planets in the habitable zone of long-lived main sequence stars (spectral type A5 and later) can be estimated as follows. The mean star–planet separation of the habitable zone for a star of luminosity L has been estimated as $a \sim (L/L_\odot)^{1/2}$ AU (Kasting et al., 1993). Over this mass range, $\sim 0.2-2M_\odot$, $L \propto M_\star^{4.5}$ (Torres et al., 2010a), so that $\alpha \propto M_\star^{2.25}$ and hence

$$\alpha \simeq 3\,\mu{\rm AU}\left(\frac{M_{\rm p}}{M_\oplus}\right)\left(\frac{M_\star}{M_\odot}\right)^{1.25}. \quad (3.3)$$

Accordingly, for a $1\,M_\oplus$ planet orbiting a main sequence star, α ranges from about $2.3\,\mu$AU for spectral type K0V,

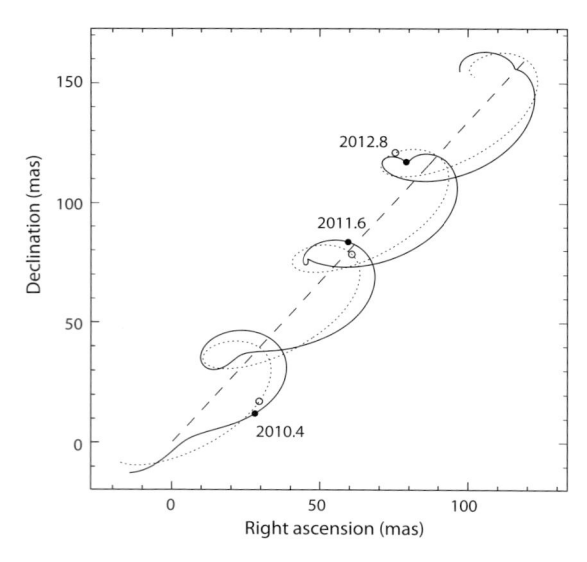

Figure 3.1: Schematic of the path on the sky of a star at $d = 50\,pc$, with a proper motion of $50\,mas\,yr^{-1}$, and orbited by a planet of $M_p = 15\,M_J$, $e = 0.2$, and $a = 0.6\,AU$. The straight dashed line shows the system's barycentric motion viewed from the solar system barycentre. The dotted line shows the effect of parallax (the Earth's orbital motion around the Sun, with a period of 1 year). The solid line shows the motion of the star as a result of the orbiting planet, the effect magnified by ×30 for visibility. Labels indicate (arbitrary) times in years.

Table 3.1: Representative astrometric signatures for a $1\,M_\odot$ host star, and a variety of star distance d, planet mass M_p, and semi-major axis a.

Type	d (pc)	M_p	a (AU)	α (μas)
Jupiter	10	$1\,M_J$	5	500.
"	100	"	"	50.
Hot Jupiter	10	$1\,M_J$	5	1.
"	100	"	"	0.1
Earth	10	$1\,M_\oplus$	1	0.3
"	100	"	"	0.03

up to about $7\,\mu$AU for A5V. Gould et al. (2003a) similarly argue that A and F stars constitute the majority of viable targets for such astrometric searches.

3.2 Astrometric accuracy from ground

Single aperture The theoretical photon-noise limited positional error for a monolithic telescope is given by (Lindegren, 1978)

$$\sigma_{ph} = \frac{\lambda}{4\pi D} \frac{1}{S/N}, \qquad (3.4)$$

where D is the telescope diameter, λ the wavelength, and S/N the signal-to-noise. For $V = 15$ mag, $\lambda = 600$ nm, $D = 10$ m, a system throughput of $\epsilon = 0.4$, and an integration time of 1 h, the resulting positional error due to photon noise alone is $\sigma_{ph} \sim 30\,\mu$as. Mainly as a result of atmospheric turbulence and differential chromatic refraction, however, even the best astrometric accuracies achieved on the ground currently fall far short of this theoretical performance.

The understanding of the accuracy limits of narrow-angle astrometry have evolved considerably over the last three decades. Until the 1970s, long-focus telescopes equipped with photographic plates were used for accurate narrow-angle astrometric measurements. Analyses of image centroids yielded measurements at the 20–30 mas level (Harrington & Dahn, 1980). The introduction of CCDs, and the Ronchi ruling photometer (Gatewood, 1987), eliminated plate-specific error sources to the point that atmospheric turbulence became a dominant source of error.

For small angular separations, $\lesssim 1$ arcmin, the time-averaged precision with which the angle between two stars near the zenith can be measured is reasonably consistent with atmospheric turbulence models (Lindegren, 1980, eqn 35a)

$$\sigma_\delta \simeq 540\, D^{-\frac{2}{3}} \theta\, t^{-\frac{1}{2}} \text{ arcsec}, \qquad (3.5)$$

where θ is their angular separation in rad, D the telescope diameter in m, and t the integration time in s; in these units, the formula is valid for $\theta \ll d/4300$. The numerical factor is obtained from the (assumed Hufnagel) phase-structure function describing the atmospheric (Kolmogorov-like) turbulence for the atmospheric seeing and wind-speed profiles typical of a good site. Values of $\theta = 1$ arcmin, $D = 1$ m, and $t = 1$ h lead to $\sigma_\delta \simeq 3$ mas, in reasonable agreement with the accuracies reported for long-focus telescopes with star separations up to a few arcmin (e.g. Gatewood, 1987; Han, 1989; Monet et al., 1992).

With several reference stars, theoretical estimates give a further improvement, variously described by

$$\sigma_\delta \propto D^{-1} \theta^{\frac{4}{3}} t^{-\frac{1}{2}} \qquad \text{Lindegren (1980),} \qquad (3.6)$$

$$\sigma_\delta \propto D^{-\frac{3}{2}} \theta^{\frac{11}{6}} t^{-\frac{1}{2}} \qquad \text{Lazorenko (2002)}. \qquad (3.7)$$

Such predictions are substantially confirmed by the results of, e.g., Pravdo & Shaklan (1996).

Further innovative approaches to observation and measurement continue to be developed. Lazorenko & Lazorenko (2004) considered apodisation of the entrance pupil, and an improved symmetry treatment of the reference field (by assigning specific weights to each reference star), leading to their modified and more general theoretical expression for the differential astrometric error due to the atmosphere

$$\sigma_\delta \propto D^{-\frac{k}{2}+\frac{1}{3}} \theta^{\frac{k\mu}{2}} t^{-\frac{1}{2}}. \qquad (3.8)$$

Here, $2 \leq k \leq (8N+1)^{1/2}-1$ is determined by the number of reference objects N, and $\mu \leq 1$ is a term dependent on

3.2 Astrometric accuracy from ground

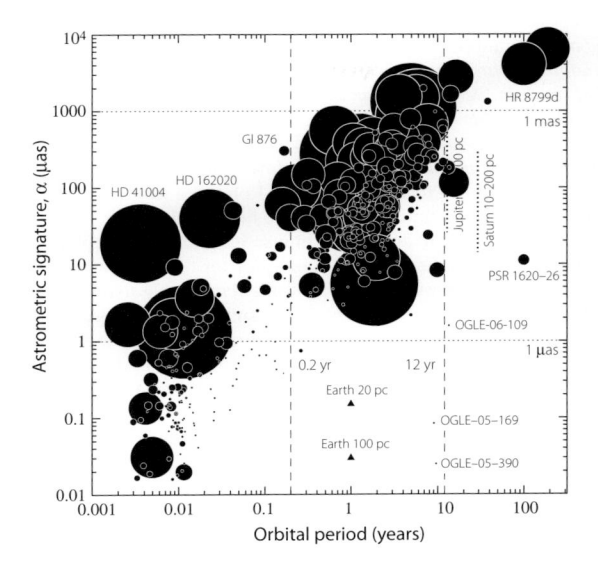

Figure 3.2: Astrometric signature, α, versus period, calculated for the known planetary systems using Equation 3.1. Circles are proportional to M_p or $M_p \sin i$ (e.g. HD 41004b = 18M_J, OGLE TR–113 = 1M_J; but are of constant size at 0.4M_J below this value). Some specific planets, and the effects of Earth, Jupiter, and Saturn at various distances, are also shown. Vertical lines at 0.2 and 12 yr mark period limits for Gaia. Horizontal lines show astrometric signatures of 1 mas and 1 μas; the lower limit for Gaia is ~30 μas. Data are from exoplanet.eu, 2010–11–01 (for 424 planets with appropriate data, out of 494 known).

k and the magnitude and distribution of the reference stars; the expression reduces to Equation 3.5 for $N = 1$ (i.e. $k = 2$). The reference field forms a virtual net filter which attenuates the image motion spectrum.

Formally, this leads to performances below 100 μas for 10-m class telescopes with very good seeing and $t \sim$ 600 s. Narrow-field imagers on Palomar and VLT, including the use of adaptive optics, have demonstrated short-term 100 – 300 μas precision (Neuhäuser et al., 2007; Lazorenko, 2006; Lazorenko et al., 2007; Röll et al., 2008; Cameron et al., 2009). Lazorenko (2006) reported VLT–FORS2 positional accuracies $\sim$ 300 μas (for $R = 16$ mag and $t = 17$ s with 0.55 arcsec seeing), dominated by photon centroiding errors of $\sim$ 250 μas. Under certain circumstances, a systematic positional precision as small as 25 μas over six years has been reported (Lazorenko et al., 2009).

Interferometry For larger angular separations, $\sim 0°.5$, both theory (Lindegren, 1980, eqn 35b) and empirical data had shown that the turbulence-limited accuracy is $\propto \theta^{1/3}$ and, to first order, is independent of telescope size. Given this dependency, long-baseline interferometry was therefore not immediately viewed as a technique that offered a significant advantage in relative positional accuracy for narrow-field measurements.

For two stars with an angular separation of $0°.5$, their light paths are separated by $\sim$100 m at the top of the turbulent atmosphere, at a height of $\sim$10 km. In conventional astrometry with long-focus telescopes, the telescope is much smaller than the beam separation, and the accuracy is independent of telescope diameter. However with interferometry, baselines can be larger than the beam separation, and a qualitative change in the behaviour of the atmospheric errors results.

Turbulence models applied to differential observations give the mean square error for an astrometric measurement, in arcsec (Shao & Colavita, 1992, eqn 2)

$$\sigma_\delta^2 \simeq 5.25\, B^{-4/3} \theta^2 \int h^2 C_n^2(h)(Vt)^{-1}\, dh\,, \qquad (3.9)$$

where B is the baseline in m, θ the star separation in arcmin, $V(h)$ is the wind speed as a function of height, and t the integration time in sec. $C_n^2(h)$ is a complicated function describing the vertical profile of the atmosphere's refractivity power spectrum, which drives atmospheric turbulence through temperature fluctuations. The equation applies in the very narrow-angle regime, $\theta h \ll B$, and for long integrations, $t \gg B/V$.

For the atmospheric conditions measured at Mauna Kea, Shao & Colavita (1992, eqn 4) found

$$\sigma_\delta \simeq 300\, B^{-\frac{2}{3}} \theta\, t^{-\frac{1}{2}} \text{ arcsec}\,. \qquad (3.10)$$

The form is equivalent to the single aperture (Figure 3.5), with B replacing D, and the smaller numerical factor reflecting the (improved) site conditions at Mauna Kea. Accordingly, interferometric accuracy depends linearly on star separation, and improves with increasing baseline. For $\theta = 20$ arcsec, $B = 200$ m, and $t = 1$ h, this implies $\sigma_\delta \simeq 10\,\mu$as (Shao & Colavita, 1992).

Antarctic The integrated amplitude of C_n^2 determines the seeing, usually quantified through the Fried parameter, r_0. At optical wavelengths, Antarctic seeing is typically $\sim$1–3 arcsec. But the determining factor for accurate interferometric astrometry is tilt anisoplanatism, the second moment of the C_n^2 profile in Equation 3.9 which, due to the h^2 factor, is dominated by high-altitude turbulence. At the south pole, the $C_n^2(h)$ profile is far less affected by the jet streams, trade winds, and high-altitude synoptic wind shear that govern the behaviour at mid-latitudes (Marks et al., 1999).

Measurements from Dome C confirm long atmospheric coherence times, and large ($\sim$30 arcsec) isoplanatic angles over which the paths are temporally coherent (e.g. Lawrence et al., 2004; Kenyon et al., 2006; Kenyon & Storey, 2006). Lloyd et al. (2002) even reported seeing Jupiter scintillating, and their studies indicate that a 100 m interferometer at Dome C could achieve 10 μas differential astrometry 300 times faster than a comparable interferometer at a good mid-latitude site.

> **Early investigations:** Of early investigations of the astrometric manifestations of planets, Holmberg (1938) was surely one of the first. The abstract reads *'In the present paper modern trigonometric parallax observations are investigated. It appears that the residuals of these observations are not always distributed at random. Many parallax stars show periodic displacements. These effects probably are to be explained as perturbations caused by invisible companions. Since the amplitudes of the orbital motions are very small, the masses of the companions will generally be very small, too. Thus Proxima Centauri probably has a companion, the mass of which is only some few times larger than the mass of Jupiter. A preliminary investigation gives the result that 25% of the total number of parallax stars may have invisible companions.'* Presumably, all that Holmberg saw was noise, but it nevertheless shows how, already in the 1930s, some understood the issues and made efforts to find exoplanets.
>
> Two discoveries of planet-like companions from astrometric measurements of long-term time-series photographic plates were announced in 1943: companions of $10M_J$ for 70 Oph by Reuyl & Holmberg (1943), and $16M_J$ for 61 Cyg by Strand (1943). Strand was unequivocal: *'The only solution which will satisfy the observed motions gives the remarkably small mass of... 16 times that of Jupiter... Thus planetary motion has been found outside the solar system.'* The results were interpreted as supporting theories of the origin of the solar system (Alfvén, 1943) and speculations on the frequency of planetary systems (Jeans, 1943). Struve (1952) mentioned the merits of planet searches using radial-velocities, transit photometry, and astrometry, stating that *'one of the burning questions of astronomy deals with the frequency of planet-like bodies in the Galaxy which belong to stars other than the Sun'*.
>
> The presence of planets around 61 Cyg and 70 Oph was excluded by Heintz (1978b). But lengthy disputes surrounded the extensive ground-based observations of Barnard's star, for which two planetary mass bodies (0.7 and 0.5 M_J) with periods of 12 and 20 years respectively were proposed (e.g. van de Kamp, 1963, 1982; Gatewood & Eichhorn, 1973; Croswell, 1988) and Lalande 21185 (e.g. Lippincott, 1960; Hershey & Lippincott, 1982; Gatewood et al., 1992; Gatewood, 1996). Early discussions of ground-based optical observations related to planet detection are given by Black & Scargle (1982) and Gatewood (1987).

3.3 Microarcsec astrometry

With μas precision on the horizon, both from ground and space, and necessary for the astrometric detection of Earth-mass planets, a number of higher-order physical effects modify the apparent position of the target star. Their consideration is crucial for the global measurements by Gaia, although less so for differential measurements over small angles. Although their treatment is delicate, it should be possible to model their contributions accurately.

Light deflection In terms of the angle ψ subtended at the observer between the Sun and a given star, the (post-Newtonian) light deflection angle due to the spherically symmetric part of the gravitational field of the Sun is (e.g. Will, 1993; Klioner, 2003, eqn 64)

$$\alpha_{pN} = \frac{(1+\gamma)GM_\odot}{c^2 r_0} \cot\frac{\psi}{2}, \qquad (3.11)$$

where G is the gravitational constant, and r_0 is the distance between the observer and the Sun. γ is a variable in the *parameterised post-Newtonian* (ppN) formalism, with $\gamma = 1$ in general relativity. With $r_0 = 1$ AU, $\gamma = 1$, and for a star at 90° to the ecliptic ($\psi = 90°$), $\alpha_{pN} \simeq 4$ mas. A star at $\delta\psi = 1$ arcmin then experiences a differential light deflection $\delta\alpha_{pN} \simeq 1\,\mu$as.

Data analysis at the μas level for Gaia is being formulated directly within the framework of general relativity, either using a non-perturbative approach (de Felice et al., 2004), or through the ppN formulation which is the present baseline (Klioner, 2003). Light deflection due to the solar quadrupole moment, and for observations near to Jupiter, will also be included. In this framework, even the basic concepts of parallax, proper motion, and radial velocity have to be refined (Klioner, 2004).

Aberration Displacement of an object's observed position resulting from the observer's motion with respect to the solar system barycentre is referred to as *aberration*. If θ is the angle between the direction towards the source and the direction of the observer's velocity, v, then the angular shift $\delta\theta$ toward the apex of the observer's motion is given by (Klioner, 2003, eqn 17)

$$\begin{aligned}\delta\theta =& \frac{v}{c}\sin\theta\left[1 + \frac{1}{c^2}(1+\gamma)w(x_0) + \frac{1}{4}\frac{v^2}{c^2}\right] \\ & - \frac{1}{4}\frac{v^2}{c^2}\sin 2\theta + \frac{1}{12}\frac{v^3}{c^3}\sin 3\theta + O(c^{-4}),\end{aligned} \quad (3.12)$$

where γ is the ppN parameter, and $w(x_0)$ is the gravitational potential of the solar system at the point of observation. For a satellite with barycentric velocity $v \sim 40\,\mathrm{km\,s^{-1}}$, the first-order (classical) aberration is ~ 28 arcsec, the second-order term may reach 3.6 mas, and third-order effects are $\sim 1\,\mu$as. Because of the size of the first-order term, an accuracy of $1\,\mu$as furthermore requires that the observer's (barycentric coordinate) velocity should be known to $\sim 10^{-3}\,\mathrm{m\,s^{-1}}$.

Source motion A star's velocity through space leads to a secular change in its observed proper motion $\dot{\mu}$, an effect referred to as *perspective acceleration*, while the radial component of its motion (its radial velocity) leads to a secular change in its trigonometric parallax $\dot{\varpi}$. These can be written (Dravins et al., 1999)

$$\dot{\mu} = -\frac{v_r}{A}2\varpi\mu \qquad (3.13)$$

$$\dot{\varpi} = -\frac{v_r}{A}\varpi^2, \qquad (3.14)$$

where A is the astronomical unit. In common units, $\dot{\mu}$ and $\dot{\varpi}$ are in arcsec yr^{-2} and arcsec yr^{-1} respectively, if

3.4 Astrophysical limits

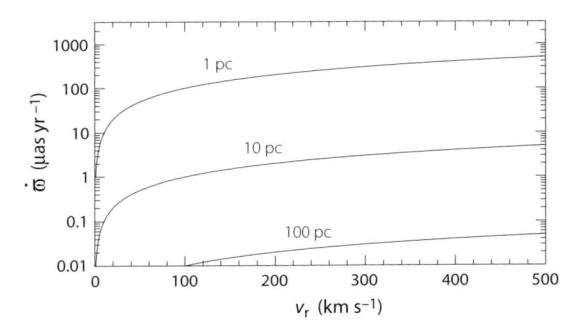

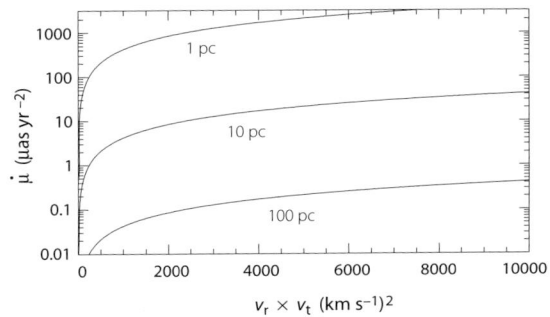

Figure 3.3: Effects of source motion. Top: $\dot{\varpi}$ (μas yr^{-1}) as a function of radial velocity v_r; and bottom: $\dot{\mu}$ (μas yr^{-2}) as a function of the product of the radial and tangential velocities, $v_r \times v_t$. The effect is shown for three representative stellar distances. Adapted from Sozzetti (2005, Figure 1).

the radial velocity v_r is in km s^{-1}, ϖ is in arcsec, μ is in arcsec yr^{-1}, and the astronomical unit is expressed as $A = 9.778 \times 10^5$ arcsec km yr s^{-1}.

Figure 3.3 shows $\dot{\varpi}$ as a function of v_r, and $\dot{\mu}$ as a function of $v_r \times v_t$, where the tangential velocity $v_t = 4.74\mu/\varpi$ km s^{-1} for ϖ, μ in arcsec(yr^{-1}). At μas accuracy levels, the effect of perspective acceleration remains significant out to tens of parsec for stars with large v_r and/or large v_t, while the secular change in parallax may be significant for nearby stars with large v_r.

3.4 Astrophysical limits

At μas astrometric accuracy, several astrophysical noise sources may contribute. Below 1 μas, stellar surface structure (Eriksson & Lindegren, 2007), and relativistic modeling (e.g. Anglada-Escudé et al., 2007) may introduce significant barriers, while the effects of interstellar and interplanetary scintillation in the optical, and stochastic gravitational wave noise, will still be negligible (Perryman et al., 2001).

3.4.1 Surface structure jitter

Limits to the intrinsic accuracy of astrometric measurements may come from 'jitter' induced by stellar surface structure (star spots, plages, granulation, and non-radial oscillations) likely to produce fluctuations in the observed photocentre. Various studies have been made (Bastian & Hefele, 2005; Reffert et al., 2005; Catanzarite et al., 2008; Lanza et al., 2008; Makarov et al., 2009). Svensson & Ludwig (2005) and Ludwig (2006) computed hydrodynamical model atmospheres for a range of stellar types, predicting both photometric and astrometric jitter caused by granulation, and finding that the latter is almost entirely determined by the surface gravity g, being proportional to g^{-1} for a wide range of models.

Eriksson & Lindegren (2007) constructed relatively well-defined statistical relations between variations in the photocentre, total flux, and radial velocity for a wide range of possible surface phenomena, and used the studies of surface structure carried out for radial velocity accuracy assessment (e.g. Saar & Donahue, 1997; Queloz et al., 2001; Henry et al., 2002; Saar et al., 2003; Hatzes, 2002; Paulson et al., 2004b) to predict the astrometric jitter in various star types, without detailed knowledge of their actual surface structures.

If the radial velocity variations are caused mainly by rotational modulation of the spotted surface, they find

$$\sigma_{\text{pos}} = \frac{R_\star \sigma_{v_R}}{v \sin i} \simeq 0.49 R_\star \sigma_m \quad (3.15)$$

$$\sigma_{v_R} \simeq 0.376 \, v \sin i \, \sigma_m \simeq 0.43 R_\star \left(\frac{2\pi}{P_{\text{rot}}}\right) \sigma_m \quad (3.16)$$

where σ_{pos}, σ_{v_R} and σ_m are the rms scatter in astrometry, radial velocity, and magnitude, $R_\star$ is the stellar radius, $v \sin i$ the projected rotational velocity, and P_{rot} the stellar rotation period. The approximate forms were derived from Monte Carlo simulations.

For most stellar types, they predict that effects due to stellar surface structures will be of order $\Delta \sim 10 \, \mu$AU or greater (or, equivalently, $10/d$ μas where d is in pc). While this results in a negligible impact on the detection of large (Jupiter-size) planets, and on the determination of stellar parallax and proper motion, the effect exceeds the typical astrometric displacement caused by an Earth-size exoplanet in the habitable zone (§3.1). Only for stars with extremely low photometric variability ($\lesssim 0.5$ mmag) and low magnetic activity, comparable to that of the Sun, will the astrometric jitter be of order 1 μAU, sufficient for the astrometric detection of an Earth-sized planet in the habitable zone.

Effects of disk instabilities Rice et al. (2003a) used high-resolution numerical simulations to study whether gravitational instabilities within circumstellar disks can produce a detectable motion of the central star. They found a dependency on the efficiency of disk cooling, and predict astrometric displacements of up to 10 – 100 μas on decade time scales at $d = 100$ pc. Effects of density waves on the motion of the disk's photocentre were also considered by Takeuchi et al. (2005b).

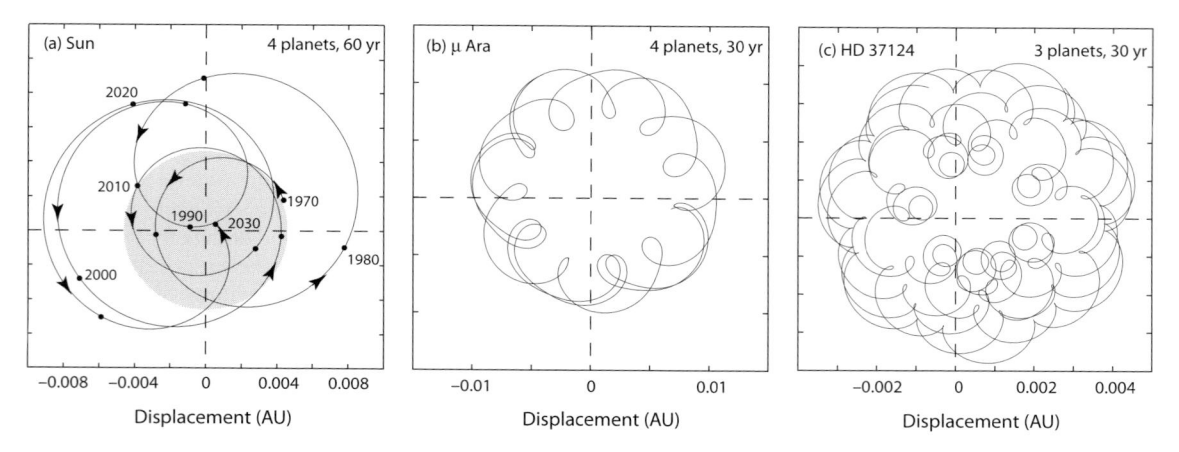

Figure 3.4: Example multiple-planet host star motions with respect to their system barycentre. (a) Path in the ecliptic plane of the Sun's reflex motion about the solar system barycentre, 1970–2030, due to the four giant planets. The scale of the displacement is indicated by the grey circle which shows the solar diameter ($R_\odot = 6.96 \times 10^8$ m, or 0.00465 AU) positioned at the barycentre. Close approaches between Sun and barycentre, or 'peribacs' (Fairbridge & Shirley, 1987), recur at mean intervals of 19.86 yr. Around 1990 (solid circle near centre) the Sun had a retrograde motion relative to the barycentre. (b–c) Barycentric motions of the multiple exoplanet host stars μ Ara and HD 37124 over the 30-year interval 2000–2030. From Perryman & Schulze-Hartung (2011, Figure 2).

3.5 Multiple planets and mandalas

The dynamical effect of multiple planets contributes to the total astrometric signature in the same way as the Sun's path over decades reflects the combined gravitational effects of all solar system objects (Figure 3.4a). As noted in the first discussions of the Sun's orbital revolution about the solar system barycentre by Newton (as quoted by Cajori, 1934) the actual motion of the Sun is rather complex *'since that centre of gravity is continually at rest, the Sun, according to the various positions of the planets, must continuously move every way, but will never recede far from that centre.'*

Perryman & Schulze-Hartung (2011) determined the barycentric motion of the host stars of known multiple exoplanet systems. Examples of the complex star paths in their orbit plane that result from the linear superposition of the reflex motions due to the Keplerian orbit of each individual planet around that star–planet barycentre (Equation 3.1) are shown in Figure 3.4(b–c).

A curve similar to the symmetric part of Figure 3.4b, illustrating the orbit of Mars viewed from Earth, appeared in Kepler's 1609 *Astronomia Nova*. These families of curves have been named *planet mandalas*, after the Sanskrit for circle (Wolfram, 2010).

Connection with the solar cycle   Solar axial rotation plays a fundamental role in the two main hypotheses for a mechanism underlying the solar cycle: attributed to a turbulent dynamo operating in or below the convection envelope, or to a large-scale oscillation superimposed on a fossil magnetic field in the radiative core.

However, the precise nature of the dynamo, and many of the details of the associated solar activity (such as the details of the sun spot cycles, or the reasons for the prolonged Maunder-type solar minima) remain unexplained (e.g. Charbonneau, 2005; Choudhuri, 2007; Weiss & Thompson, 2009; Jones et al., 2010), although certain features may arise naturally in some models (e.g. Charbonneau et al., 2007).

Empirical investigations have long pointed to a link between the Sun's barycentric motion and various solar variability indices (e.g. Wolf, 1859; Brown, 1900; Schuster, 1911; Jose, 1965; Ferris, 1969). Specifically, acceleration in the Sun's motion, or the change of its orbital angular momentum, have been linked empirically to phenomena such as the Wolf sun spot number counts (Wood & Wood, 1965), climatic changes (Mörth & Schlamminger, 1979; Scafetta, 2010), the 80–90-yr secular Gleissberg cycles (Landscheidt, 1981, 1999), the prolonged Maunder-type solar minima (Fairbridge & Shirley, 1987; Charvátová, 1990, 2000), short-term variations in solar luminosity (Sperber et al., 1990), sun spot extrema (Landscheidt, 1999), the 2400-yr cycle seen in ^{14}C tree-ring proxies (Charvátová, 2000), hemispheric sun spot asymmetry (Juckett, 2000), torsional oscillations in long-term sun spot clustering (Juckett, 2003), and violations of the Gnevishev–Ohl sun spot rule (Javaraiah, 2005).

A specific curiosity of the Sun's barycentric motion is evident in Figure 3.4a. Around 1990 (and before that, in 1811 and 1632) the Sun had a retrograde motion relative to the barycentre, i.e. its angular momentum with respect to the centre of mass was weakly negative (Jose, 1965). Javaraiah (2005) has argued that epochs violating the Gnevishev–Ohl 'sun spot rule', which states that the sum of sun spot numbers over an odd-numbered cycle exceeds that of its preceding even-numbered cycle (Gnevishev & Ohl, 1948; Komitov & Bonev, 2001; Nagovitsyn et al., 2009), are close to these intervals of the Sun's retrograde orbital motion.

Attempts to identify a spin–orbit coupling mechanism between the solar axial rotation and orbital revolution, most recently by Zaqarashvili (1997) and Juckett (2000), have been contested by Shirley (2006), who nevertheless acknowledged that a physical picture relating the Sun's rotation and revolution remains unclear. An implied connection between the Sun's equatorial rotation rate and the barycentric motion is postulated by Wilson et al. (2008b).

Perryman & Schulze-Hartung (2011) showed that behaviour cited as being correlated with the Sun's activity, for example large changes in orbital angular momentum, dL/dt, and intervals of negative orbital angular momentum, are common – but more extreme – in exoplanet systems. HD 168443 and HD 74156 in particular, with dL/dt exceeding that of the Sun by more than five orders of magnitude, offer an independent opportunity to corroborate the hypothesised link between the Sun's barycentric motion and the many manifestations of solar activity. In particular, they offer the possibility of independently testing any theories of spin–orbit coupling which are advanced in the case of the Sun.

3.6 Modeling planetary systems

Proper motion and parallax The astrometric wobble is a second-order perturbation of the observable (the position), apparent only after accounting for the much larger effects of proper motion and parallax present in the same observations.

In the absence of an orbiting companion there are five astrometric observables which describe a star's angular position on the sky: the equatorial coordinates in right ascension and declination, α_0, δ_0 [given at a specified epoch, typically J2000.0, and within a specified reference system, now conventionally the International Celestial Reference System ICRS, which replaces the earlier reference system equinox/epoch J2000, and the even earlier equinox/epoch B1950]; the corresponding orthogonal components of proper motion $\mu_{\alpha*}, \mu_\delta$ [with $\mu_{\alpha*} = \mu_\alpha \cos\delta$ denoting an angle in great-circle measure]; and the parallax ϖ [a designation often adopted in preference to π to avoid confusion with the mathematical constant].

Keplerian elements A planetary orbit determined astrometrically is characterised by the seven Keplerian parameters (§2.1): $a, e, t_p, i, \Omega, \omega, P$ (observations of the semi-major axis in angular measure, a'', can be converted into linear measure using the star distance). Orbit fitting for a system with n_p planets therefore requires a total of $5 + 7 \times n_p$ parameters.

In a simplistic picture, as for orbit reconstruction in the case of radial velocity observations, each planet would have its own orbital frequency, and Fourier analysis of the total signal would reveal the various planets. In practice, a planet in an eccentric orbit might have harmonics which could be interpreted as distinct planets, while a long-period planet observed over a short time might have noise generated at many frequencies due to the difficulty of distinguishing an orbital arc from the star's proper motion.

Numerous studies have investigated the sensitivity of astrometric measurements for the detection of exoplanets and for the measurement of their orbital elements and masses (e.g. Black & Scargle, 1982; Eisner & Kulkarni, 2001a, 2002; Konacki et al., 2002; Pourbaix, 2002; Sozzetti, 2005). For multiple planet systems detectable by Gaia and SIM, studies of the fitting procedures have shown them, as in the case of radial velocity modeling for multiple systems, to be highly non-linear (Casertano et al., 2008; Wright & Howard, 2009; Traub et al., 2010).

Given this complexity, various algorithms for single- and multiple-component orbital fits are being developed for use with the Gaia and SIM-type data (alone or together with radial velocity data), either based on frequency decomposition (Konacki et al., 2002), using minimisation techniques to optimally search the orbital parameter space, such as Levenberg–Marquardt or Markov Chain Monte Carlo analysis (Ford & Gregory, 2007), or using partial linearisation of the multi-body Kepler problem.

Mass and orbit inclination Unlike radial velocity (Doppler shift) measurements, which have an intrinsic $M_p \sin i$ degeneracy due to the fact that the orbit inclination, i, is undetermined, astrometric measurements provide a and i separately. With $M_\star$ estimated from its spectral type or evolutionary models, then the astrometric displacement yields M_p directly.

For multiple planet systems, astrometry can also (in principle) determine the relative inclinations between pairs of orbits (van de Kamp, 1981, eqn 16.5)

$$\cos \Delta i = \cos i_1 \cos i_2 + \sin i_1 \sin i_2 \cos(\Omega_1 - \Omega_2) . \quad (3.17)$$

One objective of such studies is to establish the occurrence of co-planar orbits, an important ingredient for formation theories and dynamical stability analyses.

Planet–planet interactions In numerous multi-planet systems, e.g. GJ 876 (Bean & Seifahrt, 2009), planet–planet interactions can significantly alter the radial velocity and/or astrometric signature of the system, and a full dynamical (Newtonian) fit involving an N-body code may be needed to ensure the short- and long-term stability of the solution.

Linearisation The four elements a, ω, i, Ω can be replaced by the Thiele–Innes constants (Thiele, 1883; Binnendijk, 1960; Heintz, 1978a) which are better suited to orbital description in rectangular coordinates

$$\begin{aligned}
A &= a(+\cos\omega\cos\Omega - \sin\omega\sin\Omega\cos i) \\
B &= a(+\cos\omega\sin\Omega + \sin\omega\cos\Omega\cos i) \\
F &= a(-\sin\omega\cos\Omega - \cos\omega\sin\Omega\cos i) \\
G &= a(-\sin\omega\sin\Omega + \cos\omega\cos\Omega\cos i) \, .
\end{aligned} \quad (3.18)$$

Together with $C = a\sin\omega\sin i$ and $H = a\cos\omega\sin i$, these are related to the direction cosines of the major and minor axes of the orbit in the xy tangent plane, $(A/a, B/a, C/a)$ and $(F/a, G/a, H/a)$ respectively. Analogous constants had already appeared in Gauss' treatment of planetary orbits.

Tangential coordinates in the plane of the sky are then given by (Heintz 1978a, eqn 15–16; van de Kamp 1981, eqn 10.6)

$$\begin{aligned}
x &= AX + FY \\
y &= BX + GY \, ,
\end{aligned} \quad (3.19)$$

where XY are elliptical rectangular coordinates in the orbit plane defined as

$$\begin{aligned}
X &= \cos E(t) - e \\
Y &= \sqrt{1-e^2}\sin E(t) \, ,
\end{aligned} \quad (3.20)$$

and $E(t)$ is the eccentric anomaly.

The astrometric displacement of a star in equatorial coordinates, at time t, due to parallax, proper motion, and a system of n_p unseen planets is then readily generalised as (Wright & Howard, 2009, eqn 49–50)

$$\Delta\alpha(t)\cos\delta = \sum_{j=1}^{n_\mathrm{p}} \left[B_j X_{j,t} + G_j Y_{j,t} \right] +$$
$$\Delta\alpha_0\cos\delta + \varpi\Pi_{\alpha,t} + \mu_\alpha(t-t_0) \quad (3.21)$$

$$\Delta\delta(t) = \sum_{j=1}^{n_\mathrm{p}} \left[A_j X_{j,t} + F_j Y_{j,t} \right] +$$
$$\Delta\delta_0 + \varpi\Pi_{\delta,t} + \mu_\delta(t-t_0) \, , \quad (3.22)$$

where $(\Delta\alpha_0\cos\delta, \Delta\delta_0)$ accounts for an adjustment to the nominal star position. $(\Pi_{\alpha,t}, \Pi_{\delta,t})$ are the orthogonal components of the displacements due to parallax in the coordinate directions (Green, 1985; Seidelmann, 1992)

$$\begin{aligned}
\Pi_{\alpha,t} &= r_x(t)\sin\alpha - r_y(t)\cos\alpha \\
\Pi_{\delta,t} &= \left[r_x(t)\cos\alpha + r_y(t)\sin\alpha\right]\sin\delta - r_z(t)\cos\delta \, . \quad (3.23)
\end{aligned}$$

(r_x, r_y, r_z) are the Cartesian components of the observatory position with respect to the solar system barycentre.

Combining astrometry and radial velocities Four orbital elements are in common between astrometric and spectroscopic solutions: a, e, ω, t_p. Combined observations therefore further constrain the three-dimensional orbit, and can yield individual component masses without the ambiguity of the orbital inclination.

The approach used for the combination of HST–FGS astrometry and radial velocity orbits is described by Benedict et al. (2002). This proceeds by first constraining all 'plate constants' (image scale, rotation, offsets, radial terms, and parallax factors) to those determined from their astrometry-only solution, then determining the orbital elements K, e, P, ω from the available radial velocity data, and finally constraining the orbit by minimising the astrometric and radial velocity residuals according to the identity given by Equation 2.22 (as proposed by Pourbaix & Jorissen, 2000, their Equation 12)

$$\frac{a''\sin i}{\varpi} = \frac{PK\sqrt{1-e^2}}{2\pi} \, . \quad (3.24)$$

Here, the left-hand side contains quantities derived only from the astrometric solution (parallax ϖ, angular semi-major axis a'', and orbital inclination i), and the right-hand side are those from the radial velocity orbit (period P, radial velocity semi-amplitude of the visible component K, and eccentricity e).

Further considerations of the linear formulation of the combined astrometric and radial velocity solution is given by (Wright & Howard, 2009).

3.7 Astrometric measurements from ground

Results from interferometers equipped with adaptive optics at Keck-I and VLTI have confirmed that 10 – 100 μas narrow-field astrometry should be achievable. This has led to various astrometric planets search programmes being initiated from the ground.

Palomar: STEPS Accuracy limits measured at the 5-m Palomar telescope (Pravdo & Shaklan, 1996) prompted the start of the STEllar Planet Survey (STEPS), an astrometric survey for giant planet and brown dwarf companions around some 30 nearby M-dwarfs (Pravdo et al., 2005; Pravdo & Shaklan, 2009a). The survey has detected several brown dwarf companions over its 10-year programme.

Their first proposed planet mass detection around the nearby (6 pc) low-mass ($0.08 M_\odot$) cool M8 V star, VB 10, was reported by Pravdo & Shaklan (2009b), the most recent of the claimed astrometric discoveries (see box on page 64). From a simultaneous fit of the astrometry and low-precision radial velocity data, they derived $M_\mathrm{p} \sim 6.4 M_\mathrm{J}$, $P = 0.744$ yr, $a = 0.36$ AU (62 mas), $i = 96°\!.9$, and a resulting astrometric signature $\alpha \sim 5$ mas. The existence of the planet was weakly supported by 300 m s^{-1}

precision near-infrared radial velocities from Keck II–NIRSPEC (Zapatero Osorio et al., 2009), but such a high-inclination orbit was ruled out using $10\,\mathrm{m\,s^{-1}}$ near-infrared measurements from VLT–CRIRES (Bean et al., 2010b), and probably from independent $200\,\mathrm{m\,s^{-1}}$ accuracy data (Anglada-Escudé et al., 2010b).

Palomar PTI: PHASES The Palomar Testbed Interferometer is a near-infrared instrument with a maximum baseline of 100 m, which ceased normal operations in 2008. The PHASES programme (Palomar High-precision Astrometric Search for Exoplanet Systems, Muterspaugh et al., 2010), used phase-referencing imaging to achieve $\sim 100\,\mu$as accuracy for $\sim$30 arcsec binaries (Lane et al., 2000) and $20-50\,\mu$as for sub-arcsec binaries (Lane & Muterspaugh, 2004). Observations have excluded tertiary companions of a few M_J with $a < 2\,\mathrm{AU}$ in several binary systems (Muterspaugh et al., 2006).

VLTI–PRIMA: ESPRI ESO's VLTI (Very Large Telescope Interferometer) consists of four fixed 8.2 m Unit Telescopes (UTs), four moveable 1.8-m Auxiliary Telescopes (ATs) and six long-stroke delay lines, providing baselines up to 200 m and operations from $1-13\,\mu$m. The PRIMA facility (Phase-Referenced Imaging and Microarcsecond Astrometry) implements a dual-feed capability, initially for two UTs or ATs, to allow simultaneous interferometric observations of two objects separated by up to 1 arcmin. It targets an accuracy of $10-50\,\mu$as in its narrow-field differential astrometry mode (Delplancke, 2008; Launhardt et al., 2008).

The ESPRI consortium (Exoplanet Search with PRIMA, Launhardt et al., 2008) targets the astrometric detection of low-mass planets around nearby stars of any spectral type within 15 pc, the search for massive planets orbiting young stars with ages in the range $5-300$ Myr within 100 pc, and the astrometric characterisation of known radial velocity planets within 200 pc. The initial target list includes ~900 stars, but availability of suitable reference stars will likely result in ~100 stars observed during the 5-yr survey (Launhardt, 2009).

Keck: ASTRA The Keck interferometer combines light from the two 10-m diameter primary mirrrors, separated by 85 m, for near-infrared fringe visibility and mid-infared nulling observations (Keck–I, Ragland et al., 2008). The ASTRA (ASTrometric and phase-Referenced Astronomy) upgrade (Pott et al., 2009; Woillez et al., 2010) includes dual-star capability to carry out narrow-angle astrometry at the $100\,\mu$as level between pairs of objects separated by $\lesssim 20-30$ arcsec.

Las Campanas: CAPS The Carnegie Astrometric Planet Search started in 2007 using the 2.5-m du Pont telescope at Las Campanas, Chile (Boss et al., 2009). It is optimised to follow about 100 very nearby ($\lesssim 10\,\mathrm{pc}$) low-mass stars, principally late M, L, and T dwarfs (generally too faint and red to be included in ground-based Doppler planet surveys), for 10 yr. Astrometric accuracies, of $0.3\,\mathrm{mas\,hr^{-1}}$, should detect a $1\,M_J$ companion orbiting 1 AU from a late M dwarf at 10 pc.

3.8 Astrometric measurements from space

Above the effects of the Earth's atmosphere, the theoretical accuracies given by Equation 3.4 are essentially achievable, assuming the appropriate control of relevant instrumental error sources (such as attitude jitter and time-varying optical distortions) and their calibration.

In what must have been amongst the earliest discussions of space astrometry, Couteau & Pecker (1964) considered both double stars, and the search for planetary systems. An astrometric search for Jupiter-like companions to nearby stars using the Hipparcos data was suggested by Gliese (1982).

3.8.1 Hipparcos

Wide angle, global astrometric measurements have been made from space by the Hipparcos satellite, which operated between 1989–93. This provided 1 mas level accuracy for about 120 000 stars (Perryman, 1997; Perryman et al., 1997), further improved by a more recent re-reduction (van Leeuwen, 2007), along with lower accuracy data for the Tycho and Tycho 2 catalogues (Høg et al., 1997, 2000).

The Hipparcos catalogue was constructed by fitting the five parameters of the standard model $(\alpha, \delta, \mu_{\alpha*}, \mu_\delta, \varpi)$ to the 100 or so measurements for each of the 118 000 stars. Higher-order fits were made for 2622 stars with significant acceleration terms (either quadratic or cubic polynomials of time), and orbital fits for 235 stars with significant known or measured orbital motion attributable to a companion of stellar mass (Lindegren et al., 1997; Perryman, 2009, Chapter 3). Orbital perturbations due to planets are barely recognisable at the milliarcsec level, and attempts to quantify astrometric motion due to orbiting planets has been carried out only in a few specific cases following the catalogue publication, as described below.

Intermediate astrometry In addition, the catalogue includes the *intermediate astrometric data*. These are the 100 or so one-dimensional coordinates on the measurement great-circles made for each star during the observing period. Files contain the mid-epochs and poles of the reference great circle for each orbit and, within each orbit, the residuals between the observed abscissae and those calculated from the set of reference astrometric parameters given in the main catalogue.

This allows the reconstruction of the photocentric motion on the sky, as shown schematically in Figure 3.1. It also allows the Hipparcos data to be merged rigorously with other (astrometric or radial velocity) data, or to be fit to models other than the basic fitting used to construct the Hipparcos catalogue. In the present context, periodic motion of a star due to an invisible companion whose orbital period is known, may be detectable at levels below which the star is securely identifiable as double in the absence of such prior knowledge.

In equatorial coordinates, partial derivatives of the star abscissa can be expressed with respect to the five parameters of the standard model as

$$d_1 = \partial a_i/\partial \alpha_*, \; d_2 = \partial a_i/\partial \delta, \; d_3 = \partial a_i/\partial \varpi,$$
$$d_4 = \partial a_i/\partial \mu_{\alpha*}, \; d_5 = \partial a_i/\partial \mu_\delta, \quad (3.25)$$

where a_i is the abscissa of the ith observation of a given star. In the small-angle approximation, a linearized equation for the observed abscissa difference $\Delta a_i = a_{\text{obs}} - a_{\text{calc}}$, can then be written as (Volume 1, Section 2.8 of Perryman, 1997; Goldin & Makarov, 2006)

$$\Delta a_i = d_1 \Delta x + d_2 \Delta y + d_3 \Delta \pi + d_4 \Delta \mu_x + d_5 \Delta \mu_y$$
$$+ d_1 \sum_j \frac{\partial x}{\partial \epsilon_j} \Delta \epsilon_j + d_2 \sum_j \frac{\partial y}{\partial \epsilon_j} \Delta \epsilon_j, \quad (3.26)$$

where ϵ_j are the vector components of the seven orbital elements, $\epsilon = [P, a'', e, i, \omega, \Omega, t_p]$, and $x \equiv \alpha_*, y \equiv \delta$.

Results Perryman et al. (1996) derived weak upper limits on M_p for 47 UMa ($< 7 M_J$), 70 Vir ($< 38 M_J$), and 51 Peg ($< 500 M_J$) based on adjustment of the orbital elements using a large number of trial periods, including those of the known planets. Comparable upper limits for 47 UMa were later given by Zucker & Mazeh (2001) and, following the discovery of a second planet orbiting the same system, by Fischer et al. (2002).

Mazeh et al. (1999) used a similar approach to derive a mass for the outer companion of the triple planetary system υ And, of $M_p = 10.1^{+4.7}_{-4.6} M_J$, compared with an $M_p \sin i$ from radial velocity measurements of $4.1 M_J$. They also derived estimates of the mass of the two inner planets, on the assumption that the orbits of all three are co-aligned, of $1.8 \pm 0.8 M_J$ and $4.9 \pm 2.3 M_J$ respectively.

A number of studies began to suggest rather high masses for some of the proposed planetary companions. For HD 10697 the companion was inferred to be a brown dwarf (Zucker & Mazeh, 2000). For ρ CrB, Gatewood et al. (2001) combined the Hipparcos intermediate astrometry with MAP data to derive a mass some 100 times larger than the minimum mass from radial velocity studies, although this was questioned by subsequent high-resolution infrared spectroscopy (Bender et al., 2005). Han et al. (2001b) used the intermediate astrometric data for all known planetary candidates with $P > 10$ d to conclude that spectroscopic programmes can be biased to small values of $\sin i$, leading in turn to masses much in excess of the associated minimum masses.

Possible bias in the mass estimates obtained in this way subsequently became the subject of some debate. Pourbaix (2001) and Pourbaix & Arenou (2001) re-analysed the same data, and argued instead that the trend to low inclinations is an artefact of the adopted fitting. This was supported by McGrath et al. (2002), who used HST FGS observations, formally at the level of 0.3 mas, to examine the motion of 55 Cnc. They placed 3σ upper limits of 0.3 mas on the semi-major axis of the reflex motion, ruling out the 1.15 mas perturbation proposed by Han et al. (2001b), and placing an upper limit of about $30 M_J$ on the planetary mass.

In conclusion, some inferences about the properties of systems discovered by radial velocity measurements have been possible from the Hipparcos data, but it remains evident from Equation 3.1 (and Figure 3.2) that its milliarcsec astrometry can contribute only marginally to exoplanet detection and orbit characterisation, the more so since the elementary data have individual accuracies somewhat poorer than those of the five primary astrometric parameters derived from them.

3.8.2 HST–Fine Guidance Sensor

The Fine Guidance Sensors on HST provide targeted, narrow angle, relative astrometry (Benedict et al., 1999, 2000). Within the field of view (the FGS pickle), the single-measurement precision for $V \lesssim 16$ mag is $\sim$1–2 mas, primarily limited by spacecraft jitter. From multiple measurements, astrometric signatures down to

---

Other Hipparcos limits: Mass limits for other planet candidates discovered from radial velocity observations include: HD 179949 and HD 164427 (Tinney et al., 2001); ι Dra (Frink et al., 2002); and HD 11977 (Setiawan et al., 2005). Zucker & Mazeh (2001) analysed Hipparcos astrometry for 47 planet and 14 brown dwarf secondaries, finding that the lowest derived upper limit is for 47 UMa at $0.014 M_\odot$, similar to the limits given by Perryman et al. (1996). For 13 other candidates, upper limits exclude a stellar companion although brown dwarf secondaries are still an option, further supporting the idea that most exoplanets are not merely disguised stellar secondaries.

The intermediate astrometric data were used to constrain, or in three cases to solve for, the orbital inclinations and masses for a further ten low-mass companions from the Keck precision radial velocity survey (Vogt et al., 2002). Astrometric orbital solutions for a few Doppler-detected systems containing companions with minimum masses close to and slight above the dividing line between planets and brown dwarfs have been obtained by Reffert & Quirrenbach (2006) and Sozzetti & Desidera (2010).

Various studies illustrate the rigorous use of combined radial velocity and astrometric data. Hauser & Marcy (1999) concluded that their combined results were consistent with the possibility that perturbations from 16 Cyg A cause the eccentricity in the planet around 16 Cyg B. Torres (2007) used a combination of radial velocity measurements and Hipparcos astrometry in a detailed study of the orbit of the bright K1 III–IV star γ Cep. The minimum mass of the companion is $M_p \sin i = 1.43 \pm 0.13 M_J$ from the radial velocity data alone, while the Hipparcos astrometric data place a dynamical upper limit on this mass of $13.3 M_J$.

3.9 Future observations from space

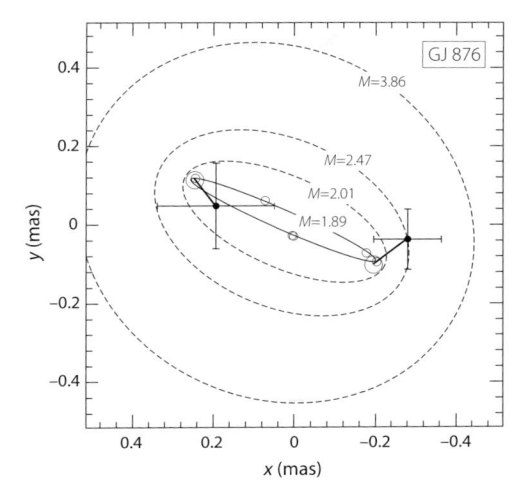

Figure 3.5: HST–FGS astrometric observations of GJ 876, showing four possible primary star orbits due to the perturbation by the longest period planet GJ 876 b, labeled according to inferred mass (M_J). The densest sets of observations, at pericentre and apocentre, are shown by large circles. The astrometry-only residual normal points at phases 0.26 (pericentre, lower right) and 0.72 are shown as large crosses, connected to the derived orbit by their residual vectors. From Benedict et al. (2002, Figure 3), reproduced by permission of the AAS.

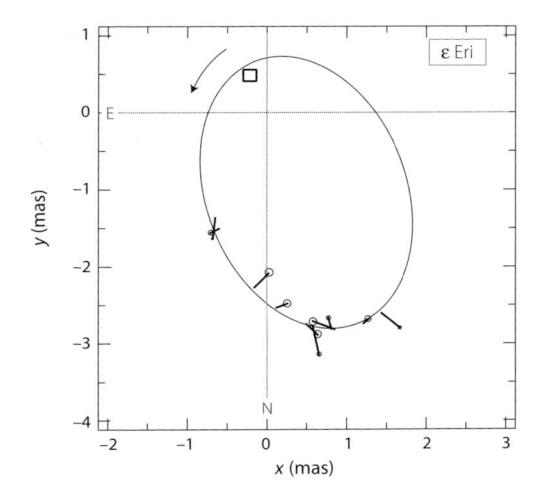

Figure 3.6: HST–FGS astrometric observations of ϵ Eri. Vectors connect the normal point residuals to their predicted position at each observation epoch. Circle sizes are proportional to the number of observations forming the normal point. The square marks pericentre passage, $T_0 = 2007.3$, and the arrow indicates the direction of the star motion. From Benedict et al. (2006, Figure 10), reproduced by permission of the AAS.

around 0.25 mas can be detected (McArthur et al., 2010, section 4.2). Various astrometric results on exoplanet candidates have been published.

55 Cnc McGrath et al. (2002) gave an upper mass limit for the first planet in the 55 Cnc system, 55 Cnc b, of $\sim 30 M_J$. McArthur et al. (2004) determined $M_p = 17.7 \pm 5 M_\oplus$ for the inner planet, 55 Cnc e, under the assumption of coplanarity.

GJ 876 Benedict et al. (2002) measured an astrometric signature $\alpha = 0.25 \pm 0.06$ mas for GJ 876 b, from a combined fit to HST–FGS astrometry and high-precision radial velocities (Figure 3.5). Assuming $M_\star = 0.32 M_\odot$, they determined a preferred solution $M_p = 1.89 \pm 0.34 M_J$, $i = 84 \pm 6°$, $\alpha = 0.25 \pm 0.06$ mas, and $\varpi = 214.6 \pm 0.2$ mas.

Bean & Seifahrt (2009) constrained the coplanarity of planets b and c using a combination of Doppler measurements, HST–FGS astrometry, and dynamical considerations. They determined a mutual inclination $\Delta i_{bc} = 5.0^{+3.9}_{-2.3}°$, concluding that the planets probably formed in a circumstellar disk, and that their subsequent dynamical evolution into a 2:1 mean motion resonance only led to excitation of a small mutual inclination.

ϵ Eri Benedict et al. (2006) measured an astrometric displacement due the long-period planet in ϵ Eri of $\alpha = 1.88 \pm 0.20$ mas, and a resulting mass $M_p = 1.55 \pm 0.24 M_J$ (Figure 3.6).

υ And The results for υ And (McArthur et al., 2010) illuminate the prospects of sub-milliarcsec astrometry (Figure 3.7). They determined the masses of υ And c ($13.98^{+2.3}_{-5.3} M_J$) and υ And d ($10.25^{+0.7}_{-3.3} M_J$), and their mutual inclination ($\Delta i_{cd} = 29.9° \pm 1°$), representing the first direct determination of relative orbit inclinations.

Other By determining M_p rather than $M_p \sin i$ other studies combining HST–FGS astrometry and high-precision radial velocities have identified companions as brown dwarfs or M dwarfs rather than massive planets: HD 33636 (Bean et al., 2007) and HD 136118 (Martioli et al., 2010).

Other HST–FGS data have been acquired and are being analysed in order to determine the actual masses of the Doppler-detected planets HD 47536 b, HD 136118 b, HD 168443 c, HD 145675 b, and HD 38529 c (Benedict et al., 2008). Observations are also being collected for other multiple systems, including HD 128311, HD 202206, μ Ara, and γ Cep, with the aim of measuring directly their coplanarity.

3.9 Future observations from space

Large-scale acquisition of microarcsec astrometric measurements promises at least three important developments. First, measurements significantly below 0.1 mas offer planet detection possibilities well below the Jupiter mass limit, out to 50–200 pc. Second, in combination with spectroscopic measurements and an estimated host star mass, they provide direct determination of the planet mass, independent of the orbital inclination. Third, the relative orbital inclination of multi-planet systems can be determined.

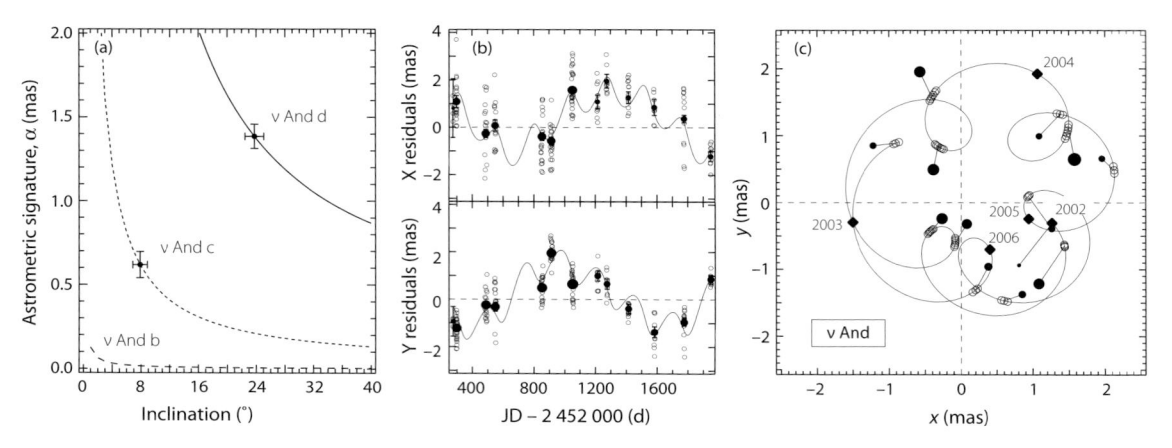

Figure 3.7: υ And observed with HST–FGS: (a) astrometric signature versus orbit inclination, with the HST–FGS determinations for planets c and d indicated; (b) components of the stellar reflex motion due to planets b and c. Filled circles indicate normal points, with size proportional to the number of individual measurements at that epoch (open circles); (c) astrometric reflex motion of υ And due to planets c and d. The solid curve shows the modeled motion (cf. Figure 3.4). Open circles lying on the curve show the times of observations, and filled circles indicate the associated normal points. From McArthur et al. (2010, Figures 7, 9, 10), reproduced by permission of the AAS.

Gaia Gaia is an ESA project in the final stages of development, and due for launch in 2012 (Perryman et al., 2001; Lindegren, 2009). Gaia will survey approximately a billion stars to $V \sim 20$ mag as part of a census of the Galactic stellar population. Around 80 distinct measurements will be made for each star, with a magnitude-dependent single-measurement accuracy determined by the image centroid accuracy derived from a focal plane scan. The Gaia accuracies for bright stars require location accuracies of ~0.001 of a CCD pixel, and the corresponding capability to calibrate attitude motion and optical distortions, including chromatic terms.

Early estimates of 10–30 000 exoplanet systems detectable by Gaia (Lattanzi et al., 2000; Perryman et al., 2001; Quist, 2001; Sozzetti et al., 2001) were made on the basis of target accuracies at the time of mission acceptance, and are no longer applicable.

More recent evaluations have used the latest mission accuracy estimates, a single-measurement error for bright stars $\sigma_\psi = 8\,\mu$as, a mission duration of five years, and a more realistic double-blind protocol for assessing the significance of the solutions (Casertano et al., 2008). Simulations show that planets with astrometric signature $\alpha \gtrsim 3\sigma_\psi$ and $P < 5$ years could be detected reliably with a small number of false positives. At twice this limit, uncertainties in orbital parameters and masses are typically 15–20%. Over 70% of two-planet systems with well-separated periods in the range $0.2 \le P \le 9$ yr, and eccentricity $e \le 0.6$, are correctly identified. For favourable configurations of two-planet systems with well separated periods under 4 yr, it should be possible to carry out meaningful coplanarity tests, with typical uncertainties on the mutual inclination angle of $\lesssim 10°$.

Extrapolating from the statistical properties of the known exoplanet sample, Gaia should discover and measure several thousand giant planets with semi-major axes $a = 3 - 4$ AU out to 200 pc, and will characterise hundreds of multiple-planet systems, including meaningful tests of coplanarity.

SIM, SIM PlanetQuest, and SIM Lite NASA's Space Interferometry Mission (SIM) is a pointed (Michelson) interferometer, originally with a 10 m baseline, which started development in 1996, based on concept studies made several years earlier (Shao, 1993). In a revised configuration, SIM PlanetQuest (Unwin et al., 2008), it targeted parallaxes at about $4\,\mu$as for $V \lesssim 20$, and a differential accuracy of $0.6\,\mu$as on bright stars.

A further redesign, SIM Lite, had a shorter baseline (6 m), and consisted of two Michelson stellar interferometers and a precision telescope working as a star tracker (Goullioud et al., 2008). It targeted sub-$1\,\mu$as narrow-angle astrometry in 1.5 hr integrations on bright targets ($V \le 6$) linked to moderately fainter reference stars ($V \simeq 9 - 10$) in a two degree field.

An accuracy on the position of the delay lines of a few tens of pm, and a positional stability of internal optical path lengths of ~10 nm, is required to maintain the necessary fringe visibility (Zhai et al., 2008). During a five year mission, it would search 65 nearby stars for planets down to $1 M_\oplus$ in the habitable zone. In pointed mode, it would reach $8\,\mu$as accuracy at $V = 19$ in 100 hr of integration.

The pointed capability means that it would be applicable for detailed orbit determinations of targeted planetary systems, including those known from previous radial velocity or transit experiments. The possibility of

3.9 Future observations from space

deriving relative inclinations would follow the principles already demonstrated by HST–FGS (cf. Figure 3.7).

Various detailed assessments of its planet detection and characterisation capability have been given (Sozzetti et al., 2002, 2003; Ford & Tremaine, 2003; Ford, 2004a, 2006a; Catanzarite et al., 2006; Unwin et al., 2008; Brown, 2009a; Traub et al., 2010).

Amongst these, Traub et al. (2010) used detailed double-blind simulations to evaluate planet detection reliability and completeness, highlighting how different detection and orbit fitting algorithms, using the same data, can perform in measurably different ways. Simulated systems included solar system analogues, and single terrestrial habitable zone planets. With 250 visits to a given star over the mission lifetime, each of differential accuracy $\sigma_\psi = 1.4\,\mu$as along one axis, and the addition of 15 years of radial velocity observations to assist characterisation of any long-period planets in the system, the study demonstrated that Earth-like planets could be detected, even in the presence of other orbiting planets.

A further optical interferometer study, Planet Hunter, has been carried out at JPL as a NASA astrophysics strategic mission concept (Marcy, 2009b).

Interferometric space astrometry does not appear, however, as a high priority in the 2010 US Decadal Survey report (Blandford et al., 2010).

Other planned space missions JASMINE (Japan Astrometry Satellite Mission for INfrared Exploration) is a sky-scanning mission with a target accuracy of $10\,\mu$as. It will operate in the infrared, and has a projected launch date around 2016 (Gouda et al., 2008). Nano-JASMINE is a smaller sky-scanning precursor experiment operating in the z-band, aiming for accuracies of around 1 mas (Kobayashi et al., 2008).

JMAPS (Joint Milliarcsecond Pathfinder Survey) is a US Department of Navy bright star astrometric all-sky survey scheduled for launch around 2012, targeting some 20 million stars to $V = 14$ mag, also with accuracy ~1 mas (Dorland et al., 2009; Hennessy et al., 2010).

Earlier concepts Various other national space astrometry programmes were studied and proposed over the past decade which included planet detection amongst their scientific objectives, but which are no longer under consideration. These included the German DIVA project targeting 0.2 mas accuracy to 15 mag (Röser, 1999), the USNO sky-scanning projects FAME targeting 10 million stars to 14 mag (Johnston, 2003), AMEX and OBSS reaching to 23 mag (Johnston et al., 2006b), and MAPS operating in a step-stare mode (Zacharias & Dorland, 2006). Russian studies included OSIRIS and LIDA (Bagrov, 2006).

4

Timing

An orbiting planet is accompanied by the periodic oscillation of the position of the host star about the system barycentre, recognisable through changes in the radial velocity and astrometric position of the primary.

If the primary also possesses periodic time signatures, then these can provide an alternative route to the dynamical detection of orbiting planets through the change in measured period due to light travel time. This has an amplitude related to the displacement of the primary along the line-of-sight

$$\tau_\mathrm{p} = \frac{1}{c} \frac{a \sin i \, M_\mathrm{p}}{M_\star}, \qquad (4.1)$$

where c is the velocity of light.

There are three classes of object, in addition to transiting planets, which offer this possibility: radio pulsars, pulsating stars, and eclipsing binaries.

Pulsars provide short period and extremely stable timing signals, and their careful monitoring led to the first exoplanet discovery, around the millisecond pulsar PSR B1257+12, in 1992. Two planet-hosting pulsars are now known, and they include the lowest-mass planet discovered to date, at just $0.02 M_\oplus$.

The realisation that planets could exist around stars in their final evolutionary stages has led to numerous, but as yet unsuccessful (or at least, unconfirmed), searches for planets around (pulsating) white dwarfs. More recently, planets have been inferred orbiting a pulsating extreme horizontal branch star, and around three eclipsing binary systems.

As of the end of 2010, a total of 10 planets around six objects have been discovered by timing techniques (Table 4.2).

4.1 Pulsars

Pulsars are rapidly spinning highly-magnetised neutron stars, formed during the core collapse of massive stars ($\sim 8-40 M_\odot$) in a supernova explosion. They emit narrow beams of radio emission parallel to their magnetic dipole axis, seen as pulses at the object's spin frequency due to a misalignment of the magnetic and spin axes.

There are two broad classes: 'normal' pulsars, with spin periods around 1 s; and the msec pulsars, 'recycled' old ($\sim 10^9$ yr) neutron stars that have been spun-up to very short periods during mass and angular momentum transfer from a binary companion; most of the known msec pulsars still have (non-accreting) binary companions, either white dwarfs or neutron stars.

Millisecond pulsars are extremely accurate frequency standards, with periods normally changing only through a tiny spin-down at a rate $\sim 10^{-19}$ s s^{-1} due to their low magnetic field strength (Bailes, 1996). Pulse arrival time residuals can be measured with an accuracy of order μs, with one of the most stable, PSR J0437–4715, showing residuals of 130 ns (van Straten et al., 2001).

The known pulsar population exceeds 1700, with more than 80 millisecond pulsars associated with the disk of the Galaxy, and some 130 known in Galactic globular clusters of which Terzan 5 counts 33, and 47 Tuc 22. The fastest spin-period known is the 716-Hz (1.4-ms) eclipsing binary pulsar PSR J1748–2446ad in Terzan 5.

The high accuracy of pulsar timing allows low mass bodies orbiting the pulsar to be detected from changes in pulse arrival times. For a circular edge-on orbit of period P, and assuming a canonical pulsar mass of $1.35 M_\odot$, Equation 4.1 can be written (Wolszczan, 1997)

$$\tau_\mathrm{p} \simeq 1.2 \left(\frac{M_\mathrm{p}}{M_\oplus}\right) \left(\frac{P}{1\,\mathrm{yr}}\right)^{2/3} \mathrm{ms}. \qquad (4.2)$$

Jovian or terrestrial planets should be detectable around 'normal' slow pulsars, while substantially lower masses, down to that of the Moon and largest asteroids, could be recognised in millisec pulsar timing residuals.

4.1.1 PSR B1257+12

The first planet discovered around an object other than the Sun was found around the 6.2-ms pulsar PSR B1257+12[1] at $d \sim 300$ pc (Wolszczan & Frail, 1992).

---

[1] Pulsar designations signify the object's right ascension and declination. 'B' indicates discoveries in the earlier reference system of equinox and epoch B1950.0. More recent discoveries are designated 'J' for the epoch/equinox J2000 (now ICRS).

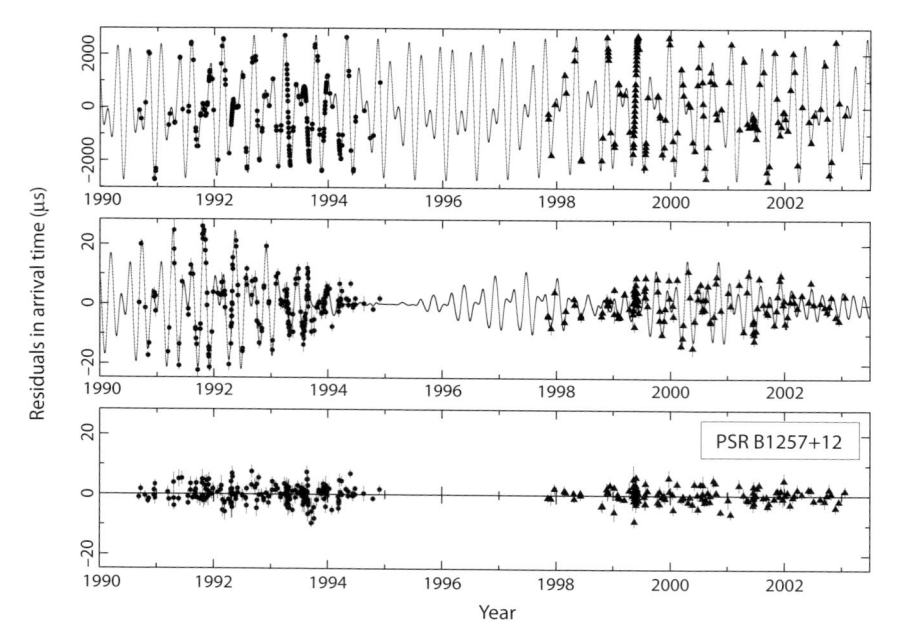

Figure 4.1: Daily-averaged arrival time residuals for PSR B1257+12 observed at 430 MHz with the 305-m Arecibo radio telescope between 1990–2003. Top: residuals for a model without planets. Middle: residuals for a Keplerian model for all three planets, dominated by perturbations between the outer planets. Bottom: residuals after inclusion of planet–planet perturbations. From Konacki & Wolszczan (2003, Figure 1), reproduced by permission of the AAS.

At least two terrestrial-mass companions were inferred from the timing residuals, with masses of $M \sin i \simeq 2.8$ and $3.4 M_\oplus$. Their orbits were almost circular, with $a = 0.47$ and 0.36 AU. Of importance for the subsequent characterisation of the system is the fact that their periods ($P = 98.22$ and 66.54 d respectively) are close to a 3:2 (mean motion) resonance.

Confirmation and other planets Although a number of alternatives to explain the observed timing residuals were examined (Phillips & Thorsett, 1994), the planet hypothesis could be rigorously verified: the semi-major axes of the orbits are sufficiently similar that the two planets would perturb one another significantly during their orbital close encounters, every $\simeq 200$ d, with resulting three-body effects leading to departures from a simple non-interacting Keplerian model which would grow with time (Rasio et al., 1992; Malhotra et al., 1992; Malhotra, 1993b; Peale, 1994).

Continued monitoring provided confirmation of the predicted mutual gravitational perturbations, described by osculating terms in which the Keplerian elements evolve with time. Modeling of the orbit evolution allowed masses to be derived without *a priori* knowledge of their orbital inclination, $\sin i$. They also provided evidence for a third planet with $P = 25.34$ d and $M_p = 0.02 M_\oplus$, roughly that of the Moon (Wolszczan, 1994a,b; Konacki et al., 1999b; Wolszczan, 2008).

That the third planet might have been an artefact of temporal changes of heliospheric electron density at the solar rotation rate was discussed (Scherer et al., 1997), but later discounted. Suggestions of a possible fourth planet with $P \sim 170$ yr (Joshi & Rasio, 1997; Wolszczan, 1997) are unconfirmed.

Konacki & Wolszczan (2003) presented updated results from observations made between 1990–2003 and analysed according to the detailed timing model of Konacki et al. (2000, see Figure 4.1), leading to the physical parameters given in Table 4.1.

System stability Early dynamical simulations indicated that the orbit configuration of the two outer planets should be stable over some hundreds of thousands of years (Gladman, 1993). This was confirmed by later studies using the more accurate initial conditions determined by Konacki & Wolszczan (2003).

Goździewski et al. (2005b) assessed stability using MEGNO (§2.6, Cincotta & Simó, 2000), a fast indicator that aims to distinguish between regular and chaotic motions by integration over some 10^4 orbital periods of the outermost planet, thereby rapidly examining a large number of initial conditions. Direct integration over 1 Gyr further demonstrated that no secular changes of a, e or i appeared during this interval, and that the relative inclination of the lowest mass planet to the mean orbital plane of the other two must be rather small.

Other studies of the stability of the PSR B1257+12 system were reported by Callegari et al. (2006).

Table 4.1: Orbital and physical parameters of the PSR B1257+12 planets. Planet designations follow those given in Wolszczan (1994a) and Konacki & Wolszczan (2003).

Parameter	Planet A	Planet B	Planet C
Projected semi-major axis (ms)	0.0030	1.3106	1.4134
Eccentricity, e	0.0	0.0186	0.0252
Orbital period, P (d)	25.262	66.5419	98.2114
Mass ($M_\oplus$)	0.020	4.3	3.9
Inclination, i (°)	–	53	47
Semi-major axis, a (AU)	0.19	0.36	0.46

Formation Early models to explain the existence of planets around pulsars hypothesised three formation mechanisms (Phinney & Hansen, 1993; Podsiadlowski, 1993; Banit et al., 1993; Phillips & Thorsett, 1994). In the first, a planet formed around a normal star, the pulsar progenitor, its present existence then implying that it must have survived the supernova explosion. A second possibility is that it was formed around another star, before being captured by the pulsar through dynamical interaction. Alternatively, the planet was only formed after the supernova explosion which created the neutron star.

In this latter scenario, the supernova would need to retain some residual material that could fall back to form a debris disk around the young neutron star. Subsequent fragmentation into planets would then follow the standard model of planet formation. Difficulties in modeling planets which survive the supernova explosion may favour the 'fallback' accretion disk model. At the same time, the process cannot be common: only two planet systems are known from the precision timing of Galactic millisecond pulsars (Lorimer, 2005).

A quark nova formation scenario is described by Keränen & Ouyed (2003).

Dust disks The simulations of Goździewski et al. (2005b) showed that, for PSR B1257+12, the zone beyond 1 AU is also stable, which in turn means that any residual outer dust disk or Kuiper-type belt should also be stable. More direct evidence for the formation and survival of dust disks around neutron stars has also come from the tentative identification of a cool $10M_\oplus$ disk around the young X-ray pulsar 4U 0142+61 (Wang et al., 2006).

Foster & Fischer (1996) quantified the observability of a pulsar disk by assuming that a fraction f of the pulsar spin-down luminosity L heats N dust grains of size a to a temperature T. The expected infrared flux can then be estimated from $fL \sim 4\pi a^2 N\sigma T^4$, along with the pulsar distance estimated from its dispersion measure.

Several attempts to detect dust disks around millisecond pulsars have been made (e.g. Lazio & Fischer, 2004; Löhmer et al., 2004; Wolszczan, 2008). The most stringent limits for PSR B1257+12 come from Spitzer–MIPS observations at 24 and 70 μm (Bryden et al., 2006a) which gave limits on a 100–200 K dust disk of below 3×10^{-5} of the pulsar's spin-down luminosity.

Indirect constraints on the origin of the debris which formed the planets around PSR B1257+12 has come from modeling by Currie & Hansen (2007). They showed that a progenitor disk produced by supernova fallback could supply the material needed to form the known planets. In contrast, tidal disruption of a companion star would underproduce solids within 1 AU, while overproducing solids where no body of lunar mass or greater now exists.

The fact that the orbits are nearly coplanar might further support standard models of planet formation. Residual disk material could also have provided the means to circularise the orbits and bring them close to the observed 3:2 resonance (Ruden, 1993).

4.1.2 PSR B1620–26

Shortly after its discovery, the 11-ms pulsar PSR B1620–26 in the globular cluster M4 was found to have a binary companion, a $0.3M_\odot$ white dwarf in a 191-d low eccentricity orbit (Lyne et al., 1988; McKenna & Lyne, 1988). Inadequacy of the published pulse timing models to properly predict the pulsar period and phase at later epochs led to the hypothesis (Backer, 1993) and subsequent probable confirmation (Backer et al., 1993; Thorsett et al., 1993) that the observed second period derivative of the pulsar could be attributed to a $10M_J$ object orbiting the pulsar–white dwarf binary pair. The planetary mass object is the lightest and most distant member of a hierarchical triple system, in an orbit with $a \sim 35$ AU and $P \sim 100$ yr (Thorsett et al., 1999).

The survival of this outer companion in such a wide orbit in the cluster environment was immediately recognised as problematic because of its low binding energy, and the frequent likely encounters with other cluster stars (Backer et al., 1993). More recent studies of survivability of planets in wide orbits do suggest that such a planet can undergo a number of encounters without unbinding it from the system (Woolfson, 2004).

Data acquired since 2001 with the 100-m NRAO Green Bank telescope place the planet mass at around $1 - 2M_J$, with an orbital period of a few decades, a high orbital inclination relative to the plane of the inner pulsar–white dwarf binary, and with modest orbital eccentricity. Signatures of the Newtonian interaction between the planet and the white dwarf are also observed (Sigurdsson et al., 2008).

Formation Three formation scenarios for this hierarchical triple system have been proposed (Figure 4.2). In the first (A; Sigurdsson, 1992, 1993, 1995; Ford et al., 2000a; Sigurdsson et al., 2003, 2008) the planet forms around a main sequence star, then the system migrates towards the cluster core where it encounters a neutron star binary. One neutron star captures the star and planet, and ejects its original companion. The main sequence star then evolves into a red giant (and eventually

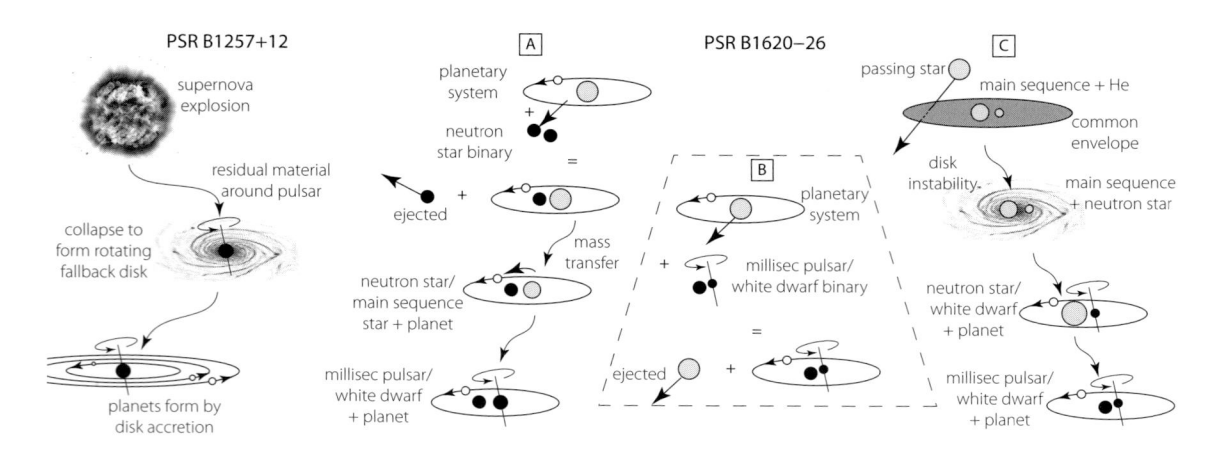

Figure 4.2: Schematic of the formation of the pulsar planets. Left: for PSR B1257+12, the planet is surmised to have formed from a 'fallback disk' created after the supernova explosion. For PSR B1620−26, three options are described in the text. A and B involve different capture-and-exchange scenarios of a pre-existing main sequence star–planet system, while C involves planet formation via disk instability after a passing star disrupts the common envelope of a main sequence/giant binary.

a white dwarf), transferring mass and so spinning up the neutron star to its final millisecond pulsar status.

A second alternative, also involving exchange (B; Joshi & Rasio, 1997; Fregeau et al., 2006) is that a primordial main sequence planetary system encountered a pre-existing binary millisecond pulsar, in which the planetary system is disrupted, with the main sequence star being ejected and the planet captured.

Since both these exchange models require the original planet formation to take place in the low-metallicity globular cluster environment, which appears to occur with low probability, a third possibility (C; Beer et al., 2004b) is that the planet only formed as a result of gravitational instability as a passing star perturbed the common-envelope of a main sequence/evolved binary. The main sequence star then transferred mass and so spun-up the neutron star to its present millisecond pulsar status, before evolving to a white dwarf.

These models each have their own difficulties. With only one exemplar of such a hierarchical triple, and with such a very long period planetary orbit, it is presently difficult to decide between them.

4.1.3 Other considerations

False alarms and unconfirmed planets Before the announcement of PSR B1275+12 in 1992, there had been two pulsar–planet false alarms. Evidence for a long-period planet around the slow pulsar PSR B0329+54 (spin period 0.71 s) had been reported by Demianski & Proszynski (1979), based on large second time derivatives of the spin frequency, although alternative explanations were also given. The planetary interpretation was supported by Shabanova (1995), who gave $P = 16.9$ yr, $M_p > 2 M_\oplus$, $e = 0.23$ and $a = 7.3$ AU. Both groups reported an additional 3 yr periodicity in pulse arrival times. The planetary hypothesis was questioned by Konacki et al. (1999a) based on further observations, with variations in the timing residuals for this relatively young neutron star attributed to spin irregularities or precession of the pulsar spin axis.

For PSR B1829−10, Bailes et al. (1991) announced a possible $10 M_\oplus$ companion which was also subsequently retracted (Lyne & Bailes, 1992).

For the radio quiet pulsar Geminga, Mattox et al. (1998) reported evidence for a companion ($a = 3.3$ AU and $M_p \sin i = 1.7 M_\odot$) from γ ray observations, although this may have been an artefact of the spin period.

For PSR B1828−11, Liu et al. (2007) attributed correlated periodic variations in pulse shape and slow-down rate to forced precession by a quark composition planet.

Satellites of pulsar planets In the same way that transit time variations for the transiting planets can be used to place constraints on the presence of planetary satellites, the pulsar time-of-arrival residuals can be subjected to a similar analysis. Lewis et al. (2008) considered systems in which the satellite–planet and planet–pulsar orbits are circular and coplanar (cf. the study of radial velocity variations of a binary star on a distant companion by Schneider & Cabrera 2006), and showed that plausible configurations with detectable signals might exist in the case of PSR B1620−26.

Detection limits Detection limits for pulsar planets or their satellites through time-of-arrival analyses are affected by their own statistical noise properties governing 'clock purity' (Cordes, 1993), notably phase jitter due to pulse-to-pulse profile variations, and red timing noise in

4.2 Pulsating stars

Table 4.2: Planets discovered from an analysis of timing signals, corresponding to objects listed at exoplanets.eu as of 1 November 2010. For PSR B1257+12 designations, see Table 4.1. For NN Ser designations, see Beuermann et al. (2010).

Name	Type	V	$d_\star$ (pc)	M_p (M_J)	R_p (R_J)	P_{orb} (d)	a (AU)	$M_\star$ ($M_\odot$)	$R_\star$ ($R_\odot$)	Discovery reference
PSR B1257+12 B	ms pulsar	–	300	0.013	–	66.5419	0.36	–	–	Wolszczan & Frail (1992)
PSR B1257+12 C	"	"	"	0.012	–	98.2114	0.46	–	–	"
PSR B1257+12 A	"	"	"	7×10^{-5}	–	25.262	0.19	–	–	Wolszczan (1994a)
PSR B1620−26	pulsar	24	3800	2.5	–	~36000	23	1.35	–	Thorsett et al. (1993)
V391 Peg b	pulsating	14.6	1400	3.2	–	1170	1.7	0.5	0.23	Silvotti et al. (2007)
HW Vir b	eclipsing	10.9	181	19.2	–	5767	5.3	0.48/0.14	0.18/0.17	Lee et al. (2009)
HW Vir c	"	"	"	8.5	–	3321	3.6	"	"	"
DP Leo b	eclipsing	17.5	–	6.3	–	8693	8.6	0.69	–	Qian et al. (2010b)
NN Ser (ab) c	eclipsing	16.6	500	6.9	–	5660	5.38	0.65	–	Beuermann et al. (2010)
NN Ser (ab) d	"	"	"	2.2	–	2830	3.39	"	"	"

which neighbouring residuals are correlated, attributed to inhomogeneous angular momentum transport.

Other limits may arise from planet–planet interactions in the same system (cf. Laughlin & Chambers, 2001), pulsar precession (e.g. Stairs et al., 2000; Akgün et al., 2006; Haberl et al., 2006; Liu et al., 2007), periodic structural variations in the interstellar medium (Scherer et al., 1997; You et al., 2007), and stochastic gravitational wave background (e.g. Detweiler, 1979; Hobbs et al., 2009; Pshirkov et al., 2010).

4.2 Pulsating stars

The existence of planetary systems around stars in advanced stages of stellar evolution is of interest for two reasons. It provides an insight into the future of the solar system planets in general and Earth in particular. Furthermore if survivability is robust, and if such planets could be detected, it would provide another route to characterising their frequency and distribution.

With few planets known around stars in the late stages of stellar evolution, survivability beyond the main sequence is currently a topic somewhat restricted to theoretical investigation (Livio & Soker, 1984; Sackmann et al., 1993; Soker, 1994, 1996; Duncan & Lissauer, 1998; Soker, 1999; Debes & Sigurdsson, 2002; Burleigh et al., 2002; Hansen, 2004; Villaver & Livio, 2007).

At the same time, some of the scenarios investigated have observable consequences: in tracing the evolution of Jupiter's orbit as the Sun evolves to form a planetary nebula, Soker (1994) found that Jupiter is likely to deposit a substantial fraction of its orbital angular momentum in spinning-up the Sun, leading to a degree of axisymmetric mass loss from the Sun, and resulting in an elliptical planetary nebula. The observed fraction of elliptical planetary nebulae may therefore correlate with the distribution of massive planets and brown dwarfs.

4.2.1 White dwarfs

White dwarfs represent the end point of stellar evolution for stars up to about $8M_\odot$, and are common in the solar neighbourhood, with more than a hundred known within 20 pc (Sion et al., 2009). Whether planetary companions to white dwarfs exist, having survived the red giant branch and asymptotic giant branch phases, will depend on the initial orbit separation, the stellar mass-loss rate and total mass loss, tidal forces, and the details of any interaction with ejected material.

Planets in initial orbits within the final extent of the red giant envelope will be engulfed and migrate inwards (Livio & Soker, 1984). Planets further out will have a greater chance of survival, migrating outwards as mass is lost from the central star. If significant mass is lost instantaneously, then the planet will escape the system if its orbital velocity exceeds the instantaneous escape velocity. In practice, the dynamical time scale for a planet to react to even very significant mass-loss rates is measured in decades, small compared to even the shortest evolutionary phases (Burleigh et al., 2002, eqn 3).

Accordingly, planetary orbits should simply expand adiabatically, with a $1M_\odot$ progenitor evolving to become a $0.5M_\odot$ white dwarf leading to orbits expanding by a factor two. In the process, some stable orbits might nevertheless become unstable (Debes & Sigurdsson, 2002), while the system may be disrupted if velocity kicks during the red giant phase are asymmetric (Heyl, 2007).

As a result, orbits with initial semi-major axes > 0.7 AU will remain larger than the primary star radius at all stages of its evolution (see box), and white dwarfs could therefore potentially host surviving planets with $P \gtrsim 2.4$ yr. Massive white dwarfs could host even shorter period planets if the primary is the product of a gravitationally driven merger which leaves behind a remnant debris disk (Livio et al., 1992).

Various programmes are attempting to quantify the occurrence of planets around white dwarfs. They follow two principal observational routes: pulsation timing signatures (considered here), and direct imaging (§7.5).

Pulsating white dwarfs As a white dwarf cools through certain temperature ranges, C/O (at $\sim 10^5$ K), He (at $\sim 2.5 \times 10^4$ K) and H (at $\sim 10^4$ K) in its photosphere

> **Survival of the solar system planets:** In a study of the Sun's future beyond its H-burning main sequence, Sackmann et al. (1993) [see also Jorgensen 1991] predicted the following. While still on the main sequence, its luminosity will reach $1.1L_\odot$ in 1.1 Gyr, and $1.4L_\odot$ in 3.5 Gyr, at which points 'moist greenhouse' and 'runaway greenhouse' catastrophes are predicted by cloud-free climate models (Kasting, 1988). It reaches $2300L_\odot$ and $R \sim 170R_\odot$ on the red giant branch (about 7.8 Gyr hence), engulfing Mercury.
>
> After core helium burning, it climbs the asymptotic giant branch reaching its largest radial extent of $213R_\odot$ (0.9 AU) at the first thermal pulse. By this point its mass has reduced to $0.6M_\odot$, and the orbits of Venus and Earth have consequently moved out to 1.22 and 1.69 AU respectively. It reaches its peak luminosity of $5200L_\odot$ at the fourth thermal pulse, finally reaching a white dwarf mass of $0.54M_\odot$, by which time Venus and Earth will have moved outwards to 1.34 and 1.85 AU respectively.
>
> Further studies, including the effects of tidal interactions, have been reported by Rybicki & Denis (2001). The future evolution of the Sun–Jupiter system as the Sun evolves to form a planetary nebula has been investigated by Soker (1994). The more recent analysis by Schröder & Connon Smith (2008) suggests that the Earth will not survive engulfment, despite the effect of the solar mass loss.

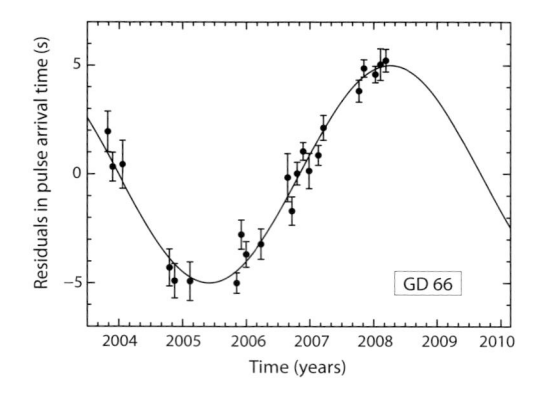

Figure 4.3: *Residuals in pulse arrival time for the 302-s mode of the white dwarf GD 66. The timing residuals are consistent with a $2M_J$ planet in a 4.5-yr orbit. From Mullally et al. (2009, Figure 5) [updated from Mullally et al. 2008, Figure 2], reproduced by permission of the AAS.*

progressively become partially ionised, driving multi-periodic non-radial g-mode pulsations. These pulsating white dwarfs are classified as GW Vir, DBV, and DAV (or ZZ Ceti) stars respectively, and have pulsation periods in the range 100–1000 s.

They include some of the most stable pulsators known, both in amplitude and phase. As a consequence, timing methods similar to those used to search for orbiting planets around radio pulsars can be applied.

G117–B15A The pulsation period of the $V = 15.5$ mag DAV white dwarf G117–B15A is $P \simeq 215.197$ s. Data extending back to 1975 were originally acquired as a probe of white dwarf interiors (Kepler et al., 1991, 2000). Observations between 1975–2005 give a period derivative (Kepler et al., 2005)

$$\dot{P} = (3.57 \pm 0.82) \times 10^{-15} \text{ s s}^{-1} . \qquad (4.3)$$

Measuring clock stability by $P/\dot{P}$, the stability of G117–B15A is comparable to that of the most stable ms pulsars having $\dot{P} = 10^{-20}$ s s^{-1}.

Similar to the timing of the pulsar planets, the effects of an orbiting planet would result in a periodic change in the measured pulse arrival time with an amplitude given by Equation 4.1. As for the astrometric signature to which this timing amplitude is related, the signal increases with both a and M_p. No periodicity in the pulse arrival time residuals has been observed, which excludes the existence of planets around G117–B15A with masses $\sim 0.1 - 10M_J$ and orbital radii in the range 10–0.1 AU respectively (Kepler et al., 2005, their figure 4).

A planet with an orbital period much longer than the observational baseline could still give rise to an apparently linear change in pulsation period, $\dot{P}$, given by (Kepler et al., 1991)

$$\dot{P} = \frac{P}{c} \frac{GM_p \sin i}{a^2} , \qquad (4.4)$$

where P is the white dwarf pulsation period. The question is then whether the observed period derivative is the consequence of a very long orbital period companion, or has some other origin.

Three other possibilities have been considered (Kepler et al., 2005). The first arises from the white dwarf cooling, estimated to result in $\dot{P} \simeq 10^{-15}$ s s^{-1}. A second contribution, estimated at $\dot{P} \leq 3.8 \times 10^{-15}$ s s^{-1}, could arise from a known wide-separation proper motion companion to the white dwarf which may or may not be in orbit around it. A third arises from the star's radial velocity and proper motion (Pajdosz, 1995; Kepler et al., 2000). If all of these possibilities could be excluded, the observed period change could signal a planet with $M_p \simeq 1M_J$ at $a \simeq 30$ AU ($P \simeq 200$ yr), or to a smaller planet in a correspondingly closer orbit.

G29–38 G29–38 is another very stable white dwarf pulsator for which there have been various claims for companions. All of these were shown to be spurious by more careful long-term pulsation timing (Kleinman et al., 1994; Provencal, 1997). The combination of timing studies, 2MASS photometry, and high contrast imaging using HST and Gemini places limits on the existence of any planet at below $12M_J$ for separations between 1–5 AU and ages between 1–3 Gyr (Debes et al., 2005a).

GD 66 Mullally et al. (2008) are monitoring 15 white dwarfs for planetary timing signatures. Twelve of these have been observed since 2003 with the 2.1-m telescope of the McDonald Observatory, with a further three objects (including G117–B15A) included in much longer-term timing campaigns. While for most objects, companion masses down to a few M_J at 5 AU can be excluded, one star, GD 66, shows a variation in arrival time consistent with a possible $2M_J$ planet in a 4.5 yr orbit (Figure 4.3). Imaging observations with Spitzer–IRAC currently place an upper limit on its mass of $5-7M_J$ for an assumed age of 1.2–1.7 Gyr (Mullally et al., 2009).

4.2.2 Hot subdwarfs

V391 Peg Silvotti et al. (2007) reported the discovery of a $M_p \sin i = 3.2 M_J$ planet orbiting the extreme horizontal branch (also known as sdB-type) star V391 Peg (HS 2201+2610), with $a = 1.7$ AU and $P = 3.2$-yr.

The star is in a rare late (post red giant) evolutionary stage. Originally with a mass similar to the Sun, V391 Peg is one of a small fraction of stars which experienced a catastrophic mass loss at the end of the red giant phase, losing its hydrogen-rich envelope in an outflowing wind, and resulting in a dense, B-type subdwarf, powered by He fusion, with $T_{eff} \sim 30\,000$ K (Han et al., 2002).

Like some 40 other stars of its class, V391 Peg shows stable short-period p-mode pulsations (Kilkenny, 2007; Lutz et al., 2009), with four or five periods between 342–354 s. Superimposed on a change of the main period at a rate of $\dot{P} = (1.46 \pm 0.07) \times 10^{-12}$ s s^{-1} is a 3.2-yr periodicity, seen in both the main and second pulsation frequencies (Figure 4.4), and taken as signatures of an orbiting planet.

V391 Peg b probably had $a \simeq 1$ AU (comparable to Earth) when it was still on the main sequence, the planet moving outwards adiabatically to its present position as the central star lost mass. Unlike planets surrounding white dwarfs, evolution to the present location has only been affected by the red giant phase (and not the asymptotic giant branch and thermal pulses).

There may be no causal link between the planet orbiting V391 Peg, and the extreme mass loss which led to this specific pulsating hot subdwarf. But one possibility is that a massive orbiting companion could deposit angular momentum and energy onto the star's hydrogen envelope, enhancing its mass loss (Soker, 1998).

4.3 Eclipsing binaries

A planet orbiting both components of an eclipsing binary will result in periodically varying eclipse times as the binary itself circles the combined system barycentre.

That the majority of solar-type stars appear to reside in binary or multiple systems has been known for some time (Duquennoy & Mayor, 1991). That circumbinary planets can form and survive over long time scales, most likely coplanar with their host binaries, has also been demonstrated by theoretical and simulation studies (e.g. Bonnell & Bate, 1994; Holman & Wiegert, 1999; David et al., 2003; Moriwaki & Nakagawa, 2004; Quintana & Lissauer, 2006; Pierens & Nelson, 2008b). Observational

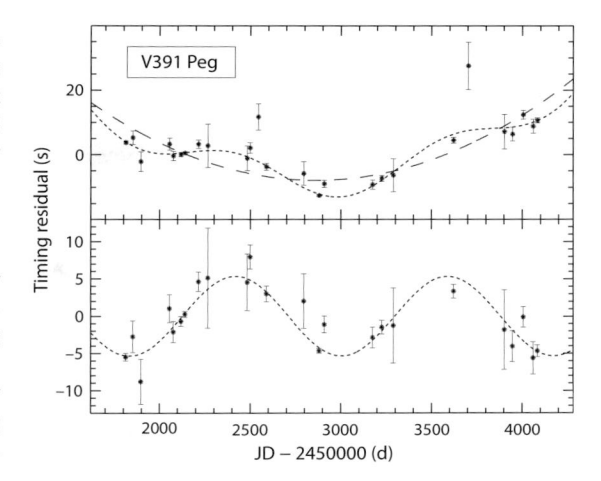

Figure 4.4: V391 Peg, showing residuals in the pulse arrival times for the main pulsation frequency over six years. Models are for a linear period derivative (top figure, dashed curve), and one including an additional sinusoidal component with $P = 3.2$ yr (top figure, dotted). Only the sinusoidal component is shown below. From Silvotti et al. (2007, Figure 1), by permission from Macmillan Publishers Ltd, Nature, ©2007.

evidence for debris surrounding main sequence binary systems is given by, e.g., Trilling et al. (2007).

In the simulations of circumbinary disks around close binary star systems with stellar separations $0.05 - 0.4$ AU and binary eccentricities in the range $e = 0 - 0.8$, Quintana & Lissauer (2006), for example, found that planetary systems formed around binaries with apocentre distances $a(1 + e) \lesssim 0.2$ AU are very similar to those around single stars, whereas those with larger maximum separations tend to be sparser, with fewer planets, especially interior to 1 AU.[2]

With timing accuracies of about 10 s for selected eclipsing binaries showing sharp eclipses, it should be possible (according to Equation 4.2) to detect circumbinary planets of $\sim 10 M_J$ in 10–20 yr orbits.

4.3.1 Confirmed planets

HW Vir The very short-period eclipsing binary HW Vir is a $P_{orb} = 2.8$ h orbital system, comprising an sdB and a main sequence M star. It is the first eclipsing binary

---

[2] For planetary orbits around one component of a binary system, Quintana et al. (2007) showed that sufficiently wide binaries leave the planet formation process largely unaffected. Binary stars with pericentre $q_B \equiv a_B(1 - e_B) > 10$ AU have a minimal effect on terrestrial planet formation within about 2 AU of the primary, whereas binary stars with $q_B \lesssim 5$ AU restrict terrestrial planet formation to within about 1 AU of the primary. Given the observed distribution of binary orbital elements for solar-type primaries, they estimate that about 40–50% of the binary population is wide enough to allow terrestrial planet formation to take place unimpeded.

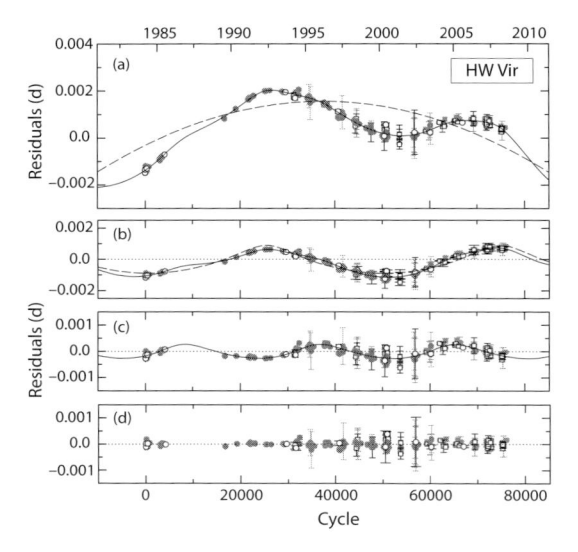

Figure 4.5: HW Vir, and the residuals of the eclipse timings measured since the early 1980s with respect to the linear terms of the Ibanoğlu et al. (2004) ephemeris: (a) the parabolic curve corresponds to a linear period decrease of $\dot{P} = -8.3 \times 10^{-9}\,\mathrm{d\,yr^{-1}}$, which might arise from magnetic stellar wind breaking; (b) residuals from the quadratic form, dependent on $\ddot{P}$; (c) residuals after including the effect of the 15.8 yr planet; (d) residuals with respect to the two-planet model. From Lee et al. (2009, Figure 5), reproduced by permission of the AAS.

system for which the hypothesis of orbiting planets as an explanation of the eclipse time variations is considered to be reasonably secure. Studied since its discovery in 1980, both Kilkenny et al. (2003) and Ibanoğlu et al. (2004) already concluded that the cause of the measured period change was a light travel time effect caused by a third body with $M \gtrsim 0.02 M_\odot$ and $P = 18-20$ yr.

Lee et al. (2009) added further observations between 2000–08, and from the timing residuals (Figure 4.5) inferred the presence of two planets, with $M_\mathrm{p} \sin i = 19.2 M_\mathrm{J}$, $P = 15.8$ yr, $a = 5.3$ AU, $e = 0.46$ and $M_\mathrm{p} \sin i = 8.5 M_\mathrm{J}$, $P = 9.1$ yr, $a = 3.6$ AU, $e = 0.31$ respectively, with a system inclination of $i \simeq 81°$. The periods suggest 5:3 or 2:1 resonant captures, and the relatively high eccentricities are in line with theoretical results on planet–planet interaction and stability (Kley et al., 2004; Ford & Rasio, 2008; Pierens & Nelson, 2008a).

Assuming Bond albedos of 0.34 as for Jupiter, the planet equilibrium temperatures are 230 K and 270 K respectively, dominated by the flux from the sdB primary.

As the primary is almost totally eclipsed, every 2.8 hr, the temperatures at the top of the planetary atmospheres would drop to about 30 K over about 10 min, recovering over a similar period at eclipse egress.

DP Leo DP Leo is a member of the class of AM Her stars or polars, strongly magnetised cataclysmic binaries, and the first such example of the class discovered to be eclipsing. Qian et al. (2010b) have shown that a significant sinusoidal light travel time component, with semi-amplitude 31.5 s, can be explained by a $P = 23.8$-yr planet ($a = 8.6$ AU) of mass $6.4 M_\mathrm{J}$. This assumes that the planet is coplanar with the eclipsing binary, i.e. with an orbital inclination $i = 79°.5$, and that the total binary mass is $0.69 M_\odot$.

NN Ser NN Ser is a $P = 3.12$-hr eclipsing binary with a white dwarf primary and red dwarf secondary (Parsons et al., 2010b). Periodic eclipse timing residuals attributed to a $15 M_\mathrm{J}$, $P = 7.56$-yr planet were first reported by Qian et al. (2009). Observations extending over 22 yr (1988–2010) indicate the presence of two planets, NN Ser (ab) c and d, with masses $6.9 M_\mathrm{J}$ and $2.2 M_\mathrm{J}$ respectively, periods of 15.5 yr and 7.7 yr respectively, and eccentricities of ~ 0 and ~ 0.2 respectively, orbiting in a 2:1 mean motion resonance (Beuermann et al., 2010).

4.3.2 Unconfirmed planets

CM Dra From eclipse timings of CM Dra dating back to 1977, Deeg et al. (2008) have proposed the existence of a third body of a few M_J and $P = 18.5 \pm 4.5$ yr, or an object in the range $1.5 M_\mathrm{J} - 0.1 M_\odot$ with a period of hundreds to thousands of years. The interpretation is not yet confirmed (Ofir, 2008).

QS Vir Qian et al. (2010a) reported a $6.4 M_\mathrm{J}$ planet in a $P = 7.86$-yr ($a = 4.2$ AU) orbit around the 'hibernating' cataclysmic variable QS Vir, with sinusoidal residuals of amplitude 12 s. QS Vir is a 3.618-h orbital period binary comprising a white dwarf primary and red dwarf (M4) secondary, in which the red dwarf is temporarily and marginally detached from its Roche lobe, such that no mass transfer onto the primary is ongoing.

An orbiting planet would provide evidence that a giant planet can survive the red giant phase of the primary star's evolutionary history. The orbital parameters are, however, considered to be incompatible with subsequent ULTRACAM observations (Parsons et al., 2010a).

5
Microlensing

By the end of 2010, eleven exoplanets in ten systems had been discovered through gravitational microlensing (Table 5.1). The first unambiguous detection of a $4M_J$ planet with a projected separation of ~4 AU was reported in 2004, and the discovery of a $5M_\oplus$ planet in 2006. With the characterisation of a two-planet system somewhat analogous to Jupiter and Saturn in 2008, in which the orbital motion of the outer planet could be detected and measured during the lensing event, these discoveries marked the emergence of the technique as a powerful and independent exoplanet probe over an important region of planetary mass and orbital radius.

5.1 Introduction

Gravitational lensing In general relativity, the presence of matter (energy density) distorts spacetime, and the path of electromagnetic radiation is deflected as a result. Under certain conditions, light rays from a distant background object (the source) are bent by the gravitational potential of a foreground object (the lens) to create images of the source which are distorted (and possibly multiple), and which may be highly focused and hence significantly amplified. Its manifestation depends upon the fortuitous alignment of the background source, the intervening lens, and the observer.

Different regimes are generally recognised for gravitational lensing, depending on whether effects are discernible at an individual object level (*strong lensing*), or only in a statistical sense (*weak lensing*). Strong lensing can be further divided, somewhat subjectively according to telescope resolution, into *macrolensing* (resulting in either multiple resolved images, or in 'arcs' in which the source is distorted, both sheared and magnified) and *microlensing* (in which discrete multiple images are essentially unresolved).

Relative motion between source, lens and observer leads to time-varying amplification of the images, which may occur over time scales of hours, months or many years, depending on the nature of the source and lens. If the foreground lensing object is itself of complex gravitational morphology, whether a cluster of galaxies, or a star orbited by one or more planets, then the background source may show a more complex time-dependent amplified light curve resulting from the time-varying alignment geometry.

Microlensing In the microlensing regime, a term introduced by Paczyński (1986a), discrete images of the source are unresolved at typical telescope resolutions. The term embraces phenomena operating on galaxy scales (quasar microlensing) and on stellar mass scales and below (relevant for exoplanets). In the latter case, the primary lens is a single point mass of order $1M_\odot$, and the two images of a background source have a separation of order only 1 mas, well below typical ground-based instrumental resolution.

In the domain of interest for exoplanet detection, the exoplanet system (host star and planet) acts as a multiple lens, and a more distant star within the Galaxy acts as the probing source. The changing magnification of the sub-images due to the time-varying alignment geometry of the observer–host star–background source may lead to a significant intensity variation of the sum of the multiply lensed images over time scales of weeks. This changing intensity with time allows the event to be recognised as a microlensing event. Careful monitoring of the light curve as the alignment changes over several hours allows the additional lensing effects of an accompanying planet to be identified.

Microlensing provides a powerful route both for the detection, and for the characterisation, of planetary systems. The basic experimental challenge is that significant magnification (brightening) requires extremely precise alignment of observer, source and lens, to within the angular Einstein radius, or around ~1 mas.

Such an alignment probability for any given source star in the Galaxy is so low that very large numbers of potential sources have to be monitored, frequently and simultaneously, to have any chance of observing even a single favourable configuration. The Galactic bulge region, with its high stellar surface density, is therefore the target monitoring region of choice.

> **Historical background:** The first observational confirmation of general relativistic light bending, based on the 1919 solar eclipse observed in Brazil, was reported by Dyson et al. (1920). The term 'lensing' in the context of light deflection was used, pejoratively, by Lodge (1919), who argued that *'it is not permissible to say that the solar gravitational field acts like a lens, for it has no focal length.'* The term has nevertheless persisted as a description of the phenomenon.
> The possibility that gravitational lensing by a foreground object could result in two distinct images of a background star was pointed out by Eddington (1920). A qualitative description, and the possibility of a ring-shaped image, was suggested in a short communication by Russian physicist Orest Chwolson (1924) who noted *'Whether the case of a fictitious double star actually occurs, I cannot judge.'* The problem was first considered quantitatively by Einstein (1936) and Link (1936). Einstein's paper starts *'Some time ago, R.W. Mandl paid me a visit and asked me to publish the results of a little calculation, which I had made at his request'*. Later he comments *'Of course, there is no hope of observing this phenomenon directly. First, we shall scarcely ever approach closely enough to such a central line. Second, [the angles] will defy the resolving power of our instruments.'* Prescient papers by Zwicky (1937a,b) later argued that *'extragalactic nebulae offer a much better chance than stars for the observation of gravitational lens effects.'*
> After a lapse of almost three decades, the subject was reopened with the independent work of Klimov (1963), Liebes (1964), and Refsdal (1964). Liebes (1964) first considered gravitational lensing as a method to detect planets around other stars, concluding that the primary effect would be to *'slightly perturb the lens action of these stars'*. He also considered the detectability of unbound planet-sized bodies *'floating about the Galaxy'*, but also concluded that the *'associated pulses would be so weak and infrequent and of such fleeting duration – perhaps a few hours – as to defy detection'*.
> Walsh et al. (1979) discovered the first case of strong lensing, a double image of the distant quasar Q0957+561. The discovery marked the start of a substantial body of both theoretical and observational work, and more than a hundred multiple images of galaxy-lensed systems are now known. Arc-like images of extended galaxies were first reported by Lynds & Petrosian (1986) and Soucail et al. (1987). Mainly through subsequent HST observations, many examples are now known.
> The first incomplete Einstein ring was reported by Hewitt et al. (1988), and a complete example, a little less than 1 arcsec in diameter around the radio source B1938+666, in which both lens and source are galaxies, was imaged in the near-infrared by HST–NICMOS (King et al., 1998). Again, dozens of examples of more-or-less complete Einstein rings are now known, both in the optical, in the near-infrared, and in the radio. Microlensing studies of Galactic structure, and the associated search for planets, was launched by the work of Paczyński (1986a,b, 1991) and Mao & Paczyński (1991).

For configurations of relevance, a typical source lies at a distance of ~8 kpc. In the unlensed state, it is generally faint or even invisible, and to first order is considered as a point source of light. Microlensing most effectively probes lens systems some half way to the source, where the host star (and any accompanying planets) may also be invisible. It is the time-varying behaviour of the magnified background star that probes the geometry and mass distribution of the foreground system.

In practice, a characteristically rising light curve of a given star, out of a large number (of order 10^7 – 10^8) being simultaneously monitored, indicates that a favourable alignment may be developing. In this case, more intensive photometric monitoring must be initiated to properly sample the crucial diagnostic period of a possible exoplanet alignment.

The light curve of the background star is the observational signature of microlensing, and encodes the geometry and mass distribution of the lens system.

Further background The treatment here focuses on the concepts necessary to understand the latest microlensing detections of exoplanets. A wider treatment of gravitational lensing is given in Schneider et al. (2006c) and, within that volume, on microlensing by Wambsganss (2006). Reviews on microlensing in general include Paczyński (1996) and Mollerach & Roulet (2002). Reviews on the planetary aspects have been given as the subject has developed (e.g. Sackett, 2004; Wambsganss, 2004; Gould, 2005; Gaudi, 2008).

5.2 Description

Light bending Formulae to analyze gravitational lensing were derived by Refsdal (1964), and are subsequently found in various forms throughout the literature. For the geometry shown in Figure 5.1, the deflection angle $\alpha_{\rm GR}$ for a light ray propagating past a lensing mass $M_{\rm L}$ with impact parameter b is (e.g. Will, 1993)

$$\alpha_{\rm GR} = \frac{4GM_{\rm L}}{c^2 b} = \frac{2R_{\rm S}}{b}, \qquad (5.1)$$

on condition that $b \gg R_{\rm S}$, where $R_{\rm S}$ is the corresponding Schwarzschild radius

$$R_{\rm S} = 2GM_{\rm L}/c^2, \qquad (5.2)$$

where G is the gravitational constant and c is the speed of light. A rigorous solution of the general relativistic equations of motion for the coupled spacetime and matter is not required, because the bending of spacetime by exoplanet systems is always small.

The condition on $\alpha_{\rm GR}$ that a deflected light ray from the source reaches the observer follows purely from the trigonometry of Figure 5.1. In the usual approximation of small angles, with $D_{\rm L}$ and $D_{\rm S}$ signifying the distances to the lens and source, and $D_{\rm LS} = D_{\rm S} - D_{\rm L}$ (applicable for a Euclidean metric, i.e. for local events, but not for lensing over cosmological distances), the angle between source and lens is

$$\theta_{\rm S} D_{\rm S} = \frac{D_{\rm S}}{D_{\rm L}} b - \alpha_{\rm GR} D_{\rm LS} = \frac{D_{\rm S}}{D_{\rm L}} b - \frac{2R_{\rm S}}{b} D_{\rm LS}. \qquad (5.3)$$

5.2 Description

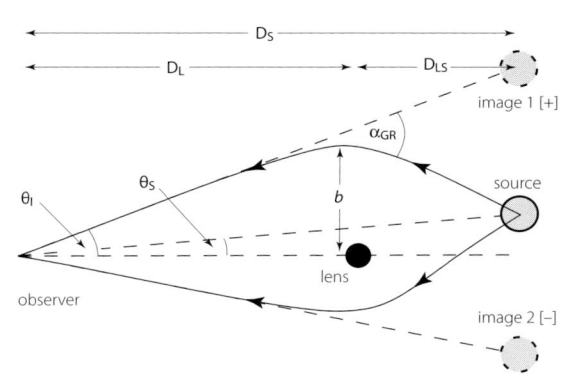

Figure 5.1: Lensing schematic for a point mass lens M_L at distance D_L, offset by the small angle θ_S from the direct line from observer to source. A light ray from the source, passing the lens at distance b, is deflected by an angle α_{GR}. An observer sees one image of the source displaced to angular position $\theta_1 = b/D_L$ on the same side as the source, and a second image on the other.

With $\theta_I = b/D_L$ denoting the angle between the deflecting mass and the deflected ray, this can be written

$$\theta_S = \theta_I - 2R_S \frac{D_{LS}}{D_L D_S} \frac{1}{\theta_I} . \tag{5.4}$$

This is a form of the *lens equation*, which describes the mapping, in the lens plane, from an image position θ_I to the source position θ_S. Being quadratic in θ_I, it has two (image) solutions.

Einstein radius It is convenient to define a characteristic angle θ_E, the *angular Einstein radius*, and a characteristic length scale in the lens plane R_E, the (linear) *Einstein radius*, given by

$$\theta_E = \left[2R_S \frac{D_{LS}}{D_L D_S} \right]^{1/2} = \left[\frac{4GM_L}{c^2} \frac{D_{LS}}{D_L D_S} \right]^{1/2} \tag{5.5}$$

$$R_E = \theta_E D_L = \left[2R_S \frac{D_L D_{LS}}{D_S} \right]^{1/2} . \tag{5.6}$$

Then the lens equation (Equation 5.4) can be written

$$\theta_I^2 - \theta_S \theta_I - \theta_E^2 = 0 . \tag{5.7}$$

This has two solutions

$$\theta_{+,-} = \frac{1}{2} \left(\theta_S \pm \sqrt{\theta_S^2 + 4\theta_E^2} \right) , \tag{5.8}$$

showing that the source has two images, one on each side of the lens (a negative value of θ meaning that the image is on the other side of the lens from the source). The angular separation between the two images is

$$\Delta\theta \equiv \theta_+ - \theta_- = \sqrt{\theta_S^2 + 4\theta_E^2} . \tag{5.9}$$

In the hypothetical case of perfect observer–lens–source alignment ($\theta_S = 0$), the configuration is rotationally symmetric about the line of sight to the lens, and the two

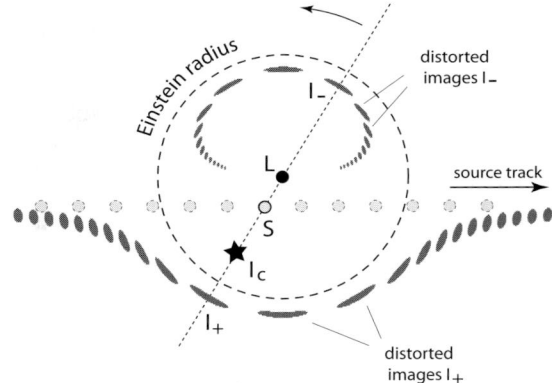

Figure 5.2: Schematic of (astrometric and photometric) microlensing. A background source (small grey circles, S) moves behind a foreground lens (L). Here L, and its Einstein radius, are stationary. S and L along with the two distorted images ($I_\pm$, inside and outside the Einstein radius), and their centroid (I_c) remain colinear. As the source moves past the lens, this connecting line rotates by almost $180°$ (here counterclockwise), and the image centroid follows a correspondingly non-linear path. Adapted from Paczyński (1996, Figure 3).

images merge to form a ring (an *Einstein ring*) of angular radius θ_E (the Einstein radius). For all other source positions, one image lies inside θ_E and one lies outside (Figure 5.2).

Introducing relevant numerical quantities provides an estimate of the Einstein radius typical for exoplanet investigations. Equations 5.5 and 5.6 can be written

$$\theta_E \simeq 1.0 \left(\frac{M_L}{M_\odot} \right)^{\frac{1}{2}} \left(\frac{D_L}{8 \text{ kpc}} \right)^{-\frac{1}{2}} \left(\frac{D_{LS}}{D_S} \right)^{\frac{1}{2}} \text{ mas} , \tag{5.10}$$

$$R_E \simeq 8.1 \left(\frac{M_L}{M_\odot} \right)^{\frac{1}{2}} \left(\frac{D_S}{8 \text{ kpc}} \right)^{\frac{1}{2}} \left(\frac{D_L D_{LS}}{D_S} \right)^{\frac{1}{2}} \text{ AU} . \tag{5.11}$$

The former shows that for a source roughly at the distance of the Galactic centre (assumed to lie at 8 kpc), and a lens with $M_L = M_\odot$ half way to the source, typical image separations are of order the Einstein angular radius $\theta_E \sim 1$ mas. This is well below the angular resolution of most ground-based instruments; consequently the image separation, and its variation with time, are generally undetectable. The latter shows that the Einstein radius for a typical host star is $R_E \simeq 4$ AU. Being of order of the orbital radius of planets in the solar system, this is a particularly fortuitous scale length for probing exoplanets.

Since general relativistic light bending is wavelength independent, the microlensing light curve is correspondingly achromatic.

Magnification Light deflection in the gravity field changes both the direction of a light ray and the cross-section of a bundle of rays. Brightening of the two images occurs because the flux from each is the product of

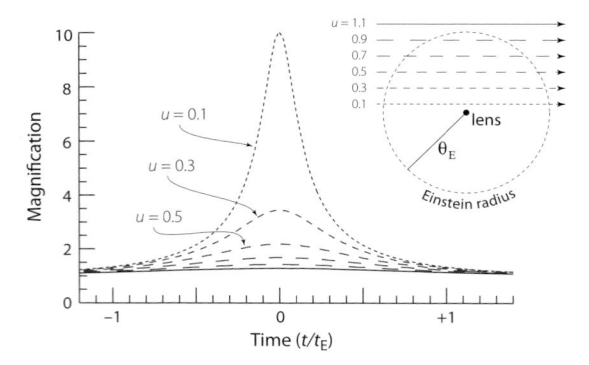

Figure 5.3: Theoretical microlensing light curves for a point source and single lens, with a constant relative transverse velocity between them. For the geometry depicted in the inset, the Einstein radius is shown dashed, and a series of trajectories are shown with their corresponding values of $u = \theta_S/\theta_E$. Microlensing events passing close to the projected lens position are then highly amplified (Equation 5.15), with the magnification becoming rapidly more pronounced with decreasing u. Adapted from Paczyński (1996, Figures 4–5).

the (constant) surface brightness of the source and the (enlarged) solid angle subtended by the distorted image. The *magnification* for each image is given by the ratio of the image area to the source area

$$A_\pm = \left| \frac{u_\pm}{u} \frac{du_\pm}{du} \right|. \quad (5.12)$$

For a Schwarzschild (point mass) lens, and writing the projected lens–source separation in units of the Einstein radius

$$u \equiv \frac{\theta_S}{\theta_E}, \quad (5.13)$$

the two images have individual magnifications

$$A_\pm = \frac{1}{2}\left(\frac{u^2+2}{u\sqrt{(u^2+4)}} \pm 1 \right), \quad (5.14)$$

although only the total magnification is observable

$$A \equiv A_+ + A_- = \frac{u^2+2}{u\sqrt{u^2+4}} \quad (5.15)$$
$$\simeq u^{-1} \quad \text{for } u \ll 1 \quad (5.16)$$
$$\simeq 1 \quad \text{for } u \gg 1. \quad (5.17)$$

For perfect observer–lens–source alignment $u \to 0$, $A \to \infty$, and the magnification is then formally infinite. For the realistic case of a finite source size, and partly because this diverging condition arises from the simplified treatment of geometrical optics, the magnification is in practice always finite.

Several microlensing events with peak values of $A \gtrsim 800$ have been reported, implying that, experimentally, a background source of (say) $I = 20$ mag would be temporarily magnified to reach $I \simeq 13$ mag. The highest magnification reported to date is ~ 3000 in the case of OGLE–2004–BLG–343 (Dong et al., 2006).

The total magnification varies as a function of time due to the relative transverse motion between source, lens, and observer (Figure 5.3). For a given relative transverse velocity between source and lens, $v_\perp$, a typical time scale for a lensing event is given by the Einstein radius crossing time

$$t_E = R_E/v_\perp \simeq 70 \left(\frac{M_L}{M_\odot}\right)^{\frac{1}{2}} \left(\frac{D_S}{8\,\text{kpc}}\right)^{\frac{1}{2}} \left(\frac{D_L D_{LS}}{D_S}\right)^{\frac{1}{2}}$$
$$\left(\frac{v_\perp}{200\,\text{km s}^{-1}}\right)^{-1} \text{days}. \quad (5.18)$$

Again, for a source in the bulge (at 8 kpc), and a lens of $1 M_\odot$ half way to the source, the Einstein time scale for the resulting microlensing event is $\simeq 35$ d.

Equivalently, in angular measure, the Einstein time scale is related to the (unknown) lens–source relative

---

Early microlensing surveys: The first microlensing surveys were motivated by the search for evidence of dark matter in galaxy halos probed by quasars (Gott, 1981; Paczyński, 1986b). Even for such 'normal' microlensing – that is, before accounting for the still smaller probabilities of detecting planetary perturbations – the alignment required for a detectable brightening is so precise that the chance of substantial microlensing magnification is extremely small. It is of order $\sim 10^{-6}$ for background stars even in the denser directions of the Galactic bulge, nearby Magellanic Clouds, or nearby spiral galaxy M31, even if all the unseen Galactic dark matter were composed of compact macroscopic objects capable of lensing.

Only since 1993, when massive observational programmes capable of surveying millions of stars got underway, was photometric microlensing observed by the EROS (Expérience de Recherche d'Objets Sombres, Aubourg et al. 1993), OGLE (Optical Gravitational Microlensing, Udalski et al. 1993), MACHO (Massive Compact Halo Objects, Alcock et al. 1993), DUO (Disk Unseen Objects, Alard 1996), and MOA (Microlensing Observations in Astrophysics, Muraki et al. 1999) projects. Early reviews of these results were given by Paczyński (1996) and Gould (1996).

Using the achromatic nature of the microlensing events to assist distinguishing them from intrinsic source variability, several thousand microlensing events have now been detected in the Galaxy (with some 10% showing binary lens structure), and many individual events and statistical results have been published.

These impressively vast monitoring programmes have demonstrated that the excess microlensing seen towards the Large Magellanic Cloud by the MACHO group (Alcock et al., 2000) requires at most 20% of the Galaxy's dark matter in the form of stellar mass objects, while the results of the EROS group (Tisserand et al., 2007) suggest that much of the excess may be caused by stars within the LMC itself.

With the microlensing constraints on dark matter largely resolved, the emphasis of observations over the last decade has focused on the detection of exoplanets.

5.3 Caustics and critical curves

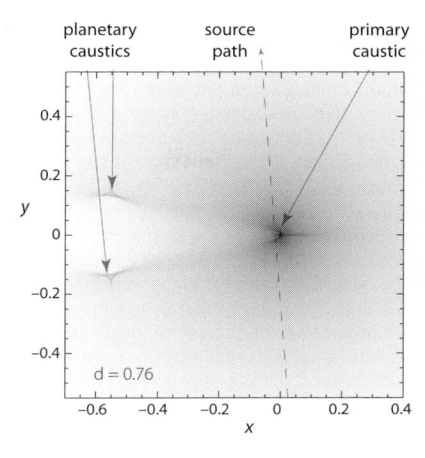

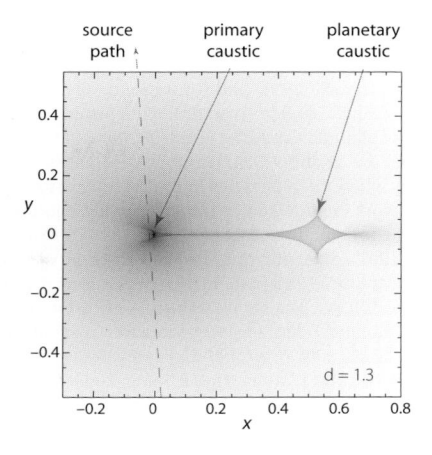

Figure 5.4: Caustics for a planet-to-lens mass ratio $q = 0.006$ (~ $6M_J/1M_\odot$), generated by inverse ray shooting in the context of models for OGLE–2005–BLG–71. The primary lens lies at the origin, with the planet along the x-axis with separation d (in units of θ_E). Left: $d = 0.76$; right: for the dual position at $d = 1/0.76 = 1.3$. Intensity is proportional to logarithmic magnification. Dashed lines show the model source star track, passing close to the central stellar caustic. Figures provided by Daniel Kubas.

proper motion, μ_{LS}, by

$$t_E = \frac{\theta_E}{\mu_{LS}} . \qquad (5.19)$$

Single lens Writing the projected lens–source separation resulting from uniform rectilinear motion as

$$u = \left(\left[\frac{t - t_0}{t_E} \right]^2 + u_0^2 \right)^{1/2} , \qquad (5.20)$$

shows that a microlensing light curve for a single lensing event depends upon three parameters (in addition to the unlensed flux of the background star, which may or may not be detectable in the absence of microlensing): the Einstein radius crossing time t_E, the time of peak magnification t_0 (at which u is a minimum, and A is a maximum), and the minimum separation u_0, which determines the peak magnification and also the specific form of the light curve (Figure 5.3).

Normal (single lens) microlensing events yield only one physically relevant parameter, the Einstein time scale, derived from the event duration. As evident from Equation 5.18 this is a degenerate combination of M_L, D_L (or, more strictly for a source at finite distance, the lens–source relative parallax), and the source–lens relative transverse velocity $v_\perp$. If the lens is unseen, its mass and distance can be determined only in particularly favourable circumstances (§5.4).

Binary lens In a binary lens system with projected component separation a, the second lens introduces three further parameters (in addition to t_E, t_0 and u_0): the mass ratio of the two components, $q = M_p/M_\star$; the projected star–planet separation at the time of the lensing event in units of the Einstein radius for the total mass, $d = a/R_E$; and the angle of the source trajectory relative to the binary axis, α. The combination leads to a very wide variety of binary lens light curves. The general properties of the two-point lens system was explored in detail by Schneider & Weiss (1986), and the first examples of binary microlensing light curves were presented by Mao & Paczyński (1991).

Compared to the smooth light curves of single-lens microlensing, the existence of a planet can result in additional short-duration peaks depending on the way in which the source path crosses the lens projection (Figure 5.6). The duration of planetary events typically scale with $q^{0.5}$ (Equation 5.18), and last typically less than a day, compared with a typical primary lens event duration of 30–40 d. As q decreases, the peak signals become rarer and briefer: for Earth-mass planets, typical time scales are only 3–5 h.

5.3 Caustics and critical curves

Regions in the lens plane where the determinant of the Jacobian matrix of the coordinate transformation from image to source plane vanishes, and the magnification according to Equation 5.12 is formally infinite, are termed *critical curves*. Corresponding points in the source plane (found from the mapping given by the appropriate lens equation), are termed *caustics*.[1] For a single point lens, the caustic for any source distance is

---

[1] In optics, a caustic is the envelope of light rays reflected or refracted by a curved surface or object, resulting in concentrations of light. Familiar examples include light shining through a wine glass onto a tablecloth, or rippling caustics formed when light shines through waves on the surface of a swimming pool. The term originates from the Greek for burning, recalling that such concentrations of sunlight can burn.

> **Development of microlensing planet studies:** A large body of theoretical and simulation studies has been carried out as part of the development of the framework for microlensing studies, both before and after the first planet detections. Chronologically Mao & Paczyński (1991) and Gould & Loeb (1992) investigated lensing when one or more planets orbit the primary lens, finding that detectable fine structure in the photometric signature of the background object occurs relatively frequently, even for low-mass planets. Gould & Loeb (1992) found that the probability of detecting such fine-structure is about 17% for a Jupiter-like planet (i.e. at about 5 AU from the central star), and 3% for a Saturn-like system; these relatively high probabilities occur specifically when the planet lies in the 'lensing zone', between about 0.6–1.6 R_E.
>
> Subsequent work included determination of detection probabilities (Bolatto & Falco, 1994), extension to Earth-mass planets including the effects of finite source size (Bennett & Rhie, 1996; Wambsganss, 1997), determination of physical parameters (Gaudi & Gould, 1997), detection rates for realistic observational programmes (Peale, 1997; Gaudi & Sackett, 2000), distinguishing between binary source and planetary perturbations (Gaudi, 1998), detection rates for high-magnification events (Griest & Safizadeh, 1998), multiple planets in high-magnification events (Gaudi et al., 1998), repeating events due to a multiple planetary system (Di Stefano & Scalzo, 1999), caustic-crossing configurations (Graff & Gaudi, 2000; Bozza, 2000), multiple planet anomalies (Han et al., 2001a), probability rates (Peale, 2001), simulations of high magnification events (Rattenbury et al., 2002), effects of wide orbit planets (Han & Kang, 2003; Han et al., 2005), effects of multiple planets (Han, 2005a), properties of central caustics (Chung et al., 2005; Han, 2006a), sensitivity to Earth-mass planets (Park et al., 2006), and double-peaked high magnification events (Han & Gaudi, 2008).

the single point behind the lens, and the critical curve (the positions of the images of these caustics) is the Einstein ring. High-magnification microlensing events occur when the source comes near to a caustic, with the peak magnification occurring at the projected distance of closest approach.

The caustics of binary lenses play a crucial role in the interpretation of exoplanet microlensing light curves. When the lens consists of two point-like objects, specifically a star and an orbiting planet, the caustic positions and shapes can be formulated in terms of the planet-to-star mass ratio $q = M_p/M_\star$, and the angular star–planet separation d (in units of θ_E). For arbitrary distances and mass ratios the caustic structures are extended and complicated in shape, but for small values of q, and for $d \neq 1$, the picture is more straightforward.

Magnification maps In the example magnification maps given in Figure 5.4, and following Griest & Safizadeh (1998), the primary (star) lens is assumed to reside at the origin, and the planet is along the positive x-axis at d. The point-like single lens *central caustic* becomes a tiny wedge-like feature, still located near $x = 0$, while one or two new caustics appear depending on the planet position. For $d > 1$ there is a new diamond-shaped *planetary caustic* located on the same side of the lens as the planet with a predominantly excess magnification, while for $d < 1$ two small triangular-shaped caustics appear close together on the opposite side of the lens, now with a pronounced magnification deficit between them (Cassan, 2008; Wambsganss, 1997). The light curve of a moving source tracks a slice across the magnification contours in this xy plane.

There are essentially two broad classes of light curves which result: those which intersect the caustic structure (caustic crossing events, giving light curves with a complex structure) and those which do not (which give a smoother light curve more reminiscent of a single lens).

The position of the planetary caustics can be reasoned as follows (Griest & Safizadeh, 1998, Appendix A). Since a planet mass is much smaller than the primary lens mass, its area of influence is small when measured in units of R_E. Thus to first approximation the planet can have a large effect only when its position is near one of the main images. The relation between planet and caustic positions should then corresponds to the relation between image and source positions (Equation 5.8). Inverting this equation (again, expressed in units of R_E) gives the expected position of the planetary caustic, which therefore lies along the x-axis at

$$x_c \simeq (d^2 - 1)/d. \qquad (5.21)$$

The mass ratio q, but not M_p itself, can generally be determined from the duration of the planetary perturbation. Determination of the star–planet separation (in units of R_E, which itself is typically unknown), can be made from the light curve, and a first-order estimate of d can be found as follows. If a planetary deviation is observed at a time when the best-fit single lens light curve results in a certain magnification A, the corresponding lens–source separation u_c can be estimated from Equation 5.15. Then the planet position is given by the solution to Equation 5.21,

$$d = \frac{1}{2}\left(u_c \pm \sqrt{u_c^2 + 4}\right). \qquad (5.22)$$

From Figure 5.4 it can be seen that the two solutions for a given u_c have similar magnification patterns in the vicinity of the central caustic, although precise photometry may allow distinguishing between the two. Light curves with a caustic crossing, or those resulting from a close approach to a cusp, may reveal other important effects, including features attributable to the finite size of the source star (§5.4).

High-magnification events Events of high magnification, $A \gtrsim 100$, are proving to be of particular importance

5.3 Caustics and critical curves

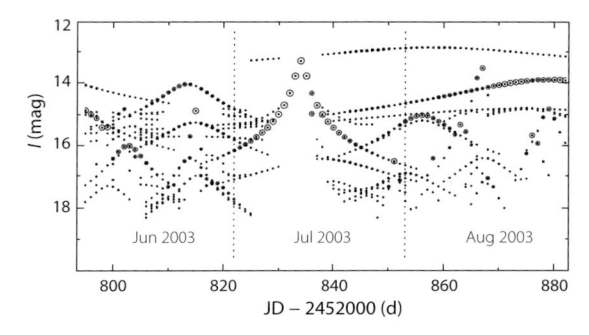

Figure 5.5: Sampling of 443 OGLE events between 2003 June–August, optimised for a 2-m telescope with total observing time 1.5 hr per night. Circles are proportional to the allocated observing time, chosen to maximise total planet detectability. Too many targets implies excessive telescope slew time, while too few is inefficient because the detection zone areas grow as $t^{1/2}$. On some nights, just one high-magnification event may be the only priority. From Horne et al. (2009, Figure 12), © John Wiley & Sons, Inc.

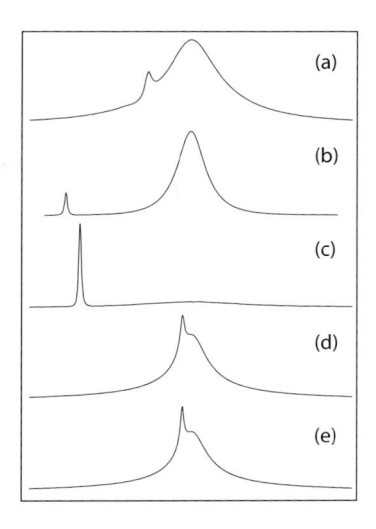

Figure 5.6: Illustrative classification of perturbation types. (a) projected star–planet separation similar to R_E (of the primary); (b) projected separation substantially larger than R_E, and the source trajectory passes the effective magnification regions of both primary and planet; (c) as (b), but the trajectory only passes the effective magnification region of the planet; (d) wide-separation planet, when the trajectory passes close to the primary; (e) the star–planet separation is much smaller than R_E, and the trajectory passes close to the primary. From Han (2007b, Figure 1), reproduced by permission of the AAS.

in the discovery and interpretation of exoplanets. In early studies of planetary microlensing, it was expected that the identification and analysis of magnification patterns of the planetary caustics, of the sort evident in Figure 5.4, would offer the best prospects of detection planetary signals (e.g. Gould & Loeb, 1992). But such events are of typically rather low magnification, and therefore less easy to detect, and less easy to observe photometrically with adequate signal-to-noise.

Griest & Safizadeh (1998) demonstrated that there were, instead, practical advantages in concentrating searches in the region of the central stellar caustic: large deviations from the single lens light curve can still occur due to the presence of a planet, and the high magnification (and the more accurate photometry therefore possible) should make these changes in the central caustic structure detectable.

This can be visualised more easily with reference to Figure 5.2: near the peak of high-magnification events, the two images created by the primary (star) lens are highly magnified and distorted, the combined image pattern reaching its closest manifestation of an Einstein ring. A planetary companion to the primary star, and lying reasonably near to the Einstein ring, will distort its symmetry. As the host star passes close to the line-of-sight to the source, the images sweep around the Einstein ring, thus probing its distortion.

Observing high-magnification events Although high magnification events are rare, the combined OGLE and MOA surveys (§5.8) are now discovering some 600–1000 microlensing events towards the bulge each year, with a few very high magnification events amongst them. Focusing on the high magnification events has a number of practical advantages: only relatively sparse temporal monitoring by the survey groups (one or two times per night) is required to predict their occurrence well in advance of maximum, thereby allowing observations to be concentrated on potentially important events as they unfold (see Figure 5.5). They are particularly well suited for the detection of multiple planets, since they automatically probe the region of the central caustic, where even low-mass planets sufficiently near the Einstein ring will further perturb the light curve (Bond et al., 2002a; Rattenbury et al., 2002; Abe et al., 2004).

Their brightness also means that they are accessible to observation by small telescopes (to which amateurs are now contributing), and since the sources are consequently typically very faint, it is easier to characterise the planetary host star with follow-up observations less perturbed by source blending.

Ray shooting For a binary or multiple lens system, the lens equation is a generalisation of Equation 5.4, in which the observer, source, and lens system no longer necessarily lie in a single plane. Although mapping from the image to source plane for an arbitrary lensing geometry can be simply formulated (e.g. in complex coordinates: Bourassa et al., 1973; Witt, 1990; Bennett & Rhie, 1996; Griest & Safizadeh, 1998), solving directly for all possible images and their total magnification for realistic configurations is generally impractical.

An alternative is to exploit the simplicity of the mapping from image to source plane in a technique originally developed by Kayser et al. (1986) and Schneider & Weiss (1987) to study the effects of microlensing in the variability of active galactic nuclei, and referred to as *ray shooting* (or equivalently as *inverse ray shooting*, because in practice the photon paths are modeled 'backwards' from the observer in the direction of the lens, and hence to the source plane).

In the approach described by Wambsganss (1997), representative binary lens magnification maps are generated for trial values of q and d by propagating a uniform bundle of light rays backwards from the observer, through the lens plane containing star and planet, and collected in the source plane. The density of rays at a particular location in the source plane is then proportional to the magnification at that point. Light curves at arbitrary trajectory angles α can be obtained from appropriate one-dimensional cuts through such a magnification pattern, convolved with the source profile.

Early examples of magnification maps and light curves were given by Gould & Loeb (1992), Wambsganss (1997), and Gaudi & Sackett (2000). Conceptually, an observed microlensing event could then be interpreted by comparison with a (huge) library of such magnification maps and light curves to infer the lensing geometry.

Classification A qualitative classification of perturbation types, given by Han (2007b), is shown in Figure 5.6. A more quantitative parameterisation according to the three different binary lens caustic topologies (close, intermediate, wide) is described by Cassan (2008), and illustrated in Figure 5.7.

Light curve modeling Practical model fitting tends to follow a rather brute force numerical approach to establish the detailed lensing geometry (e.g. Gould et al., 2006b). Present analyses aim to characterise a given exoplanetary light curve on the basis of seven numerical parameters: the three single-lens geometric parameters (t_0, u_0, t_E), the three binary lens parameters (q, d, α), and a seventh parameter ($\rho \equiv \theta_\star/\theta_E$) which is required whenever the angular radius of the source $\theta_\star$ plays a significant role in the fine structure of the light curve, in particular when the source crosses a caustic (§5.4).

For planetary lenses with (high magnification) caustic crossings there are seven observable light curve features that directly constrain the seven model parameters, up to the two-fold degeneracy $d \leftrightarrow d^{-1}$ (Dominik, 1999). Three of these, the epoch, peak and duration of the primary lensing event strongly constrain t_0, u_0 and t_E respectively, and these values can be used as seeds. The entry and exit caustic crossing times, and the height and duration of (one of) the caustic crossings, then constrain (d, q, α, ρ). Model values are extracted by a numerical grid search of parameter space, typically by holding (d, α, q) fixed at a set of values while minimising χ^2 over the remaining four parameters.

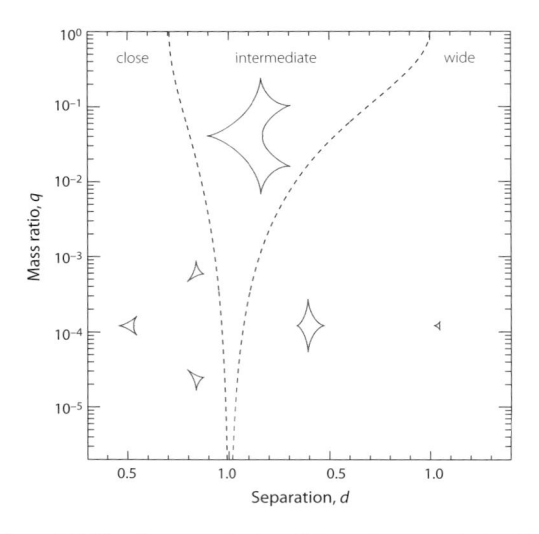

Figure 5.7: The three topologies of binary lenses as derived by Cassan (2008): the close binary lens (left), which involves a central caustic plus two off-axis small secondary caustics (plotted for $d = 0.8$ and $q = 10^{-2}$), the intermediate binary (middle, assuming $d = 1$ and $q = 10^{-2}$) with a single caustic, and the wide binary (right, $d = 1.6$ and $q = 10^{-2}$) with a central caustic and an isolated secondary caustic. When the mass ratio tends to zero, the extension of the intermediate domain also tends to zero. From Cassan (2008, Figure 1), reproduced with permission © ESO.

The modeling is computationally intensive, with the globally-best model often hard to find because of the many local minima. Interpretation of the more complex microlensing light curves often proceeds with collaborating groups using independent methods to corroborate the results. Gould (2008) cites events requiring tens of thousands of processor hours to model, and describes the hexadecapole approximation which can be orders of magnitude faster than full finite source calculations.

5.4 Other light curve effects

A number of further effects, relevant for the modeling and interpretation of planetary microlensing light curves, are described in this section. The important effects of microlens parallax and non-linear motion of the observer are described in §5.5.

Finite source size During high magnification events with small values of the projected lens–source separation, the finite angular diameter of the source means that different regions of its stellar disk are magnified at different times during the event and, for well-sampled light curves, distinctive finite source size effects may be evident. Since at high magnification $A \simeq 1/u$ (Equation 5.15), taking representative values of $\theta_E \sim 1$ mas (Equation 5.5) and $\theta_\star \sim 1$ μas, suggests that such effects should be expected for $A_{\max} \gtrsim 1000$.

5.4 Other light curve effects

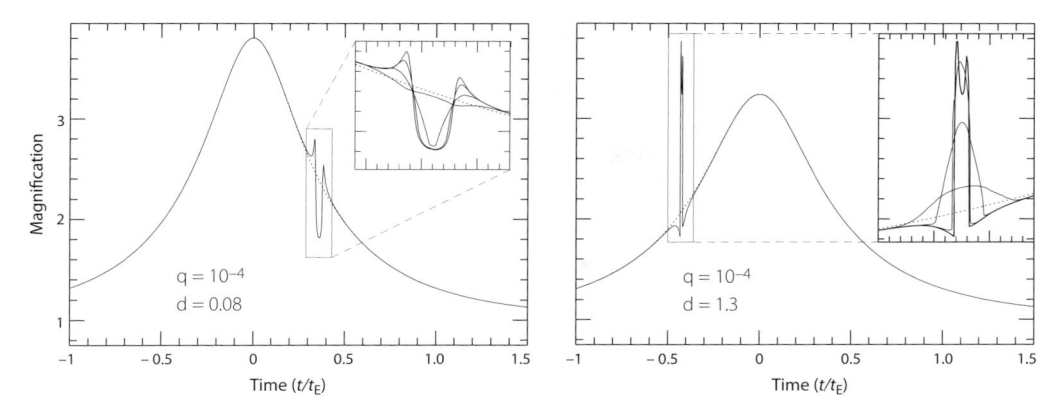

Figure 5.8: Theoretical microlensing light curves showing planetary deviations for a mass ratio $q = 10^{-4}$ and two values of d. Dashed curves are for an unperturbed single lens. The main plots are for a stellar radius $\rho = 0.003$, with the insets showing the progressively smaller amplitudes for larger stellar radii ($\rho = 0.006, 0.013, 0.03$). From Bennett & Rhie (1996, Figure 1), reproduced by permission of the AAS.

Detailed effects were calculated by Bennett & Rhie (1996), and examples for a planet mass ratio $q = 10^{-4}$, and for a range of values of the parameter $\rho = \theta_\star/\theta_E$ are shown in Figure 5.8. Planet-induced deviations become progressively less pronounced for larger $\theta_\star$, reaching the limits of detectability at $\rho \sim 0.03$. Accordingly, planets down to $M_p \sim 1 M_\oplus$ should be detectable when lensed against main sequence source stars, while even planets of much higher mass $M_p \sim 10 M_\oplus$ are less easily detectable when lensed against giant source stars. It follows that microlensing searches for Earth-mass planets should preferably focus on main sequence source stars.

Good precision can be achieved in practice. Dong et al. (2009a) determined $\rho = (3.29 \pm 0.08) \times 10^{-3}$ for the planetary event MOA–2007–BLG–400.

If the source star is visible and unblended, its angular radius $\theta_\star$ can be estimated from the angular size–colour relation based on its (extinction-corrected) magnitude and colour (Yoo et al., 2004b). With the source radius crossing time, $t_\star$, derived from the microlensing light curve, the angular Einstein radius then follows from

$$\theta_E = \frac{\theta_\star t_E}{t_\star} . \quad (5.23)$$

The dependency of M_L on D_L (whether known or not) follows from Equation 5.6, while the linear projected star–planet separation, and hence a lower limit on the size of the exoplanet orbit, follows from the modeled value of d.

Limb darkening and star spots Associated with the effects of finite source size is the possibility of a detailed modeling of limb darkening for the source star, since in practice the surface brightness of its stellar disk varies with radius. Modeling of ρ may proceed on the assumption of a uniform stellar disk, or by using a more physically appropriate radial surface brightness profile. An improved measurement of stellar limb darkening may be obtained as a by-product of a more detailed model of these finite source effects.

Limb darkening was first modeled for the non-planetary microlensing events MACHO–97–BLG–28 by Albrow et al. (1999), and subsequently for MACHO–97–BLG–41 (Albrow et al., 2000a), OGLE 99–BLG–23 (Albrow et al., 2001b), and others. A detailed profile, inconsistent with current atmospheric models, was obtained for the K giant event EROS–2000–BLG–5 by Fields et al. (2003). Such an analysis for the H-band photometry in the planetary microlensing event MOA–2007–BLG–400 (Dong et al., 2009a) nevertheless showed good agreement with the models of Claret (2000).

The possibility of detecting star spots on the source star was investigated by Chang & Han (2002).

Orbital motion of a binary lens Although not necessarily of direct interest for exoplanet events, effects attributed to an exoplanet must exclude other time-dependent phenomena as a cause of any fine structure in the light curve. Dominik (1998) investigated short-period binaries, and discussed three scenarios: a rotating binary lens, a rotating binary source, and a rotating observer (Earth orbiting the Sun), finding that the most dramatic effects are for a rotating binary lens, because the caustic structure itself changes with time. This rotation introduces five additional parameters: two rotation angles, the binary period, the eccentricity and phase.

The orbital motion of binary lenses where both components have stellar masses has been reported in a few cases where the light curve contains well-measured caustic crossings that establish key times in the orbit to $\sim 10^{-5} t_E$. Indeed, the first claimed planetary microlensing event MACHO–97–BLG–41 (Bennett et al., 1999) was revised when an improved fit was found invoking a binary with an orbital period of 1.5 yr (Albrow et al., 2000a). Modeling of EROS–2000–BLG–5 required inclusion of finite source effects, microlens parallax, and a binary orbital motion (An et al., 2002).

Orbital motion of a star–planet lens The possibility of detecting orbital motion of a planet around the host star was discussed by Bennett (2008), who reasoned that the effect could be detected in principle for planetary events lasting $\Delta t = 1-10$ d, if shifts in the lens position with respect to the source could be detected at a fraction of the finite source effect, $\sim 0.1 \theta_\star/\theta_E$.

The effect is seen in the data for the outer planet in the first multiple planetary microlensing event, OGLE–2006–BLG–109L (Gaudi et al., 2008). As discussed further in §5.9, orbital motion changes the projection geometry of the planetary caustic, leading to both its rotation and change of shape during the event. Although orbital periods are likely to be long (several years), and the peak microlensing events last only several days, the caustic size is $\propto 1/|d-1|$. For OGLE–2006–BLG–109, $d = 1.04$, such that a change in projected orbital separation of 0.5% over the eight days between the first cusp crossing and the peak gave a measurable 10% change in caustic size. The components of the projected velocity of the planet relative to the primary star, together with the stellar mass, determine the outer planet's orbit (including inclination) on the assumption that it is circular.

Blending and lens–source displacement The very dense star fields in the direction of the Galactic bulge selected for microlensing monitoring have typical angular separations much smaller than the seeing disk, and the flux of other stars frequently contributes to the measured light curve. Such *blending* can be due to a physical companion of the source star, due to the lens itself, or due to the superposition of another (non-lensing) object along the line-of-sight. This can introduce further complexity, and degeneracies, in the fitting procedures (Alard, 1997; Wozniak & Paczyński, 1997; Alard & Lupton, 1998; Gould & An, 2002).

Separation between light from the source star and the (normally unrecognisable) blended contribution of light from the lens may be possible once the angular separation between source and lens has increased to a few milliarcsec, perhaps several years after the microlensing event occurred (Bennett et al., 2007). Any difference in colour between source and lens will result in a (small) displacement in the image centroid as a function of wavelength. Measurement of the centroid shift from multicolour high-angular resolution imaging may then provide an estimate of the host star spectral type and, with some further assumptions on the underlying stellar population, an estimate of D_L, and hence a complete solution of the lens equation. The effect was detected for the first microlensing planet OGLE–2003–BLG–235 in

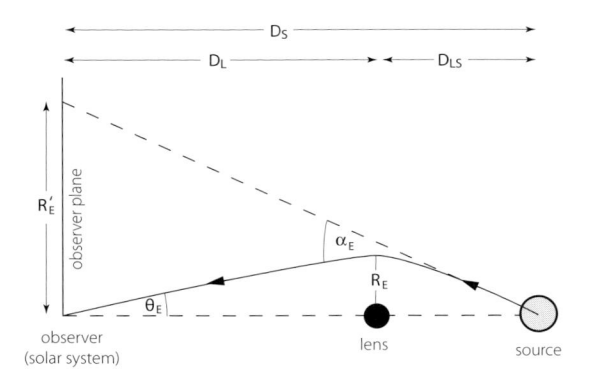

Figure 5.9: Schematic of light bending for impact parameter $b = R_E$. The linear Einstein radius R_E in the lens plane subtended at the source, projects to a scaled value at the solar system R'_E. Determination of the microlens parallax can be made from measurements of the microlensing light curve from widely spaced locations in the observer plane.

HST observations made 1.8 years after peak magnification, when the source–lens separation was ~ 6 mas (Bennett et al., 2006).

The relative source–lens transverse motion, $v_\perp$, which drives the rapidly changing magnification during a microlensing event, cannot be determined uniquely from the microlensing light curve alone. However, long after the microlensing event, small changes in the elongation of the image of the source may be detectable, e.g. using the very stable point-spread function of HST. Measurement of the relative angular proper motion between source and lens, μ_{LS}, either from the image elongation or from the colour-dependent centroid shift, can then establish the angular Einstein radius via Equation 5.19, and hence M_L as a function of D_L from Equation 5.6.

5.5 Microlens parallax and lens mass

Studies of the first microlensed exoplanets have demonstrated that a direct determination of the primary lens mass (the host star) can be in fact made. Since stellar mass determination is normally only feasible for (particular types of) binary stars, this possibility is interesting in its own right. More importantly in the present context it follows that, once determined and combined with the star–planet mass ratio q, the planet mass can be determined. The following summary, based on Gould (2009), gives the framework for an understanding of the microlensing planet mass determinations made since 2004.

In Figure 5.9, corresponding to the geometry of Figure 5.1, the impact parameter is set to the physical size of the (linear) Einstein radius, such that the corresponding deflection angle $\alpha_E = 4GM_L/(c^2 R_E)$. The undeviated ray is propagated onwards to the observer plane to define a projected value of the Einstein radius, R'_E. From trigonometry (and for small angles), $R'_E = \alpha_E D_L$

5.5 Microlens parallax and lens mass

and $R_E = \theta_E D_L$ which, using the deflection expression for α_E gives

$$\theta_E R'_E = \frac{4GM_L}{c^2}. \tag{5.24}$$

Also, from angular equalities

$$\theta_E = \frac{R'_E}{D_L} - \frac{R'_E}{D_S} = R'_E \left(\frac{1}{D_L} - \frac{1}{D_S}\right) = \frac{\varpi_{\text{rel}}}{\text{AU}} R'_E, \tag{5.25}$$

where the relative parallax, ϖ_{rel}, has an amplitude

$$\varpi_{\text{rel}} = \text{AU} \left(\frac{1}{D_L} - \frac{1}{D_S}\right) = \text{AU} \left(\frac{D_{LS}}{D_L D_S}\right), \tag{5.26}$$

where ϖ_{rel} is in radians, and D (defined in §5.2) and the AU (astronomical unit) are in common units.

By analogy with the usual trigonometric parallax $\varpi = \text{AU}/d$ (where ϖ is in rad for a distance d in AU or, alternatively, AU and d are in common units), the *microlens parallax* is now introduced. It expresses the size of the Earth orbit relative to the Einstein radius of the microlensing event projected onto the observer plane

$$\varpi_E = \frac{\text{AU}}{R'_E}. \tag{5.27}$$

Equations 5.24–5.27 can be rearranged to give

$$M_L = \frac{\theta_E}{\kappa \varpi_E}, \tag{5.28}$$

$$\varpi_E = \sqrt{\frac{\varpi_{\text{rel}}}{\kappa M_L}}, \tag{5.29}$$

$$\theta_E = \sqrt{\kappa M_L \varpi_{\text{rel}}}, \text{ and} \tag{5.30}$$

$$\varpi_{\text{rel}} = \theta_E \varpi_E, \tag{5.31}$$

where $\kappa = 4G/(c^2 \text{AU}) \sim (8.1/M_\odot)$ mas.

Application The importance of these relations is as follows. For microlensing events the (measured) Einstein time is related to the (unknown) lens–source relative proper motion according to Equation 5.19. With θ_E given by 5.30, it follows that determination of ϖ_E would eliminate uncertainty arising from the unknown relative proper motion, and would give a direct relation between M_L and ϖ_{rel} through Equation 5.29.

If both θ_E and ϖ_E (or equivalently R'_E) can be determined, then the lens (host star) mass can be established from Equation 5.28 (and the planet mass thereafter from the mass ratio, q), and the lens–source relative parallax from Equation 5.31. This provides a distance to the event, and hence the linear size of the projected star–planet separation, if the distance to the background source can be estimated.

For planetary microlensing events, estimates of θ_E can be made through the study of finite source effects during the event, or from estimates of the lens–source proper motion made months or years afterwards. Measurements of ϖ_E are being made from observations of the microlensing light curve over an extended observer baseline or, in a related manner, by including the effects of the non-linear motion of the Earth's orbit (see below).

The combination of finite source effects and Earth's orbital motion was used to determine the lens mass for the microlensing binary EROS–2000–BLG–5 by An et al. (2002). The method was first applied to an exoplanet event to determine the mass of the host star in the case of OGLE–2006–BLG–109 by Gaudi et al. (2008).

Non-linear motion of the observer Experimentally, the simplest method to measure ϖ_E results from the acceleration of the Earth's orbital motion around the Sun. The departure from rectilinear relative motion between observer–lens–source during microlensing events with suitably long Einstein crossing times results in a distorted light curve as the Earth's motion progressively deviates from a straight line over the duration of the event.

This way to estimate ϖ_E was proposed in the context of MACHO microlensing searches by Gould (1992), and developed by Smith et al. (2003). Alcock et al. (1995) used it to place the first constraints on a lens mass and distance, and the same principle has been applied to other events subsequently. Nevertheless the effect remains only occasionally detectable: Smith et al. (2002) found just one in a search of 512 microlensing candidates during 1997–99, while Poindexter et al. (2005) found just 22 events (out of ~ 3000) for which inclusion of the effect significantly improved the light curve fit.

Notwithstanding the effect's relative rarity, high magnification planetary microlensing events appear to offer a particularly favourable configuration, and the effect has been clearly measured for the planetary event OGLE–2006–BLG–109 (Gaudi et al., 2008), and perhaps also for OGLE–2005–BLG–071.

Extended baseline For a microlensing event observed simultaneously at two locations separated by a distance of order $R_E \sim 1$ AU, both the impact parameter and the time of maximum magnification will differ. Comparison of the light curves of these different alignment geometries allows a more direct estimate of ϖ_E to be made.

Refsdal (1966) proposed that the effect could be detected, and the primary lens mass determined, by observations made simultaneously from Earth and from a distant space observatory, ideas which have been further developed subsequently (Gould, 1992; Boutreux & Gould, 1996; Gould et al., 2003b). In the first practical application, Dong et al. (2007) combined ground-based and Spitzer space telescope observations to measure the microlens parallax of OGLE–2005–SMC–001. Their solution defined a lens geometry with an eightfold degeneracy (two each for parallax, binary separation, and cusp approaches) but led to a parallax measurement (Figure 5.10) yielding a projected transverse lens velocity of

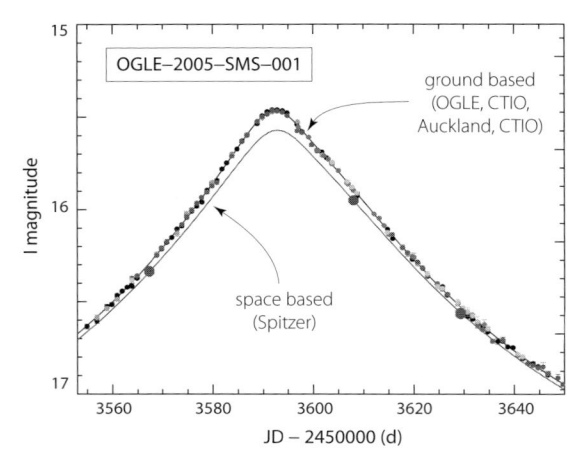

Figure 5.10: Light curve of the microlensing event OGLE–2005–SMS–001 observed simultaneously from ground-based observatories (upper curve), with a separate fit using photometric data from the Spitzer space telescope (lower curve). The model includes two additional parameters to account for binary rotation, and two further parameters to account for the microlens parallax. From Dong et al. (2007, Figure 4), reproduced by permission of the AAS.

$230\,\mathrm{km\,s^{-1}}$, a typical value expected for halo lenses, but excluding the possibility that the lens was in the Small Magellanic Cloud itself.

Terrestrial parallax From the consideration of photon statistics, Holz & Wald (1996) argued that microlens parallaxes could theoretically be measured even from a much smaller baseline, using two widely-spaced observatories on Earth. Gould (1997) showed that for events of very high magnification, and for observers separated by $\sim 1 R_\oplus$, the size of the effect is $\sim A_{\max} R_\oplus / R'_E$, and can be of order 1%. Such observations were made for the first time for the thick disk brown dwarf OGLE–2007–BLG–224 (Gould et al., 2009). This terrestrial parallax was also observed for the $A = 1600$ event OGLE–2008–BLG–279, placing mass limits of $\simeq 0.2 M_\oplus$ for projected separations close to the Einstein ring (Yee et al., 2009).

5.6 Astrometric microlensing

In addition to the photometric manifestations, the time-varying magnification of the unresolved microlensed images also leads to a small motion of their photocentre, typically by a fraction of a milliarcsec (Høg et al., 1995; Miyamoto & Yoshii, 1995; Walker, 1995; Paczyński, 1996; Miralda-Escudé, 1996; Boden et al., 1998).

The effect is shown schematically in Figure 5.2. The astrometric deflection of the centroid, from Equation 5.15, can be written

$$\delta\theta = \frac{A_+ \theta_+ + A_- \theta_-}{A_+ + A_-} - \theta_S = \frac{u\,\theta_E}{u^2 + 2}, \quad (5.32)$$

where, again, $u = \theta_S / \theta_E$. This reaches a maximum deflection angle, at $u = \sqrt{2}$, of

$$\delta\theta_{\max} = 8^{-1/2} \theta_E \simeq 0.35\,\theta_E. \quad (5.33)$$

For a typical bulge lens, with $\theta_E \sim 300\,\mu\mathrm{as}$, the shift is ~ 0.1 mas.

Although the astrometric displacement is small, the astrometric cross-section is substantially larger than the photometric, falling off as b^{-1} (Equation 5.1), meaning that the effect extends over larger angular scales. Furthermore, the normal degeneracy with regard to the mass of the lens is removed (Gaudi & Gould, 1997). This follows from Equation 5.30, where ϖ_{rel} is now the (measurable) relative parallax of the lens source assuming linear motion between the source and lens. Numerically

$$\frac{M_L}{M_\odot} = 0.123 \frac{\theta_E^2}{\varpi_{\mathrm{rel}}}, \quad (5.34)$$

where θ_E and ϖ_{rel} are in mas.

Astrometric planet detection Astrometric microlensing related specifically to planet detection has been investigated theoretically, although not yet observed (Mao & Paczyński, 1991; Safizadeh et al., 1999; Han & Chang, 1999; Dominik & Sahu, 2000; Han & Lee, 2002; Han, 2002; Asada, 2002; Han & Chang, 2003). Figure 5.11 shows predictions from Safizadeh et al. (1999). Although the planet's astrometric perturbing effect is short in duration, the amplitude can still be large for Jovian planets, with the time above 10 μas being typically of order days.

Astrometric signatures have not yet been measured, being too small for present ground-based observations. In the future, narrow-field astrometric interferometers, notably VLTI–PRIMA (Launhardt et al., 2008), and microarcsec-class space experiments, should be able to measure the effects. Concepts such as the NASA space interferometer SIM PlanetQuest (Unwin et al., 2008) would allow the astrometric study of photometric events alerted from ground (Paczyński, 1998). Scanning missions offer somewhat less favourable observational conditions, although Gaia could detect astrometric lensing independently of photometric signatures (Dominik & Sahu, 2000).

High proper motion lenses A distinct approach to identifying potential astrometric microlensing events is to select on potential lenses instead of on sources (Paczyński, 1995, 1998; Di Stefano, 2008a,b). Stars with high proper motion are preferentially nearby, with a relatively large θ_E and hence large astrometric cross-section (Equation 5.33), and they therefore cross large areas of sky over a relatively short time.

For Barnard's star, a background source at an angular separation of 9 arcsec would be displaced by 100 μas.

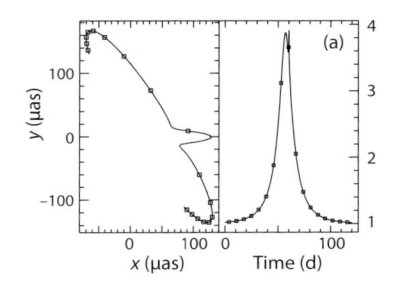

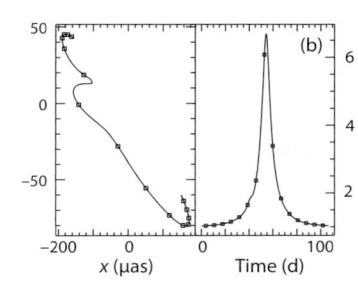

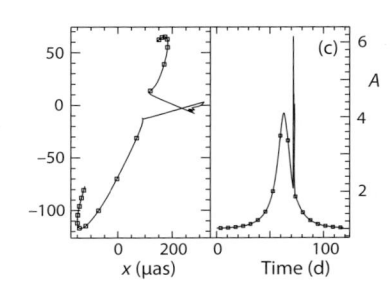

Figure 5.11: Predicted planet astrometric (leftmost plots of each pair, scale at left) and photometric (rightmost plots, scale at right) lensing curves. All assume a planet-to-lens mass ratio $q = 10^{-3}$, with a primary lens Einstein radius $\theta_E = 550\,\mu as$, corresponding to a Saturn-mass planet. Data points (squares) are plotted one per week: (a) $d = 1.3$; (b) $d = 0.7$; (c) a caustic crossing event with $d = 1.3$ (d is the projected planet–lens separation in units of the Einstein radius, θ_E). Both astrometry and photometry show a smooth perturbation due to the primary lens, with the short planetary event in addition. From Safizadeh et al. (1999, Figure 2), reproduced by permission of the AAS.

Over three years, with a proper motion of ~ 10 arcsec yr^{-1}, this star alone would probe an area of $\sim 30 \times 18$ arcsec^2. Hipparcos and Gaia stars with accurate high proper motions would allow suitable events to be predicted well in advance. Again, the absence of μas astrometry means that such events cannot yet be studied.

Light travel time A somewhat related consideration is noted for completeness. For a lens producing two or more images of a single source, the light travel times along the distinct paths are different. Two effects contribute: geometric time delay due to the different geometric path lengths, and relativistic (Shapiro) time delay due to gravitational time dilation (the divergence of coordinate time and proper time) in the gravity field (Schneider, 1985).

Although the time delay would be an important observable setting the scale of the system (image separations and brightness ratios being scale invariant), the effect (and markers for its measurement) are not (yet) within reach of exoplanet research.

5.7 Other configurations

This section summarises other source/lens configurations which might be observed in the future.

Planet orbiting the source star Essentially all discussions of microlensing focus on planets around the (foreground) lens star. While giving statistics about the planetary population, physical knowledge is restricted to planet masses and orbit parameters. In contrast, the reflected light from a planet contains physical information such as the presence of satellites or rings, and details of its atmospheric composition.

Planets might be detected in the source plane as they cross the caustics of the foreground lens and become highly magnified, while crescent-like sources can undergo substantially higher magnification than a uniformly illuminated disk (Heyrovský & Loeb, 1997; Graff & Gaudi, 2000; Lewis & Ibata, 2000; Ashton & Lewis, 2001; Gaudi et al., 2003). Lewis & Ibata (2000) also noted that corresponding polarisation fluctuations would probe atmospheric properties. Spiegel et al. (2005) calculated geometries and integration times, and concluded that CH$_4$, H$_2$O, and Na/K could be inferred from 0.6–1.4 μm spectroscopy.

Satellite orbiting a planet The detection of satellites (moons) surrounding an exoplanet, due to the changing magnification patterns, is generally considered problematic due to finite source effects (Han & Han, 2002). For the specific case of Earth-mass satellites orbiting ice-giant planets, Han (2008a) found that non-negligible satellite signals might occur when the planet-satellite separation is similar to or greater than the Einstein radius of the planet, thus for projected separations of 0.05–0.24 AU for a Jupiter-mass planet.

Liebig & Wambsganss (2010) made extensive triple-lens simulations over a range of two-dimensional projections of the three body configurations, and also concluded that the detection of planetary satellites is feasible under favourable configurations (Figure 5.12).

Planet orbiting a binary system Planets orbiting binary stars might be detectable (Han, 2008b). For $M_p \sim 1 M_J$, high detection efficiency is expected for orbits in the range 1–5 AU around binary stars with separations 0.15–0.5 AU. Lee et al. (2008) showed that in configurations where a planet orbits one of the binary components (and the other binary component is located at a large distance), both planet and secondary produce perturbations in a common region around the planet-hosting component. Signatures of both planet and binary companion may be detectable in the light curves of high-magnification events.

Free-floating planets Microlensing has the potential of detecting free-floating planets. Their occurrence rate

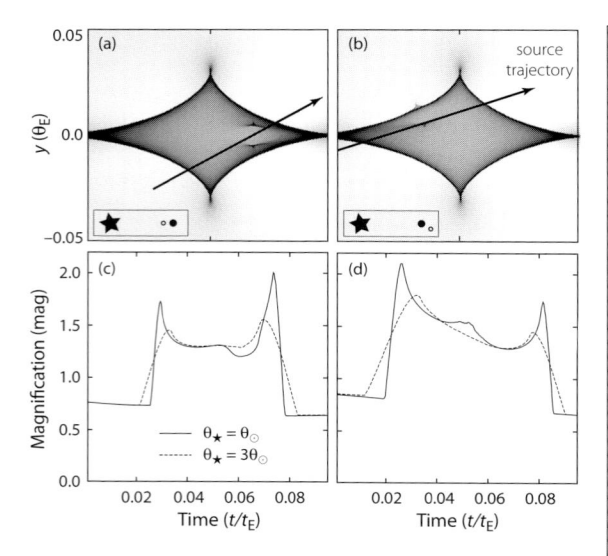

Figure 5.12: Microlensing detectability of planetary satellites, showing two configurations of triple-lens geometries, in which the satellite has moved in its orbit by 210° between the two. Mass ratios are $q = 10^{-3}$ and 10^{-2} for the planet–star and satellite–planet. Angular separations are $1.3\theta_E$ (star) and $1.3\theta_E$ (planet), respectively. Top pair: magnification maps (darker regions are higher magnification), with the geometrical configurations shown at bottom left. Bottom pair: light curves for the source trajectories indicated. Solid and dashed curves correspond to angular source sizes of $\theta_\odot$ and $3\theta_\odot$ respectively. From Liebig & Wambsganss (2010, Figures 2–3a,h), reproduced with permission © ESO.

would be important in placing constraints on their ejection by planet–planet interactions. Han et al. (2004) discussed the determination of masses using extended baselines (§5.5). Han (2006b) showed that a degeneracy between isolated events and wide-separation planets could be distinguished through astrometric microlensing follow-up, given its larger cross-section (§5.6).

Constraints on the mass of free-floating planets in the globular cluster M22 were given by Sahu et al. (2001) but the events were questioned (Gaudi, 2002) and later retracted (Sahu et al., 2002). Gil-Merino & Lewis (2005) investigated suggestions that uncorrelated variability in the gravitationally-lensed quasar QSO 2237+0305 (the Einstein Cross) could be attributed to a microlensing population of free-floating planets, and argued that such conclusions are flawed.

Microlensed transiting planets   Lewis (2001) showed that deviations in the light curve due to a transiting planet during a microlensing event would be substantial, although the probability of detecting such a transiting planet is prohibitively small, $\sim 10^{-6}$. Agol (2002) discussed microlensing during transits in binaries.

Planetesimal disks   Heng & Keeton (2009) and Hundertmark et al. (2009) have considered gravitational

Upper limits and uncertain detections: Between 1997 and the first unambiguous microlensing planet in 2004, early results from the exoplanet microlensing surveys gave a number of unconfirmed or uncertain planet detections, and the first limits on the existence of planets based on the lack of detection of planetary signals.

Bennett et al. (1997), based on limited photometric coverage, suggested that MACHO–94–BLG–4 could be an M-dwarf with a companion gas giant of mass $\sim 5 M_J$ at $a \sim 1$ AU, and that MACHO–95–BLG–3 could be an isolated object of mass $\sim 2 M_J$ or a planet more than 5–10 AU from its parent star.

Bennett et al. (1999) claimed the first unambiguous detection of a Jupiter-mass planet orbiting a binary star MACHO–97–BLG–41, but the independent PLANET data for the same event favoured a rotating binary with an orbital period of $P \sim 1.5$ yr (Albrow et al., 2000a).

Gaudi & Sackett (2000) analysed some 100 events monitored by PLANET, finding that more than 20 had sensitivity to perturbations that would be caused by a Jovian-mass companion to the primary lens. No unambiguous signatures were detected, indicating that Jupiter-mass planets with $a = 1.5 - 3$ AU occur in less than one third of systems, with a similar limit applying to planets of mass $3 M_J$ at $a = 1 - 4$ AU.

Rhie et al. (2000) gave limits on Earth-mass planets from the high-magnification ($A \sim 80$) event MACHO–98–BLG–35. Albrow et al. (2000b) gave limits for a lower magnification event OGLE–1998–BLG–14. Albrow et al. (2001a) and Gaudi et al. (2002) analysed five years of null detections, suggesting that less than 33% of the lens stars in the inner Galactic disk and bulge could have companions of $\gtrsim 1 M_J$ between 1.5–4 AU. Bond et al. (2002b) showed that the data for MACHO–98–BLG–35 were consistent with the possible detection of a terrestrial planet.

Jaroszynski & Paczyński (2002) proposed that the event OGLE–2002–BLG–55 had a signal consistent with a detected planet, but Gaudi & Han (2004) noted that there were other possible explanations. Tsapras et al. (2003) derived upper limits from three years of OGLE data (1998–2000). Snodgrass et al. (2004) derived upper limits from 389 OGLE observations from 2002. Yoo et al. (2004a) derived constraints from the high magnification event OGLE–2003–BLG–423.

lensing by planetesimal disks around nearby stars, for which the light curve may exhibit short-term low-amplitude residuals caused by planetesimals several orders of magnitude below Earth mass.

5.8   Microlensing observations in practice

The large-scale observing programmes that have detected the first microlensing planets focus on Baade's Window in the central Galactic bulge, visible from May to September each year. The high star densities in the bulge (Figure 5.13) maximise the chances of detecting rare microlensing events; the inescapable disadvantage of observing in this region of high stellar surface density are the effects of crowding and blending on the photometric analysis of the background sources.

5.8 Microlensing observations in practice

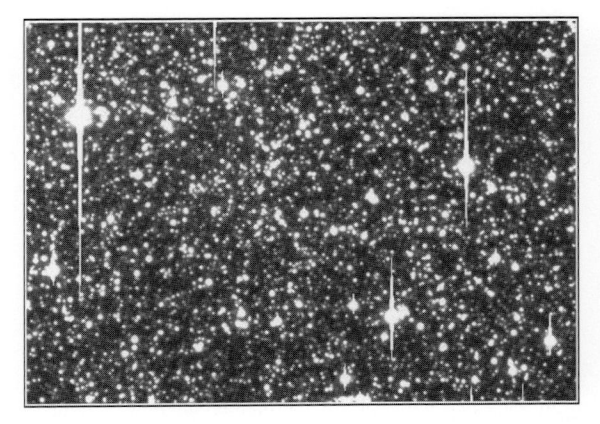

Figure 5.13: A small part of a MOA image of a field in the Galactic centre, obtained at Mt John University Observatory, New Zealand, showing a large density of faint stars. Image provided by the New Zealand–Japan MOA group.

Two-step mode In the absence of observing techniques that can observe tens of square degrees of sky several times per hour, current efforts focused on the detection of planetary microlensing events typically operate in a two-step mode (Gould & Loeb, 1992; Han & Kim, 2001; Han, 2007a). A wide-angle survey telescope detects the early stages of a microlensing event using a relatively coarse time sampling over such a large area. Once a deviation indicative of such an ongoing event is detected and issued as an alert, an array of smaller, follow-up narrow-angle telescopes distributed in Earth longitude follow the events with high-precision photometry and a much denser time coverage.

Monitoring Two monitoring teams, the MOA and OGLE collaborations, are currently each operating single dedicated telescopes, together gathering detailed photometric information about lensing events sifted from the vast survey data streams, and generating alerts for several hundred ongoing microlensing events per year, based on difference-imaging photometry and real-time event detection (Bond et al., 2002a).

The problem of the early identification of high-magnification events is severe, given the typically faint sources with attendant poor photometric accuracy; efforts to optimise this have resulted in the OGLE Early Warning System (EWS) and the Early Early Warning System (EEWS) (Udalski et al., 2005), as well as anomaly feedback based on results from the follow-up teams in the case of RoboNet–1.0 (Dominik et al., 2007).

MOA (Microlensing Observations in Astrophysics, Bond et al. 2001, Abe et al. 2004) is a NZ/Japan collaboration. Phase I (2000–2005) employed a 0.6-m telescope on Mt. John (NZ), with phase II, from 2006, using a 2-m telescope. Images are acquired in a broad I-centred band, using a 4k×6k pixel camera with exposures of 180 s. They cover some 20 square degrees a few times per night, and are geared to high magnification events, detecting typically ten events per season with $A_{\max} > 100$.

OGLE (Optical Gravitational Lens Experiment, Udalski 2003) is a Polish/US collaboration operating a 1.3-m telescope on Las Campanas (Chile), and regularly monitoring some 200 million stars. It has developed from OGLE–I (the pilot phase, 1992–1995), OGLE–II (1996–2000), OGLE–III (also devoted to transiting planets, 2001–2009), and OGLE–IV (starting in 2009). During phase III, images were acquired in the I band, using an 8k×8k camera with exposures of 120 s. OGLE–IV, focused on increasing the number of planet detections using microlensing, employs a 32-chip CCD camera.

Follow-up Three collaborations currently pursue possible planetary microlensing alerts: PLANET/RoboNet, MicroFUN, and MiNDSTEp. [Previous monitoring groups GMAN (Global Microlensing Alert Network, Pratt et al. 1996) subsequently placed their emphasis on non-planetary microlensing, while MPS (Microlensing Planet Search, Rhie et al. 1999) which began in 1997, merged with PLANET in 2004.] In addition, both survey teams, OGLE and MOA, can switch from survey to follow-up mode for confirmed high-magnification events.

Since 2005, PLANET/RoboNET (Beaulieu et al., 2007) has been a joint venture of the former PLANET collaboration (Probing Lensing Anomalies NETwork, Albrow et al. 1998, Dominik et al. 2002, Sackett et al. 2004), with its various telescopes in Australia, South Africa, and South America, together with RoboNet-1.0 (Burgdorf et al., 2007), the UK-operated robotic network comprising the 2-m telescopes of Liverpool (La Palma, Spain), the Faulkes North (Maui, Hawaii), and the Faulkes South (Siding Spring, Australia).

As an example of the dense follow-up sampling now possible, the PLANET team's caustic crossing data of EROS–2000–BLG–005 (An et al., 2002) yields a light curve consisting of more than 1000 data points over several days with photometric accuracy of order 1%.

As of 2008, the two Faulkes telescopes are owned and operated by the Las Cumbres Observatory Global Telescope network, as part of their global deployment of robotic telescopes, which is due to be expanded by 2011 with the introduction of 18 new 1-m and 24 0.4-m telescopes (Hidas et al., 2008). RoboNet–II, meanwhile, uses the same telescope resources as RoboNet–1.0, along with improved software for prioritising and reducing the observations (Tsapras et al., 2009). An optimised follow-up scheme taking account of telescope and other observing resources, and the predicted nature of the targets being observed, has been described by Horne et al. (2009, see also Figure 5.5).

MicroFUN (Microlensing Follow-Up Network, Yoo et al. 2004b) is a consortium of observers, coordinated by Ohio State University, which focuses almost entirely on high-magnification events for the reasons described in §5.3. The collaboration represents some 40 (professional and amateur) observers from 10 countries (including New Zealand, South Korea, Israel, and South Africa), typically using 0.5-m class telescopes.

MiNDSTEp (Microlensing Network for the Detection of Small Terrestrial Exoplanets, Dominik et al. 2010) is a large European/Eurasian effort also established to exploit the dense monitoring of ongoing gravitational microlensing events.

5.9 Exoplanet results

5.9.1 Individual objects

Planets discovered by microlensing through to the end of 2010 are given in Table 5.1. Comments about these specific systems follow.

OGLE–2003–BLG–235 The first confirmed planet detected by mirolensing was reported by Bond et al. (2004), and the observational data and model results are shown in Figure 5.14 (the event is also identified as MOA–2003–BLG–53). The long-term behaviour is typical for a point mass microlensing event, while the two sharp spikes correspond to the caustic entry and exit, their short duration indicating an extreme mass-ratio binary system. Bond et al. (2004) derived a mass ratio $q = 0.0039$. Various assumptions led to their estimates of $D_L = 5.2^{+0.2}_{-2.9}$ kpc, $M_p = 1.5^{+0.1}_{-1.2} M_J$, and a transverse separation $a = 3.0^{+0.1}_{-1.7}$ AU. The probable detection of the planetary host star using multicolour HST observations (Bennett et al., 2006) provides an estimate of the lens star spectral type (indicating a K-dwarf), an estimate of its distance and linear transverse motion, and thereafter a complete solution of the lens system, yielding the parameters given in Table 5.1.

OGLE–2005–BLG–71 Udalski et al. (2005) reported the second planet discovery (Figure 5.15). Their preferred model gives $t_E = 70.9 \pm 3.3$ d, and a mass ratio $q = (7.1 \pm 0.3) \times 10^{-3}$. Subsequent HST observations by Dong et al. (2009b) provided an estimate of the distance to the lens, and suggest that it is an M dwarf with disk kinematics, $M_\star = 0.46 \pm 0.04 M_\odot$ at a distance $D_L = 3.2 \pm 0.4$ kpc. The best-fit model leads to $M_p = 3.8 \pm 0.4 M_J$, a projected separation 3.6 ± 0.2 AU from the host star, leading to an equilibrium temperature $T_{eq} \sim 55$ K, similar to Neptune.

OGLE–2005–BLG–390 The first low-mass microlensing planet was reported by Beaulieu et al. (2006). The prominence of the planet peak was strongly reduced by the significant angular extent of the source star. The microlensing fit gave a mass ratio $q = (7.6 \pm 0.7) \times 10^{-5}$, and a projected planet–star separation $d = 1.610 \pm 0.008 \theta_E$. The source was identified as a relatively bright clump giant, with photometry giving the source star's angular radius of $\theta_\star = 5.25 \pm 0.73 \mu$as for

> **Naming convention:** Microlensing events are named after the team who first reported them, so that while OGLE–2003–BLG–235 and MOA–2003–BLG–53 refer to the same event, the former takes precedence. The lens and source can be specified with the suffixes 'L' and 'S' respectively. Additional capital or lower case letters designate companions of stellar or planetary mass respectively.
>
> Accordingly OGLE–2006–BLG–109LA, OGLE–2006–BLG–109Lb, and OGLE–2006–BLG–109Lc designate specifically the star, and the two known planets of this multiple lens system. Similarly ...Sb would refer to a planetary companion to the source star (a configuration that has not been observed to date, cf. §5.7).

an estimated bulge distance of 8.5 kpc. From estimates of the Einstein radius crossing time of 11.03 ± 0.11 d, a crossing time of the source star radius of 0.282 ± 0.010 d, and a probability estimate of Galactic distributions, they derived $M_\star = 0.22^{+0.21}_{-0.11}$ (suggesting an M dwarf host star), $M_p = 5.5^{+5.5}_{-2.7} M_\oplus$, at a distance of $D_L = 6.6 \pm 1$ kpc. The projected star–planet separation of $2.6^{+1.5}_{-0.6}$ AU yields a probable orbital period of 9^{+9}_{-3} years.

Ehrenreich et al. (2006) modeled the likely rock-to-ice mass ratio, included the effects of (internal) radiogenic heating, and argued that liquid water might have been present below an icy surface when the planet was ≲5 Gyr old, but is now likely to be entirely frozen. Kubas et al. (2008) showed that the existence of other Neptune/Jupiter mass planets is not ruled out.

OGLE–2005–BLG–169 Gould et al. (2006b) reported this extremely high magnification event, $A \sim 800$. It was initially flagged on 2005 April 21 as a faint ($I = 19.4$ mag) but brightening bulge source, reaching $I \sim 13$ mag at maximum. The light curve gave $q = 8^{-3}_{+2} \times 10^{-5}$ and a component separation $d = 1.0 \pm 0.02 \theta_E$. In an analysis somewhat along the lines of that for OGLE–2005–BLG–390, they derived $M_p = 13^{+4}_{-5} M_\oplus$, along with the other parameters given in Table 5.1.

OGLE–2006–BLG–109 Gaudi et al. (2008) reported the detection of the first multiple-planet microlensing system. Although the caustics of the two planets merge to form a single caustic curve, their effects were largely independent, such that parts of the caustic associated with the individual planets could be identified.

Using the finite source size during caustic exit provided the source radius relative to the Einstein radius $\rho = \theta_\star/\theta_E$, and hence θ_E based on an estimated $\theta_\star$ from the source's spectral type. Other subtle distortions in the light curve could be attributed to acceleration of the Earth in its orbit around the Sun, leading to an estimate of the physical size of the Einstein radius projected onto the plane of the observer R'_E (§5.4). These two measures of the Einstein radius allowed triangulation of the event, and hence the host star distance, yielding $D_L = 1.49 \pm 0.13$ kpc and $M_\star = 0.50 \pm 0.05 M_\odot$.

5.9 Exoplanet results

Table 5.1: *Exoplanets discovered by microlensing, chronologically through to the end of 2010. Mass ratios q are from the discovery papers. Derived physical parameters (lens distance D_L, host star mass $M_\star$, planet mass M_p, and projected star–planet separation a) are based on higher-order modeling, probability arguments, or subsequent observations, as discussed in the text.*

#	Event	q	D_L (kpc)	$M_\star$ ($M_\odot$)	M_p	a (AU)	Discovery reference
1	OGLE–2003–BLG–235Lb	3.9×10^{-3}	5.8 ± 0.6	$0.63^{+0.07}_{-0.09}$	$2.6^{+0.8}_{-0.6} M_J$	$4.3^{+2.5}_{-0.8}$	Bond et al. (2004)
2	OGLE–2005–BLG–71Lb	7.1×10^{-3}	3.2 ± 0.4	$0.46^{+0.04}_{-0.04}$	$3.8^{+0.4}_{-0.4} M_J$	$3.6^{+0.2}_{-0.2}$	Udalski et al. (2005)
3	OGLE–2005–BLG–390Lb	7.6×10^{-5}	6.6 ± 1.0	$0.22^{+0.21}_{-0.11}$	$5.5^{+5.5}_{-2.7} M_\oplus$	$2.6^{+1.5}_{-0.6}$	Beaulieu et al. (2006)
4	OGLE–2005–BLG–169Lb	8.0×10^{-5}	2.7 ± 1.4	$0.49^{+0.23}_{-0.29}$	$13^{+4.0}_{-5.0} M_\oplus$	$2.7^{+1.5}_{-1.0}$	Gould et al. (2006b)
5	OGLE–2006–BLG–109Lb	1.3×10^{-3}	1.5 ± 0.1	$0.50^{+0.05}_{-0.05}$	$0.71^{+0.08}_{-0.08} M_J$	$2.3^{+0.2}_{-0.2}$	Gaudi et al. (2008)
	OGLE–2006–BLG–109Lc	4.9×10^{-4}	"	"	$0.27^{+0.03}_{-0.03} M_J$	$4.6^{+0.5}_{-0.5}$	"
6	MOA–2007–BLG–192Lb	2.0×10^{-4}	1.0 ± 0.4	$0.06^{+0.02}_{-0.02}$	$3.3^{+4.9}_{-1.6} M_\oplus$	$1.0^{+0.4}_{-0.2}$	Bennett et al. (2008)
7	MOA–2007–BLG–400Lb	2.5×10^{-3}	5.8 ± 0.7	$0.30^{+0.19}_{-0.12}$	$0.83^{+0.49}_{-0.31}$	0.72 or 6.5	Dong et al. (2009a)
8	OGLE–2007–BLG–368Lb	9.5×10^{-5}	5.9 ± 1.2	$0.64^{+0.21}_{-0.26}$	$20^{+7.0}_{-8.0} M_\oplus$	$3.3^{+1.4}_{-0.8}$	Sumi et al. (2010)
9	MOA–2008–BLG–310Lb	3.3×10^{-4}	≥ 6.0	$0.67^{+0.14}_{-0.14}$	$74^{+17.0}_{-17.0} M_\oplus$	$1.25^{+0.1}_{-0.1}$	Janczak et al. (2010)
10	MOA–2009–BLG–319Lb	3.9×10^{-4}	6.1 ± 1.1	$0.38^{+0.34}_{-0.18}$	$0.16^{+0.14}_{-0.08} M_J$	2.0 ± 0.4	Miyake et al. (2011)

The orbital motion of the outer planet (planet c) could also be identified in the data: orbital motion changes the projection geometry of the associated planetary caustic, leading to both a rotation and change of shape of the caustic during the event (Figure 5.16). They could then constrain the two components of the projected velocity of the planet relative to the primary star which, together with the stellar mass, completely determine the outer planet's orbit (including inclination) on the assumption that it is circular. The orbital parameters of planet b followed from assumptions of circularity and coplanarity. Resulting planet masses are 0.71 ± 0.08 and $0.27 \pm 0.03 M_J$, at (three-dimensional) orbital separations 2.3 ± 0.02 and 4.6 ± 0.05 AU respectively.

The system resembles a scaled solar system. Although the primary star mass is only half solar, the mass ratios of the planets and their host $M_b/M_\star = 1.35 \times 10^{-3}$, $M_c/M_b = 0.36$ are similar to $M_J/M_\odot = 0.96 \times 10^{-3}$ and $M_S/M_J = 0.30$ for Jupiter, Saturn and the Sun. The ratio of projected separations $a_b/a_c = 0.60$ is also similar to the Jupiter/Saturn value $a_J/a_S = 0.55$. Estimated equilibrium temperatures $T_b \sim 81 \pm 12$ K and $T_c \sim 59 \pm 7$ K are some 30% smaller than those of Jupiter and Saturn. Bennett et al. (2010) determined an orbital eccentricity $0.15^{+0.17}_{-0.10}$ and an orbital inclination $i = 64^{+4°}_{-7}$.

Malhotra & Minton (2008) explored the system's habitability. The two planets detected would resonantly excite large orbital eccentricities on a putative Earth-type planet, driving it out of the habitable zone. The existence of an additional inner planet of $\gtrsim 0.3 M_\oplus$ at $\lesssim 0.1$ AU would suppress the eccentricity perturbation and greatly improve the prospects for habitability. Thus, the planetary architecture of a potentially habitable system, with two terrestrial planets and two Jovian planets, could bear close resemblance to our own. Wang et al. (2009) studied the system's dynamical stability and possible formation history.

MOA–2007–BLG–192 Bennett et al. (2008) derived a best-fit light curve including both the parallax and finite source effects, giving the parameters listed in Table 5.1. Adaptive optics images taken with VLT–NACO are consistent with a lens star that is either a brown dwarf or a star at the bottom of the main sequence.

MOA–2007–BLG–400 Dong et al. (2009a) discovered this planet in a high-magnification event with peak magnification $A_{max} \sim 600$. The small primary lens/source impact parameter resulted in a strong smoothing of the peak of the event. The angular extent of the region of perturbation due to the planet is significantly smaller than the angular size of the source, and as a result the planetary signature is also smoothed out by the finite source size. Thus, the deviation from a single-lens fit is broad and relatively weak. The derived mass ratio is $q = 2.5 \times 10^{-3}$, but the planet–star projected separation is subject to a strong degeneracy, leading to two indistinguishable solutions that differ in separation by a factor of ~ 8.5.

OGLE–2007–BLG–368 Sumi et al. (2010) reported this event, and derived the parameters given in Table 5.1.

MOA–2008–BLG–310 Janczak et al. (2010) modeled this event with a large lens distance, $D_L \gtrsim 6$ kpc which, if the primary lens is a star, would make it the first planet found within the distant Galactic bulge.

MOA–2009–BLG–319 Miyake et al. (2011) discovered this event from the high-cadence monitoring of the MOA–II survey, allowing its identification as a high magnification event 24 h prior to its peak. The planetary signal was observed by 20 different telescopes.

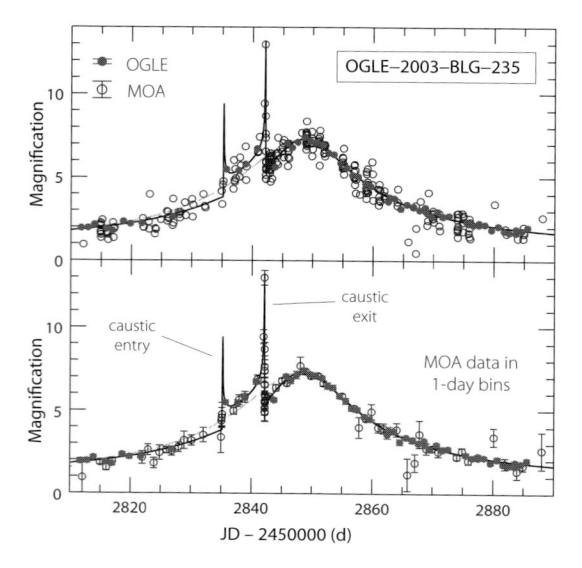

Figure 5.14: The first microlensing planet system, OGLE–2003–BLG–235, with data from OGLE and MOA over a period of about 80 days during 2003. The bottom panel shows binned data. Caustic crossings (entry and exit) occur on days 2835 and 2842. The binary and single lens fits are shown by the solid and fainter (lower) lines respectively. From Bond et al. (2004, Figure 1), reproduced by permission of the AAS.

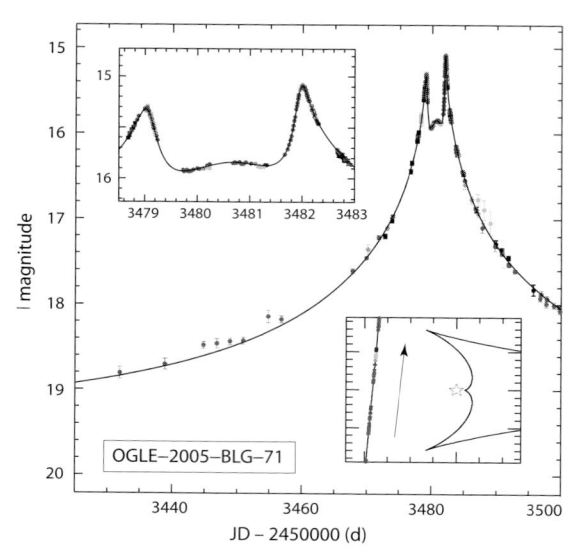

Figure 5.15: Light curve for the second microlensing planet system, OGLE–2005–BLG–71. The upper inset shows an enlargement of the planetary anomaly near peak magnification. The triple peak indicates that the source passed three cusps of a caustic (lower inset), the middle one being relatively weak. From Udalski et al. (2005, Figure 1), reproduced by permission of the AAS.

5.9.2 Statistical results

Although the number of planets detected by microlensing is currently small, the systematic approach to the detection and follow-up of high-magnification events provides the basis of an unbiased sample from which the absolute planet frequency can be inferred over the relevant ranges of planet/star mass ratio and projected separation.

Gould et al. (2010) constructed such a sample based on six planets detected from intensive follow-up observations of high-magnification ($A > 200$) microlensing events during 2005–2008. The sampled host stars have a typical mass $M_\star \sim 0.5 M_\odot$, with maximum planet detection sensitivity corresponding to deprojected separations roughly three times that of the 'snow line'. At the mean mass ratio $q = 5 \times 10^{-4}$ they derived

$$\frac{d^2 N_p}{d\log q \, d\log s} = (0.36 \pm 0.15) \, \text{dex}^{-2} \,, \tag{5.35}$$

with no significant deviation from a flat (Öpik's law) distribution in log projected separation s.

Their inferred planet frequency is a factor seven larger than that derived from Doppler studies at a factor ~25 smaller star–planet separations (i.e., $P \sim 2 - 2000$ d). The difference is nevertheless reasonably consistent with the extrapolated gradient derived from Doppler studies, suggesting a universal separation distribution across a factor 100 in star–planet distance, 100 in mass ratio, and factor two in host mass.

If all planetary systems were analogues of the solar system, their sample would have yielded 18.2 planets (11.4 Jupiters, 6.4 Saturns, 0.3 Uranus, 0.2 Neptunes) including 6.1 systems with two or more planets. This compares with six planets including one two-planet system in the observational sample. This in turn implies a first estimate of 1/6 for the frequency of solar-like systems.

5.10 Summary of limitations and strengths

Limitations In the early days of microlensing planet searches, the rarity and non-repeatability of the microlensing phenomenon were seen as something of a disadvantage. Without question, the probability of an individual planet lensing event occurring for an arbitrary background star, even in the Galactic bulge, is very small, of order 10^{-8} or less, such that hundreds of millions of stars must be monitored to detect the few that might be microlensing at any time. Even then, planetary deviations in the light curves are short-lived, of order hours or days, and the most crucial features easily missed due to observational constraints. Nevertheless, current survey and follow-up programmes are now delivering significant numbers of planetary events.

A planet, once found, will typically be very distant (of order a few kpc). The distance itself may be difficult to estimate without additional constraints such as the microlens parallax, or unless the host star is visible. Planet parameters, such as mass and orbital radius, scale with

5.10 Summary of limitations and strengths

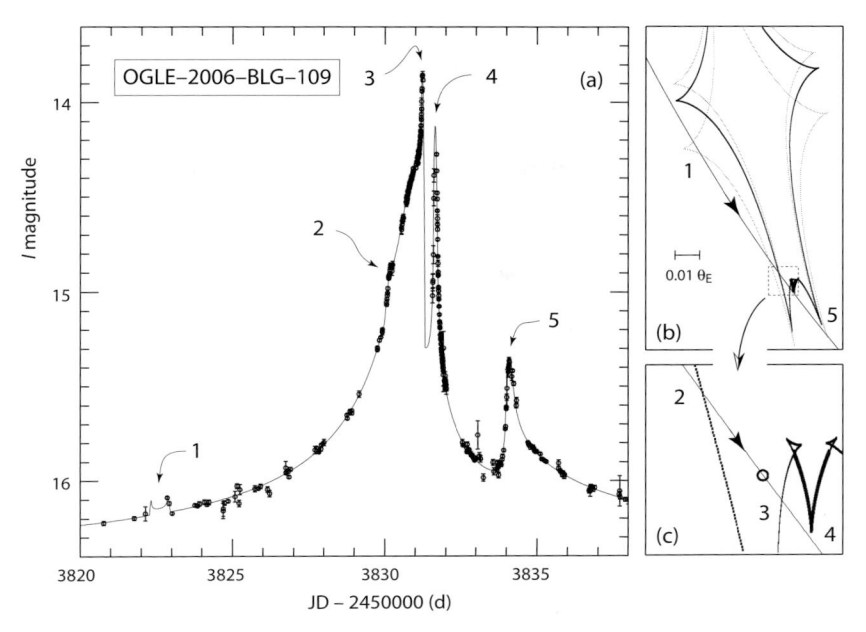

Figure 5.16: The two planet system OGLE–2006–BLG–109: (a) the light curve, with all data points from the OGLE, MicroFUN, MOA, PLANET, and RoboNet telescopes; (b) and inset (c): the reconstructed path of the source, whose size is indicated by the circle in (c), is shown by the arrowed lines. The heavier line shows the position of the combined two-planet caustic at the peak of the event (feature 3); the lighter diamond-shaped curves show the combined caustic at two other epochs corresponding to features 1 and 5 (the three span about 10 days). The five numbered features are caused by the source approaching or crossing the combined caustic, whose shape and orientation change mainly because of the orbital motion of the outer planet. The majority of the caustic is due to the outer planet. The additional cusp (the thick line in c) corresponds to feature 4, and is due to the inner planet. The horizontal bar in (b) shows the angular scale of $0.01\,\theta_E \sim 15\,\mu as$. Adapted from Gaudi et al. (2008, Figure 1).

the (possibly uncertain) properties of the host star, and some parameter degeneracy may also be present.

Single measurement epoch While planetary deviations captured in microlensing light curves never repeat, at least on any relevant time scale due to the rarity of suitable alignments, this is not a barrier to interpretation. Many other astrophysical phenomena (supernovae, γ-ray bursts, detailed accretion events) never repeat in the same object, and plausibility is more a matter of signal-to-noise and repeatability as a class. The wealth of diagnostic data evident from the high-magnification events, in particular, should ensure that the technique will grow further in importance.

Strengths Given suitably high monitoring frequency and high photometric accuracy, the strengths of this detection method, which will be further augmented if space-based measurements are carried out in the future, include its sensitivity to low-mass planets, in principle down to $M_\oplus$ and below (precise practical lower limits depend on source structure and various geometrical alignment details). To first order, the amplitude of the light curve deviation (although not the detection probability) is independent of the planet mass, remaining large even for low planet masses, and limited only by finite source size effects.

The method's greatest sensitivity covers the lensing zone, roughly between 0.6–1.6 AU. For low-mass main sequence stars, this separation coincidentally implies that microlens-detected planets provide particularly important constraints on the frequency of planets in the habitable zone. Multiple planetary systems are also detectable for the same microlensing event, either via very high magnification events with sufficiently small impact parameters passing the central caustic and thus carrying the signature of all planets in the system, or through the chance passage through two or more planetary caustics.

Searches are sensitive to lens systems anywhere along the sight line to the Galactic bulge, with a maximum sensitivity for a lens position roughly halfway to the source. It is a technique largely unbiased in terms of host star properties, whether in terms of spectral type or activity: planets and their host stars should therefore be found in proportion to their actual frequency in the Galaxy disk. As a search technique it remains effective out to very large (kpc) distances, with the capability of detecting planets in M31 being considered (Baltz & Gondolo, 2001; Chung et al., 2006; Ingrosso et al., 2009; Calchi Novati et al., 2010).

The method should also detect free-floating planet mass bodies. Since these could include planets ejected from their host system by planet–planet interactions, statistics on their occurrence should provide further constraints on planet formation and evolution.

5.11 Future developments

Ground-based Next-generation microlensing experiments expect to operate on different principles from the present two-step mode (surveys with alerts and a follow-up search mode). The goal is for wide-field ($2° \times 2°$) cameras on 2-m telescopes spaced around the Earth to monitor several square degrees of the Galactic bulge once every 10 min, without interruption. This higher cadence is expected to find 6000 events per year, a factor 10 increase over the present rates. All will automatically be monitored for planetary perturbations by the search survey itself, in contrast to the roughly 50 events monitored per year at present. Together, these improvements will yield a 100-fold increase in the number of events probed and, correspondingly, in the number of planet detections (Gould et al., 2007; Han, 2007b). Wide separation planets, and free-floating planets, could also be detected and distinguished more efficiently (Han, 2009).

The rationale for a small (2-m) telescope in the Antarctic, providing more continuous target monitoring under improved atmospheric conditions, especially important for the fleeting passages of the smallest Earth-like planets, has also been discussed in the general framework of astronomy from the Antarctic (Yock, 2006).

Space based A space mission would circumvent two key problems for ground observations. The first is the extreme crowding of main sequence source stars in the Galactic bulge, for which blending would be strongly reduced in the absence of atmospheric seeing (although not entirely suppressed, Han 2005b).

Space-based observations would also give uninterrupted photometric time coverage over hours or days, which is difficult and frequently impossible from the ground (Peale, 2003).

Microlensing Planet Finder (MPF) was a proposed Discovery-class mission (Bennett et al., 2004), succeeding the earlier concept of the Galactic Exoplanet Survey Telescope (GEST, Bennett et al., 2003). With a mirror diameter 1.1 m, pointing stability 24 mas, spectral range 600–1600 nm in two bands, and mission duration 3.7 years, it would continuously view two 1.3 square degree fields in the Galactic bulge. Experiment goals were the detection and measurement of $90f$ planets where f is the average number of planets per star with the same mass ratio as Earth ($q = 3 \times 10^{-6}$) at separations 1–2.5 AU; $200g$ planets where g is the corresponding number around M dwarfs ($q = 10^{-5}$); $2800j$ Jupiters and $500s$ Saturns; $25n$ free-floating planets where n is the number of free-floating Earth mass planets for every star in the Galaxy; and $66z$ terrestrial and $3300z$ gas giants if a fraction z have similar mass ratios and separations as in the solar system. Detailed simulations were reported by Bennett & Rhie (2002).

The 2010 US Decadal Survey Report (Blandford et al., 2010) proposes, as its highest priority for new large space activities, a Wide Field Infrared Survey Telescope (WFIRST). This aims to combine two distinct scientific objectives within one instrument: a dark energy probe targeting cosmic shear measurements based on the JDEM–Omega proposal (Gehrels, 2010), and a microlens exoplanet survey based on MPF. The report notes that '*A 2013 new start should enable launch in 2020*'.

A similar ESA dark energy proposal, the 1.2-m Euclid mission, is also considering a 3–12 month microlensing programme (Beaulieu et al., 2010). Euclid is competing for one of two medium-class launch opportunities scheduled for 2017–2018.

6

Transits

After the first transiting exoplanet was observed in 1999, other systems discovered from radial velocity surveys were also monitored to look for possible transits. Surveys from the ground and from space were quickly set up to carry out 'blind searches' for new planets from their periodic transit signatures alone.

Transiting planets are of particular importance because their light curves provide an estimate of their radii. Densities follow from their mass, which in turn gives a first estimate of their composition. Further probing of the planet's structural and atmospheric properties are accessible from photometry and spectroscopy during the transit, and during the secondary eclipse when the planet passes behind the star.[1]

By the end of 2010, some 70 transiting planets were known, searches from ground and space were still intensifying, and new properties of the planetary population – both individual and statistical – were rapidly unfolding.

This chapter covers sequentially the search for transiting planets, their characterisation from their light curves, the search for reflected light, second-order effects on the light curves and transit times, and follow-up photometry and spectroscopy during the transit and secondary eclipse. A high proportion of transiting planets turn out to be close-in 'hot Jupiters', and a review of their general properties is also covered.[2]

---

[1] An eclipse is the (partial) obscuration of one celestial body by another. When of very different angular size, the term *transit* refers to the smaller (here the planet) moving in front of the larger (the star), at the time of inferior conjunction; an occultation, or *secondary eclipse*, refers to the planet passing behind the star (superior conjunction). 'First contact' marks the start of the transit, when the projected outer rim of the planet makes contact with the projected outer rim of the star, and similarly for fourth contact on exit. Second and third contact are the times when the projected planet lies just inside the projected rim of the star, on ingress and egress. Grazing transits and secondary eclipses occur when the projections never fully overlap.

[2] *Hot Jupiters* are loosely defined as Jupiter-mass planets at $a \lesssim 0.1$ AU, significantly heated by stellar irradiation. *Very hot Jupiters* have been defined as those with $P < 3$ d (Beatty & Gaudi, 2008), or $a \lesssim 0.025$ AU (Ragozzine & Wolf, 2009).

6.1 Introduction

Given a suitable alignment geometry, light from the host star is attenuated by the transit of a planet across its disk, with the effect repeating at the orbital period. The probability of observing such a transit for any given star, seen from a random direction and at a random time, is extremely small. The effect being sought is also small: a planet with $R \sim R_J$ transiting a star of $1 R_\odot$ results in a drop of the star flux of $(\Delta F/F) \simeq 1.1 \times 10^{-2}$, or around 0.01 mag. For planets of Earth or Mars radius, $\Delta F \simeq 8.4 \times 10^{-5}$ and 3×10^{-5} respectively. Depths of up to 7% might occur for M dwarfs (Haghighipour et al., 2010), and significantly more for planets around white dwarfs (Drake et al., 2010; Agol, 2011; Faedi et al., 2011).

The first exoplanet transit, HD 209458, was observed by Henry et al. (1999, 2000) and independently by Charbonneau et al. (2000). The latter observed two transits (Figure 6.1), of duration 2.5 h and a depth of 1.5%, at an interval consistent with the known orbit. It gave the first confirmation that Jupiter-mass planets in close orbits about their host stars have radii and densities comparable to the gas-giants of our own solar system.

Since then, the number of known transiting planets has grown. A few of the brightest have been found in the same way as HD 209458b, by photometric follow-up of known Doppler planets at times of inferior conjunction as estimated from the spectroscopic orbit (Kane, 2007; Kane et al., 2009). More are now being found using small-aperture, wide-field imaging systems based on commercial optics of modest cost, surveying the entire sky for prominent transits of the brightest stars.

Ground-based searches are able to discover transits with depths up to about $(\Delta F/F) \simeq 1\%$, revealing gas-giant planets around stars frequently bright enough for radial-velocity confirmation and mass measurements with 2m-class telescopes, or for study of their atmospheric transmission and emission spectra from space-based observations, notably using HST and Spitzer.

Surveys from space, beyond the effects of atmospheric seeing and scintillation, are discovering planets with transit depths of a few times 10^{-4}, extending detectable exoplanet masses down to just a few $M_\oplus$.

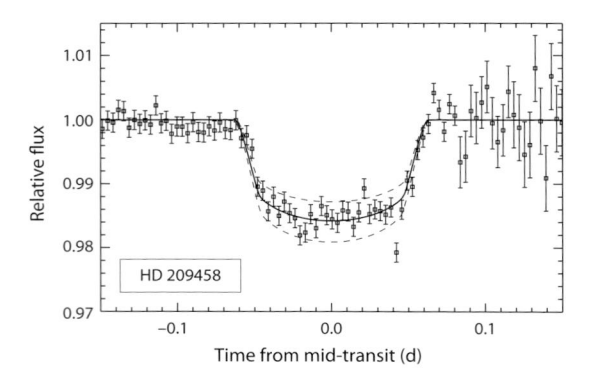

Figure 6.1: *The first detected transiting exoplanet, HD 209458, showing the measured flux versus time. Measurement noise increases to the right due to increasing atmospheric air mass. From Charbonneau et al. (2000, Figure 2), reproduced by permission of the AAS.*

6.2 Transit searches

The increasing majority of transiting planets are being found from dedicated wide-angle searches. Since there is little to indicate *a priori* which stars may have planets, which of those that do might be oriented favourably for a transit to be observed, and when or how frequently such transits may occur, surveys simply monitor large numbers of stars, simultaneously and for long periods of time, searching for the tiny periodic drops in intensity that might be due to transiting planets.

Initially, there was considerable optimism of finding large numbers of exoplanets from dedicated transit searches (e.g. Horne, 2001; Gillon et al., 2005). But the stringent observational requirements quickly became evident: the need for dedicated telescopes, the highest photometric precision, stable instrumentation with low systematic noise, and procedures for handling large number of false positive detections, either due to isolated or blended eclipsing binary systems, or to noise (Brown, 2003). This in turn called for optimal follow-up strategies, with access to suitable high-accuracy spectroscopic instruments (Tingley & Sackett, 2005).

A rise in the discovery rate since 2005 followed as these techniques were mastered: after just three discoveries in total to the end of 2003, eight were found in 2004–05, 26 in 2006–07, and 27 in 2008–09 (a comparable behaviour over time was seen for the early radial velocity discoveries, cf. Pont 2008). Search efficiency is now such that objects are even found independently from different surveys (WASP–11 b ≡ HAT–P–10 b; HAT–P–14 b ≡ WASP–27 b).

Survey yields Predicting survey yields, taking account of telescope size, exposure, wavelength, window function, planet frequencies, stellar distributions and interstellar extinction, has been tackled in various studies (most recently Beatty & Gaudi, 2008; Heller et al., 2009;

> **Early studies:** Detection of an exoplanet by measuring the photometric signature of the planetary transit across the face of the star was mentioned by Struve (1952). The possibility was developed by Rosenblatt (1971), who proposed detecting the event's colour signature as a result of limb darkening, and who considered the effects of stellar noise sources (intrinsic stellar variations, flares, coronal effects, sun spots, etc.) and Earth atmospheric effects (air mass, absorption bands, seeing, and scintillation). Further developments were brought by Borucki & Summers (1984) and Borucki et al. (1985).
>
> Even before the detection of the first exoplanet in 1995, and before the detection of the first transiting planet in 1999, the method was considered as one of the most promising means of detecting planets with masses significantly below that of Jupiter, with the detection of Earth-class (and hence habitable) planets quickly seen as being within its capabilities (Schneider & Chevreton, 1990; Hale & Doyle, 1994; Schneider, 1994; Heacox, 1996; Janes, 1996; Schneider, 1996; Deeg, 1998; Sartoretti & Schneider, 1999).

von Braun et al., 2009b), including a consideration of the gains to be made by combining the various survey results (Fleming et al., 2008).

Practical searches vary in their details of execution, but follow broadly similar principles: tens of thousands of relatively bright stars ($V \lesssim 13$ mag) are monitored over weeks or months, possible transits identified in the photometric time series, stellar binaries eliminated, and confirmation of the planetary nature of the companion made only after radial velocity observations.

Outline of the data processing In the first reduction step of ground surveys such as HAT and SuperWASP, automatic field recognition matches observed stars to a reference catalogue (Tycho 2 or 2MASS), providing an astrometric solution to 1–3 arcsec (Pál & Bakos, 2006).

Magnitudes at each epoch are based on the optimised techniques of aperture photometry (Gilliland & Brown, 1988; Kjeldsen & Frandsen, 1992; Everett & Howell, 2001), augmented by difference imaging (Alard & Lupton, 1998). Error sources include photon noise, atmospheric effects including scintillation (§6.3), and differential extinction. Photometric zero-points are determined with reference to USNO–B1.0 or 2MASS.

Correct handling of non-Gaussian noise (trends and other systematics) is paramount for the detection of transits where the depth of the light drop may be only a few mmag. Typically at the end of each observing season, correlated systematic errors are removed from the entire data set (e.g. using SysRem, Tamuz et al., 2005; Smith et al., 2006), along with other trends, e.g. using the trend-fitting algorithm (TFA, Kovács et al., 2005) or external parameter decorrelation (EPD, Bakos et al., 2010b). Furthermore, an empirically-determined variance may be added to match observed noise properties to facilitate a suitable statistical treatment (Collier Cameron et al., 2006).

6.2 Transit searches

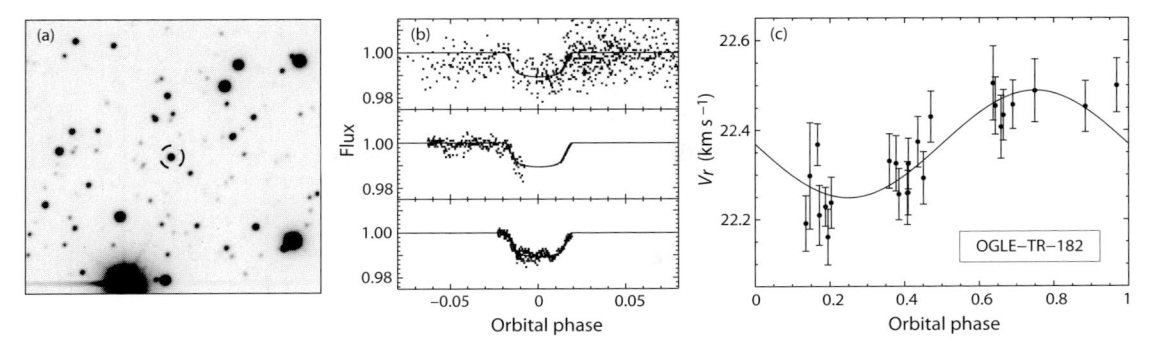

Figure 6.2: OGLE–TR–182 ($P = 3.98$ d, $I = 15.9$ mag), representative of the state-of-the-art in OGLE transit searches: (a) VLT I-band image (50 arcmin square field); (b) light curves from the OGLE telescope (top), and VLT–FORS1 (bottom two); (c) radial velocity from VLT–FLAMES/UVES. From Pont et al. (2008b, Figures 1–3), reproduced with permission © ESO.

Candidate identification Candidate searches typically adopt a modified box-least squares algorithm (Kovács et al., 2002) with a coarse search grid to identify epoch, period, depth and duration of the strongest signals, which can then be refined using analytically-differentiable transit profile models (Protopapas et al., 2005). Literature on search techniques and their comparison includes, e.g., Aigrain & Favata 2002; Tingley 2003; Aigrain & Irwin 2004; Tingley 2004; Moutou et al. 2005; Weldrake & Sackett 2005; Collier Cameron et al. 2006; Schwarzenberg-Czerny & Beaulieu 2006; Régulo et al. 2007; Carpano & Fridlund 2008; Ford et al. 2008b; Carter & Winn 2009.

The identification of possible transiting planets is complicated by the presence of much larger numbers of stellar mass binaries (Willems et al., 2006). Ellipsoidal (binary star) variables are rejected from the out-of-transit light curves, and probable giants can be eliminated on the basis of their reduced proper motions (e.g. using 2MASS, Collier Cameron et al., 2007b). A first estimate of stellar mass and radius, and planet radius and impact parameter, are then derived from light curve models using an appropriate limb darkening description. Possible transiting objects are then subjected to more careful photometry on the ground to exclude faint background eclipsing binaries.

Candidate confirmation Only the most promising candidates are finally subjected to radial velocity follow-up measurements: a small radial velocity amplitude, combined with the transit signature implying that the inclination $i \simeq 90°$, together ruling out stellar mass eclipsing binaries with small inclination. Double-lined (stellar mass) binaries and fast rotators can typically be rejected after a single observation, narrow single-lined stellar binaries after a second, while objects with planet-like radial velocity amplitudes are observed until ten or so measurements confirm their sub-stellar mass.

Star and planet parameters Host star parameters can be determined from the radial velocity observations, e.g. deriving T_{eff} and $\log g$ from diagnostic spectral lines (such as Hα, Na I D and Mg I b), and thereafter $M_\star$ and $R_\star$ from appropriate stellar evolutionary models. Planetary radii can then be estimated by a combined χ^2 fit to the photometric and radial velocity measurements together. Details depend on orbital eccentricity, and the adopted model for limb darkening.

Exoplanet transit data bases Various web-based databases collate results of exoplanet transit measurements. These include the NASA/IPAC/NExScI Star and Exoplanet Database, NStES (von Braun et al., 2009a) at nsted.ipac.caltech.edu; the Exoplanet Transit Database of the Czech Astronomical Society, ETD (Poddaný et al., 2010) at var.astro.cz/ETD; the Exoplanet Transit Parameters data base, ETP, at www.inscience.ch/transits, and the Amateur Exoplanet Archive (AXA, brucegary.net/AXA/x.htm) for data obtained by amateur observers.

6.2.1 Large-field searches from the ground

The main features of the major transit survey experiments are given in this section (which does not however list all such efforts ongoing).

Appendix D lists the discoveries to date. Although new transiting systems are now being announced continuously, the table is included to provide a synopsis of experiment successes, typical values of V and $d_\star$, and the range of parameters ($M_\mathrm{p}, R_\mathrm{p}, P_{\mathrm{orb}}, a$) probed.

HAT/HATNet The Hungarian Automated Telescope project started in 2003 with a single telescope. HAT-north is now a network of six automated northern hemisphere telescopes, four stationed at the Whipple Observatory (Arizona) and two on Mauna Kea (Bakos et al., 2002, 2004). Equipped with 2k×2k CCDs, they provide an 8°×8° field with a pixel size of 14 arcsec. They achieve a precision of 3–10 mmag at $I \sim 8-11$.

From late 2009, six further telescopes at three southern sites (Las Campanas, Chile; Siding Spring, Australia; HESS γ-ray site, Namibia) form the operational HAT-south network, together permitting near continuous monitoring of selected fields (Bakos et al., 2009a). Its sensitivity, to $R \simeq 14$ across a $128°$ field, leads to a predicted detection rate of around 25 per year.

Details of the data processing steps applied to the HAT data are given by, e.g., Bakos et al. (2010b), while procedures to follow up candidates with photometry and spectroscopy are described by Latham et al. (2009).

Their 26 discoveries to date (HAT-P-1 to 26, Appendix D) comprise a rich diversity of systems. These include HAT-P-2 b, their highest eccentricity ($e = 0.52$; Bakos et al., 2007a); HAT-P-7 b, in which the host star shows tidally-distorted ellipsoidal light-curve variations (Welsh et al., 2010); HAT-P-13 (Szabó et al., 2010) and HAT-P-17 (Howard et al., 2010), systems comprising both inner transiting and outer non-transiting planets; HAT-P-14 b in a retrograde orbit (Winn et al., 2011); HAT-P-23 b, their shortest orbital period ($P = 1.2$ d, $a = 0.02$ AU; Bakos et al., 2010a); and HAT-P-26 b, their lowest mass ($0.06 M_J$; Hartman et al., 2011a).

OGLE Starting with the third phase of the Optical Gravitational Lensing Experiment (OGLE–III, §5.8), a search for planetary transits was initiated using their 1.3-m telesope at Las Campanas (Udalski et al., 2002a). In 2001, photometric observations of three fields towards the Galactic centre (800 epochs per field) were collected on 32 nights over 45 days. Out of 5 million stars monitored, 52 000 with photometry better than 1.5% were analysed for flat-bottomed eclipses with depth $\lesssim 0.08$ mag. Altogether, 46 low-luminosity transiting objects were detected. For 42 (185 transits in total), multiple transits were observed, allowing a determination of the orbit period. A total of six campaigns were made between 2001–06 (Udalski et al., 2002b, 2004; Udalski, 2007). Planet candidates were subject to radial velocity follow-up (Dreizler et al., 2002, 2003; Konacki et al., 2003b; Bouchy et al., 2005b; Pont et al., 2005).

These resulted in the first of the confirmed OGLE transiting exoplanets: the hot (1900 K) 1.2-d period $\sim 1 M_J$ mass OGLE–TR–56 (Konacki et al., 2003a; Sasselov, 2003); its secondary eclipse of 0.036% was measured with VLT and Magellan by Sing & López-Morales (2009). Over subsequent years, improved procedures to distinguish between stellar and planet companions have been developed (e.g., Silva & Cruz, 2006), and further transiting planets discovered (Appendix D): chronologically, OGLE–TR–113 and OGLE–TR–132 (Bouchy et al., 2004), OGLE–TR–111 (Pont et al., 2004), OGLE–TR–10 (Konacki et al., 2005), OGLE–TR–211 (Udalski et al., 2008) and OGLE–TR–182 (Pont et al., 2008b, see also Figure 6.2). By 2006, of just six transiting planets known in total, five had been detected by OGLE.

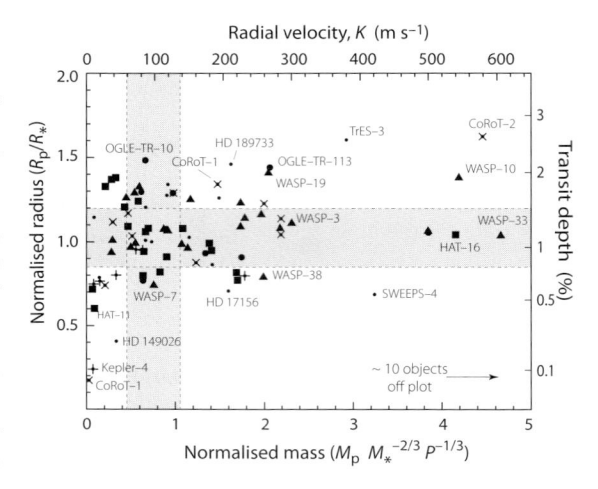

Figure 6.3: Mass–radius diagram for transiting planets, as of 2010 November 1 (Appendix D). Symbols: ● OGLE, ▲ WASP, ■ HAT, × CoRoT, + Kepler, · other; some outlying objects are labeled. The horizontal axis is $M_p M_\star^{-2/3} P^{-1/3}$ (scaled to $M_J M_\odot^{-2/3} \, 4\,\mathrm{days}^{-1/3}$), proportional to radial-velocity semi-amplitude K (top axis). The vertical axis is $R_p/R_\star$ normalised to $R_J/R_\odot$ (left), related to the transit depth (right axis). Shaded regions show near-threshold zones for photometric detection (horizontal) and spectroscopic confirmation (vertical) for OGLE. Their intersection is especially problematic, requiring large telescopes for confirmation. Updated from Pont et al. (2008b, Figure 4).

A search of the earlier OGLE–II data (1996–2000), also using SysRem for detrending and a parameter search using box least-squares (Snellen et al., 2007), resulted in the discovery of OGLE2–TR–L9 (Snellen et al., 2009b). Many follow-up observations, both photometric and radial velocity, have also been made to improve knowledge of the planet and host star properties.

Detailed analysis of the three final seasons of the OGLE transit survey is still ongoing. Meanwhile, the radial velocity follow-up to date has shown that the vast majority of transit candidates were eclipsing binaries, with a typical rate of one planet per 10–20 eclipsing binaries. A higher rate of planets is found near the detection threshold, although at the expense of more false positives (Figure 6.3).

Comparison of the number of OGLE detections with the number expected on the basis of radial velocity surveys still suffers from small number statistics, but they appear broadly compatible (Gould et al., 2006a; Fressin et al., 2007a).

MACHO/MOA Searches have also been made in the MACHO microlensing project photometry data base (Drake & Cook, 2004; Hügelmeyer et al., 2007), and the 2000–05 MOA observations of 7 million stars (Abe et al., 2005; Fukui et al., 2009), with only MAESTRO–1 b currently of possible planet status (Setiawan et al., 2008c).

6.2 Transit searches

TrES The Trans-Atlantic Exoplanet Survey network employs three small-aperture (0.1 m), wide-field (6°), CCD-based systems with an angular resolution of 11 arcsec per pixel. The emphasis is on bright star transits to facilitate follow-up observations, although eliminating false positives is still a major issue (O'Donovan et al., 2006b; Alonso et al., 2007). The telescopes used are STARE (STellar Astrophysics and Research on Exoplanets, Brown & Charbonneau 2000; Rabus et al. 2007), Tenerife, Canary Islands; PSST (Planet Search Survey Telescope, Dunham et al. 2004), Lowell Observatory, Arizona; and Sleuth, Mount Palomar, California).

Observations are made in Johnson R or Sloan r, together observing the same field almost continuously for 2 month intervals. Images are acquired every 2 min, and binned to 9 min resolution. Photometric precision ranges from $\lesssim 2$ mmag for the brightest non-saturated stars at $R \simeq 8$ to better than 10 mmag at $R \simeq 12.5$.

Discoveries to date are TrES–1 to TrES–4 (Alonso et al., 2004; O'Donovan et al., 2006a, 2007; Mandushev et al., 2007, respectively).

Amongst subsequent investigations of TrES–1, the secondary eclipse has been observed with Spitzer (Charbonneau et al., 2005), and the first inferences about atmospheric composition, temperature, and global circulation have been made (Burrows et al., 2005; Fortney et al., 2005). Transit times have been used to place limits on other planets (Steffen & Agol, 2005; Narita et al., 2007; Rabus et al., 2009a), and the Rossiter–McLaughlin effect has been measured (Table 6.1, Narita et al., 2007).

WASP/SuperWASP The WASP consortium (Wide-Angle Search for Planets) operates two wide-field camera arrays in the northern (La Palma, Canary Islands) and southern (Sutherland, S. Africa) hemispheres (Pollacco et al., 2006; Collier Cameron et al., 2009). The prototype WASP0 monitored 35 000 stars in Draco for two months (Kane et al., 2004, 2005a; Christian et al., 2006), with interruptions to view a field in Pegasus when transits of HD 209458 occurred (Kane et al., 2005b).

Each SuperWASP telescope uses eight 2k×2k CCD cameras on a robotic equatorial mount, forming a mosaic with a field of view of 15° × 30° (RA, dec), and a pixel size of 14 arcsec. During the first two years of full-time operation, starting in 2006 May, the search fields have been primarily located at $\delta \simeq \pm 30°$. Fields are observed if accessible for at least 4 h each night, and the observing season for a given object typically spans 120–150 nights, with data rates up to 100 Gbyte per night. The first 6 months of SuperWASP-north produced light curves of 6.7 million objects with 12.9 billion data points (Christian et al., 2006; Clarkson et al., 2007; Lister et al., 2007; Street et al., 2007; Kane et al., 2008). Long-term rms scatter for non-variable stars, after pipeline treatment, is 4 mmag at $V = 9.5$, degrading to 10 mmag at $V = 12$. Details of the WASP data processing are given by, e.g.,

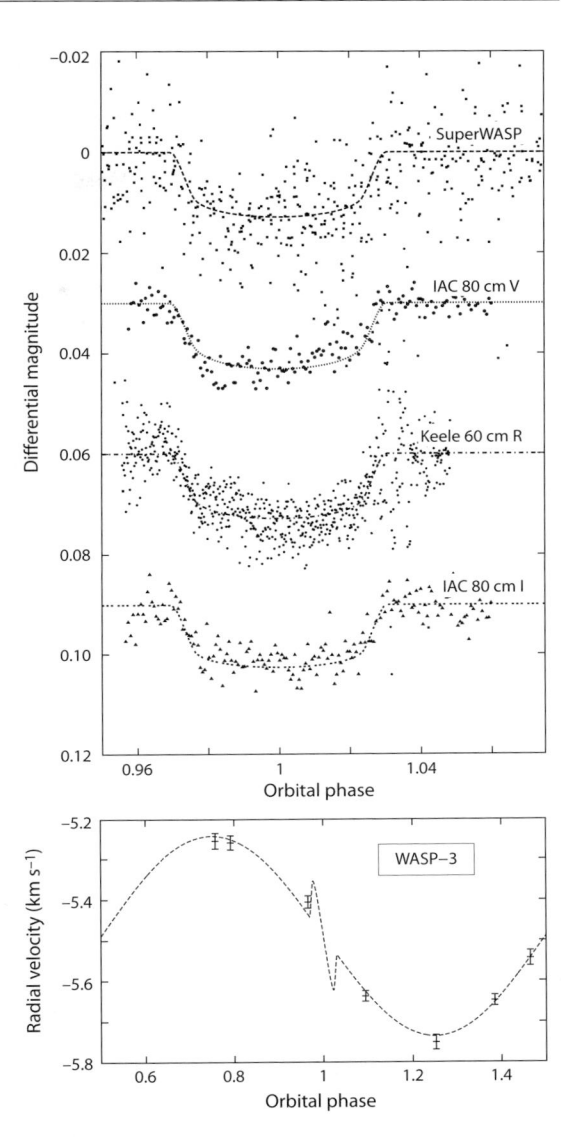

Figure 6.4: WASP–3: simultaneous solution to the SuperWASP-N, IAC 80 V I and Keele R photometry. The orbit is assumed to be circular. Lower panel: fit to the radial velocity data, including the predicted Rossiter–McLaughlin effect (§6.4.11) for the star's $v \sin i = 13.4$ km s^{-1}. From Pollacco et al. (2008, Figure 3), © John Wiley & Sons, Inc.

Pollacco et al. (2006); Collier Cameron et al. (2009). Public release details are described by Butters et al. (2010).

SuperWasp report a success rate of one confirmed planet per five or six selected for radial velocity follow-up (typically using SOPHIE at the 1.9-m OHP telescope, and CORALIE at the 1.2-m Euler telescope at La Silla).

Their 37 discoveries to date (WASP–1 to 38, excluding the false positive WASP–9 b) comprise a rich diversity of systems. These include WASP–8 b, their highest eccentricity planet ($e = 0.31$), moving on a retrograde orbit with respect to the rotation of its host star (Queloz et al., 2010); WASP–17 b, with the lowest density of all known planets ($\rho = 0.09\,\mathrm{Mg\,m^{-3}}$; Anderson et al., 2010b); WASP–19 b, their shortest orbital period ($P = 0.79$ d, $a = 0.016$ AU Hebb et al., 2010); WASP–21 b, with a probable thick disk host star (Bouchy et al., 2010); WASP–29 b, their lowest mass ($0.24 M_J$ Hellier et al., 2010); and WASP–33 b, the first planet orbiting a δ Scu host star (Herrero et al., 2011), in which general relativistic precession (§6.4.13) exceeds that of the Sun–Mercury system by a factor $\sim 10^{10}$ (Iorio, 2011).

Example light curve results, for WASP–3 and WASP–17, are shown in Figures 6.4 and 6.5.

XO Also aiming to detect bright star transits, XO has observed since 2003 concentrating on a large-area survey using two 0.11-m telescopes located on the summit of Haleakala, Maui (McCullough et al., 2005; Burke et al., 2007). The 1k×1k CCD system operates in drift-scan mode with an instantaneous field of $7° \times 7°$. In its first year of operation, for example, XO observed 7% of the sky in $7°$ wide strips, from $0°$ to $+63°$ in dec and centred every 4 h in RA, providing photometry of 100 000 stars at more than 1000 epochs per star, and a precision of 10 mmag per measurement for $V < 12$ mag.

Discoveries to date are XO–1 to XO–5 (McCullough et al., 2006; Burke et al., 2007; Johns-Krull et al., 2008; McCullough et al., 2008; Burke et al., 2008, respectively).

6.2.2 Other searches from the ground

This summary is restricted to programmes that have detected planets to date.

MEarth project The MEarth project uses eight identical 0.4-m automated telescopes (two operational since January 2008) in a single enclosure at Mt Hopkins, to monitor 2000 nearby M dwarfs with masses between $0.10 - 0.35 M_\odot$, selected from nearby stars with large proper motion (Nutzman & Charbonneau, 2008; Irwin et al., 2009); the low mass of M dwarfs means that a survey should detect transits down to a few $M_\oplus$ in their habitable zones. It has resulted in the detection of a $6.5 M_\oplus$ transiting planet around the 13 pc distant M dwarf GJ 1214 (Charbonneau et al., 2009).

Siding Spring Observatory WFI The eight CCD Wide Field Imager (WFI) of the Siding Spring Observatory, Australia, was used for a 52×52 arcmin^2 transit survey towards Lupus in 2005–06 June. The survey resulted in the discovery of Lupus–TR–3 (Weldrake et al., 2008a); of their other candidates, Lupus–TR–1 and 4 are blended systems, while Lupus–TR–2 is a binary (Bayliss et al., 2009b). A deeper survey, SuperLupus, doubling the number of images of the same field from 1700 to 3400, was completed in 2008 (Bayliss et al., 2009a).

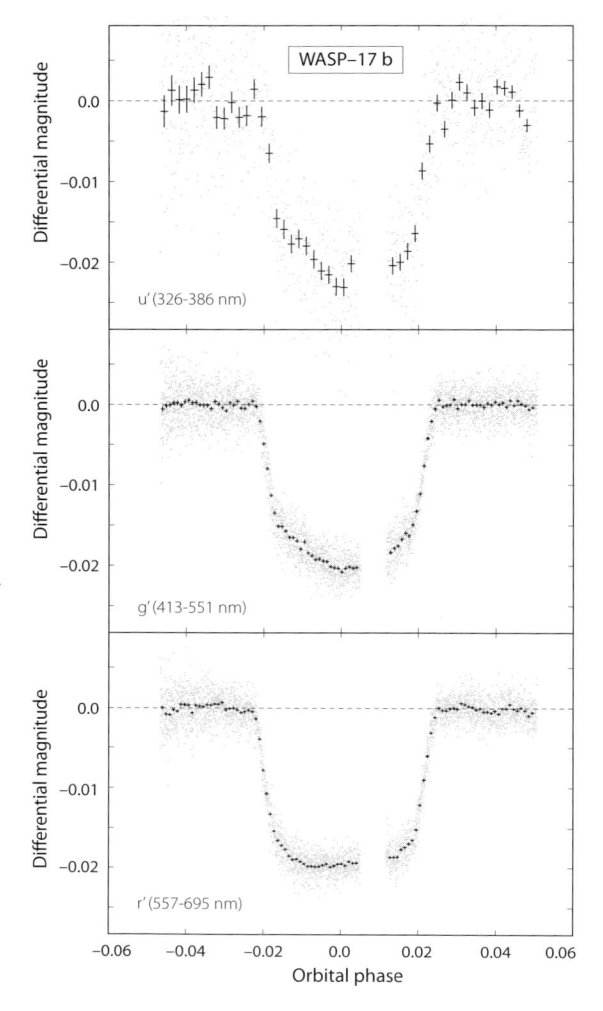

Figure 6.5: WASP–17 ($V = 11.6$ mag, $P = 3.74$ d) observed with ULTRACAM at the 3.5-m NTT on 2010 April 26. The instrument observes simultaneously in three bands. The observations (grey points) were acquired at 7.8-s integration in g and r, and at 31-s in u, binned here (solid crosses with 1σ standard errors) at 311 s in g and r, and 623 s in u. Data gaps are due to zenith-pointing constraints. WASP–17 is the lowest density of all known planets, $\rho = 0.09\,\mathrm{Mg\,m^{-3}}$ or some factor of 10 lower than that of Jupiter, giving it the largest atmospheric scale height, and therefore a good target for transmission spectroscopy. From Bento et al. (2011, in preparation).

Radial velocity discoveries Several transiting planets (Appendix D) were first detected from radial velocity observations, and at least two radial velocity programmes have subsequently focused on short-period giant planets, which are more likely to transit their host star (da Silva et al., 2006). The ELODIE metallicity-biased programme combines high-precision radial velocities with the 1.93-m telescope of OHP, and high-precision photometry from the 1.20-m telescope, and resulted in discovery of the transiting HD 189733 b (Bouchy et al., 2005c). Its short period of 2.219 d, and its large transit depth of 3% ($M \simeq 1.15 M_J, R \simeq 1.26 R_J$) make it one of the prime targets for atmospheric studies with HST/Spitzer. The N2K programme has also discovered one transiting planet using a similar approach, HD 149026 b (Sato et al., 2005a).

Long-period planets Planets with $P \gtrsim 10$ d are not easily followed photometrically after radial velocity detection because the geometric transit probability becomes very small, and ground-based surveys which rely on photometry folding rapidly become incomplete.

Since 2002, the Transitsearch.org network has monitored these cases, using small telescopes and amateur participation, based on possible transit predictions (Seagroves et al., 2003; Shankland et al., 2006; Kane, 2007; Kane et al., 2009). The network observed a transit of the high-eccentricity 21-d orbital period HD 17156 b on 2007 September 9–10 (Barbieri et al., 2007), for which the favourable geometry from the radial velocity orbit had given an *a priori* transit probability of 13%, and a transit time predicted to within a few hours.

The 111-d orbital period of HD 80606 b is even more extreme. It was discovered from its radial velocity signature as a giant planet on an extremely eccentric orbit ($4 M_J, e = 0.93$; Naef et al., 2001). The secondary transit around pericentre was discovered from Spitzer 8 μm observations (Laughlin et al., 2009), and the two data sets together implied an orbit inclination $i \simeq 90°$, and an associated 15% probability that the planet would transit. With a corresponding transit window predicted for 2009 February 14, three groups independently observed and published a transit signature (Moutou et al., 2009; Fossey et al., 2009; Garcia-Melendo & McCullough, 2009; Hidas et al., 2010).

6.2.3 Searches in open and globular clusters

A very significant observational effort has been devoted to transit searches in Galactic open clusters and, to a lesser extent, globular clusters.

The scientific motivation includes establishing the effects of stellar density and metallicity on planet occurrence (e.g. Ida et al., 2000; Kobayashi & Ida, 2001; Kenyon & Bromley, 2002a), determining time scales of planet formation and migration, and modeling the effects of stellar encounters on formation and disruption (Fregeau et al., 2006; Malmberg et al., 2007b; Weldrake, 2008; Fragner & Nelson, 2009). No transiting planet has yet been confirmed in either cluster type.

Some of the search programmes operate under their own acronyms: EXPLORE–OC (von Braun et al., 2005); Monitor (Hodgkin et al., 2006); PISCES (Mochejska et al., 2004); and STEPSS (Burke et al., 2004). Typical photometric accuracies are in the range 3–10 mmag.

Open clusters Several groups have targeted Galactic open clusters in their search for transiting planets.[3] The absence of planets found to date is probably not unexpected, given the relatively few stars per cluster (Weldrake, 2008). Pepper & Gaudi (2005, 2006) argued that surveys of the nearest and richest open clusters, with moderate instruments over more than 20 nights (e.g. a Pan-STARRS survey of the Hyades and Praesepe), still have the potential to detect transiting hot Neptunes and hot Earths around low-mass stars. Limits arising from the existence of correlated noise were studied by Aigrain & Pont (2007).

Globular clusters The central core region of 47 Tuc was surveyed in an 8-d observation by HST, which yielded no candidates (Gilliland et al., 2000). Ground-based surveys of the outer regions of two clusters have been made with the ANU 1-m telescope: for 47 Tuc (22 000 stars over 33 nights; Weldrake et al. 2005) and for ω Cen (31 000 stars over 25 nights; Weldrake et al. 2008b).

Based on known planet frequencies, more than 20 might have been expected in the combined (core and halo) surveys of 47 Tuc, with around five expected for ω Cen. The null results suggest either that planet formation has a significant dependency on metallicity, or that the migration or survival of close-in planets as a result of dynamical disruption may be significantly altered in the high star-density cluster environment.

Simulations for 47 Tuc have shown that planets in the dense core region with $a = 1$ AU would survive disruptions by stellar encounters for 10^8 yr, and significantly longer for the short-period planets to which transit surveys are most sensitive (Davies & Sigurdsson, 2001). In the outer regions, planets with $a = 10$ AU should still be intact, with planets at 0.04 AU unaffected by the cluster dynamics (Bonnell et al., 2001).

---

[3] These include NGC 188: 87 h over 45 nights at the Whipple 1.2-m (Mochejska et al., 2008); NGC 1245: 19 nights at the Hiltner 2.4-m (Burke et al., 2006); NGC 2099 (M37): 20 nights at the MMT 6.5-m (Hartman et al., 2008a,b); NGC 2158: 20 nights at the Whipple 1.2-m (Mochejska et al., 2004, 2006); NGC 2301: 14 nights at the Univ. Hawaii 2.2-m (Tonry et al., 2005; Howell et al., 2005); NGC 2660 and 6208 (von Braun et al., 2005); NGC 6633: at the 0.5-m UNSW APT (Hidas et al., 2005); NGC 6791: 8 nights at the NOT 2.5-m (Bruntt et al., 2003), 10 nights at the CFHT 3.6-m and others (Montalto et al., 2007), and 300 h over 84 nights at the Whipple 1.2-m (Mochejska et al., 2002, 2005); NGC 6819: 38 000 stars at the INT 2.5-m (Street et al., 2003); NGC 6940: 50 000 stars over 18 nights at the INT 2.5-m (Hood et al., 2005); NGC 7086: 1000 stars over 12 nights at the DAO 1.8-m (Rosvick & Robb, 2006); NGC 7789: 30 nights monitoring 33 000 stars at 900 epochs at the INT 2.5-m (Bramich et al., 2005; Bramich & Horne, 2006); and the Monitor programme targeting ten star-forming regions and open clusters with ages between 1–200 Myr which started in 2004 using various 2–4-m telescopes worldwide (Hodgkin et al., 2006; Aigrain et al., 2007; Miller et al., 2008).

If short-period gas giant planets formed in globular clusters with comparable frequencies as in the solar neighbourhood, they should be detectable in reasonable numbers in transit surveys. Since metallicity plays a key role in the frequency of gas giants, the globular cluster results suggest that this metallicity trend continues to much lower values than those probed by radial velocity surveys. Soker & Hershenhorn (2007) have argued that planets might be present, but at larger separations typically inaccessible to transit surveys, because of the correlation between metallicity and the quantity $I_e = M_p[a(1-e)]^2$ which concentrates at the higher range of I_e for low metallicity systems.

6.2.4 Future searches from the ground

Other surveys under development or planned, and described in the literature, include:

– SkyMapper Telescope transit search: a 5.7 deg^2 (32 2k×4k) survey for the 1.35-m survey telescope at Siding Spring Observatory, building on experience from the SuperLupus search (Bayliss & Sackett, 2007);

– Pan–Planets will use the 1.8-m Pan–STARRS telescope to cover 21 sq. degrees over four years to $i = 16 - 17$ mag. With 120 h of observations per year and 6 min time sampling, some 100 hot Jupiters could be found (Afonso & Henning, 2007; Koppenhoefer et al., 2009). The more sparsely time-sampled Pan–STARRS–1 multiband 3π area survey may detect single epoch transits from Jupiters transiting M dwarfs (Dupuy & Liu, 2009);

– Subaru Suprime–Cam monitored 100 000 stars in the Galactic plane to assess its suitability for fainter transit surveys to beyond 18 mag (Urakawa et al., 2006);

– the Alsubai project (QA) is planning deployment along the lines of SuperWASP, based on a 5-telescope prototype camera operating in New Mexico.

Antarctic Transit observations from Antarctica would assist two of the typical limitations of ground-based surveys: limited (diurnal-cycle) time sampling, and systematic photometric variations on time scales of hours induced by the atmosphere. Instrument tests and site characterisation, along with simulations, have assessed the suitability of the Antarctic plateau, including Dome C, for photometric time series observations in general, and for planet transit detection in particular (e.g. Viotti et al., 2003; Caldwell et al., 2004; Pont & Bouchy, 2005; Crouzet et al., 2009, 2010b).

Rauer et al. (2008) compared performances with a site in Chile, and a lower-latitude three-site network, concluding that Dome C (at latitude $-75°$), is a prime site for long-duration observations during winter, nonetheless best operated as part of a network of sites.

Of various proposals and site-testing projects, ASTEP (Antarctica Search for Transiting Extrasolar Planets, with leading institutes from F/CH) has conducted first transit observations with a 0.4-m automatic telescope at Dome C, which targets observations to $V = 16$ (Fressin et al., 2005, 2007b; Crouzet et al., 2010a; Daban et al., 2010). ICE–T (International Concordia Explorer Telescope, with leading institutes from D/I/E) proposes use of a dual 0.6-m wide-field Schmidt telescope (Strassmeier et al., 2007).

At the highest point on the Antarctic plateau, Dome A, the Chinese Small Telescope ARray (CSTAR) was installed in 2008, and has observed, automatically, over three Antarctic winters (Yuan et al., 2010). A triple 0.5-m Schmidt telescope system, AST3, is scheduled for operation in 2011.

6.2.5 Searches from space

Ground-based photometry to better than about 0.1% accuracy is complicated by variable atmospheric extinction, while scintillation, the rapidly varying turbulent refocusing of rays passing through the atmosphere, imposes limits at about 0.01%. Extension of the transit method to space experiments, where very long uninterrupted observations can be made above the Earth's atmosphere, are therefore particularly important.

CoRoT The CoRoT satellite, led by CNES (F) with several European partners, was launched on 2006 December 27 into a 900 km altitude, 103-min polar orbit, and is dedicated to both stellar seismology and the search for transiting exoplanets (Auvergne et al., 2009). The telescope has an effective diameter 0.27 m, the field of view is $2°\!.7 \times 3°\!.0$, and the focal plane comprises four 2k×2k CCDs. The measured attitude stability is ~ 0.15 arcsec rms. The polar orbit gives two preferred viewing directions separated by 180°, in the directions of the Galactic centre and anti-centre, each of which can be observed continuously for up to 150 days, with at least five different fields planned over the mission lifetime. In the planet transit search mode, some 12 000 target stars are observed simultaneously.

Stars are sampled with a cadence of either 512 s or 32 s, and the photometric signal is extracted on board using optimised apertures selected for the target stars. A dispersing prism allows brighter stars to be measured in three channels (red, green, blue), assisting discrimination of planet transits from the more chromatic stellar variability. For the remainder, samples are combined into a single white light channel.

The first long observation run, LRc01 towards the Galactic centre in the direction of Aquila, lasted from 2007 May–October (Cabrera et al., 2009). Measured photometric precision is 7.1×10^{-4} at $R = 15$ mag for 512 s sampling (for details of noise properties and related calibration issues, see Aigrain et al., 2009; Alapini & Aigrain, 2009; Mazeh et al., 2009a). Photometric analysis resulted in 42 candidates, of which 26 were settled by mid-2009,

6.2 Transit searches

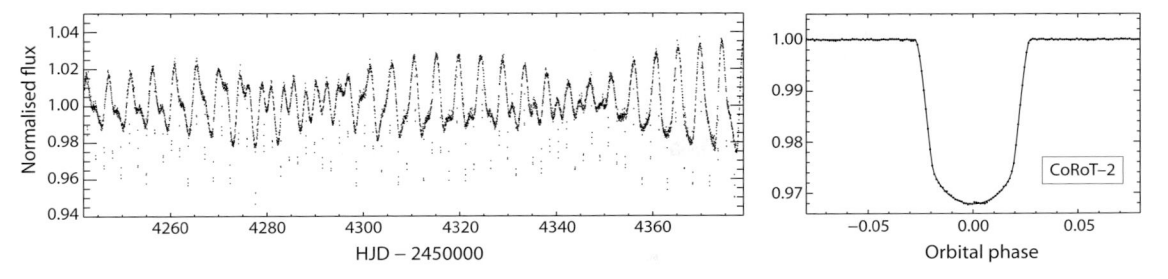

Figure 6.6: CoRoT–2. Left: normalized flux over 150 days, showing 78 orbital periods with a resolution of 34 min. The prominent low-frequency modulation is due to spots on the stellar surface. Right: phase-folded light curve of the 78 transits with a time resolution of 2.5 min. The planetary parameters are $P = 1.742996$ d, $i = 87°\!.84$ and (in combination with radial velocity data) $M_p = 3.31 M_J$, $R_p = 1.465 R_J$, $\rho = 1.31$ Mg m^{-3}. From Alonso et al. (2008a, Figures 1–2), reproduced with permission © ESO.

leading to a subset of confirmed planets and 16 open cases; the non-planets are either binaries, grazing binaries, or contaminating systems where light from an eclipsing binary is blended with light from a nearby star to produce a planet transit-like signal (Almenara et al., 2009). The first long run in the Galactic anti-centre direction, LRa01, ran from 2007 October to 2008 March.

The precision with which planetary parameters can be measured from the photometry relies on the precision of stellar parameters, of which limb darkening is partially degenerate with the orbit inclination. The limb darkening is therefore preferably established independently (Torres et al., 2008), for example using $BVRI$, and JHK_s from 2MASS, which provides the temperature and spectral class, and hence model limb-darkening coefficients. Their Exo-Data data base, providing prior knowledge and other complementary information, is described by Deleuil et al. (2009). Public access details are described by Solano et al. (2009).

As for ground-based programmes (§6.2.1) extensive photometric and radial velocity ground-based follow-up observations are required. CoRoT routinely uses (at least) ESO, OHP (F), IAC (E), McDonald (US), Wise (ISR), HARPS and Leonard Euler time of the Geneva Observatory (CH), the low-resolution spectrograph of the Thüringer Landessternwarte Tautenburg (D), BEST II (D), and ESA's optical ground station in Tenerife (E). Stellar characterisation for the predominantly faint targets employs the 8-m VLT–UVES/GIRAFFE spectrographs (Loeillet et al., 2008a). Transit duration, planet size, and orbital eccentricity can be significantly improved using follow-up photometry on the ground (Colón & Ford, 2009; Deeg et al., 2009).

As of 2010 November (Appendix D), CoRoT has discovered a number of confirmed planets, CoRoT–1 b to CoRoT–17 b (as of 2009 March, its discoveries are no longer designated CoRoT–Exo–n). An example light curve, for CoRoT–2 b, is shown in Figure 6.6. This is representative of a number of systems which show a clear transit signal on top of a significant semi-periodic rotational variation of stellar flux due to star spots. CoRoT–3 b, with $M_p \simeq 22 M_J$, is presumably a brown dwarf.

For CoRoT–7, with a particularly shallow transit depth $\Delta F/F \simeq 3.4 \times 10^{-4}$, radial velocity observations confirm the existence of the transiting planet CoRoT–7 b, with $P = 0.853$ d and $M_p \simeq 0.02 M_J \simeq 4.8 M_\oplus$, and reveal the presence of a second (but non-transiting) planet CoRoT–7 c, with $P = 3.69$ d and $M_p \simeq 8.4 M_\oplus$ assuming coplanarity (Queloz et al., 2009; Lanza et al., 2010). A third planet with $P \simeq 9$ d may also be present (Hatzes et al., 2010).

The secondary eclipses for at least CoRoT–1 (Snellen et al., 2009a; Alonso et al., 2009a) and CoRoT–2 (Alonso et al., 2009b) have also been recognised in the satellite data and, in the case of CoRoT–1, also from the ground (Rogers et al., 2009). From these, atmospheric temperatures, albedos, and energy redistribution parameters can be inferred (§6.5). Many other results on the CoRoT discoveries have now been published.

Kepler NASA's Kepler satellite was launched on 2009 March 6 into an Earth-trailing heliocentric orbit. It comprises a 0.95 m aperture modified Schmidt telescope, and 42 2k×1k CCDs covering a wide (115 square degree) field observed over the wavelength range 430–890 nm. It concentrates on monitoring 150 000 main sequence stars (8–15 mag) in the Cygnus region, with four measurements per hour, continuously over four years (Borucki et al., 2010a; Koch et al., 2010a). Kepler evolved from an earlier proposal, FRESIP, which was unsuccessful in its initial bid for selection as part of NASA's Discovery Program (Koch et al., 1996).

From the 33-d (Q1) observations starting in 2009 May, 156 000 stars were selected to maximise the number both bright and small enough to show detectable signals for small planets in and near the habitable zone. Prior to launch, the stellar characteristics of all stars in the field with $K < 14.5$ were determined (Koch et al., 2006). Analysis includes removal of false positives, and radial velocity follow-up (Gautier et al., 2007; Latham, 2008a; Yee & Gaudi, 2008).

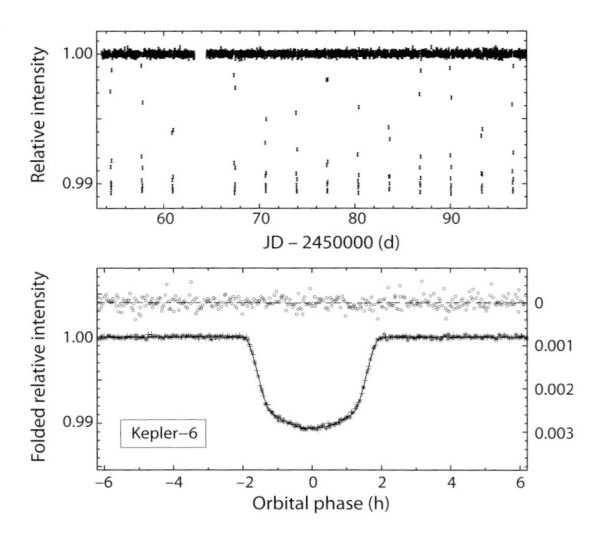

Figure 6.7: Kepler–6 b. Top: detrended light curve. Bottom: photometry folded at the planet period, with the model fit to the primary transit (lower curve, left scale), and the best fit to the secondary eclipse (upper curve, right scale). The planetary parameters are $P = 3.234723$ d, $i = 86°.8$ and, combined with radial velocity data, give $M_p = 0.67 M_J$, $R_p = 1.32 R_J$, $\rho = 0.35$ Mg m^{-3}. From Dunham et al. (2010, Figure 1), reproduced by permission of the AAS.

Simulations suggest that a population of hot super-Earths should be detectable (Schlaufman et al., 2010). In addition to transits, some 100 reflected light detections have been predicted for planets with orbital periods up to 7 days (Jenkins & Doyle, 2003).

The Kepler-discovered planet systems confirmed as of 2010 November 1 are Kepler–4 to 9 (Appendix D), the designations 1–3 b being used to refer to previously known planets in the Kepler field (TrES–2 b, HAT–P–7 b, and HAT–P–11 b respectively). An example light curve, for Kepler–6 b, is shown in Figure 6.7.

These first discoveries include the double transiting system Kepler–9 b/c (Holman et al., 2010), the first host star for which multiple transiting planets have been observed, and which show prominent transit time variations due to their proximity to a 2:1 orbital resonance (§6.4.13).

As of 2010 November 1, the Kepler project also lists 312 candidates for which follow-up observations are still required to confirm or reject the transiting planet nature of the observed light curve variations (Table 1.3).

SWEEPS The Advanced Camera for Surveys on HST was used to monitor 180 000 stars in the dense Sagittarius I window in the Galactic bulge, to look for transiting planets around F, G, K, and and M dwarfs (Sagittarius Window Eclipsing Extrasolar Planet Search, Sahu et al., 2006). From a 7-d survey in 2004 February, they discovered 16 candidates with periods of 0.6–4.2 d. Radial velocity observations of the two brightest ($V \sim$ 19 mag) confirmed the planetary nature of SWEEPS–4 and SWEEPS–11.

6.2.6 Follow-up observations from space

The Hubble Space Telescope (HST) and Spitzer Space Telescope have provided high signal-to-noise observations of light curves and spectra of transiting exoplanets, providing detailed characterisation of several transiting systems discovered from the ground. They are described only briefly in this section, with results summarised in §6.5. Hipparcos and MOST have made other contributions to the observations of planetary transits.

Hubble Space Telescope Launched in 1990, the 2.4-m aperture NASA/ESA Hubble Space Telescope has been used to observe known transit events in considerable detail. Observations have been made with the spectrographic mode of Space Telescope Imaging Spectrograph (STIS, installed in 1997, failed in 2004, and replaced in May 2009), and with the Advanced Camera for Surveys (ACS) in HRC mode with grisms (installed in 2002, with reduced functionality after 2006). With STIS (e.g. for HD 209458, Brown et al. 2001) light from the bright host star can be dispersed over many pixels, with the charge-handling capacity of 2.5×10^8 photons per readout, and a summation over all pixels, leading to photometric signal-to-noise ratio of more than 10^4. The readout time of the 1024×64 pixel subarray is about 20 s allowing an 80 s sampling cadence with a duty cycle of 75%. ACS observations of three transits of HD 189733 by Pont et al. (2007) resulted in a very-high accuracy light curve, with signal-to-noise of 15 000 on individual measurements, and 35 000 over 10-min averages.

Spitzer Space Telescope NASA's Spitzer Space Telescope, launched on 2003 August 25, has a 0.85-m primary mirror, cooled to 5.5 K until exhaustion of the liquid helium coolant in 2009 May. It has three instruments: IRAC (Infrared Array Camera), which operates simultaneously at four wavelengths (3.6, 4.5, 5.8, and 8 μm); IRS (Infrared Spectrograph) with four submodules operating at wavelengths 5.3–14 and 14–40 μm (low resolution) and 10–19.5 and 19–37 μm (high resolution); and MIPS (Multiband Imaging Photometer for Spitzer), three detector arrays for the far infrared at 24, 70, and 160 μm.

The first exoplanet observations were of HD 209458 (Deming et al., 2005b) and TrES–1 (Charbonneau et al., 2005). Subsequently, Spitzer has made numerous measurements using the four IRAC bandpasses, the IRS array at 16 μm, and the MIPS array at 24 μm. The instrumental effects most relevant to exoplanet eclipse studies (ramp, pixel phase, and slit loss) are summarised by Deming (2009). The shortest wavelength 3.6 and 4.5 μm IRAC channels continue to operate at full efficiency in the warm mission phase because the instrument remains radiatively cooled at $\sim$ 35 K.

Hipparcos detections While primarily an astrometric mission, the broad optical pass band of the Hipparcos detection system, combined with the 100 or so distinct measurement epochs between 1990–93 for 118 000 stars, provides a well-calibrated multi-epoch global system of photometric measurements.

Once photometric transits had been reported for HD 209458 by Charbonneau et al. (2000) and Henry et al. (2000), the period transit signal was discovered *a posteriori* in the Hipparcos epoch photometry data (Söderhjelm et al., 1999; Robichon & Arenou, 2000). Hipparcos observed the star on 89 occasions, of which five corresponded to planetary transits. The orbital period of about 3.5 d, and a transit duration of about 0.1 d, imply a 3% probability of observing a transit at any given epoch. The Hipparcos median magnitude is $Hp = 7.7719 \pm 0.002$, and the transits resulted in a 2.3 ± 0.4% mean decrease in flux. From the temporal baseline of more than eight years, or nearly 1000 periods, between the Hipparcos observations in the early 1990s and the first ground-based detection in the late 1990s, the period was improved 20-fold, from 3.524 47 d determined on ground (Mazeh et al., 2000) to 3.524 739(14) d when combined with the Hipparcos data (Robichon & Arenou, 2000; Castellano et al., 2000).

The second transiting star bright enough to appear in the Hipparcos catalogue, HD 149026, has not been detected in the Hipparcos photometric data.

Bouchy et al. (2005c) and Hébrard & Lecavelier des Etangs (2006) both reported transits for the ninth transiting planet to be discovered, and the third bright enough to be in the catalogue, HD 189733. It has $P = 2.219$ d, a transit depth of around 3%, and a transit duration of around 1.6 hr. Hébrard & Lecavelier des Etangs (2006) showed that around 3% of randomly chosen observations would be expected to fall during a transit, such that out of the 176 available Hipparcos observations, some five corresponding to transit periods would be expected. Folding of the Hipparcos light curve indeed shows that such a number of data points fall at the time of transits. The combination of Hipparcos and the ground-based observations of Bouchy et al. (2005c) span a period of 15 years, from which $P = 2.218574(8)$ d, with a resulting accuracy in the orbital period of ~1 s.

Hipparcos searches Other transit events probably remain buried in the Hipparcos epoch photometry data base, although they remain problematic to identify because of the large and unconstrained search space, and the complexity of dealing with the resulting large number of false positives. Hébrard et al. (2006) systematically searched the data for periods compatible with known planetary transits, and constructed a ranked list for follow-up measurements; 194 were observed with HARPS, but revealed only active stars.

Koen & Lombard (2002) described procedures for identifying transit events in photometric time series, applying them to 10 820 bright ($Hp < 7$ mag) non-variable Hipparcos stars. Laughlin (2000) defined a sample of 206 metal-rich stars of spectral type FGK which have an enhanced probability of harbouring short-period planets, and searched for Hipparcos transits: various candidate transit periods were identified. Jenkins et al. (2002) presented empirical methods for setting detection thresholds and for establishing confidence levels, applicable to the sparse Hipparcos data.

A corresponding, but improved analysis should also be possible from the Gaia photometric time series (Robichon, 2002; Eyer & Mignard, 2005).

MOST The Canadian MOST satellite (Microvariability and Oscillations of Stars) was a precursor of CoRoT, designed for studies of stellar oscillations but with applicability to parallel planetary transit studies. Launched on 2003 June 30, it comprises a 0.15-m aperture telescope feeding two CCDs, one for attitude tracking and one for science, with a single 300 nm wide filter centred at 525 nm. The polar Sun-synchronous low-Earth orbit allows stars between $-19° < \delta < +36°$ to be observed continuously for up to 60 days (Walker et al., 2003a).

Observations of the known transiting system HD 209458 over 14 days in 2004 and 44 days in 2005 placed upper limits on the planet albedo ruling out bright reflective clouds in its atmosphere (Rowe et al., 2006, 2008), revealed no other transit periods (Croll et al., 2007a), and placed upper limits on transit time deviations for the known planet thus constraining the perturbing effects of others (Miller-Ricci et al., 2008b). Absence of transit time deviations for HD 189733 excluded super-Earths in orbital resonance (Croll et al., 2007b; Miller-Ricci et al., 2008a). Observations of τ Boo detected stellar variability possibly induced by its planetary companion (Walker et al., 2008)

EPOXI–EPOCh NASA's Deep Impact cometary flyby spacecraft, launched in 2005 January, was re-oriented in 2008 (and renamed as EPOXI) for the fly-by of comet Hartley 2. Before this, as the EPOCh phase (Extrasolar Planet Observation and Characterisation) phase, the satellite was reconfigured to make two distinct observations relevant to exoplanets. Between 2008 January–August its larger telescope, the 0.3-m High Resolution Instrument, observed transits of seven exoplanets, each observed continuously for several weeks (Ballard et al., 2008): XO–2, GJ 436, HAT–P–4, TrES–3, WASP–3, TrES–2, and HAT–P–7 (e.g. Christiansen et al., 2010).

EPOXI also observed the Earth's disk, at 0.18 AU and 57° phase angle, and at 0.34 AU and 77° phase angle in 2008 June (Livengood et al., 2008). Observations with various filters and time sampling over a range of wavelengths from 350 nm to 4.8 μm, provided data intended as a template for characterising Earth-type exoplanets.

Its infrared spectroscopy shows the signatures of H_2O and CO_2, demonstrates significant time-variability of the optical spectrum associated with various terrestrial phenomena, and quantifies the suitability of the 'red edge' as a signature of chlorophyll-based life.

6.2.7 Future observations from space

Amongst the confirmed or proposed experiments, some are devoted to more efficient searches (PLATO, TESS), with others partly focused on (JWST, SPICA) or fully devoted to (THESIS) follow-up observations of known transiting systems.

PLATO PLATO (PLAnetary Transits and Oscillations of stars) is a proposed transit (and asteroseismology) mission under consideration by ESA, for launch not before 2018 (Catala, 2009a,b). With a significantly wider field than CoRoT and Kepler, it will focus on brighter stars, for improved accuracy and to facilitate post-detection investigation. It aims to detect significantly smaller planets through short-cadence (30 s) uninterrupted monitoring, over three years, of 100 000 bright stars ($V \leq 11$), and a further 400 000 to $V = 13$. The instrument has 28 identical small, very wide-field telescopes, each with its own CCD focal plane, and all directed at the same $26°$ diameter field. [Two earlier asteroseismology/transit detection missions were studied by ESA: STARS, with $0.5\,m^2$ collecting area and $1°\!.5$ field which was not accepted (Schneider, 1996), and a revised concept Eddington (Favata, 2004), with an enlarged telescope of $1\,m^2$ collecting area and $6\,deg^2$ field of view, which was accepted in 2000 but subsequently cancelled.]

TESS TESS (Transiting Exoplanet Survey Satellite) is a SMEX-class proposal to NASA for a two-year all-sky survey of some two million solar-type (mid-F through to K) stars down to $I = 12$ mag, using an array of CCDs covering 2000 sq. deg. in a single pointing. Its emphasis is also on nearby and typically brighter candidates than those targeted by CoRoT and Kepler, and is partly intended as a rapid mission to provide targets for JWST (Deming et al., 2009). Some 10 000 M dwarfs within 30 pc will also be included (Latham, 2008b).

JWST The NASA/ESA James Webb Space Telescope (JWST), due for launch in 2014, has a 6-m primary mirror, and will provide high signal-to-noise light curves and spectra of transiting planets. The imaging instruments NIRCam ($0.6-5\,\mu m$) and MIRI ($5-29\,\mu m$) should provide light curves of the primary and secondary eclipses of planets down to $1M_\oplus$, with filters, and the Tunable Filter Imager (TFI)TFI, providing limited spectral diagnostics. For spectroscopy, NIRCam ($2.4-5\,\mu m$), NIRSpec ($1-5\,\mu m$), and MIRI ($5-29\,\mu m$) provide a variety of wavelength coverage and resolution. Instrument configurations for exoplanet applications are tabulated by Clampin (2009), with exoplanet objectives detailed by Seager et al. (2009).

SPICA SPICA (Space Infra-Red Telescope for Cosmology and Astrophysics) is a proposed high-sensitivity mid- to far-infrared Japanese mission, with a 3.5-m mirror operating from $3.5-200\,\mu m$, and a planned launch in 2017 (Goicoechea et al., 2008). Mechanical cryocoolers cool the mirror to 4.5 K (cf. 80 K of Herschel) providing improved sensitivity at $10-100\,\mu m$. Instrumentation includes a $30-200\,\mu m$ imaging spectrometer (SAFARI, Goicoechea & Swinyard, 2010), a $4-40\,\mu m$ high-resolution spectrograph, a $10-100\,\mu m$ low-resolution spectrograph, and a $5-20\,\mu m$ imaging coronagraph.

THESIS THESIS (Transiting Habitable-zone Exoplanet Spectroscopy Infrared Spacecraft), is a MIDEX-class proposal to NASA for the spectroscopic investigation of a large population of planets including non-transiting planets and super-Earths (Swain et al., 2009a, 2010; Deroo et al., 2010). It comprises a 1.4-m primary mirror, and a moderate resolution spectrometer ($R = 2000$), to provide spectra from $2-14\,\mu m$, possibly extending to the near-visible. Simultaneity of this extended spectral range is a prerequisite, to probe planetary atmospheres over a wide range of pressures, to allow some molecules to be identified in multiple wavelength bands, and to permit temporal variability to be attributed to either changes in temperature or changes in composition.

6.2.8 Searches around specific stellar types

Circumbinary planets One early approach to selecting stars with an increased prospect of having transiting planets was to observe eclipsing binary systems, whose geometry implies $i \sim 90°$, under the assumption that the planetary and binary orbital planes are co-aligned (Doyle, 1988; Schneider & Chevreton, 1990; Hale, 1994). Configurations in which a planet orbits one component of a widely-separated binary (in which case transits are improbable due to the arbitrary long planetary period), or is in orbit around a close binary, both result in large domains of dynamical stability (Harrington, 1977; Heppenheimer, 1978; Benest, 1998).

Observations of the short-period eclipsing binary, CM Dra with a period of 1.28 days, was pursued by the TEP (Transits of Extrasolar Planets) network since 1994, using six telescopes located around the world (Doyle et al., 1996; Deeg et al., 1998; Doyle et al., 2000). Timing of the binary eclipse minima provided a planet detection sensitivity down to about $1\,M_J$ (Deeg et al., 2000).

The commonly used 'box least squares' algorithm (Kovács et al., 2002) is a three-dimensional search algorithm which assumes the periodicity, and the uniformity of the transit depth, of a transiting planet. Neither of these assumptions apply in the case of circumbinary planets. Ofir (2008) developed a generalisation of the method to include an underlying binary model (around which the planet orbits), determining for each data point whether the planet was transiting the primary or the secondary, and assessing the transit depth accordingly. Blind testing applied to the CoRoT data is described by Ofir et al. (2009).

Transiting planet parameters, as well as the theories of planetary formation, depend on whether the host star is single or a binary. Daemgen et al. (2009) carried out a survey of 14 transiting planet host stars using 'lucky imaging' to minimise effects of atmospheric seeing. They confirmed WASP–2 as a binary, and discovered previously unknown binarity for TrES–2 and TrES–4.

M dwarfs Several planetary companions to M dwarfs are known from radial velocity surveys. The specific interest of transit surveys focused on M dwarfs (along with those of L spectral type, collectively also referred to as ultracool dwarfs) is that, because of their small radii, signals from an Earth-size planet should be readily distinguishable, although their intrinsic variability may be a limiting factor (Caballero & Rebolo, 2002; Caballero et al., 2003; Gould et al., 2003c; Snellen, 2005; Nutzman & Charbonneau, 2008; Plavchan et al., 2008). Owing to partial electron degeneracy, objects of $100 M_J$ and older than 10^8 years have radii $\simeq 1 R_J$ (Burrows et al., 2001).

Additionally, their habitable zone lies much closer to the host star, thus improving their geometric transit probability: for an M5 star, a planet receiving the same stellar flux as the Earth would lie only 0.074 AU from the star, presenting a 1.6% geometric transit probability. Such planets may spend billions of years in the habitable zone (Andreeshchev & Scalo, 2004; Tarter et al., 2007).

Although more than 2000 ultracool dwarfs are known, many from the 2MASS survey, their relatively low surface density on the sky requires observing them individually. Nutzman & Charbonneau (2008) established design requirements for a network of ten 0.3-m survey telescopes. The MEarth project implemented eight identical 0.4-m telescopes in a single enclosure at Mt Hopkins, and resulted in the first such transiting planet, of $6.5 M_\oplus$ around the 13 pc distant M dwarf GJ 1214 (Charbonneau et al., 2009, see also §6.2.2).

Blake et al. (2008) carried out a study for transit detection in the near infrared, based on 13 ultracool dwarfs observed with the 1.3-m automatic Peters telescope at Mt Hopkins, aiming at ~0.01 mag photometry.

The 2MASS calibration fields, which were observed between 562–3692 times during the four years of the Two-Micron All Sky Survey project, have been searched for transiting M dwarfs (Plavchan et al., 2008). From 7554 sources in the range $K_s = 5.6 - 16.1$ mag, three M dwarf eclipsing systems have been reported, including two transiting planet candidates.

Giant stars Planets have been detected from Doppler surveys around evolved stars with $R_\star \gtrsim 2.5 R_\odot$, but none have been detected and further characterised from transit surveys because of their small transit depths ($\Delta F \simeq 10^{-4}$) and long transit durations $t_T \gtrsim 50$ hr. Assef et al. (2009) proposed to improve detection prospects using narrow-band measurements to isolate the thin ring of chromospheric emission at the limb of the giant stars.

6.3 Noise limits

Limitations from space The fundamental limits on high-accuracy time-series transit photometry, obtained above the Earth's atmosphere, are imposed by the presence of stellar surface structure (star spots, plages, granulation, and non-radial oscillations), which are strong functions of spectral type. It is the same surface structure which sets limits on achievable radial velocity and astrometric accuracies; the relationships between them, parameterised by Eriksson & Lindegren (2007), are detailed in §3.4.

Aigrain et al. (2004) used the solar irradiance power spectrum from SOHO, and relationships between chromospheric activity, rotation period, and colour index, to predict the variability power on various time scales, and the impact of microvariability on the detection capability of transit searches as a function of stellar spectral type and age. For Kepler, at $V = 12$ and with 10 min sampling, the smallest detectable planetary radii for 4.5 Gyr old G2, K0 and K5 stars, given a total of three or four transits in the light curves, were found to be 1.5, 1.0 and $0.8 R_\oplus$ respectively. Despite their higher variability, K-stars are promising transit candidates due to their small radius (see also Jenkins, 2002; Bonomo & Lanza, 2008).

Svensson & Ludwig (2005) and Ludwig (2006) computed hydrodynamical model atmospheres for a range of stellar types, predicting both photometric and astrometric jitter caused by granulation.

Limitations from the ground Nearby exoplanet host stars are typically comparatively bright, and telescope apertures can be very large. For surveys of the brightest stars, the contribution of photon noise to photometric transit measurements is accordingly generally negligible. Limitations on the smallest transit depths ΔF, and hence the smallest planet masses arise, instead, from a combination of atmospheric transparency variations, atmospheric scintillation noise, and detector granularity. The latter is addressed by high-precision autoguiding, perhaps combined with telescope defocusing of order 10–20 arcsec (e.g. Southworth et al., 2009a,b, 2010).

Very stable sites in terms of atmospheric transparency fluctuations are required before scintillation dominates. If H_2O drives the transparency fluctuations, telluric-line absorption variations may dominate longward of ~600 nm. The effects have been quantified by the statistical treatment of error propagation in the presence of red noise (e.g. Pont et al., 2006).

Atmospheric influences on stellar images: A wavefront incident on a telescope pupil experiences both phase and intensity fluctuations, both caused by refractive index variations within the turbulent atmosphere.

Phase fluctuations are responsible for atmospheric 'seeing' and image motion. This first-order effect (affecting the wavefront angle) results from the integrated effect of light propagating through the atmosphere's full vertical depth, although it may be dominated in practice by the contributions from tolerably well-identifiable layers, both at high altitude as well as close to the ground. Rays are effectively redirected across the telescope pupil, resulting in a change in angle of the arriving wavefront, and degrading the image resolution. The phase effects can be (partially) corrected using adaptive optics.

Scintillation is a second-order effect (affecting the wavefront curvature). It results in the spatially and temporally varying intensity of starlight reaching the ground. Arising from the refocusing through turbulence, it is perceptible only at large distances from the disturbance, and is consequently dominated by high-altitude turbulence. Since the flux intercepted by the telescope undergoes rapid changes (due to photons falling outside of the telescope pupil when they should have been collected, and *vice versa*), the apparent brightness of the star flickers. The naked eye perceives the effect as 'twinkling'. Adaptive optics correction may provide a diffraction-limited image, but its intensity still changes due to scintillation.

The mechanisms causing scintillation are described further by Dravins et al. (1997a). Effects decrease to longer wavelength (Dravins et al., 1997b), and improve with the telescope diameter as $D^{-2/3}$ (Reiger, 1963; Young, 1967).

Image intensity is also affected by atmospheric transparency variations, which can be largely corrected, in long-exposure imaging, through differential photometry with respect to comparison stars nearby on the sky. However, differential photometry corrects neither for phase fluctuations (unless the comparison star is within the isoplanatic patch), nor scintillation effects. On the contrary, because of the high-altitude origin of scintillation, and the resulting small angular coherence scale of ~1 arcsec in the optical, calibration with respect to a comparison star tends to increase scintillation noise (Osborn et al., 2011).

In part because it is the phase component which dominates astronomical image resolution, and in part because of the restricted domains of astrophysical investigation which demand high-accuracy high-time resolution photometry (notably asteroseismology), efforts to correct for phase variations have been intensive and highly successful, while scintillation noise has received less attention.

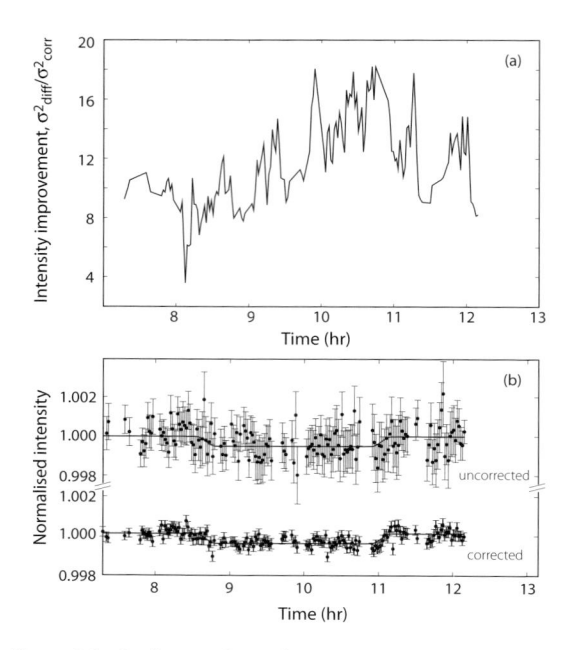

Figure 6.8: Conjugate-plane photometry: (a) expected (modeled) improvement factor in intensity variance as a function of time for data acquired from San Pedro Mártir on 2000 May 19; (b) simulated light curves of a secondary eclipse of an exoplanet with a transit depth of 0.05%, uncorrected (above) and corrected (below) for scintillation noise, assuming a 2-m telescope, and 30 s exposures in the V-band for a host star of V = 11 mag. From Osborn et al. (2011, Figures 11–12).

Fundamental noise limits using fast photometry from the ground evidently have several contributions, but appear to be poorly quantified and not particularly well understood. An evaluation by Mary (2006) suggests that once transparency variations are suppressed, the overall noise is of a mixed Poisson nature, arising from photon noise mixed by scintillation.

Broadly, ground-based transit photometry is limited to accuracies of a few mmag over relevant integration times. Ground-based telescopes are, consequently, presently able to discover exoplanet transits with depths up to about $(\Delta F/F) \simeq 1\%$ only, and are generally unable to observe secondary transits. It is the contributions other than photon noise that leads to the majority of transit surveys being conducted from relatively small aperture telescopes.

Conjugate-plane photometry Osborn et al. (2011) have proposed the concept of *conjugate-plane photometry* to improve scintillation noise (see box). The principle involves conjugating the telescope pupil to the dominant high atmospheric turbulent layer, then apodising it before calibration with a comparison star.

The method requires that the atmospheric inhomogeneities are concentrated within some relatively discrete layer(s), so that it is possible to define the limiting telescope entrance pupil corresponding to that altitude. Conceptually, an 'oversized' telescope mirror then collects the light that is sometimes refracted outside its nominal diameter.

Simulations of an atmosphere dominated by high-altitude turbulence suggest a reduction in intensity variance by a factor of 30, while models based on measurements at San Pedro Mártir suggest that a factor 10 improvement in variance may be achievable (Figure 6.8). Its efficiency in improving ground-based transit light curves remains to be demonstrated in practice.

6.4 Transit light curves

Figure 6.9 is a schematic of an orbiting planet showing, progressively, the transit of the planet as it passes in front of the star, the subsequent rise in flux as the planet's illuminated surface comes into view, and the drop in flux during the secondary eclipse as the planet passes behind the star.

As the field has progressed over the past decade, observation and interpretation has moved rapidly beyond simply the photometric detection of the transit, to include multicolour photometry and spectroscopy during the transit and secondary eclipse phases, and searches for detailed photometric structure, transit time variations, and reflected light during day-side illumination.

6.4.1 Observables

There are four principle observables which characterise the duration and profile of the primary transit: the period P, the transit depth ΔF, the interval between the first and fourth contacts t_T, and the interval between the second and third contacts t_F (see Figure 6.9).

From these, three geometrical equations together describe the principle features of the transit light curve (Seager & Mallén-Ornelas, 2003, eqn 1–3): the transit depth ΔF itself, the total transit duration t_T, and the transit shape specified by the ratio of the flat (fully occulted) part to the total transit duration t_F/t_T which, for circular orbits, are

$$\Delta F \simeq \left(\frac{R_p}{R_\star}\right)^2 \tag{6.1}$$

$$\sin(t_T \pi/P) = \frac{R_\star}{a}\left\{\frac{[1+(R_p/R_\star)]^2 - [(a/R_\star)\cos i]^2}{1-\cos^2 i}\right\}^{1/2} \tag{6.2}$$

$$\frac{\sin(t_F \pi/P)}{\sin(t_T \pi/P)} = \left\{\frac{[1-(R_p/R_\star)]^2 - [(a/R_\star)\cos i]^2}{[1+(R_p/R_\star)]^2 - [(a/R_\star)\cos i]^2}\right\}^{1/2} \tag{6.3}$$

The first follows from the ratio of the areas of the projected disks of the planet and star. The total transit time follows from the fraction of the orbital period P during which the projected distance between the centres of the star and planet is less than the sum of their radii (see Figure 6.9). The transit shape is derived similarly.

In the simplest interpretation, if the radius of the star can be estimated from, say, spectral classification, then R_p can be estimated from Equation 6.1. Setting $i = 90°$, $b = 0$, the duration of the transit for a circular orbit is numerically

$$t_T \simeq 13 \left(\frac{M_\star}{M_\odot}\right)^{-1/2} \left(\frac{a}{1\,\mathrm{AU}}\right)^{1/2} \left(\frac{R_\star}{R_\odot}\right) \text{ hours}, \tag{6.4}$$

giving a transit period of about 25 h for a Jupiter-type orbit and 13 h for an Earth-type. The minimum inclination

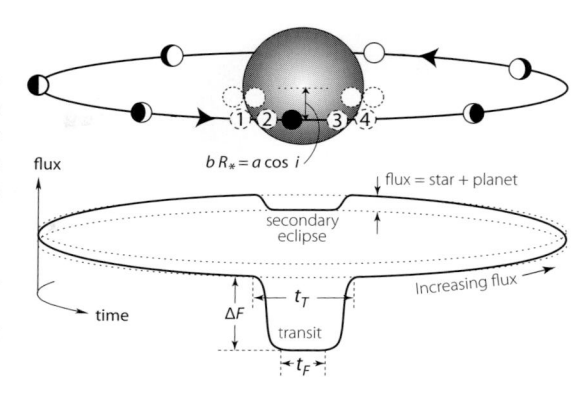

Figure 6.9: Schematic of a transit. During the transit, the planet blocks a fraction of the star light. After the transit, the planet's brighter day-side progressively comes into view, and the total flux rises. It drops again during the secondary eclipse as the planet passes behind the star. Dashed circles show the first to fourth contact points; those for smaller impact parameter (dotted) are more closely separated in time, and the ingress/egress slopes correspondingly steeper. The total transit duration t_T is between first and fourth contact, while t_F is timed between second and third contact. After Winn (2009, Figure 1).

where transits can occur is given by $\cos i_{\min} = (R_\star/a)$, while grazing incidence transits occur for $a \cos i = (R_\star \pm R_p)$.

The probability for a randomly-oriented planet on a circular orbit to be favourably aligned for a transit, or secondary eclipse, is (Borucki & Summers, 1984)

$$p = \frac{R_\star}{a} \simeq 0.005 \left(\frac{R_\star}{R_\odot}\right) \left(\frac{a}{1\,\mathrm{AU}}\right)^{-1}, \tag{6.5}$$

given by the solid angle on the sphere swept out by a planet's shadow (see the schematic of Figure 6.14 for the case of an eccentric orbit). Evaluation of i and p for realistic cases demonstrates that transits only occur for $i \simeq 90°$, while p is very small. The transit probability is independent of star distance, but the corresponding photometric accuracy decreases.

6.4.2 Theoretical light curves

The problem of constructing theoretical transit light curves, and using them to infer properties of the transiting planet–star system, appears simple enough in principle. One projected sphere (that of the planet) passes across another (the star) and the light from the star is attenuated according to the fraction of the two surfaces which overlap. In practice, algebraic treatment of the basic problem of two overlapping circles is unwieldy. Including the effects of stellar limb darkening, light reflected from the planet, blending due to background objects, and effects due to orbit eccentricity and other higher-order terms, conspire to make the problem a challenging one, yet rich in the physical information that the transit light curves convey.

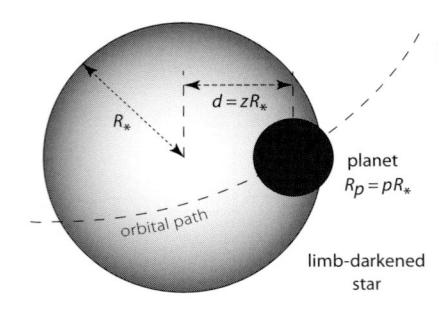

Figure 6.10: Schematic for the derivation of the theoretical light curve, showing the instantaneous transit geometry as viewed by the observer, and the definition of the parameters $p = R_p/R_\star$ and $z = d/R_\star$. Limb-darkening blurs the boundary between transit ingress and egress. After Mandel & Agol (2002, Figure 1), reproduced by permission of the AAS.

Limb darkening Limb darkening refers to the drop of intensity in a stellar image moving from the centre to its limb, and results from the combined effects of optical depth with the decreasing star density and temperature with radius. It is typically represented by functions of $\mu = \cos\theta$, where θ is the angle between the normal to the stellar surface and the line of sight to the observer.

The limb darkening functions of Claret (2000) provide suitable prescriptions and are widely used. In the non-linear law, the radial dependence of specific intensity is approximated by the fourth-order Taylor series

$$I(r) = 1 - \sum_{n=1}^{4} c_n (1 - \mu^{n/2}), \quad (6.6)$$

with $I(0) = 1$, and coefficients c_i dependent on effective temperature, luminosity class, and metallicity. The quadratic law is a limiting case with $c_1 = c_3 = 0$.

The effect is very significant for planetary transits, giving rise to a small colour change with λ already inferred by Rosenblatt (1971), although it diminishes with increasing wavelength. In the mid-infrared, where increasing numbers of transit and secondary eclipse measurements are being made, stellar disks are of rather uniform brightness, an approximation reasonably valid already in the near infrared. Accurate transit modeling of HD 209458 by Deeg et al. (2001) was the first time that a limb-darkening sequence could be determined for a single late main sequence star beyond the Sun.

More completely, a transit light curve results from the attenuation of the stellar photosphere by the planet and its own atmosphere (its own limb darkening), an effect which is ignored in this section.

Geometric formulation The formulation of Mandel & Agol (2002) in terms of the geometry of overlapping circles provides an insight into observed transit light curves. They develop an expression for the ratio of obscured to unobscured flux, $F(p, z) = 1 - \lambda(p, z)$, in terms of $p = R_p/R_\star$ and $z = d/R_\star$, where R_p is the planet radius, $R_\star$ is the stellar radius, and d is the distance between star and planet centres (Figure 6.10).

For a uniform source, i.e. without limb darkening, there are three geometrical regimes, handled separately. Outside of transit $(1 + p < z)$ there is no attenuation and $\lambda(p, z) = 0$. Within the fully occulted region $(z \le 1 - p)$ the attenuation is simply the ratio of the projected areas, $\lambda(p, z) = p^2$. For the partially overlapping region, and determined only by geometry,

$$\lambda(p,z) = \frac{1}{\pi}\left[p^2\kappa_0 + \kappa_1 - \sqrt{\frac{4z^2 - (1+z^2-p^2)^2}{4}}\right], \quad (6.7)$$

where $\kappa_0 = \cos^{-1}[(p^2 + z^2 - 1)/2pz]$,
and $\kappa_1 = \cos^{-1}[(1 - p^2 + z^2)/2z]$.

The limb-darkened light curve is given by

$$F(p,z) = \left[\int_0^1 I(r) 2r\,dr\right]^{-1} \int_0^1 I(r)\,dr\, \frac{d\left[F\left(\frac{p}{r},\frac{z}{r}\right)r^2\right]}{dr}. \quad (6.8)$$

Expressions for $F(p, z)$ in the presence of limb darkening depend on the region of (p, z) parameter space, for example whether the planet is crossing the limb, the centre of the disk, or (in theory) both. Theoretical light curves which include limb darkening (e.g. Mandel & Agol, 2002; Seager & Mallén-Ornelas, 2003) show that its effects are threefold: changing the transit depth ΔF as a function of impact parameter; making the bottom rounder (and hence the flat part shorter, reducing t_F); and blurring the boundary between ingress/egress and the flat bottom (Figure 6.11).

For small R_p, the description of the fully occulted part of the light curve can be simplified by assuming that the surface brightness of the star is constant under the disk of the planet. For $p \lesssim 0.1$ and $1 - p < z < 1 + p$ an approximation, accurate to ~ 2%, is then (their Equation 8)

$$F = 1 - \frac{1}{4\Omega(1-a)} \int_{z-p}^1 I(r) 2r\,dr \left(p^2 \cos^{-1}\left[\frac{z-1}{p}\right] \right. \\ \left. -(z-1)\sqrt{p^2 - (z-1)^2}\right), \quad (6.9)$$

where $\Omega = \sum c_n(n+4)^{-1}$. Finally, to determine z as a function of time requires the planetary orbital parameters, which for zero eccentricity is given by $z = aR_\star^{-1}[(\sin\omega t)^2 + (\cos i \cos\omega t)^2]^{1/2}$, where ω is the orbital frequency.

Their routines for computing these transit light curves, for an arbitrary limb-darkening function (for example according to the spectral type of the star), are frequently used for transit modeling. Pál (2008) derived partial derivatives of these functions for the case of quadratic limb darkening, useful for model fitting. He also compared the statistical errors in the light curve parameters as a function of the observing bandpass, from the near-ultraviolet to mid-infrared.

6.4 Transit light curves

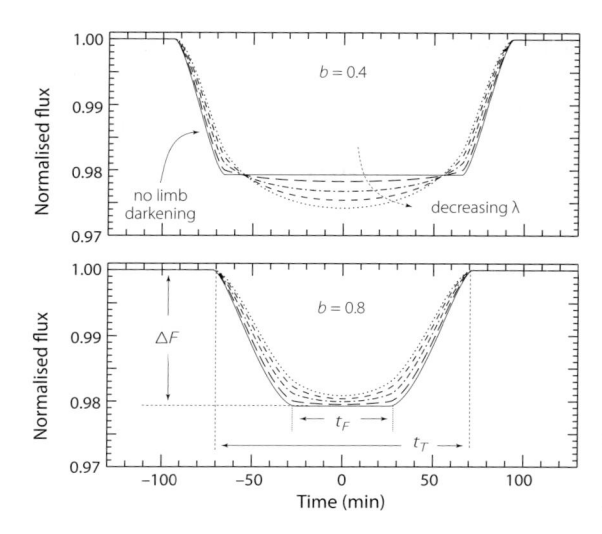

Figure 6.11: Theoretical transit light curves for two different impact parameters b, without (solid curves) and with (solar-type) limb darkening at 3, 0.8, 0.55 and 0.45 µm (the effects increase towards shorter wavelength). Model parameters are: $R_p = 1.4 R_J$, $a = 0.05$ AU, $R_\star = R_\odot$, $M_\star = M_\odot$. From Seager & Mallén-Ornelas (2003, Figure 11), reproduced by permission of the AAS.

Methods from eclipsing binaries Synthetic light curve codes, developed for the analysis of eclipsing binaries, have been adapted and used to study planetary transits. EBOP (Etzel, 1993) is based on a geometrical solution for biaxial ellipsoids (so accounting for tidal deformations from spherical shape), but originally considered only a linear law of limb darkening.

Modifications to take account of non-linear limb darkening were made by Giménez & Diaz-Cordovés (1993) and, more recently, developed for the analysis of exoplanet light curves, including error analysis, by Southworth et al. (2004, 2007).

A somewhat similar numerical approach building on the sequence of discrete geometries and specific limb darkening introduced by Mandel & Agol (2002) was developed by Giménez (2006a). This is based on the work of Kopal (1977) who, also in the context of close binary systems, used the mathematics of physical optics to express the loss of light during mutual eclipses as a cross-correlation of two circular apertures representing the eclipsing and eclipsed disks. The approach can be generalised to eccentric orbits, arbitrary limb-darkening expressions, and non-zero luminosity of the planet.

The fractional loss of light as a function of orbital phase is first expressed as

$$\alpha(\theta) = \sum_{n=0}^{N} C_n \alpha_n, \quad (6.10)$$

where the coefficients C_n are related to the terms in a general law of limb darkening (cf. Equation 6.6)

$$I(\mu) = 1 - \sum_{n=1}^{N} u_n(1 - \mu^n), \quad (6.11)$$

such that the geometrical and limb darkening parameters are decoupled.

The α_n are formulated, as in Mandel & Agol (2002), in terms of the apparent separation between the centres of the projected disks of the star and planet d as

$$\alpha_n(b,c) = \frac{b^2(1-c^2)^{\nu+1}}{\nu \, \Gamma(\nu+1)} \sum_{j=0}^{\infty} (-1)^j (2j+\nu+2) \frac{\Gamma(\nu+j+1)}{\Gamma(j+2)}$$
$$\times \left\{ G_j(\nu+2, \nu+1; 1-b) \right\}^2 G_j(\nu+2, 1; c^2), \quad (6.12)$$

where $b \equiv R_p/(R_\star + R_p)$, $c \equiv d/(R_\star + R_p)$, and $\nu \equiv (n+2)/2$; Γ is the gamma function, and $G_n(p, q; x)$ are the Jacobi–Gegenbauer polynomials.

Generating the light curve thus involves defining $I(\mu)$, stepping through orbit phase to determine the instantaneous star–planet projected separation d through numerical integration of Kepler's equation, then evaluating the α_n using this convergent series to the desired accuracy (e.g. to $j \simeq 20$ for a geometric precision better than 5×10^{-5}).

The light curve synthesis code ELC, making use of the NextGen model atmospheres (Orosz & Hauschildt, 2000), has been applied, amongst others, to the ellipsoidal variations observed in the Kepler light curve of HAT–P–7 (§6.4.12; Welsh et al., 2010).

Light curve fitting The inverse problem, that of establishing the appropriate physical parameters (and limb darkening coefficients) for a given light curve, proceeds through a χ^2 search for the unknowns through some appropriate region of parameter space, seeded by first-order estimates of the relative planetary radius ($R_p/R_\star$), the impact parameter ($b \equiv a \cos i/R_\star$), and the normalised separation between star and planet ($a/R_\star$).

Efficient numerical minimisation schemes, such as the Bayesian-based Markov Chain Monte Carlo algorithm, which also provides the full multi-dimensional joint probability distribution for all the parameters, or the Levenberg–Marquardt acceleration (Press et al., 2007), can be adopted, essentially providing confidence levels determined by surfaces of constant χ^2. For practical examples see, e.g., Winn et al. (2007).

Example fits Figure 6.12 shows a number of examples of transit light curves, illustrating the actual ratios of planet and star radii, and a schematic of the impact parameter derived from the light curve analysis.

6.4.3 Circular orbits

Various physical parameters can be extracted from the light curves, according to the accuracy and type of observation (e.g. whether the secondary eclipse is observed), the assumptions made (e.g. of a circular orbit,

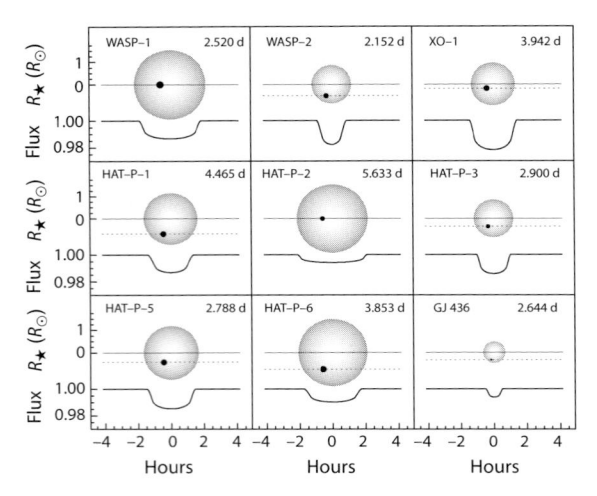

Figure 6.12: Examples of transit light curves, on a uniform time and relative flux scale. Planet and star sizes are also shown to scale, with the transit depth proportional to the ratio of their projected areas. Planet trajectories are shown as dotted lines, according to their estimated impact parameters. From Torres et al. (2008, Figure 8), reproduced by permission of the AAS.

and use of Kepler's law), information about the orbit available from other sources (e.g. radial velocity or astrometry), and astrophysical assumptions (such as the mass–luminosity relation for the primary star).

There is a unique and exact solution of the planet and star parameters which can be obtained from a transit light curve with two or more transits, under the assumptions of a circular orbit, for observations in a (long wavelength) bandpass where limb darkening is negligible, and ignoring possible contributions of contaminating (blended) sources. This outline follows the development of Seager & Mallén-Ornelas (2003).

From the three geometrical equations describing the light curve (Equations 6.1–6.3), three dimensionless combinations can be constructed: the planet–star radius ratio directly from Equation 6.1

$$\frac{R_p}{R_\star} = \sqrt{\Delta F} \; ; \quad (6.13)$$

the impact parameter b, defined geometrically as the projected distance between the planet and star centres during mid-transit in units of $R_\star$ (Figure 6.9)

$$b \equiv \frac{a}{R_\star} \cos i$$

$$= \left\{ \frac{(1-\sqrt{\Delta F})^2 - \frac{\sin^2(t_F \pi/P)}{\sin^2(t_T \pi/P)}(1+\sqrt{\Delta F})^2}{1 - [\sin^2(t_F \pi/P)/\sin^2(t_T \pi/P)]} \right\}^{1/2} ; \quad (6.14)$$

and the ratio $a/R_\star$

$$\frac{a}{R_\star} = \left\{ \frac{(1+\sqrt{\Delta F})^2 - b^2[1 - \sin^2(t_T \pi/P)]}{\sin^2(t_T \pi/P)} \right\}^{1/2}. \quad (6.15)$$

Additionally invoking Kepler's third law

$$P^2 = \frac{4\pi^2 a^3}{G(M_\star + M_p)} \quad (6.16)$$

to set the physical length scale, an expression for the stellar density $\rho_\star$ can be derived from Equation 6.15, assuming $M_p \ll M_\star$,

$$\rho_\star \equiv \frac{M_\star}{R_\star^3}$$

$$= \left(\frac{4\pi^2}{P^2 G}\right) \left\{ \frac{(1+\sqrt{\Delta F})^2 - b^2[1 - \sin^2(t_T \pi/P)]}{\sin^2(t_T \pi/P)} \right\}^{3/2}. \quad (6.17)$$

A unique solution can be imposed on the dimensionless ratios by additionally invoking the stellar mass–radius relation

$$R_\star = k M_\star^x, \quad (6.18)$$

where k is a constant, distinct for main sequence or giants, and x is the corresponding power law.

The five physical parameters $R_\star$, $M_\star$, i, a, and R_p can then be derived (Seager & Mallén-Ornelas, 2003, eqn 10–14)

$$\frac{M_\star}{M_\odot} = \left(k^3 \frac{\rho_\star}{\rho_\odot}\right)^{1/(1-3x)}, \quad (6.19)$$

$$\frac{R_\star}{R_\odot} = k\left(\frac{M_\star}{M_\odot}\right)^x = \left(k^{1/x} \frac{\rho_\star}{\rho_\odot}\right)^{x/(1-3x)}, \quad (6.20)$$

$$a = \left(\frac{P^2 G M_\star}{4\pi^2}\right)^{1/3}, \quad (6.21)$$

$$i = \cos^{-1}\left(b\frac{R_\star}{a}\right), \quad (6.22)$$

$$\frac{R_p}{R_\odot} = \frac{R_\star}{R_\odot}\sqrt{\Delta F} = \left(k^{1/x} \frac{\rho_\star}{\rho_\odot}\right)^{x/(1-3x)} \sqrt{\Delta F}. \quad (6.23)$$

For main sequence stars $k = 1$ and $x \approx 0.8$ (Cox, 2000, pp355–357), and $(R_p/R_\odot) = (\rho_\star/\rho_\odot)^{-0.57}\sqrt{\Delta F}$.

With the further approximation $R_\star \ll a$, equivalent to $t_T \pi/P \ll 1$, the expressions for $b, a, \rho_\star$ simplify to

$$b = \left\{\frac{(1-\sqrt{\Delta F})^2 - (t_F/t_T)^2(1+\sqrt{\Delta F})^2}{1-(t_F/t_T)^2}\right\}^{1/2}, \quad (6.24)$$

$$\frac{a}{R_\star} = \frac{2P}{\pi}\Delta F^{1/4}\left(t_T^2 - t_F^2\right)^{-1/2}, \quad (6.25)$$

$$\rho_\star = \frac{32 P}{G\pi}\Delta F^{3/4}\left(t_T^2 - t_F^2\right)^{-3/2}. \quad (6.26)$$

Uniqueness of this solution allows first-order physical parameter estimates to be made from the transit light curve measurements alone, including $\rho_\star$ (Equation 6.17) and hence R_p (Equation 6.23). If $\rho_\star$ implies that the star is a giant, the companion may immediately be identified as a more massive object of lesser interest.

If $M_\star$ and $R_\star$ are assumed known from the spectral type, then the problem is over-constrained, and Equation 6.26 can be re-arranged to give an expression for the orbital period, even if only a single transit is observed

$$P = \frac{G\pi}{32} \frac{M_\star}{R_\star^3} \frac{\left(t_T^2 - t_F^2\right)^{3/2}}{\Delta F^{3/4}}. \quad (6.27)$$

For P in days, the first term on the right hand side is $G\pi/32 = 288.73$. Seager & Mallén-Ornelas (2003) estimate that the period can be determined in this way to $\sim 15-20\%$ for $\delta t < 5$ min and $\sigma \sim 0.0025$ mag.

Carter et al. (2008) used a similar parameterisation, and derived analytic approximations for the corresponding uncertainties and covariances.

6.4.4 Eccentric orbits

Some 25% of transiting planets have significant eccentricities as determined from radial velocity measurements. The assumption of a circular orbit, which underlies the formulation of Seager & Mallén-Ornelas (2003), is therefore frequently an oversimplification.

Light curves Some lengthy algebra leads to the equations defining a general elliptical orbit in three dimensions, and thereafter to prescriptions for the corresponding light curve features, P, $R_p/R_\star$, t_T, and t_F (Kipping, 2008, Appendix A). These reduce to the results of Seager & Mallén-Ornelas (2003) for $e = 0$ (cf. the models of Tingley & Sackett 2005 and Ford et al. 2008a). Of the seven variables which can be adjusted to produce a light curve with the same four primary features, the orbital eccentricity e and argument of pericentre ω (as well as the stellar mass $M_\star$) cannot be determined from the transit light curve alone.

Transit duration Under the same assumption that $R_p \ll R_\star \ll a$, the transit duration is much less than the orbital period, and the planet–star separation during the transit is nearly constant. Ford et al. (2008a) derived an expression for the total transit duration for an eccentric orbit (cf. Equation 6.2)

$$\frac{t_T}{P} \simeq \frac{R_\star}{\pi a \sqrt{1-e^2}} \left\{ \left(1 + \frac{R_p}{R_\star}\right)^2 - b^2 \right\}^{1/2} \left(\frac{r_t}{a}\right), \quad (6.28)$$

where r_t is the planet–star separation at time of mid-transit

$$r_t = a(1 - e\cos E_t) = \frac{a(1-e^2)}{1 + e\cos v_t} = \frac{a(1-e^2)}{1 + e\cos\omega}, \quad (6.29)$$

where v_t and E_t are the true and eccentric anomalies at the time of transit, and ω the argument of pericentre. The ratio of the transit duration to that of a planet on an equivalent circular orbit is then $(1 + e\cos\omega)/(1-e^2)^{1/2}$. This is dependent on e and ω, being less than unity for planets that transit near pericentre, and greater than unity for those transiting near apocentre. For highly eccentric orbits (say, $e = 0.9$), the ratio can range between $\sim 0.2 - 4$. It follows that estimates of R_p will be biased if scaled to $R_\star$ which is itself derived from the transit light curve on the erroneous assumption of a circular orbit.

A specific example of such a bias was given by Kipping (2008): from the accurate Spitzer timing measurements for HD 209458 (Richardson et al., 2006), and an estimate of $e = 0.014(5)$ (Winn et al., 2005), he found a 1% larger value for R_p from the transit timing analysis than under the assumption $e = 0$. A more robust procedure for light-curve fitting is therefore to use measurements independent of the transit to determine $M_\star$, M_p, e, and ω (and to invoke stellar evolution if no measure of $M_\star$ is available) and then use the transit observations to determine P, $R_\star$, and finally R_p once e is known.

Light curve asymmetry Barnes (2007) and Kipping (2008) quantified the light curve asymmetry as a function of e. On an elliptical orbit, and unless the planet is at pericentre or apocentre at the time of mid transit, the planet's azimuthal velocity will change between ingress and egress: if the transit occurs after pericentre and before apocentre, the ingress time will be shorter than egress, and *vice versa*. The ingress and egress durations are given by (Barnes, 2007, eqn 19)

$$\tau = \frac{R_p}{v_v \cos(\sin^{-1} b)}, \quad (6.30)$$

where $v_v = v_0 \frac{1 + e\cos f}{\sqrt{1-e^2}}. \quad (6.31)$

Here, v_v is the planet's azimuthal velocity at the position in the orbit corresponding to the true anomaly v, v_0 the equivalent velocity for a circular orbit with the same semi-major axis, and b is the impact parameter.

HAT–P–2 b (HD 147506) has a particularly favourable geometry with true anomaly at mid-transit close to $v = \pi/2$, and yields $\tau_{\text{ingress}} = 636$ s and $\tau_{\text{egress}} = 517$ s (Figure 6.13). In general, while mid-transit times can be shifted from seconds to minutes depending on e, it will be a challenge to distinguish between different eccentricities from the difference in ingress and egress shapes, or from the difference between ingress and egress durations, even in the infrared where the effects of limb darkening are minimised.

Transit probability Although an eccentric planet spends the majority of its time at distances from the star larger than its semi-major axis, the majority of its true anomaly is spent at smaller distances. This results in a larger fraction of the celestial sphere being intercepted by the planet's shadow, and a higher probability that the planet will transit than for a corresponding circular orbit, although the time spent in transit at these locations will be shorter (Figure 6.14).

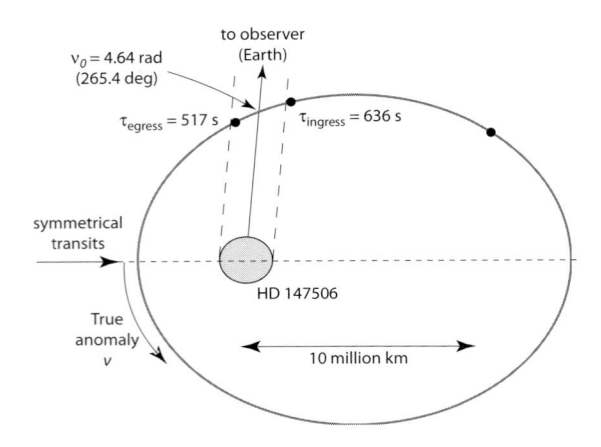

Figure 6.13: Schematic of the orbit of HD 147506. As viewed from Earth, the transit occurs at a value of the true anomaly $v_0 = 4.64$ rad (close to $-90°$), such that the asymmetry between ingress and egress is close to its maximum. If the transit could be viewed at $v = 0, \pi$ (i.e. from the left or right in the figure), the transit would be symmetrical. From Barnes (2007, Figure 5), reproduced by permission of University of Chicago Press.

The transit probability is a function of both the true anomaly, v, and the polar angle from the orbit plane. From the expression for the star–planet distance as a function of v, and integrating over the extent of the planet's shadow for all values of ω (cf. Barnes, 2007, eqn 1–8) leads to the result that for planets on eccentric orbits

$$p = \left(\frac{R_\star \pm R_p}{a}\right)\left(\frac{1}{1-e^2}\right), \qquad (6.32)$$

where inclusion of the small term $\pm R_p$ excludes or includes grazing transits. This reduces to Equation 6.5 for $e = 0$, and shows that planets on eccentric orbits are more likely to transit than those in circular orbits with the same semi-major axis, by a factor $(1-e^2)^{-1}$. For a circular orbit, the geometric conditions for transits and secondary eclipses are identical, while for eccentric orbits an observer may see transits without a secondary eclipse, and *vice versa*.

Implications for transit surveys The implication for transit surveys is two-fold (Burke, 2008): an increase in the probability for the planet to generate transits near pericentre, accompanied by a corresponding reduction in the detectability due to a shorter transit duration. For an eccentricity distribution matching known planets with $P > 10$ d, the probability of transits is $\simeq 1.25$ times higher than the equivalent circular orbit and the average transit duration is $\simeq 0.88$ times shorter. These two opposing effects nearly cancel for a realistic transit survey.

There is a similar consequence for efforts to detect transits amongst planets discovered by radial velocity measurements: the transit probability for any given planet is not a strictly declining function of semi-major axis, but depends on a favourable combination of e and ω (Seagroves et al., 2003; Kane & von Braun, 2008).

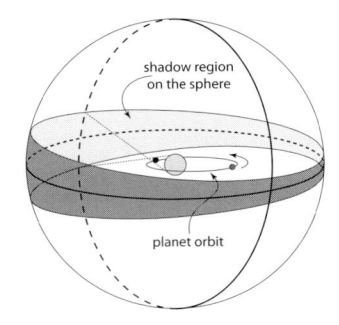

Figure 6.14: Geometry of an eccentric planetary orbit, and the area of the celestial sphere onto which the transit is projected. At pericentre (planet at left in the figure) the region is larger than when the planet is at apocentre. Similar considerations for a circular orbit lead to the transit probability given by Equation 6.5. Adapted from Barnes (2007, Figure 1).

One of the highest *a priori* transit probabilities at the time of the study by Seagroves et al. (2003) belongs not to one of the short-period hot Jupiters, which tend to average transit probabilities of ~12%, but rather to the $P = 550$-d planet around HIP 75458 (ι Dra). Here, the large semi-major axis (1.34 AU) is compensated by the large stellar radius, $12.8 R_\odot$, and favourable orbital geometry ($e = 0.7$, $\omega = 94°$), leading to an *a priori* transit probability of 15%.

Although this planet does not, in fact, transit, similar arguments led to the search for and successful detection of transits for the $P = 21$-d HD 17156 b and the $P = 111$-d HD 80606 b (p109).

Multiple transiting planets Kepler–9 is the first host star for which multiple transiting planets have been observed (Holman et al., 2010). From seven months of Kepler observations, two Saturn-size planets, with periods of 19.2 and 38.9 d, are presently experiencing periods increasing and decreasing at respective average rates of 4 and 39 min per orbit. The transit times of the inner body also display an alternating variation of smaller amplitude. These signatures are characteristic of gravitational interaction of two planets near a 2:1 orbital resonance. An additional transiting super-Earth-size planet with a period of 1.6 d is also inferred.[4]

Simultaneous transits Planets with orbits of terrestrial-size will be a challenge to detect from their periodic photometric signal alone, due to their very long orbital periods. Simultaneous transits of two such planets will presumably be improbably rare.

---

[4] Note added in press: six coplanar transiting planets have been discovered around Kepler–11, with orbital periods between 10–118 d (Lissauer et al., 2011).

6.4 Transit light curves

Compounded by the motion of the orbiting Earth, transits of Mercury occur only 13 or 14 times per century. Transits of Venus occur in a pattern that repeats every 243 years, with pairs eight years apart, separated by gaps of more than a century. The last pair were in December 1874 and December 1882, and the most recent on 2004 June 8 and 2012 June 6.

According to Meeus & Vitagliano (2004) simultaneous transits of Mercury and Venus seen from Earth last occurred on 373 173 September 22 BCE, and will next occur on 69 163 July 26, and thereafter only in 224 508 March. A simultaneous transit of Venus and Earth will be visible from Mars in the year 571 741 CE.

6.4.5 Physical quantities

Stellar density The stellar density $\rho_\star$ estimate which can be derived from the light curve alone (Equation 6.26) provides a direct constraint on $R_\star$, and is a sensitive indicator of its evolutionary state. A more prescriptive form, dependent on the derived quantity $(a/R_\star)$, is given by (Sozzetti et al., 2007)

$$\frac{M_\star}{R_\star^3} = \frac{4\pi^2}{GP^2}\left(\frac{a}{R_\star}\right)^3 - \frac{M_p}{R_\star^3}. \quad (6.33)$$

The first term on the right is entirely determined from measurable quantities while the second, although unknown until $M_\star$ is determined, is two or more magnitudes smaller and can be ignored, at least for a first iteration (and included thereafter). As an example, for TrES–2, with $P = 2.47063(1)$ d (O'Donovan et al., 2006a), and $a/R_\star = 7.63 \pm 0.12$ (Holman et al., 2007), the uncertainty on $\rho_\star$ is only 1.6%.

A uniform re-analysis of transiting systems to establish $\rho_\star$ has been made for 23 systems by Torres et al. (2008), and for 14 systems by Southworth (2008).

Planet surface gravity The planet's surface gravity, g_p, can be derived by combining the transit observations with parameters determined directly from the radial velocity reflex motion (Southworth et al., 2007). This follows from the mass function of a spectroscopic binary (cf. Equation 2.24)

$$\mathcal{M} = \frac{(1-e^2)^{3/2}K_\star^3 P}{2\pi G} = \frac{M_p^3 \sin^3 i}{(M_\star + M_p)^2}, \quad (6.34)$$

where $K_\star$ is the radial velocity semi-amplitude. Including Kepler's third law (Equation 6.16) and solving for $(M_\star + M_p)$ gives

$$g_p \equiv \frac{GM_p}{R_p} = \frac{2\pi}{P}\frac{(1-e^2)^{1/2}K_\star}{(R_p/a)^2 \sin i}. \quad (6.35)$$

P can be obtained from the radial velocity, or the light curve, and is typically determined rather precisely. The radial velocities also give e and $K_\star$, while i and (R_p/a) are found directly from the transit light curve. An equivalent expression directly in terms of the light curve observable is given by (Torres et al., 2008, eqn 6)

$$\begin{aligned}\log g_p &= -4.1383 - \log P + \log K_\star \\ &\quad -\frac{1}{2}\log\left(1-\left[\frac{b}{a/R_\star}\frac{1-e^2}{1+e\sin\omega}\right]^2\right) \\ &\quad +2\log\left(\frac{a/R_\star}{R_p/R_\star}\right) + \frac{1}{2}\log(1-e^2),\end{aligned} \quad (6.36)$$

where the numerical constant is such that g_p is in m s^{-2} when P is in days and $K_\star$ is in m s^{-1}.

Errors on the relevant light curve parameters show that it can provide a superior gravity indicator than the traditional estimate of $\log g$ based on the widths of pressure-sensitive absorption lines. For HD 209458 b, for example, Southworth et al. (2007) determined $g_p = 9.28 \pm 0.15$ m s^{-1}, an order of magnitude more precise than that obtained through measurement of its mass, radius and density. It confirmed that the planet has a lower surface gravity that that predicted by models of gas giant planets at the time.

In general, the immunity of g_p to systematic errors in stellar properties suggests that theoretical models of planet structure should be based on the comparison between theoretical and observed surface gravities, rather than theoretical and observed planet radii.

6.4.6 Interferometric observations

Interferometric observations can provide a direct measurement of stellar angular diameters $\theta_\star$, which can then be transformed into $R_\star$ on the basis of a trigonometric parallax, independently of stellar evolution models. For example, CHARA angular diameter measurements of 24 planet hosting stars (mostly non-transiting) are given by Baines et al. (2008).

Of the transiting planets, the star with the largest angular diameter is HD 189733 with $\theta_\star = 376 \pm 31\,\mu$as determined using the CHARA array (Baines et al., 2007). The angular diameters of GJ 436, HD 149026, and HD 209458 are also in the range 170–250 μas.

Uniquely, interferometers like CHARA or VLTI should also be capable of directly determining the position angle of the planet's orbital plane, because of the significant asymmetry in the source brightness introduced by the transiting planet (van Belle, 2008). R_p and b are measurement by-products which, as a minimum, provide independent checks on these parameters.

At any instant, an interferometer measures one Fourier component of the source's brightness distribution. In aperture synthesis, the overall source structure is then reconstructed from the different telescope combinations, whose projected baseline separations also vary as the Earth rotates.

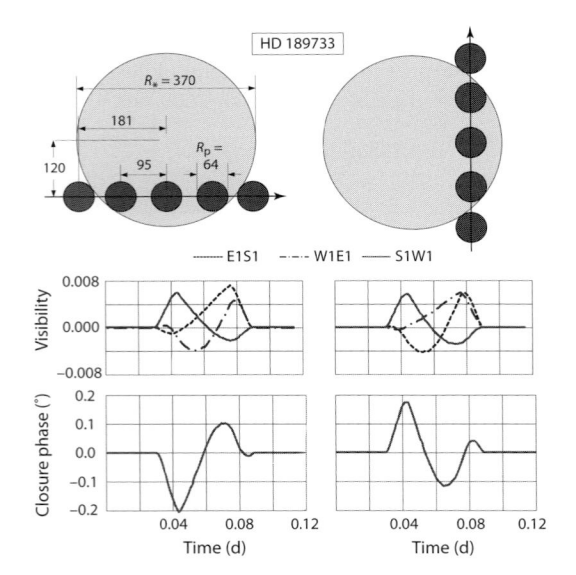

Figure 6.15: Simulated interferometric observations of HD 189733. The figure shows two different possible transit geometries (top), the corresponding expected visibility amplitude for each of three CHARA baselines (middle), and the visibility phase for the array's S1W1 baseline (bottom). In the top panels, the dimensions are the angles in microarcsec inferred for the system from the discovery parameters (Bouchy et al., 2005c). From van Belle (2008, Figure 2), reproduced by permission of University of Chicago Press.

Ignoring the absolute phase term, the complex interferometric visibility of a transiting star–planet system is given by (van Belle, 2008, eqn 2)

$$V = \frac{V_\star + rV_p e^{-2\pi i \mathbf{B} \cdot \Delta \mathbf{s}/\lambda}}{1+r}, \qquad (6.37)$$

where $V_\star, V_p$ are the visibility functions of the star and planet, $V = 2J_1(x)/x$, where J_1 is the Bessel function, $x = \theta_{ud}\pi B/\lambda$, θ_{ud} the angular diameters of the relevant uniform disks, λ the observing wavelength, $\mathbf{B}$ and $\mathbf{s}$ are the baseline and relative separation vectors, and the brightness ratio $r = -(R_p/R_\star)^2$.

In practice, visibility phases are corrupted by atmospheric turbulence. But for three or more telescopes, the closure phase Φ, given by the sum of the visibility phases around a closed loop of baselines, and long used in radio astronomy, cancels many of the atmospheric and instrument effects, and provides a highly sensitive probe for interferometric image reconstruction (Pearson & Readhead, 1984; Monnier, 2007).

The CHARA interferometer consists of six 1-m telescopes in a Y-configuration, with two telescopes per arm, and with the three longest baselines $B > 300$ m. The Michigan Infrared Combiner (MIRC) instrument combines currently four (and potentially six) telescopes simultaneously, providing three (and potentially 10) independent closure phase measurements with a precision of $\sigma_\Phi \simeq 0.°03$ (Monnier et al., 2006). Preliminary studies on υ And were reported by Zhao et al. (2008).

VLTI–AMBER can also determine closure phases, but at a somewhat lower accuracy level of a few degrees (Weigelt et al., 2007). Beyond VLTI–AMBER, PIONIER is a precursor for the interferometric instruments VSI and GRAVITY, and employs more stable 'integrated optics on a chip'. Due for installation at the VLTI in late 2010, it will perform four-way telescope combination providing three precision closure phase estimates.

Figure 6.15 illustrates the dimensions for HD 189733, and shows how the transit geometry (top panel) affects the complex visibilities (middle panels) and more particularly the closure phases (bottom panels). Interferometric closure phases should provide an important probe of the transit geometry in the future, particularly valuable for higher-order transit light curve effects.

6.4.7 Reflected light

Principles Outside of the transit and secondary eclipse, and indeed even in the absence of transits, an orbiting planet should reveal its presence by the orbital modulation of reflected star light (Figure 6.9). The presence of reflected light is implicit in the existence of a secondary eclipse, and has been measured explicitly in transiting systems, for example in the case of CoRoT–1 (Snellen et al., 2009a, Figure 6.26). This section refers specifically to searches in non-transiting systems.

Bromley (1992) already proposed to use reflected light during bright stellar flares, noting that *'possible planetary configurations that may be probed by this method are limited to Jupiter-size objects in tight orbits about the parent star'*. Other considerations (Charbonneau et al., 1998; Seager & Sasselov, 1998; Seager et al., 2000) led to the first direct searches.

A planet of radius R_p at distance a from the star intercepts a fraction $(R_p/2a)^2$ of the stellar luminosity. For $R_p \ll R_\star \ll a$ the planet/star flux ratio, ε, can be written (Collier Cameron et al., 1999; Leigh et al., 2003a)

$$\varepsilon(\alpha,\lambda) \equiv \frac{f_p(\alpha,\lambda)}{f_\star(\lambda)} = p(\lambda)\left(\frac{R_p}{a}\right)^2 g(\alpha), \qquad (6.38)$$

where the geometric albedo $p(\lambda)$ is the ratio of brightness at zero phase (i.e., seen from the star) to that of a fully reflecting, diffusely scattering (Lambertian) flat disk with the same cross-section. For a Lambert sphere, $p(\lambda) = 2/3$.

The relevant phase function is given by (Sobolev, 1975; Charbonneau et al., 1999)

$$g(\alpha) = (\sin\alpha + (\pi-\alpha)\cos\alpha)/\pi, \qquad (6.39)$$

$$\text{with } \cos\alpha = -\sin i \sin(2\pi\phi),$$

where $\alpha \in [0,\pi]$ is the angle between star and observer subtended at the planet, $i \in [0,\pi/2]$ is the orbit inclina-

6.4 Transit light curves

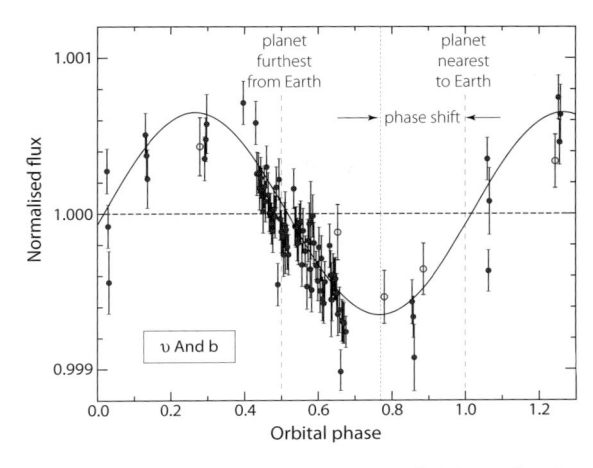

Figure 6.16: υ And b, observed at 24 μm with Spitzer, showing reflected light modulated at the 4.6-d orbital period. The best-fit sinusoid (solid curve) exhibits a phase offset of ~80°, indicating a planet 'hot spot' advected almost to the planet's day–night terminator. Circles represent earlier Spitzer data from Harrington et al. (2006). From Crossfield et al. (2010, Figure 3), reproduced by permission of the AAS.

tion, and $\phi \in [0, 1]$ is the orbital phase, with $\phi = 0$ at the time of radial velocity maximum.

The relative amplitude of the reflected light decreases with the square of the planet–star distance (Equation 6.38), with no modulation for face-on systems, $i = 0°$. The proximity of hot Jupiters to their host stars makes them optimum targets for reflected-light searches. At optical wavelengths, starlight scattered off the planet's atmosphere is expected to dominate over thermal emission. As noted by Charbonneau et al. (1999) a photometric detection would yield i, and hence the planetary mass, breaking the $M_p \sin i$ degeneracy implicit in radial velocity measurements.

Results Charbonneau et al. (1999) selected τ Boo for study, with $P = 3.3$ d and $a = 0.046$ AU. Although the star–planet separation is at most 3 mas, the reflected light component is $L_p/L_\star \sim 10^{-4}$, some $10^4 - 10^5$ times higher than for Jupiter.

The observed spectrum should contain a planetary component, reaching a maximum amplitude of $K_\star [(M_\star + M_p)/M_p] \simeq 152$ km s^{-1}. The similarity between the orbital period and the stellar rotation period suggest that the τ Boo system is tidally locked, in which case the planet should reflect a non-rotationally broadened stellar spectrum, yielding relatively narrow planetary lines superimposed on the broader stellar lines. From Keck–HIRES observations over three nights, no evidence for a highly reflective planet was found, yielding a geometric albedo $p \lesssim 0.3$.

Reports of a detected modulation at the orbital period (Collier Cameron et al., 1999), and a weak CH_4 signature in the infrared spectrum (Wiedemann et al.,

2001), were followed by more stringent limits from 75 h of échelle spectroscopy at the 4.2-m WHT telescope over 17 nights which revised the claim (Leigh et al., 2003b). Searches giving only upper limits were also conducted for the $P = 4.6$-d υ And (Collier Cameron et al., 2002), and the 3.5-d orbital period HD 75289 (Leigh et al., 2003c; Rodler et al., 2008).

These optical albedo limits may still be some way from the true values of $p(\lambda)$. The upper limit of $p = 0.038 \pm 0.045$ from MOST observations of HD 209458 b (Rowe et al., 2008), is compatible with the very low albedos predicted by the cloudless models of Burrows et al. (2008b).

Success, in the mid-infrared at least, came with observations of υ And at 24 μm with Spitzer (Harrington et al., 2006). They measured the flux at five epochs over the 4.6-d period, achieving a S/N of 6300 at each. With Spitzer secondary eclipse measurements of HD 209458, TrES–1, and HD 189733 suggesting that only some 0.1% of the total infrared light from the system is emitted by the planet, their S/N nevertheless allowed them to determine the planet/star flux ratio of $(2.9 \pm 0.7) \times 10^{-3}$. Their detection of such a modulation demonstrated that there is a significant illumination difference between the planet's day- and night-sides.

A more recent determination, also from 24 μm Spitzer–MIPS data (Crossfield et al., 2010), is shown in Figure 6.16. A large phase offset of ~80° with respect to the radial velocity curve suggests a planet 'hot spot' advected almost to the planet's day–night terminator (§6.5). They determined a 'hot side' temperature of ~1800 K, implying a hemisphere-averaged temperature contrast of ~900 K, and an orbital inclination $i \gtrsim 28°$.

Attempts to measure the reflected light from υ And based on closure phase measurements using the CHARA–MIRC interferometer (see also §6.4.6) are described by Zhao et al. (2008).

Prospects for Kepler Brown (2009b) introduced the formulation of the photometric orbit as a Keplerian entity, i.e. an observational set governed by the dynamical motion of two bodies interacting by the force of gravity, corresponding to the classical treatment of both radial velocity and astrometric orbit determinations. The six photometric Keplerian parameters are a, e, i, the mean anomaly, the argument of pericentre, and an effective planet radius defined as $\sqrt(p(\lambda) R_p)$.

Simulation of the Kepler satellite observations, assuming a precision of 2×10^{-5} throughout three years of observation, showed that orbital solutions might be possible for many of the 100–760 hot Jupiters that it might discover (Jenkins & Doyle, 2003). Radial velocity observations combined with Kepler photometry could mean that M_p is then determined without the $\sin i$ ambiguity, although uncertainties about their phase functions, albedos, and instrument systematics remain.

6.4.8 Doppler variability

A non-zero spectral index of the stellar spectrum means that there is a flux variability associated with any Doppler reflex motion. Loeb & Gaudi (2003) showed, at nonrelativistic velocities, that this amounts to

$$\frac{\Delta F}{F} = (3 - \alpha)\frac{K}{c}, \qquad (6.40)$$

where K is the radial velocity semi-amplitude, and $\alpha = d\ln(F_\nu)/d\ln\nu$ is the logarithmic slope of the source spectral flux in the observed frequency band, with $\alpha \simeq -1.3$ for solar-type stars observed in the optical band. For close-in planets with $P \lesssim 0.2$ yr, the resulting periodic Doppler variability, of order μmag, is significant relative to variability caused by reflected light from the planetary companion, and dominates for $P \gtrsim 0.2$ yr. The effect is relevant for almost any orbital inclination.

Loeb & Gaudi (2003) estimated that Kepler should have the photometric sensitivity to detect all planets with minimum mass $M_p \sin i \gtrsim 5 M_J$ and $P \lesssim 0.1$ yr around the 10^4 main sequence stars with spectral types A–K and $V < 12$ in its field of view.

The extension to higher, stellar-mass binary companions was considered by Zucker et al. (2007), effects which should be more easily detectable by Kepler.

6.4.9 Polarisation

Motivation Light from solar-like stars has a very small overall polarisation. The integrated solar disk, for example, has a linear polarisation of less than 2×10^{-7} in V (Kemp et al., 1987). The resolved solar disk nevertheless shows a polarisation variation from centre to limb, resulting from scattering opacity (Leroy, 2000).

There are two distinct manifestations of polarisation that are receiving attention. In the first, the transiting planet breaks the symmetry integrated over the stellar disk, resulting in a non-zero linear polarisation which varies as the transit progresses. Simulations by Carciofi & Magalhães (2005) for transits of GKM dwarfs suggest that the effect may be observable even for Earth-like planets, while asymmetry around mid-transit may further constrain e.

The second approach is to search for polarised light reflected from the planet outside of the transit or secondary eclipse phases, either from the integrated system light, or directly from the planet itself if this can be resolved in the future (Seager et al., 2000; Tamburini et al., 2002; Baba & Murakami, 2003; Hough et al., 2003; Saar & Seager, 2003; Stam et al., 2004, 2006; Stam, 2008; Buenzli & Schmid, 2009; Fluri & Berdyugina, 2010). Ignoring polarisation will, in contrast, affect the interpretation of reflected light studies (Stam & Hovenier, 2005), especially for elliptical orbits, or if the planet is oblate (Sengupta & Maiti, 2006).

Light scattered in the planetary atmosphere is linearly polarised perpendicular to the scattering plane, and characterized by the Stokes parameters Q and U. These parameters will vary as the scattering angle changes, and should show two peaks per orbital period. Polarisation variability would constrain the nature of scattering particles in the planetary atmosphere, and allow determination of P, e, and i; the latter, in combination with radial velocity measurements, would resolve the $M_p \sin i$ degeneracy.

In the case of an unresolved star–planet system, any net polarisation will be very small, at 10^{-5} or less, largely dictated by the ratio of reflected to total light, and with a strong dependency on the structure of the planetary atmosphere (Stam, 2004). Despite the small effect, a high (relative) accuracy is possible on the ground.

In the more distant future, circular polarisation spectroscopy could provide a remote-sensing technique in the search for life. This consideration is based on *homochirality*, in which biogenic molecules are presumed likely to possess the same sense of chirality (enantiomorphism), with the same sense of circular polarisation when viewed in scattered light (Sparks et al., 2009).

Results Berdyugina et al. (2008) claimed the first detection of polarised light from HD 189733 b. B band measurements with DIPol on the 0.6-m KVA telescope on La Palma, distributed over orbital phase, gave two polarisation peaks near maximum predicted elongation with an amplitude of 2×10^{-4}. Fixing P, a, R_p, and T_0, they found i and e in agreement with radial velocity-derived values, and a (Lambert) radius of the scattering atmosphere 30% larger than that of the opaque body inferred from transits. While a possible explanation was quickly forthcoming (Sengupta, 2008), the measurements have not been confirmed (Wiktorowicz, 2009).

Lucas et al. (2009) reported observations of 55 CnC and τ Boo at the WHT 4.2-m telescope with Planetpol, an APD-based polarimeter targeting a sensitivity better than 10^{-6}, and specifically designed to detect unresolved planets (Hough et al., 2003, 2006). Their measured nightly standard deviation on Q and U was 2.2×10^{-6}, but they derived only upper limits on the polarised flux for 55 Cnc e and τ Boo b. This stands in contrast with the much higher polarisation claimed for HD 189733 b.

Other experiments ESO's VLT–SPHERE 'planet finder' (§7.3.1), due to be operational in 2012, combines adaptive optics, spectroscopy and coronagraphy with an imaging polarimeter (ZIMPOL). This targets an accuracy of 10^{-5} for exoplanet imaging (Thalmann et al., 2008).

6.4.10 Secondary eclipse

The total light just before the time of the secondary eclipse is the sum of the stellar flux and that of the star.

During the secondary eclipse the total light is that of the star alone. The difference therefore corresponds to the flux of the planet's day-side region. By measuring the relative depths of the secondary eclipse in multiple bandpasses, a low-resolution spectrum of the planet's flux can therefore be reconstructed.

Infrared photometry of the secondary eclipse also provides an estimate of the planetary temperature. In the Rayleigh–Jeans limit the depth of the secondary eclipse is given by (Charbonneau, 2003)

$$\Delta F \simeq \frac{T_p}{T_\star} \left(\frac{R_p}{R_\star}\right)^2 . \qquad (6.41)$$

For $R_p = 1 R_J$, $R_\star = 1 R_\odot$, and planet/star temperatures of 1500 K and 6000 K, $\Delta F \simeq 4$ mmag.

At optical wavelengths, the planet/star flux ratio is at least two orders of magnitude smaller than in the infrared, due to the planet's lower thermal emission at shorter wavelengths, and the low optical reflectance expected from cloud-free models (Figure 6.25). While detection of the secondary eclipse is therefore more difficult at optical wavelengths, it also provides important atmospheric diagnostics.

Results from the observations of secondary eclipses are described further in §6.5.

Orbital dependence The probability of an observable secondary eclipse depends on the orbital parameters of the planet, and particularly on its eccentricity and argument of pericentre. Kane & von Braun (2009) provide analytical expressions for the probabilities, calculate their values for the known transiting systems, and discuss constraints on the existence and observability of primary transits if a secondary eclipse is observed.

The timing of the secondary eclipse with respect to the primary transit is also related to the orbital eccentricity, and specifically to $e \cos \omega$. For example, the $\simeq 1$ h secondary eclipse of GJ 436 b, with $e = 0.15$, occurs at orbit phase 0.587, more than five hours after the mid-point between transits (Deming et al., 2007a).

6.4.11 Rossiter–McLaughlin effect

In addition to the photometric signal of a planet transit, a spectroscopic signal, on top of the basic orbital Doppler shift, is also present. The phenomenon was originally seen as a rotational effect in eclipsing binary systems, independently reported by R.A. Rossiter and D.B. McLaughlin in 1924.

As suggested by Schneider (2000), the effect should be seen during exoplanetary transits as a small positive or negative anomaly in the radial velocity curve, caused by the progressive occultation of the rotating stellar disk (Figure 6.17). First reported for HD 209458 by Queloz et al. (2000a), measurements are being made for a growing number of systems (e.g. Winn, 2008; Winn et al., 2008a; Dreizler et al., 2009; Fabrycky & Winn, 2009).

Background of the Rossiter–McLaughlin Effect: Back-to-back papers in the Astrophysical Journal in 1924 reported anomalies in the velocity curves of the binary systems β Lyr (Rossiter, 1924) and Algol (McLaughlin, 1924). For the 12.9 day β Lyr system, a secondary oscillation in an interval of ± 1.6 d on either side of the principal minimum was apparent from 442 spectra taken over 1911–21.

Appealing to the rotation of the eclipsed star, normally symmetrically broadened by the rotation of one limb away from the observer and the other toward, Rossiter explained the effect as *'When the star is entering eclipse, the receding limb is visible and the approaching limb is covered [and vice versa]'*, leading to an asymmetry in the velocity curve. Described by them as the *'rotational effect'*, its magnitude amounted to 26 and 35 km in β Lyr and Algol respectively. Now described in the exoplanet literature as the Rossiter–McLaughlin effect the phenomenon, and its possible rotational origin, had in actuality previously been noted by Schlesinger in the cases of both δ Lib (Schlesinger, 1910, p134) and λ Tau (Schlesinger, 1916, p28).

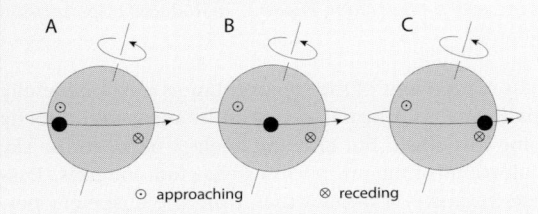

As the exoplanet transits its rotating host star, it blocks out more of the star's rotationally blue-shifted light at A, and more of its red-shifted light at C, causing changes in the radial velocity profile according to orbital phase. The precise form of the radial velocity deviations from a Keplerian fit provides information on the star's spin axis orientation with respect to the planet's orbital plane.

Relevance to formation and migration Although in the solar system the orbit planes of all eight planets are well aligned with the solar equator, such alignment cannot be assumed for exoplanets in general. Of the theories proposed to explain the existence of the close-in giant planets, disk migration acting alone may largely preserve the initial spin-orbit alignment (Lin et al., 1996; Ward, 1997a; Murray et al., 1998), while additional planet–planet scattering or Kozai migration would produce at least occasionally large misalignments (Ford et al., 2001a; Yu & Tremaine, 2001; Papaloizou & Terquem, 2001; Terquem & Papaloizou, 2002; Marzari & Weidenschilling, 2002; Thommes & Lissauer, 2003; Wu et al., 2007; Fabrycky & Tremaine, 2007; Wu et al., 2007; Chatterjee et al., 2008; Jurić & Tremaine, 2008; Nagasawa et al., 2008). Tidal forces are not expected to have a strong effect (Winn et al., 2005).

Measurement of the angle ϕ between a planet's orbital axis and that of its host star therefore provides diagnostics of theories of planet formation and subsequent evolution. Yet while the inclination of the planet orbit $i \simeq 90°$ is measured via transit photometry, the line-of-

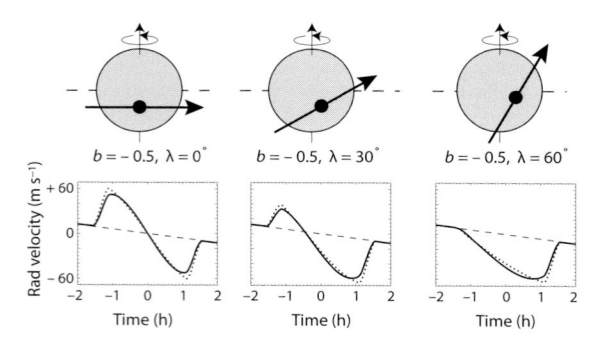

Figure 6.17: Three different theoretical transit trajectories, with the same impact parameter and therefore the same light curves, showing the dependence of the Rossiter–McLaughlin radial velocity signature on λ, computed with the formulae of Ohta et al. (2005). The long-dashed lines show the star's radial velocity in the absence of the Rossiter–McLaughlin effect. Solid and dotted lines are with and without some limb darkening. Adapted from Gaudi & Winn (2007, Figure 2), reproduced by permission of the AAS.

sight component of the stellar rotation axis is generally unknown. As a result, ϕ cannot be measured directly using this effect, but only the angle λ between the sky projections of the orbital and stellar rotation axes. Positive (negative) λ implies that, from the observer's perspective, the projected stellar spin axis is rotated clockwise (counterclockwise) with respect to the projected orbit normal. Values $|\lambda| > (\pi - i)$ correspond to retrograde orbits (Fabrycky & Winn, 2009).

The maximum amplitude of the anomaly is

$$\Delta V \simeq (R_p/R_\star)^2 \sqrt{(1-b^2)}\, v\sin i_\star\,, \qquad (6.42)$$

where $v\sin i_\star$ is the projected stellar equatorial rotation velocity. For a Sun-like star, $v\sin i \simeq 2\,\mathrm{km\,s^{-1}}$, and the maximum size of the effect is around $20\,\mathrm{m\,s^{-1}}$ for a Jupiter-like planet, and around $0.2\,\mathrm{m\,s^{-1}}$ for an Earth. Because the transit observations give a precise and independent measure of $(R_p/R_\star)^2$, spectroscopic monitoring of ΔV tracks the planet's trajectory referred to the sky-projected stellar rotation axis. Different approaches to the analysis use either a modeled stellar photosphere (Queloz et al., 2000a; Winn et al., 2005), or analytic expressions depending on the assumed form of limb darkening (Ohta et al., 2005; Giménez, 2006b).

Constraints on rotation axis and planet orbit A constraint on the true angle between the stellar rotation axis and the planet orbit can be found by combining estimates of $R_\star$, the stellar rotation period $P_{\rm rot}$, and the projected rotational velocity $v\sin i_\star$, which together determine $\sin i_\star$, and hence (Winn et al., 2007, eqn 7–8)

$$\cos\phi = \cos i_\star \cos i + \sin i_\star \sin i \cos \lambda\,. \qquad (6.43)$$

For HD 189733, variability attributed to spots gave an estimate of the stellar rotation period, and an upper bound

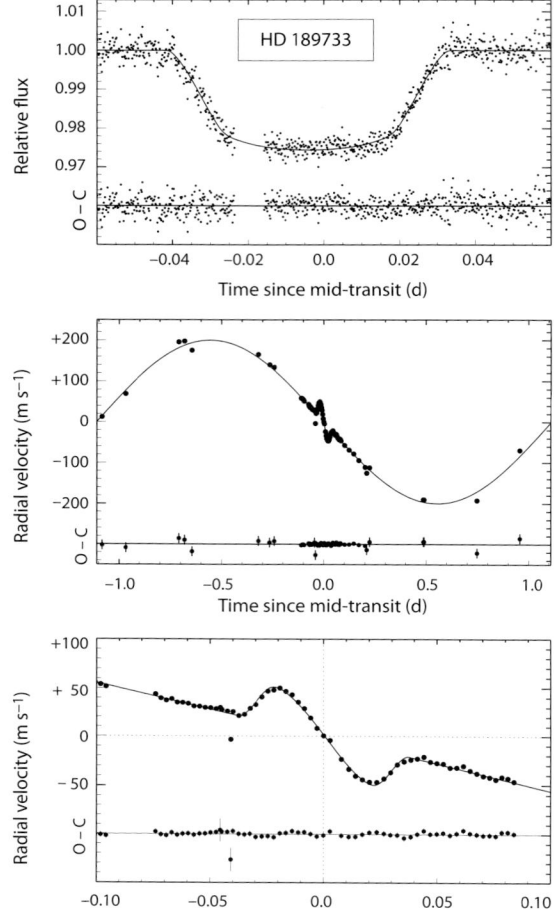

Figure 6.18: Rossiter–McLaughlin effect for HD 189733. Top: transit photometry in the z-band, with the best-fitting model (solid line). Middle: radial velocities as a function of orbital phase, along with the model (solid line). Bottom: close-up near the mid-transit time (residuals below). The abscissa scale is different in each case. From Winn et al. (2006, Figure 1), reproduced by permission of the AAS.

of $\phi < 27°$ (Winn et al., 2007). An alternative constraint on the stellar rotation axis may be provided by astero-seismology (e.g. Gizon & Solanki, 2003).

Results Figure 6.18 shows the detailed measurements in the example case of HD 189733 (see also Figure 6.4). Table 6.1 summarises the transiting planets for which the Rossiter–McLaughlin effect has been reported, as of late 2010.

Retrograde orbits The existence of retrograde orbits, in which the planet's component of orbital angular momentum is counter to that of the stellar rotation, had been predicted by planet–planet scattering or Kozai migration models. The possibility of chaotic transitions of

6.4 Transit light curves

Table 6.1: Exoplanets in which the Rossiter–McLaughlin spin–orbit effect has been reported (as of 2010 November 1). Star and planet properties (V, spectral type, d, R_p) are from exoplanet.eu, and indicate the range of systems over which the effect is currently measured. λ is the angle between the projections of the axes of stellar rotation and planet orbit. Errors are only quoted for λ. Typical errors on i are $0.5-1°$, while typical errors on $v\sin i$ are $0.5-1$ km s^{-1}. Instruments used are: C = Euler+CORALIE, E = OHP+ELODIE, H = ESO 3.6+HARPS, K = Keck+HIRES, N = NOT+FIES, O = OHP+SOPHIE, S = Subaru+HDS. In the case of multiple studies of the same object, the reference given is generally the most recent.

System	V	ST	d (pc)	R_p (R_J)	i (°)	λ °	$v\sin i$ km s^{-1}	Instr.	Reference
CoRoT–1	13.6	G0V	460	1.49	85.1	-77 ± 11	5.2	K	Pont et al. (2010)
CoRoT–2	12.6	K0V	300	1.46	87.8	7.2 ± 4.5	11.5	H/O	Bouchy et al. (2008)
CoRoT–3	13.3	F3V	680	1.01	86.1	-37.6 ± 15	35.8	H	Triaud et al. (2009)
CoRoT–11	12.9	F6V	560	1.43	83.2	$\sim 0\pm5$	40.0	O/H/K	Gandolfi et al. (2010)
HAT–P–1	10.4	G0V	139	1.23	86.3	3.7 ± 2.1	3.8	K/S	Johnson et al. (2008a)
HAT–P–2	8.7	F8	118	1.16	90.0	0 ± 12	22.9	O	Loeillet et al. (2008b)
HAT–P–4	11.2	F	310	1.27	88.8	-4.9 ± 11.9	5.8	K	Winn et al. (2011)
HAT–P–7	10.5	–	320	1.42	80.8	182.5 ± 9.4	4.9	K/S	Winn et al. (2009)
HAT–P–11	9.6	K4	38	0.45	89.2	103 ± 15	1.3	K	Winn et al. (2010c)
HAT–P–13	10.6	G4	214	1.28	83.4	1.9 ± 8.6	1.7	K	Winn et al. (2010b)
HAT–P–14	10.0	F	205	1.20	83.4	187.8 ± 4.4	7.3	K	Winn et al. (2011)
HD 17156	8.2	G0	78	1.02	85.4	10 ± 51	6.3	S	Narita et al. (2009a)
HD 80606	8.9	G5	58	0.92	89.6	50 ± 40	2.2	O	Pont et al. (2009)
HD 149026	8.2	G0IV	79	0.65	86.1	-12 ± 15	6.2	O	Wolf et al. (2007)
HD 189733	7.7	K1–2	19	1.14	85.5	-1.4 ± 1.1	3.3	K	Winn et al. (2006)
HD 209458	7.7	G0V	47	1.32	86.6	-3.9 ± 1.6	4.7	E	Queloz et al. (2000a)
Kepler–8	13.9	–	1330	1.42	84.1	-26.9 ± 4.6	10.5	K	Jenkins et al. (2010)
TrES–1	11.8	K0V	157	1.08	88.4	30 ± 21	1.3	S	Narita et al. (2007)
TrES–2	11.4	G0V	220	1.27	83.6	-9 ± 12	1.0	K	Winn et al. (2008b)
TrES–4	11.6	F	440	1.81	82.8	7.3 ± 4.6	8.3	S	Narita et al. (2010)
WASP–2	12.0	K1	144	1.04	84.7	-153 ± 13	1.0	C/H	Triaud et al. (2010)
WASP–3	10.6	F7V	223	1.45	84.9	3.3 ± 3	14.1	K	Tripathi et al. (2010)
WASP–4	12.6	G8	300	1.42	89.5	4 ± 40	2.1	C/H	Triaud et al. (2010)
WASP–5	12.3	G5	297	1.17	86.1	12.1 ± 10	3.2	C/H	Triaud et al. (2010)
WASP–6	12.4	G8	307	1.22	88.5	-11 ± 16	1.6	H	Gillon et al. (2009)
WASP–8	9.9	G8	87	1.04	88.5	-123.3 ± 4	1.6	C/H	Queloz et al. (2010)
WASP–14	9.8	F5V	160	1.26	84.3	-14 ± 17	2.9	N/O	Joshi et al. (2009)
WASP–15	10.9	F7	308	1.43	86.0	-139.6 ± 5	4.3	C/H	Triaud et al. (2010)
WASP–17	11.6	F6	–	1.74	86.6	-147 ± 15	9.9	H/others	Anderson et al. (2010b)
WASP–18	9.3	F6	100	1.16	80.6	4.0 ± 5	14.6	C/H	Triaud et al. (2010)
XO–3	9.8	F5V	260	1.22	82.5	70 ± 15	18.3	O	Hébrard et al. (2008)

a solar system planet under tidal interactions (Beletskii et al., 1996; Laskar, 2003) has been extended to the influence of tidal and magnetic perturbations for short-period exoplanets by Gusev & Kitiashvili (2006).

The observations of HAT–P–7 by Winn et al. (2009) were the first to reveal a probable polar, or even retrograde, planet orbit. Its retrograde nature, with $\lambda \simeq -132°$, was confirmed by Narita et al. (2009b). Other examples of such highly perturbed orbits are now being found (Table 6.1).

Implications for migration models Many of the measured systems have small values of λ, but a significant number are highly misaligned. The earliest results were interpreted as providing possible evidence for two mechanisms for inserting hot Jupiters into their close-in orbits. In the first, type II migration, disk–planet tidal interactions cause the planet to migrate inwards, largely preserving their initial spin–orbit alignment. In the second, some of the close-in giant planets may have arrived at their current locations through gravitational perturbations from other massive bodies, possibly arising in part from mutually inclined binary systems (Wu et al., 2007), in either case followed by tidal dissipation and circularisation (Triaud et al., 2010; Winn et al., 2010a; Matsumura et al., 2010).

Another possibility to explain large values of λ is a misalignment between the rotation of the host star and the protoplanetary disk (Bate et al., 2010; Lai et al., 2010). Other processes which may adjust planet orbits after formation are discussed in §12.6. For terrestrial-mass planets, the giant impact stage of protoplanetary collisions is presumed to determine their initial spin state (e.g. Kokubo & Ida, 2007).

Statistical results Triaud et al. (2010) used a sample of 26 measured spin–orbit projections to derive the underlying distribution of ϕ (Equation 6.43), finding that between 45–85% of hot Jupiters have $\phi > 30°$. They concluded that the results are consistent with predictions of Kozai resonance scattering, provisionally concluding that most hot Jupiters are formed from a dynamical and tidal origin without the need for migration to drive the planets to very small orbital radii. On the contrary, they found that disk migration cannot explain the observations without invoking additional processes.

The principal component and clustering analysis of Marchi et al. (2009), meanwhile, found strong indications for two types of hot Jupiter, with parameters reflecting the physical mechanisms of type II migration and scattering.

Host star dependence Based both on these radial velocity observations (Winn et al., 2010a, 2011), as well as on the line-of-sight stellar rotation velocities of transit hosts (Schlaufman, 2010), it appears that misaligned systems tend to be associated with hotter host stars ($T_{\rm eff} \gtrsim 6250$ K, or $M_\star \gtrsim 1.2 M_\odot$). If confirmed, this trend may indicate either that planet formation and migration are different for low- and high-mass stars, or that the subsequent tidal evolution is different, perhaps due to different outer convection zones and consequently slower rates of tidal dissipation (Winn et al., 2010a).

Line-profile tomography Derivation of the inclination and obliquity of the orbital plane has also been determined by *line-profile tomography* (Collier Cameron et al., 2010a,b). The method is based on modeling the global shape of the stellar cross-correlation function as the convolution of a limb-darkened rotation profile and a Gaussian representing the Doppler core of the average photospheric line profile. The light blocked by the planet during the transit is a Gaussian of the same intrinsic width, whose trajectory across the line profile then provides a measure of the misalignment angle, and an independent measure of the projected stellar rotational velocity, $v \sin i$.

The travelling Doppler 'shadow' cast by the planet creates an identifiable distortion in the line profiles, which has been shown to yield self-consistent measures of the projected stellar rotation rate, the intrinsic width of the mean local photospheric line profile, the projected spin–orbit misalignment angle, and the system's centre-of-mass velocity.

Results for HD 189733 b, $\lambda = -0°.4 \pm 0°.2$ (Collier Cameron et al., 2010a), are very close to those obtained from the Rossiter–McLaughlin effect. Applied to WASP–33 b, Collier Cameron et al. (2010b) derived $\lambda = -251°.6 \pm 0°.7$.

6.4.12 Higher-order photometric effects

Additional contributions to the light from an isolated planet may result in more complex light curves. Predictions suggest that some of these will be apparent with photometric accuracy close to that achieved at present.

Planetary satellites A planet's transit light curve will be distorted by the presence of an accompanying satellite (Sartoretti & Schneider, 1999; Barnes & O'Brien, 2002; Domingos et al., 2006; Cabrera & Schneider, 2007; Sato & Asada, 2009), which can be assessed as follows.

The orbital radius of a satellite around a planet must lie, to first order, somewhere between the Roche limit (within which it will disintegrate due to the planet's tidal forces exceeding its gravitational self-attraction) and the Hill radius (the gravitational sphere of influence of the planet in the proximity of the more massive host star, page 43). For rigid bodies the Roche limit is given by

$$R_{\rm R} = R_{\rm p} \left(\frac{2\rho_{\rm p}}{\rho_s} \right)^{1/3}, \qquad (6.44)$$

where ρ the density, and subscripts denote planet and satellite. In reality, the Roche limit depends on the body's internal viscosity and tensile strength.

For HD 209458 b (with $M_{\rm p} \sin i = 0.69 M_{\rm J}, M_\star = 1.01 M_\odot, a = 0.0468$ AU) $R_{\rm H} \simeq 4.2 \times 10^5$ km $\simeq 5.9 R_{\rm J}$. With the planet's orbital speed of ~ 140 km s^{-1}, transit features of a gravitationally bound object could lie as much as 49 min before or after the corresponding planet features, compared to the full transit duration of 184 min. A satellite might therefore be detectable from its own photometric transit signature significantly before or after the main transit. The transit duration could also differ from that of the planetary transit if it has significant orbital motion, or due to its different transit projection.

Such effects have been searched for without success in HST observations of HD 209458 b (Brown et al., 2001). Similar observations of HD 189733 b have ruled out the existence of $1 M_\oplus$ satellites around this $1.13 M_{\rm J}$ planet (Pont et al., 2007).

Rings and comets If a transiting planet is surrounded by rings with significant opacity, or associated toroidal atmospheres, they would cause distortions of the light curve relative to that of a spherical body, with dips in the light curve before first and after fourth contact (Schneider et al., 1998; Schneider, 1999; Arnold & Schneider, 2004; Barnes & Fortney, 2004; Johnson & Huggins, 2006).

For HD 209458 b, Brown et al. (2001) placed limits on any ring radius of $1.8 R_{\rm p}$, slightly smaller than the radius of Saturn's ring system when measured in units of Saturn's radius. The low sensitivity is a consequence of the assumption that rings must lie in the planet's orbital plane, and hence nearly edge-on as seen from Earth.

Simulations using a geometrical model of the dust distribution, and optical properties of cometary grains, such as those inferred from the photometric variations for β Pic (Lecavelier des Etangs et al., 1997, 1999), suggest that the detection of comets might be possible with transit accuracies of 10^{-4}; they are predicted to give a

6.4 Transit light curves

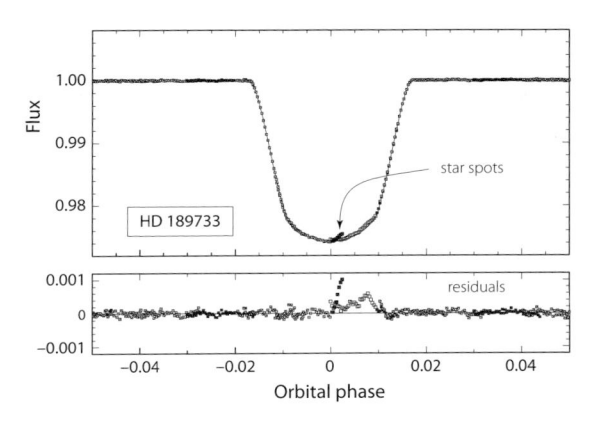

Figure 6.19: HST–ACS light curve for HD 189733 b, showing the effects of star spots on the detailed structure of the transit light curve (top), with residuals around the best-fit transit model (bottom). Residuals outside the spot complexes are at the $\pm 10^{-4}$ level. From Pont et al. (2007, Figure 1), reproduced with permission © ESO.

'rounded triangular' shape, largely achromatic, as a typical transit signature (Lecavelier des Etangs et al., 1999).

Planetary oblateness, rotation and weather Significant planetary oblateness (see box on page 134) will lead to a slightly different light curve from that of a spherical planet of the same projected area, an effect which can be quantified by calculating, at any moment across the transit, the intersection between the (elliptical) projection of the oblate planet and the (circular) projection of the star. The effect on the light curve is most pronounced during the ingress/egress phases for orbital inclination different from 90° and a non-zero projected obliquity (Seager & Hui, 2002; Barnes & Fortney, 2003), where the differences may reach 1.5×10^{-4} for rotational flattening similar to Jupiter and Saturn. Barnes et al. (2009a) used dynamical models of HD 209458 b and HD 189733 b to show that planet shapes resulting from rotation are unlikely to introduce detectable light curve deviations (below 10^{-5} of the host star). Carter & Winn (2010) used Spitzer photometry of seven transits of HD 189733 to place a limit on its oblateness roughly equal to that of Saturn. The effect of oblateness on the transit timing is considered in §6.4.13.

Atmospheric refraction of starlight from points in the planet's shadow further modifies the ingress and egress shapes (Hui & Seager, 2002), as do changes in light propagation delay times from the orbiting planet, with ingress advancing at a higher rate than egress by $\sim 10^{-4} - 10^{-3}$ for hot Jupiters (Loeb, 2005). The large spatial scales of moving atmospheric structures could generate significant photospheric variability at a level detectable with Spitzer (Rauscher et al., 2007a).

Light scattered by an Earth-like planet will vary in intensity and colour as the planet rotates, such that the global planetary environment might be inferred from photometric transit observations. Studies have been made in the context of imaging missions, but may also be applicable to transit observations. Models predict diurnal variations of a factor several, depending on ice and cloud cover, seasonal variations, surface composition (e.g. ocean versus land composition), atmospheric structures, and zonal winds (e.g. Ford et al., 2001b; Gaidos & Williams, 2004; Pallé et al., 2008a; Williams & Gaidos, 2008; Barnes et al., 2009a; Oakley & Cash, 2009).

Star spots Structure in high signal-to-noise transit curves has been attributed to flares or star spots in the case of HD 189733 b (Pont et al., 2007, Figure 6.19), HD 209458 (Silva, 2008), TrES–1 (Dittmann et al., 2009; Rabus et al., 2009a), and OGLE–TR–10 b (Bentley et al., 2009). Such spots can also be inferred from photometric (rotational) variability, e.g. as demonstrated for GJ 436 by Demory et al. (2007).

If successive transits pass across the same spots (as HD 189733 and HD 209458), the stellar rotation period may be determined independently from rotationally-broadened spectral lines (Silva, 2003). For a suitably-aligned star, the spot structure will recur in successive transits for as long as the spot is on the visible hemisphere, steadily advancing in phase due to the star's rotation between transits. Silva (2008) used structure in the four transits of HD 209458 b observed with HST by Brown et al. (2001) to derive a stellar rotation period of 9.9 or 11.4 d.

The nature of the spot structure across successive transits in principle constrains the angle between the planet orbit and the star's rotation axis (Winn et al., 2010c). For stars like HAT–P–11, with the planet orbit almost orthogonal to the stellar rotation axis ($\lambda \sim 90°$, Table 6.1) star spot events will not recur across successive transits, because the star's rotation moves the spot away from the transit chord. A spot must complete a full rotation before returning to the transit chord, and even then, the planet will miss it unless it has also completed an integral number of orbits.

Loeb (2009) quantified the small effect on the transit duration caused by variations in photospheric radius of the host star attributable to changes in its magnetic activity. Effects would be larger for M dwarfs. Specific studies for CoRoT 2 were reported by Czesla et al. (2009).

Rapid stellar rotation Barnes (2009) demonstrated how rapid stellar rotation, which leads to photospheric temperatures hotter at the pole than the equator by several thousand K (von Zeipel, 1924) should lead to distinctive light curves from which the relative alignment of the stellar rotation pole and the planet orbit normal can be derived. It is analogous to the Rossiter–McLaughlin effect on the radial velocities, and could similarly be used to constrain theories of planet formation and migration, especially for rapid rotators where radial velocity measures are difficult.

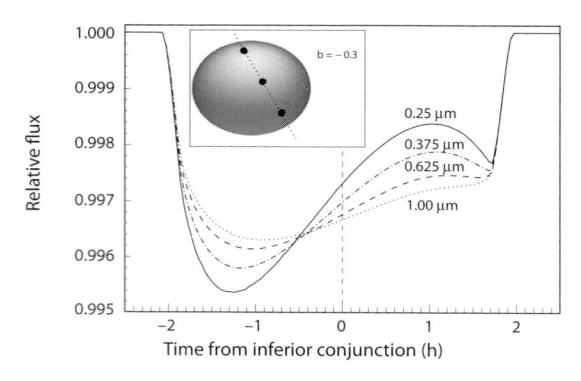

Figure 6.20: Model light curves for a $1R_J$ planet in a 0.05 AU orbit around an Altair-like star with a 60° obliquity (inset). Curves correspond to a transit impact parameter $b = -0.3 R_{pole}$ at different wavelengths, assuming identical limb darkening. The contrast is greatest at the short-wavelength (Wien) side of the blackbody curve, and least at the long-wavelength (Rayleigh–Jeans) side. From Barnes (2009, Figure 9), reproduced by permission of the AAS.

For planets which are spin–orbit aligned, the von Zeipel effect will cause systematic errors in the radii determined both for the star and planet, especially for high impact parameters, and will lead to broad-band colour variations during the transit. For significant misalignments, highly asymmetric transit light curves can result, constraining both the stellar spin pole direction and the spin–orbit alignment, but which may not be immediately recognisable as planetary transits (Figure 6.20). Barnes (2009) estimates that some 5–10% of the Kepler target stars should be rapid rotators.

Ellipsoidal variations The assumption of a spherical star for transit modeling also breaks down in the case of a close orbiting planet, which can tidally distort the star, and consequently leads to a flux modulation twice per orbit. The resulting *ellipsoidal variations* in the light curve of the luminous fluid body are well-known in studies of close stellar binaries, and the models which have been developed for them can similarly be applied to the resulting phase varying light from a star and close-orbiting planet (Wilson, 1990; Orosz & Hauschildt, 2000; Wittenmyer et al., 2005).

The presence of ellipsoidal variations in exoplanet systems was anticipated by Loeb & Gaudi (2003) and Drake (2003). Pfahl et al. (2008) presented a detailed theoretical investigation, and estimated that ellipsoidal oscillations on the star, induced by tidal forcing by an orbiting giant planet, could be detected in approximately 100 of the Kepler satellite targets.

Significant ellipsoidal variations were first reported by Welsh et al. (2010) from the Kepler light curve of the transiting system HAT–P–7. The planet is separated by $\sim 4R_\star$, and leads to the anticipated flux modulation twice per orbit (Figure 6.21).

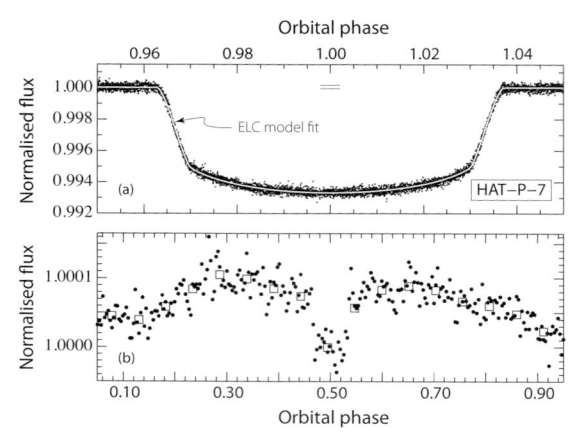

Figure 6.21: The Kepler light curve of HAT–P–7: (a) phase-folded light curve across the transit at 1 min cadence, with the fit from the ELC model code of Orosz & Hauschildt (2000) superimposed (white line); (b) the light curve over the full orbital period averaged in 5 min (filled circles) and 75 min (open squares) bins. The double-hump profile is due to the ellipsoidal variations of the star plus reflected light from the planet. The vertical extent of the lower panel is indicated by the central horizontal lines in (a). From Welsh et al. (2010, Figure 1), reproduced by permission of the AAS.

Early transit ingress and bow shocks Wavelength-dependent transit depth variations probe the outer regions of the planet's atmosphere (§6.5). An early transit ingress observed in the near ultraviolet spectrum of the highly-irradiated WASP–12 b has been attributed to the presence of a disk of previously stripped material (Fossati et al., 2010).

In an alternative explanation, Vidotto et al. (2010) suggested that the early ingress is caused by a bow shock ahead of the planet. They derived an upper limit to the magnetic field of WASP–12 b of $B_p \lesssim 2.4 \times 10^{-3}$ T, and concluded that shock formation leading to an observable early ultraviolet ingress is likely to be a common feature of transiting systems, placing constraints on planetary magnetic field strengths, and their associated radio emission (§7.6.2).

6.4.13 Higher-order timing effects

To first order, a transiting planet may be considered as having a constant orbital period, which can be determined from a linear fit to the observed transit times after correction for the observer's motion around the solar system barycentre, including light travel time.

Various effects may lead to changes in the transit time duration, in the interval between successive transits, in the interval between transits and the secondary eclipse, and in the form of the light curve including differences in the ingress and egress shapes and times.

Such transit time variations include perturbations due to other gravitating bodies, tidal forces, and rela-

6.4 Transit light curves

tivistic precession, as well as apparent effects due to changes in geometrical projection (proper motion and parallax) as viewed by the observer. Other effects noted in the literature but not considered further here include orbit decay (Sasselov, 2003), magnetic breaking (Lee et al., 2009; Barker & Ogilvie, 2009), and non-gravitational (Yarkovsky) effects (Fabrycky, 2008).

Timing precision Even when transit observations during ingress and egress are well sampled, the standard error on the determination of the mid-transit time cannot surpass that due to photon statistics alone (Holman & Murray, 2005, eqn 3)

$$\frac{\sigma_{t_c}}{t_T} \sim (S\, t_T)^{-1/2} \left(\frac{R_p}{R_\star}\right), \quad (6.45)$$

where t_T is the transit duration and S the photon count rate of the star. For the 0.95-m aperture Kepler, $S \simeq 7.8 \times 10^8\, 10^{-0.4(V-12)}\, \mathrm{hr}^{-1}$, such that a Jupiter-mass planet in a 1 AU orbit around a $1 M_\odot$ star at $V = 12$ ($t_T \simeq 13$ hr, $R_p/R_\odot \simeq 0.1$) gives $\sigma_{t_c} \simeq 20$ s, while for a terrestrial-size planet $\sigma_{t_c} \simeq 500$ s. Scaled to an 8-m telescope gives limits of 0.3 s and 10 s respectively.

Apsidal precession Spherical masses obey a r^{-2} force law, and execute closed elliptical orbits as a result. In reality, the centrifugal potential of spinning bodies causes rotational flattening, while the tidal potential of a nearby mass raises tidal bulges. Both effects create gravitational quadrupole fields, with an r^{-3} dependency, that result in orbit precession. Other orbiting planets, and the spacetime metric of general relativity, also result in precessional motion.

Observational consequences for the planet orbit can be split into two components: *apsidal precession* in which the orbit ellipse rotates in its own plane, and *nodal precession*, out of the orbit plane, in which the orbit normal precesses about the total angular momentum vector. For eccentric orbits, both will result in long-term variations of the transit times, and of the transit duration. Miralda-Escudé (2002) and Ragozzine & Wolf (2009) derived expressions for various combinations of these effects. Typically, the apsidal precession component is more dominant.

Irrespective of the cause, contributions add linearly to first order. The total apsidal precession is then

$$\dot\omega_\mathrm{tot} = \dot\omega_\mathrm{tid,p} + \dot\omega_\mathrm{gr} + \dot\omega_\mathrm{rot,p} + \dot\omega_\mathrm{rot,\star} + \dot\omega_\mathrm{tid,\star} + \dot\omega_\mathrm{p2}\,, \quad (6.46)$$

with the terms broadly in order of their importance for an isolated hot Jupiter (Ragozzine & Wolf, 2009, eqn 13), along with a final term due to the possible presence of a second planet.

The general relativistic precession rate is given, again to first order, by (e.g. Will, 1993)

$$\dot\omega_\mathrm{gr} = \frac{3 G M_\star n}{a c^2 (1-e^2)}\,, \quad (6.47)$$

Instrumentation for transit time determinations: The importance of accurate transit time measurements has brought a renewed focus to instrumental techniques.

The fast readout camera RISE at the 2-m Liverpool telescope, La Palma, currently reaches ~10 s timing accuracy for the best transits (Steele et al., 2008; Gibson et al., 2010, and references).

Other high-speed instrumental photometry measurements of transit events have been reported. Three-colour optical imaging at frame rates up to 500 Hz have been made using ULTRACAM (Dhillon et al., 2007; Bentley et al., 2009), at the 4.2-m WHT at La Palma, and the 8.2-m VLT and the 3.5-m NTT in Chile (Figure 6.5), although the 5 arcmin field of view provides a limited probability of finding a comparison star of suitable brightness.

Orthogonal transfer array CCDs shift accumulated charge both horizontally and vertically during an exposure to produce broad, stable point-spread functions. Observations of WASP–10 and TrES–3 with a photometric precision of 0.7 mmag per minute have been reported by Johnson et al. (2009).

The possibility of measuring the energy of an individual optical photon directly, using superconducting tunnel junctions, and without recourse to filters or dispersive devices, was first proposed theoretically by Perryman et al. (1993), and demonstrated by Peacock et al. (1996). Similar technological implementations have since been reported (e.g. Cabrera et al., 1998). These devices have very high intrinsic quantum efficiency, approaching 100% from the ultraviolet to the near infrared, along with μs-level arrival time and spatial position of each photon (e.g. Perryman et al., 1999; de Bruijne et al., 2002; Reynolds et al., 2003).

STJ-based transit observations of HD 209458, HD 189733, and TrES–1, using a 12×10 superconducting pixel array, have been reported by Stankov et al. (2007). The 2 μm interpixel gaps lead to a variable seeing-induced photometric light loss which limits current applications of the technology.

where $n = (GM_\star/a^3)^{1/2}$ is the mean motion, in rad s^{-1}. For $P = 10$ d, $a \sim 0.1$ AU, and $e \simeq 0$, this gives $\dot\omega_\mathrm{gr} \simeq 4 \times 10^{-7} n \simeq 10^{-4}$ yr^{-1}. This general relativistic precession component has also been considered by Pál & Kocsis (2008) and Rafikov (2009b), and its observability for hot Jupiters by Iorio (2006a), Adams & Laughlin (2006b), and Jordán & Bakos (2008).

$\dot\omega_\mathrm{p2}$ is the precession rate due to a second planet, derived using the epicycle approximation for small e as (Miralda-Escudé, 2002, eqn 17)

$$\dot\omega_\mathrm{p2} = \frac{3 M_2 a^3}{4 M_\star a_2^3}\, n\,. \quad (6.48)$$

For an Earth-like planet with $M_2/M_\star \simeq 3 \times 10^{-6}$, and $a_2 \simeq 2a$, the precession rate is $\dot\omega_\mathrm{p2} \simeq 3 \times 10^{-7} n$, comparable to that due to relativistic precession.

$\dot\omega_\mathrm{rot,\star}$ and $\dot\omega_\mathrm{rot,p}$ are the precession rates due to the quadrupole field resulting from rotational flattening. For the star (Miralda-Escudé, 2002, eqn 16)

$$\dot\omega_\mathrm{rot,\star} = \frac{3 J_2 R_\star^2}{2 a^2}\, n\,. \quad (6.49)$$

For $J_2 \simeq 10^{-6}$ and $R_\star \lesssim 0.1a$, $\dot\omega_{\rm rot,\star} \lesssim 10^{-8} n$, much smaller than $\dot\omega_{\rm gr}$. For a hot Jupiter tidally locked and synchronously rotating with the star, spin periods are somewhat longer than that of Jupiter, $\simeq 10\,{\rm h}$. The rotational flattening scales with the inverse square of the spin period, with the consequence that $\dot\omega_{\rm rot,p}$ is also small. The quadrupole term has been considered explicitly for HD 209458 by Iorio (2006b).

$\dot\omega_{\rm tid,\star}, \dot\omega_{\rm tid,p}$ are the precession rates due to the tidal potentials of the star and planet, given approximately by (Ragozzine & Wolf, 2009, eqn 6–7)

$$\dot\omega_{\rm tid,\star} \simeq \frac{15}{2} k_{2\star} \left(\frac{R_\star}{a}\right)^5 \frac{M_{\rm p}}{M_\star}\left(1 + \frac{13}{2} e^2\right) n \qquad (6.50)$$

for the star, and analogously for the planet. k_2 is the Love number (see box). Using Jupiter and solar values for the planet and star, their ratio is

$$\frac{\dot\omega_{\rm tid,p}}{\dot\omega_{\rm tid,\star}} = \frac{k_{2\rm p}}{k_{2\star}} \left(\frac{R_{\rm p}}{R_\star}\right)^5 \left(\frac{M_\star}{M_{\rm p}}\right)^2 \simeq 100 \, , \qquad (6.51)$$

assuming $k_{2\rm p}/k_{2\star} \simeq 10$. In consequence, the star raises a significant tidal bulge on the planet, which dominates the total precession (Equation 6.46).

As a result, apsidal precession for hot Jupiters is expected to be primarily determined by the contribution from the planetary tidal bulge which, for small e, can be written (Ragozzine & Wolf, 2009, eqn 14)

$$\dot\omega_{\rm tid,p} \simeq 0.59 \left(\frac{k_{2\rm p}}{0.3}\right) \left(\frac{M_\star}{M_\odot}\right)^{3/2} \left(\frac{M_{\rm p}}{M_{\rm J}}\right)^{-1} \left(\frac{R_{\rm p}}{R_{\rm J}}\right)^5 \times \left(\frac{a}{0.025\,{\rm AU}}\right)^{-13/2} \text{degrees yr}^{-1} . \qquad (6.52)$$

The precession rate reaches a few degrees per year for the known planets (Ragozzine & Wolf, 2009, Table 1), and up to $19°\!.9\,{\rm yr}^{-1}$ for WASP–12 b. Its measurement, through its effect on the measured light curves, would directly probe the interior of the orbiting planet. Detailed studies for HAT–P–13 show how improved measurement of the orbital eccentricity will place improved limits on $k_{2\rm p}$ and consequently on the planetary core mass (Batygin et al., 2009).

Models which calculate oblateness for hot Jupiters assuming different core masses are given by, e.g., Barnes & Fortney (2003), and constraints on oblateness from transit light curves (but not the secular time evolution) have been determined by Carter & Winn (2010).

Nodal precession The discovery of transiting planets with almost polar orbits (§6.4.11) leads to the possibility of measuring transit time variations in systems in which nodal precession dominates. An appropriate candidate is WASP–33, a fast rotating main sequence star which hosts a hot Jupiter with $a = 0.02\,{\rm AU}$ and $e \sim 0$, moving along a retrograde and almost polar orbit (Collier Cameron et al., 2010b; Herrero et al., 2011).

Rotation, J_2, and Love numbers: The rotation of stars and planets leads to a flattening of their polar regions. The resulting obslateness is quantified by the flattening parameter, $(R_+ - R_-)/R_-$, where R_+ and R_- are the equatorial and polar radii, which can be determined empirically from imaging. They are related to the rotational period through contours of constant (gravitational and centrifugal) potential, expressed in terms of the spherical mass moments, J_n.

For the Sun, the quadrupole moment $J_2 = (C - A)/M_\odot R_\odot^2$ (where C and A are the moments of inertia about the body's rotation and equatorial axes respectively) results from the rotation of the stellar interior, and its mass distribution as a function of radius. Observations of the solar diameter indicate an oblateness ranging from 8.8×10^{-6} from stratospheric balloon observations (Lydon & Sofia, 1996), 1.1×10^{-5} measured from the ground (Bursa, 1986), and 9.8×10^{-6} from the MDI instrument on SOHO (Kuhn et al., 1998). The resulting difference in apparent diameter from equator to pole lies between 17–22 mas. Combined with a model of the solar interior and of the differential rotation constrained by helioseismology, these give model-dependent estimates of $J_2 = (2.2 \pm 0.1) \times 10^{-7}$ (Mecheri et al., 2004). Empirical estimates of $J_2 \sim 3 \times 10^{-6}$ are found from the effect of the solar quadrupole on lunar libration and the precession of planetary orbits.

In the solar system, oblateness reaches 0.098 for Saturn and 0.065 for Jupiter (Murray & Dermott, 2000).

A planet's Love number, k_2, is the constant of proportionality between an applied second-degree potential, and the resulting field that it induces at the planet's surface

$$V_2^{\rm ind}(R_{\rm p}) \equiv k_{2\rm p} V_2^{\rm app}(R_{\rm p}) , \qquad (6.53)$$

and is a measure of how the redistribution of mass caused by an external potential affects the external gravity field of the planet. Its importance for exoplanets is that it depends on the planet's internal mass distribution, including the presence or absence of a solid core. For main sequence stars $k_2 \simeq 0.03$ (Claret, 1995) implies that their low-mass outer envelopes have little effect on the gravity field, while higher values for Saturn (0.32) and Jupiter (0.49) reflect their more uniform density distributions albeit with differing central condensation.

The classical and relativistic nodal precession rates for WASP–33 b have been determined by Iorio (2011). As a consequence of the close-in orbit and rapid stellar rotation, the quadrupole moment and angular momentum of the star are 1900 and 400 times larger than those of the Sun respectively, resulting in substantial classical and general relativistic non-Keplerian orbital effects. The resulting nodal precession rate is 9×10^9 times larger than that induced by the Sun's oblateness on the orbit of Mercury, while the general relativistic gravitomagnetic nodal precession is 3×10^5 times larger than the Lense–Thirring effect on Mercury due to the Sun's rotation.

The magnitudes of the resulting rate of change in the transit duration are of order 3×10^{-6}, 2×10^{-7}, and 8×10^{-9} for the stellar J_2, the planet's rotational oblateness, and general relativity, respectively, suggesting that the effects may be measurable over several years.

6.4 Transit light curves

Other orbiting planets In addition to the precession treatment, transit time variations due to additional gravitating bodies has been studied in some depth (Miralda-Escudé, 2002; Agol et al., 2005; Holman & Murray, 2005; Heyl & Gladman, 2007; Nesvorný & Morbidelli, 2008; Nesvorný, 2009; Holman, 2010). Effects are particularly sensitive to bodies in resonant orbits, where mid-transit timing accuracies of 10 s would allow the detection of Earth-mass planets in low-order mean-motion resonance, or more massive planets out of resonance (Gibson et al., 2010).

Transit time studies have been made for CoRoT–1 (Bean, 2009); GJ 436 (Alonso et al., 2008b; Bean et al., 2008a; Coughlin et al., 2008; Cáceres et al., 2009); HAT–P–3 (Gibson et al., 2010); HD 189733 (Miller-Ricci et al., 2008a; Hrudková et al., 2010); HD 209458 (Brown et al., 2001; Agol & Steffen, 2007; Miller-Ricci et al., 2008b); OGLE–TR–111 b (Díaz et al., 2008; Adams et al., 2010a); OGLE–TR–113 b (Adams et al., 2010b); TrES–1 (Steffen & Agol, 2005; Rabus et al., 2009b,c); TrES–2 (Mislis & Schmitt, 2009; Rabus et al., 2009c); TrES–3 (Gibson et al., 2009); WASP–3 (Gibson et al., 2008; Maciejewski et al., 2010); and XO–1 (Cáceres et al., 2009).

Changes in transit duration for TrES–2, ~3 min over two years, were reported by Mislis & Schmitt (2009). Similar variations reported for OGLE–TR–111 b by Díaz et al. (2008) were not confirmed by Adams et al. (2010a).

Transit time variations of ~1–2 min for WASP–3 b were reported by Maciejewski et al. (2010). Their favoured explanation was a second planet of $15 M_\oplus$ located close to the outer 2:1 mean motion resonance.

Significant transit time variations of the two-planet transiting system Kepler–9 have been measured from seven months of Kepler observations (Holman et al., 2010). The 19.2- and 38.9-d orbital periods of the two planets are increasing and decreasing at respective average rates of 4 and 39 min per orbit, explicable by gravitational interaction of two planets near a 2:1 orbital resonance (Figure 6.22). The transit times of the inner body display an additional variation of smaller amplitude.

Satellites Various studies have considered the detectability of satellites or exomoons from transit timing variations (Sartoretti & Schneider, 1999; Arnold, 2005; Schneider, 2005; Szabó et al., 2006; Simon et al., 2007; Kipping, 2009a,b; Kipping et al., 2009).

For a circular satellite orbit, the displacement of the planet with respect to the planet–satellite barycentre is

$$\delta a = a_\mathrm{s} \frac{M_\mathrm{s}}{M_\mathrm{p}}, \qquad (6.54)$$

where $a_\mathrm{s}, M_\mathrm{s}$ are the satellite's semi-major axis and mass. For circular coplanar orbits, the peak-to-peak time difference between the mid-transit point for the planet and system barycentre is then (Sartoretti &

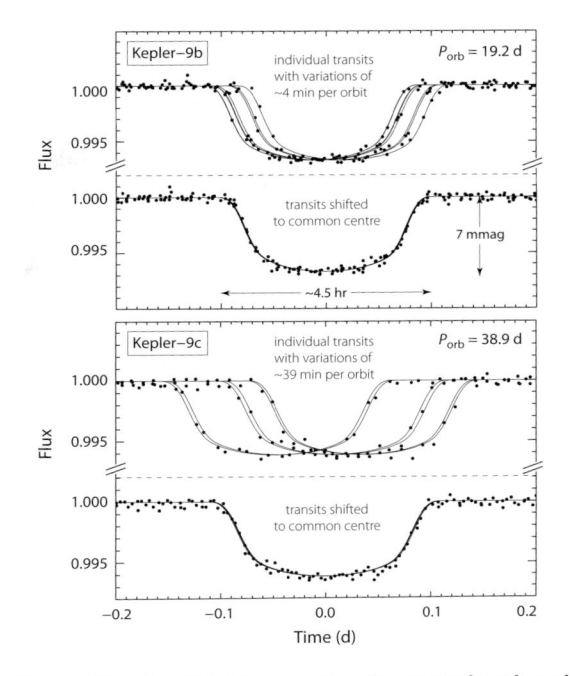

Figure 6.22: Transit light curves for planets Kepler–9 b and Kepler–9 c. In both panels, the top curve shows the data folded with the best-fit period. Significant displacements between different transits are due to transit time variations arising from the gravitational interactions between the planets (in the original colour figure, data points from different transits are distinguishable). The bottom curves show the transits shifted to a common centre. Adapted from Holman et al. (2010, Figure 5).

Schneider, 1999; Kipping, 2009a)

$$\Delta t \simeq 2 a_\mathrm{s} \frac{M_\mathrm{s}}{M_\mathrm{p}} \frac{P}{2\pi a}, \qquad (6.55)$$

where P, a refer, as usual, to the period and orbital radius of the planet. For a $1 M_\oplus$ satellite orbiting HD 209458 b at a maximum distance of the Hill radius, R_H, the amplitude of the timing excursion about the mean orbital phase is 13 s, comparable to the present standard error on the central time of a single transit (Brown et al., 2001).

Since the displacement is proportional to the product $a_\mathrm{s} M_\mathrm{s}$, determining M_s requires an independent knowledge of a_s (Ford & Holman, 2007). Estimating the satellite orbit period uniquely is impeded by the Nyquist-limited sampling imposed by the fundamental (transit) sampling rate of $0.5 P^{-1}$ (Kipping, 2009a).

Kipping (2009a) included orbital eccentricity in the expression for the rms transit time variation. He also studied the variation of transit duration, which has a different dependency being $\propto M_\mathrm{s} a_\mathrm{s}^{-1/2}$. Together, the transit time variation and the transit time duration allow M_s and a_s to be determined separately. For the three most favourable, rms transit times and transit durations are respectively {13.7, 12.6 s} for GJ 436 b, {7.7, 9.3 s} for CoRoT–4 b, and {4.6, 6.8 s} for HAT–P–1 b.

The more general case of non-coplanar orbits was studied by Kipping (2009b). Results applied to Kepler-class photometry suggest that habitable zone satellites down to $0.2 M_\oplus$ could be detected (Kipping et al., 2009).

Trojan planets Trojan planets reside at the Lagrange L4/L5 points of a planet's orbit, as are found accompanying Mars, Jupiter and Neptune in the solar system. They may be a frequent by-product of planet formation and evolution, either formed *in situ* from the proto-planetary disk and surviving through inward migration (Laughlin & Chambers, 2002; Chiang & Lithwick, 2005), or from various capture mechanisms (Morbidelli et al., 2005; Chiang & Lithwick, 2005). In theory they may reach Trojan/planet mass ratios much larger than the $\lesssim 7 \times 10^{-9}$ found in the solar system, perhaps as large as unity (Laughlin & Chambers, 2002). Terrestrial mass Trojans of giant planets in the habitable zone have been hypothesised (Ji et al., 2005; Schwarz et al., 2005).

Ford & Gaudi (2006) showed that, given a large-enough libration amplitude, Trojans could be detected from dynamical measurements (radial velocity or astrometry), as well as from transit photometry or transit-timing measurements. They could also be detected from a (systematic) difference between the transit mid-time and that calculated from the radial velocity data alone, an effect that recurs at every transit with amplitude

$$\Delta t \simeq \pm 37.5 \left(\frac{P}{3\,\text{days}}\right)\left(\frac{M_T}{10 M_\oplus}\right)\left(\frac{0.5 M_J}{M_p + M_T}\right) \text{ min}, \quad (6.56)$$

where M_T is the Trojan mass.

Their studies ruled out Trojan companions to HD 209458 b and HD 149026 b more massive than 13 and $25 M_\oplus$ respectively. Madhusudhan & Winn (2009) extended the search to 25 systems, finding the most constraining limit of $2.8 M_\oplus$ for GJ 436. Ford & Holman (2007) extended the idea to search for Trojans from variations in the transit time using photometric data alone.

Effects of parallax and space motion The timing and duration of a transit has a periodic dependency on the observer's position due to trigonometric parallax, and an apparent secular evolution due to the star's space motion (Figure 6.23); both effects have been considered by Scharf (2007) and Rafikov (2009b).

For an exoplanet orbit coplanar with that of the Earth–Sun system, observers at the two extremes of the Earth's orbit would register a given alignment phenomenon displaced in time by

$$\Delta t = \frac{A_m P}{\pi d_\star}, \quad (6.57)$$

where A_m is the mean orbital radius of the Earth (1 AU), P the orbital period, and $d_\star$ the distance to the star. For $P = 400$ d and $d_\star = 10$ pc, $\Delta t \simeq 5$ sec; the effect decreases linearly with increasing $d_\star$ and decreasing P, and so

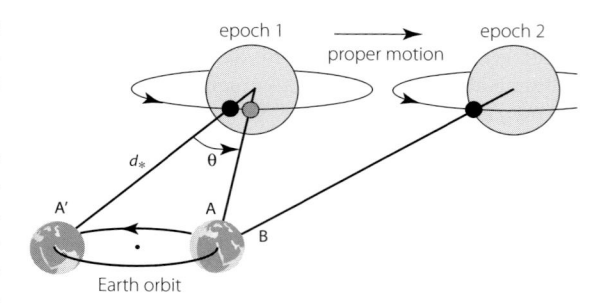

Figure 6.23: From different points in the Earth's annual orbit around the Sun (AA') the observer registers a different time for a given contact due to parallax. As a distinct effect, the star's proper motion through space leads to a change in the exoplanet's projected orbital plane viewed by the observer (AB), leading to a secular change in the measured transit times.

should be irrelevant for hot Jupiters. For different alignment geometries, the observer's orbital motion results in a changing effective orbital inclination i, and hence a change in the apparent transit duration (Equation 6.2).

Change in alignment geometry due to the star's proper motion also leads to changes in the measured transit time and transit duration (Rafikov, 2009b, eqn 11). For the coplanar alignment geometry, with $d_\star = 100$ pc, $P = 400$ d, and a proper motion corresponding to a transverse stellar velocity of $20\,\text{km s}^{-1}$, the secular change in transit time is 10 s over 10 yr. For an orthogonal alignment corresponding to a drift in the systems's apparent orbital inclination, the effect reaches $\sim 20-200$ s (Scharf, 2007).

Rafikov (2009b) also gives an expression for the analogue of the Shklovskii effect in pulsar timing (Shklovskii, 1970) due to the special relativistic contribution to the orbital period arising from a combination of the star's radial motion and the finite velocity of light. For large systemic radial velocities, this term may be comparable to the contribution from general relativistic precession.

6.4.14 Higher-order spectroscopic effects

In addition to the basic orbital Doppler shift, and the Rossiter–McLaughlin effect due to the selective blocking of the stellar light by the orbiting planet, other more subtle spectroscopic effects have been studied (Seager & Sasselov, 2000; Brown, 2001). These include the wavelength dependence of atmospheric opacity (leading to the possibility of characterising the atmospheric composition, considered further in §6.5), effects due to the planet's (possibly tidally locked) rotation, and atmospheric winds. For HD 209458, Spiegel et al. (2007) estimated the centroid shift due to rotation as $0.6\,\text{m s}^{-1}$ on the stellar absorption lines, making the possibility of distinguishing between a rotating and non-rotating planet impossible according to current capabilities.

6.5 Transmission and emission spectroscopy

6.5.1 Background

The transit and secondary eclipse represent two distinct but related configurations which can be used to probe an exoplanet's atmosphere. In both situations, observations are made in the combined light of the star–planet system, and conditions in the planetary atmosphere are deduced from the differences in flux as the planet moves in front of, or behind the star (Figure 6.24). Both effects are at the limit of what can be studied effectively from the ground, even for the brightest systems.

Observations from space, notably from HST and Spitzer, avoid atmospheric effects and in particular scintillation. Spitzer in particular gives access to the mid-infrared and beyond where limb darkening is less significant, the planet–star contrast ratio is significantly improved (from $10^{-5} - 10^{-6}$ in the optical to $10^{-3} - 10^{-4}$ in the infrared; see Figure 6.25), and where a number of strong molecular absorption bands, including H_2O, CO and CH_4, are present (Tinetti et al., 2007a).

Transmission spectroscopy When the planet transits in front of the star, a fraction of the stellar light passes through the narrow annulus of the planet's atmosphere surrounding its limb: at wavelengths where the planet atmosphere absorbs the grazing star light more strongly, the transit depth increases. The wavelength-dependent *transmission spectrum* probes the outer regions of the atmosphere, and conveys information about the atoms, molecules, and condensates present in it (Brown, 2001).

The area of the planetary atmosphere intercepted is approximately an annulus of radial dimension $\simeq 5H$ (Seager et al., 2009), where

$$H = \frac{kT}{\mu_m g_p} \qquad (6.58)$$

is the atmospheric scale height, k is Boltzmann's constant, T the atmospheric temperature, μ_m the mean molecular weight, and g_p is the planet's surface gravity. The fractional contribution of the transmission signal is given by the ratio of the annular to stellar areas

$$\delta \simeq 5 \times \frac{2R_p H}{R_\star^2} \; . \qquad (6.59)$$

For a hot Jupiter dominated by an atmosphere of hydrogen ($\mu_m = 2$) $\delta \simeq 10^{-4}$. Since the effect scales with H, and therefore $\propto \mu_m^{-1}$, detection of the atmosphere of a (super)-Earth planet by means of CO_2 absorption, for example, is proportionally more of a challenge (Miller-Ricci et al., 2009b).

Emission spectroscopy Observations spanning the secondary eclipse, when the planet passes out of view behind the star, provides a measure of the planet's thermal emission and associated spectral features. This is

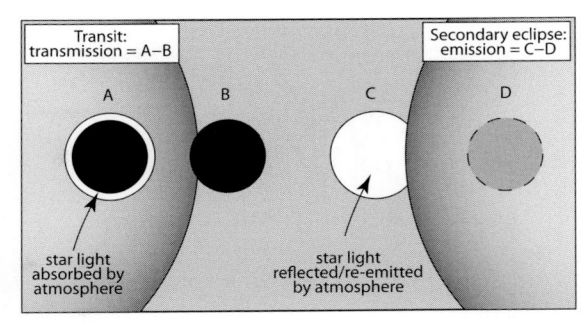

Figure 6.24: Geometry of transmission (left) and emission spectroscopy (right). During the transit, part of the background star light passes through the (annular) atmosphere of the planet. During the secondary eclipse, there is a switching off of the light reflected or emitted from the day-side surface of the planet.

estimated from the difference between the total light just before or after eclipse (star+planet day-side) and that during the eclipse (star only). The *emission spectrum* contains information about the atmosphere's temperature and gradient. The presence of absorption lines indicates a temperature profile decreasing with height, while emission lines signify the converse. An isothermal profile would produce a featureless emission spectrum.

CoRoT–1, as observed by the satellite itself, illustrates the type of complete orbit light curve that has been obtained for a few transiting systems (Snellen et al., 2009a). The satellite data over 55 days, and phase folded at the orbital period, clearly show the transits, followed by an increasing flux as the day-side hemisphere rotates into view, followed by the secondary eclipse by the star, before the planet rotates out of view again (Figure 6.26). The observations are well reproduced by a day-side hemisphere of uniform surface brightness with a ratio of planet to stellar flux (from the transit depth and the phase variation) of 1.26×10^{-4}, and a night-side ratio (from the secondary eclipse depth of 0.016% and the phase variation) of $< 3 \times 10^{-5}$, consistent with the night-side hemisphere of the planet being entirely black.

Tidal locking and asymmetric heating Hot Jupiters are expected to be tidally locked to their host stars, such that the stellar flux is always incident on the same illuminated hemisphere. This highly asymmetric heating drives strong atmospheric circulation which in turn may transport significant amounts of energy to the night-side (Showman & Guillot, 2002; Cho et al., 2003). Observations of the total flux as a function of orbital phase can establish whether the resulting planet temperatures are extremely high on the illuminated side and very low on the other, or whether atmospheric circulation efficiently redistributes the absorbed stellar radiation from the illuminated to the non-illuminated hemispheres.

If the stellar radiation is absorbed high in the atmosphere, reradiation is expected to dominate over radiation transport around the atmosphere due to bulk flow

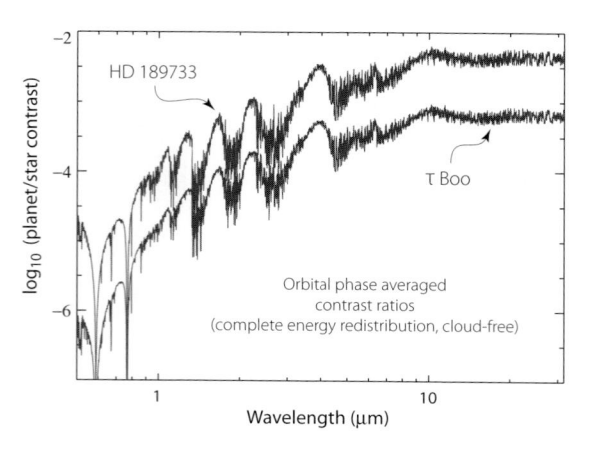

Figure 6.25: Theoretical planet–star contrast ratios versus wavelength for two of the closest giant planets, HD 189733 and τ Boo. HD 209458, TrES–1, and 51 Peg lie between these two cases. The models assume complete energy redistribution and ignore cloud effects. From Burrows et al. (2006, Figure 1), reproduced by permission of the AAS.

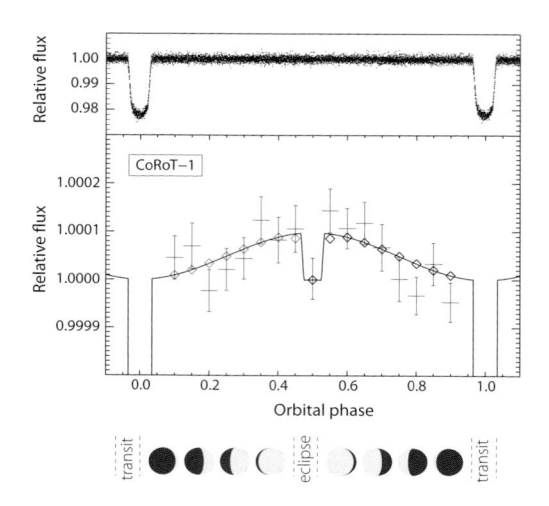

Figure 6.26: CoRoT–1 b observed over 55 days (36 orbits), and phase-folded at the orbital period P = 1.5089557 d. In the centre panel the data are binned in phase intervals of 0.05. The data are consistent with the day-side hemisphere rotating into view, being eclipsed by the star, and rotating out of view again (bottom schematic). The solid curve is a model assuming uniform (but distinct) brightness for the day and night hemispheres. From Snellen et al. (2009a, Figure 1), by permission from Macmillan Publishers Ltd, Nature, ©2009.

(advection), resulting in a high temperature contrast between the illuminated and non-illuminated hemispheres. If, in contrast, the stellar radiation penetrates deep into the atmosphere before it is absorbed, advection can dominate, and heat may be transported around the planet resulting in a smaller temperature contrast (Showman et al., 2008b).

Reconstructing the longitudinally resolved brightness map of the day-side photospheric emission has been developed based on eclipse-mapping techniques (e.g. Williams et al., 2006; Rauscher et al., 2007b; Knutson et al., 2007a; Cowan & Agol, 2008).

Equilibrium temperatures The equilibrium temperature of the planet results from a balance between the incident radiation from the host star, and that absorbed by the planet or its atmosphere. Ignoring additional heat sources (such as tidal deformation, radiogenic decay, and the greenhouse effect), and considering the incident energy intercepted by the planet's disk, the Stefan–Boltzmann law can be written

$$T_{eq} = T_\star \left(\frac{R_\star}{2a}\right)^{1/2} [f(1-A_B)]^{1/4}, \qquad (6.60)$$

where $T_\star$ is the stellar effective temperature and a is the planet's semi-major axis. A_B is the Bond albedo (the fraction of incident radiation, over all wavelengths, which is scattered). While Jupiter and the other gas giants have (geometric and Bond) albedos in the range 0.3–0.5, hot Jupiters appear to have much lower albedos; high-precision optical photometry of HD 209458 with MOST gave a geometric albedo of only 4% (Rowe et al., 2008). An upper limit on T_{eq} can be derived from Equation 6.60 by setting $A_B = 0$.

The factor f describes the effectiveness of atmospheric circulation, and the degree to which the energy absorbed is transferred from the planet's day to nightsides. If the planet is tidally locked, but the incident energy is redistributed (for example, by advection) to give an isotropic reemission and uniform equilibrium temperature over both hemispheres, then $f = 1$. If the dayside alone reradiates the incident energy (and the nightside remains cold), its higher resulting equilibrium temperature is given by $f = 2$. Further adjustments to f are used to account for the fact that the angle of incidence of the insolating stellar flux decreases from the substellar point to the terminator, both in theoretical models (Burrows et al., 2003a), and in their interpretation (e.g. Harrington et al., 2006; Léger et al., 2009).

Theoretical models Various model atmospheres have been constructed to interpret the transit and secondary eclipse fluxes (see, for example, Burrows et al., 2005; Barman et al., 2005; Fortney, 2005; Fortney et al., 2005; Seager et al., 2005a; Burrows et al., 2006; Cooper & Showman, 2006; Showman et al., 2006; Burrows et al., 2007a; Chabrier & Baraffe, 2007; Koskinen et al., 2007b,a; Langton & Laughlin, 2007; Cho et al., 2008; Dobbs-Dixon & Lin, 2008; Hansen, 2008; Showman et al., 2008a,b, 2009). They variously calculate planet/star flux ratios during secondary eclipse, include parameterisation for the redistribution of heat to the planet's night-side, and calculate average day-side and night-side atmospheric temperature/pressure profiles (§11.7).

6.5 Transmission and emission spectroscopy

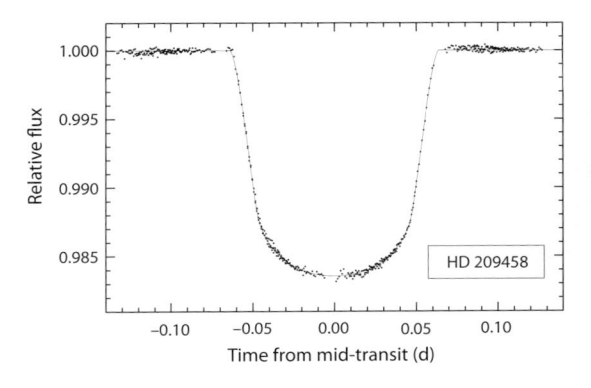

Figure 6.27: HD 209458, observed over four transits with the HST Imaging Spectrograph, and co-phased assuming a period P = 3.5247 d. From Brown et al. (2001, Figure 3), reproduced by permission of the AAS.

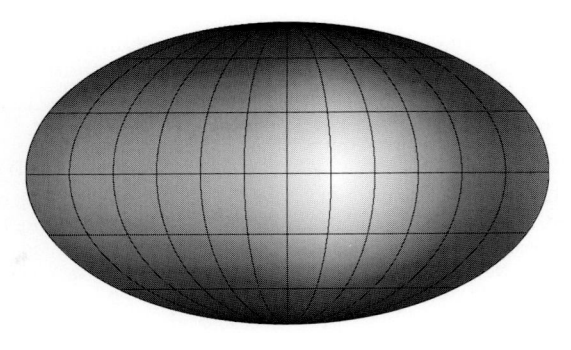

Figure 6.28: HD 189733, observed with Spitzer–IRAC at 8 µm, showing brightness estimates for twelve longitudinal strips on the planet surface. The planet is assumed tidally locked, edge-on, and with no limb-darkening. A sinusoidal dependence on latitude has been added. From Knutson et al. (2007a, Figure 3), by permission from Macmillan Publishers Ltd, Nature, ©2007.

Of particular importance in the interpretation of the transmission and emission spectroscopy features are the effects of opacity of TiO and VO influencing the global atmospheric temperature profiles, and the presence of strong molecular absorption bands in the infrared between 1–10 µm, which can be used as diagnostics for the presence of H_2O, CO and CH_4. For a hot-Jupiter atmosphere in thermochemical equilibrium, for example, the dominant carbon-bearing molecule is expected to be CO at higher temperatures ($T > 1200$ K) and CH_4 at lower temperatures ($T < 800$ K).

6.5.2 Observations

In this section, a summary of the transit and secondary eclipse observations for the two brightest transiting objects, HD 209458 and HD 189733, is followed by reference to other systems for which emission spectroscopy is being undertaken, concluding with the present picture of their large-scale atmospheric properties.

HD 209458 HD 209458 is a bright ($V = 7.7$) G0V star at $d = 47$ pc, and one of the most intensively observed transiting planets on account both of its brightness and large transit depth of almost 2% (Table 6.2). From radial velocity observations the planet, with $M_p = 0.69 M_J$, is known to orbit the $M_\star = 1.01 M_\odot$ star at a distance of 0.047 AU. Its accurate orbital period of 3.524 748 59 d was determined after *a posteriori* detection in the Hipparcos epoch photometry (Robichon & Arenou, 2000). It was the first known transiting planet, initially observed at modest signal-to-noise with small-diameter discovery instruments by Henry et al. (1999, 2000) and independently by Charbonneau et al. (2000, see Figure 6.1). It is the availability of both a planet radius and mass, an age estimate from stellar evolution theory, and a luminosity from its trigonometric parallax, which combine to make the system so valuable for theoretical study.

HST gave a much improved transit light curve (Figure 6.27), from which Brown et al. (2001) derived $R = 1.35 R_J$ from the transit decrement. The large radius for such a small mass (discussed further in §6.6) showed that these massive close-orbit planets were not rocky cores, but large, Jupiter-sized objects. Its density $\rho_p = 0.35$ Mg m^{-3}, surface gravity $g_p = 9.43$ m s^{-2}, and escape velocity $v_e = 43$ km s^{-1}, also indicate that the planet is stable against disruption by tidal forces, thermal evaporation, or mass stripping by the stellar wind.

Estimates of $M_\star, R_\star, M_p, R_p$ have been revised as improved stellar models, and improved transit light curves and orbit analyses, have been applied. The mass–radius relation continues to be a topic of investigation.

From the wavelength dependence of the estimated transit radius, the presence of neutral Na in the atmosphere was inferred (Charbonneau et al., 2002). Similar limits on CO (Brown et al., 2002), H_2O (Richardson et al., 2003a), and CH_4 (Richardson et al., 2003b) followed.

The existence of an extended and escaping gaseous envelope, or exosphere, resulting from high temperatures in its upper atmosphere, was invoked to explain strong atomic hydrogen absorption (Vidal-Madjar et al., 2003). The resulting mass loss is, nevertheless, only 0.1% over the age of the system. Carbon and oxygen are also present in the exosphere (Vidal-Madjar et al., 2004), but since both elements are too heavy to be lost by Jean's escape, their presence was attributed to hydrodynamic loss in which the heavier elements are carried by the outflowing hydrogen.

More recent HST and Spitzer observations have been interpreted, through atmospheric models, as indicating the presence of silicate clouds (Richardson et al., 2007), H_2O (Barman, 2007), TiO and VO absorption (Désert et al., 2008), CH_4, H_2O and CO_2 (Swain et al., 2009b), and the existence of an atmospheric temperature inversion resulting from high-altitude absorbers

Table 6.2: Observations of HD 209458 (top) and HD 189733 (bottom), ordered by publication date, through to the end of 2009. Under 'Type' of observation is listed the number of transits (T), or eclipses (E) observed.

Instrument	Type	Wavelength	Results	Reference
			HD 209458	
APT 0.8-m	1T	by	$R_p = 1.42 R_J$, $\rho = 0.27\,\mathrm{Mg\,m^{-3}}$	Henry et al. (1999, 2000)
STARE 0.1-m	2T	R	$R_p = 1.27 R_J$, $i = 87°\!.1$, $g_p = 9.7\,\mathrm{m\,s^{-2}}$	Charbonneau et al. (2000)
Hawaii 2.2-m	1T	400–800 nm	$R_p = 1.55 R_J$, $i = 85°\!.9$	Jha et al. (2000)
Sierra Nevada 0.9-m	1T	$uvby$	derivation of limb darkening	Deeg et al. (2001)
HST–STIS	4T	582–638 nm	$R_p = 1.347 R_J$, $\rho = 0.35\,\mathrm{Mg\,m^{-3}}$, $i = 86°\!.6$	Brown et al. (2001)
VLT–UVES	1T	328–699 nm	limit on exosphere	Moutou et al. (2001)
HST–STIS	4T	589.3 nm	detection of Na, $\delta\Delta = 2.3 \times 10^{-4}$ (published data)	Charbonneau et al. (2002)
Keck–NIRSPEC	1T	2–2.5 μm	limit on CO	Brown et al. (2002)
VLT–ISAAC	2E	3.6 μm	limit on CH_4	Richardson et al. (2003b)
IRTF–SpeX	2E	1.9–4.2 μm	limits on CO and H_2O	Richardson et al. (2003a)
HST–STIS	3T	121.5 nm	detection of H Ly-α in an extended exosphere	Vidal-Madjar et al. (2003)
VLT–ISAAC	1T	1080 nm	limit on He I	Moutou et al. (2003)
HST–STIS	4T	118–171 nm	H I + O I, C II due to hydrodynamic escape	Vidal-Madjar et al. (2004)
Subaru–HDS	1T	410–680 nm	limit on Hα	Winn et al. (2004)
Spitzer–MPIS	1E	24 μm	thermal emission $T \simeq 1130$ K, implies irradiation	Deming et al. (2005b)
Keck–NIRSPEC	3T	2 μm	limit on CO	Deming et al. (2005a)
Subaru–HDS	2T	410–680 nm	limit on Na, Li, H$\alpha/\beta/\gamma$, Fe, Ca	Narita et al. (2005)
Lick/Keck Doppler	–	–	probable eccentricity, $e = 0.014 \pm 0.009$	Laughlin et al. (2005b)
UKIRT 3.8-m	2E	2.2 μm	limit on secondary eclipse	Snellen (2005)
Various	–	–	improved orbit: $R_p = 1.35 R_J$, $M_p = 0.66 M_J$	Wittenmyer et al. (2005)
Spitzer–MPIS	2T	24 μm	$R_p = 1.26 R_J$ (for $M_\star = 1.17 R_\odot$, $R_\star = 1.06 R_\odot$)	Richardson et al. (2006)
HST–STIS	4T	290–1030 nm	limb darkening, $R_p = 1.32 R_J$, transit time limits	Knutson et al. (2007b)
HST–STIS	1T	300–550 nm	hot H: dense atmosphere of 5000 K at 8000 km	Ballester et al. (2007)
Spitzer–IRS	2E	7.5–13.2 μm	emission at 7.8/9.6 μm, possibly silicate clouds	Richardson et al. (2007)
MOST	15T	optical	limit on other planets and Trojans at $\simeq 2 - 4 M_\oplus$	Croll et al. (2007a)
IRTF	2E	3.8 μm	limit on secondary eclipse from the ground	Deming et al. (2007b)
HST–STIS	4T	290–1030 nm	water absorption from models (published data)	Barman (2007)
HST–STIS	4T	582–638 nm	surface gravity $g_p = 9.28\,\mathrm{m\,s^{-2}}$ (published data)	Southworth et al. (2007)
Spitzer–IRAC	1E	3.6–8 μm	atmospheric temperature inversion	Knutson et al. (2008)
Spitzer–IRS	2E	7.5–13.2 μm	emission at 7.5–8.5 μm (published data)	Swain et al. (2008a)
HST–ACS	2T	Ly-α	confirmation of exospheric H	Ehrenreich et al. (2008)
Subaru–HDS	1T	410–680 nm	confirmation of Na from HST (published data)	Snellen et al. (2008)
Spitzer	2T	24 μm	$R_p = 1.275 R_J$ using $e = 0.014$ (published data)	Kipping (2008)
HST–STIS	5T	290–1030 nm	TiO and VO from models (published data)	Désert et al. (2008)
MOST	15T	optical	albedo $p < 0.08$	Rowe et al. (2008)
HST–NICMOS	1E	1.5–2.5 μm	CH_4, H_2O, CO_2; temperature inversion	Swain et al. (2009b)
			HD 189733	
Spitzer–IRS	1T	7.5–14.7 μm	limit on H_2O, CH_4; efficient heat distribution	Grillmair et al. (2007)
various	8T	optical	$R_p = 1.154 R_J$, $\rho = 0.91\,\mathrm{Mg\,m^{-3}}$, $g_p = 9.28\,\mathrm{m\,s^{-2}}$	Winn et al. (2007)
Spitzer–IRAC	1T/1E	8 μm	small day/night temperature range, 973–1212 K	Knutson et al. (2007a)
CHARA	–	1.67 μm	$R_p = 1.19 R_J$ directly from $R_\star = 0.779 R_\odot$	Baines et al. (2007)
Spitzer–IRAC	2T	3.6, 5.8, 8 μm	absorption by H_2O inferred (published data)	Tinetti et al. (2007b)
Spitzer–IRAC	1T	3.5, 5.8 μm	limit on H_2O	Ehrenreich et al. (2007)
Keck–NIRSPEC	1T	2.0–2.4 μm	limit on H_2O and CO absorption	Barnes et al. (2007)
HST–ACS	3T	550–1050 nm	$R_p = 1.154 R_J$, $i = 85°\!.68$, spots, no moons	Pont et al. (2007)
HET–HRS	11T	500–900 nm	Na I doublet lines	Redfield et al. (2008)
Spitzer	–	–	H_2O from models; day–night transfer = 43%	Barman (2008)
HST–ACS	3T	550–1050 nm	featureless spectrum suggests hazy clouds	Pont et al. (2008a)
HST–NICMOS	1T	1.5–2.5 μm	CH_4 at 2.2 μm, resolved H_2O at 1.9 μm	Swain et al. (2008b)
Spitzer–IRAC	1T	3.6, 5.8 μm	H_2O in upper atmosphere	Beaulieu et al. (2008)
HST–ACS	3T	550–1050 nm	$MgSiO_3$ grains of 0.1–0.01 μm (published data)	Lecavelier et al. (2008)
Spitzer–IRAC	2E	3.6–24 μm	4 μm low-opacity emission, no inversion	Charbonneau et al. (2008)
Spitzer–IRS	10E	3.6–24 μm	H_2O vibration (6–6.5 μm), possible CO (4.5 μm)	Grillmair et al. (2008)
Spitzer–IRAC	2T	3.6–8 μm	improved $R_p/R_\star$ ratios, no H_2O at 5.8 μm	Désert et al. (2009)
HST–NICMOS	5T	1.66/1.87 μm	Raleigh scattering by sub- μm haze particles	Sing et al. (2009)

6.5 Transmission and emission spectroscopy

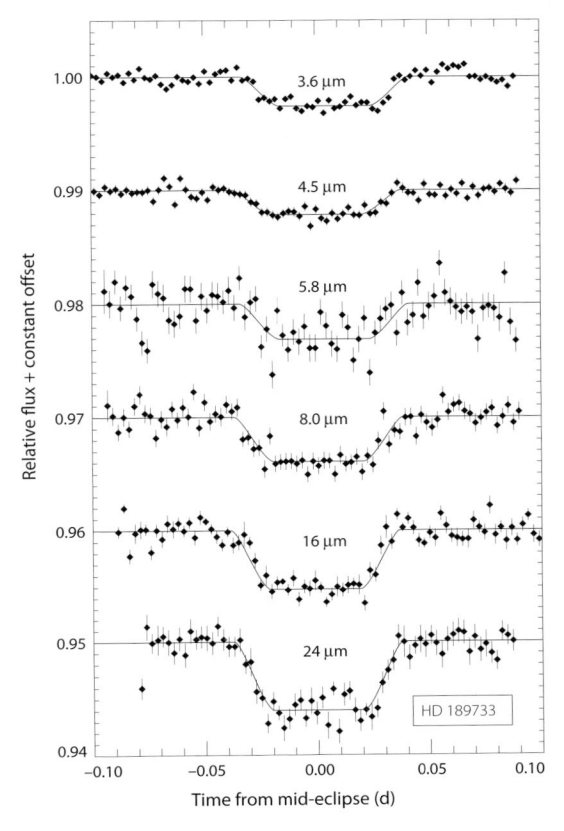

Figure 6.29: HD 189733: Spitzer time series observations over 3.6–24 µm, including best-fit eclipse curves. Each time series is binned in 3.5-min intervals, normalised, and plotted with a distinct constant offset. From Charbonneau et al. (2008, Figure 2), reproduced by permission of the AAS.

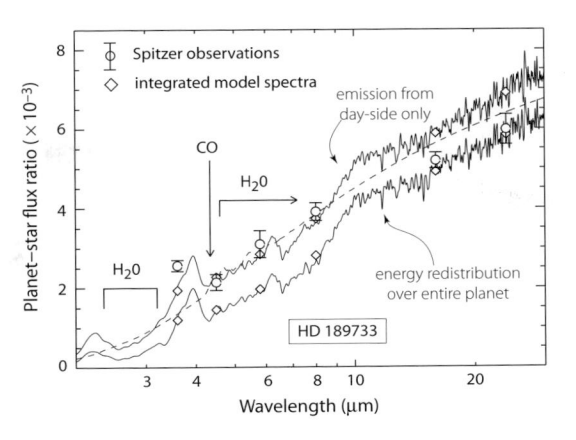

Figure 6.30: HD 189733: eclipse depths observed with Spitzer over 3.6–24 µm (circles). Models assume either that the emission of the absorbed stellar flux is constrained to the day-side only (upper curve), or uniformly redistributed over the entire planet (lower curve). Diamonds indicate the synthetic spectra integrated over the Spitzer bandpasses. The flux ratio expected for a black-body emission spectrum with T = 1292 K is shown dashed. From Charbonneau et al. (2008, Figure 3), reproduced by permission of the AAS.

(Knutson et al., 2008; Swain et al., 2009b).[5]

HD 189733 HD 189733 is a bright ($V = 7.7$) K0 star at $d = 19$ pc, orbited by a transiting giant planet. From radial velocity observations, the planet ($M_p = 1.13 M_J$, $R_p = 1.14 R_J$), orbits the $M_\star = 0.80 M_\odot$, $R_\star = 0.79 R_\odot$ star at a distance of 0.031 AU. A period $P = 2.218574(8)$ d was determined from its *a posteriori* transit detection in the Hipparcos epoch photometry (Hébrard & Lecavelier des Etangs, 2006). HD 189733 is the other most intensively observed transiting planet on account of both its brightness, its large transit depth

---

[5]A parallel can be drawn with the Earth's atmosphere, which is heated from below by solar radiation absorbed at its surface. Air above is heated by convection, with the troposphere characterised by a decreasing temperature with height. The tropopause marks the start of the Earth's temperature inversion layer, the stratosphere, which is caused by the high-level ozone layer that absorbs solar ultraviolet radiation and provides a source of heating from above. The absence of turbulence in the stratosphere is a consequence of the inverse temperature gradient, and the resulting inhibition of convection.

($\sim 2.5\%$), its proximity, and its short orbital period. The host star appears to be one component of a binary star system, the secondary being an M dwarf at a projected distance of 216 AU and with an orbital period of about 3200 years (Bakos et al., 2006).

Knutson et al. (2007a) followed the flux over over half an orbital period, and constructed a longitudinal map of the temperature distribution as the day-side of the planet rotated into view. They estimated minimum and maximum brightness temperatures of 973 ± 33 K and 1212 ± 11 K at a wavelength of 8 µm, a relatively small difference suggesting that energy from the irradiated day-side is efficiently redistributed throughout the atmosphere. Their data indicate that the peak hemisphere-integrated brightness occurs $16 \pm 6°$ before opposition (Figure 6.28).

Figure 6.29 shows the Spitzer infrared measurements of the secondary eclipse depths over 3.6–24 µm. These measurements suggest the presence of both H_2O and CO in the planet atmosphere (Charbonneau et al., 2008, Figure 6.30). Observations over 1.5–2.5 µm by Swain et al. (2008b), suggest both the presence of H_2O and CH_4 in the atmosphere (Figure 6.31).

WASP–12 An extensive set of wavelength-dependent transit-depth features has been observed, using HST–COS, for the highly-irradiated transiting exoplanet WASP–12 b (Fossati et al., 2010). The spectra cover three distinct wavelength ranges in the near ultraviolet: NUVA (253.9–258.0 nm), NUVB (265.5–269.6 nm), and NUVC (277.0–281.1 nm), and the light curves at these wavelengths imply effective radii of $2.69 \pm 0.24 R_J$, $2.18 \pm$

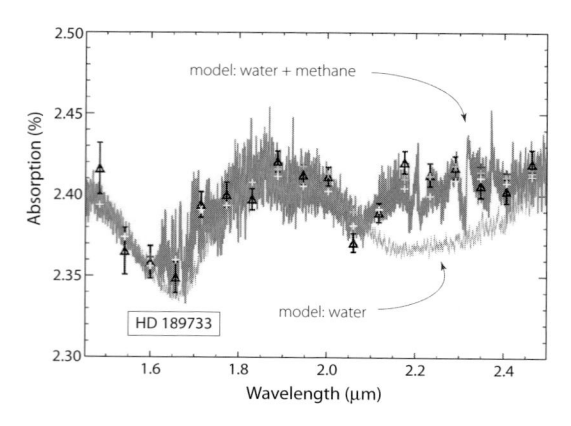

Figure 6.31: HD 189733: HST–NICMOS transit observations over 1.5–2.5 µm (▲). Two theoretical spectra of the predominantly H_2 atmosphere showing the effects of water at an abundance of 5×10^{-4} (lower), and the improved fits at 1.7–1.8 µm and at 2.15–2.4 µm with an additional methane abundance of 5×10^{-5} (upper). [The model spectrum can be improved further with the addition of small amounts of either ammonia or carbon monoxide.] From Swain et al. (2008b, Figure 2), by permission from Macmillan Publishers Ltd, Nature, ©2008.

$0.18 R_J$, and $2.66 \pm 0.22 R_J$ respectively, suggesting that the planet is surrounded by an extended absorbing cloud which overfills the Roche lobe. They detected enhanced transit depths at the wavelengths of resonance lines of neutral sodium, tin, and manganese, and at singly-ionised ytterbium, scandium, manganese, aluminium, vanadium, and magnesium.

The pM and pL classes of hot Jupiters Although each planet presumably has its own unique atmosphere, interior structure, and accretion history, the Spitzer measurements suggest that the atmospheres of hot Jupiters fall into two reasonably distinct categories, characterised by the overall level of incident flux (Figure 6.32), and the effect that this has on the opacity of TiO and VO (Hubeny et al., 2003). They are referred to as 'pM-class' and 'pL-class' in analogy with the M- and L-type brown dwarfs (Burrows et al., 2007b; Fortney et al., 2008a).

The hotter pM-class are characterised by high levels of incident stellar flux, resulting in gas-phase TiO/VO (Burrows et al., 2007b; Fortney et al., 2008a; Burrows et al., 2008a) or non-equilibrium products of photochemistry, as seen in solar system bodies (Burrows et al., 2008a; Zahnle et al., 2009). The additional absorber in the upper atmosphere generates the temperature inversion. These planets are expected to appear bright in the infrared, and to exhibit molecular bands in emission. The fact that the absorbed energy is reradiated before it can be transported to the night-side results in large day-side/night-side temperature contrasts and negligible phase shifts in their thermal emission light curves.

Examples are HD 209458 b (Deming et al., 2005b; Knutson et al., 2008); HD 149026 (Harrington et al., 2007); TrES–4 b (Knutson et al., 2009a); XO–2 b (Machalek et al., 2009); and HAT–P–7 (Christiansen et al., 2010). Large day–night contrasts have also been determined for υ And b (Harrington et al., 2006) and HD 179949 b (Cowan et al., 2007), although both are non-transiting, and their orbit inclinations and radii therefore unknown.

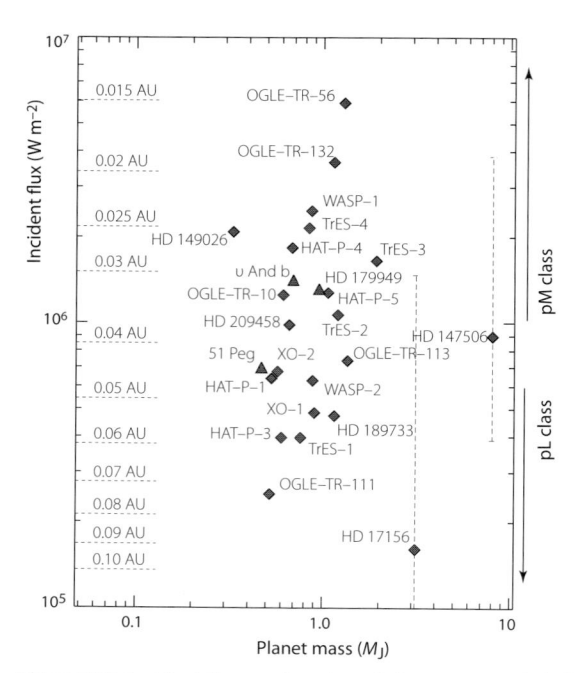

Figure 6.32: Incident flux as a function of planet mass. Labeled lines at left indicate the distance from the Sun at which the planet would intercept this same flux. ♦ = transiting, ▲ = non-transiting. Error bars for HD 147506 and HD 17156 indicate the flux range experienced over their eccentric orbit. Separations around ~ 0.04 – 0.05 AU represent the predicted transition region between the classes. From Fortney et al. (2008a, Figure 1), reproduced by permission of the AAS.

The cooler pL-class are characterised by atmospheres that receive less stellar flux. Titanium and vanadium are condensed out, the incident flux is absorbed deeper in the atmosphere, and no temperature inversion is generated. The absorption of incident flux deeper in the atmosphere leads to a greater redistribution of absorbed energy, in turn resulting in cooler day-sides, warmer night-sides, and strong advecting jet flows leading to significant phase shifts in their thermal emission light curves.

Examples are HD 189733 b, characterised by a day-side/night-side temperature differences of only 240 K at both 8 and 24 µm, and accompanied by a phase shift of 20–30° (Deming et al., 2006; Knutson et al., 2007a; Grillmair et al., 2008; Knutson et al., 2009b); TrES–1, the first of the secondary eclipses observed by Spitzer (Charbonneau et al., 2005); and GJ 436 (Deming et al., 2007a; Demory et al., 2007; Bean et al., 2008a).

6.6 Properties of transiting planets

Secondary eclipse measurements remain challenging. They have also been observed for CoRoT–1, from the satellite (Snellen et al., 2009a; Alonso et al., 2009a), and from the ground (Rogers et al., 2009); CoRoT–2 with a secondary eclipse of 0.006% (Alonso et al., 2009b); HAT–P–1 (Todorov et al., 2010a); and OGLE–TR–56 with a secondary eclipse of 0.036%, observed with VLT and Magellan (Sing & López-Morales, 2009).

6.6 Properties of transiting planets

Most transiting planets orbit stars with masses close to $1 M_\odot$ (Appendix D), with extremes ranging from $0.16 M_\odot$ (GJ 1214) to $1.5 M_\odot$ (WASP–33 and OGLE2–TR–L9): stars of spectral type later than K are too dim to be found in large numbers by wide-field transit surveys, while those earlier than F have rotationally broadened spectral lines and inherent stellar noise that may restrict radial velocity follow-up.

The basic properties of the ensemble of transiting planets nevertheless vary widely: masses range from $4.8 M_\oplus$ (CoRoT–7 b) to some $10 M_J$ (WASP–18, XO–3). Some are significantly larger for their mass than expected from gas giant models (HD 209458, WASP–2, TrES–4), while others are significantly smaller (HD 149026, HAT–P–3).

Bulk densities range from as little as $0.09\,\mathrm{Mg\,m^{-3}}$ (WASP–17), through the 0.6–$1.3\,\mathrm{Mg\,m^{-3}}$ more typical of the solar system gas giants, to $\sim 5.5\,\mathrm{Mg\,m^{-3}}$ indicating rocky planets with densities comparable to the Earth (CoRoT–7), and extending to the $26.4\,\mathrm{Mg\,m^{-3}}$ of the brown dwarf CoRoT–3.

Orbital periods are typically below 10 d. They range from below one day (with WASP–19 the current shortest at 0.79 d) upwards through the other close-in hot Jupiters, to the 95-d orbit of CoRoT–9 b, and the 111-d orbit of HD 80606 (discovered from Doppler measurements), with a corresponding range of stellar irradiation. WASP–12 b is an extreme example: its equilibrium temperature is $\simeq 2510\,\mathrm{K}$, and the stellar disk would subtend an angle of $35°$ from its surface. Its inferred tidal bulge implies a predicted apsidal precession of $19°\!.9\,\mathrm{yr}^{-1}$ (Ragozzine & Wolf, 2009).

Some planets follow highly eccentric orbits (notably HD 80606 with $e \simeq 0.93$, with HAT–P–2, CoRoT–10, and HD 17156 having $e \simeq 0.5 - 0.7$). Some have an orbit highly inclined to the stellar rotation axis, with at least two (HAT–P–7 and WASP–17) appearing to be on retrograde orbits. Orbit inclinations with respect to the line of sight are generally close to the $90°$ expected for transiting systems, with CoRoT–7 b, CoRoT–14 b, CoRoT–17 b, and OGLE–TR–56 as small as $i \simeq 78-80°$.

Maximum angular star–planet separations are $\lesssim 2\,\mathrm{mas}$, reaching 3 mas for the nearby GJ 436 b, and 8 mas for the long-period HD 80606.

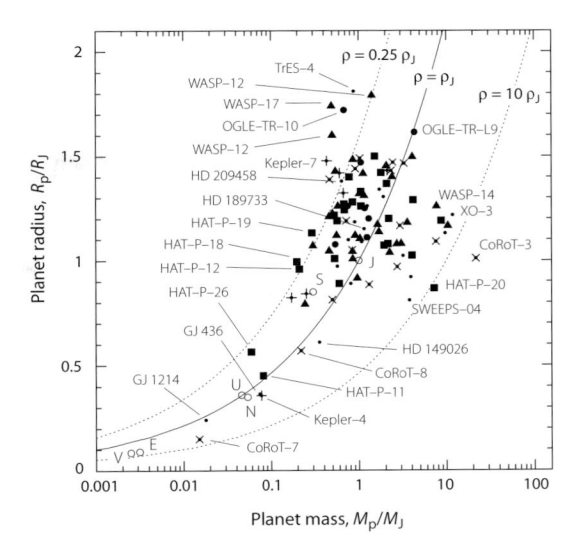

Figure 6.33: Mass–radius diagram for transiting exoplanets (● OGLE, ▲ WASP, ■ HAT, × CoRoT, + Kepler, · other). Data are from exoplanet.eu, 2010 November 1 (Appendix D). V, E, J, S, U, N indicate positions of Venus, Earth, Jupiter, Saturn, Uranus, Neptune. Dashed lines are of constant density in Jovian units $(0.25, 1, 10\rho_J)$. CoRoT–3 b $(22 M_J)$ is a brown dwarf.

6.6.1 Mass–radius relation

Overall features The mass–radius diagram for transiting planets is shown in Figure 6.33. It shows a clustering in the range $0.5 - 1 M_J$ and $0.9 - 1.5 R_J$, with few objects with densities below $\sim 0.25\rho_J$, and few above $\sim 10\rho_J$.

Figure 6.34 shows the transiting planets in a diagram covering more than three orders of magnitude in mass, extending from the stellar and sub-stellar regime down to the hot gaseous planets. The two lines are theoretical models of solar composition, and for two isochronal ages. That the broad observational features are reasonably replicated by theory suggests that the overall description of their internal structure, composition, and heat content is tolerably well understood.

The detailed structure of the diagram is discussed by Chabrier et al. (2009). The general behaviour is reflected in the polytropic mass–radius relation $R \propto M^{(1-n)/(3-n)}$ (Burrows & Liebert, 1993; Chabrier & Baraffe, 2000). For low-mass stars with large radiative cores $n \simeq 3$, decreasing to $n = 3/2$ below $0.4 M_\odot$ when the star becomes fully convective. From the bottom of the main sequence, below the hydrogen-burning minimum mass, brown dwarfs are supported primarily by electron degeneracy. As mass decreases the increasing electrostatic contribution (Coulomb pressure) leads to a decreasing density, and a decrease in n to a value $n = 1$ at a few M_J. In terms of an equation of state relating pressure and density, $P \propto \rho^\gamma$ with $\gamma = 1 + 1/n$, the decreasing polytropic index from stars to planets broadly corresponds to progressively less compressible interiors.

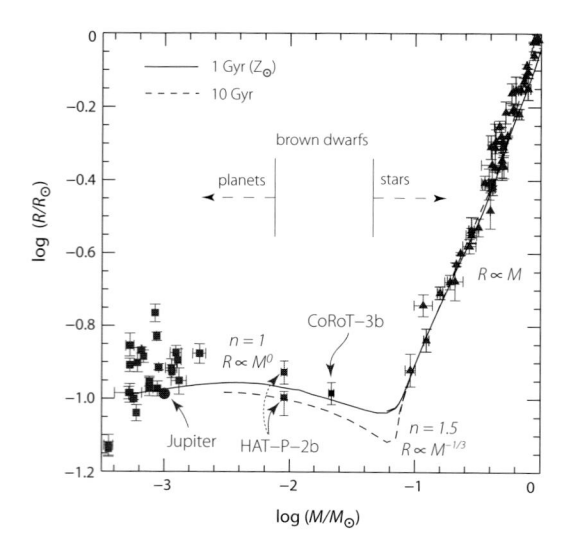

Figure 6.34: Mass–radius relation for stars and planets. Solid and short-dashed lines are models with solar composition for two isochrones, and n indicates the relevant polytropic index. Lines demarcating planets and stars are somewhat arbitrary. The CoRoT–3 b radius is from Deleuil et al. (2008). For HAT-P-2 b, both the original determination (lower point, Bakos et al., 2007a) and a later revision (upper point, Pál et al., 2010) are shown. From Chabrier et al. (2009, Figure 2).

Theoretical models Planets from terrestrial mass upwards are expected to be primarily composed of four successive layers in differing proportions (§11.2): an iron/nickel core, a silicate (rock) layer, an 'ice' layer, and a H/He envelope. The mass–radius relation follows from the relevant (pressure versus density) equation of state, along with equations describing mass conservation, energy conservation, and hydrostatic equilibrium.

Amongst more recent models for giant planets (§11.3.3) and super-Earths (§11.5), Fortney et al. (2007) computed radii for masses in the range $0.01 M_\oplus - 10 M_J$ for pure iron, rock, water, and H/He, as well as various mixtures. They included a dependency on orbital distance in the range 0.02–10 AU, coupling planetary evolution to stellar irradiation through a nongray radiative-convective equilibrium atmosphere model. Baraffe et al. (2008) presented similar grids of planetary evolution models from $10 M_\oplus - 10 M_J$, with various fractions of heavy elements.

Planet composition Such models show that a planet's position in the mass–radius diagram provides an indication of its overall composition, with temperature playing a relatively minor role. GJ 436 b, for example, is a $23 M_\oplus$ planet comparable to Neptune in mass and radius, orbiting an M dwarf (Gillon et al., 2007). Its position in the model grid of Fortney et al. (2007) suggests that it is an ice giant like Uranus and Neptune, composed largely of water ice (Figure 6.35). Its high equi-

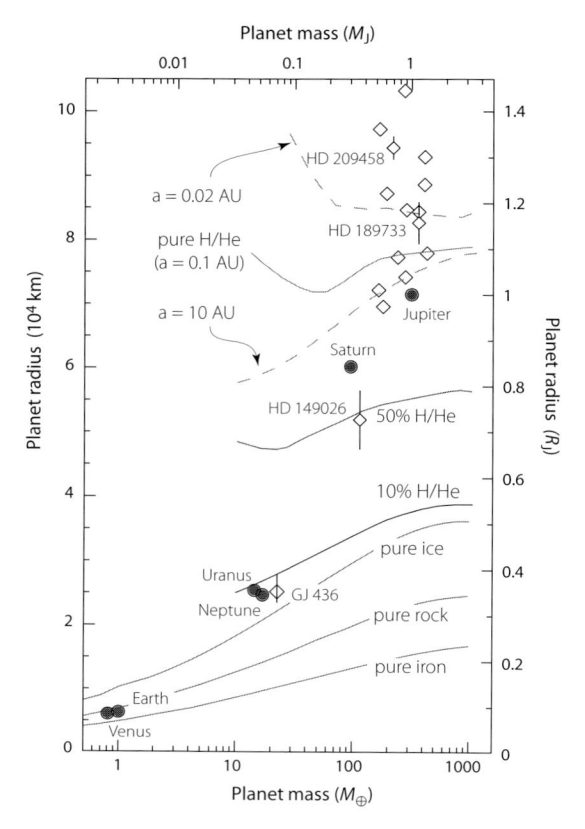

Figure 6.35: Mass–radius diagram from the Fortney et al. (2007) models, showing the position of the first transiting hot Jupiters (diamonds) and solar system planets. Models are for pure iron, silicate, and water ice, along with 10%, 50% and 100% H/He atmospheres irradiated at 0.1 AU from a solar-type star. Dotted lines are for a cold (a = 10 AU) and hot (a = 0.02 AU) pure H/He gas giant. From Gillon et al. (2007, Figure 3), reproduced with permission © ESO.

librium surface temperature, between 520–620 K, would imply a steam atmosphere. It may possess a H/He envelope, of an extent which depends on the presence of an iron/rock core, and which could be retained over long time scales, despite evaporation, because of the small size and low temperature of the primary star (Lecavelier des Etangs, 2007). *In situ* formation of an ice giant so close to its parent star is considered implausible. It presumably formed at a large orbital radius, beyond the 'snow line' where the protoplanetary disk is cool enough for water to condense, before migrating inwards to its present position.

Highest mass transiting planets Values of R_p for the two highest mass transiting objects provide the first constraints on the mass–radius relation for this cool, dense, and partially degenerate sub-stellar regime.

While there is broad agreement between observation and theory (Figure 6.34), details remain unclear.

6.6 Properties of transiting planets

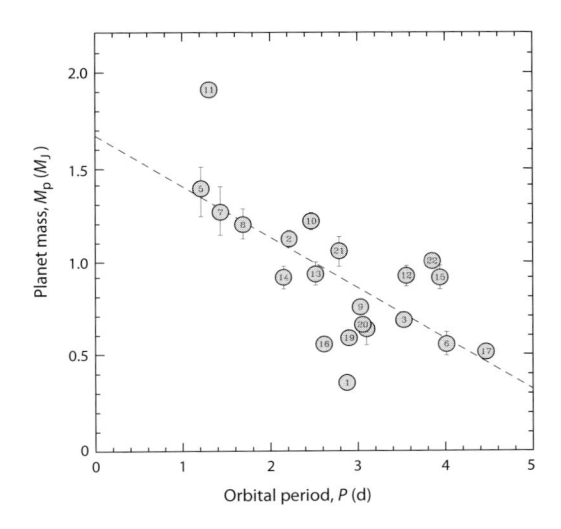

Figure 6.36: *Planet mass M_p as a function of orbit period. The dashed line is a linear fit. From Torres et al. (2008, Figure 9), reproduced by permission of the AAS.*

CoRoT–3 b, 'the first secure inhabitant of the brown-dwarf desert', could either be a brown dwarf of solar composition and age 2 Gyr, or an irradiated and inflated planet with a rocky core (Deleuil et al., 2008). From their detailed models with different metallicities, Baraffe et al. (2008) have argued that, for its inferred age of 2–3 Gyr, HAT–P–2 b is too dense to be a brown dwarf. This shows, in turn, that planets can form with masses at least up to $9M_J$. At the same time, it requires at least $200M_\oplus$ of heavy material in its interior, a requirement at the limit of current predictions if the planet was formed by core-accretion. Alternatively, it may have formed from collisions between one or more other massive planets.

Inflated radii of hot Jupiters Although the earliest transit measurements of HD 209458 b showed that its radius, $1.35R_J$, is broadly consistent with it being a gas giant composed primarily of hydrogen, its radius was nevertheless some 10% larger than theory had predicted (Guillot et al., 1996). This led Burrows et al. (2000a) to conclude that it must have migrated inward very early on in its lifetime, such that the incident irradiation from the star inhibited the further convective cooling and contraction expected for an object of its age. But studies incorporating realistic atmospheric temperature profiles showed that models could replicate the observed radius only if the deep atmosphere is unrealistically hot (Bodenheimer et al., 2001; Guillot & Showman, 2002; Bodenheimer et al., 2003; Baraffe et al., 2003).

While models taking into account the stellar irradiation on the internal heat content can successfully reproduce the radii of hot Jupiters in many cases (e.g. Barman et al., 2001; Chabrier et al., 2004; Baraffe et al., 2005, 2008; Fortney et al., 2006), a number, including HD 209458 b, TrES–4 b, and various WASP objects (Ibgui et al., 2010), have radii larger than most theoretical predictions. Various resolutions to this discrepancy have been proposed.

Bodenheimer et al. (2003) invoked tidal heating through ongoing orbital circularisation resulting from perturbations due to a second planetary companion. This was ruled out for HD 209458 b by Laughlin et al. (2005b). Tidal heating due to orbit circularisation by the host stars has also been studied (Gu et al., 2004; Mardling, 2007; Ibgui & Burrows, 2009; Miller et al., 2009; Ibgui et al., 2010).

Miller et al. (2009), for example, computed a grid of cooling and contraction paths for 45 transiting systems, starting from a large phase space of initial semi-major axes and eccentricities. Although matches can be found for a large fraction of planets with anomalously large radii, they found that orbit circularisation can be preceded by long periods when the semi-major axis is only slowly decreasing. This explanation would require that some of the systems are being viewed at privileged times of tidal evolution and hence radius inflation.

Winn & Holman (2005) proposed obliquity tides, normally damped through tidal dissipation, but which may persist if the planet is in a Cassini state (a resonance between spin precession and orbital precession). Later work showed that a Cassini resonance is unlikely for short-period planets (Fabrycky et al., 2007; Levrard et al., 2007; Peale, 2008).

Guillot & Showman (2002) and Showman & Guillot (2002) proposed that extra heating due to strong insolation-driven weather patterns on the planet could lead to the conversion of kinetic wind energy into thermal energy at pressures of tens of bars. Hansen & Barman (2007) suggested that if irradiation-driven evaporation preferentially removes helium, the consequent reduction in mean molecular weight may result in anomalously large radii for a given mass.

Burrows et al. (2003a) argued that the size discrepancy stems from an improper interpretation of the transit radius, and that the measured radius lies higher in the planetary atmosphere than generally assumed. Burrows et al. (2007a) invoked enhanced atmospheric opacities that retain the internal heat.

Gaudi (2005) drew attention to a Malmquist-like selection effect, whereby the number of planets with radius R_p detected in a signal-to-noise limited transit survey, is $\propto R_p^\alpha$, with $\alpha \sim 4-6$. For a dispersion in the intrinsic distribution of planetary radii σ, this leads to detected planets being larger on average by a fractional amount $\alpha(\sigma/\langle R_p \rangle)^2$ relative to the mean radius $\langle R_p \rangle$ of the underlying distribution.

Small hot Jupiters HD 149026 b was the first of several transiting giant planets found to have a radius significantly smaller than predicted by standard theories (Sato et al., 2005a). One explanation is that it could have

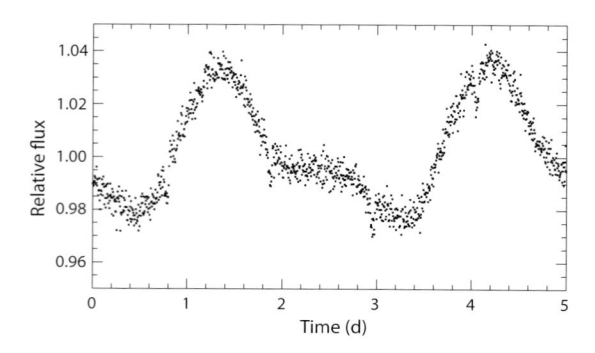

Figure 6.37: Transit light curve of an imaginary planet with $P \sim 0.9$ d and $M \sim 2M_J$, illustrating the very different light curve that may result from a planet which has spiraled inwards imparting significant angular momentum to a solar-type host star. Data points with 6 min sampling have been extrapolated from the light curve of CoRoT–2. The simulation is an extreme example of the form seen in Figure 6.21. From Pont (2009, Figure 7), © John Wiley & Sons, Inc.

some $70M_\oplus$ (2/3 of its total) in heavy elements, assumed to be in a core, as inferred for Saturn and Jupiter (Burrows et al., 2007a). Another, also assumed to be heavy-element rich, is HAT–P–3 b (Torres et al., 2007).

6.6.2 Observed correlations

Despite the wide range of properties of the transiting planets, presumably reflecting their own atmosphere, interior structure, and accretion and migration history, a number of correlations have been reported. The following discussion is based on two published subsets of transiting planets which have been subjected to a uniform re-analysis. Torres et al. (2008) derived the host star properties $M_\star$ and $R_\star$ based on stellar evolution models, incorporating the transit constraint on the stellar density, $\rho_\star$ (§6.4.5). Southworth et al. (2007) and Southworth (2008) determined a uniform set of surface gravities for 14 of the objects from their transit light curves and their stellar spectroscopic orbits (§6.4.5). These various correlations, along with a possible correlation between stellar $T_{\rm eff}$ and $R_{\rm p}$, have also been investigated in a wide simulation of the CoRoT observations (Fressin et al., 2009). The effects of metallicity are considered in more detail in §8.4.

Mass versus period Zucker & Mazeh (2002) pointed out a correlation between planet mass and orbital period for the 70 or so planets known at the time, manifested as a paucity of massive planets with short orbital periods. Over the much narrower domain of mass and radius occupied by the transiting planets, a decreasing linear relation between mass and period was found by Mazeh et al. (2005), and subsequently confirmed by others (e.g. Hansen & Barman, 2007; Torres et al., 2008; Southworth, 2009). The relation for the subset analysed by Torres et al. (2008) is shown in Figure 6.36.

They found a further and probable causal dependency: planets around metal-poor stars are more massive than those around metal-rich stars at a given orbital period.

Two possible explanations have been proposed: the trend for larger masses at shorter orbital periods could be related to the mechanism that halts migration, with larger planets able to migrate further in. The metallicity dependence would then imply that planets in metal-poor systems must be more massive to migrate to the same inward point compared with more metal-rich planets. Sozzetti et al. (2006) suggested that a dependency of migration on metallicity could arise from slower migration rates in metal-poor disks (Livio & Pringle, 2003; Boss, 2005), or through longer time scales for giant planet formation around metal-poor stars, which would reduce the migration efficiency before the disk dissipates (Ida & Lin, 2004b; Alibert et al., 2005a).

Alternatively, the mass–period relation may reflect the survival prospects close to the star, due to thermal evaporation driven by the ultraviolet flux (Baraffe et al., 2004; Mazeh et al., 2005; Davis & Wheatley, 2009; Lammer et al., 2009b). In this picture, planets with initial masses below some critical value expand as they evaporate, speeding up the evaporation process. If metal-rich planets are more likely to develop a rocky core (Pollack et al., 1996; Guillot et al., 2006; Burrows et al., 2007a), and if the presence of such a core slows down evaporation (Baraffe et al., 2004; Lecavelier des Etangs et al., 2004), then survival at a given period would have a dependence on metallicity.

Tidal effects may also result in detection biases. For $P \lesssim 3-5$ d, Pont (2009) found that, in order of increasing mass, close-in planets will be tidally unaffected ($M \ll M_J$), circularised ($M \simeq M_J$), spiraling in ($M \simeq 1-2M_J$), destroyed ($M \simeq 2-3M_J$), and synchronised ($M \gtrsim 3M_J$). Host stars which are rapidly rotating as a result of tidal spin-up may be dropped from Doppler searches due to their broadened spectral lines, and due to the expected correlation between spin rate and photospheric activity.

Rotation-induced variability coupled with very short-period transits may also lead to transit light curves qualitatively different from those assumed in standard transit searches (Figure 6.37), resulting in a detection bias against massive close-in planets for transit surveys also.

Surface gravity versus period A related correlation is that of planetary surface gravity versus orbital period (Southworth et al., 2007; Hansen & Barman, 2007; Southworth, 2008). This has been confirmed from the uniform treatment of the 22 systems considered by Torres et al. (2008, Figure 6.38a), and the 14 systems considered by Southworth (2009). Since $g_{\rm p}$ can be derived from transit measurements independently of the mass or radius of the host star (§6.4.5), it should be free of possible systematics arising from stellar evolutionary models.

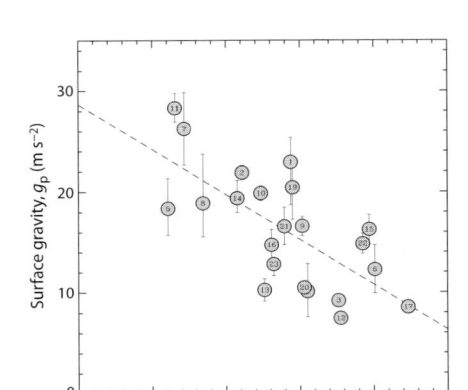

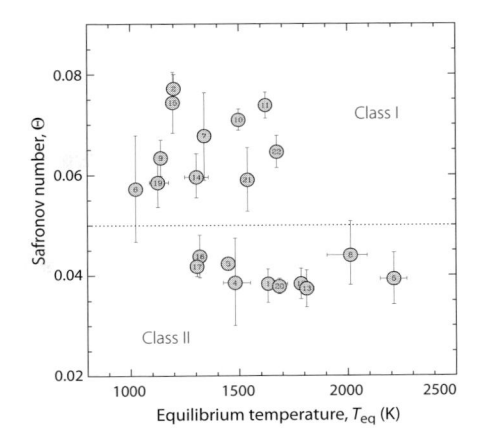

6.6 Properties of transiting planets

Figure 6.38: Left: surface gravity g_p versus orbital period for the transiting planets considered by Torres et al. (2008), except for HAT–P–2 b which is off scale at $g_p \sim 234 \, \mathrm{m \, s^{-2}}$. A linear fit is shown. Right: Safronov number, Θ, versus equilibrium temperature, T_{eq}. Planet classes are labeled according to the proposal by Hansen & Barman (2007), with a dividing line shown at $\Theta = 0.05$. From Torres et al. (2008, Figures 10–11), reproduced by permission of the AAS.

The $g_p - P$ correlation may point more directly to the underlying effect responsible for the $M_p - P$ correlation, since surface gravity is a fundamental parameter entering into the evaporation of planetary atmospheres. If metallicity influences planetary radii, and given that $g_p \propto M_p/R_p^2$, metallicity must also introduce some scatter into the $g_p - P$ correlation. Fressin et al. (2009) attributed part of the effect to the decreasing transit probability, and decreasing detection efficiency, for larger P and higher g_p.

Safronov number The *Safronov number* is defined in terms of the escape velocity from the surface of the planet, $V_{\mathrm{esc}}^2 = 2GM_p/R_p$, and the orbital velocity of the planet about its host star, $V_{\mathrm{orb}}^2 = GM_\star^2/a(M_\star + M_p)$, as

$$\Theta = \frac{1}{2}\left(\frac{V_{\mathrm{esc}}}{V_{\mathrm{orb}}}\right)^2 = \left(\frac{a}{R_p}\right)\left(\frac{M_p}{M_\star}\right). \quad (6.61)$$

It provides a measure of the ability of a planet to gravitationally scatter other bodies in the same system (Safronov, 1972).

Hansen & Barman (2007) identified a distinction between two classes of hot Jupiter, based on the relation between their Safronov number and the zero-albedo equilibrium temperature, T_{eq} (Equation 6.60). The known transiting planets fell into two groups, differing by almost a factor two in their values of Θ at fixed T_{eq}. Generally, the Class II objects orbit the hotter host stars, and appear to orbit stars of higher mass. They surmised that it may reflect the influence of planetary scattering in determining when migration stops. Another possibility is that if evaporation preferentially removes helium, the consequent reduction in the mean molecular weight may lead to some planets having anomalously large radii.

Torres et al. (2008) confirmed the separation of transiting planets into two classes according to their Safronov number (Figure 6.38b), and found evidence that the systems with small Safronov numbers are more metal-rich on average. Both Southworth (2009) and Fressin et al. (2009) found the evidence for the existence of two distinct classes to be less compelling.

7
Imaging

7.1 Introduction

Imaging generally refers to the detection of a point source image of the exoplanet. This may be either in the reflected light from the parent star (in the visible), or through its own thermal emission (in the infrared). This is to be distinguished from spatial resolution of an exoplanet surface. It is also distinguished here from the detection of an increased flux of the combined star–planet system as a result of light reflected from the planet, a decrease in intensity of the combined light as the planet enters a secondary eclipse, or a modified stellar spectrum in absorption or emission by a transiting planet, all of which are discussed in Chapter 6.

The scientific interest of obtaining exoplanet images is to confirm the accumulating but generally indirect evidence for their existence, as a precursor to more extensive spectroscopic investigations, and as a first step towards a far future goal of obtaining resolved spatial imaging of an exoplanet surface.

Star–planet brightness The ratio of the planet to stellar brightness depends on the stellar spectral type and luminosity class, the planet's proximity to the star given by its orbital semi-major axis (a) and instantaneous projected separation, on the planet's mass, composition, radius (R_p) and age, on the scattering properties of its atmosphere, and on the observation wavelength. For reflected light of wavelength λ, the planet/star flux ratio can be written

$$\frac{f_p(a,\lambda)}{f_\star(\lambda)} = p(\lambda)\left(\frac{R_p}{a}\right)^2 g(\alpha), \quad (7.1)$$

where $p(\lambda)$ is the geometric albedo, and $g(\alpha)$ is a phase-dependent function (§6.4.7). The formula, including the phase dependence, is modified if the planet's thermal emission is significant.

The ratio $f_p/f_\star$ is very small. For the Jupiter–Sun system, it is $\sim 10^{-9}$ at maximum elongation, with an angular separation of 0.5 arcsec at 10 pc. For the Earth–Sun it is $\sim 10^{-10}$. Values for exoplanets of primary interest are expected to range from 10^{-5} in the infrared to 10^{-10} in the optical (Figure 7.1).

Angular separation The star–planet angular separation depends on the orbital parameters as well as on the stellar distance. Exoplanets of interest typically lie very close in angular terms to the host star, within 0.1–0.5 arcsec, and swamped by the bright stellar glare. From a ground-based telescope, the planet signal is immersed within the star's 'seeing' profile, of order 0.3–1 arcsec, arising from turbulent atmospheric refraction.

Even when corrected using adaptive optics, or eliminated by observations from space, two further sources of light from the host star make direct exoplanet detection problematic: diffracted light from the telescope and supporting structures, and scattered light from wavefront aberrations which results in a residual intensity in the focal plane in the form of instrumental 'speckles'.

The technical challenge Under these conditions elementary signal-to-noise calculations imply that obtaining a direct image of a planet is not feasible, and some means of removing or attenuating the star light is required in order to improve the signal-to-noise ratio at the position of the planet.

The technical challenges are considerable. Ground-based instruments targeting exoplanet imaging, for a favourable subset of star–planet properties and ages, have been under development for a number of years, and some major instruments are soon to become operational. From space, ambitious plans for space interferometers and coronagraphs by NASA and ESA grew from an early phase of optimism in the late 1990s, but are currently not being pursued with the same urgency.

Observations in the infrared are facilitated by the simultaneous decrease in emission from the star and increased thermal emission from the planet, and by the implementation advantages of adaptive optics at longer wavelengths. They are exacerbated by decreasing diffraction-limited angular resolution, and engineering complications of observing in the thermal infrared.

Typical targets To date, the first images of massive, widely separated exoplanets have been acquired, although none so far resemble those of the solar system. The detection of exo-Earths remains out of reach of any

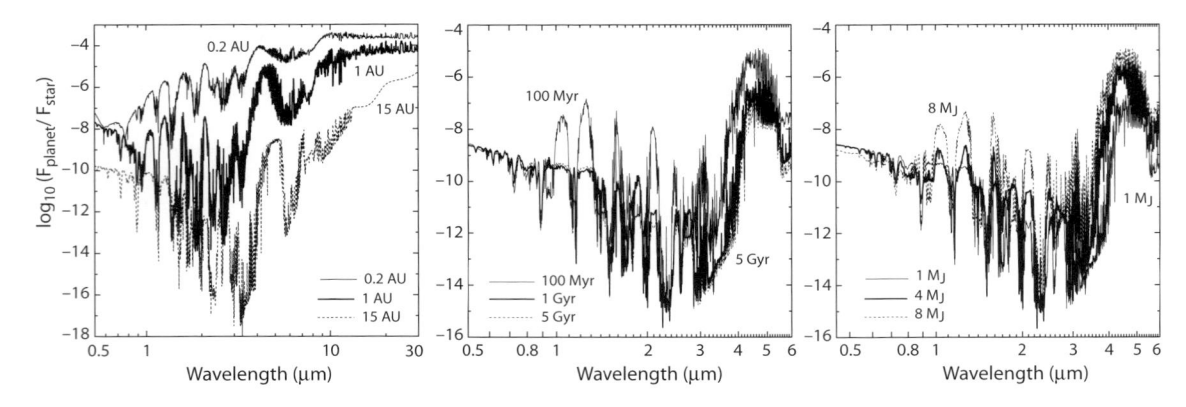

Figure 7.1: Predicted planet/star flux ratios versus wavelength. Models are for giant planets orbiting a G2V star of solar metallicity (phase-averaged, zero eccentricity, zero orbital inclination, effects of H_2O and NH_3 clouds included): (a) for a $1 M_J$ planet with an age of 5 Gyr as a function of orbital distance. The flux ratio is dominated by reflection in the optical (Rayleigh scattering and clouds), and by emission in the infrared; the transition occurs between 0.8–3 μm depending on separation; (b) for a $1 M_J$ planet at 4 AU as a function of age; (c) for a 5 Gyr planet at 4 AU as a function of planet mass. From Burrows et al. (2004b, Figures 3, 6, 7), reproduced by permission of the AAS.

of the imaging facilities currently under development. Planned facilities (ground-based 'extreme adaptive optics' imagers, extremely large telescopes, or JWST) will instead focus on exoplanet categories where imaging prospects are highest, for example:

(a) young stars (10–100 Myr, $d < 100$ pc), around which the planets are still young, warm, and hence self-luminous. Evolutionary models predict exoplanet luminosities higher than for mature planets by several orders of magnitude depending on mass and age (Burrows et al., 1997; Chabrier et al., 2000; Burrows et al., 2004b);

(b) stars with known planets, for which the increasing temporal baseline of radial velocity surveys are identifying long-period trends suggestive of the existence of high-mass giant planets in wide-separation orbits;

(c) nearby stars, $d < 5$ pc, in which a shorter-period giant planet with significant reflected light might be detected because of its relatively large angular separation from the star due to its proximity. For $a = 0.05 - 1$ AU, the flux ratio at 20 μm can reach 10^{-3} (Burrows et al., 2004b).

7.2 Techniques

Various techniques are being applied in attempts to image exoplanets at the highest angular resolution and contrast: the use of large apertures to improve signal-to-noise and resolution; adaptive optics to minimise the effects of atmospheric turbulence, or imaging from space to eliminate its effects altogether; coronagraphic masks to suppress light from the host star; post-processing to treat residual aberrations; interferometers to improve angular resolution, with nulling interferometry to eliminate the stellar light; and improving contrast between the planet and star by observing at longer wavelengths.

7.2.1 Active optics

Telescope mirrors were originally designed to retain their shape by virtue of their thickness and intrinsic stiffness, a condition which limited maximum primary mirror diameters to 5-m or so, as typified by the Palomar Observatory's Hale Telescope of 1949.

Since the demonstration of ESO's New Technology Telescope (NTT) in 1989 (Wilson, 1991), large telescopes have made use of thin-mirror technologies, in which the low-order figure of the primary is maintained through the use of *active optics*. By monitoring image quality over seconds to minutes, coupled with an array of actuators operating to keep the mirror in an optimal shape, the effects of gravity, wind, and telescope alignment can be minimised (e.g., for LBT, see Hill, 2010).

The same principle of active mirror control is used in the replication of large aperture sizes built up through the use of smaller segmented mirrors. This technique was used for the original MMT, is used for current segmented telescopes such as Keck and HET, and will be used for the extremely large telescopes E–ELT and TMT.

7.2.2 Adaptive optics

The related but more complex technique of *adaptive optics* operates on shorter time scales of order 1 ms, and aims to compensate for atmospheric phase fluctuations across the telescope pupil to achieve diffraction-limited resolution ($\lambda/D \simeq 0.01$ arcsec at 500 nm for a 10-m telescope), using either a small corrective mirror or an adjustable secondary. Originally proposed by Babcock (1953), with first applications reported by Buffington et al. (1977), it has been under intensive development since the early 1990s (e.g. Golimowski et al., 1992;

7.2 Techniques

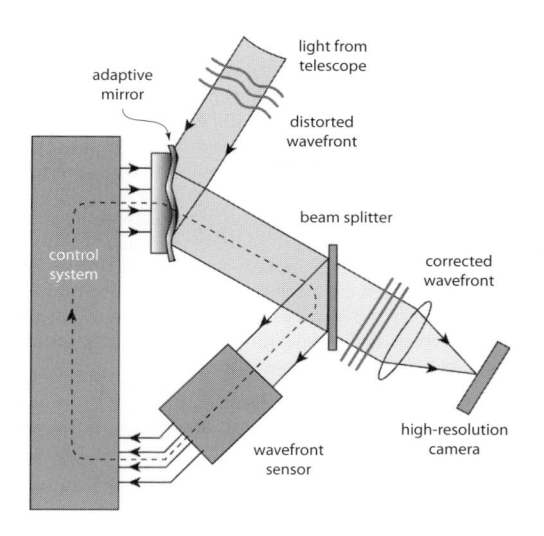

Figure 7.2: In adaptive optics, a wavefront distorted by the atmosphere is reflected from a deformable mirror with hundreds of actuators glued to its rear. The system operates in closed-loop, measuring the residual wavefront error after reflection using a wavefront sensor. Corrections are applied such as to leave the wavefront flat, updated thousands of times per second to match the rapidly changing effects of atmospheric turbulence. Courtesy: Claire Max, Center for Adaptive Optics, UCSC.

Beckers, 1993; Hubin & Noethe, 1993; Angel, 1994; Stahl & Sandler, 1995; Hardy, 1998; Wizinowich et al., 2000).

In its basic form, measurement of the atmospheric phase fluctuations affecting a target star are made via the continuous measurement of the wavefront of a bright reference star nearby on the sky. An equal but opposite wavefront correction is then imposed using a deformable mirror containing voltage-responsive actuators distributed across its surface (Figure 7.2).

Wavefront sensing measurements, at frequencies of order 1 kHz, are made over pupil sub-apertures down to size $\simeq r_0$, where the Fried parameter r_0 is a measure of the coherence length of the atmospheric wavefront errors (Fried, 1965, 1966). Typical values at a good site are $r_0 \sim 0.15 - 0.2$ m at visible wavelengths, increasing to ~ 1 m at $2\,\mu$m. Although the number of sensors and actuators required is ideally of order $(D/r_0)^2$ (Angel, 1994), a significantly smaller number still produces images with a sharp core (Wang & Markey, 1978; Hardy, 1982; Roddier et al., 1991). Current systems employ of order 200 actuators, extreme adaptive optics systems under development employ of order 40×40, while systems of order 10^4 are planned for the future ELTs.

Wavefront sensing In practice, various wavefront-sensing schemes are in use: the Shack–Hartmann sensor measures the displacement of an array of sub-aperture images (Rousset et al., 2003); the curvature sensor introduces a spherical phase aberration into the focal plane then transforms phase aberrations into light intensity modulations in the pupil plane (Roddier, 1988); and the pyramid sensor divides the focal plane into four quadrants which are then reimaged in a separate pupil plane (Ragazzoni, 1996). Comparative studies of various schemes have been reported (e.g. Guyon, 2005; Vérinaud et al., 2005), while newer variants continue to be developed (e.g. Le Roux & Ragazzoni, 2005; Oti et al., 2005a; Butterley et al., 2006; Guyon et al., 2009; Guyon, 2010; Peter et al., 2010).

Stars bright enough to serve as a natural wavefront reference must lie close in angular terms to the source being observed, within of order the *isoplanatic patch* (over which the paths are temporally coherent): about 3 arcsec for visible light, and some 30 arcsec in the infrared. As one example of site characterisation, Masciadri et al. (2010b) present statistics for Mt Graham (LBT) covering seeing, wavefront coherence time, and isoplanatic angle.

Residual effects Adaptive optics systems cannot correct for atmospheric effects on spatial scales smaller than the interactuator spacing in the pupil plane. Power at higher spatial frequencies in the pupil plane transforms into noise on large angular scales in the image plane. The final image therefore comprises a diffraction-limited core surrounded by an extended halo whose form reflects the number of actuators, the characteristics of the atmospheric turbulence, and residual optical aberrations. Present adaptive optics systems nevertheless routinely deliver a *Strehl ratio*, the ratio of the peak intensity of the image to that of a perfect imaging system operating at the diffraction limit, of order 70%.

Laser guide stars The fraction of sky with suitably bright natural guide stars is only of order a few per cent. The use of artificial laser guide stars was developed, in both the military and astronomical communities, as a way of extending the applicability of adaptive optics to arbitrary locations on the sky (e.g. Beckers, 1993; Hubin & Noethe, 1993; Lloyd-Hart et al., 1998; Ageorges & Dainty, 2000). The principle is to use a laser beam pulsed on the ground and reflected from the higher atmosphere, to mimic a bright artificial source of light subject to the same wavefront distortions.

Sodium beacon guide stars use (yellow) lasers tuned to the 589 nm Na D lines to excite non-ionized sodium atoms naturally present in a ~ 5 km thick mesospheric layer at an altitude of around 80–100 km (Chapman, 1939). The fluorescing region appears as an artificial star at the targeted location. The mesospheric metal layer originates from the continuous ablation of meteors (e.g. Kane & Gardner, 1993); below it, Na is normally bound.

Rayleigh beacons employ shorter wavelength (green) lasers, and rely on molecular Rayleigh scattering within the lower atmosphere, at altitudes of typically 10–15 km. Pulsed at around 10 kHz, they can be time-gated to select the atmospheric layer over which the phase fluctuations are preferentially compensated.

GLAO, MCAO, and XAO In *ground-layer adaptive optics*, the goal is to correct the dominant boundary-layer turbulence over a wide field of order a few arcmin, improving the angular isoplanatism associated with single conjugated adaptive optics at the expense of diffraction-limited resolution (Rigaut et al., 2000; Hubin et al., 2005; Andersen et al., 2006).

In *multi-conjugate adaptive optics*, the goal is to increase the corrected field of view and reduce point-spread function variation over the field. Multiple wave-front sensors point at different locations, using several deformable mirrors optically conjugated at different altitudes to provide an optimum correction for the deformable mirrors (Goncharov et al., 2005; Vernet-Viard et al., 2005; Fusco et al., 2006a). These systems are being built into the designs of the new extremely large telescopes, both for the TMT (Gilles et al., 2006), and for the European–ELT (MAORY, Diolaiti et al., 2010).

In *extreme adaptive optics*, Strehl ratios of $\gtrsim 90\%$ are targeted in the near infrared, using more sub-apertures (of order 40×40) and higher control rates (of order 2 kHz). With coronagraphy and image post processing, systems such as VLT–SPHERE (Petit et al., 2008) and Gemini–GPI (Macintosh et al., 2008) target contrast ratios of order $10^{-6} - 10^{-7}$ at 0.5 arcsec separation.

Operational systems The routine use of laser guide star adaptive optics started at the Lick, Palomar and Keck observatories around 2006 (Wizinowich et al., 2006b), and at the VLT using NACO and SINFONI around 2007 (Rousset et al., 2003; Bonaccini Calia et al., 2003). Other systems are operational or under development at all large telescopes, including Subaru–AO188 with 188 actuators (Watanabe et al., 2004), Gemini–ALTAIR with 177 (Tracy et al., 2004; Lafrenière et al., 2008), and LBT–ARGOS with 672 (Rabien et al., 2008; Hill, 2010). In 2013, four laser guide star systems will be available at the ESO VLT as part of the Adaptive Optics Facility (Arsenault et al., 2006; Bonaccini Calia et al., 2010).

7.2.3 Coronagraphic masks

For the reasons noted above, the point-spread function of a telescope equipped with adaptive optics typically consists of a bright diffraction-limited core, with several Airy rings superimposed on a wide scattered light halo containing several percent of the total flux. The improved image quality provides access to a region within a few times the telescope's diffraction width, with the dynamic range limited by the stellar halo and the bright Airy rings, rather than by atmospheric seeing.

Further significant enhancement in contrast can be achieved by suppressing the noise associated with the stellar light by rejecting it from the area of interest in

---

The Lyot coronagraph for solar observation: Coronagraphy was originally developed for studies of the solar corona outside periods of total eclipse, and included early attempts by Hale (1895). With the coronal brightness at 2 arcmin from the Sun's edge being a factor 10^6 less than that of the solar disk, the system had to overcome bright points of light diffracted by small bubbles in the glass, hollows and scratches on its surface, dust particles and, most severely, light diffracted by the edge of the lens.

The technical problems were considerable, and results were first achieved only in July 1931 from the Pic du Midi Observatory (Lyot, 1939). Glass free from inclusions, highly polished lenses, and a high altitude site to minimise atmospheric pollution were all essential.

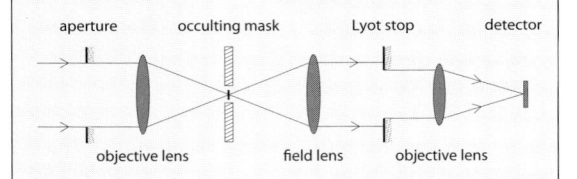

In the Lyot coronagraph as implemented for observations of the solar corona, the first objective lens forms an image of the disk and corona, and an occulting mask blocks the image of the disk. If the occulting mask were the only blocking element, diffracted light would still swamp faint off-axis structure. Consequently, a field lens re-images the objective lens and its diffraction pattern, and the 'Lyot stop' intercepts the diffraction ring while allowing most of the light from surrounding structure to pass. The second objective lens relays the resulting image onto the detector plane. In a perfect Lyot coronagraph some 50% of the light from a nearby source might be lost, compared to the suppression of some 99% of the stellar light.

A similar sequence forms the basis of all coronagraphic concepts, with the occulting mask replaced by the relevant coronagraphic mask.

---

the focal plane. The technique, employing some form of mask in the telescope focal plane, is referred to as *coronagraphy*, after its early development to observe the solar corona. The original Lyot coronagraph used an amplitude mask, physically blocking the central stellar light, while more recent coronagraphs also employ phase masks to cancel light through self-interference.

As a result, the combination of a coronagraph with an adaptive optics system aims to block the core of the image of an on-axis point source, suppress the bright diffraction rings and halo, remove light that would otherwise reduce the dynamic range, and improve the prospects of imaging faint off-axis structure.

Coronagraphy without adaptive wave front correction, or with only low-order tip–tilt correction, has been used to access angular scales close to the central star and to prevent detector saturation (e.g. Golimowski et al., 1992; Nakajima et al., 1995), but is of limited relevance for exoplanet imaging from the ground where the very highest contrast ratios are now required.

Table 7.1: *Coronagraphs theoretically able to achieve 10^{10} contrast within $5\,\lambda/d$, from Guyon et al. (2006b, Table 1). Designs post-dating that compilation, or of some other specific nature, are shown without abbreviation in column 2.*

Coronagraph	Abbrev.	Reference
Interferometric coronagraphs:		
Achromatic interferometric coronagraph	AIC	Baudoz et al. (2000)
Common-path achromatic interferometer-coronagraph	CPAIC	Tavrov et al. (2005)
Visible nulling coronagraph, $x - y$ shear (fourth-order null)	VNC	Mennesson et al. (2003)
Pupil swapping coronagraph	PSC	Guyon & Shao (2006)
Pupil apodisation:		
Conventional pupil apodisation and shaped-pupil	CPA	Debes et al. (2002), Kasdin et al. (2003)
Achromatic pupil phase apodisation	PPA	Yang & Kostinski (2004)
Phase induced amplitude apodisation coronagraph	PIAAC	Guyon (2003), Guyon et al. (2005)
Phase induced zonal Zernike apodisation	PIZZA	Martinache (2004)
Spiderweb/star-shaped mask	–	Vanderbei et al. (2003b), Vanderbei et al. (2003a)
Checkerboard mask	–	Vanderbei et al. (2004)
Improved Lyot concept with amplitude masks:		
Apodised pupil Lyot coronagraph	APLC	Soummer et al. (2003a,b)
Apodised pupil Lyot coronagraph, multi-stage (N steps)	APLCN	Aime & Soummer (2004)
Band limited, fourth-order	BL4	Kuchner & Traub (2002)
Band limited, eighth-order	BL8	Kuchner et al. (2005), Shaklan & Green (2005)
Band limited, notch-filter	–	Kuchner & Spergel (2003), Debes et al. (2004)
Achromatic prolate apodised Lyot coronagraph	–	Aime (2005)
Binary apodisation	–	Cady et al. (2009)
Improved Lyot concept with phase masks:		
Phase mask	PM	Roddier & Roddier (1997)
Four quadrant phase mask	4QPM	Rouan et al. (2000), Riaud et al. (2001)
Eight octant phase mask	–	Murakami et al. (2008)
Achromatic phase knife coronagraph	APKC	Abe et al. (2001)
Optical vortex coronagraph, with topological charge m	OVCm	Foo et al. (2005), Palacios (2005)
Annular groove phase mask coronagraph	AGPMC	Mawet et al. (2005)
Optical differentiation	ODC	Oti et al. (2005b)
Achromatic chessboard	–	Rouan & Pelat (2008)

Lyot coronagraphy and adaptive optics Ideally, a coronagraph coupled to an adaptive optics system would perform as if placed above the atmosphere. The image of a point source in the image plane would be a pure Airy disk, and the size of the Lyot stop could be chosen using Fourier theory. Malbet (1996) and Sivaramakrishnan et al. (2001) provide further details of the basic Lyot coronagraph combined with adaptive optics, with the latter demonstrating how the Lyot stop must be optimised for a given telescope according to specific atmospheric conditions.

Classical Lyot coronagraphs are used, for example, with Gemini South–NICI (Chun et al., 2008), Subaru–CIAO (Tamura et al., 2000), and with VLT–NACO, where a focal spot of 0.7 or 1.4 arcsec is available (Boccaletti et al., 2009). Application to segmented mirrors was evaluated by Sivaramakrishnan & Yaitskova (2005). Other comparisons have been made by Crepp et al. (2007).

For a perfect coronagraph, and an entrance pupil with no phase aberrations, there would be zero light outside of some specified angle. In practice, and as considered further in §7.2.4, two effects limit the efficiency of all coronagraphs in the imaging of exoplanets in which the host stars are point-like sources and the images are dominated by diffraction: residual effects due to imperfect atmospheric correction, and imperfect optics.

Post-Lyot coronagraphs Alternative concepts for high-rejection coronagraphs have been stimulated by the prospects of exoplanet imaging, and a large literature on optimised designs and laboratory tests has appeared in the last ten years (for a broad perspective, see e.g. Ferrari et al., 2007; Guyon, 2007; Oppenheimer & Hinkley, 2009). New and optimised designs are still being discovered, some exploiting new basic physics.

Coronagraphic surveys are already underway in a number of ground-based telescopes, and coronagraphs have been built into space-based telescopes notably HST and JWST. Designs include modified opaque disks to suppress the amplitude of the starlight, as in the original Lyot design, as well as systems which modify its phase to create self-destructive interference (Gay & Rabbia, 1996). The results of an international meeting held in 2004 with the goal of optimising coronagraphic designs for exoplanet detection, and in particular within the context of JWST and TPF–C, has been made available by Quirrenbach (2005).

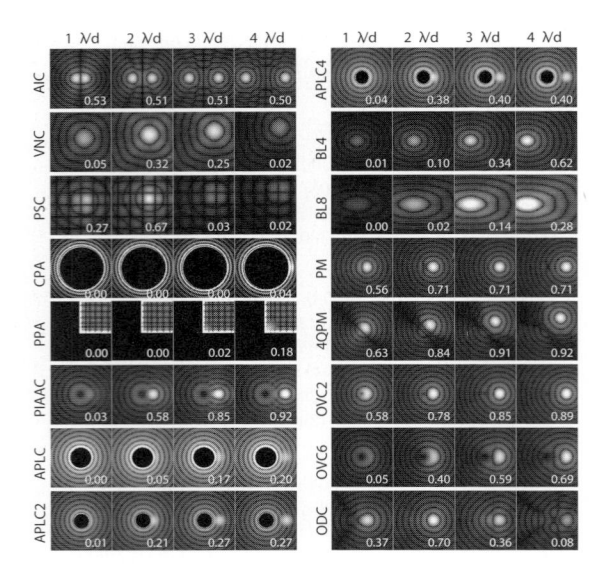

Figure 7.3: *Simulated monochromatic point-source images of a 10^{10} contrast system, at angular separations λ/d to $4\lambda/d$, for the coronagraphs in Table 7.1. The number in each image is the throughput for the off-axis source. The pixel scale is the same in all images, but the brightness scale is not. The companion is moving on a diagonal (rather than horizontal) line for VNC, PSC, PPA and 4QPM. From Guyon et al. (2006b, Figure 1), reproduced by permission of the AAS.*

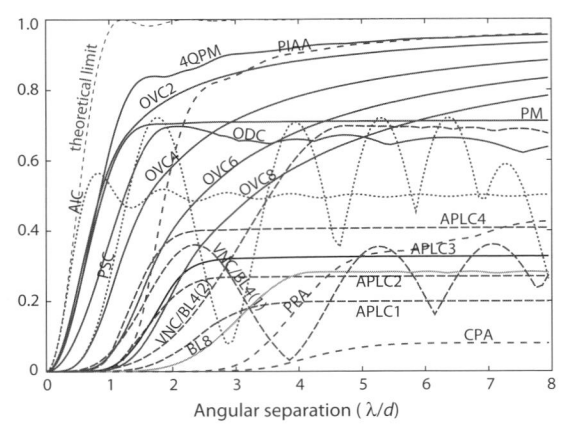

Figure 7.4: *Throughput, at the 10^{10} contrast level, of the coronagraphs listed in Table 7.1 as a function of angular separation, for a monochromatic point-like source. For coronagraphs with preferential directions (BL4, BL8, 4QPM, ODC, VNC, PSC), the peak throughput is shown, assuming that the telescope orientation is optimal. From Guyon et al. (2006b, Figure 3), reproduced by permission of the AAS.*

Key performance parameters are the *inner working angle*, scaled to λ/D and loosely defined as the angular offset at which the star flux suppression matches the Sun–Earth contrast, and the fraction of the azimuthal space that may be simultaneously searched (which determines the coronagraph's efficiency).

Guyon et al. (2006b) made a synthesis of the performance of various designs relevant for unobstructed circular pupils, known to provide a theoretically achievable 10^{10} point-spread function contrast within $5\lambda/D$ of the central source. Their synthesis is summarised in Table 7.1, and their performances illustrated in Figures 7.3–7.4. Some of these concepts are known to provide very high attenuation, but only under specific conditions which are hardly achievable in practice.

A further complication is that the inner working angle can be very sensitive to the source size. The highest performance coronagraphs, those with the largest throughput and smallest inner working angle, are often those most sensitive to a finite stellar diameter, thus presumably ruling them out for terrestrial planet imaging (Guyon et al., 2006b).

Polarisation dependencies are reviewed by Breckinridge & Oppenheimer (2004) and, in the context of nulling interferometry, Spronck & Pereira (2009).

Classification of coronagraph concepts Guyon et al. (2006b) divided the existing concepts into four categories. These are summarised as follows (with references given in Table 7.1), with two highly performing examples – the four-quadrant phase mask and the optical vortex coronagraph – described in more detail.

Interferometric coronagraphs: akin to nulling interferometers, this class relies on interferometric combination of discrete beams derived from the entrance pupil. Examples include the *achromatic interferometric coronagraph*, which uses a beam splitter to destructively combine two copies of the entrance pupil, one of them π-phase shifted and flipped. The *common-path achromatic interferometric coronagraph* uses the same principles, with a common-path interferometer. The *visible-nulling coronagraph* is the coronagraphic equivalent of a double-Bracewell nulling interferometer, in which two successive shears in perpendicular directions produce four beams which, when combined, yield a fourth-order null in the pupil plane, thus combining a deep null with imaging capability. In the *pupil swapping coronagraph* parts of the pupil are geometrically swapped prior to destructive interferometric combination, thus avoiding the throughput loss due to the shear in the visible-nulling coronagraph.

Pupil apodisation coronagraphs: these designs are all characterised by a modification of the pupil complex amplitude, yielding a point-spread function suitable for high-contrast imaging. Apodisation can be performed by a pupil plane amplitude mask which can be continuous or binary, or by a phase mask.

The phase-induced amplitude apodisation coronagraph of Guyon et al. (2005), for example, achieves an inner working angle of $1.5\lambda/D$, a radial field of $100\lambda/D$, and nominal detectability of an exo-Earth at 10 pc with a

4-m space telescope in 30 s. The same technique is also referred to as *pupil mapping* (Traub & Vanderbei, 2003; Vanderbei & Traub, 2005), and as *intrinsic apodisation* (Goncharov et al., 2002).

Improved Lyot coronagraphs with amplitude masks: improved performance of the basic Lyot design can be obtained by better matching the Lyot stop to the light distribution in the reimaged pupil. This can be by adapting the pupil to the hard-edge of the focal plane mask using apodisation (non-uniform transmissivity), as in the *apodised pupil Lyot coronagraph*, or by adapting the mask to the telescope pupil by using *band-limited masks* in the focal plane.

Apodised Lyot masks are used for JWST–NIRCam, and Krist et al. (2007) illustrates how the corresponding Lyot stops are matched to the (segmented) telescope pupil pattern. Martinez et al. (2008) considered the apodised pupil Lyot coronagraph as a suitable baseline for the 30–40 m extremely large telescopes.

Improved Lyot coronagraphs with phase masks: a phase mask can be used to introduce phase shifts in the focal plane to create self-destructive interference, rather than employing an opaque disk to block the stellar light. Examples include the *four-quadrant phase mask coronagraph*, and the related *achromatic phase knife coronagraph*. Variants include the *optical vortex coronagraph* in which the phase-shift varies azimuthally around the centre, and the *vector vortex coronagraph* which rotates the angle of polarisation.

Four-quadrant phase mask (4QPM) Figure 7.5 shows the successive steps in the image generation in the case of the *four-quadrant phase mask* coronagraph (Rouan et al., 2000). This is based on the principle of a phase mask originally proposed by Roddier & Roddier (1997), but one which is less sensitive to atmospheric turbulence and misalignment.

The focal plane is divided into four equal-area quadrants, centred on the optical axis. A π phase shift is applied to opposing quadrants, which results in destructive interference for a bright star located at its geometric centre. A Lyot stop is placed in the exit pupil to remove diffracted starlight, as in the original Lyot coronagraph. Implementation is based on the precision mounting of four half-wave plates (Riaud et al., 2003). In practice, implementation of the π phase shift tends to introduce chromatic terms, while the phase transitions between adjacent quadrants introduces large dead zones (Riaud et al., 2001).

Such a mask has been used with VLT–NACO, providing central light attenuation ∼10 for long exposures (Boccaletti et al., 2004), limited by terms uncorrected by the adaptive optics system. Scientific results are reported by, e.g., Gratadour et al. (2005) and Riaud et al. (2006). It will be one of the coronagraphs used in the

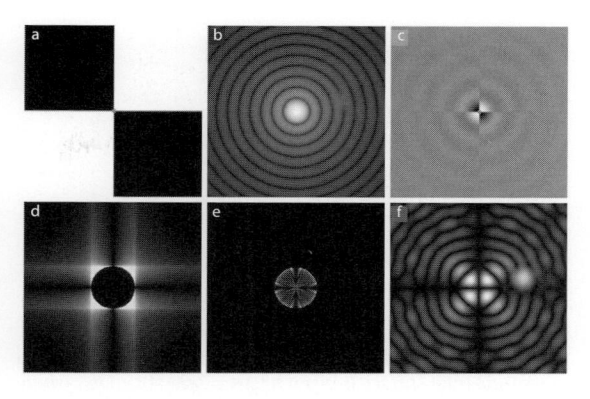

Figure 7.5: Simulation of the four-quadrant phase mask coronagraph and the detection of a companion 15 mag fainter at angular distance $2.1\lambda/d$: (a) the phase mask, with white/black for $0/\pi$ phase shifts, which takes the place of the occulting mask in the Lyot design; (b) the Airy pattern; (c) the complex amplitude of the star phase-shifted by the mask; (d) the exit pupil; (e) the exit pupil viewed through the Lyot stop; (f) final coronagraphic image, with companion. From Rouan et al. (2000, Figure 2), reproduced by permission of University of Chicago Press.

exoplanet imaging instrument VLT–SPHERE (Boccaletti et al., 2008), and in the JWST instrument MIRI (Baudoz et al., 2006).

Optical vortex coronagraph (OVCm) This is a type of phase mask based on optical vortex spatial filtering (Khonina et al., 1992), itself based on the orbital angular momentum properties of light (Allen et al., 1992; Harwit, 2003). The technique was proposed for exoplanet coronagraphy as an *optical vortex coronagraph* (Swartzlander, 2001; Foo et al., 2005) and, based on similar principles, as an *annular groove phase mask* (Mawet et al., 2005), and as a *quantum coronagraphic mask* (Tamburini et al., 2006).

The vortex phase mask, located at the focus of a circularly symmetric optical system, introduces a phase proportional to $m\theta$, where θ is an azimuthal coordinate in the focal plane, and m is an integer called the winding number or *topological charge* (Figure 7.6). The resulting intensity profile of the diffracted vortex beam is characterised by a dark central hole and annular rings, referred to as an *optical vortex* (Figure 7.7). For $m = 2$, zero-intensity values for the on-axis starlight results over the entire exit pupil, a phenomenon demonstrate analytically by Foo et al. (2005). For perfect optics, the vortex mask with a simple Lyot stop will therefore extinguish all light from an on-axis source.

The optical vortex coronagraph has high performance (Guyon et al., 2006b), can be designed to be achromatic (Swartzlander, 2006), has small sensitivity to low-order aberrations (Palacios & Hunyadi, 2006), and performs respectably in the presence of central telescope obscuration and atmospheric turbulence (Jenkins, 2008).

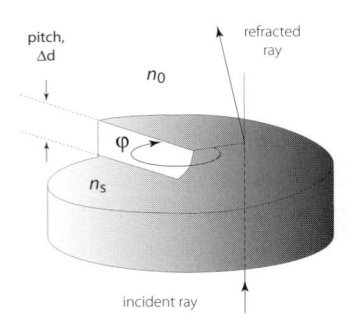

Figure 7.6: Schematic of a vortex lens phase mask. The pitch of the substrate surface is given by $\Delta d = m\lambda_0/(n_s - n_0)$ where λ_0 is the design wavelength, m is the topological charge, and n_s and n_0 are the refractive indices of the substrate and superstrate (Swartzlander, 2006, eqn 3). For $m = 2$ and an air–glass system with $n_s - n_0 \sim 0.5$, $\Delta d \sim 4\lambda_0$. Adapted from Swartzlander (2006, Figure 1).

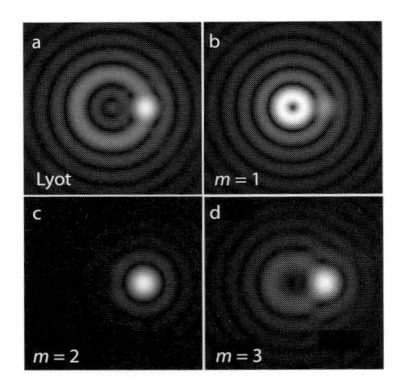

Figure 7.7: Simulation of the optical vortex coronagraph with star–planet intensity ratio 100, and the planet at the angular separation of the $m = 1$ ring: (a) performance of the Lyot coronagraph; (b–d) performance of the optical vortex coronagraph with $m = 1, 2, 3$ respectively. For $m = 2$ the starlight is essentially eliminated. From Foo et al. (2005, Figure 3).

From a manufacturing perspective, (scalar) vortex phase masks were originally developed in discrete steps by etching the phase ramp in a dielectric material such as fused silica. Nearly continuous helical surfaces can now be made out of liquid crystal polymers (Mawet et al., 2009a). [A further class, the *vectorial optical vortex*, relies on the manipulation of the transverse polarisation state of the light].

Astronomical tests (Swartzlander et al., 2008; Mawet et al., 2010) were followed by use with a high-performant 1.5-m sub-aperture of the Hale 5-m telescope (Serabyn et al., 2010), which successfully observed the three point source images now known to surround HR 8799 (Marois et al., 2008b), including the innermost planet separated by $2\lambda/D$ from the central star (§7.5).

Pupil replication A somewhat distinct technique for increasing contrast is that of *pupil replication*. This optically rearranges the incoming wavefront to decrease the diameter of the image of a star on or very near to the optical axis, while preserving (subject to some chromatic defect) the angular position of an off-axis source (Greenaway et al., 2005).

The technique transforms the telescope into a pseudo-interferometer, and the field of view properties are analogous to those in pupil densification schemes (§7.4.3). Effects of realistic instrumental defects, such as surface error, chromatic smearing, and pupil shift, may limit the gains expected theoretically (Riaud et al., 2005; Spaan & Greenaway, 2007).

Pupil filtering and remapping A variation of the interferometric coronagraphs divides the pupil into coherent subapertures, feeding each into a single-mode optical fibre, and remapping the exit pupil to allow non-redundant interference of all sub-apertures. A diffraction-limited image, with high dynamic range and free of speckle noise, is reconstructed from the fringe pattern (Perrin et al., 2006; Lacour et al., 2007; Kotani et al., 2009; Serabyn, 2009).

Free-flying occulters A distinct coronagraphic implementation could be achieved in principle, although with no confirmed plans yet to do so in practice, by placing an occulting screen between star and telescope in the form of a suitably equipped free-flying satellite. Proposed concepts have included UMBRAS, designed around a 4-m telescope and a 10-m occulter (Schultz et al., 2003), the much larger BOSS, the New Worlds Observer, and Planetscope which places the free-flying coronagraph on a stratospheric balloon platform (Traub et al., 2008). In all cases, control of diffracted light is still paramount.

The BOSS (Big Occulting Steerable Satellite) concept comprised a large occulting mask, nominally a 70×70 m^2 transparent square with a 35 m radius, and a radially-dependent, circular transmission function inscribed (Copi & Starkman, 2000). It would be supported by a framework of inflatable or deployable struts, and aligned with a ground- or space-based telescope. In combination with JWST, for example, both would be in a Lissajous-type orbit around the Sun–Earth Lagrange point L2, with the mask steered to observe a selected object using a combination of solar sailing and ion or chemical propulsion. Attenuation of 4×10^{-5} of the stellar light is predicted at 1 μm. Their simulations suggested that planets separated by 0.1–0.2 arcsec from the host star of 8 mag could be seen down to a relative intensity of 1×10^{-9}. For systems like the solar system, Earth and Venus would be visible to 5 pc, with Jupiter and Saturn to 20 pc.

The New Worlds Observer proposal (Cash et al., 2005; Cash, 2006; Vanderbei et al., 2007) calls for a 50 m occulter separated from a 4 m telescope by a distance of 80 000 km. The occulter could also be used in combination with a ground-based telescope (Janson, 2007).

A variant of the separated occulter uses a Fresnel lens. An array of sub-apertures on one spacecraft focuses the first diffraction order onto a field telescope some km distant (Koechlin et al., 2005). The design can be combined with coronagraphs and apodisation.

Lunar occultation The use of the lunar limb as an external occulter for exoplanet detection was considered already by Elliot (1978), who proposed a space telescope in an orbit yielding a stationary lunar occultation of any star lasting two hours, using the black limb of the moon as an occulting edge. The idea was re-examined by Richichi (2003), who estimated that in the K and L bands (2.2 and 3.6 μm) 8–10 m telescopes could thereby detect companions 5–11 mag fainter at separations down to 0.01 arcsec. Although insufficient to detect hot Jupiters, further gains would come with the 30–40 m ELTs.

Coronagraphic discovery space Factors such as the wavelength dependence of the inner working angle and contrast dependencies due to phase errors, the expected exoplanet radii and albedos, and the contribution of zodiacal and exo-zodiacal light, can be taken into account in the optimisation of coronagraphic searches (Brown, 2005; Agol, 2007).

Agol (2007) found that for blind planet surveys the optimum search wavelength is between 400–500 nm for Earth-like planets and 420–580 nm for Jovian planets; that target stars should be ranked by their luminosity divided by distance to the sixth or eighth power depending on the dominant noise source; and that stars with different metallicities should be assigned exposure times that vary with the cube of the metallicity.

7.2.4 Speckle noise

The contrast detection limit within the wings of the point-spread function in adaptive optics imaging, with or without coronagraphs, is determined by two contributions: photon noise, and *speckle noise*. In the presence of photon noise alone, a 30–100 m telescope could detect an exo-Earth around very nearby stars (Angel, 2003b; Hawarden et al., 2003). In reality, exoplanet detection limits are currently dictated by how well speckle noise can be suppressed or calibrated.

Speckle noise arises from the essentially random intensity pattern produced by the mutual interference of a set of wavefronts. The term is used to describe effects due to both rapid atmospheric phase fluctuations, which evolve on time scales of order 1–10 ms, and those due to instrumental imperfections, which evolve on a variety of intermediate to long time scales. Even extremely small drifts in the wavefront change the speckle's intensity noticeably: for a speckle intensity 10^{-5} of the host star intensity to be stable to within 1%, the corresponding spatial frequency would need to be stable to $2.5 \times 10^{-6} \lambda$, or 4 pm at 1.6 μm (Guyon, 2005).

Speckle patterns arising from atmospheric fluctuations formed the basis of the short-exposure technique of speckle interferometry (Labeyrie, 1970). In the related *dark speckle* technique (Labeyrie, 1995; Boccaletti et al., 1998), rapid changes in optical path length due to the atmospheric turbulence are exploited, with the goal of detecting the planet in very short exposures (~ 1 ms) when, by chance, the star light interferes destructively at the planet location. Atmospheric speckles are strongly attenuated in adaptive optics imaging (and totally eliminated in space imaging), and this technique is accordingly now of only historical interest.

More relevant for exoplanet imaging, quasi-static speckle patterns are seen in a star image observed through imperfect optics. At separations $\gtrsim$ 0.5 arcsec, the main source of long-exposure adaptive optics speckles results from surface errors on the telescope mirrors and instrument optics. The resulting noise can be an order-of-magnitude larger than the residual photon noise (Racine et al., 1999; Marois et al., 2000; Perrin et al., 2003; Bloemhof, 2004; Soummer et al., 2007; Marois et al., 2008a).

Various methods are now in use, or under development, to reduce its effects, either through suppression (preferable since it eliminates the underlying noise contribution), or through calibration.

Speckle suppression Speckle suppression techniques are presently largely developmental. Measurements of residual wavefront errors may be possible in the coronagraph image plane using a 'post-coronagraph wavefront sensor', with a further deformable mirror to apply additional corrections. Various technical implementations have been proposed, and some have been developed as laboratory prototypes (Angel, 2003b; Bloemhof, 2003; Codona & Angel, 2004; Give'on et al., 2005; Bordé & Traub, 2006; Trauger & Traub, 2007; Galicher et al., 2008; Sivaramakrishnan et al., 2008). It is an approach considered in the context of the TMT Planet Formation Imager (PFI) by Vasisht et al. (2006).

Interferometric subtraction makes use of the fact that, unlike the light from the nearby companion, speckles arising from the light of the central source are coherent with it, whether the speckles are evolving more rapidly as on the ground, or are essentially stationary as in space (Guyon, 2004; Galicher et al., 2008). Post-coronagraphic phase-shifted holographic suppression has also been proposed (Labeyrie & Le Coroller, 2004; Ricci et al., 2009).

Sivaramakrishnan et al. (2002) showed that the widely-cited technique of *speckle decorrelation*, introducing many independent realisations of additional phase error into a wave front during one speckle lifetime, thus changing the instantaneous speckle pattern, was not effective in reducing the speckle noise.

Speckle calibration An alternative to speckle suppression is speckle calibration, which aims to disentangle the stellar speckles from the exoplanet signal by virtue of specific characteristics present in the planetary light but not in the starlight, for example based on some spectral (Racine et al., 1999) or polarimetric (Kuhn et al., 2001; Tamura et al., 2006) signal.

Automatic recognition of possible exoplanets in a speckle-dominated image has been developed using wavelet analysis, essentially aiming to identify structures of a given spatial scale (Masciadri & Raga, 2004), and using matched filtering and Bayesian techniques in the context of TPF (Kasdin & Braems, 2006; Marsh et al., 2006b).

In non-coronagraphic imaging with adaptive optics systems on large telescopes, the point-spread function wings in long-exposure images are relatively smooth, and with appropriate calibration, a point-source detection limit some 10 times fainter than the background intensity level can be achieved within the central arcsec. Other techniques can lead to further gains.

Angular difference imaging exploits the intrinsic field-of-view rotation of altitude/azimuth telescopes, by keeping the telescope pupil fixed on the science camera, and allowing the field-of-view to rotate slowly with time around the star, such that patterns associated with optical aberrations are cancelled (Marois et al., 2006; Lafrenière et al., 2007b). Limits of $6 M_J$ around GJ 450 were reported using MMT–Clio (Heinze et al., 2006), and quasi-Gaussian residuals have been reported using Gemini–ALTAIR–NIRI (Marois et al., 2008a).

The technique was used in the imaging of the three companions around HR 8799 using both the Keck and Gemini telescopes (Marois et al., 2008b, §7.5). As *roll subtraction*, it is baselined for the coronagraphic observations with, e.g., JWST–NIRCam (Krist et al., 2007).

Simultaneous differential imaging aims to suppress the speckle pattern by isolating light from a strong absorption or emission feature, for example by separating the contribution of methane in a companion from the stellar spectrum (Racine et al., 1999; Marois et al., 2000, 2004, 2005). This specific choice of spectral feature exploits the fact that cool ($T_{eff} < 1200$ K) substellar objects have strong CH_4 absorption longward of 1.62 μm (Burrows et al., 2001, 2003b).

The VLT–NACO–SDI simultaneous differential imager (Lenzen et al., 2004) provides four simultaneous images through three narrowband filters, with two inside (1.625 μm) and two outside (1.575/1.600 μm) the CH_4 features. The point-spread function, including residual aberrations and speckles, are largely identical. The mode was designed to search for methane-rich objects near to very bright stars, with contrasts of 50 000 accessible. It was used to detect a companion at 1.17 arcsec from the host star SCR 1845 (Biller et al., 2006).

Similar systems are implemented in MMT–ARIES–MEDI (Freed et al., 2003), in CFHT–TRIDENT (Marois et al., 2005, 2008a), and in Gemini South–NICI (Near-Infrared Coronagraphic Imager) which includes an 85-element adaptive optics system, a Lyot coronagraph, and a dual-channel imager for spectral difference imaging which can also be used together with angular differential imaging (Artigau et al., 2008; Chun et al., 2008). Other wavelength bands, for example [Fe II], may also be used (Tamura et al., 2006).

Integral field spectroscopy This is a general technique for acquiring a low-resolution spectrum of each spatial element in a two-dimensional field, which has been used extensively for mapping velocity structures in external galaxies (e.g. Allington-Smith, 2006). It has been applied to host star spectroscopy for the transiting systems HD 209458 and HD 189733 using VLT–SINFONI (Angerhausen et al., 2006), WHT–INTEGRAL (Arribas et al., 2007), and Keck–OSIRIS (Angerhausen et al., 2009).

For resolved exoplanet imaging, integral field spectroscopy does not suppress the speckles, but it can assist distinguishing exoplanets from artefact speckles, as well as characterising *bona fide* planets spectroscopically (Sparks & Ford, 2002; Berton et al., 2006a,b; Vigan et al., 2007, 2008; Antichi et al., 2009). In the technique of *spectral deconvolution* described by Sparks & Ford (2002), recognition of the planetary spectrum within the multiple spectra into which the image is divided is achieved using pattern recognition, the process aided by the time-varying velocity of the orbiting planet.

Integral field spectroscopy is baselined in both the VLT–SPHERE and Gemini–GPI exoplanet imaging instruments (§7.3.1).

7.3 Ground-based imaging instruments

Some of the ground-based telescope–instrument (and coronagraph) combinations targeting exoplanet imaging have been mentioned in the previous sections, including CFHT–TRIDENT, Gemini–ALTAIR–NIRI, Gemini–NICI, Hale–OVC, Keck–AO, LBT–ARGOS, Subaru–AO188–CIAO, VLT–NACO, and WHT–NAOMI–OSCA. But in the absence of extreme adaptive optics or high-performant coronagraphs, their search space is restricted to massive exoplanets, either young and luminous, and/or wide-orbiting with $a \gtrsim 25$ AU.

7.3.1 Extreme adaptive optics instruments

Two major instruments which are being developed specifically for exoplanet detection and characterisation at 8-m telescopes will come on-line in 2011–2012: SPHERE at the VLT, and GPI at Gemini South. SPHERE developed from ESO's call for proposals for an exoplanet imaging instrument in 2001, illustrating the decade-long development time typical of these complex instruments. Both have dedicated extreme adaptive optics, employing of order 2000 actuators, and targeting Strehl ratios of 90% in the K-band. Both employ optimised coronagraphs.

In addition, Subaru–HiCIAO will be upgraded to a more performant adaptive optics system on the same schedule.

The number of exoplanet detections is difficult to predict. Signal-to-noise drops rapidly with distance, due to the combined effects of inverse-square brightness decrease, and the reduced star–planet angular separation.

VLT–SPHERE SPHERE (Spectro-Polarimetric High-contrast Exoplanet Research, Beuzit et al., 2008) is a second-generation instrument for the VLT, building on the experience of NACO and aiming for an order-of-magnitude imaging performance improvement. It is due for first light in 2012. The instrument, a synthesis of two earlier concept studies (CHEOPS and Planet Finder), includes an extreme adaptive optics system with a 41 × 41 actuator deformable mirror, and a 40 × 40 lenslet Shack–Hartmann wavefront sensor covering 450–950 nm (Fusco et al., 2006b; Petit et al., 2008).

A modular coronagraphic package includes an (achromatic) four-quadrant phase mask, a classical Lyot coronagraph, and an apodised Lyot coronagraph (Boccaletti et al., 2008).

Both evolved and young planetary systems will be observed, the former primarily through their reflected light using a visual differential polarimetry channel (ZIMPOL), and the latter through their intrinsic infrared emission using differential imaging and integral field spectroscopy at $R = 30$ (Claudi et al., 2006; Vigan et al., 2007, 2008). The design goals include access to angular scales of 0.1–3 arcsec from the host star, and the detection of giant planets at contrasts of 10^{-6} at $J = 6$ mag, and at an angular separation of 0.5 arcsec.

The SPHERE survey, over several hundred VLT nights, will include nearby young associations; young active FK dwarfs; stars within 20 pc; stars with known planets; and candidates from astrometric surveys. As an indication, SPHERE should be able to detect a young 10 Myr, $5M_J$ mass planet at 40 pc. For mature planets, the most promising targets lie within 5–10 pc, with $a \gtrsim 3-5$ AU.

Gemini Planet Imager GPI (Macintosh et al., 2008) shares a number of central design features with SPHERE. It includes its own extreme adaptive optics system, with 1800 actuators, 0.18 m sub-apertures, and a control rate of 2 kHz. It uses an apodised Lyot coronagraph to target a planet contrast of 10^{-7} at $I = 6$ mag, and an inner working angle of 0.13 arcsec (Soummer et al., 2006). The wavelength coverage is 1–2.4 μm. Lenslet-based integral field spectroscopy at $R = 45$ will be used for exoplanet confirmation and characterisation. The instrument is due for first light on Gemini South in early 2011.

Subaru–AO188–HiCIAO Subaru, with its AO188 adaptive optics system with 188 actuators, does not achieve the same high Strehl ratios targeted by SPHERE and GPI, but it is also equipped with a High Contrast Coronagraphic Imager for Adaptive Optics (HiCIAO), with first light reported in December 2008. It uses a classical Lyot coronagraph with mask diameter $4.8\lambda/D$ as baseline, a polarisation channel, and a 2024 × 2024 HgCdTe array with a pixel scale of 0.01 arcsec operating in the infrared JHK bands, giving contrasts of 4×10^{-6}, and an inner working angle of 0.1 arcsec (Tamura et al., 2006; Tamura, 2009).

Post-processing is based on angular difference imaging, simultaneous differential imaging using CH_4 and [Fe II] bands, and polarisation differential imaging. The HiCIAO survey, using 120 nights of Subaru time over 5 years, started in October 2009. An upgrade to a 1024 actuator adaptive optics system, AO1024, is targeted for early 2011.

7.3.2 Extremely large telescopes

Exoplanet imaging and spectroscopy are important science goals for the very large optical aperture telescopes now under development, notably the 42-m European Extremely Large Telescope (E–ELT), the Giant Magellan Telescope (GMT, Johns, 2008b), and the Thirty Metre Telescope (TMT, Crampton et al., 2009). All are targeting first light around 2018.

Detailed exoplanet imaging studies were also carried out for the 100-m OWL (Overwhelmingly Large Telescope), which would have provided $\lambda/D \sim 1$ mas in the V-band, with S/N = 10 at 35 mag in 1 hr for imaging, and at 30 mag in 3 hr for $R = 1000$ spectroscopy (Vérinaud et al., 2006; Lenzen et al., 2006; Fusco et al., 2006c). Apertures of 100 m are no longer under active consideration.

Exoplanet detection challenges of the extremely large telescopes are technically similar to those for the 10-m class. Results will depend sensitively on reaching very high Strehl ratios, and on the success of diffraction and speckle suppression techniques to deliver the highest contrasts (Cavarroc et al., 2006; Hainaut et al., 2007).

Image structure For a Strehl ratio S, simulations show a central image spike resembling a diffraction-limited image containing some S % of the total light, less a modest fraction diverted into wider components of the telescope point-spread function. The spike is surrounded by a halo containing the residual light distributed like an unmodified seeing disk, i.e. with a generally Lorentzian distribution. This halo overlies, and within the seeing disk generally dominates, the fainter wings of the telescope point-spread function.

Table 7.2: V magnitude and separation α (arcsec) relevant for ELT observations. Modeled planet/star contrasts are in the range $\sim 10^{-8}$ to 10^{-9}. From Perryman et al. (2005, Table 6).

Distance (pc)	Star (mag)		Hot Jupiter 0.2 AU	Earth 1 AU	Jupiter 5 AU
10	4.8	V	24.1	27.9	25.8
		α	0.020	0.100	0.500
25	6.8	V	26.1	29.9	27.8
		α	0.008	0.040	0.200
50	8.3	V	27.6	31.4	29.3
		α	0.004	0.020	0.100
100	9.8	V	29.1	32.9	30.8
		α	0.002	0.010	0.050

Table 7.3: Indicative distance limits and numbers of stars as a function of exoplanet mass and primary mirror diameter, and based on simplistic assumptions, for both imaging (Im) and spectroscopy (Sp). From Perryman et al. (2005, Table 7).

D (m)		Earth-type		Jupiter-type	
		Im	Sp	Im	Sp
30	d (pc)	10	–	70	5
	$n(\star)$	22	0	6800	3
60	d (pc)	22	–	120	18
	$n(\star)$	210	0	35 000	170
100	d (pc)	40	15	500	35
	$n(\star)$	1200	67	2 500 000	860

The wings combine the light diffused by the small-scale imperfections of the optics and any accumulated dust, as well as the light diffracted by the central obstruction and geometric edges of the mirrors and supporting structures. The diffracted light strongly dominated the uncorrected seeing light.

For adaptive optics-corrected images, the detailed structure of the halos is more complex. At large radii they are composed of the rapidly varying 'classical' speckles, with size $\sim \lambda/D$. Less well-understood 'super-speckles' occur closer to the first 2–3 diffraction rings; they are larger, brighter, less rapidly variable and therefore less likely to average out on a long-exposure image. At the wavelengths of interest they should not, however, occur more than 10 mas from the image centre, corresponding to 0.1 AU at a distance of 10 pc. While techniques such as simultaneous differential imaging may cancel these super-speckles, their noise contribution remains even after subtraction, and is typically the strongest noise source in the 5–15 λ/D region. This can only be reduced by improving the Strehl ratio of the adaptive optics system.

Exoplanet detection Detection of a terrestrial-like planet is made possible in principle by its relatively large angular separation from the central diffraction peak of its host star. At separations of 100 mas (1 AU at 10 pc) only the sky background, the wider scattering components of the intrinsic point-spread function, and the adaptive optics halo contribute to the background.

The orbital radius at which an exoplanet is detectable is then bounded by two effects. At the inner extreme, the bright inner structures of the stellar image reduce sensitivity below 10–20 mas (1 AU at 50–100 pc), so that within these angles only young self-luminous exoplanets could be detected. At the outer extremes, reflected starlight falls off with increasing orbital distance. Some relevant angular scales are shown in Table 7.2.

Assuming that a planet is detected only beyond an angular distance of $5\lambda/D$ from the host star, the volume of space accessible is proportional to D^3. The numbers for G stars range from about about 900 for 100 m aper-tures to 25 for 30 m. For background-limited measurements, and a planet lying within the uncorrected stellar light, the time to achieve a given S/N scales as $(S/N)^{-2}$, i.e. $S \propto D^2$, background $\propto D^2$, and diffraction-limited pixel size $\propto D^{-2}$, so that $S/N \propto D^2$ and $t \propto D^{-4}$. Thus a 30 m aperture would require 120 times longer than a 100 m to observe a planet accessible to both.

With specific assumptions about the form of the instrumental aberrations, Chelli (2005) derived expressions for the signal-to-noise ratio on the planet flux, for direct and differential imaging, in the presence of speckle noise and photon noise due to the residual stellar halo. For an Earth at 10 pc to be detected with S/N = 5 in an integration time of 12 hr, a 100-m telescope would require the intensity of the incoherent halo at 0.1 arcsec radius to be reduced from that of the central source by a factor 1.2×10^6. For a 30-m telescope, the corresponding factor is 1.3×10^7.

Distance limits The limiting distances at which particular observations can be performed depend sensitively on these various assumptions. Estimates of the numbers of single G and early K stars that would be suitable for the imaging and spectroscopy of Earth- and Jupiter-like planets were made in the context of an ESA–ESO Working Group in 2005, and are given in Table 7.3.

According to the study made at that time, 30–40-m ELTs will not be capable of imaging Earth-like planets, and will be challenged to provide useful spectroscopy even for very nearby Jupiter-type planets. More massive and more widely separated exoplanets will be measurable, and form the focus of specific instrumentation efforts now underway.

Instrumentation for exoplanet imaging The proposed multi-conjugate adaptive optics module for the E-ELT, MAORY (Diolaiti et al., 2010), is foreseen to provide average Strehl ratios of ~ 0.5 over a corrected field of view up to 2 arcmin at 0.8–2.4 μm. It will operate with six Na laser guide stars, and three levels of wavefront correction comprising the telescope adaptive mirror, and two post-focal deformable mirrors conjugated at heights of 4 and 12.7 km.

7.3 Ground-based imaging instruments

Two dedicated exoplanet imaging instruments are under study for E–ELT. EPICS (Exoplanet Imaging Camera and Spectrograph) aims for near-infrared, high-contrast (10^8–10^9), high angular resolution (~10 mas) down to 30 mas from the host star (Vérinaud et al., 2006; Kasper et al., 2008). METIS (Mid-Infrared ELT Imager and Spectrograph) targets the mid-infrared, 3–20 μm (Brandl et al., 2008).

For the TMT, the adaptive optics module foresees a 2–4 kHz pyramid wavefront sensor, driving a 10^4 (128 × 128, d = 0.24 m) actuator deformable mirror, and targeting Strehl ratios of 0.9 (Vasisht et al., 2006).

The Planet Formation Imager (PFI, Macintosh et al., 2006; Macintosh, 2007) for TMT has similar contrast and angular scale objectives as EPICS. It includes a dedicated integral field spectrograph, with $R \sim 70$ and a Nyquist-sampled 2 × 2 arcsec^2 field. The role of a coronagraphic mask (and diffraction suppression) is implemented as a (dual-stage shearing) nulling interferometer, providing a small inner working angle $\sim 3\lambda/D$. Further contrast enhancement, by suppression of the quasi-static speckle pattern resulting from other wavefront errors, is under study, employing a post-coronagraphic wavefront sensor driving a dedicated deformable mirror.

For the Giant Magellan Telescope (GMT), planet detection programmes are described by Codona (2004), Angel et al. (2006), and Close (2007).

7.3.3 Imaging from the Antarctic

The Antarctic Plateau provides excellent astronomical conditions in terms of telescope acuity and sensitivity across optical to submillimeter wavelengths. High atmospheric transparency and low sky emission result from the extremely cold and dry air. It yields one long observing 'night' per year, with complete darkness and 80% clear skies in June–July, decreasing to 7 hr darkness during March and October, although with an extended 'twilight' period.

Science currently conducted from the Antarctic includes optical, infrared, terahertz and sub-mm astronomy, solar astronomy, measurement of cosmic microwave background anisotropies, and high-energy astrophysical measurements of cosmic rays, gamma rays, and neutrinos (Burton, 2010).

Astronomical activities are underway at four plateau sites: the Amundsen–Scott station at the South Pole (US), the Concordia station at Dome C (France/Italy), the Kunlun station at Dome A (China), and the Fuji station at Dome F (Japan).

Dome C (elevation 3250 m, latitude $-75°$) was opened for winter operations in 2005 after a decade-long construction phase (Figure 7.8), and is the polar site best characterised astronomically (e.g. Aristidi et al., 2003; Storey et al., 2003; Lawrence et al., 2004). The ambient temperature ranges from 195–235 K, resulting in low and stable thermal emission. Very low surface winds, with a median of 2.7 m s^{-1} and below 5 m s^{-1} for more than 90% of the time, result in low free-air turbulence, with a quasi-absence of jet streams inside the polar vortex. Future plans for Dome C have been formulated within the EU-funded ARENA consortium, a network programme run from 2006–09 which concluded its activities with a development roadmap (Epchtein, 2010).

Even better conditions may exist at Dome A and Dome F (Lawrence, 2004b), with gains of up to three orders of magnitude in various performance indicators compared to the best mid-latitude sites such as Mauna Kea and Chajnantor. Dome A is the highest location on the Antarctic plateau, at 4083 m, and first visited only in 2005. China began construction of the Kunlun station at the site in 2009. Various astronomical observations are ongoing using an Australian robotic observatory, and there are plans for several major astronomical facilities now underway (Gong et al., 2010).

Performance gains Broadly, Antarctica offers the following performance gains compared to excellent temperate latitude sites (Storey, 2009; Burton, 2010): (i) atmospheric seeing a factor 2–3× better (above a 10–40 m boundary layer), providing better spatial resolution and improved point-source sensitivity; (ii) an isoplanatic angle some 2–3× larger, from the 2.9 arcsec normalised value for Paranal, to some 5.9 arcsec, providing better adaptive optics sky coverage using natural guide stars; (iii) improved atmospheric coherence time, some 2.5× longer, resulting in increased sensitivity for adaptive optics and interferometry; (iv) reduced scintillation noise, by a factor 3–4, resulting in more precise photometry; (v) reduced infrared sky background, by a factor 20–100, resulting in increased infrared sensitivity and improved photometric stability; (vi) extremely dry air, a factor 3–5× lower with 250 μm precipitable water vapour typical, resulting in enlarged and improved infrared and sub-mm transmission windows; and (vii) dust aerosols some factor of 50 lower, resulting in better infrared windows and improved sky stability.

The combination of coldness and dryness for the atmosphere results in infrared photometric gains that peak at about a factor of 25 in the K and L bands, i.e. an Antarctic 1.8-m telescope is more efficient than an 8-m telescope at a temperate site. In the H and N bands the gain (of order 3) is significant although less dramatic.

Exoplanet imaging Of specific interest for exoplanet imaging is the confinement of most of the atmospheric turbulence to a thin layer just above the ice. This creates conditions that are particularly favourable for adaptive optics wavefront correction, as well as providing telescopes with excellent seeing if raised above the boundary layer. Specific studies have been made of the performance of such ground-layer adaptive optics systems

Figure 7.8: The French–Italian Concordia Station at Dome C, and the experiments of the Concordiastro site characterisation programme in 2008: PAIX (extinction), GSM and DIMM (seeing), MOSP (turbulence outer scale), SBM (sky brightness), ASTEP Sud (photometry), ASTEP 400 and LUCAS (spectra of Earth-shine off the Moon). The twin towers of the station are seen at the rear. Courtesy Karim Agabi.

(e.g. Lawrence, 2004a; Travouillon et al., 2009; Carbillet et al., 2010; Masciadri et al., 2010a).

There are no confirmed plans for exoplanet imaging experiments from the Antarctic (although transit searches are ongoing there, §6.2.4). Various studies have been made, in particular based on nulling interferometry (Swain et al., 2003; Lawrence et al., 2006; Valat et al., 2006; Lloyd, 2006). An ambitious proposal for an interferometer at Dome C, KEOPS, called for an array of 39 1–2 m telescopes spread over kilometric baselines and operated in the thermal infrared, aimed at characterising all exo-Earths within 1 kpc (Vakili et al., 2005).

7.3.4 Ground-based interferometry

By virtue of their long optical baselines and high angular resolution, all major interferometers have embraced exoplanet imaging as part of their scientific goals (for a review of principles and applications of interferometry see, e.g., Quirrenbach, 2001; Monnier, 2003).

VLT–I, Keck–I, LBT–I Three large-aperture systems have been designed to operate with interferometric combination of the individual telescope beams:

(a) the ESO Very Large Telescope Interferometer (VLTI) at Cerro Paranal, Chile. This has four 8-m and four auxiliary 1.8-m telescopes over a 200-m baseline (Glindemann et al., 2003). Exoplanet detection has been considered in phase-closure mode for the first generation interferometric instrument AMBER (Joergens & Quirrenbach, 2004; Millour et al., 2006), and using colour-differential photometry (Vannier et al., 2006). It is planned in narrow-angle astrometric mode with PRIMA (ESPRIT, Launhardt et al., 2008), and in imaging mode with the second-generation VLTI Spectro-Imager instrument (VSI, Renard et al., 2008). VSI targets exoplanet imaging using 4–8 telescopes and low spectral resolution reaching angular scales down to 1.1 mas in the near-infrared;

> **Long-baseline small-aperture interferometers:** Operational instruments are: the 30-m baseline 2-element Mitaka Optical and Infrared Array (MIRA, Yoshizawa et al., 2006; Ohishi et al., 2008); the 75-m baseline 3-element 1.65-m telescope Infrared Spatial Interferometer (ISI, Townes & Wishnow, 2008); the 160-m baseline Sydney University Stellar Interferometer (SUSI, Davis et al., 2006); the 200-m 6-element 1-m telescope Y-configuration CHARA interferometer of Georgia State University at Mount Wilson (ten Brummelaar et al., 2003); the 340-m baseline 10-element 1.4-m telescope Y-configuration Magdalena Ridge Observatory Interferometer, with first light due in late 2010 (MROI, Buscher et al., 2006a; Creech-Eakman et al., 2008); and the 437-m baseline 6-element US Navy Prototype Optical Interferometer (NPOI, Johnston et al., 2006a).
>
> Interferometers no longer operational include the 32-m Mark III interferometer at Mount Wilson, operational 1987–1992 (Shao et al., 1988); the 38-m Infrared-Optical Telescope Array at Mount Hopkins, operational 1993–2006 (IOTA, Schloerb et al., 2006); the 65-m GI2T of the Observatoire de la Côte d'Azur, closed in 2006 (Mourard et al., 2006); the 100-m Cambridge Optical Aperture Synthesis Telescope, operational 1993–2005 (COAST, Buscher et al., 2006b); and the 110-m baseline Palomar Testbed Interferometer, operational 1995–2009 (PTI, Akeson, 2006).

(b) the Keck Observatory Interferometer in Hawaii (Keck–I), has two 10-m and four auxiliary 2-m telescopes over an 85-m baseline (Wizinowich et al., 2006a);

(c) the Large Binocular Telescope, Mt Graham, Arizona, has two 8.4-m telescopes on a common mount with a 14.4-m baseline. Binocular imaging began in early 2008, separate adaptive optics imaging for the two telescopes is planned for 2010, and mirror cophasing is foreseen for 2012 (Hill, 2010).

Small-aperture long-baseline interferometers The discovery space for small-aperture interferometers is limited by the faint exoplanet flux. A list of operational systems is given in the accompanying box. Nulling has not been developed for any of these systems.

Stratospheric observations Bridging the technological and environmental conditions between ground and space, stratospheric observations have been proposed. A coronagraphic camera with active optics at an altitude of 35 km, Artemis/UHST, was considered by Ford et al. (2002). Traub et al. (2008) showed that free-atmospheric and locally-generated turbulence should not limit observations at the 10^{-9} contrast levels for planet–star separations of 0.5 arcsec.

7.4 Space-based imaging

7.4.1 Existing telescopes

The contribution of HST, Spitzer, and JWST to transit photometry and spectroscopy, and a short description of each mission, is given in §6.2.6 and §6.2.7. This sec-

7.4 Space-based imaging

Nulling interferometry: *Nulling* is one mode of interferometric operation considered for exoplanet imaging, first proposed by Bracewell (1978) and Bracewell & MacPhie (1979). It introduces destructive interference between the pupils of two telescopes, or the subapertures of a single telescope, for an on-axis star.

Identical path lengths through the two beams leads to an interference maximum for an on-axis source. Off-axis rays traverse different optical path lengths, resulting in circular interference fringes with a bright central maximum. Introducing a phase difference of π rad in one of the paths suppresses the central maximum. The transmission pattern on the sky, for monochromatic light, is a sequence of fringes given by $T(\theta) = \sin^2(\pi\theta D/\lambda)$. By varying the baseline D, a range of constructive interference angles can be examined for the presence of an off-axis source.

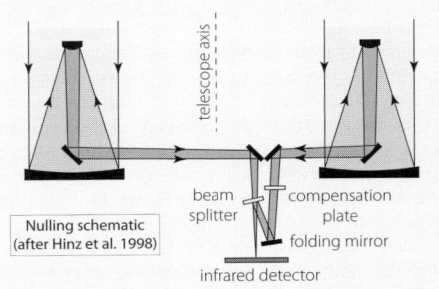

On the ground, nulling requires adaptive optics and operation at mid-infrared wavelengths, around 5–10 μm, because rapid atmospheric path length differences of order a few μm otherwise shift the interference randomly between constructive and destructive states.

Various implementations have been described (e.g. Hinz et al., 1998; Wallace et al., 2000; Hinz et al., 2001). Constraints on errors in the required π rad phase difference can be relaxed if more than two beams are recombined (Mieremet & Braat, 2003; Lane et al., 2006). Closure phase measurements for a nulling interferometer with three or more telescopes may also be possible (Danchi et al., 2006).

A nulling prototype for Darwin, GENIE, was considered but not implemented for VLTI (Gondoin et al., 2008). Nulling has been implemented at the MMT using two subapertures (BLINC, Hinz et al., 2000), and for Keck–I (KIN, Mennesson et al., 2006; Barry et al., 2008b; Colavita et al., 2009). A nulling camera is under development for LBT–I (NIC, Hinz et al., 2008). Nulling operation is also foreseen for GMT (Angel et al., 2006).

Direct detection prospects for known exoplanets have been evaluated in the case of nulling operations for Keck–I, VLTI–I, and LBT–I, with predicted integration times of minutes to hours (Langlois et al., 2006).

tion refers only to imaging-related considerations, and specifically to their coronagraphic capabilities.

HST NICMOS (near-infrared Camera and Multi-Object Spectrometer), operational from 1997–99 and 2002–08, provided an occulting spot of 0.6 arcsec diameter. It was used to confirm the discovery of 2M J1207 b (Song et al., 2006, §7.5).

ACS (Advanced Camera for Surveys) High Resolution Channel, operational from 2002–06, included a coronagraphic mask with two occulting spots of diameter 1.8 and 3 arcsec (Ford et al., 1998). This mode was used for the detection of Fomalhaut b (Kalas et al., 2008, §7.5).

STIS (Space Telescope Imaging Spectrograph (STIS) operational from 1997–2004, included focal plane wedges and a Lyot stop providing white-light coronagraphy between 0.2–1.0 μm (Grady et al., 2003).

Spitzer The imaging capability of Spitzer (launched 2003) is determined by the diffraction limit of the relatively small 0.85-m telescope (1.5 arcsec at 6.5 μm). It therefore has limited application for direct exoplanet imaging, although it has been widely used for the detection and imaging of circumstellar and debris disks. Its sensitivity corresponds to the detection of dust around a solar-type star at 30 pc at below that inferred to exist in small grains in the Kuiper Belt (6×10^{19} kg).

JWST Three of the four instruments on the 6-m JWST (launch 2014) will have coronagraphs:

NIRCam, the near-infrared camera, provides coronagraphic imaging from 1–5 μm using a Lyot coronagraph with five apodised occulting masks at the telescope focus (three circular and two wedge-shaped), and matching aperture masks (Lyot stops) at subsequent images of the telescope pupil (Krist et al., 2009). At around 4.6 μm, and after angular (roll) difference imaging, contrast levels of $10^{-5} - 10^{-6}$ at separations above 0.6 arcsec are predicted. Monte Carlo simulations have been used to estimate the numbers of stars around which planets of various masses would be detected at various projected distances, broadly corresponding to $2M_J$ at ages of 1 Gyr and less than $1M_J$ for young objects with ages less than 300 Myr (Krist et al., 2007, Table 1).

MIRI, the mid-infrared camera, provides coronagraphic imaging from 5–28 μm using three four-quadrant phase-mask coronagraphs and a classical Lyot coronagraph. Each has its own wavelength filter (at 10.6, 11.4, 15.5, and 23 μm respectively), with the coronagraphic stops deposited onto the filters (Boccaletti et al., 2005; Baudoz et al., 2006). Contrasts of $10^4 - 10^5$ are predicted between 0.5–1.0 arcsec, corresponding to the signal of a $5M_J$ planet at an age of 1 Gyr.

TFI, the Tuneable Filter Imager, is implemented within the JWST Fine Guidance Sensor, and provides coronagraphic imaging with four occulting spots, from 1.6–4.9 μm at a resolving power $R \sim 100$ (Doyon et al., 2008). Using spectral deconvolution, it is expected to reach contrasts of 10^5 at 1 arcsec separations.

7.4.2 Space interferometry

HST, Spitzer, and JWST do not have the angular resolution, nor were they designed, for the task of exoplanet imaging. More targeted ideas nevertheless began to emerge already in the 1970s (see box on page 164).

Requirements for high angular resolution imaging developed into the infrared nulling interferometry concepts which formed the basis both of ESA's Darwin study, and of NASA's original Terrestrial Planet Finder (TPF) study, which have been carried out over the last decade. In both cases, several large-aperture telescopes would fly in formation, and their light combined in a central hub using precisely controlled path delays. In both concepts, direct detection in the infrared would be accompanied by low-resolution spectroscopy.

The prospect of a realistic search for life on terrestrial exoplanets lay at the foundation of both projects. Their goals were the detection of exo-Earths, their physical and chemical characterisation, and the identification of possible biosignatures in their atmospheres.

Search space for exo-Earths Basing the premise of the search for life on the principles of organic chemistry, there are a number of arguments that have been used to constrain the search space. While these arguments are certainly imprecise, and possibly incorrect, they serve as a basis for discussion and proceed broadly as follows (§11.9).

If life requires liquid water, then the search for life should focus on the habitable zone in which liquid water is present on the planet's surface or, more strictly, on the continuously habitable zone in which liquid water could have existed over billions of years as the star evolved in luminosity on the main sequence. The precise boundary of such a habitable zone, in terms of semi-major axis, eccentricity, albedo, greenhouse effect, and others is uncertain. Accordingly, a primary search zone can be considered to extend from 0.9–1.1 AU, with a desirable search zone from 0.7–2 AU. For host star spectral types other than solar, the habitable zone scales as $L_\star^{1/2}$.

Plate tectonic activity has been considered necessary to sequester and recycle CO_2. The minimum planetary mass required to sustain tectonic motion is unknown, so therefore is the minimum diameter of a terrestrial exoplanet, and its associated reflecting area. A minimum has been proposed at around $R_p = 0.4 R_\oplus$. Planets with $M_p > 10 M_\oplus$, i.e. $R_p > 2 R_\oplus$ for rocky planets and $R_p > 3 R_\oplus$ for water planets, would likely accrete hydrogen and evolve into gas giants.

Exoplanet surface albedos are also expected to vary widely, from $A_B = 0.05$ for an ocean or a Moon-like surface to values of 0.4 or higher, introducing a large uncertainty in the amount of reflected light expected in the visible. But for terrestrial planets massive enough to retain an atmosphere, the total albedo is likely to be dominated by atmospheric scattering, and albedos towards the upper range are to be expected.

Optical versus infrared: technical As the TPF and Darwin studies developed, there was a continuing debate about the relative merits of detecting and characterising Earth-like planets in the visible stellar light re-

---

Early ideas for exoplanet imaging from space: Bonneau et al. (1975) considered Lyot filtering to decrease the brightness of the Airy rings in what was then the Large Space Telescope, while KenKnight (1977) suggested an analogue of phase-contrast microscopy to attenuate scattered light arising from the imperfect figure of a 2-m space telescope. Elliot (1978) proposed a space telescope in an orbit yielding a stationary lunar occultation of any star lasting two hours, using the black limb of the moon as an occulting edge to reduce the background light from the planet's star.

Bracewell (1978) and Bracewell & MacPhie (1979) noted that with Sun/Jupiter temperatures of 6000 K and 128 K, detection of thermal emission in the Rayleigh–Jeans regime longward of $\sim 20\,\mu m$ (where the emission from the planet is strongest) would result in a factor of 10^5 improvement in contrast. They also introduced the principle of nulling interferometry to enhance the planet/star signal.

Ideas for improved space missions (Angel et al., 1986; Korechoff et al., 1994) or balloon experiments above altitudes of 30 km (Terrile & Ftaclas, 1997) were subsequently developed. It was shown that multi-element arrays can provide a deep central null with high-resolution fringes that can be used for mapping. These were predicted to yield full constructive interference for a close-in planet even in the presence of a resolved stellar disk or dust cloud (Angel & Woolf, 1996, 1997; Woolf & Angel, 1997).

Nulling interferometry has remained the baseline for TPF–I and Darwin, although different approaches, such as pupil apodisation (Nisenson & Papaliolios, 2001), and pupil densification (Riaud et al., 2002), have been proposed as alternatives.

---

flected by the planet, or in the thermal infrared where the planet/star contrast is greatly improved. The technical solution to the former has generally been considered to be via coronagraphy associated with a single telescope (although coronagraphy could equally well be implemented with an interferometric array), with the technical solution in the infrared being generally associated with the concept of a free-flying interferometer (e.g. Beckwith, 2008).

There has been no definitive resolution of this issue. Technologically, various advantages accrue from searches at visible wavelengths. For a given angular resolution λ/D the required aperture scales with λ, resulting in mirror sizes in the optical of order 6–12 m, which are still substantial but perhaps more tractable than interferometric baselines in the thermal infrared a factor three larger, i.e. at around 20–40 m.

At visible wavelengths the telescope does not contribute to the radiation background, meaning that the telescope can be operated at around 300 K, offering significant engineering advantages over a thermal infrared instrument which would need to be cooled to ~40 K.

In the thermal infrared, background radiation from solar system zodiacal light is a further complication; the emission from solar system dust (arising from comets and asteroids) itself peaks around 10–20 μm. If the instrument could be placed at 4–5 AU from the Sun, the corresponding zodiacal background contribution would be reduced by about 100 (Léger et al., 1998; Landgraf & Jehn, 2001; Gurfil et al., 2002).

Although a visible light system could be significantly smaller than a comparable infrared interferometer, significant advances in optical mirror technology would be required. The TPF studies concluded that the mirrors would need to be 'ultra-smooth' to minimise scattered light, and that active optics would be needed to maintain low and mid-spatial frequency mirror structure.

Infrared interferometry would require large boom technology or formation flying, with separation accuracies of around 0.01 m and short internal delay lines.

Optical versus infrared: scientific Scientifically, there are arguments for both spectral regions (e.g. Schneider, 2002b,a; Arnold et al., 2002; Schneider, 2003).

In the visible, spectral features and associated planetary diagnostics are numerous (Figure 11.22). They include the absorption bands of H_2O, O_2, and O_3 (including the Chappuis bands at 520–580 nm); Rayleigh scattering from the columnar abundance of gases and clouds; vegetation 'red edge' (which occurs at 725 nm for terrestrial vegetation); overall exoplanet colour (e.g. whether similar to Venus, Earth, Mars or Jupiter); brightness (indicating planet size and albedo); brightness variations (indicating rotation rate, weather patterns, and the presence of oceans or land masses); and polarisation (characterising the molecular atmosphere).

Measured values can further be used to infer properties such as temperature, diameter, mass, surface gravity, and atmospheric pressure. If the planet more closely resembles the primitive Earth (Figure 12.7), then CH_4 and CO_2 may also be significant.

A reference optical wavelength range could be set at 500–800 nm, although extensions to 300 nm would include the Huggins O_3 bands at 330 nm, while extension to 1.3 µm would include additional H_2O and CO_2 bands.

In the infrared, the important molecular bands of CH_4, CO_2 H_2O, and O_3 have been considered as particularly valuable diagnostics for the presence of life (§11.9.3). However, while the O_3 band at 9.6 µm may be the signature of an O_2-rich atmosphere produced by photosynthetic life forms, O_2, and hence O_3, may also be built up without life from the photolysis of H_2O and CO_2. The analysis of Selsis et al. (2002) suggests that, while the detection of O_2 may provide only an ambiguous indicator of life, the triple signature of CO_2, H_2O, and O_3 may provide a robust discriminator between photochemical O_2 production and biological photosynthesis.

In either case, a spectral resolution of $R \sim 70$ is deemed necessary to quantify the presence and equivalent widths of the various spectral features.

Achievable signal-to-noise An order-of-magnitude estimate of the achievable signal-to-noise for exoplanet

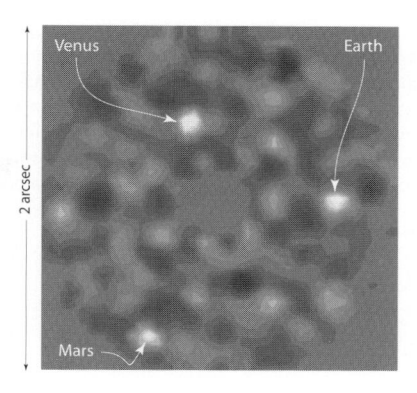

Figure 7.9: Simulation of a 60-hour exposure of the original Darwin interferometer, with six 1.5-m telescopes in a 1 AU orbit and 50-m baseline observing between 6–17 µm. It covers a 2×2 arcsec^2 field, and simulates a solar-type star at 10 pc, with solar-level zodiacal dust, and three inner planets yielding terrestrial-level flux at locations corresponding to Venus, Earth and Mars. The stellar light, at the image centre, has been nulled.

detection has been given by Quirrenbach (2005, Section 3.5), which underlines its challenging nature. For a G star of $V = 5$ mag, an Earth in the habitable zone with $V = 30$ mag over the range 400–800 nm contributes $N = 100 \epsilon A$ photons hr^{-1}, where ϵ is the total detection efficiency and A is the collecting area in m^2.

Assuming $\epsilon = 5\%$ and $A = 3.5 \times 7$ m^2, then $N = 100$ photons hr^{-1}. If the photon noise dominates both speckle noise and detector noise, then for S/N = 7, the detection times for different spectral features (Rayleigh scattering with $R = 3$, vegetation red edge, CH_4, H_2O) range from 2–80 hr. The detection of albedo variations of 30% at the same S/N consequently requires exposure times of up to 400 hr.

Darwin ESA embarked on the Darwin Infrared Space Interferometer study as a high-priority but longer-term programme in 1996. It was proposed as a nulling interferometer, operating in the infrared, which could search for Earth-like planets around 100–200 stars out to distances of 15–20 pc using direct imaging, and analyse their atmospheres for the chemical signature of life (Léger et al., 1996; Mariotti et al., 1997; Mennesson & Mariotti, 1997; Léger et al., 1998; Fridlund, 2000; Fridlund et al., 2003; Fridlund, 2004; Mennesson et al., 2005; Ollivier, 2007; Cockell et al., 2009a,b).

It was originally conceived as a set of eight spacecraft, placed at L2, with six free-flying 1.5-m class telescopes, one beam combination unit, and one communication unit. It would be passively-cooled to 40 K, and provide baselines up to 50 m. It would operate between 6–17 µm to cover spectral lines including H_2O, CH_4, O_3 and CO_2 (Figure 7.9).

Later studies adopted a configuration employing four telescopes separated by up to 50–100 m operated in a 'dual-Bracewell' configuration, requiring a dual Soyuz–

Table 7.4: Integration times for Darwin for the detection of Earth-like planets at S/N = 5, and spectroscopy at S/N = 7 in the faintest part of the spectrum (in hours). Assumptions are: spectral range 6–17 μm, spectral resolving power $R = 20$; $A_{total} = 40\ m^2$; $T_{optics} = 40\ K$; effective planetary temperature 265 K (signal scales as T_{eff}^4); $R = R_\oplus$, with 10 times the zodiacal dust than in the inner solar system; Spitzer Si–As detectors.

Stellar type	10 pc	20 pc	30 pc
G2V	18–33	28–54	109–173
G5V	12–22	27–46	105–166
K2V	4–9	26–37	104–157
K5V	4–6	26–35	249–155

Fregat launch, and with a target launch date of 2015. The mission scheduled a detection phase of two years, and a spectroscopic phase of three years.

Table 7.4 summarises the estimated Darwin integration times for various stellar types, using a 90% confidence level for a non-detection based on three observations. Two ±45° sky caps near the ecliptic poles are inaccessible. The study identified some 500 F5–K9 stars out to 25 pc, of which some 285 are single. There are 2–5 G0V–G2V stars within 10 pc, and 21 single, non-variable G0V–G4V stars within 15 pc. Darwin targeted a survey of all 285 single FGK stars out to 25 pc, with a spectrum of 30 planets in the case of a planet prevalence of ~10%.

Terrestrial Planet Finder NASA's Terrestrial Planet Finder (TPF) advanced in parallel with the Darwin study, with discussions taking place between the two teams. TPF was conceived to take the form of either a coronagraph operating at visible wavelengths or a 75–100-m baseline interferometer operating in the infrared (Beichman, 1996; Thronson, 1997; Beichman, 1998, 2003; Lawson et al., 2006, 2008).

In May 2002, two concepts were selected for further evaluation: an infrared interferometer (multiple small telescopes on a fixed structure or on separated spacecraft flying in precision formation and utilising nulling), and a visible light coronagraph (a large optical telescope, with a mirror three to four times larger and at least 10 times more precise in wave-front error than HST).

In April 2004, NASA announced that it would embark on a 6 × 3.5 m² visual coronagraph in 2014, designated TPF–C. It would operate over 0.6–1.06 μm, and would target a full search of 32 nearby stars and an incomplete search for 80–130 stars. For the detection of ozone at distances of 15 pc and S/N~25, apertures of about 40 m², and observing times of 2–8 weeks per object, were indicated. A free-flying interferometer, with ESA, could then be considered before 2020 (designated TPF–I).

Design work has continued, with three concepts representing a possibly phased implementation (TPF–C, an external occulter TPF–O, and TPF–I now being refined (Traub et al., 2007). TPF–C was re-baselined as a probe-class mission in 2006 with less ambitious objectives, and in which context concepts such as ACCESS (Trauger et al., 2008), EPIC (Clampin, 2007), and PECO (Guyon et al., 2008) are being studied.

Precursors or alternatives The 2000 Decadal Survey Committee (McKee & Taylor, 2000) qualified its endorsement of the TPF mission with the condition that the abundance of Earth-size planets be determined prior to any start. In addition, issues of solar system zodiacal emission, and the effects of extra-solar zodiacal emission on exoplanet detection probabilities (Angel, 1998), meant that precursor missions with less-ambitious goals were also proposed (Malbet et al., 1995).

Many ideas for technological precursors or replacements for TPF were subsequently put forward to NASA, typically involving 2-m class telescopes with combinations of adaptive wavefront correction, coronagraphs and apodisation, or larger interferometers and free-flying occulters. These included, lexically, ACCESS (Trauger et al., 2008); BOSS (Copi & Starkman, 2000); Eclipse (Trauger, 2007); EPIC (Clampin et al., 2004, 2006; Clampin, 2007); ESPI (Lyon et al., 2003); ExPO (Gezari et al., 2003); FKSI (Danchi et al., 2003; Danchi & Lopez, 2007; Barry et al., 2008a; Danchi et al., 2008); OPD (Mennesson et al., 2003); PECO (Guyon et al., 2008); TOPS (Guyon et al., 2006a, 2007); UMBRAS (Schultz et al., 2003); and 4mTPF (Brown et al., 2003).

NASA's Space Interferometry Mission, SIM, primarily designed for optical astrometry at microarcsec level with maximum baselines of about 10 m, was also originally considered as a technological precursor for TPF, capable of imaging and nulling albeit at much lower angular resolution (Böker & Allen, 1999).

On the ground, GENIE was proposed as a nulling interferometer for VLTI, to test the Darwin technology (den Hartog et al., 2006; Gondoin et al., 2008).

On the European side, a specific space interferometer precursor for Darwin, PEGASE, has also been studied, utilising three free-flying satellites comprising two 0.4-m siderostats and one beam combiner (Ollivier et al., 2006, 2009). A pre-cursor coronagraphic mission, SEE–COAST (Super-Earth Explorer Coronagraphic Off Axis Space Telescope), with polarimetric and spectroscopic capability, was submitted to ESA's Cosmic Vision programme in 2007 (Schneider et al., 2006b, 2009b).

Current status After decade-long studies, the futures of Darwin and TPF in the ESA and NASA programmes remain uncertain. Contributing factors are their technical challenges and costs. Current system performances still give only modest signal-to-noise for the detection of Earth-like planets within 10 pc, an objective resting on the unproven assumption that such planets exist.

New technical studies in the framework of the New Worlds Observer have been given high priority in the 2010 US Decadal Survey (McKee & Taylor, 2000).

7.4 Space-based imaging

Table 7.5: *Detection capabilities for an Earth-type planet at 10 pc, from Angel (2003a), assuming $\Delta\theta = 0.1$ arcsec, $t_{int} = 24$ hr, QE = 0.2, $\Delta\lambda/\lambda = 0.2$. Mode N corresponds to a nulling system, C to a coronagraph. Ground-based results assume fast atmospheric correction, and that long-term averaging is realistic. Optical spectroscopy only becomes feasible for $D \sim 100$ m. An Antarctic 100-m telescope gives comparable signal-to-noise to Darwin/TPF at 11 μm, although O_3 at 9.6 μm and CO_2 at 15 μm are inaccessible due to atmospheric opacity.*

Telescope	D (m)	λ (μm)	Mode	S/N
Darwin/TPF–I	4 × 2	11	N	8
TPF–C	3.5	0.5	C	11
"	7	0.8	C	5–34
Antarctic	21	11	N	0.5
"		0.8	C	6
TMT, GMT	30	11	N	0.3
"		0.8	C	4
OWL	100	11	C	4
"		0.8	C	46
Antarctic OWL	100	11	C	17
"		0.8	C	90

Overall comparison Table 7.5 indicates the signal-to-noise that could be expected from a 24 hr observation of an Earth at 10 pc, in the optical and near-infrared, for various large telescope concepts. A 100-m aperture, and one sited in the Antarctic, are included for performance comparison, although neither is currently under serious consideration. Even if Earths at distances as close as 10 pc exist, detection and spectroscopy will be challenging, even with 100-m apertures or space interferometers.

7.4.3 The future: resolved imaging

Within NASA's Origins Program HST, Spitzer and others were originally referred to as 'precursor missions'. SIM and JWST were referred to as 'First Generation Missions', employing either large, lightweight optics or collections of small telescopes providing images equivalent to those obtainable with a single, much larger instrument. The first generation missions were to serve as technological stepping stones to the second generation missions, including Terrestrial Planet Finder (TPF), and the third generation missions, including Life Finder and Planet Imager. Short descriptions of these yet more ambitious programmes illustrate the complexity that will be involved in going beyond TPF/Darwin.

Life Finder TPF targeted the detection of the nearest planetary systems, with low-quality spectra foreseen as a realistic by-product. Life Finder was only to have been considered after the Darwin/TPF results became available, and once O_2 or O_3 had been discovered. Life Finder would then target confirmation of the presence of life, searching for an atmosphere significantly out of chemical equilibrium, for example through the simultaneous presence of CO_2, H_2O, and O_3.

Some pointers to the technology requirements and complexity of Life Finder have been described by Woolf (2001). They considered that the light collecting area of Life Finder would have to be substantially larger than that of TPF, and adopted targets of 500–5000 m^2. One of the primary technical challenges would be to produce such a collecting area at affordable cost and mass. According to their study, a 'mini-Life Finder' might be a 50 × 10 m^2 telescope, made with 12 segments of 8.3 × 5 m^2, made of 5 kg m^{-2} glass, piezo-electric controlled adaptive optics, and a total mass (optics and structure) of about 10^4 kg.

To detect the 7.6 μm methane feature, the required collecting area rises from 220 m^2 (four or five 8-m telescopes) for a planet at 3.5 pc, to 4000 m^2 (eighty 8-m telescopes) at 15 pc. An outline technology development plan for Life Finder indicated costs as ≫$2 billion.

Planet Imager The imaging of planetary systems referred to so far in this chapter concerns exoplanet detection in which the planet appears only as a point source.

Resolving the surface of a planet is, in contrast, a far future goal requiring substantial technology development that is not yet even in planning. Much longer baselines, from tens to hundreds of km, would be needed. Formation flying will require technology development well beyond that of Darwin/TPF, including complex control systems, ranging and metrology, wavefront sensing, optical control and on-board computing.

A plausible Planet Imager would demand 50–100 Life Finder telescopes used together in an interferometric array. Woolf (2001) concluded that *'the scientific benefit from this monstrously difficult task does not seem commensurate with the difficulty'*.

Bender & Stebbins (1996) also undertook a partial design of a separated spacecraft interferometer which could achieve visible light images with 10 × 10 resolution elements across an Earth-like planet at 10 pc. This called for 15–25 telescopes of 10-m aperture, spread over 200 km baselines. Reaching 100 × 100 resolution elements would require 150–200 spacecraft distributed over 2000 km baselines, and an observation time of 10 years per planet. The effects of planetary rotation on the time variability of the spectral features complicates the imaging task, while more erratic time variability (climatic, cloud coverage, etc.) will further exacerbate any imaging attempts. They noted that the resources identified would dwarf those of the Apollo Program or the Space Station, concluding that it was *'difficult to see how such a program could be justified'*.

Hypertelescopes Parallel to the Planet Imager studies in the US, resolved imaging of exoplanets using 'hypertelescopes' on Earth and in space have also been considered in Europe (Labeyrie, 1996; Pedretti et al., 2000; Labeyrie, 2002; Riaud et al., 2002; Gillet et al., 2003; Labeyrie, 2003; Labeyrie et al., 2003; Labeyrie &

Table 7.6: Companion objects detected by direct imaging, divided into brown dwarfs (an incomplete compilation included for comparison), objects bordering brown dwarfs and high-mass planets, and probable planets (chronological within each category). A selection is shown in Figure 7.10 to illustrate relevant scales of mass and separation (note that the two 2M host objects are themselves brown dwarfs). Companions indicated ⋆ are listed as planets in exoplanet.eu (June 2010).

	Object	$M_\star$ ($M_\odot$)	$d_\star$ (pc)	M_p (M_J)	a'' (arcsec)	a (AU)	Discovery instrument	Reference
	Brown dwarfs:							
	GJ 229 B	0.6	6	20–50	7.8	45	Palomar–AOC	Nakajima et al. (1995)
	DH Tau B	–	144?	30–50	2.3	330	Subaru–CIAO	Itoh et al. (2005)
	HD 203030 B	1.0	41	24	11.9	487	Palomar–PALAO	Metchev & Hillenbrand (2006)
	GJ 758 B	0.97	15	10–40	1.9	29	Subaru–HiCIAO	Thalmann et al. (2009a)
	Borderline objects:							
⋆	GQ Lup b	0.7	140	21	0.7	100	VLT–NACO	Neuhäuser et al. (2005)
⋆	AB Pic b	0.83	47	14	5.5	248	VLT–NACO	Chauvin et al. (2005b)
	CHRX 73 b	–	161	12	1.3	210	HST–ACS	Luhman et al. (2006)
	HN Peg b	–	18	16	43.0	795	Spitzer–IRAC	Luhman et al. (2007)
⋆	USco108 b	0.06	145	14	4.6	670	2MASS	Béjar et al. (2008)
	EK 60 b	–	125	14	8.7	1100	Subaru–CIAO	Kuzuhara et al. (2009)
	Planetary mass:							
⋆	2M J1207 b	0.03	70	4	0.8	46	VLT–NACO	Chauvin et al. (2005a)
⋆	SCR 1845 b	–	4	8	1.2	5	VLT–NACO–SDI	Biller et al. (2006)
	1RSXJ1609 b	0.85	150	8	2.2	330	Gemini–ALTAIR	Lafrenière et al. (2008)
⋆	Fomalhaut b	2.1	8	3	15.0	119	HST–ACS	Kalas et al. (2008)
⋆	HR 8799 b	1.5	39	7	1.7	68	Keck/Gemini	Marois et al. (2008b)
⋆	" c	"	"	10	1.0	38	"	"
⋆	" d	"	"	10	0.6	24	"	"
⋆	CT Cha b	–	165	17	2.7	440	VLT–NACO	Schmidt et al. (2008)
⋆	β Pic b	1.8	19	8	0.4	12	VLT–NACO	Lagrange et al. (2009b)
⋆	2M J0441 b	0.02	140	7	0.1	15	HST–WFPC2/Gemini	Todorov et al. (2010b)

Le Coroller, 2004; Le Coroller et al., 2004; Martinache & Lardière, 2006; Reynaud & Delage, 2007; Aime, 2008). The goal is to design an imaging interferometer which generates usable images at the combined focus of a diluted interferometric array. The basic hypertelescope design is based on a dilute array of smaller apertures with a 'densified' exit pupil, meaning that the exit pupil has sub-pupils having a larger relative size than the corresponding sub-apertures in the entrance pupil (Pedretti et al., 2000, their figure 1).

Riaud et al. (2002) combined the hypertelescope with a phase-mask coronagraph to yield attenuations of 10^{-8}. Simulations of 37 telescopes of 0.6 m aperture over a baseline of 80 m in the infrared, observing the 389 Hipparcos M5–F0 stars out to 25 pc, and including the simulated contributions from zodiacal and exo-zodiacal background, yielded 10-hour snapshot images in which an Earth-like planet was detectable around 73% of the stars. Gains of a factor 20–30 with respect to a simple Bracewell nulling interferometer were reported.

In space, the hypertelescope would comprise a flotilla of dozens or hundreds of small elements, deployed in the form of a large dilute mosaic mirror. Pointing would be achieved by globally rotating the array, using solar sails attached to each element. Somewhat paralleling the conclusions of the Life Finder and Planet Imager studies, an exo-Earth discoverer would require a 100–1000 m baseline hypertelescope, while an exo-Earth imager would require baselines of order 150 km. A 30-min exposure employing 150 3-m diameter mirrors with separations up to 150 km, would be sufficient to detect 'green' spots similar to the Earth's Amazon basin on a planet at 3 pc (Labeyrie, 2002).

Superlenses The diffraction limit inherent in conventional optics is explained by the Huygens–Fresnel principle of secondary wavefront generation and mutual interference, and its mathematical formulation by Kirchhoff and others (e.g. Born & Wolf, 1999, Section 8.1). Combined with the preceding considerations, it is the diffraction limit which renders resolved exoplanet imaging implausible for the foreseeable future.

In the formulation of diffraction based on Maxwell's equations, evanescent waves carry structure at subwavelength scales. In media with positive permittivity ϵ and permeability μ, these waves decay exponentially and transfer no energy to the image plane (e.g. Born & Wolf, 1999, Section 11.4.2).

Research in *metamaterials* (those engineered to provide properties not found in nature) is actively pursuing the development of negative refractive index materials in which both ϵ and μ are negative (Veselago, 1968). This has led to the conjecture of *superlenses* in which imaging at sub-wavelength resolution, by enhancing evanescent

7.5 Imaging results

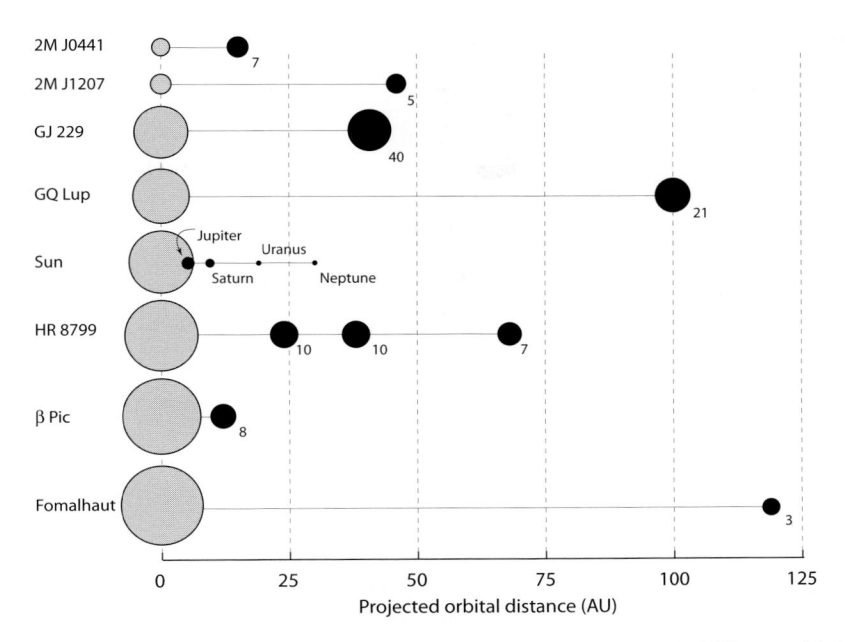

Figure 7.10: Schematic comparison of $M_\star$, M_p, a for a subset of the imaged systems given in Table 7.6, with the Sun and outer solar systems planets included for comparison. The host star (grey) is shown with size proportional to $R_\star \propto \sqrt[3]{M_\star}$. Companions (black), with masses indicated in units of M_J, are shown at a distinct scale with size proportional to $R_p \propto \sqrt[3]{M_p}$, and at their appropriate projected orbital distance, in AU.

waves, might be feasible (Pendry, 2000; Ramakrishna, 2005). Sub-diffraction limited resolution was reported at 1.06 GHz by Grbic & Eleftheriades (2004), and at 365 nm with $\lambda/6$ resolution by Fang et al. (2005).

Practical relevance, notably for distant sources in which evanescent fields are absent, can be questioned. Nevertheless, applied to astronomical telescope optics, May & Jennetti (2004) postulated that evanescent fields created by reflection off the primary could be amplified and modified to generate fields that sharpen the focus, hinting that techniques might be developed that will improve telescope resolution to below the diffraction limit.

7.5 Imaging results

The preceding sections have set out the techniques which are presently directed at exoplanet imaging, and the variety of instruments, on ground and in space, that are being applied to such searches.

Although many search programmes are now underway, the prospects of imaging an Earth-mass planet in an Earth-like orbit appear improbable with current telescopes and instruments, while even the imaging of a Jupiter-like planet in a Jupiter-like orbit will represent a difficult challenge. But exoplanets come in a more extended range of masses and orbits, and the first detections of more massive and more widely spaced planets have recently been announced.

The basic step in establishing a faint object close to the host star as a gravitationally-bound companion, rather than a chance alignment of a background star, is through multi-epoch measurements demonstrating that the star and companion are comoving, i.e. sharing the same common proper motions. Small relative image shifts may then represent true orbital motion.

Uncertainties in classification Companion objects are now being imaged in increasing numbers. However, discrepancies between the on-line compilation at exoplanet.eu, and other syntheses (e.g. Schmidt et al., 2009), in part result from two considerations.

The first is due to the sensitive boundary between objects that are definitively brown dwarfs on the one hand, or most probably of planetary mass on the other. Estimated companion masses are dependent on evolutionary models, as a consequence of the difficulties of precise age determination (Figure 7.12), and objects such as AB Pic b are difficult to classify (Chauvin et al., 2005b).

Similar uncertainty surrounds the formation mechanism of some of the low-mass objects being discovered, particularly those in wide orbits surrounding very low-mass stars and brown dwarfs. Whether the companion has formed by one-step gravitational collapse (out of the same gas cloud as the host star), or by a two-step process involving core accretion, may prove to be an important criterion for accepting a low-mass object as a planet.

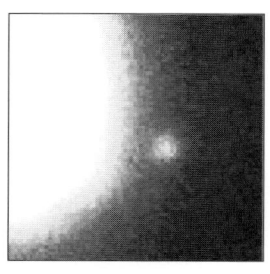

Figure 7.11: *The brown dwarf GJ 229 B imaged with an early adaptive optics/coronagraph system at the Palomar Observatory 60-inch telescope (left) and with HST (right). The brown dwarf is 7 arcsec from the bright star GJ 229A. The brightness ratio is ~5000, and the separation corresponds to that of Sun–Pluto. A Jupiter mass planet at a distance of 10 pc would be 14 times closer to its parent star, and roughly 200 000 times dimmer than GJ 229 B. From Nakajima et al. (1995, Figure 1), by permission from Macmillan Publishers Ltd, Nature, ©1995.*

Discoveries to date Discoveries to date are summarised in Table 7.6, and a subset are shown schematically in Figure 7.10 to illustrate the relevant scales of mass and separation. Despite the potential of adaptive optics with coronagraphy, few discoveries are currently of planetary mass within 1–2 arcsec, although the panorama is changing rapidly.

7.5.1 Searches around nearby stars

Early results on the path to planet imaging included discovery of the brown dwarf GJ 229 B (Figure 7.11), imaged from the ground with adaptive optics, and from space by HST (Nakajima et al., 1995).

Other coronagraphic searches for brown dwarfs and massive planets have since been undertaken (e.g. Oppenheimer et al., 2001; Liu et al., 2002; McCarthy & Zuckerman, 2004). From an infrared coronagraphic survey of 178 stars at Steward and Lick, with detection thresholds $> 30 M_J$ between 140–1200 AU, and of 102 stars at Keck, with detection thresholds $> 10 M_J$ between 75–300 AU, McCarthy & Zuckerman (2004) detected one brown dwarf companion and no planets. In a search using VLT–NACO for 28 stars, Masciadri et al. (2005) excluded the existence of $5 M_J$ planets beyond 14 AU, and $10 M_J$ planets beyond 8.5 AU for 50% of the targets.

Of the reported low-mass companions detected to date (Table 7.6), two example images are shown, although in both cases the primaries are considered as brown dwarfs rather than stars.

2M J1207 b, with a mass of $5-8 M_J$, was discovered using VLT–NACO (Chauvin et al., 2005a), and confirmed using HST–NICMOS (Song et al., 2006, Figure 7.13). The nature of 2M J1207 b continues to be debated (e.g. Mohanty et al., 2007; Mamajek & Meyer, 2007).

SCR 1845 b was discovered using VLT–NACO with simultaneous differential imaging (Biller et al., 2006). It has a mass of $9-65 M_J$, and lies at a projected separation of 4.5 AU (1.17 arcsec) from the M8.5 host star, itself only 3.85 pc from the Sun (Figure 7.14).

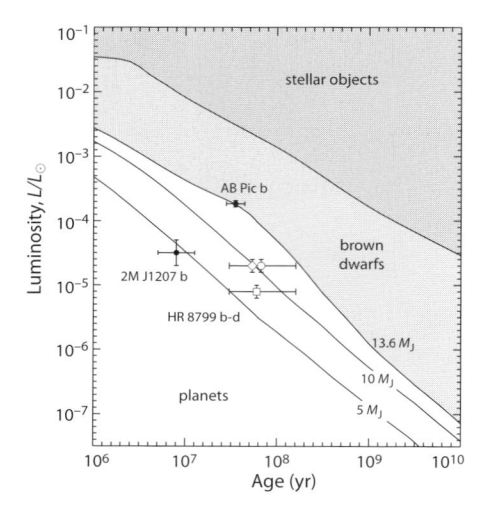

Figure 7.12: *Luminosity versus time for a variety of masses, from the evolutionary models of Baraffe et al. (2003), and the locations of various companion objects detected by direct imaging: the three coeval points of HR 8799 b (square), c (diamond), and d (circle), the latter two displaced horizontally for clarity; the low-mass object AB Pic b, on the planet/brown dwarf dividing line; and the planetary mass companion to the brown dwarf, 2M J1207 b. The deuterium burning limit, ~ 13.6 M_J, separates planets from brown dwarfs, while the boundary between stars and brown dwarfs is set by stable hydrogen burning. In all cases, estimated companion masses depend on the assumed age of the host star. Based on Marois et al. (2008b, Figure 4).*

7.5.2 Searches around exoplanet host stars

Various deep adaptive optics imaging searches have been made around stars known to possess orbiting planets, currently without success. These include:

– a Keck search around 25 stars (Luhman & Jayawardhana, 2002), and a VLT–NACO and CFHT–PUEO–KIR search around 26 stars, six with long-term radial velocity drifts (Chauvin et al., 2006);

– a VLT–NACO and MMT search around 54 stars using (methane-band) simultaneous differential imaging (Biller et al., 2007). With H-band contrasts of 9–10 mag, the survey was sensitive to a $7 M_J$ planet at 15 AU from a 70 Myr K1 star at 15 pc or a $7.8 M_J$ planet at 2 AU from a 12 Myr M star at 10 pc;

– a Gemini–ALTAIR search around 85 nearby young stars using angular difference imaging (Lafrenière et al., 2007a), with sensitivity sufficient to reach $2 M_J$ in the separation range 40–200 AU.

Coronagraphic searches around individual objects include: 55 Cnc using HST–NICMOS (Schneider et al., 2001) which failed to confirm the disk reported by Trilling & Brown (1998); GJ 86 using VLT–NACO (Lagrange et al., 2006); and ϵ Eri using VLT–NACO–SDI (Janson et al., 2007).

7.5 Imaging results

Figure 7.13: Confirmation of 2M J1207 b. HST–NICMOS imaging at 0.9, 1.1 and 1.6 μm (left to right). The star, centred in the 0.2 arcsec radius circle, is subtracted from each using a second image acquired at a different celestial orientation. The companion lies 0.77 arcsec distant. From Song et al. (2006, Figure 1), reproduced by permission of the AAS.

7.5.3 Searches in systems with debris disks

In 2008, imaging observations identified planets orbiting two stars surrounded by dusty debris disks (§10.3.4). Both stars – Fomalhaut, an A3V star at $d = 7.7$ pc, and HR 8799, an A5V star at $d = 39.4$ pc – are younger, brighter, and more massive than the Sun. As noted above, masses of planets detected by imaging are inferred from their brightness, and since they fade as they radiate away the heat of their formation, mass estimates are correspondingly sensitive to their assumed ages.

Fomalhaut Fomalhaut is surrounded by a belt of cold dust, whose structure is consistent with gravitational clearing by an orbiting planet: specifically, a 15 AU offset between the star and geometric centre of the belt, and a sharp truncation of its inner edge (Stapelfeldt et al., 2004; Kalas et al., 2005). The eccentricity and sharpness of the disk's inner edge led Quillen (2006a) to propose the existence of a planet just interior to it, with a mass between that of Neptune and Saturn, and with $a \sim 119$ AU and $e \sim 0.1$.

From HST–ACS coronagraphic observations in 2004, Kalas et al. (2005) had imaged the dust belt, and also detected several faint sources near to the star. Keck II observations in 2005, and further HST–ACS coronagraphic observations in 2006, showed that Fomalhaut b shared its proper motion with that of the central star, but with an additional offset of 0.184 ± 0.022 arcsec corresponding to a 0.82 ± 0.10 AU yr^{-1} projected motion relative to the host star (Kalas et al., 2008). The observations are consistent with counterclockwise orbital motion, and a projected distance of 119 AU from the star (Figure 7.15). At such a distance, and for $M_\star = 2 M_\odot$, the orbital period is ~ 870 yr. Although the planet mass is not determined directly, dynamical models of the interaction between the planet and the disk imply $M_p \lesssim 3 M_J$; a higher mass would lead to gravitational disruption of the disk.

Atmospheric models indicate that Fomalhaut b is a cooling Jovian-mass planet with an age of 100–300 Myr. Although faint, it is still some 100 times brighter than from reflected light from a Jupiter-like planet at that orbital radius.

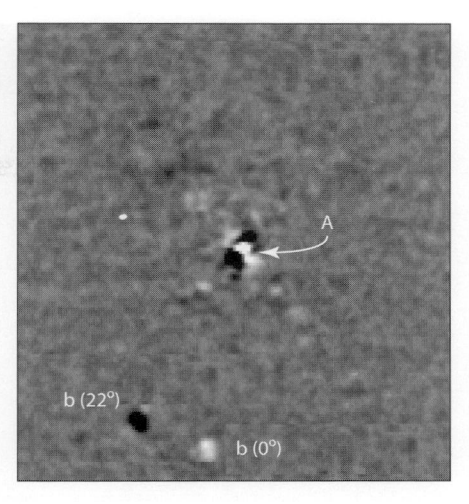

Figure 7.14: SCR 1845 observed with VLT–NACO–SDI. 30-min observations at position angles 0° and 22° were made in each of the four SDI filters. Subtraction removes most of the speckles from the primary (A), and reveals the sub-stellar methane-rich mid-T dwarf (SCR 1845 b) separated by 1.17 arcsec. From Biller et al. (2006, Figure 2), reproduced by permission of the AAS.

HR 8799 High-contrast observations with Keck and Gemini revealed three planets orbiting the star HR 8799, with projected separations of 24, 38, and 68 AU (Marois et al., 2008b, Figure 7.16). Multi-epoch data show counter-clockwise orbital motion for all three planets. From theoretical models of giant planet evolution, the luminosity of the companions measured in three infrared bands (J, H, and Ks), and the age of the system (estimated based on four different methods as 30–160 Myr), imply planetary masses between $5 - 13 M_J$. The system resembles a scaled-up version of the outer portion of the solar system.

The three objects were subsequently detected using a high-performant 1.5-m sub-aperture of the Hale 5-m telescope equipped with an optical vortex coronagraph (Serabyn et al., 2010). Spatially-resolved spectroscopy of HR 8799 c using VLT–NACO at 4 μm was reported by Janson et al. (2010).

The companion masses are sensitively dependent on the estimated age of the central star, being determined from the theoretical age–luminosity relation derived from brown dwarf/giant planet evolutionary tracks (Figure 7.12). The asteroseismology analysis of Moya et al. (2010), with its own uncertainties due to the unknown orientation of the rotation axis, implies an age for HR 8799 closer to 1 Gyr, and consequently companions of brown dwarf rather than of planetary mass.

β Pic β Pic (§10.3.4) was the third prominent debris disk system in which a probable planet has been detected. VLT–NACO observations by Lagrange et al. (2009b) revealed a $M \sim 8 M_J$ object at ~ 8 AU from the central star (Figure 10.4).

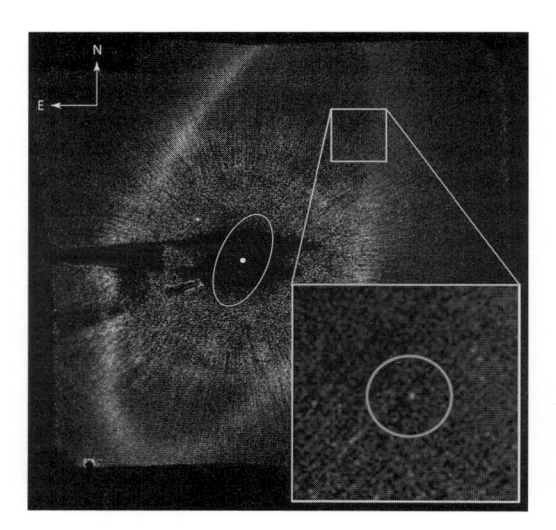

Figure 7.15: HST–ACS coronagraphic image of Fomalhaut at 0.6 µm showing the location of object b (expanded as inset) 12.7 arcsec from the star (about 115 AU), and just within the inner boundary of the dust belt. Other objects are either background stars or false-positives. The central circle marks the location of the star behind the occulting spot. The central ellipse has a semi-major axis of 30 AU (3.9 arcsec), corresponding to the orbit of Neptune. Originally published as Kalas et al. (2008, Figure 1), this version is courtesy: Paul Kalas, UC Berkeley/NASA/ESA.

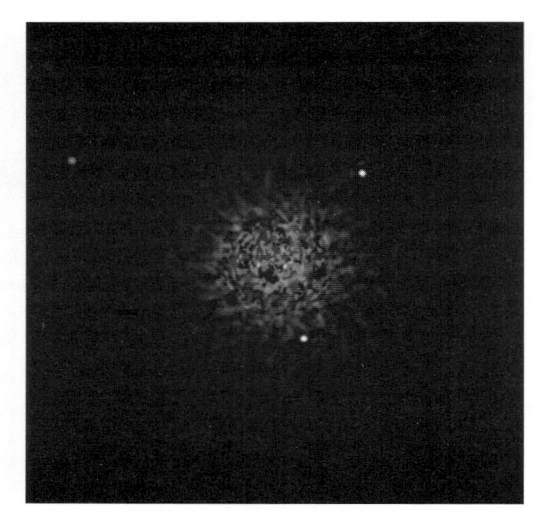

Figure 7.16: Combined J-, H-, and Ks-band image of HR 8799 from Keck in 2008 July (H) and September (J and Ks). Light from the host star has been removed by angular difference imaging (and left unmasked to show the speckle noise). Planet b is to the top left, c to top right, and d is just below the central star. The inner part of the H-band image has been rotated by 1° to compensate for the orbital motion of d between July–September. The scale is such that the distance between the central star and object d is 0.6 arcsec, or 24 AU. Originally published as Marois et al. (2008b, Figure 1), this version is courtesy: NRC–HIA, C. Marois & Keck Observatory.

7.5.4 Searches around white dwarfs

Giant planets that survive the demise of their parent star are interesting candidates for direct imaging because of the favourable planet–white dwarf flux ratio (Ignace, 2001; Burleigh et al., 2002; Livio et al., 2005; Gould & Kilic, 2008).

The ratio of temperatures and bolometric luminosities of the planet and white dwarf in a circular orbit are given by (Ignace, 2001)

$$\frac{T_\mathrm{p}}{T_\mathrm{wd}} = (1-A_\mathrm{B})\left(\frac{R_\mathrm{wd}}{2a}\right)^{(1/2)}, \qquad (7.2)$$

$$\frac{L_\mathrm{p}}{L_\mathrm{wd}} = \frac{R_\mathrm{p}^2 \, T_\mathrm{p}^4}{R_\mathrm{wd}^2 \, T_\mathrm{wd}^4} = \left(\frac{R_\mathrm{p}}{2a}\right)^2 (1-A_\mathrm{B}), \qquad (7.3)$$

where T_p, T_wd are the planet's equilibrium temperature and white dwarf effective temperature respectively, R are the radii, a is the orbital radius, and A_B is the planet's Bond albedo. The bolometric brightness contrast between the planet and star is therefore a function only of R_p, a, and A_B.

White dwarf luminosities are some 1000 times smaller than solar-type stars (because of their smaller sizes), and the ratio of bolometric luminosities may be a factor 10^3-10^4 more favourable for imaging compared to the situation for a late-type main-sequence star. For hot white dwarfs, the atmosphere of a Jovian planet will also be photoionised, leading to the emission of hydrogen recombination lines which may be detectable spectroscopically (Chu et al., 2001).

Wavelength choice Imaging prospects are best in the infrared, where the ratio between the thermal emission from the planet and that of the rapidly cooling white dwarf is largest. For a white dwarf with $T_\mathrm{wd}=10\,000$ K, orbited by a Jupiter-size planet with $P \simeq 100$ d, the 10 µm flux density at its Wien peak may be comparable to the emission of the white dwarf well into its Rayleigh–Jeans tail (Ignace, 2001). At the same time, the Roche limit imposes a maximum temperature and hence brightness of an orbiting planet. For $M_\mathrm{wd}=1M_\odot$ and $R_\mathrm{wd}=1R_\oplus$, a Jupiter-like planet cannot orbit closer than $\sim 24R_\mathrm{J}$, so that its temperature is limited to $T_\mathrm{p} \lesssim 0.04 T_\mathrm{wd}$.

A number of imaging searches aiming to detect planetary mass companions around nearby white dwarfs have been undertaken, all unsuccessful,[1] although

---

[1] Clarke & Burleigh (2004) and Burleigh et al. (2006) searched for common proper motion companions using adaptive optics imaging to $J=24$ using Gemini and WHT. Debes et al. (2005a) used HST–NICMOS and Gemini–Altair–NIRI imaging of G29–38, placing limits of $16M_\mathrm{J}$ and $6M_\mathrm{J}$ at separations 3–12 AU and > 12 AU respectively; further Spitzer observations are reported by Reach et al. (2005). Debes et al. (2005b) used HST–NICMOS to search seven DAZ white dwarfs within 50 pc to lim-

white dwarfs are known as wide companions to stars hosting exoplanets.[2] Searches for extended dust disks have been more successful (e.g. Jura, 2003; Becklin et al., 2005; Kilic et al., 2005; Gänsicke et al., 2006; Jura et al., 2007; Farihi et al., 2009; Reach et al., 2009; Klein et al., 2010). Their own origin is attributed to tidal disruption of either comets (Debes & Sigurdsson, 2002) or asteroids (Jura, 2003, 2006, 2008).

In summary, while the most sensitive searches to date could detect companions with $M_p \gtrsim 5M_J$, no such objects orbiting white dwarfs have been found. Their absence is consistent with the conclusions of McCarthy & Zuckerman (2004), whose infrared coronagraphic search of 178 stars at Steward and Lick observatories, found that no more than 3% of main-sequence stars harbour $5-10M_J$ planets orbiting between 75–300 AU.

7.6 Observations at radio wavelengths

There are two routes to the possible detection of exoplanets at radio wavelengths, neither of which have yet yielded positive results: astrometry, and direct imaging.

7.6.1 Astrometry

For stars that are themselves significant and compact sources of radio emission, planet detection may be possible using the measured astrometric displacements of the host star. As in optical astrometry, sub-mas (and preferable μas) positional accuracy is required at a number of suitably spaced measurement epochs. From them, the five astrometric parameters (two positional components, two proper motion components, and parallax) as well as the star's reflex motion due to the orbiting planet, can in principle be derived.

Very long baseline interferometry (VLBI), which employs baselines of up to several thousands km combined with phase referencing techniques, is required to achieve such sub-mas accuracies. The Very Long Baseline Array (VLBA), operational since 1993 and with baselines up to 8000 km, routinely achieves astrometric accuracies of $\sim 100\,\mu$as at a single epoch (Reid, 2008). The European VLBI Network (EVN), formed in 1980, achieves comparable performances. Theoretical accuracies of order $50\,\mu$as are achievable for both (Pradel et al., 2006).

Targets for VLBI must have a sufficiently high brightness temperature, $T_b > 10^7 - 10^8$ K, in order to be detectable by a high-resolution radio interferometer. This precludes the observation of sources of thermal emission at the longest baselines, and requires sources with significant non-thermal flux to achieve the highest astrometric accuracies.

Non-thermal radio emission has been detected from various stellar types (Güdel, 2002), including brown dwarfs (Berger, 2006), proto-stars (Bower et al., 2003), massive stars with winds (Dougherty et al., 2005), late-type stars (Gary & Linsky, 1981), and close binaries including RS CVn-type, Algol-type, and X-ray binaries.

RS CVn binaries In RS CVn binaries, quiescent and flaring gyro-synchrotron radio emission is generated from MeV electrons in magnetic structures related to the intra-stellar region and stellar photosphere respectively. For the close binary σ^2 CrB, observations since 1987 yielded post-fit rms residuals of 0.20 mas (Lestrade et al., 1996, 1999). This corresponds, at its distance $d = 21$ pc, to the displacement expected for a Jupiter-like planet around the binary system. No residuals attributable to orbiting planets were apparent.

M dwarfs Radio emission in late-type stars originates from cyclotron radiation due to non-relativistic electrons in the coronal plasma. M dwarfs are sufficiently bright and numerous for large-scale astrometric exoplanet searches to be undertaken at radio wavelengths, complementing the radial velocity technique which is particularly challenging for later spectral types.

Bower et al. (2009) surveyed 172 active M dwarfs within 10 pc with the VLA, detecting 29 with flux densities exceeding 100 μJy. Seven of these were then observed with the Very Long Baseline Array (VLBA) providing a detection threshhold of 500 μJy. Four stars were detected at just two or three epochs. Although such a small number of measurements is insufficient for the detection of planetary perturbations directly, their residuals from the optically determined apparent motions were used to exclude planetary companions more massive than $3-6M_J$ at orbital separations of around 1 AU.

7.6.2 Direct imaging

More direct detection of exoplanets at radio wavelengths should be possible through emission arising from their magnetospheric interaction with their host stellar wind, as observed for the radio-emitting planets in the solar

---

its of $\sim 10-18M_J$. Friedrich et al. (2005) used HST–NICMOS to search seven Hyades white dwarfs, where adiabatic expansion is expected to have resulted in orbital distances of 25 AU and magnitudes $JH \simeq 20.5 - 23.3$ mag. Hansen et al. (2006) used Spitzer–IRAC to observe 14 hot white dwarfs. Mullally et al. (2007) used Spitzer–IRAC to observe 124 bright white dwarfs, placing limits of $5M_J$ for eight stars, and $10M_J$ for 23 stars. Kilic et al. (2009) used Spitzer–IRAC to observe 14 white dwarfs of high-mass stellar remnants, giving limits of $5-10M_J$. Mullally et al. (2009) used Spitzer–IRAC to place an upper limit of $5-7M_J$ on a possible companion inferred from pulse timing variations.

[2]Mugrauer & Neuhäuser (2005) used VLT–NACO to establish that a stellar mass companion discovered by Els et al. (2001) to be orbiting the star GJ 86A is a cool white dwarf. There are also white dwarf companions to HD 147513A (Mayor et al., 2004) and HD 27442A (Mugrauer et al., 2007a). These systems provide some evidence that planets can survive post-main sequence evolution, albeit at distances of $\gtrsim 10-20$ AU.

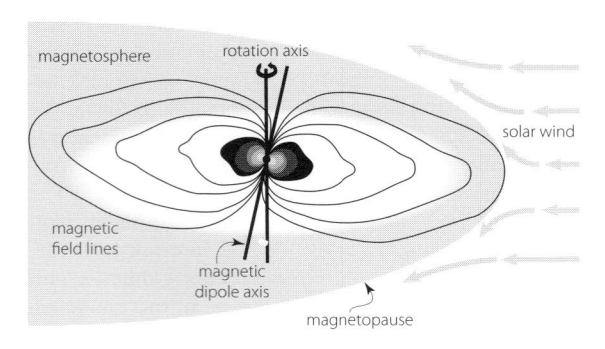

Figure 7.17: Schematic of the interaction between the solar wind, and the magnetosphere of a planet with a significant magnetic dipole moment. An offset between the dipole and rotation axes leads to rotational modulation of the radio cyclotron emission, analogous to that seen in radio pulsars.

system. This is somewhat analogous to direct optical imaging, but facilitated by the typical absence of significant non-thermal radio emission from the closely-separated host star.

Solar system decametric emission Of the various sources of radio emission in the solar system (see, e.g., Kraus, 1966, Chapter 8), Jupiter was first detected at radio wavelengths using the Mills Cross antenna at 22.2 MHz (Burke & Franklin, 1955). It has since been known as a bright and variable source at decametric wavelengths (3–40 MHz).

Five of the solar system planets with a dynamo-driven magnetic field are now known to produce non-thermal low-frequency radio emission as a result of interaction between the solar wind and their planetary magnetospheres (the cavity created in the flow of the solar wind by the planet's internally generated magnetic field, Figure 7.17): Earth, Jupiter, Saturn, Uranus and Neptune. In addition, radio bursts originate from electrodynamic interactions between Jupiter and its satellite Io (Zarka et al., 2001; Hess et al., 2008).

The radio emission from the gas giants other than Jupiter falls off below the Earth's ionospheric cut-off at around 10 MHz (Figure 7.18). The magnetospheres responsible for their radio emission were instead directly detected and measured by the fly-by missions: first by Pioneer 10 for Jupiter in 1973, by Pioneer 11 for Saturn in 1979, and by Voyager 2 for Uranus in 1986 (Ness et al., 1986) and Neptune in 1989 (Ness et al., 1989). Cassini, the fourth probe to visit Saturn, and the first to enter orbit around it, has made *in situ* radio measurements since 2004 (Wang et al., 2010).

Analytic and numerical calculations indicate that the magnetic fields in giant planets are generated by a dynamo process arising through convective motions in the electrically-conducting metallic hydrogen circulating in their outer cores (Connerney, 1993; Guillot et al.,

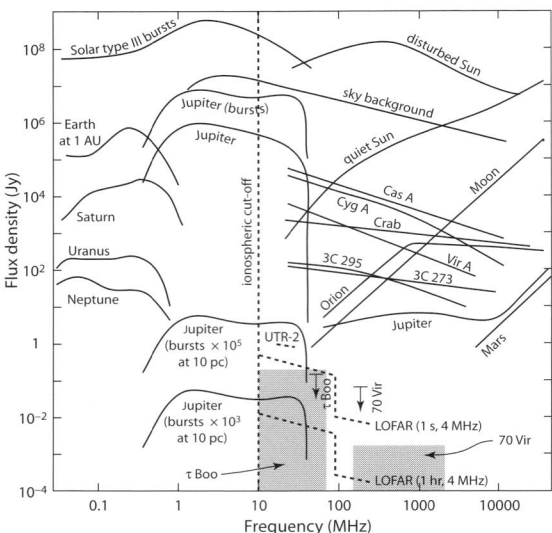

Figure 7.18: Flux versus wavelength for astronomical radio sources ($1\,\mathrm{Jy}=10^{-26}\,W\,m^{-2}\,Hz^{-1}$). Galactic, extragalactic and solar spectra are from Kraus (1966, Figure 8.6a). Planetary spectra corresponding to auroral emission are from Zarka (1992). Shaded boxes are predictions for τ Boo and 70 Vir from Farrell et al. (2004b), with corresponding illustrative upper limits from recent observations. The Earth's ionospheric cut-off is shown at 10 MHz. Adapted from Zarka (2004, Figure 1).

2004). The solar system planetary dynamos are nevertheless highly diverse, with Uranus and Neptune having displaced and significantly tilted fields (Appendix A, Table A2), with quadrupole components, and magnetic field lines distorted by orbiting satellites (Russell, 1993). Detailed modeling encounters non-linear chaotic processes, which prove difficult to solve numerically even for the solar system giants with their abundance of observational data (Stevenson, 2003).

Although other effects contribute (e.g. Blanc et al., 2005; Zarka & Kurth, 2005), the dominant mechanism for low-frequency radio waves from the solar system planets is cyclotron maser emission from energetic (keV) electrons interacting with the planetary magnetic field. Characteristic frequencies are determined by the magnetic field strength, with a sharp cut-off above ~ 40 MHz in the case of Jupiter.

The associated radio emission is not isotropic, but emitted over a certain solid angle, some 1.6 sr in the case of Jupiter (Zarka et al., 2004). Intensities are furthermore time varying, modulated by solar coronal mass ejections, planetary rotation, and satellite interactions (Higgins et al., 1997; Gurnett et al., 2002). At higher (GHz) frequencies, lower-level synchrotron emission dominates.

Radio Bode's law The interaction between the solar wind and the planetary magnetosphere controls the auroral radio emission of the five magnetised solar system planets. Various forms of an empirical scaling law, the

7.6 Observations at radio wavelengths

radio Bode's law, have been constructed from the observed relation between their average radio power and the incident kinetic power due to the solar-wind ram pressure on their magnetospheric cross-sectional area (Desch & Kaiser, 1984; Zarka, 1992). The average planetary radio emission has been expressed as (Zarka et al., 2001, eqn 1)

$$P_{\text{radio}} \propto \left(\frac{N_0}{a^2}\right) V_W R_{\text{mp}}, \qquad (7.4)$$

where N_0 is the average stellar wind density at 1 AU, a is the planet–star separation, V_W is the solar wind speed, and R_{mp} is the dayside magnetopause distance.

An alternative form has been given as (Farrell et al., 2004a, eqn 1)

$$P_{\text{radio}} = 4 \times 10^{11} \,\text{W} \left(\frac{\omega}{10\,\text{hr}}\right)^{0.79} \left(\frac{M_p}{M_J}\right)^{1.33} \left(\frac{a}{5\,\text{AU}}\right)^{-1.6}, \qquad (7.5)$$

where ω is the planetary rotation period, and a is the planet–star separation. The quantities ω and M are proxies for the planetary magnetic moment.

Relevance for exoplanets By analogy with the solar system planets, radio emission should accompany extrasolar planets which possess dynamo-driven magnetic fields combined with a source of energetic magnetospheric electrons. The latter may arise from auroral processes, or from magnetic planet–satellite coupling.

Detection of exoplanet radio emission would provide information on the planet's magnetic field strength, its rotation period, and perhaps the presence of moons. Polarisation may be a further diagnostic.

In the context of the development of life, and analogous with the situation on Earth, magnetic fields may be important in providing protection from the energetic particles arising from stellar winds, stellar flares, and cosmic rays. Accordingly, the presence of radio emission may be one possible proxy for habitability.

Exoplanet magnetic fields: theory The existence of an intrinsic magnetic field in giant exoplanets is expected from theoretical models of their internal structure. These models predict the existence of an internal energy source driving convection (which increases with mass, and decreases with time as the planet cools), and the existence of an electrically-conducting liquid-hydrogen metallic layer formed by the dissociation and ionisation of molecular hydrogen under extreme pressures (Burrows et al., 1997; Hubbard et al., 2002).

Sánchez-Lavega (2004) derived estimates of the magnetic field and magnetic dipole moment as a function of exoplanet mass, age, and rotation rate, for masses in the range $0.3 - 10 M_J$, and with rotation periods ranging from tidal synchronism (1–4 d) to centrifugal breakup (2–3 hr). Based on the heat generation given by the evolutionary models of Burrows et al. (1997),

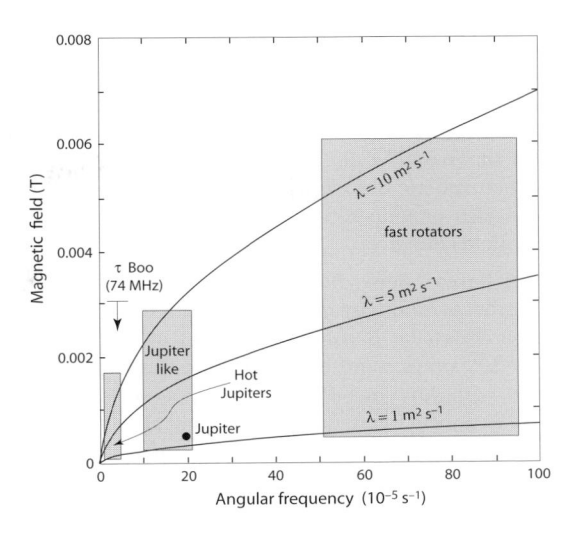

Figure 7.19: Predicted magnetic field intensity in giant exoplanets, at the top of the metallic-hydrogen layer, as a function of angular rotation frequency, for three values of the magnetic diffusivity, λ (Nellis, 2000). Boxes delineate plausible values for the magnetic field for three rotation periods: hot Jupiters (periods of 1–4 d, assuming spin-orbit synchronism), Jupiter-like planets (periods of 10–20 hr), and fast rotators (periods of 2–3 hr). The surface value for Jupiter is indicated. The upper limit for τ Boo b was derived from radio observations at 74 MHz (Lazio & Farrell, 2007). From Sánchez-Lavega (2004, Figure 1), reproduced by permission of the AAS.

Sánchez-Lavega (2004) showed that convective motions develop in the metallic region, and derived estimates of the resulting magnetic field strengths from relevant scaling laws. They concluded that strong magnetic fields, $30-60 \times 10^{-4}$ T, are likely to occur in young, massive and rapidly rotating planets, falling to $\sim 10^{-4}$ T in the case of older or spin-orbit synchronised planets (Figure 7.19). A comparable law with corresponding predictions is given by Christensen et al. (2009).

Exoplanet magnetic fields: evidence Indirect evidence for the existence of exoplanet magnetic fields includes the modulation of activity indicators of their host stars (such as Ca emission-line intensity or X-ray activity) phased with the planetary orbits. Such variations have been attributed to magnetic reconnection events between the stellar and planetary fields (Cuntz et al., 2000; Cuntz & Shkolnik, 2002; Preusse et al., 2006; Lanza, 2008, 2009). Specific studies have been reported in the case of HD 179949 (Shkolnik et al., 2003; McIvor et al., 2006; Saar et al., 2008) and υ And (Shkolnik et al., 2005).

Combining both archival and targeted surveys, Kashyap et al. (2008) found that out of the 230 stars then identified as possessing planets, roughly one-third were detected in X-rays, with stars accompanied by close-in giant planets being on average more X-ray active by a factor of about four than those with planets that are more distant. This has also been taken to suggest that giant planets in close proximity to their host stars influence the stellar magnetic activity.

Early ingress in the ultraviolet transit light curve for the highly-irradiated transiting planet WASP–12 b (§6.4.12; Fossati et al., 2010) has been attributed to a magnetospheric bow shock (Vidotto et al., 2010). This could also provide constraints on the planetary magnetic field strength.

Super-flares Super-flares have been observed around a number of otherwise normal F–G main sequence stars. These are stellar flares with energies of $10^{26} - 10^{31}$ J, $10^2 - 10^7$ times more energetic than the largest solar flares, with durations of hours to days and visible from X-ray to optical frequencies (Schaefer et al., 2000).

Rubenstein & Schaefer (2000) proposed that super-flares are caused by magnetic reconnection between fields of the primary star and a close-in Jovian planet, in close analogy with the flaring observed in RS CVn binary systems. If the companion has a magnetic dipole moment of adequate strength, the magnetic field lines connecting the pair will be wrapped by orbital motion. Interaction of specific field loops with the passing planet will initiate reconnection events. At the time of that study, only one super-flare star, κ Ceti, had been the target of a specific planet search, and the presence of a Saturn-mass planet had not been excluded.

Super-flares on the Sun, which would be catastrophic for life on Earth, are not expected since the solar system does not have a planet with a large magnetic dipole moment in a close orbit.

Radio flux predictions For radio emission originating from a stellar wind incident on the planetary magnetosphere, various expressions for the expected radio flux, P_{radio}, have been formulated in terms of the physical conditions at the planet (Zarka et al., 2001; Lazio et al., 2004a; Stevens, 2005; Grießmeier et al., 2005; Zarka & Halbwachs, 2006; Grießmeier et al., 2007a,b; Zarka, 2007; Jardine & Collier Cameron, 2008). George & Stevens (2007, eqn 1) suggest the form

$$P_{radio} \propto \dot{M}_\star^{2/3} V_w^{5/3} \mu_p^{2/3} a^{-4/3} d^{-2} , \qquad (7.6)$$

where $\dot{M}_\star$ is the stellar mass-loss rate, V_w is the stellar wind speed at the magnetosphere, μ_p is the planetary magnetic moment, a is the planet–star separation, and d is the stellar distance.

Specific dependencies can then be invoked to establish predicted radio fluxes tied to particular observables. Thus the stellar mass-loss rate is related to the stellar X-ray luminosity (Wood et al., 2002a; Cranmer, 2008). The planetary magnetic moment is likely to scale with the planetary mass. In the solar system the relationship, known as Blackett's law (Blackett, 1947), is variously given as $\mu_p \propto M_p$, or $\mu_p \propto \omega M_p^{5/3}$, where μ_p is the planetary magnetic moment, M_p the planetary mass, and ω the planetary angular rotation frequency.

Lazio et al. (2004a) estimated that most of the known extrasolar planets should emit in the frequency range 10–1000 MHz with flux densities of order 1 mJy. Magnetised hot Jupiters might emit radio emission several orders of magnitude stronger than Jovian levels (Zarka et al., 2001). Stevens (2005) compiled a potential target list, and estimated that τ Boo, GJ 86, υ And, HD 1237, and HD 179949 are the most promising, with peak frequencies in the range 8–48 MHz. For nearby stars with surface field strengths above $10^{-4} - 10^{-3}$ T, Jardine & Collier Cameron (2008) predicted radio fluxes of tens of mJy for exoplanets close to their host stars.

There are further complications. Close-in planets with short orbital periods may have more complex magnetospheric geometries and interactions (Ip et al., 2004). But tidal locking may modify the planetary dynamo, perhaps resulting in a weaker magnetic field and a smaller magnetic moment (Grießmeier et al., 2004).

Magnetic white dwarfs Corresponding electron–cyclotron radio emission from terrestrial planets in close orbits around magnetic white dwarfs may also render even low-mass planets detectable in the radio (Willes & Wu, 2004, 2005).

Measurements and upper limits Even before detection of the first exoplanets, a number of radio searches had been conducted. Yantis et al. (1977) reported a search for Jovian-like exoplanets using the Clark Lake Radio Observatory at 26.3 MHz, aiming to distinguish planetary bursts from stellar bursts by the presence of a high-frequency cut-off, and possibly modulation associated with planetary rotation. Winglee et al. (1986) used the VLA at 1.4 and 0.33 GHz to search six nearby stars.

Subsequent low-frequency radio surveys, many of which have been applied to exoplanet searches, includes the Cambridge 6C (151 MHz, Hales et al., 1993), 7C (151 MHz, Riley et al., 1999) and 8C (38 MHz, Hales et al., 1995) surveys, the Ukrainian UTR–2 (Braude et al., 2006), the VLA Low-Frequency Sky Survey VLSS (74 MHz, Lane et al., 2008) and the Giant Metrewave Radio Telescope GMRT (Ananthakrishnan, 1995).

Since the first exoplanet discoveries, radio detections have been attempted numerous times, with various telescopes, and at different (mostly low) frequencies. Although there are significant uncertainties in predicting which exoplanets are most likely to be the strongest radio emitters, searches to date have focused on the short-period planets, but with more speculative observations of wider orbital separations also being pursued.

Reported work, all so far without detections, includes: use of the VLA targeting seven known exoplanets at 1.4 GHz, 0.33 GHz, and 74 MHz, reaching 50 mJy at the lowest frequencies (Bastian et al., 2000); use of the Green Bank 100 m at 330 MHz to survey 20 exoplanets (Langston et al., 2002); use of the UTR–2 to survey 20 exoplanets at 10–25 MHz down to 1.6 Jy (Ryabov et al.,

2003); use of the VLA at 74 MHz to observe τ Boo, with upper limits in the range 135–300 mJy (Lazio & Farrell, 2007); and use of the GMRT at 150 MHz to observe the short-period exoplanets τ Boo, 70 Vir, υ And, GJ 876, HD 162020, HD 179949 (Majid et al., 2006), and the longer-period exoplanets ε Eri and HD 128311, with upper limits in the range 3–6 mJy (George & Stevens, 2007).

The bright, nearby hot-Jupiter transiting system HD 189733 has been observed with GMRT at 244 and 614 MHz, with upper limits of 2 mJy and 160 μJy respectively (Lecavelier des Etangs et al., 2009). It has also been observed at 327–347 MHz with the Green Bank telescope with an upper limit of 47 mJy (Smith et al., 2009a), and specifically during the period of secondary eclipse (Smith et al., 2009b).

Future radio surveys Under development is the Low-Frequency Array, LOFAR (Pradel, 2009), which operates over the frequency range 10–250 MHz, with design sensitivity of 2 mJy in a 15 min integration. It was officially opened, in incomplete form, in June 2010. The final configuration foresees up to 10 000 dipole antennae geographically sited in NL, D, S, UK, and F, with baselines up to 1500 km. The sensitivity and wide field of LOFAR should be particularly important in the next phase of the search for exoplanet radio emission (Cairns, 2004; Farrell et al., 2004b; Zarka, 2004).

The next major development after LOFAR, the Square Kilometre Array, SKA (Lazio et al., 2004b; Tarter, 2004; Lazio, 2008), is not expected to come online before 2017 at the earliest.

7.7 Observations at mm/sub-mm wavelength

The Atacama Large Millimeter Array (ALMA) is a mm and sub-mm interferometer consisting of 64 × 12-m antennae under construction in the Atacama desert in Northern Chile at an altitude of 5050 m. The antennae can be spaced from a compact configuration with a minimum separation of 150 m, to a maximum spacing of 16 km, providing a resolution of 10 mas at shortest wavelengths. The receivers will cover the atmospheric windows in the 35–1000 GHz range (350 μm – 7 mm) with a bandwidth of 8 GHz in two polarisations.

Astrometry Analogous to astrometric searches in the optical, the possibility of indirectly detecting exoplanets based on the astrometric displacement of the thermal emission from the stellar photosphere has also been considered (Lestrade, 2003, 2008). At 345 GHz, which optimises the combination of expected detector noise, object spectrum, and site characteristics, the theoretical astrometric precision of ALMA is of order 0.1 mas, and minimum planetary masses of $0.1 M_J$ could be established astrometrically for some 400 nearby stars.

Direct imaging ALMA will be able to image disks around young stars out to several hundred parsecs, providing density and temperature profiles through measurements of thermal dust emission, and providing information on disk dynamics and chemistry through measurements of spectral lines. In the case of protoplanetary disks, ALMA may image gaps, warps and holes caused by protoplanets (Butler et al., 2004; Wolf & D'Angelo, 2005; Wolf, 2008), a prospect of particular interest in the context of type II migration (§10.8.2).

For direct exoplanet detection, ALMA's capabilities are more limited. At its highest resolution, the system could in principle resolve a Jupiter-like planet from its host star out to distances of 100–150 pc. The main limitation comes from the planetary flux. At 345 GHz, the flux density of a mature giant exoplanet is given by

$$F_{345} = 6.10^{-8} \, T \left(\frac{R_p}{R_J}\right)^2 \left(\frac{1}{d}\right)^2 , \qquad (7.7)$$

where d is the distance in pc, and T is the temperature in K. Together with the expected sensitivity of ALMA, a mature Jupiter will be detectable only out to about 1 pc. For a 'hot Jupiter' with $R = 1.5 R_J$ and $T = 1000$ K, this limit extends only to a few pc. A proto-Jupiter, with $R = 30 R_J$ and $T = 2500$ K, would be detectable in minutes to hours out to tens of parsecs (Butler et al., 2004).

7.8 Miscellaneous signatures

7.8.1 Planetary and proto-planet collisions

Stern (1994) considered the detectability of giant impacts occurring during the late stages of planetary formation. Massive colliding and accreting pairs, with $M \sim 0.1 M_\oplus$, would deposit sufficient energy to turn their surfaces molten and temporarily render them much more luminous at infrared wavelengths, with ocean vaporisation or energy radiation to space dominating according to mass. This occurs even for low-velocity approaches (i.e. for co-planar, zero-eccentricity orbits at 1 AU), the gravitational attraction leading to impact velocities of around $10 \, \mathrm{km \, s^{-1}}$. A luminous 1500–2500 K photosphere persisting for some 10^3 yr could be created for a terrestrial-mass planet (impacts on giant planets would be more luminous but shorter lived) leading to an estimate of 1/250 young stars affected.

Planet–planet collisions are also possible in quasi-mature planetary systems as a consequence of planet–planet perturbations. Zhang & Sigurdsson (2003) investigated the electromagnetic signals accompanied with planetary collisions and their event rate. An Earth–Jupiter collision would give rise to a prompt extreme ultraviolet/soft X-ray flash lasting for hours, and a bright infrared afterglow lasting for thousands of years or more. Signals would be above the X-ray detectability limits of

Chandra and XMM, and possibly above the photometric detection abilities of Gaia. Other estimates of detection prospects have been given (Kenyon & Bromley, 2005; Anic et al., 2007; Miller-Ricci et al., 2009a).

2M J1207 b The imaged system 2M J1207 b (§7.5.1, Mohanty et al., 2007) has been proposed as an example of such a collisional process. Its apparently underluminous nature has been explained as a hot proto-planet collision afterglow, with specific predictions which follow for its surface gravity (Mamajek & Meyer, 2007).

7.8.2 Collisional debris

Fine grains and dust are also expected as a by-product of protoplanetary collisions. In the solar system, the slow but persistent collisions between asteroids generates a tenuous cloud of dust, recognisable through the reflection of solar light as the *zodiacal light* (e.g. Ishiguro & Ueno, 2003; Müller et al., 2005). In the young solar system, collisions and dust production rates were presumably many times larger.

Dusty circumstellar disks were discovered by IRAS in 1983, and are now known around hundreds of main sequence stars (Zuckerman, 2001). The dust is generally cold, presumably primordial, and originates in the outer regions analogous to the solar system Edgeworth–Kuiper belt beyond Neptune (Luu & Jewitt, 2002).

A relatively small number of main sequence stars also possess warm (> 120 K) dust analogous to the zodiacal dust seen in the solar system. The dust temperature provides an indication of the distance between the dust and the star. Some of these warm dust clouds have been attributed to collisional events, such as ζ Lep (Chen & Jura, 2001), β Pic (Telesco et al., 2005), and BD+20 307 (Song et al., 2005).

BD+20 307 For BD+20 307, for example, arguments leading to the hypothesis of recent or ongoing collisional processes proceed as follows. Hipparcos, Keck, Gemini, 2MASS and IRAS data spanning 0.5–25 μm identify a large amount of warm silicate dust particles, which must be smaller than a few μm to account for the prominent spectral peak at around 10–12 μm (Song et al., 2005). The fraction of the stellar luminosity reradiated by the dust is large, $\tau \sim 0.04$. Such a quantity of dust translates into a collisional grinding time scale of a few hundred years (Kenyon & Bromley, 2005). Grains fractured to sub-μm dimensions will be cleared from the stellar vicinity by the radiation pressure from the host star, with the Poynting–Robertson drag time scale for μm-sized particles being of order 1000 yr (Wyatt & Whipple, 1950). As a result, the warm dust observed must either be the result of ongoing collisional processes, or of a recent collisional event. Song et al. (2005) argue that a 300-km sized asteroid would have to have been recently pulverised to account for the infrared excess observed for BD+20 307.

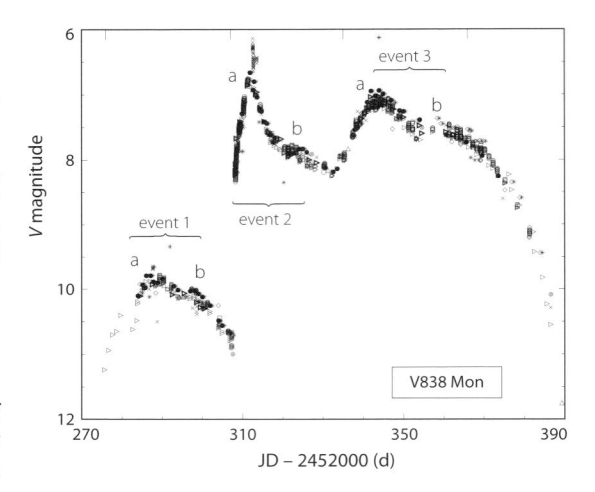

Figure 7.20: The light curve of V838 Mon in 2002, over a period of 120 days, showing the three similar 'double outburst' events, each marked by a major peak (a) and accompanied by a smaller peak (b). The bursts have been attributed to a planet progressively being slowed in three separate density-driven stages, and finally being engulfed in the envelope of a massive B star. From Retter et al. (2006, Figure 1), © John Wiley & Sons, Inc.

7.8.3 Accretion onto the central star

One terminal stage of an inspiraling planet may be its eventual accretion onto the star. Modeling has been carried out for stars in their final evolutionary stages as the outer stellar envelope expands, notably for asymptotic giant branch stars (Siess & Livio, 1999a) and for solar-mass stars on the red giant branch (Siess & Livio, 1999b).

In the latter case observational signatures that accompany the engulfing of the planet include ejection of a shell and a subsequent phase of infrared emission, increase in the ^{7}Li surface abundance, spin-up of the star because of the deposition of orbital angular momentum, and the possible generation of magnetic fields and the related X-ray activity caused by the development of shears at the base of the convective envelope.

Infrared excess and high Li abundances are observed in 4–8% of G and K stars, and Siess & Livio (1999b) postulated that these signatures might originate from the accretion of a giant planet, a brown dwarf, or a very low-mass stars. The discovery of ^{6}Li in the atmosphere of the metal-rich solar-type star HD 82943, which is known to have an orbiting giant planet, may be evidence for a planet (or planets) having been engulfed by the parent star (Israelian et al., 2001).

Direct evidence for planetary accretion is limited, but at least two systems are contenders.

V838 Mon V838 Mon has been considered as the prototype of a new class of star which undergoes pronounced outbursts repeated at short intervals. The class comprises at least two other objects, M31RV and V4332 Sgr, with interpretations including nova-type out-

7.8 Miscellaneous signatures

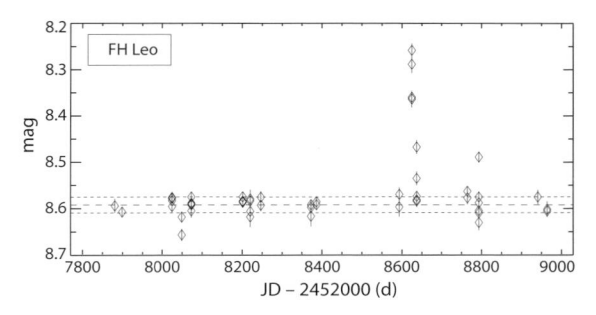

Figure 7.21: Hipparcos light curve for FH Leo. The dashed line is the mean quiescent magnitude (dotted lines are at $\pm 1\sigma$). Error bars are the Hipparcos standard errors. The outburst is possibly a planetary accretion event. From Dall et al. (2005, Figure 1), reproduced with permission © ESO.

bursts, thermal pulses, and stellar mergers (Retter & Marom, 2003, and references).

V838 Mon underwent an unusual outburst in 2003, showing at least three peaks over a period of 100 days. Imaging revealed a light echo around the object, providing evidence for a dust shell which was emitted several thousand years ago and is now reflecting light from the eruption (Bond et al., 2003). Spectral analysis suggests that the object was relatively cold throughout the event, which was characterised by an expansion to extremely large radii. Retter & Marom (2003) suggested that the outburst was caused by the expansion of a red giant, followed by the successive swallowing of three relatively massive planets in close orbits.

Retter et al. (2006) reasoned that the similarity of the three events, each accompanied by secondary shallower peaks (Figure 7.20), rather argues for a three-step process in the accretion of a single planet into a giant stellar envelope: the captured planet reaches some critical density in the stellar envelope, triggers an initial super-Eddington event that causes the star to expand and the density of material surrounding the planet to decrease.

The descent of the planet continues, and the process is repeated, until it finally reaches the nuclear burning shell at $\sim 1 R_\odot$, where it dissolves or evaporates (Livio & Soker, 1984; Soker, 1998; Siess & Livio, 1999a,b).

FH Leo The common proper motion system FH Leo is a wide visual binary of separation 8.31 arcsec, observed together by Hipparcos as HIP 54268. It was classified as a nova-like variable due to an optical outburst observed in the Hipparcos photometry (Figure 7.21). From spectroscopy, and a study of the elemental abundances including Li and α-elements, Dall et al. (2005) concluded that the component stars, HD 96273 and BD+07° 2411 B, do constitute a physical binary, being of normal late-F and early-G type. At $d = 117$ pc, the lower limit for the physical separation is 936 AU.

Dall et al. (2005) concluded that the rise-time seen in the Hipparcos light curve might have been very fast, while the decay probably lasted at least 13 days, with a possible second event about 170 days later. Their favoured explanation is a planetary accretion event which resulted in an energy outburst and a polluted stellar atmosphere. From the magnitude rise, they estimated a mass of the accreted matter as 5×10^{20} kg, about the mass of a large asteroid like Pallas or Vesta. A scenario where BD+07° 2411 B accreted such a companion could explain the Hipparcos outbursts, the overabundances found in BD+07° 2411 B with respect to HD 96273, and the presence of lithium in BD+07° 2411 B.

Further analysis by Vogt (2006) has argued that FH Leo may still be a dwarf nova in a triple system, but the debate is still open.

7.8.4 Gravitational wave modulation

Planets around compact binaries that are strong sources of gravitational radiation should modulate the phase of the gravitational waves. LISA could place limits $\gtrsim 4 M_J$ around 3000 Galactic double white dwarfs (Seto, 2008).

8
Host stars

Properties of the host stars of exoplanets are derived from a combination of astrometric, photometric, and spectroscopic observations, interpreted primarily within the context of stellar evolutionary models.

Planets are now known to exist around a wide variety of stellar types: not only around main sequence stars like the Sun, but around M dwarfs (§2.4.4) and as lower mass companions to brown dwarfs (§7.5), around pulsating stars including hot subdwarfs (§4.2.2) and δ Scu variables (WASP–33), around giants (§2.4.5), and around objects in the terminal stages of evolution including pulsars (§4.1) and probably white dwarfs (§4.2.1). They are found in binary systems (§4.3), around stars of the thick disk (e.g. WASP–21), around stars of low-metallicity both from radial velocity surveys (§2.4.6) and from transit surveys (e.g. WASP–37), in open clusters (§2.4.6), and perhaps in the bulge (MOA–2008–BLG–310).

8.1 Knowledge from astrometry

8.1.1 Hipparcos distances and proper motions

The majority of stars monitored for radial velocity or photometric transit observations are bright and consequently relatively nearby ($d < 50 - 70$ pc). Distances and proper motions are therefore generally well determined from the Hipparcos satellite measurements. Operated between 1989–93, Hipparcos provided 1 mas accuracy in positions, parallaxes and annual proper motions for about 120 000 stars (Perryman, 1997; Perryman et al., 1997), subsequently improved through an enhanced satellite attitude solution (van Leeuwen, 2007). These precise distances (Figure 8.1) translate into improved determination of the host star properties.

The Tycho 2 catalogue (Høg et al., 2000), fully superseding the original Tycho catalogue (Høg et al., 1997), is a further source of high-accuracy astrometric data. Together, the Hipparcos and Tycho astrometric catalogues provide the underlying reference frame for all ground-based astrometry (Perryman, 2009, Chapter 2).

Completeness The magnitude completeness limit of the Hipparcos catalogue ranges between $V = 7.3 - 9$ mag, depending on Galactic latitude and spectral type (Perryman, 1997, Volume 3, Section 3.2). Although specific efforts were made during the input catalogue compilation to ensure that all known nearby stars down to the satellite observability limit of $V \sim 12$ were included, the catalogue is incomplete fainter than $V \sim 7.3$ mag.

Use in target lists The Hipparcos data, notably the distances and derived luminosities, have been extensively used in establishing target lists for radial velocity and other planetary search programmes, for example for Keck (Vogt et al., 2002), CORALIE/HARPS (Tamuz et al., 2008) and many of the other Doppler programmes listed in Table 2.4, for the Darwin interferometer mission study (Kaltenegger et al., 2006), and more generally for studies of multiplicity of known exoplanetary systems (Raghavan et al., 2006). The Tycho 2 catalogue was used, for example, to select targets for N2K, the large-scale radial velocity survey of 2000 stars (Fischer et al., 2005).

8.1.2 Nearby star census

The definition of the nearby stellar population figures in various areas of exoplanet research, ranging from evaluations of their statistical occurrence to the identification of nearby candidates for imaging or spectroscopy.

It remains, however, a difficult task to establish a complete census of stars within the immediate solar neighbourhood, even out to distances of only 10–20 pc. The earliest ground-based parallax surveys were very successful in identifying nearby bright stars, but problems persist at the faint end of the luminosity function, $M_V \gtrsim 15$, where a complete parallax survey even out to only 10 pc remains impossible.

Surveys searching for high-proper motion stars (e.g. Luyten, 1979) were efficient at detecting nearby candidate stars which were then added to such parallax programmes (including the Hipparcos Input Catalogue in the early 1980s), but they implied a strong bias towards high-velocity halo objects. For this reason, early nearby star compilations used spectroscopic and photometric distance estimates to identify additional nearby candidates. The advent of accurate all-sky multi-colour

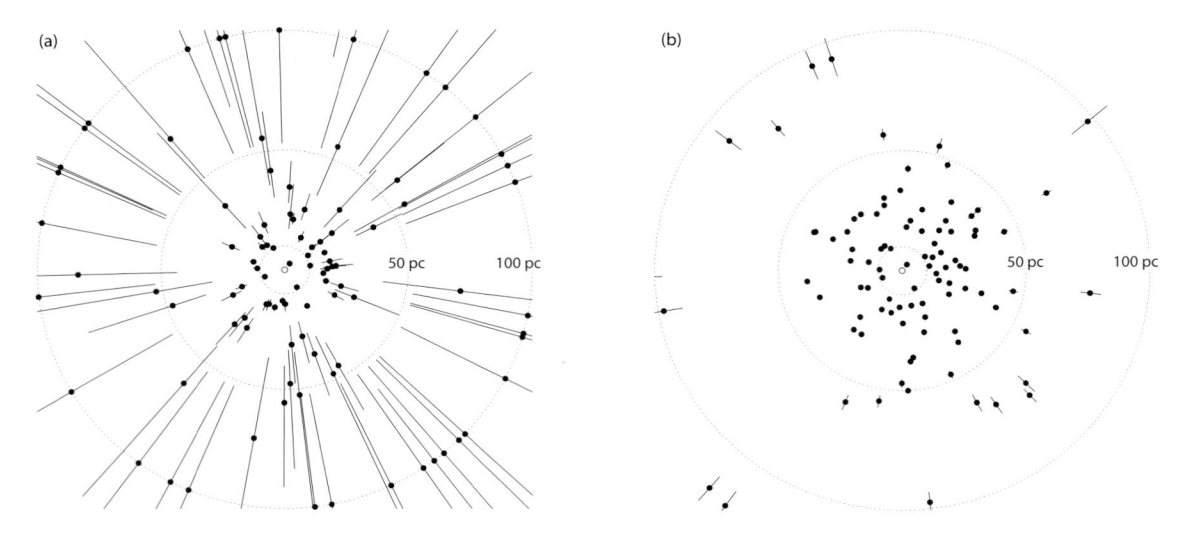

Figure 8.1: Improvement in the knowledge of exoplanet host star distances by Hipparcos. For the 100 brightest stars with exoplanets known from radial velocity measurements at the end of 2010 (V < 7.2 mag), estimated distances and standard errors are shown from: (a) the ground-based compilation of van Altena et al. (1995), and (b) from Hipparcos (Perryman, 1997). Azimuthal coordinates correspond to right ascension, independent of declination. Distances undetermined in (a) are arbitrarily assigned a parallax of 10 ± 9 mas. Hipparcos substantially improved both parallax standard errors, and their systematics.

surveys further facilitated the search for nearby, low-luminosity stars.

Catalogue of Nearby Stars The *'Catalogue of Stars within Twenty-Five Parsecs of the Sun'* (Woolley, 1970) was one of the first attempts to compile a census of known stars in the solar neighbourhood, largely based on trigonometric parallaxes.

An evolving compilation has been maintained by the Astronomisches Rechen-Institut in Heidelberg over the last 50 years. Gliese (1957) published the *'Katalog der Sterne näher als 20 Parsek für 1950.0'*, containing 915 single stars and systems within 20 pc (1094 components altogether), with probable parallax errors of 9.2 mas. Gliese (1969) published the updated *'Catalogue of Nearby Stars'*, or CNS2, with a slightly enlarged distance limit of 22.5 pc ($\pi \geq 0.045$ arcsec). It contained 1049 stars or systems within 20 pc, and the standard errors were estimated as 7.6 mas. In both compilations trigonometric, photometric, and spectroscopic parallax estimates were employed.

The *'Third Catalogue of Nearby Stars'*, or CNS3, was only published in preliminary form (Gliese & Jahreiß, 1991). This extended the census to some 1700 stars nearer or apparently nearer than 25 pc (the trigonometric parallax limit is actually 0.0390 arcsec), and was based on the latest edition of the General Catalogue of Trigonometric Stellar Parallaxes. Information includes spectral types from a variety of sources, broad-band *UBVRI* photometric data, photometric parallaxes, parallaxes based on luminosity and space-velocity components, spectral type–luminosity and colour–luminosity relations, positions, and proper motions. Contrary to the CNS2, trigonometric parallaxes and photometric or spectroscopic parallaxes were not combined. The resulting parallax is the trigonometric parallax if $\sigma_\pi/\pi < 0.14$, or the photometric or spectroscopic parallax if the trigonometric parallax was less accurate or if it was not available.

The *'Fourth Catalogue of Nearby Stars'*, or CNS4, incorporates data from the Hipparcos Catalogue, and accordingly provides a major development in the comprehensive inventory of the solar neighbourhood up to a distance of 25 pc from the Sun (Jahreiß & Wielen, 1997). It is not published as of late 2010. The binary star content of CNS4 is discussed by Jahreiß & Wielen (2000).

Other compilations There are other compilations of nearby stars. Northern Arizona University 'NStars Database', dating from 1998, maintains a compilation of all stellar systems within 25 pc, and at the end of 2010 listed around 2600 objects.

Georgia State University's 'Research Consortium on Nearby Stars', RECONS (Henry et al. 2006 and references, Henry 2009) aims to discover and characterise 'missing' stars within 10 pc, via astrometric, photometric, and spectroscopic techniques. On 2009 January 1, their compilation listed 354 objects in 249 systems within 10 pc, compared with 182 Hipparcos entries out to 10 pc. They are currently extending survey work to 25 pc.

Andronova (2000) has separately compiled a catalogue of about 5000 stellar systems considered to be within 25 pc, containing data from the Hipparcos, Tycho 2, Washington Double Star, MSC, digital sky surveys and other sources.

Additional nearby candidates are emerging from the large-scale high proper motion surveys, themselves benefiting from an astrometric re-calibration of multi-epoch Schmidt plates. For example, Lépine (2005) has established, in the northern hemisphere, a list of 539 new candidate stars within 25 pc of the Sun, including 63 estimated to be within only 15 pc. He estimates that some 18% of nuclear-burning stars within 25 pc of the Sun remain to be located.

8.1.3 Galactic coordinates

The Hipparcos and Tycho catalogues, in common with other astrometric catalogues, provide positions and proper motions within the equatorial system. Various studies of the host stars make use of Galactic coordinates, or space velocities derived from proper motions and radial velocities. The two relevant transformations are given hereafter (Perryman, 1997, Volume 1, Section 1).

The basis vectors in the equatorial system are denoted [**x y z**], with **x** being the unit vector towards $(\alpha,\delta) = (0,0)$, **y** the unit vector towards $(\alpha,\delta) = (+90°,0)$, and **z** the unit vector towards $\delta = +90°$. Denoting the basis vectors in Galactic coordinates as [$\mathbf{x}_G\ \mathbf{y}_G\ \mathbf{z}_G$], an arbitrary direction **u** may be written in terms of equatorial or Galactic coordinates as

$$\mathbf{u} = [\mathbf{x\ y\ z}] \begin{pmatrix} \cos\delta\cos\alpha \\ \cos\delta\sin\alpha \\ \sin\delta \end{pmatrix} = [\mathbf{x}_G\ \mathbf{y}_G\ \mathbf{z}_G] \begin{pmatrix} \cos b\cos l \\ \cos b\sin l \\ \sin b \end{pmatrix}. \quad (8.1)$$

The transformation between them is given by

$$[\mathbf{x}_G\ \mathbf{y}_G\ \mathbf{z}_G] = [\mathbf{x\ y\ z}]\mathbf{A}_G, \quad (8.2)$$

where the matrix $\mathbf{A}_G$ relates to the definition of the Galactic pole and centre in the International Celestial Reference System (ICRS). In the published Hipparcos catalogue, the following definitions in the ICRS were adopted for the north Galactic pole (α_G,δ_G), and for the origin of Galactic longitude defined by the longitude of the ascending node of the Galactic plane on the equator of ICRS, l_Ω

$$\alpha_G = 192°85948, \quad (8.3)$$
$$\delta_G = +27°12825, \quad (8.4)$$
$$l_\Omega = 32°93192. \quad (8.5)$$

The angles α_G, δ_G and l_Ω, considered as exact quantities, are consistent with the previous (1960) definition of Galactic coordinates to a level set by the quality of optical reference frames prior to Hipparcos. The transformation matrix A_G may be computed to any desired accuracy; to 10 decimal places the result is

$$\mathbf{A}_G = \begin{pmatrix} -0.0548755604 & +0.4941094279 & -0.8676661490 \\ -0.8734370902 & -0.4448296300 & -0.1980763734 \\ -0.4838350155 & +0.7469822445 & +0.4559837762 \end{pmatrix}. \quad (8.6)$$

The star's Galactic longitude and latitude can then be computed from

$$\begin{pmatrix} \cos b\cos l \\ \cos b\sin l \\ \sin b \end{pmatrix} = \mathbf{A}'_G \begin{pmatrix} \cos\delta\cos\alpha \\ \cos\delta\sin\alpha \\ \sin\delta \end{pmatrix}. \quad (8.7)$$

Space coordinates and velocity The position of a star with respect to the solar system barycentre, **b**, measured in pc, and its barycentric space velocity, **v**, measured in km s^{-1}, are given in equatorial components by

$$\begin{pmatrix} b_x \\ b_y \\ b_z \end{pmatrix} = \mathbf{R} \begin{pmatrix} 0 \\ 0 \\ A_p/\varpi \end{pmatrix}, \quad (8.8)$$

$$\begin{pmatrix} v_x \\ v_y \\ v_z \end{pmatrix} = \mathbf{R} \begin{pmatrix} k\mu_{\alpha*}A_v/\varpi \\ k\mu_\delta A_v/\varpi \\ kv_r \end{pmatrix}, \quad (8.9)$$

with $\mu_{\alpha*} = \mu_\alpha\cos\delta$, and

$$\mathbf{R} = \begin{pmatrix} -\sin\alpha & -\sin\delta\cos\alpha & \cos\delta\cos\alpha \\ \cos\alpha & -\sin\delta\sin\alpha & \cos\delta\sin\alpha \\ 0 & \cos\delta & \sin\delta \end{pmatrix}. \quad (8.10)$$

$A_p = 1000$ mas pc and $A_v = 4.74047...$ km yr s^{-1} designate the astronomical unit expressed in the appropriate units, v_r is the radial velocity, and $k = (1 - v_r/c)^{-1}$ is the Doppler factor.

Galactic components of **b** and **v** can be obtained through pre-multiplication by $\mathbf{A}'_G$, as in Equation 8.7.

Kinematic properties There is little evidence that the kinematic properties of stars with or without planets in the solar neighbourhood differ significantly. Barbieri & Gratton (2002) reconstructed the Galactic orbits of the host stars known at the time, and found no kinematic differences with the stars studied by Edvardsson et al. (1993). Similar results are reported by, e.g., Luck & Heiter (2006).

Ecuvillon et al. (2007) studied the kinematics of metal-rich stars with and without planets, using Hipparcos astrometry and radial velocities from CORALIE, and examined their relation to the Hyades, Sirius, and Hercules dynamical streams. Showing that the planet host targets have a kinematic behaviour similar to that of the metal-rich comparison sub-sample, and appealing to the scenarios proposed for the origin of the dynamical streams, they argued that systems with giant planets could have formed more easily in the metal-rich inner Galaxy, and then been brought into the solar neighbourhood by these dynamical streams.

The question of the Galactic origin of planet host stars is considered further in §8.4.3.

8.2 Photometry and spectroscopy

Bolometric magnitudes The total energy integrated over all wavelengths is referred to as the *bolometric magnitude*, $M_{\rm bol}$ (e.g. Gray, 2000). In terms of the V band by way of illustration, the quantity required to transform an apparent magnitude in a particular bandpass (m_V) to an absolute magnitude in that bandpass (M_V) is the star's distance

$$m - M \equiv 5\log\left(\frac{d}{10}\right) = -5\log\varpi - 5, \quad (8.11)$$

where d is in pc, or the parallax ϖ is in arcsec (and where correction due to interstellar extinction is ignored). The quantity required to transform the absolute magnitude to the bolometric magnitude, $M_{\rm bol}$, is referred to as the *bolometric correction*, BC, for that magnitude system

$$\mathrm{BC}(V) \equiv M_{\rm bol} - M_V = 2.5\log\frac{\int F_\nu S_\nu d\nu}{\int F_\nu d\nu} + C, \quad (8.12)$$

where F_ν is the measured flux, and S_ν is the detector response for the relevant bandpass. BC is a function of the bandpass and the underlying stellar energy distribution of the star, the latter itself a function of $T_{\rm eff}$, $\log g$, metallicity, etc. The luminosity of the star is then given by

$$\log\left(\frac{L_\star}{L_\odot}\right) = -0.4\,[M_V + \mathrm{BC} - (M_{V,\odot} + \mathrm{BC}_\odot)]. \quad (8.13)$$

Bolometric corrections are derived from empirical calibrations or from model atmospheres. The zero point of the bolometric magnitude scale is set by reference to the Sun: Cayrel de Strobel (1996) gave $M_{\rm bol\odot} = 4.75$ and $\mathrm{BC}(V)_\odot = -0.08$ yielding $M_V = 4.83$, while Cox (2000) gives $M_{\rm bol\odot} = 4.74$, $\mathrm{BC}(V)_\odot = -0.08$, and $M_V = 4.82$ (see also Bessell et al., 1998); $M_{\rm bol\odot} = 4.75$ was adopted by IAU Commission 36 at the IAU General Assembly in 1997 (Andersen, 1999, p141).

Bolometric corrections are usually tabulated versus spectral type or colour index, and their dependence on $B-V$ is only moderately sensitive to luminosity class. For hot or cool stars, bolometric corrections in V are large, since most of the flux lies outside of the V band.

Bolometric corrections for Hipparcos photometry Whenever available, there are three compelling reasons to use the Hipparcos magnitudes Hp (van Leeuwen et al., 1997), rather than say Johnson V magnitudes as a basis for the construction of uniform bolometric magnitudes for a sample of stars: their high accuracy ($\sigma \sim 0.0015$ mag); their small systematic errors and excellent uniformity independent of position, magnitude, and colour index; and the fact that the photometric band is very wide and therefore a better observational approximation to the total flux.

Until recently, a restriction in the wider use of the Hp magnitudes has been the absence of associated bolometric corrections, a shortcoming now rectified. Bessell et al. (1998) used synthetic spectra derived from ATLAS9 and NMARCS to produce broad-band colours and bolometric corrections for a wide range of $T_{\rm eff}$, g and [Fe/H]. The same synthetic spectra were then used to construct a grid of $T_{\rm eff}$, $\log g$ and Z giving corresponding values of the following quantities: $\mathrm{BC}(V)$, $\mathrm{BC}(Hp)$, $V-Hp$, $V-V_T$, $B-B_T$, B_T-V_T, $B-V$, $V-R$, and $V-I$ (where B_T and V_T are magnitudes in the Tycho photometric bands). The bolometric corrections are given in tabular form by Bessell (2007). Examples are shown in Figure 8.2.

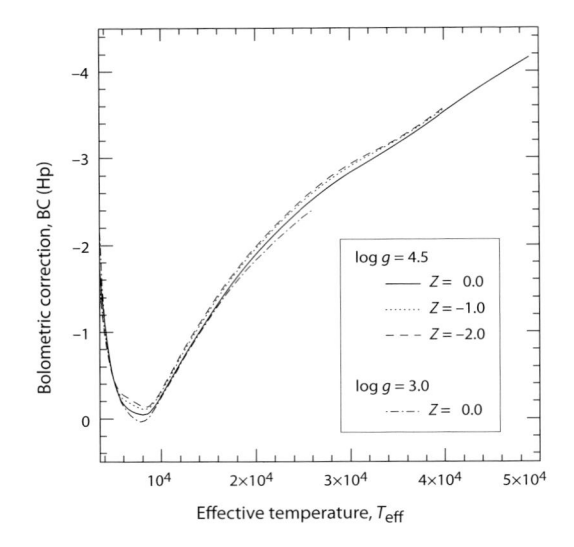

Figure 8.2: Bolometric corrections for the Hipparcos photometric passband, BC(Hp), versus effective temperature $T_{\rm eff}$, for a subset of values of $\log g$ and Z. The determinations, by M.S. Bessell, are also available in tabular form (Bessell, 2007).

Effective temperature The effective temperature of a star, $T_{\rm eff}$, is defined as the temperature of a blackbody radiator with the same radius and same luminosity (total energy output), related via the Stefan–Boltzmann law

$$L_\star = 4\pi\sigma R_\star^2 T_{\rm eff}^4 \quad (8.14)$$

where $L_\star$ is the luminosity, $R_\star$ is the radius, and σ is the Stefan–Boltzmann constant. Determination of $T_{\rm eff}$ based on this definition therefore requires knowledge of both $R_\star$ (itself requiring knowledge of the distance d and angular diameter θ) as well as $L_\star$ (itself requiring knowledge of the star's bolometric magnitude, according to Equation 8.13).

Angular diameters In addition to linear diameters estimated from theoretical evolutionary models, direct angular size measurements have been made in a few cases using interferometry. Baines et al. (2008) determined 24 exoplanet host star diameters using the CHARA array (Figure 8.3). Combining these with the Hipparcos distances yields the linear radii which, amongst other applications, provides an independent estimate of their

8.2 Photometry and spectroscopy

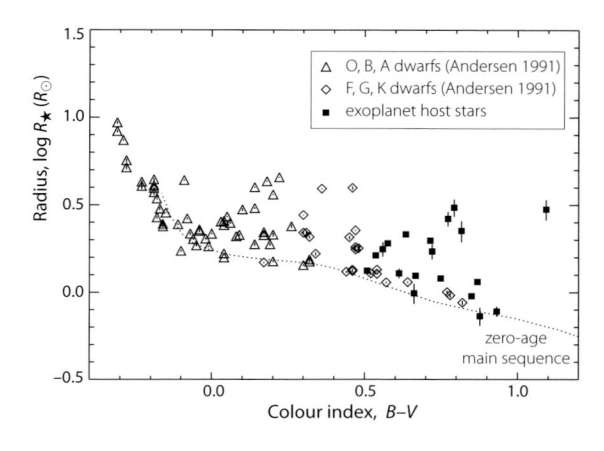

Figure 8.3: Host star stellar radii derived from interferometric angular diameter measurements with the CHARA array. Triangles represent O, B, and A dwarfs; diamonds represent F, G, and K dwarfs. Both sets of measurements are from Andersen (1991). Filled squares represent exoplanet host stars diameters with errors < 15%. The dotted line indicates the zero-age main sequence from Girardi et al. (2000) for stars with masses between $0.15 - 5.0 M_\odot$. From Baines et al. (2008, Figure 3), reproduced by permission of the AAS.

evolutionary state. Similar observations were reported using the Palomar Testbed Interferometer (von Braun & van Belle, 2008).

Parameters from spectroscopy In practice, spectroscopic analysis provides robust estimates of a number of basic stellar quantities. In most recent work, the measured equivalent widths of some 50 Fe I and Fe II absorption lines are used to estimate the four basic stellar parameters which influence the relative line strengths and line profiles: effective temperature $T_{\rm eff}$, surface gravity $\log g$, microturbulence velocity, and metallicity [Fe/H]. Typical observational requirements call for 2–3-m class telescopes, with high-dispersion échelle spectroscopy ($R \sim 50 - 70\,000$), covering most of the optical spectrum at high signal-to-noise ratios of several hundred per resolution element.

Analysis makes use of atmospheric models, typically assuming a plane-parallel geometry (e.g. Kurucz, 1993) under the assumption of local thermodynamic equilibrium (LTE), along with basic atomic data, most critically the line oscillator strengths. Differential analysis with respect to the solar flux spectrum can employ reflectors such as the Moon (e.g. Fuhrmann et al., 1998) or Callisto (e.g. Luck & Heiter, 2006).

Further details are given in a number of the references cited below (e.g. Santos et al., 2004c; Luck & Heiter, 2006; Sousa et al., 2008). A somewhat different (spectrum-matching) technique was used by Valenti & Fischer (2005). In either case, typical uncertainties are in the range 0.02–0.5 dex for [Fe/H], and 40–70 K for $T_{\rm eff}$ (Figure 8.4), although higher formal precision is often

achieved (Kovtyukh et al., 2003).

Gray (2000) described the inaccuracy of the present knowledge of the effective temperature scale as one of the greatest barriers to the advancement of stellar evolutionary theory. Its importance in the study of exoplanets is through its role in abundance determinations, which are sensitive to the assumed $T_{\rm eff}$. Calibration of the effective temperature scale via the Stefan–Boltzmann law (e.g. using the infrared flux method) has given some concordant results (e.g. Ribas et al., 2003; Sousa et al., 2008) and other discordant results (e.g. Ramírez & Meléndez, 2004, 2005a,b; Casagrande et al., 2006; Gonzalez, 2006a). Although procedures are clearly converging, uncertainties in individual value of $T_{\rm eff}$ are still sometimes considered to be nearer to 100 K (Luck & Heiter, 2006).

Sousa et al. (2008) derived an empirical fit as a function of $B - V$ given in the Hipparcos catalogue and [Fe/H]

$$T_{\rm eff} = 9114 - 6827(B-V) + 2638(B-V)^2 + 368 [{\rm Fe/H}], \qquad (8.15)$$

with a standard deviation of 47 K over the intervals $4500\,{\rm K} < T_{\rm eff} < 6400\,{\rm K}$, $-0.85 < [{\rm Fe/H}] < 0.40$, and $0.51 < B - V < 1.20$.

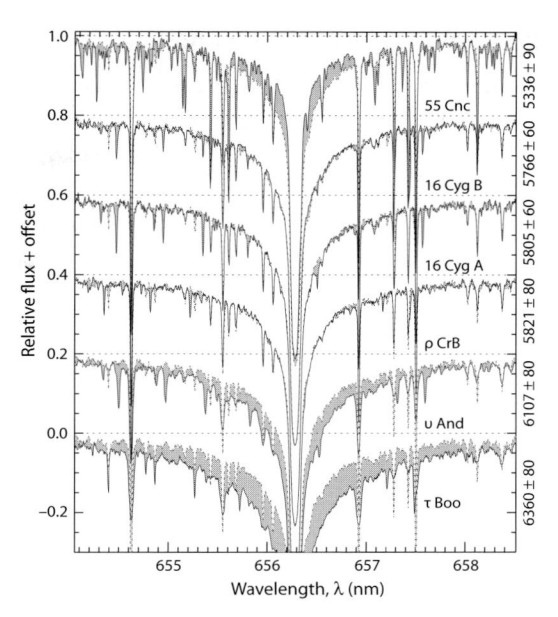

Figure 8.4: Hα line profiles of 55 Cnc, 16 Cyg B, 16 Cyg A, ρ CrB, υ And, and τ Boo. The sequence of increasing $T_{\rm eff}$ (given at right) is superimposed on the spectrum of the Moon for determining the instrumental profile (dotted), with differences depicted by shading. From Fuhrmann et al. (1998, Figure 1), reproduced with permission © ESO.

Abundances of other elements Once the principle stellar parameters have been estimated, the abundances of other elements are derived from individual spectral lines using either measured equivalent widths or comparison with synthetic spectra. Abundance comparisons for stars with and without planets are usually made relative to Fe, as [x/Fe], which removes the first-order difference in [Fe/H] between the two sample types.

Light elements such as Li, C, N, O, Na, Al, Mg, and S have relatively few spectral lines in solar-type stars, and high-quality spectra are required for reliable abundances. Some observational considerations for other specific lines are discussed by Gonzalez (2006a).

Rotation velocities In an early study, Barnes (2001) compared the rotation periods of 11 exoplanet host stars, derived through photometric or spectroscopic variability, with comparison stars, and results from evolutionary stellar models. The analysis suggested no statistical difference compared with other main sequence stars, although the preferential selection of radial velocity surveys according to low rotation or low activity has inevitably resulted in samples typically more biased towards older stars.

Additional determinations of rotation periods of M dwarfs was given by Jenkins et al. (2009b), and for other host stars by Simpson et al. (2010a).

Stellar ages are sometimes estimated from their rotational velocities, based on the following arguments: during the pre-main sequence phase, angular momentum evolution is mainly driven by the magnetic coupling between the star and its disk, and solar-type stars arrive on the zero age main sequence with a large spread in rotation (e.g. Königl, 1991). From this point on, angular momentum evolution is dictated by stellar winds whose intensity is itself a function of magnetic activity and rotation (e.g. Kawaler, 1988). This creates a regulation mechanism leading to a steady rotational decrease, and to the convergence of the rotation rate at about 1 Gyr.

Studies of Li abundances (§8.4.7) suggests that angular momentum transfer arising from long-duration star–disk interactions during the pre-main sequence phase may have resulted in yet systematically slower stellar rotation rates at the beginning of the star's main sequence evolution for some planet host stars.

In a few individual cases of close-in planets, there is some evidence for tidal spin-up of the star as a result of angular momentum deposition by inwardly-migrating massive planets (§10.9.3). Anomalous rotation rates may also be expected as a result of giant planet ingestion during the expanding star's red giant, horizontal branch and early asymptotic branch phases (Massarotti, 2008).

Determination of the line-of-sight inclinations of the stellar rotation axes, based on the projected equatorial velocity ($v \sin i$), the stellar rotation period ($P_{\rm rot}$), and the stellar radius ($R_\star$), can be used to estimate masses of Doppler detected planets by removing the $\sin i$ discrepancy, but only under the assumption that the stellar rotation axis is aligned with the planet's orbital axis (e.g. Simpson et al., 2010a; Watson et al., 2010).

Spectroscopic notation used in this section:
- [x/H] ≡ $\log_{10}(N_{\rm x}/N_{\rm H}) - \log_{10}(N_{\rm x}/N_{\rm H})_\odot$, where $N_{\rm x}$ is the number of atoms of element x per unit volume; thus
- [Fe/H] ≡ log number abundance of Fe/H relative to solar
- [α/Fe] ≡ $\log_{10}(N_\alpha/N_{\rm Fe}) - \log_{10}(N_\alpha/N_{\rm Fe})_\odot$, the α-element abundance relative to solar
- 'metals': in astronomical usage, generally taken to embrace all elements heavier than He
- dex: a contraction of 'decimal exponent', with n dex meaning 10^{-n}, and 1 dex corresponding to a factor 10

8.3 Evolutionary models

Masses, radii, and ages Reliable stellar mass determinations are important in establishing dynamical orbits of planetary systems. Accurate radii are important for setting the linear scale of transiting planets. Ages provide information on the star's evolutionary state and potentially their Galactic origin, provide temporal constraints on the processes and efficiency of resonances and tidal interactions for orbital synchronisation and circularisation, and will be important in optimising searches for exo-Earths.

These properties are (typically) not primary observables, and are generally inferred by comparison with theoretical evolutionary models. Other routes to mass estimation are via orbital dynamics for binary stars, or using spectroscopic gravity indicators.

Accurate stellar distances are essential in locating the position of a star in the observational Hertzsprung–Russell diagram, often presented as absolute magnitude versus colour index (Figure 8.5). Theoretical evolutionary models, in contrast, are usually characterised by the bolometric magnitude ($M_{\rm bol}$) and the stellar effective temperature ($T_{\rm eff}$), and predict a star's position in the theoretical Hertzsprung–Russell diagram on the basis of its initial chemical composition, initial mass, and assumed age.

Relating theoretical to observed quantities, and hence inferring stellar properties such as mass and age, is however subject to numerous complications. First is the choice of theoretical model, of which various well-validated examples are available and in use, and which differ in the detailed treatment of physical processes (such as line opacity, convective overshooting, and He diffusion), chemistry (such as the presence of α elements), and astrophysical properties (such as rotation). To convert an observed magnitude to an absolute magnitude requires knowledge of the star's distance (Equation 8.11), and to convert an absolute magnitude to a bolometric magnitude requires an estimate of the bolometric correction (Equation 8.12).

A compilation of recent stellar evolutionary models, and further details in the context of the Hipparcos literature, is given by Perryman (2009, Chapter 7). An important consideration is that ages from isochrones typi-

8.3 Evolutionary models

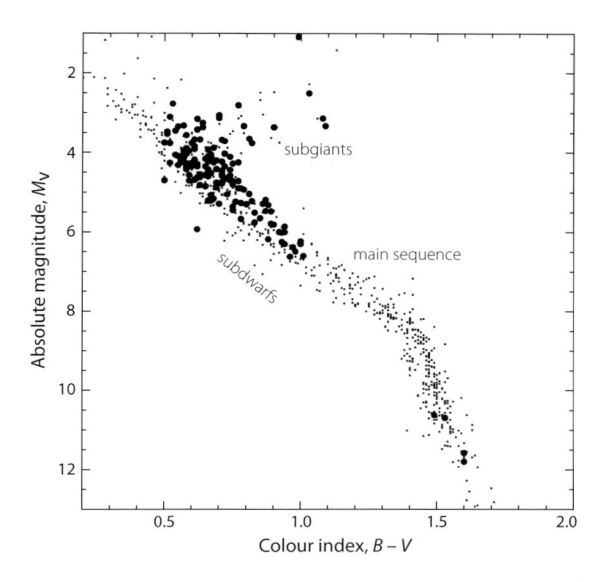

Figure 8.5: The Hipparcos-based Hertzsprung–Russell diagram for stars within 25 pc from the sample of Reid et al. (2002), with systems identified with planetary companions shown as filled circles. Adapted from Hawley & Reid (2003, Figure 3), to reflect the status as of mid-2007, by Neill Reid.

cally have a precision that varies significantly according to location in the Hertzsprung–Russell diagram (e.g. Jørgensen & Lindegren, 2005).

Example derivations As two early examples of the procedures that have been frequently applied subsequently in the field of exoplanets, Fuhrmann et al. (1997) used the Hipparcos parallaxes to estimate absolute magnitudes, and hence masses and ages from stellar evolutionary tracks, for 51 Peg and 47 UMa, deriving masses of $1.12 \pm 0.06\,M_\odot$ and $1.03 \pm 0.05\,M_\odot$ respectively, and ages of 4.0 ± 2.5 Gyr and 7.3 ± 1.9 Gyr respectively. Fuhrmann et al. (1998) made a similar analysis for other F and G-type stars with planetary companions: υ And, 55 Cnc, τ Boo, 16 Cyg and ρ CrB. Figure 8.6, for the case of υ And, typifies the difficulty of establishing metallicity unambiguously, and the improvement brought to stellar modeling by the Hipparcos results.

As an example of procedures on a larger scale, Valenti & Fischer (2005) presented a spectroscopic compilation of 1040 FGK dwarfs from the Keck, Lick, and AAT planet search programmes. They use a spectrum-matching technique that generates differences with respect to the observed spectrum that are then minimised over parameter space to simultaneously determine effective temperature $T_{\rm eff}$, surface gravity $\log g$, projected rotational velocity $V_{\rm rot}$, and metallicity [Fe/H], along with abundances of Na, the iron-peak elements Fe and Ni, and the α-process elements Si and Ti. The Hipparcos parallaxes, combined with V-band photometry and corresponding bolometric corrections, were used to determine bolometric luminosities. Interpolating theoretical Yonsei–Yale isochrones (Demarque et al., 2004) with respect to $L_\star$, $T_{\rm eff}$, [Fe/H], and [α/Fe] for each star then yielded a theoretical mass, radius, gravity, and age for most of their sample. Estimated precision was 44 K in $T_{\rm eff}$, 0.03 dex in [Fe/H], 0.06 dex in $\log g$, and $0.5\,{\rm km\,s^{-1}}$ in $V_{\rm rot}$.

A similar procedure was used by Robinson et al. (2007) to estimate the atmospheric parameters for 1907 metal-rich stars in the N2K spectroscopic survey, based on the estimates of temperatures and metallicities for more than 100 000 FGK dwarfs from Tycho 2 (Ammons et al., 2006) or from the Geneva–Copenhagen survey (Nordström et al., 2004b).

Age distribution As a class, exoplanet host stars appear to show no very strong departures from the typical ages of stars in the solar neighbourhood.

Stellar ages can be estimated using a variety of techniques, preferably using isochrone fitting when suitably constrained (e.g. Jørgensen & Lindegren, 2005). Lower limits are sometimes inferred from lithium abundances as a function of effective temperature based on calibrations using clusters of different ages (Martin, 1997; Jeffries et al., 2002; Sestito & Randich, 2005); from the Ca II activity–age relation (Henry et al., 1996; Pace & Pasquini, 2004; Wright et al., 2004); or from the stellar rotational velocity (e.g. Melo et al., 2006).

Saffe et al. (2005) estimated ages from chromospheric activity indicators and isochrones to derive median ages of 49 exoplanet host stars of 5.2 ± 4 Gyr and 7.4 ± 4 Gyr respectively.

Reid (2002) used a control sample of 486 Hipparcos FGK stars to show that the planetary host stars exhibit velocities with a lower dispersion than the field star sample, suggesting that their average age is only some 60% that of a representative subset of the disk. This perhaps reflects the higher proportion of metal-rich stars in the planet host sample.

Reid et al. (2007) used the data set described by Valenti & Fischer (2005) to demonstrate that, while the age distribution of nearby FGK dwarfs is broadly consistent with a uniform star-formation rate over the 10 Gyr history of the Galactic disk, most stars known to have giant planetary companions are, in contrast, younger than 5 Gyr. Systems with star–planet separations < 0.4 AU have a significantly flatter age distribution, suggesting that they are dynamically stable on time scales of many Gyr. If the frequency of terrestrial planets is furthermore correlated with stellar metallicity (§8.4), then the median age of such planetary systems is likely to be around 3 Gyr.

Melo et al. (2006) examined the ages of five stars hosting very short orbital period ($P < 3$ d) transiting planets, to establish whether their non-detection in radial velocity surveys is due, as a class, to them be-

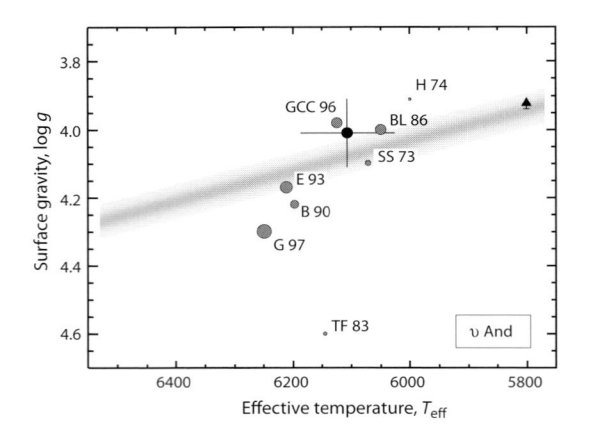

Figure 8.6: $\log g$ versus T_{eff} for υ And. The black circle, with error bars, marks the spectroscopically derived values. The inclined greyscale bar indicates the most probable parameter space allowed by the accurate Hipparcos distance. Grey circles represent results from earlier work, with diameters proportional to derived metallicity. A systematic shift of the bar as a result of different metallicity scales is indicated by the vertical arrow for a decrease in [Fe/H] by 0.1 dex. From Fuhrmann et al. (1998, Figure 2), reproduced with permission © ESO.

Table 8.1: An incomplete list of some of the larger compilations of [Fe/H] relevant to exoplanet investigations, mostly for FGK dwarfs. The compilations often provide other parameters such as $\log g$ and T_{eff}. $N_\star$ is the total sample size, N_{p} is the number of planet stars hosts when quoted. References are for the latest descriptions in the case of progressively enlarged samples.

$N_\star$	N_{p}	Reference
3356	–	Cayrel de Strobel et al. (2001)
14 000	–	Nordström et al. (2004b)
1040	99	Valenti & Fischer (2005)
160	27	Takeda & Honda (2005)
100 000	–	Ammons et al. (2006)
216	55	Luck & Heiter (2006)
1907	–	Robinson et al. (2007)
118	28	Bond et al. (2008)
451	–	Sousa et al. (2008)

ing rapidly evaporated due to a high ultraviolet flux of their (presumably young and hot) host stars. From their chromospheric activity indices, none were found to be younger than 0.5 Gyr.

8.4 Element abundances

Chemical abundance analysis, using high-resolution high signal-to-noise spectroscopy, provides a fundamental diagnostic of host star properties, and an important if indirect probe of planetary formation and subsequent evolution.

8.4.1 Metallicity

An important aspect of a star's chemical composition is the fraction of *metals* (i.e. elements heavier than He in astronomy usage). Iron abundance, expressed as [Fe/H], is frequently used as the reference element for exoplanet host star studies (e.g. Fuhrmann et al., 1997, 1998; Gonzalez, 1997, 1998; Gonzalez et al., 1999; Gonzalez & Laws, 2000; Giménez, 2000; Murray et al., 2001; Santos et al., 2001; Murray & Chaboyer, 2002; Laws et al., 2003; Santos et al., 2003, 2004c, 2005, and others).

The abundances of other elements are providing increasingly valuable diagnostics (e.g. Gonzalez, 1998; Gonzalez & Vanture, 1998; Gonzalez & Laws, 2000; Santos et al., 2000; Gonzalez et al., 2001b; Smith et al., 2001; Sadakane et al., 2002; Zhao et al., 2002; Bodaghee et al., 2003; Ecuvillon et al., 2004a,b, and others).

Already from the earliest studies of just four systems (51 Peg, 55 Cnc, υ And, and τ Boo) it appeared that stars hosting planets have significantly higher metal content than the average solar-type star in the solar neighbourhood (Gonzalez, 1997). While the Sun and other nearby solar-type dwarfs have [Fe/H] ~ 0 (Reid, 2002), typical exoplanet host stars have [Fe/H] ≳ 0.15. Values of [Fe/H] = +0.45 for two early discoveries, 55 Cnc and 14 Her, placed them amongst the most metal-rich stars in the solar neighbourhood (Gonzalez & Laws, 2000).

Although planets around even very low-metallicity stars have since been found, an overall correlation between metallicity and planet occurrence has been confirmed by subsequent work, using different samples and different analysis procedures.

Comparison stars Consistent agreement in determining effective temperatures and metallicities has proven notoriously difficult. Published results for a given star frequently formally disagree, as a combined result of differing spectral resolution, the choice of spectral lines, analysis procedures, and the adopted scales of metallicity and T_{eff}. For the case of υ And shown in Figure 8.6, for example, nine publications pre-2000 give values spanning the range [Fe/H]= −0.23, T_{eff} = 6000 K (Hearnshaw, 1974) to [Fe/H]= +0.17, T_{eff} = 6250 K (Gonzalez, 1997).

To establish statistical differences between stars with and without planets at the level of 0.1–0.2 dex, a secure sample of comparison stars is required. The comparison sample should be demonstrably companion-free, and analyses for both samples should preferably be based on the same sets of spectroscopic lines, observed and analysed in the same way. A number of such spectroscopic host star samples and comparison sets have now been constructed and investigated.

Santos et al. (2001) first presented a spectroscopic study of a volume-defined set of 43 F8–M1 stars within 50 pc included in the CORALIE programme, and for which constant radial velocities over a long time interval provided evidence that the comparison stars are planet-free. A further 54 comparison stars were added by Santos et al. (2005), yielding two large and uniform samples of

8.4 Element abundances

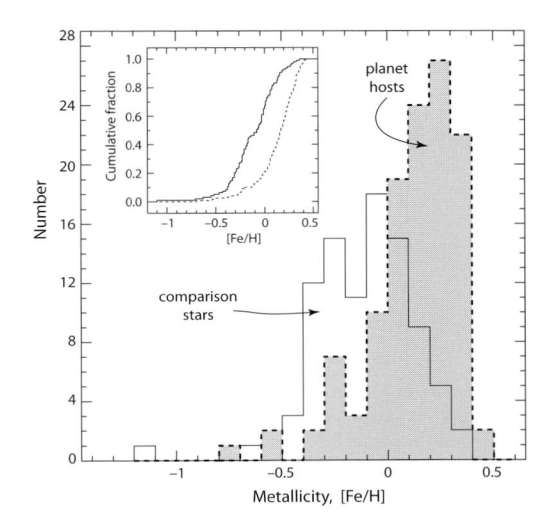

Figure 8.7: Metallicity distribution for 119 planet-host stars (shown as the dashed line, shaded), and for a volume-limited comparison sample of 94 stars with no known planets (continuous line, unshaded). The average metallicity difference of the two samples is 0.24 dex. Inset: cumulative distribution functions. A statistical Kolmogorov–Smirnov test shows that the probability that both distributions belong to the same population is $\sim 10^{-12}$. From Santos et al. (2005, Figure 1), reproduced with permission © ESO.

119 planet-host stars, and 94 stars without known planets, all of which have accurate stellar parameters and [Fe/H] estimates. These samples have been the basis of various subsequent abundance analyses (Santos et al., 2003; Bodaghee et al., 2003; Santos et al., 2004c; Israelian et al., 2004; Gilli et al., 2006). A further 64 comparison stars were added by Sousa et al. (2006).

Other large uniform spectroscopic surveys of exoplanet host stars and comparison stars include (see also Table 8.1): the 99 planet host stars from the 1040 FGK dwarfs of the Keck, Lick, and AAT programme, selected according to magnitude, colour, and luminosity (Valenti & Fischer, 2005); the 27 planet host star and 133 comparison stars observed at Okayama (Takeda et al., 2005; Takeda & Honda, 2005); the 28 planet host stars and 90 comparison stars from the Anglo–Australian planet search programme (Bond et al., 2006, 2008); and the 216 star sample of the 'nearby stars project' (Heiter & Luck, 2003; Luck & Heiter, 2005, 2006).

Luck & Heiter (2006) detail the overlap between these and other samples, including the extensive Geneva–Copenhagen spectroscopic and kinematic survey (Nordström et al., 2004a,b).

Occurrence versus metallicity A comparison based on the host star and control samples of Santos et al. (2005) confirmed previous indications that the frequency of giant planets rises as a function of [Fe/H] (Figure 8.7). Similar results were found by Fischer & Valenti (2005), who gave the incidence of Doppler-detected giant planets as < 3% for [Fe/H] < −0.5, and 25% for [Fe/H] > +0.5. Over the range −0.5 < [Fe/H] < 0.5, and for FGK-type main-sequence stars, they expressed the probability of formation of a gas giant planet, with orbital period shorter than 4 yr and $K > 30\,\mathrm{m\,s^{-1}}$, as

$$P(\mathrm{planet}) = 0.03 \times 10^{2.0[\mathrm{Fe/H}]}$$
$$= 0.03 \left[\frac{N_{\mathrm{Fe}}/N_{\mathrm{H}}}{(N_{\mathrm{Fe}}/N_{\mathrm{H}})_\odot} \right]^2, \qquad (8.16)$$

the second expression following from the definition of [Fe/H] (box on page 186).

As discussed further below, the correlation between occurrence and metallicity may not extend to giant stars, to stars of intermediate metallicity, to M dwarfs, or to the occurrence of low-mass planets.

Transiting planets The correlation between metallicity and occurrence appears to extend to the close-in giant planets discovered by transit photometry.

To explain the observed radius anomalies for transiting planets known at the time (including HD 209458 and OGLE–TR–10 considered to be anomalously large, and HD 149026 considered to be anomalously small), Guillot et al. (2006) suggested an exoplanet composition/evolution model which included an additional internal energy source equal to 0.5% of the incoming stellar luminosity. This additional heat source acts to slow the cooling of the planet.

With this adjustment to bring the radii into better consistency with theoretical models, they showed that for the nine transiting planets known at the time, the amount of heavy elements that had to be added to match their observed radii was a steep function of the host star metallicity: from less than $20M_\oplus$ of heavy elements around stars of solar composition, to up to $100M_\oplus$ for stars with three times the solar metallicity (Figure 8.8). These results add to the picture that heavy elements play a key role in the formation of close-in giant planets.

A uniform determination of spectroscopic parameters for 13 host stars of transiting planets was made by Ammler-von Eiff et al. (2009), and supplemented by a compilation of results for a total of 50 transit host stars. A systematic offset in the abundance scale was found for the TrES and HAT objects.

Giant stars Pasquini et al. (2007) found that planet occurrence around a sample of 14 giant stars does not correlate with increasing metallicity, in contrast with main sequence stars. While they favoured an explanation based on the accretion of metal-rich material (§8.4.3), other interpretations are also possible, perhaps due to differences in migration, or to the presence of a dual-formation mechanism (Matsuo et al., 2007) with a metal-independent mechanism more effective for large

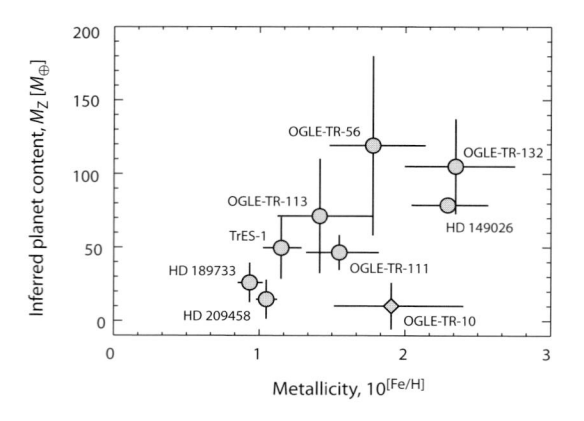

Figure 8.8: Mass of heavy elements in the transiting planets as a function of the metal content of the host star. The mass of heavy elements is that required to fit the observed radii, calculated on the basis of evolution models including an additional heat source assumed equal to 0.5% of the incoming stellar heat flux. From Guillot et al. (2006, Figure 3), reproduced with permission © ESO.

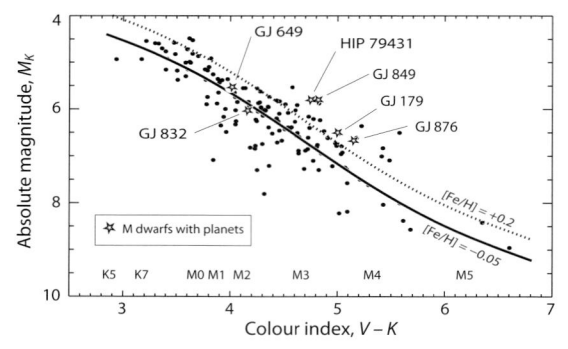

Figure 8.9: Nearby low-mass stars from the Keck sample in the M_{K_S} versus $V - K_S$ plane, with the corresponding spectral types shown at bottom. Small black circles indicate a volume-limited sample of single K dwarfs ($d < 20$ pc) and M dwarfs ($d < 10$ pc). The solid line is a fifth-order polynomial fit to the mean main sequence, and corresponds to roughly solar metallicity. Open symbols indicate all M dwarfs known to harbour at least one giant planet. From Johnson et al. (2010c, Figure 1), reproduced by permission of University of Chicago Press.

stellar masses. The choice of metallicity scale may also be a contributing factor (Santos et al., 2009). A similar trend was found for 322 late-G giants, including 10 planet host stars, by Takeda et al. (2008b).

Stars of intermediate metallicity Haywood (2008) used α-element abundances, and the Galactic velocity components (notably the component V in the direction of Galactic rotation), to classify the fourteen exoplanet host stars in the metallicity range -0.7 to -0.2 dex according to their membership of the thin disk or thick disk populations. All but one belong to the thick disk, and just one to the metal-poor tail of the thin disk. A similar result for older, lower metallicity host stars with enhanced [α/Fe] had been noted by Reid et al. (2007).

The classification by population is possible because, at these intermediate metallicities, stars in the solar vicinity fall into two main groups: the thin disk being solar in α-elements and rotating faster than the *local standard of rest* (viz. the velocity of a hypothetical group of stars in strictly circular orbits at the solar position), while the thick disk is enriched in α-elements, [α/Fe] >0.1 dex, and lags the local standard of rest.

The distinct properties of the thin and thick disk in terms of α-element enrichment as a function of metallicity are illustrated in the studies of e.g. Fuhrmann (1998), and Reddy et al. (2003, 2006).

M dwarfs The complex spectra of low-mass M dwarfs precludes the use of standard LTE spectroscopic modeling, and the knowledge of their metallicity distribution has been based until recently on photometric calibration (Bonfils et al., 2005a). This calibration originally suggested that M dwarfs in the solar neighbourhood, including those with known planets, are systematically metal poor compared to their higher-mass counterparts (Bean et al., 2006).

Johnson & Apps (2009) derived a revised metallicity calibration based on $V - K_S$ photometry of a volume-limited sample with common proper motion companions and found that, in contrast, M dwarfs with planets appear to be systematically metal rich. The mean metallicity for their M dwarf sample with planets is Fe/H = +0.16, compared with +0.15 for the FGK dwarfs with planets. The result brings the M dwarfs into a consistent pattern of metallicity excess being correlated with planet occurrence. There is still a systematically lower fraction of Jovian planets around M dwarfs than FGK dwarfs at any given metallicity, a result likely to be a reflection of their lower stellar masses, rather than an effect of metallicity.

By late 2010, seven Doppler-detected giant planets were known around six M dwarfs (Figure 8.9). From their volume-limited Keck sample, Johnson et al. (2010c) estimate that $3.4^{+2.2}_{-0.9}\%$ of stars with $M_\star < 0.6 M_\odot$ host planets with $M \sin i > 0.3 M_J$ and $a < 2.5$ AU. Restricted to metal-rich stars with [Fe/H] > +0.2, the occurrence rate rises to $10.7^{+5.9}_{-4.2}\%$.

Neptune-mass planets The first Doppler detections of low-mass planets already suggested that their occurrence might show a different dependence on the host star metallicity than the case for giant planets (e.g. Udry et al., 2006).

Sousa et al. (2008) derived [Fe/H] for 451 stars from the HARPS high-precision sample. They found that, in contrast to the giant Jupiter-mass planets, Neptune-like planets do not form preferentially around metal-rich stars, with the ratio of Jupiter to Neptune mass planets being an increasing function of metallicity. The number of Neptune mass planets is currently only small, and most hosts with only Neptune-mass planets are in any case M dwarfs.

8.4.2 Possible biases

That the planetary occurrence trend versus metallicity might arise from a variety of observational biases has been considered in some detail. Specific sample selection criteria certainly affect the resulting statistics, although the studies cited below suggest that none of the possible effects identified is likely to explain the observed probability–metallicity correlation.

Selection bias Most radial velocity surveys are biased against very young stars with their higher rotational velocities or higher chromospheric activities, effects quantified in this context by Paulson & Yelda (2006). They are also biased against multiple star systems, and against metal poor stars, although the latter are in any case only poorly represented in the solar neighbourhood.

Magnitude-limited and volume-limited surveys result in different representations of stars of different spectral types, as discussed by Fischer & Valenti (2005). But even the distance-limited criterion of Santos et al. (2005) is not a volume-limited survey, since their sample of F8–M1 stars within 50 pc does not include all late G, K, and M dwarfs within that distance.

A more subtle bias against low-mass high-metallicity stars results from a magnitude cut-off, while a colour cut-off gives the opposite (Murray & Chaboyer, 2002).

Orbital period bias Radial velocity surveys are themselves biased in detecting planets with shorter orbital periods and larger masses. The observed metallicity correlation could therefore reflect a dependence of orbital radius on metallicity, perhaps signaling a dependence of migration rate on metallicity (Gonzalez, 2003; Sozzetti, 2004; Gonzalez, 2006a).

Observationally, the weak correlation between metallicity and orbital period first reported by Sozzetti (2004) has not been confirmed. In later studies with larger samples, there is no evident correlation between metallicity and semi-major axis (Fischer & Valenti 2005, their figure 15; Jones et al. 2008b, their figure 6.8), nor indeed with eccentricity or planet mass (Bond et al., 2008, their figure 4).

Neither is there any theoretical expectation yet invoked of a link between metallicity and migration rate (Livio & Pringle, 2003). Pinotti et al. (2005) have proposed a model that correlates the metallicity of the host star with the semi-major axis of its most massive planet prior to migration. Further out, the formation of debris-generating planetesimals at tens of AU may still be independent of the metallicity of the primordial disk (Greaves et al., 2006).

In summary, there remains the possibility that the majority of giant planets currently known have experienced significant migration, and therefore that current Doppler surveys still sample only a specific subset of the overall exoplanet population.

8.4.3 Origin of the metallicity difference

Excluding the possibility of selection bias, two principal hypotheses have been put forward to explain the connection between high metallicity and the presence of massive planets: either causative as a consequence of higher primordial abundances facilitating accretion, or by self-enrichment as a result of the capture of metallicity-enhanced material. Based on the additional evidence described below, the current consensus is that while some host stars may be polluted to some degree by material capture, the dominant effect is likely to be primordial, with planets simply more likely to form around metal-rich stars.

However, the anomalies noted above – the absence of a similar correlation for giant stars, for low-mass planets, and for stars of intermediate metallicity – point to the picture being incomplete. A collective explanation may lie in the possible migration of giant planet hosts from the inner regions of the Galaxy. These various possibilities are considered further in this section.

Primordial occurrence According to this hypothesis, the high metallicity observed in certain hosts is a bulk property of the star, and represents the original composition out of which the protostellar and protoplanetary molecular clouds formed.

In this picture, the higher the metallicity of the primordial cloud, the higher the proportion of dust to gas in the protoplanetary disk. This facilitates condensation and accelerates protoplanetary accretion before the gas disk is lost (Pollack et al., 1996). Giant planets are subsequently formed by runaway accretion of gas onto such rocky cores with $M \sim 10 M_\oplus$, rather than by gravitational instabilities in a gaseous disk which predicts formation much less sensitive to metallicity (Boss, 2002). The cut-off in the metallicity distribution for host stars at [Fe/H] $\gtrsim 0.5$ (Figure 8.7) then represents the upper limit to metallicities in the solar neighbourhood.

Observationally, the probability of forming a giant planet appears to be proportional to the square of the number of Fe atoms (Equation 8.16). Since particle collision rates are similarly proportional to the square of the number of particles, this result has been further used to argue a physical link between dust particle collisions in the primordial disk and the formation rate of giant planets. Based on the core accretion model, Kornet et al. (2005), Wyatt et al. (2007), and Ida & Lin (2004b, 2005a) were able to reproduce the distribution of giant planets with host star metallicity. The latter model also predicts that short-period giant planets should be rare around M dwarfs, but that Neptune mass ice-giant planets might be common, a trend which is broadly apparent in M dwarf Doppler surveys (Endl et al., 2006).

If metallicity determines the time scale for giant planet formation, there should be a correlation between planet mass and metallicity. Rice & Armitage (2005) showed some evidence for this correlation, although the same trend is not evident in the studies of, e.g., Bond et al. (2008, their Figure 4).

Self-enrichment An alternative explanation is that the high metallicity is a phenomenon restricted to the surface region of the star, arising from the capture of metal-rich material, and the resulting 'pollution' of its outer convective envelope. This might be the result of the terminal inward migration of a planet onto the star as a result of dynamical friction (Laughlin & Adams, 1997; Gonzalez, 1998; Siess & Livio, 1999b; Israelian et al., 2001; Sandquist et al., 2002; Israelian et al., 2003), self-pollution due to the transfer of gas-depleted, metal-rich material from the disk to the star as a result of migration (Goldreich & Tremaine, 1980; Lin et al., 1996; Laughlin, 2000; Gonzalez et al., 2001b), or to the break-up and infall of planets or planetesimals onto the star due to gravitational interactions with other companions (Rasio & Ford, 1996; Queloz et al., 2000b; Quillen & Holman, 2000; Quillen, 2002).

A planet added to a fully convective star would be folded into the entire stellar mass, and would lead to a negligible overall metallicity enhancement. However, main sequence solar mass stars like the Sun have radiative cores with relatively small outer convection zones comprising only a few percent of the stellar mass. At ages of $\gtrsim 10^8$ yr the Sun's outer convection zone had reduced to $\sim 0.03 M_\odot$ (Ford et al., 1999, and Figure 8.10). For higher mass stars, the convection zone is smaller still ($0.006 M_\odot$ at $1.2 M_\odot$), so that the sensitivity of surface metallicity to accreted matter rises steeply for stars more massive than the Sun.

Under these conditions, planet capture could significantly enhance the heavy element content in the convective zone, resulting in elemental abundances deviating from the underlying trends resulting from Galactic chemical evolution (Figure 8.10). Planet hosts having the shallowest convection zones would be expected to have the highest metallicities (Laughlin & Adams, 1997; Ford et al., 1999; Siess & Livio, 1999b; Pinsonneault et al., 2001; Li et al., 2008b).

Studies have pursued various implications of the self-enrichment model. Murray & Chaboyer (2002) estimated that some $5 M_\oplus$ of iron would have to be slowly accreted over some 50 Myr to explain the observed mass–metallicity and age–metallicity relations. Sandquist et al. (1998) showed that a capture event may also influence the further orbital migration of other

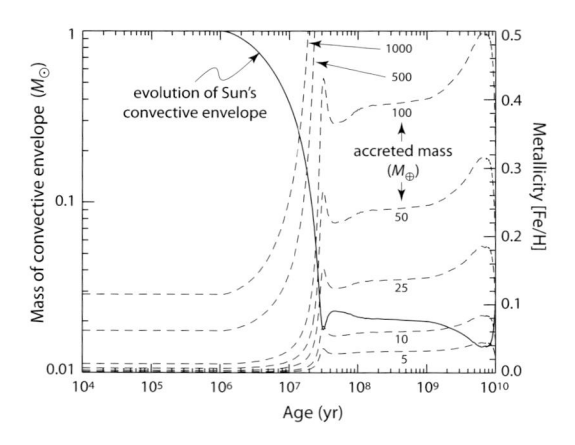

Figure 8.10: High metallicity arising from pollution. The solid line (left axis) shows the evolution of the Sun's convective envelope. Dashed lines (right axis) indicate the surface metallicity that would result from the instantaneous accretion of rocky material onto the star at each time in the Sun's past (masses in $M_\oplus$), assuming that the accreted material is mixed throughout the convective envelope. Producing high surface metallicities of 0.2–0.3 dex is possible with the accretion of $\sim 10-25 M_\oplus$ of rocky material after 10^7 yr. From Ford et al. (1999, Figure 5), reproduced by permission of the AAS.

planets in the system due to changes in angular momentum or magnetic field. Sandquist et al. (2002) argued that an infalling planet could penetrate the convection zone. Cody & Sasselov (2005) developed a stellar evolution code to model stars with non-uniform metallicity distributions, motivated by the phenomenon.

One important test of the enrichment hypothesis makes use of the fact that when a star leaves the main sequence, its convection zone deepens significantly. This would lead to strong dilution if the high metallicity is a result of surface pollution. Fischer & Valenti (2005) used their sample of 1040 nearby FGK dwarfs (Valenti & Fischer, 2005), including 99 planet hosts, to show that there is no correlation between planet host metallicity and the convection zone depth, either while the star is on the main sequence, or after it evolves to the subgiant branch and its convection zone deepens (Figure 8.11). This result has been taken to support the primordial basis of the observed correlation.

Another test of self-enrichment is to search for compositional differences between common proper motion pairs (i.e. binary components with large separations), which presumably formed together out of the same molecular cloud with the same chemical composition. Accretion has been invoked to explain abundance anomalies in the binary 16 Cyg A and 16 Cyg B (where the planet orbits the B component), which have very different Fe abundances and Li content (Gonzalez, 1998; Laws & Gonzalez, 2001). Similar arguments were given for HD 219542 (Gratton et al., 2001), although the exoplanet host status of the latter was subsequently re-

8.4 Element abundances

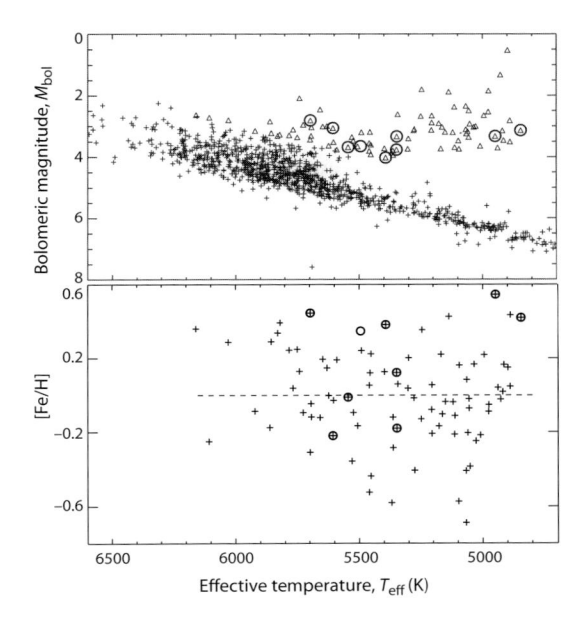

Figure 8.11: Top: Hertzsprung–Russell diagram for stars from the Keck, Lick, and AAT planet search projects, as of 2005. Key: + = main sequence stars; △ = subgiants (M_{bol} > 1.5 mag above the lower main sequence boundary); circled symbols: subgiants with detected planets. Bottom: subgiants, with detected planets circled. If the high metallicity correlated with the presence of Jovian planets is limited to the convective envelope of main sequence stars, subgiants with planets should show progressively lower metallicity as a result of dilution as they evolve across the subgiant branch. No such gradient is observed. From Fischer & Valenti (2005, Figures 11–12), reproduced by permission of the AAS.

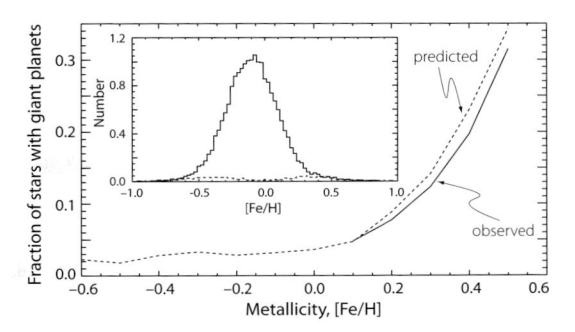

Figure 8.12: Inset: the simulated 'local' metallicity distribution (main histogram). The contributions of the metal-rich and metal-poor components assumed to have come to the solar neighbourhood by radial migration are shown by the lower dashed line. Main: the predicted fraction of stars with giant planets obtained assuming the metallicity distribution and intrinsic giant planet proportion of 0% in the metal-poor component, 5% locally, and 25% in the metal rich component (dashed line). The solid line is the fraction of planet hosts versus stellar metallicity according to Udry & Santos (2007). From Haywood (2009, Figure 4b), reproduced by permission of the AAS.

tracted (Desidera et al., 2003). Fifty common proper motion pairs were studied by Desidera et al. (2004) and Desidera et al. (2006), and nine pairs by Luck & Heiter (2006). The implications for the reported differences remains unclear.

Other evidence for accretion has been attributed to the presence of ^{6}Li, discussed further in §8.4.7, in the metal-rich dwarf HD 82943 (Israelian et al., 2001, 2003), in 59 Vir (Fuhrmann, 2004), in non-planet hosting stars including the super-lithium-rich F dwarf J37 in NGC 6633 (Laws & Gonzalez, 2003; Ashwell et al., 2005), and for various white dwarfs (Jura, 2006).

Notwithstanding what is now a general consensus that the occurrence–metallicity correlation is essentially primordial, material accretion would appear to be an inevitable by-product of planet formation and evolution. Certainly, the Sun continues to accrete cometary material today: accretion of up to $100 M_\oplus$ of metal-rich material was proposed as a partial resolution of the solar neutrino problem (Jeffery et al., 1997), but such a mass is probably ruled out by solar models, where the differences between solar photospheric and meteoritic abundances display a weak but significant trend with con-

densation temperature, suggesting that the metallicity of the Sun's envelope has been enriched relative to its interior by about 0.07 dex (Gonzalez, 2006b).

Different Galactic origins As noted above, the correlation between stellar metallicity and the occurrence of giant planets appears to break down for giant stars, and for stars of intermediate metallicity for which giant planets are found preferentially orbiting thick disk stars. Haywood (2008, 2009) has suggested that the explanation underlying the giant planet occurrence–metallicity correlation is a dynamical manifestation related to the migration of stars in the Galactic disk. Giant planet formation is then hypothesised to correlate with Galactocentric distance, rather than being primarily linked to metallicity, according to the following picture.

Most metal-rich stars (Fe/H > +0.25 dex) found in the solar neighbourhood, including those hosting planets, are considered to have migrated from the inner disk, i.e. from within the solar Galactocentric radius, by the effect of radial mixing (Sellwood & Binney, 2002). Given a Galactic radial metallicity gradient of 0.07–0.1 dex kpc^{-1} (e.g. Edvardsson et al. 1993; Wielen et al. 1996 and references; Maciel & Costa 2009), stars with a mean metallicity [Fe/H] = +0.35 will have originated at about 3–5 kpc from the Sun in the direction of the Galactic centre. If 25% of stars at this location systematically host giant planets, independent of metallicity, then the observed correlation between occurrence and metallicity follows from dilution introduced by radial mixing.

The origin of the Sun is not inconsistent with this picture. Wielen et al. (1996) inferred its birthplace at a Galactocentric radius of $R_{i,\odot}$ = 6.6 ± 0.9 kpc, based on its metallicity which is larger by 0.17 ± 0.04 dex than the av-

Thin and thick disk populations: While dominated by thin disk stars, the local stellar population contains approximately 5–10% of thick disk stars, and some 0.1–0.5% of halo stars. Any studies of the local disk population are therefore intimately connected with the determination and segregation of the ages, chemical composition, and velocity structures of the various component populations. Significant sub-structure in the local disk is also present in the form of open clusters, moving groups, and associations, including the Gould Belt.

The recognition of the existence of stellar populations differing in age, chemical composition, spatial distribution, and kinematic properties, represented a breakthrough in the knowledge of Galactic structure, and underpins the basis for recent models of galaxy formation and evolution. An extensive and complex literature now exists on the topic of the Galaxy's disk: its separation into thin and thick disk components, whether they represent discrete or continuous populations, their respective scale heights, their kinematic, metallicity, and age properties, the relationship between them, and their origin.

The vertical distribution of the different populations is frequently described either in terms of a *characteristic thickness*, defined as the ratio of the surface density (integrated over disk thickness) to its volume density at the Galactic plane, or in terms of a *scale height*, z_h, defined by $\exp(-z/z_h)$ for an exponential distribution. The thin and thick disks have characteristic thicknesses of 180–200 pc and 700–1000 pc respectively, with the interstellar medium having a scale height of about 40 pc (Dehnen & Binney, 1998). Even for the thin disk, however, its scale height is different for different classes of stars, with old stars found at greater distances from the plane partly as a result of disk heating in which the irregular gravitational field of spiral arms and molecular clouds gradually increases their random velocities over time.

Characteristic Galactic rotational velocities and velocity dispersions are of order $\langle V_{\rm rot} \rangle = 205, 180 \pm 50, 20$ and $\sigma_{uvw} = 20, 50, 100$ for the thin, thick and halo components respectively (Reid, 1998), where the velocity components uvw are taken conventionally towards the Galactic centre, in the direction of Galactic rotation, and towards the north Galactic pole, respectively. Numerous determinations can be found in the literature (e.g. Soubiran et al., 2003; Bensby et al., 2004).

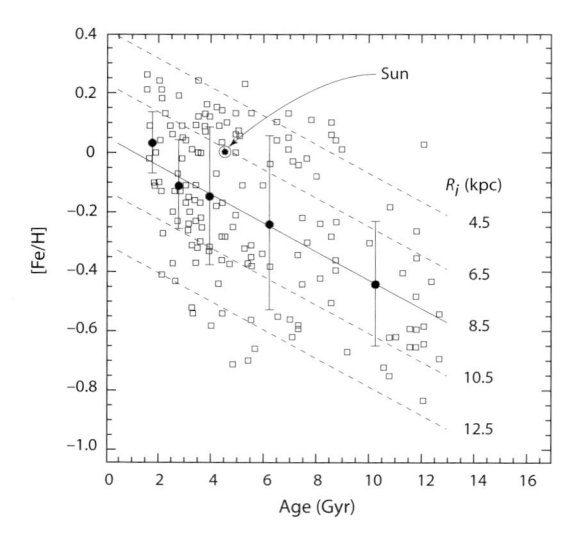

Figure 8.13: Places of formation of stars which are now nearby. Metallicity [Fe/H] is plotted against age for stars from the sample of Edvardsson et al. (1993). The position of the Sun is indicated. Inclined lines are of constant Galactocentric distance for their derived age–metallicity relation, such that the place of formation, R_i, can be determined for each star, including the Sun. Mean metallicities as a function of age are shown as filled circles. From Wielen et al. (1996, Figure 3), reproduced with permission © ESO.

erage of nearby stars of solar age, combined with a similar radial Galactic metallicity gradient. This is also consistent with its space motion, and has similar implications for the origin of nearby high-metallicity stars (Figure 8.13).

Haywood (2009) has suggested that planet formation is related, not to metallicity, but to the presence of molecular hydrogen in the form of the Galaxy's molecular ring (Clemens et al., 1988; Jackson et al., 2006). This provides a large reservoir of H_2, itself considered to be directly linked to star formation (e.g. Kennicutt, 2008). At its maximum density of 2–5 times its local density, 3–5 kpc from the Sun, planets are then expected to form preferentially. In this picture, the region of enhanced giant planet formation 'happens' to correspond to the metallicity range of 0.3–0.5 dex. Combined with radial mixing, this model offers consistency with the proportion of giant planets found around metal-rich and solar-metallicity stars (Figure 8.12).

In this picture, the giant star giant-planet hosts contain only a limited bias towards metal-rich objects because they are typically younger than the dwarfs, with ages of < 1 Gyr (Takeda et al., 2008b). Their relative youth implies that they are therefore less contaminated in metallicity by radial mixing.

The hypothesis has a further observational consequence: stars hosting non-giant planets, i.e. of Neptune or Earth mass, may form in less dense H_2 environments, such that a predominance of metal-rich stars among the Neptune/super-Earth hosts is not expected. Such behaviour has been confirmed in the HARPS results noted previously (Sousa et al., 2008).

8.4.4 Refractory and volatile elements

Planet formation involves *condensation*, the change from gaseous phase into the liquid or solid phase of the same element or chemical species. This involves the loss of kinetic energy by collision, or by adsorption onto an existing, colder, condensation centre.

In planetary science, elements and compounds with high equilibrium condensation temperatures are re-

8.4 Element abundances

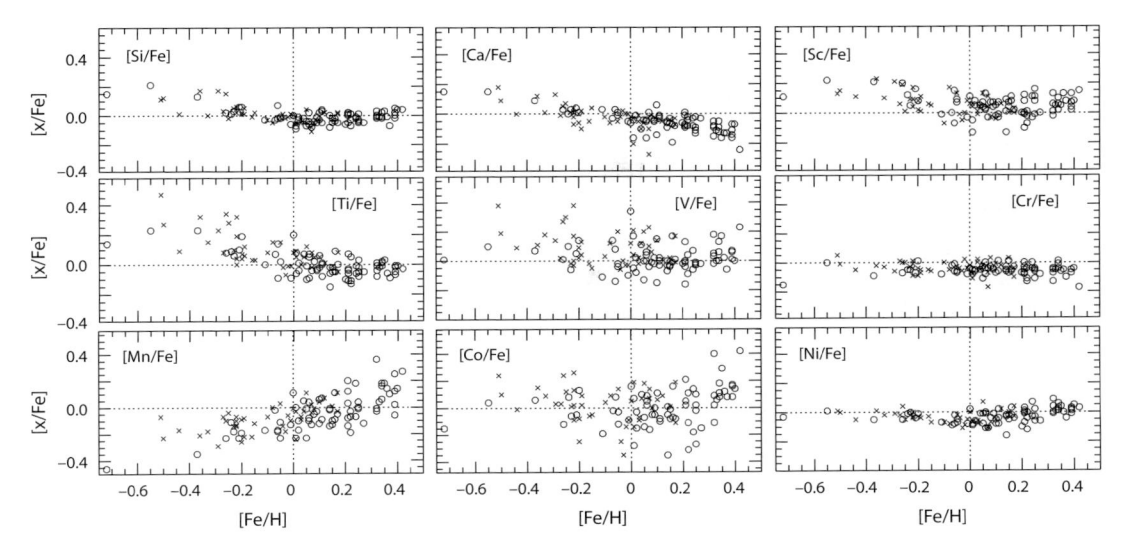

Figure 8.14: [x/Fe] versus [Fe/H] for nine (moderately) refractory elements. Open circles denote the planet-host stars, and crosses represent comparison stars. Dashed lines represent values for the Sun. From Bodaghee et al. (2003, Figure 2), reproduced with permission © ESO.

ferred to as *refractory*, while those with low condensation temperatures are referred to as *volatile*.[1]

Of the elements which have been most studied in the context of planet host stars, the (moderately) refractory include Al, Ca, Ti, and V; those with intermediate condensation temperatures include Co, Fe, Mg, Ni, and Si; while the (moderately) volatile include C, Cu, N, Na, O, S, and Zn.

If the correlation between occurrence and metallicity originates from primordial abundances, then similar occurrence trends may be expected for metals other than iron. Self-enrichment should, in contrast, result in an overabundance of refractory elements deposited in the stellar photosphere, and a reduced abundance of volatile elements due to preferential evaporation (Smith et al., 2001).

Refractory elements For metals in general, values of [x/H] typically show enhancements for stars with planets, as in the case of [Fe/H]. Studies of the relative abundances of refractory or moderately refractory elements have conveyed a less clear-cut picture when comparing values of [x/Fe] for planet hosts to those of comparison stars of the same [Fe/H].

Takeda et al. (2001) and Beirão et al. (2005) reported no significant trends in the various elements which they studied. Gonzalez et al. (2001b) found smaller values of Mg and Al (as well as Na), with no significant differences for Si, Ca, and Ti, but with less significant differences found in later studies (Gonzalez, 2006b). Sadakane et al. (2002) reported some host stars with the volatile elements C and O underabundant with respect to the refractories Si and Ti. Huang et al. (2005) reported a possible Mg enhancement.

A comparison for 77 planet hosts and 42 comparison stars was made by Bodaghee et al. (2003), with results for Si, Ca, Sc, Ti, V, Cr, Mn, Co and Ni shown in Figure 8.14. Their conclusion is that the abundance trends of the refractory elements, [x/Fe] versus [Fe/H], for planet hosts are largely identical to those for the comparison stars at the corresponding (high) values of [Fe/H]. The work was extended to 101 stars with and 93 without known planetary companions by Gilli et al. (2006).

Some evidence for Si and Ni enrichment of planet host stars compared with the general metal-rich population was reported by Robinson et al. (2006a) using the 1040 FGK dwarf sample of Valenti & Fischer (2005), but these results were not confirmed by Gonzalez & Laws (2007). The latter studied 18 elements in 31 host

---

[1] More details can be found in, e.g., Larimer (1988); Cowley (1995); Taylor (2001), and the topic is considered further in §11.2.2. Briefly, the demarcation is set by the condensation temperature of the 'common elements' Mg, Si, and Fe, while condensation temperatures are themselves (weakly) pressure dependent. Further sub-classification, is somewhat arbitrary and defined variously. For example, Taylor (2001) uses very volatile (condensation temperatures < 700 K), volatile (700–1100 K), moderately volatile (1100–1300 K), moderately refractory (1300–1500 K), refractory (1500–1700 K), and super-refractory (> 1700 K). A slightly different classification is suggested by Lodders (2003), who uses: highly volatile (< 371 K), volatile (371–704 K), moderately volatile (704–1290 K), intermediate (1290–1360 K), refractory (1360–1500 K), highly refractory (1500–1650 K), and ultra-refractory (> 1650 K).

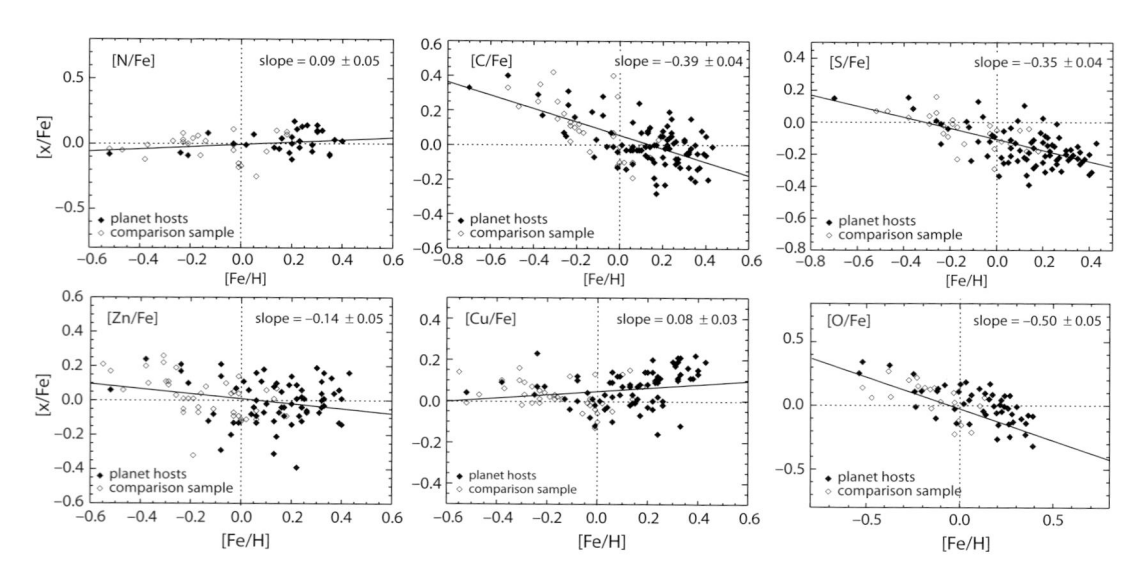

Figure 8.15: [x/Fe] versus [Fe/H] for the (moderately) volatile elements N, C, S, Zn, Cu, O. Planet host stars are shown as filled diamonds, and comparison stars as open symbols. Linear least-squares slopes apply to all objects. From Ecuvillon et al. (2004b, Figure 7b) for N; Ecuvillon et al. (2004a, Figures 8–11) for C, S, Zn, Cu; and Ecuvillon et al. (2006b, Figure 10e) for O, reproduced with permission © ESO.

stars, finding some differences between their comparison sample in the case of Al, Si, and Ti, but also demonstrating some inconsistencies with previous results.

Notwithstanding some small and possibly important abundance differences, the broad conclusion is that planet-hosting stars are largely indistinguishable from other Population I stars in their enrichment histories of refractory elements. One implication of this result is that no extraordinary chemical events, such as a nearby supernova, are necessary to stimulate planet formation.

Volatile elements C and O are significant opacity sources in stars, and important in the chemistry of protoplanetary disks. Their forbidden lines are preferentially used as abundance indicators, but they are weak and blended, and demand high-quality spectroscopy.

Both Gonzalez & Laws (2000) and Santos et al. (2000) reported sub-solar values for [C/Fe] and [Na/Fe] in planet host stars, while subsequent investigations using inhomogeneous comparison samples gave a more uncertain picture of the trends for [C/Fe], [O/Fe], and [N/Fe] (Gonzalez et al., 2001b; Takeda et al., 2001; Sadakane et al., 2002; Takeda & Honda, 2005).

The uniform sample of planet hosts and comparison stars defined by Santos et al. (2001) was used in a series of studies (Ecuvillon et al., 2004a,b, 2006a,b) to derive abundances of N, C, S, Zn, Cu and O (Figure 8.15). While most elements show a decreasing trend of [x/Fe] with increasing [Fe/H], the overall conclusion is again that the abundance trends for planet host stars for the (moderately) volatile elements, [x/Fe] versus [Fe/H], are largely identical to those for the comparison stars at the corresponding (high) values of [Fe/H]. Some evidence for S enhancement was reported by Huang et al. (2005). No significant trends were found by Gonzalez (2006b).

With no compelling evidence of preferential volatile depletion amongst stars with planets, most investigators have concluded that the metal enrichment of planet hosts is primordial, and that stars hosting giant planets form preferentially in metal-rich molecular clouds.

Implications for terrestrial planet formation Gonzalez et al. (2001a) estimated that a metallicity at least half that of the Sun is required to build a habitable terrestrial planet, as dictated by heat loss, volatile element inventory, and atmospheric loss. The concentrations (with respect to Fe) of the radiogenic isotopes ^{40}K, ^{235}U, ^{238}U and ^{232}Th will affect the efficiency of plate tectonics (Urey, 1955), which may be an important recycling process providing feedback to stabilise temperatures on planets with oceans and atmospheres, while the relative abundances of Si and Mg (with respect to Fe) affect the mass of the core relative to the mantle. Since all these elements and isotopes vary with time and location within the Galaxy as a result of star formation activity, even planetary systems with the same metallicity as the Sun will not necessarily form habitable Earth-like planets. Regions of the Galaxy accordingly least likely to contain Earth-mass planets are the halo, the thick disk, and the outer thin disk. The bulge should contain Earth-mass planets, but with a different mix of elements compared to the Sun.

8.4 Element abundances

The r-process, s-process, and α elements: The elements are considered to have been broadly formed as follows (Burbidge et al., 1957; Thielemann, 2002): H in the Big Bang; He in the Big Bang and in stars; C and O in low- and high-mass stars; Ne–Fe ($Z = 10 - 26$) in high-mass stars; Co–Bi ($Z = 27 - 84$) in the s- and r-processes; and Po–U ($Z = 84 - 92$) in the r-process. The s- and r-processes (for slow and rapid neutron capture, with respect to the β-decay rate of the nuclei, respectively) are the two principal paths leading to the 'trans-Fe' elements.

The *s-process* involves neutrons liberated during core and shell He-burning being captured by a nucleus, with the neutron subsequently undergoing β-decay to produce a proton. It being easier to add the chargeless neutron to a nucleus than it is to add a proton directly, this results in the progressive build-up of elements up to Pb and Bi, starting on existing heavy nuclei around Fe. It is thought to occur mostly in asymptotic giant branch stars during the thermal pulse stage (e.g. Reddy et al., 2003).

The *r-process* involves the rapid addition of many neutrons to existing nuclei, which again decay into protons, producing the heavier elements. The r-process is a subset of explosive Si-burning, which differs strongly from its hydrostatic counterpart, and is thought to occur only in supernovae, and mostly in those of Type II, the end points of massive star evolution, rather than those of Type Ia, resulting from binary systems (Qian, 2003). Observational evidence is based on the existence of elements like Au, and the fact that in some of the oldest stars in the Galaxy, which were formed after only a small number of Type II supernovae had enriched the interstellar medium, the abundance of Fe is very low, while the abundances of r-process elements are anomalously high.

α-*elements* are those whose most abundant isotopes are integral multiples of the He nuclei or α-particle: (C, N, O), Ne, Mg, Si, S, Ar, Ca, and Ti. Type Ia supernovae predominantly produce elements of the iron peak (V, Cr, Mn, Fe, Co and Ni) as a result of normal freeze-out of charged particle reactions during cooling. Low-density freeze-outs, most pronounced in Type II supernovae, leave a large α abundance, and a higher proportion of α-elements. This includes O, so that O enhancement is well correlated with the presence of α-elements. C and N, as well as O, are sometimes included within the class since they are synthesized by nuclear α-capture reactions, although the enrichment of the interstellar medium by C and N is not due to Type II supernova explosions but due to stellar winds of the more massive asymptotic giant branch stars.

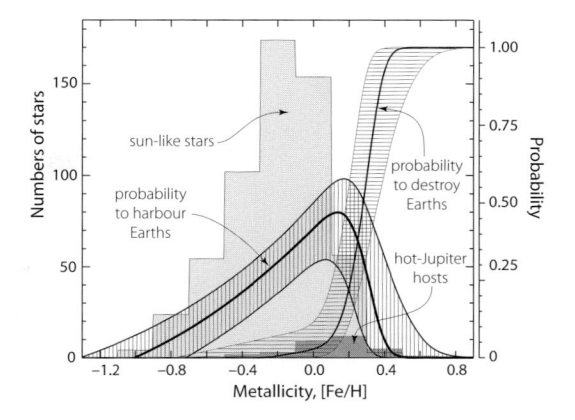

Figure 8.16: For a given metallicity, the probability of destroying Earths is taken to be the ratio of the number of hot Jupiter hosts to the number of stars surveyed. The probability of harbouring Earth-like planets is based on the assumption that the production of Earths is linearly proportional to metallicity, but is cut off at high metallicity by the increasing probability to destroy Earths. From Lineweaver (2001, Figure 1), with permission from Elsevier.

If stellar abundances reflect planetary abundances, spectroscopic host star studies may also represent a step in identifying terrestrial planet composition and evolution. Stars with different Mg/Si ratios may have terrestrial planets with differing compositions of the pyroxene-silicate mineral series (Mg,Fe)SiO3, or of the olivine series (Mg,Fe)$_2$SiO$_4$, compositions likely to affect volcanic activity and plate tectonics.

Lineweaver (2001) assumed that the probability of forming Earths is proportional to metallicity. With very low metallicity, Earths are unable to form, but with a very high metallicity, giant planets would dominate and destroy any planets of terrestrial mass. The resulting probability for a stellar system to harbour an Earth-like planet is shown in Figure 8.16. Combined with estimates of the star formation rate, and the gradual build-up of metals, Lineweaver (2001) derived an estimate of the age-distribution of Earth-like planets in the Galaxy. This analysis indicates that three quarters of Earth-like planets are older than Earth, with an average age 1.8 ± 0.9 billion years older.

Models of planet formation involve the concept of a snow line (§11.2.2), the distance from the central protostar beyond which it is cool enough for hydrogen compounds such as H$_2$O, NH$_3$, and CH$_4$ to condense into solid ice grains, demarcating the regions in which terrestrial and Jovian planets can form. Depending on density, the temperature of the snow line is estimated to be about 150 K, and it lies at around 2.7 AU in the present solar system. Lower temperatures beyond the snow line makes more solid grains available for accretion into planetesimals and eventually into planets.

The short-lived isotope ^{26}Al ($\tau_{1/2} \sim 7 \times 10^5$ yr), produced from Ar by cosmic-ray spallation, has long been recognised as a possible heat source responsible for differentiation of early planetesimals (Urey, 1955). Such a phase is invoked to explain the thermal processing evident in certain meteorites. Heating by ^{26}Al may also impose a *snow moment*, with planetesimals forming earlier than this being depleted of volatiles (Gilmour & Middleton, 2009).

Implications for SETI observers Something of an aside relates to the possibility of searching for developed civilisations in parallel with host star spectroscopic characterisation. Both Drake (1965) and Shklovskii

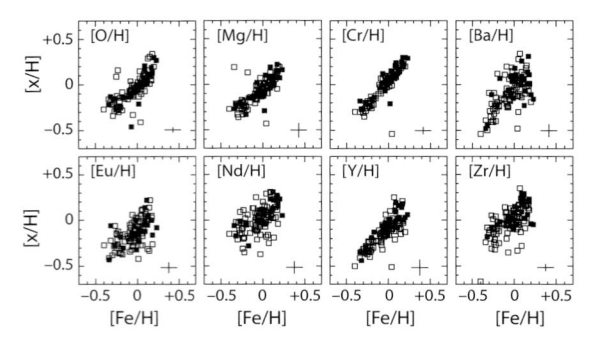

Figure 8.17: [x/H] versus [Fe/H] for the planet hosts and comparison stars from the Anglo–Australian planet search programme. Filled squares represent host stars, and open squares represent non-host stars. Typical error bars are shown at the bottom right of each panel. The r- and s-process elements are Ba, Eu, Nd, Y, Zr. From Bond et al. (2008, Figure 1), reproduced by permission of the AAS.

& Sagan (1966) independently suggested that extraterrestrial civilisations could announce their presence by adding a short-lived isotope into their stellar atmosphere, such that external observers could detect its absorption lines, and recognise its artificial origin.

8.4.5 The r- and s-process elements

The very different origins of the r- and s- process elements means that their abundances provide information on the history of the material incorporated into planets and their host stars, as well as on overall models of Galactic chemical evolution. Various studies of their occurrence in stars without planets have been reported (e.g. Edvardsson et al., 1993; Allende Prieto et al., 2004; Bensby et al., 2005; Reddy et al., 2006).

Studies of these elements are now also being made for exoplanet host stars. Huang et al. (2005) included the s-process Ba in their broader study of abundances of 22 host stars, finding [Ba/Fe] typically solar. Gonzalez & Laws (2007) included the r-process Eu amongst their study of 18 elements in 31 host stars.

A more extensive study of 28 host stars and 90 comparison stars from the Anglo–Australian planet search programme was reported by Bond et al. (2008). In addition to the elements C, Si, O, Mg, and Cr, they determined abundances for Eu (r-process), Ba, Y, and Zr (s-process), and Nd (arising from both processes). Their results show that the abundances in host stars are different from both the standard solar abundances and the abundances in non-host stars in all elements studied, with enrichments over non-host stars ranging from 0.06 dex for O, to 0.11 dex for Cr and Y (Figure 8.17).

The results provide further evidence that metal enhancement observed in planetary host stars is the result of normal Galactic chemical evolution processes, and that the observed chemical traits of planetary host stars are primordial in origin.

Origin and depletion of lithium: Lithium provides an important if complex diagnostic of both primordial nucleosynthesis and of (pre-) main sequence stellar evolution. Essentially, ^{7}Li is produced in the Big Bang, along with ^{2}H (deuterium), ^{3}He, and ^{4}He, and its primordial abundance therefore provides an important test of Big Bang nucleosynthesis (Burbidge et al., 1957). ^{6}Li is produced primarily through cosmic-ray fusion reactions with the interstellar gas (e.g., Fields & Olive, 1999).

Both isotopes are destroyed at relatively low temperatures due to their low binding energies: ^{6}Li at $T \gtrsim 2 \times 10^6$ K, and ^{7}Li at $T \gtrsim 2.5 - 3 \times 10^6$ K, the precise values of T depending on the time scale of the transport mechanism below the convective zone as a result of the accelerating reaction rates with temperature (Lumer et al., 1990; Montalbán & Rebolo, 2002). At typical main sequence densities, ^{7}Li survives only in the outer 2–3% of the stellar mass. Otherwise, when surface material is mixed down to depths where $T \gtrsim 2.5 \times 10^6$ K, Li burning occurs, primarily through the reaction ^{7}Li$(p,\alpha)^4$He (Caughlan & Fowler, 1988), and Li depleted material is returned to the surface. In giants, surface Li decreases as a consequence of dilution.

Li abundance therefore reflects the mixing of matter, element diffusion, and angular momentum evolution throughout the star's evolutionary history (Pinsonneault et al., 1992; Pinsonneault, 1997; Stephens et al., 1997; Montalbán & Rebolo, 2002), providing a sensitive but complex diagnostic. The *lithium test*, for example, uses the presence or absence of Li to distinguish candidate brown dwarfs from low-mass stars (Rebolo et al., 1992, 1996).

High lithium abundance combined with high chromospheric activity often indicates stellar youth, although alone it is not considered to be a reliable tracer of age for solar-type stars (e.g., Pasquini et al., 1994; Mallik, 1999).

A complication in understanding the distribution of lithium in the Galaxy is the existence of the 'Spite plateau' (Spite & Spite, 1982a,b), the uniform ^{7}Li abundance in halo dwarfs spanning a wide range of T_{eff} and metallicity, but at a factor of ten below the abundances found in young Population I objects (Charbonnel & Primas, 2005).

8.4.6 The alpha elements

Israelian (2008) has drawn attention to three unexpected trends which appear in the behaviour of the α elements, whose abundances primarily reflect their production in Type II supernovae (see box).

The reasonably uniform star formation history over the 10 Gyr history of the Galactic disk, leads to the expectation that the slope of $[\alpha/\text{Fe}]$ versus [Fe/H] should be rather constant (Tsujimoto et al., 1995a). In contrast, present evidence suggests that the α element ratios [Si/Fe], [Ti/Fe], and [Mg/Fe] show a rather abrupt change at [Fe/H] ∼ 0, becoming rather flat (Figure 8.14), while [Ca/Fe] decreases with increasing [Fe/H] (Gonzalez et al., 2001b; Sadakane et al., 2002; Bodaghee et al., 2003; Beirão et al., 2005; Gilli et al., 2006).

The abundance ratio [O/Fe] continues to decrease at [Fe/H] > 0 (Ecuvillon et al., 2006b, and Figure 8.15), without showing the flattening seen in the F–G field dwarf study of Nissen & Edvardsson (1992).

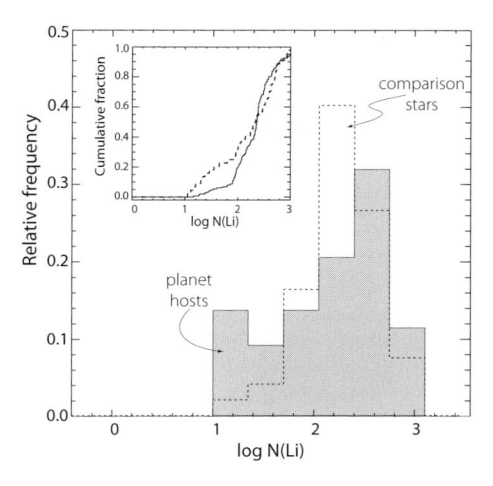

Figure 8.18: Lithium distribution for stars with planets (solid line/shaded). The distribution for comparison stars without planets (dotted line), from Chen et al. (2001), is significantly different. Inset: cumulative distribution functions. From Israelian et al. (2004, Figure 4), reproduced with permission © ESO.

The behaviour of [C/Fe] and [N/Fe] (Ecuvillon et al., 2004a,b, and Figure 8.15) are qualitatively different, despite the fact that Galactic chemical evolution models predict similar trends for all the α elements down to [Fe/H] = 0 (Tsujimoto et al., 1995b).

Israelian (2008) suggested three explanations for these anomalies: that Galactic evolution models are uncertain at high metallicity, that the presence of planets affects abundance trends in metal-rich stars, or that the abundance analysis of metal-rich stars is unreliable.

A different explanation is that the host stars are representative of different Galactic (thin or thick disk) populations, a possibility considered further in §8.4.3.

8.4.7 Lithium

As a consequence of its primordial genesis and relative ease of destruction in stars (see box on page 198), lithium depletion is known to reflect the turbulent mixing of matter, element diffusion, and angular momentum evolution throughout a star's evolutionary history. Due to additional mixing processes which might causally ensue, observed Li abundance anomalies may provide information on the past accretion history of protoplanetary disk material, planets and planetesimals (Alexander, 1967; Gonzalez, 1998; Montalbán & Rebolo, 2002), and on tidal interactions between the host star and close-orbiting planets due to resulting changes in its rotation rate (Chen & Zhao, 2006; Takeda et al., 2007b).

Observations Only one line, of Li I near 670.8 nm, is available as a Li diagnostic in solar-type stars. Several stars now known to have planetary systems were recognised as having low Li abundances prior to the discovery of their planetary companions, including ρ CrB (Lambert et al., 1991). The large Li difference between 16 Cyg A and B had also been noted, despite their similar temperatures (Friel et al., 1993).

Following the confirmation of the first exoplanetary systems, King et al. (1997b) examined 16 Cyg A and B further, and commented on six other systems, but considered that the sample size was too small to draw conclusions. Based on eight planet host stars, Gonzalez & Laws (2000) suggested that stars with planets tend to have smaller Li abundances when corrected for differences in $T_{\rm eff}$, [Fe/H], and chromospheric emission. Ryan (2000) reached a different conclusion based on 17 stars, using comparison stars drawn from open clusters and the field. He eliminated young chromospherically-active stars and subgiants as being at different evolutionary stages, and therefore with systematically different Li abundances. Evidence at the time suggested that planet-host and comparison stars of the same age, temperature and composition were indistinguishable.

From a larger sample of 79 planet hosts, and 157 comparison stars from the work of Chen et al. (2001), Israelian et al. (2004) found a difference in the two distributions (Figure 8.18), with an excess of Li depletion in planet host stars with effective temperatures in the range 5600–5850 K, but with no significant differences at higher temperatures (Figure 8.19).

These results were confirmed in an independent study based on 160 F-K disk dwarfs and subgiants, including 27 planet-host stars, by Takeda & Kawanomoto (2005), although not by Luck & Heiter (2006). A uniform treatment of Li abundances was compiled by Gonzalez (2008), who assembled data (including upper limits) for 37 planet host stars and 147 comparison stars from previous publications (Israelian et al., 2004; Takeda & Kawanomoto, 2005; Luck & Heiter, 2006; Gonzalez & Laws, 2007; Takeda et al., 2007b). This study confirmed the smaller Li abundances for planet hosts near 5800 K, and presented new evidence that planet hosts around 6100 K have an excess Li.

From 451 stars from the HARPS programme, 70 with planets, Israelian et al. (2009) found that half the solar analogues without detected planets have some 10% of the primordial Li abundance, while the planet bearing subset have less than 1%, with neither age nor metallicity correlated with the excess depletion (Figure 8.20). The presence of planets indeed appears to increase the amount of mixing, and the resulting Li depletion.

Lithium as a probe of disk interaction The general trend for stars at the lower temperatures to destroy Li more effectively is consistent with their deeper convection zones, while for stars more massive than the Sun the convective layers do not extend to Li-burning temperatures, and they therefore preserve a large fraction of their original Li.

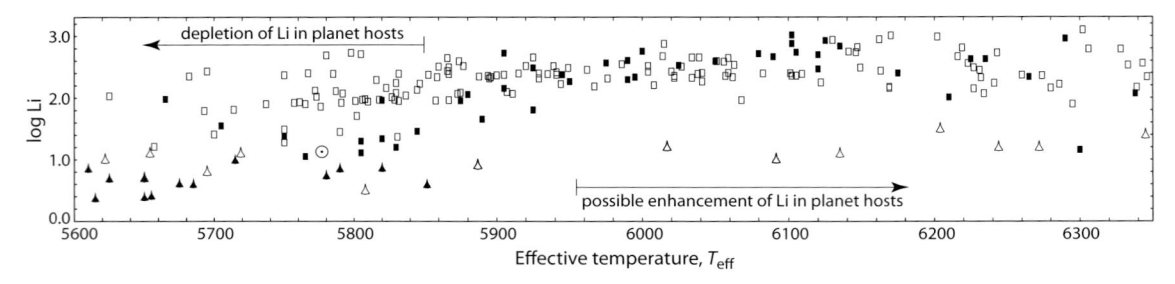

Figure 8.19: Lithium versus effective temperature for stars with planets (filled squares) and the comparison sample of Chen et al. (2001) (empty squares). Upper limits are indicated as filled triangles for planet hosts, and open triangles for the comparison sample. ⊙ indicates the position of the Sun. A regime of possible Li enhancement is indicated according to the subsequent results of Gonzalez (2008). From Israelian et al. (2004, Figure 5), reproduced with permission © ESO.

But an additional cause of Li depletion in exoplanet host stars is necessary to explain any preferential depletion compared with non-host stars. One possibility (Israelian et al., 2004; Chen & Zhao, 2006; Israelian et al., 2009) is that angular momentum transfer arising from star–disk interactions during the pre-main sequence phase resulted in systematically slower stellar rotation at the beginning of its main sequence evolution.

Some stellar evolution models which include rotation (e.g. Bouvier et al., 1997; Allain, 1998) predict that while fast rotators evolve with little decoupling between core and envelope, slower rotators develop a high degree of differential rotation. This strong differential rotation at the base of the convective envelope may then result in enhanced lithium depletion in the slower rotators (Bouvier, 2008; Castro et al., 2008, 2009). Not all theoretical models concur with this picture: Pinsonneault et al. (1989) found a rotational shear at the base of the convective envelope which scales with surface velocity.

Findings by Gonzalez (2008) that the observed Li abundance anomalies are further correlated with rotation ($v \sin i$), as well as chromospheric activity ($R'_{\rm HK}$), nevertheless suggest that planet formation processes have indeed altered the rotational history and Li abundances of stars that host Doppler-detected planets.

If confirmed, this would bring to five the number of stellar parameters that correlate with the presence of Doppler-detected planets: metallicity, mass, lithium abundance, rotation, and chromospheric activity.

A corollary of this explanation is that long-lived disks (> 5 Myr) may be a necessary condition for massive planet formation and migration (Bouvier, 2008). Theoretical work may point in the same direction (Goodman & Rafikov, 2001; Sari & Goldreich, 2004), while some relation between disk mass and a star's rotation history is rather well established (Edwards et al., 1993; Stassun et al., 1999; Barnes et al., 2001; Rebull, 2001; Rebull et al., 2002; Hartmann, 2002; Wolff et al., 2004). Accretion of planetary mass bodies onto a star can also lead to angular momentum spin-up (Siess & Livio, 1999b).

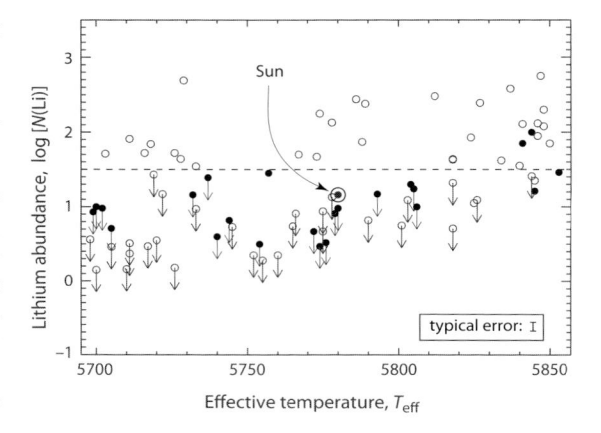

Figure 8.20: Li abundances for the planet-hosting (filled circles) and single comparison stars (open circles) from the HARPS sample study. The dashed line matches the upper envelope of the lower limits corresponding to a minimum signal-to-noise ratio of 200 in a typical solar twin. The two planet-hosting stars with the highest Li abundance also have nearly the highest effective temperatures and therefore thinner convective zones, which help to preserve Li. Other than in these stars, $\log[N(Li)] = 1.5$ is the highest value found in a planet-hosting star. Errors in $\log[N(Li)]$ (bar in bottom right corner) include uncertainties in $T_{\rm eff}$ and equivalent width measurement. From Israelian et al. (2009, Figure 1), by permission from Macmillan Publishers Ltd, Nature, ©2009.

A further corollary is that the anomalously low Li abundance of the Sun may itself result from the presence of the solar system planets, and their impact on the early evolution of the solar system.

The presence of ^{6}Li As a result of the different temperatures at which ^{6}Li and ^{7}Li are destroyed, stellar evolution models can predict whether ^{6}Li, ^{7}Li, or both, are consumed at particular stages of their evolution, as a function of mass and metallicity (Montalbán & Rebolo, 2002). Thus standard models (e.g. Forestini, 1994) predict that ^{6}Li cannot survive pre-main sequence mixing in metal-rich solar-type stars.

Accordingly, ^{6}Li should not be present in the atmosphere of a normal solar-type star, but could be present in a star that has accreted planetary matter. Its presence could conceivably provide some discrimination between different giant planet formation scenarios (Sandquist et al., 2002).

The presence of ^{6}Li is manifested as a red asymmetry in the Li I, but its inferred contribution must be decoupled from similar effects due to convective motions (Ghezzi et al., 2009). It has been reported in just one planet host star, HD 82943 by Israelian et al. (2001), attributed to the infall of rocky material (see also Sandquist et al., 2002), with the infrared excess reported for the system by Beichman et al. (2005b) perhaps implying that recent collisions have provided suitable accreting material. The detection was subsequently contested by Reddy et al. (2002), confirmed by Israelian et al. (2003), but further questioned by Ghezzi et al. (2009).

Unsuccessful searches were reported in eight planet-hosting stars by Reddy et al. (2002), in a further two by Mandell et al. (2004), and in five more by Ghezzi et al. (2009). Whatever the nature of HD 82943, these negative results suggest that post-main sequence accretion of planets or planetary material that is undepleted in lithium is uncommon.

8.4.8 Beryllium

Beryllium (^{9}Be) is, like lithium, used as a tracer of the internal structure and (pre-) main sequence evolution of solar type stars, with a higher thermonuclear destruction temperature of $\sim 3.6 \times 10^6$ K (Caughlan & Fowler, 1988; Lumer et al., 1990; Boesgaard et al., 2004).

Early inconclusive studies were made for 55 Cnc and 16 Cyg A/B (King et al., 1997a; Garcia Lopez & Perez de Taoro, 1998; Deliyannis et al., 2000). Santos et al. (2002a) reported Be abundances for 29 planet hosts and six single stars, finding several Be-depleted stars at 5200 K.

Studies for a larger sample of 41 planet hosts and 29 stars without known planets (Santos et al., 2004b,d) suggests that planet hosts have normal Be abundances, again supporting a primordial origin for the metallicity excess of planet host stars. Santos et al. (2004b) also found a small number of planet-host late-F and early-G dwarfs that might have higher than average Be abundances, perhaps related to the engulfment of planetary material, but more likely attributable to Galactic chemical evolution, or to stellar-mass differences for stars of similar temperature.

8.5 Asteroseismology

8.5.1 Principles

Asteroseismology, the study of stellar oscillations, provides independent constraints on fundamental stellar

---

Asteroseismology: Non-radial oscillations were invoked more than 50 years ago to explain puzzling spectral characteristics of β Cephei stars. This led to the search and discovery of the 5-min oscillations in the Sun, and the development of *helioseismology* to probe its internal structure (Gough & Toomre, 1991; Christensen-Dalsgaard, 2002).

As applied to other stars, *asteroseismology* uses stellar oscillations, typically excited by convection or by the κ (opacity-driven) mechanism, to study their internal structure, dynamics, and evolutionary state (e.g. Christensen-Dalsgaard, 1984; Brown & Gilliland, 1994; Christensen-Dalsgaard, 2004; Aerts et al., 2010).

Each *stellar oscillation mode* is characterised by three integers: the radial order n, the harmonic degree l, and the azimuthal order $-l \le m \le l$. Integers l and m determine the spherical harmonic component describing the properties of the mode as a function of co-latitude, θ, and longitude, ϕ, in a spherical polar coordinate system. The radial component of velocity, for example, can be expressed as (Christensen-Dalsgaard, 2004, eqn 1)

$$V = \mathscr{R}\left[v_r(r) \, Y_l^m(\theta,\phi) \, \exp(-i\omega t) \right], \tag{8.17}$$

where $Y_l^m(\theta,\phi) = c_{lm} P_l^m(\cos\theta) \exp(im\phi)$ is a spherical harmonic, P_l^m the associated Legendre function, c_{lm} a normalisation constant, and $v_r(r)$ an amplitude function which depends only on radial distance r.

Since the surface of stars other than the Sun are (typically) unresolved, signal averaging over the stellar surface suppresses information from all but the modes of lowest degree, with $l \le 3$. These involve the sound travel time across the stellar diameter, and conditions near the stellar surface and the convective boundary layer.

Frequencies generally scale as the inverse dynamical time scale (Christensen-Dalsgaard, 2004, eqn 2)

$$\omega_{\rm dyn} = \left(\frac{GM_\star}{R_\star^3}\right)^{1/2}. \tag{8.18}$$

Detailed mode properties are determined by the dominant restoring force: either pressure caused by compression and rarefaction (p-modes), or buoyancy caused by density differences affected by gravity (g-modes). Waves propagate (or are damped) depending on specific local characteristic frequencies, viz. the acoustic (or Lamb) frequency S_l, and the buoyancy (or Brunt–Väisälä) frequency N.

---

parameters such as mass, density, radius, age, rotation period, and chemical composition. This is achieved by comparing patterns of observed oscillation frequencies with theoretical predictions based on corresponding models of stellar evolution. Such information is of value since stars with the same externally observable parameters ($L_\star$, $T_{\rm eff}$, and [Fe/H]) can have very different interiors as a result of their evolutionary histories.

Stellar oscillations are predicted and observed over a wide range of mass and evolutionary states, with solar-like objects, high-amplitude δ Scuti variables, β Cephei variables, and rapidly oscillating Ap stars having received particular seismological attention.

In solar-like stars, oscillation modes are excited

stochastically by turbulent convection (e.g. Christensen-Dalsgaard, 2004, Section 2.2). Qualitatively, standing pressure (or acoustic) waves are then trapped between the density decrease toward the surface, and the increasing sound speed toward the centre which refracts the downward propagating wave back to the surface. Similar convection-driven oscillations are expected for all stars with significant surface convection, i.e. those on the cool side of the δ Scuti instability strip.

Precise mode frequencies, typically in the range of 0.1–10 mHz (periods of order 1 min to a few hours), are detectable either by precision photometry (intensity changes) or by spectroscopy (radial velocity or equivalent width changes). Under favourable circumstances, individual mode frequencies can be determined with very high accuracy, despite their very small amplitudes, being at maximum some $0.2\,\mathrm{m\,s^{-1}}$ in radial velocity, or around 5×10^{-6} in broad-band photometric intensity.

Photometric observations from the ground are hampered significantly by the atmosphere, while the long and uninterrupted observing periods necessary for the highest accuracy in frequency determinations are also limited by observing constraints. From space, MOST (Matthews, 2006), CoRoT (Michel et al., 2006) and Kepler (Christensen-Dalsgaard et al., 2008; Borucki et al., 2010a) were all designed with dedicated photometric asteroseismology capabilities, in addition to their exoplanet transit detection modes (§6.2.5).

Modeling the oscillations Amplitudes and phases of the stellar oscillations are largely controlled by the near-surface layers. Frequencies are determined by the bulk sound speed, and the internal stellar density profile. Convective motions within the differentially rotating outer convective zone modify the star's temperature, density and velocity structure, while *convective overshooting*, caused by the momentum of cool sinking material into the deeper stable radiative regions, alters the structure of the *tachocline*, the transition region between the two. Stellar rotation, introduced as a rotation profile in the code solving for the adiabatic oscillation frequencies, further influences the fine structure of the frequency spectrum through its variation with stellar radius, and because of the resulting flows and instabilities which are also a function of the star's evolutionary state (Gough & Thompson, 1990).

Given these various complexities, the interpretation of an observed frequency spectrum proceeds via a comparison with theoretical stellar models, essentially providing a description of the stellar interiors via tests of the physical models used in their construction.

Various numerical codes are used to model stellar evolution according to initial mass, chemical composition, rotation, etc. Two which have been routinely applied to seismology targets are the Aarhus code (ASTEC, Christensen-Dalsgaard, 2008b) and the

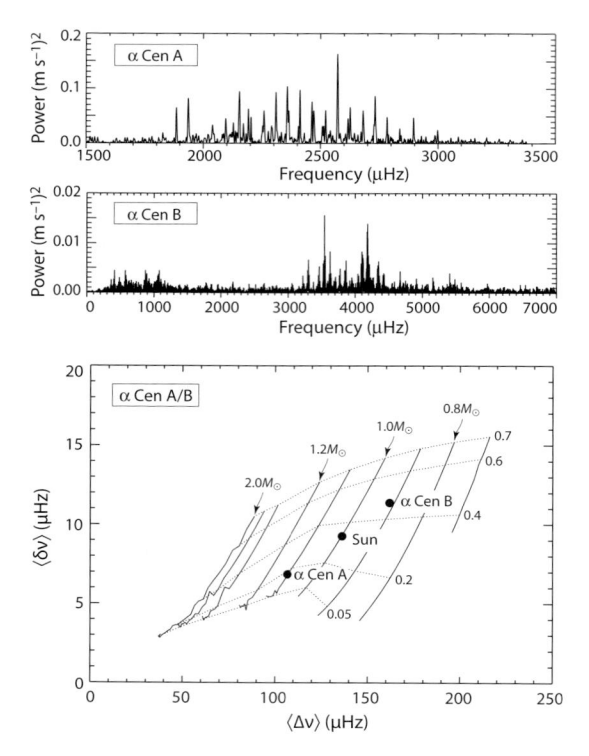

Figure 8.21: Top two: asteroseismic power spectra for α Cen A and α Cen B (on different frequency scales). Bottom: the corresponding asteroseismic Hertzsprung–Russell diagram. Axes are the average small and large frequency separations, $\langle \delta \nu \rangle$ and $\langle \Delta \nu \rangle$ respectively. Solid lines are evolutionary tracks for fixed mass, and dashed lines show constant core-H content, related to age. α Cen A is more massive and more evolved than the Sun, while α Cen B is less massive and less evolved, results agreeing with traditional models of stellar evolution. From Bedding et al. (2004, Figure 1), Kjeldsen et al. (2005, Figure 4a), and Christensen-Dalsgaard (2004, Figure 3) respectively.

Toulouse–Geneva evolution code (TGEC, Vauclair, 2010). Both use similar foundations: the OPAL equation of state (Rogers et al., 1996), updated OPAL opacities (Iglesias & Rogers, 1996), and the NACRE nuclear reaction parameters (Angulo et al., 1999). Diffusion and settling of helium can be included. Convection is treated according to the widely-adopted mixing-length formulation of Böhm-Vitense (1958). Convective overshooting is parameterised to occur over a distance of α_{ov} pressure scale heights, i.e. with $\alpha_{\mathrm{ov}} = 0$ describing its absence.

Corresponding codes for calculating oscillation frequencies include ADIPLS (Christensen-Dalsgaard, 2008a), and PULSE (Brassard et al., 1992). The former details the equations of *adiabatic stellar oscillations* (i.e. those for which the pulsation period is significantly less than the thermal time scale), the adopted approximations, and their numerical solution.

While the precise frequencies of stellar oscillations

depend on the detailed structure of the star, and on the physical properties of the gas, *asymptotic theory* (Christensen-Dalsgaard, 2004) leads to simplified expressions which are used for many seismological analyses in practice.

For a non-rotating star, asymptotic analysis expresses the mode frequencies as (Christensen-Dalsgaard, 2004, eqn 7)

$$\nu_{n,l} = \Delta\nu(n + \frac{l}{2} + \alpha) , \qquad (8.19)$$

where n is the radial order and l is the angular degree, α represents a phase shift sensitive to the properties of the near-surface layers, and where

$$\Delta\nu \simeq \left[2\int_0^R \frac{1}{c}\,dr\right]^{-1} \qquad (8.20)$$

is the inverse of the sound travel time across the stellar diameter, i.e. dependent on the mean stellar density. Hence the spectrum is uniformly spaced in radial order, with a spacing referred to as the *large separation*

$$\Delta\nu_{n,l} = \nu_{n,l} - \nu_{n-1,l} \sim \Delta\nu . \qquad (8.21)$$

An inspection of Equation 8.19 implies that there is a frequency degeneracy in modes having the same value of $n + l/2$. A more detailed analysis shows that this implied degeneracy is not perfect, and that mode frequencies are further separated by the *small separation* determined by adjacent modes with $n \to n-1$ and $l \to l+2$, viz.

$$\delta\nu_{n,l} = \nu_{n,l} - \nu_{n-1,l+2} . \qquad (8.22)$$

Rotation, and other departures from spherical symmetry, introduce a further dependence of the oscillation frequencies on m. For a slowly rotating star (Christensen-Dalsgaard, 2004, eqn 11)

$$\omega_{n,l,m} \equiv 2\pi\nu_{n,l,m} = \omega_{n,l,0} + m\langle\Omega\rangle , \qquad (8.23)$$

where m is the azimuthal order, and Ω is the angular rotational velocity. This has been found to be a good approximation in the solar case, and provides for the possibility of determining stellar rotation periods from oscillation frequencies alone.

The inclination of the rotation axis affects the amplitudes of the different azimuthal degree modes, and can also be determined in principle (Gizon & Solanki, 2003). For rapidly rotating stars, higher-order effects further complicate the frequency spectrum (Gough & Thompson, 1990).

The asteroseismic Hertzsprung–Russell diagram relates the average small separation, $\langle\delta\nu\rangle$, to the average large separation, $\langle\Delta\nu\rangle$. This can be used to estimate stellar parameters such as mass and, through central hydrogen abundance as a proxy, age (Figure 8.21).

Model frequency structures are often presented in the form of an échelle diagram (e.g. Figure 8.22), in which the frequency is expressed as

$$\nu_{n,l} = \nu_0 + k\Delta\nu + \tilde{\nu}_{n,l} , \qquad (8.24)$$

where ν_0 is some reference frequency, and k is an integer such that $0 < \tilde{\nu}_{n,l} < \Delta\nu$. The spectrum is then divided into segments of length $\Delta\nu$, and $\tilde{\nu}_{n,l}$ plotted against $\nu_0 + k\Delta\nu$. The frequencies would then line up vertically if the asymptotic expression given by Equation 8.19 were precisely valid.

Accuracies of stellar parameters In the context of the CoRoT observations, Mulet-Marquis et al. (2009) generated models to derive the uncertainties of fundamental stellar parameters for a given accuracy of oscillation frequency determination. They estimated accuracies of 1–7% in $M_\star$, 1–3% in $R_\star$, and 2–10% in $L_\star$.

Simulations reported by Kjeldsen et al. (2009) give comparable uncertainties of 5% in $M_\star$, 2–3% in $R_\star$, 1% in mean densities, and 5–10% for ages relative to the total time on the main sequence.

8.5.2 Application to exoplanet host stars

Applications of asteroseismology of relevance to exoplanet host stars have been reported as follows.

Primordial or self-enriched metallicities Oscillation frequencies and amplitudes can be compared with stellar evolutionary models to determine whether the observed metallicity is a bulk or surface property, thus providing independent evidence as to whether the high metallicity of exoplanet host stars is primordial, or a result of self-enrichment.

The four-planet system μ Ara (Pepe et al., 2007) shows the common attribute of high metallicity compared to stars without detected planets. From a Fourier transform of the HARPS radial velocity time series, Bouchy et al. (2005a) identified 43 oscillation modes, with harmonic degrees $l = 0-3$. To first order, $L_\star$ is constrained by the Hipparcos parallax of 65.5 ± 0.8 mas.

Evolutionary simulations Bazot & Vauclair (2004); Bazot et al. (2005) suggest that the metal-rich models have a convective core, while the accreting models do not. These internal structural differences lead to specific predicted signatures for the $l = 0$ and $l = 2$ modes. Although the seismic analysis provides precise and independent constraints on $L_\star$ and T_{eff} (Figure 8.22), the crucial frequency region from 2.5–2.7 mHz had only low signal-to-noise, and interpretation in terms of either primordial or accretion models could not be resolved unambiguously. Further analysis has been reported by Soriano & Vauclair (2010).

For ι Hor, some 25 solar-type oscillation modes were detected with HARPS (Laymand & Vauclair, 2007). The metallicity, helium abundance, and age derived from the

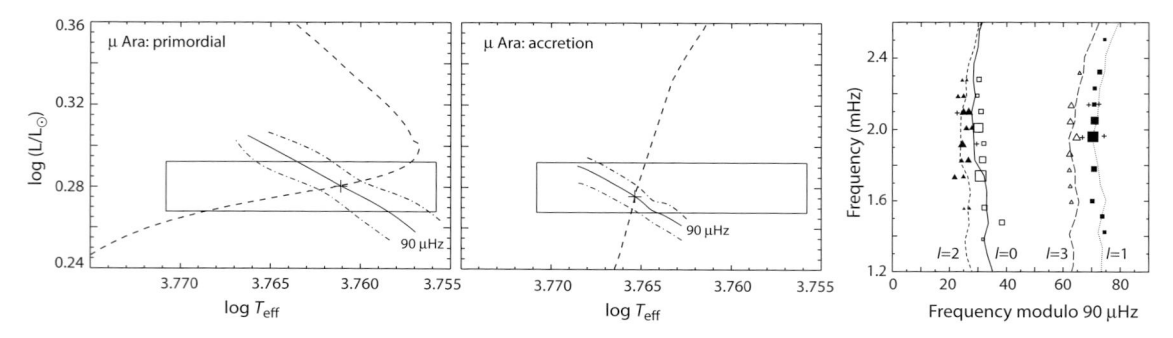

Figure 8.22: Asteroseismology of the Doppler-planet host μ Ara. Left and middle: evolutionary tracks (dashed lines) for assumed primordial overmetallic and self-enrichment/accretion scenarios, respectively. The three oblique lines correspond to asteroseismic separations of 90 μHz (as observed, solid line), as well as 89 μHz (higher curve) and 91 μHz (lower curve), and the cross corresponds to the favoured model. Error boxes correspond to the Hipparcos-based luminosity and adopted $T_{\rm eff}$. Right: échelle diagram for the overmetallic model. Lines represent the theoretical modes indicated. Symbols represent their observational counterparts, with sizes proportional to signal-to-noise. From Bazot et al. (2005, Figures 1–3), reproduced with permission © ESO.

seismic analysis were found to be characteristic of the Hyades cluster. Vauclair et al. (2008) concluded that the metallicity is therefore primordial.

Detailed seismological modeling in advance of the planned CoRoT observations of HD 52265 were reported by Soriano et al. (2007).

Radii of transiting planets Large uncertainties on the determination of transiting planet radii can result from the uncertainties on stellar radii to which they are normalised. Christensen-Dalsgaard et al. (2010) reported Kepler asteroseismology results for the three exoplanet host stars identified by ground-based transit surveys in the Kepler field: HAT–P–7, HAT–P–11, and TrES–2, all expected to show solar-like oscillations at observable amplitudes. The frequency spectrum for HAT–P–7 was used in models for each of three assumed values of the overshoot parameter $\alpha_{\rm ov}$ (Table 8.2, and Figure 8.23). The best fit is given by $\alpha_{\rm ov} = 0.2$, resulting in $R_\star = 1.981 R_\odot$ from the asteroseismic analysis.

Masses of giant host stars The determination of stellar masses for giant stars is complicated by the fact that evolutionary tracks over a wide range of masses converge to the same region of the Hertzsprung–Russell diagram. Hatzes & Zechmeister (2008) measured stellar oscillations in three planet-hosting giant stars: HD 13189, β Gem, and ι Dra. They showed that their periods (in the range 0.1–5 d) and radial velocity amplitudes (in the range 5–85 m s^{-1}) are consistent with solar-like p-mode oscillations, and derived stellar mass estimates consistent with, but independently of, isochrone models.

Masses of imaged companions The masses of companions detected by direct imaging are estimated from their brightness. Since giant planets fade as they radiate away the heat of their formation, mass estimates are correspondingly sensitive to the estimated age of the host star. For HR 8799 (§7.5.3), Marois et al. (2008b) estimated the age of the host star as 30–160 Myr using a number of indirect lines of argument, giving implied companion masses in the range $7 - 10 M_{\rm J}$.

The host star is a γ Doradus-type pulsational variable. This is a class of main sequence star in the lower part of the classical instability strip with periods around 1 d. Moya et al. (2010) used the g-mode pulsation spectrum to estimate its age using asteroseismology, where its metallicity and age both have a significant influence on the corresponding buoyancy-type restoring force. They estimated a much higher age of around 1 Gyr, albeit dependent on the unknown inclination of the stellar rotation axis. This age revision would raise the companion masses to the brown dwarf range.

Excitation of g-modes A planet orbiting a solar-type star at a radius such that the Keplerian orbital frequency, $\omega_{\rm k} = \sqrt{[G(M_\star + M_{\rm p})/a^3]}$, is half the frequency of a given resonant mode of the star, may excite g-modes as a result of their gravitational interaction. Dynamical tides raised by the planet on the star are dissipated due to turbulent viscosity in the convective zone and to radiative damping in the radiative core, as modeled in coalescing binary stars (Kokkotas & Schafer, 1995; Alexander, 1987). Even for close-in planets the effects are generally small

Table 8.2: Asteroseismology analysis for the Kepler field transit host star HAT–P–7. For three values of the overshoot parameter, $\alpha_{\rm ov}$, the model fit gives the stellar parameters which minimise χ^2_ν along the evolutionary tracks shown in Figure 8.23. Data are from Christensen-Dalsgaard et al. (2010, Table 2).

$\alpha_{\rm ov}$	$L_\star$ ($L_\odot$)	$M_\star$ ($M_\odot$)	$R_\star$ ($M_\odot$)	Age (Gyr)	$T_{\rm eff}$ (K)	χ^2_ν
0.0	5.91	1.53	1.994	1.758	6379	1.08
0.1	5.81	1.52	1.992	1.875	6355	1.04
0.2	5.87	1.50	1.981	2.009	6389	1.00

8.6 Activity and X-ray emission

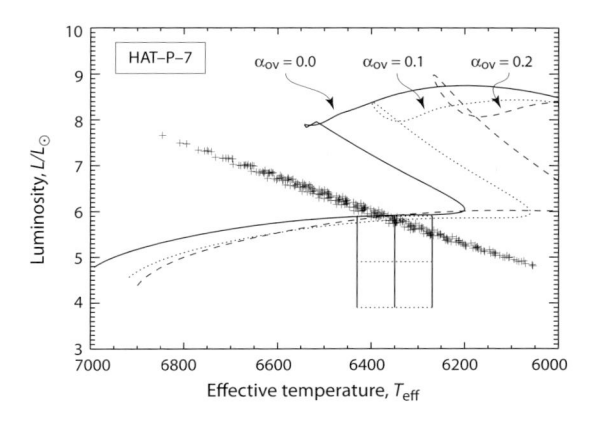

Figure 8.23: Theoretical Hertzsprung–Russell diagram for the transit host star HAT–P–7, with evolutionary tracks providing fits to three assumed values of the overshoot parameter α_{ov}. Models indicated '+' minimise the difference between the computed and observed frequencies. The box is centred on the $L_\star$ and T_{eff} values given by Pál et al. (2008), with a size matching their errors. From Christensen-Dalsgaard et al. (2010, Figure 3), reproduced by permission of the AAS.

(Terquem et al., 1998), but certain configurations may be observable (Berti & Ferrari, 2001).

8.6 Activity and X-ray emission

8.6.1 Magnetic and chromospheric activity

Exoplanet host stars display a variety of chromospheric and magnetic activity which is, to first order, dependent on the spectral type rather than on whether the star hosts planets or not. Such activity can be quantified statistically in terms of jitter in radial velocity (§2.2.6), astrometry (§3.4), and photometry (§6.3), as a function of spectral type and luminosity class.

Observations also indicate that some close-in 'hot Jupiters' undergo episodes of periodic or enhanced stellar activity, most probably linked to the presence of the planet through magnetic or tidal interactions. The effects are expected to be greatest in the outermost layers of the star (in the chromosphere, transition region, and corona). The broad, deep photospheric absorption lines of Ca II H and K provide an important diagnostic since they allow the chromospheric emission to be observed and monitored at the highest contrasts.

Progress in understanding this type of enhanced interaction-driven activity has built on studies of binary stars, which are known to be generally more active, for example in X-rays, than single stars of the same type and rotation rate (e.g. Zaqarashvili et al., 2002, and references), and where the effect has generally been attributed to tidal or magnetic interactions. These ideas were extended to the case of stars with giant planets and brown dwarfs by Cuntz et al. (2000).

Energy dissipation Tidal or magnetic interactions would be expected to display different periodicities with respect to the planet's orbital rotation. Tidal interactions should result in two tidal bulges on the star, and an enhanced activity which should be modulated at half the planet's orbital period. Cuntz et al. (2000) suggested that the energy generation due to tidal interactions is proportional to the gravitational perturbation

$$\frac{\Delta g_\star}{g_\star} = \frac{M_p}{M_\star} \frac{2R_\star^3}{(a_p - R_\star)^3} \,, \quad (8.25)$$

where the height of the tidal bulge

$$h_{tide} \propto \frac{\Delta g_\star}{g_\star} R_\star \propto a_p^{-3} \,. \quad (8.26)$$

For energy released via reconnection during interaction between the planetary magnetosphere and the stellar magnetic field, the effect should be modulated at the planet's orbital period. Saar et al. (2004) estimated that the associated energy release

$$E \propto B_\star B_p v_{rel} a_p^{-n} \propto a_p^{-n} \,, \quad (8.27)$$

where $B_\star$ and B_p are the stellar and planetary magnetic fields, v_{rel} is the relative velocity between them which produces the shear in the magnetic fields leading to reconnection. They gave $n = 3$ very close to the star and $n = 2$ farther away.

Observations Given that tidal and magnetic effects are expected to depend strongly on the star–planet separation, observations have been focused on planets with the smallest semi-major axes. Of the stars so far monitored, periodic activity has been reported in Ca II H and K and/or Balmer line emission in the case of HD 179949 (Shkolnik et al., 2003) and v And (Shkolnik et al., 2005). It has also been inferred from optical brightness variations in the case of τ Boo (Walker et al., 2008).

In these cases, activity is modulated at the stellar rotation rate, suggesting magnetically-driven interactions, while the location of maximum chromospheric heating is found to be phase shifted with respect to orbital conjunction (Shkolnik et al., 2005), leading the subplanetary point by typically 60–70°, and up to 170° in the case of v And (Figure 8.24). The activity may be intermittent over the longer term (Shkolnik et al., 2008).

The stellar magnetic field is inferred to play a key role in these interactions, either by triggering activity directly, or by tracing it indirectly, perhaps through tidal interactions influencing the magnetic field by enhancing the shear at the base of the convective envelope. The phase lag between orbital conjunction and epochs of enhanced activity observed for HD 179949 and v And has been attributed to an offset between the magnetic and rotation axes (McIvor et al., 2006), to the propagation of Alfvén waves (Preusse et al., 2006), or to other magnetic field configurations (Lanza, 2008).

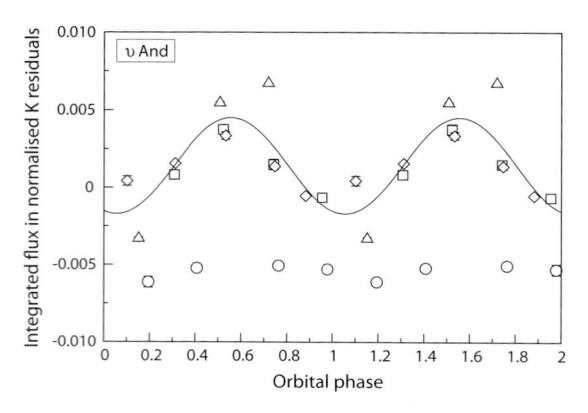

Figure 8.24: Integrated flux of the Ca II K line residuals from a normalised mean spectrum of υ And as a function of orbital phase. Data were acquired in 2001 Aug (circles), 2002 Jul (squares), 2002 Aug (triangles), and 2003 Sep (diamonds). The solid line is the best-fit hot-spot model fitted to the 2002 and 2003 data. The Ca II emission is 169° out of phase with the subplanetary point. From Shkolnik et al. (2005, Figure 8), reproduced by permission of the AAS.

In the case of CoRoT-7, a maximum entropy spot model of the light curve identifies three active regions at three discrete longitudes that appear to migrate at different rates, probably as a consequence of surface differential rotation (Lanza et al., 2010).

The magnetic cycles of τ Boo A $4.1 M_J$ planet orbits the F7 star τ Boo in a 3.31 d period with a semi-major axis of 0.048 AU (Butler et al., 2006). The star has a moderate-intrinsic activity, and a relatively shallow convective envelope of about $0.5 M_J$ (Shkolnik et al., 2005, 2008). It has a weak magnetic field, determined by Zeeman Doppler imaging, of $1-3 \times 10^{-4}$ T (Catala et al., 2007), which underwent a global magnetic polarity reversal between 2006–07 (Donati et al., 2008), and again between 2007–08 (Fares et al., 2009), with switches between a predominantly poloidal and a dominantly toroidal nature at different epochs. The observations suggest a magnetic cycle of ~ 2 yr, much shorter than that of the Sun, at 22 yr.

The star has strong differential rotation which does not change with the magnetic cycle: the equator rotates in 3 d, while its pole rotates in 3.9 d. This means that at latitude $\sim 40°$, its rotation is synchronised with the orbital period of the planet. Fares et al. (2009) hypothesise that the short magnetic cycle may be the result of the star's very shallow convection zone, or it may be connected to the presence of the close-orbiting planet, for example by tidally synchronising the outer convective envelope, and thereby enhancing the shear at the tachocline.

8.6.2 X-ray emission

A first systematic study of the X-ray emission from the host stars of Doppler-detected planets, also aimed at understanding their coronal activity and possible star–planet interactions, was made by Kashyap et al. (2008).

They searched for X-ray counterparts in archival data from the ASCA, EXOSAT, Einstein, ROSAT, XMM–Newton and Chandra space observatories, complemented by dedicated XMM–Newton targeted surveys. Out of some 230 host stars investigated, 70 were detected serendipitously or in pointed observations.

Taken as a whole, the set of main sequence stars with giant planets was found to be broadly similar in X-ray characteristics to the field star sample.

For the subset of close-in giant planets with small star–planet separations, a_p, tidal or magnetic interactions may nevertheless result in enhanced X-ray activity. In the model of Cohen et al. (2009), a close-in planetary magnetosphere restricts the expansion of the stellar magnetic field and the acceleration of the stellar wind, causing a higher plasma density in a coronal hot spot. From their study, this can lead to variations of the X-ray flux of $\sim 30\%$ as the hot spot rotates in and out of view, and to an overall X-ray flux enhanced by a factor of ~ 1.5 for a stellar dipole field, and by a factor ~ 15 for a Sun-like magnetic field at maximum activity.

After accounting for their sample biases, Kashyap et al. (2008) reported a possible difference between the X-ray properties of host stars with $a_p < 0.15$ AU ($\langle L_X \rangle = 10^{28.49 \pm 0.09}$ erg s^{-1}) and those with larger separations $a_p > 1.5$ AU ($\langle L_X \rangle = 10^{27.85 \pm 0.18}$ erg s^{-1}), hinting that stars with close-in giant planets may indeed be more active compared with those with larger separations.

Poppenhaeger et al. (2010) subsequently investigated the stellar X-ray activity of the 72 planet-bearing stars within 30 pc known at the end of 2009, using ROSAT and XMM–Newton data, but found no significant correlations of X-ray luminosity with planet parameters such as mass or semi-major axis. They concluded that while coronal star–planet interactions are evidently important for a few individual targets, they do not result in a major effect on planet-hosting stars in general.

As suggested by the magnetohydrodynamic model of Cohen et al. (2010), significant star–planet interactions require not only that the star–planet separation be small, but that the Alfvén surfaces of the stellar and planetary magnetospheres must also be interacting.

8.7 Stellar multiplicity

Background Stellar systems composed of two or more stars (binaries or multiples) exist in a wide range of configurations. They range from tight binaries with orbital periods of hours or days, and separations down to less than $10 R_\odot$ (~ 0.1 AU), to marginally-bound wide bina-

8.7 Stellar multiplicity

ries with orbital periods $> 10^5$ yr and separations up to 20 000 AU (0.1 pc) or more. The scientific interest of multiple stars ranges from statistics of their physical properties (periods, mass ratios, eccentricities) providing constraints on theoretical models of star formation, various aspects of stellar evolution and mass transfer, and their central role for stellar mass determination.

Binaries are extremely common in the Galaxy, with various estimates suggesting that some 50% or more of all stars occur in gravitationally bound groups having a multiplicity of two or higher. The true binary frequency is hard to establish for any given population due to the different techniques required to probe different separation ranges, and to the inherent difficulties of detecting binaries with very small or very large angular separations, and/or large magnitude differences. Binary frequency appears to correlate well with stellar ages: for example, low-mass pre-main-sequence stars in the star-forming region Taurus–Auriga have an (inferred) binary frequency as high as 80–100% (Leinert et al., 1993; Kohler & Leinert, 1998), values confirmed by studies of other star-forming regions.

Studies of component mass ratios give different results according to observational technique, sample population, and correction for selection effects. A bimodal distribution is often reported, with one peak arising from equal-mass components, and another at a mass ratio of around 0.2–0.3. The distribution in orbital periods P has been estimated over several orders of magnitude, from days to millions of years (Duquennoy & Mayor, 1991; Fischer & Marcy, 1992).

Orbital period distributions of spectroscopic binaries point to a bell-shaped distribution of $\log P$ with a maximum in the visual binary range at around 180 yr (e.g. Goldberg et al., 2003). Eccentricities are correlated with periods: circular orbits dominate for periods below 10 d, with longer-period systems having eccentricities ranging from 0.1 to nearly 1.0, and with an absence of circular orbits.

The formation of binary stars is broadly considered as occurring in two phases: fragmentation of molecular clouds producing the condensing cores, followed by subsequent accretion and migration which will fix the final component masses and orbital parameters (e.g. Halbwachs et al., 2003).

Multiplicity of planet hosts Assessing the multiplicity of exoplanet host stars is approached from two different directions.

Binary or multiple stars can themselves be surveyed for the presence of planets, an approach which is described separately for radial velocity observations (§2.7), and for photometric transit observations (§6.2.8). Although a large fraction of stars are known to occur in binary or multiple systems, radial velocity observations have tended to avoid close binaries in view of additional complexities of the spectral line calibration and analysis.

The other approach is to scrutinise stars around which planets have already been detected, in order to determine whether that star has other widely separated or fainter stellar mass companions (e.g. Patience et al., 2002; Raghavan et al., 2006). Observations include high angular resolution imaging to probe separations out to a few arcsec, or the imaging search for more widely separated companions out to 100 arcsec or more.

In either case, to ensure their apparent physical association is real, the classification of candidate companions requires confirmation that the components share the same proper motion, i.e. that they form a *common proper motion pair*.

Various examples of such surveys (e.g. Eggenberger et al., 2007a; Thalmann et al., 2009b) and companion detections (e.g. Mugrauer et al., 2004, 2005, 2007a,b) are found in the exoplanet literature. Companions are recorded in the various on-line catalogue compilations.

A detailed statistical picture of the occurrence of planets around binaries, and their dynamical stability, is currently highly incomplete.

9

Brown dwarfs and free-floating planets

9.1 Brown dwarfs

Brown dwarfs are sub-stellar objects, too low in mass to sustain stable hydrogen fusion, but in which lower threshold nuclear reactions can still occur. Spanning the mass range $13-80M_J$, they occupy the domain between planets and stars. Originally termed black dwarfs, the term 'brown dwarf' was introduced by Tarter (1976) and has been used thereafter.

This section summarises their characteristics, and places them in the context of exoplanet studies. The connection is particularly important in understanding the nature of massive orbiting planets with $M \gtrsim 13M_J$, and of the 'free-floating planets' found in young open clusters and star-forming regions.

9.1.1 The role of fusion

Some thirty years before their discovery, theories had identified a stellar/sub-stellar boundary at $\sim 75-80M_J$, depending on chemical composition (Kumar, 1963). In the normal process of star formation, gravitational collapse releases energy which leads to increasing temperature and density, and to cores which become partially degenerate. Collapse halts when the sum of the normal gas pressure, and the free electron degeneracy pressure arising from the Pauli exclusion principle which acts as an energy sink, balances the gravitational potential.

At solar metallicity, objects above $\sim 78M_J$ reach the 3×10^7 K core temperatures necessary to initiate hydrogen fusion, and become stars. Below $\sim 74M_J$, core temperatures never rise to the levels necessary for sustained hydrogen fusion, and these 'failed stars' are termed brown dwarfs. Transition objects in the range $\sim 74-78M_J$ sustain core fusion for the first 10^9-10^{10} yr of their lifetime, before increasing degeneracy and falling temperatures transform them to brown dwarfs (D'Antona & Mazzitelli, 1985).

Decreasing metallicity leads to an increase in the mass of the stellar/sub-stellar boundary, and to higher luminosities and effective temperatures at the H-burning limit. As given by Burrows et al. (2001), the boundary for solar composition is $M \sim 0.07-0.074M_\odot \equiv$ $73-78M_J$, $T_{\rm eff} \sim 1700-1750$ K, and $L \sim 6 \times 10^{-5}L_\odot$; while at zero metallicity $M \sim 0.092M_\odot \equiv 96M_J$, $T_{\rm eff} \sim 3600$ K, and $L \sim 1.3 \times 10^{-3}L_\odot$. Uncertainties in silicate grain physics result in further ambiguities, with a boundary as low as $T_{\rm eff} \sim 1600$ K seemingly possible (Chabrier & Baraffe, 2000; Chabrier et al., 2000).

Below the H burning limit, lower threshold nuclear reactions may still occur: lithium fusion above $\sim 63-65M_J$ via the reaction ^{7}Li(p,α)^{4}He, and deuterium fusion above $\sim 13M_J$ via the reaction ^{2}H(p,γ)^{3}He, these mass estimates being applicable for objects of solar metallicity. Below the deuterium-burning limit, which is again sensitive to chemical composition, and which also depends on structural uniformity, nuclear processes, and the role of dust (Tinney, 1999), objects cannot sustain any species of nuclear burning in their interiors (Grossman & Graboske, 1973; Burrows et al., 1993; Saumon et al., 1996; Burrows et al., 1997, 2001).

Although not a directly observable quantity, the deuterium-fusion threshold of $\sim 13M_J$ therefore sets a lower mass limit for objects classified as brown dwarfs. Below that mass, the terms planet, or sub-brown dwarf, will be used depending on their formation mechanism (see box on page 210).

9.1.2 Detection

The detection and characterisation of brown dwarfs in the solar neighbourhood, and in nearby young clusters and star-forming regions, has roughly paralleled that for exoplanets over the years 1995–2010. Even two decades after their predicted existence, finding brown dwarfs remained problematic because of their low luminosities and temperatures – even at their brightest in the near infrared they are still more than 8–10 mag fainter than solar-type stars. Searches in the 1980s and early 1990s failed to detect them, or at least were unable to confirm their sub-stellar nature (e.g. Zuckerman & Becklin, 1987a; Stevenson, 1991; Basri, 2000b).

Ever since their discovery, characterising them by mass, temperature and atmospheric composition has

Nomenclature: This work will use a simplified nomenclature for distinguishing planets from brown dwarfs, based on their mass and their formation mechanism. Objects orbiting a star, and formed from a circumstellar gas disk, whether by core accretion (Lissauer, 1993) or by gravitational instabilities (Boss, 2002), are termed *planets* below the $\sim 13 M_J$ deuterium-fusion mass limit, and *brown dwarfs* (or brown dwarf planets) above it. Objects formed in the same way as stars, via rapid core collapse from molecular gas clouds (e.g. Bodenheimer et al., 1980b; Padoan & Nordlund, 2004), will again be termed *brown dwarfs* above this mass limit, but as *sub-brown dwarfs* below (Boss, 2001).

For the class of object referred to by their discoverers as 'free-floating planets', current evidence suggests that these have probably formed as the low-mass tail of the star formation process, and there is no evidence to suggest (neither is it claimed) that they have been ejected from planetary systems. For them, the term 'sub-brown dwarf' will be used. Planets gravitationally detached from their original host star presumably exist as a result of scattering processes, although no such cases are known, and the term *free-floating planet* or *ejected planet* will be used.

This nomenclature essentially follows that proposed by Boss et al. (2003), although it presupposes just two basic routes to the formation of planetary mass objects. Other issues are of relevance in classifying objects in the solar system, such as the type of orbit occupied (thus the Moon would have been considered a planet if orbiting the Sun directly, while Mercury would have been considered a satellite if orbiting Jupiter), the shape of the body (whether it is gravitationally relaxed), and whether it has cleared its region of planetesimals. A detailed discussion can be found in Basri & Brown (2006). Current definitions will almost certainly require adjustment as the variety of possible formation mechanisms becomes better understood, and as the quality of exoplanet observations improves.

similarly remained a challenge (e.g. Tinney, 1999; Basri, 2000a; Reid & Metchev, 2008; Luhman, 2010).

Early brown dwarf candidates included companions to Van Biesbroeck 8 (McCarthy et al., 1985), unconfirmed by speckle interferometry (Perrier & Mariotti, 1987); to G29–38 (Zuckerman & Becklin, 1987b) much debated but finally shown to be a circumstellar disk (Greenstein, 1988; Graham et al., 1990; Kuchner et al., 1998); to HD 114762 (Latham et al., 1989) confirmed as a radial velocity companion but with somewhat uncertain mass; and to the white dwarf GD 165 (Becklin & Zuckerman, 1988).

Confirmation of demonstrably sub-stellar mass came with the near-simultaneous discovery of two objects: the first isolated exemplar, the M8-type Teide 1 in the Pleiades (Rebolo et al., 1995), and the first companion to a normal star, the T6-type GJ 229 B at a distance of only 6 pc (Nakajima et al., 1995). With a 10 mag difference compared to its primary host star, GJ 229 B has an effective temperature of 1200 K, significantly below that of any H burning star, and a spectrum displaying strong CH_4 absorption, reminiscent of the solar system gas giants (Oppenheimer et al., 1995; Marley et al., 1996).

Large area digital detectors sensitive in the near infrared (1–2 μm) changed the panorama of brown dwarf discoveries. Following the Two Micron Sky Survey (TMSS, Neugebauer & Leighton, 1969), which scanned 70% of the sky and detected 5700 infrared sources, the late 1990s saw the advent of significantly deeper, wide-field infrared sky surveys.

The Deep near-infrared Southern Sky Survey (DENIS, Epchtein et al., 1997) surveyed the southern sky between 1995–2001 in the I, J, and K_S bands using the ESO 1-m telescope at La Silla, generating a catalogue of 355 million point sources (third release, 2005). The Two Micron All Sky Survey (2MASS, Skrutskie et al., 2006) surveyed the entire sky between 1997–2001 from Mt Hopkins and Cerro Tololo in the J (1.25 μm), H (1.65 μm), and K_S (2.17 μm) bands, generating the 2003 final-release catalogue of 300 million point sources with limiting magnitudes fainter than 15.8, 15.1, and 14.3 mag respectively.

Among the scientific goals of these surveys was the discovery of brown dwarfs which, by virtue of their intrinsic faintness and unknown properties, were at the time leading candidates to solve the problem of the Universe's missing mass. Their early results on space densities, however, swiftly ruled out brown dwarfs as the solution to the dark matter problem.

Discoveries with 2MASS (e.g. Kirkpatrick et al., 1999; Burgasser et al., 2003b) and DENIS (e.g. Delfosse et al., 1999b; Delfosse & Forveille, 2001; Kendall et al., 2004; Martín et al., 2004; Phan-Bao et al., 2008) along with the optical/far-red Sloan Digital Sky Survey, SDSS (e.g. Geballe et al., 2002; Kirkpatrick, 2005) now account for the majority of known brown dwarfs.

These wide-area imaging surveys detect primarily isolated brown dwarfs. Although they can detect wide-separation components beyond ~1000 AU (Gizis et al., 2001), they are insensitive to close-in companions of normal main sequence stars due to the stellar glare.

One isolated brown dwarf at about 100 pc, MACHO–179–A, was discovered through microlensing surveys (Poindexter et al., 2005).

Current census As of mid-2010, the online compilation of Gelino et al. (2010) lists more than 500 M dwarfs (representing an incomplete listing of all known M dwarfs, only some of which are brown dwarfs), 600 L dwarfs (many but not all of which are brown dwarfs, e.g. Reid et al., 1999), and 200 T dwarfs (all of which are brown dwarfs).[1] The majority of L and

---

[1] That some objects of class M and L are stars and some are brown dwarfs arises because some of the youngest (low-mass) brown dwarfs can have the same T_{eff} as the higher mass stars, and therefore similar spectral characteristics. Equivalently, the atmospheric gas and cloud chemistry in brown dwarfs is similar to that in gas giant exoplanets for corresponding values of T_{eff}.

T dwarfs are isolated field objects, although some 15–20% occur as binaries, both as L dwarfs (Reid et al., 2008) and T dwarfs (Burgasser et al., 2003c).

Amongst the diversity of scientific findings are metal-poor brown dwarfs with halo kinematics (Burgasser et al., 2003a), and brown dwarfs surrounded by circumstellar disks and planetary-mass objects. For some, light curve modulation in the infrared has been attributed to rotation, the existence of grain-free and cooler grain-bearing cloudy regions, and the implied existence of extreme wind and weather patterns (Artigau et al., 2009; Goldman et al., 2008), analogous to those observed on Jupiter and, through secondary eclipse observations, of transiting exoplanets (§6.5).

Ongoing and future surveys For the coolest brown dwarfs, the census of even the nearest solar neighbourhood, within 10–20 pc, remains highly incomplete because of their low luminosity. Deeper sky surveys with brown dwarf searches high in their scientific priorities, are now underway or planned.

The UK Infrared Digital Sky Survey (UKIDSS, Lawrence et al., 2007) has been operational at the 3.8-m UKIRT since 2005. It is some 3 mag more sensitive than 2MASS. The first L and T dwarf discoveries have been reported (Kendall et al., 2007a,b; Lodieu et al., 2007; Warren et al., 2007; Lodieu et al., 2009; Goldman et al., 2010), the nearest being a T dwarf companion to an L dwarf primary at 8 pc (Scholz, 2010).

The Canada–France Brown Dwarf Survey (CFBDS, Delorme et al., 2008a,b) is using CFHT–MegaCam to find ultracool dwarfs. The survey is expected to discover 500 L or M dwarfs later than M8, and 100 T dwarfs.

The Wide-field Infrared Survey Explorer (WISE, Wright, 2008, 2010a) is a NASA MIDEX space mission, with a 0.4-m aperture, mapping the sky in four infrared bands (3.3, 4.7, 12, 23 μm), and launched in December 2009. The all-sky survey started in January 2010, and the first data release is due in April 2011.

The Visible and Infrared Survey Telescope for Astronomy (VISTA, Emerson & Sutherland, 2010) is a 4-m aperture, 1°.6 field infrared telescope at ESO Paranal, operating between 0.85–2.3 μm with broad-band filters at Z, Y, J, H, K_S. Surveys commenced in early 2010.

The first of the Panoramic Survey Telescope & Rapid Response System (Pan-STARRS, Kaiser, 2007) 1.8-m 3° telescopes began surveys from Hawaii in May 2010.

The Large Synoptic Survey Telescope (LSST, Ivezic et al., 2008), with an 8.4-m aperture and 3°.5 field, targets first light in 2015. From Cerro Pachón, it aims to image the sky once every three nights. Faint, very nearby, fast-moving brown dwarfs should be detected.

9.1.3 Luminosity and age

Once their nuclear energy source is exhausted, brown dwarfs cool and fade quickly, with density playing an im-

> **MK classification:** The MK classification itself was built on the original Harvard sequence of spectral types, through a progressive revision of the list of defining standards (Morgan et al., 1943; Johnson & Morgan, 1953; Morgan & Keenan, 1973; Keenan, 1985). A convenient summary of its broader nomenclature is given by Garrison (2000).
>
> As for stars, the progressive brown dwarf spectral types are broadly a decreasing function of temperature, but with classification in practice based on their spectral features. As described by Morgan & Keenan (1973) '*The MK system is a phenomenology of spectral lines, blends, and bands, based on a general progression of colour index and luminosity. It is defined by an array of standard stars, located on the two-dimensional spectral type versus luminosity class diagram. These standard reference points do not depend on values of any specific line intensities or ratios of intensities; they have come to be defined by the appearance of the totality of lines, blends, and bands in the ordinary photographic region. The definition of a reference point, then, is the appearance of the spectrum 'as in' the standard star.*'

portant role (Chabrier et al., 2009). At the upper mass range, $\sim 75-80 M_J$, the density of a brown dwarf is set by electron degeneracy pressure as in white dwarfs, while at the lower end, $10-20 M_J$, it is set primarily by Coulomb pressure, as in planets. The result is that the radii of all brown dwarfs are approximately $1 R_J$, varying by only 10–15% over $\sim 10-80 M_J$ (Figure 6.34).

Solar-metallicity power-law relations which characterise the evolution of older sub-stellar mass objects (Burrows & Liebert, 1993; Marley et al., 1996) are summarised by Burrows et al. (2001, eqn 1–5) as

$$L = 4 \times 10^{-5} L_\odot \left(\frac{10^9 \text{ yr}}{\tau}\right)^{1.3} \left(\frac{M}{0.05 M_\odot}\right)^{2.64} \left(10^3 \kappa_R\right)^{0.35}$$

$$T_{\text{eff}} = 1550 \text{ K} \left(\frac{10^9 \text{ yr}}{\tau}\right)^{0.32} \left(\frac{M}{0.05 M_\odot}\right)^{0.83} \left(10^3 \kappa_R\right)^{0.088}$$

$$M = 35 M_J \left(\frac{g}{10^3}\right)^{0.64} \left(\frac{T_{\text{eff}}}{1000 \text{ K}}\right)^{0.23}$$

$$\tau = 1.0 \text{ Gyr} \left(\frac{g}{10^3}\right)^{1.7} \left(\frac{1000 \text{ K}}{T_{\text{eff}}}\right)^{2.8}$$

$$R = 6.7 \times 10^4 \text{ km} \left(\frac{g}{10^3}\right)^{-0.18} \left(\frac{T_{\text{eff}}}{1000 \text{ K}}\right)^{0.11}, \quad (9.1)$$

where τ is the age, g the surface gravity (m s^{-2}), and κ_R is an average Rosseland mean opacity (m^2 kg^{-1}).

The weak dependence of radius on mass and temperature means that all brown dwarfs follow a similar trajectory in the (L, T_{eff}) plane. Deriving the mass from its L or T_{eff} therefore requires an estimate of its age. For example, a brown dwarf with $L \sim 10^{-4} L_\odot$ could be a $70 M_J$ object at $\tau = 10^9$ yr, or a $10 M_J$ sub-brown dwarf at $\tau = 10^7$ yr (e.g. Burrows et al., 2001, their figure 1).

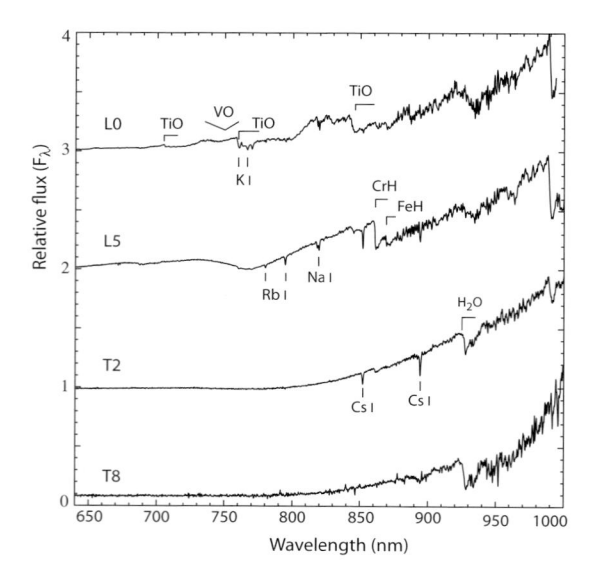

Figure 9.1: Representative optical spectra of L and T-type brown dwarfs. T_{eff} ranges from ~2100 K at L0 to ~900 K at T5. The most prominent spectral features are labeled. From Reid & Metchev (2008, Figure 3).

9.1.4 Classification

Brown dwarfs are classified spectroscopically through an extension of the MK stellar spectral type sequence, viz. OBAFGKM(SRN), as M, L, T and possibly Y (Figure 9.1). Collectively, late-M, L and T dwarfs are also known as *ultracool dwarfs*.

Atmospheric temperatures range from 2000–3000 K down to 500–600 K, so that their thermal emission lies primarily beyond 1 μm, e.g. peaking at 1.5 μm for a T_{eff} = 2000 K black body. Low and decreasing temperatures through the late M, L, and T sequence result in a wide variety of near-infrared spectral features as brown dwarfs age and cool, from relatively narrow lines of neutral atoms to broad molecular bands, all of which may have different dependencies on T_{eff}, $\log g$, and [Fe/H].

The low temperature conditions accelerate condensation out of the gas state and into the formation of grains, such that the effects of condensates, clouds, molecular abundances, and atomic opacities are all important in interpreting their spectral properties. In the case of orbiting rather than isolated brown dwarfs, supplementary irradiation by the central host star further affects their reflection spectra and albedos. Despite these complexities, evolutionary models broadly explain the systematic features of sub-stellar mass objects over three orders in mass and age, and some factor 30 in temperature (e.g. Burrows et al., 2001).

Spectral features Between 2000–3000 K, the defining characteristics of spectral class M are an optical spectrum dominated by the metal-oxide absorption bands of TiO and VO, with metal hydride (MgH, CaH) absorption in the optical and H_2O bands in the near infrared becoming more prominent in the later spectral types.

Below 2000–2100 K, L dwarfs are characterised by strong metal hydride bands (FeH, CrH, MgH, CaH), prominent alkali metal lines (Na I, K I, Cs I, Rb I), and strong near-infrared absorption of H_2O and CO (Kirkpatrick et al., 1999; Basri et al., 2000). Below 1700 K, CH_4 forms in the outer atmosphere.

Below 1200–1300 K, T dwarfs are characterised by prominent CH_4 absorption (Burgasser et al., 2002). Some of the coolest known, of spectral class T9, have T_{eff} ~ 700 K (Leggett et al., 2009).

Below 500–600 K, NH_3 is predicted to make a significant contribution to the near- and mid-infrared spectrum (Kirkpatrick, 2005). An additional Y class, with prototypes and possible distinguishing spectral features, has been proposed to encompass such objects (Burrows et al., 2003b; Burningham et al., 2008; Delorme et al., 2008a; Eisenhardt et al., 2010). Example theoretical predictions are shown in Figures 9.2 and 9.3.

The longer cooling time of the more massive objects suggests that most field brown dwarfs should lie at the upper end of the allowed mass range. Relatively small numbers show spectral features suggestive of low surface gravities (strong VO absorption, weaker alkali metal lines, and weaker hydride bands), implying lower masses and younger ages (Kirkpatrick et al., 2008; Cruz et al., 2009). Cruz et al. (2009) accordingly proposed an extension of the spectral classification for L0 to L5-type dwarfs to include three gravity classes.

9.1.5 Recognising brown dwarfs

Distinguishing high-mass brown dwarfs from low-mass stars primarily relies on accurate estimates of temperature and luminosity. Atmospheric CH_4 can be accumulated by older brown dwarfs, as for the T dwarf prototype GJ 229 B (Nakajima et al., 1995), and its spectral presence rules out an object of stellar mass. The presence of lithium may also suggest that the object is substellar, although very young stars may contain residual unburned lithium, while more massive objects like the Sun can retain lithium in their cooler convective atmospheres. Conversely, the more massive brown dwarfs, ≳ 65M_J, may be hot enough to deplete their lithium when young. A weakening at lower gravities is also predicted from model atmospheres (Kirkpatrick et al., 2008).

The lower temperature range of the L dwarfs, 700–2000 K overlaps with the upper range of the 'hot Jupiters', whose predicted temperatures are expected to reach 1200–1500 K. In consequence, there are expected to be similarities in their spectral appearance, as well as in their atmospheric temperature/density profiles.

The weak mass–radius dependence means that only density rather than radius can distinguish low-mass brown dwarfs from high-mass planets (Luhman, 2008).

9.1 Brown dwarfs

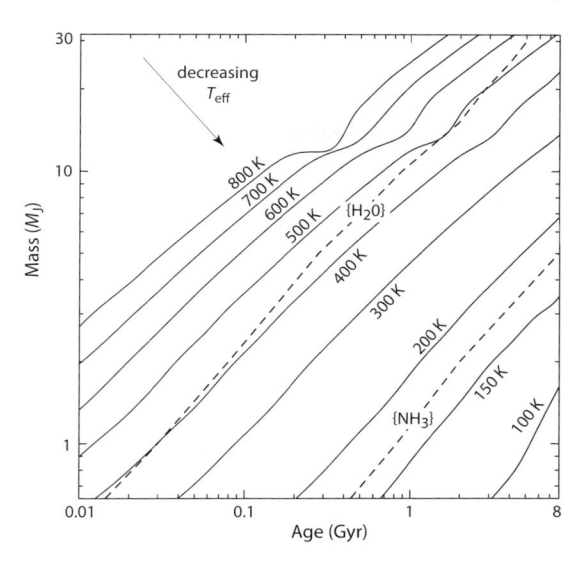

Figure 9.2: Brown dwarf mass versus age for $T_{eff} < 800$ K. At constant mass, increasing age corresponds to a declining T_{eff}. The condensation curves for H_2O and NH_3 (dashed lines) indicate ages at which condensation first occur in the outer atmosphere. The kink near $13 M_J$ is a consequence of deuterium burning. From Burrows et al. (2003b, Figure 2), reproduced by permission of the AAS.

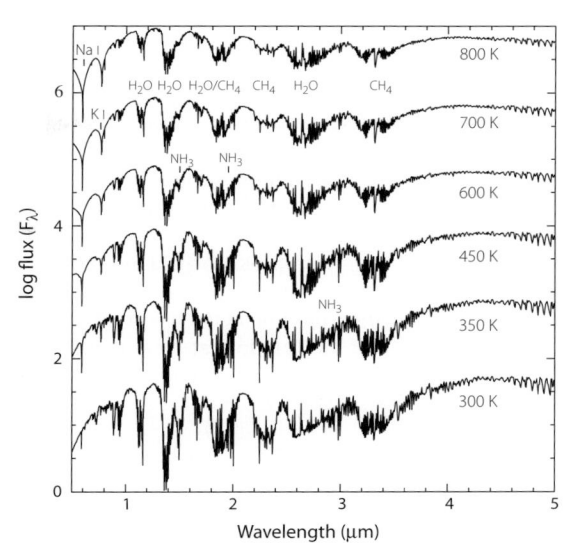

Figure 9.3: Expected spectroscopic characteristics of cool brown dwarfs in the range covered by JWST–NIRSpec, based on the model calculations by Burrows et al. (2003b). The main spectral features expected at the various temperatures are identified. NH_3 appears below ~600 K, and the alkali metal lines (Na I, K I), shortward of 1 μm, weaken below ~450 K. From Reid & Metchev (2008, Figure 13).

9.1.6 Other properties

X-ray and radio emission Convection combined with rapid rotation leads to the development of tangled magnetic fields near their surface, reaching 0.01–0.1 T (Berger, 2006). Quiescent X-ray emission, first detected for Cha Hα1 (Neuhäuser & Comeron, 1998), as well as X-ray flares, first detected for LP 944–20 (Rutledge et al., 2000), are now known for a number of objects.

As for M dwarfs (§7.6.1), brown dwarfs may also possess detectable radio emission, attributed to cyclotron radiation due to non-relativistic electrons in the coronal plasma. Radio emission was also first detected from LP 944–20 (Berger et al., 2001). An overall detection rate of around 10% for 90 objects of spectral type M5–T8 was given by Berger (2006).

Occurrence as binary companions Of brown dwarfs detected in the imaging surveys such as DENIS, 2MASS, and SDSS, the majority are isolated objects. A small subset, around 5% (Gizis et al., 2001, 2002), have also been detected in these surveys as wide common proper motion companions of nearby main sequence stars.

Probing their occurrence as companions to normal stars at smaller angular separations has been addressed by direct imaging, radial velocity, and astrometric surveys – including as by-products in the search for planets. One brown dwarf–brown dwarf eclipsing binary pair has been reported (Stassun et al., 2006).

Imaging surveys have targeted mainly nearby objects ($\lesssim$ 50 pc) to maximise the angular separation, and/or young stars ($\lesssim$ 300 Myr) to enhance the prospects of finding more luminous companions. Surveys have used a combination of coronagraphy (e.g. Oppenheimer et al., 2001), HST (e.g. Brandner et al., 2000; Lowrance et al., 2005), and ground-based adaptive optics (e.g. McCarthy & Zuckerman, 2004; Neuhäuser & Guenther, 2004; Carson et al., 2005).

Overall, for the orbital separation range 0–3 AU, these surveys have confirmed the 'brown dwarf desert' originally found from radial velocity observations, where only $\lesssim$ 0.5% of FGK dwarfs are found to have brown dwarf companions (Marcy & Butler, 2000). At orbital separations 10–1000 AU the occurrence rate rises by a factor 5–10 to reach a fraction of 2–4% (Neuhäuser & Guenther, 2004). Beyond separations of 1000 AU, the frequency of brown dwarf companions to 2MASS stars is consistent with the frequency of stellar secondaries, at around 18%.

Taken together with the results of radial velocity surveys for stellar binaries, there is a strong preference for near-equal mass stellar systems to exist at small physical separations, presumably reflecting the effects of competitive accretion during the star formation process. Wide systems, $\gtrsim$ 100 AU, originate from the gravitational association of independent stellar cores, and consequently span a greater range of mass ratios.

Disks and planets Planet formation in protoplanetary disks, which is marked by the growth and crystallisation of sub-μm dust grains accompanied by dust settling towards the disk mid-plane, also appears to take place in a similar manner in brown dwarf disks (Apai et al., 2005; Payne & Lodato, 2007). Paralleling the instances of stars orbited by circumstellar disks, brown dwarfs and planets, there are growing numbers of examples of brown dwarfs themselves known to be orbited by dust disks (e.g. Zapatero Osorio et al., 2007; Luhman & Muench, 2008, and references) and planets, amongst them 2M J1207 (Chauvin et al., 2005a) and 2M J0441 (Todorov et al., 2010b), both detected by direct imaging (§7.5). Even some of the very lowest mass objects, the sub-brown dwarfs, are now known to be surrounded by disks (Luhman et al., 2005b).

9.1.7 Formation

Brown dwarfs are roughly twice as numerous as main-sequence stars, although contributing no more than ~15% to the total local disk mass. This conclusion was reached by Reid et al. (1999), who represented the density of L dwarfs as a power-law mass function $\psi(M) \propto M^{-\alpha}$, with $1 < \alpha < 2$. With their results favouring a value nearer to the lower limit, adopting $\alpha = 1.3$ implies that the local space density of $10 - 78 M_J$ brown dwarfs is of order $10\,\mathrm{pc}^{-3}$.

That brown dwarfs are comparable in number to hydrogen-burning stars implies that a robust theory of star formation should also explain their origin. The problem with models attributing their formation to the same process has been that the mass of brown dwarfs is some two orders of magnitude smaller than the average Jeans mass in star-forming clouds. If the average Jeans mass sets an approximate lower limit to the stellar mass (Larson, 1992; Adams & Fatuzzo, 1996; Padoan et al., 1997; Elmegreen, 1999), the existence of a significant population of brown dwarfs becomes problematic.

Turbulent fragmentation   Padoan & Nordlund (2002, 2004) showed that supersonic turbulence in molecular clouds can generate a complex density field with very large density contrasts, with higher densities resulting in a lower Jeans mass (Equation 10.1). In simulations, a fraction of brown dwarf mass cores formed by turbulent flow are dense enough to be gravitationally unstable.

Based on the density, temperature, and Mach number typical of cluster-forming regions, Padoan & Nordlund (2004) propose that star formation by turbulent fragmentation can also account for the observed brown dwarf and sub-brown dwarf abundances. Specifically, they found a good match to the distribution of stellar and substellar members of the young cluster IC 348 determined by Luhman et al. (2003).

Other processes   Other mechanisms which have been considered to explain the presence of brown dwarfs rely either on forming a high gas density (and thus a low

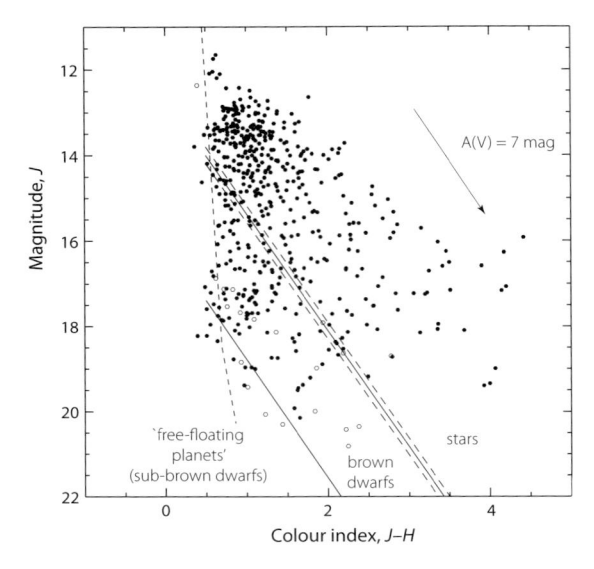

Figure 9.4: Young brown dwarfs and sub-brown dwarfs ('free-floating planets') in Orion, in the colour–magnitude diagram (J versus $J - H$). Open circles are uncertain data points. The dotted line is an approximate zero-reddening track. Solid lines, parallel to the $A(V) = 7$ reddening vector, divide the population into stars, brown dwarfs and planet mass, using the predictions of Burrows et al. (1997) and an age of 1 Myr. Dashed lines correspond to 0.3 Myr and 2 Myr, indicating effects of an age spread on the classification. The effect is similar at the planetary boundary. From Lucas & Roche (2000, Figure 1a), © John Wiley & Sons, Inc.

Jeans mass) in the pre-stellar cores (as in turbulent fragmentation), or on the halting of accretion once a low-mass fragment has formed. Multiple cloud fragmentation during collapse (Burkert et al., 1997), or core accretion followed by the capture of gas in protostellar disks (Bodenheimer, 1998), tend to form a gravitationally bound multiple system, requiring efficient mechanisms to rapidly disrupt the resulting orbits.

Plausible processes have been reviewed by Whitworth et al. (2007), and include (a) hierarchical fragmentation (Boss et al., 2000; Whitehouse & Bate, 2006); (b) disk fragmentation in which a massive circumstellar disk is unstable to gravitational fragmentation, potentially induced by a stellar fly-by (Bate et al., 2002; Matzner & Levin, 2005; Whitworth & Stamatellos, 2006; Clarke et al., 2007; Goodwin & Whitworth, 2007; Stamatellos et al., 2007; Whitworth et al., 2007), with Goodwin & Whitworth (2007) suggesting that brown dwarfs are formed in disks beyond 100 AU, with the resulting binaries being disrupted by passing stars; (c) the related mechanism of gravitational fragmentation of infalling filamentary gas into stellar clusters (Bate et al., 2002; Bonnell et al., 2008); (d) the ejection of newly-formed fragments or stellar embryos in multiple systems or circumbinary disks, halting post-formation accretion

9.2 Free-floating objects of planetary mass

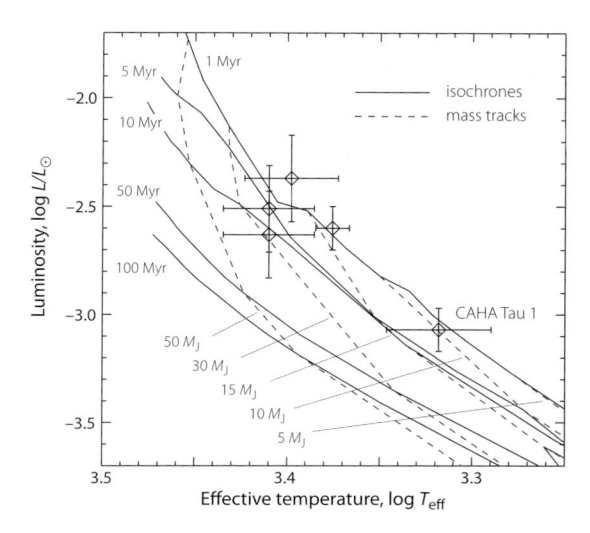

Figure 9.5: Hertzsprung–Russell diagram illustrating the location of some of the low-mass members of the Taurus star forming region, all with spectral types ≥ M9.0. Luminosities are derived using bolometric corrections, and a reddening $A_V = 2.3$ mag. Isochrones and mass tracks are from Chabrier et al. (2000). Assuming that the candidates are cluster rather than background objects, CAHA Tau 1 falls between the 1–5 Myr isochrones and, depending on age, its properties translate into a mass between $5-15 M_J$. From Quanz et al. (2010, Figure 10), reproduced by permission of the AAS.

such that the fragment retains a low mass (Boss, 2001; Reipurth & Clarke, 2001; Smith & Bonnell, 2001; Bate et al., 2002; Kroupa & Bouvier, 2003; Jiang et al., 2004; Bate & Bonnell, 2005); and (e) the photo-evaporation of a collapsing pre-stellar core, for example through the wind of nearby supernovae or OB stars (Whitworth & Zinnecker, 2004; Gahm et al., 2007).

9.2 Free-floating objects of planetary mass

Although the lower mass limit for bodies formed by the fragmentation of collapsing interstellar gas clouds remains somewhat uncertain, the mass function observed in Orion and other young star-forming regions already gave the first indications that isolated objects even down to planetary mass could form as a by-product of the star formation process (Tamura et al., 1998; Béjar et al., 1999; Oasa et al., 1999; Lucas & Roche, 2000; Hillenbrand & Carpenter, 2000; Najita et al., 2000).

Searches for isolated objects of (sub-)brown dwarf or planetary mass have focused much attention on star-forming regions and young clusters. These offer the advantage that, as a result of their passive cooling, isolated sub-stellar objects at an age of only a few Myr are some three orders of magnitude more luminous than those with ages of a few Gyr. Once detected photometrically, masses can be derived from a comparison of the observed luminosity, or observed temperature, with theo-

Table 9.1: Candidate cluster or 'free-floating planets', ordered by discovery chronology. Only those objects that appeared in the `exoplanet.eu` *compilation on 2010 November 1 are given. Objects lie at a distance of about 100 pc (ρ Oph 4450), 160 pc (Cha 110913) and 440 pc (S Ori 68–70).*

Name	$M(M_J)$	T (K)	Reference
S Ori 68	5	–	Zapatero Osorio et al. (2000)
S Ori 70	3	1100	Zapatero Osorio et al. (2002)
Cha 110913	8	1350	Luhman et al. (2005a)
CAHA Tau 1	10	2080	Quanz et al. (2010)
CAHA Tau 2	11.5	2280	Quanz et al. (2010)
ρ Oph 4450	2–3	1400	Marsh et al. (2010)

retical evolutionary tracks (e.g. Burrows et al., 1997), although in star forming regions, photometric or spectroscopic calibration is complicated by extinction.

Reports of free-floating objects of planetary mass have come from deep optical and near-infrared imaging of the Chamaeleon I dark cloud (Tamura et al., 1998; Oasa et al., 1999; Luhman et al., 2008); of the star cluster around σ Ori (Zapatero Osorio et al., 2000; Martín et al., 2001; Bihain et al., 2009); of the Trapezium cluster in Orion (Lucas & Roche, 2000; Lucas et al., 2001, 2005, 2006; Weights et al., 2009); and of the Taurus star forming region (Quanz et al., 2010).

The continuum of properties between stars, brown dwarfs, and the suggested 'free-floating planets' are evident in Figure 9.4. An example of the Hertzsprung–Russell diagram (L versus $T_{\rm eff}$) used to estimate masses from theoretical evolutionary models is shown in Figure 9.5, although whether these particular objects are members of the Taurus star-forming region, or background sources, remains to be confirmed.

These observations suggest that the formation of brown dwarf and planetary mass objects are relatively abundant at least in some young clusters (e.g. Lodieu et al., 2008; Quanz et al., 2010, and references), although not in all (Lyo et al., 2006). They also imply that the formation of unbound planetary mass objects is completed within a few Myr. Masses may extend down to $\sim 3 M_J$ in the case of ρ Oph 4450 (Marsh et al., 2010) and S Ori 70 (Zapatero Osorio et al., 2002; Martín & Osorio, 2003), although the cluster membership and hence the mass of the latter were challenged by Burgasser et al. (2004).

Whether these objects are truly isolated was probed by a survey of the σ Ori cluster by Caballero et al. (2006). They discovered a M5–6 brown dwarf of $45 M_J$ at a separation of 4.6 arcsec from S Ori 68, or ~1700 AU. Such a wide separation is reminiscent of the brown dwarf–planet pairs 2M J1207 at 55 AU (Chauvin et al., 2005a), GQ Lup at 100 AU (Neuhäuser et al., 2005), and AB Pic at 250 AU (Chauvin et al., 2005b).

Objects classified as confirmed 'cluster planets' or 'free-floating planets' by Schneider (2010) are listed in Table 9.1. According to the nomenclature above, these should rather be referred to as sub-brown dwarfs.

10

Formation and evolution

10.1 Overview

Planetary systems, the solar system amongst them, are believed to form as inevitable and common by-products of star formation. For orientation, an overview of the processes described in this chapter is as follows.

The present paradigm starts with star formation in molecular clouds. Brown dwarfs are formed as the low-mass tail of this process, although some may be formed as a high-mass tail of planet formation. Gas and dust in the collapsing molecular cloud which does not fall directly onto the protostar resides in a relatively long-lived accretion disk which provides the environment for the subsequent stages of planet formation. Terrestrial-mass planets are formed within the disk through the progressive agglomeration of material denoted, as it grows in size, as dust, rocks, planetesimals and protoplanets. A similar process typically occurring further out in the disk results in the cores of giant planets, which then gravitationally accumulate their mantles of ice and/or gas.

As the planet-forming bodies grow in mass, growth and dynamics become more dominated by gravitational interactions. Towards the final phases, and before the remaining gas is lost through accretion or dispersal, the gas provides a viscous medium at least partially responsible for planetary migration. Some migration also occurs during these later stages as a result of gravitational scattering between the (proto-)planets and the residual sea of planetesimals. The final structural stabilisation of the planetary system may be affected by planet–planet interactions, until a configuration emerges which may be dynamically stable over billions of years.

The current observational data for exoplanet systems is broadly compatible with this overall picture. Other constraints are available from a substantial body of detailed observations of the solar system.

10.2 Star formation

The macrophysics of star formation deals with the formation of systems of stars, ranging from clusters to galaxies. It is the microphysics of star formation that is of most relevance to planet formation: how protostars acquire their mass via gravitational collapse, how the infalling gas loses its magnetic flux and angular momentum, and how the resulting stellar properties are determined by the medium from which they form.

The accepted paradigm for star formation is that gravitational instabilities in molecular clouds of gas and dust grains lead to gravitational collapse, fragmentation, and accretion (Shu et al., 1987; McKee & Ostriker, 2007), while for relatively massive stars, *competitive accretion* is an alternative (Zinnecker, 1982; Bonnell et al., 1997). Only the aspects most relevant for planet formation are outlined here, while a more comprehensive summary of 17 key steps in the current picture are listed by McKee & Ostriker (2007, Section 5).

Molecular clouds Molecular clouds are complex structures with wide density and composition variations, whose nature and scales are defined by turbulence. The gas, whose composition can be established from spectroscopy, consists primarily of H_2, along with H and He atoms and numerous other molecules (CO, CO_2, CH_4, H_2O, HCO, CH_3OH, NH_3, ...), with more than 150 species detected in the interstellar medium to date (Semenov et al., 2010).

The dust grains comprise sub-micron amorphous and crystalline silicates (both the olivine, Mg_2SiO_4–Fe_2SiO_4, and pyroxene, $MgSiO_3$–$FeSiO_3$, solid solution series), amorphous carbon, and many other species (Gail & Hoppe, 2010). Their size distribution is established from the extinction, scattering, and polarisation properties of the cloud as a function of wavelength. At low temperatures, the more volatile molecular gases condense onto the dust grains as icy mantles. Rising temperatures in the collapsing molecular clouds lead to the sublimation of the icy mantles and, in the innermost regions of the forming protoplanetary disk, the less volatile and more refractory dust grains.

Protostars and protostellar collapse Star formation starts with some process which initiates gravitational collapse in the molecular cloud. This process is uncertain, but may be related to highly supersonic turbulent

> **Protostellar, protoplanetary and debris disks:** During the early stages of star formation, the star-forming cloud, the accretion disk feeding the stellar embryo, and the stellar embryo itself, are usually referred to as *protostellar*. In the final phases of protostar evolution, as astronomical interest shifts to planet formation within the residual gas and dust, the disk is referred to as *protoplanetary*. There is, however, no specific point at which the disk becomes unambiguously 'protoplanetary', and the distinction between *protostellar disks* and *protoplanetary disks* is accordingly somewhat arbitrary.
>
> Much later in the process of planet formation, as planetesimals and protoplanets have formed and start a new phase of collisional attrition, the resulting circumstellar disks are referred to as *debris disks*.

flow (McKee & Ostriker, 2007).

Whatever the trigger, local density enhancements due to the compression of finite volumes within the flow become gravitationally unstable. According to the *Jeans instability criterion* (Jeans, 1902), this occurs if their size λ exceeds the critical Jeans length

$$\lambda_J = \left(\frac{\pi c_s^2}{G\rho}\right)^{1/2}, \qquad (10.1)$$

where $c_s^2 = kT/\mu$ is the isothermal sound velocity, μ is the mean molecular weight ($\sim 2m_H$ for a gas predominantly composed of H_2), $T \sim 10$ K is the gas temperature, and ρ is the local density. The corresponding Jeans mass is $M_{\rm Jeans} \equiv (4\pi/3)(\lambda_J/2)^3 \rho$.

If $\lambda > \lambda_J$ thermal pressure cannot resist self-gravity, and runaway collapse follows. This leads to an isothermal near free-fall contraction, during which the density increases by some 15 orders of magnitude. Contraction continues until the central region becomes opaque, compressional heating starts to exceed radiative cooling of the gas–dust mixture, and the temperature rises.

The gravitational time scale is usually expressed in terms of the *free-fall time*, the time taken for a pressure-free, spherical cloud to collapse to a point owing to its self-gravity (McKee & Ostriker, 2007, eqn 14)

$$t_{\rm ff} = \left(\frac{3\pi}{32 G\rho}\right)^{1/2} \simeq 6.4 \times 10^4 \;{\rm yr} \; \left(\frac{M}{M_\odot}\right)\left(\frac{T}{10\,{\rm K}}\right)^{-3/2}. \quad (10.2)$$

Collapse proceeds through the dissociation of H_2 and the ionisation of H, leading to the formation of a protostellar embryo, which continues to grow in mass by accreting material from its local environment.

The time scales for the early stages of spherically symmetrical collapse of low-mass stars are short: $\sim 1\,t_{\rm ff}$ from initial collapse to the protostellar embryo, and $\sim 4\,t_{\rm ff}$ for the subsequent 80% of the remaining stellar mass. Accretion of the final 20% of mass occurs on a much longer time scale of about 0.5 Myr for a $1M_\odot$ star (Wuchterl & Tscharnuter, 2003).

The resulting stars shine by thermonuclear fusion, with stable hydrogen burning occurring for masses above $\sim 0.07 - 0.09 M_\odot$ ($\sim 75 - 95 M_J$), depending on chemical composition, when the central temperature triggers nucleosynthesis. The most massive stars thereafter evolve rapidly, creating new elements by nucleosynthesis, and dispersing them through gaseous outflows or supernova explosions. Part of the chemically-enriched material remains in the gas phase, while part condenses into solid dust grains, together providing material for subsequent generations of star formation.

Within this broad picture, at least two aspects of the microphysics of star formation remain uncertain: the process by which an unstable region fragments into the many sub-units necessary to create star clusters, and the process which leads to the significant fraction of binary and multiple stars (McKee & Ostriker, 2007).

Brown dwarfs, which occupy the mass range of about 13–80 M_J, are broadly considered to have been formed by the same physical process, and may experience partial hydrogen burning, and deuterium and/or lithium burning according to mass (§9.1.1). Below about $12-13 M_J$, objects derive no luminosity from thermonuclear fusion at any stage in their evolutionary lifetime (Burrows et al., 1993; Saumon et al., 1996). The discovery of even lower mass sub-stellar objects in young clusters, down to $\sim 1 M_J$ or below, may still be consistent with formation through the same process (§9.1.7).

10.3 Disk formation

The study of protostellar and protoplanetary disks is an extensive field, with a substantial body of both observational and theoretical support. Only an outline of the phases of disk evolution most relevant for planet formation are given here. Recent reviews give further details and additional references to the relevant literature (e.g. Ciesla & Dullemond, 2010).

10.3.1 Initial collapse

The material within a gravitational unstable region of a molecular cloud inevitably carries some angular momentum, which prevents the collapsing material from falling directly onto the stellar embryo. Instead it falls, by symmetry, onto a flat rotating disk in the plane through the centre of mass and orthogonal to the total angular momentum of the cloud.

The fraction of the gas and dust falling onto the disk rather than directly onto the star depends on the angular momentum of the collapsing cloud, but may exceed 90%. For high disk/protostar mass ratios the disk is gravitationally unstable, spiral waves develop, and rapid mass accretion onto the star continues until the mass ratio falls below the Toomre instability limit. For a thin axisymmetric collisionless disk, Toomre (1964) showed that a non-axisymmetric perturbation will be gravitationally

10.3 Disk formation

unstable if the parameter

$$Q = \frac{c_s \Omega}{\pi G \Sigma} < 1, \quad (10.3)$$

with $Q \simeq 1$ implying marginal (in)stability. Here, $c_s = (\gamma P/\rho)^{1/2}$ is the sound speed, Ω is the angular velocity, and Σ is the disk *surface density*, i.e. the projected mass per unit area at a given radial distance.

The angular momentum of the residual disk material prevents its rapid inspiraling onto the star. Any subsequent inward migration of mass requires an outward transfer of angular momentum. Such a process only occurs on the time scale of viscous disk accretion, which is much longer than the formation time of the initial stellar embryo, and it is the inefficiency of this (incompletely understood) accretion process which makes planet formation possible.

Disk evolution, proceeding from the massive accretion disks to more tenuous protoplanetary disks, is determined by viscosity, accretion onto the central star, grain coagulation, and photoevaporation.

A plausible overall evolutionary chronology is shown in Figure 10.1, where ages are reckoned from the time of initial collapse. Massive accretion disks appear typical for stars with ages ≲ 1 Myr. These evolve to lower mass protoplanetary disks, with little or no ongoing accretion, which usually fall in the age range 1–10 Myr. Disks older than ~10 Myr are generally non-accreting debris disks, i.e. secondary disks generated by collisional process at later evolutionary stages (§10.3.4).

10.3.2 Young stellar objects

Within the newly-formed accretion disks, young stellar objects mark the sites of the earliest stages of disk and planet formation. They are generally characterised by an infrared excess due to hot dust in the disk, and possibly also an ultraviolet excess attributed to hot spots resulting from material accreting onto the stellar surface. Young stellar objects show evidence of stellar winds and outflows, including the Herbig–Haro type collimated jets (Figure 10.2), phenomena which accompany mass accretion that interacts with magnetic fields and rotation (Konigl & Pudritz, 2000; Pudritz et al., 2007).

Classification Young stellar objects are frequently assigned to one of four classes (André et al., 2000), according to their spectral index over the wavelength region 2.5–10 μm and beyond

$$\alpha_{IR} = \frac{d \log \lambda F_\lambda}{d \log \lambda}, \quad (10.4)$$

and which also reflects their successive evolutionary stages (e.g. Adams et al., 1987; Gail & Hoppe, 2010):

Class 0: sources with a very faint protostar in the optical and near infrared, and with a spectral energy distribution peaking in the far infrared or sub-mm region.

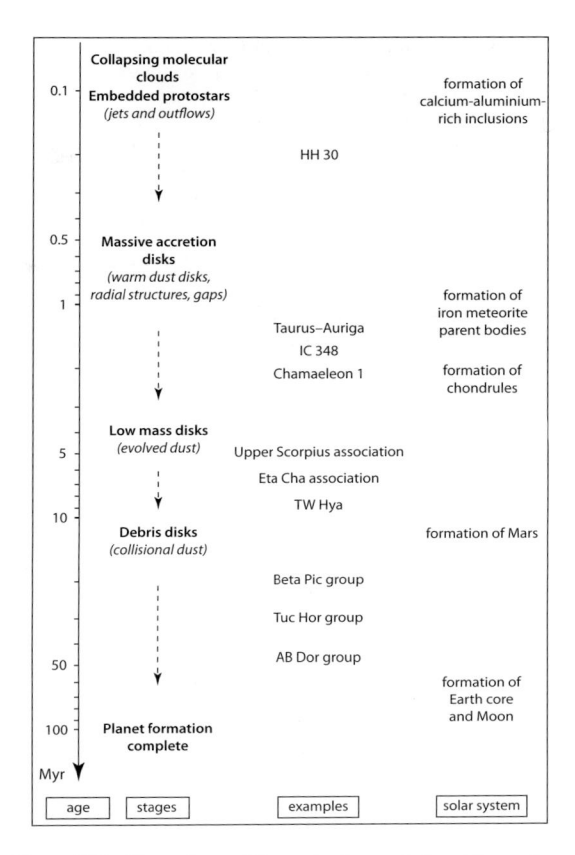

Figure 10.1: Chronology of the early stages of planet formation, from an arbitrary initial time, showing some of the main evolutionary stages, some representative astronomical examples, and some specific epochs relevant for the solar system (see also §12.5). The figure is an adaptation of a more detailed chronology given by Apai & Lauretta (2010, Figure 1.3).

These are sources in which the molecular cloud is starting to collapse, and in which the stellar embryo and its associated disk are starting to become established;

Class I: sources with a spectral energy distribution flat or rising into the mid-infrared, $\alpha_{IR} > 0$. These are protostars with circumstellar disks and envelopes. During this phase, stars accrete most of their final mass from the disk and the surrounding gas and disk envelope;

Class II: sources with a spectral energy distribution declining into the mid-infrared, $-1.5 < \alpha_{IR} < 0$. These are the pre-main sequence stars with observable accretion disks (*classical T Tauri stars*);

Class III: sources with little or no infrared excess, $\alpha_{IR} < -1.5$. These are pre-main sequence stars without detectable accretion, in which the initial disk has been largely cleared (*weak-lined T Tauri stars*).

The early evolutionary stages (classes 0 and I) are optically hidden by the dust of the collapsing envelope, and are only observable in the far-infrared. The optically visible class II–III sources reach some 90% of their final

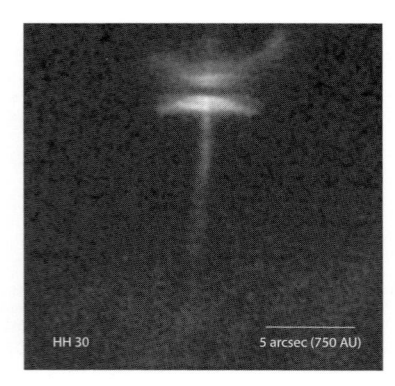

Figure 10.2: The Herbig–Haro object HH30, at a distance of ~150 pc, observed by HST. Two thin jets flow outwards from the young stellar object in the centre. The two lenticular regions are scattered light from dust in the disk. The dark central lane is the accretion disk observed edge-on (courtesy NASA/ESA/STScI).

mass at the transition between the classical and weak-lined T Tauri phases, which occurs for solar-mass stars some 2×10^5 yr after the onset of collapse.

Understanding the chemical and isotopic composition and evolution of protoplanetry disks during these different phases is limited by the sensitivity and spatial resolution of the necessary (sub)mm interferometric observations. Nevertheless, radio-interferometric studies, Spitzer and Herschel observations, laboratory experiments, and meteoritic measurements are all contributing to a developing picture of the relevant physical and chemical processes (Semenov et al., 2010; Min, 2010).

10.3.3 Protoplanetary disks

Over the next few million years, most of the remaining disk mass is accreted, the residual gas is dispersed, and the star proceeds on its final evolution to the main sequence as a class III object. This final phase of protostar evolution is loosely referred to as protoplanetary, and it is within the *protoplanetary disk* that the processes of planet formation are developing. As noted at the start of this chapter, there is no specific starting point at which the disk becomes unambiguously protoplanetary, and the distinction between protostellar disks and protoplanetary disks is somewhat arbitrary.

At early epochs, when the disk mass is largest, *active circumstellar disks* are characterised by the conversion of gravitational energy into thermal radiation as gas flows inwards and accretes onto the star (Pringle, 1981; Frank et al., 2002). Towards the end of the accretion and gas clearing phase, *passive circumstellar disks* derive most of their luminosity from reprocessed starlight.

With certain assumptions, for example that the disk absorbs all incident stellar radiation and re-emits it as a single temperature blackbody, the shape and radial temperature profile of a passive disk can be estimated (Adams & Shu, 1986; Kenyon & Hartmann, 1987; Chiang & Goldreich, 1997; Armitage, 2007a). The vertical structure of the disk, determined by hydrostatic equilibrium, may flare, i.e. become thicker with increasing radius, in which case the outer disk regions will intercept a larger fraction of the stellar photons, leading to higher temperature regions at larger radial distances.

The spectral energy distribution of such a disk can be derived by integrating the temperature of the disk at each radius, weighted by the contributing annular area. Mid-plane temperature profiles vary from $T \propto r^{-1/2}$ for a flared disk, to the more rapidly declining $T \propto r^{-3/4}$ for a flat disk (Kenyon & Hartmann, 1987). The highest temperatures in a static disk are reached at its innermost edge, closest to the star.

Chiang & Goldreich (1997) showed that a passive disk model comprising a hot surface dust layer that directly re-radiates half of the stellar flux, combined with a cooler disk interior that absorbs and re-emits the other half as thermal radiation, reproduces most observed spectral energy distributions when allowing for a flaring geometry (see also Dullemond et al., 2007).

Minimum mass solar nebula Attempts to infer the general structure of protoplanetary disks were made by Weidenschilling (1977b) and Hayashi (1981). They reconstructed the solar nebula by taking the mass of solid material presently in the form of planets and asteroids, distributing it across its 'feeding zone' to form a smooth distribution, adding the amount of H and He at each location to produce a solar composition, and arguing that the resulting structure represented the least amount of initial material needed to form the solar system. This *minimum mass solar nebula* was described by a radial surface density profile, $\Sigma(r)$, and temperature profile, $T(r)$ given by

$$\begin{aligned}\Sigma(r) &= 1.7 \times 10^4 \, r_{\text{AU}}^{-3/2} \, \text{kg m}^{-2}\,, \\ T(r) &= 280 \, r_{\text{AU}}^{-1/2} \, \text{K}\,, \end{aligned} \quad (10.5)$$

where r_{AU} is the radial distance from the Sun in AU. This structure produced a disk with a mass between $0.01 - 0.07 M_\odot$ out to 40 AU. Although subject to uncertainties, the concept of a minimum mass solar nebula remains a point of reference in quantifying the mass contained both within protoplanetary disks, and within more mature planetary systems (Kuchner, 2004).

Estimating the mass of real protoplanetary disks is not straightforward. The largely molecular hydrogen gas cannot be measured directly, so that mass estimates are based on mm observations which trace the distribution of solid bodies up to ~0.01 m in size, and futher assumptions about the size distribution and dust-to-gas ratio (Piétu et al., 2007; Andrews & Williams, 2007). Mass estimates obtained in this way are reasonably consistent with the range given by the minimum mass solar nebula hypothesis (Williams et al., 2005), scaling roughly linearly with the mass of the central star (Klein et al., 2003).

10.3 Disk formation

Disk viscosity and turbulence Inferred accretion rates for the active circumstellar disks observed around classical T Tauri stars are estimated at $10^{-9} - 10^{-7} M_\odot \, \mathrm{yr}^{-1}$ (Gullbring et al., 1998). But it has been a challenge to understand how or why disk material should flow inwards at these high rates. To do so, material must lose significant angular momentum, which has been problematic to explain theoretically.

The standard picture is that some form of frictional forces act to redistribute the matter in the disk, allowing most of the matter to move inwards, while pushing some outwards to absorb the excess angular momentum (e.g. Shakura & Sunyaev, 1973; Lynden-Bell & Pringle, 1974; Hartmann et al., 1998). Over time, the disk is therefore expected to grow in radius while decreasing in mass, resulting in a decreasing accretion rate. This appears to be consistent with the statistics of accretion rate for sources of different ages (Sicilia-Aguilar et al., 2004). Alternatively, angular momentum could be lost, for example in a magnetically-driven wind (Blandford & Payne, 1982).

The terms 'viscosity', and 'viscous heating' are frequently used in connection with the redistribution of angular momentum within the disk. Classical molecular viscosity is, however, negligible in protoplanetary disks. For a gas, the molecular kinematic viscosity is given by (Armitage, 2007a, eqn 66)

$$\nu \sim \lambda c_s, \qquad (10.6)$$

where λ is the mean free path, and c_s is the sound speed. For a gaseous disk of surface density $\Sigma = 10^4 \, \mathrm{kg \, m^{-2}}$, and aspect ratio $h/r = 0.05$ at $1 \, \mathrm{AU}^1$, Armitage (2007a) gives $\lambda \sim 0.025 \, \mathrm{m}$, $\nu \sim 40 \, \mathrm{m^2 \, s^{-1}}$, and a resulting disk evolution time scale

$$\tau \sim r^2/\nu \sim 10^{13} \, \mathrm{yr}. \qquad (10.7)$$

This is some 10^6 times too slow to account for the type of disk evolution observed.

The most widely accepted view is that instabilities within the disk drive turbulence, which in turn increases the effective viscosity of the gas (e.g. Klahr et al., 2006). The effect was identified in the context of black hole accretion disks by Shakura & Sunyaev (1973), who adopted the parameterisation

$$\nu = \alpha c_s h, \qquad (10.8)$$

where the maximum scale of turbulence is of the order of the scale height of the disk h, and the maximum turbulent velocity is comparable to the sound speed c_s. In these so-called α-disks, α is a dimensionless constant which measures the efficiency of turbulence in creating angular momentum transport. Observations of protoplanetary disks suggest $\alpha \sim 0.01$ (Hartmann et al., 1998), and time-dependent calculations of the structure and evolution of the primordial solar nebula during the viscous diffusion stage suggest a similar value (Ruden & Lin, 1986). In β-disks, shear instabilities provide the corresponding viscosity (Richard & Zahn, 1999).

Other effects which have been considered in generating the turbulence needed to provide adequate viscous drag are non-linear hydrodynamic instabilities (Balbus et al., 1996; Godon & Livio, 1999a; Ioannou & Kakouris, 2001; Afshordi et al., 2005; Balbus, 2006; Balbus & Hawley, 2006; Ji et al., 2006; Shen et al., 2006); vortices (Godon & Livio, 1999b; Chavanis, 2000; Barranco & Marcus, 2005; Johnson & Gammie, 2005; Klahr & Bodenheimer, 2006); and convection (Ryu & Goodman, 1992; Stone & Balbus, 1996).

Self-gravitational instability leading to the development of trailing spiral arms, which act to transport angular momentum outwards (Toomre, 1964), may play a role in massive protoplanetary disks at early epochs.

Magnetorotational instability The presence of a magnetic field transforms the disk's dynamical stability. Even with a very weak magnetic field, the resulting magnetohydrodynamic flow leads to an unstable disk with perturbations growing exponentially. The resulting *magnetorotational instability* (Hawley & Balbus, 1991; Balbus & Hawley, 1991, 1998) leads to self-sustaining turbulence (Brandenburg et al., 1995; Stone et al., 1996), an outward transfer of angular momentum, an inward mass flow with the necessary transport efficiency, and perhaps to conditions favourable for the aggregation of 50–100 m size bodies into those of 10–100 km (Edgeworth–Kuiper belt) size (Carballido et al., 2006).

Layered accretion disks Angular momentum transport via magnetorotational instability is believed to operate in accretion disks around white dwarfs, neutron stars, and black holes. But the low ionisation in protoplanetary disks limits conductivity, prevents coupling between the gas and the magnetic field, suppresses the turbulence, and inhibits accretion (Blaes & Balbus, 1994; Gammie, 1996; Desch, 2004; Salmeron & Wardle, 2005).

Ionisation necessary to support magnetorotational instability, and hence accretion, may arise through collisional excitation of alkali metals at the higher temperatures and densities of the innermost parts of the disk (Umebayashi, 1983). X-rays or cosmic rays may provide an ionisation source throughout the lower column densities of the outer parts of the disk, or in the surface layers elsewhere (Umebayashi & Nakano, 1981).

It has been argued, but not verified observationally, that this combination of ionising sources, shown schematically in Figure 10.3, results in a layered disk with an inner turbulent-free 'dead zone' within the outer

---

[1] Such a surface density, typical of that considered for protoplanetary disks, may be compared with that of the local mean Galactic disk surface density, e.g. within 1100 pc, of $\Sigma(1100 \, \mathrm{pc}) \sim 52.8 \, M_\odot \, \mathrm{pc}^{-2}$ (Holmberg & Flynn, 2004), or $\sim 0.1 \, \mathrm{kg \, m^{-2}}$.

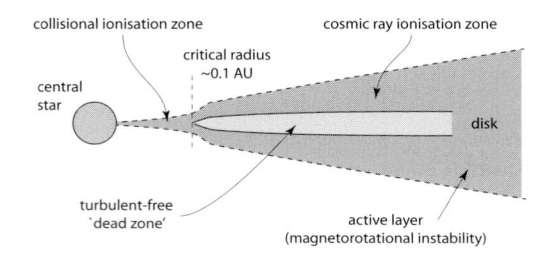

Figure 10.3: Schematic of a layered accretion disk. Inside 0.1 AU, where $T \sim 10^3$ K, collisional ionisation ensures magnetorotational instability, and enables accretion. At large radii, cosmic rays penetrate the entire disk. At intermediate radii, they ionise a layer of thickness $\sim 10^3$ kg m^{-2} on either side. Between the active layers is a 'dead zone' where turbulence and accretion are inhibited. Adapted from Gammie (1996, Figure 1).

turbulent structure which fuels the accretion (Gammie, 1996; Matsumura & Pudritz, 2006). Such a geometry would provide a quiescent environment interior to the turbulent zone in which the settling of dust, the subsequent growth of planetesimals, the structured formation of terrestrial mass planets and ice giants, and the migration of low-mass planets might operate (Armitage et al., 2001; Matsumura & Pudritz, 2003, 2005, 2006; Matsumura et al., 2007; Oishi et al., 2007; Turner et al., 2007).

The model of Gammie (1996) predicts a stellar accretion rate of $\sim 10^{-8} M_\odot$ yr^{-1} independent of the rate of material infall onto the disk, with non-accreted matter accumulating in the inner few AU of the disk at about $10^{-7} M_\odot$ yr^{-1}. Matsumura & Pudritz (2005, 2006) find that the dead zone typically extends out to 12–15 AU, acts to slow down migration due to its lower viscosity, and encourages the growth of eccentricities through planet–disk interactions.

Disk dispersal The lifetime of protoplanetary disks determines the time available for planet formation. As the original dusty disk is lost, no raw material remains to form planetesimals, terrestrial planets, or giant planets. Ages of dust disks are generally estimated from the presence of the excess thermal emission emerging from small, warm dust grains. The declining fraction of stars with dust disks suggests a disk half-life of 3–5 Myr (e.g. from Spitzer, Hernández et al., 2007), at least for the fine dust, and probably progressing radially outwards.

As the accretion rate falls, extreme ultraviolet radiation ($h\nu > 13.6$ eV) from the central star, or possibly from an external source (Adams et al., 2004, 2006), irradiates gas remaining in the outer disk, and removes it from the system by Poynting–Robertson drag and photoevaporation. Gas in the surface layer is heated to $T \sim 10^4$ K, ionised, and flows away from the star as a thermal wind no longer gravitationally bound to the disk (Bally & Scoville, 1982; Shu et al., 1993; Hollenbach et al., 1994; Johnstone et al., 1998; Font et al., 2004; Richling et al.,

2006). An example of such an process ongoing is seen in the HST observations of the core of the Trapezium cluster in Orion (O'Dell et al., 1993).

The time scale for the final stages of disk clearing by photoevaporation, corresponding to the transition between optically thick to optically thin disks in the thermal infrared, is very short. Estimates of the order of 10^5 yr are inferred, both observationally from the population of transition objects intermediate between the classical and weak-line T Tauri stars (Simon & Prato, 1995; Wolk & Walter, 1996), from numerical simulations (Clarke et al., 2001; Alexander et al., 2005, 2006a,b), and from modeling of the protosolar nebula from the properties of meteorites, asteroids, and planets (Pascucci & Tachibana, 2010).

10.3.4 Debris disks

Circumstellar dust rings were first identified observationally through the infrared excess of α Lyr (Vega), observed by the IRAS satellite (Aumann et al., 1984). During the course of the mission, three other bright main sequence stars were found to exhibit a similar infrared excess: Fomalhaut, ϵ Eri, and β Pic. Together these are regarded as prototypes of the 'Vega-type' phenomenon.

Their ages range from some 10 Myr (e.g. β Pic), to around $2-4 \times 10^8$ yr in the case of Vega and Fomalhaut, and up to ~ 1 Gyr in the case of ϵ Eri and τ Ceti. These ages, and the short dispersal time expected for the original dust disk, suggest that the disks observed in many of these older stars originate less from the remnant protostellar gas and dust, and more from gas-poor *debris disks* arising from collisions of planetesimals at a much later evolutionary stage (Kenyon & Bromley, 2002b). The youngest examples may be a combination of the remnant protostellar disk and the secondary debris disk.

Various examples of dust disks are now known around nearby main sequence stars within about 50 pc (Laureijs et al., 2002; Mann et al., 2006). The disks are physically large, perhaps extending out to ~ 500–1000 AU from the star, and the surface area of the small particles which make up the disk is many orders of magnitude larger than that of a planet (Beckwith, 1996). They therefore emit and reflect light very well, and are particularly prominent at infrared wavelengths, between about 2 μm and 1 mm. They display a distinctive spectrum much broader than any single-temperature black body, which originates from thermal emission over a wide range of temperatures, from ~ 1000 K close to the star to ~ 10–30 K near the outer edges of the disk.

Figure 10.4 shows three examples. HD 141569, at $d \sim 100$ pc, has an age of about 1–10 Myr, and the disk may comprise both protostellar and debris dust (Weinberger et al., 1999). A bright inner region is separated from a fainter outer region by a dark band, resembling the Cassini division in Saturn's rings. The disk extends to

10.3 Disk formation

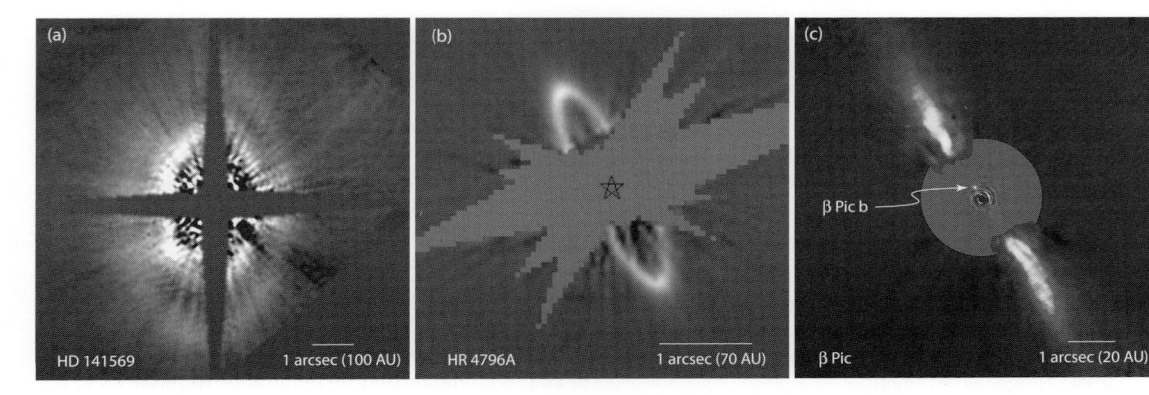

Figure 10.4: *Examples of imaged debris disks: (a) HD 141569 observed at 1.1 μm with HST–NICMOS (Weinberger et al., 1999, Figure 1). (b) HR 4796A, observed in the optical with HST–STIS (Schneider et al., 2009a, Figure 2; this version courtesy G. Schneider). (c) β Pic, from combined ESO 3.6-m ADONIS imaging in 1996 (outer region), and 3.6 μm observations with VLT–NACO (inner region) revealing the probably planet, β Pic b (Lagrange et al., 2009b, this version courtesy A.M. Lagrange, D. Ehrenreich, and ESO). In all cases, the geometric central structures are artefacts of the coronagraphic imaging.*

~400 AU, with the gap at 250 AU. An unseen planet may have carved out the gap, in which case its mass can be estimated at ~ $1.3\,M_J$, and its orbital period as 2600 yr. If it takes ~ 300 periods to clear such material (Bryden et al., 1999) then the gap could be opened in ~ 8×10^5 yr, consistent with the age of the star.

In the case of HR 4796A, with an age of 8 ± 3 Myr, the prominent dust ring had been previously imaged in the infrared with HST–NICMOS (Schneider et al., 1999). Particles with a size of a few μm are indicated larger than typical interstellar grains. Debes et al. (2008) have inferred the presence of tholins (complex, radiationally-evolved molecules) within the ring.

Significant structure, and perhaps planets themselves, are now inferred or detected in a number of debris disks (Faber & Quillen, 2007; Krivov et al., 2007; Edgar & Quillen, 2008; Reche et al., 2008; Stark & Kuchner, 2008; Quillen, 2010; Kalas, 2010), although known planetary systems possessing prominent disk systems are not common. Collisional models are having some success in reproducing the spectral energy distributions of various disks (e.g. Krivov et al., 2008)

Some other examples are given hereafter.

ε Eri ε Eri is a main sequence star, actually at a distance of only 3 pc, with both a known planetary companion ε Eri b, detected by radial velocity (Hatzes et al., 2000) and astrometry (Benedict et al., 2006), and a well-resolved dust disk. Numerous observations of the disk have been reported (Greaves et al., 1998; Hatzes et al., 2000; Quillen & Thorndike, 2002; Liou et al., 2002; Macintosh et al., 2003; Schütz et al., 2004b; Proffitt et al., 2004; Moran et al., 2004; Greaves et al., 2005; Benedict et al., 2006; Croll et al., 2006; Poulton et al., 2006; Janson et al., 2007, 2008; Heinze et al., 2008; Brogi et al., 2009; Backman et al., 2009; Marengo et al., 2009).

Fomalhaut A debris disk, and inferred collisional processes, are also well observed in the case of Fomalhaut (Wyatt & Dent, 2002; Holland et al., 2003; Stapelfeldt et al., 2004; Chiang et al., 2009; Marengo et al., 2009; Le Bouquin et al., 2009; Absil et al., 2009; Marengo et al., 2009). These disk observations led to the prediction (Kalas et al., 2005; Quillen, 2006a), and eventual imaging (Kalas et al., 2008) of a planet-mass body orbiting at 119 AU (§7.5.3). The large orbital separation has been interpreted in the context of outward migration of a pair of planets, leading to the prediction that a second planet should be orbiting at 75 AU (Crida et al., 2009b).

HR 8799 A well-developed debris disk is also observed around HR 8799 (§7.5.3), in which three planets are immersed (Marois et al., 2008b; Reidemeister et al., 2009; Lafrenière et al., 2009; Fukagawa et al., 2009; Su et al., 2009; Metchev et al., 2009; Goździewski & Migaszewski, 2009; Fabrycky & Murray-Clay, 2010; Janson et al., 2010; Moya et al., 2010).

β Pic β Pic, at a distance of 19.4 pc, has an estimated age of only ~10 Myr (Smith & Terrile, 1984). The inner region appears to be devoid of gas and dust, and ring-like structures are clearly observed (Artymowicz, 1997; Beust et al., 1998; Kalas et al., 2000; Augereau et al., 2001; Wahhaj et al., 2003; Weinberger et al., 2003; Brandeker et al., 2004; Telesco et al., 2005; Chen et al., 2007; Ahmic et al., 2009; Boccaletti et al., 2009).

Comet-like bodies (Beust et al., 1994; Roberge et al., 2006b), effects of stellar encounters (Kalas et al., 2001; Larwood & Kalas, 2001), collisional debris (Beust & Morbidelli, 2000; Thébault & Beust, 2001; Thébault et al., 2003; Karmann et al., 2003; Telesco et al., 2005; Okamoto et al., 2004), and effects of a purported planet (Lecavelier des Etangs et al., 1996, 1997; Crossley & Haghighipour, 2004; Gorkavyi et al., 2004; Galland et al., 2006a;

Golimowski et al., 2006; Roberge et al., 2006a; Freistetter et al., 2007; Quillen et al., 2007) have all been reported.

A candidate planet, evident in Figure 10.4c, is also suggested (Lagrange et al., 2009a,b; Lecavelier des Etangs & Vidal-Madjar, 2009; Fitzgerald et al., 2009). Provisionally, its properties are estimated as $a \sim 8\,\mathrm{AU}$, $M \sim 8 M_\mathrm{J}$, and $T \sim 1500\,\mathrm{K}$ (Lagrange et al., 2009b).

Vega The disk around Vega has also been detected and further characterised in various studies (Dominik et al., 1998; Wilner et al., 2002; Metchev et al., 2003; Macintosh et al., 2003; Dominik & Decin, 2003; Decin et al., 2003; Wyatt, 2003; Su et al., 2005; Marsh et al., 2006a; Hinz et al., 2006; Heinze et al., 2008).

Others Structure in young protostellar disks, including evidence for planetesimal infall, or gas disk clearing, have been observed around numerous stars (e.g. Chakraborty et al., 2004; Greaves, 2004; Kalas et al., 2004; Liu et al., 2004; Oppenheimer et al., 2008).

Debris disks, with a qualitative similarity to the asteroid/Kuiper belts and zodiacal dust cloud in the solar system, have been observed in various other more evolved systems, including: HD 12039 (Hines et al., 2006), HD 15745 (Kalas et al., 2007), HD 32297 (Schneider et al., 2005; Fitzgerald et al., 2007; Mawet et al., 2009b), HD 53143 (Kalas et al., 2006), HD 69830 (Beichman et al., 2005a; Bryden et al., 2006b; Smith et al., 2009c), HD 98800 (Verrier & Evans, 2008), HD 100546 (Grady et al., 1997; Liu et al., 2003; Grady et al., 2005), HD 107146 (Ardila et al., 2004), HD 141569 (Weinberger et al., 1999; Marsh et al., 2002; Augereau & Papaloizou, 2004), HD 145263 (Honda et al., 2004), HD 181327 (Schneider et al., 2006a), HR 4796A (Schneider et al., 1999; Wahhaj et al., 2005; Debes et al., 2008), γ Oph (Su et al., 2008), and τ Cet (Greaves et al., 2004b).

Various studies have evaluated the evidence for planets or planetesimals within these disks, or carried out related surveys (e.g. Laureijs et al., 2002; Wolf et al., 2002; Wood et al., 2002b; Macintosh et al., 2003; Kenyon & Bromley, 2004b,a; Meyer et al., 2004; Zuckerman & Song, 2004; Moro-Martín et al., 2005; Rieke et al., 2005; Wolf & D'Angelo, 2005; Acke & van den Ancker, 2006; Gorlova et al., 2006; Varnière et al., 2006; Matthews et al., 2007; Apai et al., 2008; Ricci et al., 2008; Bryden et al., 2009; Lawler et al., 2009).

Apart from the cases noted above, early searches for dust disks around stars with known planets tended to yield null or only marginal results (e.g. Greaves et al., 2004a; Saffe & Gómez, 2004; Schütz et al., 2004a; Beichman et al., 2005b; Liseau et al., 2008; Shankland et al., 2008). Report of the detection of a disk around 55 Cnc were subsequently contested (Dominik et al., 1998; Trilling & Brown, 1998; Jayawardhana et al., 2002). A more recent Spitzer–MIPS survey at 24 and 70 μm identified 10 new debris disks around 150 stars with known planets (Kóspál et al., 2009).

10.4 Terrestrial planet formation

10.4.1 The context

Today, the most widely considered *solar nebula theory*[2] holds that planet formation in the solar system, and by inference in other exoplanet systems, follows on from the process of star formation and accretion disk formation, through the agglomeration of residual material as the protoplanetary disk collapses and evolves.

In the present paradigm, planet formation proceeds by a 'bottom-up' process, with bodies of ever-increasing size being produced. It proceeds through a number of stages characterised by qualitative differences in the respective particle interactions. Together these extend over more than 14 orders of magnitude in size, from the sub-micron dust grains to the giant planets.

Historical background Scientific theories of the formation of the solar system had been under consideration since well before the discovery of exoplanets and circumstellar disks, and even before current theories underlying star formation and protoplanetary disks had been developed. Different theories date back to the works of Emanuel Swedenborg in 1734, the Compte de Buffon in 1749, Immanuel Kant in 1755, and Pierre-Simon Laplace in 1796.

Laplace (1796) considered that gaseous clouds, or nebulae, slowly rotate, gradually collapse and flatten due to gravity, and eventually form stars and planets. In this model, the latter formed by detachment of a discrete system of gaseous rings, within which clumps and planets grew by spontaneous gravitational collapse. Laplace's nebula model was pre-eminent in the 19th century. But it was largely abandoned in the early 20th century because it was unclear how material could be spontaneously partitioned such that the Sun retained 99.86% of the mass of the solar system while the planets ended up with 99.5% of the total angular momentum in their orbital motion.

Various alternatives were subsequently proposed. These included (Figure 10.5) the idea that planetesimals formed out of material from an erupting Sun (Chamberlin, 1901; Moulton, 1905), a tidal model involving interaction between the Sun and a massive star (Jeans, 1917), the star–Sun collision theory (Jeffreys, 1929a,b), the Schmidt–Lyttleton cloud accretion model (Schmidt, 1944; Lyttleton, 1961), the turbulence-collision driven protoplanet theory (McCrea, 1960, 1988), the tidal capture theory involving the Sun and a cool low-mass pro-

---

[2] The *solar nebula* refers to the gas and dust disk left over from the Sun's formation. In cosmogenic theories of the solar system, theories of formation are divided into *monistic* or *dualistic* depending on whether the Sun and planets were produced by the same or different processes. The present solar nebula theory is considered monistic.

10.4 Terrestrial planet formation

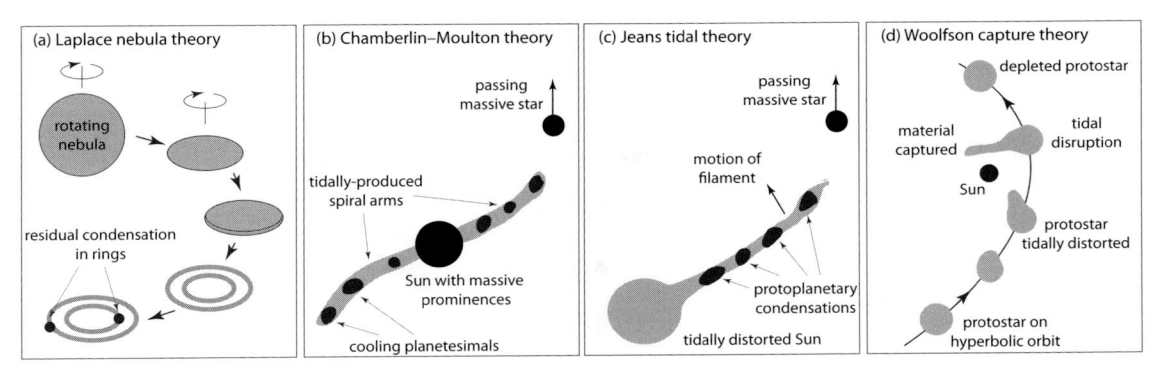

Figure 10.5: Schematic of alternative mechanisms that have been proposed for the formation of the solar system: (a) Laplace nebula theory, in which the rotating nebula collapses, leaving rings of material in which condensations form; (b) Chamberlin–Moulton planetesimal theory, where high-density regions come from solar prominences drawn out by tidal disruption; (c) Jeans tidal theory, where condensations form in ejected filaments from the tidally-distorted Sun; (d) Woolfson capture theory, showing successive configurations of a disrupted protostar leading to captured material. Adapted from Woolfson (1993, Figures 1–4).

tostar (Woolfson, 1964; Williams & Woolfson, 1983; Dormand & Woolfson, 1989), and a modern form of the Laplacian hypothesis (Prentice, 1978a,b).

These and other aspects of earlier and alternative ideas of the formation of the solar system, and a perspective of the difficulties faced by the current solar nebula theory, are discussed further by Woolfson (1993, 2000a,b, 2003, 2007). A review of theories developed during 1956–85 is given by Brush (1990). The possibility that the formation of most exoplanet systems differs from that of the solar system was considered by Beer et al. (2004a).

10.4.2 Stages in the formation of terrestrial planets

Many of the basic ideas central to the current picture of terrestrial planet formation were presented by Safronov (1969), and in its English translation (Safronov, 1972), and have been refined subsequently (e.g. Goldreich & Ward, 1973; Cameron, 1973; Wetherill, 1990; Lissauer, 1993, 1995; Wetherill, 1996; Bodenheimer, 2006).

In this 'bottom-up' picture, planet formation occurs in a number of successive stages (Figure 10.6): (i) dust settles into the mid-plane of the system; (ii) collisional processes or gravitational instabilities in the dust disk build up larger bodies of km size and larger called planetesimals; (iii) planetesimals collect together to form (eventually) either terrestrial planets or, further out in the disk, more massive cores of the giant planets; (iv) the giant planet cores acquire a gaseous envelope. Description of these stages are given hereafter, while a number of recent reviews of formation and migration provide more extensive details (e.g. Thommes & Lissauer, 2005; Masset & Kley, 2006; Papaloizou & Terquem, 2006; Thommes, 2007; Raymond, 2010; Youdin, 2010).

A number of the stages and their associated time scales remain uncertain, amongst them the coagulation of dust and other particles over several orders of magnitude in size and mass, aspects of the angular momentum transport, the role of magnetorotational turbulence and the relevance of dead zones, and the contribution of gravitational fragmentation in the build up of planetesimals and in the growth of giant planets.

Dust to rocks: sub-micron to 10 m *Dust* is a significant constituent of the Universe, comprising some 1% of the mass of the interstellar medium, and characterised by condensed particles with sizes ranging from sub-micron and upwards. Interstellar dust has a diverse origin, the constituent matter having either condensed in the winds of evolved stars and in the ejecta of nova and supernova explosions, or having formed in dense interstellar clouds (Apai & Lauretta, 2010; Gail & Hoppe, 2010). The process of planet formation starts with the presence of dust in collapsing molecular clouds, some existing dust being carried inwards by the infalling gas, with additional dust condensing from the gas phase.

Qualitatively, the dust grains are presumed to settle into a dense layer in the mid-plane of the disk, where they begin to stick together through a combination of electrostatic forces and collisional impacts, growing as they collide (Kusaka et al., 1970). But unmodified micron-sized nebula dust would probably take too long to settle into the disk mid-plane, even in the absence of turbulence (Dullemond & Dominik, 2004; Johansen & Klahr, 2005; Carballido et al., 2005; Fromang & Papaloizou, 2006; Turner et al., 2006, 2007). Weidenschilling et al. (1989) accordingly constructed models in which the dust particles stuck together to form larger entities that would settle more quickly. Their models suggested that the dust structures so formed would be looser than the dense-packed $m \propto d^3$, and with their predicted $m \propto d^2$ structures, a settling time scale of $10^5 - 10^6$ yr was estimated.

Quantitatively, detailed collisional and coagulation models in these earliest phases are partially founded on laboratory studies (e.g. Supulver et al., 1997; Blum, 2000; Wurm & Blum, 2006; Zsom et al., 2010; Brucato & Nuth, 2010), including the Space Shuttle experiment CODAG (Blum et al., 2000), but they remain subject to numerous uncertainties arising from the complex physics (Weidenschilling, 1980; Dominik & Tielens, 1997; Sekiya, 1998; Blum & Wurm, 2000; Goodman & Pindor, 2000; Dullemond & Dominik, 2005; Henning et al., 2006; Johansen et al., 2006; Trieloff & Palme, 2006; Johansen et al., 2007; Blum & Wurm, 2008; Johansen et al., 2008; Youdin, 2010). Formation of a dust disk within the lifetime of the nebula disk is, nevertheless, a pre-requisite of the model. The inference of mm-sized grains in the terrestrial zone of the 3 Myr old protoplanetary disk of KH 15D provides the type of observational support necessary to corroborate such theories (Herbst et al., 2008).

Progressive aggregation is believed to form macroscopic objects, or 'rocks', with sizes of order 0.01–10 m, probably analogous to the few hundred meter thick rings around Saturn (Zebker et al., 1985; Esposito, 2002; Porco et al., 2005; Nicholson et al., 2008).

Radial forces are significant at different scale sizes for different reasons. The dust grains and smaller bodies are well-coupled to the circulating gas, and orbit the protostar in the same direction and in the same plane. They are subject to aerodynamic drag by the gas, which affects both the vertical motion and radial drift, with a resulting force (Armitage, 2007a, eqn 90–94)

$$f_d = -\frac{1}{2} C_d (\pi a^2)(\rho v^2), \qquad (10.9)$$

where a is the particle radius (assumed spherical), and the gas of density ρ moves with velocity v. The drag coefficient, C_d, depends on the particle size compared to the mean free path λ of the gas molecules (Weidenschilling, 1977a). The resulting force is termed *Stokes drag* for particles with $a > 9\lambda/4$, i.e. above ~0.01 m in size (Stokes, 1851), and *Epstein drag* for smaller particles (Epstein, 1924), and both result in an inward particle drift.

Larger rocks are less strongly coupled to the gas, and move with the slightly larger Keplerian disk velocity. The result is a loss of orbital angular momentum, and again a net inward particle drift (Takeuchi & Lin, 2002; Armitage, 2007a, eqn 105–16). The peak inspiraling rate, at ~1 AU, occurs for particle sizes ~ 0.01 – 10 m, leading to rapid radial drift time scales of order 100 yr.

Such a rapid radial drift implies that particle growth up to 10 m or more in size must itself be very rapid, with consequences for the local ratio of solids to gas as a function of time and radius (Takeuchi et al., 2005a; Youdin & Chiang, 2004). In the presence of turbulence, local pressure maxima may assist rapid growth by providing sites where solids might concentrate (Haghighipour & Boss, 2003; Rice et al., 2004; Durisen et al., 2005).

Rocks to planetesimals: 10 m to 10 km Over the next $10^4 - 10^5$ years, further collisions lead to the formation of *planetesimals*, objects of size ~1 km and above.[3] This is generally considered to occur through a continuation of the same type of pairwise collisional growth as at smaller size scales, gravitational interactions between individual bodies remaining very weak until, by definition, the minimum planetesimal mass is achieved.

An alternative, and currently less-favoured, hypothesis is that planetesimals are formed rapidly by gravitational fragmentation of a dense particle sub-disk near the gas mid-plane. This *Goldreich–Ward mechanism* (Goldreich & Ward, 1973) has the merit of a short formation time scale, and one which would essentially bypass the size scales that are most susceptible to radial drift. But it appears to fail for standard gas to dust ratios, of order 100, once the turbulent effects of Kelvin–Helmholtz instabilities are included (Cuzzi et al., 1993). Large factors of local dust enrichment, of order 10–100, would be required for the mechanism to work (Garaud & Lin, 2004), perhaps via radial pile-up (Youdin & Chiang, 2004) or photoevaporation (Throop & Bally, 2005).

Once of planetesimal size, the bodies are largely decoupled from the gas (by definition), and their evolution can subsequently be modeled analytically or by N-body simulations. In this regime, Safronov (1972) showed that if the random relative velocity between planetesimals is less than the escape speed from the largest of the them, then that body will grow at the expense of the others, accreting all other bodies that collide with it.

Newly-formed planetesimals move on elliptical orbits, and with gravitational interactions between them, equivalent to elastic collisions, increasing their random motions. Eventually, this increase in relative velocities and orbital eccentricities increases the probability of inelastic conditions, which damp the random component of the motions. A balance between the two effects occurs when the mean random velocity is of the order of the escape velocity from the largest planetesimal.

The protoplanet then grows at a rate given by (Pollack et al., 1996, eqn 1)

$$\frac{dM_e}{dt} = \pi R_c \Sigma_p \Omega F_g, \qquad (10.10)$$

where M_e is the embryo mass, R_c is its effective or capture radius, Σ_p is the surface mass density of plantesimals, Ω is the orbital frequency, and F_g is the gravita-

---

[3] Planetesimals are loosely defined as solid objects arising during the accumulation of planets whose internal strength is dominated by self-gravity and whose orbital dynamics are not significantly affected by gas drag. This corresponds to objects larger than approximately 1 km in the solar nebula. The description is retained for objects up to 100–1000 km or so in size, when dominant *embryos* or protoplanets start their accelerated growth towards potential planets.

10.4 Terrestrial planet formation

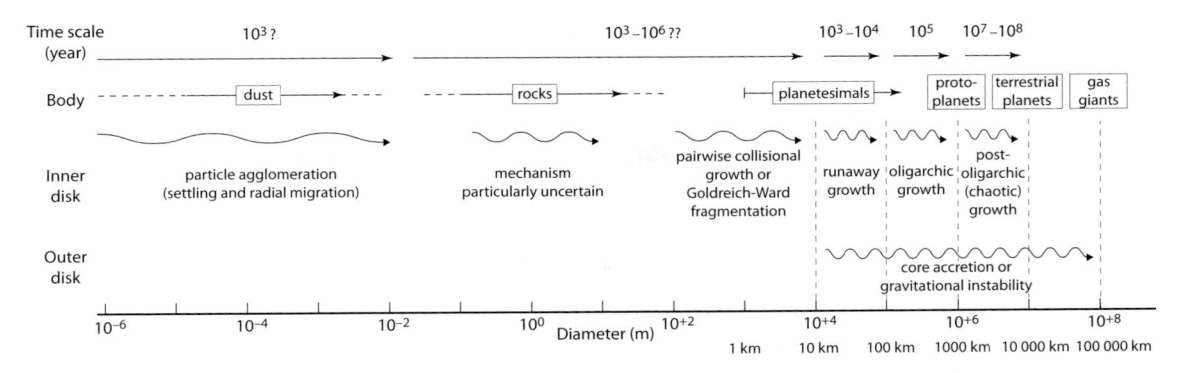

Figure 10.6: Schematic of the growth of planets, starting with sub-micron dust, and extending up to the terrestrial planets in the inner disk, and the gas giants in the outer disk. Some indicative time scales are given, although some intervals, especially around the meter-size barrier, remain highly uncertain.

tional enhancement factor given by the ratio of the gravitational to geometric cross sections. For Keplerian motion, $\Omega \propto r^{-3/2}$, and the time scale for planet growth increases steeply with increasing r as a consequence of the dependence of Ω on r, and because Σ_p is expected to decrease with increasing r.

For plausible models of disk mass of order $0.1 M_\odot$, and a surface density varying as r^{-1}, total formation times estimated from such models would be of order 4×10^6 yr for the Earth ($r = 1$ AU, $\Sigma \sim 10^3$ kg m^{-2}); 5×10^8 yr for a $10 M_\oplus$ Jupiter core ($\Sigma \sim 200$ kg m^{-2}); and 3×10^{10} yr for Neptune ($\Sigma \sim 30$ kg m^{-2}), the latter greatly exceeding the age of the solar system. Since the lifetimes of nebula disks are of order a few million years at most, local enhancements of the surface density, or a reduction in the relative planetesimal velocities, are required to reduce the formation time scales to plausible durations.

In numerical simulations, protoplanetary growth proceeds in several different regimes, depending on the dynamical state of the planetesimal disk. If the growth of the velocity dispersion is dominated by the embryos (objects which are growing in size at the expense of others, and which will eventually grow to form planets), and assuming that gravitational focusing is weak, then (e.g. Rafikov, 2003a, eqn 3)

$$\frac{1}{M_e} \frac{dM_e}{dt} \propto M_e^{-1/3}. \qquad (10.11)$$

This dependency of growth rate on M_e shows that the embryo's growth slows as its mass increases. This type of growth law (which may or may not occur in practice) is known as *orderly growth*, and it implies that many embryos grow at roughly the same rate. Orderly growth is formally characterised by very long time scales, of order $10^8 - 10^9$ yr (Safronov, 1972), although in the simulations of Rafikov (2003a), for example, such orderly growth never develops.

Time scale for growth to 10 km Growth over the *meter-size barrier* of 0.01–10 m, perhaps extending up to 10^4 m, remains particularly uncertain due to the combined effects of fragmentation and radial drift (Rice et al., 2006; Brauer et al., 2008; Johansen et al., 2008). The associated formation time scales are correspondingly highly uncertain. Growth to 0.01 m may require as little as $\sim 10^3$ yr, with growth from $0.01 - 10^4$ m requiring somewhere in the range $10^3 - 10^6$ yr. Inwardly drifting dusty bodies up to this size, approaching closer than ~ 0.4 AU to the star, may also be rapidly eroded by *thermophoresis* as a result of internal thermal gradients (Wurm, 2007).

Runaway growth: 10 km to 100 km Mutual gravitational interactions between planetesimals become progressively more important as their mass increases, since their effective collisional cross-section is much larger than their physical cross-section. As a result, object deflection, or *gravitational focusing*, starts to have a significant effect on collisional probabilities (Greenzweig & Lissauer, 1990; Armitage, 2007a, eqn 151–155). At the same time, dynamical friction develops as an important mechanism for transferring kinetic energy from the larger planetesimals to the smaller ones (Stewart & Wetherill, 1988), keeping the velocity dispersion of the most massive bodies relatively low, and reducing the formation time scale.

Neglecting the embryo's effect on growth of the velocity dispersion in the disk, and assuming gravitational focusing is strong, then (Wetherill & Stewart, 1989; Rafikov, 2003b,a, eqn 4)

$$\frac{1}{M_e} \frac{dM_e}{dt} \propto M_e^{1/3}, \qquad (10.12)$$

and the embryo growth accelerates as its mass increases. This corresponds to a phase of *runaway growth* (Stewart & Wetherill, 1988; Wetherill & Stewart, 1989), in which a few bodies grow rapidly at the expense of the others.

Runaway growth is considered to lead to the formation of $\gtrsim 100$ km-sized bodies at 1 AU in some 10^4 yr (Kokubo & Ida, 1996; Weidenschilling et al., 1997; Rafikov, 2003a).

Detailed modeling must consider whether collisions lead to accretion growth, or to fragmentation. Colliding bodies may (i) bounce off each other elastically and remain unbound, (ii) dissipate sufficient kinetic energy to become gravitationally bound, with or without fragmentation, or (iii) disrupt or shatter one or both bodies, creating a number of unbound fragments. The overall outcome is determined by mass, impact velocity, and more uncertain parameters such as their intrinsic strength (Benz & Asphaug, 1999; Leinhardt & Richardson, 2002; Ryan & Melosh, 1998; Barnes et al., 2009c).

With the progressive formation of massive planets changing the velocity distribution through gravitational interactions, conditions are believed to develop from those more favourable for accretion into those more favourable for disruption. In the solar system, constraints come from collisional processes inferred in the evolution of the Kuiper belt (Stern & Colwell, 1997a,b), and from present-day collisions in the asteroid belt where impact velocities can be large (Bottke et al., 1994), leading to the fragmentation and formation of asteroid families (Nesvorný et al., 2002).

A consideration of three-body *shear dominated* dynamics, affected by the gravitational influence of the Sun, leads to a reduction in collisional rate compared with the expectations from two-body *dispersion dominated* dynamics (Rafikov, 2003a; Goldreich et al., 2004b).

Oligarchic growth: 100 km to 1000 km The size of a growing planet's *feeding zone* is set by the maximum distance over which its gravity is able to perturb other orbits sufficiently to allow collisions. The feeding zone itself scales with the Hill radius, the radius within which the gravity of the planet dominates that of other bodies within the system (see box on page 43). As the mass of the largest planetesimals grow, they perturb the velocities of smaller planetesimals in their vicinity, reducing the number on roughly circular orbits which are available for accretion.

This leads to a slowdown in the accretion rate once a certain *isolation mass* is reached, and runaway growth up to this point gives way to a phase of slower *oligarchic growth*. The largest objects still grow at the expense of smaller bodies, but more slowly, and all roughly at the same rate.

This slowdown occurs because, although the feeding zone expands as the planet mass grows, the number and mass of available planetesimals in the enlarged feeding zone rises more slowly than linearly. Armitage (2007a, eqn 172–177) gives for the isolation mass

$$M_{iso} \propto M_\star^{-1/2} \Sigma_p^{3/2} r^3 , \qquad (10.13)$$

where Σ_p is the local surface density of planetesimals at radial distance r. For $\Sigma_p = 100\,\mathrm{kg\,m^{-2}}$, conditions in the terrestrial planet region (Lissauer, 1993) and gas giant core regimes (Pollack et al., 1996) suggest isolation masses of order $0.07 M_\oplus$ and $9 M_\oplus$ respectively. The latter is comparable to estimates of the Jovian core mass (Guillot, 2005).

A more quantitative approach to describing the size evolution of a number of merging bodies is based on the *coagulation equation* (Smoluchowski, 1916), which describes the evolution of the mass spectrum of a collection of particles due to successive mergers. In discrete form, the bodies are represented as integer multiples of a small mass m_1, with n_k bodies of mass $m_k = km_1$ at time t. Neglecting fragmentation

$$\frac{dn_k}{dt} = \frac{1}{2} \sum_{i+j=k} A_{ij} n_i n_j - n_k \sum_{i=1}^{\infty} A_{ki} n_i , \qquad (10.14)$$

where A_{ij} is the rate of mergers between bodies of mass m_i and m_j, including effects such as gravitational focusing. The first term represents the increase in the number of bodies of mass m_k due to collisions of all pairs of bodies whose masses m_i, m_j sum to m_k, while the second term describes the loss of masses m_k due to their incorporation into larger bodies.

Implementation of the coagulation equation in planet-formation studies can incorporate gas drag, Poynting–Robertson drag, radiation pressure, and collisional fragmentation to treat the growth of planetesimals into oligarchs, with perturbations of eccentricity and inclination and explicit N-body calculations introduced to follow the evolution of oligarchs into planets (Safronov, 1969; Safronov, 1972; Wetherill & Stewart, 1993; Kenyon & Luu, 1998; Lee, 2000; Inaba et al., 2001; Bromley & Kenyon, 2006).

Simulations results differ in detail, and do not necessarily display distinct sequences of orderly growth, runaway growth, and oligarchic growth (e.g. Rafikov, 2003a). Some illustrative results are shown in Figure 10.7.

In typical simulations, runaway growth may occur after $\sim 10^5$ yr when the relative velocities of the larger bodies temporarily fall into a low-velocity regime. After $\sim 10^6$ yr, results point to a number of relatively isolated protoplanets (or planetary embryos) distributed throughout the disk.

Conceptually, the resulting picture is of some 100 objects around 1000 km or more in size, with masses of the order of $10^{22} - 10^{23}$ kg (Moon- to Mars-sized), accompanied by a swarm of billions of 1–10 km planetesimals (Kokubo & Ida, 1998, 2000, 2002; Rafikov, 2003a; Thommes et al., 2003; Fogg & Nelson, 2005; Chambers, 2006b; Cresswell & Nelson, 2006; Kenyon & Bromley, 2006; Ford & Chiang, 2007; Fortier et al., 2007; Zhou et al., 2007; Brunini & Benvenuto, 2008; Chambers, 2008; Fortier et al., 2009).

10.4 Terrestrial planet formation

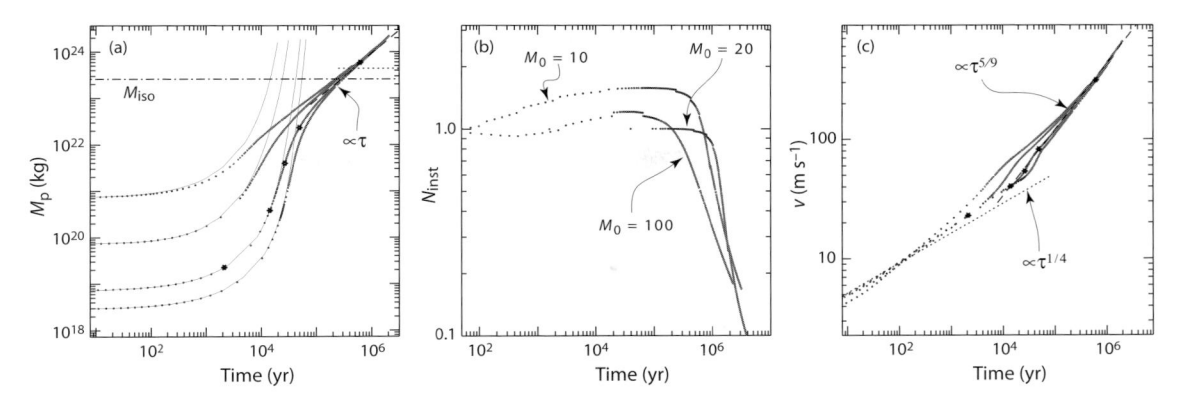

Figure 10.7: Simulated growth of objects in a planetesimal disk: (a) embryo mass as a function of time (solid or dashed lines) for $a = 0.9$ AU and $m = 7.9 \times 10^{17}$ kg, for different initial conditions. Thin solid lines indicate runaway curves obtained by neglecting the embryo's dynamical effects. The dot–dashed line is the isolation mass. The $M_p \propto \tau$ line shows the asymptotic behaviour; (b) instantaneous surface density of planetesimals which can be accreted by the embryo N_{inst} as a function of time, for three different values of (dimensionless) initial embryo mass, M_0. There is a sharp drop of N_{inst} after the embryos start to dominate the disk dynamics; (c) growth of the planetesimal velocity dispersion as a function of time, for different initial conditions. The dotted line with slope 1/4 shows the initial growth due to planetesimal–planetesimal scattering alone. The dashed line with slope 5/9 is the asymptotic behaviour. From Rafikov (2003a, Figures 7, 10, 11), reproduced by permission of the AAS.

Post-oligarchic growth: 1000 km to 10 000 km When the mass in these relatively isolated protoplanets, or oligarchs, is roughly comparable to the mass in planetesimals, strong dynamical interactions lead to a more chaotic collision and merger rate. The orbits of neighbouring embryos begin to cross one another, and their eccentricities and inclinations increase rapidly. Gravitational focusing becomes weaker, and the growth rate drops significantly.

This *post-oligarchic* or *chaotic growth phase* is characterised by large-scale, stochastic mixing of the planetary embryos and catastrophic collisions, leading to the formation of several terrestrial planets, and ultimately defining the final architecture of the planetary system. The growing protoplanetary embryos, above ~3000 km in size, are also accompanied by their internal melting and progressively differentiated interiors, as a result of radiogenic, gravitational, or impact heating.

Head-on collisions lead to mergers with little mass loss, with large impacts causing extensive heating and the formation of magma oceans (Tonks & Melosh, 1992). Oblique collisions may separate and escape, with some exchange of material (Agnor & Asphaug, 2004).

Mergers proceed until the orbit spacing becomes large enough that the configuration is quasi-stable, with the final assembly of the terrestrial planets taking ~10–100 Myr (Chambers & Wetherill, 1998; Goldreich et al., 2004a; Moorhead & Adams, 2005; Kokubo et al., 2006; Kenyon & Bromley, 2006; Ogihara & Ida, 2009).

The final configuration of the terrestrial planets in terms of ultimate size and spacing depends on numerous factors, amongst them the initial conditions (Chambers, 2001; O'Brien et al., 2006), the (unknown) viscosity of the protoplanetary disk (Lin & Papaloizou, 1979), the surface density of planetesimals (Raymond et al., 2005b; Kenyon & Bromley, 2006), and the existence and migration of giant planets (Levison & Agnor, 2003). Simulations show that, in the solar system, the oligarchic and post-oligarchic growth stages are strongly influenced by the presence of the gas giants, Jupiter and Saturn.

Modern theories of planetary growth do not yield deterministic 'Bode's Law' formulae for planetary orbits (see box on page 48), but characteristic spacings do exist, suggesting that gaps develop broadly in proportion to orbital radius (Hayes & Tremaine, 1998; Laskar, 2000).

Applied to the solar system, numerical simulations suggest that the Earth reached half its current mass in 10–30 Myr, and its present mass in ~ 100 Myr, with the growth rate declining exponentially with time (Chambers, 2001; Raymond et al., 2005b; O'Brien et al., 2006). Such estimates are consistent with those derived from radiometric data, and from cratering records (Grieve & Pesonen, 1996; Stothers, 1998).

The protracted duration of the final stage of accretion implies considerable radial mixing, with material acquired from different regions of the inner disk. This may explain the slightly different oxygen isotope ratios found on Earth and Mars (Clayton & Mayeda, 1996).

Earth and Venus each perhaps experienced some 10 giant impacts with other embryos during this phase. Mercury and Mars may be embryos that failed to grow. Mercury's high density may indicate that it lost much of its mantle during a high-speed collisional-stripping impact with another embryo (Benz et al., 1988).

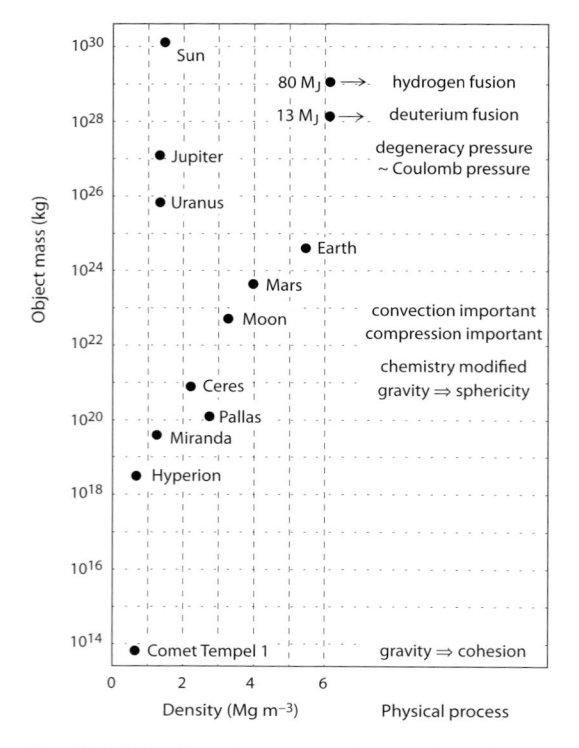

Figure 10.8: The effect of gravity on shape and structure as a function of mass. Some solar system objects are given as examples. The radii of brown dwarfs are approximately constant above $\sim 13 M_J$, and their densities fall off the plot to the right.

10.5 Size, shape, and internal structure

As bodies grow progressively in size, there is a developing influence of gravity on their overall shape, and on their internal structure. The relevant effects have been outlined by Basri & Brown (2006), and are illustrated schematically in Figure 10.8, with some solar system examples in Figure 10.9.

Rocks to planetesimals The bulk volume of small and intermediate sized rocks is determined by (atomic and molecular) bound electron degeneracy pressure, and gravity plays no role in their size or shape. As mass increases, gravity reaches the level necessary to hold together a rock or rubble pile. This depends on density and rotation, but starts to operate for small asteroids, as well as comets such as Shoemaker–Levy 9 (which impacted Jupiter in 1994), and Tempel 1 (imaged by Deep Impact in 2005), both with $D \sim 5$ km and $M \sim 8 \times 10^{13}$ kg.

At $D \sim 500 - 1000$ km, gravity overcomes an object's material strength, forcing it (in the absence of rotation) to take the equipotential figure of a sphere. With a dependence on mass and density, solar system objects become gravitationally relaxed, and spherical, at around the size and mass of Ceres ($D = 940$ km, $M = 9.4 \times 10^{20}$ km, $\rho = 2.1$ Mg m^{-3}; Thomas et al., 2005), while the smaller asteroids Pallas (608 km) and Vesta (538 km) are slightly elongated in HST images (e.g. McFadden et al., 2007). The largest known substantially non-spherical ice body is Saturn's satellite Hyperion ($D \sim 100 - 180$ km, $M = 0.6 \times 10^{19}$ kg, $\rho = 0.6$ Mg m^{-3}).

Planetesimals to planets The mass threshold for an object to develop a solid-state convective interior, and thereafter to behave as a fluid over geological time scales, depends on the Rayleigh number in the interior (Turcotte & Schubert, 2002). Scaling as D^5, solid-state convection rapidly becomes important somewhere between the size and mass of the Moon ($D \sim 3500$ km, $M \sim 7.3 \times 10^{22}$ kg) and that of Mars ($D \sim 6500$ km, $M \sim 6.4 \times 10^{23}$ kg).

At a similar mass ($D \sim 3000$ km for a terrestrial body) the gravitational energy per atom exceeds 1 eV. This corresponds to the typical energy of chemical reactions, so that self-gravity starts to modify the chemical composition of the initial material.

At $D \sim 6000$ km for rocky material and ~ 1000 km for ices, the central pressure due to gravity exceeds the *bulk modulus* of the material, $K = -V \partial P / \partial V$ ($\sim 100 - 1$ GPa, respectively), and significant compression and density increase occur. For rocky materials, this also corresponds approximately to the mass of Mars.

Brown dwarfs to stars The next significant change occurs only at $\sim 1 - 2 M_J$. For such gas giants, core temperatures are large due to the large pressures, and electrons are freed, resulting in a pressure term due to Fermi exclusion. At $\sim 2 M_J$, free electron degeneracy pressure in the cores becomes comparable to normal Coulomb pressure, and above $2 - 5 M_J$ an increase in mass leads to an increase in density (i.e. a decrease in radius).

The resulting maximum size, at $2 - 5 M_J$, is modified if the non-degenerate outer layers are hot, i.e. if the giant planet or brown dwarf is young or irradiated, in which case the object can be somewhat more bloated.

Differentiation Gravitational pressure, combined with collisional impact and radioactive heating, facilitate convection and may lead to melting, enhancing *physical differentiation* as materials with higher density sink. Thus iron, the most common element forming a dense molten phase, sinks towards planetary interiors, while the less dense materials rise upwards. Meteoritic compositions indicate that differentiation occurred even in some asteroids, and gravitational settling is also inferred for the Sun.

In the case of the Earth, the result is a series of compositionally-distinct layers: a dense Fe-rich core, a less-dense magnesium silicate-rich mantle, a thin crust dominated by lighter silicates (Al, Na, Ca, K), a liquid water hydrosphere, and a gaseous N-rich atmosphere.

Chemical differentiation is further modified by reactive affinities, determined by the chemical bonds that

10.6 Giant planet formation

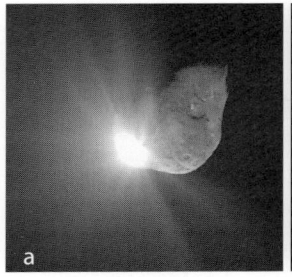

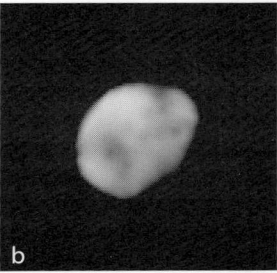

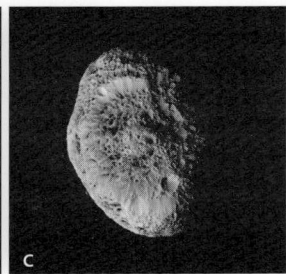

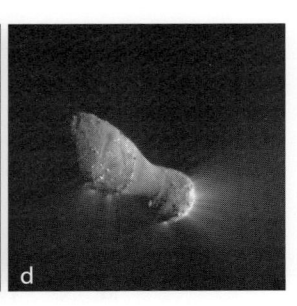

Figure 10.9: Examples of non-spherical shapes of low-mass solar system bodies. (a) Comet Tempel 1 imaged 67 s after the collision with the Deep Impact probe in 2005 (NASA). (b) Asteroid Vesta imaged by HST in 2007 (NASA, L. McFadden). (c) Saturn's icy satellite Hyperion imaged during the Cassini flyby in 2005 (NASA); (d) Comet Hartley 2 imaged from 700 km by EPOXI in 2010 (NASA/JPL–Caltech/UMD).

the element forms. The geochemical *Goldschmidt classification* (Figure 10.10), proposed in the 1920s by Victor Goldschmidt, denotes preferred host phases, viz. *lithophile* (literally, 'rock' or 'silicate loving', but also O-loving in practice), *siderophile* ('iron loving'), *chalcophile* ('sulphur loving'), and *atmophile* ('gas loving'), dictated by predominantly ionic, metallic, covalent and van de Waals bonding respectively. Thus iron is the host phase for siderophiles such as Ni, Co, or Au, which therefore tend to concentrate in the Earth's core, while silicates provide the host phase for lithophiles such as K, Na, U and W, which tend to concentrate in the crust.

The combination of physical and chemical differentiation on Earth results in an average density of $5.5\,\mathrm{Mg\,m^{-3}}$, a mantle density of $3.4\,\mathrm{Mg\,m^{-3}}$, and a crustal density of $2.7\,\mathrm{Mg\,m^{-3}}$. Seismic measurements suggest that the core itself comprises a solid inner core of radius ~1200 km and a liquid outer core extending to a radius of ~3400 km.

10.6 Giant planet formation

The terms *giant planets*, *gas giants*, or simply *Jupiters*, refer to large planets, typically $\gtrsim 10 M_\oplus$, that are not composed primarily of rock or other solid matter. When orbiting close to the host star they are referred to as *hot Jupiters* or *very hot Jupiters*. Planets in the mass range $\sim 3-10 M_\oplus$ are frequently called super-Earths or occasionally, for the higher mass examples, *gas dwarfs*.

An extension of the mechanism for terrestrial planet formation requires some adjustments in explaining the existence of the giant planets: at small orbital distances the supply of gas and dust required for stable accretion is inadequate, while at large orbital distances the long accretion time scales are potentially problematic.

An understanding of the formation of giant planets is therefore driven by identifying conditions under which a sufficiently massive accreting core can form, and which can in turn accumulate the necessary quantity of gas before the protoplanetary gas disk disperses.

The existence of these substantial gas envelopes provides a strong temporal constraint on their formation: they must form rapidly, before the gas in the protoplanetary disk dissipates, viz. within ~5–10 Myr.

There are two theories for the formation of giant exoplanets under active consideration: core accretion, and gravitational disk instability. Model developments have been strongly guided by the general structure (mass and density), elemental composition, and orbital characteristics of the solar system giants (§12.2).

10.6.1 Formation by core accretion

The *core accretion model* of giant planet formation is the most widely considered of two processes proposed to explain their existence. The same mechanism is considered to apply to the solar system gas and ice giants, and to the exoplanet giants, at least to those orbiting within $r \lesssim 10-50$ AU.

Sometimes described as 'bottom-up', it is broadly conceptualised as a two-stage process. The first resembles that of terrestrial planet formation described previously, leading to (and requiring) the formation of a massive planet, or core, of some $5-20 M_\oplus$. The second stage is characterised by a progressively more rapid accretion of gas onto the resulting core, along with the continued accretion of some planetesimals.

Qualitatively, the scenario can explain a wide range of planetary architectures. Cores of adequate mass for large-scale gas accretion only form under certain conditions within the protoplanetary disk. Further, the processes of massive core formation, and gas dispersal with the disk, operate on comparable time scales of order 5–10 Myr. Consequently, if the disk conditions are not suitable for a massive core to form, or the massive core cannot form rapidly enough, the gas disk may be dispersed before it can accrete, and no giants are formed. For smaller core masses, or longer core formation time scales, the residual gas can be dispersed before accretion runs its full course, and ice giants rather than the gas-rich gas giants may result.

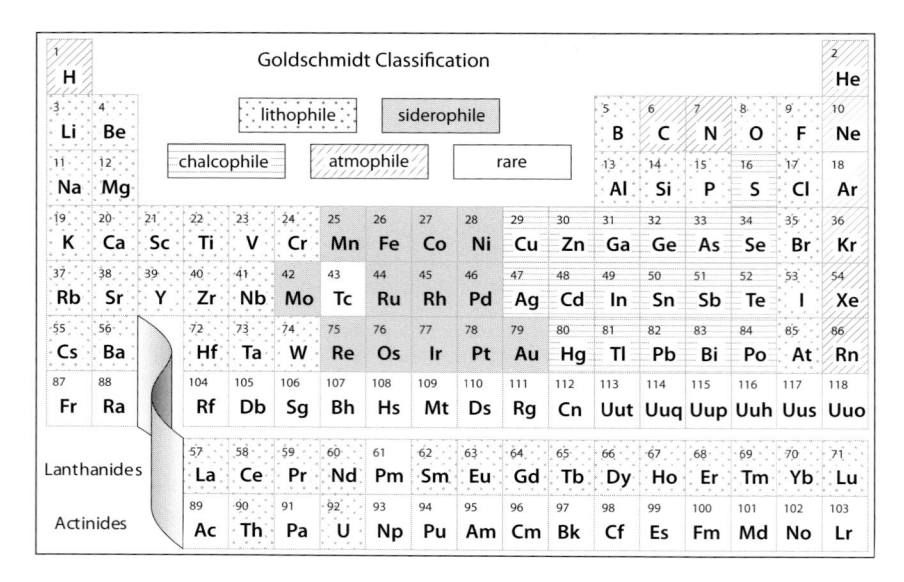

Figure 10.10: The periodic table according to the Goldschmidt classification, which groups the elements according to their preferred geochemical host phases, and provides a framework for describing the structure of the Earth from material present in the early solar system. Elements can be assigned to more than one group, such that the scheme indicates general trends only.

Mizuno et al. (1978) and Mizuno (1980) originally showed that this model could account for the relative amounts of high- and low-Z materials in the giant planets. Early ideas were developed subsequently by numerous workers (e.g. Stevenson, 1982; Pollack, 1984; Bodenheimer & Pollack, 1986; Pollack et al., 1986; Podolak et al., 1988; Pollack et al., 1996; Papaloizou & Terquem, 1999; Kokubo & Ida, 2000; Bodenheimer et al., 2000).

Various recent reviews have been devoted to the field (e.g. Weidenschilling, 2005; Hubickyj, 2006; Thommes & Duncan, 2006; Mordasini et al., 2008; Hubickyj, 2010).

Core accretion in outline It is a prerequisite of the model that one or more massive cores of $\sim 5-20 M_\oplus$ form in the protoplanetary disk, presumably as a result of continued planetesimal collisions. Simulations indicate that there are generally insufficient solids in the inner disk to allow such massive cores to develop, but the situation is quantitatively different beyond the snow line, where the temperature is low enough for the formation of water (and other) ices, thus substantially augmenting the surface density of condensates (§11.2.2). This enhancement in potential accretion material, combined with the reduced gravitational dominance of the central star, allows large solid cores to form more easily in the outer disk regions.

Over a period of a few million years, the progressively growing core eventually reaches a *critical core mass* (Bodenheimer & Pollack, 1986), also referred to as the *crossover mass* (Pollack et al., 1996), beyond which the gas accretion rate exceeds that of planetesimals by an amount which grows exponentially with time.

Once the envelope mass exceeds a few percent of $M_\oplus$, a combination of gas drag, evaporation, and dynamical pressure makes it increasingly difficult for planetesimals to arrive intact at the core boundary, and a significant mass fraction dissolves in their gaseous envelopes, enriching them in high-Z elements. Such a scenario broadly accounts for the enhancement of some high-Z elements in the atmospheres of the solar system giants, and for their progressive enrichment from Jupiter to Saturn to Uranus/Neptune (Podolak et al., 1988; Simonelli et al., 1989). Planetesimal accretion might be inhibited at certain stages through the protective mechanism of mean motion resonances (Zhou & Lin, 2007), while late-stage dust accretion may be limited to particles of very small size (Paardekooper, 2007).

The accretion process ends when the planetesimal and gas supplies terminate, either due to the opening of a disk gap (assuming that this is unbridged by accretion streams), or because the gas disk dissipates (e.g. Tanigawa & Ikoma, 2007). The competing time scales of core formation and disk dispersal further suggests that the resulting architecture of individual systems remains sensitive to initial conditions.

Early models faced a number of difficulties, notably: reproducing the detailed low- and high-Z abundances of Uranus and Neptune, and their partitioning between core and envelope (Pollack et al., 1996); the possibility of dynamical instabilities leading to ejection of the gaseous envelope (Wuchterl, 1990, 1991a,b); and the general problem of the long time scales required for planetary accretion far out in the disk (Lissauer, 1987).

10.6 Giant planet formation

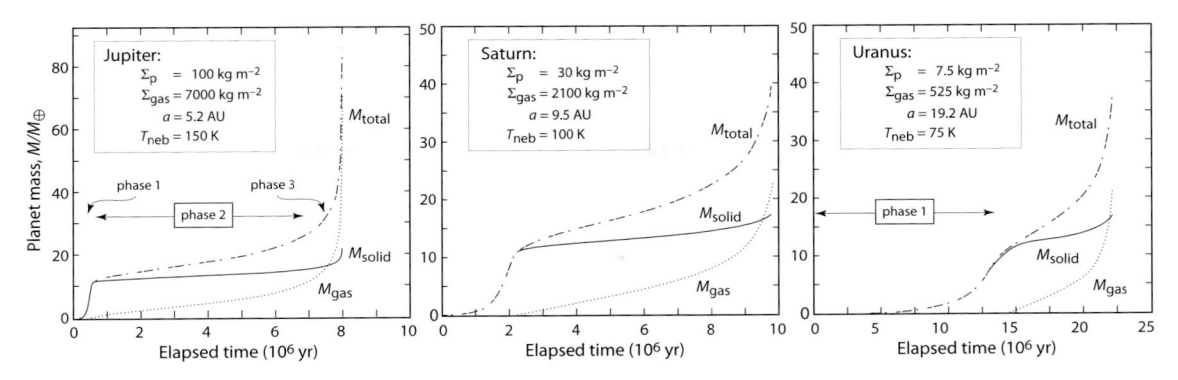

Figure 10.11: Giant planet formation by core accretion. Simulations correspond to the cases of Jupiter, Saturn, and Uranus (models J1, S1, and U1 from Table III of Pollack et al. 1996). Values shown are the initial conditions for the planetesimal and gas surface densities Σ_p and Σ_{gas}, the initial embryo semi-major axis a, and nebula temperature T_{neb}. Simulations begin with an embryo mass comparable to that of Mars, with almost all its mass in a high-Z core. Planetesimals have a radius of 100 km. The results show, as a function of time, the total planet mass M_{total}, and the corresponding contributions from accumulated solids M_{solid}, and accumulated gas M_{gas} (dashed, solid, and dotted lines respectively). From Pollack et al. (1996, Figures 1, 4, 5).

Example simulations Pollack et al. (1996) described simulations which still serve to demonstrate the physical basis of the core accretion model, and which are outlined here. They combined three numerical steps: (i) accretion by an isolated embryo immersed in a sea of planetesimals, (ii) interaction and dissolution of accreted planetesimals with the growing gaseous envelope, and (ii) gas accretion onto the core based on quasi-hydrostatic core–envelope models.

For the first step, the growing planet, with an initial embryo mass corresponding to that of Mars, and with almost all its mass in a high-Z core, is assumed to be embedded in a disk of gas and planetesimals (of radius 100 km), with a locally uniform initial surface mass density. The embryo accretion rate is derived from three-body (Sun, protoplanet, and planetesimal) orbital integrations, with a suitable distribution of planetesimal eccentricities and inclinations.

For the second step, the growing importance of the gaseous envelope around the high-Z core enhances the capture radius R_c (Equation 10.10), leading to the deposition of mass and energy within the envelope once the incoming planetesimal intercepts a mass of gas comparable to its own mass (Pollack et al., 1986).

For the third step, the rate of gas accretion is estimated from the mass and radius of the core at that time, assuming $\rho_{core} = 3.2 \, \text{Mg m}^{-3}$, and based on equations of state and opacity coefficients using a solar mixture of elements, normal stellar opacities for $T > 3000$ K, and opacity contributions from H_2O, silicates, Fe, and TiO at lower temperatures.

Giant planet formation time scales consistent with the lifetime of the solar nebula (cf. Equation 10.10) are met for planetesimal surface densities Σ_p somewhat larger than those given by the minimum mass solar nebula (§10.3.3), and if core growth occurs mainly during runaway accretion when the effects of gravitational focusing can be very large, $F_g \sim 10^4$ (Lissauer, 1987).

Their initial conditions assume specific values for the semi-major axis of the initial embryo orbit a, the nebula temperature T, the initial planetesimal surface density Σ_p, and initial gas surface density Σ_{gas}. Illustrative results are shown in Figure 10.11 for their (baseline) models for Jupiter, Saturn and Uranus. In each case, the results show, as a function of time, the total planet mass M_{total}, and the corresponding contributions from accumulated solids M_{solid}, and accumulated gas M_{gas}.

Three phases of growth For Jupiter, growth actually occurs in three distinct phases. During phase 1, the first 5×10^5 yr, the embryo accumulates solids by runaway planetesimal accretion, a process which ends with depletion of its feeding zone. During most of phase 2, $\sim 7 \times 10^6$ yr, the accretion rates of gas and solids are reasonably constant. The planet's growth accelerates towards the end of this phase, and runaway accumulation of gas (and to a lesser extent solids) ensues during phase 3. The process ends when the supplies of planetesimals and gas terminate.

For Saturn, phase 1 lasts four times longer, while the overall duration of phase 2 is similar. For Uranus, phase 1 is a further factor of eight longer, while the overall duration is a factor 2–3 longer than for Jupiter.

Further developments These results identify a plausible mechanism by which Jupiter and Saturn, and by inference other giant exoplanets, can undergo rapid gas accretion within an elapsed time of a few million years, i.e. within the estimated lifetime of the solar nebula. The bulk compositions of Uranus and Neptune are explained as the result of the dissipation of the nebula gas while the planets were still in their long-lived phase 2.

Within this model, detailed results depend on the precise planetesimal accretion scenario, including the allowance for stirring by nearby competing planetesimals, the impact of dissolved planetesimals on the envelope properties, and the effects of occasional very massive planetesimals.

In the simulations of Bodenheimer & Pollack (1986), critical core masses lie in the range $10-30M_\oplus$. More recent estimates are rather in the range $5-20M_\oplus$, and with a dependency on accretion rate, $\dot{M}_{core}$, and grain opacity, κ. Ikoma et al. (2000, eqn A2) give an analytical fit to simulation results

$$M_{core}^{crit} \sim 7M_\oplus \left(\frac{\dot{M}_{core}}{1\times 10^{-7} M_\oplus\, yr^{-1}}\right)^{q'} \left(\frac{\kappa}{0.1\, m^2\, kg^{-1}}\right)^s, (10.15)$$

with exponents q' and s in the range 0.2–0.3. This dependency indicates that faster growth can be achieved, but at the expense of a larger core mass.

Effects of migration Simulations of the effect of (generally inward) migration (§10.8.2) indicate that it significantly modifies the time scale and outcome of the core formation process (Ward, 1989; Papaloizou & Terquem, 1999; Tanaka & Ida, 1999; Alibert et al., 2004, 2005a; Chambers, 2006a; Kornet & Wolf, 2006; Kornet et al., 2006; Thommes & Murray, 2006; Benvenuto et al., 2007; Crida et al., 2007; Edgar, 2007; Ogihara et al., 2007; Thommes et al., 2007; Chambers, 2008; D'Angelo & Lubow, 2008; Crida et al., 2009a; Kley et al., 2009; Paardekooper & Papaloizou, 2009).

For stochastic migration due to turbulent fluctuations in the disk, Rice & Armitage (2003) showed that the formation of Jupiter can be accelerated by almost an order of magnitude if the growing core executes a random walk with an amplitude of ~ 0.5 AU. At the same time, predicted migration rates place overall constraints on the plausible mass of the protoplanetary disk and, in turn, on the likely parameters of the minimum mass solar nebula (§10.3.3) in the case of the solar system (Desch, 2007; Crida, 2009).

Ongoing investigations Areas of ongoing investigation or uncertainty include:
(1) the first stage demands the assembly of a massive core, a process which is subject to the same uncertainties of basic physics as for terrestrial planet formation;
(2) the time scales for core assembly lie in the middle of the age range at which young solar-type stars lose their gaseous disks. If the gas disk has dissipated by the time that the massive cores form, then the creation of Jupiter-mass planets and heavier will be inhibited;
(3) the required formation time scales for the solar system ice giants and other giant exoplanets orbiting at large separations may be excessive if they formed at their current locations. Outward migration as a result of planetesimal disk interactions appears to be a plausible mechanism which would remove this concern (§10.8.3);
(4) the opacity of the upper envelope regions remain uncertain, with lower opacities resulting in shorter formation time scales (Podolak, 2003; Hubickyj et al., 2005).

Meanwhile, any further downward revision of the estimated core masses for Jupiter and Saturn (Guillot et al., 1997; Guillot, 1999; Guillot et al., 2004; Helled & Schubert, 2008) might formally diminish the appeal of core accretion, unless the cores dissipate following runaway gas accretion.

Conversely, while the observed correlation between planetary frequency and host star metallicity has been taken as providing support for the core accretion model, a full theoretical understanding of the underlying correlation currently remains elusive (§8.4.3).

The case of HD 69830 It is a requirement of the core accretion model that it should be able to replicate the properties of multiple planet systems, using a consistent coupling between the mechanisms of core accretion and inward migration. Alibert et al. (2006) have proposed a possible formation scenario for the three-Neptune mass system HD 69830, in which the planets have masses of 10, 12, and $18M_\oplus$, with semi-major axes $a = 0.08$ AU, 0.19 AU, and 0.63 AU respectively.

Their model predicts that the innermost planet formed from an embryo which started inside the snow line, and is consequently composed of a rocky core surrounded by a small gaseous envelope. The two outermost planets started their formation beyond the snow line and, in consequence, accreted a substantial amount of H_2O-ice during their formation and subsequent migration. The current thermodynamical conditions inside the two outer planets suggests that they are made of a rocky core surrounded by a shell of fluid H_2O, contained within a more extended gaseous envelope. McNeil & Nelson (2010) were, on the other hand, unable to reproduce the observed distributions.

The solar system The low eccentricities and low inclinations of the terrestrial planets in the solar system provide a particular challenge for the core accretion model. The 'dynamical shake-up' scenario of Nagasawa et al. (2005) and Thommes et al. (2008c) has been proposed as a supplemental mechanism for producing terrestrial planets with the nearly circular and coplanar orbits of Earth and Venus.

Their scenario invokes a dissipating gas disk, within which the net precession of each protoplanet is determined by the summed contributions from Jupiter, Saturn, and the disk itself. When the total precession rate of any given protoplanet matches that of one of the giant planets, a secular resonance between the two bodies occurs, resulting in rapid pumping of the smaller body's eccentricity. As the gas disk is depleted, the location of the inner secular resonance with Jupiter sweeps inward, transiting the asteroid belt and then the terrestrial region. One after another, the protoplanets are hit

10.6 Giant planet formation

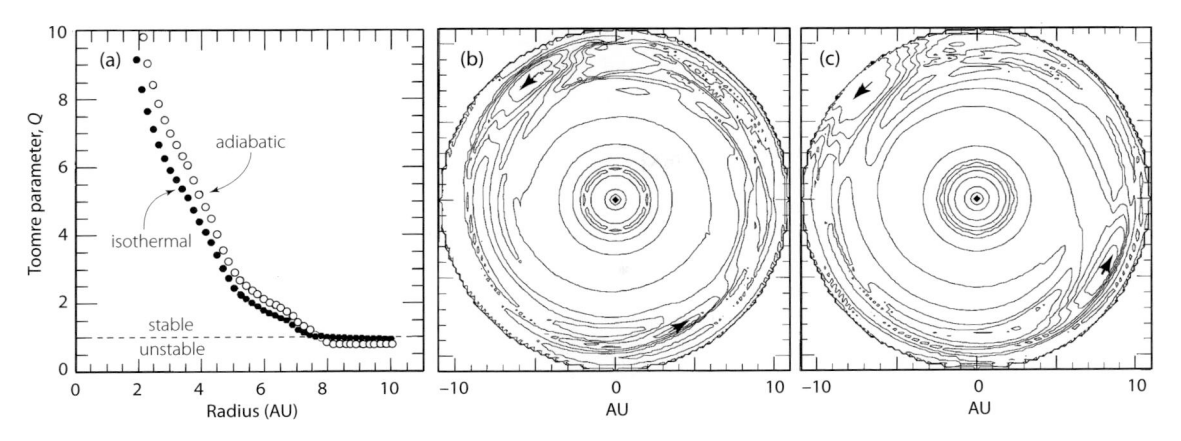

Figure 10.12: Giant planet formation by gravitational instability. (a) the Toomre stability parameter Q as a function of radius, at the start of the simulations, for both locally isothermal and locally adiabatic models. The horizontal dashed line shows the critical value, $Q = 1$, for instability to the growth of non-axisymmetric perturbations. Right pair: equatorial density contours in the three-dimensional disk showing the outcome of (b) the locally isothermal model after 430 yr, and (c) the locally adiabatic model after 550 yr. Each disk forms two giant gaseous protoplanets, of unequal mass, with the arrows denoting their centres and direction of motion. Each protoplanet is trailed by a low-density region, and thin spiral arms. Adapted from Boss (1997, Figures 1–2).

by the resonance, resulting in an inward-passing wave of orbit crossing, and in numerous collisions between protoplanets. The final stage of terrestrial planet formation is therefore initiated as the disk dissipates, rather than afterwards. If the whole process completes while sufficient gas remains, the end products are then damped to low eccentricities and inclinations.

The model offers two further features of note. The final assembly of the planets is completed within several tens of Myr, a factor 10 faster than many simulations without the sweeping resonance, but in good agreement with the time scales from cosmochemical data. Furthermore, a significant delivery of H_2O-rich material from the outer asteroid belt is a natural by-product (§12.5).

An alternative process of rapid assembly of the terrestrial planets matching the observed low eccentricities and inclinations has been modeled with all of the mass initially confined to a narrow annulus between 0.7–1.0 AU (Hansen, 2009).

10.6.2 Formation by gravitational disk instability

An alternative mechanism that may be responsible for the formation of (some) giant planets, is that of *gravitational disk instability*.

Sometimes described as 'top-down', this envisages instead a rapid, primarily single-step collapse in a massive, gravitationally (marginally) unstable protoplanetary gas disk (Kuiper, 1951; Cameron, 1978a; Bodenheimer et al., 1980a; Adams & Benz, 1992; Boss, 1995, 1997, 1998, 2003, 2006a,c,d,e; Michikoshi et al., 2007; Boss, 2008; Mayer, 2010).

The model is founded on the premise that at least some protoplanetary gas disks, during some phase of their evolution, are likely to be gravitationally unstable. Under such conditions, self-gravity in the disk alters its structure and evolution. Density perturbations then grow on a dynamic (or freefall) time scale, developing into spiral arms that produce efficient outward transfer of angular momentum, and inward transfer of mass through gravitational torques.

The central issue is whether such gravitational instabilities, which are not contested, can result in *fragmentation*, in which the disk breaks up into bound self-gravitating clumps of gas and dust. Such clumps are referred to in this model as *giant gaseous protoplanets*.

If such a mechanism operates, it would circumvent the uncertainties inherent in the progressive accumulation of solids, from sub-μm scales upwards, and would side-step the lengthier formation time scale demanded by core accretion.

Relevant disk models are characterised by their Toomre Q parameters (Equation 10.3), with $Q > 1$ indicating stability, and $Q \sim 1$ indicating (marginal) instability. For a disk with an aspect ratio $h/r = 0.05$ at 10 AU around a $1 M_\odot$ star, $Q = 1$ requires a disk surface density $\Sigma \sim 10^4 \, \mathrm{kg\,m}^{-2}$. This is a factor of two larger than that given by the minimum mass solar nebula at the same radial distance (Equation 10.5), suggesting that the mechanism operates, if at all, preferentially at early epochs when the disk mass is still high.

Early interest in the model wavered because of its failure to explain the large and similar values of the estimated core masses of the solar system giants (Pollack, 1984; Stevenson, 1982), although in view of the present uncertainties in their internal structures, these particular constraints appear as less of an obstacle.

Example simulations The essence of the gravitational instability model is captured in the three-dimensional hydrodynamic simulations described by Boss (1997). The model starts with an axisymmetric disk of mass $140 M_J$ ($0.13 M_\odot$), orbiting a star of $1 M_\odot$, and extending between 1–10 AU (rotation period 1–28 yr). The disk is seeded with bar-like density perturbations, biasing the disk towards a two-armed spiral structure.

The disk is assumed to be either locally isothermal ($\gamma = 1$ in the pressure–density relation $P \propto \rho^\gamma$), or locally adiabatic ($\gamma = 7/5$, implying that the density enhancements are unable to cool radiatively during their growth phase). Over some 15 rotation periods of the outer disk, the inner disk regions remain stable ($Q \gg 1$), becoming unstable ($Q \sim 1$) at around 7.5 AU in both cases (Figure 10.12). In both models, the initial bar-shaped perturbation winds up into a two-arm spiral. Self-gravity overcomes the thermal pressure, and each disk breaks up into two giant gaseous protoplanets (the number reflecting the form of the initial perturbations), with masses in the range $1 - 8 M_J$. The Jeans mass, of around $0.2 - 1 M_J$, is exceeded, and further collapse follows.

Boss (1997) argued that, at solar composition, a $1 M_J$ protoplanet region contains $\sim 6 M_\oplus$ of elements heavier than H/He. During the subsequent 10^5 yr of contraction to planetary densities, slowed by the gas pressure, dust grains grow by collisional coagulation and sedimentation, along the lines expected for terrestrial planet formation, forming a substantial rock and ice core.

Effects of cooling Early studies of gravitational instabilities had suggested that marginally unstable disks would evolve through the formation of spiral density waves (Laughlin & Bodenheimer, 1994), which would transport angular momentum, thereby lowering the surface density and enhancing stability (Lin & Pringle, 1990), but also leading to dissipation and heating, thereby raising the sound speed, and further resisting fragmentation (Cassen et al., 1981). Fragmentation then depends crucially on whether the disk self-gravity can overcome the thermal pressure in the disk, which acts to damp the growth of perturbations.

Gammie (2001) further quantified the cooling time as the control parameter driving the outcome of instabilities, showing that the disk fragments for

$$\tau_c \lesssim 3 \Omega^{-1}, \qquad (10.16)$$

otherwise reaching a steady 'gravito-turbulent' state in which $Q \sim 1$, and cooling is balanced by heating due to the dissipation of turbulence. The same fragmentation boundary was confirmed in three-dimensional hydrodynamic simulations by Rice et al. (2003b), and extended to other equations of state by Rice et al. (2005).

Conflicting results In recent simulations, some groups have confirmed that fragmentation occurs, perhaps over specific regimes of orbital radius or opacity, while others have not. For example, Rafikov (2005) found that fragmentation may occur for very massive planets at 50–100 AU, but would be difficult at significantly smaller orbital radii. Boley et al. (2006) used three-dimensional radiative hydrodynamics simulations with realistic opacities for a gravitationally unstable $0.07 M_\odot$ disk around a $0.5 M_\odot$ star, finding cooling times too long to permit fragmentation at all radii. Other results continue to suggest that fragmentation is unlikely, or plausible only at very large radii (e.g. Rafikov, 2007; Boley et al., 2007; Stamatellos & Whitworth, 2008; Rafikov, 2009a; Cai et al., 2010).

Of work suggesting that fragmentation can occur, Boss (2005) demonstrated robust self-gravitating protoplanets forming within spiral arms based on a high-resolution numerical treatment of disk thermodynamics and radiative transfer. Mayer et al. (2007) used three-dimensional smoothed-particle hydrodynamics simulations, finding that gravitationally bound clumps with masses close to $1 M_J$ can occur for disks with masses $> 0.12 M_\odot$ and for mean molecular weights comparable to or higher than solar metallicity. In their simulations, fragmentation is driven by vertical convective-like motions which transport heat from the disk mid-plane to its surface on a time scale of ~ 40 yr at 10 AU.

Other work continues to suggest that fragmentation can occur, with some focus given to the large separation planets around HR 8799 and Fomalhaut (e.g. Boss, 2006b, 2007b; Clarke et al., 2007; Durisen et al., 2008; Dodson-Robinson et al., 2009; Nero & Bjorkman, 2009; Clarke, 2009; Rice & Armitage, 2009; Cossins et al., 2010; Meru & Bate, 2010; Rice et al., 2010; Vorobyov & Basu, 2010).

Dual mode or hybrid mechanisms With results divided as to the regimes in which giant planet formation by gravitational instabilities may occur, the possibilities of dual or hybrid modes have been suggested. Matsuo et al. (2007) have argued, for example, that 90% of the planets known at the time are consistent with core accretion, with 10% demanding disk instability.

Durisen et al. (2007) described hybrid scenarios where gravitational instabilities facilitate core accretion.

Boss (2008) argued that if the same mechanism occurs in a disk being photoevaporated by ultraviolet radiation from nearby massive stars, then these collapsing protoplanets would also be stripped of their gaseous envelopes, resulting in ice giants, or cold super-Earths.

Boley (2009) noted that if giant planets form by disk instability at $r \gtrsim 100$ AU, with core accretion dominating for $r \lesssim 100$ AU, then a bimodal distribution of gas giant semi-major axes should result. Further, if core accretion is indeed less efficient in low-metallicity systems, the ratio of gas giants at large radii to lower mass planets at small radii should increase with decreasing metallicity.

Font-Ribera et al. (2009) suggested that formation at large radii by fragmentation could be followed by inward migration as a result of dynamical friction.

Observational signatures Jang-Condell & Boss (2007) considered the observability of ongoing planet formation by gravitational instability by simulating scattered light images of such a circumstellar disk. Density structures at high disk altitudes, and photometric variations on time scales much shorter than the orbital period of the planet, may be observable by future extremely large telescopes.

10.7 Formation of planetary satellites

The detailed formation history of satellite systems remains uncertain (Scharf, 2008), with a variety of processes likely to be involved (Stevenson et al., 1986). Some, like the Moon, appear to have formed via an impact event (§12.6), while some of the irregular satellites of Jupiter and other giants may have originated via capture (Astakhov & Farrelly, 2004). Orbits may have been subsequently affected by satellite–satellite interactions including resonances (Nesvorný et al., 2003).

Present theories generally assume that the initial planet formation process, whether by core accretion or gravitational instability, is accompanied by the collapse of its gaseous envelope into a sub-disk, forming satellites by coagulation processes analogous to that occurring in the protoplanetary disk itself (Lubow et al., 1999; Alibert et al., 2005b; Lubow & D'Angelo, 2006; Canup & Ward, 2006).

In modeling the satellites of the outer solar system planets, some theories distinguish between a young sub-disk, fed by radial and vertical infall from the protoplanetary disk of gas and dust, and a late sub-disk in which the gas has dispersed (Canup & Ward, 2002; Alibert et al., 2005b; Canup & Ward, 2006). Other models argue for an entirely late-stage gas-poor planetesimal capture mechanism (Estrada & Mosqueira, 2006).

Again, and in analogy with current planet formation and migration theories, satellites which form during the young disk phase may migrate inwards and be accreted by the planet. Surviving satellites may then simply be the last objects to form in the disk, a process which may lead to a general scaling law between the host planet mass and the total satellite mass (Canup & Ward, 2006). The latter find that their models consistently produce satellites totaling a few 10^{-4} of the host planet mass.

Dynamical stability analyses suggest that around stars of $M_\star > 0.15 M_\odot$, satellites are only stable over ~ 5 Gyr for planetary orbits with $a \gtrsim 0.6$ AU (Ward & Reid, 1973; Barnes & O'Brien, 2002). Within this radius, stellar tides are effective in removing the satellites, suggesting that the hot Jupiters within ~ 0.1 AU are unlikely to be accompanied by satellites. Current limits on transit time variations appear to be consistent with this conclusion. A more detailed study of the survivability of Earth-mass satellites around hot Jupiters as a function of orbital radius, including effects of tidal heating and thermal evaporation, is given by Cassidy et al. (2009).

10.8 Orbital migration

10.8.1 Evidence for migration

A robust formation mechanism must explain not only the observed properties of the solar system, but also the exoplanet frequency and mass distribution, their frequently large eccentricities, and the large number of semi-major axes $a < 0.2$ AU, where both high temperature and relatively small amount of protostellar matter available would preclude accretion *in situ* (Boss, 1995; Bodenheimer et al., 2000).

Various mechanisms have been proposed to explain the location of planets at small orbital distances. These include gravitational interactions between two or more Jupiter-mass planets leading to orbit crossing, and to the ejection of one planet while the other is left in a more compact orbit (Rasio & Ford, 1996; Weidenschilling & Marzari, 1996); the effects of a bound companion star (Holman et al., 1997; Mazeh et al., 1997a; Marcy et al., 1999); and the effects of stellar encounters, perhaps in a young star cluster (de la Fuente Marcos & de La Fuente Marcos, 1997; Laughlin & Adams, 1998; Malmberg & Davies, 2009). Malmberg & Davies (2009) find that their simulations result in an eccentricity distribution similar to that observed in the range of separation $1 \lesssim a \lesssim 6$ AU.

A further mechanism widely favoured to explain the relatively large number of very short-period planets, and the overall architecture of many systems, is migration (e.g. Trilling et al., 2002).

Hot Jupiters The existence of close-in hot Jupiters has focused attention on some earlier predictions that Jupiter-mass planets (gas giants) could be formed further from the star, followed by non-destructive migration inwards. This inward migration could be driven by tidal interactions with the protoplanetary disk (Goldreich & Tremaine, 1979, 1980; Lin & Papaloizou, 1986a,b; Ward & Hourigan, 1989), with the orbits subject to tidal circularisation in the process (Terquem et al., 1998; Ford et al., 1999).

But the existence of many exoplanets with small orbital radius provides a complication for migration models: the orbital migration time scale decreases with decreasing orbital period, which should lead to rapid orbital decay for successively smaller orbits. This suggests that any inward migration must somehow be halted at some point, at least until the disk evaporates. Theories and numerical simulations have therefore examined orbital migration, halting mechanisms, and the origin of orbital eccentricities as part of the overall formation scenario (Lin et al., 1996; Rasio et al., 1996; Mazeh et al., 1997b; Ward, 1997a,b; Murray et al., 1998; Trilling et al., 1998; Ward, 1998; Ruden, 1999).

At least three mechanisms are now believed to lead to substantial orbital evolution once the planets have formed: interactions between the planets and the residual gaseous protoplanetary disk (gas disk migration), interactions between the planets and the remnant planetesimals (planetesimal disk migration), and planet–planet scattering. The physical basis for each of these processes is now reasonably well understood, and summarised in the following sections. More comprehensive reviews are given by Papaloizou & Terquem (2006) and Armitage (2010).

Hot Earths, hot super-Earths and hot Neptunes The more recent discoveries of significantly lower-mass planets in short-period orbits, the hot Earths and super-Earths ($\sim 1-10 M_\oplus$) and the hot Neptunes ($\sim 10-20 M_\oplus$), have a somewhat wider range of plausible origins than the planet–planet scattering and migration possibilities for the more massive gas giants.

Models include (Raymond et al., 2008c, and references) (1) *in situ* accretion; (2) formation at larger orbital radii followed by inward type I migration, (3) formation from material being shepherded inward by a migrating gas giant, (4) formation from material being shepherded by moving secular resonances during disk dispersal, (5) tidal circularisation of eccentric terrestrial planets with small pericentres, (6) mass-loss of a close-in gas or ice giant by photoevaporation, (7) mass-loss by collisional stripping (Marcus et al., 2009).

Raymond et al. (2008c) argue that the planetary system architecture, and the bulk composition of the transiting close-in planet, may together provide sufficient diagnostics to distinguish between these different models. The degree of evaporation is expected to decrease with decreasing stellar mass (Kennedy & Kenyon, 2008a).

10.8.2 Gas disk migration

The runaway, oligarchic, and early post-oligarchic stages of planet formation are accompanied by the continuing presence of residual gas within the disk. A planet moving through the gas disk excites spiral density waves within it. Orbital migration, eccentricity evolution, and gap opening may all result from an interaction between the planet and the perturbed disk (e.g. Goldreich & Tremaine, 1979, 1980; Lin & Papaloizou, 1993; Lin et al., 2000; Nelson et al., 2000; Terquem et al., 2000; Papaloizou & Nelson, 2003; Nelson & Papaloizou, 2003, 2004; Papaloizou & Terquem, 2006).

The excited density waves are pronounced at the locations of the Lindblad and corotation resonances (Binney & Tremaine, 2008), and the planet experiences a torque exerted by each of them. For a planet on a circular orbit with angular frequency Ω_p, the *corotation resonance* exists at the radius in the disk where

$$\Omega = \Omega_p . \quad (10.17)$$

The multiple inner and outer *Lindblad resonances* occur where

$$m(\Omega - \Omega_p) = \pm \kappa_0 , \quad (10.18)$$

where m is an integer, and κ_0 is the epicyclic frequency. For a Keplerian gas disk, $\kappa_0 = \Omega$, and the radii of the Lindblad resonances fall at

$$r_L = r_p \left(1 \pm \frac{1}{m}\right)^{2/3} , \quad (10.19)$$

where r_p is the orbital radius of the planet.

Goldreich & Tremaine (1979, 1980) derived expressions for the torques on the planet embedded in a two-dimensional disk. The net torque, mainly resulting from the positive torques exerted by the Lindblad resonances located interior to the planet and the negative torques exerted by the exterior Lindblad resonances, causes the radial orbit migration of the planet. For the more general case of eccentric orbits, (non-coorbital) corotation resonances still exist, but now at harmonics of the mean motion angular frequency (Artymowicz & Lubow, 1994; Papaloizou & Larwood, 2000).

The net differential Lindblad torque in both two-dimensional disks (e.g. Hourigan & Ward, 1984; Ward, 1986; Korycansky & Pollack, 1993) and in three-dimensional disks (e.g. Takeuchi & Miyama, 1998; Lubow & Ogilvie, 1998; Miyoshi et al., 1999; Tanaka et al., 2002) results in an inward migration time for an Earth-sized planet at 5 AU of order $1 - 10 \times 10^5$ yr.

Compared with the much longer time required for the formation of giant planets by core accretion, of order 10^7 yr (§10.6.1), these results imply that the massive core falls to the Sun before gas accretion can complete. This disparity in time scales results in one of the difficulties in the standard theory of planet formation. As noted above, migration may shorten the accretion time by an order of magnitude, making the two time scales more comparable, but still leaving planet migration as an important factor in planet formation.

Type I migration Evolution of the planet's orbit by wave excitation under the condition that the gas surface density profile remains approximately unperturbed, such that the disk response can be considered using linear analysis, is referred to as *type I migration* (Ward, 1997a). Analytical and numerical simulations show that this condition holds for $M_p \lesssim 10 M_\oplus$. The overall torque on the planet is found by summing the individual torques exerted by each resonance (Figure 10.13). The migration time varies inversely with M_p, and type I migration is consequently most rapid for the largest body

10.8 Orbital migration

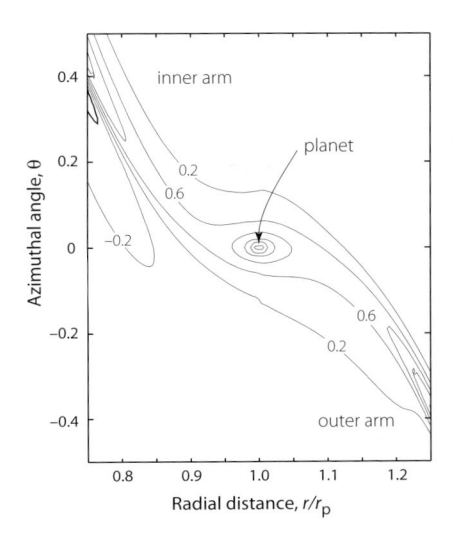

Figure 10.13: Simulations of surface density perturbations excited by a planet in a three-dimensional disk, in the (r,θ) plane. High surface density arms are formed inside and outside the planet orbit, the inner exerting a positive torque on the planet, and vice versa. Their asymmetry causes the net Lindblad torque on the planet. Similar forms are found from hydrodynamical simulations (e.g. Miyoshi et al., 1999). From Tanaka et al. (2002, Figure 6a), reproduced by permission of the AAS.

for which the assumptions that the gas disk remains unaffected by the planet remains valid. Type I migration is illustrated in Figure 10.14a.

For a three-dimensional isothermal gas disk, and including the effects of both corotation and Lindblad resonances, Tanaka et al. (2002) found that, for a disk surface profile $\Sigma(r) \propto r^{-\gamma}$, the migration time scale is given by (their equation 70)

$$\tau_{\rm I} = \frac{1}{(2.1+1.1\gamma)} \frac{M_\star}{M_{\rm p}} \frac{M_\star}{r_{\rm p}^2 \Sigma(r_{\rm p})} \left(\frac{c_{\rm s}}{r_{\rm p}\Omega_{\rm p}}\right)^2 \frac{1}{\Omega_{\rm p}}, \quad (10.20)$$

where $\Sigma(r_{\rm p})$, $c_{\rm s}$, and $\Omega_{\rm p}$ are respectively the gas surface density, sound speed, and angular velocity at the location of a planet orbiting at distance $r_{\rm p}$. As in the two-dimensional case, the migration speed is proportional to the planet mass and the disk surface density, although longer than in the two-dimensional case by a factor 2–3.

Generally, although not exclusively, the net torque due to Lindblad resonances leads to (rapid) inward migration. For a typical low-mass protoplanetary disk with $\Sigma(r) = 1500(r/5\,{\rm AU})^{-3/2}\,{\rm kg\,m}^{-2}$ at $T = 130$ K, an Earth-mass planet at 5 AU has an (inward) migration lifetime of $\tau_{\rm I} \sim 8\times 10^5$ yr (Tanaka et al., 2002), significantly shorter than the formation time scale of gas giants, and that of typical protoplanetary disks (Haisch et al., 2001). The inward migration of $0.1-1 M_\oplus$ planets is accompanied by orbit crossings, mergers, and resonance captures, with multiple hot super-Earths or Neptunes as a result (Terquem & Papaloizou, 2007).

As the gas disperses, migration slows, possibly to the point that the planet can survive for the remaining life of the disk. Accretion may be inefficient as a result (McNeil et al., 2005), and planet formation may be most successful in low-mass disks where migration is less rapid.

Migration rates may be modified by various effects: the presence of a significant toroidal magnetic field (Terquem, 2003); changes in the vertical distribution of gas due to the migrating planet (Jang-Condell & Sasselov, 2005); radiative transfer effects in the inner disk regions (Paardekooper & Mellema, 2006); corotation torques which grow with increasing mass, and which may slow or reverse migration (Masset et al., 2006); and additional torques arising from overdense regions in the presence of magnetorotational instabilities (Laughlin et al., 2004b). These torques are expected to vary with time, randomising the migration, and leading to a reduced dynamical lifetime, although a small fraction of planets may survive for much longer times (Johnson et al., 2006a; Adams & Bloch, 2009; Adams et al., 2009).

Rapid inward type I migration during the early protoplanetary growth phase would lead to the loss of a significant fraction of cores into the star. The problem may be eased in models of non-isothermal disks employing radiative diffusion for realistic opacities, which show significant outward migration for low planetary masses (Kley & Crida, 2008). A steep surface density in the inner disk, expected at the boundary between an active disk and a dead zone, may also inhibit core loss onto the star (Morbidelli et al., 2008).

Type II migration Type I migration is characterised by an azimuthally-averaged gas surface density profile which is largely unperturbed by the planet. For larger planet masses, the surface density profile is modified by the presence of the planet. Under these circumstances, the planet's angular momentum dominates the disk's viscous forces, gas is repelled from the vicinity of the planet, and an annular gap in the gas opens up at the orbital radius of the planet. The resulting orbital evolution, in which angular momentum exchange occurs through a combination of wave excitation and shock dissipation, is referred to as *type II migration* (Figure 10.14b).

Two conditions are necessary for gap formation (Armitage, 2007a, eqn 212–216). The first is that the radius of the planet's Hill sphere is larger than the thickness of the gas disk. This corresponds to the condition

$$R_{\rm H} \equiv \left(\frac{M_{\rm p}}{3M_\star}\right)^{1/3} r \gtrsim h, \quad (10.21)$$

which requires

$$\frac{M_{\rm p}}{M_\star} \gtrsim 3\left(\frac{h}{r}\right)_{\rm p}^3. \quad (10.22)$$

For typical protoplanetary disks with aspect ratio $h/r \sim$

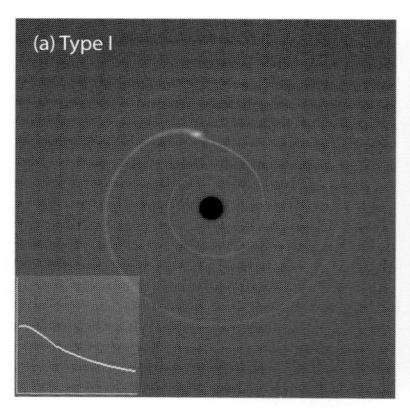

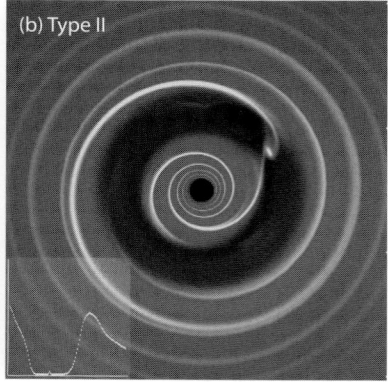

Figure 10.14: *Type I and type II migration. Simulations of the interaction between a planet on a circular orbit with a laminar (non-turbulent) protoplanetary disk, computed from a two-dimensional isothermal hydrodynamic code with a constant kinematic viscosity: (a) in type I migration, a relatively low-mass planet excites a wave in the gas disk, but does not significantly perturb the azimuthally-averaged surface density profile (inset); (b) in type II migration, a more massive planet (here of $10M_J$) clears an annular gap, within which the surface density is a small fraction of its unperturbed value. As the disk evolves, the planet follows the motion of the gas (either inward or outward) while remaining within the gap. From Armitage & Rice (2005, Figure 1).*

0.05, this is satisfied for $M_p/M_\star \sim 10^{-4}$, i.e. for planets between the mass of Saturn and Jupiter.

The second condition is that maintaining a persistent gap requires that tidal torques must remove gas from the gap faster than viscosity will allow it to be replenished (Goldreich & Tremaine, 1980; Lin & Papaloizou, 1980; Papaloizou & Lin, 1984; Takeuchi et al., 1996). Armitage (2007a, eqn 216) gives for this condition

$$\frac{M_p}{M_\star} \gtrsim \left(\frac{c_s}{r_p \Omega_p}\right)_p^2 \alpha^{1/2} . \quad (10.23)$$

For α-disks with $\alpha \sim 0.01$, this condition for gap opening implies $M_p/M_\star \gtrsim 10^{-4}$.

For the two conditions taken together, Jupiter mass planets should be massive enough to force gap opening, while Saturn-type planets may be only marginally so. A more massive planet is required to open a comparable gap when the planet is on an eccentric orbit (Hosseinbor et al., 2007).

Once a gap is opened, orbital evolution is expected to occur on the same time scale as the viscous time scale of the protoplanetary disk (Shakura & Sunyaev, 1973; Ward, 1997a), giving a type II migration time scale (Papaloizou & Terquem, 2006, eqn 75)

$$\tau_{II} = \frac{1}{3\alpha}\left(\frac{r_p}{h}\right)^2 \frac{1}{\Omega_p} . \quad (10.24)$$

For an α-disk with $\alpha = 0.01$ and $h/r = 0.1$, this gives $\tau_{II} \sim 10^3 - 10^4$ yr at $r_p = 1 - 5$ AU, very much shorter than the disk lifetime or planetary formation time scale.

More realistic models include modifications in the disk surface density (Syer & Clarke, 1995); the finite disk accretion rate (Bell et al., 1997); tidal streams that bridge the gap for marginal gap-opening masses (Lubow et al., 1999; Lubow & D'Angelo, 2006); magnetic activity and gas disk dispersal (Armitage, 2002); the global evolution of the disk (Crida & Morbidelli, 2007); the minimum planet mass for gap opening (Edgar et al., 2007); and spiral groove modes (Meschiari & Laughlin, 2008).

There is currently no direct observational confirmation of the existence of such annular gaps, but ALMA (§7.7) may have the sensitivity and angular resolution to detect them (Wolf et al., 2002; Wolf & D'Angelo, 2005).

Type III migration For a smooth initial density profile in the disk, and a small density gradient at the corotation radius, the corotation torques are less pronounced than those arising from the Lindblad resonances. Under certain circumstances, however, the corotation resonance can modify the type I migration rate substantially (Masset & Papaloizou, 2003; Artymowicz, 2004a,b; Masset et al., 2006; Paardekooper & Mellema, 2006), possibly leading to particularly rapid migration referred to as *type III migration*, or *runaway migration*.

The effect depends strongly on the mass accumulation rate, and on the structure of the gas flow in the coorbital region and, like type I migration in principle, does not have a predetermined direction. Two- and three-dimensional simulations have been carried out, covering both inward and outward migration (Masset & Papaloizou, 2003; D'Angelo et al., 2005; Papaloizou, 2005; Pepliński et al., 2008a,b,c).

Simulations show a transition from the fast to a slow regime, which ends type III migration well before the planet reaches the star. In the fast regime the migration rate and induced eccentricity are lower for less massive disks, but are reasonably independent of planet mass. Eccentricity is damped on the migration time scale.

10.8 Orbital migration

Migration and the brown dwarf desert Armitage & Bonnell (2002) proposed that the dearth of brown dwarfs in short-period orbits around solar mass stars, the brown dwarf desert (Grether & Lineweaver, 2006), is a consequence of inward migration in the presence of an evolving protoplanetary disk. Brown dwarf secondaries forming at the same time as the primary star have masses which are comparable to the initial mass of the protoplanetary disk, and are destroyed via merger with the star as a result of inward migration (Matzner & Levin, 2005). Massive planets forming at a later epoch, when the disk is largely dispersed, survive. According to this model, the brown dwarf desert arises because the mass at the H-burning limit is coincidentally comparable to the initial disk mass for a solar mass star.

Migration and the existence of terrestrial planets Dynamical simulations by Mandell & Sigurdsson (2003) suggest that a significant fraction of existing terrestrial planets could survive the inward migration of a giant planet, possibly returning to near circular orbits relatively close to their original positions. Although the migrating giant moves through various orbital resonances with the inner planets, and may excite large eccentricities, the dynamics are chaotic, and the ultimate fate of specific planets is highly dependent on initial conditions. But once the giant planet has moved sufficiently close to the star, it is effectively decoupled from any remaining bound terrestrial planets, and the remaining planets settle into quasi-stable orbits. A fraction of the final orbits are in the habitable zone, suggesting that planetary systems with close-in giant planets remain viable targets for searches for Earth-like habitable planets (Jones et al., 2006a). A population of planetesimals of very high eccentricity and inclination may persist as signatures of such events (Lufkin et al., 2006; Veras & Armitage, 2006).

If a giant planet forms and migrates quickly, the planetesimal population may have time to regenerate within the lifetime of the disk, and terrestrial planets may still be able to form thereafter, albeit with a reduced efficiency (Armitage, 2003). Potentially habitable planets may only be able to form for small final orbital radii of the hot Jupiter, $a \lesssim 0.25$ AU, while for $a \gtrsim 0.5$ AU their formation appears to be suppressed (Raymond et al., 2005a; Raymond, 2006). Nevertheless, water-rich terrestrial planets should be able to form in the habitable zones of hot Jupiter systems, along with hot Earths and hot Neptunes, even for modest giant planet orbital eccentricities (Fogg & Nelson, 2005; Mandell et al., 2007; Fogg & Nelson, 2007b,a, 2009).

Halting migration Conditions which drive inward migration would require some halting mechanism if the planet is not to fall terminally onto the host star.

Various mechanisms have been hypothesised which might bring about such a halting, none of which have been fully embraced. These include: Roche lobe overflow (Trilling et al., 1998; Hansen & Barman, 2007), possibly as a result of inflation due to ohmic heating caused by the planet's response to the star's tilted magnetic field (Laine et al., 2008); tidal friction (Trilling et al., 1998; Lin et al., 2000); inner disk truncation by the stellar magnetosphere (Lin et al., 1996; Shu et al., 2000; Romanova & Lovelace, 2006; Rice et al., 2008; Adams et al., 2009), stellar magnetospheric winds (Lovelace et al., 2008), or photoevaporation (Matsuyama et al., 2003); holes in the protoplanetary disk arising from magnetorotational instabilities (Kuchner & Lecar, 2002; Matsumura & Pudritz, 2005; Papaloizou et al., 2007); magnetic resonance torques in the disk (Terquem, 2003); magnetohydrodynamic turbulence (Laughlin et al., 2004b; Nelson & Papaloizou, 2004); perturbations from a nearby star (Mal'Nev et al., 2006); twisted magnetic toroidal–poloidal torques linking the star and planet (Fleck, 2008); resonant trapping (considered further below), leading to a reversal of type II migration.

Migration would also be stopped at any radius when the disk dissipates (Trilling et al., 1998, 2002; Lecar & Sasselov, 2003), although this appears to require very fine tuning of the disk mass and disk lifetime to explain the observed population. In a more causal variant of this hypothesis, a giant planet could survive if, after formation, not enough material was left in the disk for significant migration to occur. In this scenario, a number of giant planets could assemble and fall onto the star (Gonzalez, 1997; Laughlin & Adams, 1997), with disk material left over to allow a further planet to form but not migrate (Lin, 1997).

A slightly different picture would follow if the close-in planets reached their present configurations through tidal circularisation (§10.9) following scattering into eccentric orbits with small pericentres (e.g. Marzari & Weidenschilling, 2002), as suggested by their spin–orbit alignments revealed by the Rossiter–McLaughlin effect (§6.4.11). For close-in planets, tidal decay continues, eventually to destruction, even after the orbits have circularised. The hot and very hot Jupiters observed might simply be those next in line for tidal destruction, after the many that have already spiraled down into their host stars (Jackson et al., 2009). The short life expectancies inferred for some of the shortest-period planets, such as WASP–12 b, CoRoT 7 b, and OGLE–TR–56 b, would be a consequence.

Indirect evidence for this mechanism may come from the chemical pollution (§8.4.3) or spin-up (§10.9) of host stars which have consumed a close-in planet. Tidal destruction could be halted, or at least delayed, if spin-up of the host star leads to tidal locking in which the rotation period of the host star becomes equal to that of the orbital period as the planet spirals inward.

More direct evidence for this scenario will come

from improved statistics of spin–orbit alignment and semi-major axis distribution. For a population still dominated by inward migration, the inner edge of the mass–period distribution should correspond to the Roche limit. For a population resulting from initially eccentric orbits subsequently tidally circularised, the inner edge should correspond to twice the Roche limit (Pätzold & Rauer, 2002; Ford & Rasio, 2006).

Modification of eccentricities The same disk torques which cause migration also alter the orbital eccentricities and inclinations of the migrating planet. Whether eccentricities undergo growth or decay depends on the relative strengths of the Lindblad resonances which increase eccentricity, and (non-coorbital) corotation resonances which act to damp it (Goldreich & Tremaine, 1980; Artymowicz, 1993; Goldreich & Sari, 2003; Moorhead & Adams, 2008).

Some analytic estimates suggest that, for low-mass planets, e and i can be damped on time scales $10^2 - 10^3$ times shorter than the migration time (e.g. Tanaka & Ward, 2004), and this damping may play a significant role in orbit circularisation for terrestrial planets (Kominami & Ida, 2002), at least for the 10 Myr until disk dispersal (Pascucci et al., 2006).

If the planet is of sufficient mass to clear a gap in the disk, the efficiency of eccentricity damping by coorbital Lindblad resonances is reduced, and eccentricity excitation may dominate (Goldreich & Sari, 2003; Ogilvie & Lubow, 2003; Sari & Goldreich, 2004; Masset & Ogilvie, 2004). While such eccentricity growth is not predicted for Jovian mass planets, the effect becomes important for brown dwarf masses and above (Artymowicz et al., 1991; Papaloizou et al., 2001).

Resonant migration The discovery, through imaging, of exoplanets orbiting at distances out to $a \sim 120\,\mathrm{AU}$, notably those surrounding Fomalhaut and HR 8799 (§7.5.3), provides another challenge to formation theories: at these distances, typical dynamical and accretion time scales are expected to be too long to allow their formation *in situ*. Three possibilities have been considered to explain their existence.

In the first, planet formation by gravitational instability remains effective in the outer parts of the disk beyond $\sim 50 - 100\,\mathrm{AU}$, in regions where the cooling time relative to the dynamical time becomes short, and where the Toomre Q parameter may be smaller (§10.6.2). Alternatively, the planets may have formed in the $5-20\,\mathrm{AU}$ region, and their current locations with their large semi-major axes may have resulted from scattering with other giant planets (§10.8.4), perhaps producing a population of giant planets as distant as $\sim 100 - 10^5\,\mathrm{AU}$ (Veras et al., 2009).

A third possibility combines type II migration with the mechanism by which two migrating planets can become trapped in mean motion resonance (Snellgrove

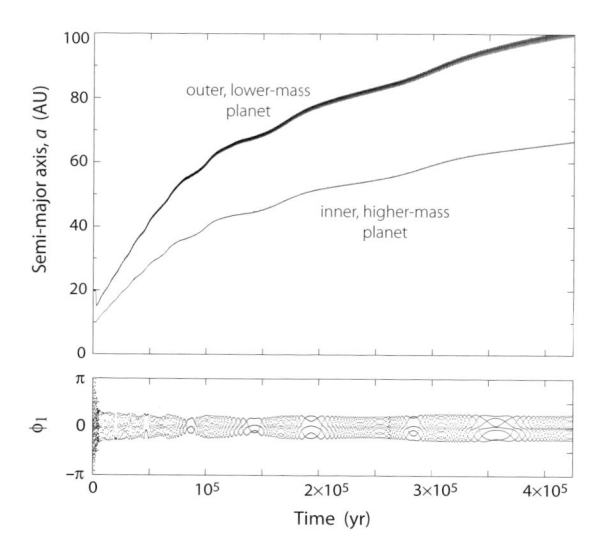

Figure 10.15: Resonant migration for two planets initially at $a = 10$ and $20\,\mathrm{AU}$, orbiting a $M_\star = 2M_\odot$ star, within a disk of surface density $\Sigma = 2666\,\mathrm{kg\,m^{-2}}$ at $10\,\mathrm{AU}$. Top: semi-major axes as a function of time. Bottom: resonant angle $\phi_1 = -\lambda_1 + 2\lambda_2 - \varpi_1$ (Equation 2.52) as a function of time. From Crida et al. (2009b, Figure 1), reproduced by permission of the AAS.

et al., 2001; Murray et al., 2002; Papaloizou, 2003; Ferraz-Mello et al., 2003; Wyatt, 2003; Kley et al., 2004; Veras & Armitage, 2004b; Kley & Sándor, 2007; Martin et al., 2007; Pierens & Nelson, 2008a; Crida et al., 2008). Masset & Snellgrove (2001) found that, under certain conditions during type II migration, resonant trapping can reverse the normal inward migration, resulting in an outward migration of both planets.

Specifically, if two planets open up overlapping gaps, then the inner planet experiences a positive torque from the inner disk, but a reduced torque from the disk's outer parts. Similarly, the outer planet experiences a mostly negative torque from the outer disk. If the inner planet is more massive, the total torque applied to the pair is positive, and a resonantly trapped pair then migrate outwards together.

The importance of the effect depends on a number of factors. To continue over a long period, material lying outside of the common gap must be funneled towards the inner disk, and the inner disk must also be replenished (Crida et al., 2009b). The drift rate is also a decreasing function of the disk's aspect ratio, with the one-sided Lindblad torque being proportional to $(h/r)^{-3}$ (Masset & Snellgrove, 2001; Morbidelli & Crida, 2007).

Simulation results from Crida et al. (2009b) are shown in Figure 10.15. This models the evolution of the semi-major axes for a pair of outwardly migrating planets, initially at $r = 10$ and $20\,\mathrm{AU}$, for a disk surface density $\Sigma = 2666\,\mathrm{kg\,m^{-2}}$ at $10\,\mathrm{AU}$ (6.6 times heavier than the minimum mass solar nebula, §10.3.3), and with $M_\star =$

10.8 Orbital migration

$2M_\odot$, corresponding to the case of Fomalhaut. The outer planet migrates to 100 AU in 4×10^5 yr, reaching the same distance in 4×10^6 yr in a disk a factor of ten less massive.

The model of Crida et al. (2009b) predicts that, in the Fomalhaut system with Fomalhaut b at 115 AU (Kalas et al., 2008), a second planet, with $M_p \sim 1-10 M_J$, should now be orbiting at ~ 75 AU.

10.8.3 Planetesimal disk migration

After formation of the terrestrial and gas/ice giants, and after dispersal of the gas disk, some planet migration is still expected as a result of gravitational scattering between the planets and any remaining planetesimals.

For a planetesimal of mass δm interacting with a planet of mass M_p at orbital radius r then, by conservation of angular momentum, an inward scattering of the planetesimal will be accompanied by an outward movement of the planet, and *vice versa*, leading to a planet displacement of order (Malhotra, 1995; Armitage, 2007a)

$$\frac{\delta r}{r} \sim \pm \frac{\delta m}{M_p} \qquad (10.25)$$

respectively. For significant migration, the total mass of scatterable planetesimals should be at least comparable to that of the planet. Even so, outwardly scattered planetesimals may be lost from the system while, to first order, a single planet may be expected to undergo only limited overall migration as a result of comparable numbers of inward and outward scatterings.

Relatively small migrations nevertheless appear to have had significant effects on the architecture of the solar system, and in particular over its outer regions (e.g. Gomes et al., 2004; Tsiganis et al., 2005; Murray-Clay & Chiang, 2006; Morbidelli & Levison, 2008). The detailed consequences of planetesimal disk migration for the solar system, for which observational constraints are both numerous and detailed, including perhaps the precipitation of the lunar late heavy bombardment, are considered further in §12.7. These considerations provide additional support for the paradigm of planet formation according to the solar nebula theory.

In the simulations of Thommes et al. (2008a) young planetary systems with closely spaced orbits may have arrived in stable mean motion resonance configurations through interaction with the gas disk. Following gas dispersal, the subsequent effects of planetesimal migration can lead to further large-scale dynamical instability, with more-or-less cataclysmic results. Planetesimal disk migration may therefore be an evolutionary step in many planetary systems. Some outcomes may resemble the architecture of the solar system, while others end up with the type of high-eccentricity orbits commonly observed in exoplanet systems.

10.8.4 Planet–planet scattering

Planet–planet scattering resulting from gravitational interactions with other orbiting planets in the same system can continue after both the gas and planetesimal disks have been lost or depleted (Barnes, 2010; Marzari, 2010). Evolution of an initially unstable multiple planetary system can result in the ejection of one or more planets (typically the lightest), in an increased orbital separation tending towards a more stable configuration, or in planet–planet or planet–star collisions. The relative probability of these events is determined by the initial orbital radii, masses, and orbital eccentricities, and N-body experiments are needed to study the evolution of any given system.

Scattering may be accompanied or stimulated by gas disk migration, and may be accompanied by Kozai resonance enhancing eccentricity growth.

Typically, the end result is a modest inward migration of the surviving planets, often with a significant gain in eccentricity. High-eccentricity planets with very large apocentric distances, $\gtrsim 10\,000$ AU, also appear in these simulations, representing a transient phase on the path to ejection and the formation of free-floating planets (Veras et al., 2009; Scharf & Menou, 2009). The detection of free-floating planets may be possible through microlensing (§5.7), or perhaps through more exotic impact signatures on the horizontal branch in dense globular clusters (Soker et al., 2001).

Such planet–planet scattering is the leading candidate for explaining the occurrence of non-circular orbits, with simulations successfully reproducing the incidence and distribution of high orbital eccentricities amongst the most massive (Weidenschilling & Marzari, 1996; Rasio & Ford, 1996; Lin & Ida, 1997; Levison et al., 1998; Marzari & Weidenschilling, 2002; Adams & Laughlin, 2003; Ford et al., 2003; Adams, 2004; Ford et al., 2005; Moorhead & Adams, 2005; Veras & Armitage, 2005; Barnes & Greenberg, 2007a; Chatterjee et al., 2008; Ford & Rasio, 2008; Jurić & Tremaine, 2008; Moeckel et al., 2008; Nagasawa et al., 2008; Raymond et al., 2008a; Thommes et al., 2008b; Veras et al., 2009; Scharf & Menou, 2009; Raymond et al., 2009b,a, 2010).

Planet–planet with planetesimal scattering Specific simulations which include the effects of both planet–planet scattering, and planetesimals disk interactions at larger orbital radii, are also being carried out (Raymond et al., 2009a, 2010). Their simulations follow the evolution of three massive planets, initially formed in marginally-unstable configurations (specifically, unstable over $10^5 - 10^6$ yr), randomly separated by 4–5 mutual Hill radii, and orbiting up to 10 AU. A planetesimal disk with a total mass of $50 M_\oplus$, corresponding to that used in the Nice model (§12.7), orbits between 10–20 AU. Each of 5000 long-duration N-body simulations were in-

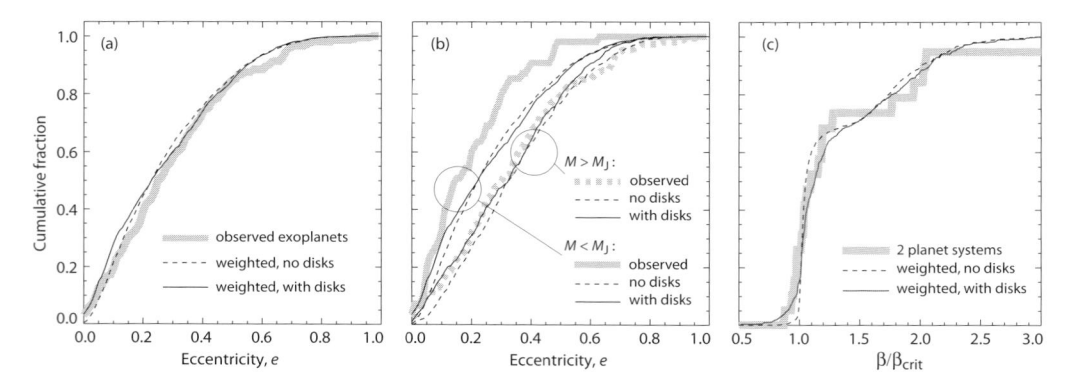

Figure 10.16: Simulations versus observed distributions for numerical experiments based on both planet–planet and planetesimal disk scattering, showing: (a) the cumulative eccentricity distribution, (b) the eccentricity distribution subdivided by mass, and (c) the measure of the proximity of a pair of orbits to the Hill stability limit, $\beta/\beta_{\rm crit}$. Observed distributions are shown by the thick lines, while the simulations, with and without planetesimal disks, are shown by solid and dashed lines, respectively. From Raymond et al. (2010, Figures 1 and 21), reproduced by permission of the AAS.

tegrated over 100 Myr with a 20-day time step. A selection of the results is shown in Figure 10.16.

The picture which emerges from these and other numerical simulations is that the broad eccentricity distribution observed for large exoplanet masses ($M_{\rm p} \gtrsim 0.3 M_{\rm J}$), which is primarily derived from planets at $a \lesssim 3$ AU, is consistent with the distribution arising from isolated planet–planet scattering. For lower mass planets, the presence of the planetesimal disk results in a wide range of evolutionary behaviour, including strong orbital scattering, sudden jumps in eccentricity due to resonance crossings driven by divergent migration, and the strong damping of eccentricities for low-mass planets scattered to large orbital radii.

These simulations lend support to the hypothesis that the lower eccentricities observed for low-mass planets at large orbital radii reflect the past dynamical effects of residual planetesimals disks. Planetary inclinations with respect to the initial orbital plane, and the mutual inclination in multiple planetary systems, are damped in the same way as eccentricity.

The simulations also show that scattering leads to dynamically-packed systems, in which the final separations of two-planet systems cluster close to their Hill stability boundary. A measure of the proximity of a pair of orbits to the Hill stability boundary is given by the quantity $\beta/\beta_{\rm crit}$ (Barnes & Greenberg, 2006c; Raymond et al., 2010) for which the results of scattering simulations, and the observed values for known two-planet systems, are shown in Figure 10.16c.

Scattering into resonances Planet–planet scattering can also populate a variety of orbital resonances, including high-order mean motion resonances, up to 11th order in the simulations of Raymond et al. (2008a). Addition of the planetesimal disk acts as a damping force on the planetary orbits, further inducing mean motion resonances and even resonance chains (e.g. analogous to the 4:2:1 Laplace resonance in Jupiter's Galilean satellites) with high efficiency (Raymond et al., 2009a).

For systems formed in more stable configurations, and not resulting in subsequent close planet–planet encounters, damping by the planetesimal disk also acts to align the orbits into mean motion resonances through convergent migration. A large fraction of outer high-mass giant planets might therefore be expected to orbit in resonance, especially if their orbits display the low-eccentricities expected of more stable systems.

Although the observed orbital arcs are limited in duration, stability analysis suggests that the wide-orbiting triple planet system HR 8799 may be the first example of such a resonant chain (Reidemeister et al., 2009; Fabrycky & Murray-Clay, 2010).

10.9 Tidal effects

10.9.1 Tidal evolution of close-in planets

Exoplanets orbiting within $a \sim 0.2$ AU of the host star, however they might arrive there, will experience significant tidal forces as a result of their proximity (Jackson, 2010). Tides are generated both on the planet due to the potential of the star, and on the star due to the planet.

For close-in planets, tidal forces are generally expected to lead to alignment of their rotation axes, synchronisation of their rotation and orbital periods, a reduction in orbital ellipticity (*tidal circularisation*), an accompanying reduction in semi-major axis, and a conversion of orbital energy into tidal heating of the planet. The associated time scales are believed to be very long, of order 1 Gyr, but effects on the planet's orbit and its thermal heating can nonetheless be very significant.

10.9 Tidal effects

Observationally, the distribution of exoplanet eccentricities as a function of semi-major axis (Figure 2.26) shows that, for $a \gtrsim 0.2$ AU eccentricities are relatively large, averaging $e \sim 0.3$, and broadly distributed up to values close to unity, while for $a \lesssim 0.2$, eccentricities tend to be smaller, averaging $e \sim 0.09$.

It is possible that the combination of mechanisms responsible for generating large orbital eccentricities is less effective for close-in systems. But the current view is that the observed decrease of eccentricities with decreasing a arises partly as a result of damping accompanying inward migration during the few Myr before disk dispersal (§10.8.2), with tides raised between close-in planets and their host stars further contributing to circularisation for systems within $a \lesssim 0.2$ over the subsequent few Gyr (Rasio et al., 1996).

Theoretical time scales for tidal evolution have been estimated for binary star systems (e.g. Zahn, 1977; Hut, 1981) and in a number of studies of exoplanet systems (e.g. Rasio & Ford, 1996; Mardling & Lin, 2002; Dobbs-Dixon et al., 2004; Adams & Laughlin, 2006a). Hut (1980) demonstrated that a binary system can be in tidal equilibrium only if coplanarity, circularity and corotation have been established, and gave an analysis of equilibrium stability against general perturbations. Levrard et al. (2009) argued that essentially none of the transiting exoplanet systems has a tidal equilibrium state, implying their eventual collision with their host star, and suggesting that the nearly circular orbits of transiting planets and the alignment between the stellar spin axis and the planetary orbit are unlikely to be due to tidal dissipation alone. Under these conditions circularisation and synchronisation time scales lose their conventional meaning, and detailed numerical simulations are required to establish a system's tidal evolution.

Formulation The departure of a tidally-distorted body from perfect elasticity or fluidity is described by the dimensionless *specific dissipation function*, Q, defined by (e.g. Goldreich & Soter, 1966, eqn 1)

$$Q^{-1} = \frac{1}{2\pi E_0} \oint \left(-\frac{dE}{dt}\right) dt, \qquad (10.26)$$

where E_0 is the maximum energy stored in the tidal distortion, and the integral over $-dE/dt$, the rate of dissipation, is the energy lost during one orbit. The uncertainty in the knowledge of Q is discussed further below.

Dissipative processes (tidal friction) in each body result in a phase lag between the tidal forcing potential and the body's deformation. If the planet's orbital period is longer than the star's rotation period, the lagging tide is carried ahead of the planet by an angle ϵ given by (MacDonald, 1964, eqn 130)

$$Q^{-1} = \tan 2\epsilon, \qquad (10.27)$$

or, since Q is generally large, $Q^{-1} \simeq 2\epsilon$. It is the asymmetrical alignment of the tidal bulges with respect to the line

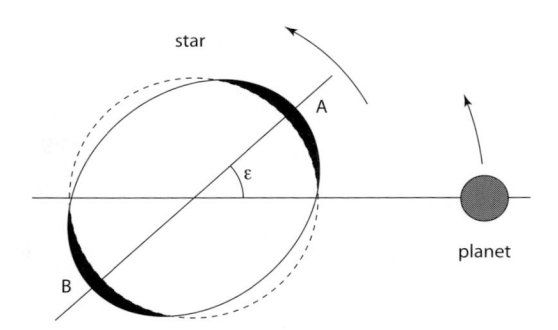

Figure 10.17: The force of attraction between an orbiting planet and the nearer tidal bulge on the star A exceeds that between the planet and B; the net torque retards the rotation of the star and accelerates the planet in its orbit, transferring angular momentum and energy from the star's rotation into the planet's orbital revolution. For a planet in an eccentric orbit, the tidal torque is larger at pericentre, but whether the net change in orbital eccentricity is positive or negative depends on the balance of tidal torques and radial forces, which in turn depends on the stellar rotation compared with the planet's orbital rotation. Adding the effects of tides raised on the planet by the star typically results in a decrease in orbital eccentricity.

of centres (Figure 10.17) which introduces a net torque between the two bodies. As observed in close binaries (e.g. Mazeh, 2008), these torques act to align the rotation axes, synchronise their rotation and orbital periods, and circularise their orbits. The final effect on a planet's orbital eccentricity, however, depends on the balance of tidal torques and radial forces, and the combined effects of the separate tides raised on the two bodies (for a detailed exposition, see Goldreich & Soter, 1966).

In the solar system, tides are raised on the Sun by the planets and their satellites, on the planets by their satellites and the Sun, and on the satellites by the planets and the Sun. Darwin (1908) developed the tidal disturbing function into Fourier components, and derived expressions for the corresponding orbital evolution. Jeffreys (1961) indicated how tides raised on planets by satellites would usually cause a secular *increase* in orbital eccentricity. Goldreich (1963) showed that tides raised on satellites tend to decrease eccentricity, and may often counteract the effect of tides raised on the planet.

10.9.2 Orbital evolution

Expressions for the resulting evolution of a and e (Goldreich & Soter, 1966; Kaula, 1968) have been given in a form appropriate for close-in exoplanets by Jackson et al. (2008b, eqn 1–2) as

$$\frac{1}{e}\frac{de}{dt} = \qquad (10.28)$$
$$-\left[\frac{63}{4}(GM_\star^3)^{1/2}\frac{R_p^5}{Q_p M_p} + \frac{171}{16}\left(\frac{G}{M_\star}\right)^{1/2}\frac{R_\star^5 M_p}{Q_\star}\right]a^{-13/2},$$

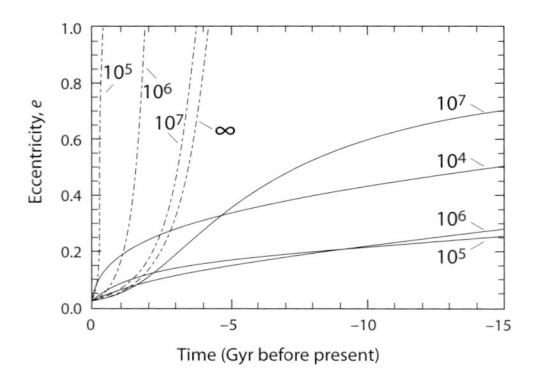

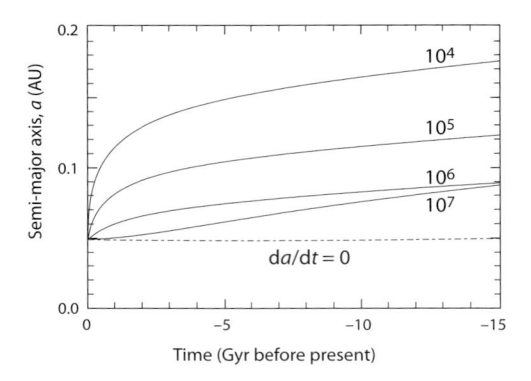

Figure 10.18: *Tidal evolution of the eccentricity, e, and the semi-major axis, a, for τ Boo b. Orbits are traced backwards in time from their current values of e = 0.023, and a = 0.0595 AU. For this example, Q_p is fixed at 10^5, and results are shown for various assumed values of $Q_\star$. Solid lines correspond to the numerical integration of Equations 10.28 and 10.29. Dashed lines show the very different exponential solutions if a is (inappropriately) assumed to be constant. From Jackson et al. (2008b, Figure 3), reproduced by permission of the AAS.*

$$\frac{1}{a}\frac{da}{dt} = \qquad (10.29)$$
$$-\left[\frac{63}{2}(GM_\star^3)^{1/2}\frac{R_p^5}{Q_p M_p}e^2 + \frac{9}{2}\left(\frac{G}{M_\star}\right)^{1/2}\frac{R_\star^5 M_p}{Q_\star}\right]a^{-13/2},$$

where Q_p and $Q_\star$ are the dissipation functions for the planet and star respectively.[4] These equations describe the orbit-averaged effects of the tides, and they include both the effect of the tide raised on the planet by the star, i.e. the *planetary tide* described by Q_p, and the effect of the tide raised on the star by the planet, i.e. the *stellar tide* described by $Q_\star$. Ignoring the stellar tide is equivalent to setting $Q_\star = \infty$.

A number of assumptions underlie these expressions (Jackson et al., 2008b): (1) that Q is independent of frequency, although alternative assumptions and the nature of the body's resulting response have been investigated (Hubbard, 1974; Goldreich & Nicholson, 1977; Hut, 1981; Eggleton et al., 1998; Ogilvie & Lin, 2004, 2007; Ferraz-Mello et al., 2008; Goodman & Lackner, 2009; Greenberg, 2009); (2) that e is small, such that terms of higher order than e^2 are negligible, although more complex tidal models for arbitrary e have also been investigated (Zahn, 1977; Hut, 1981; Eggleton et al., 1998; Mardling & Lin, 2002; Ivanov & Papaloizou, 2004b,a; Ogilvie & Lin, 2004; Efroimsky & Lainey, 2007; Ivanov & Papaloizou, 2007); (3) that the planet's orbital period is small with respect to the stellar rotation period (Goldreich & Soter, 1966; Dobbs-Dixon et al., 2004); and (4) that

---

[4]In this form, the coefficients Q include a contribution from the (unknown) tidal Love number, k, for each distorted body (see box on page 134). This depends on the body's rigidity and radial density distribution. Thus Q used by Jackson et al. (2008b), and here, is equivalent to the $Q' = 3Q/2k$ of Goldreich & Soter (1966), with numerical coefficients corresponding to Love number $k = 3/2$.

other planets on eccentric orbits are not influencing its motion (Mardling & Lin, 2004).

Spin–orbit resonance and tidal locking A further assumption inherent in Equations 10.28–10.29 is that the planet is rotating nearly synchronously with its orbit. One effect of the tidal torques is that a close-in planet should have spun down from any primordial rotation rate to near synchronous rotation over very short time intervals, perhaps measurable in years to decades (e.g. Léger et al., 2009), but certainly on time scales less than ~1 Myr (Goldreich & Soter, 1966; Goldreich & Peale, 1966; Peale, 1977; Rasio et al., 1996; Murray & Dermott, 2000). This is to be compared to the Gyr time scales expected for tidal evolution.

If the torque at a given a is of sufficient magnitude, and synchronisation is achieved, the orbiting planet is said to be *tidally locked*, as in the case of the Earth–Moon system. Hot Jupiters are similarly expected to be tidally locked to their host stars, with one hemisphere in permanent day and the other in permanent night.

Non-synchronous spin-orbit resonances are also possible (Goldreich & Peale, 1966; Winn & Holman, 2005; Correia et al., 2008). The 3:2 spin–orbit resonance of Mercury around the Sun is one such example, discovered by radar observations (Pettengill & Dyce, 1965), and explained in terms of its large eccentricity ($e = 0.21$) combined with its limited rigidity (Peale & Gold, 1965). In the general case of exoplanets with $M_p \lesssim 12 M_\oplus$, dense atmospheres, and of moderate eccentricity and low obliquity, the final equilibrium rotation state may differ from synchronous motion, with up to four distinct equilibrium possibilities, one of which can be retrograde (Correia et al., 2008).

Constraints on planet mass and radius For close-in planets, observed values of e can provide constraints on

10.9 Tidal effects

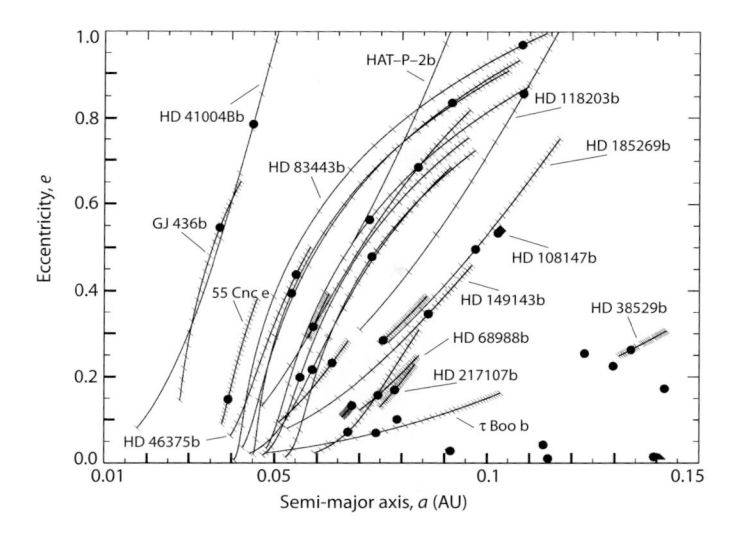

Figure 10.19: Tidal evolution of e and a for the sample of known close-in extrasolar planets using best-fit values of $Q_p = 10^{6.5}$ and $Q_\star = 10^{5.5}$ obtained from a match between predicted initial eccentricities, and the distribution of eccentricities of more distant orbiting planets. Solid curves represent the trajectories of orbital evolution from current orbits (lower left end of each curve) backward in time (toward the upper right). Tick marks are spaced every 500 Myr to indicate the rate of tidal evolution. Tidal integrations were performed for 15 Gyr for all planets, but the filled circles indicate the initial values of orbital elements at the beginning of each planet's life. From Jackson et al. (2008b, Figure 7), reproduced by permission of the AAS.

the planet's mass and radius, although such inferences rely on an assumed value for Q_p. From Equation 10.28, the smaller the value of M_p, the more rapidly the tide raised on the planet can circularise its orbit, an argument used by Trilling (2000) to provide lower mass limits for a number of planets with significant eccentricities.

Bodenheimer et al. (2003) used a similar argument to constrain planetary radii, given that the rate of circularisation increases with increasing R_p.

Constraints on tidal dissipation parameters In the solar system, values of Q separate into two distinct groups: for the terrestrial planets and satellites of the major planets $Q = 10 - 500$, while for the major planets $Q > 6 \times 10^4$ (Goldreich & Soter, 1966). Values of Q_p for exoplanet giants are currently highly uncertain, and have tended to rely either on estimates derived from models of the tidal evolution of the Galilean satellites (e.g. Yoder & Peale, 1981; Greenberg, 1982, 1989; Aksnes & Franklin, 2001), or on models of internal dissipation (e.g. Goldreich & Nicholson, 1977; Ogilvie & Lin, 2004).

Thus, values of $Q_p = 10^6$ for giant exoplanets and $Q_\star = 10^5$ for host stars have been adopted in some early studies of tidal evolution, while for terrestrial exoplanets, values of $Q_p = 100$ may be more appropriate (Goldreich & Soter, 1966; Lambeck, 1977; Dickey et al., 1994; Mardling & Lin, 2002; Barnes et al., 2008b). A continuing uncertainty on Q_p follows from the existence of close-in planets with non-zero eccentricity (Matsumura et al., 2008).

Figure 10.18 illustrates the dependence of $\dot{e}$ and $\dot{a}$ on various values of $Q_\star$ for an assumed $Q_p = 10^5$ in the case of τ Boo b. Incorporating the tidal effects on both bodies leads to a complex relation between $\dot{e}$ and $Q_\star$, as $Q_\star$ increases progressively from 10^4 to 10^7.

Jackson et al. (2008b) examined the hypothesis that tides are responsible for damping eccentricities by numerically integrating Equations 10.28–10.29 for exoplanets closer than 0.2 AU. Their integrations began at the current a and e, and proceeded backwards in time to the point where tidal evolution began to dominate the orbital evolution, i.e. around the epoch that the protoplanetary disk was cleared, and collisional effects diminished. Both tides raised on the planet and on the star are included, as is the coupling between a and e. It is further assumed that, although initially unknown, the same value of Q_p applies to all planets, and the same value of $Q_\star$ applies to all stars.

For each pair of assumed Q_p and $Q_\star$, integration yields a distribution of initial eccentricities. A comparison between these computed distributions, and the observed distributions of e for planets with larger a, leads to their best-estimate of $Q_p = 10^{6.5}$ and $Q_\star = 10^{5.5}$. Although somewhat larger than those frequently adopted in earlier studies, their plausibility provides further support to the hypothesis of tidal circularisation.

Still larger values of $Q_p > 10^7$ have been inferred for some hot Jupiters, for example OGLE–TR–56 b (Carone & Pätzold, 2007), and CoRoT–7 b (Léger et al., 2009).

Application to known exoplanet orbits Applying the resulting tidal models with the derived values of Q_p and $Q_\star$ to known close-in systems Jackson et al. (2008b) demonstrated that, in several cases, stellar and planetary tides have significantly reduced their semi-major axes after the planets formed and gas migration ceased, leading to a subsequent reduction of a factor two or so in their semi-major axes since formation (Figure 10.19). While tidal evolution for many planets is evidently slower than in the past, some, such as HD 41004B b, are still undergoing rapid changes in a and e.

Reliable estimates of a circularisation time scale require the coupled treatment of the evolution of a and e. Much shorter circularisation time scales are inferred if the concurrent changes in a are ignored, as evident from the dashed curves in Figure 10.18a. Such rapid tidal circularisation time scales estimated for HD 209458 b have led to other mechanisms for maintaining the planet's eccentricity being proposed (such as an additional planet by Mardling, 2007), and to tidal heating being excluded as an explanation for its anomalously large radius. In contrast, the coupled equations suggests that e can remain fairly large, with significant tidal heating operating over the past billion years (Jackson et al., 2008b,c).

Effects on planet shape The shape of a planet distorted both by rotation and tidal potential is a triaxial Roche ellipsoid (Chandrasekhar, 1969). The longest of the axes is directed, to within the very small angle ϵ (Equation 10.27), to the star, and the shortest along its rotation axis. For a homogeneous distribution of mass, the tidal bulge or equatorial prolateness, f_+, and the polar flattening, f_- are given by

$$f_+ = \frac{15}{4} \frac{M_\star}{M_p} \left(\frac{R_p}{a}\right)^3,$$

$$f_- = \frac{25}{8} \frac{M_\star}{M_p} \left(\frac{R_p}{a}\right)^3. \qquad (10.30)$$

These tidal bulges, while significant, are typically too small to be relevant in deriving, e.g., volume or density estimates for transiting planets. For example, for CoRoT–7 b ($R_p = 1.68 R_\oplus$), Léger et al. (2009) estimated $f_+ < 0.016$ and $f_- < 0.013$, limits which correspond to a tidal bulge of ±70 km, and a polar radius ~120 km smaller than its mean equatorial radius.

10.9.3 Spin-up of host stars

Tidal effects In addition to evidence for tidal interactions provided by the distribution of orbital eccentricities, Pont (2009) inferred excess rotation of the host stars of a number of transiting planetary systems, attributed to the tendency towards spin–orbit tidal synchronisation, an effect also studied by Mardling & Lin (2002).

Excessive stellar rotation for a given spectral type, derived from radial velocity or photometric observations, was inferred for HD 189733, CoRoT–2, HD 147506,

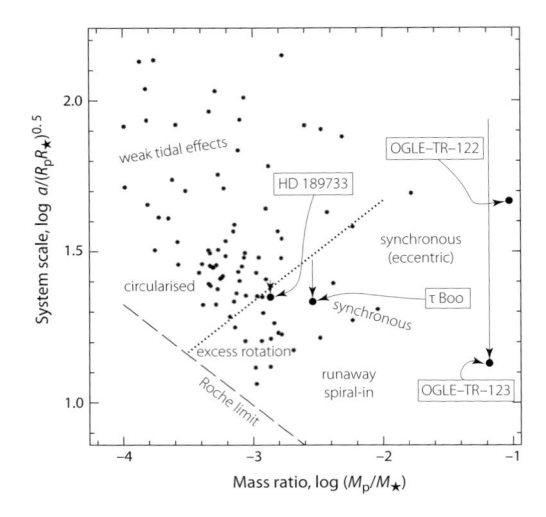

Figure 10.20: Tidal mass–period diagram for transiting planets (indicated as asterisks). The system scale (ordinate) is given by the semi-major axis in joint units of $R_\star$ and R_p. The dotted line marks the possible transition to detectable spin-up and orbit decay. Four objects are labeled, including the two transiting stars with the lowest masses (OGLE–TR–122 and OGLE–TR–123): arrows show the planet's original semi-major axis if the present excess angular momentum of the star is placed back in the planetary orbit. From Pont (2009, Figure 6), © John Wiley & Sons, Inc.

and XO–3, with four others lying above the mean relation (HAT–P–1, WASP–4, WASP–5, and CoRoT–1). Tidal spin-up had previously been excluded for HD 189733 (Bouchy et al., 2005c) and CoRoT–2 (Alonso et al., 2008a). Tidal spin-up has been studied for the OGLE transiting planets (Sasselov, 2003; Pätzold et al., 2004), and has been inferred for a number of non-transiting systems, including τ Boo, with spin rates consistent with synchronous rotation in the case of τ Boo and HD 73256.

Mass–period correlation The corresponding angular momentum exchange implies that some planets must have spiraled towards their star by substantial amounts since dissipation of the protoplanetary disk. The initial planet orbital distance before tidal infall can be estimated by assigning the excess stellar rotation to the planetary orbit. Based on these arguments, Pont (2009) constructed a tidal mass–period diagram which divides up the mass–period plane according to the planet's sensitivity to tidal evolution (Figure 10.20). Below a certain separation corresponding to orbital periods of ~3–5 d, the orbital properties (e, a, and stellar rotation) are controlled by tidal effects, and can be very different from their initial values.

Pont (2009) concluded that, in increasing order of mass, close-in planets would be expected to be tidally unaffected ($M \ll M_J$), circularised ($M \simeq M_J$), spiraling in ($M \simeq 1 - 2 M_J$), destroyed ($M \simeq 2 - 3 M_J$), and synchronised ($M \gtrsim 3 M_J$).

Rapidly rotating main-sequence stars continue to be found hosting planets, some of the most rapid amongst them being 30Ari B b (Guenther et al., 2009), OGLE2–TR–L9 (Snellen et al., 2009b), WASP–33 (Collier Cameron et al., 2010b), and CoRoT–11 (Gandolfi et al., 2010).

Magnetic field effects The time scales involved in tidal synchronisation may be too long for tidal interactions to be responsible for the observed excess in stellar angular momentum. Lanza (2010) proposed that part of the angular momentum excess may be due to star–planet magnetic interactions, leading to a reduction of magnetic breaking, and to a reduction in the angular momentum loss to the stellar wind (Weber & Davis, 1967).

The detailed magnetohydrodynamic model of Cohen et al. (2010) confirms that once the stellar and planetary Alfvén surfaces interact with each other, the stellar wind topology in the hemisphere facing the planet changes and the angular momentum loss to the wind decreases. Rather than the host stars being spun up, the effect amounts rather to a reduction in the stellar angular momentum lost.

Which of the two (tidal or magnetic) mechanisms dominates depends on the stellar type and the individual star-planet parameters.

Implications for radial velocity and transit surveys The various effects of tidal evolution have implications for radial velocity and transit searches for very close-in planets. Strong tidal spin-up of the host star could result in a bias in radial velocity surveys, in which stars with rapid rotation may be dropped from consideration either due to their broadened spectral lines, or due to the anticipated correlation between rotation rate and photospheric activity.

Rotation-induced stellar variability coupled with very short-period transits may also lead to transit light curves qualitatively different from those assumed in standard transit searches (Figure 6.37), resulting in a detection bias against massive close-in planets for these surveys also.

More generally, the probability of observing a transit increases for smaller a, but decreases as orbits become more circular (§6.4.4). These two aspects of tidal orbit evolution would imply that the probability of an observable transit has a dependence on the stellar age, which in turn would have implications for transit surveys, for example in young open clusters (§6.2.3).

10.9.4 Tidal heating

As the planetary orbit evolves through tidal coupling, orbital energy is converted into tidal energy, and substantial internal heating of the planet can result. Since the orbital energy depends only on the semi-major axis, a, the term in $\dot{a}$ corresponding to the tides raised on the

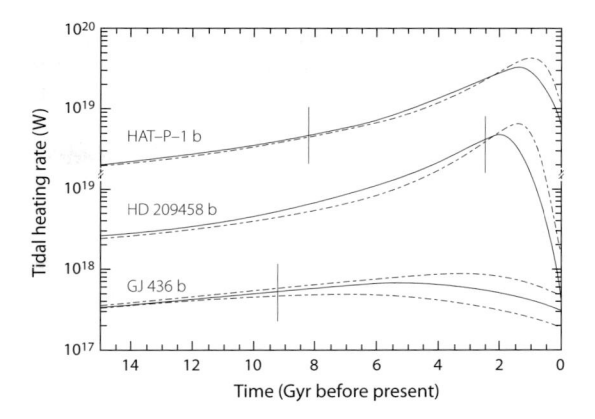

Figure 10.21: Tidal heating for HAT–P–1 b, HD 209458 b, and GJ 436 b as a function of time before present (the vertical scale is shifted for HAT–P–1 b). Vertical lines indicate estimated formation epochs. Solid curves are based on the planet's current nominal eccentricity. Dashed curves are the maximum and minimum heating consistent with observational uncertainty in the orbital elements. For HAT–P–1 b and HD 209458 b, observations cannot exclude zero eccentricity, so the lower bound on heating rates is formally zero, and only the upper limit is shown. From Jackson et al. (2008c, Figure 1), reproduced by permission of the AAS.

planet by the star (Equation 10.29) also gives the tidal heating rate for the planet, H. Jackson et al. (2008c) give

$$H = \frac{63}{4} \frac{(GM_\star)^{3/2} M_\star R_p^5}{Q_p'} a^{-15/2} e^2 , \quad (10.31)$$

$$\simeq 4 \times 10^{10} \, \text{W} \left(\frac{M_\star}{M_\odot}\right)^{\frac{5}{2}} \left(\frac{R_p}{R_J}\right)^5 \left(\frac{Q_p'}{10^{6.5}}\right)^{-1} \left(\frac{a}{1\,\text{AU}}\right)^{-\frac{15}{2}} e^2 ,$$

where $Q_p' = 3Q_p/2k$, and k is the Love number.

Tidal heating increases as the planet moves inward toward its star, i.e. as a decreases, then decreases as its orbit circularises and e decreases. Plausible heating histories for planets with measured radii have been derived by Jackson et al. (2008c), using the same tidal parameters for the star and planet estimated above ($Q_p = 10^{6.5}$, $Q_\star = 10^{5.5}$). Examples are shown in Figure 10.21.

HD 209458 b and HAT–P–1 b are accordingly inferred to have undergone substantial tidal heating of up to $3 - 4 \times 10^{19}$ W as recently as 1 Gyr ago, of the order estimated by Burrows et al. (2007a) to be necessary to explain their anomalously large radii. As an example of planets with radii more consistent with theoretical models, GJ 436 b has also been subject to much smaller tidal heating.

The same models show, for example, that GJ 876 d ($0.02 M_J$) is likely to have experienced substantial tidal heating. At its surface, the predicted tidal heating of $10^{19} - 10^{20}$ W (assuming $M_\star = 0.32 M_\odot$, $R_p = 0.143 R_J$, $a = 0.0208$ AU, $e = 0.01$, $Q_p = 100$, and $k = 0.3$) compares with the 7×10^{17} W estimated to induce substantial melting of its mantle (Valencia et al., 2007a,c), such that it may not, in consequence, be a solid rocky planet.[5]

Other studies of tidal heating, and constraints on Q, are reported by Matsumura et al. (2008), and for various transiting systems such as WASP–12 (Hebb et al., 2009), WASP–14 (Joshi et al., 2009), WASP–18 (Hellier et al., 2009a), and COROT–7 (Léger et al., 2009).

10.9.5 Tidal heating and habitability

Effects of tidal heating in rocky or terrestrial planets and satellites may have significant implications for habitability (Jackson et al., 2008a).

Relevance in the solar system The Jovian satellite Europa is a rocky body covered by 150 km of H_2O, for which tidal heating may maintain a subsurface water ocean and which, scaling from the tidal heating on Io, may have a surface ice layer only a few km thick (Greenberg, 2005). Tidal heating of an icy exoplanet may similarly generate a subsurface ocean suitable for life, even in the absence of an atmosphere (Reynolds et al., 1983; Chyba, 2000; Greenberg, 2005; Vance et al., 2007).

For Io, the chemicals required could be generated by energetic charged particles trapped in Jupiter's magnetosphere (Johnson et al., 2004), and transported into the subsurface ocean by impacts (Chyba & Phillips, 2002), or local melting (O'Brien et al., 2002). Tidal heating might also give rise to underwater volcanic vents (O'Brien et al., 2002), which could support life directly with thermal energy, by analogy with the deep ocean thermophilic bacteria found on Earth (Baross, 1983).

Contribution to tectonic activity For the Earth, tidal heating is negligible, but internal energy from radionuclide decay and the residual heat of formation drives convection in the mantle, contributing to plate tectonic motion (O'Neill & Lenardic, 2007), and helping to stabilise the atmosphere and surface temperature over hundreds of millions of years (Walker et al., 1981). If a stable surface temperature is a prerequisite for life, plate tectonics may be required for habitability (Regenauer-Lieb et al., 2001). When Mars was last tectonically active, its radiogenic heat flux was $\sim 0.04\,\mathrm{W\,m^{-2}}$ (Williams et al., 1997), perhaps a minimum amount required for tectonics in a rocky planet. For terrestrial planets with $M > 1 M_\oplus$, O'Neill & Lenardic (2007) have argued that plate tectonics may be less likely, while Valencia et al. (2007a) concluded the opposite.

M dwarf host stars are considered promising candidates for the discovery of Earth-like habitable planets, as a consequence of the smaller mass ratio required for radial velocity searches for any given planetary mass. Their lower masses implies that their habitable zones would lie closer to the host star than for more massive stars (Selsis et al., 2007; Tarter et al., 2007). However, if rocky planets around M stars have masses much less than $1 M_\oplus$ (Raymond et al., 2007b), then such planets may have too little radiogenic heating to drive long-lived plate tectonics (Williams et al., 1997).

In these circumstances, tidal heating may provide a significant alternative heat source, and may therefore be critical in determining habitability. At the opposite extreme, in analogy with the highly active Jovian satellite Io, terrestrial planets with even moderate eccentricities may have very high tidal heating, perhaps ruling out habitability as a result of excessive volcanic activity (Barnes et al., 2009b).

Contribution to atmospheric replenishment Planet atmospheres are in general believed to originate from a combination of accretion and outgassing (§11.8.1). Tidal heating can enhance the outgassing of volatiles in the mantle by increasing internal convection, and can supply or replenish atmospheres that might otherwise be lost by planetesimal erosion, thermal (Jeans) escape, and hydrodynamic escape (§11.8.2), further promoted by photolytic dissociation (§11.7).

With such scenarios in mind, Jackson et al. (2008a) modeled the tidal heating and evolution of a range of hypothetical exoplanets. They integrated the tidal equations over a range of stellar and planetary masses, assuming a planetary radius based on the geophysical models of Sotin et al. (2007), viz. $R_\mathrm{p} = M_\mathrm{p}^{0.27}$ for $1 < M_\mathrm{p} < 10$ and $R_\mathrm{p} = M_\mathrm{p}^{0.3}$ for $0.1 < M_\mathrm{p} < 1$. Each planet was assigned an initial semi-major axis in the middle of its star's habitable zone (§11.9.1).

Assuming that a heat flux between $\sim 0.04 - 2\,\mathrm{W\,m^{-2}}$ drives plate tectonics without unacceptable volcanism, the epochs at which habitability is possible can then be estimated. Thus around stars of $M_\star = 0.2 M_\odot$, planets of $M_\mathrm{p} = 1 - 10 M_\oplus$ can experience acceptable levels of tidal heating for $> 10-100$ Gyr over a wide range of initial eccentricities, $e_0 \sim 0.2 - 0.8$.

Habitability of exoplanet satellites The habitability of exoplanet satellites will be similarly affected by tidal heating (Williams et al., 1997; Scharf, 2006, 2008). As well as tidal dissipation due to eccentricity damping, tidal heating can occur as a result of other resonances, such as mean motion resonances (as in the Laplacian resonance of Io, Europa, and Ganymede), and spin–orbit libration as in the case of Enceladus (Wisdom, 2004). Again, rapid atmospheric particle loss may be diminished if the satellite possesses an intrinsic magnetic field, as in the case of Ganymede (Kivelson et al., 1998).

---

[5] For an assumed $R_\mathrm{p} = 10\,000$ km, this corresponds to a surface heat flux of $\sim 10^4 - 10^5\,\mathrm{W\,m^{-2}}$. This can be compared to the surface heat flux of the most volcanically-active body in the solar system, Io of $\sim 3\,\mathrm{W\,m^{-2}}$, dominated by tides (Peale et al., 1979; Yoder, 1979; Yoder & Peale, 1981; McEwen et al., 1992), and that of the Earth, $\sim 0.08\,\mathrm{W\,m^{-2}}$, dominated by radiogenic heat (Davies, 1999).

10.10 Population synthesis

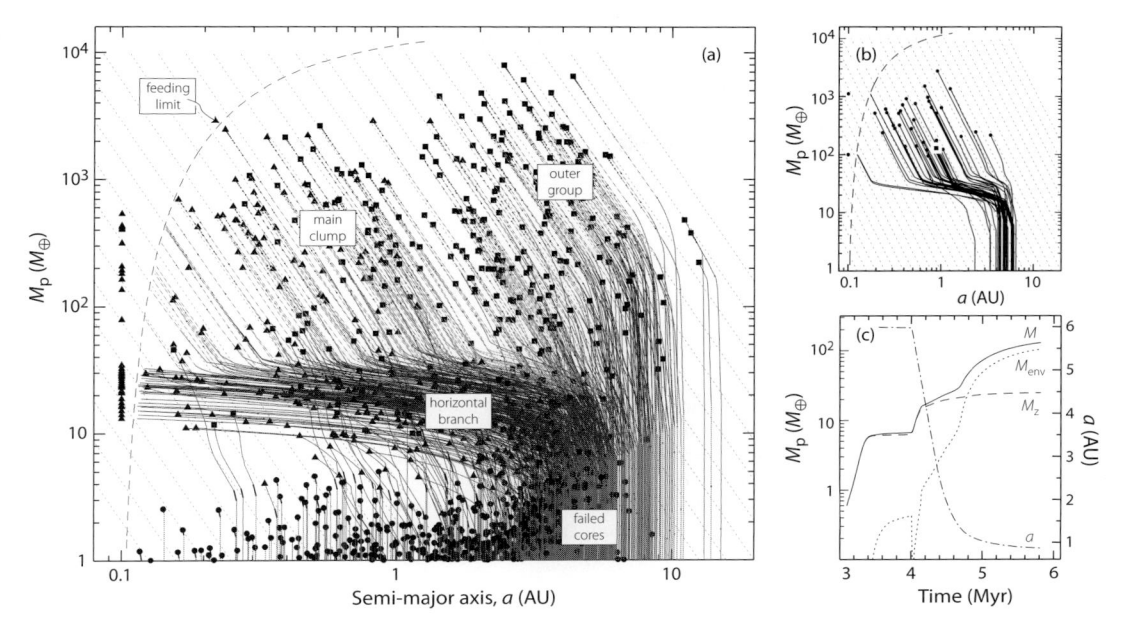

Figure 10.22: Results of population synthesis modeling: (a) planet formation tracks in the (M_p, a) plane for 1500 Monte Carlo runs. Large black symbols show final planet positions, where growth and migration stop. Planets reaching the 'feeding limit' are set arbitrarily to a final value $a = 0.1\,AU$. Short dashed lines have a slope of $-\pi$. Four main groupings are indicated (colour-coded according to migration mode in the original), and small black dots are plotted on the tracks every 0.2 Myr to indicate each planet's temporal evolution; (b–c) formation tracks for a subset of planets passing through the 'horizontal branch' to become members of the 'main clump'. The typical embryo highlighted in (b) as a black-on-white square, whose temporal evolution is shown in (c), eventually leads to a Saturn-mass planet at $a = 0.9\,AU$. In (c), curves are labeled (left ordinate) with total mass M, mass of accreted solids M_Z, mass of the envelope $M_{\rm env}$, and (right ordinate) semi-major axis a. From Mordasini et al. (2009a, Figures 8 and 11), reproduced with permission © ESO.

Scharf (2008) concluded, for example, that a Mars-sized ($0.1 M_\oplus$) satellite retaining a terrestrial-like atmosphere with an orbital radius and eccentricity similar to that of Europa around any of the known exoplanets capable of retaining satellites, could readily attain a tidally-driven habitable surface temperature (i.e. 273 K< T <373 K) over a period of several Gyr, if the necessary driving conditions (e.g. orbital resonance) are maintained. Tidal heating of ejected planet–satellite pairs has also been considered (§11.8.3).

10.10 Population synthesis

Predicting the properties of individual planets, whether solar system planets or exoplanets, is a necessary but insufficient condition for validating models of exoplanet formation and evolution. The number of model parameters is large, and they are potentially only poorly constrained by the properties of either a single system, or a small number of systems.

A large number of exoplanets are now known, with distinct properties becoming classifiable through principal component or hierarchical clustering analyses (e.g. Marchi, 2007). A *population synthesis* approach is therefore now feasible, in which Monte Carlo models are used to establish the properties of an ensemble of hypothetical planetary systems. Various combinations of plausible initial conditions can be combined with a realistic probability of occurrence to predict the final outcome of the formation process, along with their accompanying probabilities. These can then be compared with the observed planetary population in order to verify that the most important statistical properties are reproduced, and furthermore to constrain formation models, and to establish probability distributions for the most important initial conditions.

Such an approach forms the basis of the development of a deterministic model of planet formation by Ida & Lin (2004a,b, 2005b, 2008a,b, 2010), the assessment of silicon enrichment by Robinson et al. (2006a), and other similar analyses of planetary radii, masses and orbital distance (e.g. Kornet & Wolf, 2006; Thommes et al., 2008b; Broeg, 2009).

Such models can also be used to make predictions about individual planets or system architectures that have not yet been observed, either because they are created only rarely, or because observational accuracies are so far insufficient to detect them.

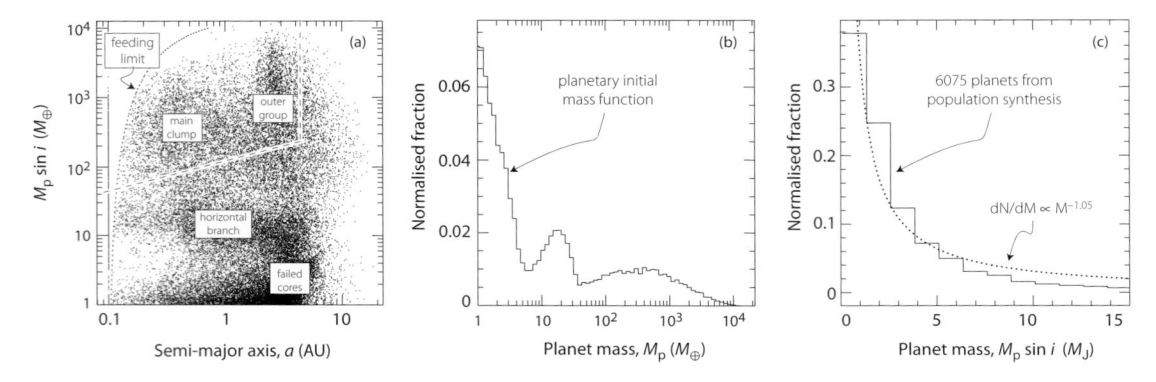

Figure 10.23: Results of population synthesis modeling: (a) distribution of the 70 000 point synthetic population in the (projected) ($M_p \sin i, a$) plane. The various sub-populations of 'failed cores', 'horizontal branch', 'main clump', and 'outer group' planets are again identified. The dashed line indicates the approximate boundary of the region currently most robustly probed by radial velocity surveys; (b) planetary initial mass function, corresponding to the moment in time when the gaseous protoplanetary disk disappears; (c) histogram of the predicted $M_p \sin i$ for the 6075 detectable simulated planets compared to the observationally inferred $dN/dM \propto M^{-1.05}$ power law of Marcy et al. (2005a). From Mordasini et al. (2009b, Figures 2, 3 and 5), reproduced with permission © ESO.

Monte Carlo models Extensive studies along these lines have been carried out by Benz et al. (2008) and Mordasini et al. (2009a,b). Their modeling is based on the principles of giant planet formation by core accretion (§10.6.1) using the models of Alibert et al. (2005a), and implicitly including type I and type II disk migration (§10.8.2). Evolution is terminated when inward migration takes the planet's feeding zone to a separation below 0.1 AU, and tidal effects are not included.

They used four Monte Carlo variables to describe a range of initial conditions: the initial dust-to-gas ratio of the protoplanetary disk (ranging between 0.013–0.13), the initial gas surface density Σ_0 at 5.2 AU (500 – 10^4 kg m^{-2}), the photoevaporation rate ($5 \times 10^{-10} - 3 \times 10^{-8} M_\odot$ yr^{-1}), and the initial semi-major axis of a seed embryo within the disk (0.1–20 AU). Each of these is accompanied by corresponding probability distributions.

Sub-populations from evolutionary tracks Some illustrative results of 1500 Monte Carlo runs from Mordasini et al. (2009a) are shown in Figure 10.22. They identified four main classes of tracks:

(a) 'failed cores': these are low-mass planets occupying the lower-right region of the diagram, for which evolution has ceased at low mass since the initial conditions do not allow the formation of more massive planets during the disk's lifetime. Embryos undergo type I migration, but only slowly, such that their evolutionary tracks essentially follow constant a. As expected from the isolation mass, failed cores can reach larger masses at larger distances. Failed-core planets are qualitatively different from terrestrial planets, which may occupy a similar region of the (M_p, a) plane, but which form from multiple giant impacts after gas disk dispersal, on much longer time scales (and not modeled by Mordasini et al., 2009a);

(b) 'horizontal branch planets': these were so-designated by Mordasini et al. (2009a) as a consequence of their morphology in the (M_p, a) plane. Type II migration dominates, and occurs on time scales much shorter than the underlying accretion. The transition from type I to type II migration can be accompanied by a rapid growth in mass outside of the snow line, followed by a crossing of the snow line on the planet's inward migration (Figure 10.22c). Mordasini et al. (2009a) describe this 'horizontal branch' population as *'the conveyor belt by which Neptune-like planets are transported close to the star'*. [The hot Neptunes are the sub-population of lower-mass exoplanets which, starting with the HARPS results reported by Lovis et al. (2006), high-precision radial velocity surveys are now finding in ever increasing numbers];

(c) 'main clump planets': these are a population with $M_p \sim 0.3 - 3 M_J$, and with $a \sim 0.3 - 2$ AU. For these, the core grows to a size that triggers runaway gas accretion, while also collecting solids as it passes through the horizontal branch region. In this third phase of formation, evolutionary tracks lie on straight lines in the (M_p, a) plane, with slope d log M_p/d log $a = -\pi$;

(d) 'outer group planets': these are planets with starting positions $a \gtrsim 4 - 7$ AU. The amount of solid material available for accretion is large, and they can grow to a supercritical mass almost *in situ*, without the need to collect additional material by migration. Their evolution tracks are nearly vertical in the (M_p, a) plane, and they do not pass through the horizontal branch phase, although there is a continuous transition between them and the main clump planets.

Other model features In addition to these four sub-populations, Mordasini et al. (2009a) have drawn at-

10.10 Population synthesis

tention to the more complex behaviour of type I and type II migration evident from their Monte Carlo modeling. They also identify an absence of massive planets ($M_p \gtrsim 10 M_J$) both close to ($a \lesssim 0.5$ AU) and very distant from ($a \gtrsim 10$ AU) the host star, a dearth of planets in the mass range $30-100 M_\oplus$, a number of high-mass, deuterium-burning planets ($M_p \gtrsim 20 M_J$), and an absence of Neptune analogues in terms of both M_p and a (but with many Neptune-mass planets at smaller orbital separations).

Comparison with observations Mordasini et al. (2009b) extended their analysis to a comparison between their population synthesis results, and observed distributions. Figure 10.23a shows results for 70 000 simulations extending over their various initial conditions (they note that this takes several days on a 50 CPU cluster, with most of the time devoted to solving the planetary envelope structure equations). The various sub-populations are again indicated.

The dashed line indicates the approximate boundary of the region currently most robustly probed by radial velocity surveys, corresponding to ~ 10 m s^{-1} Doppler accuracies, and a 4.6 AU semi-major axis cut-off corresponding to a 10 yr period baseline. The region currently accessible observationally lies to the upper left of the delineated region.

This approximate cut-off in the observed radial velocity distribution implies that only a small fraction of the underlying planet population can have been detected to date, perhaps corresponding to only some 10% of the population predicted by this particular population synthesis model.

Large numbers of low-mass planets with $M_p \lesssim 100 M_\oplus$, and high-mass planets with $M_p \gtrsim 20 M_J$ and $P \gtrsim 2000$ d ($a \gtrsim 3$ AU) would appear to await discovery. Evidence for the existence of a significant population of such massive long-period planets has been given by, e.g., Cumming et al. (2008, their figure 5).

Figure 10.23b shows the derived *initial mass function* for planets, corresponding to the moment in time when the gas disk disperses. Unlike the corresponding stellar power-law distribution, the planetary initial mass function has a complex structure with several minima and maxima. Several mechanisms can subsequently modify this initial mass distribution, with the largest changes expected to occur below $\sim 10-20 M_\oplus$.

Figure 10.23c shows the mass distribution of the detectable sub-population of 6075 planets, compared with the power-law fit determined by Marcy et al. (2005a). There is, encouragingly, a broad consistency between the present observational distribution and the population synthesis predictions.

Mordasini et al. (2009b) provide other statistical indicators of the fits between current observational data and the population synthesis models, including details of the semi-major axis and metallicity distributions.

At the present time, a wide range of planetary systems is predicted by these models, largely reflecting the wide diversity of conditions in the protoplanetary disks. In the future, the technique of population synthesis offers opportunities to constrain formation and evolution models still further and, at the same time, provide guidance on the observational domains in which planet detection and characterisation will develop.

11

Interiors and atmospheres

11.1 Introduction

A significant advance in the knowledge of exoplanet interiors and atmospheres has been made possible with the discovery of transiting exoplanets. Densities from their masses and radii are providing the first indications of their interior structure and composition, while broadband brightness temperature measurements from transit photometry and spectroscopy are providing the first insights into their atmospheric composition and thermal transport processes. All of this has been substantially facilitated by the knowledge of interiors and irradiated atmospheres of solar system planets and satellites acquired over the last half century.

Physical models of exoplanets span two extreme classes of object, and potentially much in between: the low-mass high-density 'solid' planets dominated by metallic cores and silicate-rich and/or ice-rich mantles, and the high-mass low-density gas giants dominated by their massive accreted H/He envelopes.

For the gas-rich giants, models of their interiors and models of their atmospheres are closely connected, and the most recent atmospheric models couple their emergent flux with their assembly by core accretion. Models of their interiors predict bulk properties such as the pressure–temperature relation, and their radii as a function of mass. For close-in, highly irradiated gas giants, the additional external heat source has a significant effect on the pressure–temperature structure of the outer atmosphere. Combined with inferences on their probable bulk chemical composition, atmospheric models also predict broad-band colours and spectral features arising from specific atomic and molecular species.

Interior models of low-mass planets without massive gaseous envelopes aim to determine the mass–radius relation for a given internal composition and, by application of such models, the most likely internal composition given a specific planet's measured mass and radius. For terrestrial-type planets, a somewhat distinct problem is establishing the nature of their atmospheres: whether they might have acquired a (modest) gaseous envelope either by impact accretion or by outgassing, their likely composition, and whether such an atmosphere might have been retained or eroded over its evolutionary lifetime.

Constraints on internal structure Mass estimates, or lower limits, can be made for exoplanets detected by Doppler measurement, astrometry, microlensing, or timing. Photometric transits also provide an estimate of their radii which, combined with a mass estimate, yield their mean densities. The latter, combined with theoretical models, provides a diagnostic of their composition.

In practice, a unique internal physical and chemical composition cannot be inferred from an object's mass, volume, and density alone. Neither is it possible to determine directly the presence of a central core, the core's state (whether liquid or solid), or whether such a core is structurally differentiated or partially mixed with the envelope. But by appeal to the primary constituents observed and inferred in the solar system planets (e.g. iron, rock, ices, and H/He gas), supported by thermal equilibrium calculations to predict the likely occurrence of particular species, along with their equations of state under assumed internal and external physical conditions, plausible models of their interiors are emerging. Structural inferences are, in turn, being used to provide constraints on their formation.

11.2 Planetary constituents

11.2.1 Gas, rock, and ice

Thermal equilibrium and condensation calculations for approximately solar nebula elemental compositions (§11.2.2) suggest the formation of rather similar minerals and thermal stability sequences over a wide range of total pressures of relevance to the formation of protoplanetary disks ($P \lesssim 10^2$ Pa).

Thus the most thermally stable solid containing iron is an Fe–Ni metal alloy, for magnesium it is the Mg-silicate forsterite (Mg_2SiO_4), and for sulphur it is troilite (FeS). As a result of such considerations, it is believed that the planetary constituents forming out of different

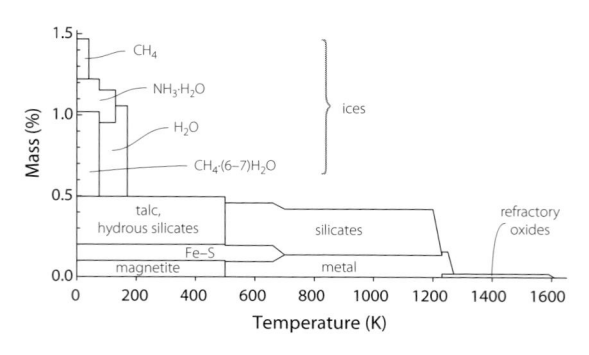

Figure 11.1: The distribution of solid phases (ice and rock) expected in a planetary accretion disk of solar composition at a total pressure of 1 Pa. The calculations assume thermodynamic equilibrium, which is limited by slow gas–gas and gas–solid reactions at low temperatures. From Lodders (2010, Figure 2a).

Table 11.1: Typical minerals (rocks) made of the more abundant elements. From Lodders (2010, Table 1.1).

Mineral group	Endmember	Formula
Olivine	Forsterite	Mg_2SiO_4
$(Mg,Fe)_2SiO_4$	Fayalite	Fe_2SiO_4
Pyroxene	Enstatite	$MgSiO_3$
$(Mg,Fe,Ca)SiO_3$	Ferrosilite	$FeSiO_3$
	Wollastonite	$CaSiO_3$
Feldspar	Anorthite	$CaAl_2Si_2O_8$
	Albite	$NaAlSi_3O_8$
	Orthoclase	$KAlSi_3O_8$
Metal alloys	Iron-nickel	$FeNi$
Sulphides	Troilite	FeS
	Pyrrhotite	$Fe_{1-x}S$
Oxides	Magnetite	Fe_3O_4
Hydrous silicates	Talc	$Mg_3(Si_4O_{10})(OH)_2$

disks should be comprised of similar oxides, silicates, and sulphides. The relative quantities are determined by the elemental composition of the accretion disk, while their radial distribution is determined by the disk temperature and pressure profiles.

The three principal types of compounds available in accretion disks are loosely referred to as gas, rock (non-volatile condensates), and ice (volatile condensates), and planets form by accreting various amounts of each. Subsequent evolution in temperature and pressure conditions within the planet can modify the phases and redistribute their radial profile through chemical and gravitational differentiation. Though still referred to as comprising 'ices' and 'rocks', the physical state of these basic constituents in a mature planet may accordingly be very different from that during the accretion phase.

The proportions of gas, rock, and ice determine the planet's mass and bulk composition (Figure 11.1), while the equations of state determine which phases will be present, along with the size of the resulting planet.

Gas The gas component of gas giants is primarily H_2 and He, along with whatever species are left over after the rocky and icy condensates have formed (including N_2 and the noble gases), as well as other secondary photochemical pollutants. H and He comprise ~ 98.5% of the mass in the solar nebula, leaving typically ~1.5% of the total mass bound up in rocks and ices.

Rock Geologically, *rock* is defined in the Oxford English Dictionary as *'Any natural material, hard or soft (e.g. clay), having a distinctive mineral composition'*. In the context of exoplanet chemistry, Lodders (2010) suggests as a working definition *'a component containing the approximate elemental abundances, and consisting of the mineral phases, that are observed in undifferentiated meteorites called chondrites'*.

All rocky compounds are relatively refractory, and require high temperatures for evaporation, and condense out of a gas of solar system composition at relatively high temperatures. Being the last material to evaporate, and the first to condense, rock is therefore always expected to be present amongst solids in an accretion disk. By analogy with the solar system abundances, the amount of rocky material is expected to be ~0.5% of all mass in an accretion disk.

Major minerals commonly encountered in planetary science are listed in Table 11.1. This includes elements like Si, Mg, Ca, Al, and Ti that form oxides and silicates, as well as Fe–Ni metal alloys and FeS. Lodders (2010) gives further details of the expected percentages of various rocky and icy substances under equilibrium and non-equilibrium conditions, and of expected condensates and gases in a solar composition gas giant planet.

Of these minerals, olivine is one of the most common in the Earth's crust and upper mantle and, being found also in meteorites, the Moon, and Mars, is frequently taken as representative of generic 'rock'. It exists as a solid-solution series, $(Mg,Fe)_2SiO_4$, ranging from the Mg-rich end-member forsterite, Mg_2SiO_4 ($\rho = 3.2$), to the Fe-rich end-member fayalite, Fe_2SiO_4 ($\rho = 4.4$). Phase transitions to various high-pressure polymorphs occur at $P \gtrsim 10-15$ GPa, depending on temperature and iron content, corresponding to depths of 400–500 km on Earth. This yields distinctive changes in seismic records, and marks the transition from the upper to lower mantle (Deer et al., 1996).

Ice In common usage, the word 'ice' portrays water in its cold and solid state. None of these three characteristics are necessarily implied when used in the context of exoplanets, where the solid phases of H_2O are termed 'ice' even though, at sufficiently high pressure, various solid phases of H_2O can exist at $T \gg 273$ K (§11.5.2).

Furthermore in planetary science, volatiles with a melting point between ~100–200 K are commonly referred to as 'ices', even when in a hot and/or liquid form. In this context, ice encompasses various sub-

11.2 Planetary constituents

11.2.2 Chemical composition and condensation

In the considerations of the formation of terrestrial planets (§10.4) and giant planets (§10.6), the detailed chemical composition of the various bodies is, in current models, largely ignored.

Issues of composition as a function of temperature and pressure are, however, becoming increasingly important considerations in various areas of exoplanet science, amongst them: (a) in the composition of the solar nebula as a starting point for formation models (§12.4); (b) for the detailed chemical studies of dust agglomeration during the earliest phases of planet formation (§10.4.2); (c) for interpreting the qualitative differences of accretion processes within and beyond the snow line (this section); (d) for modeling the interiors and bulk properties of giant planets (§11.3) and super-Earths (§11.5); (e) for modeling the atmospheres of the gas giants formed by accretion (§11.7), and of terrestrial planets formed by outgassing (§11.8.1); and (f) in interpreting the photospheres of exoplanet host stars, for example, for evidence of accreted material (§8.4.3). These considerations fall within the field of *cosmochemistry*, the study of the chemical composition of the universe and the processes that produced them (e.g. McSween & Huss, 2010).

Assuming adequate time for the relevant chemical reactions to reach equilibrium, a gas starting out with some initial elemental composition, at a specified temperature (and to a less sensitive degree, pressure), is transformed into a thermodynamically stable mix of elements and compounds in their various permitted elemental, mineralogical, gas, or ice forms. Thermochemical equilibrium calculations can predict the gas phase, gas–grain, and solid phase reactions which occur as a function of pressure and temperature, thereby estimating which gases form, which elements or compounds condense, and in what proportions.

The condensation temperatures of the constituents of a gas of solar composition are widely used as a diagnostic of chemical fractionation in various branches of astronomy, planetary science, and meteoritics. Early thermochemical computations for gas chemistry in the Sun and in cool stars, taking condensation into account, were made by Wildt (1933) and Russell (1934).

Subsequently, numerous studies have been dedicated to establishing volatility trends, which are determined by the condensation temperature of each element and its compounds (Lord, 1965; Larimer, 1967; Grossman, 1972; Grossman & Larimer, 1974; Boynton, 1975; Wai & Wasson, 1977, 1979; Sears, 1978; Fegley & Lewis, 1980; Saxena & Eriksson, 1983; Fegley & Palme, 1985; Kornacki & Fegley, 1986; Palme & Fegley, 1990; Gail, 2002, 2004; Lodders, 2003). Different dust-to-gas nebula ratios can also be considered (Ebel & Grossman, 2000, and references).

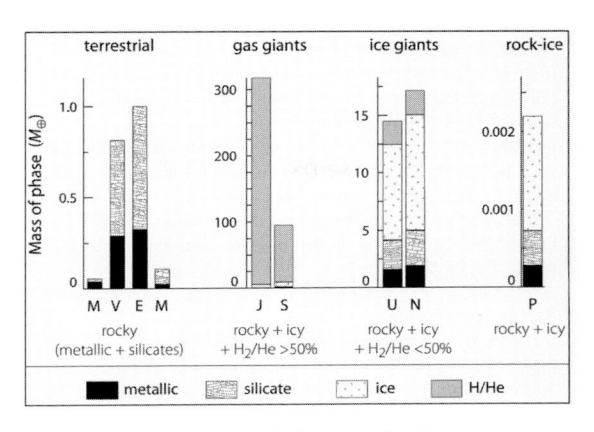

Figure 11.2: Classification of solar system bodies into four compositional types, represented by the terrestrial planets (Mercury, Venus, Earth, Mars), the gas-rich giant planets (Jupiter, Saturn), the ice giants (Uranus, Neptune), and the dwarf planets (typified by Pluto). After Lodders (2010, Figure 1).

stances containing C, N, and O. Water ice is the most important, in part because O is the third most abundant element in the solar system, but also because it condenses at the highest temperature. Other possibly abundant ices are CH_4, CO, CO_2, N_2, NH_3, and $NH_3 \cdot H_2O$. As for H_2O, the former four also occur as *clathrates*, in which a lattice of one type of molecule traps and contains another type of molecule. For water, these are represented by $X \cdot nH_2O$, where $n = 5 - 7$ is the number of water molecules that characterises a unit structure into which another molecule 'X' can be accommodated.

Composition of the solar system planets Four main compositional types are observed for planets and dwarf planets in the solar system (Figure 11.2).

Terrestrial-type composition, mainly comprising rocky materials (metals and silicates) are represented by Mercury, Venus, Earth and Mars, as well as (1) Ceres and other asteroids. Atmospheres, when present, represent a marginal contribution to their total mass.

In the gas-rich giant planets, represented by Jupiter and Saturn, H and He dominate, and their overall elemental composition is closest to that of the Sun.

Uranus and Neptune accreted only some 10–20% of their mass as H and He gas, leaving them as 'gas-poor' giants, or ice giants.

Bodies composed predominantly of rock and ice are of relatively low mass in the solar system. They are represented by the IAU-designated dwarf planets (such as Pluto, Eris, and Haumea), with some of the asteroids in the outer asteroid belt, and most objects beyond Neptune, likely to belong to the class. Tenuous atmospheres of icy evaporates may exist for the more massive. More massive exoplanets of similar composition, 'ocean planets', (§11.5.2) have been hypothesised.

Condensates and condensation temperature Given a starting prescription of elemental abundances, the resulting composition and condensation fraction can be predicted thermodynamically. The procedure is general, but in the present context can be used, for example, to estimate the composition, in elements and compounds, solids and gases, of the solar nebula.

The following description, based on Lodders (2003), outlines how such constituents can be inferred from the initial elemental abundances. Gail (2002, 2004) reports a similar series of large-scale simulations predicting the abundance of minerals formed in the early solar system, and the most important dust components.

Thermodynamically, the condensation temperature of each specific constituent is determined by the total gas pressure, the elemental abundances that determine the partial pressures, the distribution of an element between different gases and condensates, and the vapour pressure of the element. Condensation temperatures of relevance to the solar system, including those determined by Lodders (2003) and the numerous previous works cited above, are typically calculated at a total pressure of 10 Pa (10^{-4} bar), which is considered to be representative of the total pressure near 1 AU in the original solar nebula (Fegley, 2000).

In the approach described by Lodders (2003), the numerical code CONDOR (Lodders & Fegley, 1993; Fegley & Lodders, 1994) is used to determine the chemical equilibrium condensation temperatures. It simultaneously solves for the mass balance and chemical equilibrium for 2000 gases (including molecules, radicals, atoms, and ions) and 1600 condensates of all naturally occurring elements. The underlying physics can be conceptualised in two steps, which are in practice coupled and solved iteratively.

In the first step, the abundances of all 2000 gases accounted for in the code are determined as follows, taking Al as an example (Lodders, 2003, section 3.1). For a total abundance in all forms, $n(\text{Al})$, the total mole fraction is

$$X_{\Sigma\text{Al}} = \frac{n(\text{Al})}{n(\text{H} + \text{H}_2 + \text{He} + \ldots)}, \tag{11.1}$$

where the denominator is the sum of the dominant H, He, and all other gas abundances, with the temperature-dependent equilibrium ratios of H/H_2 (and others) taken into account. Multiplying by the total pressure gives the pressure of Al in all forms, which is equal to the partial pressure sum for Al. This is expressed as

$$P_{\Sigma\text{Al}} = a_{\text{Al}} \, [K_{\text{Al}} + K_{\text{AlO}} f_{\text{O}_2}^{0.5} + K_{\text{AlOH}} (f_{\text{O}_2} f_{\text{H}_2})^{0.5} + \ldots], \tag{11.2}$$

where the K_i are the equilibrium constants appropriate for forming the relevant gases from the constituent elements in their reference states. The *thermodynamic activity coefficient*, a, and the *fugacity* of each of the other gases combining with the primary element, f, account

---

> **Refractory and volatile elements:** condensation temperatures give the volatility groupings for the elements. Lodders (2003) subdivides the 50% condensation temperatures, for solar system abundances at 10 Pa, as:
> - ultra-refractory, $T > 1650$ K: the metals Os, Re, W; the lithophiles Al, Hf, Sc, Th, Y, Zr; and the heavy rare earth elements Gd, Tb, Dy, Ho, Er, Tm, Lu;
> - highly refractory, 1650–1500 K: the metals Ir, Mo, Ru; the lithophiles Ca, Nb, Ta, Ti, U; and the light rare earth elements La, Pr, Nd, Sm;
> - refractory, 1500–1360 K: the metals Pt, Rh; the lithophiles Ba, Be, Ce, Sr, V, Yb;
> - the common elements, 1360–1290 K: Mg, Si, Fe; also the metals Co, Cr, Ni, Pd; and the lithophile Eu;
> - moderately volatile, 1290–704 K: the siderophiles Ag, As, Au, Bi, Cu, Ga, Ge, P, Pb, Sb, Te; lithophiles Cs, B, K, Li, Na, Mn, Rb, Zn; the halogens Cl, F;
> - volatile, 704–371 K: the chalcophiles Cd, In, S, Se, Tl; the siderophile Sn; the halogens Br, I;
> - highly volatile, < 371 K (if at all): C, N, O; noble gases; Hg.
>
> The more abundant rock-forming elements are: (highly) refractory (Ca, Al, Ti); common (Fe, Si, Mg); moderately volatile (P); volatile (S); and highly volatile (H, C, N, O).

for deviations from ideal thermodynamic behaviour in the specified interactions. Analogous forms are written for each element in the code, with the appropriate K_i given in the thermodynamic literature, notably the NIST–JANAF thermochemical tables (Chase, 1998).

Equation 11.2 shows that the chemistry of the elements is evidently interdependent. The mass balance equations from the set of coupled, nonlinear equations are therefore solved iteratively. An initial estimate can be used for the thermodynamic activity or fugacity of each element, but this can be optimised if the major gas for each element is known. The solution gives the abundance of all resulting gases, along with the thermodynamic activity or fugacity for each combination.

In the second step, condensate stabilities are computed considering the compounds which can be formed from the elements in their respective reference states. For example, the condensation of corundum (Al_2O_3), described by the reaction 2Al (gas) + $1.5\,\text{O}_2$ (gas) = Al_2O_3 (solid), occurs when the thermodynamic activity of Al_2O_3 reaches unity. This is calculated from

$$a_{\text{Al}_2\text{O}_3} = a_{\text{Al}}^2 \, f_{\text{O}_2}^{1.5} \, K_{\text{Al}_2\text{O}_3}, \tag{11.3}$$

where $K_{\text{Al}_2\text{O}_3}$ is the temperature-dependent equilibrium constant for corundum, and a and f are taken from the gas-phase equilibrium calculations.

At the *appearance condensation temperature* at which the thermodynamic activity of a pure phase (e.g. Fe-metal, FeS, or Al_2O_3) reaches unity, the compound or element begins to condense from the gas phase. The gas phase abundance is thereby reduced by the fraction condensed. The activities of all other possible condensates are computed in a similar manner. The gas-

11.2 Planetary constituents

Table 11.2: *A selection of illustrative condensation temperatures, for solar nebula composition, and a total pressure of 10 Pa. From Lodders (2003, Table 7).*

Mineral	Name	T (K)
Al_2O_3	corundum	1677
$CaTiO_3$	perovskite	1593
$MgAl_2O_4$	spinel	1397
$CaAl_2Si_2O_8$	anorthite	1387
Fe	Fe alloy	1357
Mg_2SiO_4	forsterite	1354
$CaMgSi_2O_6$	diopside	1347
$MgSiO_3$	enstatite	1316
FeS	troilite	704
Fe_3O_4	magnetite	371
H_2O	water ice	182

phase and gas-solid chemical equilibria are coupled and solved iteratively.

Minor and trace elements The condensation temperatures of minor and trace elements that do not condense as pure phases are computed differently. Many minor and trace elements condense by forming (solid) solutions with host phases made of major elements. An understanding of these host phases for trace elements is constrained observationally by elemental analyses of minerals in meteorites, and further guided by the geochemical Goldschmidt classifiers denoting preferred host phases (§10.5).

The formation of solid solutions starts when a host phase begins to condense, and when the minor and trace elements have the same condensation temperatures as the host phase. The 50% condensation temperature, at which half the element is in the gas phase and the other half is combined into condensates, provides a more robust indication of the extent of condensation or volatility of minor and trace elements.

Table 11.2 lists a selection of condensation temperatures. Table 11.3 summarises the resulting condensate distributions for three separate abundance distributions, illustrating the sensitivity of the modeled condensates to the initial elemental composition.

As a caveat, chemical reactions proceed at different rates, which generally decrease exponentially with decreasing temperature. In consequence, there may not have been sufficient time for the predicted low-temperature equilibrium chemistry to have taken place before the local environment cooled significantly, or before the gaseous solar nebula was dispersed. Specific consequences are explored by Fegley (2000).

Refractory and volatile elements Of the numerous results concerning condensation sequences and their interdependencies described by Lodders (2003), the following points are noteworthy in gaining an overview.

Condensation of the major elements (Al, Ca, Mg, Si, Fe) and other important rock-forming elements (Ti, S, P) form the bulk of rocky material, and their condensates serve as host phases for minor and trace elements. Condensation of the three most abundant rock-forming elements, Mg, Si, and Fe (often called the *common elements*; Larimer 1988; Cowley 1995; Taylor 2001), occurs together over a relatively small (100 K) temperature interval. These elements, and their condensation temperatures, are used to define the *refractory elements*, elements condensing at higher temperatures, and the *volatile elements*, those condensing at lower temperatures (see box on page 258).

Several ultra-refractory transition elements condense before any major condensates form. The transition metals W, Re, Os, Ir, Mo, Pt, Rh, and Ru condense before iron metal.

Condensation of the highly volatile elements C, N, and O at low temperatures produces the major fraction of solids, as the high-temperature condensates of all other elements (except the noble gases) only make up about one-third of all condensable material. However, the condensation temperatures and types of condensates of C and N, and to some extent of O, are also affected by kinetic effects.

Water ice condensation sets the reference temperature for the appearance of volatile ices. Some oxygen removal from the gas has already taken place with the formation of silicates and oxides. About 23% of all oxygen is bound to rocky elements (Al, Ca, Mg, Si, and Ti) before water ice condenses at 182 K.

Condensation of carbon compounds depends on the gas-phase equilibrium between CO and CH_4

$$CO + 3H_2 \rightleftharpoons CH_4 + H_2O \,, \tag{11.4}$$

as a function of temperature. Under equilibrium conditions, CO is the major carbon-bearing gas at high temperature, but is replaced by CH_4 gas with decreasing temperature. At 10 Pa, this occurs below 650 K. Pure CH_4 ice then condenses at 41 K.

Nitrogen condensation is tied to the gas equilibrium

$$N_2 + 3H_2 \rightleftharpoons 2NH_3 \,, \tag{11.5}$$

favouring NH_3 gas formation at low temperatures (and high pressures). At 10 Pa, NH_3 is the major nitrogen-bearing gas below 325 K.

More detailed models of the atmospheric chemistry in giant planets are given for C, N, O by Lodders & Fegley (2002), for S, P by Visscher et al. (2006), and for Fe, Mg, Si by Visscher et al. (2010).

The snow line The radius in the protoplanetary disk beyond which water ice can be present is referred to as the *snow line*, and it is believed to play an important role in the architecture of planetary systems. From the picture above, water ice forms at $T \lesssim 182$ K, although this may be modified slightly by effects such as the non-equilibrium chemistry of carbon.

Table 11.3: Condensate distribution from the numerical simulations of Lodders (2003), as given in their Table 11. Values are percent by mass, assuming equilibrium condensation, for three model compositions: (1) their inferred bulk solar system composition representative of the solar nebula, (2) their revised estimate of the solar photospheric composition, and (3) the literature value from Anders & Grevesse (1989).

Condensate	(1)	(2)	(3)
silicates and oxides (a)	0.303	0.269	0.264
metal + FeS (b)	0.186	0.165	0.176
H_2O ice (c)	0.571	0.506	0.920
CH_4 ice (d)	0.330	0.292	0.408
NH_3 ice (e)	0.097	0.086	0.134
total rock (a+b)	0.489	0.434	0.440
total ices (c+d+e)	0.998	0.884	1.462
total condensates	1.487	1.318	1.903
rock among condensates	32.89	32.89	23.14
ices among condensates	67.11	67.11	76.84
H_2O ice/rock	1.17	1.17	2.09
total oxygen in rock	22.75	22.75	14.42

Within the solar system, H_2O-rich (C-class) asteroids are found predominantly in the outer asteroid belt (Morbidelli et al., 2000), suggesting that the snow line in the solar nebula lay at around 2.7 AU. This is somewhat larger than predictions based on a minimum-mass disk characterised by conventional opacities and a mass accretion rate of $10^{-8} M_\odot \text{yr}^{-1}$, which place the snow line more at 1.6–1.8 AU (Lecar et al., 2006). Passive protoplanetary disks appear to have snow lines at still smaller radii of around 1 AU, while those for classical T Tauri accreting disks may lie at ~3 AU (Lecar et al., 2006; Garaud & Lin, 2007).

As evident from Table 11.3, water and other ices increase the surface density of total condensates over that of rocks alone by a factor of 3–4. The mass of the fastest growing planetesimal (the isolation mass) is expected to scale as the surface density to the 3/2 power (Equation 10.13). Accordingly, changes in the efficiency or outcome of planet formation, perhaps the demarcation between terrestrial and gas giants planets, may be identified with the changes in the surface density and composition of solids that occur beyond the snow line (Marboeuf et al., 2008).

As a low-mass host star contracts at constant effective temperature during the pre-main sequence phase, the snow line moves broadly inwards, allowing the rapid formation of icy protoplanets that may collide and merge into super-Earths before the star reaches the main sequence (Kennedy et al., 2006, 2007). More detailed models of protostellar disks around T Tauri stars suggest a non-monotonic evolution with time (Garaud & Lin, 2007): evolving from outside 10 AU during FU Ori-type outbursts, to 2 AU during the quasi-steady accretion phase, to 0.7 AU when the accretion rate drops, finally expanding again to beyond 2.2 AU during the transition to the debris disk phase. This should have observable consequences for the distribution of H_2O, and its isotopic composition, within the solar system and beyond (§12.5).

Further detailed calculations of the location of the snow line as a function of stellar mass and age have been made by Kennedy & Kenyon (2008b) and Kenyon & Bromley (2008). Thus stars more massive than $3M_\odot$ evolve quickly to the main sequence, pushing the snow line to 10–15 AU before protoplanets form, and presumably restricting the range of disk masses that can form giant planet cores.

11.3 Models of giant planet interiors

Various models of planetary interiors of relevance to solar system planets and exoplanets have been developed over the past 40 years (Guillot, 2005).

Amongst them, Zapolsky & Salpeter (1969) determined radii at zero-temperature for bodies composed of H, He, C, Fe, and Mg over the mass range $0.3M_\oplus - 10M_\odot$. Stevenson (1982) determined radii for zero-temperature and warm H, H/He, ice, and rock planets over the range $1 - 1000 M_\oplus$. Saumon et al. (1996) determined radii of giant planets with and without cores. Guillot et al. (1996) extended the determinations to include compositions ranging from H/He to rock. Bodenheimer et al. (2003) determined radii for H/He planets with and without cores for a range of orbital separations, including effects of *insolation* (*in*cident *sol*ar radi*ation*) from the host star. Numerous studies of the effects of irradiation have been made subsequently (e.g. Guillot & Showman, 2002; Laughlin et al., 2005c; Arras & Bildsten, 2006).

Fortney et al. (2007) published model radii for planets composed of water ice, rock, and iron. Their results are generally consistent with preceding work in their regions of overlap, but they include mixed compositions, a wide range of core masses, computations over a planet mass range $0.01 M_\oplus - 10 M_J$, and incorporating the effects of insolation from the host star over orbital separations ranging from 0.02–10 AU.

Models of terrestrial exoplanets in the mass range $1 - 10 M_\oplus$ and above have been used to predict their likely composition, and their resulting mass-radius relations (e.g. Seager et al., 2007; Sotin et al., 2007; Valencia et al., 2007b; Elkins-Tanton & Seager, 2008a; Grasset et al., 2009). Models have even considered variations of lithospheric thickness and different interior temperature structures (Valencia et al., 2006; Ehrenreich et al., 2006), although effects on radii are at the few per cent level only, and unlikely to be discernible from current photometric transit data.

In this section, the focus is on predictions of the overall properties of giant planet interiors, including the pressure–temperature relations, and the resulting mass–

11.3 Models of giant planet interiors

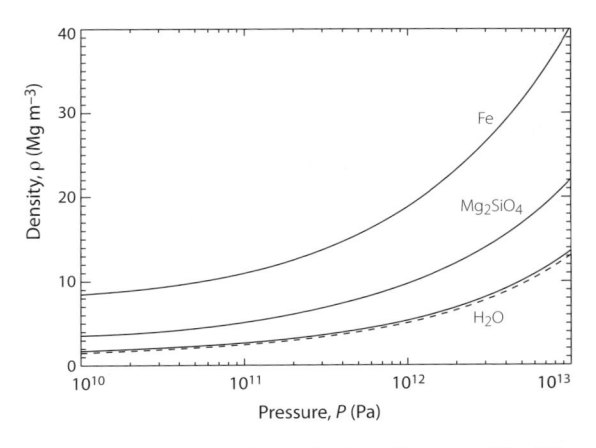

Figure 11.3: Equation of state for iron (Fe), rock (Mg_2SiO_4, olivine/forsterite), and water ice. Solid lines show the zero temperature pressure–density relations. The dashed line for H_2O indicates the effect of thermal corrections. From Fortney et al. (2007, Figure 1), reproduced by permission of the AAS.

radius relation. Some of these considerations, such as the effects of stellar irradiation, are also relevant for predicting the detailed properties of their outer atmospheres, including their colours and spectra. These properties are considered separately in §11.7.

11.3.1 Equations of state

Many astrophysical problems, exoplanet internal structures amongst them, require detailed knowledge of the physical properties of the constituent matter. An *equation of state* is a thermodynamic relation describing the equilibrium state of matter of a given composition under a specified set of physical conditions. It relates two or more 'state variables', i.e. quantities which depend only on the system's actual state (and not how the body achieved that state) but which are not necessarily independent. State variables include temperature, pressure, volume, density, internal energy, and entropy.

For a star or planet, thermodynamic relations for the pressure $P(\rho, T)$ and entropy $S(\rho, T)$ as a function of density ρ and temperature T dictate the mechanical and thermal equilibrium respectively. According to context, 'the' equation of state for a planet frequently implies the relation $P(\rho, T)$, which quantifies the compressibility of the interior, and depends on chemical composition.

Equations of state can be derived for elements, compounds or minerals either experimentally, or by invoking relevant theoretical models over different regions of density and temperature (e.g. Thomas–Fermi–Dirac theory to characterise the interactions between electrons and nuclei, rigid-rotator and harmonic-oscillator methods for molecular rotation and vibration terms). More elaborate models may account for partial dissociation and ionisation caused by both pressure and tempera-

ture. Various domains of the equation of state relevant to stars is illustrated in Figure 1 of Pols et al. (1995).

Details of the historical developments and underlying physical techniques in the determination of equations of state can be found in, e.g., Hummer & Mihalas (1988) and Saumon et al. (1995). Tabular equations of state from numerical calculations made at the Lawrence Livermore and Los Alamos National Laboratories, amongst others, have been used extensively in astrophysical applications over many years.

The planetary models of Fortney et al. (2007), for example, use the Los Alamos SESAME 2140 equation of state for Fe (Lyon & Johnson 1992; an alternative is given by Anderson & Ahrens 1994), the Sandia ANEOS (analytic equation of state) models for H_2O and olivine (Thompson, 1990), and the prescriptions given by Saumon et al. (1995) for H and He. The equations of state for H and H_2O are considered further in §11.3.2. For interior models of super-Earths (§11.5), the details of the equation of state for the rocky cores becomes more crucial (e.g. Seager et al., 2007; Sotin et al., 2007).

Limitations in the current equations of state representing the gaseous interiors of giant exoplanets are noted by Guillot (2005, Section 2.3).

Temperature dependence A non-zero internal temperature modifies the pressure–temperature equation of state through the introduction of a thermal pressure term, P_{th}, given by

$$P = P_0 + P_{th}, \qquad (11.6)$$

where P_0 is the pressure at $T = 0$ K. Thermal effects are small for high-Z species, with zero-T equations of state believed to be accurate to 1–2% for interior models of Uranus and Neptune (Hubbard & Macfarlane, 1980; Hubbard, 1984). For H_2O, thermal pressure effects amount to ~10%, with models for Uranus and Neptune (Hubbard & Macfarlane, 1980) giving

$$P_{th} = 3.59 \times 10^6 \rho T, \qquad (11.7)$$

with P_{th} in Pa, ρ in Mg m^{-3}, and T in K.

The models of Fortney et al. (2007), for example, accordingly ignore thermal corrections for rock and iron. Within the ice, they assume that the interior temperatures follow the Uranus/Neptune adiabat of Guillot (2005), viz. reaching ~550 K at 10^8 Pa and 4000 K at 10^{11} Pa. They then add the thermal pressure correction of Equation 11.7 at every P_0 (Figure 11.3).

11.3.2 Hydrogen and water

Phase diagram for hydrogen An understanding of the deep interiors of giant planets comprising H and He has developed over the last three decades (e.g. Stevenson, 1975; Stevenson & Salpeter, 1977b; Hubbard & DeWitt,

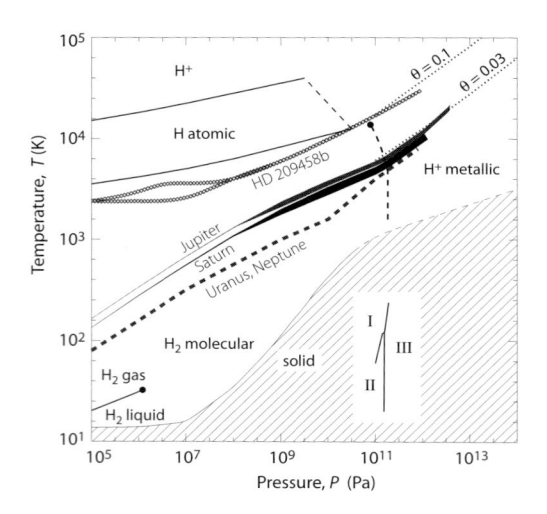

Figure 11.4: Phase diagram for hydrogen. The dashed line near 10^{11} Pa marks the molecular to metallic transition. Hydrogen is solid in the hatched area, with its three phases indicated. Values of the degeneracy parameter $\theta = T/T_F$ (where T_F is the Fermi temperature) are shown as dotted lines. Temperature–pressure profiles are shown for Jupiter, Saturn, Uranus, Neptune, and HD 209458 b. Adapted from Guillot (2005, Figure 1).

1985; Marley & Hubbard, 1988; Saumon et al., 1995; Guillot, 2005; Chabrier et al., 2007).

The behaviour of hydrogen at the relevant temperatures and pressures is complex, but fundamental in predicting densities and heat transport. In the gas giants, molecules, atoms and ions may coexist, in a fluid that is partially degenerate (i.e. the free electron energies are determined by both thermal and quantum effects) and partially coupled through Coulomb interactions.

The temperature–pressure phase diagram of hydrogen (Figure 11.4) shows a number of regions relevant to exoplanet atmospheres (Guillot, 2005): in their relatively cool low-pressure photospheres ($T \sim 50 - 3000$ K, $P \sim 10^4 - 10^6$ Pa), hydrogen is in molecular form. Deeper in their interiors, H (and He) becomes progressively fluid. At $P \gtrsim 10^{11}$ Pa extreme pressures result in ionisation, and the H becomes electrically conductive or 'metallic' (Wigner & Huntington, 1935; Loubeyre et al., 1996; Weir et al., 1996; Chabrier et al., 1992; Saumon et al., 2000).

At low temperatures and high pressures, hydrogen can solidify (Datchi et al., 2000; Gregoryanz et al., 2003), with three phases recognised (Mao & Hemley, 1994). But as noted by Hubbard (1968), and as evident from Figure 11.4, the interiors of the H/He giant planets remain fluid, rather than solid, whatever their age: an isolated Jupiter should begin partial solidification only after $\gtrsim 10^3$ Gyr of cooling.

Suggestions for high-pressure laboratory experiments on H and He to further the understanding of the interiors of the solar system and exoplanet giants have been made by Fortney (2007).

Phase diagram for water The structure and other characteristics of H_2O (in solid, liquid, or gas form) are dictated by the corresponding phase diagram and associated equations of state (Figure 11.5). Many of its properties, as collated by the International Association for the Properties of Water and Steam (Wagner & Pruß, 2002), are therefore relevant to the predicted characteristics of the ice giants, icy moons, and 'ocean planets'.

The second most common molecule in the Universe after H_2, the physical and chemical properties of water are nevertheless extremely complex (e.g. Hobbs, 1974; Brovchenko & Oleinikova, 2008), in large part due to its versatile intra-molecular hydrogen bonding. Chaplin (2010), for example, lists 67 anomalous properties as compared with other liquids.

The *triple point* of water, the temperature and pressure at which the solid, liquid and gas phases[1] coexist in thermodynamic equilibrium, occurs at $T_{tp} \equiv 273.16$ K $\equiv 0.01$ C (by definition) and at a partial vapour pressure of $P_{tp} = 611.657$ Pa (Guildner et al., 1976). [More generally, triple points occur where any three phase lines join, and H_2O has many other triple points corresponding to its many solid phases].

The *critical point* ($T_{crit} \sim 647$ K, $P_{crit} \sim 22.064$ MPa) occurs at the end of the liquid–gas phase line where the properties of these two phases become indistinguishable. Beyond the critical point, only a single homogeneous supercritical fluid phase exists.

Currently, water has 19 known solid phases (Chaplin, 2010): three non-crystalline *polyamorphs* (LDA, HDA, VHDA, designating low-density, high-density, and very high-density amorphous), and 16 crystalline *polymorphs*, distinguished by their structure, ordering and density. These are designated I h (hexagonal, the predominant form in the biosphere), I c (cubic), and II–XV (variously displaying cubic, hexagonal, monoclinic, orthorhombic, rhombohedral or tetragonal forms, and in which the H-bonding may be ordered or disordered). Ice IV and XII are metastable, with IV, XII–XIV occurring in the phase space of ice V.

Densities are $\rho < 1$ Mg m^{-3} only for ice I h/I c (0.92) and LDA. Densities of other well-characterised phases exceed 1.14, reaching 2.51 Mg m^{-3} for ice X.

Ice XI refers to both the low-temperature (ferroelectric) polymorph of ice I h, but also to a theoretical high-pressure form (also referred to as ice XIII, a name, confusingly, also assigned to another form). The high-pressure phase is expected to be a dense (~ 11 Mg m^{-3}) fluid plasma under conditions relevant for Jupiter's core

---

[1] The terms 'gas' and 'vapour' are frequently used as synonyms (as here) although, more strictly, vapour is a gaseous substance below T_{crit}, and which can therefore be liquefied by pressure alone. 'Steam' is commonly used synonymously for water vapour, especially when above the boiling point.

11.3 Models of giant planet interiors

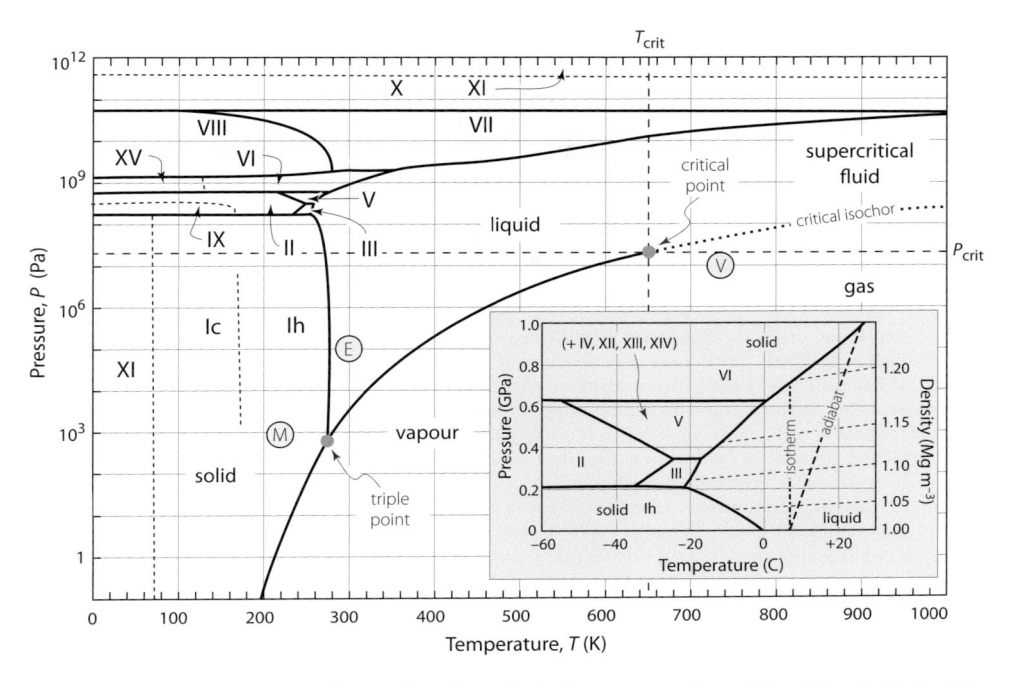

Figure 11.5: Phase diagram for water. Heavy lines indicate boundaries between two phases. The triple point is the intersection of the solid/liquid, liquid/gas, and gas/solid phase boundaries, and where these three phases can accordingly coexist ($T_{\rm tp}$ = 273.16 K, $P_{\rm tp}$ = 611.657 Pa). The critical point is the location at which the gas–liquid phases merge into a single supercritical fluid ($T_{\rm crit}$ ~ 647 K, $P_{\rm crit}$ ~ 22.064 MPa). Roman numerals designate the various solid phases. Mean surface conditions are indicated for (M)ars, (E)arth, and (V)enus. Inset: enlargement of the central phase space region, showing isodensity lines for the liquid phase (dashed). The dot-dash (isothermal) and long-dash (adiabatic) lines are relevant to ocean depths (§11.5.2). Adapted from Chaplin (2010).

of T ~ 20 000 K and P ~ 5×10^{12} Pa, and possibly superionic within Saturn and Neptune (French et al., 2009).

Other determinations of the phase diagrams of H$_2$O (and NH$_3$) of relevance for the middle ice layers of Neptune and Uranus (P = 30 – 300 GPa, T = 300 – 7000 K) were reported by Cavazzoni et al. (1999), with melting curves at high pressure studied by Datchi et al. (2000).

New and complex properties of water, notably at the more extreme conditions of temperature and pressure, are still being discovered. For example, ice XII was discovered only in 1996, ices XIII–XIV in 2006, and ice XV only in 2009 (Salzmann et al., 2009). Meanwhile, the latest models of icy exoplanet interiors (e.g. Valencia et al., 2007b) typically take into account at least the high-pressure phases ice VII (Fei et al., 1993; Frank et al., 2004) and ice X (Pruzan et al., 2003; Caracas, 2008; Aarestad et al., 2008). Although the phase structure may have limited effect on radii (Valencia et al., 2006) it does impose conditions on the compositional phases that are likely to occur towards the planet surface.

11.3.3 Structural models

The structure and evolution of a spherically symmetric giant planet are assumed to be governed by the equations of hydrostatic and thermodynamic equilibrium, and mass and energy conservation. These can be expressed as (e.g. Guillot 2005, equations 1–4)

$$\frac{\partial P}{\partial r} = -\rho g \qquad (11.8)$$

$$\frac{\partial T}{\partial r} = \frac{\partial P}{\partial r} \frac{T}{P} \frac{\mathrm{d} \ln T}{\mathrm{d} \ln P} \qquad (11.9)$$

$$\frac{\partial M}{\partial r} = 4\pi r^2 \rho \qquad (11.10)$$

$$\frac{\partial L}{\partial r} = 4\pi r^2 \rho \left(\dot{\epsilon} - T \frac{\partial S}{\partial t} \right), \qquad (11.11)$$

respectively, where P is the pressure, ρ the density, L the intrinsic luminosity, S the entropy per unit mass, and $g = GM/r^2$ the gravity. $\dot{\epsilon}$ accounts for energy produced by nuclear (above ~ $13 M_{\rm J}$), radiogenic, or tidal heating. Equivalent expressions are given by Fortney et al. (2007, eqn 1–3). The resulting models of cooling and contracting adiabatic planets, described further by Fortney & Hubbard (2003, 2004), have been applied to evolutionary models of Jupiter and Saturn (Fortney & Hubbard, 2003), cool giant exoplanets (Fortney & Hubbard, 2004), and hot Jupiters (Fortney et al., 2006).

General models of giant planets assume a core of an unknown mass, comprising an unknown generic combination of iron, refractory material ('rock') and more

Opacities: Half a century ago, in the context of stellar structure and evolution, Schwarzschild (1958) considered the determination of opacities to be *'by far the most bothersome factor in the entire theory'*. He took account of three contributing processes: photoionisation (bound–free transitions), inverse bremsstrahlung (free–free transitions) and electron scattering. Later work showed that spectral lines (bound–bound transitions) also contribute significantly, eventually leading to the widely-used Los Alamos opacities (Huebner et al., 1977). By the 1980s, certain discrepancies between observation and theory, notably for Cepheid pulsations (Simon, 1982) and for solar oscillations (Christensen-Dalsgaard et al., 1985), led to the suggestion that the Los Alamos tables were missing important opacity sources.

Combined with the need for opacities of low-Z species, this led to the re-examination of stellar opacities by two groups: the Opacity Project (Seaton et al., 1994) and the OPAL group at Livermore (Rogers & Iglesias, 1992). They adopted independent approaches, and both duly reported opacities generally higher than the previous Los Alamos opacities, reaching factors of 2–3 in stellar envelopes with $T \sim 10^5 - 10^6$ K. Opacity Project and OPAL opacities are in reasonable agreement (Seaton et al., 1994; Iglesias & Rogers, 1996); and good agreement between OPAL opacities, and those of Alexander & Ferguson (1994) or Kurucz (1991), is also found in the domains where they overlap. Subsequent comparisons have been made using revised Opacity Project data (Badnell et al., 2005). With these new opacities, a number of long-standing problems in stellar evolution have been addressed and, in turn, finer tests of stellar structure have been made possible.

Efforts are now being directed at the derivation of opacities relevant at lower temperature, including the contributions from millions of molecular and atomic lines that are fundamental for the calculation of the thermal structure and radiative transfer within the atmospheres of cool stars, brown dwarfs, and the gas giants (Sharp & Burrows, 2007). Results include compilations for a zero-metallicity (pure H/He) gas (Lenzuni et al., 1991), solar composition including grains but not including alkali metals or TiO absorption (Alexander & Ferguson, 1994), grain-free and alkali-free tables at low-temperatures (Guillot, 1999), effects of grains and alkali metals (Allard et al., 2001), and the inclusion of revised optical constants and the presence of high-temperature condensates such as Al_2O_3 and $CaTiO_3$ (Ferguson et al., 2005). The latter, with computations ranging over $T = 500 - 30\,000$ K and gas densities from $10^{-4} - 10^{-19}$ Mg m^{-3}, also provides comparisons with the OPAL and Opacity Project opacities which are, again, in good agreement at the higher temperatures.

Freedman et al. (2008) provide the most recent tables of mean opacities relevant for giant exoplanet atmospheres, in part derived from the HITRAN (High Resolution Transmission) molecular spectroscopic database (Rothman et al., 2009). Their results cover three metallicities, [M/H]= −0.3, 0.0, +0.3, and sample a grid of 324 points spanning $T = 75 - 4000$ K and $P = 30 - 3 \times 10^7$ Pa. They give both Planck mean opacities, relevant for optically thin regions, and Rosseland mean opacities, most relevant for optically thick regions (Mihalas, 1978). The relevant thermochemical equilibrium abundances used in the opacity models were computed with the CONDOR code (§11.2.2).

Freedman et al. (2008) account for the following opacity sources. Molecular species, also accounting for relevant isotopes, include H_2O (2.9×10^8 lines), CH_4 (2×10^8), NH_3 (34 000), TiO (1.7×10^8), VO (3.1×10^6), CO, H_2S, PH_3, FeH, and CrH. Significant opacity contributions also arise from the alkali atoms Na, K, Cs, Rb, and Li (Burrows et al., 2000b; Allard et al., 2003; Burrows & Volobuyev, 2003). Collision-induced absorption includes interactions between H_2–H_2, H_2–He, and H_2–H (Borysow, 2002, and references). Other opacity sources include bound–free absorption by H and H$^-$, free–free absorption by H, H$^-$, H_2, and H_2^-, Rayleigh scattering from H_2, and Thomson scattering (Lenzuni et al., 1991).

---

volatile species ('ices', such as H_2O, CH_4, and NH_3), and a fluid envelope mostly comprising H/He. In contrast to terrestrial planets, effects of viscosity, rotation and magnetic fields are currently neglected. The models of Fortney et al. (2007) also explicitly ignore tidal heating, the heat content of the core and its unknown heat transport properties (Fortney et al., 2006), and changes in radius over time for the zones of water/rock/iron (i.e. ignoring $\partial S/\partial t$ in Equation 11.11). They also ignore helium phase separation, which will act to increase the radius of cold gas giants, but which is therefore considered unlikely to affect irradiated giants within a few AU of their host stars (Fortney & Hubbard, 2004).

Heat transport and opacities The internal temperature gradient, $d \ln T/d \ln P$ in Equation 11.9, depends on the process by which the internal heat is transported from the interior regions to the surface. That (isolated) giant planet interiors are essentially convective, rather than transporting heat by radiation or conduction, follows from the rapid increase in opacity with increasing pressure and temperature (Hubbard et al., 2002).

At $P \gtrsim 10^5$ Pa and $T \lesssim 1000$ K, the dominant opacity sources are H_2O, CH_4, and collision-induced H_2 absorption (Guillot, 2005).

At intermediate temperatures, $T \sim 1000 - 2000$ K, an effect of decreasing H/He opacity with increasing temperature was predicted by Guillot et al. (1994a,b) which, it was argued, could result in the presence of a radiative shell deep in the interiors of Jupiter, Saturn, and Uranus. This possibility was revised by Guillot et al. (2004) after the identification of additional sources of (Na/K) opacity in brown dwarfs in the spectral regions where H, He, H_2O, CH_4, and NH_3 are relatively transparent (Burrows et al., 2000b). These sources are also confirmed in more recent opacity determinations (Freedman et al., 2008).

At $T \gtrsim 1500 - 2000$ K two additional opacity sources contribute: absorption by H_2^- and H$^-$, and the presence of TiO, which is a strong absorber at visible wavelengths, and which results from the vaporisation of $CaTiO_3$.

The thin radiative atmosphere which exists above the otherwise fully convective region controls the ultimate cooling of the giant planet, and hence the con-

11.4 Predictions of interior models

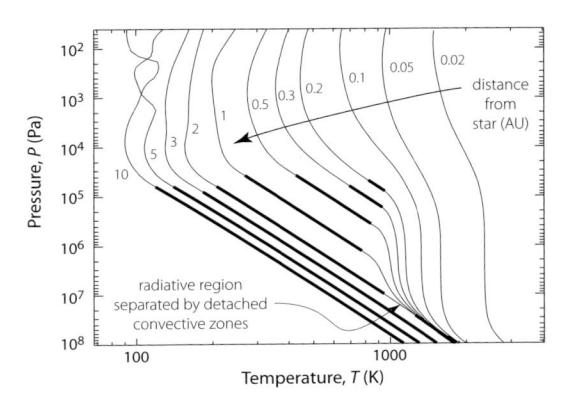

Figure 11.6: Pressure–temperature profiles for Jupiter-like planets of age ~4.5 Gyr ($g = 25\,m\,s^{-2}$, $T_{int} = 100\,K$) from 0.02–10 AU from the Sun. Thin lines indicate radiative regions, and thick lines convective regions. At small separations, a radiative region extends deep into the atmosphere, and convection does not begin until $P > 10^8$ Pa. From Fortney et al. (2007, Figure 3), reproduced by permission of the AAS.

traction of its interior.[2] The fact that giant planet atmospheres have numerous atomic and molecular absorbers (including H_2O, CH_4, NH_3, Na, and K) means that their atmospheres are distinctly non-blackbody (Burrows et al., 1997; Marley et al., 1999; Sudarsky et al., 2000). As a result, self-consistent 'non-grey' atmospheric models must be used to determine T_{eff}, and hence the external boundary conditions for the evolutionary models (Baraffe et al., 2003; Marley et al., 2007b).

Practical construction of equilibrium structure models cover a grid of gravities, T_{eff}, and incident fluxes. Thus Fortney et al. (2007) used recent low-temperature mean opacities (Freedman et al., 2008), assumed solar metallicity atmospheres (Lodders, 2003, §12.4), and included the effects of condensation and gravitational settling ('rainout') on the atmospheric composition (Lodders & Fegley, 2002; Lodders, 2003, §11.7).

Effects of stellar irradiation In addition to internal heat sources (radiogenic or tidal) which will change the rate of cooling, any incident stellar radiation further modifies the thermal properties of an otherwise isolated planetary atmosphere. These effects must therefore be incorporated into realistic model atmospheric grids.

The effects are significant for hot Jupiters at small orbital separations (Guillot & Showman, 2002; Baraffe et al., 2003; Burrows et al., 2004a; Marley et al., 2007a), but also for older low-mass planets at larger orbital separations for which their diminishing intrinsic effective temperature, T_{int}, would otherwise fall below its equilibrium temperature, T_{eq}. The latter is itself determined by the incident stellar flux. By definition (Fortney et al., 2007, eqn 6)

$$T_{eff}^4 = T_{int}^4 + T_{eq}^4 \,. \qquad (11.12)$$

For $T_{int} \ll T_{eq}$, the incident stellar flux dominates, and a more extensive radiative zone develops.

Resulting pressure–temperature profiles for a Jupiter-like planet from 0.02–10 AU from the Sun are shown in Figure 11.6. The adopted intrinsic temperature, $T_{int} = 100$ K, is close to Jupiter's current value (Table 12.2). Bond albedos range from $A_B \sim 0.05 - 0.1$ within 0.1 AU, to larger values of $A_B \sim 0.3 - 0.4$ from 1–10 AU. At larger separations, cooler temperatures result in the strong alkali metal absorbers Na and K condensing into clouds below the visible atmosphere.

A deep external radiative zone appears in the most highly irradiated models, and for planets at < 0.05 AU, convection does not begin until $P \gtrsim 10^8$ Pa. From 0.1–2 AU, the presence of two detached convective zones has an important consequence: heat flow from the central regions is inhibited, and stellar irradiation over this separation range has roughly the same retarding effect on cooling and contraction, even though the incident fluxes vary by a factor 400.

11.4 Predictions of interior models

Structural models make no predictions as to what might constitute a plausible planet, neither in terms of mass or composition, and they do not take account of formation mechanisms. Rather, specifying a mass, and some possible combination of ice, rock and iron leads, from the assumed equations of state, to its predicted radius. These models of mass as a function of composition and radius can be compared with knowledge of solar system planets to establish their plausibility, then used to provide the first estimates of possible exoplanet compositions on the basis of their measured mass and, in the case of transiting planets, their measured radius.

The following results on ice–rock–iron planets, and H/He dominated giants, are from the work of Fortney et al. (2007).

11.4.1 Dependence on composition

Results for planets with $M_p = 0.01 - 1000 M_\oplus$ composed of pure iron, pure rock, and pure ice, and various ice/rock and rock/iron mixtures, are shown in Figure 11.7. It is assumed that any equilibrium structure will have differentiated gravitationally (as observed for solid bodies in the solar system), with ice overlaying rock, and rock overlaying iron.

In all cases, at low masses, planetary radii grow initially with mass as $M_p^{1/3}$. At higher mass, a larger fraction of the material becomes pressure ionised, the material behaves progressively more like a Fermi gas, and

---

[2] In the case of Jupiter, the resulting cooling rate estimated by Hubbard (1977) is $\Delta T_{eff} = -6.9\,K\,Gyr^{-1}$.

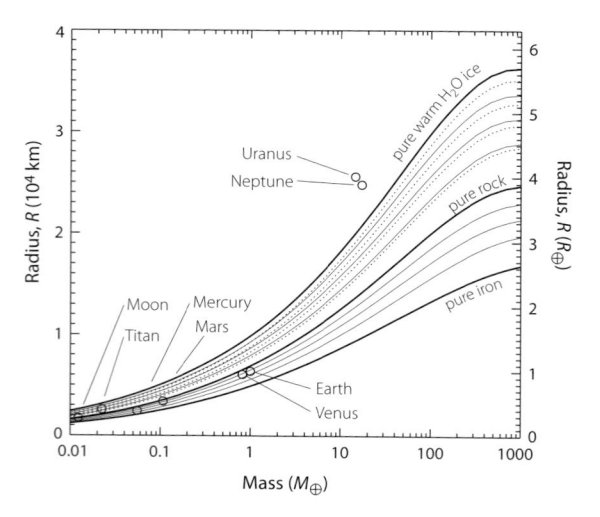

Figure 11.7: Mass versus radius for planets composed of H_2O ice, rock (Mg_2SiO_4), and iron. The thin curves between pure ice and pure rock, are from top to bottom, 75% ice/25% rock, 50/50, and 25/75 (inner layer rock, outer later ice). Dotted lines between pure warm ice and rock are for zero-temperature ice. The thin curves between pure rock and iron, are from top to bottom, 75% rock/25% iron, 50/50, and 25/75 (inner layer iron, outer later rock). Solar system objects are shown as open circles. From Fortney et al. (2007, Figure 4), reproduced by permission of the AAS.

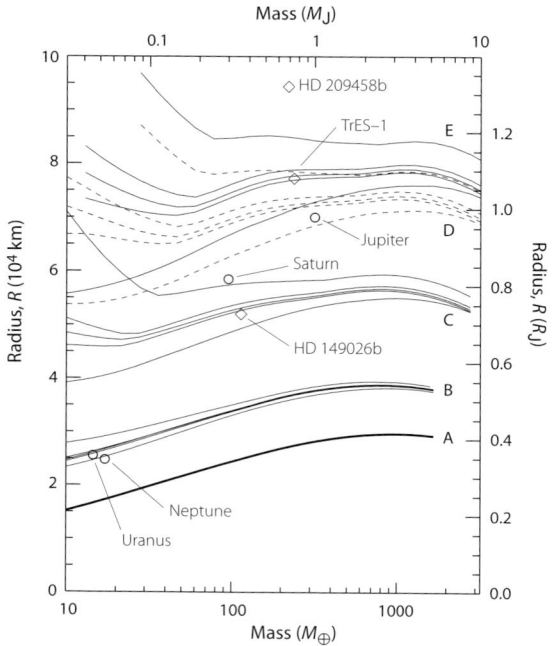

Figure 11.8: Theoretical planetary radii at 4.5 Gyr as a function of mass for various compositions. For B–E, sets of five curves are calculated at 0.02, 0.045, 0.1, 1.0, and 9.5 AU distance from the host star (top to bottom, respectively). Compositions are: (A) 50% rock/50% ice; (B) 90% heavy elements; (C) 50% heavy elements; (D, dotted) 10% heavy elements; (E) pure H/He. Open circles are solar system planets, diamonds are selected exoplanets. From Fortney et al. (2007, Figure 7), reproduced by permission of the AAS.

the mass–radius curves flatten near $1000 M_⊕$. Beyond this, the radius decreases as mass increases, eventually falling as $M_p^{-1/3}$ (§6.6.1).

The positions of the solar system terrestrial planets, ice giants, and the Moon and Titan, are also indicated in Figure 11.7. Their known internal structures are in reasonable agreement with these theoretical models.

Determination of an unambiguous internal composition, based on measurements of M_p and R_p alone, will not be straightforward. For planets around $1 M_⊕$, for example, a mass determined with an accuracy of 50%, combined with a radius accurate to $0.25 R_⊕$, would not be sufficient to discriminate between an internal composition of pure iron, or one comprised of an equal mixture of ice and rock.

11.4.2 H/He dominated gas giants

Predicted radii as a function of mass for H/He-dominated planets of age 4.5 Gyr from the models of Fortney et al. (2007) are shown in Figure 11.8. The models include the effects of stellar heating at five orbital separations from 0.02–9.5 AU, and for a range of core masses (ranging from 0–90% of the planet mass) and compositions. In general, planets with larger cores have smaller radii, and planets closer to their host stars have larger radii as a result of the additional thermal heating. The largest radii occur for core-free planets comprising pure H/He.

The solar system giants are shown in the same figure. From these models, Jupiter's position corresponds to a plausible 10% heavy element composition, with Saturn's position suggesting a higher heavy element abundance. The positions of Uranus and Neptune lie close to the 90% heavy element curves, although they probably contain more ice than rock (Podolak et al., 1995).

Knowledge about the more detailed interior composition of the solar system giants, including their rotation, core composition, ages, and possible H/He demixing, is reviewed by Guillot (2005).

The example of HD 149026 b While the massive close-in transiting planets detected to date are inferred to be gas giants, their very different densities indicate significant variations in heavy element composition.

As one example, HD 149026 b has an estimated mass $M_p = (0.36 ± 0.04) M_J$ and an estimated radius $R_p = (0.725 ± 0.05) R_J$, yielding $ρ ∼ 1.17$ Mg m^{-3}, or $1.7 ρ_{Saturn}$ (Sato et al., 2005a). For such a mass, very different radii would follow for an assumed uniform solar composition ($1.14 R_J$), a $20 M_⊕$ core of constant-density $ρ = 5.5$ Mg m^{-3} ($0.97 R_J$), or an object surmised to consist entirely of water ice ($0.43 R_J$), or of olivine ($0.28 R_J$).

11.5 Super-Earths

The results for HD 149026 b have rather been taken to imply the presence of a $60 - 93 M_⊕$ core composed of elements heavier than H/He, corresponding to ~2/3 heavy elements by mass, i.e. a substantial enrichment above solar composition (Sato et al., 2005a; Fortney et al., 2006; Ikoma et al., 2006).

11.5 Super-Earths

The discovery of the first exoplanets with $M \lesssim 10 M_⊕$, loosely termed *super-Earths*, began with the detections of GJ 876 d, a Doppler-detected $7.5 M_⊕$ planet at 0.02 AU (Rivera et al., 2005), OGLE–2005–BLG–390L b, a microlensed $5 M_⊕$ planet at 5 AU (Beaulieu et al., 2006), and HD 69830 b, a Doppler-detected $10.8 M_⊕$ planet at 0.08 AU (Lovis et al., 2006). Although so far without measured radii and therefore with unknown densities, the prospect of further discoveries from radial velocity, microlensing, and space transit surveys has stimulated the more detailed modeling of the interior composition of lower-mass exoplanets.

Compositional extremes Early discussions of terrestrial-mass planets, and in particular those that might exist in the habitable zone, generally assumed that planets of terrestrial mass, $\sim 0.2 - 5 M_⊕$, would be of roughly terrestrial composition, i.e. made up mostly of silicates and iron-peak elements like Earth (Figure 11.2).

These views have changed as formation theories have advanced, and it is now considered that planets of more exotic composition or structure should exist. These might include super-Earth planets formed beyond the snow line, with a significant mass fraction in H_2O (Kuchner, 2003; Léger et al., 2004), the so-called 'ocean worlds' considered further in §11.5.2.

Planets formed in systems with $C/O > 1$ might result in carbon-dominated terrestrial planets formed substantially from SiC and other carbon compounds (Kuchner & Seager, 2005). Coreless terrestrial planets, without a differentiated metallic core (essentially a giant silicate mantle into which iron is bound in the form of iron oxide), could result from different oxidation states during planetary accretion and solidification (Elkins-Tanton & Seager, 2008a). Na and K clouds could surround hot super-Earths in which volatile elements have been lost from the planet (Terquem & Papaloizou, 2008; Schaefer & Fegley, 2009).

11.5.1 General models

The predicted bulk properties of these and other low-mass exoplanets, and the broad range of compositions that might exist, has been considered in various work (e.g. Ehrenreich et al., 2006; Valencia et al., 2006; Seager et al., 2007; Sotin et al., 2007; Valencia et al., 2007b,c; Elkins-Tanton & Seager, 2008a,b; Grasset et al., 2009; Schaefer & Fegley, 2009).

Equation of state The information required to determine the mass–radius relation (and ignoring the details of heat transport) is contained in the equations for hydrostatic equilibrium and mass conservation (Equations 11.8 and 11.10), and in the equation of state giving the thermodynamic relation between pressure and density $P(\rho, T)$. Some published models have been computed without temperature dependence (e.g. Seager et al., 2007), and others with (e.g. Valencia et al., 2006, 2007c; Sotin et al., 2007).

The adopted equation of state may rest partly on empirical fits to experimental data obtained at pressures $P \lesssim 200$ GPa, such as the formulations of Vinet or Burch–Murnaghan or, at high pressures $P \gtrsim 10^4$ GPa as the electron degeneracy pressure becomes increasingly important, on the Thomas–Fermi–Dirac theoretical formulation. Further details are given by, e.g., Seager et al. (2007) and Sotin et al. (2007, their Appendix A).

By way of illustration, the simplest form adopted by Seager et al. (2007) is that of Vinet et al. (1989)

$$P = 3K_0 \eta^{2/3} \left[1 - \eta^{-\frac{1}{3}}\right] \exp\left(\frac{3}{2}(K_0' - 1)\left[1 - \eta^{-\frac{1}{3}}\right]\right), \quad (11.13)$$

where $\eta = \rho/\rho_0$ is the compression ratio, $K_0 = -V(\partial P/\partial V)_T$ is the bulk modulus of the material, and K_0' is the pressure derivative. Similar dependencies are shown in Figure 11.3. Seager et al. (2007) have noted that the constant temperature equations of state for the solid materials they considered can be approximated by

$$\rho(P) = \rho_0 + cP^n \, . \quad (11.14)$$

They list best-fit parameters ρ_0, c, and n for various materials over the range $P < 10^{16}$ Pa, with these approximations valid to some 1–10% depending on pressure.

Example results and ternary diagrams Valencia et al. (2007b) modeled a general low-mass exoplanet comprising concentric differentiated shells of homogeneous composition, and solved the differential equations for density, pressure, mass, and gravity structure under hydrostatic equilibrium within each layer. The regions are: (1) the core, mainly Fe, divided into inner solid and outer liquid layers if the temperature profile crosses the Fe melting curve; (2) an upper and lower mantle comprising various Mg-Fe silicates and oxides; and (3) a H_2O layer divided into a water ocean over an icy shell comprising the high-pressure ices VII and X. The thicknesses of the regions depend on the assumed ice mass fraction and core mass fraction for a given planet.

Their results, of which a selection are shown in Figure 11.9, are displayed in the form of *ternary diagrams*. In these, the data of a three-component system are plotted in a triangle whose sides depict the three compositional axes. Each vertex represents 100% of the given component, with the opposite side corresponding to 0%

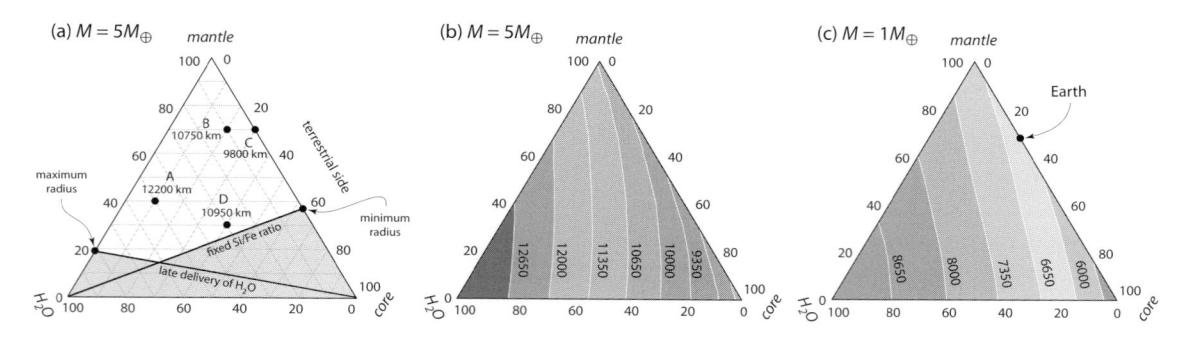

Figure 11.9: Ternary diagrams with radius predictions for super-Earth models based on arbitrary compositions of a Fe-core/silicate mantle/H_2O-layer planet. In (a), for $M = 5M_\oplus$, four different compositions are labeled (CMF = core mass fraction, IMF = ice mass fraction): (A) CMF=10%, IMF=50%, R = 12 200 km; (B) CMF=20%, IMF=10%, R = 10 750 km; (C) CMF=30%, IMF=0%, R = 9800 km; and (D) CMF=40%, IMF=30%, R = 10 950 km. The shaded regions show improbable compositions from solar nebula abundances and accretion processes. In (b,c), contours of equal radius are shown as a function of the three-component composition for total planet masses of $M = 5M_\oplus$ and $M = 1M_\oplus$ respectively. The Earth has a negligible mass in the form of water, and appears on the side opposite the H_2O vertex accordingly. From Valencia et al. (2007b, Figures 1–2), reproduced by permission of the AAS.

of that component, and lines parallel to that side indicating the percentage of that component. The radius of a planet is then determined by the model for any combination of components and for a given mass. Different ternary diagrams show the radius (or other properties) for a specific planet mass.

Certain regions of the overall composition space are likely to be excluded on the basis of plausible nebula abundances and accretion mechanisms, and these are shown shaded in Figure 11.9a. The minimum planet radius, for pure Fe or a heavy Fe+Ni alloy, occurs at the lower right vertex, and correspondingly for the maximum radius assuming a pure H_2O composition at the lower left. Allowing for excluded regions, the minimum and maximum radii are unlikely to fall at the vertices.

The isoradius diagrams of Figures 11.9b–c show that the same value of the radius is obtained for different combinations of core, mantle, and water content, illustrating the degeneracy in composition which results from a knowledge only of the density. But some constraints on composition follow from the fact that there will be a threshold radius such that larger values imply that the planet is ocean-dominated. This is given by the largest isoradius curve that intersects the 'terrestrial side' (zero water content) of the ternary diagram. For 1, 2.5, 5, 7.5, and $10M_\oplus$ planets, the threshold radius is 6600, 8600, 10 400, and 11 600 km respectively. These arguments have been applied to the $7.5M_\oplus$ GJ 876 d by Valencia et al. (2007c).

Mass–radius relation From these models of super-Earths, various mass–radius relations have been derived (e.g. Léger et al., 2004; Valencia et al., 2006; Seager et al., 2007; Sotin et al., 2007; Valencia et al., 2007b; Grasset et al., 2009). These relations aim to quantify the extent to which a planet deviates from the $R \propto M^{1/3}$ relation expected if the interiors were homogeneous and incompressible. Expressed as a simple power-law

$$R_p = a\, R_\oplus \left(\frac{M_p}{M_\oplus}\right)^b, \qquad (11.15)$$

Sotin et al. (2007) found $(a, b) = (1, 0.306)$ and $(1, 0.274)$ for terrestrial planets of solar composition (and minimal H_2O content) in the ranges $0.01 - 1 M_\oplus$ and $1 - 10 M_\oplus$ respectively. For 50% water content, corresponding to the class of 'ocean planets', they found $(a, b) = (1.258, 0.302)$ and $(1.262, 0.275)$ for the same mass ranges. The essentially identical exponent means that, independent of mass, a 50% H_2O planet is predicted to be 25% larger than a terrestrial planet (and the surface gravity correspondingly smaller).

Valencia et al. (2007b) gave a more general power-law expression as a function of the H_2O-ice mass fraction, under the further assumption that the mantle-to-core ratio remains constant (such that the Fe/Si ratio is itself approximately constant, and consistent with the solar nebula composition). Then

$$R_p = (1 + 0.56\,i)\, R_\oplus \left(\frac{M_p}{M_\oplus}\right)^{0.262(1-0.138\,i)}, \qquad (11.16)$$

where i denotes the ice mass fraction in percent.

With a particular scaling of mass and radius, Seager et al. (2007) derived an expression for $M_p \lesssim 20 M_\oplus$ more related to their underlying approximation to the equation of state (Equation 11.14), viz.

$$\log R'_p = k_1 + (1/3)\log(M'_p) - k_2 M'^{k_3}_p, \qquad (11.17)$$

where M'_p and R'_p are scaled mass and radius values, and k_1, k_2, k_3 are obtained from the model fits.

Grasset et al. (2009) derived the parameterised form

$$\log\left(\frac{R_p}{R_\oplus}\right) = \log\alpha + \left[\beta + \gamma \frac{M_p}{M_\oplus} + \epsilon\left(\frac{M_p}{M_\oplus}\right)^2\right]\log\left(\frac{M_p}{M_\oplus}\right), \quad(11.18)$$

11.5 Super-Earths

> **Nomenclature for super-Earths:** Valencia et al. (2007b) proposed the following nomenclature. A super-Earth is a planet in the range $M = 1 - 10 M_\oplus$ (and likely to have large amounts of solid material). The lower-bound corresponds to the mass of the Earth, and the upper, set somewhat arbitrarily, demarcates planets likely to acquire large envelopes of H/He during their formation.
>
> Sub-classes include both super-Earths of terrestrial-type composition, as well as 'ocean planets' with a mass fraction > 10% in the form of H_2O (in either solid or liquid phase). They term planets with more massive Fe or Fe–Ni cores as 'super-Mercuries', and they may be terrestrial or have large oceans. The name 'ice giants', as exemplified by Uranus and Neptune, is used for objects of $M \gtrsim 10 M_\oplus$ with some ($\lesssim 10\%$) gas.

where $\alpha, \beta, \gamma, \epsilon$ are quadratic functions of the mass fraction of water. Over $1 - 100 M_\oplus$, they estimate 1% accuracy for compositions ranging from Earth-like (essentially zero H_2O content) to pure ice.

11.5.2 Ocean planets

Three classes of solar system object have a higher fraction of ices than either the terrestrial planets or the gas giants: comets, the ice giants Uranus and Neptune, and the icy satellites of the outer planets (Johnson, 2005).

The interior structures and cooling histories of Uranus (Podolak et al., 1991) or Neptune (Hubbard et al., 1995) are still held to be less well understood than those of Jupiter or Saturn. But they are of too low a density to be comprised solely or predominantly of ices (Figure 11.7), and models for their interiors suggest H/He envelopes around massive protoplanetary cores of rock and ice (Figure 11.2). Ices in their interiors are expected to be in a fluid phase (Cavazzoni et al., 1999). The existence of H_2O oceans deep in their interiors has been considered by Wiktorowicz & Ingersoll (2007).

Icy satellites such as J III Ganymede, J IV Callisto, S V Rhea and S VIII Iapetus are of much lower masses, typically $\lesssim 0.02 - 0.001 M_\oplus$, but have significantly higher ice fractions. Their inferred compositions, mostly ices and silicates, are consistent with formation near the snow line. They are believed to have formed very rapidly, on time scales of 100–10 000 years in the case of Ganymede and Callisto (Lunine & Stevenson, 1982). During their assembly, optically thick gaseous envelopes would have prevented them from dissipating the heat of accretion by radiation, and they are presumed to have begun life with their ice mostly melted, before freezing again in the outer solar system (Lunine & Stevenson, 1982).

Hypothetical existence Kuchner (2003) and Léger et al. (2004) hypothesised that terrestrial- or super-Earth mass planets could exist, with compositions dominated by volatiles such as H_2O ice. According to current understanding, these objects, which have been termed *ocean planets* or 'water worlds', could form at large orbital distances, beyond the snow line. Although a rising fraction of more volatile ices (NH_3, CH_4, CO, N_2, CO_2) would be expected for objects forming further out, the longer times required to reach chemical equilibrium at lower temperatures and densities (§11.2.2), makes predictions of specific compositions difficult.

Nevertheless, such objects in the range $1 - 10 M_\oplus$, and thus below the presumed core masses required for giant planet formation, would not be expected to have accreted a large H/He envelope. A fraction could thereafter have migrated inwards, some plausibly to reside in the habitable zone. Objects towards the upper mass range would also be feasible targets for detection by radial velocity or transit techniques.

Atmospheres Kuchner (2003) argued that a H_2O-dominated planet arriving at 1 AU should quickly develop a gaseous atmosphere of $\gtrsim 10^7$ Pa, under conditions similar to those hypothesised during Earth's accretion history (Abe & Matsui, 1985). In this scenario, shock impact degassing of volatile materials results in the formation of an impact-generated proto-atmosphere, and the planet's surface temperature thereafter increases with subsequent planetesimal impact due to the atmosphere's blanketing effect.

If accretion-driven impact energy at its base exceeds a certain value, the atmosphere can become a runaway greenhouse. Otherwise, the hot atmosphere provides stable greenhouse heating up to a basal temperature (Matsui & Abe, 1986; Kuchner, 2003)

$$T_G = \left(\frac{\lambda + \sqrt{1-\omega}}{1 + \sqrt{1-\omega}} \right)^{1/4} T_{BB} \approx \lambda^{1/4} T_{BB}, \quad (11.19)$$

where λ is the ratio of infrared to optical opacities, ω is the *single scattering albedo* at short wavelengths, and T_{BB} is the local blackbody temperature

$$T_{BB} = 278 \text{K} \left(\frac{a}{1 \text{AU}} \right)^{-1/2} \left(\frac{L_\star}{L_\odot} \right)^{1/4}. \quad (11.20)$$

For $\lambda \sim 10 - 100$ (~31 for Venus), and $\omega \sim 1$, $T_G \sim 2.4 T_{BB}$.

When $T_G < T_{crit} = 647$ K, the *critical temperature*, water condenses out of a vapour atmosphere in equilibrium with the stellar radiation, leaving a hot ocean on the planet's surface (Matsui & Abe, 1986; Zahnle et al., 1988). Again assuming $\lambda = 31$, $\omega = 1$, the planet's vaporisation halts for $a \gtrsim 1.03$ AU $(L_\star/L_\odot)^{1/2}$, and the planet can survive indefinitely, so long as it retains its atmosphere[3]. Closer in, the planet will slowly vaporise.

---

[3] Matsui & Abe (1986) used such arguments to infer the temperature of the first rain on Earth of ~600 K. They considered this to be consistent with the wide temperature limits of the early Archean (~3.5 Ga) ocean of ~ 273 – 419 K as estimated from the ^{18}O isotope composition of metamorphosed sedimentary rock by Oskvarek & Perry (1976).

Ocean depth Amongst other investigations, Léger et al. (2004) considered a representative $6M_\oplus$ ocean planet, comprising $3M_\oplus$ of metals and silicates and $3M_\oplus$ of H_2O. Depending on orbital radius and thermal properties, the outer H_2O envelope could exist as a hot gaseous phase for which, at $P_{\rm crit}$, $T > T_{\rm crit}$, such that no liquid surface exists. At progressively lower temperatures, it would have either a liquid or ice surface.

For a planet located in the habitable zone with a liquid water surface, Léger et al. (2004) posed the question: '*how deep is the ocean?*'. Because of the nature of the ice phase space, this apparently loosely-constrained question has, in fact, a rather well-defined answer. For any given surface water temperature, and depending on the amount of internal planetary heat being transferred outwards, the increasing temperature as a function of depth has its maximum gradient for an adiabatic dependency, a zero gradient for a perfectly isothermal convection-driven transfer, and even a negative gradient as a result of large-scale currents as in the case of the Earth's ocean.

The temperature–pressure relations for the adiabatic and isothermal dependencies are shown in the inset of Figure 11.5 for an ocean surface at $T = 7°C$. The bottom of the ocean is encountered at the pressure at which the water solidifies, i.e. the intercept of the isotherm or adiabat with the liquid–solid (actually ice VI) phase boundary.

For an adiabatic dependency, and an ocean surface at 0, 7, 30°C (273, 280, 303 K) the resulting ocean depth is 60, 72, and 133 km respectively. For an isothermal ocean at the same surface temperatures, the ocean depth is 40, 45, and 65 km respectively. Similar considerations, and similar results, were given by Sotin et al. (2007).

Candidate ocean planets GJ 1214 b, at 13 pc, was discovered as a transiting planet by Charbonneau et al. (2009). With $M = 6.55 M_\oplus$ and $R = 2.68 M_\oplus$, its low density ($\rho = 1.9 \,{\rm Mg\,m^{-3}}$), could be modeled by a dense iron–silicate core, combined with a H–He envelope. But its small mass argues against the accretion of an extended gaseous envelope, and suggest instead an internal composition comprising a Fe–Ni core, a silicate mantle, some 50% of its mass in H_2O, and possibly a thin outer H–He envelope (Marcy, 2009a). A deep liquid H_2O ocean, and a hot vapour atmosphere, are suggested by its equilibrium surface temperature of ~460 K (~190 C).

A cold ocean planet model was introduced for the low-temperature microlensed planet OGLE–2005–BLG–390L b (Ehrenreich et al., 2006).

Given the compositional range that can match a given bulk density (at least for smaller radii planets), the secure identification of ocean planets based on the mass–radius relation alone may still be questionable unless a significant gas layer can be excluded by other means (Adams et al., 2008).

11.6 Diagnostics from rotation

The limited information on interior structure which is conveyed by the mass and radius of an exoplanet is augmented, at least in principle, by the fact that the planet is rotating (Zharkov & Trubitsyn, 1971, 1974, 1978).

In hydrostatic equilibrium, the external gravitational potential at coordinate position (r,θ) is given by (Zharkov & Trubitsyn, 1974, eqn 1)

$$V(r,\theta) = \frac{GM_{\rm p}}{r}\left[1 - \sum_{n=1}^{\infty}\left(\frac{R_+}{r}\right)^{2n} J_{2n}\, P_{2n}(\theta)\right], \quad (11.21)$$

where R_+ is the equatorial radius, J_{2n} are the planet's gravitational moments, P_{2n} are Legendre polynomials, and θ is the colatitude. The external gravitational field is then fully specified by the set of gravitational multipole moments (Zharkov & Trubitsyn, 1974, eqn 2)

$$J_{2n} = -\frac{1}{M_{\rm p} R_+^{2n}} \int \rho(r)\, r^{2n}\, P_{2n}(\theta)\, {\rm d}\tau\,, \quad (11.22)$$

where τ is a volume element, and the integration extends over the planet's entire volume.

In the case of the solar system giants (§12.2 and Table 12.2), masses are derived from the motions of their natural satellites, radii are derived from radio occultations, and rotation periods are inferred from the time variation of their magnetic fields. All four are relatively fast rotators, with periods of 10–11 h for Jupiter and Saturn, and 16–17 h for Uranus and Neptune. Estimates of J_2 and J_4, and hence some constraints on their interior density profiles, have been established from the Pioneer and Voyager spacecraft flybys (see Table 12.2 for references).

For exoplanets, where *in situ* measure of the J_{2n} is not possible, the oblateness may be accessible from transit observations (Seager & Hui, 2002; Barnes & Fortney, 2003) (§6.4.12). The oblateness is given in terms of the equatorial and polar radii by (Guillot, 2005, eqn 15)

$$\frac{R_+}{R_+ - R_-} \simeq \left(\frac{3}{2}\Lambda_2 + \frac{1}{2}\right), \quad (11.23)$$

where $\Lambda_2 = (J_2/q) \sim 0.1 - 0.2$, and where

$$q = \frac{\omega^2 R_+^3}{GM}, \quad (11.24)$$

the ratio of centrifugal acceleration to gravity (ω is the rotation rate), provides a measure of the rotation rate. It may therefore be plausible to obtain the rotation rate from an oblateness measure or, given a rotation rate determined independently, some constraint on their interior structure.

Galilean satellites Gravity measurements from the Galileo mission have similarly been used to construct interior structure and composition models for Jupiter's

11.7 Atmospheres of gas giants

Galilean satellites (Sohl et al., 2002). These models provide constraints on the core sizes, and on the mantle densities and possible mineralogies, for example consistent with an olivine-dominated composition in the case of Io and Europa.

For Ganymede, the thickness of the silicate mantle is between 900–1100 km, with an outermost ice shell 900 km thick, further subdivided by pressure-induced phase transitions into ice I, ice III, ice V, and ice VI layers. Bulk Fe/Si ratios suggest a difference in fractionation between the inner (Io, Europa) and outer (Ganymede, Callisto) satellite pairs during formation.

Equilibrium models of heat transfer suggest that Europa, and possibly Ganymede and Callisto, may have oceans underneath their ice shells tens to hundreds of km deep, depending on ice phase, and depending on the degree to which radiogenic is supplemented by tidal heating (Spohn & Schubert, 2003).

11.7 Atmospheres of gas giants

Bulk densities for transiting giant planets indicate massive H/He-rich envelopes surrounding a denser core of metals, silicates, and ices in various proportions. Atmospheric modeling aims to derive the properties of these outer gaseous envelopes, notably the pressure–temperature profiles of their outer regions, day–night temperature gradients and circulation patterns, their chemical and condensate composition, and their resulting spectral properties including broad-band colours as well as more detailed predictions of the presence of molecular bands and atomic lines in the optical and infrared spectral regions (Seager & Deming, 2010; Seager, 2010). Coupling of such atmosphere models to interior thermal evolution models further provides a more self-consistent insight into their combined interior and atmospheric structures.

Numerous factors influence the temperature–pressure structure of the outer atmosphere. These include the age of the planet combined with its cooling rate, its internal (radiogenic and tidal) heat sources, external heating due to stellar irradiation, and the physical and chemical properties of the atmospheric constituents, notably abundances, gas opacities (see box on page 264), and the specific condensed phases present at the relevant temperatures and pressures.

Elemental composition and enrichment An initial assumption for modeling might be that giant exoplanet atmospheres have elemental compositions similar to that of their host stars, viz. typically close to solar, but with some enrichment of heavy elements (Fortney et al., 2008b). Models frequently use elemental abundance ratios inferred for the solar nebula (§12.4), thereafter computing equilibrium compositions at various metallicities (Fegley & Lodders, 1994; Lodders & Fegley, 2002).

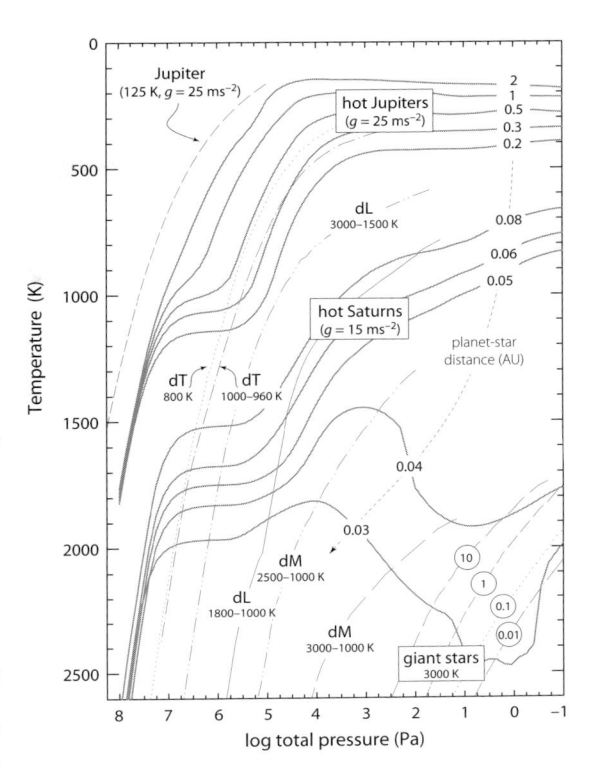

Figure 11.10: Pressure–temperature relations from model atmospheres. Thin lines, labeled with T_{eff}, show relations for (upper left to lower right) Jupiter; T, L, M dwarfs; and giant stars, for which circled numbers indicate surface gravity, g, in $m\,s^{-2}$. Thick lines are relations for Jupiter- and Saturn- mass planets at different distances from their primary stars, labeled in AU. Temperature and pressure increase to the lower left, as if towards the interiors. From Lodders (2010, Figure 4).

The solar system gas giant atmospheres are, however, known to be enriched in at least some heavy elements with respect to the Sun, by a factor of ~2–5. Estimates for a number of elements for Jupiter's atmosphere (O, C, N, S, and various noble gases) were obtained from measurements with the Galileo entry probe (Niemann et al., 1998; Atreya et al., 2003), with measurements of C and P for Saturn from Cassini and ISO spectroscopy (Flasar et al., 2005; Visscher & Fegley, 2005).

Processes which caused such enrichment are uncertain, and whether they operate to a similar degree in exoplanet atmospheres is unknown. Nevertheless, various enrichment mechanisms have been suggested for Jupiter and Saturn, including planetesimal bombardment and accumulation during their formation (Owen et al., 1999; Gautier et al., 2001a,b; Alibert et al., 2005d), erosion of the heavy-element core (Stevenson, 1985), accretion of a metal-rich disk gas (Guillot & Hueso, 2006), and chemical fractionation within the planet (Stevenson & Salpeter, 1977a; Lodders, 2004b).

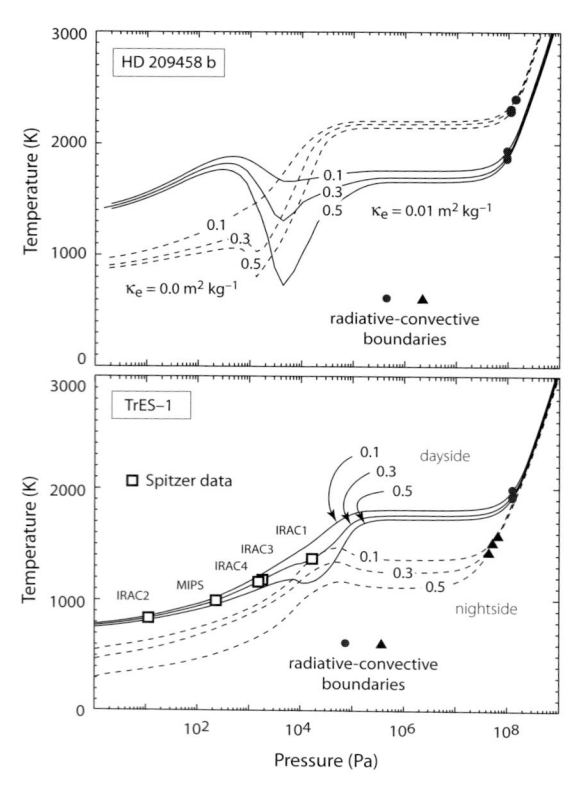

Figure 11.11: *Example temperature–pressure profiles for two of the six close-in planets modeled by Burrows et al. (2008a). Dayside profiles incorporate stellar irradiation. The numbers attached to each curve are the values of P_n which characterise the efficiency of irradiated energy transfer from dayside to nightside, ranging from 0 (no energy redistribution) to 0.5 (full redistribution). The models for HD 209458 b include an extra upper-atmosphere absorber, characterised by κ_e. When $\kappa_e \neq 0$, the pressure–temperature profiles show distinct thermal inversions. From Burrows et al. (2008a, Figures 1a,c), reproduced by permission of the AAS.*

Atmospheric models For an ideal gas in hydrostatic equilibrium, the number density of molecules per unit area above a specified altitude z is given by

$$N \sim \frac{p(z)}{\mu g(z)} = n(z) H, \qquad (11.25)$$

where μ is the mean molecular weight, H is the atmospheric scale height, $p(z)$ is the pressure, and $g(z)$ is the gravity. The equality, exact for an isothermal gas, provides an insight into a number of the basic characteristics of atmospheres (Chamberlain & Hunten, 1987; Houghton, 2002), including that of the Earth.

However, the complexity of the transfer of radiation through an atmosphere whose absorption properties depends on the atomic and molecular composition, temperature and pressure, means that numerical models are required to predict their detailed (temperature–pressure) structures. These are based on one-dimensional radiative transfer codes (e.g. Toon et al., 1989). Condensation processes, or 'rainout', are now also taken into account. Since these are, in turn, dependent on temperature and pressure, the construction of atmospheric models is essentially an iterative process.

The model described by Fortney et al. (2008b), for example, computes radiative–convective equilibrium pressure–temperature profiles and low-resolution spectra, with high-resolution spectra computed using a line-by-line radiative transfer code. Most recent applications use the solar nebula abundances derived by Lodders (2003), and the opacity database of Freedman et al. (2008). Such models have been applied to the atmospheres of a variety of planetary and substellar objects, including Titan (McKay et al., 1989) and Uranus (Marley & McKay, 1999); brown dwarfs and extrasolar giant planets in general (Marley et al., 1999; Chabrier & Baraffe, 2000); ultracool brown dwarfs (Marley et al., 2002; Saumon & Marley, 2008); specific T dwarfs including GJ 229 B (Marley et al., 1996) and others (Saumon et al., 2006, 2007); and a number of individual giant exoplanets, including TrES–1 and HD 209458 b (Fortney et al., 2005; Baraffe et al., 2003), HD 149026 b (Fortney et al., 2006), and HD 209458 b and HD 189733 b observed by Spitzer (Fortney & Marley, 2007).

Another comprehensive atmospheric modeling programme widely applied to brown dwarf and (irradiated) giant exoplanet atmospheres is described by Burrows et al. (2008a). Model atmospheres are computed using the code CoolTLUSTY (Sudarsky et al., 2003; Hubeny et al., 2003; Burrows et al., 2006), a variant of the general atmosphere code TLUSTY (Hubeny, 1988; Hubeny & Lanz, 1995). Molecular and atomic opacities are taken from Sharp & Burrows (2007). Chemical abundances, including condensate rainout, are derived using thermochemical modeling (Burrows & Sharp, 1999; Burrows et al., 2001; Sharp & Burrows, 2007).

These models have been applied to brown dwarfs and exoplanets (Burrows et al., 1997; Sudarsky et al., 2000; Burrows et al., 2001; Sudarsky et al., 2003; Burrows et al., 2004b; Sudarsky et al., 2005); irradiated planets (Burrows et al., 2000a; Hubeny et al., 2003; Burrows et al., 2007a); and secondary eclipses of transiting systems (Burrows et al., 2005, 2006, 2007b).

In the standard *hot start* models (a term introduced by Marley et al., 2007b), the initial state is considered to be an arbitrary large, hot, non-rotating, adiabatic sphere, from which the object thereafter cools and contracts. To first order, details of the starting conditions are considered irrelevant since the initial cooling and contraction are very fast.

Pressure–temperature relations Examples of pressure and temperature relations are given in Figures 11.10 and 11.11. Both figures partly anticipate descriptions of the effects of external heating (below).

11.7 Atmospheres of gas giants

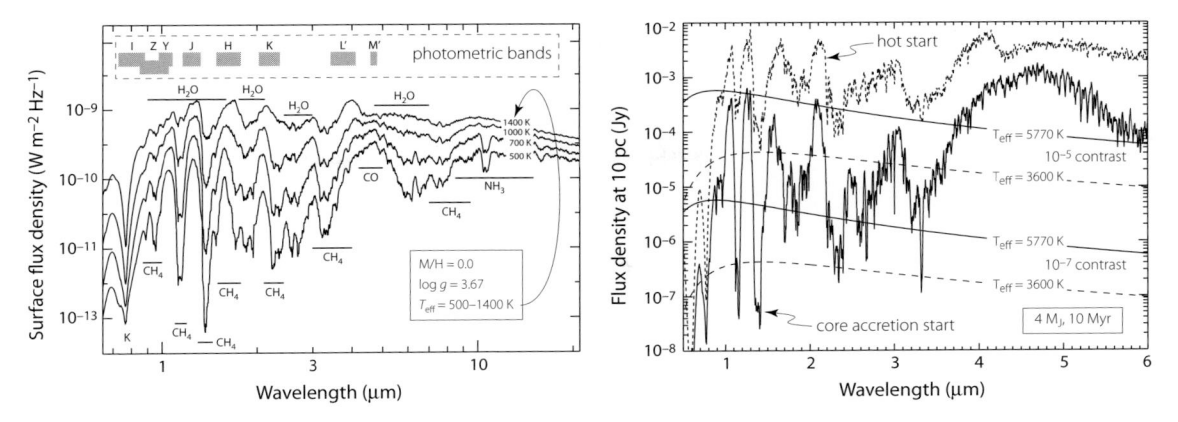

Figure 11.12: Dependence of the predictions of atmospheric models on starting conditions. Left (a): Predicted spectra for cool CH_4-dominated atmospheres, of solar metallicity, in the range $T_{\rm eff} = 500-1400$ K, expected for young gas-giant planets based on 'core accretion start' models. Right (b): the flux density at 10 pc illustrates the contrast ratio needed to detect a $4M_J$ planet according to the predictions of the 'core accretion start' models (solid lines) compared with the more luminous predictions from the 'hot start' models. Contrast ratios of 10^{-5} and 10^{-7} are shown for solar-type dwarfs with $T_{\rm eff} = 5770$ K (solid curve) and M2 dwarfs with $T_{\rm eff} = 3600$ K (dashed curve). From Fortney et al. (2008b, Figures 3–4), reproduced by permission of the AAS.

Figure 11.10, from Lodders (2010), shows model predictions for Jupiter, for a selection of T, L and M dwarfs, and for giant stars with $T_{\rm eff} = 3000$ K and a range of surface gravities. The 800 K T dwarf corresponds to GJ 570 D, while the 1000–960 K T dwarf corresponds to GJ 229 B.

The effective temperatures of gas giant planets of a given mass which are not externally heated decrease with age, as for brown dwarfs after the completion of deuterium burning (Burrows et al., 2001, 2003b, see also Figure 9.2). Correspondingly, in the absence of any external heating, the chemistry of their outer atmospheres will also change with age.

Figure 11.11 shows results for two of the six close-in planets modeled by Burrows et al. (2008a). These models include an *ad hoc* upper atmosphere absorber with constant opacity, κ_e, consistent with the expected abundances for TiO and VO. The effects of energy transport from dayside to nightside, deep in the atmosphere, is characterised by the parameter P_n, ranging from 0 (no energy redistribution) to 0.5 (full redistribution).

Effects of external heating By modifying the pressure–temperature relation in its outer regions, an external heat source in the form of stellar irradiation affects the mass–radius of a giant planet, as considered in §11.3.3. The same effects modify the emission spectrum of the planet, by altering the pressure–temperature structure, and thereby also modifying the conditions for condensation of atomic and molecular species.

Figure 11.10 also shows the consequences of external heating on the pressure–temperature relations for Jupiter- and Saturn-mass exoplanets over a range of separations (0.03–2 AU) from their host star. Planets of Jupiter mass, with $T_{\rm eff} \sim 100$ K in the absence of stellar heating (upper lines), display modified pressure–temperature profiles at smaller orbital separations which resemble those of isolated T dwarfs (Fortney et al., 2008b). However, their different surface gravities result in different opacities at a given atmospheric pressure and temperature, resulting in different $T_{\rm eff}$ for giant planets and brown dwarfs at a given location in the pressure–temperature diagram.

At smaller orbital separations, the atmospheres of planets such as TrES–1 b and HD 209458 b reach temperatures of 1200–1400 K or higher, resulting in atmospheric chemistry more comparable to that of isolated brown dwarfs near the L–T dwarf transition.

At the closest separations of 0.04–0.03 AU are the very short-period transiting planets such as TrES–3 b. Occupying the lower-right region of Figure 11.10, temperatures in the outer regions are high and pressures are low, corresponding to conditions in giant and supergiant stellar atmospheres. Stratospheric-type temperature inversions are predicted by these models.

Atmospheric circulation The existence of the detached convective zones for the close-in, highly irradiated atmospheres (Figure 11.6), results in a deep internal heat flow distinct from the thermalised incident radiation. The global temperature distribution is no longer relatively homogeneous, and equator-to-pole and day–night temperature gradients can be large. This is in contrast to the internal heatflow in Jupiter, which is less-strongly irradiated, and in which convective transfer results in a relatively isothermal planet at the convective–radiative boundary (Ingersoll & Porco, 1978).

The resulting atmospheric circulation in hot Jupiters

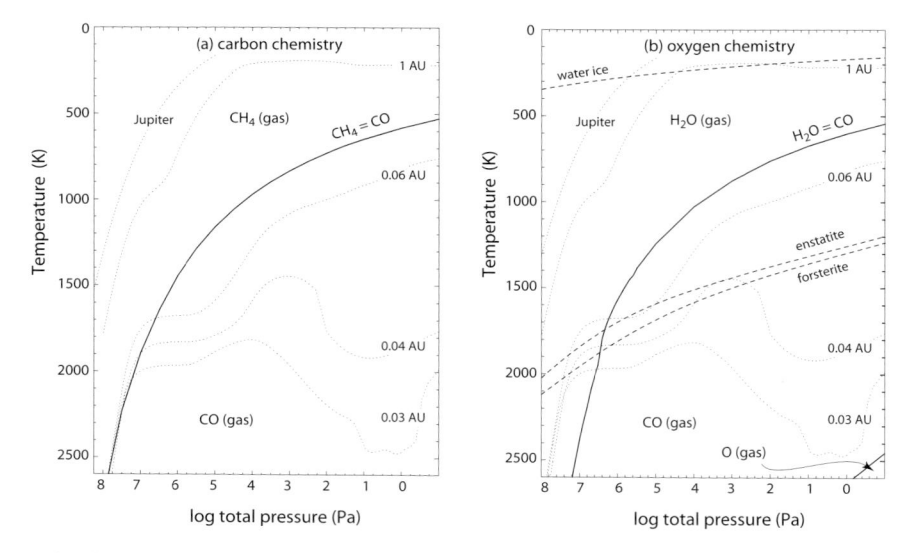

Figure 11.13: (a) The chemistry of carbon in low-mass objects is dominated by CO (at high temperatures and low pressures) and CH_4 (conversely). The solid line is the equal abundance curve for CO and CH_4, described by Equation 11.26. Dotted lines are pressure–temperature profiles for Jupiter and Jupiter-mass objects at various separations from the host star. (b) Corresponding diagram for the chemistry of oxygen. Temperature and pressure increase to the lower left. From Lodders (2010, Figures 5–6).

results in a redistribution of energy by possibly high-velocity atmospheric winds, and thermal emission distributions over the planet surface which depends on the incident energy flux, the atmospheric scale height, the planet rotation rate, and whether it is tidally locked to the star. Observations are discussed further in §6.5.

Dependence on initial conditions The starting conditions for many atmospheric evolution models begins with an arbitrarily large and hot, non-rotating, adiabatic sphere. Initially this cools and contracts rapidly and, over a period of several million years, is assumed to lose all memory of the starting conditions. This has been termed the 'hot start' model (Marley et al., 2007b).

More reliable predictions of the atmospheres of young gas giants should be possible by coupling spectral calculations to thermal evolution models that include details of giant planet formation by the paradigm model of core-accretion and gas capture. Such models, based on initial conditions given by the (one-dimensional) core accretion models of Hubickyj et al. (2005), are described by Marley et al. (2007b) and Fortney et al. (2008b). Predictions of the emergent flux density for a range of T_{eff} are shown in Figure 11.12a.

As a result of the treatment of gas accretion, the 'core accretion start' models begin their lives significantly smaller and colder than the 'hot start' models, with the result that atmospheric models based initial conditions representative of core accretion are initially less luminous by a factor 10–100. Significantly, some memory of the initial conditions appears to persist over evolution times as long as ~ 0.1–1 Gyr.

In consequence, such cool young Jupiters would be 1–6 mag fainter than predicted by standard cooling tracks (Figure 11.12a), with observations therefore providing some diagnostics of the underlying formation mechanism. Even higher contrast ratios would be required to detect these young objects, with the resulting H_2O and NH_3 bands being extremely dark.

Chemical composition Thermodynamic calculations, of the form described in §11.2.2, are used to establish the equilibrium abundances of the elements and chemical species expected to be present in a gas of given elemental composition, temperature, and pressure. Subsequent conditions then determine which condensates form, and which gases are thereby removed such that their spectral signatures disappear.

A table of the principal condensates and gases expected in giants planets of solar composition is given in (Lodders, 2010, Table 2). Determination of the atmospheric composition of Jupiter and Saturn, using these procedures, are described by Fegley & Lodders (1994).

The atmospheric chemistry for giant exoplanets, including the condensation of particular species into clouds, is likely to be broadly similar to that for brown dwarfs of corresponding T_{eff}. One difference is the expected presence of deuterated gases (e.g., CH_3D, HDO, NH_2D) in the lower-mass planets, which is depleted by D-burning for the higher mass ($\gtrsim 13 M_J$) brown dwarfs.

The following outline, from Lodders (2010), provides some orientation amongst the detailed predictions of such calculations.

The reaction pathways of the more abundant reac-

11.7 Atmospheres of gas giants

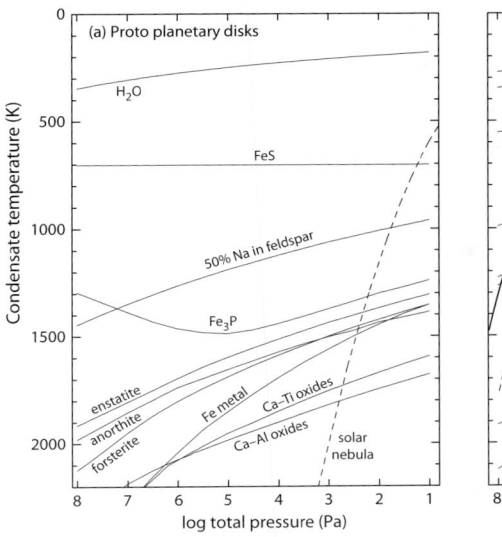

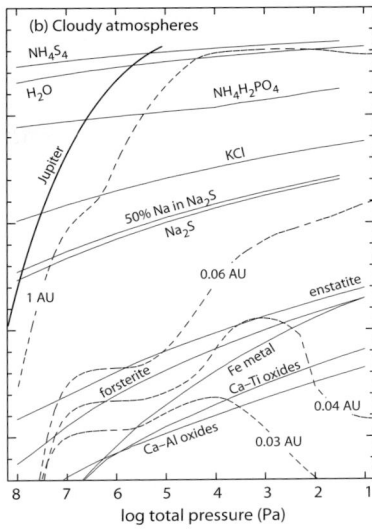

Figure 11.14: Condensates predicted according to the importance of gravitational settling: (a) in low-gravity environments such as protoplanetary disks, high-temperature condensates (e.g. F) remain dispersed within the gas, and can continue to react with it to form secondary condensates at lower temperature (e.g. FeS); (b) in high-gravity environments such as planetary atmospheres, the condensates settle into cloud layers and are removed from further reactions. There are no secondary condensates, and other species condense at lower temperature. From Lodders (2010, Figure 7).

tive elements (notably C, N, and O) influences the chemistry of the less abundant elements. The latter may form various oxides in cooling gases, depending on the availability of oxygen. The formation of many molecular gases and condensates is therefore strongly regulated by the C and O chemistry (Lodders & Fegley, 2002).

Carbon chemistry The equilibrium distribution of C, over a large range of temperature and pressure, is controlled by the reaction between CO and CH_4 through

$$CO + 3H_2 \rightleftharpoons CH_4 + H_2O. \qquad (11.26)$$

As illustrated in Figure 11.13a, at high temperatures and low pressures (e.g. $T \gtrsim 1500$ K, $P \lesssim 10^5$ Pa), CO is the major C-bearing gas, while at lower temperatures and higher pressures, CH_4 dominates. The latter conditions apply for unirradiated Jupiter-mass planets (including Jupiter), and for separations down $a \gtrsim 0.1$ AU, while the former conditions apply at smaller separations.

These predictions are modified for lower temperature formation conditions when chemical equilibrium is not reached. This results in quenching of the CO to CH_4 conversion, and an overabundance of CO compared to the expected equilibrium values. The effect is seen in Jupiter, Saturn, and Neptune (Fegley & Lodders, 1994), as well as in various T dwarfs (Fegley & Lodders, 1996; Noll et al., 1997a; Saumon et al., 2007).

Oxygen chemistry The major O bearing gases are CO and H_2O. With the presence of SiO and other metal oxides reducing the H_2O abundance, CO becomes the most abundant gas, and the conversion from CO to CH_4, which produces equal molar amounts of H_2O, therefore controls the abundance of H_2O.

As a result, the curve of equal abundances of CO and H_2O (Figure 11.13b) is at a similar location in the pressure–temperature diagram to that of the equal CO and CH_4 abundance (Figure 11.13a). In consequence, CO again dominates at higher temperatures and lower pressures, while H_2O dominates at lower temperature and higher pressures, as applicable for unirradiated Jupiter-mass planets.

The abundances of O-bearing gases are further affected by the formation of atmospheric condensates. At high temperatures, oxides and silicates lock up some 20% of the available oxygen, most importantly in forsterite (Mg_2SiO_4) and enstatite ($MgSiO_3$). The remaining oxygen at lower temperatures is mainly in H_2O vapour, which condenses into clouds at 200–300 K depending on the pressure–temperature profile.

For a $1 M_J$ planet at 1 AU, for example, the pressure–temperature profile coincides with the condensation curve of H_2O ice over a wide range (Figure 11.13b), and spectral signatures of H_2O would be expected. For Jupiter at 5 AU in contrast, H_2O clouds lie too deep in the atmosphere to influence the emergent spectra.

C/O abundance ratio Variations in the C/O ratio of a planet, compared to the value of ~0.5 in an otherwise solar composition gas, would have important consequences for the chemistry of gas giant atmospheres because the equilibrium balance between CO, CH_4, and H_2O modifies the interior evolution, strongly affecting resulting spectra (e.g. Fortney et al., 2005, 2006).

Processes which might lead to enhanced exoplanet C/O ratios (e.g., formation in H_2O-depleted or carbonaceous-enriched regions), or diminished C/O ratios (e.g., preferential accretion of H_2O ice), are considered further by Lodders (2010).

Condensate clouds and rainout The abundances of species predicted by thermodynamic equilibrium calculations are affected by the formation of condensates within the atmosphere. For a gas parcel rising and cooling adiabatically from a hot deep interior in which all gases present are well mixed, a point is reached where a condensible species condenses, and a cloud base is formed. The first species to condense, at the lowest layers, are the highly refractory oxides such as perovskite ($CaTiO_3$) and corundum (Al_2O_3), followed by various magnesium silicates, and so on through species of progressively lower volatility. As the more refractory species condense out with decreasing temperature, those species are depleted, and at least partially removed from further gaseous reactions.

Detailed results of the condensation process differ according to whether the medium is strongly gravitationally stratified (such as in exoplanet atmospheres), or exists instead as a low-gravity environment (such as in protoplanetary disks).

In the former case, condensates settle due to gravity and form cloud layers, and thus are no longer available for reaction with the cooler gases at higher altitudes. In the latter case, the condensates can remain dispersed throughout the gas, with secondary reactions still possible (Figure 11.14).

The consideration of condensing cloud layers, or *rainout*, was originally developed for the modeling of the solar system giants (Lewis, 1969; Barshay & Lewis, 1978). It was subsequently applied to models of cool dwarfs (Lodders, 1999), and to brown dwarfs and giant exoplanets (Burrows & Sharp, 1999). Inclusion of rainout is required for the accurate spectral modeling of brown dwarfs and the solar system giant planets (Fegley & Lodders, 1994; Burrows et al., 2002; Marley et al., 2002).

Expected cloud compositions for the gravitationally-settled models follow from a consideration of the abundances and volatility of the various species. The broad range of estimated gas giant equilibrium temperatures, from around 200 K in the case of 47 UMa to around 1500 K in the case of τ Boo, signifies planetary atmospheres very different from each other, and from the planetary atmospheres in the solar system, with consequences for their composition and their spectral appearance (Figure 11.15). The following outline is, again, from the more extensive description by Lodders (2010).

Condensates versus temperature The most abundant rock-forming elements Mg, Si, and Fe produce clouds of forsterite (Mg_2SiO_4), enstatite ($MgSiO_3$), and liquid iron respectively. Less abundant rock-forming elements (Ca, Al, Ti, Mn, Cr) result in less massive clouds. Lower volatility condensates composed of O, S, N and C form at $T \lesssim 200$ K.

Even for giant planets with low-temperature upper atmospheres, which are expected to have massive ice-cloud layers of H_2O (solid or liquid), NH_3, and CH_4, their deeper atmospheric layers will host clouds covering the full sequence of more refractory species. Although these deep layers are not seen directly in Jupiter, for example, the presence of iron clouds is inferred from the presence of FeS gas (Niemann et al., 1998).

Heated gas giants are dominated by TiO and VO bands in the optical, and H_2O, CO, and FeH bands in the infrared. Refractory ceramics such as corundum (Al_2O_3), Ca-aluminates, and Ca-titanates appear near 1800–2200 K, leading to the disappearance of the bands TiO and VO. For the closest (hottest) planets, temperatures are too high for the condensation of the refractory ceramics, and TiO and VO remain.

At ~1200–1500 K, depending on pressure, Fe-metal condensation, Cr/Cr_2O_3 condensation, and Mg-silicate (Mg_2SiO_4/$MgSiO_3$) condensation consume the Fe, Cr, Mg, and Si gases, removing all of the major rock-forming elements from the atmosphere above the $MgSiO_3$ cloud. Lines of monatomic alkali metals (K, Na, Cs, Rb) will persist to lower temperatures, gradually converting to NaCl, KCl, oxides, hydroxides, and hydrides, before being removed as sulphide and halide condensates.

Decreasing temperatures, leading to the disappearance of CO and the appearance of CH_4 (Figure 11.13), results in CH_4 absorption bands at 1.6, 2.2, and 3.3 μm.

Thereafter, CH_4, H_2O, and NH_3 characterise the cooler objects until the condensation of H_2O into liquid or solid form, and the condensation of NH_3 into solid form, leaves only CH_4 in an otherwise H/He-rich atmosphere. Condensation of CH_4 only occurs at very low temperatures, $T \lesssim 50$ K. It is the high-level NH_3 clouds, along with various by-products of photolytic dissociation, that reflect sunlight back from Jupiter.

Temperature inversions or stratospheres Under externally irradiated conditions where more incident energy is absorbed in the outer atmosphere than can be emitted by the cooler isothermal profile below, the absorbing layer heats up until its thermal emission balances the absorbed incident energy. This creates a *temperature inversion layer* or *stratosphere*. The *tropopause* occurs at the altitude where the warmer radiative upper layer meets the cooler convective troposphere, with its decreasing temperature with height. Ozone absorption is responsible for the Earth's stratosphere, while CH_4 and other gases are responsible for the stratospheres of

11.7 Atmospheres of gas giants

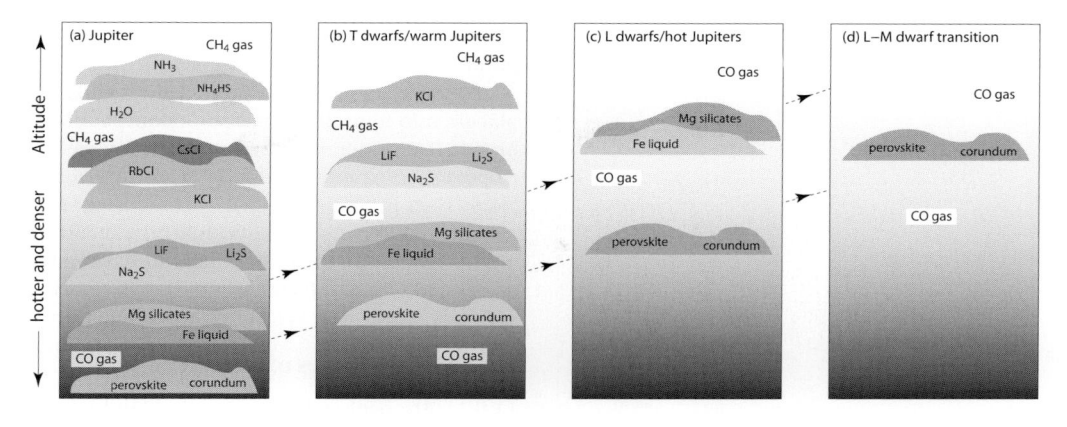

Figure 11.15: Schematic of cloud-like condensations versus altitude expected in exoplanets and brown dwarfs of various temperatures: (a) Jupiter; (b) T dwarfs and Jupiters too warm for H_2O clouds to form; (c) L dwarfs and hot Jupiters; (d) L–M dwarf transition objects. The overall structures are similar, but cloud levels are higher in the warmer atmospheres for the refractory species, while the more volatile species may not condense at all as the temperature increases. From Lodders (2004a, Figure 1).

the solar system giants (Figure 11.16).

As predicted by atmospheric modeling, and as evident for separations $\lesssim 0.04$ AU in Figure 11.10, such inversions have been identified through photometry and transmission spectroscopy for a number of strongly irradiated gas giants, e.g. for HD 209458 b and TrES–4 b (discussed further in §6.5). TiO and VO are generally considered to be the species providing the high-altitude opacity necessary to generate this temperature inversion (Hubeny et al., 2003; Fortney et al., 2008a), although this has been questioned (Spiegel et al., 2009b).

Photochemistry and photolysis The term *photochemistry* refers to all aspects of the interaction between light and atoms or molecules, including ionisation and isomerisation.[4] *Photolysis* (or photolytic dissociation) concerns more specifically the breakdown of molecules as a result of photon interactions. Atmospheric photochemistry is driven by the more energetic stellar photons (ultraviolet, extreme ultraviolet and, in principle, X-rays and γ-rays) and, in a more liberal use of the term, energetic electrons.

Whether a particular photochemical reaction is important in a given planet or exoplanet atmosphere depends on the penetration depth of the energetic photon, and whether it reaches the relevant gaseous or condensable species. Atmospheric dissociation by the absorption of ultraviolet radiation is usually therefore of relevance only high in the atmosphere before the bulk of the incident radiation is absorbed or scattered back to space

(Yung & Demore, 1999). As examples

$$H_2 + h\nu \; (\lambda < 84.5\,\text{nm}) \; \rightarrow \; H + H$$
$$CH_4 + h\nu \; (\lambda < 145\,\text{nm}) \; \rightarrow \; CH_3 + H$$
$$NH_3 + h\nu \; (\lambda < 230\,\text{nm}) \; \rightarrow \; NH_2 + H$$
$$O_2 + h\nu \; (\lambda < 240\,\text{nm}) \; \rightarrow \; O + O \qquad (11.27)$$

with the latter being the first step in the formation of ozone (O_3) in the Earth's stratosphere. More details of the complex reaction pathways, photochemical products, and circumstances of relevance in the different solar system atmospheres are given by Strobel (2005).

More generally, photolytic dissociation results in photochemical products which can contribute to atmospheric radiative transfer, participate in further chemical reactions, and accelerate hydrodynamic escape. The latter will, in turn, facilitate detectability of a planetary atmosphere which will, in consequence, vary – in some cases very significantly – according to stellar type and evolutionary phase (Rybicki, 2006).

In Jupiter, molecular species such as H_2O, H_2S, and NH_3 are condensed below other cloud layers, and so are unavailable for photodissociation. In hot Jupiter atmospheres, in contrast, these species are gaseous, and no longer protected (Marley, 1998; Liang et al., 2004). While photochemical hydrocarbon aerosols have been shown to be insignificant (Liang et al., 2004), S and N compounds may be important in hot Jupiter photochemistry and perhaps in haze production (Marley et al., 2007a). Photolytic compounds including CO_2, HCN, and C_2H_6 may be present in GJ 229 B and, by analogy, in H_2O-bearing gas giant atmospheres (Troyer et al., 2007).

Photolytic dissociation may be a significant complication in characterising terrestrial atmospheres according to their colours (e.g. Pallé et al., 2008b). An Earth-like planet with a higher abundance of CH_4 may be en-

---

[4] An example of fundamental importance, although not to planet formation, is the torsional isomerisation of rhodopsin. This first stage of photon reception in the human eye, and all other eyes, is a process which relaxes in $\lesssim 200$ fs (Schoenlein et al., 1991).

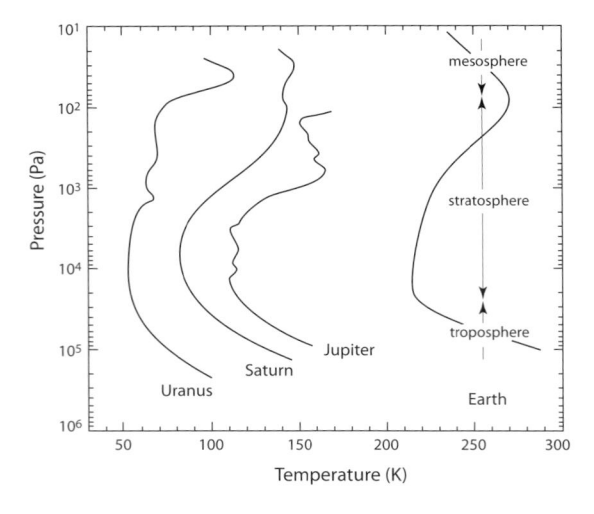

Figure 11.16: Atmospheric pressure–temperature profiles for Uranus, Saturn, Jupiter and Earth. All have deep adiabatic tropospheres and stratospheres, characterised by temperature inversions (for Earth, between ~ 10^2–10^4 Pa). Uranus and Jupiter data are from the Voyager radio science occultation experiments, and the Earth profile is from the 1976 US Standard Atmosphere (NASA, 1976). Adapted from Marley (2010, Figure 1), with the Saturn profile provided by Marley (2010, priv. comm.).

veloped in a photochemical haze, and unrecognisable as an ocean- or vegetation-dominated planet (Zahnle, 2008). On the other hand, photodissociation of H_2O may make water in terrestrial or ocean planet atmospheres detectable through ultraviolet absorption in the Lyman lines of atomic hydrogen (Jura, 2004).

Albedos The *albedo* measures the reflectivity of the planet's atmosphere resulting from the totality of absorption and scattering processes taking place within it. The two variants of most importance in exoplanet studies are the *geometric albedo* and the *Bond albedo*.

The geometric albedo, $p(\lambda)$, is the ratio of brightness at zero phase (i.e., seen from the star) to that of a fully reflecting, diffusively scattering (Lambertian) flat disk with the same cross-section. It is a function of wavelength, relevant for the detection of reflected light, and described further in §6.4.7.

The Bond albedo, A_B, measures the fraction of incident radiation, over all wavelengths, which is scattered. It appears in the expression for the planet's equilibrium temperature and is described further, along with present observational constraints, in §6.5.

Albedos are strongly influenced by the presence of clouds and *aerosols* (fine solid particles or liquid droplets in suspension). Planets with significant cloud cover are expected to reflect stellar light more strongly (high albedo), hence reducing their infrared luminosity, but increasing the amount of reflected stellar radiation. When only rare species condense, more light will be absorbed (low albedo), and the effective temperature and infrared luminosity will be higher.

Sudarsky et al. (2000) proposed dividing the giant planets into four albedo classes, corresponding to four broad effective temperature ranges: a 'Jovian' class with tropospheric ammonia clouds ($T_{\rm eff} \lesssim 150$ K); a 'water cloud' class primarily affected by condensed H_2O ($T_{\rm eff} \sim 250$ K); a clear class lacking cloud ($T_{\rm eff} \gtrsim 350$ K); and a high-temperature class for which alkali metal absorption predominates ($T_{\rm eff} \gtrsim 900$ K).

11.8 Atmospheres of terrestrial planets

11.8.1 Atmospheric formation

Three sources have been identified as the possible origin of the atmospheres of the terrestrial planets: the capture of nebular volatiles, outgassing during accretion, and outgassing from later tectonic activity. The specific instance of the Earth's atmosphere is considered further in §12.8.

Capture of nebular gases In the core accretion model of gas giant formation, the capture of nebular gases is a fundamental process underlying the later stages of their growth. While nebula gas accretion has been considered as, at least, a partial contribution to the atmospheres of the terrestrial planets (e.g. Hayashi et al., 1979; Pollack & Black, 1982; Cameron, 1983; Ikoma & Genda, 2006), its overall significance remains unproven (Elkins-Tanton & Seager, 2008b; Zahnle et al., 2010). The argument underlying this view is that low-mass terrestrial planets are unable to capture and retain nebula gases during their early accretion phase, while nebular gases may have largely dissipated from the inner solar system by the time of final stages of planetary accretion.

Outgassing through accretion The terrestrial planets are believed to have formed within the solar nebula's snow line, where H_2O was not condensed. Theories of the origin of water on Earth (§12.5) consider instead that H_2O was acquired through impacts of comets and carbonaceous asteroids (e.g. Abe & Matsui, 1985; Matsui & Abe, 1986; Chyba, 1990), along with other surface volatiles such as CH_4 (Butterworth et al., 2004; Court & Sephton, 2009; Sephton & Court, 2010). In the primitive class of chondritic meteorites (box on page 297) 'water' is mostly in the form of OH within silicate minerals – it is referred to as water since it most probably existed as H_2O when the mineral was formed, and is released as H_2O if the mineral melts.

On a given body, the efficiency of impact accretion and H_2O outgassing depends on the competition between impact delivery of new volatiles, and the impact erosion of those already present. For the inner solar system planets, Chyba (1990) argued that the net accumulation of planetary oceans was strongly favoured.

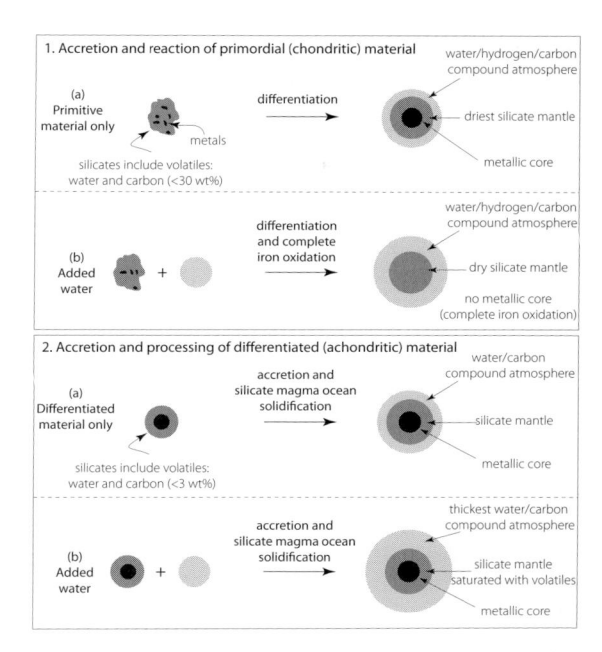

Figure 11.17: Schematic of the outgassing models of Elkins-Tanton & Seager (2008b). (1a) metallic Fe reacts with H_2O to form H and FeO_X, until the metallic Fe or H_2O is exhausted, resulting in an atmosphere of H and C compounds only. (1b) H_2O is added from an external source until all metallic Fe is oxidised, resulting a H_2O/H mixture, and C compounds. (2a) a molten achondritic silicate mantle solidifies, partitioning volatiles between silicates, liquids, and an atmosphere of H_2O and C compounds. (2b) H_2O is added, leading to solidification of the magma ocean, and an atmosphere of H_2O and C compounds. In all cases, traces of He and N are found in the atmosphere. Adapted from Elkins-Tanton & Seager (2008b, Figure 1), reproduced by permission of the AAS.

Elkins-Tanton & Seager (2008b) modeled outgassing during accretion in order to estimate the range of atmospheric mass and composition likely to exist on exoplanets ranging from $1 - 30 M_\oplus$. They assumed that the material being accreted is either primitive (chondritic), or differentiated (achondritic). In each of the two cases, additional water and other volatiles is, or is not, accreted at the same time, giving four model configurations (Figure 11.17). This results in oxidising conditions during accretion, or reducing conditions during accretion, respectively. The latter can produce a planet with a metallic core and water-rich atmosphere, while the former can produce planets without a metallic core but with a hydrogen-rich atmosphere.

The models of Elkins-Tanton & Seager (2008a) predict two types of planet not present in the solar system: those consisting of silicate rock with no metallic core, and those with a deep water surface layer resulting directly from its initial solidification.

Their models suggest that initial atmospheres range from below 1% to over 20% of the planet's total mass,
with H_2O and C compounds dominating. H amounts to less than 6% of the planet mass, while He is found only in traces. Nitrogen, although in very low concentration in the accretion material, appears in sufficient quantity to build the N_2-based atmosphere of the Earth.

Outgassing through tectonic activity Later in the life of the planet, internal convection induced by heating (either primordial, radiogenic, or tidal) can augment the outgassing of volatiles locked up in the mantle. Papuc & Davies (2008) estimated that, over a wide range of planetary masses, volcanic activity outgassing may provide an atmosphere as massive as the Earth's in a few Gyr.[5]

Kite et al. (2008) estimated volcanism versus time for planets with Earth-like composition and masses up to $25 M_\oplus$, as a contribution toward predicting atmospheric mass on terrestrial exoplanets. They found that (1) volcanism is likely to proceed on massive planets with plate tectonics over the main-sequence lifetime of the host star; (2) plate tectonics may not operate on high-mass planets because of their buoyant crust which is difficult to subduct; and (3) melting is necessary but insufficient for efficient volcanic degassing; volatiles partition into the earliest, deepest melts, which may be denser than the residue and sink to the base of the mantle on young, massive planets. Magma must also crystalise at or near the surface, and the pressure of overlying volatiles must be fairly low, if volatiles are to reach the surface.

11.8.2 Atmospheric erosion

Various processes can erode an existing atmosphere.

Planetesimal erosion The same type of accreting impact responsible for the build-up of volatiles on the terrestrial planets can also erode an existing atmosphere. Of most importance is the effect of the late stages of residual planetesimal impacts, which may result in significant or even catastrophic ablation, either through multiple small impacts, or one or more larger impacts. The efficiency of this atmospheric ejection is related to

---

[5]There is a connection between theories of planetary formation, and more speculative theories of the origin of oil and 'natural gas' (primarily CH_4) in the Earth's crust. In the Russian–Ukrainian theory of deep abiogenic hydrocarbon formation (originating with Mendeleev 1877, and Kudryavtsev 1951), petroleum is considered not as a fossil fuel, but as a primordial material erupted at the Earth's surface from great depth. The *deep gas theory* (Gold, 1979, 1985, 1993; Gold & Soter, 1980), which differs in details rather than in principles, invokes deep faults as the pathway for the continuous migration of primordial CH_4 and other gases (including He) from the Earth's upper mantle to the surface, where they are converted into higher hydrocarbons (oil and gas) in the upper crust. A review of both theories is given by Glasby (2006). While a biogenic origin for both is widely favoured, ongoing research suggests that at least some abiogenic formation is plausible (e.g. Scott et al., 2004; Proskurowski et al., 2008; Kolesnikov et al., 2009).

the mass of the impactor and the atmospheric scale height (O'Keefe & Ahrens, 1982; Ahrens, 1993; Newman et al., 1999).

Prescriptions for the impactor flux in the early solar system, based on observed cratering on the surface of Mars and the Moon, have been given by Tremaine & Dones (1993) and Melosh & Vickery (1989). The latter are based on the Schmidt–Holsapple scaling law for silicate projectiles (Schmidt & Housen, 1987), which equates the impact flux to a crater density assuming an average impact velocity of $20\,\text{km}\,\text{s}^{-1}$, mean impact angle of $45°$, and a rim-to-rim diameter 25% larger than the apparent transient crater diameter (Figure 11.18).

In the form given by Kuchner (2003, eqn 5), and based on the solar system model, the number of projectiles per year of mass greater than m_p that fall on a planet of (effective gravitational) radius R_p is

$$N = 0.25\,\text{yr}^{-1}\left(\frac{R_\text{p}}{R_\oplus}\right)^2 \left(1 + 2300\,e^{-t/2.2\times 10^8\,\text{yr}}\right)$$
$$\times \left(\frac{m_\text{p}}{1\,\text{kg}}\right)^{-0.47}. \qquad (11.28)$$

Considering impactors with masses between 10^{15} kg and $0.01 M_\text{Moon}$ on a planet with $R_\text{p} = R_\oplus$, Kuchner (2003) derived a total impact mass of $< 0.003 M_\oplus$, allowing the survival of even an $0.1 M_\oplus$ volatile-rich planet.

This size distribution model, however, fails for very high-mass impactors (Tremaine & Dones, 1993). Such giant impacts probably created the Earth's Moon (§12.6), may have spun up Mars (Dones & Tremaine, 1993a), stripped Mercury's silicate mantle (Benz et al., 1988), and largely removed (in the case of Mars) or diminished (in the case of Venus) the primordial atmospheres of some of the terrestrial planets (Cameron, 1983).

Further evidence for collisional atmospheric erosion may come from the radiogenic and primordial noble gas content of the Venus, Earth and Mars atmospheres. These show a several order-of-magnitude decrease in ^{20}Ne and ^{36}Ar (but a nearly constant ratio) in progressing from Venus to Mars. For example, the abundances of ^{36}Ar in the atmospheres are, in relative mass (kg per kg of planet) 2.5×10^{-9} for Venus, 3.5×10^{-11} for Earth, and 2.1×10^{-13} for Mars.

These and other noble gas trends were originally attributed to grain accretion, and were assumed to reflect conditions in the solar nebula at the time of formation of the planetary atmospheres, either related to the composition of the nebula, or the composition of the trapped gases in small nebular bodies (Pollack & Black, 1979; Wetherill, 1980; McElroy & Prather, 1981; Pollack & Black, 1982). But Cameron (1983) argued that the Earth lost essentially all of its primordial Ar due to the collisional Moon-forming event, and that the amount now present represents that brought in by late accretion.

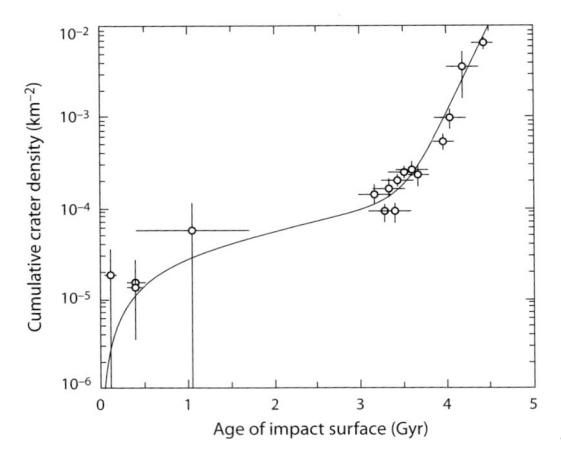

Figure 11.18: Comparison between the lunar impact crater density, and that predicted by the impactor model of Melosh & Vickery (1989). The model yields a cumulative density for craters exceeding 4 km diameter of $N(>4\,\text{km}) = 2.68\times 10^5 [T + 4.57\times 10^{-7}(e^{\lambda T} - 1)]$, where T is the age of the cratered surface, and $\lambda = 4.53\,\text{Gyr}^{-1}$. Crater density data is from Kaula et al. (1981, Table 8.4.2). From Melosh & Vickery (1989, Figure 2), by permission from Macmillan Publishers Ltd, Nature, ©1989.

Jeans escape In *Jeans escape*, or Jeans evaporation, individual molecules from the high–velocity tail of the thermal Maxwell–Boltzmann distribution may reach escape velocity and overcome the planet's gravity field.

The mechanism defines the uppermost layer of the atmosphere, the *exosphere*. Its lower boundary, the *exobase* or thermopause, is the height at which the molecular mean free path is equal to one pressure scale height, such that upward traveling particles are on a ballistic trajectory. The exosphere is the transition zone between the atmosphere and interplanetary space, with the altitude of the Earth's exobase ranging from 250–500 km depending on solar activity.

The relevant dimensionless parameter describing the atmospheric loss is the Jeans parameter (which is proportional to the mass of the atom or molecule)

$$\lambda_\text{J} = \frac{v_\text{esc}^2}{v_\text{p}^2}, \qquad (11.29)$$

where v_p is the most probable velocity of the distribution, and v_esc is the escape velocity at the exobase. Thermal escape becomes important for $\lambda_\text{J} \lesssim 10$.

For the high-gravity cold solar system gas giants, Jeans escape is insignificant, with λ_J at their exobases of 480, 420, 50, and 120 for Jupiter, Saturn, Uranus and Neptune (Strobel, 2002; Hunten, 2002). For planets of smaller mass and higher temperature, such as Earth, only light atoms escape (e.g. Irwin, 2006, Table 3.1).

Soon after the discovery of the first hot Jupiter, 51 Peg, orbiting 100 times closer than Jupiter, Guillot et al. (1996) showed that the planet is stable to classical Jeans escape, and also to photodissociation and EUV-induced mass loss, even under extreme irradiation conditions assuming tidally locking. They estimated a mass loss rate of $10^{-16} M_\odot \text{ yr}^{-1}$, or $10^{-4} M_J$ over 1 Gyr.

Effects of X-ray irradiation on mass loss have been considered separately for G dwarfs (Penz et al., 2008), and M dwarfs (Penz & Micela, 2008).

Hydrodynamic escape In *hydrodynamic escape*, a thermally-driven Jeans escape of light atoms or molecules can carry heavier species along with it through collisional drag (e.g. Shizgal & Arkos, 1996). The mechanism requires a continuous energy source at altitude to maintain the high-velocity tail.

Thermally sustained by planetesimal accretion during early formation, hydrodynamic escape has also been invoked to explain the anomalous noble gas isotope ratios of Venus, Earth, and Mars (Zahnle et al., 1990).

The problem is compounded by EUV heating, conduction, and tidal forces (Watson et al., 1981; Lammer et al., 2003; Lecavelier des Etangs et al., 2004; Lammer et al., 2004; Yelle, 2004; Lammer et al., 2008; Yelle et al., 2008; Tian, 2009).

Irradiation may be relevant even at large orbital separations. Over Pluto's lifetime, the solar EUV-driven wind may have led to the erosion of several km of Pluto's surface ice (Watson et al., 1981; Hunten & Watson, 1982; Trafton et al., 1997), and a small fraction of its total mass through N_2 escape (Tian & Toon, 2005).

Kuchner (2003) evaluated the survival time against volatile atmospheric escape as a function of planetary mass and orbital distance, and found that an Earth-mass volatile-rich planet even at 0.3 AU from a Sun-like star can retain its volatiles for the age of the solar system. Nevertheless, the steady atmospheric erosion means that such ocean planets should be more common around young stars, and stars with lower time-averaged EUV luminosity. Compared to the FGK type main sequence stars, M stars would be particularly hospitable to close-in volatile Earth-mass planets.

Photolytic dissociation The loss of volatiles by Jeans or hydrodynamic escape may be further accelerated by photolytic dissociation under irradiation by ultraviolet light (§11.7). For example, NH_3 is expected to be converted into N_2 and H_2 in less than 2 Gyr for a planet located at 1 AU from a G2V star (Léger et al., 2004). Dissociation of H_2O in a transiting terrestrial or ocean planet may be detectable through ultraviolet absorption in the Lyman lines of atomic hydrogen (Jura, 2004).

Vigorous stellar activity adds to the depletion, with coronal mass ejections in M stars possibly removing atmospheres in the habitable zone (Lammer et al., 2007; West et al., 2008). With its low surface gravity, impact erosion by the solar wind may have reduced the atmosphere of Mars from an initial pressure of 10^5 Pa to its current CO_2-dominated mean surface level pressure of ~600 Pa over the lifetime of the solar system (Melosh & Vickery, 1989). Without sufficient shielding by a magnetosphere, an Earth-like planet with $a < 0.2$ AU could lose its atmosphere in ~1 Gyr (Lammer et al., 2007).

Stripping of hot Jupiters and Neptunes For close orbiting systems in the range 0.01–0.1 AU, significant gas escape through hydrodynamic escape is predicted, although at rates that may still leave atmospheres stable for several Gyr (Guillot et al., 1996; Yelle, 2004; Tian et al., 2005). Observations of HD 209458 b, for example, indicate the escape of H I (Vidal-Madjar et al., 2003; Ehrenreich et al., 2008) as well as O I and C II (Vidal-Madjar et al., 2004), while still leaving the strongly irradiated planet stable against hydrodynamic escape.

Significant atmospheric evaporation may be compounded by inward migration and Roche lobe overflow (Trilling et al., 1998, 2002; Jaritz et al., 2005; Erkaev et al., 2007), with a resulting inflation of the outer atmospheric layers (Baraffe et al., 2004; Hubbard et al., 2007).

Strong evaporation and/or tidal stripping may be responsible for the formation of hot-Neptunes such as GJ 436 b and 55 Cnc e, and it is possible that the more gaseous hot Jupiters and the higher-density hot Neptunes may share the same origin and evolution history (Baraffe et al., 2005, 2006).

CoRoT–7 b is another example of a giant planet that may have been largely stripped of its H/He envelope due to the proximity to its host star. The remaining rocky or metallic core would resemble a terrestrial planet, and the hypothetical class of stripped giants has been termed a *chthonian planet* (Hébrard et al., 2004).

11.8.3 Atmospheres of ejected planets

Although there is no observational evidence for the existence of true free-floating planets, present models suggest that rock and ice embryos, including those with $M_p \sim 1 M_\oplus$, may be ejected from a young planetary system by gravitational scattering. Even in more mature architectures like the solar system, terrestrial-type planets may still be ejected (§12.6). Studies have addressed the atmospheric properties of such ejecta, and in particular whether they might support life.

Isolated planets Stevenson (1999) has reasoned that planets ejected into interstellar space may survive, with atmospheres, for billions of years.

An Earth-mass body develops an atmosphere with $M_{atm}/M_p \sim 0.01$ assuming a pressure-induced opacity of hydrogen (Birnbaum et al., 1996), or $M_{atm}/M_p \sim 0.001$ for more opaque models (Stevenson, 1982). The atmospheric escape time after nebula clearing may be as short as 1 Myr at 1 AU, but several Gyr in the interstellar medium if *sputtering*, collisions with high-velocity interstellar atomic or molecular hydrogen, is also small.

An interstellar Earth-like planet 4.6 Gyr after formation would have a long-lived radionuclide-driven luminosity of $4 \times 10^{13} (M_p/M_\oplus)$ W (Stacey & Davis, 2008).

Table 11.4: *A comparison of various planet heat sources. Earth's internal heat comes from a combination of residual heat from planetary accretion (about 20%), the balance largely from radioactive decay (Turcotte & Schubert, 2002, p136). Currently* ^{40}K, ^{238}U, ^{235}U, *and* ^{232}Th *are the primary radiogenic heat sources in the Earth (Stacey & Davis, 2008); heat generation now is 2–3 times less than in the Archean era (3.8–2.0 Gyr). For the ejected systems, energies are typical tidal heat dissipations from the simulations of Debes & Sigurdsson (2007).*

Heat source	Energy (W)
Solar insolation (Earth, present day)	1.5×10^{17}
Radiogenic heating (Earth, present day)	4×10^{13}
Radiogenic heating (Earth, Archean era)	1×10^{14}
Tidal heating on Io (Veeder et al., 1994)	1×10^{14}
Ejected Earth–Moon (tidal, rocky)	4.2×10^{15}
Ejected Earth–Moon (tidal, icy)	2.1×10^{16}

From consideration of the atmospheric pressure and opacity, Stevenson (1999) derived a surface temperature

$$T \sim 425 \left(\frac{M_p}{M_\oplus}\right)^{1/12} \left(\frac{(M_{atm}/M_p)}{0.001}\right)^{0.36} \text{K}, \quad (11.30)$$

with $T > 273$ K for basal pressures of 10^8 Pa, implying that bodies with liquid oceans are possible in interstellar space. They may have volcanic cores, and dynamo-driven magnetic fields with large magnetospheres.

Planet–satellite pairs In subsequent simulations, Debes & Sigurdsson (2007) found that interactions between giant planets and terrestrial-sized protoplanets with lunar-sized satellites led to 3.3% cases in which only the Earth-mass planet is ejected, while 4.6% ended in the ejection of a bound Earth–Moon type system. In most cases the eccentricity of the bound pair increases in the process, thereafter supplying heating of the planet through tidal circularisation, and augmenting that originating from interior radionuclides.

Similar tidal heating has been investigated in relation to the thermal history of the Moon (Peale & Cassen, 1978), vulcanism on Io (Peale et al., 1979), the heating of Europa (Cassen et al., 1979; Carr et al., 1998), and in assessing the long-term fate of the solar system (Laughlin & Adams, 2000).

Debes & Sigurdsson (2007) give the following expressions for tidal heating arising from circularisation of the satellite's eccentric orbit, and from synchronisation between the planet's spin and the satellite's mean motion, respectively

$$\dot{E}_{circ} = -\frac{63 e^2 n}{4 \tilde{\mu} Q_p} \left(\frac{R_p}{a}\right)^5 \frac{G M_s^2}{a}, \quad (11.31)$$

$$\dot{E}_{sync} = -(\omega - n) \frac{3 k_{2p}}{Q_p} \frac{M_s^2}{M_s + M_p} \left(\frac{R_p}{a}\right)^5 n^2 a^2 (\omega - n), (11.32)$$

where n is the mean motion, M_s is the satellite mass, M_p is the planet mass, $\tilde{\mu}$ is the ratio of elastic to gravitational forces (the effective rigidity of the planet), k_{2p} is the planet's Love number, ω its orbital rotation, and Q_p its specific dissipation function. For a rocky planet they assume $k_{2p} = 0.299$, $\tilde{\mu} = 4$, $Q_p = 12$. For an icy planet they assume $k_{2p} = 0.7$, $\tilde{\mu} = 1$, $Q_p = 100$. A comparison of the various resulting heat sources is given in Table 11.4.

With various coarse assumptions, they estimate a local space density of such ejected pairs as $\sim 10^{-3}$ pc^{-3}. Their detection would be difficult based on thermal radiation expected to peak at around 80 μm, or on their non-thermal radio emission, but more plausible based on gravitational microlensing.

The tidal heat production declines with time. Debes & Sigurdsson (2007) estimate that it drops to radiogenic levels after 140 Myr for a rocky planet, and after 246 Myr for an icy planet, perhaps time enough *'for life to arise and adapt to the decreasing temperatures'*.

11.9 Habitability

The search for other planets is partly motivated by efforts to understand their formation and, by analogy, to gain an improved understanding of the formation of the solar system. Search accuracies are expected to improve to the point that the detection of terrestrial planets in the 'habitable zone' will become more feasible if not routine, and there is presently no reason to assume that such planets do not exist in very large numbers.

Improvements in spectroscopy, from the ground or from space, and developments of atmospheric modeling, will presumably lead to searches for planets which are progressively habitable, inhabited by micro-organisms, and ultimately by intelligent life (these searches may or may not prove fruitless). Search strategies will be assisted by improved understanding of the conditions required for development of life on Earth (e.g. Gogarten, 1998; Des Marais, 1998; McKay, 1998) combined with observational feasibility (e.g. Schneider, 1994; Mariotti et al., 1997; Léger, 1999).

The new discipline of *astrobiology*, the study of the origin, evolution, distribution, and future of life in the Universe, encompasses the search for habitable environments in the solar system and beyond, and for biospheres that might be very different from Earth's. It is a cross-disciplinary effort, with knowledge being assimilated from astronomy, biology, chemistry, geography, geology, physics, and planetary science (Chyba & Hand, 2005; Impey, 2010).

The discovery of exoplanets catalysed a number of new astrobiology initiatives (e.g. Cowan et al., 1999; Des Marais et al., 2003, 2008), along with various conferences on the search for life (e.g. Cosmovici et al., 1997; Des Marais, 1997; Woodward et al., 1998), beginning to quantify philosophical debate that has been ongoing for centuries (Crowe, 1986; Dick, 1996).

This section touches only on some of the very broad-

11.9 Habitability

> **NASA and ESA astrobiology roadmaps:** NASA's astrobiology roadmap addresses three basic questions: how does life begin and evolve, does life exist elsewhere in the Universe, and what is the future of life on Earth and beyond? The key domains of investigation are summarised by Des Marais et al. (2008) as (listed verbatim):
>
> (1) understand the nature and distribution of habitable environments in the Universe. Determine the potential for habitable planets beyond the solar system, and characterise those that are observable;
>
> (2) determine any past or present habitable environments, prebiotic chemistry, and signs of life elsewhere in our solar system. Determine the history of any environments having liquid water, chemical ingredients, and energy sources that might have sustained living systems.
>
> (3) understand how life emerges from cosmic and planetary precursors. Perform observational, experimental, and theoretical investigations to understand the general physical and chemical principles underlying the origins of life;
>
> (4) understand how life on Earth and its planetary environment have co-evolved through geological time. Investigate the evolving relationships between Earth and its biota by integrating evidence from the geosciences and biosciences that shows how life evolved, responded to environmental change, and modified environmental conditions on a planetary scale;
>
> (5) understand the evolutionary mechanisms and environmental limits of life. Determine the molecular, genetic, and biochemical mechanisms that control and limit evolution, metabolic diversity, and acclimatisation of life;
>
> (6) understand the principles that will shape the future of life, both on Earth and beyond. Elucidate the drivers and effects of microbial ecosystem change as a basis for forecasting future changes on time scales ranging from decades to millions of years, and explore the potential for microbial life to survive and evolve in environments beyond Earth, especially regarding aspects relevant to US space policy;
>
> (7) determine how to recognise signatures of life on other worlds and on early Earth. Identify biosignatures that can reveal and characterise past or present life in ancient samples from Earth, extraterrestrial samples measured *in situ* or returned to Earth, and remotely measured planetary atmospheres and surfaces. Identify biosignatures of distant technologies.
>
> ESA's roadmap for the detection and characterisation of other Earths is described by Fridlund et al. (2010a).

est considerations. A recent review of the wide variety of considerations for planet habitability is given by Lammer et al. (2009a).

11.9.1 The habitable zone

Assessment of the suitability of a planet for supporting life is largely based on the knowledge of life on Earth. With the general consensus among biologists that carbon-based life requires water for its self-sustaining chemical reactions (Owen, 1980), the search for habitable planets has focused on identifying environments in which liquid water is stable over billions of years. At the same time, while carbon is central to life on Earth, and one of the most abundant elements in the solar system, speculation that other life may exist has long been posited (e.g. Cuntz & Williams, 2006).

So guided, the *habitable zone* is loosely defined by the range of distances from a star where liquid water can exist on the planet's surface (Huang, 1959, 1960; Dole, 1964; Hart, 1979; Kasting et al., 1993; Kasting, 1996; Williams et al., 1997; Kasting, 2008). Primarily controlled by the star–planet separation (Figure 11.19), it is also affected by orbital eccentricity, planet rotation, heat sources other than stellar irradiation, and atmospheric properties including circulation.

The latter point merits particular emphasis, since a planet's equilibrium temperature defined on the basis of the incidence of stellar radiation will be increased, often significantly, by the trapping of radiation within its atmosphere. This includes contributions both from the so-called 'greenhouse gases' and, in the case of gas giants, from the pressure-induced far-infrared opacity of H_2, where even for effective temperatures as low as 30 K, atmospheric basal temperatures can exceed the melting point of water (Stevenson, 1999)

The habitable zone will be extended outwards by other atmospheric effects including stellar X-ray heating due to photoionisation and Compton scattering (Cecchi-Pestellini et al., 2006), and additional internal heat sources including long-lived radionuclides (U^{235}, U^{238}, K^{40} etc., as on Earth, cf. Heppenheimer 1978), or tidal heating due to gravitational interactions as in the case of Jupiter's moon Io.

For Earth-like planets orbiting main-sequence stars, Kasting (1988) found an almost constant inner boundary defined by water loss and the runaway greenhouse effect, as exemplified by the CO_2-rich atmosphere and resulting temperature of Venus (Rasool & de Bergh, 1970), and a more extended outer boundary determined by the onset of CO_2 condensation, increasing planetary albedo, and the ensuing runaway glaciation (Figure 11.20).

These considerations result, for a $1\,M_\odot$ star, in an inner habitability boundary at about 0.75–0.95 AU and an outer boundary at around 1.37–1.77 AU (Kasting et al., 1993).

Continuously habitable zone The habitable zone evolves outwards with time, and according to stellar mass, because of the increasing stellar luminosity with age. This results in a narrower width of the *continuously habitable zone*, again loosely defined as the range of orbital distances over which liquid water could have existed continuously over the ill-defined period of time necessary for the evolution of life (Hart, 1978, 1979; Underwood et al., 2003).

In the case of the Sun, the continuously habitable zone over the past ~4 Gyr is around 0.95–1.15 AU. Positive feedback due to the greenhouse effect and plane-

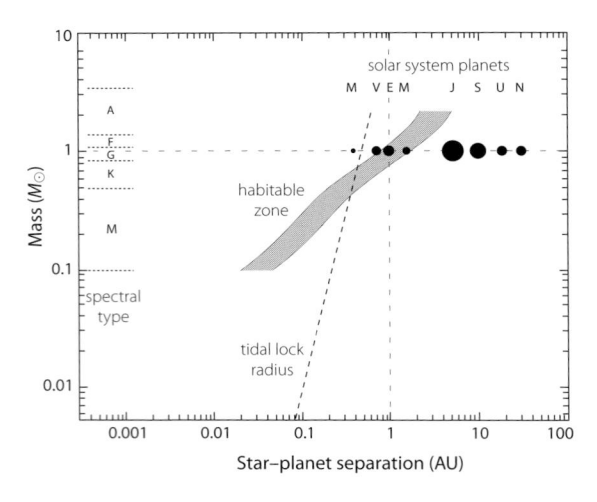

Figure 11.19: *The habitable zone, within which a planet can maintain liquid water on its surface as a consequence of the star's (zero-age main sequence) mass and luminosity. Earth sits within the habitable zone, while Venus and Mars are located at its edges. The dotted line indicates the distance at which an Earth-like planet in a circular orbit would be in synchronous rotation within 4.5 Gyr as a result of tidal damping. Earth-like planets within the habitable zone of an M star would be within this tidal-lock radius. From Kasting et al. (1993, Figure 16), with permission from Elsevier.*

tary albedo variations, and negative feedback due to the link between atmospheric CO_2 level and surface temperature, via the carbonate–silicate cycle, modify these boundaries further (Kasting, 1988; Kasting et al., 1993; Kasting, 1996).

While Earth's habitability over early geological time is complex and incompletely understood (§12.8), its climate has remained conducive to life for the past 3.5 Gyr years or more, perhaps since as early as 10–20 Myr after the Moon-forming impact (Hart, 1978; Martin et al., 2006a,b; Zahnle et al., 2007), despite a large increase in solar luminosity over this period. Previous higher concentrations of CO_2 and/or CH_4, combined with negative feedback loops, may have helped to stabilise the climate (Kasting & Catling, 2003, see also §11.8.1). It is believed that such effects were sufficient to accommodate the 30% increase in the Sun's luminosity over the last 4.6 Gyr needed in order to sustain the presence of liquid water evident from geological records (Kasting, 1996; Lean & Rind, 1998; Lunine, 1999a,b).

In the future, the Sun will increase to roughly three times its present luminosity by the time it leaves the main sequence, in about 5 Gyr. Over the forthcoming 1.1 Gyr, the Sun's luminosity will increase by a further 10%, and the Earth may leave the continuously habitable zone some 500–900 Myr from now (Caldeira & Kasting, 1992).

Searches for water Various searches for H_2O in other exoplanet systems can in principle be undertaken. An unsuccessful search for H_2O masers towards 18 exoplanets using the Australia Telescope Compact Array at 12 mm was reported by Minier & Lineweaver (2006).

Spectrally-dispersed near-infrared interferometry has spatially and spectrally resolved the presence of H_2O interior to 1 AU of the young star MWC 480 (Eisner, 2007), attributed to the sublimation of migrating icy bodies. Spatially resolving and discerning the snow line in circumstellar disks directly may be possible by observing the 3 μm H_2O-ice feature in scattered light (Inoue et al., 2008).

Ultraviolet radiation zone Ultraviolet radiation, in particular between 200–300 nm, is damaging to most terrestrial biological systems, inhibiting photosynthesis, destroying DNA, and damaging a wide variety of proteins and lipids. Ultraviolet radiation is nevertheless believed to have been an important energy source on the primitive Earth, underlying the synthesis of many biochemical compounds and, therefore, essential for several biogenic processes. In the Archean Earth, without its atmospheric protection, radiation spanning the UVB (280–315 nm) and UVC (190–280 nm) regions would, for example, have probably reached the Earth's surface with little attenuation (Sagan, 1973).

Buccino et al. (2002, 2004, 2006) used these considerations to define the boundaries of a plausible *ultraviolet habitable zone* as a function of stellar spectral type. They derived an inner limit determined by the levels of ultraviolet radiation tolerable by DNA, and an outer limit characterised by the minimum levels needed in biogenic processes. They applied these various criteria to the exoplanet systems whose host stars had been observed by the International Ultraviolet Explorer satellite, IUE.

Further studies around M dwarfs, which are generally favourable for studies of the habitable-zone, but which may emit significant ultraviolet radiation during flares, were reported by Buccino et al. (2007).

Orbital dependence Williams & Pollard (2002) used a three-dimensional general-circulation climate model and a one-dimensional energy-balance model to examine highly elliptical orbits near the habitable zone, arguing that long-term climate stability depends primarily on the average stellar flux received over an entire orbit, not the length of the time spent within the habitable zone. Selsis et al. (2007) related a star's habitable zone to its mass, the planet's semi-major axis, and its cloud cover. Barnes et al. (2008b) further included the effects of eccentricity on the range of acceptable semi-major axes. The complex effects of climate dynamics and feedback are starting to be considered (Mandell, 2008; Spiegel et al., 2008), also as a function of the planet's obliquity (Spiegel et al., 2009a).

Tajika (2008) examined the conditions of a planet which is covered with ice, but which has an internal ocean for the time scale of planetary evolution owing to

11.9 Habitability

geothermal heat flow from the planetary interior. Liquid water can exist if the planetary mass and the water abundance are comparable to the Earth, although a planet with a mass $\lesssim 0.4 M_\oplus$ would not be able to maintain the internal ocean. Liquid water would be stable for a planet with a mass $M \gtrsim 0.4 M_\oplus$ either on its surface or beneath the ice, irrespective of planetary orbit and luminosity of the central star.

Models for tidally locked, synchronous rotators (Joshi & Haberle, 1997; Joshi et al., 1997; Joshi, 2003), suggest that synchronously rotating planets within the circumstellar habitable zones of M dwarfs should be habitable (Tarter et al., 2007).

Lopez et al. (2000, 2005) considered the migration or development of the habitable zone at much larger distances, 5–50 AU, during the short period of post-main-sequence evolution corresponding to the subgiant and red giant phases.

System stability and architecture The overall planetary system architecture will determine the likelihood of planets in multiple planet systems remaining within the habitable zone over extended periods (e.g. Jones et al., 2001; Goździewski, 2002; Jones & Sleep, 2002; Noble et al., 2002; Dvorak et al., 2003a,b; Menou & Tabachnik, 2003; Asghari et al., 2004; Érdi et al., 2004; Jones et al., 2005; Fatuzzo et al., 2006; Haghighipour, 2006; Jones et al., 2006a; Laakso et al., 2006; Sándor et al., 2007b; von Bloh et al., 2007b; Hinse et al., 2008; Ji et al., 2008; Pilat-Lohinger et al., 2008a,b; Cuntz & Yeager, 2009; Smith & Lissauer, 2009).

As examples of this considerable body of work, Menou & Tabachnik (2003) and Jones et al. (2006a) determined the borders of the zero-age main sequence and present habitable zones, respectively. Sándor et al. (2007b) determined a series of stability maps (§2.6.9), which can help to establish where Earth-like planets could exist in planetary systems having one giant planet.

The presence of an eccentric Jupiter may reduce the water content of planets formed inward of it (Raymond et al., 2007a). In contrast, very water-rich planets may form in the wake of migrating Jupiters (Raymond et al., 2005a, 2006b; Mandell et al., 2007).

Planet radius Within the ~1 AU habitability zone, Earth- (or super-Earth) class planets can be considered as those with masses between about 0.5–10 $M_\oplus$ or, equivalently, radii between 0.8–2.2 $R_\oplus$. Planets below this mass are likely to lose their life-supporting atmospheres because of their low gravity and lack of plate tectonics, while more massive systems are unlikely to be habitable because they can attract a H–He atmosphere and become gas giants (Huang, 1960).

Stellar spectral type Habitability is also likely to be governed by the range of stellar types for which life has enough time to evolve, i.e. stars not more massive

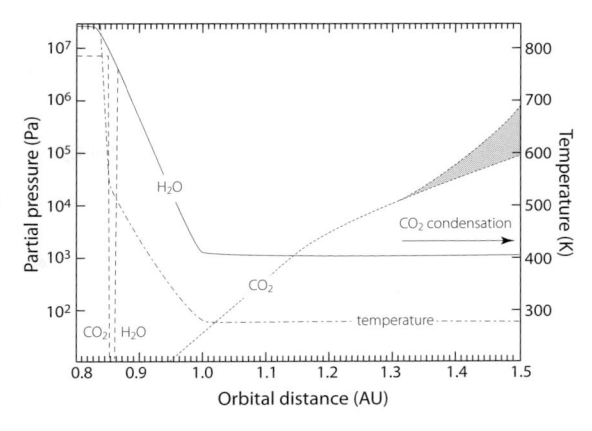

Figure 11.20: The partial pressures of CO_2 and H_2O (left axis) and mean surface temperature (right axis) as a function of orbital distance for a habitable planet within the habitable zone. From Kaltenegger & Selsis (2007, Figure 2).

than spectral type A. However, even F stars have somewhat narrower continuously habitable zones because they evolve more rapidly.

Late K and M dwarfs may not be the most favourable to host habitable planets, in part because the planets can become trapped in synchronous rotation due to tidal damping (also resulting in relatively weak intrinsic planetary magnetic moments), and in part because of the high incidence of coronal mass ejections, together resulting in little protection from high-energy particle radiation (Khodachenko et al., 2007). They may also be strong intermittent emitters of ultraviolet radiation during flares (Buccino et al., 2007). A further restriction may emerge from detailed formation studies: even if planets are formed at suitable orbital separations, they may be underabundant in required volatiles (Lissauer, 2007). In a detailed evaluation of the advantages and disadvantages of M stars as targets in searches for terrestrial-zone habitable planets, Scalo et al. (2007) reasoned that their planets *'must survive a number of early trials in order to enjoy their many Gyr of dynamical stability.'*

Mid- to early-K and G stars may therefore be optimal for the development of life (Kasting et al., 1993). Early simulations of the formation of habitable systems, within the framework of models of planet formation in general and the solar system in particular, are given for example by Wetherill (1996) and Lissauer (1997).

Habitability may be further confined within a narrow range of [Fe/H] of the parent star (Gonzalez, 1999b). If the occurrence of gas giants decreases at lower metallicities, their shielding of inner planets in the habitable zone from frequent cometary impacts, as occurs in our solar system, would also be diminished (Wetherill, 1994). At higher metallicity, asteroid and cometary debris left over from planetary formation may be more plentiful, enhancing impact probabilities.

The importance of the Moon The Earth's obliquity appears to have been stabilised by the Moon (Laskar & Robutel, 1993; Laskar et al., 1993; Neron de Surgy & Laskar, 1997). In its absence, considered to be an accident of accretion (§12.6), an Earth-like planet may undergo large-amplitude chaotic obliquity fluctuations on time scales of order 10 Myr, depending also on the planet's land-sea distribution. These motions would result in large local temperature excursions (Ward, 1974; Laskar & Robutel, 1993; Laskar et al., 1993; Williams & Kasting, 1997; Atobe et al., 2004) although perhaps not precluding the accompanying migration of life.

Galactic habitable zone Gonzalez (1999a) investigated the habitability implications of the anomalously small motion of the Sun with respect to the local standard of rest, both in terms of its pseudo-elliptical component within the Galactic plane, and its vertical excursion with respect to the mid-plane. Such an orbit could provide effective shielding from high-energy ionising photons and cosmic rays from nearby supernovae, from the X-ray background by neutral hydrogen in the Galactic plane, and from temporary increases in the perturbed Oort comet impact rate. If chiral asymmetry is also a prerequisite for life (e.g. Bailey, 2000, 2004), habitability may further depend on the polarisation environment of the star forming region.

Prantzos (2008) concluded that the physical processes underlying the concept of a Galactic habitable zone are *'hard to identify and even harder to quantify'*. No significant conclusions have been drawn about the extent of such a zone. It may be possible that the entire Galaxy disk is broadly suitable for complex life (Lineweaver et al., 2004; Bounama et al., 2007), perhaps with the outer disk regions somewhat favoured (Ćirković & Bradbury, 2006), but perhaps only sparsely populated (Cole, 2006).

The anthropic principle Many other conditions have been hypothesised as necessary or desirable for the development of life, largely based on the single life form known in the Universe, namely that on the Earth. Hypothetical conditions rapidly become tied to the more philosophical discussions of the *anthropic principle*, which broadly states that the physical Universe must be compatible with the conscious life that observes it.

In the definitions proposed by Barrow & Tipler (1988), the *weak anthropic principle* holds that *'The observed values of all physical and cosmological quantities are not equally probable but they take on values restricted by the requirement that there exist sites where carbon-based life can evolve and by the requirement that the Universe be old enough for it to have already done so'*.

The *strong anthropic principle* assets that *'The Universe must have those properties which allow life to develop within it at some stage in its history'*. This extensive topic will not be considered further here.

Solar twins In the search for habitable planets, attention has partly focused on *solar twins*, stars which are, by definition, non-binary stars identical to the Sun in all of their astrophysical parameters: mass, age, luminosity, chemical composition, temperature, surface gravity, magnetic field, rotation velocity, and chromospheric activity (*solar analogues* are, similarly, stars that looked in the past, or will look in the future, very similar to the Sun, thus providing a look at the Sun at some other point in its evolution).

Solar twins may be considered as being those most likely to possess planetary systems similar to the solar system, and best-suited to host life forms based on carbon chemistry and water oceans.

The systematic search for and study of solar twins and analogues started with the work of Hardorp (1978), who surveyed the near ultraviolet (360–410 nm) spectra of 77 solar-type stars in parts of the northern and southern hemispheres, finding no G2V star which matches the properties of the Sun.

The pre-Hipparcos status of subsequent searches is reviewed by Cayrel de Strobel (1996). Starting with 109 photometric solar-like candidates, they were also unable to identify a 'perfect' twin although two of the first three exoplanetary systems discovered, 51 Peg and 47 UMa, were on the initial list. The G2 star HD 146233 (HR 6060 = HIP 79672 = 18 Sco) at 14 pc comes very close to being such a twin, although with slightly higher luminosity and age (Porto de Mello & da Silva, 1997). The two G components of the binary 16 Cyg A/B were considered to be the next closest twins.

Further efforts to identify solar twins and other astrobiologically-interesting stars have been based on the Hipparcos data (Porto de Mello et al., 2000; Pinho et al., 2003; Pinho & Porto de Mello, 2003; Galeev et al., 2004; King et al., 2005; Porto de Mello et al., 2006). These suggest that HR 6060 (18 Sco) and HIP 78399 are the most promising solar twins, with essentially all parameters coinciding within the errors. Of stars with planets on the Keck, Lick, and AAT Doppler planet-search programme at that time, Gray et al. (2006) found that only HD 186427 has properties close, rather than very close, to those of the Sun.

11.9.2 Exoplanets in the habitable zone

GJ 581 At present, one planetary system, GJ 581, is considered to offer the best prospects for planets in the habitable zone. The host star is a nearby (6.27 pc) M3V dwarf. Radial velocity data, the latest published set comprising 119 measurements over 4.3 yr from HARPS, and 122 over 11 yr from HIRES (Bonfils et al., 2005b; Udry et al., 2007; Mayor et al., 2009a; Vogt et al., 2010a) suggest the presence of up to six planets (not all of which have been independently confirmed to date), listed in Table 11.5, and illustrated schematically in Figure 11.21.

11.9 Habitability

Table 11.5: The planets of GJ 581, ordered by increasing semi-major axis. Data are from Vogt et al. (2010a).

Planet	$M_p \sin i$ ($M_\oplus$)	a (AU)	P (d)
e	1.7	0.028	3.15
b	15.6	0.041	5.37
c	5.6	0.073	12.92
g	3.1	0.146	36.56
d	5.6	0.218	66.87
f	7.0	0.758	433

Dynamical stability (Beust et al., 2008), issues of aliasing (Dawson & Fabrycky, 2010), and the possibility of hidden and stable orbits (Zollinger & Armstrong, 2009; Anglada-Escudé et al., 2010a) have been studied.

Udry et al. (2007) originally suggested that planet c ($M_p \sin i = 5.1 M_\oplus$, $P = 12.9$ d) lies close to the inner edge of the habitable zone, while planet d ($M_p \sin i = 8.3 M_\oplus$, $P = 83.4$ d) lies close to its outer edge.

The original estimate of the equilibrium temperature of planet c ($T_{eq} \sim 320$ K, assuming a Bond albedo $A_B \sim 0.5$) was revised by later studies, but with the emerging consensus that planets c and d could nevertheless support habitability depending on their atmospheres, and whether they are tidally locked (Selsis et al., 2007; von Bloh et al., 2007a; Chylek & Perez, 2007).

The most recent data and analysis by Vogt et al. (2010a) suggests that a proposed sixth planet, GJ 581 g ($M_p \sin i = 3.1 M_\oplus$, $P = 36.6$ d, $a = 0.146$ AU, $e = 0$) could place it in the middle of the star's habitable zone. The radius of GJ 581 g is estimated as $1.3 - 1.5 R_\oplus$ if primarily composed of magnesium silicates, $1.7 - 2 R_\oplus$ if predominantly H_2O-ice, and with all radii smaller by 20% if it is significantly differentiated. The mass and radius together imply a surface gravity $\sim 1.1 - 1.7 g_\oplus$.

Its estimated $T_{eq} \sim 209 - 228$ K, depending on A_B, satisfies the conditions on the equilibrium temperature necessary for habitability of $T_{eq} \lesssim 270$ K derived by Selsis et al. (2007). The actual surface temperature is expected to be higher than T_{eq}. For Earth ($T_{eq} \sim 255$ K), atmospheric greenhouse heating results in a mean surface temperature $T_s \sim 288$ K. For a comparable or more massive atmosphere, greenhouse heating, combined with its probable synchronous-rotation (tidal locking), will result in higher surface temperature, plausibly implying the presence of liquid water.

HIP 5750 A Saturn-mass planet orbiting the nearby M4 dwarf HIP 57050 with $P = 41.4$ d was reported by Haghighipour et al. (2010), placing it in the habitable zone with $T_{eq} \sim 230$ K. Although too massive to be terrestrial, and *a priori* unlikely to be transiting, it would have an expected transit depth of 7%. This suggests that habitable zone planets with transit depths offering opportunities for high-quality atmospheric studies may soon be found.

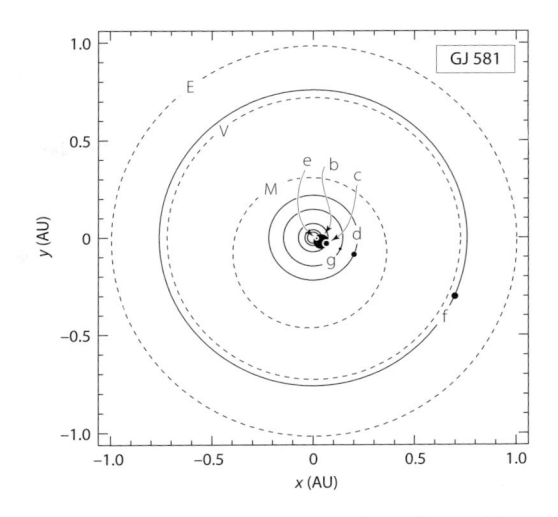

Figure 11.21: View of the GJ 581 system from above, with coordinates in AU. Planets, shown with a circle size proportional to mass, are from inner to outer: e, b, c, g, d, f, of orbital period 3.1, 5.4, 12.9, 36.6, 66.9, 433 d respectively, and mass ($M_p \sin i$) 1.7, 15.6, 5.6, 3.1, 5.6, 7.0 $M_\oplus$ respectively. The orbits of M(ercury), V(enus), and E(arth) are shown to the same scale. From Vogt et al. (2010a, Figure 6), reproduced by permission of the AAS.

Implied frequency of habitable planets From the statistics of solar-type or later stars within the distance horizon of GJ 581, and the statistics of radial velocity measurements and detections to date, Vogt et al. (2010a) argued that the fraction of stars with potentially habitable planets, $\eta_\oplus$, could be substantial, perhaps of the order of a few tens of percent.

11.9.3 Spectroscopic indicators of life

General considerations In the development of the Darwin and Terrestrial Planet Finder space imaging studies (§7.4.2), an important consideration was the spectral range demanded (optical, infrared, or both), and which spectral lines might be observed with the best prospects of indicating the presence of life. The arguments proceeded broadly as follows. They may be criticised as being founded on C-based life as presently known, and the identification of more complex chemical pathways may modify the conclusions (Saar et al., 2008; Schneider, 2008a,b).[6]

---

[6] Despite the immense accumulation of empirical data, there is no generally accepted definition of life. In biology, life is considered to be a characteristic of organisms that display all, or most of, a set of phenomena including metabolism, growth, organisation, reproduction, adaptation, homeostasis, and response to stimuli. At another extreme, Erwin Schrödinger suggested that the characteristic feature of a living organism is one that takes negative entropy from its environment to delay its decay into thermodynamic equilibrium, while Karl Popper pointed out that such a definition describing both an oil-fired

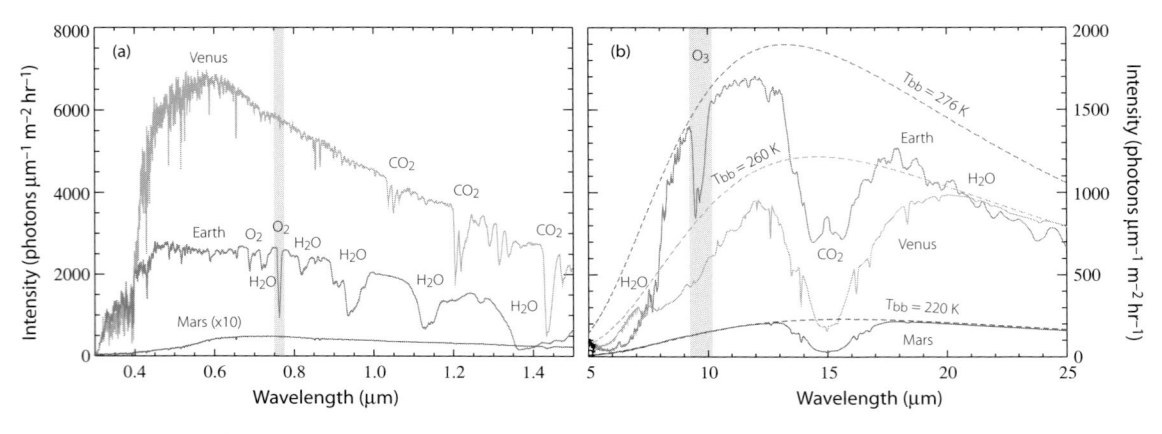

Figure 11.22: Spectra of Venus, Earth, and Mars: (a) reflection spectra (which mix solar and planetary lines) at a resolution ~100 over the ultraviolet–visible–near infrared; (b) mid-infrared thermal emission at the same resolution, with the black-body emission of a planet of the same radius (dashed). For Venus and Mars, only CO_2 is observable at low-resolution. The Earth also shows the deep and sharp 9.6 μm O_3 and 0.76 μm O_2 bands due to life, and H_2O bands. The flux corresponds to a solar system analogue at 10 pc. Spectra are model results validated by observations. From Selsis et al. (2008a, Figure 1).

Owen (1980) already argued that large-scale biological activity on a telluric planet necessarily produces a large quantity of O_2. Photosynthesis builds organic molecules from CO_2, with the help of H^+ ions which can be provided from different sources. In the case of oxygenic bacteria on Earth, H^+ ions are provided by the photodissociation of H_2O, in which case O_2 is produced as a by-product. However, this is not the case for anoxygenic bacteria, and thus O_2 is considered as a possible but not a necessary by-product of life. Indeed, Earth's atmosphere was O_2-free until about 2 Ga, suppressed for more than 1.5 Gyr after life originated (Kasting, 1996; Bekker et al., 2004).

Owen (1980) noted the possibility, quantified by Schneider (1994), of using the 760 nm band of O_2 as a spectroscopic tracer of life since, being highly reactive with reducing rocks and volcanic gases, it would disappear in a short time in the absence of a continuous production mechanism. Plate tectonics and volcanic activity provide a sink for free O_2, and are the result of internal radiogenic heating and of silicate fluidity, both of which are expected to be generic whenever the mass of the planet is sufficient and when liquid H_2O is present. For small enough planet masses, volcanic activity disappears some time after planet formation, as do the associated oxygen sinks.

Ozone Angel et al. (1986) showed that O_3 is itself a tracer of O_2 and, with a prominent spectral signature at 9.6 μm in the infrared where the planet/star contrast is significantly larger than in the optical, should be easier to detect. These considerations motivated the consideration of the Darwin and TPF space interferometers observing in the infrared for the study of lines such as H_2O at 6–8 μm, CH_4 at 7.7 μm, O_3 at 9.6 μm and CO_2 at 15 μm (Woolf & Angel, 1997). Higher resolution studies could reveal the presence of CH_4, its presence on Earth resulting from a balance between anaerobic decomposition of organic matter and its interaction with atmospheric oxygen; its highly disequilibrium co-existence with O_2 could be strong evidence for the existence of life (Margulis & Lovelock, 1974; Kasting & Donahue, 1980; Léger et al., 1994a,b; Kasting, 1996; Segura et al., 2003; Selsis et al., 2008b).

Other potential biosignatures, especially when seen in the presence of O_2, could include the products of biomass burning, such as CH_3Cl and N_2O, which may be even more prominent for planets around M dwarfs (Segura et al., 2005; Grenfell et al., 2007).

The possibility that O_3 is not an unambiguous identification of Earth-like biology but rather a result of abiotic processes (Owen, 1980; Kasting, 1996; Noll et al., 1997b; Schneider, 1999; Segura et al., 2007) was examined in detail by Léger et al. (1999). They considered various production processes such as abiotic photodissociation of CO_2 and H_2O, followed by the preferential escape of hydrogen from the atmosphere.

In addition, cometary bombardment could bring O_2 and O_3 sputtered from H_2O by energetic particles (cf. the ultraviolet spectral signature of O_3 in the satellites of Saturn, Rhea and Dione; Noll et al. 1997b), according to temperature, greenhouse blanketing, and the presence of volcanic activity. Léger et al. (1999) concluded that a

---

boiler and a self-winding watch (Popper, 1976). The review by Tsokolov (2009) analyses three approaches to defining life, and concludes that '... *all three are problematic in that they attempt to define life with undefined terms, confuse a description with a definition, or define life arbitrarily in terms of minimal living systems.*'

11.9 Habitability

simultaneous detection of significant amounts of H_2O and O_3 in the atmosphere of a planet in the habitable zone stands as a criterion for large-scale photosynthetic activity on the planet. Whether these would correspond to a true signature of a biological process is less clear.

An absence of O_2, on the contrary, would not necessarily imply the absence of life. However, an atmosphere rich in O_2 provides the largest feasible source of chemical energy, and abundant O_2 may be necessary for the high-energy demands of complex life anywhere, viz. for actively mobile organisms of 0.1–1 m size scale with specialised, differentiated anatomy (Catling et al., 2005). Specifically, Catling et al. (2005) have argued that the partial pressure of atmospheric O_2 must exceed $\sim 10^3$ Pa to allow organisms that rely on O_2 diffusion to evolve to a size ~ 1 mm, while $\sim 10^3 - 10^4$ Pa is required to exceed a size-threshold of 0.01 m for complex life with circulatory physiology.

Characterising Earth-like atmospheres Assuming adequate spectral resolution, the temperature and pressure environment of specific molecular species can be derived from the shape and width of the spectral line, notably as a result of Doppler broadening dominating at lower pressures (i.e. higher altitudes), pressure-induced Lorentz broadening dominating at higher pressures, and the temperature distribution with altitude. Disentangling the effects of temperature structure and vertical distribution is best carried out using a species evenly mixed within the atmosphere, such as CO_2 on Earth, rather than more vertically inhomogeneous species such as O_3, H_2O, or CH_4 (Meadows, 2008).

Earth's main atmospheric constituent, N_2, has no strong spectral features redward of the far ultraviolet. O_2 has a triplet between 0.6–0.76 μm, and is a significant spectral feature for the Earth's atmosphere, but not for those of Venus and Mars. CO_2 is a trace gas in the Earth's atmosphere, but a bulk gas on Venus and Mars, and produces strong spectral features in the mid-infrared and, for sufficient partial pressures, also in the near infrared between 1.0–1.7 μm (Des Marais et al., 2002; Meadows, 2008; Selsis et al., 2008a; Vázquez et al., 2010). Modeled atmospheric spectra for Earth, and comparison with those of Venus and Mars, are shown in Figure 11.22.

H_2O vapour absorbs strongly throughout the visible, near and mid-infrared, and its presence may indicate a surface ocean. Other indicators of oceans may be glints, rainbows, and polarisation (Ford et al., 2001b; Schmid et al., 2006; Stam et al., 2006; Bailey, 2007; Cowan et al., 2009).

Other distinctive spectral signatures at ultraviolet, visible, and infrared wavelengths can be used to establish molecular abundances, infer the presence of greenhouse gases, and estimate surface temperatures, and in particular to determine whether H_2O is likely to exist in liquid form (Schindler & Kasting, 2000; Woolf et al.,

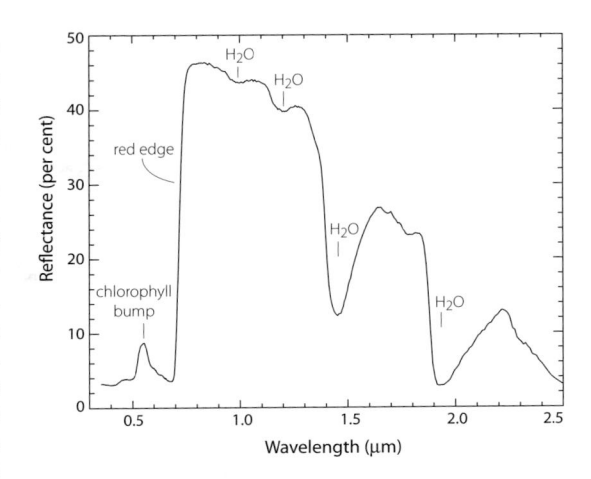

Figure 11.23: Reflection spectrum of a deciduous leaf. The bump near 550 nm is a result of chlorophyll absorption (at 450 nm and 680 nm), which gives plants their green colour. The sharp rise between 700–800 nm, the red edge, is due to the contrast between the strong absorption of chlorophyll and the otherwise reflective leaf. From Seager et al. (2005b, Figure 1).

2002; Turnbull et al., 2006; Tinetti et al., 2006a,b; Selsis et al., 2008a; Langford et al., 2009; Pallé et al., 2009; Vidal-Madjar et al., 2010).

Signatures of plant life and photosynthesis The Earth displays specific spectral signatures due to plant life (Figure 11.23). Chlorophyll has a varying absorption with wavelength, with stronger absorption in the ultraviolet/blue and in the red as compared to $\lambda \sim 550$ nm. Plants also show a sharp increase in reflectivity at $\lambda \gtrsim 0.7\,\mu$m, known as the *red edge*. It was observed as a distinguishing feature in the Earth's spectrum by the Galileo spacecraft during Earth flyby (Sagan et al., 1993), and is used as the basis of vegetative remote-sensing.

The 'red edge' has been considered as a possible biosignature for extraterrestrial vegetation (Arnold et al., 2002; Seager et al., 2005b; Tinetti et al., 2006c; Arnold et al., 2009), although it is only rather weakly visible in the Earth's global spectrum (Montañés-Rodríguez et al., 2006; Hamdani et al., 2006), and not observed in Earth light reflected from the Moon (Montañés-Rodriguez et al., 2005). A comparable sharp spectral signature may be stronger on other planets with different forms of plant life, or may appear at different wavelengths for planets around stars of different spectral type (Franck et al., 2001; Wolstencroft & Raven, 2002; Raven & Cockell, 2006; Tinetti et al., 2006b; Kiang et al., 2007a,b).

Small temporal variations in CO_2 and CH_4 also arise on Earth as a result of seasonal changes in patterns of photosynthesis (Tucker et al., 1986).

Earth's spectrum over geological history The Earth's atmosphere has evolved significantly over geological time (§11.8.1). To model the observable spectra of an

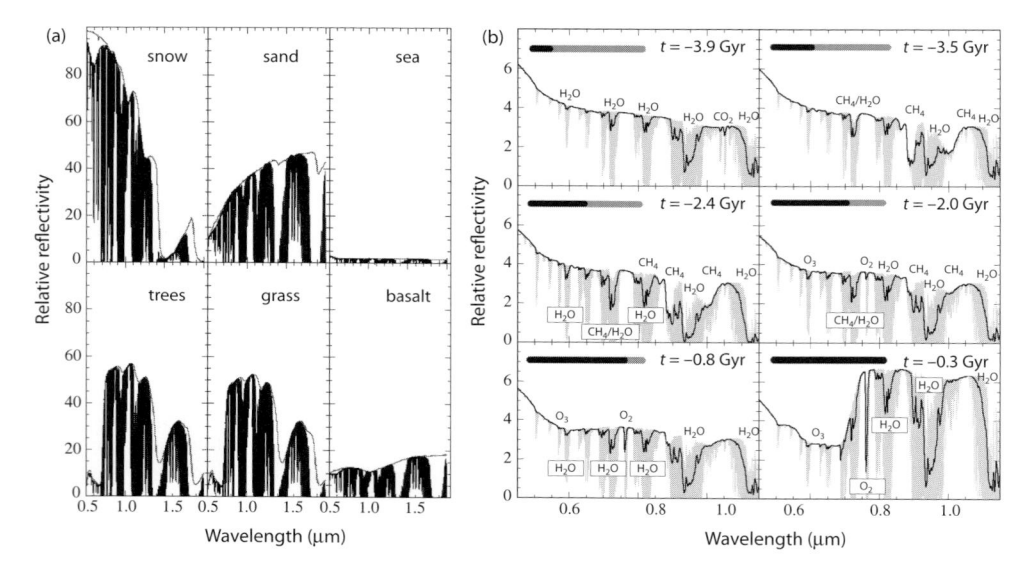

Figure 11.24: Spectra of an Earth-like planet over geological history: (a) the contribution to a present-day atmosphere of distinct surface features, assuming a clear atmosphere in the absence of clouds; (b) visible and near-infrared spectra of an Earth-like planet at six distinct geological epochs, again in the absence of clouds. The spectral lines (grey) change significantly as the planet evolves from CO_2-rich (−3.9 Gyr), through CO_2/CH_4-rich (−2.4 Gyr), to a present-day atmosphere (lower right). Solid curves show a spectral resolution of 70, comparable to the proposed TPF–C mission concept. From Kaltenegger et al. (2007, Figures 1 and 9), reproduced by permission of the AAS.

Earth-like planet throughout its history, and to guide the interpretation of an observed Earth-like planet by an instrument such as Darwin or Terrestrial Planet Finder, Kaltenegger et al. (2007) characterised the atmosphere at six epochs spanning a period extending from 3.9 Gyr ago to the present day.

Based on reflectance spectra of specific compositional features such as snow, sand, sea and vegetation (Figure 11.24a), they constructed visible and near-infrared spectra based on estimated surface abundances of H_2O, CO_2, CH_4, O_2, O_3, and N_2O at the various epochs. Example results, for an Earth-like planet in the absence of clouds, are shown in Figure 11.24b.

11.9.4 SETI

Considerations of habitability may well imply that the fraction of habitable planets is small. The *search for extra-terrestrial intelligence* (SETI), is nevertheless motivated by the belief that life, and intelligence, is likely to emerge under conditions resembling those on the early Earth (e.g. Cocconi & Morrison, 1959; Drake, 1961; von Hoerner, 1961, 1973; Townes, 1997; Leigh & Horowitz, 1997; Bhathal, 2000; Tarter, 2001; Drake, 2008).[7]

Quantifying the probability that intelligent life exists elsewhere in the Galaxy, for example through consideration of the Drake equation (e.g. Chyba, 1997; Ćirković, 2004), or resolving the *Fermi paradox*, viz. *'If other advanced civilisations exist, where are they?'* (e.g. Tipler, 2003), may lie far in the future. Success or persistent failure in such searches would be of considerable significance (cf. footnote on page 28).

Radio/microwave surveys NASA's High Resolution Microwave Survey, a programme to search for continuous and pulsed radio signals generated by extra-solar technological civilisations, consisted of an all-sky survey between 1–10 GHz, as well as a targeted search of 1000 nearby stars at higher spectral resolution and sensitivity in the 1–3 GHz range. Although Congress terminated the survey in 1993, the SETI Institute raised private funds to continue the targeted portion of the search as Project Phoenix, which now carries out observations at the Arecibo Observatory in conjunction with simultaneous observations from the Lovell Telescope at Jodrell Bank (Morison, 2006).

---

[7] The essence of this so-called 'principle of mediocrity', that life on Earth and our technology are about average is, naturally, not universally shared. The primitive single-celled Archaea, found in a broad range of habitats and not only as extremophiles, are the simplest known living organisms, and widely used by astrobiologists as models of possible extraterrestrial life. Howland (2000) cautioned that, given precursors such as amino acids and believing that a living organism would inevitably arise through self-assembly is *'... similar to a tornado descending on a junkyard and producing, by self-assembly, a jet airliner'.*

11.9 Habitability

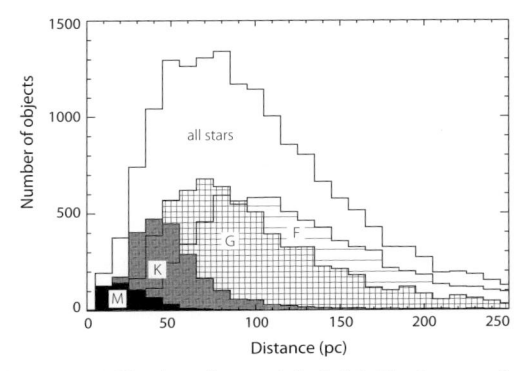

Figure 11.25: Number of potentially habitable stars as a function of distance and spectral type in the SETI Institute's Catalogue of Nearby Habitable Systems. Transmitted power comparable to the Arecibo planetary radar will be detectable by the Allen Telescope Array out to ~300 pc. From Turnbull & Tarter (2003a, Figure 11), reproduced by permission of the AAS.

In a joint effort by the SETI Institute and the University of California, Berkeley, the Allen Telescope Array is being constructed at Hat Creek Observatory, northern California (Tarter, 2006). The array consists of 350 dishes, each 6.1 m in diameter, a bandwidth covering 0.5–11 GHz, and a capability of observing some 10 000 target stars per year.

Optical surveys The search for optical signals from extraterrestrial civilisations was considered already by Schwartz & Townes (1961). Pilot programmes have been undertaken at a number of observatories (Kingsley, 2001; Shuch, 2001). The principle is based on intercepting targeted interstellar laser communications, and current searches aim to detect a brief (few ns), intense light pulse with fast photon detectors. Its feasibility is founded on the fact that current laser technology on Earth allows the generation of directional pulses that outshine the broad-band visible light of the Sun by a factor of at least 10^4.

A group at University of California, Berkeley, has been searching at optical wavelengths since 1997. More recently, they have searched for narrow-band coherent signals in the spectra taken as part of their search for planets at Lick, Keck, and the AAT observatories (Werthimer et al., 2001).

An optical search has also been underway at Harvard since 1998 (Horowitz et al., 2001), more recently in synchronised mode with Princeton. A search for repetitive ns-pulsed signals from a list of 13 000 Sun-like stars has been operational since 2006 (Howard et al., 2004, 2007).

Lick Observatory's optical SETI programme started observations in 2000. The results of a search of 4605 Hipparcos stars within ~60 pc were reported by Stone et al. (2005). A two-telescope dual channel system has been used for a similar programme at the University of Western Sydney (Bhathal, 2001).

Another approach is to use ground-based gamma-ray Cherenkov detectors (Holder, 2005). The Solar Tower Atmospheric Cherenkov Effect Experiment (STACEE) has been used to search for ns blue-green laser pulses from 187 nearby stars that might host habitable planets (Hanna et al., 2009).

Catalogue of Nearby Habitable Systems The SETI Institute's Catalogue of Nearby Habitable Systems (Turnbull & Tarter, 2003a,b) was created for Project Phoenix from the Hipparcos Catalogue. Information on distances (for signal propagation), variability (for climate stability), multiplicity (for orbital stability), kinematics (as a metallicity indicator), and spectral classification, was complemented by data on X-ray luminosity, Ca II H and K activity, rotation, spectral types, kinematics, metallicity, and Strömgren photometry.

Combined with theoretical studies on habitable zones, stellar evolutionary tracks, and three-body orbital stability, unsuitable stars were removed, leaving a list of targets that, according to present knowledge, are the best candidates for potentially habitable hosts for complex life. The analysis resulted in 17 129 Hipparcos habitable star candidates in the vicinity of the Sun, of which 75% lie within 140 pc, and some 2200 of which are known or suspected to be members of binary or triple systems (Figure 11.25).

The catalogue is intended to provide at least three target stars, within the large primary field of view of the Allen Telescope Array, during routine observations. The algorithm for prioritising objects in the full target list includes scoring based on the category of each target, and its proximity to the Sun.

Who will speak for Earth? In 2001, the International Academy of Astronautics established a 'Permanent SETI Study Group'. Under it, a 'Post-Detection Task Group' is mandated to *'prepare, reflect on, manage, advise, and consult in preparation for and upon the discovery of a putative signal of extraterrestrial intelligent origin'*.

Evidently, concerns such as 'who will speak for Earth' when or if contact is made (Goldsmith, 1988), are some way from today's scientific mainstream, but perhaps not as far as they were 20 years ago.

12

The solar system

Theories of exoplanet formation and migration can be confronted with a wealth of diverse observational constraints from the solar system, of which this chapter provides an incomplete and selective summary.

Amongst them are the orbital motions of the planets (including spacings, eccentricities and inclinations, dynamical stability and resonances), planetary masses and rotation (and their angular momentum distribution), the existence of planetary satellites and rings, the occurrence of other minor bodies (comets, asteroids and meteorites, including the presence of the Oort Cloud and the Edgeworth–Kuiper belt), bulk and isotopic compositions, radiogenic isotope ages, and cratering records.

The publications listed in Table 12.1 provide more detailed perspectives, with a recent synthesis given by de Pater & Lissauer (2010). Early theories of the formation of the solar system are summarised in §10.4.1.

12.1 Birth in clusters

Most stars are born in molecular clouds as members of stellar clusters although, rarely, some stars might be born in isolation (Lada & Lada, 2003). Whether the Sun was formed in isolation or in a cluster remains uncertain (Adams, 2010). Evidence for the latter includes the dynamical structure of the Edgeworth–Kuiper belt which suggests a nearby encounter with another star (Morbidelli & Levison, 2004), and short-lived radionuclides and their decay products in the proto-solar nebula (Hester et al., 2004; Hester & Desch, 2005; Gounelle & Meibom, 2008), explicable in terms of a supernova explosion within 1–2 pc of the young Sun (Looney et al., 2006).

Nevertheless, no other star sharing the Sun's age, metallicity, and Galactic kinematics is yet known (Portegies Zwart, 2009).

12.2 The solar system giants

There are four giant planets in the solar system: the *gas giants* Jupiter and Saturn, along with Uranus and Neptune which are also classified as gas giants, but whose

Table 12.1: Selected publications in the University of Arizona Space Science Series on solar system bodies.

Title	Reference
Protostars and Planets V	Reipurth et al. (2007)
Meteorites and the Early Solar System II	Lauretta et al. (2006)
Comets II	Festou et al. (2004)
Asteroids III	Bottke et al. (2002)
Origin and Evolution of Planetary and Satellite Atmospheres	Atreya et al. (1989)
Satellites	Burns & Matthews (1986)
Planetary Rings	Greenberg et al. (1984)
Mercury	Vilas et al. (1988)
Venus II	Bougher et al. (1997)
Origin of the Earth and Moon	Canup & Righter (2000)
Mars	Kieffer et al. (1992)
Jupiter	Gehrels (1976)
Satellites of Jupiter	Morrison (1982)
Saturn	Gehrels et al. (1984)
Uranus	Bergstralh et al. (1991)
Neptune and Triton	Cruikshank et al. (1995)
Solar System beyond Neptune	Barucci et al. (2008)
Pluto and Charon	Stern & Tholen (1997)

less massive gas envelopes and distinct compositions also justifies their alternative designation as *ice giants*.

Masses and radii Detailed constraints on masses, radii, and internal structure are available (Table 12.2). Masses are derived from the motions of their natural satellites, radii are derived from radio occultations, and rotation periods are inferred from the time variation of their magnetic fields. All are relatively fast rotators, with periods of 10–11 h for Jupiter and Saturn, and 16–17 h for Uranus and Neptune.

Estimates of J_2 and J_4, and constraints on their interior density profiles, have been established from the Pioneer and Voyager spacecraft flybys. Schematics of their inferred internal structures are shown in Figures 12.1.

Jupiter and Saturn Jupiter and Saturn consist mostly of H and He. Jupiter has $R_J \sim 70000$ km ($11 R_\oplus$), $M_J \sim 318 M_\oplus$, and an average density $\rho_J = 1.33$ Mg m^{-3}, with elements heavier than He representing some $11-45 M_\oplus$ of its total mass (e.g. Zharkov & Gudkova, 1991; Chabrier

Table 12.2: Mass, radii (equatorial R_+, polar R_-, and mean $\bar{R}$), mean density, gravity fields (J_2, J_4), and rotation periods (P_ω) of the solar system giants, taken from the compilation of Guillot (2005, Table 1). Numbers in parentheses are the uncertainty in the final digits. Primary references are: (a) Campbell & Synnott (1985), (b) Campbell & Anderson (1989), (c) Anderson et al. (1987), (d) Tyler et al. (1989), (e) Lindal et al. (1981), (f) Lindal et al. (1985), (g) Lindal (1992), (h) from fourth-order figure theory (Zharkov & Trubitsyn, 1971), (i) $(2R_+ + R_-)/3$, (j) Davies et al. (1986), (k) Warwick et al. (1986), (l) Warwick et al. (1989). The Bond albedo, A_B, and effective temperature, T_{eff}, are determinations from Voyager IRIS data, as compiled by Guillot (2005, Table 2).

	Jupiter	Saturn	Uranus	Neptune
$M_p \times 10^{-26}$ [kg]	18.986 112(15)[a]	5.684 640(30)[b]	0.868 320 5(34)[c]	1.024 354 2(31)[d]
$R_+ \times 10^{-7}$ [m]	7.149 2(4)[e]	6.026 8(4)[f]	2.555 9(4)[g]	2.476 6(15)[g]
$R_- \times 10^{-7}$ [m]	6.685 4(10)[e]	5.436 4(10)[f]	2.497 3(20)[g]	2.434 2(30)[g]
$\bar{R} \times 10^{-7}$ [m]	6.989 4(6)[h]	5.821 0(6)[h]	2.536 4(10)[i]	2.462 5(20)[i]
$\bar{\rho}$ [Mg m^{-3}]	1.327 5(4)	0.688 0(2)	1.270 4(15)	1.637 7(40)
$J_2 \times 10^2$	1.469 7(1)[a]	1.633 2(10)[b]	0.351 60(32)[c]	0.353 9(10)[d]
$J_4 \times 10^4$	−5.84(5)[a]	−9.19(40)[b]	−0.354(41)[c]	−0.28(22)[d]
$P_\omega \times 10^{-4}$ [s]	3.572 97(41)[j]	3.835 77(47)[j]	6.206(4)[k]	5.800(20)[l]
P_ω [h]	9.92	10.65	17.24	16.11
q (Equation 11.24)	0.089 23(5)	0.154 91(10)	0.029 51(5)	0.026 09(23)
Bond albedo, A_B	0.343(32)	0.342(30)	0.300(49)	0.290(67)
T_{eff} [K]	124.4(3)	95.0(4)	59.1(3)	59.3(8)

et al., 1992). Values for Saturn are $R_S \sim 60\,000$ km ($9.4R_\oplus$), $M_S \sim 95M_\oplus$, and $\rho_S = 0.69$ Mg m^{-3}.

Structurally, they are believed to comprise small rocky cores at high temperature and pressure ($\sim 30\,000$ K and ~ 3000 GPa at the core boundary for Jupiter) but with poorly understood composition and properties. The surmised existence of cores is guided by plausible formation models, as well as by gravity measurements (which suggest a core mass of $12-45M_\oplus$ for Jupiter), although they may now be diluted or even absent (Guillot et al., 1997; Guillot, 1999; Guillot et al., 2004).

Surrounding the core is a thick dense layer of liquid *metallic hydrogen*, where extreme pressures result in unbound electrons, and the H is consequently electrically conductive (Wigner & Huntington, 1935; Loubeyre et al., 1996; Weir et al., 1996; Chabrier et al., 1992; Saumon et al., 2000).

Above this is an outer layer comprising some 90% H and 10% He by volume, or approximately 75% H and 25% He by mass (close to the theoretical composition of the primordial solar nebula), with traces of CH$_4$, H$_2$O, NH$_3$, Si-based compounds, and others.

Uranus and Neptune Uranus ($R_U \sim 25\,000$ km or $\sim 4R_\oplus$, $M_U \sim 14.5M_\oplus$, $\rho_U = 1.27$ Mg m^{-3}) is believed to comprise a small rocky core of uncertain mass consisting of Fe, Ni and silicates ($0.5-3.7M_\oplus$, $\rho \sim 9$ Mg m^{-3}, $P \sim 800$ GPa, $T \sim 5000$ K), an ice mantle of $9.3-13.5M_\oplus$ composed primarily of H$_2$O, CH$_4$ and NH$_3$, and an outer gaseous H/He envelope of $0.5-1.5M_\oplus$ extending from about 80–85% of its radius out to the cloudtops. However, different rock/ice/gas compositions also satisfy the observations (Bergstralh et al., 1991; Lunine, 1993; Podolak et al., 1995).

The 'ice mantle' is not composed of conventional ice, but is rather a hot dense fluid of high electrical conductivity.[1] This gradually merges into the outer gaseous atmosphere, with a 'surface' at about $R \sim 25\,000$ km designated as the radius corresponding to an atmospheric pressure of 100 kPa (1 bar).

Neptune ($R_N \sim 25\,000$ km or $\sim 3.9R_\oplus$, $M_N \sim 17.1M_\oplus$, $\rho_N = 1.64$ Mg m^{-3}) has an internal structure resembling but denser than that of Uranus, with a rocky core of $\sim 1.2M_\oplus$, an ice mantle of $10-15M_\oplus$, and an outer atmosphere of $1-2M_\oplus$ (Lunine, 1993; Podolak et al., 1995).

Noble gas enrichment The origin of the noble gas enrichment of Jupiter's atmosphere, in particular Ar which condenses only at very low temperatures (~ 30 K), remains uncertain. Hypotheses include *clathration* (lattice trapping) of noble gases in ices (Gautier et al., 2001a; Hersant et al., 2004; Alibert et al., 2005c), or the delivery of planetesimals formed at very low temperatures (Owen et al., 1999).

The enrichment of C and possibly N of the atmospheres of Uranus and Neptune may similarly indicate that a significant planetesimal mass ($\gtrsim 0.1M_\oplus$) impacted the planets after the bulk of their H/He envelopes had been captured.

Noble gases in the lunar regolith are primarily implanted by the solar wind. Such species include apparently orphan radiogenic ^{40}Ar and ^{129}Xe in excess of the primordial solar origin, and therefore generally attributed instead to lunar degassing. In analogy with exoplanet host star pollution caused by captured planets,

---

[1] The terms 'gas' and 'ice' are planetary-science designations for volatiles with exceptionally low melting points, or melting points above about 100 K, respectively. The terms are applied irrespective of whether the compounds are solids, liquids or gases. In the case of the gas and ice giants, the majority of the 'gas' and 'ice' in their interiors is, in fact, hot and liquid.

Ozima et al. (2004) considered that pollution of the Sun by a $2M_\oplus$ planet capture could account for orphan Xe in the Moon.

12.3 Minor bodies in the solar system

Planetesimals and protoplanets In the solar system, planetesimals which grew to modest size without forming larger objects, as well as post-collisional debris, are represented by meteoroids, asteroids, and comets.

Meteoroids are mainly 'rock' (a combination of iron- and magnesium-bearing silicates and metallic iron) with irregular orbits and between 100 μm and some 10–50 m in size (Millman, 1961; Beech & Steel, 1995).[2]

Asteroids are taken to be larger bodies which formed inside the orbit of Jupiter rather than in the outer solar system. Examples of protoplanets which have survived more-or-less intact are (dwarf planet) 1 Ceres and (asteroids) 2 Pallas and 4 Vesta in the inner solar system, and perhaps the Kuiper belt dwarf planets further out.

Comets Comets, as 'dirty snowballs', comprise frozen ice (H_2O, CO_2), dust grains, and small rocky particles, and range from a few hundred meters to tens of kilometers across (Hahn & Malhotra, 1999). There are two broad families, with both classes providing constraints on the early evolution of the outer solar system (Malhotra, 1993a), and on collisional models for planet formation (Kenyon, 2002).

The *Edgeworth–Kuiper belt*, as a subset of all trans-Neptunian objects (i.e. those orbiting beyond Neptune Schulz, 2002), extends from 50 AU (beyond Pluto) to hundreds or a few thousand AU, and forms a vast system possibly identified with the remnant of the Sun's protoplanetary disk (Jewitt & Luu, 1993; Williams, 1997; Jewitt, 1999; Chiang et al., 2007). These objects are classified in several distinct groupings: a large population of *resonant Kuiper belt objects* or *plutinos*, in Pluto-like orbits, and in 3:2 mean motion resonance with Neptune; the *Centaurs*, which are non-resonant objects with perihelion distances interior to Neptune; *classical Kuiper belt objects*, orbiting beyond and little influenced by Neptune; and *scattered disk Kuiper belt objects*, other objects with perihelia beyond Neptune.

The *Oort cloud* consists of some 10^{12} comets of all orbital orientations, extending out to tens of thousands of AU, but with a total mass of only a few $M_\oplus$. It originated from the gravitational scattering of planetesimals early in the history of the solar system by Uranus and Neptune, when objects were sent outwards at close to the solar system escape velocity, and were perturbed into long-lived orbits by nearby stars or the Galactic tidal field. Bodies scattered from Jupiter and Saturn, deeper within the Sun's gravitational potential well, would have been lost from the solar system. The Oort cloud's innermost extension is perhaps represented by (90377) Sedna, a 1000–1500 km diameter object with semi-major axis $a = 480 \pm 40$ AU, large eccentricity $e = 0.84 \pm 0.01$, and large inclination $i \sim 12°$ (Brown et al., 2004; Kenyon & Bromley, 2004c).

Collisional debris The smallest solar system objects, swarms of sand to small meteoroids, are considered to be impact debris from the final stage of its formation, of which the Moon's impact craters provide evidence. Even now, at intervals of several tens of Myr, a small planetesimal or comet in an Earth-crossing orbit strikes the Earth. Such impacts may have been responsible for mass extinctions such as at the Cretaceous/Tertiary boundary 65 Myr ago (e.g. Stothers, 1998). Interplanetary dust found by Pioneer beyond Saturn is similarly explained (Landgraf et al., 2002).

12.4 Solar nebula abundances

The products of the thermochemical equilibrium processes described in §11.2.2 are sensitive to the initial elemental composition of the gas. To infer the gas and mineralogical composition of the solar nebula out of which the solar system planets formed, therefore requires an accurate estimate of its initial elemental abundances.

Such an estimate can be made from a combined consideration of solar photospheric and meteoritic abundances, although the process is not straightforward. It needs to address the extent to which observed solar photospheric abundances reflect the Sun's true bulk composition, and the extent to which the Sun's bulk composition, and the wide variety of observed meteoritic compositions, properly reflect the conditions in the original solar nebula.

Elemental photospheric and meteoritic abundances published by Anders & Grevesse (1989) are still widely referenced, although various revisions in abundances have since been proposed (e.g. Grevesse & Noels, 1993; Grevesse et al., 1996; Grevesse & Sauval, 1998, 2002). In these cited works, the abundances derived from photospheric and meteoritic data are generally taken as representative of the bulk solar system composition.[3]

---

[2]Murray et al. (2004) have argued that particles larger than ~ 10 μm can propagate for tens of pc through the interstellar medium, and that such particles have probably been detected by ground-based radar at Arecibo and New Zealand, as well as by satellites.

[3]On the logarithmic astronomical scale, the numerical abundance of element 'x' is designated $A(x) = \log \epsilon(x)$. On this scale, the number of H atoms is set to $A(H) = \log n(H) = 12$, so that $A(x) = \log[n(x)/n(H)] + 12$. On the cosmochemical scale, abundances, designated as $N(x)$, are normalised to the number of silicon atoms of $N(Si) = 10^6$.

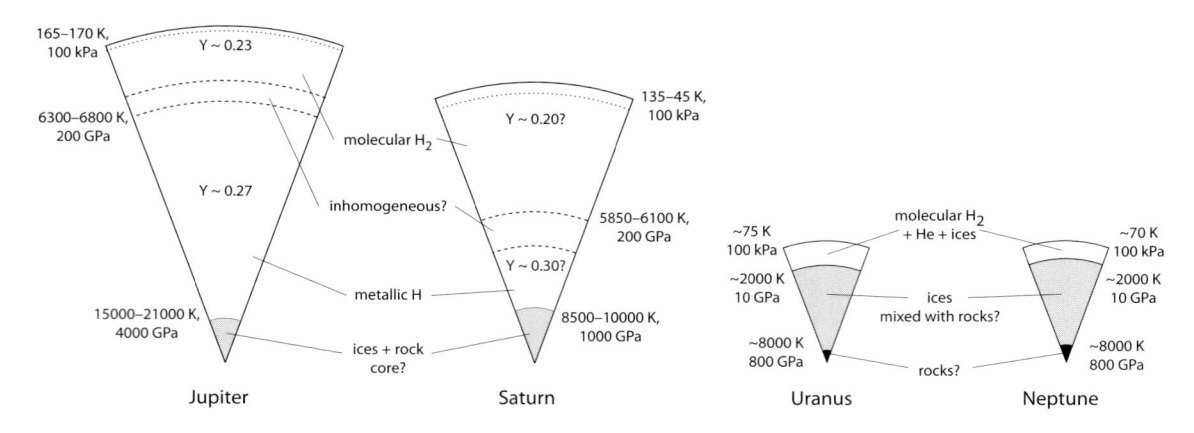

Figure 12.1: *Schematic scaled representation of the interiors of the solar system giants. For Jupiter and Saturn, He mass mixing ratios Y are indicated. The size of their central rock and ice cores are very uncertain and, in the case of Saturn, the inhomogeneous region may extend down to the core. Adapted from Guillot (2005, Figures 6 and 9).*

For meteorite abundances, the small class of CI (Ivuna type) carbonaceous chondrites lack chondrules and refractory inclusions, have the most 'primitive' compositions (essentially identical to that of the Sun, excluding gaseous elements like H and He), and are therefore used as a standard for assessing the degree of chemical fractionation experienced by materials formed throughout the solar system. Significant importance has traditionally been given to the intensively-studied 14 kg Orgueil meteorite, the most massive of the CI chondrites, which fell in France in 1864.

Abundances determined the Sun's photospheric spectrum and from CI-type meteorites agree reasonably well (Anders & Grevesse, 1989), but there are differences. Compared to the solar photosphere, CI-meteorites are depleted in noble gases, and in H, C, N, and O which readily form gaseous compounds, while they are enriched in those elements (notably Li) that are processed in the Sun. Taking account of such caveats, the usually more precise analyses of CI chondrites can then be used to refine photospheric abundance estimates, preferably based on the mean composition of the CI group (Lodders, 2003).

Estimates of present solar system abundances have also been further revised as a result of improved understanding of various physical processes. Models of the Sun's evolution and interior show that observed photospheric abundances (relative to hydrogen) are lower than those of the proto-Sun, because helium and other heavy elements have settled toward the Sun's interior over its evolutionary lifetime (e.g. Pinsonneault, 1997; Talon, 2008).

Additionally, Allende Prieto et al. (2001, 2002) presented downward revisions of the solar abundances of oxygen and carbon, partly as a result of detailed modeling of solar granulation, and partly by accounting for departures from local thermodynamical equilibrium. This is important because these two abundant elements govern much of the chemistry of other less abundant elements. A lower absolute oxygen abundance lowers the condensation temperatures of O-bearing compounds, while the C/O ratio further influences many condensation temperatures.

Reference abundances Lodders (2003) determined a reference set of both photospheric and protosolar abundances, derived from such an evaluation of solar and meteoritic abundance determinations. Table 1 of Lodders (2003) gives their estimated photospheric and meteoritic abundances for all 83 naturally occurring elements. Their Table 2 gives the corresponding solar system abundances, i.e. the inferred protosolar values, unfractionated with respect to hydrogen. Their Table 6 gives isotopic abundances in the solar system.

12.5 Constraints on formation

Age and early chronology The Sun's age is loosely defined as the time elapsed since it arrived on the zero-age main sequence. This in turn may be defined as the point in time when nuclear reactions just began to dominate gravitation as the primary energy source. It is estimated from the latest of a sequence of helioseismology studies to be 4.57 ± 0.11 Gyr (Guenther & Demarque, 1997; Dziembowski et al., 1999; Bonanno et al., 2002). The same value, 4.57 ± 0.07 Gyr, was derived from meteoritic dating by Bahcall et al. (1995), although the latter authors state that '... *at what stage of solar evolution this narrow time band represents is not obvious*'.

Absolute ages, primarily from the coupled U–Pb decay routes, and relative ages, based on the ^{26}Al→^{26}Mg

12.5 Constraints on formation

Meteorites: *Meteorites* are fragments of (proto)planetary material that have survived their passage through the atmosphere to land on the Earth. All known meteorites are fragments of either asteroids, the Moon, or Mars, with the former dominating. They display a large diversity of texture and mineralogy. Most are ancient, dating from the first 10 Myr of the solar system formation. Some $10^5 - 10^6$ kg of meteoritic material falls on Earth each day.

Modern meteorite taxonomy is complex (Krot et al., 2007). Some 86% are *chondrites*, which retain some record of geochemical process in the solar neighbourhood. Some 8% are *achondrites*, which underwent subsequent melting and differentiation. Around 5% are iron meteorites which, from their estimated cooling rates of $1-10$ K Myr^{-1} inferred from the Thomson–Widmänstatten structures of Ni-rich taenite and Ni-poor kamacite, are considered to have originated from the cores of once-molten asteroids. The remaining 1% are the stony–iron meteorites, a mixture of iron–nickel metal and silicate minerals.

Chondrites were originally defined as meteorites containing chondrules; now the term signifies a bulk chemical composition, excluding the most volatile elements, similar to that of the Sun. *Carbonaceous chondrites* are those with bulk refractory lithophile abundances, normalised to Si, equal to or greater than that of the solar photosphere. They generally contain carbon compounds, water (as OH within silicate minerals), and other volatiles although, despite the name, they can be relatively poor in carbon.

Chondritic meteorites comprise *chondrules*, spherical bodies that existed prior to their incorporation into meteorites and which show evidence for partial or complete melting; *calcium-aluminium-rich inclusions (CAIs)* composed largely of Ca, Al, and Ti oxides; and a matrix of varying composition and origin, including presolar grains, nebular condensates, and CAI and chondrule fragments.

Meteorites trace key stages underlying the solar system evolution, viz. the initial interstellar medium, protostellar collapse, disk formation, dust condensation, dust coagulation, thermal processing, and planetesimal and planet formation. Their isotopic anomalies and range of oxidation states provide extensive constraints on detailed models of the solar system formation (Apai & Lauretta, 2010).

and ^{182}Hf→^{182}W chronometers amongst others, are providing an increasingly reliable and consistent sequence of timings for events in the early solar system deduced from the meteoritic records (Figure 10.1).

The mm to cm-sized refractory calcium-aluminum-rich inclusions in chondritic meteorites are believed to have been the earliest solids to form, with absolute U–Pb dating indicating formation at 4567.2 ± 0.7 Myr (Allègre et al., 1995; Amelin, 2006; Amelin & Krot, 2007; Wadhwa et al., 2007). Thereafter, relative dating indicates that an interval of up to 3 Myr preceded the formation of the chondrules (Allègre et al., 1995; Amelin et al., 2002; Kita et al., 2005), while the parent bodies of iron meteorites may have formed just after, or a little before, the chondrules (Kleine et al., 2002).

Other meteoritic events (e.g. Kleine et al., 2005a; Scott, 2007; Chambers et al., 2010), the differentiation histories of Earth (e.g. Kleine et al., 2002, 2004; Yin et al., 2002; Halliday, 2004), Mars (e.g. Borg & Drake, 2005; Nimmo & Kleine, 2007), and the Moon (e.g. Kleine et al., 2005b; Touboul et al., 2007, 2009), the formation of Earth's atmosphere (Allègre et al., 1995; Halliday, 2004), and the spatial heterogeneity of short-lived radionuclides (Boss, 2007a), are all now being similarly dated.

Attempts to estimate the age of the Earth itself has a long and varied history (Dalrymple, 1991). In a recent contribution, zircon (ZrSiO$_4$) from Jack Hills in Western Australia has yielded U–Pb ages up to 4.404 Ga, making them the oldest minerals dated on Earth (Wilde et al., 2001). The age of the Earth, corresponding to the end of its accretion, its early differentiation, core formation, and atmospheric outgassing is, as a result, dated at some 100 Myr after the formation of the first meteorites (Allègre et al., 1995).

Transient heating events One of the factors underlying acceptance of the solar nebular theory in the 1970s was the realisation that many meteoritic features could be understood in terms of their condensation from a hot vapour. The first theoretical studies of the condensation of material of solar system composition were published at around that time (e.g. Larimer, 1967; Grossman, 1972), and these reinforced the idea that material in the early solar system had been in a hot gaseous form. But as concluded by Cameron (1978b) *'At no time, anywhere in the solar nebula, anywhere outwards from the orbit of Mercury, is the temperature in the unperturbed solar nebula ever high enough to evaporate completely the solid materials contained in interstellar grains'.*

Nevertheless, as deduced from meteoritic crystalline silicates, chondrules and refractory inclusions, some form of intense and repeated transient heating must have been important in the early solar system. Such heating is not part of the current models for planetary system formation, and the origin of these high-temperature products has been the subject of prolonged investigation (e.g. Boss, 1996; Shu et al., 1996; Rubin, 2000; Jones et al., 2000b; Ghosh et al., 2004; Connolly et al., 2006; Lauretta et al., 2006; Scott, 2007; Wooden et al., 2007; Apai et al., 2010).

While the heating mechanism remains unknown, contenders are plentiful, including lightning-type discharges caused by turbulence in the solar nebula (Cameron, 1966; Whipple, 1966; Pilipp et al., 1992, 1998; Love et al., 1995; Desch & Cuzzi, 2000), the hypothetical X-wind from an early active Sun (Shu et al., 1996, 1997, 2000; Shu, 2001), the decay of ^{26}Al and other short-lived radionuclides (Bennett & McSween, 1996), shock waves (Ciesla & Hood, 2002; Desch & Connolly, 2002; Boss & Durisen, 2005; Desch et al., 2005), X-ray flares (Feigelson & Montmerle, 1999; Joung et al., 2004; Nakamoto et al., 2005), γ-ray bursts and associated discharges (McBreen & Hanlon, 1999; Duggan et al., 2003; McBreen et al.,

> **Geochronology:** *Geochronology*, the scientific determination of absolute and relative ages of the geological record, exploits a variety of radiometric (and stratigraphic) dating techniques, generally based on comparison between the observed abundance of a naturally occurring radioactive isotope (or *radionuclide*) and its decay or daughter products (referred to as *radiogenic nuclides*, whether they are stable or not), using known decay rates (e.g. Dickin, 2005).
>
> One of the oldest and most refined methods is the uranium–lead scheme, used for one of the earliest accurate estimates of the age of the Earth (Patterson, 1956). It achieves a precision of ~1 Myr in *absolute ages* over the entire geological record. In its most general form, it exploits the two parallel U–Pb decay routes: $^{238}U \rightarrow {}^{206}Pb$ ($t_{0.5} \sim 4.5 \times 10^9$ yr) and $^{235}U \rightarrow {}^{207}Pb$ ($t_{0.5} \sim 0.7 \times 10^9$ yr) to determine two independent 'concordant' ages. Discrepancies in the two estimates provides constraints on sample contamination or radiogenic loss. The U–Pb isochron dating variant uses a single decay route, usually $^{238}U \rightarrow {}^{206}Pb$, while the Pb isotope or lead–lead dating method makes use of the Pb isotope ratios alone.
>
> The method is frequently applied to the mineral zircon ($ZrSiO_4$), which incorporates U into its crystalline structure but strongly rejects Pb, such that the entire measured lead content is considered to be of radiogenic origin.
>
> Such absolute radiometric dating requires a measurable fraction of the parent radionuclide to remain in the sample. For bodies dating back to the beginning of the solar system, this requires extremely long-lived parent isotopes, which in turn implies that the derived absolute ages are inevitably imprecise.
>
> Accurate *relative ages* can be determined from the decay products of short-lived isotopes that are themselves no longer present. From their measured decay products, it is inferred that various short-lived radionuclides, including ^{10}Be, ^{26}Al, ^{53}Mn, ^{60}Fe, ^{129}I and ^{182}Hf, were injected into the early solar nebula, possibly from a supernova explosion (e.g. Meyer & Zinner, 2006; Bizzarro et al., 2007).
>
> The $^{26}Al \rightarrow {}^{26}Mg$ chronometer is widely used to estimate the relative ages of chondrules. With $t_{0.5} \sim 720\,000$ yr, the parent nuclide is no longer extant, and dating exploits the deviation from the natural abundance of ^{26}Mg in comparison with the ratio of the stable isotopes ^{27}Al and ^{24}Mg (e.g. Krot et al., 2005b). Relative ages can be similarly estimated from the (short-lived decay) chronometers $^{53}Mn \rightarrow {}^{53}Cr$, $^{60}Fe \rightarrow {}^{60}Ni$, and $^{182}Hf \rightarrow {}^{182}W$.

2005), and impact heating by partially molten planetesimals (Hutchison & Graham, 1975; Hutchison et al., 1988; Rubin, 1995; Krot et al., 2005a).

The mass of Mars Time-dependent calculations of the structure and evolution of the primordial solar nebula during the viscous diffusion stage suggest that the value of the α parameter of the protoplanetary disk (§10.3.3) was about 0.01 (Ruden & Lin, 1986).

The assumption of a corresponding surface density distribution of the solar nebula which is smooth and monotonic, provides a reasonable match to the mass distribution in the solar system, which shows an overall radial trend from Uranus to Neptune approximated by $r^{-3/2}$ (Weidenschilling, 1977b).

In comparison with this monotonic trend, however, the zones of Mercury, Mars and the asteroids, appear strongly depleted in mass (Weidenschilling, 1975, 1977b). Processes which might take mass out of the Mars region have been proposed. Thus Safronov (1969) and Weidenschilling (1975) suggested that planetesimals from the region of Jupiter would be perturbed into eccentric orbits, passing through the Mars/asteroid region, colliding and disrupting the planetesimals that had formed there.

In planet accretion models in which α is a function of both radius and time, since both T and Σ are functions of radius and time, Jin et al. (2008) found a sustained mass decrement to be present in the surface density profile at the radius corresponding to Mars. They suggest that this leaves Mars gaining little mass during the final stages of chaotic growth.

The origin of water on Earth The location of the snow line at ~3 AU in the solar nebula (§11.2.2) suggests that water in the early solar system was probably restricted to its outer regions: the outer asteroid belt, the giant planet regions, and the Edgeworth–Kuiper belt. The question then arises as to the origin of the water on Earth (Drake et al., 2004).

Chyba (1990) argued that the terrestrial planets acquired their water oceans, and other surface volatiles, through late-accretion of comets and carbonaceous asteroids during the 'late heavy bombardment' about 4 Gyr ago (§12.7). He considered the competition between impact delivery of new volatiles and impact erosion of those already present (§11.8.2) and argued that, for the inner solar system planets, the net accumulation of water was favoured.

From dynamical models of the primordial evolution of solar system bodies, Morbidelli et al. (2000) found it plausible that the Earth has accreted water throughout its formation history, from the earliest phases when the solar nebula was still present, extending to the late stages of gas-free sweep-up of scattered planetesimals.

They concluded that asteroids and comets from the Jupiter–Saturn region were the first water deliverers, when the Earth was less than half its present mass. The bulk of the water presently on Earth was subsequently carried by a few planetary embryos, originally formed in the outer asteroid belt and accreted by the Earth during the final stages of its formation.

Finally in their chronology, and accounting for $\lesssim$ 10% of the present mass of H_2O, was the accretion of water-bearing comets from the Uranus–Neptune region and from the Kuiper belt. Simulations by Raymond et al. (2004, 2007a) gave somewhat comparable contributions from a few planetary embryos arriving stochastically, and from millions of planetesimals arriving in a more statistically robust manner.

Raymond et al. (2005a) found that the presence of

an outer giant planet such as Jupiter does not enhance the H_2O content of the terrestrial planets, but rather decreases both their formation and their H_2O delivery timescale. The volume delivered is sensitive to the adopted orbital eccentricities of Jupiter and Saturn (O'Brien et al., 2006): the present eccentricities yield almost none, but with large amounts delivered for the initially circular orbits predicted by the Nice model (§12.7). Significant H_2O-delivery is also a by-product of the 'dynamical shake-up' model of Thommes et al. (2008c). Similar considerations apply to exoplanets and, presumably, their H_2O-based habitability (Raymond, 2006).

The net result is that the water on Earth had essentially the deuterium/protium ratio typical of the water condensed in the outer asteroid belt. This is in agreement with the observation that the $^2H/^1H$ ratio in the oceans is very close to the mean value of that of the H_2O inclusions in carbonaceous chondrites (Lunine, 2006).

The evolution of the snow line during the early stages of the solar system formation was not monotonic (Garaud & Lin, 2007). This would have consequences for the relative amounts of hydration, and isotopic compositional differences, for Venus, Earth and Mars.

The isotopic composition of some of the Western Australian zircons indicates that there was already water on the surface of the Earth, and therefore a hydrosphere interacting with the crust, at around 4.3–4.4 Ga (Wilde et al., 2001; Mojzsis et al., 2001; Nemchin et al., 2006; Ushikubo et al., 2008).

12.6 Orbit considerations

Resonances The various types of orbital resonances known in the solar system are summarised in the box on page 300. Examples of all types are already known, or inferred, in various multiple exoplanet systems (§2.6), including the first example of a three-body Laplace resonance in the case of GJ 876 (Figure 2.38).

Orbital stability and chaos General considerations on orbital integrations for multiple planetary system, including the concepts of stability and chaos, have been noted in §2.6.1.

The more rapid the planet's orbital movement, the finer the time step needed to integrate its motion. Limited by computational power, until 1991 the only numerical integration of a realistic model of the full solar system was the JPL ephemeris DE102, spanning the interval 1411 BC to 3002 AD (Newhall et al., 1983).

Although very long-term numerical integration of planetary orbits in the solar system can be used to identify the general domains of stability and chaos, detailed results are very sensitive to the knowledge of initial conditions and masses. Certain broad features were already identified in early studies, including the 100 Myr of the LONGSTOP (outer planets) project (Nobili, 1988; Roy et al., 1988; Milani et al., 1989); the 200 Myr of the MIT project (Applegate et al., 1986); the 845 Myr of the Digital Orrery project (Sussman & Wisdom, 1988); the 1 Gyr of the first symplectic integration (or N-body mapping, Wisdom & Holman, 1991); and up to 5 Gyr in more global terms (Laskar, 1994, 1996; Laskar & Robutel, 2001).

LONGSTOP, for example, identified long-periodic variations in the semi-major axes of the outer planets with periods of the order of 1 Myr (Nobili, 1988). In the case of Pluto, Milani et al. (1989) used a 100 Myr integration to confirm the 3:2 mean motion resonance with Neptune, a 19 900-year longitude libration, and a 3.78 Myr libration of its argument of pericentre. Its orbit is chaotic, with a Lyapunov exponent of 1/(20 Myr) (Sussman & Wisdom, 1988), although its small mass means that its chaotic orbit has negligible consequences on the dynamics of the other planets.

Both Sussman & Wisdom (1992) and Laskar (1994, 1996) found that, over several Gyr, the inner solar system is chaotic, showing large orbital diffusion, and a Lyapunov exponent of 1/(5 Myr). The chaotic behaviour originates from two secular resonances, one related to the perihelia and nodes of Mars and Earth, the other related to those of Mercury, Venus and Jupiter. Variations of the eccentricities and inclinations are clearly visible over periods of a few million years (Figure 12.2), behaviour known qualitatively to Laplace and Leverrier (as reported by Laskar, 1992).

In the simulations of Laskar (1994, 1996), Mercury's eccentricity reaches $e \sim 0.5$ at some epochs. While insufficient to cross the orbit of Venus, even small changes in the initial position assumed for the Earth, of ~150 m, can lead to orbits in which Mercury escapes from the solar system or collides with Venus. Similar escape solutions exist for Mars. Although such escapes occur with only low probability, consistent with the continued presence of these bodies in the solar system, their existence demonstrates that the solar system is not strictly stable.

Extreme sensitivity to the initial conditions in dynamical phase space extends to the introduction of higher-order effects. Varadi et al. (2003) and Laskar & Gastineau (2009) carried out long-term numerical integration of the equations of motion which include contributions from the Moon, and adjustments due to the effects of general relativity. While still resulting in some solutions in which Mercury collides with Venus, the latter simulations also reveal solutions in which a decrease in the eccentricity of Mercury induces a change of angular momentum from the giant planets that destabilises all of the terrestrial planets ~3.34 Gyr from now, with possible collisions of Mercury, Mars, or Venus with the Earth.

At the same time, the fact that strong instabilities (collision or escape) occur only on time scales comparable to its age, i.e. of order 5 Gyr, implies that the solar system can be considered as marginally stable. The

> **Solar system resonances:** A substantial advance in the knowledge of the solar system over recent years has revealed more than 300 planetary satellites (counting some 200 for Saturn), many tens of thousands of asteroids with established orbits (and of various classes), as well as comets and Edgeworth–Kuiper belt objects. Complex ring systems are now known to accompany all of the giant planets (Esposito, 2006). Their collective physical structure and dynamical behaviour are determined by gravitational forces, and sculpted by the complex phenomenon of resonance (Peale, 1976). Some examples follow.
>
> *Spin–orbit resonance:* most major satellites are in a 1:1 (or synchronous) spin–orbit resonance, in which the orbital period is equal to the rotation period. It is the end result of retarding torques on a circular orbit, and results (for example) in the same face of the Moon being always directed towards the Earth. Other states are possible: the unexpected 3:2 spin–orbit resonance of Mercury around the Sun was discovered by radar observations (Pettengill & Dyce, 1965), and explained in terms of its unusually large eccentricity ($e = 0.21$) combined with its limited rigidity (Peale & Gold, 1965).
>
> *Mean motion resonance:* this important class of orbit–orbit coupling occurs when two bodies have orbital periods (or mean motions, as defined by Equation 2.8) in a simple integer ratio. Qualitatively, stable orbits result when the two bodies never closely approach. Examples in the solar system include the stable orbit of Pluto and the Plutinos, despite crossing the orbit of Neptune, because of the 3:2 Neptune–Pluto resonance; and the Trojan asteroids, preserved by a 1:1 resonance with Jupiter. Examples of unstable resonances include: (a) the largely empty lanes in the asteroid belt, the Kirkwood gaps; these correspond to mean-motion resonances with Jupiter (most prominently at the 4:1, 3:1, 5:2, and 2:1 Jovian resonances), in which almost all asteroids have been ejected (while there are asteroid concentrations at the 3:2 and 1:1 resonances); (b) resonances between Saturn and its inner satellites, giving rise to gaps in the rings of Saturn; (c) the 1:1 resonances between bodies with similar orbital radii causing large solar system bodies to 'clear out' the region around their orbits by ejecting nearly everything else around them; this effect was incorporated into the IAU 2006 definition of a planet (§1.5).
>
> *Laplace resonance:* this is a specific case of mean motion resonance, known to Laplace, which occurs when three (or more) orbiting bodies have a simple integer ratio between their orbital periods. Jupiter's inner satellites Ganymede, Europa, and Io are in a 1:2:4 Laplace resonance, with Ganymede completing 1 orbit in the time that Europa makes 2, and Io makes 4. Their mean motions (Equation 2.8) are related by $n_I - 3n_E + 2n_G = 0$, satisfied to nine significant figures (Peale, 1976).
>
> *Secular resonance:* a secular resonance occurs when one of the eigenfrequencies of the perturbing system is in a simple numerical ratio with the rate of change of the (proper) longitude of pericentre or the (proper) longitude of the ascending node (Murray & Dermott, 2000, Section 7.11). The consequence is that a small body in secular resonance with a much larger one will precess at the same rate, steadily changing its eccentricity and inclination. An example is the secular resonance between asteroids and Saturn. Asteroids which approach it have their eccentricity slowly increased until they become Mars-crossers, at which point they are usually ejected from the asteroid belt due to a close encounter with Mars.
>
> *Kozai resonance:* this is experienced by bodies on highly inclined orbits, in which inclination and eccentricity oscillate synchronously, i.e. increasing one at the expense of the other, whilst conserving (the Delaunay quantity) $H_K = \sqrt{(1-e^2)} \cos i$ (Kozai, 1962). An important consequence is the absence of bodies following highly-inclined orbits, since their growing eccentricity would result in a small pericentre, typically leading to collision or destruction by tidal forces.

same simulations also show that the inner solar system is dynamically full, in the sense that any test objects introduced escape on time scales $\ll 5$ Gyr. This dynamical filling has recently been noted for many multiple exoplanet systems (§2.6).

The outer solar system is very stable (Figure 12.2), and also dynamically full (Gladman & Duncan, 1990; Holman & Wisdom, 1993; Levison & Duncan, 1993).

One general conclusion from these types of simulation is that the solar system, at the end of its formation, was most probably significantly different from its present configuration. Laskar (1994, 1996) already concluded that exoplanet systems with only one or two planets are improbable. If they do exist, they are likely to be crowded with the original planetesimals not ejected by planetary perturbations.

Multiple systems with large outer planets and small inner planets should be subject to instabilities similar to those evident in these solar system simulations, including instabilities of planet obliquities along with the attendant varying 'climate' implications (Ward, 1973; Laskar & Robutel, 1993).

Planetary and solar obliquities The spins of the solar system planets are neither perfectly aligned nor randomly oriented. Their *obliquities*, the angle between their spin and orbital angular momentum vectors (Cox, 2000), range from $0°.03$ (Mercury), $3°.1$ (Jupiter), $23°.4$ (Earth), $25°.2$ (Mars), $27°$ (Saturn), $30°$ (Neptune), $98°$ (Uranus), $118°$ (Pluto), and $177°$ (Venus). The solar obliquity is $7°$ with respect to the mean orbital plane, and $6°$ with respect to the total angular momentum, or *invariable plane*.

The wide range of obliquities are demonstrably inconsistent with an isotropic distribution, with too many small values (Tremaine, 1991), and a degree of prograde rotation perhaps indicative of semi-collisional accretion (Schlichting & Sari, 2007). If the planets had grown smoothly and continuously by the accretion of small bodies and gas in an isolated flat disk, their obliquities should be zero. Accretion of a relatively small number of massive bodies, with a nonzero velocity dispersion normal to the disk, would impart the growing protoplanet with a randomly directed component of spin angular momentum (Kokubo & Ida, 2007), but would also lead to

12.6 Orbit considerations

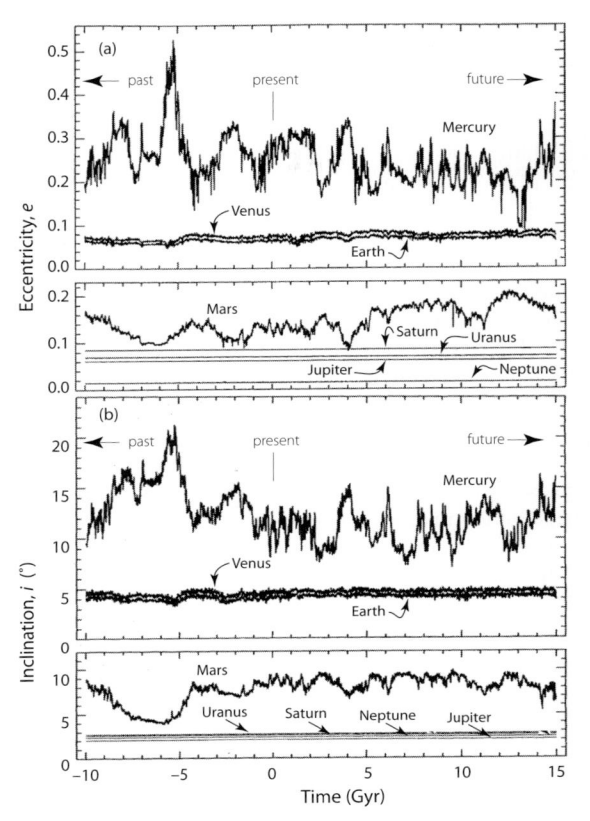

Figure 12.2: Numerical integration of the orbits of the solar system planets, 10 Gyr into the past and 15 Gyr into the future, showing (a) eccentricities, and (b) inclinations with respect to the fixed ecliptic at J2000, each separated into two panels for clarity. For each planet, values plotted are the maxima over 10 Myr. The inner planets show large and irregular variations, especially for Venus and Mars, while the orbits of the giant planets (Jupiter, Saturn, Uranus, Neptune) are regular, and their curves of maximum eccentricity and inclination appear as straight lines. From Laskar (1994, Figure 1), reproduced with permission © ESO.

orbital eccentricities and inclinations larger than those observed (Harris & Ward, 1982). As a result, a combination of effects is likely to be responsible for the observed distribution.

Planetesimal or protoplanet collisions may be a contributing factor, and one implying stochastic rather than smooth accretion (Lissauer & Kary, 1991; Lissauer & Safronov, 1991; Dones & Tremaine, 1993b), also in the case of the Earth's spin state (Dones & Tremaine, 1993a).

Excitation of the obliquities of the giant planets also arises if their spin axis precession frequencies passed through resonance with their orbital precession frequencies during migration (Harris & Ward, 1982). For Uranus, both tilt due to collision (Bergstralh et al., 1991) and tilt during migration (Boué & Laskar, 2010) have been proposed. A Jupiter–Saturn resonance crossing during divergent migration has been proposed as an explanation of both the small obliquity of Jupiter and the large obliquity of Saturn (Mosqueira & Estrada, 2006; Brunini, 2006; Lee et al., 2007). The origin of Saturn's obliquity has also been attributed to gravitational perturbations by Neptune (Ward & Hamilton, 2004; Hamilton & Ward, 2004).

None of the obliquities of the terrestrial planets are believed to be primordial. Secular orbit perturbations are likely to have stimulated large and chaotic variations in the past. Tidal dissipation has since stabilised the obliquities of Mercury and Venus, while that of Mars, currently 25°, is still in a largely chaotic regime ranging from 0–60° (Ward, 1973; Laskar & Robutel, 1993).

Earth's obliquity may have been stabilised by the Moon (Laskar & Robutel, 1993; Laskar et al., 1993; Neron de Surgy & Laskar, 1997; Atobe & Ida, 2007). It may have had a large primordial obliquity of ~70° as a result of the Moon-forming impact, with a subsequent secular decrease attributable to core–mantle torques, with climatic consequences during the Proterozoic–Phanerozoic resulting from passages through critical values (Williams, 1993).[4]

Further contributions to the observed obliquities may arise from external torques, causing the total angular momentum of the planetary system to change after formation. This may be due to passing stars or molecular clouds (Heller, 1993), tidal interactions with other protostars (Larson, 1984; Wesson, 1984), orbiting satellites (Atobe & Ida, 2007), or a distant outer planet (Gomes et al., 2006), or the varying angular momentum of infalling material during protostellar collapse due to cloud core inhomogeneities.

The effects of inhomogeneous infall on the obliquities of the (outer) planets was assessed by Tremaine (1991), in a model which also may account for the

---

[4] At the present epoch, and as a consequence of N-body interactions, Earth's obliquity oscillates with an amplitude of 2°.4 (22°.1–24°.5) over a period of ~41 000 yr. The collective effects on the Earth's climate of long-term variations in this axial tilt, along with changes in orbital eccentricity and precession, are usually referred to as *Milankovitch cycles*, after the contributions of Milankovitch (1941) [in English translation, Milankovitch 1969]. In the absence of secular changes in the semi-major axis a, orbital effects on the Earth's climate are described by two independent parameters, the obliquity ϵ, and the *climatic precession index*, $\Delta e \sin \bar{\omega}$ (where e is the eccentricity, and $\bar{\omega}$ is the longitude of perihelion). Although any dependency was largely discarded in the 1960s, geological data from deep ocean cores led to the theory's revival in the 1970s. Imbrie (1982) reviews the development of the astronomical theory of the Pleistocene ice ages, the role of the Earth's orbit, and the key publications in astronomy, climatology, and geology, pre-1982 (e.g. Berger, 1976, 1977a,b, 1980; Hays et al., 1976; Weertman, 1976; Schneider, 1979; Imbrie & Imbrie, 1980; Ghil & Le Treut, 1981; Kutzbach, 1981). For an introduction to the additional effects of variations in solar irradiance, see, e.g. Keller (2004).

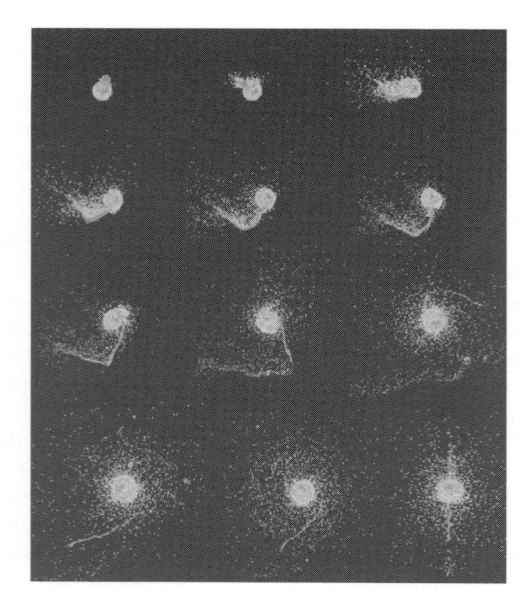

Figure 12.3: Hydrodynamic simulations of the Moon's formation as a result of an oblique giant impact between the Earth and a 6×10^{23} kg protoplanet. Results are projected onto the impact plane at times t = 0.3, 0.7, 1.4, 1.9, 3, 3.9, 5, 7.1, 11.6, 17 and 23 h after impact (left to right); the final frame is t = 23 h viewed edge-on. From Canup & Asphaug (2001, Figure 1), by permission from Macmillan Publishers Ltd, Nature, ©2001.

present obliquity of the Sun. With a similar premise of turbulent and chaotic infall, especially when truncated by stellar encounters, significant misalignment between the final stellar rotation axis and disk spin axis was also found by Bate et al. (2010).

Orbits of the terrestrial planets The low eccentricities and low inclinations of the terrestrial planets provide a specific constraint for the core accretion model of the gas giants, possibly explicable in terms of the 'dynamical shake-up' model of Nagasawa et al. (2005) and Thommes et al. (2008c). This is described further in §10.6.1.

Planetary satellites Most of the planets possess satellites, with more than 200 classified in the case of Saturn. They provide evidence that the giant planets were themselves probably surrounded by disks early in the formation of the solar system, although some appear to have been captured (§10.7), while an impact origin has been proposed in the case of the Moon.

Origin of the Moon It was originally hypothesised that the Moon had formed by centrifugal spin-off from the Earth (Darwin, 1898; Binder, 1974). This was later replaced by the paradigm of an impact-induced event (Hartmann & Davis, 1975; Cameron & Ward, 1976; Cameron & Benz, 1991; Canup & Righter, 2000).

Hydrodynamic simulations provide a reasonably consistent dynamical account of such an impact origin, positing an oblique and late-stage giant collision between the Earth and a 6×10^{23} kg protoplanet, i.e. a Mars-mass object, which has been named 'Theia' (Canup et al., 1999; Canup & Asphaug, 2001; Canup, 2004a,b; Halliday, 2008).

The simulations of Canup & Asphaug (2001) show that the cores of the impacting bodies rapidly coalesce (Figure 12.3), while the orbital liquid and vapour debris disk solidifies and accretes into a single large satellite in ~1000 yr (Kokubo et al., 2000; Pahlevan & Stevenson, 2007). Core formation models suggest that both the Earth and the impactor were already differentiated by the time of the impact (Tonks & Melosh, 1992).

Evidence for an impact origin includes lunar samples indicating that its surface was once molten. The low Fe abundance in the lunar mantle suggests that the impact happened near the end of the Earth's accumulation and differentiation (§10.5), although questions concerning volatiles and isotope ratios remain (Halliday, 2000). These are perhaps circumvented if the impactor formed by accretion at the same radial distance in a Lagrange L4 or L5-type orbit (Belbruno & Gott, 2005). The lack of a clear ^{182}W excess in lunar samples (box on page 298) implies that the impact took place > 50 Myr after the formation of the solar system (Touboul et al., 2007).

The common occurrence of low-velocity oblique impacts in simulations of post-oligarchic stages of terrestrial planet formation suggests that planets like Earth and Venus are likely to experience at least one such impact during formation. The consequence is that large impact-generated satellites may be common by-products of exoplanet formation (Agnor et al., 1999).

12.7 Planetesimal migration in the solar system

An argument in support of the existence of planetesimals, and their role in planet formation, is that most bodies in the solar system have evidently been bombarded by objects of comparable size, while some planetesimal-like bodies still remain in the form of asteroids and comets.

Further circumstantial evidence appears in the form of observational consequences of migration within a planetesimal disk. The simplified picture of such migration (§10.8.3) suggests that the effects on the overall structure of the solar system might be relatively minor, at least in terms of the general spacing of the major planets. Small orbital migration may nevertheless have had significant consequences in shaping the detailed architecture and phenomena observed in the solar system, and particularly in its outer parts.

Evidence for this began to emerge from numerical simulations carried out by Fernandez & Ip (1984), who found that during an ensemble of such scattering events Jupiter, as the main ejector of planetesimals,

12.7 Planetesimal migration in the solar system

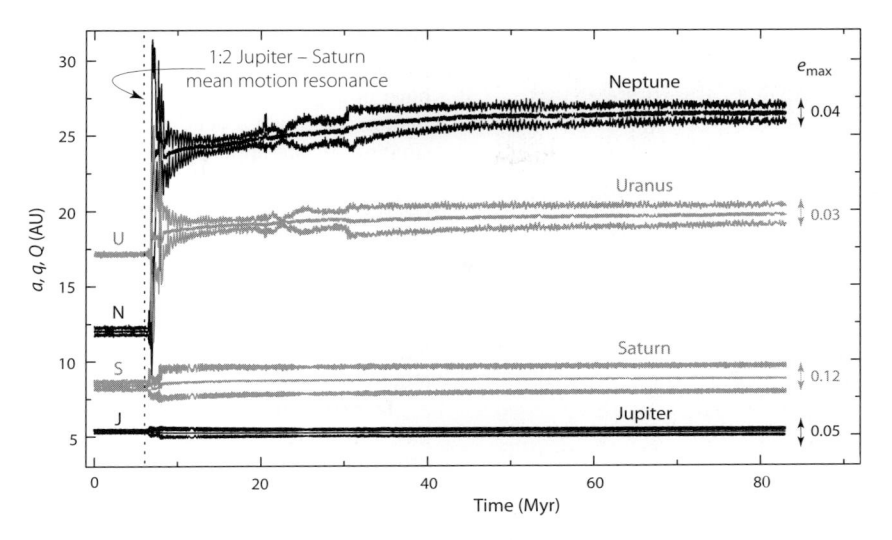

Figure 12.4: The starting configuration of the outer solar system planets given by the Nice model, with their subsequent orbital evolution from an N-body simulation with $35M_\oplus$ of disk planetesimals in 3500 particles out to 30 AU. The three curves for each planet indicate the semi-major axis a, and their minimum (q) and maximum (Q) heliocentric distances. The vertical dashed line marks the epoch of the 1:2 Jupiter–Saturn mean motion resonance capture. During subsequent dynamical interactions, the eccentricities of Uranus and Neptune can exceed 0.5, and in 50% of simulation runs (including this), they exchange orbits. The maximum eccentricity over the last 2 Myr of evolution is indicated. From Tsiganis et al. (2005, Figure 1), by permission from Macmillan Publishers Ltd, Nature, ©2005.

loses orbital angular momentum and so moves slightly inwards. On average, Neptune, Uranus, and to some extent Saturn, gain orbital angular momentum in their interactions with planetesimals, and they are displaced more significantly outwards.

A similar migratory effect was proposed for planetary cores which evolved into orbits resembling those of Uranus and Neptune (Thommes et al., 1999).

This type of radial migration could then be responsible for the eccentric orbit of Pluto as a result of resonance capture (Malhotra, 1993a, 1995). As Neptune migrated outwards, it could have captured Pluto and a number of other small Kuiper belt objects into mean motion resonances (§2.6), with the eccentricities of the captured bodies increasing thereafter as Neptune continued its outward migration. For a body captured into resonance, integration of the first-order perturbation equations implies that its eccentricity increases as Neptune's orbit expands as (Malhotra, 1995, eqn 5)

$$e_{\text{final}}^2 \simeq e_{\text{initial}}^2 + \frac{1}{j+1} \ln\left(\frac{a_{\text{N,final}}}{a_{\text{N,initial}}}\right), \quad (12.1)$$

where $a_{\text{N,initial}}$ is the semi-major axis of Neptune's orbit at the time of resonance capture, $a_{\text{N,final}}$ is its final value, and the particle is assumed locked into an exterior $j+1:j$ orbital resonance.

Accordingly, if Pluto was on an initially circular orbit, and was subsequently captured into the 3:2 Neptune resonance, its current $e = 0.25$ implies that Neptune was at $a_\text{N} \sim 25$ AU and Pluto at 33 AU at the point of capture, compared with their current locations at 30.1 AU and 39.5 AU respectively.

The large population of Kuiper belt objects in 3:2 resonance with Neptune appears consistent with this same outward migration (Hahn & Malhotra, 2005; Murray-Clay & Chiang, 2005, 2006).

The Nice model More generally, Tsiganis et al. (2005) have proposed a specific starting configuration for the outer solar system planets, which they hypothesise existed before an extended phase of planetesimal scattering altered their orbits. In this so-called *Nice model*, the giant planets all formed on circular and coplanar orbits, Jupiter started marginally further out from the Sun (at 5.45 AU, compared to 5.2 AU at present), the other giant planets started in a more compact configuration within about 15–17 AU, and a sea of planetesimals of total mass $30-50M_\oplus$ resided interior to 30 AU with a density falling off linearly with heliocentric distance (Figure 12.4).

Provided that Jupiter and Saturn crossed their 1:2 resonance on their migratory paths as they interacted with the planetesimal disk, N-body simulations can reproduce a number of primary characteristics of the giant planet orbits, namely their present semi-major axes and eccentricities, and their mutual inclinations with respect to the mean orbital plane of Jupiter of some 2°. The dynamics of the Jupiter Trojans would be significantly influenced by such a resonant crossing (Marzari & Scholl, 2007).

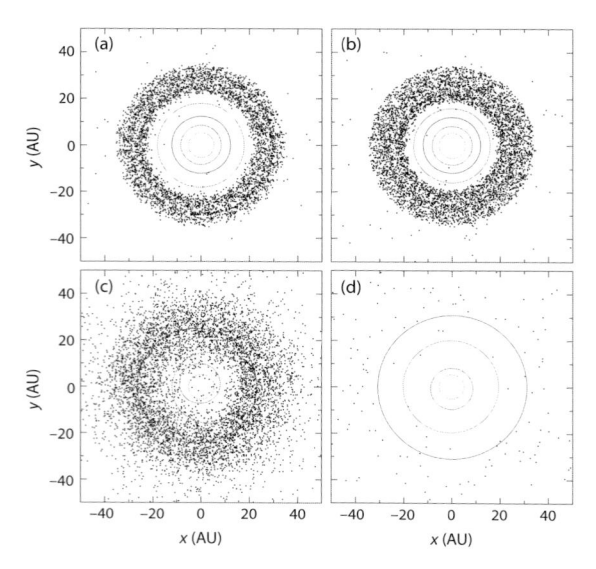

Figure 12.5: Numerical simulations of the lunar heavy bombardment. Planetary orbits and disk planetesimals are projected on the initial mean orbital plane. The four giant planets were initially on nearly circular coplanar orbits, with $a = 5.45$, 8.18, 11.5 and $14.2\,AU$, and the planetesimal disk, of total mass $35 M_\oplus$, extended from 15.5–$34\,AU$. Panels represent the system at (a) the start of planetary migration (100 Myr); (b) just before the start of the lunar heavy bombardment (879 Myr); (c) just after its start (882 Myr); and (d) 200 Myr later when only 3% of the initial disk mass remains, and the planets have achieved their final orbits. From Gomes et al. (2005, Figure 2), by permission from Macmillan Publishers Ltd, Nature, ©2005.

Lunar heavy bombardment One of the implications of the increasing eccentricities of the ice giants as Saturn crossed the 1:2 Jupiter resonance would have been the penetration of their orbits into the outer planetesimal disk, resulting in a brief phase of rapid and enhanced planetesimal scattering (Figure 12.5).

Gomes et al. (2005) associated this phenomenon with the *late heavy bombardment* of the Moon. Also known as the *lunar cataclysm*, this is a period around 4.1–3.8 Gyr ago during which a large number of impact craters are believed to have formed on the Moon, and by inference also on Mercury, Venus, Earth, and Mars.

Evidence for the impact clustering comes mainly, but not exclusively, from radiometric ages of impact melt rocks collected during the Apollo 15–17 lunar missions (Tera et al., 1974), with the subsequent literature being substantial albeit somewhat discordant (e.g. Hartmann et al., 2000; Morbidelli et al., 2001; Strom et al., 2005; Bottke et al., 2007; Thommes et al., 2008a; Ćuk et al., 2010).

With this association, the lunar impact chronology then fixes the date of the inferred resonant crossing to ~700 Myr after the solar system formation. Jupiter's Trojan asteroids may have been captured into their current orbits at about the same time (Morbidelli et al., 2005).

Dynamical capture of irregular satellites The distribution and detailed properties of planetary satellites have not figured prominently in models of planet formation, although a need for additional sources of dissipation at late epochs has been identified (Raymond et al., 2006c). The irregular satellites of the giant planets, dormant comet-like objects that reside on both stable prograde and retrograde orbits, provide one specific testing ground which has been examined. Their size distributions and total numbers are comparable to one another, with the observed populations at Jupiter, Saturn, and Uranus having shallow power-law slopes for objects larger than 8–10 km in diameter.

One recent scenario suggests that they were dynamically captured during the same planetesimal-dominated orbital migration phase of the outer planets, in which multiple close encounters between the giant planets allowed scattered comets near the encounters to be captured via three-body reactions. This suggests that the irregular satellites should be related to other dormant comet-like populations produced at the same time from the same disk.

Bottke et al. (2010) explained the similarity of their overall size distribution, and their differences compared with those of the Trojan asteroids, Kuiper belt, and scattered disk objects, by collisional evolution of a captured population of Trojan-like asteroids. The simulated collisions produced some 0.001 lunar masses of dark dust at each giant planet as a by-product, reminiscent of the layer of dark carbonaceous chondrite-like material seen on numerous outer planet satellites such as Callisto, Titan, Iapetus, Oberon, and Titania.

Other implications The Nice model, and the accompanying orbital evolution of the giant planets, has been adopted as a reference for various other investigations of the detailed solar system structure.

These studies include formation of the giant planets (Desch, 2007; Benvenuto et al., 2009) and the migration of Neptune (Gomes et al., 2004), constraints on outer solar system chronology (Johnson et al., 2008b), formation of the terrestrial planets (O'Brien et al., 2006; Brasser et al., 2009), the delivery of H_2O to the Earth (O'Brien et al., 2006), implied encounters between Saturn and Uranus or Neptune (Morbidelli et al., 2009), accretion in the inner solar system (Raymond et al., 2009c), resonances in the formation of the Kirkwood gaps (Minton & Malhotra, 2009), the resonant structure of Jupiter's Trojan asteroids (Robutel & Bodossian, 2009), formation of the Kuiper belt (Booth et al., 2009; Batygin & Brown, 2010), and formation and dynamical evolution of Neptune's Trojan asteroids (Lykawka et al., 2010).

Similar planetesimal disk contributions have been adopted in exoplanet studies, to assess the extent and importance of the planetesimal contribution to migration. These are considered further in §10.8.4.

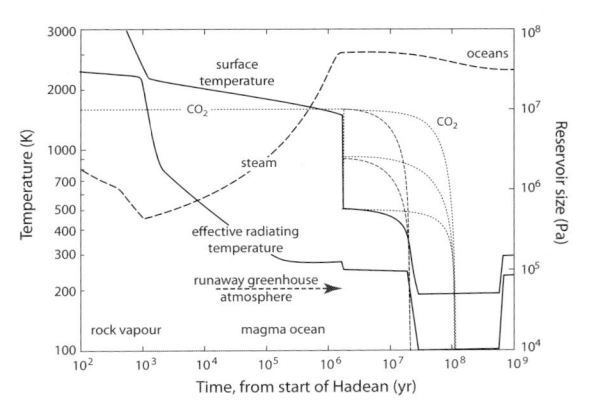

Figure 12.6: Energetically self-consistent schematic of the Earth's early atmosphere, showing the temperature, H_2O, and CO_2 content during the Hadean, between 4.6–3.8 Gyr ago (the start of the Hadean, beginning with the Moon-forming impact, is at the left). The phase of CO_2 subduction is shown with an assumed time scale of 20 Myr (long dashes) or 100 Myr (short dashes). From Zahnle et al. (2010, Figure 3).

The broad consistency between the consequences of planetesimal-driven migration, and the numerous facets of the complex architecture of the solar system, lends further weight to the overall paradigm of planet formation according to the solar nebula hypothesis.

12.8 Atmosphere of the Earth

As an instance of terrestrial-type atmospheres (§11.8), the Earth's is considered to have originated principally as a result of accretion, with the capture of nebula volatiles probably having had a minor role (Lewis & Prinn, 1984; Zahnle et al., 2010, and references).

The Earth's atmospheric composition, and its evolution over time, have been estimated from three classes of models of relevance over successive epochs: those derived from planet formation models, those over geological time scales for which better observational constraints exist, and more recent and definitive constraints on gases and temperatures from ice-core records.

Earth's early atmosphere Zahnle et al. (2007) have presented a possible time line of the early atmosphere, starting with the Moon-forming impact about 4.6 Gyr ago, and extending over the ~0.8 Gyr of the Hadean (or pre-Archean) eon. The start of the Archean is characterised by the appearance of the geological record on Earth, about 3.8 Gyr ago.

Figure 12.6 illustrates their speculative narrative, as summarised by Zahnle et al. (2010). The earliest phase was probably marked by a hot silicate atmosphere that would have cooled and condensed over ~1000 yr. Volatile elements were partitioned between the atmosphere and the mantle according to their solubility in the silicate melt: H_2O and S predominantly entering the mantle, with CO, CO_2, and other gases entering the atmosphere. With the Moon-forming impact likely to have eroded a significant fraction of the atmosphere, but leaving much of the water (Genda & Abe, 2003, 2005), the subsequent cooling would have been controlled by the hot H_2O (steam) atmosphere. Fe sank through the mantle to the core, and the silicate/magma mantle solidified from the interior outwards over a period of ~2 Myr. At this stage, the atmospheric gas composition was mostly (in decreasing fraction) H_2, CO, H_2O, CO_2, and N_2.

After solidification (freezing), the mantle would no longer convect as easily, the geothermal heat flow would have dropped, the runaway greenhouse atmosphere inhibited, and the steam rained out to form hot H_2O oceans with a high salinity due to the dissolved atmospheric NaCl. At this point, the surface temperature had dropped to ~500 K, and the atmosphere contained $\sim 1-2 \times 10^7$ Pa of CO_2.

The resulting CO_2 atmosphere, liquid H_2O ocean, and basaltic surface crust formed an unstable combination (Urey, 1952), with the CO_2 reacting to form carbonates and thereby removed from the atmosphere (Sleep & Zahnle, 2001; Zahnle et al., 2007, their figure 11). Hydrothermal circulation through the oceanic crust on the sea floor, combined with rapid mantle turnover, may have subducted $\sim 10^7$ Pa of CO_2 from the atmosphere to the mantle in $\lesssim 10$ Myr (Sleep, 2010).

Following the subduction of CO_2, and in the absence of another greenhouse gas, the Earth's surface in the late Hadean may have been very cold, ~220 K, because of the lower luminosity of the young Sun. Thin ice-covered oceans may have had liquid water around volcanoes, with other liquid episodes following the hundreds or thousands of intermittent asteroid impacts continuing from the residual planetesimal accretion phase of the Earth's formation. These freezing pools, as sites of concentrated HCN and H_2CO, may have been favourable for the origin of life (Bada et al., 1994), perhaps accelerated by lightning (Chameides & Walker, 1981) or bolide impacts (Fegley et al., 1986; Chyba & Sagan, 1992).

Evolution of CO_2, CH_4, and O_2 A schematic of the relative concentrations of CO_2, CH_4, and O_2 over the entire history of the Earth, from Kasting (2004), is shown in Figure 12.7.

The luminosity of the Sun at around the epoch of formation of the Earth, around 4.6 Gyr ago, was significantly lower than it is today, at around $0.7 L_\odot$. Nevertheless, the fact that there is no evidence in the geological record for widespread glaciation until about 2.3 Gyr ago, suggests that the Earth was probably warmer then than it is today.

Greenhouse gases are presumed to have been responsible for resolving this *faint Sun paradox*. Early ideas that NH_3 was responsible (Sagan & Mullen, 1972) were revised when it became evident that NH_3 would be

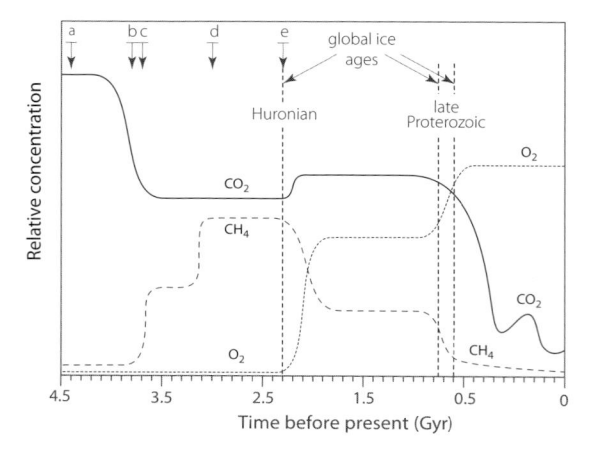

Figure 12.7: Relative concentrations of CO_2, CH_4, and O_2 over Earth's history. Key epochs are: (a) high concentrations of CO_2 compensate for the lower luminosity of the young Sun; (b) the first microscopic life begins consuming CO_2; (c) methanogens start to contribute to the atmosphere; (d) the appearance of O_2-producing bacteria; (e) the appearance of atmospheric O_2. The epoch of the origin of the CH_4-producing microbes is somewhat arbitrary. Adapted from Kasting (2004, Figure 3).

rapidly destroyed by ultraviolet radiation in an O_2-free atmosphere. In consequence, CO_2 was considered as the next most likely candidate (Wigley & Brimblecombe, 1981, and references).

This was in turn revised, in view of the absence of the mineral siderite, implying that CO_2 concentrations were far lower than would be necessary to account for the required greenhouse warming (Rye et al., 1995). Methanogens, CH_4-producing microbes feeding directly on H_2 and CO_2, have been hypothesised as being responsible for this period of greenhouse warming (Pavlov et al., 2000; Kasting & Siefert, 2002; Pavlov et al., 2003; Ueno et al., 2006), although microbial life-forms producing organosulphur-based compounds such as methanethiol (CH_3SH) have also been considered (Pilcher, 2003). An *anti-greenhouse effect* caused by a high-altitude polymerised hydrocarbon haze, may have provided negative feedback to stabilise both atmospheric temperature and composition.

Evidence for the progressive oxidation of Earth's atmosphere is well documented geologically, but the underlying causes remain uncertain. In the scenario described by Holland (2009), volcanic outgassing, in which the H_2/H_2O ratio has remained constant over time but where the CO_2/H_2O and SO_2/H_2O ratios have increased, combined with recycling of CO_2, SO_2, and H_2O, can account for many of the major steps in the oxygenation of the atmosphere (Figure 12.7). Holland (2009) details the relevance processes which may have operated during the 'great oxidation event' around 2.4 Gyr ago (Holland, 2002, 2006), and the long period of redox stability between 1–2 Gyr ago (the 'boring billion').

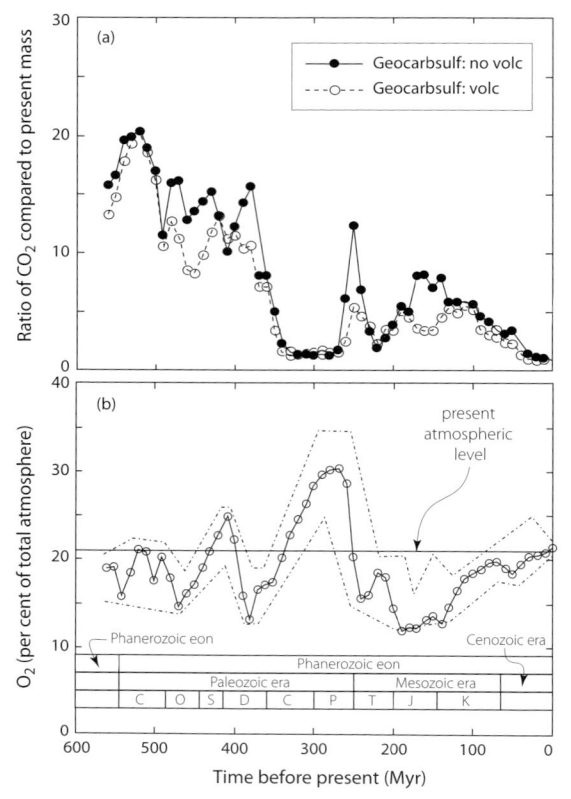

Figure 12.8: Atmospheric levels of CO_2 and O_2 during the Phanerozoic eon (from ~540 Myr ago to the present time): (a) ratio of CO_2 atmospheric mass compared to the present (the weighted mean for the past 1 Myr) from the `Geocarbsulf` *model of Berner (2006). The terms 'volc' and 'no volc' refer to the inclusion, or not, of variable volcanic weathering. (b) O_2 abundance, as a percentage of the total atmospheric mass, also from the* `Geocarbsulf` *model. Dashed lines indicate estimated error ranges. From Berner (2006, Figures 18 and 20).*

Snowball Earth Rising O_2 levels, and the emergence of primitive cyanobacteria and the destruction of CH_4 (Cloud, 1972; Farquhar et al., 2000, 2001; Pavlov et al., 2001; Falkowski, 2005; Raymond & Segrè, 2006), are believed to have led eventually to a significant drop in global temperatures and the emergence of a 'snowball Earth', firstly at around 2.3 Gyr ago (the Huronian glaciation) and, subsequently, the late Proterozoic glaciations ~750 and 600 Myr ago (e.g. Schrag et al., 2002; Hoffman & Schrag, 2002).

The Phanerozoic Better geological constraints on the CO_2 and O_2 atmospheric concentrations are available since the start of the Phanerozoic eon (e.g. Berner & Kothavala, 2001; Berner, 2001, 2006). As the current eon in geological time scale, its beginning, at around 540 Myr ago (starting with the Paleozoic era and, within that, the Cambrian period), was marked by the appearance of diverse hard-shelled animals.

12.8 Atmosphere of the Earth

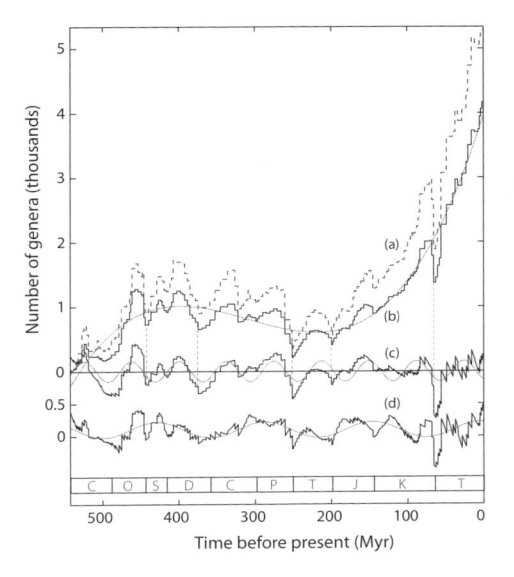

Figure 12.9: The diversity of marine genera during the Phanerozoic, from an analysis of the compendium of Sepkoski (2002) by Rohde & Muller (2005): (a) illustrates the rapid growth since the Cretaceous period ('K' in the period symbols at bottom); (b) the same data with single occurrence and poorly-dated genera removed; (c) after subtraction of the trend line shown by the smooth curve in (b), and with a 62 Myr sine wave superposed; (d) after removal of the 62 Myr cycle, and with a 140 Myr sine wave superposed. Dashed vertical lines indicate the times of five major extinctions. From Rohde & Muller (2005, Figure 1), by permission from Macmillan Publishers Ltd, Nature, ©2005.

The `Geocarb III` model of Berner & Kothavala (2001), and the `Geocarbsulf` model of Berner (2006) are widely-referenced models for the long-term cycles of C, and C+S, respectively, including effects of feedback, weathering, and degassing. The `Geocarb III` results were used, for example, for the paleoclimate synopsis in the 2007 Report of the Intergovernmental Panel on Climate Change (Solomon et al., 2007).

Resulting models demonstrate a generally decreasing concentration of CO_2 (Figure 12.8a), but with a broad minimum around the Carboniferous–Permian periods, and a peak at the Permian/Triassic boundary.

Atmospheric O_2 (Figure 12.8b) shows a broad late Paleozoic peak with a maximum value of about 30% in the Permian, a secondary peak centred near the Silurian/Devonian boundary, variation between 15–20% during the Cambrian–Ordovician, a sharp drop from 30% to 15% at the Permian/Triassic boundary, and a steady rise in O_2 from the late Triassic to the present.

Concentrations of $\gtrsim 12\%$ are considered to be the minimum necessary to support the forest fires evident in the geological record (e.g. Chaloner, 1989; Wildman et al., 2004; Scott, 2000). The O_2 curve may be a contributing factor to animal gigantism during the Permian–Triassic (Graham et al., 1995), to extinction at the Permian/Triassic boundary (Huey & Ward, 2005), and to mammalian evolution (Falkowski et al., 2005).

Ice-core records Constraints on more recent atmospheric properties and chemical compositions, via direct measurements of gases and reliable proxies of temperature, come from deep ice-core records. Antarctic samples extend back to 0.42 Myr for Vostok (e.g. Petit et al., 1999; Bender, 2002; Kawamura et al., 2007; Juselius & Kaufmann, 2009), and 0.8 Myr for the 3200 m EPICA at Dome C (Wolff et al., 2010).

Growth and patterns in biodiversity This short introduction to the Earth's atmospheric evolution will make only passing reference to the widely researched topic of the explosive biodiversity during the Phanerozoic (e.g. von Bloh et al., 2003).

The growth in the diversity of genera versus time over this interval, based on the first and last stratigraphic appearance of 36 380 marine genera compiled by Sepkoski (2002), is shown in Figure 12.9. In their analysis of these data, Rohde & Muller (2005) identified a prominent 62 ± 3 Myr cycle, possibly correlated with the five great extinctions (Raup & Sepkoski, 1982).

Numerous studies over the past few decades have proposed a causal connection between terrestrial mass extinctions, climate change, and the motion of the Sun through the Galaxy (see, e.g., Perryman, 2009, sections 10.5 and 10.6).

Proposed mechanisms include enhanced comet impacts via perturbation of the Oort cloud (e.g. García-Sánchez et al., 1999, 2001), as well as changes in cosmic rays and supernovae modulated by the passage of the Sun through the Galactic spiral arms (e.g. Frakes et al., 1992; Shaviv, 2002, 2003; Yeghikyan & Fahr, 2004a,b; Gies & Helsel, 2005; Svensmark, 2006a, 2007) or the Galactic mid-plane (Svensmark, 2006b).

Supposed periodicities in the fossil record and impact cratering dates are often cited as supporting these hypotheses (e.g. Stothers, 1998; Jetsu & Pelt, 2000, and references), but the topic remains controversial. Mechanisms, and the evidence for and against the relevance of astronomical phenomena to climate change and evolution, are reviewed by Bailer-Jones (2009).

Appendix A. Numerical quantities

Table A1 lists a number of relevant physical quantities and derived parameters at high levels of precision. Table A2 lists approximate values of some relevant properties of the solar system planets, which can be used for comparison with the emerging properties of exoplanets.

The systems of standards

Beyond a small number of basic physical quantities (notably c, G, and h) the relevant system of astronomical units is essentially defined by four numbers: the length of the day, d; the mass of the Sun, $M_\odot$, or in practice $GM_\odot$; the astronomical unit, A_m; and the Gaussian constant of gravitation, k (a discussion of the complexities and limitations can be found in Klioner 2008).

The source for these referenced standards are as follows:

- CODATA06: recommended values from the CODATA Task Group on Fundamental Constants (see www.codata.org and physics.nist.gov/cuu/Constants).
- INPOP06: self-consistent (TCB-compatible) solar system quantities from the numerical planetary ephemeris developed at the IMCCE–Observatoire de Paris (Fienga et al., 2008). Small changes in planetary masses accompany INPOP08 (Fienga et al., 2009), which have not been incorporated, and a further revision is expected in 2011. INPOP provides an alternative to the JPL development ephemeris solutions DE405 and DE414 (Konopliv et al., 2006).
- IAU(1976): the IAU (1976) system of astronomical constants.

Notes to Table A1

(1) the astronomical unit is a defining constant in INPOP06, consistent with the combination of the light travel time and the speed of light used in the JPL DE solutions.

(2) the value here is derived from the other defining quantities (CODATA06 gives $5.670\,400(40) \times 10^{-8}$).

(3) because of the limited accuracy in the knowledge of G, planetary ephemerides, and associated masses and linear dimensions, are still calculated in terms of the Gaussian constant of gravitation, $k = 4\pi^2 A_m^3/(P_\oplus^2(M_\odot + M_\oplus))$. Gauss (1857) used a numerical value of k which was subsequently incorporated into the IAU (1976) system of astronomical constants, and remains that used as a defining constant in, e.g., INPOP06.

(4) the Julian Year is an IAU definition, entering directly in the definition of the unit of proper motion, mas yr^{-1}.

(5) in high-accuracy applications, care is needed in representing quantities dependent on π due to rounding errors. Double-precision floating-point numbers following IEEE standards have 64 bits (16 significant digits) yielding, to appropriate accuracy:

- $\pi = 3.141\,592\,653\,589\,793\,238$
- 1 degree in radians: $\pi/180 = 1.745\,329\,251\,994\,33 \times 10^{-2}$ rad
- 1 mas in radians: $\pi/(180 \times 3600 \times 1000) = 4.848\,136\,811\,095\,36 \times 10^{-9}$ rad

(6) $\gamma = 1$ is fixed in INPOP06. Other terms given here can be derived numerically, e.g., Klioner (2003). $\alpha_{pN}(\perp)$ is the post-Newtonian deflection angle for an observer at 1 AU from the Sun, of a light ray arriving orthogonal to the solar direction due to the spherically symmetric part of the gravitational field of the Sun. $\alpha_{pN}(R_\odot)$ is the post-Newtonian deflection angle for an observer at 1 AU from the Sun, of a solar-limb grazing light ray due to the spherically symmetric part of the gravitational field of the Sun.

(7) accurate barycentric arrival time determinations require corrections for both geometrical (Rømer) delay associated with the observer's motion around the barycentre, and relativistic (Shapiro) delay caused by the gravitational bodies in the solar system; see, e.g., Lindegren & Dravins (2003, Section 4.3), and Will (2003, eqn 7).

(8) the value in INPOP06 is itself derived from other defining constants. Although $M_\odot$ is currently considered as a constant, the physical mass of the Sun decreases at $\sim 10^{-13} M_\odot$ yr^{-1} carried by solar radiation (Krasinsky & Brumberg, 2004). Secular acceleration in the mean longitudes of the inner planets currently places a limit of $\dot{G}/G = -2 \pm 5 \times 10^{-14}$ yr^{-1} (Pitjeva, 2005), equivalent to the same limit on the determination of $\dot{M}/M$ (Klioner, 2008).

(9) the solar apparent radius at 1 AU is that derived from astrolabe observations by Chollet & Sinceac (1999). At this accuracy level, solar oblateness is irrelevant.

(10) see Perryman (2009, Section 7.2.1) for a discussion.

(11) average of the three measurements cited.

(12) Grevesse & Noels (1993) give $Y_\odot$ and $Z_\odot$, from which $X_\odot = 1 - Y_\odot - Z_\odot$. Associated parameters relevant for the equation of state are the mean molecular weight $\mu_\odot = 1/(2X_\odot + 3Y_\odot/4 + Z_\odot/2) = 0.6092$ and the mean molecular weight per free electron $\mu_{e,\odot} = 2/(X_\odot + 1) = 1.1651$. These derivations assume complete ionisation, and that metals give ~ 0.5 particles per m_H (e.g. Karttunen, 1987, Section 11.2).

(13) here, 'Earth' includes the Earth's atmosphere, but excludes the Moon. 'Earth-system' includes the Moon. Similarly, the 'Jupiter-system' includes the contribution from its moons.

(14) the *de facto* standard Jupiter radius for exoplanet studies. Due to its oblateness, Jupiter's own mean radius is actually $0.978 R_J$.

(15) mean value at J2000.0 from a 250-year fit to JPL DE200. The orbital period is derived from mean longitude rates.

(16) Cox (2000, Section 12.1)

Table A1: Fundamental and derived quantities, and associated reference values.

Symbol	Meaning	Source/Derivation	Value
	Physical constants:		
c	Speed of light in vacuum (exact)	CODATA06	$299\,792\,458$ m s^{-1}
A_m	Astronomical unit (in m)[1]	INPOP06	$1.495\,978\,706\,910\,000 \times 10^{11}$ m
G	Newton's constant of gravitation	CODATA06	$6.674\,28(67) \times 10^{-11}$ m^3 kg^{-1} s^{-2}
h	Planck constant	CODATA06	$6.626\,068\,96(33) \times 10^{-34}$ J s
σ	Stefan–Boltzmann constant[2]	$2\pi^5 k^4/(15c^2 h^3)$	$5.670\,399 \times 10^{-8}$ W m^{-2} K^{-4}
k	Gaussian gravitational constant (exact)[3]	IAU (1976)	$0.017\,202\,098\,95$ AU$^{3/2}$ day^{-1} $M_\odot^{-1/2}$
	Time scales:		
d	Day in s (any time scale)		$86\,400$ (exactly)
y	Julian Year in days[4]		365.25 $(= 31\,557\,600$ s$)$
J2000	Julian date of standard epoch, J2000.0		$2\,451\,545.0$ JD
	Derived quantities[5]:		
A_v	Proper motion constant	$A_\mathrm{m}/(y \times d \times 1000)$	$4.740\,470\,463$ km yr s^{-1}
pc$_\mathrm{m}$	Parsec in m	$A_\mathrm{m} \times 180 \times 3600/\pi$	$3.085\,677\,581\,305\,729 \times 10^{16}$ m
pc$_\mathrm{AU}$	Parsec in AU	$180 \times 3600/\pi$	$2.062\,648\,06 \times 10^5$ AU
ly$_\mathrm{m}$	Light year in m	$y \times d \times c$	$9.460\,730\,473 \times 10^{15}$ m
ly$_\mathrm{AU}$	Light year in AU	$y \times d \times c/A_\mathrm{m}$	$6.324\,107\,708\,8 \times 10^4$ AU
ly$_\mathrm{pc}$	Light year in pc	$y \times d \times c/$pc$_\mathrm{m}$	$0.306\,601$ pc
	General Relativity:		
γ	General relativistic PPN parameter[6]	Current assumption	1
$\alpha_\mathrm{pN}(\perp)$	Deflection angle at 1 AU perp to ecliptic[6]	Klioner (2003)	4.072×10^{-3} arcsec
$\alpha_\mathrm{pN}(R_\odot)$	Deflection angle at 1 AU at solar limb[6]	Klioner (2003)	$1.750\,453$ arcsec
Δ	Shapiro time-delay constant[7]	$(1+\gamma)(GM_\odot)/c^3$	9.851×10^{-6} s
	Sun:		
$GM_\odot$	Heliocentric gravitational constant[8]	INPOP06	$1.327\,124\,420\,76 \times 10^{20}$ m^3 s^{-2}
$M_\odot$	Mass[8]	$(GM_\odot)/G$	$1.988\,4 \times 10^{30}$ kg
$\theta_\odot$	Apparent radius in arcsec[9]	Chollet & Sinceac (1999)	$959.63(8)$ arcsec
$R_\odot$	Apparent radius in m	$A_\mathrm{m} \times \pi\theta_\odot/(3600 \times 180)$	$6.959\,917\,56 \times 10^8$ m
$\rho_\odot$	Mean mass density	$M_\odot/((4/3)\pi R_\odot^3)$	1.408 Mg m^{-3}
$g_\odot$	Surface gravity	$(GM_\odot)/R_\odot^2$	274.0 m s^{-2}
$J_{2,\odot}$	Oblateness	INPOP06	$1.95(55) \times 10^{-7}$
$M_{V,\odot}$	Johnson absolute V magnitude[10]	Cayrel de Strobel (1996)	$+4.83$ mag
$M_\mathrm{bol,\odot}$	Bolometric magnitude[10]	Cayrel de Strobel (1996)	$+4.75$ mag
BC$_{V,\odot}$	Bolometric correction, Johnson V band[10]	Cayrel de Strobel (1996)	-0.08 mag
$J_\odot$	Energy flux at 1 AU[11]	Duncan et al. (1982)	1371 W m^{-2}
$L_\odot$	Luminosity	$4\pi A_\mathrm{m}^2 J_\odot$	3.856×10^{26} W
$T_\mathrm{eff,\odot}$	Effective (black-body) temperature	$(L_\odot/4\pi\sigma R_\odot^2)^{1/4}$	5781 K
$X_\odot, Y_\odot, Z_\odot$	H, He, and metal abundances by mass[12]	Grevesse & Noels (1993)	$0.716\,6, 0.265\,9, 0.017\,5$
$\alpha_\mathrm{MLT,\odot}$	Mixing-length parameter	Girardi et al. (2000)	1.68
	Earth-related:		
$GM_\oplus$	Geocentric gravitational constant[13]	INPOP06	$3.986\,004\,390\,77 \times 10^{14}$ m^3 s^{-2}
$M_\oplus$	Earth mass[13]	$(GM_\oplus)/G$	$5.972\,2 \times 10^{24}$ kg
$M_\odot/M_\oplus$	Sun/Earth mass ratio[13]	INPOP06	$3.329\,460\,508\,95 \times 10^5$
	Sun/Earth-system mass ratio[13]	INPOP06	$3.289\,005\,614\,00 \times 10^5$
	Jupiter-related:		
$M_\odot/M_\mathrm{J}$	Sun/Jupiter-system mass ratio[13]	INPOP06	$1047.348\,6$
M_J	Jupiter-system mass[13]	$M_\odot/(M_\odot/M_\mathrm{J})$	1.899×10^{27} kg $\simeq 9.5 \times 10^{-4}$ $M_\odot$
R_J	Jupiter equatorial radius at 10^5 Pa (1 bar)[14]	Cox (2000, Table 12.3)	7.1492×10^7 m
a_J	Orbital semi-major axis of Jupiter-system[15]	Seidelmann (1992)	$5.203\,363\,01$ AU
P_J	Jupiter sidereal orbital period[15]	Seidelmann (1992)	$11.862\,615$ yr
	Solar system mass composition[16]:		
	Total mass of planets		$446.6 M_\oplus$
	Total mass of satellites		$0.104 M_\oplus$
	Total mass of asteroids		$0.000\,30 M_\oplus$
	Total mass of meteors and comets		$10^{-9} M_\oplus$
	Total mass of planetary system		$0.001\,34 M_\odot$
	Total angular momentum of planetary system		3.148×10^{43} kg m^2 s^{-1}
	Local Galactic constants:		
A	Oort constant, A	Feast & Whitelock (1997)	$+14.82$ km s^{-1} kpc^{-1}
B	Oort constant, B	Feast & Whitelock (1997)	-12.37 km s^{-1} kpc^{-1}
Ω_0	Angular velocity	$A - B$	$+27.19$ km s^{-1} kpc^{-1}
R_0	Galactocentric radius of the Sun	Perryman (2009, Section 9.2.1)	8.2 kpc
V_0	Circular velocity at R_0	$R_0 \Omega_0$	223 km s^{-1}
P_rot	Galactic rotation period	$(2\pi/\Omega_0)($pc$_\mathrm{m}/y \times d)$	2.26×10^8 yr
κ_0	Epicycle frequency	$(-4B(A-B))^{0.5}$	36.7 km s^{-1} kpc^{-1}
Z_0	Sun's height from Galactic mid-plane	Perryman (2009, Section 9.2.2)	$+20$ pc
ρ_0	Oort limit (disk+dark halo)	Holmberg & Flynn (2000)	$0.102 M_\odot$ pc^{-3}
$P_\perp$	Vertical oscillation period	$(\pi/G\rho_0)^{0.5}$ s (ρ_0 in SI)	8.2×10^7 yr

Appendix A. Numerical quantities

Table A2: Properties of the solar system planets.

Property	Mercury	Venus	Earth	Mars	Jupiter	Saturn	Uranus	Neptune
Orbit:								
semi-major axis, a (10^6 km)	58	108	150	228	778	1427	2871	4498
" (AU)	0.39	0.72	1.00	1.52	5.20	9.54	19.19	30.07
siderial period (Julian years)	0.24	0.62	1.00	1.88	11.86	29.42	83.75	163.72
eccentricity, e	0.206	0.007	0.017	0.093	0.048	0.054	0.047	0.009
inclination to ecliptic, i (°)	7.0	3.4	0.0	1.9	1.3	2.5	0.8	1.8
obliquity (°)	0.0	177.3	23.4	25.2	3.1	26.7	97.9	29.6
Physical properties:								
equatorial radius, R_+ (km)	2440	6052	6378	3397	71492	60268	25559	24764
" $R_+/R_\oplus$	0.38	0.95	1	0.53	11.21	9.45	4.00	3.88
volume, $V/V_\oplus$	0.05	0.88	1	0.15	1316	755	52	44
mass, M (10^{24} kg)	0.33	4.87	5.97	0.64	1898.70	568.51	86.85	102.44
" $M/M_\oplus$	0.06	0.81	1	0.11	317.82	95.16	14.54	17.15
" M/M_J	0.0002	0.0026	0.0031	0.0003	1	0.2994	0.0457	0.0540
" $M/M_\odot$	1.66×10^{-7}	2.45×10^{-6}	3.00×10^{-6}	3.23×10^{-7}	9.55×10^{-4}	2.86×10^{-4}	4.37×10^{-5}	5.15×10^{-5}
density, ρ (Mg m^{-3})	5.43	5.24	5.52	3.94	1.33	0.70	1.30	1.76
surface gravity, g (m s^{-2})	3.70	8.87	9.80	3.71	23.12	8.96	8.69	11.00
escape velocity (km s^{-1})	4.2	10.4	11.2	5.0	59.5	35.5	21.3	23.7
oblateness, $(R_+ - R_-)/R_-$	0.0001	0.0001	0.0035	0.0052	0.0649	0.0980	0.0229	0.0171
quadrupole moment, J_2	0.0001	0.0000	0.0011	0.0020	0.0147	0.0163	0.0033	0.0034
magnetic moment (T m^3)	5.08×10^{12}	$< 6.65 \times 10^{12}$	8.04×10^{15}	$< 2.35 \times 10^{12}$	1.57×10^{20}	4.60×10^{18}	3.81×10^{17}	2.02×10^{17}
dipole tilt (°)	< 10	–	11.5	–	9.6	0.8	58.6	47.0
Satellites	0	0	1	2	63	∼ 200	27	13
Astrometric effects on the Sun:								
$a_\odot$ due to planet (km) [1]	10	264	451	74	742903	407998	125399	231732
$a_\odot$ at 10 pc (μas) [2]	0.006	0.177	0.301	0.049	497	273	84	155
Surface properties:								
visibility: solid/cloud	s	c	s/c	s	c	c	c	c
visual geometric albedo	0.11	0.65	0.37	0.15	0.52	0.47	0.51	0.41
Bond albedo, A_B	0.12	0.75	0.31	0.25	0.34	0.34	0.30	0.29
effective temperature (K)	–	∼230	∼255	∼212	124	95	59	59
mean temperature (K)	440	730	288–293	183–268	–	–	–	–
scale height, H (km)	–	15	8	11	19–25	35–50	22–29	18–22
Atmospheric composition:								
N_2	–	0.035	0.781	0.027	–	–	–	–
O_2	–	–	0.209	1.3×10^{-3}	–	–	–	–
CO_2	–	0.965	3.3×10^{-4}	0.953	–	$< 3 \times 10^{-10}$	$< 3 \times 10^{-10}$	$< 5 \times 10^{-10}$
CO	–	3×10^{-7}	2×10^{-7}	2.7×10^{-3}	1×10^{-10}	$1 - 3 \times 10^{-9}$	$< 5 \times 10^{-7}$	6×10^{-7}
CH_4	–	–	2.0×10^{-6}	–	2.4×10^{-3}	$1 - 4 \times 10^{-3}$	< 0.02	< 0.02
NH_3	–	–	4×10^{-9}	–	$< 7 \times 10^{-4}$	$< 1 \times 10^{-4}$	$< 1 \times 10^{-4}$	–
H_2O	–	2×10^{-5}	1×10^{-6}	3×10^{-4}	$< 6 \times 10^{-5}$	2×10^{-7}	$< 0.5 \times 10^{-8}$	$< 2 \times 10^{-9}$
H_2	–	–	5×10^{-7}	–	0.863	0.94	0.85	0.85
He	–	1.2×10^{-5}	5.2×10^{-6}	–	0.156	0.06	0.15	0.15

Notes to Table A2

This is a (generally low-precision) summary of some relevant properties of the solar system planets. Values are mostly from Cox (2000, Tables 12.1–12.10), supplemented by oblateness and quadrupole moment from Lang (1992) and Yoder (1995), atmospheric composition from Encrenaz et al. (2004), Bond albedos from Cole & Woolfson (2002, Topic X), and magnetic moment and dipole tilt from Lang (1992).

(1) orbital semi-major axis of a $1 M_\odot$ star due to a planet with the same M_p and a_p, i.e. $a_\odot = a_p \times (M_p/M_\odot)$.

(2) associated astrometric signature for a $1 M_\odot$ star at 10 pc, i.e. $\alpha_\odot$ (arcsec) $= (a(\mathrm{AU})/10) \times (M_p/M_\odot)$.

Appendix B. Notation

The principal notation used is summarised in the following table. Multiple use of certain symbols is inevitable if accepted usage is adhered to. When followed by a chapter number in braces [...], the notation is generally used only within that chapter.

Appendix B. Notation

a	semi-major axis		f_-	polar flattening (tidal) [10]
a	dust particle radius [10]		F_g	gravitational scattering enhancement [10]
a	activity coefficient (thermodynamic) [11]		g	stellar surface gravity [3]
a''	semi-major axis in angular measure [2]		g_p	planet surface gravity [6]
a_p	semi-major axis of planet around star [2]		G	gravitation constant
$a_\star$	semi-major axis of star around barycentre [2]		γ	systemic radial velocity [2]
a_s	planetary satellite orbital radius [6]		γ	curvature per unit mass (ppN) [3]
A	astronomical unit [2]		γ	radial exponent of disk surface density [10]
A	microlensing magnification (total) [5]		Γ	gamma function [6]
$A_\pm$	microlensing image magnifications [5]		h	height of turbulent layers in atmosphere [3]
A_B	Bond albedo		h	disk height [10]
A_m	astronomical unit (in m) [6]		h	Planck constant
A_p, A_v	astronomical unit [8]		H	atmospheric scale height
A-C, F-H	Thiele–Innes constants [3]		H	tidal heating [10]
$\mathbf{A}_G$	transformation matrix: Galactic system/ICRS [8]		i	orbit inclination ($i = 0$ is face-on)
α	astrometric signature [3]		i	ice mass fraction [11]
α	source trajectory relative to star–planet axis [5]		$I(r)$	stellar intensity as a function of radius [6]
α	exponent of power-law mass function [9]		J_2	quadrupole moment [6]
α	turbulent efficiency in α-disks [10]		$\mathscr{J}$	fast Lyapunov indicator, MEGNO [2]
α_E	microlensing deflection angle at R_E [5]		k	Boltzmann constant
α_{GR}	gravitational deflection angle [5]		k_2	Love number [6]
α_G, δ_G	north Galactic pole (ICRS) [8]		k	Doppler factor [8]
α_{IR}	spectral index in infrared [10]		K	radial velocity semi-amplitude
α_{pN}	post-Newtonian deflection angle [3]		K_0	bulk modulus [11]
b	semi-minor axis [2]		K_i	equilibrium constant (thermodynamic) [11]
b	impact parameter (light deflection) [5]		κ	grain opacity [10]
b	impact parameter (transit) [6]		κ_0	epicyclic frequency [10]
b	Galactic latitude [8]		κ_R	Rosseland mean opacity [9]
$\mathbf{b}$	star position wrt solar system barycentre [8]		l	Galactic longitude [8]
B	interferometer baseline [3]		l_Ω	origin of Galactic longitude [8]
BC	bolometric correction [8]		L	luminosity
c	speed of light		$L_\star$	stellar luminosity
c, h, v_0	linearised parameters for orbit fitting [2]		L_X	X-ray luminosity [8]
c_s	sound speed [10]		$L_1 - L_5$	Lagrange points [2]
C_d	drag coefficient [10]		λ	wavelength of light
$C_n^2(h)$	atmospheric structure function [3]		λ	mean longitude [2]
d	angular star–planet separation, in units of θ_E [5]		λ	projected angle between stellar and orbit axes [6]
d	projected star–planet separation [6]		λ	mean-free path [10]
$d, d_\star$	stellar distance		λ_J	Jeans length (gravitational collapse) [10]
$d(t)$	velocity trend parameter [2]		λ_J	Jeans parameter (atmospheric escape) [11]
D	telescope diameter		Λ	Lyapunov exponent [2]
D	object diameter [10]		m	apparent magnitude [8]
D_L	lens distance [5]		m_p	planetesimal (accretor) mass [11]
D_{LS}	lens–source distance [5]		M	absolute magnitude [8]
D_S	source distance [5]		M_{bol}	bolometric magnitude [8]
$\delta\theta$	astrometric deflection of microlens image centroid [5]		M_{atm}	atmospheric mass [9]
Δ	photocentre displacement due to star spots [3]		$M_\oplus$	Earth mass
ΔF	flux decrement during transit [6]		M_e	embryo mass [10]
$\Delta\theta$	angular separation between microlens images [5]		M_{iso}	isolation mass (embryo) [10]
e	eccentricity (orbit)		M_J	Jupiter mass
$E(t)$	eccentric anomaly		M_L	lens mass [5]
ϵ	planet/star flux ratio [6]		M_p	planet mass
$\dot\epsilon$	internal energy source [11]		M_p'	planet mass (scaled) [11]
η	compression ratio [11]		M_s	satellite (exomoon) mass
f	atmospheric circulation parameter [6]		$M_\odot$	solar mass
f	fugacity (thermodynamic) [11]		$M_\star$	stellar mass
f_p	planet flux [7]		M_T	Trojan planet mass [6]
$f_\star$	star flux [7]		$M(t)$	mean anomaly [2]
f_+	equatorial prolateness (tidal) [10]			

Appendix B. Notation

$\mathcal{M}$	mass function [2]	R_R	Roche radius [6]
μ	stellar proper motion	R_S	Schwarzschild radius [5]
μ	limb darkening parameter [6]	$R_\star$	stellar radius
μ_m	mean molecular weight	ρ	angular diameter of source star, in units of θ_E [5]
μ_p	planetary magnetic moment [7]	ρ	density
$\tilde{\mu}$	effective rigidity of body [9]	S	stellar photon count rate [6]
μ_{LS}	relative lens–source proper motion [5]	S	Strehl ratio [7]
n	mean motion	S	entropy (per unit mass) [11]
n_p	number of planets in system	σ'_{rv}	radial velocity jitter [2]
N_x	number of atoms of element x per volume [8]	σ_δ	relative angular positional error [3]
ν	spectral line frequency [2]	σ_{ph}	photon-noise limited positional error [3]
ν	molecular kinematic viscosity [10]	σ_ψ	single measurement error [3]
$\nu(t)$	true anomaly	Σ	disk surface density [10]
ω	planetary rotation period [7]	Σ_p	disk surface density in planetesimals [10]
ω	argument of pericentre	t	integration time [3]
$\tilde{\omega}$	longitude of pericentre [2]	t_0	central microlensing event time [5]
$\dot{\omega}$	precession rate [6]	$t_{0.5}$	radionuclide half-life [10]
Ω	longitude of ascending node [2]	t_E	Einstein radius crossing time [5]
Ω	disk angular velocity [10]	t_F	duration of flat part of transit [6]
		t_{ff}	free-fall time [10]
p	resonance parameter [2]	t_p	time of specified pericentre passage
p	transit probability [6]	t_T	total transit duration [6]
$\dot{p}$	pulsation period derivative [4]	T	temperature
$p(\lambda)$	geometric albedo	T_{crit}	critical temperature [11]
P	orbital period	T_{eff}	effective temperature [8]
P	(partial) pressure	T_{eq}	planet equilibrium surface temperature [6]
P_{2n}	Legendre polynomial [11]	T_p	planet temperature [4]
P_{crit}	critical pressure [11]	T_{tp}	triple point [11]
P_{rot}	stellar rotation period [3]	τ	ingress and/or egress durations [6]
P_{th}	thermal pressure contribution to EOS [11]	τ	stellar age [9]
ϕ	resonant argument [2]	τ_a	e-folding time, semi-major axis migration [2]
ϕ	angle between stellar rotation and orbit axes [6]	τ_c	cooling time (disk) [10]
Φ	interferometric closure phase [6]	τ_e	e-folding time, eccentricity migration [2]
π	mathematical constant, $\pi = 3.141592653589...$	τ_p	amplitude of light travel time from primary [4]
ϖ	stellar parallax	θ	true longitude [2]
ϖ_E	microlens parallax [5]	θ	angular separation for reference/target star [3]
ϖ_{rel}	relative source–lens parallax [5]	θ_E	Einstein radius (angular) [5]
Π_α, Π_δ	displacements due to parallax [3]	θ_I	angle between lens and microlensed image [5]
ψ	light deflection: angle between Sun and star [3]	$\theta_\pm$	angular positions of a pair of microlensed images [5]
		θ_S	angle between lens and source [5]
q	pericentre distance [2]	$\theta_\star$	angular diameter of source star [5]
q	resonance order [2]	Θ	Safronov number [6]
q	planet–star mass ratio [5]		
Q	apocentre distance [2]	u	angular lens–source separation, in units of θ_E [5]
Q	Toomre (disk) stability parameter [10]		
Q	specific dissipation function $(Q_p, Q_\star)$	v	dust particle velocity [10]
		v_{esc}	escape velocity at atmosphere exobase [11]
r	radial distance from star [10]	v_r	radial velocity
r	radial coordinate within planet [11]	v_t	transverse velocity [3]
r_p	radial distance of planet [10]	$v_\perp$	transverse velocity between source and lens [5]
r_t	star–planet separation at mid-transit [6]	$v \sin i$	stellar rotation velocity
r_0	Fried parameter (atmosphere)	$\mathbf{v}$	barycentric space velocity [8]
r_L	radius of Lindblad resonance [10]	V	interferometric visibility function [6]
R	spectroscopic resolving power [7]	$V(h)$	wind velocity as a function of atmospheric height [3]
R_c	gravitational capture radius [10]	V_w	solar/stellar wind speed [7]
R_E	linear Einstein radius in the lens plane [5]		
R'_E	projection of R_E onto observer plane [5]	$w(x_0)$	gravitational potential of solar system at x_0 [3]
R_H	Hill radius		
R_p	planet radius	XY	rectangular coordinates in the orbit plane [3]
R'_p	planet radius (scaled) [11]		
R_+, R_-	equatorial/polar radii [6]	z	coordinate along line-of-sight [2]
$\bar{R}$	mean radius [12]	z	altitude (atmosphere) [11]
		Z	atomic number [11]

Appendix C. Radial velocity planets

As of 2010 November 1, `exoplanet.eu` listed 494 exoplanet discoveries in total, of which radial velocity measurements had been made for 461 planets in 390 systems (of which 45 are multiple).

Of these 494 planets, the following table lists the 358 which were *discovered* by radial velocity measurements, as of the same date. Planets discovered by other methods, notably by transit photometry, for which radial velocity measurements have subsequently been acquired, are not included here. Such discoveries are listed in the chapters describing the relevant detection techniques.

New planets are being discovered by radial velocity measurements and reported more-or-less continuously, at the rate of approximately 50 per year during 2009 and 2010. While this table is therefore out-of-date already before publication, and although the current status can be consulted on-line, it is included to specify the observational context in which this review has been prepared.

The planets, and the associated data, have been taken from the compilation at `exoplanet.eu` as of 2010 November 1. Ordering is by object name.

Data have been rounded to a fixed number of decimal places, irrespective of accuracy. For up-to-date data at full precision, additional information for each planet and host star, and the discovery reference and source of the data, the compilation at `exoplanet.eu` should be consulted. R_p is included for congruence with Appendix D.

The columns are:
  (1) planet name
  (2) discovery year
  (3) stellar spectral type
  (4) stellar magnitude, V
  (5) stellar distance, d
  (6) planet mass, $M_p \sin i$
  (7) planet radius, R_p
  (8) orbit period, $P_{\rm orb}$
  (9) semi-major axis, a
  (10) orbit eccentricity, e
  (11) stellar mass, $M_\star$
  (12) stellar radius, $R_\star$

Table C1. Exoplanets discovered by radial velocity measurements (1/6).

(1) Name	(2) Year	(3) ST	(4) V	(5) $d_\star$ (pc)	(6) $M_p \sin i$ (M_J)	(7) R_p (R_J)	(8) P_{orb} (d)	(9) a (AU)	(10) e	(11) $M_\star$ ($M_\odot$)	(12) $R_\star$ ($R_\odot$)
14 And b	2008	K0III	5.2	76	4.80		185.84000	0.8300	0.00	2.20	11.00
υ And b	1996	F8V	4.1	13	0.69		4.61714	0.0590	0.01	1.27	1.63
c	1999	F8V	4.1	13	14.57		237.70000	0.8610	0.24	1.27	1.63
d	1999	F8V	4.1	13	10.19		1302.61000	2.5500	0.27	1.27	1.63
ξ Aql b	2008	G9IIIb	4.7	63	2.80		136.75000	0.6800	0.00	2.20	12.00
μ Ara b	2000	G3IV-V	5.2	15	1.68		643.25000	1.5000	0.13	1.08	1.25
c	2004	G3IV-V	5.2	15	0.03		9.63860	0.0909	0.17	1.08	1.25
d	2004	G3IV-V	5.2	15	0.52		310.55000	0.9210	0.07	1.08	1.25
e	2006	G3IV-V	5.2	15	1.81		4205.80000	5.2350	0.10	1.08	1.25
30 Ari B b	2009	F6V	7.1	39	9.88		335.10000	0.9950	0.29	1.13	1.13
BD−08 2823 b	2009	K3V	9.9	42	0.05		5.60000	0.0560	0.15	0.74	
c	2009	K3V	9.9	42	0.33		237.60000	0.6800	0.19	0.74	
BD−10 3166 b	2000	G4V	10.1		0.48		3.48800	0.0460	0.07	0.99	1.71
BD−17 63 b	2008	K5V	9.6	35	5.10		655.60000	1.3400	0.54	0.74	0.69
BD+14 4559 b	2009	K2V	9.6	50	1.47		268.94000	0.7770	0.29	0.86	
BD+20 2457 b	2009	K2II	9.8	200	21.42		379.63000	1.4500	0.15	2.80	
c	2009	K2II	9.8	200	12.47		621.99000	2.0100	0.18	2.80	
τ Boo b	1996	F7V	4.5	15	3.90		3.31350	0.0460	0.02	1.30	1.33
γ Cep b	2003	K2V	3.2	14	1.60		902.90000	2.0440	0.12	1.40	6.20
81 Cet b	2008	G5III	5.7	97	5.30		952.70000	2.5000	0.21	2.40	11.00
55 Cnc b	1996	G8V	6.0	13	0.82		14.65162	0.1150	0.01	1.03	1.15
c	2002	G8V	6.0	13	0.17		44.34460	0.2400	0.09	1.03	1.15
d	2002	G8V	6.0	13	3.84		5218.00000	5.7700	0.03	1.03	1.15
e	2004	G8V	6.0	13	0.02		2.81705	0.0380	0.07	1.03	1.15
f	2007	G8V	6.0	13	0.14		260.00000	0.7810	0.20	1.03	1.15
κ CrB b	2007	K1IVa	4.8	31	1.80		1191.00000	2.7000	0.19	1.80	4.71
ρ CrB b	1997	G0V/G2V	5.4	17	1.04		39.84500	0.2200	0.04	0.99	1.28
16 Cyg B b	1996	G2.5V	6.2	21	1.68		799.50000	1.6800	0.69	1.01	0.98
18 Del b	2008	G6III	5.5	73	10.30		993.30000	2.6000	0.08	2.30	8.50
42 Dra b	2009	K1.5III	4.8	97	3.88		479.10000	1.1900	0.38	0.98	22.03
ε Eri b	2000	K2V	3.7	3	1.55		2502.00000	3.3900	0.70	0.83	0.90
GJ 86 b	2000	K1V	6.2	11	4.01		15.76600	0.1100	0.05	0.79	0.86
GJ 179 b	2010	M3.5	12.0	12	0.82		2288.00000	2.4100	0.21	0.36	0.38
GJ 317 b	2007	M3.5	12.0	9	1.20		692.90000	0.9500	0.19	0.24	
GJ 436 b	2004	M2.5	10.7	10	0.07	0.37	2.64390	0.0289	0.15	0.45	0.46
GJ 581 b	2005	M3	10.6	6	0.05		5.36874	0.0410	0.00	0.31	0.38
c	2007	M3	10.6	6	0.02		12.92920	0.0700	0.17	0.31	0.38
d	2007	M3	10.6	6	0.02		66.80000	0.2200	0.38	0.31	0.38
e	2009	M3	10.6	6	0.01		3.14942	0.0300	0.00	0.31	0.38
GJ 649 b	2009	M1.5	9.7	10	0.33		598.30000	1.1350	0.30	0.54	
GJ 674 b	2007	M2.5	9.4	5	0.04		4.69380	0.0390	0.20	0.35	
GJ 676A b	2009	M0	9.6	16	4.00		1000.00000				
GJ 832 b	2008		8.7	5	0.64		3416.00000	3.4000	0.12	0.45	
GJ 849 b	2006	M3.5	10.4	9	0.82		1890.00000	2.3500	0.06	0.36	0.52
GJ 876 b	2000	M4V	10.2	5	2.28		61.11660	0.2083	0.03	0.33	0.36
c	2000	M4V	10.2	5	0.71		30.08810	0.1296	0.26	0.33	0.36
d	2005	M4V	10.2	5	0.02		1.93778	0.0208	0.21	0.33	0.36
e	2010	M4V	10.2	5	0.05		124.26000	0.3343	0.06	0.33	0.36
GJ 3021 b	2000	G6V	6.6	18	3.37		133.71000	0.4900	0.51	0.90	
HD 142 b	2001	G1IV	5.7	21	1.03		339.00000	1.0000	0.37	1.10	0.86
HD 1461 b	2009	G0V	6.6	23	0.02		5.77270	0.0634	0.14	1.08	1.10
HD 2039 b	2002	G2/G3IV-V	9.0	90	4.90		1183.00000	2.2000	0.67	0.98	1.21
HD 2638 b	2005	G5	9.4	54	0.48		3.44420	0.0440	0.00	0.93	
HD 3651 b	2003	K0V	5.8	11	0.20		62.23000	0.2840	0.63	0.79	0.95
HD 4113 b	2007	G5V	7.9	44	1.56		526.62000	1.2800	0.90	0.99	
HD 4203 b	2001	G5	8.7	78	2.07		431.88000	1.1640	0.52	1.06	1.33
HD 4208 b	2001	G5V	7.8	34	0.80		829.00000	1.7000	0.04	0.93	0.85
HD 4308 b	2005	G5V	6.5	22	0.04		15.60900	0.1180	0.27	0.83	0.92
HD 4313 b	2010	G5 D	7.8	137	2.30		356.00000	1.1900	0.04	1.72	4.90
HD 5319 b	2007	G5IV	8.1	100	1.94		675.00000	1.7500	0.12	1.56	3.26

Table C1 (continued). Exoplanets discovered by radial velocity measurements (2/6).

(1) Name	(2) Year	(3) ST	(4) V	(5) $d_\star$ (pc)	(6) $M_p \sin i$ (M_J)	(7) R_p (R_J)	(8) P_{orb} (d)	(9) a (AU)	(10) e	(11) $M_\star$ ($M_\odot$)	(12) $R_\star$ ($R_\odot$)
HD 5388 b	2009	F6V	6.8	53	1.96		777.00000	1.7600	0.40	1.21	
HD 6434 b	2000	G3IV	7.7	40	0.48		22.09000	0.1500	0.30	1.00	0.57
HD 6718 b	2009	G5V	8.5	56	1.56		2496.00000	3.5600	0.10	0.96	1.02
HD 7924 b	2009	K0V	7.2	17	0.03		5.39780	0.0570	0.17	0.83	0.78
HD 8535 b	2009	G0V	7.7	53	0.68		1313.00000	2.4500	0.15	1.13	1.19
HD 8574 b	2003	F8	7.1	44	2.11		227.55000	0.7700	0.29	1.04	1.37
HD 8673 b	2010	F7V	7.6	38	14.20		1634.50000	3.0200	0.72	1.30	
HD 9446 b	2010	G5V	8.4	53	0.70		30.05200	0.1890	0.20	1.00	1.00
c	2010	G5V	8.4	53	1.82		192.90000	0.6540	0.06	1.00	1.00
HD 9578 b	2009	G1V	8.2	57	0.62		494.00000			1.12	
HD 10180 c	2010	G1V	7.3	40	0.04		5.75979	0.0641	0.05		
d	2010	G1V	7.3	40	0.04		16.35790	0.1286	0.09		
e	2010	G1V	7.3	40	0.08		49.74500	0.2699	0.03		
f	2010	G1V	7.3	40	0.08		122.76000	0.4929	0.14		
g	2010	G1V	7.3	40	0.07		601.20000	1.4220	0.19		
h	2010	G1V	7.3	40	0.20		2222.00000	3.4000	0.08		
HD 10647 b	2003	F8V	5.5	17	0.93		1003.00000	2.0300	0.10	1.07	1.10
HD 10697 b	2000	G5IV	6.3	33	6.38		1076.40000	2.1600	0.10	1.15	1.72
HD 11506 b	2007	G0V	7.5	54	3.44		1270.00000	2.4300	0.22	1.19	1.38
c	2009	G0V	7.5	54	0.82		170.46000	0.6390	0.42	1.19	1.38
HD 11964 b	2005	G5	6.4	34	0.11		37.82000	0.2290	0.15	1.13	2.18
c	2005	G5	6.4	34	0.61		2110.00000	3.3400	0.06	1.13	2.18
HD 11977 b	2005	G8.5III	4.7	67	6.54		711.00000	1.9300	0.40	1.91	13.00
HD 12661 b	2000	G6V	7.4	37	2.30		263.60000	0.8300	0.35	1.07	1.12
c	2002	G6V	7.4	37	1.57		1708.00000	2.5600	0.03	1.07	1.12
HD 13189 b	2005	K2II	7.6	185	14.00		471.60000	1.8500	0.28	4.50	50.39
HD 13931 b	2010	G0	7.6	44	1.88		4218.00000	5.1500	0.02	1.02	1.23
HD 16141 b	2000	G5IV	6.8	36	0.23		75.56000	0.3500	0.21	1.00	1.00
HD 16175 b	2007	G0	7.3	60	4.40		990.00000	2.1000	0.59	1.35	1.87
HD 16417 b	2009	G1V	5.8	26	0.07		17.24000	0.1400	0.20	1.18	
HD 16760 b	2009	G5V	8.7	50	14.30		465.10000	1.1300	0.07	0.88	
HD 17092 b	2007		7.7	109	4.60		359.90000	1.2900	0.17	2.30	
HD 17156 b	2007	G0	8.2	78	3.19	1.10	21.21640	0.1623	0.68	1.28	1.51
HD 19994 b	2003	F8V	5.1	22	1.68		535.70000	1.4200	0.30	1.35	1.93
HD 20367 b	2002	G0	6.4	27	1.07		500.00000	1.2500	0.23	1.04	1.18
HD 20782 b	2006	G2V	7.4	36	1.90		591.90000	1.3810	0.97	1.00	
HD 20868 b	2008	K3/4IV	9.9	49	1.99		380.85000	0.9470	0.75	0.78	0.79
HD 23079 b	2001	F8/G0V	7.1	35	2.61		738.45900	1.6500	0.10	1.10	1.13
HD 23127 b	2007	G2V	8.6	89	1.50		1214.00000	2.4000	0.44	1.13	
HD 23596 b	2002	F8	7.2	52	7.19		1558.00000	2.7200	0.31	1.27	2.09
HD 27442 b	2000	K2IVa	4.4	18	1.28		423.84100	1.1800	0.07	1.20	6.60
HD 27894 b	2005	K2V	9.4	42	0.62		17.99100	0.1220	0.05	0.75	
HD 28185 b	2001	G5	7.8	39	5.70		383.00000	1.0300	0.07	1.24	1.03
HD 28254 b	2009	G5V	7.7	56	1.16		1116.00000	2.1500	0.81	1.06	
HD 30177 b	2002	G8V	8.4	55	9.17		2819.65400	3.8600	0.30	0.95	1.12
HD 30562 b	2009	F8V	5.8	27	1.29		1157.00000	2.3000	0.76	1.22	1.64
HD 32518 b	2009	K1III	6.4	117	3.04		157.54000	0.5900	0.01	1.13	10.22
HD 33283 b	2006	G3V	8.1	86	0.33		18.17900	0.1680	0.48	1.24	1.20
HD 33564 b	2005	F6V	5.1	21	9.10		388.00000	1.1000	0.34	1.25	1.10
HD 34445 b	2004	G0	7.3	47	0.79		1049.00000	2.0700	0.27	1.07	1.38
HD 37124 b	1999	G4V	7.7	33	0.64		154.46000	0.5290	0.06	0.91	0.82
c	2003	G4V	7.7	33	0.68		2295.00000	3.1900	0.20	0.91	0.82
d	2004	G4V	7.7	33	0.62		843.60000	1.6400	0.14	0.91	0.82
HD 37605 b	2004	K0V	8.7	43	2.30		55.00000	0.2500	0.68	0.80	
HD 38529 b	2000	G4IV	5.9	42	0.78		14.31040	0.1310	0.25	1.39	2.44
c	2002	G4IV	5.9	42	17.70		2134.76000	3.6950	0.36	1.39	2.44
HD 38801 b	2010	K0IV	8.3	99	10.70		696.30000	1.7000	0.00	1.36	2.53
HD 39091 b	2001	G1IV	5.7	21	10.35		2063.81800	3.2900	0.62	1.10	2.10
HD 40307 b	2008	K2.5V	7.2	13	0.01		4.31150	0.0470	0.00		
c	2008	K2.5V	7.2	13	0.02		9.62000	0.0810	0.00		

Table C1 (continued). Exoplanets discovered by radial velocity measurements (3/6).

(1) Name	(2) Year	(3) ST	(4) V	(5) $d_\star$ (pc)	(6) $M_p \sin i$ (M_J)	(7) R_p (R_J)	(8) P_{orb} (d)	(9) a (AU)	(10) e	(11) $M_\star$ ($M_\odot$)	(12) $R_\star$ ($R_\odot$)
HD 40307 d	2008	K2.5V	7.2	13	0.03		20.46000	0.1340	0.00		
HD 40979 b	2002	F8V	6.7	33	3.32		267.20000	0.8110	0.23	1.08	1.21
HD 41004 A b	2004	K1V	8.7	43	2.54		963.00000	1.6400	0.39	0.70	
HD 41004 B b	2004	M2	12.3	43	18.40		1.32830	0.0177	0.08	0.40	
HD 43197 b	2009	K0	9.0	55	0.60		327.80000	0.9200	0.83	0.96	
HD 43691 b	2007	G0IV	8.0	93	2.49		36.96000	0.2400	0.14	1.38	
HD 44219 b	2009	G5	7.7	50	0.58		472.30000	1.1900	0.61	1.00	
HD 45350 b	2004	G5IV	7.9	49	1.79		890.76000	1.9200	0.78	1.02	1.27
HD 45364 b	2009	K0V	8.1	33	0.19		226.93000	0.6813	0.17	0.82	
c	2009	K0V	8.1	33	0.66		342.85000	0.8972	0.10	0.82	
HD 45652 b	2008	G8-K0	8.1	36	0.47		43.60000	0.2300	0.38		
HD 46375 b	2000	K1IV	7.9	33	0.25		3.02400	0.0410	0.04	0.91	1.00
HD 47186 b	2008	G5V	7.8	38	0.07		4.08450	0.0500	0.04	0.99	
c	2008	G5V	7.8	38	0.35		1353.60000	2.3950	0.25	0.99	
HD 47536 b	2003	K1III	5.3	121	5.00		430.00000		0.20	0.94	23.47
c	2007	K1III	5.3	121	7.00		2500.00000			0.94	23.47
HD 48265 b	2008	G5V	8.1	87	1.16		700.00000	1.5100	0.18	0.93	
HD 49674 b	2002	G5V	8.1	41	0.12		4.94370	0.0580	0.23	1.07	0.94
HD 50499 b	2005	GIV	7.2	47	1.71		2582.70000	3.8600	0.23	1.27	1.38
HD 50554 b	2002	F8	6.9	31	4.90		1279.00000	2.3800	0.42	1.04	1.11
HD 52265 b	2000	G0V	6.3	28	1.13		118.96000	0.4900	0.29	1.20	1.25
HD 59686 b	2003	K2III	5.5	92	5.25		303.00000	0.9110	0.00		11.62
HD 60532 b	2008	F6IV-V	4.5	26	3.15		201.83000	0.7700	0.28	1.44	
c	2008	F6IV-V	4.5	26	7.46		607.06000	1.5800	0.04	1.44	
HD 62509 b	2006	K0IIIb	1.2	10	2.90		589.64000	1.6900	0.02	1.86	8.80
HD 63454 b	2005	K4V	9.4	36	0.38		2.81782	0.0360	0.00	0.80	
HD 63765 b	2009	G9V	8.1	33	0.69		356.00000			0.86	
HD 65216 b	2003	G5V	8.0	34	1.21		613.10000	1.3700	0.41	0.92	
HD 66428 b	2006	G5	8.3	55	2.82		1973.00000	3.1800	0.47	1.15	
HD 68988 b	2001	G0	8.2	58	1.90		6.27600	0.0710	0.14	1.18	1.14
HD 69830 b	2006	K0V	6.0	13	0.03		8.66700	0.0785	0.10	0.86	0.90
c	2006	K0V	6.0	13	0.04		31.56000	0.1860	0.13	0.86	0.90
d	2006	K0V	6.0	13	0.06		197.00000	0.6300	0.07	0.86	0.90
HD 70573 b	2007	G1-1.5V	8.7	46	6.10		851.80000	1.7600	0.40	1.00	
HD 70642 b	2003	G5IV-V	7.2	29	2.00		2231.00000	3.3000	0.10	1.00	0.84
HD 72659 b	2002	G0V	7.5	51	2.96		3177.40000	4.1600	0.20	0.95	1.43
HD 73256 b	2003	G8/K0	8.1	37	1.87		2.54858	0.0370	0.03	1.24	0.90
HD 73267 b	2008	G5V	8.9	55	3.06		1260.00000	2.1980	0.26	0.89	1.04
HD 73526 b	2002	G6V	9.0	99	2.90		188.30000	0.6600	0.19	1.02	1.49
c	2006	G6V	9.0	99	2.50		377.80000	1.0500	0.14	1.02	1.49
HD 73534 b	2009	G5IV	8.2	97	1.15		1800.00000	3.1500	0.05	1.29	2.65
HD 74156 b	2003	G0	7.6	65	1.88		51.65000	0.2940	0.64	1.24	1.58
c	2003	G0	7.6	65	8.03		2476.00000	3.8500	0.43	1.24	1.58
d	2007	G0	7.6	65	0.40		336.60000	1.0100	0.25	1.24	1.58
HD 75289 b	1999	G0V	6.4	29	0.42		3.51000	0.0460	0.05	1.05	1.25
HD 75898 b	2007	G0	8.0	81	2.51		418.20000	1.1900	0.10	1.28	1.60
HD 76700 b	2002	G6V	8.1	60	0.20		3.97100	0.0490	0.13	1.00	1.33
HD 80606 b	2003	G5	8.9	58	3.94	0.92	111.43637	0.4490	0.93	0.90	
HD 81040 b	2005	G2/G3	7.7	33	6.86		1001.70000	1.9400	0.53	0.96	
HD 81688 b	2008	K0III-IV	5.4	88	2.70		184.02000	0.8100	0.00	2.10	13.00
HD 82943 b	2003	G0	6.5	27	1.75		441.20000	1.1900	0.22	1.18	1.12
c	2003	G0	6.5	27	2.01		219.00000	0.7460	0.36	1.18	1.12
HD 83443 b	2002	K0V	8.2	44	0.40		2.98563	0.0406	0.01	0.79	1.04
HD 85390 b	2009	K1V	8.5	34	0.13		788.00000	1.5200	0.41	0.76	
HD 86081 b	2006	F8V	8.7	91	1.50		2.13750	0.0390	0.01	1.21	1.22
HD 86226 b	2010	G2V	7.9	42	1.50		1534.00000	2.6000	0.73	1.02	
HD 86264 b	2009	F7V	7.4	73	7.00		1475.00000	2.8600	0.70	1.42	1.88
HD 87883 b	2009	K0V	7.6	18	1.78		2754.00000	3.6000	0.53	0.82	0.76
HD 88133 b	2004	G5IV	8.0	75	0.22		3.41600	0.0470	0.13	1.20	1.93
HD 89307 b	2004	G0V	7.1	31	1.78		2157.00000	3.2700	0.24	1.03	1.05

Appendix C. Radial velocity planets

Table C1 (continued). Exoplanets discovered by radial velocity measurements (4/6).

(1) Name	(2) Year	(3) ST	(4) V	(5) $d_\star$ (pc)	(6) $M_p \sin i$ (M_J)	(7) R_p (R_J)	(8) P_{orb} (d)	(9) a (AU)	(10) e	(11) $M_\star$ ($M_\odot$)	(12) $R_\star$ ($R_\odot$)
HD 89744 b	2000	F7V	5.7	40	7.99		256.60500	0.8900	0.67	1.40	1.10
HD 90156 b	2009	G5V	7.0	40	0.06		49.77000	0.2500	0.31	0.84	
HD 92788 b	2000	G5	7.3	33	3.86		325.81000	0.9700	0.33	1.06	0.99
HD 93083 b	2005	K3V	8.3	29	0.37		143.58000	0.4770	0.14	0.70	
HD 95089 b	2010	K0D	7.9	139	1.20		507.00000	1.5100	0.16	1.58	4.90
HD 96167 b	2009	G5D	8.1	84	0.68		498.90000	1.3000	0.71	1.31	1.86
HD 99109 b	2006	K0	9.1	61	0.50		439.30000	1.1050	0.09	0.93	
HD 99492 b	2004	K2V	7.6	18	0.11		17.04310	0.1232	0.25	0.78	0.81
HD 100777 b	2007	K0	8.4	53	1.16		383.70000	1.0300	0.36	1.00	
HD 101930 b	2005	K1V	8.2	30	0.30		70.46000	0.3020	0.11	0.74	
HD 102117 b	2004	G6V	7.5	42	0.17		20.67000	0.1532	0.11	0.95	1.27
HD 102195 b	2005	K0V	8.1	29	0.45		4.11378	0.0490	0.00	0.93	0.84
HD 102272 b	2008	K0	8.7	360	5.90		127.58000	0.6140	0.05	1.90	10.10
c	2008	K0	8.7	360	2.60		520.00000	1.5700	0.68	1.90	10.10
HD 102956 b	2010	A	8.0	126	0.96		6.49500	0.0810	0.05	1.68	4.40
HD 103197 b	2009	K1V	9.4	49	0.10		47.84000	0.2490	0.00	0.90	
HD 104067 b	2009	K2V	7.9	21	0.16		55.80000			0.79	
HD 104985 b	2003	G9III	5.8	102	6.30		198.20000	0.7800	0.03	1.50	10.87
HD 106252 b	2002	G0	7.4	37	6.81		1500.00000	2.6100	0.54	1.05	1.09
HD 107148 b	2006	G5	8.0	51	0.21		48.05600	0.2690	0.05	1.12	
HD 108147 b	2002	F8/G0V	7.0	39	0.26		10.89850	0.1020	0.53	1.27	1.22
HD 108874 b	2003	G5	8.8	69	1.36		395.40000	1.0510	0.07	1.00	1.22
c	2005	G5	8.8	69	1.02		1605.80000	2.6800	0.25	1.00	1.22
HD 109246 b	2010	G0V	8.8	66	0.77		68.27000	0.3300	0.12	1.01	1.02
HD 109749 b	2005	G3IV	8.1	59	0.28		5.24000	0.0635	0.01	1.20	
HD 110014 b	2009	K2III	4.7	90	11.09		835.47700	2.1400	0.46	2.17	20.90
HD 111232 b	2004	G8V	7.6	29	6.80		1143.00000	1.9700	0.20	0.78	
HD 114386 b	2003	K3V	8.7	28	1.24		937.00000	1.6500	0.23	0.75	0.76
HD 114729 b	2002	G3V	6.7	35	0.82		1131.47800	2.0800	0.31	0.93	1.46
HD 114762 b	1989	F9V	7.3	39	11.02		83.89000	0.3000	0.34	0.84	1.24
HD 114783 b	2001	K0	7.6	22	0.99		501.00000	1.2000	0.10	0.92	0.78
HD 117207 b	2004	G8VI/V	7.3	33	2.06		2627.08000	3.7800	0.16	1.07	1.09
HD 117618 b	2004	G2V	7.2	38	0.18		25.82700	0.1760	0.42	1.05	1.19
HD 118203 b	2005	K0	8.1	89	2.13		6.13350	0.0700	0.31	1.23	
HD 121504 b	2003	G2V	7.5	44	1.22		63.33000	0.3300	0.03	1.00	
HD 122430 b	2003	K3III	5.5	135	3.71		344.95000	1.0200	0.68	1.39	22.90
HD 125595 b	2009	K4V	9.0	27	0.05		9.67000			0.76	
HD 125612 b	2007	G3V	8.3	53	3.20		502.00000	1.2000	0.39	1.10	1.05
c	2009	G3V	8.3	53	0.07		4.15470			1.10	1.05
d	2009	G3V	8.3	53	7.10		4613.00000			1.10	1.05
HD 126614 b	2010	K0	8.8	72	0.38		1244.00000	2.3500	0.41	1.15	1.09
HD 128311 b	2002	K0	7.5	17	2.18		448.60000	1.0990	0.25	0.80	0.73
c	2005	K0	7.5	17	3.21		919.00000	1.7600	0.17	0.80	0.73
HD 129445 b	2010	G8V	8.8	68	1.60		1840.00000	2.9000	0.70	0.99	
HD 130322 b	1999	K0V	8.1	30	1.08		10.72400	0.0880	0.05	0.79	0.83
HD 131664 b	2008	G3V	8.1	55	18.15		1951.00000	3.1700	0.64	1.10	1.16
HD 132406 b	2007	G0V	8.5	71	5.61		974.00000	1.9800	0.34	1.09	
HD 134987 b	1999	G5V	6.5	25	1.59		258.19000	0.8100	0.23	1.05	1.20
c	2009	G5V	6.5	25	0.82		5000.00000	5.8000	0.12	1.05	1.20
HD 136418 b	2010	G5	7.9	98	2.00		464.30000	1.3200	0.26	1.33	3.40
HD 139357 b	2009	K4III	6.0	121	9.76		1125.70000	2.3600	0.10	1.35	11.47
HD 141937 b	2002	G2/G3V	7.3	33	9.70		653.22000	1.5200	0.41	1.00	1.06
HD 142022 A b	2005	K0V	7.7	36	4.40		1923.00000	2.8000	0.57	0.99	0.71
HD 142415 b	2003	G1V	7.3	34	1.62		386.30000	1.0500	0.50	1.09	1.03
HD 143361 b	2008	G0V	9.2	59	3.12		1057.00000	2.0000	0.15	0.95	
HD 145377 b	2008	G3V	8.1	58	5.76		103.95000	0.4500	0.31	1.12	1.14
HD 145457 b	2010	K0	6.6	126	2.90		176.30000	0.7600	0.11	1.90	9.90
HD 147018 b	2009	G9V	8.3	43	2.12		44.23600	0.2388	0.47	0.93	
c	2009	G9V	8.3	43	6.56		1008.00000	1.9220	0.13	0.93	
HD 147513 b	2003	G3/G5V	5.4	13	1.00		540.40000	1.2600	0.52	0.92	1.00

Table C1 (continued). Exoplanets discovered by radial velocity measurements (5/6).

(1) Name	(2) Year	(3) ST	(4) V	(5) $d_\star$ (pc)	(6) $M_p \sin i$ (M_J)	(7) R_p (R_J)	(8) P_{orb} (d)	(9) a (AU)	(10) e	(11) $M_\star$ ($M_\odot$)	(12) $R_\star$ ($R_\odot$)
HD 148156 b	2009	G1V	7.7	53	0.85		1010.00000	2.4500	0.52	1.22	
HD 148427 b	2009	K0IV	6.9	59	0.96		331.50000	0.9300	0.16	1.45	3.22
HD 149026 b	2005	G0IV	8.2	79	0.36	0.61	2.87589	0.0429	0.00	1.30	1.50
HD 149143 b	2005	G0IV	7.9	63	1.33		4.07200	0.0530	0.02	1.21	
HD 149382 b	2009	B5 D	9.0	74	15.50		2.39100		0.00	0.41	
HD 152079 b	2010	G6V	9.2	85	3.00		2097.00000	3.2000	0.60	1.02	
HD 153950 b	2008	F8V	7.4	50	2.73		499.40000	1.2800	0.34	1.12	1.34
HD 154345 b	2006	G8V	6.7	18	0.95		3340.00000	4.1900	0.04	0.88	
HD 154672 b	2008	G3IV	8.2	66	5.02		163.91000	0.6000	0.61	1.06	1.27
HD 154857 b	2004	G5V	7.3	69	1.80		409.00000	1.2000	0.47	1.17	2.42
HD 155358 b	2007	G0	7.5	43	0.89		195.00000	0.6280	0.11	0.87	
c	2007	G0	7.5	43	0.50		530.30000	1.2240	0.18	0.87	
HD 156411 b	2009	F8	6.7	55	0.74		842.20000	1.8800	0.22	1.25	
HD 156668 b	2010	K2	8.4	24	0.01		4.64600	0.0500	0.00		
HD 156846 b	2007	G0V	6.5	49	10.45		359.51000	0.9900	0.85	1.43	
HD 159868 b	2007	G5V	7.2	53	1.70		986.00000	2.0000	0.69	1.09	
HD 162020 b	2002	K2V	9.2	31	13.75		8.42820	0.0720	0.28	0.80	0.71
HD 164604 b	2010	K2V	9.7	38	2.70		606.40000	1.1300	0.24	0.80	
HD 164922 b	2006	K0V	7.0	22	0.36		1155.00000	2.1100	0.05	0.94	0.90
HD 167042 b	2007	K1III	6.0	50	1.60		416.10000	1.3000	0.03	1.64	4.30
HD 168443 b	1998	G5	6.9	38	8.02		58.11289	0.3000	0.53	1.06	1.63
c	2001	G5	6.9	38	18.10		1765.80000	2.9100	0.21	1.06	1.63
HD 168746 b	2002	G5	8.0	43	0.23		6.40300	0.0650	0.08	0.92	1.12
HD 169830 b	2000	F8V	5.9	36	2.88		225.62000	0.8100	0.31	1.40	1.84
c	2003	F8V	5.9	36	4.04		2102.00000	3.6000	0.33	1.40	1.84
HD 170469 b	2007	G5IV	8.2	65	0.67		1145.00000	2.2400	0.11	1.14	1.22
HD 171028 b	2007	G0	8.3	90	1.98		550.00000	1.3200	0.59	0.99	1.95
HD 171238 b	2009	K0V	8.7	50	2.60		1523.00000	2.5400	0.40	0.94	
HD 173416 b	2009	G8	6.1	135	2.70		323.60000	1.1600	0.21	2.00	13.50
HD 175167 b	2010	G5IV/V	9.2	67	7.80		1290.00000	2.4000	0.54	1.10	
HD 175541 b	2007	G8IV	8.0	128	0.61		297.30000	1.0300	0.33	1.65	3.85
HD 176051 b	2010		5.2	16	1.50		1016.00000	1.7600	0.00		
HD 177830 b	1999	K0	7.2	59	1.28		391.00000	1.0000	0.43	1.48	2.99
HD 178911 B b	2001	G5	8.0	47	6.29		71.48700	0.3200	0.12	1.07	1.14
HD 179079 b	2009	G5IV	8.0	64	0.08		14.47600	0.1100	0.12	1.09	1.48
HD 179949 b	2000	F8V	6.3	27	0.95		3.09250	0.0450	0.02	1.28	1.19
HD 180902 b	2010	K0III/IV	7.8	110	1.60		479.00000	1.3900	0.09	1.52	4.10
HD 181342 b	2010	K0III	7.6	111	3.30		663.00000	1.7800	0.18	1.84	4.60
HD 181433 b	2008	K3IV	8.4	26	0.02		9.37430	0.0800	0.40	0.78	
c	2008	K3IV	8.4	26	0.64		962.00000	1.7600	0.28	0.78	
d	2008	K3IV	8.4	26	0.54		2172.00000	3.0000	0.48	0.78	
HD 181720 b	2009	G1V	7.9	56	0.37		956.00000	1.7800	0.26	0.92	
HD 183263 b	2004	G2IV	7.9	53	3.69		634.23000	1.5200	0.38	1.17	1.21
c	2008	G2IV	7.9	53	3.82		2950.00000	4.2500	0.25	1.17	1.21
HD 185269 b	2006	G0IV	6.7	47	0.94		6.83800	0.0770	0.30	1.28	1.88
HD 187085 b	2006	G0V	7.2	45	0.75		986.00000	2.0500	0.47	1.22	
HD 187123 b	1998	G5	7.9	50	0.52		3.09700	0.0420	0.03	1.06	1.17
c	1998	G5	7.9	50	1.99		3810.00000	4.8900	0.25	1.06	1.17
HD 188015 b	2004	G5IV	8.2	53	1.26		456.46000	1.1900	0.15	1.09	1.10
HD 189733 b	2005	K1-K2	7.7	19	1.15	1.15	2.21857	0.0314	0.00	0.80	0.79
HD 190360 b	2003	G6IV	5.7	16	1.50		2891.00000	3.9200	0.36	1.04	1.20
c	2005	G6IV	5.7	16	0.06		17.10000	0.1280	0.01	1.04	1.20
HD 190647 b	2007	G5	7.8	54	1.90		1038.10000	2.0700	0.18	1.10	
HD 190984 b	2009	F8V	8.8	103	3.10		4885.00000	5.5000	0.57	0.91	1.53
HD 192263 b	1999	K2V	7.8	20	0.72		24.34800	0.1500	0.00	0.81	0.75
HD 192699 b	2007	G8IV	6.4	67	2.50		351.50000	1.1600	0.15	1.68	4.25
HD 195019 b	1998	G3IV-V	6.9	37	3.70		18.20163	0.1388	0.01	1.06	1.38
HD 196050 b	2002	G3V	7.5	47	3.00		1289.00000	2.5000	0.28	1.17	1.29
HD 196885 b	2007	F8V	6.4	33	2.58		1333.00000	2.3700	0.46	1.33	1.79
HD 200964 b	2010	K0	6.6	68	1.85		613.80000	1.6010	0.04	1.44	4.30

Appendix C. Radial velocity planets

Table C1 (continued). Exoplanets discovered by radial velocity measurements (6/6).

(1) Name	(2) Year	(3) ST	(4) V	(5) $d_\star$ (pc)	(6) $M_p \sin i$ (M_J)	(7) R_p (R_J)	(8) P_{orb} (d)	(9) a (AU)	(10) e	(11) $M_\star$ ($M_\odot$)	(12) $R_\star$ ($R_\odot$)
HD 200964 c	2010	K0	6.6	68	0.90		825.00000	1.9500	0.18	1.44	4.30
HD 202206 b	2002	G6V	8.1	46	17.40		255.87000	0.8300	0.44	1.13	1.02
c	2004	G6V	8.1	46	2.44		1383.40000	2.5500	0.27	1.13	1.02
HD 204313 b	2009	G5V	8.0	47	4.05		1931.00000	3.0820	0.13	1.05	
HD 205739 b	2008	F7V	8.6	90	1.37		279.80000	0.8960	0.27	1.22	1.33
HD 206610 b	2010	K0	8.3	194	2.20		610.00000	1.6800	0.23	1.56	6.10
HD 208487 b	2004	G2V	7.5	45	0.45		123.00000	0.4900	0.32	1.30	1.15
HD 209458 b	1999	G0V	7.7	47	0.64	1.38	3.52475	0.0475	0.07	1.00	1.15
HD 210277 b	1998	G0	6.6	21	1.23		442.10000	1.1000	0.47	1.09	1.10
HD 210702 b	2007	K1III	5.9	56	2.00		341.10000	1.1700	0.15	1.85	4.72
HD 212301 b	2005	F8V	7.8	53	0.45		2.45700	0.0360	0.00	1.05	
HD 212771 b	2010	G8IV	7.6	131	2.30		373.30000	1.2200	0.11	1.15	5.00
HD 213240 b	2001	G4IV	6.8	41	4.50		951.00000	2.0300	0.45	1.22	1.50
HD 215497 b	2009	K3V	9.0	44	0.02		3.93000			0.87	
c	2009	K3V	9.0	44	0.33		567.00000			0.87	
HD 216435 b	2002	G0V	6.0	33	1.26		1311.00000	2.5600	0.07	1.25	2.00
HD 216437 b	2002	G4IV-V	6.1	27	2.10		1294.00000	2.7000	0.34	1.07	1.10
HD 216770 b	2003	K1V	8.1	38	0.65		118.45000	0.4600	0.37	0.90	1.00
HD 217107 b	1998	G8IV	6.2	20	1.33		7.12689	0.0730	0.13	1.02	1.08
c	1998	G8IV	6.2	20	2.49		4210.00000	5.2700	0.52	1.02	1.08
HD 219449 b	2003	K0III	4.2	45	2.90		182.00000	0.3000			
HD 219828 b	2007	G0IV	8.0	81	0.07		3.83350	0.0520	0.00	1.24	1.70
HD 221287 b	2007	F7V	7.8	53	3.09		456.10000	1.2500	0.08	1.25	
HD 222582 b	1999	G5	7.7	42	7.75		572.38000	1.3500	0.73	1.00	1.15
HD 224693 b	2006	G2IV	8.2	94	0.71		26.73000	0.2330	0.05	1.33	1.70
HD 231701 b	2007	F8V	9.0	108	1.08		141.60000	0.5300	0.10	1.14	1.35
HD 240210 b	2009	K3III	8.3	143	6.90		501.75000	1.3300	0.15	1.25	
HD 285968 b	2007	M2.5V	10.0	9	0.03		8.78360	0.0660	0.00	0.49	0.53
HD 290327 b	2009	G5IV	9.0	55	2.54		2443.00000	3.4300	0.08	0.90	
HD 330075 b	2004	G5	9.4	50	0.76		3.36900	0.0430	0.00	0.95	
14 Her b	2002	K0V	6.7	18	4.64		1773.40000	2.7700	0.37	0.90	0.71
HIP 5158 b	2009	K5	10.2	45	1.30		344.00000			0.78	
HIP 12961 b	2009	M0	9.7	24	0.47		57.00000				
HIP 14810 b	2006	G5	8.5	53	3.88		6.67386	0.0692	0.14	0.99	1.00
c	2006	G5	8.5	53	1.28		147.73000	0.5450	0.16	0.99	1.00
d	2009	G5	8.5	53	0.57		962.00000	1.8900	0.17	0.99	1.00
HIP 57050 b	2010	M4V	11.9	11	0.30		41.39700	0.1635	0.31	0.34	0.40
HIP 70849 b	2009	K7V	10.4	24	5.00		3000.00000			0.63	
HIP 75458 b	2002	K2III	3.3	32	8.82		510.70200	1.2750	0.71	1.05	13.50
HIP 79431 b	2010	M3V	11.3	14	2.10		111.70000	0.3600	0.29	0.49	
HR 810 b	1999	G0V pecul.	5.4	16	1.94		311.28800	0.9100	0.24	1.11	1.85
γ1 Leo b	2009	K0III	2.0	39	8.78		428.50000	1.1900	0.14	1.23	31.88
6 Lyn b	2008	K0IV	5.9	57	2.40		899.00000	2.2000	0.13	1.70	5.20
NGC 2423 3 b	2007		9.5	766	10.60		714.30000	2.1000	0.21	2.40	
NGC 4349 127 b	2007		7.4	2176	19.80		677.80000	2.3800	0.19	3.90	
51 Peg b	1995	G2IV	5.5	15	0.47		4.23077	0.0520	0.00	1.11	1.27
24 Sex b	2010	G5	7.4	75	1.99		452.80000	1.3330	0.09	1.54	4.90
c	2010	G5	7.4	75	0.86		883.00000	2.0800	0.29	1.54	4.90
ϵ Tau b	2007	K0III	3.5	45	7.60		594.90000	1.9300	0.15	2.70	13.70
4 UMa b	2007	K1III	5.8	62	7.10		269.30000	0.8700	0.43	1.23	
47 UMa b	1996	G0V	5.1	14	2.53		1078.00000	2.1000	0.03	1.03	1.24
c	2001	G0V	5.1	14	0.54		2391.00000	3.6000	0.10	1.03	1.24
d	2010	G0V	5.1	14	1.64		14002.00000	11.6000	0.16	1.03	1.24
11 UMi b	2009	K4III	5.0	120	10.50		516.22000	1.5400	0.08	1.80	24.08
61 Vir b	2009	G5V	4.7	9	0.02		4.21500	0.0502	0.12	0.95	0.94
c	2009	G5V	4.7	9	0.06		38.02100	0.2175	0.14	0.95	0.94
d	2009	G5V	4.7	9	0.07		123.01000	0.4760	0.35	0.95	0.94
b	1996	G4V	5.0	22	7.44		116.68900	0.4800	0.40	1.10	1.97

Appendix D. Transiting planets

As of 2010 November 1, exoplanet.eu listed 494 exoplanet discoveries in total, of which transits had been observed for 106 planets in 105 systems (of which one system showed multiple transits). Six others were known to be members of multiple systems in which the existence of other planets has been determined by radial velocity measurements.

The following table lists the known transiting planets, which were either discovered by transit measurements or, in a few cases, discovered from radial velocity measurements and subsequently observed to transit. The distinction is indicated in the table.

New planets are being discovered by transit measurements and reported more-or-less continuously, at the rate of approximately 30–40 per year during 2009 and 2010. While this table is therefore out-of-date already before publication, and although the current status can be consulted on-line, it is included to specify the observational context in which this review has been prepared.

The planets, and the associated data, have been taken from the compilation at exoplanet.eu as of 2010 November 1, but with the publication-pending WASP–20, 23, 34 and 35 also included.

Data have been rounded to a fixed number of decimal places, irrespective of accuracy (typical errors are ± 0.1 in M_p and $M_\star$, and ± 0.06 in R_p and $R_\star$). For up-to-date data at full precision, additional information for each planet and host star, and the source of the data, the compilation at exoplanet.eu should be consulted.

HD 80606 was independently reported by three groups. Kepler 1–3 are names reserved for previously-known transiting planets in the Kepler field.

The columns are:
(1) planet name
(2) RV: known from radial velocity; BD: brown dwarf
(3) stellar magnitude, V
(4) stellar distance, d
(5) planet mass, M_p
(6) planet radius, R_p
(7) orbit period, P_{orb}
(8) semi-major axis, a
(9) orbit eccentricity, e
(10) stellar mass, $M_\star$
(11) stellar radius, $R_\star$
(12) transit reference (with announcement year)

Table D1. Transiting exoplanets (1/2).

(1) Name	(2)	(3) V	(4) $d_\star$ (pc)	(5) M_p (M_J)	(6) R_p (R_J)	(7) P_{orb} (d)	(8) a (AU)	(9) e	(10) $M_\star$ ($M_\odot$)	(11) $R_\star$ ($R_\odot$)	(12) Reference
CoRoT–1 b		13.6	460	1.03	1.49	1.50896	0.0254	0.00	0.95	1.11	Barge et al. (2008)
CoRoT–2 b		12.6	300	3.31	1.46	1.74300	0.0281	0.00	0.97	0.90	Alonso et al. (2008a)
CoRoT–3 b	BD										Deleuil et al. (2008)
CoRoT–4 b		13.7	–	0.72	1.19	9.20205	0.0900	0.00	1.10	1.15	Aigrain et al. (2008)
CoRoT–5 b		14.0	400	0.47	1.39	4.03790	0.0495	0.09	1.00	1.19	Rauer et al. (2009)
CoRoT–6 b		13.9	–	2.96	1.17	8.88659	0.0855	0.10	1.05	1.02	Fridlund et al. (2010b)
CoRoT–7 b		11.7	150	0.02	0.15	0.85359	0.0172	0.00	0.93	0.87	Léger et al. (2009)
CoRoT–8 b		14.8	380	0.22	0.57	6.21229	0.0630	0.00	0.88	0.77	Bordé et al. (2010)
CoRoT–9 b		13.7	460	0.84	1.05	95.27380	0.4070	0.11	0.99	0.94	Deeg et al. (2010)
CoRoT–10 b		15.2	345	2.75	0.97	13.24060	0.1055	0.53	0.89	0.79	Bonomo et al. (2010)
CoRoT–11 b		12.9	560	2.33	1.43	2.99433	0.0436	0.00	1.27	1.37	Gandolfi et al. (2010)
CoRoT–12 b		15.5	1150	0.92	1.44	2.82804	0.0402	0.07	1.08	1.12	Gillon et al. (2010)
CoRoT–13 b		15.0	1310	1.31	0.88	4.03519	0.0510	0.00	1.09	1.01	Cabrera et al. (2010)
CoRoT–14 b		16.0	1340	7.60	1.09	1.51214	0.0270	0.00	1.13	1.21	Tingley et al. (2011)
CoRoT–15 b	BD										Bouchy et al. (2011)
CoRoT–16 b		15.6	–	0.50	0.81	5.35342	–	–	–	0.81	to be submitted
CoRoT–17 b		15.5	–	2.45	1.47	3.76812	–	–	–	2.00	to be submitted
GJ 436 b	RV	10.7	10	0.07	0.37	2.64390	0.0289	0.15	0.45	0.46	Gillon et al. (2007)
GJ 1214 b		14.7	13	0.02	0.24	1.58039	0.0140	0.27	0.16	0.21	Charbonneau et al. (2009)
HAT-P–1 b		10.4	139	0.52	1.22	4.46525	0.0553	0.07	1.13	1.12	Bakos et al. (2007b)
HAT-P–2 b		8.7	118	8.74	1.19	5.63347	0.0674	0.52	1.36	1.64	Bakos et al. (2007a)
HAT-P–3 b		11.9	140	0.60	0.89	2.89970	0.0389	0.00	0.94	0.82	Torres et al. (2007)
HAT-P–4 b		11.2	310	0.68	1.27	3.05654	0.0446	0.00	1.26	1.59	Kovács et al. (2007)
HAT-P–5 b		12.0	340	1.06	1.26	2.78849	0.0408	0.00	1.16	1.17	Bakos et al. (2007c)
HAT-P–6 b		10.5	200	1.06	1.33	3.85299	0.0523	0.00	1.29	1.46	Noyes et al. (2008)
HAT-P–7 b		10.5	320	1.80	1.42	2.20473	0.0379	0.00	1.47	1.84	Pál et al. (2008)
HAT-P–8 b		10.2	230	1.52	1.50	3.07632	0.0487	0.00	1.28	1.58	Latham et al. (2009)
HAT-P–9 b		–	480	0.78	1.40	3.92281	0.0530	0.00	1.28	1.32	Shporer et al. (2009)
HAT-P–10 b		≡ WASP–11 b									Bakos et al. (2009c)
HAT-P–11 b		9.6	38	0.08	0.45	4.88780	0.0530	0.20	0.81	0.75	Bakos et al. (2010b)
HAT-P–12 b		12.8	142	0.21	0.96	3.21306	0.0384	0.00	0.73	0.70	Hartman et al. (2009)
HAT-P–13 b		10.6	214	0.85	1.28	2.91629	0.0426	0.01	1.22	1.56	Bakos et al. (2009b)
HAT-P–14 b		10.0	205	2.20	1.20	4.62766	0.0594	0.09	1.39	1.47	Torres et al. (2010b)
HAT-P–15 b		12.2	190	1.95	1.07	10.86350	0.0964	0.19	1.01	1.08	Kovács et al. (2010)
HAT-P–16 b		10.8	235	4.19	1.29	2.77596	0.0413	0.04	1.22	1.24	Buchhave et al. (2010)
HAT-P–17 b		10.5	90	0.53	1.01	10.33852	0.0882	0.35	0.86	0.84	Howard et al. (2010)
HAT-P–18 b		12.8	166	0.20	1.00	5.50802	0.0559	0.08	0.77	0.75	Hartman et al. (2011b)
HAT-P–19 b		12.9	215	0.29	1.13	4.00878	0.0466	0.07	0.84	0.82	Hartman et al. (2011b)
HAT-P–20 b		11.3	70	7.25	0.87	2.87532	0.0361	0.00	0.76	0.69	Bakos et al. (2010a)
HAT-P–21 b		11.7	254	4.06	1.02	4.12446	0.0494	0.23	0.95	1.10	Bakos et al. (2010a)
HAT-P–22 b		9.7	82	2.15	1.08	3.21222	0.0414	0.00	0.92	1.04	Bakos et al. (2010a)
HAT-P–23 b		12.4	393	2.09	1.37	1.21288	0.0232	0.00	1.13	1.20	Bakos et al. (2010a)
HAT-P–24 b		11.8	306	0.69	1.24	3.35524	0.0465	0.07	1.19	1.32	Kipping et al. (2010)
HAT-P–25 b		13.2	297	0.57	1.19	3.65284	0.0466	0.03	1.01	0.96	Quinn et al. (2010)
HAT-P–26 b		11.7	134	0.06	0.56	4.23452	0.0479	0.12	0.82	0.79	Hartman et al. (2011a)
HD 17156 b	RV	8.2	78	3.21	1.02	21.21688	0.1623	0.68	1.24	1.45	Barbieri et al. (2007)
HD 80606 b	RV	8.9	58	3.94	0.92	111.43636	0.4490	0.93	0.90	–	Moutou et al. (2009)*
HD 149026 b	RV	8.2	79	0.36	0.61	2.87589	0.0429	0.00	1.30	1.50	Sato et al. (2005a)
HD 189733 b	RV	7.7	19	1.15	1.15	2.21857	0.0314	0.00	0.80	0.79	Bouchy et al. (2005c)
HD 209458 b	RV	7.7	47	0.64	1.38	3.52475	0.0475	0.07	1.00	1.15	Charbonneau et al. (2000)
Kepler–4 b		12.7	550	0.08	0.36	3.21346	0.0456	0.00	1.22	1.49	Borucki et al. (2010b)
Kepler–5 b		–	–	2.11	1.43	3.54846	0.0506	0.00	1.37	1.79	Koch et al. (2010b)
Kepler–6 b		–	–	0.67	1.32	3.23423	0.0457	0.00	1.21	1.39	Dunham et al. (2010)
Kepler–7 b		–	–	0.43	1.48	4.88553	0.0622	0.00	1.35	1.84	Latham et al. (2010)
Kepler–8 b		13.9	1330	0.60	1.42	3.52254	0.0483	0.00	1.21	1.49	Jenkins et al. (2010)
Kepler–9 b		13.9	–	0.25	0.84	19.24316	0.1400	–	1.00	1.10	Holman et al. (2010)
c		13.9	–	0.17	0.82	38.90861	0.2250	–	1.00	1.10	Holman et al. (2010)

Appendix C. Transiting planets

Table D1 (continued). Transiting exoplanets (2/2).

(1) Name	(2)	(3) V	(4) $d_\star$ (pc)	(5) M_p (M_J)	(6) R_p (R_J)	(7) P_{orb} (d)	(8) a (AU)	(9) e	(10) $M_\star$ ($M_\odot$)	(11) $R_\star$ ($R_\odot$)	(12) Reference
Lupus–TR–3 b		17.4	–	0.81	0.89	3.91405	0.0464	0.00	0.87	0.82	Weldrake et al. (2008a)
OGLE–TR–10 b		–	1500	0.68	1.72	3.10129	0.0416	0.00	1.18	1.16	Konacki et al. (2005)
OGLE–TR–56 b		16.6	1500	1.30	1.20	1.21191	0.0225	0.00	1.17	1.32	Konacki et al. (2003a)
OGLE–TR–111 b		–	1500	0.54	1.08	4.01451	0.0470	0.00	0.82	0.83	Pont et al. (2004)
OGLE–TR–113 b		–	1500	1.24	1.11	1.43248	0.0229	0.00	0.78	0.77	Bouchy et al. (2004)
OGLE–TR–132 b		–	1500	1.17	1.25	1.68987	0.0306	0.00	1.26	1.34	Bouchy et al. (2004)
OGLE–TR–182 b		16.8	–	1.06	1.47	3.97910	0.0510	0.00	1.14	1.14	Pont et al. (2008b)
OGLE–TR–211 b		–	–	0.75	1.26	3.67724	0.0510	0.00	1.33	1.64	Udalski et al. (2008)
OGLE2–TR–L9 b		–	900	4.34	1.61	2.48553	–	–	1.52	1.53	Snellen et al. (2009b)
SWEEPS–04		18.8	8500	3.80	0.81	4.20000	0.0550	–	1.24	1.18	Sahu et al. (2006)
SWEEPS–11		19.8	8500	9.70	1.13	1.79600	0.0300	–	1.10	1.45	Sahu et al. (2006)
TrES–1		11.8	157	0.76	1.10	3.03007	0.0393	0.00	0.87	0.82	Alonso et al. (2004)
TrES–2		11.4	220	1.25	1.26	2.47061	0.0356	0.00	0.98	1.00	O'Donovan et al. (2006a)
TrES–3		12.4	–	1.91	1.30	1.30619	0.0226	0.00	0.92	0.81	O'Donovan et al. (2007)
TrES–4		11.6	440	0.88	1.81	3.55395	0.0509	0.00	1.38	1.81	Mandushev et al. (2007)
WASP–1 b		11.8	–	0.86	1.48	2.51997	0.0382	0.00	1.24	1.38	Collier Cameron et al. (2007a)
WASP–2 b		12.0	144	0.85	1.04	2.15223	0.0314	0.00	0.84	0.83	Collier Cameron et al. (2007a)
WASP–3 b		10.6	223	2.06	1.45	1.84683	0.0317	0.00	1.24	1.31	Pollacco et al. (2008)
WASP–4 b		12.6	300	1.12	1.42	1.33823	0.0230	0.00	0.90	1.15	Wilson et al. (2008a)
WASP–5 b		12.3	297	1.64	1.17	1.62842	0.0273	0.00	1.02	1.08	Anderson et al. (2008)
WASP–6 b		12.4	307	0.50	1.22	3.36101	0.0421	0.05	–	–	Gillon et al. (2009)
WASP–7 b		9.5	140	0.96	0.92	4.95466	0.0618	0.00	1.28	1.24	Hellier et al. (2009b)
WASP–8 b		9.9	87	2.24	1.04	8.15872	0.0801	0.31	1.03	0.95	Queloz et al. (2010)
WASP–9 b		false positive									
WASP–10 b		12.7	90	3.06	1.08	3.09276	0.0371	0.06	0.71	0.78	Christian et al. (2009)
WASP–11 b		11.9	125	0.46	1.04	3.72247	0.0439	0.00	0.82	0.81	West et al. (2009b)
WASP–12 b		11.7	267	1.41	1.79	1.09144	0.0229	0.02	1.35	1.57	Hebb et al. (2009)
WASP–13 b		10.4	156	0.46	1.21	4.35298	0.0527	0.00	–	–	Skillen et al. (2009)
WASP–14 b		9.8	160	7.72	1.26	2.24377	0.0370	0.09	1.32	1.30	Joshi et al. (2009)
WASP–15 b		10.9	308	0.54	1.43	3.75207	0.0499	0.00	1.18	1.48	West et al. (2009a)
WASP–16 b		11.3	–	0.86	1.01	3.11860	0.0421	0.00	1.02	0.95	Lister et al. (2009)
WASP–17 b		11.6	–	0.49	1.74	3.73544	0.0510	0.13	1.20	1.38	Anderson et al. (2010b)
WASP–18 b		9.3	100	10.43	1.16	0.94145	0.0205	0.01	1.28	1.23	Hellier et al. (2009a)
WASP–19 b		12.3	–	1.15	1.31	0.78884	0.0164	0.02	0.95	0.93	Hebb et al. (2010)
WASP–20 b											to be submitted
WASP–21 b		11.6	230	0.30	1.07	4.32248	0.0520	0.00	1.01	1.06	Bouchy et al. (2010)
WASP–22 b		12.0	300	0.56	1.12	3.53269	0.0468	0.02	1.10	1.13	Maxted et al. (2010b)
WASP–23 b		12.7	–	0.87	0.96	2.94000	–	–	–	–	Triaud et al. (2011)
WASP–24 b		11.3	330	1.03	1.10	2.34121	0.0359	0.00	1.13	1.15	Street et al. (2010)
WASP–25 b		11.9	169	0.58	1.26	3.76483	0.0474	0.00	1.00	0.95	Enoch et al. (2011)
WASP–26 b		11.3	250	1.02	1.32	2.75660	0.0400	0.00	1.12	1.34	Smalley et al. (2010a)
WASP–27 b		≡ HAT–P–14 b									Simpson et al. (2010b)
WASP–28 b		12.0	334	0.91	1.12	3.40882	0.0455	0.05	1.08	1.05	to be submitted
WASP–29 b		11.3	80	0.24	0.79	3.92273	0.0457	0.03	0.82	0.85	Hellier et al. (2010)
WASP–30 b	BD										Anderson et al. (2011)
WASP–31 b		11.7	400	0.48	1.54	3.40591	0.0466	0.00	1.16	1.24	Anderson et al. (2010a)
WASP–32 b		11.3	–	3.60	1.18	2.71865	0.0394	0.02	1.10	1.11	Maxted et al. (2010a)
WASP–33 b		8.3	116	4.10	1.50	1.21987	0.0255	–	1.50	1.44	Collier Cameron et al. (2010b)
WASP–34 b		10.4	120	0.59	1.22	4.31768	0.0524	0.04	1.01	0.93	Smalley et al. (2010b)
WASP–35 b											to be submitted
WASP–36 b		–	–	2.40	1.40	1.50000	–	–	–	–	to be submitted
WASP–37 b		12.7	338	1.70	1.14	3.57747	0.0434	0.00	0.85	0.98	Simpson et al. (2011)
WASP–38 b		9.4	110	2.71	1.08	6.87182	0.0755	0.03	1.22	1.36	Barros et al. (2011)
XO–1 b		11.3	200	0.90	1.18	3.94151	0.0488	0.00	1.00	0.93	McCullough et al. (2006)
XO–2 b		11.2	149	0.57	0.97	2.61584	0.0369	0.00	0.98	0.96	Burke et al. (2007)
XO–3 b		9.8	260	11.79	1.22	3.19152	0.0454	0.26	1.21	1.38	Johns-Krull et al. (2008)
XO–4 b		10.7	293	1.72	1.34	4.12502	0.0555	0.00	1.32	1.55	McCullough et al. (2008)
XO–5 b		12.1	255	1.08	1.09	4.18775	0.0487	0.00	0.88	1.06	Burke et al. (2008)

References

Aarestad B, Frank MR, Scott H, et al., 2008, The ice VII–ice X phase transition with implications for planetary interiors. *AGU Fall Meeting Abstracts*, A1698 {263}

Abe F, Bennett DP, Bond IA, et al., 2004, Search for low-mass exoplanets by gravitational microlensing at high magnification. *Science*, 305, 1264–1267 {89, 97}

Abe F, Bond IA, Furuta Y, et al., 2005, Candidate extrasolar planet transits discovered in the microlensing observations in astrophysics. I. Galactic bulge data. *MNRAS*, 364, 325–334 {106}

Abe L, Vakili F, Boccaletti A, 2001, The achromatic phase knife coronagraph. *A&A*, 374, 1161–1168 {153}

Abe Y, Matsui T, 1985, The formation of an impact-generated H_2O atmosphere and its implications for the early thermal history of the Earth. *Lunar and Planetary Science Conference Proceedings*, volume 15, 545–59 {269, 278}

Absil O, Mennesson B, Le Bouquin J, et al., 2009, An interferometric study of the Fomalhaut inner debris disk. I. Near-infrared detection of hot dust with VLTI–VINCI. *ApJ*, 704, 150–160 {223}

Abt HA, 1979, The frequencies of binaries on the main sequence. *AJ*, 84, 1591–1597 {56}

Acke B, van den Ancker ME, 2006, Resolving the disk rotation of HD 97048 and HD 100546 in the [O I] 630 nm line: evidence for a giant planet orbiting HD 100546. *A&A*, 449, 267–279 {224}

Adams ER, López-Morales M, Elliot JL, et al., 2010a, Lack of transit timing variations of OGLE–TR–111 b: a re-analysis with six new epochs. *ApJ*, 714, 13–24 {135}

—, 2010b, Six high-precision transits of OGLE–TR–113 b. *ApJ*, 721, 1829–1834 {135}

Adams ER, Seager S, Elkins-Tanton L, 2008, Ocean planet or thick atmosphere: on the mass-radius relationship for solid exoplanets with massive atmospheres. *ApJ*, 673, 1160–1164 {270}

Adams FC, 2004, Planet migration with disk torques and planet–planet scattering. *KITP Conference: Planet Formation: Terrestrial and Extra Solar*, 1–10 {243}

—, 2010, The birth environment of the solar system. *ARA&A*, 48, 47–85 {293}

Adams FC, Benz W, 1992, Gravitational instabilities in circumstellar disks and the formation of binary companions. *IAU Colloq. 135: Complementary Approaches to Double and Multiple Star Research*, volume 32 of *ASP Conf Ser*, 185–194 {235}

Adams FC, Bloch AM, 2009, General analysis of Type I planetary migration with stochastic perturbations. *ApJ*, 701, 1381–1397 {239}

Adams FC, Cai MJ, Lizano S, 2009, Migration of extrasolar planets: effects from X-wind accretion disks. *ApJ*, 702, L182–L186 {239, 241}

Adams FC, Fatuzzo M, 1996, A theory of the initial mass function for star formation in molecular clouds. *ApJ*, 464, 256–271 {214}

Adams FC, Hollenbach D, Laughlin G, et al., 2004, Photoevaporation of circumstellar disks due to external far-ultraviolet radiation in stellar aggregates. *ApJ*, 611, 360–379 {222}

Adams FC, Lada CJ, Shu FH, 1987, Spectral evolution of young stellar objects. *ApJ*, 312, 788–806 {219}

Adams FC, Laughlin G, 2003, Migration and dynamical relaxation in crowded systems of giant planets. *Icarus*, 163, 290–306 {39, 243}

—, 2006a, Long-term evolution of close planets including the effects of secular interactions. *ApJ*, 649, 1004–1009 {245}

—, 2006b, Relativistic effects in extrasolar planetary systems. *International Journal of Modern Physics D*, 15, 2133–2140 {133}

Adams FC, Proszkow EM, Fatuzzo M, et al., 2006, Early evolution of stellar groups and clusters: environmental effects on forming planetary systems. *ApJ*, 641, 504–525 {222}

Adams FC, Shu FH, 1986, Infrared spectra of rotating protostars. *ApJ*, 308, 836–853 {220}

Aerts C, Christensen-Dalsgaard J, Kurtz DW, 2010, *Asteroseismology*. Springer {201}

Afonso C, Henning T, 2007, The Pan-Planets project. *Transiting Extrasolar Planets Workshop*, volume 366 of *ASP Conf Ser*, 326–331 {110}

Afshordi N, Mukhopadhyay B, Narayan R, 2005, Bypass to turbulence in hydrodynamic accretion: Lagrangian analysis of energy growth. *ApJ*, 629, 373–382 {221}

Ageorges N, Dainty C, 2000, *Laser Guide Star Adaptive Optics for Astronomy*. Kluwer {151}

Agnor CB, Asphaug E, 2004, Accretion efficiency during planetary collisions. *ApJ*, 613, L157–L160 {229}

Agnor CB, Canup RM, Levison HF, 1999, On the character and consequences of large impacts in the late stage of terrestrial planet formation. *Icarus*, 142, 219–237 {302}

Agol E, 2002, Occultation and microlensing. *ApJ*, 579, 430–436 {96}

—, 2007, Rounding up the wanderers: optimising coronagraphic searches for extrasolar planets. *MNRAS*, 374, 1271–1289 {157}

—, 2011, Transit surveys for Earths in the habitable zones of white dwarfs. *ArXiv e-prints* {103}

Agol E, Steffen J, Sari R, et al., 2005, On detecting terrestrial planets with timing of giant planet transits. *MNRAS*, 359, 567–579 {135}

Agol E, Steffen JH, 2007, A limit on the presence of Earth-mass planets around a Sun-like star. *MNRAS*, 374, 941–948 {135}

Ahmic M, Croll B, Artymowicz P, 2009, Dust distribution in the β Pic circumstellar disks. *ApJ*, 705, 529–542 {223}

Ahrens TJ, 1993, Impact erosion of terrestrial planetary atmospheres. *Annual Review of Earth and Planetary Sciences*, 21, 525–555 {280}

Aigrain S, Collier Cameron AC, Ollivier M, et al., 2008, Transiting exoplanets from the CoRoT space mission. IV. CoRoT–4 b: a transiting planet in a 9.2 day synchronous orbit. *A&A*, 488, L43–L46 {326}

Aigrain S, Favata F, 2002, Bayesian detection of planetary transits: a modified version of the Gregory–Loredo method for Bayesian periodic signal detection. *A&A*, 395, 625–636 {105}

Aigrain S, Favata F, Gilmore G, 2004, Characterising stellar microvariability for planetary transit searches. *A&A*, 414, 1139–1152 {115}

Aigrain S, Hodgkin S, Irwin J, et al., 2007, The Monitor project: searching for occultations in young open clusters. *MNRAS*, 375, 29–52 {109}

Aigrain S, Irwin M, 2004, Practical planet prospecting. *MNRAS*, 350, 331–345 {105}

Aigrain S, Pont F, 2007, On the potential of transit surveys in star clusters: impact of correlated noise and radial velocity follow-up. *MNRAS*, 378, 741–752 {109}

Aigrain S, Pont F, Fressin F, et al., 2009, Noise properties of the CoRoT data. A planet-finding perspective. *A&A*, 506, 425–429 {110}

Aime C, 2005, Principle of an achromatic prolate apodised Lyot coronagraph. *PASP*, 117, 1012–1119 {153}

—, 2008, Imaging with hypertelescopes: a simple modal approach. *A&A*, 483, 361–364 {168}

Aime C, Soummer R, 2004, Multiple-stage apodised pupil Lyot coronagraph for high-contrast imaging. *SPIE Conf Ser*, volume 5490, 456–461 {153}

Akeson RL, 2006, Recent progress at the Palomar Testbed Interferometer. *SPIE Conf Ser*, volume 6268, 14 {162}

Akeson RL, Koerner DW, Jensen ELN, 1998, A circumstellar dust disk around T Tauri N: subarcsecond imaging at $\lambda = 3$ mm. *ApJ*, 505, 358–362 {56}

Akgün T, Link B, Wasserman I, 2006, Precession of the isolated neutron star PSR B1828–11. *MNRAS*, 365, 653–672 {79}

Aksnes K, Franklin FA, 2001, Secular acceleration of Io derived from mutual satellite events. *AJ*, 122, 2734–2739 {247}

Alapini A, Aigrain S, 2009, An iterative filter to reconstruct planetary transit signals in the presence of stellar variability. *MNRAS*, 397, 1591–1598 {110}

Alard C, 1996, First results of the DUO programme. *Astrophysical Applications of Gravitational Lensing*, volume 173 of *IAU Symposium*, 215–220 {86}

—, 1997, Lensing of unresolved stars towards the Galactic Bulge. *A&A*, 321, 424–433 {92}

Alard C, Lupton RH, 1998, A method for optimal image subtraction. *ApJ*, 503, 325–331 {92, 104}

Albrow MD, An J, Beaulieu J, et al., 2001a, Limits on the abundance of Galactic planets from 5 years of PLANET observations. *ApJ*, 556, L113–L116 {96}

—, 2001b, PLANET observations of microlensing event OGLE–1999–BUL–23: Limb-darkening measurement of the source star. *ApJ*, 549, 759–769 {91}

Albrow MD, Beaulieu J, Birch P, et al., 1998, The 1995 pilot campaign of PLANET: searching for microlensing anomalies through precise, rapid, round-the-clock monitoring. *ApJ*, 509, 687–702 {97}

Albrow MD, Beaulieu J, Caldwell JAR, et al., 1999, Limb darkening of a K giant in the Galactic bulge: PLANET photometry of MACHO–1997–BLG–28. *ApJ*, 522, 1011–1021 {91}

—, 2000a, Detection of rotation in a binary microlens: PLANET photometry of MACHO–1997–BLG–41. *ApJ*, 534, 894–906 {91, 92, 96}

—, 2000b, Limits on stellar and planetary companions in microlensing event OGLE–1998–BUL–14. *ApJ*, 535, 176–189 {96}

Alcock C, Akerlof CW, Allsman RA, et al., 1993, Possible gravitational microlensing of a star in the Large Magellanic Cloud. *Nature*, 365, 621–623 {86}

Alcock C, Allsman RA, Alves D, et al., 1995, First observation of parallax in a gravitational microlensing event. *ApJ*, 454, L125–L128 {93}

Alcock C, Allsman RA, Alves DR, et al., 2000, The MACHO project: microlensing results from 5.7 years of LMC observations. *ApJ*, 542, 281–307 {86}

Alexander DR, Ferguson JW, 1994, Low-temperature Rosseland opacities. *ApJ*, 437, 879–891 {264}

Alexander JB, 1967, A possible source of lithium in the atmospheres of some red giants. *The Observatory*, 87, 238–240 {199}

Alexander ME, 1987, Tidal resonances in binary star systems. *MNRAS*, 227, 843–861 {204}

Alexander RD, Clarke CJ, Pringle JE, 2005, Constraints on the ionising flux emitted by T Tauri stars. *MNRAS*, 358, 283–290 {222}

—, 2006a, Photoevaporation of protoplanetary disks. I. Hydrodynamic models. *MNRAS*, 369, 216–228 {222}

—, 2006b, Photoevaporation of protoplanetary disks. II. Evolutionary models and observable properties. *MNRAS*, 369, 229–239 {222}

Alfvén H, 1943, Non-solar planets and the origin of the solar system. *Nature*, 152, 721 {64}

Ali B, Stauffer J, Carson J, 2007, The NASA Star and Exoplanet Database (NStED). *In the Spirit of Bernard Lyot: The Direct Detection of Planets and Circumstellar Disks in the 21st Century* {7}

Alibert Y, Baraffe I, Benz W, et al., 2006, Formation and structure of the three Neptune-mass planets system around HD 69830. *A&A*, 455, L25–L28 {234}

Alibert Y, Mordasini C, Benz W, 2004, Migration and giant planet formation. *A&A*, 417, L25–L28 {234}

Alibert Y, Mordasini C, Benz W, et al., 2005a, Models of giant planet formation with migration and disk evolution. *A&A*, 434, 343–353 {36, 146, 234, 252}

Alibert Y, Mousis O, Benz W, 2005b, Modeling the Jovian subnebula. I. Thermodynamic conditions and migration of proto-satellites. *A&A*, 439, 1205–1213 {237}

—, 2005c, On the volatile enrichments and composition of Jupiter. *ApJ*, 622, L145–L148 {294}

Alibert Y, Mousis O, Mordasini C, et al., 2005d, New Jupiter and Saturn formation models meet observations. *ApJ*, 626, L57–L60 {271}

Allain S, 1998, Modeling the angular momentum evolution of low-mass stars with core-envelope decoupling. *A&A*, 333, 629–643 {200}

Allard F, Hauschildt PH, Alexander DR, et al., 2001, The limiting effects of dust in brown dwarf model atmospheres. *ApJ*, 556, 357–372 {264}

Allard NF, Allard F, Hauschildt PH, et al., 2003, A new model for brown dwarf spectra including accurate unified line shape theory for the Na I and K I resonance line profiles. *A&A*, 411, L473–L476 {264}

Allègre CJ, Manhès G, Göpel C, 1995, The age of the Earth. *Geochim. Cosmochim. Acta*, 59, 1445–1456 {297}

Allen L, Beijersbergen MW, Spreeuw RJC, et al., 1992, Orbital angular momentum of light and the transformation of Laguerre-Gaussian laser modes. *Phys. Rev. A*, 45, 8185–8189 {155}

Allende Prieto C, Barklem PS, Lambert DL, et al., 2004, S^4N: a spectroscopic survey of stars in the solar neighbourhood. The nearest 15 pc. *A&A*, 420, 183–205 {198}

Allende Prieto C, Lambert DL, Asplund M, 2001, The forbidden abundance of oxygen in the Sun. *ApJ*, 556, L63–L66 {296}

—, 2002, A reappraisal of the solar photospheric C/O ratio. *ApJ*, 573, L137–L140 {296}

Allington-Smith J, 2006, Basic principles of integral field spectroscopy. *New Astronomy Reviews*, 50, 244–251 {158}

Almenara JM, Deeg HJ, Aigrain S, et al., 2009, Rate and nature of false positives in the CoRoT exoplanet search. *A&A*, 506, 337–341 {111}

Alonso R, Alapini A, Aigrain S, et al., 2009a, The secondary eclipse of CoRoT–1 b. *A&A*, 506, 353–358 {111, 143}

Alonso R, Auvergne M, Baglin A, et al., 2008a, Transiting exoplanets from the CoRoT space mission. II. CoRoT–2 b: a transiting planet around an active G star. *A&A*, 482, L21–L24 {111, 248, 326}

Alonso R, Barbieri M, Rabus M, et al., 2008b, Limits to the planet candidate GJ 436c. *A&A*, 487, L5–L8 {135}

Alonso R, Brown TM, Charbonneau D, et al., 2007, The Transatlantic Exoplanet Survey (TrES): a review. *Transiting Extrasolar Planets Workshop*, volume 366 of *ASP Conf Ser*, 13–22 {107}

Alonso R, Brown TM, Torres G, et al., 2004, TrES–1: the transiting planet of a bright K0V star. *ApJ*, 613, L153–L156 {2, 107, 327}

Alonso R, Guillot T, Mazeh T, et al., 2009b, The secondary eclipse of the transiting exoplanet CoRoT–2 b. *A&A*, 501, L23–L26 {111, 143}

Amelin Y, 2006, The prospect of high-precision Pb isotopic dating of meteorites. *Meteoritics and Planetary Science*, 41, 7–17 {297}

Amelin Y, Krot A, 2007, Pb isotopic age of the Allende chondrules. *Meteoritics and Planetary Science*, 42, 1321–1335 {297}

Amelin Y, Krot AN, Hutcheon ID, et al., 2002, Lead isotopic ages of chondrules and calcium-aluminum-rich inclusions. *Science*, 297, 1678–1683 {297}

Ammler-von Eiff M, Santos NC, Sousa SG, et al., 2009, A homogeneous spectroscopic analysis of host stars of transiting planets. *A&A*, 507, 523–530 {189}

Ammons SM, Robinson SE, Strader J, et al., 2006, The N2K consortium. IV. New temperatures and metallicities for more than 100 000 FGK dwarfs. *ApJ*, 638, 1004–1017 {29, 187, 188}

An JH, Albrow MD, Beaulieu J, et al., 2002, First microlens mass measurement: PLANET photometry of EROS–BLG–2000–5. *ApJ*, 572, 521–539 {92, 93, 97}

Ananthakrishnan S, 1995, The Giant Meterwave Radio Telescope (GMRT). *Journal of Astrophysics and Astronomy Supplement*, 16, 427–435 {176}

Anders E, Grevesse N, 1989, Abundances of the elements: meteoritic and solar. *Geochim. Cosmochim. Acta*, 53, 197–214 {260, 295, 296}

Andersen DR, Stoesz J, Morris S, et al., 2006, Performance modeling of a wide-field ground-layer adaptive optics system. *PASP*, 118, 1574–1590 {152}

Andersen J, 1991, Accurate masses and radii of normal stars. *A&A Rev.*, 3, 91–126 {185}

—, 1999, Proceedings of the Twenty-third General Assembly. *Transactions of the International Astronomical Union, Series B*, 23 {184}

Anderson DR, Collier Cameron A, Hellier C, et al., 2010a, WASP–31 b: a low-density planet transiting a late-F-type star. *ArXiv e-prints* {327}

—, 2011, WASP–30 b: a 61 Jupiter-mass brown dwarf transiting a V=12, F8 star. *ApJ*, 726, L19–22 {327}

Anderson DR, Gillon M, Hellier C, et al., 2008, WASP–5 b: a dense, very hot Jupiter transiting a 12 mag southern-hemisphere star. *MNRAS*, 387, L4–L7 {327}

References

Anderson DR, Hellier C, Gillon M, et al., 2010b, WASP-17 b: an ultra-low density planet in a probable retrograde orbit. *ApJ*, 709, 159–167 {**108, 129, 327**}

Anderson JD, Campbell JK, Jacobson RA, et al., 1987, Radio science with Voyager 2 at Uranus: results on masses and densities of the planet and five principal satellites. *J. Geophys. Res.*, 92, 14877–14883 {**294**}

Anderson WW, Ahrens TJ, 1994, An equation of state for liquid iron and implications for the Earth's core. *J. Geophys. Res.*, 99, 4273–4284 {**261**}

André P, Ward-Thompson D, Barsony M, 2000, From prestellar cores to protostars: the initial conditions of star formation. *Protostars and Planets IV*, 59–96 {**219**}

Andreeshchev A, Scalo J, 2004, Habitability of brown dwarf planets. *Bioastronomy 2002: Life Among the Stars*, volume 213 of *IAU Symposium*, 115–118 {**115**}

Andrews SM, Williams JP, 2007, High-resolution submillimeter constraints on circumstellar disk structure. *ApJ*, 659, 705–728 {**220**}

Andronova AA, 2000, The catalogue of the nearest stellar systems: NESSY. *IAU Joint Discussion*, volume 13 {**182**}

Angel JRP, 1994, Ground-based imaging of extrasolar planets using adaptive optics. *Nature*, 368, 203–207 {**151**}

—, 1998, Sensitivity of nulling interferometers to extrasolar zodiacal emission. *Exozodiacal Dust Workshop*, 209–212 {**166**}

—, 2003a, Direct detection of terrestrial exoplanets: comparing the potential for space and ground telescopes. *Earths: Darwin/TPF and the Search for Extrasolar Terrestrial Planets*, volume 539 of *ESA Special Publication*, 221–230 {**167**}

—, 2003b, Imaging extrasolar planets from the ground. *Scientific Frontiers in Research on Extrasolar Planets*, volume 294 of *ASP Conf Ser*, 543–556 {**157**}

Angel JRP, Cheng AYS, Woolf NJ, 1986, A space telescope for infrared spectroscopy of Earth-like planets. *Nature*, 322, 341–343 {**164, 288**}

Angel JRP, Codona JL, Hinz P, et al., 2006, Exoplanet imaging with the Giant Magellan Telescope. *SPIE Conf Ser*, volume 6267, 73 {**161, 163**}

Angel JRP, Woolf NJ, 1996, Searching for life on other planets. *Scientific American*, 274, 46–52 {**164**}

—, 1997, An imaging nulling interferometer to study extrasolar planets. *ApJ*, 475, 373–379 {**164**}

Angerhausen D, Krabbe A, Iserlohe C, 2006, Near-infrared integral-field spectroscopy of HD 209458 b. *SPIE Conf Ser*, volume 6269, 152 {**158**}

—, 2009, Phase-differential NIR integral field spectroscopy of transiting Hot Jupiters. *IAU Symposium*, volume 253, 552–555 {**158**}

Anglada-Escudé G, Klioner SA, Soffel M, et al., 2007, Relativistic effects on imaging by a rotating optical system. *A&A*, 462, 371–377 {**65**}

Anglada-Escudé G, López-Morales M, Chambers JE, 2010a, How eccentric orbital solutions can hide planetary systems in 2:1 resonant orbits. *ApJ*, 709, 168–178 {**287**}

Anglada-Escudé G, Shkolnik EL, Weinberger AJ, et al., 2010b, Strong constraints to the putative planet candidate around VB 10 using Doppler spectroscopy. *ApJ*, 711, L24–L29 {**69**}

Angulo C, Arnould M, Rayet M, et al., 1999, A compilation of charged-particle induced thermonuclear reaction rates. *Nuclear Physics A*, 656, 3–183 {**202**}

Anic A, Alibert Y, Benz W, 2007, Giant collisions involving young Jupiter. *A&A*, 466, 717–728 {**178**}

Antichi J, Dohlen K, Gratton RG, et al., 2009, BIGRE: a low cross-talk integral field unit tailored for extrasolar planets imaging spectroscopy. *ApJ*, 695, 1042–1057 {**158**}

Apai D, Connolly HC, Lauretta DS, 2010, Thermal processing in protoplanetary nebulae. *Protoplanetary Dust: Astrophysical and Cosmochemical Perspectives*, 230–262, Cambridge University Press {**297**}

Apai D, Janson M, Moro-Martín A, et al., 2008, A survey for massive giant planets in debris disks with evacuated inner cavities. *ApJ*, 672, 1196–1201 {**36, 224**}

Apai D, Lauretta DS, 2010, Planet formation and protoplanetary dust. *Protoplanetary Dust: Astrophysical and Cosmochemical Perspectives*, 1–26, Cambridge University Press {**219, 225, 297**}

Apai D, Pascucci I, Bouwman J, et al., 2005, The onset of planet formation in brown dwarf disks. *Science*, 310, 834–836 {**214**}

Applegate JH, Douglas MR, Gursel Y, et al., 1986, The outer solar system for 200 million years. *AJ*, 92, 176–194 {**299**}

Apps K, Clubb KI, Fischer DA, et al., 2010, M2K: I. A Jupiter-mass planet orbiting the M3V star HIP 79431. *PASP*, 122, 156–161 {**32, 33**}

Ardila DR, Golimowski DA, Krist JE, et al., 2004, A resolved debris disk around the G2V star HD 107146. *ApJ*, 617, L147–L150 {**224**}

Aristidi E, Agabi A, Vernin J, et al., 2003, Antarctic site testing: first daytime seeing monitoring at Dome C. *A&A*, 406, L19–L22 {**161**}

Armitage PJ, 2002, Magnetic activity in accretion disk boundary layers. *MNRAS*, 330, 895–900 {**240**}

—, 2003, A reduced efficiency of terrestrial planet formation following giant planet migration. *ApJ*, 582, L47–L50 {**241**}

—, 2007a, Lecture notes on the formation and early evolution of planetary systems. *ArXiv Astrophysics e-prints* {**220, 221, 226–228, 239, 240, 243**}

—, 2007b, Massive planet migration: theoretical predictions and comparison with observations. *ApJ*, 665, 1381–1390 {**36**}

—, 2010, *Astrophysics of Planet Formation*. Cambridge University Press {**238**}

Armitage PJ, Bonnell IA, 2002, The brown dwarf desert as a consequence of orbital migration. *MNRAS*, 330, L11–L14 {**241**}

Armitage PJ, Livio M, Pringle JE, 2001, Episodic accretion in magnetically layered protoplanetary disks. *MNRAS*, 324, 705–711 {**222**}

Armitage PJ, Rice WKM, 2005, Planetary migration. *ArXiv Astrophysics e-prints* {**240**}

Arnold L, 2005, Transit light-curve signatures of artificial objects. *ApJ*, 627, 534–539 {**135**}

Arnold L, Bréon F, Brewer S, 2009, The Earth as an extrasolar planet: the vegetation spectral signature today and during the last Quaternary climatic extrema. *International Journal of Astrobiology*, 8, 81–94 {**289**}

Arnold L, Gillet S, Lardière O, et al., 2002, A test for the search for life on extrasolar planets. Looking for the terrestrial vegetation signature in the Earthshine spectrum. *A&A*, 392, 231–237 {**165, 289**}

Arnold L, Schneider J, 2004, The detectability of extrasolar planet surroundings. I. Reflected-light photometry of unresolved rings. *A&A*, 420, 1153–1162 {**130**}

Arras P, Bildsten L, 2006, Thermal structure and radius evolution of irradiated gas giant planets. *ApJ*, 650, 394–407 {**260**}

Arribas S, Gilliland RL, Sparks WB, et al., 2007, The potential of integral field spectroscopy observing extrasolar planet transits. *Science Perspectives for 3D Spectroscopy*, 53–62 {**158**}

Arsenault R, Hubin N, Stroebele S, et al., 2006, The VLT Adaptive Optics Facility project: adaptive optics modules. *The Messenger*, 123, 11–15 {**152**}

Artigau É, Biller BA, Wahhaj Z, et al., 2008, NICI: combining coronagraphy, ADI, and SDI. *SPIE Conf Ser*, volume 7014, 66 {**158**}

Artigau É, Bouchard S, Doyon R, et al., 2009, Photometric variability of the T2.5 brown dwarf SIMP J013656.5+093347: evidence for evolving weather patterns. *ApJ*, 701, 1534–1539 {**211**}

Artymowicz P, 1993, Disk-satellite interaction via density waves and the eccentricity evolution of bodies embedded in disks. *ApJ*, 419, 166–180 {**242**}

—, 1997, β Pic: an early solar system? *Annual Review of Earth and Planetary Sciences*, 25, 175–219 {**223**}

—, 2004a, Dynamics of gaseous disks with planets. *Debris Disks and the Formation of Planets*, volume 324 of *ASP Conf Ser*, 39–52 {**240**}

—, 2004b, Migration Type III. *KITP Conference: Planet Formation: Terrestrial and Extra Solar* {**240**}

Artymowicz P, Clarke CJ, Lubow SH, et al., 1991, The effect of an external disk on the orbital elements of a central binary. *ApJ*, 370, L35–L38 {**242**}

Artymowicz P, Lubow SH, 1994, Dynamics of binary-disk interaction. 1. Resonances and disk gap sizes. *ApJ*, 421, 651–667 {**58, 238**}

Asada H, 2002, Perturbative approach to astrometric microlensing due to an extrasolar planet. *ApJ*, 573, 825–828 {**94**}

Asghari N, Broeg C, Carone L, et al., 2004, Stability of terrestrial planets in the habitable zone of GJ 777A, HD 72659, GJ 614, 47 UMa and HD 4208. *A&A*, 426, 353–365 {**285**}

Ashton CE, Lewis GF, 2001, Gravitational microlensing of planets: the influence of planetary phase and caustic orientation. *MNRAS*, 325, 305–311 {**95**}

Ashwell JF, Jeffries RD, Smalley B, et al., 2005, Beryllium enhancement as evidence for accretion in a lithium-rich F dwarf. *MNRAS*, 363, L81–L85 {**193**}

Assef RJ, Gaudi BS, Stanek KZ, 2009, Detecting transits of planetary companions to giant stars. *ApJ*, 701, 1616–1626 {**115**}

Astakhov SA, Farrelly D, 2004, Capture and escape in the elliptic restricted three-body problem. *MNRAS*, 354, 971–979 {**237**}

Atobe K, Ida S, 2007, Obliquity evolution of extrasolar terrestrial planets. *Icarus*, 188, 1–17 {**301**}

Atobe K, Ida S, Ito T, 2004, Obliquity variations of terrestrial planets in habitable zones. *Icarus*, 168, 223–236 {**286**}

Atreya SK, Mahaffy PR, Niemann HB, et al., 2003, Composition and origin of the atmosphere of Jupiter-an update, and implications for the extrasolar giant planets. *Planet. Space Sci.*, 51, 105–112 {**271**}

Atreya SK, Pollack JB, Matthews MS, 1989, *Origin and Evolution of Planetary and Satellite Atmospheres*. University of Arizona Press {**293**}

Aubourg E, Bareyre P, Bréhin S, et al., 1993, Evidence for gravitational microlensing by dark objects in the Galactic halo. *Nature*, 365, 623–625 {**86**}

Augereau JC, Nelson RP, Lagrange AM, et al., 2001, Dynamical modeling of large scale asymmetries in the β Pic dust disk. *A&A*, 370, 447–455 {**223**}

Augereau JC, Papaloizou JCB, 2004, Structuring the HD 141569 A circumstellar dust disk. Impact of eccentric bound stellar companions. *A&A*, 414, 1153–1164 {**224**}

Aumann HH, Beichman CA, Gillett FC, et al., 1984, Discovery of a shell around α Lyr. *ApJ*, 278, L23–L27 {**222**}

Auvergne M, Bodin P, Boisnard L, et al., 2009, The CoRoT satellite in flight: description and performance. *A&A*, 506, 411–424 {**110**}

Baba N, Murakami N, 2003, A method to image extrasolar planets with polarised light. *PASP*, 115, 1363–1366 {**126**}

Babcock HW, 1953, The possibility of compensating astronomical seeing. *PASP*, 65, 229–236 {**150**}

Backer DC, 1993, A pulsar timing tutorial and NRAO Green Bank observations of PSR 1257+12. *Planets Around Pulsars*, volume 36 of *ASP Conf Ser*, 11–18 {**77**}

Backer DC, Foster RS, Sallmen S, 1993, A second companion of the millisecond pulsar PSR B1620–26. *Nature*, 365, 817–819 {**77**}

Backman D, Marengo M, Stapelfeldt K, et al., 2009, ϵ Eri's planetary debris disk: structure and dynamics based on Spitzer and Caltech submillimeter observatory observations. *ApJ*, 690, 1522–1538 {**223**}

Bada JL, Bigham C, Miller SL, 1994, Impact melting of frozen oceans on the early Earth: implications for the origin of life. *Proceedings of the National Academy of Science*, 91, 1248–1250 {**305**}

Badnell NR, Bautista MA, Butler K, et al., 2005, Updated opacities from the Opacity Project. *MNRAS*, 360, 458–464 {**264**}

Bagrov AV, 2006, Russian projects of space missions for astrometry. *IAU Special Session*, 1 {**73**}

Bahcall JN, Pinsonneault MH, Wasserburg GJ, 1995, Solar models with helium and heavy-element diffusion. *Reviews of Modern Physics*, 67, 781–808 {**296**}

Bailer-Jones CAL, 2009, The evidence for and against astronomical impacts on climate change and mass extinctions: a review. *International Journal of Astrobiology*, 8, 213–219 {**307**}

Bailes M, 1996, Millisecond pulsar surveys. *IAU Colloq. 160: Pulsars: Problems and Progress*, volume 105 of *ASP Conf Ser*, 3–10 {**75**}

Bailes M, Lyne AG, Shemar SL, 1991, A planet orbiting the neutron star PSR 1829–10. *Nature*, 352, 311–313 {**78**}

Bailey J, 2000, Circular polarisation and the origin of biomolecular homochirality. *Bioastronomy 99*, volume 213 of *ASP Conf Ser*, 349–355 {**286**}

—, 2004, Extraterrestrial chirality. *Bioastronomy 2002: Life Among the Stars*, volume 213 of *IAU Symposium*, 139–144 {**286**}

—, 2007, Rainbows, polarisation, and the search for habitable planets. *Astrobiology*, 7, 320–332 {**289**}

Baines EK, McAlister HA, ten Brummelaar TA, et al., 2008, CHARA array measurements of the angular diameters of exoplanet host stars. *ApJ*, 680, 728–733 {**123, 184, 185**}

Baines EK, van Belle GT, ten Brummelaar TA, et al., 2007, Direct measurement of the radius and density of the transiting exoplanet HD 189733 b with the CHARA array. *ApJ*, 661, L195–L198 {**123, 140**}

Bakos GÁ, Afonso C, Henning T, et al., 2009a, HAT-South: a global network of southern hemisphere automated telescopes to detect transiting exoplanets. *IAU Symposium*, volume 253, 354–357 {**106**}

Bakos GÁ, Hartman J, Torres G, et al., 2010a, HAT-P-20 b–HAT-P-23 b: four massive transiting extrasolar planets. *ArXiv e-prints* {**106, 326**}

Bakos GÁ, Howard AW, Noyes RW, et al., 2009b, HAT-P-13 b,c: a transiting hot Jupiter with a massive outer companion on an eccentric orbit. *ApJ*, 707, 446–456 {**2, 326**}

Bakos GÁ, Kovács G, Torres G, et al., 2007a, HD 147506 b: a supermassive planet in an eccentric orbit transiting a bright star. *ApJ*, 670, 826–832 {**106, 144, 326**}

Bakos GÁ, Lázár J, Papp I, et al., 2002, System description and first light curves of the Hungarian Automated Telescope, an autonomous observatory for variability search. *PASP*, 114, 974–987 {**105**}

Bakos GÁ, Noyes RW, Kovács G, et al., 2004, Wide-field millimagnitude photometry with the HAT: a tool for extrasolar planet detection. *PASP*, 116, 266–277 {**105**}

—, 2007b, HAT-P-1 b: a large-radius, low-density exoplanet transiting one member of a stellar binary. *ApJ*, 656, 552–559 {**2, 326**}

Bakos GÁ, Pál A, Latham DW, et al., 2006, A stellar companion in the HD 189733 system with a known transiting extrasolar planet. *ApJ*, 641, L57–L60 {**141**}

Bakos GÁ, Pál A, Torres G, et al., 2009c, HAT-P-10 b: a light and moderately hot Jupiter transiting a K dwarf. *ApJ*, 696, 1950–1955 {**326**}

Bakos GÁ, Shporer A, Pál A, et al., 2007c, HAT-P-5 b: a Jupiter-like hot Jupiter transiting a bright star. *ApJ*, 671, L173–L176 {**326**}

Bakos GÁ, Torres G, Pál A, et al., 2010b, HAT-P-11 b: a super-Neptune planet transiting a bright K star in the Kepler field. *ApJ*, 710, 1724–1745 {**104, 106, 326**}

Balan ST, Lahav O, 2009, EXOFIT: orbital parameters of extrasolar planets from radial velocities. *MNRAS*, 394, 1936–1944 {**14**}

Balbus SA, 2006, Fluid dynamics: spinning disks in the lab. *Nature*, 444, 281–283 {**221**}

Balbus SA, Hawley JF, 1991, A powerful local shear instability in weakly magnetised disks. I. Linear analysis. *ApJ*, 376, 214–233 {**221**}

—, 1998, Instability, turbulence, and enhanced transport in accretion disks. *Reviews of Modern Physics*, 70, 1–53 {**221**}

—, 2006, An exact, three-dimensional, time-dependent wave solution in local Keplerian flow. *ApJ*, 652, 1020–1027 {**221**}

Balbus SA, Hawley JF, Stone JM, 1996, Nonlinear stability, hydrodynamical turbulence, and transport in disks. *ApJ*, 467, 76–86 {**221**}

Baliunas SL, Donahue RA, Soon WH, et al., 1995, Chromospheric variations in main-sequence stars. *ApJ*, 438, 269–287 {**22**}

Ballard S, Charbonneau D, A'Hearn MF, et al., 2008, Preliminary results from the NASA EPOXI mission. *AAS/Division for Planetary Sciences Meeting Abstracts*, volume 40, 01.02 {**113**}

Ballester GE, Sing DK, Herbert F, 2007, The signature of hot hydrogen in the atmosphere of the extrasolar planet HD 209458 b. *Nature*, 445, 511–514 {**140**}

Bally J, Scoville NZ, 1982, Structure and evolution of molecular clouds near H II regions. II. The disk constrained H II region, S106. *ApJ*, 255, 497–509 {**222**}

Baltz EA, Gondolo P, 2001, Binary events and extragalactic planets in pixel microlensing. *ApJ*, 559, 41–52 {**101**}

Baluev RV, 2008a, Assessing the statistical significance of periodogram peaks. *MNRAS*, 385, 1279–1285 {**13**}

—, 2008b, Optimal strategies of radial velocity observations in planet search surveys. *MNRAS*, 389, 1375–1382 {**16**}

—, 2010, Optimal planning of radial velocity observations for multi-planet extrasolar systems. *EAS Publications Series*, volume 42, 97–104 {**16**}

Banit M, Ruderman MA, Shaham J, et al., 1993, Formation of planets around pulsars. *ApJ*, 415, 779–796 {**77**}

Baraffe I, Alibert Y, Chabrier G, et al., 2006, Birth and fate of hot-Neptune planets. *A&A*, 450, 1221–1229 {**281**}

Baraffe I, Chabrier G, Barman T, 2008, Structure and evolution of super-Earth to super-Jupiter exoplanets. I. Heavy element enrichment in the interior. *A&A*, 482, 315–332 {**144, 145**}

Baraffe I, Chabrier G, Barman TS, et al., 2003, Evolutionary models for cool brown dwarfs and extrasolar giant planets: the case of HD 209458. *A&A*, 402, 701–712 {**145, 170, 265, 272**}

—, 2005, Hot-Jupiters and hot-Neptunes: a common origin? *A&A*, 436, L47–L51 {**145, 281**}

Baraffe I, Selsis F, Chabrier G, et al., 2004, The effect of evaporation on the evolution of close-in giant planets. *A&A*, 419, L13–L16 {**146, 281**}

Baranne A, Mayor M, Poncet JL, 1979, CORAVEL: a new tool for radial velocity measurements. *Vistas in Astronomy*, 23, 279–316 {**17**}

Baranne A, Queloz D, Mayor M, et al., 1996, ELODIE: a spectrograph for accurate radial velocity measurements. *A&AS*, 119, 373–390 {**17, 20, 21, 24**}

Barbieri M, Alonso R, Laughlin G, et al., 2007, HD 17156b: a transiting planet with a 21.2-day period and an eccentric orbit. *A&A*, 476, L13–L16 {**2, 109, 326**}

Barbieri M, Gratton RG, 2002, Galactic orbits of stars with planets. *A&A*, 384, 879–883 {**183**}

Barbieri M, Marzari F, Scholl H, 2002, Formation of terrestrial planets in close binary systems: the case of α Cen A. *A&A*, 396, 219–224 {**58**}

Barge P, Baglin A, Auvergne M, et al., 2008, Transiting exoplanets from the CoRoT space mission. I. CoRoT-1 b: a low-density short-period planet around a G0V star. *A&A*, 482, L17–L20 {**2, 326**}

Barker AJ, Ogilvie GI, 2009, On the tidal evolution of hot Jupiters on inclined orbits. *MNRAS*, 395, 2268–2287 {**133**}

Barman TS, 2007, Identification of absorption features in an extrasolar planet atmosphere. *ApJ*, 661, L191–L194 {**139, 140**}

—, 2008, On the presence of water and global circulation in the transiting planet HD 189733 b. *ApJ*, 676, L61–L64 {**140**}

Barman TS, Hauschildt PH, Allard F, 2001, Irradiated planets. *ApJ*, 556, 885–895 {**145**}

—, 2005, Phase-dependent properties of extrasolar planet atmospheres. *ApJ*, 632, 1132–1139 {**138**}

Barnes JR, Barman TS, Prato L, et al., 2007, Limits on the 2.2-μm contrast ratio of the close-orbiting planet HD 189733 b. *MNRAS*, 382, 473–480 {**140**}

Barnes JW, 2007, Effects of orbital eccentricity on extrasolar planet transit detectability and light curves. *PASP*, 119, 986–993 {**121, 122**}

—, 2009, Transit lightcurves of extrasolar planets orbiting rapidly rotating stars. *ApJ*, 705, 683–692 {**131, 132**}

Barnes JW, Cooper CS, Showman AP, et al., 2009a, Detecting the wind-driven shapes of extrasolar giant planets from transit photometry. *ApJ*, 706, 877–884 {**131**}

Barnes JW, Fortney JJ, 2003, Measuring the oblateness and rotation of transiting extrasolar giant planets. *ApJ*, 588, 545–556 {**131, 134, 270**}

—, 2004, Transit detectability of ring systems around extrasolar giant planets. *ApJ*, 616, 1193–1203 {**130**}

Barnes JW, O'Brien DP, 2002, Stability of satellites around close-in extrasolar giant planets. *ApJ*, 575, 1087–1093 {**130, 237**}

Barnes R, 2008, Dynamics of multiple planet systems. *Exoplanets: Detection, Formation, Properties, Habitability*, 177–208, Springer {**42–44**}

—, 2010, Planet–planet interactions. *Formation and Evolution of Exoplanets*, 49–70, Wiley {**243**}

Barnes R, Goździewski K, Raymond SN, 2008a, The successful prediction of the extrasolar planet HD 74156 d. *ApJ*, 680, L57–L60 {**48**}

Barnes R, Greenberg R, 2006a, Behaviour of apsidal orientations in planetary systems. *ApJ*, 652, L53–L56 {**42**}

—, 2006b, Extrasolar planetary systems near a secular separatrix. *ApJ*, 638, 478–487 {**44**}

—, 2006c, Stability limits in extrasolar planetary systems. *ApJ*, 647, L163–L166 {**43, 244**}

—, 2007a, Apsidal behaviour among planetary orbits: testing the planet-planet scattering model. *ApJ*, 659, L53–L56 {**42, 243**}

—, 2007b, Stability limits in resonant planetary systems. *ApJ*, 665, L67–L70 {**44**}

Barnes R, Jackson B, Greenberg R, et al., 2009b, Tidal limits to planetary habitability. *ApJ*, 700, L30–L33 {**250**}

Barnes R, Quinn T, 2001, A statistical examination of the short-term stability of the υ And planetary system. *ApJ*, 550, 884–889 {**47**}

—, 2004, The (in)stability of planetary systems. *ApJ*, 611, 494–516 {**43, 44**}

Barnes R, Quinn TR, Lissauer JJ, et al., 2009c, N-body simulations of growth from 1 km planetesimals at 0.4 AU. *Icarus*, 203, 626–643 {**228**}

Barnes R, Raymond SN, 2004, Predicting planets in known extrasolar planetary systems. I. Test particle simulations. *ApJ*, 617, 569–574 {**44**}

Barnes R, Raymond SN, Jackson B, et al., 2008b, Tides and the evolution of planetary habitability. *Astrobiology*, 8, 557–568 {**247, 284**}

Barnes SA, 2001, An assessment of the rotation rates of the host stars of extrasolar planets. *ApJ*, 561, 1095–1106 {**186**}

Barnes SA, Sofia S, Pinsonneault M, 2001, Disk locking and the presence of slow rotators among solar-type stars in young star clusters. *ApJ*, 548, 1071–1080 {**200**}

Baross JA, 1983, Growth of 'black smoker' bacteria at temperatures of at least 250 C. *Nature*, 303, 423–426 {**250**}

Barranco JA, Marcus PS, 2005, Three-dimensional vortices in stratified protoplanetary disks. *ApJ*, 623, 1157–1170 {**221**}

Barros SCC, Faedi F, Collier Cameron A, et al., 2011, WASP-38 b: a transiting exoplanet in an eccentric, 6.87-d period orbit. *A&A*, 525, A54 {**327**}

Barrow JD, Tipler FJ, 1988, *The Anthropic Cosmological Principle*. Oxford University Press {**286**}

Barry RK, Danchi WC, Rajagopal J, et al., 2008a, The Fourier–Kelvin stellar interferometer: a progress report and preliminary results from our laboratory testbed. *The Power of Optical/IR Interferometry: Recent Scientific Results and Second Generation*, 547–550 {**166**}

Barry RK, Danchi WC, Traub W, et al., 2008b, First science with the Keck interferometer nuller: high spatial resolution N-band observations of the recurrent nova RS Oph. *SPIE Conf Ser*, volume 7013, 22 {**163**}

Barshay SS, Lewis JS, 1978, Chemical structure of the deep atmosphere of Jupiter. *Icarus*, 33, 593–611 {**276**}

Barucci MA, Boehnhardt H, Cruikshank DP, et al., 2008, *The Solar System Beyond Neptune*. University of Arizona Press {**293**}

Basri G, 2000a, The discovery of brown dwarfs. *Scientific American*, 282(4), 57–63 {**210**}

—, 2000b, Observations of brown dwarfs. *ARA&A*, 38, 485–519 {**209**}

Basri G, Brown ME, 2006, Planetesimals to brown dwarfs: what is a planet? *Annual Review of Earth and Planetary Sciences*, 34, 193–216 {**7, 210, 230**}

Basri G, Mohanty S, Allard F, et al., 2000, An effective temperature scale for late-M and L dwarfs, from resonance sbsorption lines of Cs I and Rb I. *ApJ*, 538, 363–385 {**212**}

Bass RW, del Popolo A, 2005, Dynamical derivation of Bode's law. *International Journal of Modern Physics D*, 14, 153–169 {**48**}

Bastian TS, Dulk GA, Leblanc Y, 2000, A search for radio emission from extrasolar planets. *ApJ*, 545, 1058–1063 {**176**}

Bastian U, Hefele H, 2005, Astrometric limits set by surface structure, binarity, microlensing. *The Three-Dimensional Universe with Gaia*, volume 576 of *ESA Special Publication*, 215–218 {**65**}

Bate MR, Bonnell IA, 2005, The origin of the initial mass function and its dependence on the mean Jeans mass in molecular clouds. *MNRAS*, 356, 1201–1221 {**215**}

Bate MR, Bonnell IA, Bromm V, 2002, The formation mechanism of brown dwarfs. *MNRAS*, 332, L65–L68 {**214, 215**}

Bate MR, Lodato G, Pringle JE, 2010, Chaotic star formation and the alignment of stellar rotation with disk and planetary orbital axes. *MNRAS*, 401, 1505–1513 {**129, 302**}

Batygin K, Bodenheimer P, Laughlin G, 2009, Determination of the interior structure of transiting planets in multiple-planet systems. *ApJ*, 704, L49–L53 {**134**}

Batygin K, Brown ME, 2010, Early dynamical evolution of the solar system: pinning down the initial conditions of the Nice model. *ApJ*, 716, 1323–1331 {**304**}

Baudoz P, Boccaletti A, Riaud P, et al., 2006, Feasibility of the four-quadrant phase mask in the mid-infrared on the James Webb Space Telescope. *PASP*, 118, 765–773 {**155, 163**}

Baudoz P, Rabbia Y, Gay J, 2000, Achromatic interfero coronagraphy. I. Theoretical capabilities for ground-based observations. *A&AS*, 141, 319–329 {**153**}

Bayliss DDR, Sackett PD, 2007, The SkyMapper transit survey. *Transiting Extrasolar Planets Workshop*, volume 366 of *ASP Conf Ser*, 320–325 {**110**}

Bayliss DDR, Sackett PD, Weldrake DTF, 2009a, SuperLupus: a deep, long duration transit survey. *IAU Symposium*, volume 253 of *IAU Symposium*, 333–335 {**108**}

Bayliss DDR, Weldrake DTF, Sackett PD, et al., 2009b, The Lupus transit survey for hot Jupiters: results and lessons. *AJ*, 137, 4368–4376 {**108**}

Bazot M, Vauclair S, 2004, Asteroseismology of exoplanets hosts stars: tests of internal metallicity. *A&A*, 427, 965–973 {**203**}

Bazot M, Vauclair S, Bouchy F, et al., 2005, Seismic analysis of the planet-hosting star μ Ara. *A&A*, 440, 615–621 {**203, 204**}

Bean JL, 2009, An analysis of the transit times of CoRoT–1 b. *A&A*, 506, 369–375 {**135**}

Bean JL, Benedict GF, Charbonneau D, et al., 2008a, A HST transit light curve for GJ 436b. *A&A*, 486, 1039–1046 {**135, 142**}

Bean JL, Benedict GF, Endl M, 2006, Metallicities of M dwarf planet hosts from spectral synthesis. *ApJ*, 653, L65–L68 {**190**}

Bean JL, McArthur BE, Benedict GF, et al., 2007, The mass of the candidate exoplanet companion to HD 33636 from HST astrometry and high-precision radial velocities. *AJ*, 134, 749–758 {**71**}

—, 2008b, Detection of a third planet in the HD 74156 system using the Hobby–Eberly telescope. *ApJ*, 672, 1202–1208 {**44, 48**}

Bean JL, Seifahrt A, 2009, The architecture of the GJ 876 planetary system: masses and orbital coplanarity for planets b and c. *A&A*, 496, 249–257 {**67, 71**}

Bean JL, Seifahrt A, Hartman H, et al., 2010a, The CRIRES search for planets around the lowest-mass stars. I. High-precision near-infrared radial velocities with an ammonia gas cell. *ApJ*, 713, 410–422 {**20, 24**}

—, 2010b, The proposed giant planet orbiting VB 10 does not exist. *ApJ*, 711, L19–L23 {**69**}

Beatty TG, Gaudi BS, 2008, Predicting the yields of photometric surveys for transiting extrasolar planets. *ApJ*, 686, 1302–1330 {**103, 104**}

Beaugé C, Ferraz-Mello S, Michtchenko TA, 2003, Extrasolar planets in mean-motion resonance: apses alignment and asymmetric stationary solutions. *ApJ*, 593, 1124–1133 {**42**}

Beaugé C, Michtchenko TA, 2003, Modeling the high-eccentricity planetary three-body problem: application to the GJ 876 planetary system. *MNRAS*, 341, 760–770 {**49**}

Beaugé C, Michtchenko TA, Ferraz-Mello S, 2006, Planetary migration and extrasolar planets in the 2:1 mean-motion resonance. *MNRAS*, 365, 1160–1170 {**42, 49**}

Beaugé C, Sándor Z, Érdi B, et al., 2007, Co-orbital terrestrial planets in exoplanetary systems: a formation scenario. *A&A*, 463, 359–367 {**54**}

Beaulieu JP, Albrow M, Bennett D, et al., 2007, Hunting for frozen super-Earths via microlensing. *The Messenger*, 128, 33–34 {**97**}

Beaulieu JP, Bennett DP, Batista V, et al., 2010, EUCLID: dark universe probe and microlensing planet hunter. *ArXiv e-prints* {**102**}

Beaulieu JP, Bennett DP, Fouqué P, et al., 2006, Discovery of a cool planet of 5.5 Earth masses through gravitational microlensing. *Nature*, 439, 437–440 {**2, 98, 99, 267**}

Beaulieu JP, Carey S, Ribas I, et al., 2008, Primary transit of the planet HD 189733 b at 3.6 and 5.8 μm. *ApJ*, 677, 1343–1347 {**140**}

Beckers JM, 1976, Reliability of sunspots as tracers of solar surface rotation. *Nature*, 260, 227–229 {**19**}

—, 1993, Adaptive optics for astronomy: principles, performance, and applications. *ARA&A*, 31, 13–62 {**151**}

—, 2005, Sunspots, gravitational redshift and exoplanet detection. *Acta Historica Astronomiae*, 25, 285–297 {**19**}

—, 2007, Can variable meridional flows lead to false exoplanet detections? *Astronomische Nachrichten*, 328, 1084–1086 {**18**}

Becklin EE, Farihi J, Jura M, et al., 2005, A dusty disk around GD 362, a white dwarf with a uniquely high photospheric metal abundance. *ApJ*, 632, L119–L122 {**173**}

Becklin EE, Zuckerman B, 1988, A low-temperature companion to a white dwarf star. *Nature*, 336, 656–658 {**210**}

Beckwith SVW, 1996, Circumstellar disks and the search for neighbouring planetary systems. *Nature*, 383, 139–144 {**222**}

—, 2008, Detecting life-bearing extrasolar planets with space telescopes. *ApJ*, 684, 1404–1415 {**164**}

Bedding TR, Kjeldsen H, Butler RP, et al., 2004, Oscillation frequencies and mode lifetimes in α Cen A. *ApJ*, 614, 380–385 {**202**}

Beech M, Steel D, 1995, On the definition of the term meteoroid. *QJRAS*, 36, 281–284 {**295**}

Beer ME, King AR, Livio M, et al., 2004a, How special is the Solar system? *MNRAS*, 354, 763–768 {**225**}

Beer ME, King AR, Pringle JE, 2004b, The planet in M4: implications for planet formation in globular clusters. *MNRAS*, 355, 1244–1250 {**78**}

Beichman CA, 1996, A Road Map for the Exploration of Neighbouring Planetary Systems (ExNPS) 96-22. Technical report, JPL {**166**}

Beichman CA, 1998, Terrestrial Planet Finder: the search for life-bearing planets around other stars. *SPIE Conf Ser*, volume 3350, 719–723 {**166**}

—, 2003, Recommended architectures for the Terrestrial Planet Finder. *Hubble's Science Legacy: Future Optical/Ultraviolet Astronomy from Space*, volume 291 of *ASP Conf Ser*, 101–108 {**166**}

Beichman CA, Bryden G, Gautier TN, et al., 2005a, An excess due to small grains around the nearby K0V star HD 69830: asteroid or cometary debris? *ApJ*, 626, 1061–1069 {**55, 224**}

Beichman CA, Bryden G, Rieke GH, et al., 2005b, Planets and infrared excesses: preliminary results from a Spitzer–MIPS survey of solar-type stars. *ApJ*, 622, 1160–1170 {**201, 224**}

Beirão P, Santos NC, Israelian G, et al., 2005, Abundances of Na, Mg and Al in stars with giant planets. *A&A*, 438, 251–256 {**195, 198**}

Béjar VJS, Zapatero Osorio MR, Pérez-Garrido A, et al., 2008, Discovery of a wide companion near the deuterium-burning mass limit in the Upper Scorpius association. *ApJ*, 673, L185–L189 {**168**}

Béjar VJS, Zapatero Osorio MR, Rebolo R, 1999, A search for very low mass stars and brown dwarfs in the young σ Ori cluster. *ApJ*, 521, 671–681 {**215**}

Bekker A, Holland HD, Wang P, et al., 2004, Dating the rise of atmospheric oxygen. *Nature*, 427, 117–120 {**288**}

Belbruno E, Gott JR III, 2005, Where did the Moon come from? *AJ*, 129, 1724–1745 {**302**}

Beletskii VV, Pivovarov ML, Starostin EL, 1996, Regular and chaotic motions in applied dynamics of a rigid body. *Chaos*, 6, 155–166 {**129**}

Bell KR, Cassen PM, Klahr HH, et al., 1997, The structure and appearance of protostellar accretion disks: limits on disk flaring. *ApJ*, 486, 372–387 {**240**}

Bender C, Simon M, Prato L, et al., 2005, An upper bound on the 1.6 μm flux ratio of the companion to ρ CrB. *AJ*, 129, 402–408 {**70**}

Bender ML, 2002, Orbital tuning chronology for the Vostok climate record supported by trapped gas composition. *Earth and Planetary Science Letters*, 204, 275–289 {**307**}

Bender PL, Stebbins RT, 1996, Multi-resolution element imaging of extrasolar Earth-like planets. *J. Geophys. Res.*, 101, 9309–9312 {**167**}

Bender R, 1990, Unraveling the kinematics of early-type galaxies: presentation of a new method and its application to NGC 4621. *A&A*, 229, 441–451 {**17**}

Benedict GF, McArthur B, Chappell DW, et al., 1999, Interferometric astrometry of Proxima Centauri and Barnard's Star using HST–FGS3: detection limits for substellar companions. *AJ*, 118, 1086–1100 {**70**}

Benedict GF, McArthur BE, Bean JL, 2008, HST FGS astrometry: the value of fractional millisecond of arc precision. *IAU Symposium*, volume 248, 23–29 {**71**}

Benedict GF, McArthur BE, Forveille T, et al., 2002, A mass for the extrasolar planet GJ 876 b determined from HST–FGS3 astrometry and high-precision radial velocities. *ApJ*, 581, L115–L118 {**68, 71**}

Benedict GF, McArthur BE, Franz OG, et al., 2000, Interferometric astrometry of the detached white dwarf–M dwarf binary Feige 24 using HST FGS3: white dwarf radius and component mass estimates. *AJ*, 119, 2382–2390 {**70**}

Benedict GF, McArthur BE, Gatewood G, et al., 2006, The extrasolar planet ϵ Eri b: orbit and mass. *AJ*, 132, 2206–2218 {**2, 71, 223**}

Benest D, 1988, Planetary orbits in the elliptic restricted problem. I. The α Cen system. *A&A*, 206, 143–146 {**57**}

—, 1989, Planetary orbits in the elliptic restricted problem. II. The Sirius system. *A&A*, 223, 361–364 {**57**}

—, 1993, Stable planetary orbits around one component in nearby binary stars. II. *Celestial Mechanics and Dynamical Astronomy*, 56, 45–50 {**57**}

—, 1996, Planetary orbits in the elliptic restricted problem. III. The η CrB system. *A&A*, 314, 983–988 {**57**}

—, 1998, Stable planetary orbits in double star systems. *Brown Dwarfs and Extrasolar Planets*, volume 134 of *ASP Conf Ser*, 277–279 {**114**}

Bennett DP, 2008, Detection of extrasolar planets by gravitational microlensing. *Exoplanets: Detection, Formation, Properties, Habitability*, 47–88, Springer {**92**}

Bennett DP, Alcock C, Allsman RA, et al., 1997, Planetary microlensing from the MACHO project. *Planets Beyond the Solar System and the Next Generation of Space Missions*, volume 119 of *ASP Conf Ser*, 95–99 {**96**}

Bennett DP, Anderson J, Bond IA, et al., 2006, Identification of the OGLE–2003–BLG–235/MOA–2003–BLG–53 planetary host star. *ApJ*, 647, L171–L174 {**92, 98**}

Bennett DP, Anderson J, Gaudi BS, 2007, Characterisation of gravitational microlensing planetary host stars. *ApJ*, 660, 781–790 {**92**}

Bennett DP, Bally J, Bond I, et al., 2003, The Galactic Exoplanet Survey Telescope (GEST). *SPIE Conf Ser*, volume 4854, 141–157 {**102**}

Bennett DP, Bond I, Cheng E, et al., 2004, The Microlensing Planet Finder: completing the census of extrasolar planets in the Milky Way. *SPIE Conf Ser*, volume 5487, 1453–1464 {**102**}

References

Bennett DP, Bond IA, Udalski A, et al., 2008, A low-mass planet with a possible sub-stellar-mass host in microlensing event MOA–2007–BLG–192. *ApJ*, 684, 663–683 {99}

Bennett DP, Rhie SH, 1996, Detecting Earth-mass planets with gravitational microlensing. *ApJ*, 472, 660–664 {**88, 89, 91**}

—, 2002, Simulation of a space-based microlensing survey for terrestrial extrasolar planets. *ApJ*, 574, 985–1003 {**102**}

Bennett DP, Rhie SH, Becker AC, et al., 1999, Discovery of a planet orbiting a binary star system from gravitational microlensing. *Nature*, 402, 57–59 {**91, 96**}

Bennett DP, Rhie SH, Nikolaev S, et al., 2010, Masses and orbital constraints for the OGLE–2006–BLG–109L b,c Jupiter/Saturn analogue planetary system. *ApJ*, 713, 837–855 {99}

Bennett ME III, McSween HY, 1996, Revised model calculations for the thermal histories of ordinary chondrite parent bodies. *Meteoritics and Planetary Science*, 31, 783–792 {**297**}

Bensby T, Feltzing S, Lundström I, 2004, A possible age-metallicity relation in the Galactic thick disk? *A&A*, 421, 969–976 {**194**}

Bensby T, Feltzing S, Lundström I, et al., 2005, α-, r-, and s-process element trends in the Galactic thin and thick disks. *A&A*, 433, 185–203 {**198**}

Bentley SJ, Hellier C, Maxted PFL, et al., 2009, A stellar flare during the transit of the extrasolar planet OGLE–TR–10 b. *A&A*, 505, 901–902 {**131, 133**}

Benvenuto OG, Brunini A, Fortier A, 2007, Envelope instability in giant planet formation. *Icarus*, 191, 394–396 {**234**}

Benvenuto OG, Fortier A, Brunini A, 2009, Forming Jupiter, Saturn, Uranus and Neptune in few million years by core accretion. *Icarus*, 204, 752–755 {**304**}

Benz W, Asphaug E, 1999, Catastrophic disruptions revisited. *Icarus*, 142, 5–20 {**228**}

Benz W, Mayor M, 1981, A new method for determining the rotation of late spectral type stars. *A&A*, 93, 235–240 {**17**}

Benz W, Mordasini C, Alibert Y, et al., 2008, Giant planet population synthesis: comparing theory with observations. *Physica Scripta Volume T*, 130(1), 014022 {**252**}

Benz W, Slattery WL, Cameron AGW, 1988, Collisional stripping of Mercury's mantle. *Icarus*, 74, 516–528 {**229, 280**}

Berdyugina SV, Berdyugin AV, Fluri DM, et al., 2008, First detection of polarised scattered light from an exoplanetary atmosphere. *ApJ*, 673, L83–L86 {**126**}

Berger AL, 1976, Obliquity and precession for the last 5 000 000 years. *A&A*, 51, 127–135 {**301**}

—, 1977a, Long-term variations of the Earth's orbital elements. *Celestial Mechanics*, 15, 53–74 {**301**}

—, 1977b, Support for the astronomical theory of climatic change. *Nature*, 269, 44–47 {**301**}

—, 1980, The Milankovitch astronomical theory of paleoclimates: a modern review. *Vistas in Astronomy*, 24, 103–122 {**301**}

Berger E, 2006, Radio observations of a large sample of late M, L, and T dwarfs: the distribution of magnetic field strengths. *ApJ*, 648, 629–636 {**173, 213**}

Berger E, Ball S, Becker KM, et al., 2001, Discovery of radio emission from the brown dwarf LP 944–20. *Nature*, 410, 338–340 {**213**}

Bergstralh JT, Miner ED, Matthews MS, 1991, *Uranus*. University of Arizona Press {**293, 294, 301**}

Berner RA, 2001, Modeling atmospheric O_2 over Phanerozoic time. *Geochim. Cosmochim. Acta*, 65, 685–694 {**306**}

—, 2006, Geocarbsulf: a combined model for Phanerozoic atmospheric O_2 and CO_2. *Geochim. Cosmochim. Acta*, 70, 5653–5664 {**306, 307**}

Berner RA, Kothavala Z, 2001, Geocarb III: a revised model of atmospheric CO_2 over Phanerozoic time. *Am J Sci*, 301, 182–204 {**306, 307**}

Berti E, Ferrari V, 2001, Excitation of g-modes of solar-type stars by an orbiting companion. *Phys. Rev. D*, 63(6), 064031 {**205**}

Berton A, Feldt M, Gratton R, et al., 2006a, The search for extrasolar giant planets using integral field spectroscopy: simulations. *New Astronomy Reviews*, 49, 661–669 {**158**}

Berton A, Gratton RG, Feldt M, et al., 2006b, Detecting extrasolar planets with integral field spectroscopy. *PASP*, 118, 1144–1164 {**158**}

Bessell MS, 2007, Bolometric corrections to Hipparcos Hp magnitudes and UBVRI colours. *http://msowww.anu.edu.au/~bessell/* {**184**}

Bessell MS, Castelli F, Plez B, 1998, Model atmospheres broad-band colours, bolometric corrections and temperature calibrations for O–M stars. *A&A*, 333, 231–250 {**184**}

Beuermann K, Hessman FV, Dreizler S, et al., 2010, Two planets orbiting the recently formed post-common envelope binary NN Ser. *A&A*, 521, L60 {**79, 82**}

Beust H, Bonfils X, Delfosse X, et al., 2008, Dynamical evolution of the GJ 581 planetary system. *A&A*, 479, 277–282 {**287**}

Beust H, Lagrange A, Crawford IA, et al., 1998, The β Pic circumstellar disk. XXV. The Ca II absorption lines and the falling evaporating bodies model revisited using UHRF observations. *A&A*, 338, 1015–1030 {**223**}

Beust H, Morbidelli A, 2000, Falling evaporating bodies as a clue to outline the structure of the β Pic young planetary system. *Icarus*, 143, 170–188 {**223**}

Beust H, Vidal-Madjar A, Ferlet R, et al., 1994, Cometary-like bodies in the protoplanetary disk around β Pic. *Ap&SS*, 212, 147–157 {**223**}

Beuzit J, Feldt M, Dohlen K, et al., 2008, SPHERE: a planet finder instrument for the VLT. *SPIE Conf Ser*, volume 7014, 41 {**159**}

Bhathal R, 2000, The case for optical SETI. *Astronomy and Geophysics*, 41(1), 25–26 {**290**}

—, 2001, Optical SETI in Australia. *SPIE Conf Ser* (eds. Kingsley SA, Bhathal R), volume 4273, 144–152 {**291**}

Bihain G, Rebolo R, Zapatero Osorio MR, et al., 2009, Candidate free-floating super-Jupiters in the young σ Ori open cluster. *A&A*, 506, 1169–1182 {**215**}

Biller BA, Close LM, Masciadri E, et al., 2007, An imaging survey for extrasolar planets around 45 close, young stars with the simultaneous differential imager at the VLT and MMT. *ApJS*, 173, 143–165 {**170**}

Biller BA, Kasper M, Close LM, et al., 2006, Discovery of a brown dwarf very close to the Sun: a methane-rich brown dwarf companion to the low-mass star SCR 1845–6357. *ApJ*, 641, L141–L144 {**158, 168, 170, 171**}

Binder AB, 1974, On the origin of the Moon by rotational fission. *Moon*, 11, 53–76 {**302**}

Binnendijk L, 1960, *Properties of Double Stars: A Survey of Parallaxes and Orbits*. University of Pennsylvania Press, Philadelphia {**68**}

Binney J, Tremaine S, 2008, *Galactic Dynamics*. Princeton University Press, Second Edition {**238**}

Birnbaum G, Borysow A, Orton GS, 1996, Collision-induced absorption of H_2–H_2 and H_2–He in the rotational and fundamental bands for planetary applications. *Icarus*, 123, 4–22 {**281**}

Bizzarro M, Ulfbeck D, Trinquier A, et al., 2007, Evidence for a late supernova injection of ^{60}Fe into the protoplanetary disk. *Science*, 316, 1178–1181 {**298**}

Black DC, 1982, A simple criterion for determining the dynamical stability of three-body systems. *AJ*, 87, 1333–1337 {**56**}

Black DC, Scargle JD, 1982, On the detection of other planetary systems by astrometric techniques. *ApJ*, 263, 854–869 {**64, 67**}

Blackett PMS, 1947, The magnetic field of massive rotating bodies. *Nature*, 159, 658–666 {**176**}

Blaes OM, Balbus SA, 1994, Local shear instabilities in weakly ionised, weakly magnetised disks. *ApJ*, 421, 163–177 {**221**}

Blake CH, Bloom JS, Latham DW, et al., 2008, Near-infrared monitoring of ultracool dwarfs: prospects for searching for transiting companions. *PASP*, 120, 860–871 {**115**}

Blanc M, Kallenbach R, Erkaev NV, 2005, Solar system magnetospheres. *Space Sci. Rev.*, 116, 227–298 {**174**}

Blandford RD, Haynes MP, Huchra JP, et al., 2010, *New Worlds, New Horizons in Astronomy and Astrophysics: Committee for a Decadel Survey*. The National Academies Press, Washington {**73, 102**}

Blandford RD, Payne DG, 1982, Hydromagnetic flows from accretion disks and the production of radio jets. *MNRAS*, 199, 883–903 {**221**}

Bloemhof EE, 2003, Suppression of speckle noise by speckle pinning in adaptive optics. *ApJ*, 582, L59–L62 {**157**}

—, 2004, Remnant speckles in a highly corrected coronagraph. *ApJ*, 610, L69–L72 {**157**}

Blum J, 2000, Laboratory experiments on preplanetary dust aggregation. *Space Sci. Rev.*, 92, 265–278 {**226**}

Blum J, Wurm G, 2000, Experiments on sticking, restructuring, and fragmentation of preplanetary dust aggregates. *Icarus*, 143, 138–146 {**226**}

—, 2008, The growth mechanisms of macroscopic bodies in protoplanetary disks. *ARA&A*, 46, 21–56 {**226**}

Blum J, Wurm G, Kempf S, et al., 2000, Growth and form of planetary seedlings: results from a microgravity aggregation experiment. *Physical Review Letters*, 85, 2426–2429 {**226**}

Boccaletti A, Abe L, Baudrand J, et al., 2008, Prototyping coronagraphs for exoplanet characterisation with SPHERE. *SPIE Conf Ser*, volume 7015, 34 {**155, 159**}

Boccaletti A, Augereau J, Baudoz P, et al., 2009, VLT–NACO coronagraphic observations of fine structures in the disk of β Pic. *A&A*, 495, 523–535 {**153, 223**}

Boccaletti A, Baudoz P, Baudrand J, et al., 2005, Imaging exoplanets with the coronagraph of JWST–MIRI. *Advances in Space Research*, 36, 1099–1106 {**163**}

Boccaletti A, Moutou C, Labeyrie A, et al., 1998, Present performance of the dark-speckle coronagraph. *A&AS*, 133, 395–402 {**157**}

Boccaletti A, Riaud P, Baudoz P, et al., 2004, The four-quadrant phase mask coronagraph. IV. First light at the VLT. *PASP*, 116, 1061–1071 {**155**}

Bodaghee A, Santos NC, Israelian G, et al., 2003, Chemical abundances of planet-host stars. Results for alpha and Fe-group elements. *A&A*, 404, 715–727 {**188, 189, 195, 198**}

Boden AF, Shao M, van Buren D, 1998, Astrometric observation of MACHO gravitational microlensing. *ApJ*, 502, 538–549 {**94**}

Bodenheimer P, 1998, Formation of substellar objects orbiting stars. *Brown Dwarfs and Extrasolar Planets*, volume 134 of *ASP Conf Ser*, 115–127 {**214**}

—, 2006, Historical notes on planet formation. *Planet Formation*, 1–13, Cambridge University Press {**225**}

Bodenheimer P, Grossman AS, Decampli WM, et al., 1980a, Calculations of the evolution of the giant planets. *Icarus*, 41, 293–308 {**235**}

Bodenheimer P, Hubickyj O, Lissauer JJ, 2000, Models of the in situ formation of detected extrasolar giant planets. *Icarus*, 143, 2–14 {**232, 237**}

Bodenheimer P, Laughlin G, Lin DNC, 2003, On the radii of extrasolar giant planets. *ApJ*, 592, 555–563 {**145, 247, 260**}

Bodenheimer P, Lin DNC, Mardling RA, 2001, On the tidal inflation of short-period extrasolar planets. *ApJ*, 548, 466–472 {**145**}

Bodenheimer P, Pollack JB, 1986, Calculations of the accretion and evolution of giant planets: the effects of solid cores. *Icarus*, 67, 391–408 {**232, 234**}

Bodenheimer P, Tohline JE, Black DC, 1980b, Criteria for fragmentation in a collapsing rotating cloud. *ApJ*, 242, 209–218 {**210**}

Boesgaard AM, Armengaud E, King JR, et al., 2004, The correlation of lithium and beryllium in F and G field and cluster dwarf stars. *ApJ*, 613, 1202–1212 {**201**}

Böhm-Vitense E, 1958, Über die Wasserstoffkonvektionszone in Sternen verschiedener Effektivtemperaturen und Leuchtkräfte. *Zeitschrift fur Astrophysik*, 46, 108 {**202**}

Bohr J, Olsen K, 2010, Long-range order between the planets in the solar system. *MNRAS*, 403, L59–L63 {**48**}

Bois E, Kiseleva-Eggleton L, Rambaux N, et al., 2003, Conditions of dynamical stability for the HD 160691 planetary system. *ApJ*, 598, 1312–1320 {**47, 48**}

Böker T, Allen RJ, 1999, Imaging and nulling with the Space Interferometry Mission. *ApJS*, 125, 123–142 {**166**}

Bolatto AD, Falco EE, 1994, The detectability of planetary companions of compact Galactic objects from their effects on microlensed light curves of distant stars. *ApJ*, 436, 112–116 {**88**}

Boley AC, 2009, The two modes of gas giant planet formation. *ApJ*, 695, L53–L57 {**236**}

Boley AC, Durisen RH, Nordlund Å, et al., 2007, Three-dimensional radiative hydrodynamics for disk stability simulations: a proposed testing standard and new results. *ApJ*, 665, 1254–1267 {**236**}

Boley AC, Mejía AC, Durisen RH, et al., 2006, The thermal regulation of gravitational instabilities in protoplanetary disks. III. Simulations with radiative cooling and realistic opacities. *ApJ*, 651, 517–534 {**236**}

Bonaccini Calia D, Allaert E, Araujo C, et al., 2003, VLT laser guide star facility. *SPIE Conf Ser*, volume 4839, 381–392 {**152**}

Bonaccini Calia D, Feng Y, Hackenberg W, et al., 2010, Laser development for sodium laser guide stars at ESO. *The Messenger*, 139, 12–19 {**152**}

Bonanno A, Schlattl H, Paternò L, 2002, The age of the Sun and the relativistic corrections in the equation of state. *A&A*, 390, 1115–1118 {**296**}

Bonavita M, Desidera S, 2007, The frequency of planets in multiple systems. *A&A*, 468, 721–729 {**59**}

Bond HE, Henden A, Levay ZG, et al., 2003, An energetic stellar outburst accompanied by circumstellar light echoes. *Nature*, 422, 405–408 {**179**}

Bond IA, Abe F, Dodd RJ, et al., 2001, Real-time difference imaging analysis of MOA Galactic bulge observations during 2000. *MNRAS*, 327, 868–880 {**97**}

—, 2002a, Improving the prospects for detecting extrasolar planets in gravitational microlensing events in 2002. *MNRAS*, 331, L19–L23 {**89, 97**}

Bond IA, Rattenbury NJ, Skuljan J, et al., 2002b, Study by MOA of extrasolar planets in gravitational microlensing events of high magnification. *MNRAS*, 333, 71–83 {**96**}

Bond IA, Udalski A, Jaroszyński M, et al., 2004, OGLE–2003–BLG–235/MOA–2003–BLG–53: a planetary microlensing event. *ApJ*, 606, L155–L158 {**2, 98–100**}

Bond JC, Lauretta DS, Tinney CG, et al., 2008, Beyond the iron peak: r- and s-process elemental abundances in stars with planets. *ApJ*, 682, 1234–1247 {**188, 189, 191, 192, 198**}

Bond JC, Tinney CG, Butler RP, et al., 2006, The abundance distribution of stars with planets. *MNRAS*, 370, 163–173 {**189**}

Bonfils X, Delfosse X, Udry S, et al., 2004, A radial velocity survey for planets around M-dwarfs. *Spectroscopically and Spatially Resolving the Components of the Close Binary Stars*, volume 318 of *ASP Conf Ser*, 286–287 {**33**}

—, 2005a, Metallicity of M dwarfs. I. A photometric calibration and the impact on the mass-luminosity relation at the bottom of the main sequence. *A&A*, 442, 635–642 {**190**}

Bonfils X, Forveille T, Delfosse X, et al., 2005b, The HARPS search for southern extrasolar planets. VI. A Neptune-mass planet around the nearby M dwarf GJ 581. *A&A*, 443, L15–L18 {**30, 55, 286**}

Bonfils X, Mayor M, Delfosse X, et al., 2007, The HARPS search for southern extrasolar planets. X. An 11 Earth mass planet around the nearby spotted M dwarf GJ 674. *A&A*, 474, 293–299 {**30**}

Bonneau D, Josse M, Labyrie A, 1975, Lock-in image subtraction detectability of circumstellar planets with the Large Space Telescope. *Image Processing Techniques in Astronomy*, volume 54 of *Astrophysics and Space Science Library*, 403–409 {**164**}

Bonnell IA, Bate MR, 1994, Massive circumbinary disks and the formation of multiple systems. *MNRAS*, 269, L45–L48 {**81**}

Bonnell IA, Bate MR, Clarke CJ, et al., 1997, Accretion and the stellar mass spectrum in small clusters. *MNRAS*, 285, 201–208 {**217**}

Bonnell IA, Clark P, Bate MR, 2008, Gravitational fragmentation and the formation of brown dwarfs in stellar clusters. *MNRAS*, 389, 1556–1562 {**214**}

Bonnell IA, Smith KW, Davies MB, et al., 2001, Planetary dynamics in stellar clusters. *MNRAS*, 322, 859–865 {**109**}

Bonomo AS, Lanza AF, 2008, Modeling solar-like variability for the detection of Earth-like planetary transits. I. Performance of the three-spot modeling and harmonic function fitting. *A&A*, 482, 341–347 {**115**}

Bonomo AS, Santerne A, Alonso R, et al., 2010, Transiting exoplanets from the CoRoT space mission. X. CoRoT–10 b: a giant planet in a 13.24 day eccentric orbit. *A&A*, 520, A65 {**326**}

Booth M, Wyatt MC, Morbidelli A, et al., 2009, The history of the solar system's debris disk: observable properties of the Kuiper belt. *MNRAS*, 399, 385–398 {**304**}

Bordé PJ, Bouchy F, Deleuil M, et al., 2010, Transiting exoplanets from the CoRoT space mission. XI. CoRoT–8 b: a hot and dense sub-Saturn around a K1 dwarf. *A&A*, 520, A66 {**326**}

Bordé PJ, Traub WA, 2006, High-contrast imaging from space: speckle nulling in a low-aberration regime. *ApJ*, 638, 488–498 {**157**}

Borg L, Drake MJ, 2005, A review of meteorite evidence for the timing of magmatism and of surface or near-surface liquid water on Mars. *Journal of Geophysical Research (Planets)*, 110, 12–21 {**297**}

Born M, Wolf E, 1999, *Principles of Optics*. Cambridge University Press {**168**}

Borucki WJ, Koch D, Basri G, et al., 2010a, Kepler planet-detection mission: introduction and first results. *Science*, 327, 977–980 {**111, 202**}

Borucki WJ, Koch DG, Brown TM, et al., 2010b, Kepler–4 b: a hot Neptune-like planet of a G0 star near main-sequence turnoff. *ApJ*, 713, L126–L130 {**2, 326**}

Borucki WJ, Scargle JD, Hudson HS, 1985, Detectability of extrasolar

References

planetary transits. *ApJ*, 291, 852–854 {**104**}

Borucki WJ, Summers AL, 1984, The photometric method of detecting other planetary systems. *Icarus*, 58, 121–134 {**104, 117**}

Borysow A, 2002, Collision-induced absorption coefficients of H_2 pairs at temperatures from 60 K to 1000 K. *A&A*, 390, 779–782 {**264**}

Boss AP, 1995, Proximity of Jupiter-like planets to low-mass stars. *Science*, 267, 360–362 {**37, 235, 237**}

—, 1996, A concise guide to chondrule formation models. *Chondrules and the Protoplanetary Disk*, 257–263 {**297**}

—, 1997, Giant planet formation by gravitational instability. *Science*, 276, 1836–1839 {**235, 236**}

—, 1998, Evolution of the solar nebula. IV. Giant gaseous protoplanet formation. *ApJ*, 503, 923–927 {**235**}

—, 2001, Formation of planetary-mass objects by protostellar collapse and fragmentation. *ApJ*, 551, L167–L170 {**210, 215**}

—, 2002, Stellar metallicity and the formation of extrasolar gas giant planets. *ApJ*, 567, L149–L153 {**191, 210**}

—, 2003, Rapid formation of outer giant planets by disk instability. *ApJ*, 599, 577–581 {**235**}

—, 2005, Evolution of the solar nebula. VII. Formation and survival of protoplanets formed by disk instability. *ApJ*, 629, 535–548 {**146, 236**}

—, 2006a, Gas giant protoplanets formed by disk instability in binary star systems. *ApJ*, 641, 1148–1161 {**58, 59, 235**}

—, 2006b, Giant-planet formation: theories meet observations. *Planet Formation*, 192–202, Cambridge University Press {**236**}

—, 2006c, On the formation of gas giant planets on wide orbits. *ApJ*, 637, L137–L140 {**235**}

—, 2006d, Rapid formation of gas giant planets around M dwarf stars. *ApJ*, 643, 501–508 {**235**}

—, 2006e, Rapid formation of super-Earths around M dwarf stars. *ApJ*, 644, L79–L82 {**235**}

—, 2007a, Evolution of the solar nebula. VIII. Spatial and temporal heterogeneity of short-lived radioisotopes and stable oxygen isotopes. *ApJ*, 660, 1707–1714 {**297**}

—, 2007b, Testing disk instability models for giant planet formation. *ApJ*, 661, L73–L76 {**236**}

—, 2008, Rapid formation of gas giants, ice giants and super-Earths. *Physica Scripta Volume T*, 130(1), 014020 {**235, 236**}

Boss AP, Basri G, Kumar SS, et al., 2003, Nomenclature: brown dwarfs and gas giant planets. *Brown Dwarfs*, volume 211 of *IAU Symposium*, 529–537 {**210**}

Boss AP, Durisen RH, 2005, Chondrule-forming shock fronts in the solar nebula: a possible unified scenario for planet and chondrite formation. *ApJ*, 621, L137–L140 {**297**}

Boss AP, Fisher RT, Klein RI, et al., 2000, The Jeans condition and collapsing molecular cloud cores: filaments or binaries? *ApJ*, 528, 325–335 {**214**}

Boss AP, Weinberger AJ, Anglada-Escudé G, et al., 2009, The Carnegie astrometric planet search programme. *PASP*, 121, 1218–1231 {**69**}

Bottke WF, Cellino A, Paolicchi P, et al., 2002, *Asteroids III*. University of Arizona Press {**293**}

Bottke WF, Levison HF, Nesvorný D, et al., 2007, Can planetesimals left over from terrestrial planet formation produce the lunar Late Heavy Bombardment? *Icarus*, 190, 203–223 {**304**}

Bottke WF, Nesvorný D, Vokrouhlický D, et al., 2010, The irregular satellites: the most collisionally evolved populations in the solar system. *AJ*, 139, 994–1014 {**304**}

Bottke WF, Nolan MC, Greenberg R, et al., 1994, Velocity distributions among colliding asteroids. *Icarus*, 107, 255–268 {**228**}

Bouchy F, Bazot M, Santos NC, et al., 2005a, Asteroseismology of the planet-hosting star μ Ara. I. The acoustic spectrum. *A&A*, 440, 609–614 {**203**}

Bouchy F, Connes P, Bertaux JL, 1999, A new spectrograph dedicated to precise stellar radial velocities. *IAU Colloq. 170: Precise Stellar Radial Velocities*, volume 185 of *ASP Conf Ser*, 22–28 {**27**}

Bouchy F, Deleuil M, Guillot T, et al., 2011, Transiting exoplanets from the CoRoT space mission. XV. CoRoT-15 b: a brown-dwarf transiting companion. *A&A*, 525, A68 {**326**}

Bouchy F, Hebb L, Skillen I, et al., 2010, WASP-21 b: a hot-Saturn exoplanet transiting a thick disk star. *A&A*, 519, A98 {**108, 327**}

Bouchy F, Pepe F, Queloz D, 2001, Fundamental photon noise limit to radial velocity measurements. *A&A*, 374, 733–739 {**22**}

Bouchy F, Pont F, Melo C, et al., 2005b, Doppler follow-up of OGLE transiting companions in the Galactic bulge. *A&A*, 431, 1105–1121 {**106**}

Bouchy F, Pont F, Santos NC, et al., 2004, Two new very hot Jupiters among the OGLE transiting candidates. *A&A*, 421, L13–L16 {**106, 327**}

Bouchy F, Queloz D, Deleuil M, et al., 2008, Transiting exoplanets from the CoRoT space mission. III. The spectroscopic transit of CoRoT-2 b with SOPHIE and HARPS. *A&A*, 482, L25–L28 {**129**}

Bouchy F, Udry S, Mayor M, et al., 2005c, ELODIE metallicity-biased search for transiting hot Jupiters. II. A very hot Jupiter transiting the bright K star HD 189733. *A&A*, 444, L15–L19 {**109, 113, 124, 248, 326**}

Boué G, Laskar J, 2010, A collisionless scenario for Uranus tilting. *ApJ*, 712, L44–L47 {**301**}

Bougher SW, Hunten DM, Phillips RJ, 1997, *Venus II*. University of Arizona Press {**293**}

Bounama C, von Bloh W, Franck S, 2007, How rare is complex life in the Milky Way? *Astrobiology*, 7, 745–756 {**286**}

Bourassa RR, Kantowski R, Norton TD, 1973, The spheroidal gravitational lens. *ApJ*, 185, 747–756 {**89**}

Boutreux T, Gould A, 1996, Monte Carlo simulations of MACHO parallaxes from a satellite. *ApJ*, 462, 705–711 {**93**}

Bouvier J, 2008, Lithium depletion and the rotational history of exoplanet host stars. *A&A*, 489, L53–L56 {**200**}

Bouvier J, Forestini M, Allain S, 1997, The angular momentum evolution of low-mass stars. *A&A*, 326, 1023–1043 {**200**}

Bower GC, Bolatto A, Ford EB, et al., 2009, Radio interferometric planet search. I. First constraints on planetary companions for nearby, low-mass stars from radio astrometry. *ApJ*, 701, 1922–1939 {**173**}

Bower GC, Plambeck RL, Bolatto A, et al., 2003, A giant outburst at millimeter wavelengths in the Orion Nebula. *ApJ*, 598, 1140–1150 {**173**}

Boynton WV, 1975, Fractionation in the solar nebula: condensation of yttrium and the rare earth elements. *Geochim. Cosmochim. Acta*, 39, 569–584 {**257**}

Bozza V, 2000, Caustics in special multiple lenses. *A&A*, 355, 423–432 {**88**}

Bracewell RN, 1978, Detecting nonsolar planets by a spinning infrared interferometer. *Nature*, 274, 780–781 {**163, 164**}

Bracewell RN, MacPhie RH, 1979, Searching for nonsolar planets. *Icarus*, 38, 136–147 {**163, 164**}

Bramich DM, Horne K, 2006, Upper limits on the hot Jupiter fraction in the field of NGC 7789. *MNRAS*, 367, 1677–1685 {**109**}

Bramich DM, Horne K, Bond IA, et al., 2005, A survey for planetary transits in the field of NGC 7789. *MNRAS*, 359, 1096–1116 {**109**}

Brandeker A, Liseau R, Olofsson G, et al., 2004, The spatial structure of the β Pic gas disk. *A&A*, 413, 681–691 {**223**}

Brandenburg A, Nordlund A, Stein RF, et al., 1995, Dynamo-generated turbulence and large-scale magnetic fields in a Keplerian shear flow. *ApJ*, 446, 741–754 {**221**}

Brandl BR, Lenzen R, Pantin E, et al., 2008, METIS: the mid-infrared E-ELT imager and spectrograph. *SPIE Conf Ser*, volume 7014, 55 {**161**}

Brandner W, Zinnecker H, Alcalá JM, et al., 2000, Timescales of disk evolution and planet formation: HST, adaptive optics, and ISO observations of weak-line and post-T Tauri stars. *AJ*, 120, 950–962 {**213**}

Brassard P, Fontaine G, Wesemael F, et al., 1992, Adiabatic properties of pulsating DA white dwarfs. IV. An extensive survey of the period structure of evolutionary models. *ApJS*, 81, 747–794 {**202**}

Brasser R, Morbidelli A, Gomes R, et al., 2009, Constructing the secular architecture of the solar system. II. The terrestrial planets. *A&A*, 507, 1053–1065 {**304**}

Braude SY, Sidorchuk KM, Sidorchuk MA, et al., 2006, Decameter discrete sources survey of the northern sky using the UTR-2 radio telescope. *IAU Joint Discussion*, volume 12, 44 {**176**}

Brauer F, Dullemond CP, Henning T, 2008, Coagulation, fragmentation and radial motion of solid particles in protoplanetary disks. *A&A*, 480, 859–877 {**227**}

Breckinridge JB, Oppenheimer BR, 2004, Polarisation effects in reflecting coronagraphs for white-light applications in astronomy. *ApJ*, 600, 1091–1098 {**154**}

Broeg C, 2009, The full set of gas giant structures. I. On the origin of planetary masses and the planetary initial mass function. *Icarus*, 204, 15–31 {**251**}

Brogi M, Marzari F, Paolicchi P, 2009, Dynamical stability of the inner belt around ϵ Eri. *A&A*, 499, L13–L16 {**223**}

Bromley BC, 1992, Detecting faint echoes in stellar-flare light curves.

PASP, 104, 1049–1053 {**124**}

Bromley BC, Kenyon SJ, 2006, A hybrid N-body-coagulation code for planet formation. *AJ*, 131, 2737–2748 {**228**}

Broucke RA, 2001, Stable orbits of planets of a binary star system in the three-dimensional restricted problem. *Celestial Mechanics and Dynamical Astronomy*, 81, 321–341 {**57**}

Brovchenko I, Oleinikova A, 2008, Multiple phases of liquid water. *ChemPhysChem*, 9(18), 2660–2675 {**262**}

Brown EW, 1900, A possible explanation of the sunspot period. *MNRAS*, 60, 599–605 {**66**}

Brown ME, Trujillo C, Rabinowitz D, 2004, Discovery of a candidate inner Oort Cloud planetoid. *ApJ*, 617, 645–649 {**295**}

Brown RA, 2004, New information from radial velocity data sets. *ApJ*, 610, 1079–1092 {**14**}

—, 2005, Single-visit photometric and obscurational completeness. *ApJ*, 624, 1010–1024 {**157**}

—, 2009a, On the completeness of reflex astrometry on extrasolar planets near the sensitivity limit. *ApJ*, 699, 711–715 {**73**}

—, 2009b, Photometric orbits of extrasolar planets. *ApJ*, 702, 1237–1249 {**125**}

Brown RA, Burrows CJ, Casertano S, et al., 2003, The 4-m space telescope for investigating extrasolar Earth-like planets in starlight: TPF is HST2. *SPIE Conf Ser*, volume 4854, 95–107 {**166**}

Brown TM, 2001, Transmission spectra as diagnostics of extrasolar giant planet atmospheres. *ApJ*, 553, 1006–1026 {**136, 137**}

—, 2003, Expected detection and false alarm rates for transiting Jovian planets. *ApJ*, 593, L125–L128 {**104**}

Brown TM, Charbonneau D, 2000, The STARE project: a transit search for hot Jupiters. *Disks, Planetesimals, and Planets*, volume 219 of *ASP Conf Ser*, 584–589 {**107**}

Brown TM, Charbonneau D, Gilliland RL, et al., 2001, HST time-series photometry of the transiting planet of HD 209458. *ApJ*, 552, 699–709 {**112, 130, 131, 135, 139, 140**}

Brown TM, Gilliland RL, 1994, Asteroseismology. *ARA&A*, 32, 37–82 {**201**}

Brown TM, Kotak R, Horner SD, et al., 1998a, Exoplanets or dynamic atmospheres? The radial velocity and line shape variations of 51 Peg and τ Boo. *ApJS*, 117, 563–585 {**28**}

—, 1998b, A search for line shape and depth variations in 51 Peg and τ Boo. *ApJ*, 494, L85–L88 {**28**}

Brown TM, Libbrecht KG, Charbonneau D, 2002, A search for CO absorption in the transmission spectrum of HD 209458 b. *PASP*, 114, 826–832 {**139, 140**}

Brown TM, Noyes RW, Nisenson P, et al., 1994, The AFOE: a spectrograph for precise Doppler studies. *PASP*, 106, 1285–1297 {**24**}

Brucato JR, Nuth JA III, 2010, Laboratory studies of simple dust analogs in astrophysical environments. *Protoplanetary Dust: Astrophysical and Cosmochemical Perspectives*, 128–160, Cambridge University Press {**226**}

Brunini A, 2006, Origin of the obliquities of the giant planets in mutual interactions in the early solar system. *Nature*, 440, 1163–1165 {**301**}

Brunini A, Benvenuto OG, 2008, On oligarchic growth of planets in protoplanetary disks. *Icarus*, 194, 800–810 {**228**}

Bruntt H, Grundahl F, Tingley B, et al., 2003, A search for planets in the old open cluster NGC 6791. *A&A*, 410, 323–335 {**109**}

Brush SG, 1990, Theories of the origin of the solar system 1956–1985. *Reviews of Modern Physics*, 62, 43–112 {**225**}

Bryden G, Beichman CA, Carpenter JM, et al., 2009, Planets and debris disks: results from a Spitzer–MIPS search for infrared excess. *ApJ*, 705, 1226–1236 {**224**}

Bryden G, Beichman CA, Rieke GH, et al., 2006a, Spitzer–MIPS limits on asteroidal dust in the pulsar planetary system PSR B1257+12. *ApJ*, 646, 1038–1042 {**77**}

Bryden G, Beichman CA, Trilling DE, et al., 2006b, Frequency of debris disks around solar-type stars: first results from a Spitzer–MIPS survey. *ApJ*, 636, 1098–1113 {**224**}

Bryden G, Chen X, Lin DNC, et al., 1999, Tidally induced gap formation in protostellar disks: gap clearing and suppression of protoplanetary growth. *ApJ*, 514, 344–367 {**223**}

Bryden G, Rozyczka M, Lin DNC, et al., 2000, On the interaction between protoplanets and protostellar disks. *ApJ*, 540, 1091–1101 {**41**}

Buccino AP, Lemarchand GA, Mauas PJD, 2006, Ultraviolet radiation constraints around the circumstellar habitable zones. *Icarus*, 183, 491–503 {**284**}

—, 2007, Ultraviolet habitable zones around M stars. *Icarus*, 192, 582–587 {**284, 285**}

Buccino AP, Mauas PJD, Lemarchand GA, 2002, Ultraviolet radiation and habitable zones. *Origins Life Evol. Biosphere*, 32(542-548) {**284**}

—, 2004, Ultraviolet radiation in different stellar systems. *Bioastronomy 2002: Life Among the Stars*, volume 213 of *IAU Symposium*, 97–100 {**284**}

Buchhave LA, Bakos GÁ, Hartman JD, et al., 2010, HAT–P–16 b: a 4 Jupiter-mass planet transiting a bright star on an eccentric orbit. *ApJ*, 720, 1118–1125 {**326**}

Buenzli E, Schmid HM, 2009, A grid of polarisation models for Rayleigh scattering planetary atmospheres. *A&A*, 504, 259–276 {**126**}

Buffington A, Crawford FS, Muller RA, et al., 1977, First observatory results with an image-sharpening telescope. *Journal of the Optical Society of America*, 67, 304–305 {**150**}

Burbidge EM, Burbidge GR, Fowler WA, et al., 1957, Synthesis of the elements in stars. *Reviews of Modern Physics*, 29, 547–650 {**197, 198**}

Burgasser AJ, Kirkpatrick JD, Brown ME, et al., 2002, The spectra of T dwarfs. I. Near-infrared data and spectral classification. *ApJ*, 564, 421–451 {**212**}

Burgasser AJ, Kirkpatrick JD, Burrows A, et al., 2003a, The first sub-stellar subdwarf? Discovery of a metal-poor L dwarf with halo kinematics. *ApJ*, 592, 1186–1192 {**211**}

Burgasser AJ, Kirkpatrick JD, McElwain MW, et al., 2003b, The 2MASS wide-field T dwarf search. I. Discovery of a bright T dwarf within 10 pc of the Sun. *AJ*, 125, 850–857 {**210**}

Burgasser AJ, Kirkpatrick JD, McGovern MR, et al., 2004, S Ori 70: just a foreground field brown dwarf? *ApJ*, 604, 827–831 {**215**}

Burgasser AJ, Kirkpatrick JD, Reid IN, et al., 2003c, Binarity in brown dwarfs: T dwarf binaries discovered with the HST–WFPC2. *ApJ*, 586, 512–526 {**211**}

Burgdorf MJ, Bramich DM, Dominik M, et al., 2007, Exoplanet detection via microlensing with RoboNet–1.0. *Planet. Space Sci.*, 55, 582–588 {**97**}

Burke BF, Franklin KL, 1955, Observations of a variable radio source associated with the planet Jupiter. *J. Geophys. Res.*, 60, 213–217 {**174**}

Burke CJ, 2008, Impact of orbital eccentricity on the detection of transiting extrasolar planets. *ApJ*, 679, 1566–1573 {**122**}

Burke CJ, Gaudi BS, DePoy DL, et al., 2004, Survey for Transiting Extrasolar Planets in Stellar Systems (STEPSS). I. Fundamental parameters of the open cluster NGC 1245. *AJ*, 127, 2382–2397 {**109**}

—, 2006, Survey for transiting extrasolar planets in stellar systems. III. A limit on the fraction of stars with planets in the open cluster NGC 1245. *AJ*, 132, 210–230 {**109**}

Burke CJ, McCullough PR, Valenti JA, et al., 2007, XO–2 b: transiting hot Jupiter in a metal-rich common proper motion binary. *ApJ*, 671, 2115–2128 {**108, 327**}

—, 2008, XO–5 b: a transiting Jupiter-sized planet with a 4-day period. *ApJ*, 686, 1331–1340 {**108, 327**}

Burkert A, Bate MR, Bodenheimer P, 1997, Protostellar fragmentation in a power-law density distribution. *MNRAS*, 289, 497–504 {**214**}

Burleigh MR, Clarke FJ, Hodgkin ST, 2002, Imaging planets around nearby white dwarfs. *MNRAS*, 331, L41–L45 {**79, 172**}

Burleigh MR, Hogan E, Clarke F, 2006, Direct imaging searches for planets around white dwarf stars. *The Scientific Requirements for Extremely Large Telescopes*, volume 232 of *IAU Symposium*, 344–349 {**172**}

Burningham B, Pinfield DJ, Leggett SK, et al., 2008, Exploring the substellar temperature regime down to ~550 K. *MNRAS*, 391, 320–333 {**212**}

Burns JA, Matthews MS, 1986, *Satellites*. University of Arizona Press {**293**}

Burrows A, Budaj J, Hubeny I, 2008a, Theoretical spectra and light curves of close-in extrasolar giant planets and comparison with data. *ApJ*, 678, 1436–1457 {**142, 272, 273**}

Burrows A, Burgasser AJ, Kirkpatrick JD, et al., 2002, Theoretical spectral models of T dwarfs at short wavelengths and their comparison with data. *ApJ*, 573, 394–417 {**276**}

Burrows A, Guillot T, Hubbard WB, et al., 2000a, On the radii of close-in giant planets. *ApJ*, 534, L97–L100 {**145, 272**}

Burrows A, Hubbard WB, Lunine JI, et al., 2001, The theory of brown dwarfs and extrasolar giant planets. *Reviews of Modern Physics*, 73, 719–765 {**115, 158, 209, 211, 212, 272, 273**}

References

Burrows A, Hubbard WB, Saumon D, et al., 1993, An expanded set of brown dwarf and very low mass star models. *ApJ*, 406, 158–171 {**209, 218**}

Burrows A, Hubeny I, Budaj J, et al., 2007a, Possible solutions to the radius anomalies of transiting giant planets. *ApJ*, 661, 502–514 {**138, 145, 146, 249, 272**}

—, 2007b, Theoretical spectral models of the planet HD 209458 b with a thermal inversion and water emission bands. *ApJ*, 668, L171–L174 {**142, 272**}

Burrows A, Hubeny I, Hubbard WB, et al., 2004a, Theoretical radii of transiting giant planets: the case of OGLE–TR–56 b. *ApJ*, 610, L53–L56 {**265**}

Burrows A, Hubeny I, Sudarsky D, 2005, A theoretical interpretation of the measurements of the secondary eclipses of TrES–1 and HD 209458 b. *ApJ*, 625, L135–L138 {**107, 138, 272**}

Burrows A, Ibgui L, Hubeny I, 2008b, Optical albedo theory of strongly irradiated giant planets: the case of HD 209458 b. *ApJ*, 682, 1277–1282 {**125**}

Burrows A, Liebert J, 1993, The science of brown dwarfs. *Reviews of Modern Physics*, 65, 301–336 {**143, 211**}

Burrows A, Marley M, Hubbard WB, et al., 1997, A nongray theory of extrasolar giant planets and brown dwarfs. *ApJ*, 491, 856–875 {**150, 175, 209, 214, 215, 265, 272**}

Burrows A, Marley MS, Sharp CM, 2000b, The near-infrared and optical spectra of methane dwarfs and brown dwarfs. *ApJ*, 531, 438–446 {**264**}

Burrows A, Sharp CM, 1999, Chemical equilibrium abundances in brown dwarf and extrasolar giant planet atmospheres. *ApJ*, 512, 843–863 {**272, 276**}

Burrows A, Sudarsky D, Hubbard WB, 2003a, A theory for the radius of the transiting giant planet HD 209458 b. *ApJ*, 594, 545–551 {**138, 145**}

Burrows A, Sudarsky D, Hubeny I, 2004b, Spectra and diagnostics for the direct detection of wide-separation extrasolar giant planets. *ApJ*, 609, 407–416 {**150, 272**}

—, 2006, Theory for the secondary eclipse fluxes, spectra, atmospheres, and light curves of transiting extrasolar giant planets. *ApJ*, 650, 1140–1149 {**138, 272**}

Burrows A, Sudarsky D, Lunine JI, 2003b, Beyond the T dwarfs: theoretical spectra, colours, and detectability of the coolest brown dwarfs. *ApJ*, 596, 587–596 {**158, 212, 213, 273**}

Burrows A, Volobuyev M, 2003, Calculations of the far-wing line profiles of sodium and potassium in the atmospheres of substellar-mass objects. *ApJ*, 583, 985–995 {**264**}

Bursa M, 1986, The Sun's flattening and its influence on planetary orbits. *Bulletin of the Astronomical Institutes of Czechoslovakia*, 37, 312–313 {**134**}

Burton MG, 2010, Astronomy in Antarctica. *A&A Rev.*, 18, 417–469 {**161**}

Buscher DF, Bakker EJ, Coleman TA, et al., 2006a, The Magdalena Ridge Observatory Interferometer: a high-sensitivity imaging array. *SPIE Conf Ser*, volume 6307, 11 {**162**}

Buscher DF, Boysen RC, Dace R, et al., 2006b, Design and testing of an innovative delay line for the MROI. *SPIE Conf Ser*, volume 6268, 78 {**162**}

Butler BJ, Wootten A, Brown RL, 2004, Observing extrasolar planetary systems with ALMA. *Planetary Systems in the Universe*, volume 202 of *IAU Symposium*, 442–444 {**177**}

Butler RP, Marcy GW, 1996, A planet orbiting 47 UMa. *ApJ*, 464, L153–L156 {**28**}

—, 1998, The near term future of extrasolar planet searches. *Brown Dwarfs and Extrasolar Planets*, volume 134 of *ASP Conf Ser*, 162–168 {**21, 29**}

Butler RP, Marcy GW, Fischer DA, et al., 1999, Evidence for multiple companions to υ And. *ApJ*, 526, 916–927 {**14, 38, 46, 47**}

Butler RP, Marcy GW, Vogt SS, et al., 2003, Seven new Keck planets orbiting G and K dwarfs. *ApJ*, 582, 455–466 {**47, 50, 52**}

Butler RP, Marcy GW, Williams E, et al., 1996, Attaining Doppler precision of 3 m s^{-1}. *PASP*, 108, 500–509 {**19, 21**}

—, 1997, Three new 51 Peg-type planets. *ApJ*, 474, L115–L118 {**47, 48**}

Butler RP, Tinney CG, Marcy GW, et al., 2001, Two new planets from the Anglo–Australian planet search. *ApJ*, 555, 410–417 {**48**}

Butler RP, Wright JT, Marcy GW, et al., 2006, Catalogue of nearby exoplanets. *ApJ*, 646, 505–522 {**16, 21, 28, 29, 206**}

Butterley T, Love GD, Wilson RW, et al., 2006, A Shack–Hartmann wavefront sensor projected on to the sky with reduced focal anisoplanatism. *MNRAS*, 368, 837–843 {**151**}

Butters OW, West RG, Anderson DR, et al., 2010, The first WASP public data release. *A&A*, 520, L10 {**107**}

Butterworth AL, Aballain O, Chappellaz J, et al., 2004, Combined element (H and C) stable isotope ratios of methane in carbonaceous chondrites. *MNRAS*, 347, 807–812 {**278**}

Butusov KP, 1973, Svojstva simmetrii solnechnoj sistemy. *Nekotoryje Problemy Issledovanija Vselennoj (Akademya Nauk SSSR, Leningrad)* {**48**}

Caballero JA, Béjar VJS, Rebolo R, 2003, Variability of L dwarfs in the near infrared. *Brown Dwarfs*, volume 211 of *IAU Symposium*, 455–458 {**115**}

Caballero JA, Martín EL, Dobbie PD, et al., 2006, Are isolated planetary-mass objects really isolated? A brown dwarf-exoplanet system candidate in the σ Ori cluster. *A&A*, 460, 635–640 {**215**}

Caballero JA, Rebolo R, 2002, Variability in brown dwarfs: atmospheres and transits. *Stellar Structure and Habitable Planet Finding*, volume 485 of *ESA Special Publication*, 261–264 {**115**}

Cabrera B, Clarke RM, Colling P, et al., 1998, Detection of single infrared, optical, and ultraviolet photons using superconducting transition edge sensors. *Applied Physics Letters*, 73, 735–737 {**133**}

Cabrera J, Bruntt H, Ollivier M, et al., 2010, Transiting exoplanets from the CoRoT space mission. XIII. CoRoT–13 b: a dense hot Jupiter in transit around a star with solar metallicity and super-solar lithium content. *A&A*, 522, A110 {**326**}

Cabrera J, Fridlund M, Ollivier M, et al., 2009, Planetary transit candidates in CoRoT LRc01 field. *A&A*, 506, 501–517 {**110**}

Cabrera J, Schneider J, 2007, Detecting companions to extrasolar planets using mutual events. *A&A*, 464, 1133–1138 {**130**}

Cáceres C, Ivanov VD, Minniti D, et al., 2009, High cadence near infrared timing observations of extrasolar planets. I. GJ 436 b and XO–1 b. *A&A*, 507, 481–486 {**135**}

Cady E, Macintosh B, Kasdin NJ, et al., 2009, Shaped pupil design for the Gemini planet imager. *ApJ*, 698, 938–943 {**153**}

Cai K, Pickett MK, Durisen RH, et al., 2010, Giant planet formation by disk instability: a comparison simulation with an improved radiative scheme. *ApJ*, 716, L176–L180 {**236**}

Cairns IH, 2004, Solar, interplanetary, planetary, and related extrasolar system science for LOFAR. *Planet. Space Sci.*, 52, 1423–1434 {**177**}

Cajori F, 1934, *Newton's Principia, Book III, Proposition XIII*. University of California Press {**66**}

Calchi Novati S, Dall'Ora M, Gould A, et al., 2010, M31 pixel lensing event OAB–N2: a study of the lens proper motion. *ApJ*, 717, 987–994 {**101**}

Caldeira K, Kasting JF, 1992, The life span of the biosphere revisited. *Nature*, 360, 721–723 {**284**}

Caldwell DA, Borucki WJ, Showen RL, et al., 2004, Detecting extrasolar planet transits from the south pole. *Bioastronomy 2002: Life Among the Stars*, volume 213 of *IAU Symposium*, 93–96 {**110**}

Callegari N, Ferraz-Mello S, Michtchenko TA, 2006, Dynamics of two planets in the 3:2 mean-motion resonance: application to the planetary system of the pulsar PSR B1257+12. *Celestial Mechanics and Dynamical Astronomy*, 94, 381–397 {**76**}

Calvet N, D'Alessio P, Watson DM, et al., 2005, Disks in transition in the Taurus population: Spitzer IRS spectra of GM Aur and DM Tau. *ApJ*, 630, L185–L188 {**50**}

Cameron AGW, 1966, The accumulation of chondritic material. *Earth and Planetary Science Letters*, 1, 93–96 {**297**}

—, 1973, Accumulation processes in the primitive solar nebula. *Icarus*, 18, 407–450 {**225**}

—, 1978a, Physics of the primitive solar accretion disk. *Moon and Planets*, 18, 5–40 {**235**}

—, 1978b, The primitive solar accretion disk and the formation of the planets. *Origin of the Solar System*, 49–74, Wiley {**297**}

—, 1983, Origin of the atmospheres of the terrestrial planets. *Icarus*, 56, 195–201 {**278, 280**}

Cameron AGW, Benz W, 1991, The origin of the moon and the single impact hypothesis. IV. *Icarus*, 92, 204–216 {**302**}

Cameron AGW, Ward WR, 1976, The origin of the Moon. *Lunar and Planetary Institute Science Conference Abstracts*, volume 7 of *Lunar and Planetary Inst. Technical Report*, 120–122 {**302**}

Cameron PB, Britton MC, Kulkarni SR, 2009, Precision astrometry with adaptive optics. *AJ*, 137, 83–93 {**63**}

Campbell B, Walker GAH, 1979, Precision radial velocities with an absorption cell. *PASP*, 91, 540–545 {**19**}

Campbell B, Walker GAH, Yang S, 1988, A search for substellar companions to solar-type stars. *ApJ*, 331, 902–921 {**2, 24, 28**}

Campbell JK, Anderson JD, 1989, Gravity field of the Saturnian system from Pioneer and Voyager tracking data. *AJ*, 97, 1485–1495 {**294**}

Campbell JK, Synnott SP, 1985, Gravity field of the Jovian system from Pioneer and Voyager tracking data. *AJ*, 90, 364–372 {**294**}

Canup RM, 2004a, Dynamics of lunar formation. *ARA&A*, 42, 441–475 {**302**}

—, 2004b, Simulations of a late lunar-forming impact. *Icarus*, 168, 433–456 {**302**}

Canup RM, Asphaug E, 2001, Origin of the Moon in a giant impact near the end of the Earth's formation. *Nature*, 412, 708–712 {**302**}

Canup RM, Levison HF, Stewart GR, 1999, Evolution of a terrestrial multiple-moon system. *AJ*, 117, 603–620 {**302**}

Canup RM, Righter K, 2000, *Origin of the Earth and Moon*. University of Arizona Press {**293, 302**}

Canup RM, Ward WR, 2002, Formation of the Galilean satellites: conditions of accretion. *AJ*, 124, 3404–3423 {**237**}

—, 2006, A common mass scaling for satellite systems of gaseous planets. *Nature*, 441, 834–839 {**237**}

Caracas R, 2008, Dynamical instabilities of ice X. *Physical Review Letters*, 101(8), 085502 {**263**}

Carballido A, Fromang S, Papaloizou J, 2006, Mid-plane sedimentation of large solid bodies in turbulent protoplanetary disks. *MNRAS*, 373, 1633–1640 {**221**}

Carballido A, Stone JM, Pringle JE, 2005, Diffusion coefficient of a passive contaminant in a local magnetohydrodynamic model of a turbulent accretion disk. *MNRAS*, 358, 1055–1060 {**225**}

Carbillet M, Maire A, Le Roux B, et al., 2010, Adaptive optics and ground-layer adaptive optics for Dome C: numerical simulation results. *EAS Publications Series*, volume 40, 157–164 {**162**}

Carciofi AC, Magalhães AM, 2005, The polarisation signature of extrasolar planet transiting cool dwarfs. *ApJ*, 635, 570–577 {**126**}

Carone L, Pätzold M, 2007, Constraints on the tidal dissipation factor of a main sequence star: the case of OGLE–TR–56 b. *Planet. Space Sci.*, 55, 643–650 {**247**}

Carpano S, Fridlund M, 2008, Detecting transits from Earth-sized planets around Sun-like stars. *A&A*, 485, 607–613 {**105**}

Carr MH, Belton MJS, Chapman CR, et al., 1998, Evidence for a subsurface ocean on Europa. *Nature*, 391, 363–365 {**282**}

Carson JC, Eikenberry SS, Brandl BR, et al., 2005, The Cornell high-order adaptive optics survey for brown dwarfs in stellar systems. I. Observations, data reduction, and detection analyses. *AJ*, 130, 1212–1220 {**213**}

Carter JA, Winn JN, 2009, Parameter estimation from time-series data with correlated errors: a wavelet-based method and its application to transit light curves. *ApJ*, 704, 51–67 {**105**}

—, 2010, Empirical constraints on the oblateness of an exoplanet. *ApJ*, 709, 1219–1229 {**131, 134**}

Carter JA, Yee JC, Eastman J, et al., 2008, Analytic approximations for transit light-curve observables, uncertainties, and covariances. *ApJ*, 689, 499–512 {**121**}

Casagrande L, Portinari L, Flynn C, 2006, Accurate fundamental parameters for lower main-sequence stars. *MNRAS*, 373, 13–44 {**185**}

Casertano S, Lattanzi MG, Sozzetti A, et al., 2008, Double-blind test programme for astrometric planet detection with Gaia. *A&A*, 482, 699–729 {**67, 72**}

Cash W, 2006, Detection of Earth-like planets around nearby stars using a petal-shaped occulter. *Nature*, 442, 51–53 {**156**}

Cash W, Kasdin J, Seager S, et al., 2005, Direct studies of exoplanets with the New Worlds Observer. *SPIE Conf Ser*, volume 5899, 274–285 {**156**}

Cassan A, 2008, An alternative parameterisation for binary-lens caustic-crossing events. *A&A*, 491, 587–595 {**88, 90**}

Cassen P, Reynolds RT, Peale SJ, 1979, Is there liquid water on Europa. *Geophys. Res. Lett.*, 6, 731–734 {**282**}

Cassen P, Smith BF, Miller RH, et al., 1981, Numerical experiments on the stability of preplanetary disks. *Icarus*, 48, 377–392 {**236**}

Cassidy TA, Mendez R, Arras P, et al., 2009, Massive satellites of close-in gas giant exoplanets. *ApJ*, 704, 1341–1348 {**237**}

Castellano T, Jenkins J, Trilling DE, et al., 2000, Detection of planetary transits of the star HD 209458 in the Hipparcos data set. *ApJ*, 532, L51–L53 {**113**}

Castro M, Vauclair S, Richard O, et al., 2008, Lithium abundances in exoplanet-host stars. *Memorie della Societa Astronomica Italiana*, 79, 679–681 {**200**}

—, 2009, Lithium abundances in exoplanet-host stars: modeling. *A&A*, 494, 663–668 {**200**}

Catala C, 2009a, PLATO: PLAnetary Transits and Oscillations of stars. *Communications in Asteroseismology*, 158, 330 {**114**}

—, 2009b, PLATO: PLAnetary Transits and Oscillations of stars. *Experimental Astronomy*, 23, 329–356 {**114**}

Catala C, Donati J, Shkolnik E, et al., 2007, The magnetic field of the planet-hosting star τ Boo. *MNRAS*, 374, L42–L46 {**206**}

Catanzarite J, Law N, Shao M, 2008, Astrometric detection of exo-Earths in the presence of stellar noise. *SPIE Conf Ser*, volume 7013 {**65**}

Catanzarite J, Shao M, Tanner A, et al., 2006, Astrometric detection of terrestrial planets in the habitable zones of nearby stars with SIM PlanetQuest. *PASP*, 118, 1319–1339 {**73**}

Catling DC, Glein CR, Zahnle KJ, et al., 2005, Why O_2 is required by complex life on habitable planets and the concept of planetary oxygenation time. *Astrobiology*, 5, 415–438 {**289**}

Caton DB, Davis SA, Kluttz KA, 2000, A search for Trojan extrasolar planets: planets in V442 Cas and YZ Aql? *Bulletin of the American Astronomical Society*, volume 32, 1416 {**59**}

Caughlan GR, Fowler WA, 1988, Thermonuclear reaction rates. V. *Atomic Data and Nuclear Data Tables*, 40, 283–334 {**198, 201**}

Cavarroc C, Boccaletti A, Baudoz P, et al., 2006, Fundamental limitations on Earth-like planet detection with extremely large telescopes. *A&A*, 447, 397–403 {**159**}

Cavazzoni C, Chiarotti GL, Scandolo S, et al., 1999, Superionic and metallic states of water and ammonia at giant planet conditions. *Science*, 283, 44–46 {**263, 269**}

Cayrel de Strobel G, 1996, Stars resembling the Sun. *A&A Rev.*, 7, 243–288 {**184, 286, 310**}

Cayrel de Strobel G, Soubiran C, Ralite N, 2001, Catalogue of [Fe/H] determinations for FGK stars. *A&A*, 373, 159–163 {**188**}

Cecchi-Pestellini C, Ciaravella A, Micela G, 2006, Stellar X-ray heating of planet atmospheres. *A&A*, 458, L13–L16 {**283**}

Chabrier G, Baraffe I, 2000, Theory of low-mass stars and substellar objects. *ARA&A*, 38, 337–377 {**143, 209, 272**}

—, 2007, Heat transport in giant (exo)planets: a new perspective. *ApJ*, 661, L81–L84 {**138**}

Chabrier G, Baraffe I, Allard F, et al., 2000, Evolutionary models for very low-mass stars and brown dwarfs with dusty atmospheres. *ApJ*, 542, 464–472 {**150, 209, 215**}

Chabrier G, Baraffe I, Leconte J, et al., 2009, The mass-radius relationship from solar-type stars to terrestrial planets: a review. *American Institute of Physics Conference Series*, volume 1094, 102–111 {**143, 144, 211**}

Chabrier G, Barman T, Baraffe I, et al., 2004, The evolution of irradiated planets: application to transits. *ApJ*, 603, L53–L56 {**145**}

Chabrier G, Saumon D, Hubbard WB, et al., 1992, The molecular-metallic transition of hydrogen and the structure of Jupiter and Saturn. *ApJ*, 391, 817–826 {**262, 294**}

Chabrier G, Saumon D, Winisdoerffer C, 2007, Hydrogen and helium at high density and astrophysical implications. *Ap&SS*, 307, 263–267 {**262**}

Chakraborty A, 2008, Extrasolar planets. *Bulletin of the Astronomical Society of India*, 25, 28–33 {**24**}

Chakraborty A, Ge J, Mahadevan S, 2004, Evidence of planetesimal infall onto the very young Herbig Be star LkHα-234. *ApJ*, 606, L69–L72 {**224**}

Chaloner WG, 1989, Fossil charcoal as an indicator of palaeoatmospheric oxygen level. *J. Geol. Soc. London*, 14, 171–174 {**307**}

Chamberlain JW, Hunten DM, 1987, *Theory of Planetary Atmospheres: an Introduction to their Physics and Chemistry*, volume 36. Orlando Academic Press, Second Edition {**272**}

Chamberlin TC, 1901, On a possible function of disruptive approach in the formation of meteorites, comets, and nebulae. *ApJ*, 14, 17–39 {**224**}

Chambers JE, 1999, A hybrid symplectic integrator that permits close encounters between massive bodies. *MNRAS*, 304, 793–799 {**43, 44**}

References

—, 2001, Making more terrestrial planets. *Icarus*, 152, 205–224 {**229**}

—, 2006a, Planet formation with migration. *ApJ*, 652, L133–L136 {**234**}

—, 2006b, A semi-analytic model for oligarchic growth. *Icarus*, 180, 496–513 {**228**}

—, 2008, Oligarchic growth with migration and fragmentation. *Icarus*, 198, 256–273 {**228, 234**}

Chambers JE, O'Brien DP, Davis AM, 2010, Accretion of planetesimals and the formation of rocky planets. *Protoplanetary Dust: Astrophysical and Cosmochemical Perspectives*, 299–335, Cambridge University Press {**297**}

Chambers JE, Quintana EV, Duncan MJ, et al., 2002, Symplectic integrator algorithms for modeling planetary accretion in binary star systems. *AJ*, 123, 2884–2894 {**44, 56**}

Chambers JE, Wetherill GW, 1998, Making the terrestrial planets: N-body integrations of planetary embryos in three dimensions. *Icarus*, 136, 304–327 {**229**}

Chameides WL, Walker JCG, 1981, Rates of fixation by lightning of carbon and nitrogen in possible primitive atmospheres. *Origins of Life*, 11, 291–302 {**305**}

Chandrasekhar S, 1969, *Ellipsoidal Figures of Equilibrium*. The Silliman Foundation Lectures, Yale University Press {**248**}

Chang H, 2010, Titius–Bode's relation and distribution of exoplanets. *Journal of Astronomy and Space Sciences*, 27, 1–10 {**48**}

Chang HY, Han C, 2002, Variation of spot-induced anomalies in caustic-crossing binary microlensing event light curves. *MNRAS*, 335, 195–200 {**91**}

Chaplin M, 2010, Water structure and science. *www.lsbu.ac.uk/water/* {**262, 263**}

Chapman S, 1939, Notes on atmospheric sodium. *ApJ*, 90, 309–316 {**151**}

Charbonneau D, 2003, HD 209458 and the power of the dark side. *Scientific Frontiers in Research on Extrasolar Planets*, volume 294 of *ASP Conf Ser*, 449–456 {**127**}

Charbonneau D, Allen LE, Megeath ST, et al., 2005, Detection of thermal emission from an extrasolar planet. *ApJ*, 626, 523–529 {**2, 107, 112, 142**}

Charbonneau D, Berta ZK, Irwin J, et al., 2009, A super-Earth transiting a nearby low-mass star. *Nature*, 462, 891–894 {**2, 108, 115, 270, 326**}

Charbonneau D, Brown TM, Latham DW, et al., 2000, Detection of planetary transits across a Sun-like star. *ApJ*, 529, L45–L48 {**2, 103, 104, 113, 139, 140, 326**}

Charbonneau D, Brown TM, Noyes RW, et al., 2002, Detection of an extrasolar planet atmosphere. *ApJ*, 568, 377–384 {**139, 140**}

Charbonneau D, Jha S, Noyes RW, 1998, Spectral line distortions in the presence of a close-in planet. *ApJ*, 507, L153–L156 {**124**}

Charbonneau D, Knutson HA, Barman T, et al., 2008, The broad-band infrared emission spectrum of the exoplanet HD 189733 b. *ApJ*, 686, 1341–1348 {**140, 141**}

Charbonneau D, Noyes RW, Korzennik SG, et al., 1999, An upper limit on the reflected light from the planet orbiting the star τ Boo. *ApJ*, 522, L145–L148 {**124, 125**}

Charbonneau P, 1995, Genetic algorithms in astronomy and astrophysics. *ApJS*, 101, 309–334 {**14**}

—, 2005, Dynamo models of the solar cycle. *Living Reviews in Solar Physics*, 2, 2 {**66**}

Charbonneau P, Beaubien G, St-Jean C, 2007, Fluctuations in Babcock-Leighton dynamos. II. Revisiting the Gnevyshev–Ohl rule. *ApJ*, 658, 657–662 {**66**}

Charbonnel C, Primas F, 2005, The lithium content of the Galactic halo stars. *A&A*, 442, 961–992 {**198**}

Charvátová I, 1990, On the relation between solar motion and solar activity in the years 1730-80 and 1910-60 AD. *Bull. Astron. Institutes of Czechoslovakia*, 41, 200–204 {**66**}

—, 2000, Can origin of the 2400-year cycle of solar activity be caused by solar inertial motion? *Annales Geophysicae*, 18, 399–405 {**66**}

Chase MW, 1998, *NIST-JANAF Thermochemical Tables*. American Chemical Society, Fourth Edition {**258**}

Chatterjee S, Ford EB, Matsumura S, et al., 2008, Dynamical outcomes of planet–planet scattering. *ApJ*, 686, 580–602 {**127, 243**}

Chauvin G, Lagrange AM, Dumas C, et al., 2005a, Giant planet companion to 2MASSW J1207334–393254. *A&A*, 438, L25–L28 {**2, 168, 170, 214, 215**}

Chauvin G, Lagrange AM, Udry S, et al., 2006, Probing long-period companions to planetary hosts: VLT and CFHT near infrared coronagraphic imaging surveys. *A&A*, 456, 1165–1172 {**170**}

Chauvin G, Lagrange AM, Zuckerman B, et al., 2005b, A companion to AB Pic at the planet/brown dwarf boundary. *A&A*, 438, L29–L32 {**168, 169, 215**}

Chavanis PH, 2000, Trapping of dust by coherent vortices in the solar nebula. *A&A*, 356, 1089–1111 {**221**}

Chelli A, 2000, Optimising Doppler estimates for extrasolar planet detection. I. A specific algorithm for shifted spectra. *A&A*, 358, L59–L62 {**32**}

—, 2005, Imaging Earth-like planets with extremely large telescopes. *A&A*, 441, 1205–1210 {**160**}

Chen CH, Jura M, 2001, A possible massive asteroid belt around ζ Lep. *ApJ*, 560, L171–L174 {**178**}

Chen CH, Li A, Bohac C, et al., 2007, The dust and gas around β Pic. *ApJ*, 666, 466–474 {**223**}

Chen YQ, Nissen PE, Benoni T, et al., 2001, Lithium abundances for 185 main-sequence stars: Galactic evolution and stellar depletion of lithium. *A&A*, 371, 943–951 {**199, 200**}

Chen YQ, Zhao G, 2006, A comparative study on lithium abundances in solar-type stars with and without planets. *AJ*, 131, 1816–1821 {**199, 200**}

Chiang EI, Goldreich P, 1997, Spectral energy distributions of T Tauri stars with passive circumstellar disks. *ApJ*, 490, 368–376 {**220**}

Chiang EI, Kite E, Kalas P, et al., 2009, Fomalhaut's debris disk and planet: constraining the mass of Fomalhaut b from disk morphology. *ApJ*, 693, 734–749 {**223**}

Chiang EI, Lithwick Y, 2005, Neptune Trojans as a test bed for planet formation. *ApJ*, 628, 520–532 {**136**}

Chiang EI, Lithwick Y, Murray-Clay R, et al., 2007, A brief history of trans-Neptunian space. *Protostars and Planets V*, 895–911 {**295**}

Chiang EI, Murray N, 2002, Eccentricity excitation and apsidal resonance capture in the planetary system υ And. *ApJ*, 576, 473–477 {**43, 47**}

Chiang EI, Tabachnik S, Tremaine S, 2001, Apsidal alignment in υ And. *AJ*, 122, 1607–1615 {**42, 47**}

Cho JYK, Menou K, Hansen BMS, et al., 2003, The changing face of the extrasolar giant planet HD 209458 b. *ApJ*, 587, L117–L120 {**137**}

—, 2008, Atmospheric circulation of close-in extrasolar giant planets. I. Global, barotropic, adiabatic simulations. *ApJ*, 675, 817–845 {**138**}

Chollet F, Sinceac V, 1999, Analysis of solar radius determination obtained by the modern CCD astrolabe of the Calern Observatory: a new approach of the solar limb definition. *A&AS*, 139, 219–229 {**309, 310**}

Choudhuri AR, 2007, An elementary introduction to solar dynamo theory. *Kodai School on Solar Physics*, volume 919, 49–73 {**66**}

Christensen UR, Holzwarth V, Reiners A, 2009, Energy flux determines magnetic field strength of planets and stars. *Nature*, 457, 167–169 {**175**}

Christensen-Dalsgaard J, 1984, What will asteroseismology teach us? *Space Research in Stellar Activity and Variability*, 11–18 {**201**}

—, 2002, Helioseismology. *Reviews of Modern Physics*, 74, 1073–1129 {**201**}

—, 2004, Physics of solar-like oscillations. *Sol. Phys.*, 220, 137–168 {**201–203**}

—, 2008a, ADIPLS: the Aarhus adiabatic oscillation package. *Ap&SS*, 316, 113–120 {**202**}

—, 2008b, ASTEC: the Aarhus STellar Evolution Code. *Ap&SS*, 316, 13–24 {**202**}

Christensen-Dalsgaard J, Arentoft T, Brown TM, et al., 2008, The Kepler asteroseismic investigation. *Journal of Physics Conference Series*, 118(1), 012039–49 {**202**}

Christensen-Dalsgaard J, Duvall TL, Gough DO, et al., 1985, Speed of sound in the solar interior. *Nature*, 315, 378–382 {**264**}

Christensen-Dalsgaard J, Kjeldsen H, Brown TM, et al., 2010, Asteroseismic investigation of known planet hosts in the Kepler field. *ApJ*, 713, L164–L168 {**204, 205**}

Christian DJ, Gibson NP, Simpson EK, et al., 2009, WASP-10 b: a 3M_J, gas-giant planet transiting a late-type K star. *MNRAS*, 392, 1585–1590 {**327**}

Christian DJ, Pollacco DL, Skillen I, et al., 2006, The SuperWASP wide-field exoplanetary transit survey: candidates from fields 23 h<RA<03 h. *MNRAS*, 372, 1117–1128, erratum: 2007, MNRAS, 376,

1424 {107}

Christiansen JL, Ballard S, Charbonneau D, et al., 2010, Studying the atmosphere of the exoplanet HAT-P-7 b via secondary eclipse measurements with EPOXI, Spitzer, and Kepler. *ApJ*, 710, 97–104 {**113, 142**}

Chu YH, Dunne BC, Gruendl RA, et al., 2001, A search for Jovian planets around hot white dwarfs. *ApJ*, 546, L61–L64 {**172**}

Chun M, Toomey D, Wahhaj Z, et al., 2008, Performance of the near-infrared coronagraphic imager on Gemini-South. *SPIE Conf Ser*, volume 7015, 49 {**153, 158**}

Chung SJ, Han C, Park BG, et al., 2005, Properties of central caustics in planetary microlensing. *ApJ*, 630, 535–542 {**88**}

Chung SJ, Kim D, Darnley MJ, et al., 2006, The possibility of detecting planets in the Andromeda galaxy. *ApJ*, 650, 432–437 {**101**}

Chwolson O, 1924, Über eine mögliche Form fiktiver Doppelsterne. *Astronomische Nachrichten*, 221, 329 {**84**}

Chyba CF, 1990, Impact delivery and erosion of planetary oceans in the early inner solar system. *Nature*, 343, 129–133 {**278, 298**}

—, 1997, Catastrophic impacts and the Drake equation. *IAU Colloq. 161: Astronomical and Biochemical Origins and the Search for Life in the Universe*, 157–164 {**290**}

—, 2000, Energy for microbial life on Europa. *Nature*, 403, 381–382 {**250**}

Chyba CF, Hand KP, 2005, Astrobiology: the study of the living universe. *ARA&A*, 43, 31–74 {**282**}

Chyba CF, Phillips CB, 2002, Europa as an abode of life. *Origins of Life and Evolution of the Biosphere*, 32, 47–67 {**250**}

Chyba CF, Sagan C, 1992, Endogenous production, exogenous delivery and impact-shock synthesis of organic molecules: an inventory for the origins of life. *Nature*, 355, 125–132 {**305**}

Chylek P, Perez MR, 2007, Considerations for the habitable zone of super-Earth planets in GJ 581. *ArXiv e-prints* {**287**}

Ciesla FJ, Dullemond CP, 2010, Evolution of protoplanetary disk structures. *Protoplanetary Dust: Astrophysical and Cosmochemical Perspectives*, 66–96, Cambridge University Press {**218**}

Ciesla FJ, Hood LL, 2002, The nebular shock wave model for chondrule formation: shock processing in a particle-gas suspension. *Icarus*, 158, 281–293 {**297**}

Cincotta PM, Simó C, 2000, Simple tools to study global dynamics in non-axisymmetric galactic potentials. *A&AS*, 147, 205–228 {**46, 76**}

Ćirković MM, 2004, The temporal aspect of the Drake equation and SETI. *Astrobiology*, 4, 225–231 {**290**}

Ćirković MM, Bradbury RJ, 2006, Galactic gradients, postbiological evolution and the apparent failure of SETI. *New Astronomy*, 11, 628–639 {**286**}

Clampin M, 2007, Extrasolar planetary imaging coronagraph (EPIC). *In the Spirit of Bernard Lyot: The Direct Detection of Planets and Circumstellar Disks in the 21st Century*, 37 {**166**}

—, 2009, Comparative Planetology: Transiting Exoplanet Science with JWST. *Astro2010: The Astronomy and Astrophysics Decadal Survey*, Astronomy, 46–53 {**114**}

Clampin M, Melnick G, Lyon R, et al., 2006, Extrasolar planetary imaging coronagraph (EPIC). *SPIE Conf Ser*, volume 6265 {**166**}

Clampin M, Melnick GJ, Lyon RG, et al., 2004, Extrasolar Planetary Imaging Coronagraph (EPIC). *SPIE Conf Ser*, volume 5487, 1538–1544 {**166**}

Claret A, 1995, Stellar models for a wide range of initial chemical compositions until helium burning. I. From X = 0.60 to X = 0.80 for Z = 0.02. *A&AS*, 109, 441–446 {**134**}

—, 2000, A new non-linear limb-darkening law for LTE stellar atmosphere models. *A&A*, 363, 1081–1190 {**91, 118**}

Clarke CJ, 2009, Pseudo-viscous modeling of self-gravitating disks and the formation of low mass ratio binaries. *MNRAS*, 396, 1066–1074 {**236**}

Clarke CJ, Gendrin A, Sotomayor M, 2001, The dispersal of circumstellar disks: the role of the ultraviolet switch. *MNRAS*, 328, 485–491 {**222**}

Clarke CJ, Harper-Clark E, Lodato G, 2007, The response of self-gravitating protostellar disks to slow reduction in cooling time-scale: the fragmentation boundary revisited. *MNRAS*, 381, 1543–1547 {**214, 236**}

Clarke FJ, Burleigh MR, 2004, Imaging planets around white dwarfs: first results. *Extrasolar Planets: Today and Tomorrow*, volume 321 of *ASP Conf Ser*, 76–83 {**172**}

Clarkson WI, Enoch B, Haswell CA, et al., 2007, SuperWASP-north extrasolar planet candidates between $03^h < RA < 06^h$. *MNRAS*, 381, 851–864 {**107**}

Claudi RU, Turatto M, Antichi J, et al., 2006, The integral field spectrograph of SPHERE: the planet finder for VLT. *SPIE Conf Ser*, volume 6269, 93 {**159**}

Clayton RN, Mayeda TK, 1996, Oxygen isotope studies of achondrites. *Geochim. Cosmochim. Acta*, 60, 1999–2017 {**229**}

Clemens DP, Sanders DB, Scoville NZ, 1988, The large-scale distribution of molecular gas in the first Galactic quadrant. *ApJ*, 327, 139–155 {**194**}

Close L, 2007, Extrasolar planet imaging with the Giant Magellan Telescope. *In the Spirit of Bernard Lyot: The Direct Detection of Planets and Circumstellar Disks in the 21st Century* {**161**}

Cloud P, 1972, A working model of the primitive Earth. *Am J Sci*, 272, 537–548 {**306**}

Cocconi G, Morrison P, 1959, Searching for interstellar communications. *Nature*, 184, 844–846 {**290**}

Cochran WD, Endl M, Wittenmyer RA, et al., 2007, A planetary system around HD 155358: The lowest metallicity planet host star. *ApJ*, 665, 1407–1412 {**24, 34, 55**}

Cochran WD, Hatzes AP, 1994, A high-precision radial-velocity survey for other planetary systems. *Ap&SS*, 212, 281–291 {**24**}

Cochran WD, Hatzes AP, Hancock TJ, 1991, Constraints on the companion object to HD 114762. *ApJ*, 380, L35–L38 {**28**}

Cochran WD, Hatzes AP, Paulson DB, 2002, Searching for planets in the Hyades. I. The Keck radial velocity survey. *AJ*, 124, 565–571 {**33, 34**}

Cockell CS, Herbst T, Léger A, et al., 2009a, Darwin: an experimental astronomy mission to search for extrasolar planets. *Experimental Astronomy*, 23, 435–461 {**165**}

Cockell CS, Léger A, Fridlund M, et al., 2009b, Darwin: a mission to detect and search for life on exoplanets. *Astrobiology*, 9, 1–22 {**165**}

Codona JL, 2004, Exoplanet imaging with the Giant Magellan Telescope. *SPIE Conf Ser*, volume 5490, 379–388 {**161**}

Codona JL, Angel JRP, 2004, Imaging extrasolar planets by stellar halo suppression in separately corrected colour bands. *ApJ*, 604, L117–L120 {**157**}

Cody AM, Sasselov DD, 2005, Stellar evolution with enriched surface convection zones. I. General effects of planet consumption. *ApJ*, 622, 704–713 {**192**}

Cohen O, Drake JJ, Kashyap VL, et al., 2009, Interactions of the magnetospheres of stars and close-in giant planets. *ApJ*, 704, L85–L88 {**206**}

—, 2010, The impact of hot Jupiters on the spin-down of their host stars. *ApJ*, 723, L64–L67 {**206, 249**}

Colavita MM, Serabyn E, Millan-Gabet R, et al., 2009, Keck interferometer nuller data reduction and on-sky performance. *PASP*, 121, 1120–1138 {**163**}

Cole GHA, 2006, Observed exoplanets and intelligent life. *Surveys in Geophysics*, 27, 365–382 {**286**}

Cole GHA, Woolfson MM, 2002, *Planetary Science: The Science of Planets Around Stars*. Institute of Physics Publishing, UK {**9, 311**}

Collier Cameron AC, Bouchy F, Hébrard G, et al., 2007a, WASP-1 b and WASP-2 b: two new transiting exoplanets detected with SuperWASP and SOPHIE. *MNRAS*, 375, 951–957 {**2, 327**}

Collier Cameron AC, Bruce VA, Miller GRM, et al., 2010a, Line-profile tomography of exoplanet transits. I. The Doppler shadow of HD 189733 b. *MNRAS*, 403, 151–158 {**130**}

Collier Cameron AC, Guenther E, Smalley B, et al., 2010b, Line-profile tomography of exoplanet transits. II. A gas-giant planet transiting a rapidly rotating A5 star. *MNRAS*, 407, 507–514 {**130, 134, 249, 327**}

Collier Cameron AC, Horne K, Penny A, et al., 1999, Probable detection of starlight reflected from the giant planet orbiting τ Boo. *Nature*, 402, 751–755 {**124, 125**}

—, 2002, A search for starlight reflected from the υ And innermost planet. *MNRAS*, 330, 187–204 {**125**}

Collier Cameron AC, Pollacco D, Hellier C, et al., 2009, The WASP transit surveys. *IAU Symposium*, volume 253 of *IAU Symposium*, 29–35 {**107**}

Collier Cameron AC, Pollacco D, Street RA, et al., 2006, A fast hybrid algorithm for exoplanetary transit searches. *MNRAS*, 373, 799–810 {**104, 105**}

Collier Cameron AC, Wilson DM, West RG, et al., 2007b, Efficient identification of exoplanetary transit candidates from SuperWASP light curves. *MNRAS*, 380, 1230–1244 {**105**}

Colón KD, Ford EB, 2009, Benefits of ground-based photometric follow-up for transiting extrasolar planets discovered with Kepler and CoRoT. *ApJ*, 703, 1086–1095 {**111**}

Connerney JEP, 1993, Magnetic fields of the outer planets. *J. Geophys. Res.*, 98, 18659–18679 {**174**}

Connes P, 1985, Absolute astronomical accelerometry. *Ap&SS*, 110, 211–255 {**25, 27**}

—, 1994, Development of absolute accelerometry. *Ap&SS*, 212, 357–367 {**27**}

Connolly HC, Desch SJ, Ash RD, et al., 2006, Transient heating events in the protoplanetary nebula. *Meteorites and the Early Solar System II*, 383–397, University of Arizona Press {**297**}

Cooper CS, Showman AP, 2006, Dynamics and disequilibrium carbon chemistry in hot Jupiter atmospheres, with application to HD 209458 b. *ApJ*, 649, 1048–1063 {**138**}

Copi CJ, Starkman GD, 2000, The Big Occulting Steerable Satellite (BOSS). *ApJ*, 532, 581–592 {**156, 166**}

Cordes JM, 1993, The detectability of planetary companions to radio pulsars. *Planets Around Pulsars*, volume 36 of *ASP Conf Ser*, 43–60 {**78**}

Correia ACM, Levrard B, Laskar J, 2008, On the equilibrium rotation of Earth-like extrasolar planets. *A&A*, 488, L63–L66 {**246**}

Correia ACM, Udry S, Mayor M, et al., 2005, The CORALIE survey for southern extrasolar planets. XIII. A pair of planets around HD 202206 or a circumbinary planet? *A&A*, 440, 751–758 {**52, 59**}

—, 2009, The HARPS search for southern extrasolar planets. XVI. HD 45364, a pair of planets in a 3:2 mean motion resonance. *A&A*, 496, 521–526 {**52**}

Cosmovici CB, Bowyer S, Werthimer D (eds.), 1997, *Astronomical and Biochemical Origins and the Search for Life in the Universe*, volume 161 of *IAU Colloquia* {**282**}

Cossins P, Lodato G, Clarke C, 2010, The effects of opacity on gravitational stability in protoplanetary disks. *MNRAS*, 401, 2587–2598 {**236**}

Coughlin JL, Stringfellow GS, Becker AC, et al., 2008, New observations and a possible detection of parameter variations in the transits of GJ 436 b. *ApJ*, 689, L149–L152 {**135**}

Court RW, Sephton MA, 2009, Meteorite ablation products and their contribution to the atmospheres of terrestrial planets: an experimental study using pyrolysis-FTIR. *Geochim. Cosmochim. Acta*, 73, 3512–3521 {**278**}

Couteau P, Pecker JC, 1964, Space astrometry. *Bulletin d'Information de l'ADION, Nice Observatory*, 1(17) {**69**}

Cowan D, Grady M, Penny A, 1999, Astrobiology in the UK, Community Report to the British National Space Centre. Technical report, BNSC {**282**}

Cowan NB, Agol E, 2008, Inverting phase functions to map exoplanets. *ApJ*, 678, L129–L132 {**138**}

Cowan NB, Agol E, Charbonneau D, 2007, Hot nights on extrasolar planets: mid-infrared phase variations of hot Jupiters. *MNRAS*, 379, 641–646 {**142**}

Cowan NB, Agol E, Meadows VS, et al., 2009, Alien maps of an ocean-bearing world. *ApJ*, 700, 915–923 {**289**}

Cowley CR, 1995, *An Introduction to Cosmochemistry*. Cambridge University Press {**195, 259**}

Cox AN, 2000, *Allen's Astrophysical Quantities*. AIP Press; Springer, Fourth Edition {**120, 184, 300, 309–311**}

Crampton D, Simard L, Silva D, 2009, TMT science and instruments. *Science with the VLT in the ELT Era*, 279–288 {**159**}

Crane JD, Shectman SA, Butler RP, et al., 2008, The Carnegie planet finder spectrograph: a status report. *SPIE Conf Ser*, volume 7014 {**24**}

Cranmer SR, 2008, Winds of main-sequence stars: observational limits and a path to theoretical prediction. *14th Cambridge Workshop on Cool Stars, Stellar Systems, and the Sun*, volume 384 of *ASP Conf Ser*, 317–326 {**176**}

Creech-Eakman MJ, Romero V, Westpfahl D, et al., 2008, Magdalena Ridge Observatory Interferometer: progress toward first light. *SPIE Conf Ser*, volume 7013, 26 {**162**}

Crepp JR, Vanden Heuvel AD, Ge J, 2007, Comparative Lyot coronagraphy with extreme adaptive optics systems. *ApJ*, 661, 1323–1331 {**153**}

Cresswell P, Nelson RP, 2006, On the evolution of multiple protoplanets embedded in a protostellar disk. *A&A*, 450, 833–853 {**228**}

Crida A, 2009, Minimum mass solar nebulae and planetary migration. *ApJ*, 698, 606–614 {**234**}

Crida A, Baruteau C, Kley W, et al., 2009a, The dynamical role of the circumplanetary disk in planetary migration. *A&A*, 502, 679–693 {**234**}

Crida A, Masset F, Morbidelli A, 2009b, Long range outward migration of giant planets, with application to Fomalhaut b. *ApJ*, 705, L148–L152 {**223, 242, 243**}

Crida A, Morbidelli A, 2007, Cavity opening by a giant planet in a protoplanetary disk and effects on planetary migration. *MNRAS*, 377, 1324–1336 {**240**}

Crida A, Morbidelli A, Masset F, 2007, Simulating planet migration in globally evolving disks. *A&A*, 461, 1173–1183 {**234**}

Crida A, Sándor Z, Kley W, 2008, Influence of an inner disk on the orbital evolution of massive planets migrating in resonance. *A&A*, 483, 325–337 {**242**}

—, 2010, Planetary migration in resonance: the question of the eccentricities. *EAS Publications Series*, volume 41, 387–390 {**49**}

Croll B, Matthews JM, Rowe JF, et al., 2007a, Looking for giant Earths in the HD 209458 system: a search for transits in MOST space-based photometry. *ApJ*, 658, 1328–1339 {**113, 140**}

—, 2007b, Looking for super-Earths in the HD 189733 system: a search for transits in MOST space-based photometry. *ApJ*, 671, 2129–2138 {**113**}

Croll B, Walker GAH, Kuschnig R, et al., 2006, Differential rotation of ϵ Eri detected by MOST. *ApJ*, 648, 607–613 {**223**}

Crossfield IJM, Hansen BMS, Harrington J, et al., 2010, A new 24 micron phase curve for υ And b. *ApJ*, 723, 1436–1446 {**125**}

Crossley JH, Haghighipour N, 2004, On the stability of a planetary system embedded in the β Pic debris disk. *The Search for Other Worlds*, volume 713 of *American Institute of Physics Conference Series*, 265–268 {**224**}

Croswell K, 1988, Does Barnard's star have planets? *Astronomy*, 16, 6–17 {**64**}

Crouzet N, Agabi K, Blazit A, et al., 2009, ASTEP south: an Antarctic search for transiting planets around the celestial south pole. *IAU Symposium*, volume 253, 336–339 {**110**}

Crouzet N, Guillot T, Agabi A, et al., 2010a, ASTEP south: an Antarctic search for transiting exoplanets around the celestial south pole. *A&A*, 511, A36 {**110**}

—, 2010b, Photometric quality of Dome C for the winter 2008 from ASTEP south. *EAS Publications Series*, volume 40, 367–373 {**110**}

Crowe MJ, 1986, *The Extraterrestrial Life Debate 1750–1900. The Idea of a Plurality of Worlds from Kant to Lowell*. Cambridge University Press {**282**}

Cruikshank DP, Matthews MS, Schumann AM, 1995, *Neptune and Triton*. University of Arizona Press {**293**}

Cruz KL, Kirkpatrick JD, Burgasser AJ, 2009, Young L dwarfs identified in the field: a preliminary low-gravity, optical spectral sequence from L0 to L5. *AJ*, 137, 3345–3357 {**212**}

Ćuk M, Gladman BJ, Stewart ST, 2010, Constraints on the source of lunar cataclysm impactors. *Icarus*, 207, 590–594 {**304**}

Cumming A, 2004, Detectability of extrasolar planets in radial velocity surveys. *MNRAS*, 354, 1165–1176 {**13, 14**}

Cumming A, Butler RP, Marcy GW, et al., 2008, The Keck planet search: detectability and the minimum mass and orbital period distribution of extrasolar planets. *PASP*, 120, 531–554 {**15, 36, 253**}

Cumming A, Dragomir D, 2010, An integrated analysis of radial velocities in planet searches. *MNRAS*, 401, 1029–1042 {**14**}

Cumming A, Marcy GW, Butler RP, 1999, The Lick planet search: detectability and mass thresholds. *ApJ*, 526, 890–915 {**12, 24**}

Cuntz M, Eberle J, Musielak ZE, 2007, Stringent criteria for stable and unstable planetary orbits in stellar binary systems. *ApJ*, 669, L105–L108 {**57**}

Cuntz M, Saar SH, Musielak ZE, 2000, On stellar activity enhancement due to interactions with extrasolar giant planets. *ApJ*, 533, L151–L154 {**175, 205**}

Cuntz M, Shkolnik E, 2002, Chromospheres, flares and exoplanets. *Astronomische Nachrichten*, 323, 387–391 {**175**}

Cuntz M, Williams PE, 2006, Life without carbon? *Mercury*, 35(3), 030000 {**283**}

Cuntz M, Yeager KE, 2009, On the validity of the Hill radius criterion for the ejection of planets from stellar habitable zones. *ApJ*, 697, L86–L90

{43, 285}

Currie T, Hansen B, 2007, The evolution of protoplanetary disks around millisecond pulsars: the PSR B1257+12 system. *ApJ*, 666, 1232–1244 {77}

Cuzzi JN, Dobrovolskis AR, Champney JM, 1993, Particle-gas dynamics in the midplane of a protoplanetary nebula. *Icarus*, 106, 102–134 {226}

Czesla S, Huber KF, Wolter U, et al., 2009, How stellar activity affects the size estimates of extrasolar planets. *A&A*, 505, 1277–1282 {131}

da Silva R, Udry S, Bouchy F, et al., 2006, ELODIE metallicity-biased search for transiting hot Jupiters. I. Two hot Jupiters orbiting the slightly evolved stars HD 118203 and HD 149143. *A&A*, 446, 717–722 {29, 33, 108}

—, 2007, ELODIE metallicity-biased search for transiting hot Jupiters. IV. Intermediate period planets orbiting the stars HD 43691 and HD 132406. *A&A*, 473, 323–328 {29}

Daban J, Gouvret C, Guillot T, et al., 2010, ASTEP 400: a telescope designed for exoplanet transit detection from Dome C, Antarctica. *SPIE Conf Ser*, volume 7733, 151 {110}

Daemgen S, Hormuth F, Brandner W, et al., 2009, Binarity of transit host stars: implications for planetary parameters. *A&A*, 498, 567–574 {115}

D'Alessio P, Hartmann L, Calvet N, et al., 2005, The truncated disk of CoKu Tau/4. *ApJ*, 621, 461–472 {50}

Dall TH, Santos NC, Arentoft T, et al., 2006, Bisectors of the cross-correlation function applied to stellar spectra. Discriminating stellar activity, oscillations and planets. *A&A*, 454, 341–348 {23, 25}

Dall TH, Schmidtobreick L, Santos NC, et al., 2005, Outbursts on normal stars: FH Leo misclassified as a nova-like variable. *A&A*, 438, 317–324 {179}

Dalrymple GB, 1991, *The Age of the Earth*. Stanford University Press {297}

Danby JMA, 1964a, Stability of the triangular points in the elliptic restricted problem of three bodies. *AJ*, 69, 165–172 {52}

—, 1964b, The stability of the triangular Lagrangian points in the general problem of three bodies. *AJ*, 69, 294–296 {52}

Danchi WC, Barry RK, Lawson PR, et al., 2008, The Fourier–Kelvin Stellar Interferometer: a review, progress report, and update. *SPIE Conf Ser*, volume 7013, 83 {166}

Danchi WC, Deming D, Kuchner MJ, et al., 2003, Detection of close-in extrasolar giant planets using the Fourier–Kelvin stellar interferometer. *ApJ*, 597, L57–L60 {166}

Danchi WC, Lopez B, 2007, The Fourier–Kelvin Stellar Interferometer: a practical infrared space interferometer on the path to the discovery and characterisation of Earth-like planets around nearby stars. *Comptes Rendus Physique*, 8, 396–407 {166}

Danchi WC, Rajagopal J, Kuchner M, et al., 2006, The importance of phase in nulling interferometry and a three-telescope closure-phase nulling interferometer concept. *ApJ*, 645, 1554–1559 {163}

D'Angelo G, Bate MR, Lubow SH, 2005, The dependence of protoplanet migration rates on co-orbital torques. *MNRAS*, 358, 316–332 {240}

D'Angelo G, Lubow SH, 2008, Evolution of migrating planets undergoing gas accretion. *ApJ*, 685, 560–583 {234}

D'Antona F, Mazzitelli I, 1985, Evolution of very low mass stars and brown dwarfs. I. The minimum main-sequence mass and luminosity. *ApJ*, 296, 502–513 {209}

Darwin GH, 1898, *The tides and kindred phenomena in the solar system; the substance of lectures delivered in 1897 at the Lowell institute, Boston, Massachusetts*. Houghton & Mifflin {302}

—, 1908, *Scientific Papers Volume II*. Cambridge University Press {245}

Datchi F, Loubeyre P, Letoullec R, 2000, Extended and accurate determination of the melting curves of argon, helium, water ice, and hydrogen. *Phys. Rev. B*, 61, 6535–6546 {262, 263}

David EM, Quintana EV, Fatuzzo M, et al., 2003, Dynamical stability of Earth-like planetary orbits in binary systems. *PASP*, 115, 825–836 {81}

Davies GF, 1999, *Dynamic Earth*. Cambridge University Press {250}

Davies MB, Sigurdsson S, 2001, Planets in 47 Tuc. *MNRAS*, 324, 612–616 {109}

Davies ME, Abalakin VK, Bursa M, et al., 1986, Report of the IAU–IAG COSPAR working group on cartographic coordinates and rotational elements of the planets and satellites: 1985. *Celestial Mechanics*, 39, 102–113 {294}

Davis J, Ireland MJ, Jacob AP, et al., 2006, SUSI: an update on instrumental developments and science. *SPIE Conf Ser*, volume 6268, 4 {162}

Davis TA, Wheatley PJ, 2009, Evidence for a lost population of close-in exoplanets. *MNRAS*, 396, 1012–1017 {146}

Dawson RI, Fabrycky DC, 2010, Radial velocity planets de-aliased: a new, short period for super-Earth 55 Cnc e. *ApJ*, 722, 937–953 {287}

de Bruijne JHJ, Reynolds AP, Perryman MAC, et al., 2002, Direct determination of quasar redshifts. *A&A*, 381, L57–L60 {133}

de Felice F, Crosta MT, Vecchiato A, et al., 2004, A general relativistic model of light propagation in the gravitational field of the solar system: the static case. *ApJ*, 607, 580–595 {64}

de la Fuente Marcos C, de La Fuente Marcos R, 1997, Eccentric giant planets in open star clusters. *A&A*, 326, L21–L24 {237}

de Pater I, Lissauer JJ, 2010, *Planetary Sciences*. Cambridge University Press {293}

Debes JH, Ge J, Chakraborty A, 2002, First high-contrast imaging using a Gaussian aperture pupil mask. *ApJ*, 572, L165–L168 {153}

Debes JH, Ge J, Kuchner MJ, et al., 2004, Using notch-filter masks for high-contrast imaging of extrasolar planets. *ApJ*, 608, 1095–1099 {153}

Debes JH, Sigurdsson S, 2002, Are there unstable planetary systems around white dwarfs? *ApJ*, 572, 556–565 {79, 173}

—, 2007, The survival rate of ejected terrestrial planets with moons. *ApJ*, 668, L167–L170 {282}

Debes JH, Sigurdsson S, Woodgate BE, 2005a, Cool customers in the stellar graveyard. I. Limits to extrasolar planets around the white dwarf G29–38. *ApJ*, 633, 1168–1174 {80, 172}

—, 2005b, Cool customers in the stellar graveyard. II. Limits to substellar objects around nearby DAZ white dwarfs. *AJ*, 130, 1221–1230 {172}

Debes JH, Weinberger AJ, Schneider G, 2008, Complex organic materials in the circumstellar disk of HR 4796a. *ApJ*, 673, L191–L194 {223, 224}

Decin G, Dominik C, Waters LBFM, et al., 2003, Age dependence of the Vega phenomenon: observations. *ApJ*, 598, 636–644 {224}

Deeg HJ, 1998, Photometric detection of extrasolar planets by the transit-method. *Brown Dwarfs and Extrasolar Planets*, volume 134 of *ASP Conf Ser*, 216–223 {104}

Deeg HJ, Doyle LR, Kozhevnikov VP, et al., 1998, Near-term detectability of terrestrial extrasolar planets: TEP network observations of CM Dra. *A&A*, 338, 479–490 {59, 114}

—, 2000, A search for Jovian-mass planets around CM Dra using eclipse minima timing. *A&A*, 358, L5–L8 {114}

Deeg HJ, Garrido R, Claret A, 2001, Probing the stellar surface of HD 209458 from multicolour transit observations. *New Astronomy*, 6, 51–60 {118, 140}

Deeg HJ, Gillon M, Shporer A, et al., 2009, Ground-based photometry of space-based transit detections: photometric follow-up of the CoRoT mission. *A&A*, 506, 343–352 {111}

Deeg HJ, Moutou C, Erikson A, et al., 2010, A transiting giant planet with a temperature between 250 K and 430 K. *Nature*, 464, 384–387 {326}

Deeg HJ, Ocaña B, Kozhevnikov VP, et al., 2008, Extrasolar planet detection by binary stellar eclipse timing: evidence for a third body around CM Dra. *A&A*, 480, 563–571 {82}

Deer WA, Howie RA, Zussman J, 1996, *An Introduction to the Rock-Forming Minerals*. Prentice–Hall, Second Edition {256}

Dehnen W, Binney J, 1998, Mass models of the Milky Way. *MNRAS*, 294, 429–438 {194}

Dekker H, D'Odorico S, Kaufer A, et al., 2000, Design, construction, and performance of UVES. *SPIE Conf Ser*, volume 4008, 534–545 {24}

Deleuil M, Deeg HJ, Alonso R, et al., 2008, Transiting exoplanets from the CoRoT space mission. VI. CoRoT-3 b: the first secure inhabitant of the brown-dwarf desert. *A&A*, 491, 889–897 {144, 145, 326}

Deleuil M, Meunier JC, Moutou C, et al., 2009, Exo-Dat: an information system in support of the CoRoT exoplanet science. *AJ*, 138, 649–663 {111}

Delfosse X, Forveille T, 2001, Brown dwarfs and very low mass stars with DENIS. *SF2A-2001: Semaine de l'Astrophysique Francaise*, 91–94 {210}

Delfosse X, Forveille T, Beuzit J, et al., 1999a, New neighbours. I. 13 new companions to nearby M dwarfs. *A&A*, 344, 897–910 {33}

Delfosse X, Forveille T, Mayor M, et al., 1998, The closest extrasolar planet: a giant planet around the M4 dwarf GL 876. *A&A*, 338, L67–L70 {49}

Delfosse X, Tinney CG, Forveille T, et al., 1999b, Searching for very low-mass stars and brown dwarfs with DENIS. *A&AS*, 135, 41–56 {210}

References

Deliyannis CP, Cunha K, King JR, et al., 2000, Beryllium and iron abundances of the solar twins 16 Cyg A and B. *AJ*, 119, 2437–2444 {**201**}

Delorme P, Delfosse X, Albert L, et al., 2008a, CFBDS J005910.90-011401.3: reaching the T–Y brown dwarf transition? *A&A*, 482, 961–971 {**211, 212**}

Delorme P, Willott CJ, Forveille T, et al., 2008b, Finding ultracool brown dwarfs with MegaCam on CFHT: method and first results. *A&A*, 484, 469–478 {**211**}

Delplancke F, 2008, The PRIMA facility phase-referenced imaging and micro-arcsec astrometry. *New Astronomy Reviews*, 52, 199–207 {**69**}

Demarque P, Woo J, Kim Y, et al., 2004, Yonsei–Yale isochrones with an improved core overshoot treatment. *ApJS*, 155, 667–674 {**187**}

Demianski M, Proszynski M, 1979, Does PSR 0329+54 have companions. *Nature*, 282, 383–385 {**78**}

Deming D, 2009, Emergent exoplanet flux: review of the Spitzer results. *IAU Symposium*, volume 253 of *IAU Symposium*, 197–207 {**112**}

Deming D, Brown TM, Charbonneau D, et al., 2005a, A new search for carbon monoxide absorption in the transmission spectrum of the extrasolar planet HD 209458 b. *ApJ*, 622, 1149–1159 {**140**}

Deming D, Espenak F, Jennings DE, et al., 1987, On the apparent velocity of integrated sunlight. I. 1983–1985. *ApJ*, 316, 771–787 {**20**}

Deming D, Harrington J, Laughlin G, et al., 2007a, Spitzer transit and secondary eclipse photometry of GJ 436 b. *ApJ*, 667, L199–L202 {**127, 142**}

Deming D, Harrington J, Seager S, et al., 2006, Strong infrared emission from the extrasolar planet HD 189733 b. *ApJ*, 644, 560–564 {**142**}

Deming D, Richardson LJ, Harrington J, 2007b, 3.8-μm photometry during the secondary eclipse of the extrasolar planet HD 209458 b. *MNRAS*, 378, 148–152 {**140**}

Deming D, Seager S, Richardson LJ, et al., 2005b, Infrared radiation from an extrasolar planet. *Nature*, 434, 740–743 {**2, 112, 140, 142**}

Deming D, Seager S, Winn J, et al., 2009, Discovery and characterisation of transiting super Earths using an all-sky transit survey and follow-up by JWST. *PASP*, 121, 952–967 {**114**}

Demory BO, Gillon M, Barman T, et al., 2007, Characterisation of the hot Neptune GJ 436 b with Spitzer and ground-based observations. *A&A*, 475, 1125–1129 {**131, 142**}

den Hartog R, Absil O, Gondoin P, et al., 2006, The prospects of detecting exoplanets with the Ground-based European Nulling Interferometer Experiment (GENIE). *IAU Colloq. 200: Direct Imaging of Exoplanets: Science and Techniques*, 233–239 {**166**}

Dermott SF, Murray CD, 1981a, The dynamics of tadpole and horseshoe orbits. I. Theory. *Icarus*, 48, 1–11 {**52**}

—, 1981b, The dynamics of tadpole and horseshoe orbits. II. The coorbital satellites of Saturn. *Icarus*, 48, 12–22 {**52**}

Deroo P, Swain MR, Tinetti G, et al., 2010, THESIS: a combined-light mission for exoplanet molecular spectroscopy. *Bulletin of the American Astronomical Society*, volume 41, 424 {**114**}

Des Marais DJ (ed.), 1997, *The Blue Dot Workshop: spectroscopic search for life on extrasolar planets* {**282**}

Des Marais DJ, 1998, Earth's early biosphere and its environment. *Origins*, volume 148 of *ASP Conf Ser*, 415–434 {**282**}

Des Marais DJ, Allamandola LJ, Benner SA, et al., 2003, The NASA astrobiology roadmap. *Astrobiology*, 3, 219–235 {**282**}

Des Marais DJ, Harwit MO, Jucks KW, et al., 2002, Remote sensing of planetary properties and biosignatures on extrasolar terrestrial planets. *Astrobiology*, 2, 153–181 {**289**}

Des Marais DJ, Nuth JA III, Allamandola LJ, et al., 2008, The NASA astrobiology roadmap. *Astrobiology*, 8, 715–730 {**282, 283**}

Desch MD, Kaiser ML, 1984, Predictions for Uranus from a radiometric Bode's law. *Nature*, 310, 755–757 {**175**}

Desch SJ, 2004, Linear analysis of the magnetorotational instability, including ambipolar diffusion, with application to protoplanetary disks. *ApJ*, 608, 509–525 {**221**}

—, 2007, Mass distribution and planet formation in the solar nebula. *ApJ*, 671, 878–893 {**234, 304**}

Desch SJ, Ciesla FJ, Hood LL, et al., 2005, Heating of chondritic materials in solar nebula shocks. *Chondrites and the Protoplanetary Disk*, volume 341 of *ASP Conf Ser*, 849–872 {**297**}

Desch SJ, Connolly HC, 2002, A model of the thermal processing of particles in solar nebula shocks: application to the cooling rates of chondrules. *Meteoritics and Planetary Science*, 37, 183–207 {**297**}

Desch SJ, Cuzzi JN, 2000, The generation of lightning in the solar nebula. *Icarus*, 143, 87–105 {**297**}

Désert J, Lecavelier des Etangs A, Hébrard G, et al., 2009, Search for carbon monoxide in the atmosphere of the transiting exoplanet HD 189733 b. *ApJ*, 699, 478–485 {**140**}

Désert JM, Vidal-Madjar A, Lecavelier Des Etangs A, et al., 2008, TiO and VO broad band-absorption features in the optical spectrum of the atmosphere of the hot-Jupiter HD 209458 b. *A&A*, 492, 585–592 {**139, 140**}

Desidera S, Barbieri M, 2007, Properties of planets in binary systems: the role of binary separation. *A&A*, 462, 345–353 {**58, 59**}

Desidera S, Gratton RG, Endl M, et al., 2003, A search for planets in the metal-enriched binary HD 219542. *A&A*, 405, 207–221 {**193**}

Desidera S, Gratton RG, Lucatello S, et al., 2006, Abundance difference between components of wide binaries. II. The southern sample. *A&A*, 454, 581–593 {**193**}

Desidera S, Gratton RG, Scuderi S, et al., 2004, Abundance difference between components of wide binaries. *A&A*, 420, 683–697 {**193**}

Desort M, Lagrange A, Galland F, et al., 2009, Extrasolar planets and brown dwarfs around A-F type stars. VII. θ Cyg radial velocity variations: planets or stellar phenomenon? *A&A*, 506, 1469–1476 {**32, 33**}

—, 2010, Planets and brown dwarfs around A-F main-sequence stars: performances of radial-velocity surveys with HARPS and first detections. *EAS Publications Series*, volume 41, 99–102 {**32**}

Desort M, Lagrange AM, Galland F, et al., 2007, Search for exoplanets with the radial-velocity technique: quantitative diagnostics of stellar activity. *A&A*, 473, 983–993 {**21**}

—, 2008, Extrasolar planets and brown dwarfs around A-F type stars. V. A planetary system found with HARPS around the F6IV-V star HD 60532. *A&A*, 491, 883–888 {**50**}

Detweiler S, 1979, Pulsar timing measurements and the search for gravitational waves. *ApJ*, 234, 1100–1104 {**79**}

Dhillon VS, Marsh TR, Stevenson MJ, et al., 2007, ULTRACAM: an ultrafast, triple-beam CCD camera for high-speed astrophysics. *MNRAS*, 378, 825–840 {**133**}

Di Stefano R, 2008a, Mesolensing explorations of nearby masses: from planets to black holes. *ApJ*, 684, 59–67 {**94**}

—, 2008b, Mesolensing: high-probability lensing without large optical depth. *ApJ*, 684, 46–58 {**94**}

Di Stefano R, Scalzo RA, 1999, A new channel for the detection of planetary systems through microlensing. II. Repeating events. *ApJ*, 512, 579–600 {**88**}

Díaz RF, Rojo P, Melita M, et al., 2008, Detection of period variations in extrasolar transiting planet OGLE–TR–111 b. *ApJ*, 682, L49–L52 {**135**}

Dick SJ, 1996, *The Biological Universe*. Cambridge University Press {**282**}

Dickey JO, Bender PL, Faller JE, et al., 1994, Lunar laser ranging: a continuing legacy of the Apollo programme. *Science*, 265, 482–490 {**247**}

Dickin AP, 2005, *Radiogenic Isotope Geology*. Cambridge University Press, Second Edition {**298**}

Diego F, Fish AC, Barlow MJ, et al., 1995, The Ultra-High-Resolution Facility at the Anglo-Australian Telescope. *MNRAS*, 272, 323–332 {**17**}

Diolaiti E, Conan J, Foppiani I, et al., 2010, Towards the phase A review of MAORY, the multi-conjugate adaptive optics module for the E–ELT. *Adaptive Optics for Extremely Large Telescopes*, 2007 {**152, 162**}

Dittmann JA, Close LM, Green EM, et al., 2009, A tentative detection of a starspot during consecutive transits of an extrasolar planet from the ground: no evidence of a double transiting planet system around TrES–1. *ApJ*, 701, 756–763 {**131**}

do Nascimento JD, Charbonnel C, Lèbre A, et al., 2000, Lithium and rotation on the subgiant branch. II. Theoretical analysis of observations. *A&A*, 357, 931–937 {**32**}

Dobbs-Dixon I, Lin DNC, 2008, Atmospheric dynamics of short-period extrasolar gas giant planets. II. Dependence of nightside temperature on opacity. *ApJ*, 673, 513–525 {**138**}

Dobbs-Dixon I, Lin DNC, Mardling RA, 2004, Spin-orbit evolution of short-period planets. *ApJ*, 610, 464–476 {**245, 246**}

Dodson-Robinson SE, Veras D, Ford EB, et al., 2009, The formation mechanism of gas giants on wide orbits. *ApJ*, 707, 79–88 {**236**}

Dole SH, 1964, *Habitable Planets for Man*. Blaisdell, New York {**283**}

Döllinger MP, Hatzes AP, Pasquini L, et al., 2007, Discovery of a planet around the K giant star 4 UMa. *A&A*, 472, 649–652 {**2, 32, 33**}

—, 2009, Planetary companion candidates around the K giant stars

42 Dra and HD 139 357. *A&A*, 499, 935–942 {**32, 33**}

Domingo V, Fleck B, Poland AI, 1995, The SOHO mission: an overview. *Sol. Phys.*, 162, 1–37 {**52**}

Domingos RC, Winter OC, Yokoyama T, 2006, Stable satellites around extrasolar giant planets. *MNRAS*, 373, 1227–1234 {**130**}

Dominik C, Decin G, 2003, Age dependence of the Vega phenomenon: theory. *ApJ*, 598, 626–635 {**224**}

Dominik C, Laureijs RJ, Jourdain de Muizon M, et al., 1998, A Vega–like disk associated with the planetary system of ρ^1 Cnc. *A&A*, 329, L53–L56 {**224**}

Dominik C, Tielens AGGM, 1997, The physics of dust coagulation and the structure of dust aggregates in space. *ApJ*, 480, 647–673 {**226**}

Dominik M, 1998, Galactic microlensing with rotating binaries. *A&A*, 329, 361–374 {**91**}

—, 1999, The binary gravitational lens and its extreme cases. *A&A*, 349, 108–125 {**90**}

Dominik M, Albrow MD, Beaulieu JP, et al., 2002, The PLANET microlensing follow-up network: results and prospects for the detection of extrasolar planets. *Planet. Space Sci.*, 50, 299–307 {**97**}

Dominik M, Jørgensen UG, Rattenbury NJ, et al., 2010, Realisation of a fully-deterministic microlensing observing strategy for inferring planet populations. *Astronomische Nachrichten*, 331, 671–691 {**98**}

Dominik M, Rattenbury NJ, Allan A, et al., 2007, An anomaly detector with immediate feedback to hunt for planets of Earth mass and below by microlensing. *MNRAS*, 380, 792–804 {**97**}

Dominik M, Sahu KC, 2000, Astrometric microlensing of stars. *ApJ*, 534, 213–226 {**94**}

Donati JF, Moutou C, Farès R, et al., 2008, Magnetic cycles of the planet-hosting star τ Boo. *MNRAS*, 385, 1179–1185 {**206**}

Dones L, Tremaine S, 1993a, On the origin of planetary spins. *Icarus*, 103, 67–92 {**280, 301**}

—, 1993b, Why does the Earth spin forward? *Science*, 259, 350–354 {**301**}

Dong S, Bond IA, Gould A, et al., 2009a, Microlensing event MOA–2007–BLG–400: exhuming the buried signature of a cool, Jovian-mass planet. *ApJ*, 698, 1826–1837 {**91, 99**}

Dong S, DePoy DL, Gaudi BS, et al., 2006, Planetary detection efficiency of the magnification 3000 microlensing event OGLE–2004–BLG–343. *ApJ*, 642, 842–860 {**86**}

Dong S, Gould A, Udalski A, et al., 2009b, OGLE–2005–BLG–071Lb, the most massive M dwarf planetary companion? *ApJ*, 695, 970–987 {**98**}

Dong S, Udalski A, Gould A, et al., 2007, First space-based microlens parallax measurement: Spitzer observations of OGLE–2005–SMC–001. *ApJ*, 664, 862–878 {**93, 94**}

Donnison JR, 2006, The Hill stability of a binary or planetary system during encounters with a third inclined body. *MNRAS*, 369, 1267–1280 {**43**}

—, 2009, The Hill stability of inclined bound triple star and planetary systems. *Planet. Space Sci.*, 57, 771–783 {**43**}

Dorland BN, Dudik RP, Dugan Z, et al., 2009, The Joint Milli-Arcsecond Pathfinder Survey (JMAPS): mission overview and attitude sensing applications. *ArXiv e-prints* {**73**}

Dormand JR, Woolfson MM, 1989, *The Origin of the Solar System: The Capture Theory*. Ellis Horwood/Prentice Hall {**225**}

Dougherty SM, Beasley AJ, Claussen MJ, et al., 2005, High-resolution radio observations of the colliding-wind binary WR 140. *ApJ*, 623, 447–459 {**173**}

Douglas NG, 1997, Heterodyned holographic spectroscopy. *PASP*, 109, 151–165 {**25**}

Doyle LR, 1988, Progress in determining the space orientation of stars. *IAU Colloq. 99: Bioastronomy – The Next Steps*, volume 144 of *Astrophysics and Space Science Library*, 101–105 {**114**}

Doyle LR, Deeg HJ, Kozhevnikov VP, et al., 2000, Observational limits on terrestrial-sized inner planets around the CM Dra system using the photometric transit method with a matched-filter algorithm. *ApJ*, 535, 338–349 {**59, 114**}

Doyle LR, Dunham ET, Deeg H, et al., 1996, Ground-based detectability of terrestrial and Jovian extrasolar planets: observations of CM Dra at Lick Observatory. *J. Geophys. Res.*, 101, 14823–14830 {**114**}

Doyon R, Rowlands N, Hutchings J, et al., 2008, The JWST Tunable Filter Imager (TFI). *SPIE Conf Ser*, volume 7010, 30 {**163**}

Drake AJ, 2003, On the selection of photometric planetary transits. *ApJ*, 589, 1020–1026 {**132**}

Drake AJ, Beshore E, Catelan M, et al., 2010, Discovery of eclipsing white dwarf systems in a search for Earth-size companions. *ArXiv e-prints* {**103**}

Drake AJ, Cook KH, 2004, Photometric transits from the MACHO project database. *ApJ*, 604, 379–387 {**106**}

Drake FD, 1961, Project Ozma. *Physics Today*, 14, 40–46 {**290**}

—, 1965, The radio search for intelligent extraterrestrial life. *Current Aspects of Exobiology (Pergamon, New York)*, 323–345 {**197**}

—, 2008, SETI: the early days and now. *Frontiers of Astrophysics: A Celebration of NRAO's 50th Anniversary*, volume 395 of *ASP Conf Ser*, 213–224 {**290**}

Drake MJ, Stimpfl M, Lauretta DS, 2004, How did the terrestrial planets acquire their water? *Workshop on Oxygen in the Terrestrial Planets*, 3043 {**298**}

Dravins D, 1975, Physical limits to attainable accuracies in stellar radial velocities. *A&A*, 43, 45–50 {**18, 23**}

—, 1999, Stellar surface convection, line asymmetries, and wavelength shifts. *IAU Colloq. 170: Precise Stellar Radial Velocities*, volume 185 of *ASP Conf Ser*, 268–277 {**18**}

Dravins D, Lindegren L, Madsen S, 1999, Astrometric radial velocities. I. Non-spectroscopic methods for measuring stellar radial velocity. *A&A*, 348, 1040–1051 {**18, 64**}

Dravins D, Lindegren L, Mezey E, et al., 1997a, Atmospheric intensity scintillation of stars. I. Statistical distributions and temporal properties. *PASP*, 109, 173–207 {**116**}

—, 1997b, Atmospheric intensity scintillation of stars. II. Dependence on optical wavelength. *PASP*, 109, 725–737 {**116**}

Dravins D, Lindegren L, Nordlund A, 1981, Solar granulation: influence of convection on spectral line asymmetries and wavelength shifts. *A&A*, 96, 345–364 {**18, 23**}

Dreizler S, Hauschildt PH, Kley W, et al., 2003, OGLE–TR–3: a possible new transiting planet. *A&A*, 402, 791–799 {**106**}

Dreizler S, Rauch T, Hauschildt P, et al., 2002, Spectral types of planetary host star candidates: two new transiting planets? *A&A*, 391, L17–L20 {**106**}

Dreizler S, Reiners A, Homeier D, et al., 2009, On the possibility of detecting extrasolar planets' atmospheres with the Rossiter–McLaughlin effect. *A&A*, 499, 615–621 {**127**}

Duggan P, McBreen B, Carr AJ, et al., 2003, Gamma-ray bursts and X-ray melting of material to form chondrules and planets. *A&A*, 409, L9–L12 {**298**}

Dullemond CP, Dominik C, 2004, The effect of dust settling on the appearance of protoplanetary disks. *A&A*, 421, 1075–1086 {**225**}

—, 2005, Dust coagulation in protoplanetary disks: a rapid depletion of small grains. *A&A*, 434, 971–986 {**226**}

Dullemond CP, Hollenbach D, Kamp I, et al., 2007, Models of the structure and evolution of protoplanetary disks. *Protostars and Planets V*, 555–572 {**220**}

Duncan CH, Willson RC, Kendall JM, et al., 1982, Latest rocket measurements of the solar constant. *Solar Energy*, 28, 385–387 {**310**}

Duncan MJ, Levison HF, Lee MH, 1998, A multiple time step symplectic algorithm for integrating close encounters. *AJ*, 116, 2067–2077 {**44**}

Duncan MJ, Lissauer JJ, 1998, The effects of post-main-sequence solar mass loss on the stability of our planetary system. *Icarus*, 134, 303–310 {**79**}

Dunham EW, Borucki WJ, Koch DG, et al., 2010, Kepler–6 b: a transiting hot Jupiter orbiting a metal-rich star. *ApJ*, 713, L136–L139 {**112, 326**}

Dunham EW, Mandushev GI, Taylor BW, et al., 2004, PSST: The Planet Search Survey Telescope. *PASP*, 116, 1072–1080 {**107**}

Dupuy TJ, Liu MC, 2009, Detectability of transiting Jupiters and low-mass eclipsing binaries in sparsely sampled Pan–STARRS–1 survey data. *ApJ*, 704, 1519–1537 {**110**}

Duquennoy A, Mayor M, 1991, Multiplicity among solar-type stars in the solar neighbourhood. II. Distribution of the orbital elements in an unbiased sample. *A&A*, 248, 485–524 {**28, 56, 59, 81, 207**}

Durisen RH, Boss AP, Mayer L, et al., 2007, Gravitational instabilities in gaseous protoplanetary disks and implications for giant planet formation. *Protostars and Planets V*, 607–622 {**236**}

Durisen RH, Cai K, Mejía AC, et al., 2005, A hybrid scenario for gas giant planet formation in rings. *Icarus*, 173, 417–424 {**226**}

Durisen RH, Hartquist TW, Pickett MK, 2008, The formation of fragments at corotation in isothermal protoplanetary disks. *Ap&SS*, 317, 3–8

References

{236}

Dvorak R, 1982, Planetary orbits in double star systems. *Oesterreichische Akademie Wissenschaften Mathematisch naturwissenschaftliche Klasse Sitzungsberichte Abteilung*, 191, 423–437 {**56**}

—, 1984, Numerical experiments on planetary orbits in double stars. *Celestial Mechanics*, 34, 369–378 {**57**}

—, 1986, Critical orbits in the elliptic restricted three-body problem. *A&A*, 167, 379–386 {**57**}

Dvorak R, Froeschle C, Froeschle C, 1989, Stability of outer planetary orbits (P-types) in binaries. *A&A*, 226, 335–342 {**57**}

Dvorak R, Pilat-Lohinger E, Bois E, et al., 2004, Planets in double stars: the γ Cep system. *Revista Mexicana de Astronomia y Astrofisica Conference Series* (eds. Allen C, Scarfe C), volume 21, 222–226 {**57**}

Dvorak R, Pilat-Lohinger E, Funk B, et al., 2003a, Planets in habitable zones: a study of the binary γ Cep. *A&A*, 398, L1–L4 {**57, 285**}

—, 2003b, A study of the stable regions in the planetary system HD 74156: can it host Earth-like planets in habitable zones? *A&A*, 410, L13–L16 {**44, 285**}

Dyson FW, Eddington AS, Davidson C, 1920, A determination of the deflection of light by the Sun's gravitational field, from observations made at the total eclipse of 29 May 1919. *Royal Society of London Philosophical Transactions Series A*, 220, 291–333 {**84**}

Dziembowski WA, Fiorentini G, Ricci B, et al., 1999, Helioseismology and the solar age. *A&A*, 343, 990–996 {**296**}

Ebel DS, Grossman L, 2000, Condensation in dust-enriched systems. *Geochim. Cosmochim. Acta*, 64, 339–366 {**257**}

Ecuvillon A, Israelian G, Pont F, et al., 2007, Kinematics of planet-host stars and their relation to dynamical streams in the solar neighbourhood. *A&A*, 461, 171–182 {**183**}

Ecuvillon A, Israelian G, Santos NC, et al., 2004a, C, S, Zn and Cu abundances in planet-harbouring stars. *A&A*, 426, 619–630 {**188, 196, 199**}

—, 2004b, Nitrogen abundances in planet-harbouring stars. *A&A*, 418, 703–715 {**188, 196, 199**}

—, 2006a, Abundance ratios of volatile versus refractory elements in planet-harbouring stars: hints of pollution? *A&A*, 449, 809–816 {**196**}

—, 2006b, Oxygen abundances in planet-harbouring stars: comparison of different abundance indicators. *A&A*, 445, 633–645 {**196, 198**}

Eddington AS, 1920, *Space, Time and Gravitation. An Outline of the General Relativity Theory*. Cambridge University Press {**84**}

Edelstein J, Muterspaugh MW, Erskine D, et al., 2008, Dispersed interferometry for infrared exoplanet velocimetry. *SPIE Conf Ser*, volume 7014, 242–247 {**27**}

Edelstein J, Muterspaugh MW, Erskine DJ, et al., 2007, TEDI: the Triple-Spec exoplanet discovery instrument. *SPIE Conf Ser*, volume 6693 {**24, 27, 30, 33**}

Edgar RG, 2007, Giant planet migration in viscous power-law disks. *ApJ*, 663, 1325–1334 {**234**}

Edgar RG, Quillen AC, 2008, The vertical structure of planet-induced gaps in protoplanetary disks. *MNRAS*, 387, 387–396 {**223**}

Edgar RG, Quillen AC, Park J, 2007, The minimum gap-opening planet mass in an irradiated circumstellar accretion disk. *MNRAS*, 381, 1280–1286 {**240**}

Edser E, Butler CP, 1898, A simple method of reducing prismatic spectra. *Phil. Mag.*, 46(5), 207–216 {**25**}

Edvardsson B, Andersen J, Gustafsson B, et al., 1993, The chemical evolution of the Galactic disk. I. Analysis and results. *A&A*, 275, 101–152 {**183, 193, 194, 198**}

Edwards S, Strom SE, Hartigan P, et al., 1993, Angular momentum regulation in low-mass young stars surrounded by accretion disks. *AJ*, 106, 372–382 {**200**}

Efroimsky M, Lainey V, 2007, Physics of bodily tides in terrestrial planets and the appropriate scales of dynamical evolution. *Journal of Geophysical Research (Planets)*, 112, 12003 {**246**}

Eggenberger A, 2010, Detection and characterisation of planets in binary and multiple systems. *EAS Publications Series*, volume 42, 19–37 {**33, 59**}

Eggenberger A, Halbwachs J, Udry S, et al., 2004a, Statistical properties of an unbiased sample of F7-K binaries: towards the long-period systems. *Revista Mexicana de Astronomia y Astrofisica Conference Series*, volume 21, 28–32 {**56**}

Eggenberger A, Mayor M, Naef D, et al., 2006, The CORALIE survey for southern extrasolar planets. XIV. HD 142022 b: a long-period planetary companion in a wide binary. *A&A*, 447, 1159–1163 {**12, 13, 17**}

Eggenberger A, Udry S, Chauvin G, et al., 2007a, The impact of stellar duplicity on planet occurrence and properties. I. Observational results of a VLT–NACO search for stellar companions to 130 nearby stars with and without planets. *A&A*, 474, 273–291 {**58, 207**}

Eggenberger A, Udry S, Mayor M, 2004b, Statistical properties of exoplanets. III. Planet properties and stellar multiplicity. *A&A*, 417, 353–360 {**59**}

Eggenberger A, Udry S, Mazeh T, et al., 2007b, No evidence of a hot Jupiter around HD 188753 A. *A&A*, 466, 1179–1183 {**60**}

Eggleton PP, Kiseleva LG, Hut P, 1998, The equilibrium tide model for tidal friction. *ApJ*, 499, 853–870 {**246**}

Ehrenreich D, Hébrard G, Lecavelier des Etangs A, et al., 2007, A Spitzer search for water in the transiting exoplanet HD 189733 b. *ApJ*, 668, L179–L182 {**140**}

Ehrenreich D, Lecavelier des Etangs A, Beaulieu JP, et al., 2006, On the possible properties of small and cold extrasolar planets: is OGLE–2005-BLG-390L b entirely frozen? *ApJ*, 651, 535–543 {**98, 260, 267, 270**}

Ehrenreich D, Lecavelier des Etangs A, Hébrard G, et al., 2008, New observations of the extended hydrogen exosphere of the extrasolar planet HD 209458 b. *A&A*, 483, 933–937 {**140, 281**}

Einstein A, 1936, Lens-like action of a star by the deviation of light in the gravitational field. *Science*, 84, 506–507 {**84**}

Eisenhardt PRM, Griffith RL, Stern D, et al., 2010, Ultracool field brown dwarf candidates selected at 4.5 μm. *AJ*, 139, 2455–2464 {**212**}

Eisner JA, 2007, Water vapour and hydrogen in the terrestrial-planet-forming region of a protoplanetary disk. *Nature*, 447, 562–564 {**284**}

Eisner JA, Kulkarni SR, 2001a, Sensitivity of the astrometric technique in detecting outer planets. *ApJ*, 561, 1107–1115 {**67**}

—, 2001b, Sensitivity of the radial-velocity technique in detecting outer planets. *ApJ*, 550, 871–883 {**14**}

—, 2002, Detecting outer planets in edge-on orbits: combining radial velocity and astrometric techniques. *ApJ*, 574, 426–429 {**67**}

Elkins-Tanton LT, Seager S, 2008a, Coreless terrestrial exoplanets. *ApJ*, 688, 628–635 {**260, 267, 279**}

—, 2008b, Ranges of atmospheric mass and composition of super-Earth exoplanets. *ApJ*, 685, 1237–1246 {**267, 278, 279**}

Elliot JL, 1978, Direct imaging of extrasolar planets with stationary occultations viewed by a space telescope. *Icarus*, 35, 156–164 {**157, 164**}

Elmegreen BG, 1999, A prediction of brown dwarfs in ultracold molecular gas. *ApJ*, 522, 915–920 {**214**}

Els SG, Sterzik MF, Marchis F, et al., 2001, A second substellar companion in the GJ 86 system: a brown dwarf in an extrasolar planetary system. *A&A*, 370, L1–L4 {**173**}

Emerson J, Sutherland W, 2010, The Visible and Infrared Survey Telescope for Astronomy (VISTA): looking back at commissioning. *The Messenger*, 139, 2–5 {**211**}

Encrenaz T, Bibring J, Blanc M, 2004, *The Solar System*. Springer {**311**}

Endl M, Cochran WD, Kürster M, et al., 2006, Exploring the frequency of close-in Jovian planets around M dwarfs. *ApJ*, 649, 436–443 {**30, 192**}

—, 2008, New results from the McDonald Observatory and ESO-VLT planet surveys. *ASP Conf Ser*, volume 398, 51–58 {**24, 30**}

Endl M, Cochran WD, Tull RG, et al., 2003, A dedicated M dwarf planet search using the Hobby–Eberly telescope. *AJ*, 126, 3099–3107, erratum: 2006, AJ, 132, 2755 {**33**}

Endl M, Kürster M, Els S, et al., 2002, The planet search programme at the ESO Coudé Echelle spectrometer. III. The complete Long Camera survey results. *A&A*, 392, 671–690 {**33**}

Enoch B, Cameron AC, Anderson DR, et al., 2011, WASP-25 b: a 0.6 Jupiter-mass planet in the southern hemisphere. *MNRAS*, 410, 1631–1636 {**327**}

Epchtein N, 2010, *A vision for European astronomy and astrophysics at the Antarctic station Concordia, Dome C*. Prepared by Antarctic Research, a European Network for Astrophysics (ARENA) for EC–FP6 contract RICA 026150 {**161**}

Epchtein N, de Batz B, Capoani L, et al., 1997, The deep near-infrared southern sky survey (DENIS). *The Messenger*, 87, 27–34 {**210**}

Epstein PS, 1924, On the resistance experienced by spheres in their motion through gases. *Physical Review*, 23, 710–733 {**226**}

Érdi B, Dvorak R, Sándor Z, et al., 2004, The dynamical structure of the habitable zone in the HD 38529, HD 168443 and HD 169830 systems.

MNRAS, 351, 1043–1048 {47, 285}

Érdi B, Nagy I, Sándor Z, et al., 2007, Secondary resonances of co-orbital motions. MNRAS, 381, 33–40 {54}

Eriksson U, Lindegren L, 2007, Limits of ultra-high-precision optical astrometry. Stellar surface structures. A&A, 476, 1389–1400 {65, 115}

Erkaev NV, Kulikov YN, Lammer H, et al., 2007, Roche lobe effects on the atmospheric loss from hot Jupiters. A&A, 472, 329–334 {281}

Erskine DJ, 2003, An externally dispersed interferometer prototype for sensitive radial velocimetry: theory and demonstration on sunlight. PASP, 115, 255–269 {25, 26}

Erskine DJ, Edelstein J, Feuerstein WM, et al., 2003, High-resolution broad-band spectroscopy using an externally dispersed interferometer. ApJ, 592, L103–L106 {25}

Erskine DJ, Ge J, 2000, A novel interferometer spectrometer for sensitive stellar radial velocimetry. Imaging the Universe in Three Dimensions, volume 195, 501–507 {25}

Esposito L, 2002, Planetary rings. Reports on Progress in Physics, 65, 1741–1783 {226}

—, 2006, Planetary rings: structure and history. European Planetary Science Congress 2006, 196–197 {300}

Esposito M, Guenther E, Hatzes AP, et al., 2006, Planets of young stars: the TLS radial velocity survey. Tenth Anniversary of 51 Peg-b: Status of and prospects for hot Jupiter studies, 127–134 {24}

Estrada PR, Mosqueira I, 2006, A gas-poor planetesimal capture model for the formation of giant planet satellite systems. Icarus, 181, 486–509 {237}

Etzel PB, 1993, Current status of the EBOP code. IAU Commission on Close Binary Stars, 21, 113–124 {119}

Everett ME, Howell SB, 2001, A technique for ultrahigh-precision CCD photometry. PASP, 113, 1428–1435 {104}

Eyer L, Mignard F, 2005, Rate of correct detection of periodic signal with the Gaia satellite. MNRAS, 361, 1136–1144 {113}

Faber P, Quillen AC, 2007, The total number of giant planets in debris disks with central clearings. MNRAS, 382, 1823–1828 {223}

Fabrycky DC, 2008, Radiative thrusters on close-in extrasolar planets. ApJ, 677, L117–L120 {133}

Fabrycky DC, Johnson ET, Goodman J, 2007, Cassini states with dissipation: why obliquity tides cannot inflate hot Jupiters. ApJ, 665, 754–766 {145}

Fabrycky DC, Murray-Clay RA, 2010, Stability of the directly imaged multiplanet system HR 8799: resonance and masses. ApJ, 710, 1408–1421 {223, 244}

Fabrycky DC, Tremaine S, 2007, Shrinking binary and planetary orbits by Kozai cycles with tidal friction. ApJ, 669, 1298–1315 {60, 127}

Fabrycky DC, Winn JN, 2009, Exoplanetary spin-orbit alignment: results from the ensemble of Rossiter–McLaughlin observations. ArXiv e-prints {127, 128}

Faedi F, West RG, Burleigh MR, et al., 2011, Detection limits for close eclipsing and transiting substellar and planetary companions to white dwarfs in the WASP survey. MNRAS, 410, 899–911 {103}

Fairbridge RW, Shirley JH, 1987, Prolonged minima and the 179-yr cycle of the solar inertial motion. Sol. Phys., 110, 191–210 {66}

Falkowski PG, 2005, Tracing oxygen's imprint on Earth's metabolic evolution. Science, 311, 1724–1725 {306}

Falkowski PG, Katz ME, Milligan AJ, et al., 2005, The rise of oxygen over the past 205 Myr and the evolution of large placental mammals. Science, 309, 2202–2204 {307}

Fang N, Lee H, Sun C, et al., 2005, Sub-diffraction-limited optical imaging with a silver superlens. Science, 308, 534–537 {169}

Fares R, Donati J, Moutou C, et al., 2009, Magnetic cycles of the planet-hosting star τ Boo. II. A second magnetic polarity reversal. MNRAS, 398, 1383–1391 {206}

Farihi J, Jura M, Zuckerman B, 2009, Infrared signatures of disrupted minor planets at white dwarfs. ApJ, 694, 805–819 {173}

Farquhar J, Bao H, Thiemens M, 2000, Atmospheric influence of Earth's earliest sulphur cycle. Science, 289, 756–759 {306}

Farquhar J, Savarino J, Airieau S, et al., 2001, Observation of wavelength-sensitive mass-independent sulphur isotope effects during SO_2 photolysis: implications for the early atmosphere. J. Geophys. Res., 106, 32829–32840 {306}

Farrell WM, Lazio TJW, Desch MD, et al., 2004a, Radio emission from extrasolar planets. Bioastronomy 2002: Life Among the Stars, volume 213 of IAU Symposium, 73–76 {175}

Farrell WM, Lazio TJW, Zarka P, et al., 2004b, The radio search for extrasolar planets with LOFAR. Planet. Space Sci., 52, 1469–1478 {174, 175}

Fatuzzo M, Adams FC, Gauvin R, et al., 2006, A statistical stability analysis of Earth-like planetary orbits in binary systems. PASP, 118, 1510–1527 {57, 285}

Favata F, 2004, The Eddington baseline mission. Stellar Structure and Habitable Planet Finding, volume 538 of ESA Special Publication, 3–11 {114}

Feast MW, Whitelock PA, 1997, Galactic kinematics of Cepheids from Hipparcos proper motions. MNRAS, 291, 683–693 {310}

Fegley B, 2000, Kinetics of gas-grain reactions in the solar nebula. Space Sci. Rev., 92, 177–200 {258, 259}

Fegley B, Lewis JS, 1980, Volatile element chemistry in the solar nebula: Na, K, F, Cl, Br, and P. Icarus, 41, 439–455 {257}

Fegley B, Lodders K, 1994, Chemical models of the deep atmospheres of Jupiter and Saturn. Icarus, 110, 117–154 {258, 271, 274–276}

—, 1996, Atmospheric chemistry of the brown dwarf GJ 229 B: thermochemical equilibrium predictions. ApJ, 472, L37–41 {275}

Fegley B, Palme H, 1985, Evidence for oxidising conditions in the solar nebula from Mo and W depletions in refractory inclusions in carbonaceous chondrites. Earth and Planetary Science Letters, 72, 311–326 {257}

Fegley B, Prinn RG, Hartman H, et al., 1986, Chemical effects of large impacts on the Earth's primitive atmosphere. Nature, 319, 305–308 {305}

Fei Y, Mao H, Hemley RJ, 1993, Thermal expansivity, bulk modulus, and melting curve of H_2O-ice VII to 20 GPa. J. Chem. Phys., 99, 5369–5373 {263}

Feigelson ED, Montmerle T, 1999, High-energy processes in young stellar objects. ARA&A, 37, 363–408 {297}

Fellgett P, 1955, A proposal for a radial velocity photometer. Optica Acta, 2, 9–16 {17}

Ferguson JW, Alexander DR, Allard F, et al., 2005, Low-temperature opacities. ApJ, 623, 585–596 {264}

Fernandez JA, Ip W, 1984, Some dynamical aspects of the accretion of Uranus and Neptune: the exchange of orbital angular momentum with planetesimals. Icarus, 58, 109–120 {302}

Ferrari A, Soummer R, Aime C, 2007, An introduction to stellar coronagraphy. Comptes Rendus Physique, 8, 277–287 {153}

Ferraz-Mello S, Beaugé C, Michtchenko TA, 2003, Evolution of migrating planet pairs in resonance. Celestial Mechanics and Dynamical Astronomy, 87, 99–112 {242}

Ferraz-Mello S, Michtchenko TA, Beaugé C, 2005, The orbits of the extrasolar planets HD 82943c and b. ApJ, 621, 473–481 {43, 52}

Ferraz-Mello S, Rodríguez A, Hussmann H, 2008, Tidal friction in close-in satellites and exoplanets: the Darwin theory revisited. Celestial Mechanics and Dynamical Astronomy, 101, 171–201 {246}

Ferris GAJ, 1969, Planetary influences on sunspots. Journal of the British Astronomical Association, 79, 385–388 {66}

Festou MC, Keller HU, Weaver HA, 2004, Comets II. University of Arizona Press {293}

Fields BD, Olive KA, 1999, The evolution of ^6Li in standard cosmic-ray nucleosynthesis. New Astronomy, 4, 255–263 {198}

Fields DL, Albrow MD, An J, et al., 2003, High-precision limb-darkening measurement of a K3 giant using microlensing. ApJ, 596, 1305–1319 {91}

Fienga A, Laskar J, Morley T, et al., 2009, INPOP08, a 4-d planetary ephemeris: from asteroid and time-scale computations to ESA Mars Express and Venus Express contributions. A&A, 507, 1675–1686 {18, 309}

Fienga A, Manche H, Laskar J, et al., 2008, INPOP06: a new numerical planetary ephemeris. A&A, 477, 315–327 {18, 309}

Fischer DA, Laughlin G, Butler P, et al., 2005, The N2K consortium. I. A hot Saturn planet orbiting HD 88133. ApJ, 620, 481–486 {29, 33, 181}

Fischer DA, Marcy GW, 1992, Multiplicity among M dwarfs. ApJ, 396, 178–194 {207}

Fischer DA, Marcy GW, Butler RP, et al., 2002, A second planet orbiting 47 UMa. ApJ, 564, 1028–1034 {38, 70}

—, 2003, A planetary companion to HD 40979 and additional planets orbiting HD 12661 and HD 38529. ApJ, 586, 1394–1408 {55}

—, 2008, Five planets orbiting 55 Cnc. ApJ, 675, 790–801 {2, 13, 30, 31,

48, 49, 52}

Fischer DA, Valenti J, 2005, The planet-metallicity correlation. *ApJ*, 622, 1102–1117 {**189, 191–193**}

Fischer DA, Vogt SS, Marcy GW, et al., 2007, Five intermediate-period planets from the N2K sample. *ApJ*, 669, 1336–1344 {**33**}

Fitzgerald MP, Kalas PG, Graham JR, 2007, A ring of warm dust in the HD 32297 debris disk. *ApJ*, 670, 557–564 {**224**}

—, 2009, Orbital constraints on the β Pic inner planet candidate with Keck adaptive optics. *ApJ*, 706, L41–L45 {**224**}

Flasar FM, Achterberg RK, Conrath BJ, et al., 2005, Temperatures, winds, and composition in the Saturnian system. *Science*, 307, 1247–1251 {**271**}

Fleck RC, 2008, A magnetic mechanism for halting inward protoplanet migration: I. Necessary conditions and angular momentum transfer timescales. *Ap&SS*, 313, 351–356 {**241**}

Fleming SW, Ge J, Mahadevan S, et al., 2010, Discovery of a low-mass companion to a metal-rich F star with the MARVELS pilot project. *ApJ*, 718, 1186–1199 {**29**}

Fleming SW, Kane SR, McCullough PR, et al., 2008, Detecting temperate Jupiters: the prospects of searching for transiting gas giants in habitable zones. *MNRAS*, 386, 1503–1520 {**104**}

Fluri DM, Berdyugina SV, 2010, Orbital parameters of extrasolar planets derived from polarimetry. *A&A*, 512, A59 {**126**}

Fogg MJ, Nelson RP, 2005, Oligarchic and giant impact growth of terrestrial planets in the presence of gas giant planet migration. *A&A*, 441, 791–806 {**228, 241**}

—, 2007a, The effect of Type I migration on the formation of terrestrial planets in hot-Jupiter systems. *A&A*, 472, 1003–1015 {**241**}

—, 2007b, On the formation of terrestrial planets in hot-Jupiter systems. *A&A*, 461, 1195–1208 {**241**}

—, 2009, Terrestrial planet formation in low-eccentricity warm-Jupiter systems. *A&A*, 498, 575–589 {**241**}

Font AS, McCarthy IG, Johnstone D, et al., 2004, Photoevaporation of circumstellar disks around young stars. *ApJ*, 607, 890–903 {**222**}

Font-Ribera A, Miralda Escudé J, Ribas I, 2009, Protostellar cloud fragmentation and inward migration by disk capture as the origin of massive exoplanets. *ApJ*, 694, 183–191 {**237**}

Foo G, Palacios DM, Swartzlander GA, 2005, Optical vortex coronagraph. *Optics Letters*, 30, 3308–3310 {**153, 155, 156**}

Ford EB, 2004a, Choice of observing schedules for astrometric planet searches. *PASP*, 116, 1083–1092 {**15, 73**}

—, 2004b, Quantifying the uncertainty in the orbits of extrasolar planets with Markov Chain Monte Carlo. *The Search for Other Worlds*, volume 713 of *American Institute of Physics Conference Series*, 27–30 {**14**}

—, 2006a, The effects of multiple companions on the efficiency of Space Interferometry Mission planet searches. *PASP*, 118, 364–384 {**73**}

—, 2006b, Improving the efficiency of Markov Chain Monte Carlo for analyzing the orbits of extrasolar planets. *ApJ*, 642, 505–522 {**14**}

—, 2008, Adaptive scheduling algorithms for planet searches. *AJ*, 135, 1008–1020 {**15**}

Ford EB, Chiang EI, 2007, The formation of ice giants in a packed oligarchy: instability and aftermath. *ApJ*, 661, 602–615 {**228**}

Ford EB, Gaudi BS, 2006, Observational constraints on Trojans of transiting extrasolar planets. *ApJ*, 652, L137–L140 {**136**}

Ford EB, Gregory PC, 2007, Bayesian model selection and extrasolar planet detection. *Statistical Challenges in Modern Astronomy IV*, volume 371 of *ASP Conf Ser*, 189–193 {**67**}

Ford EB, Havlickova M, Rasio FA, 2001a, Dynamical instabilities in extrasolar planetary systems containing two giant planets. *Icarus*, 150, 303–313 {**127**}

Ford EB, Holman MJ, 2007, Using transit timing observations to search for Trojans of transiting extrasolar planets. *ApJ*, 664, L51–L54 {**135, 136**}

Ford EB, Joshi KJ, Rasio FA, et al., 2000a, Theoretical implications of the PSR B1620–26 triple system and its planet. *ApJ*, 528, 336–350 {**77**}

Ford EB, Kozinsky B, Rasio FA, 2000b, Secular evolution of hierarchical triple star systems. *ApJ*, 535, 385–401 {**60**}

Ford EB, Lystad V, Rasio FA, 2005, Planet-planet scattering in the υ And system. *Nature*, 434, 873–876 {**46, 47, 50, 243**}

Ford EB, Quinn SN, Veras D, 2008a, Characterizing the orbital eccentricities of transiting extrasolar planets with photometric observations. *ApJ*, 678, 1407–1418 {**121**}

Ford EB, Rasio FA, 2006, On the relation between hot Jupiters and the Roche limit. *ApJ*, 638, L45–L48 {**242**}

—, 2008, Origins of eccentric extrasolar planets: testing the planet–planet scattering model. *ApJ*, 686, 621–636 {**82, 243**}

Ford EB, Rasio FA, Sills A, 1999, Structure and evolution of nearby stars with planets. I. Short-period systems. *ApJ*, 514, 411–429 {**192, 237**}

Ford EB, Rasio FA, Yu K, 2003, Dynamical instabilities in extrasolar planetary systems. *Scientific Frontiers in Research on Extrasolar Planets*, volume 294 of *ASP Conf Ser*, 181–188 {**243**}

Ford EB, Seager S, Turner EL, 2001b, Characterisation of extrasolar terrestrial planets from diurnal photometric variability. *Nature*, 412, 885–887 {**131, 289**}

Ford EB, Tremaine S, 2003, Planet-finding prospects for the Space Interferometry Mission. *PASP*, 115, 1171–1186 {**73**}

Ford HC, Bartko F, Bely PY, et al., 1998, Advanced camera for the Hubble Space Telescope. *SPIE Conf Ser*, volume 3356, 234–248 {**163**}

Ford HC, Bhatti W, Hebb L, et al., 2008b, Detecting transits in sparsely sampled surveys. *American Institute of Physics Conference Series*, volume 1082, 275–281 {**105**}

Ford HC, Petro LD, Burrows C, et al., 2002, Artemis: a stratospheric planet finder. *Advances in Space Research*, 30, 1283–1288 {**162**}

Forestini M, 1994, Low-mass stars: pre-main sequence evolution and nucleosynthesis. *A&A*, 285, 473–488 {**200**}

Fortier A, Benvenuto OG, Brunini A, 2007, Oligarchic planetesimal accretion and giant planet formation. *A&A*, 473, 311–322 {**228**}

—, 2009, Oligarchic planetesimal accretion and giant planet formation. II. *A&A*, 500, 1249–1252 {**228**}

Fortney JJ, 2005, The effect of condensates on the characterisation of transiting planet atmospheres with transmission spectroscopy. *MNRAS*, 364, 649–653 {**138**}

—, 2007, The structure of Jupiter, Saturn, and exoplanets: key questions for high-pressure experiments. *Ap&SS*, 307, 279–283 {**262**}

Fortney JJ, Hubbard WB, 2003, Phase separation in giant planets: inhomogeneous evolution of Saturn. *Icarus*, 164, 228–243 {**263**}

—, 2004, Effects of helium phase separation on the evolution of extrasolar giant planets. *ApJ*, 608, 1039–1049 {**263, 264**}

Fortney JJ, Lodders K, Marley MS, et al., 2008a, A unified theory for the atmospheres of the hot and very hot Jupiters: two classes of irradiated atmospheres. *ApJ*, 678, 1419–1435 {**142, 277**}

Fortney JJ, Marley MS, 2007, Analysis of Spitzer spectra of irradiated planets: evidence for water vapor? *ApJ*, 666, L45–L48 {**272**}

Fortney JJ, Marley MS, Barnes JW, 2007, Planetary radii across five orders of magnitude in mass and stellar insolation: application to transits. *ApJ*, 659, 1661–1672 {**144, 260, 261, 263–266**}

Fortney JJ, Marley MS, Lodders K, et al., 2005, Comparative planetary atmospheres: models of TrES–1 and HD 209458 b. *ApJ*, 627, L69–L72 {**107, 138, 272, 276**}

Fortney JJ, Marley MS, Saumon D, et al., 2008b, Synthetic spectra and colours of young giant planet atmospheres: effects of initial conditions and atmospheric metallicity. *ApJ*, 683, 1104–1116 {**271–274**}

Fortney JJ, Saumon D, Marley MS, et al., 2006, Atmosphere, interior, and evolution of the metal-rich transiting planet HD 149026 b. *ApJ*, 642, 495–504 {**145, 263, 264, 267, 272, 276**}

Forveille T, Bonfils X, Delfosse X, et al., 2009, The HARPS search for southern extrasolar planets. XIV. Gl 176 b, a super-Earth rather than a Neptune, and at a different period. *A&A*, 493, 645–650 {**30**}

Fossati L, Haswell CA, Froning CS, et al., 2010, Metals in the exosphere of the highly irradiated planet WASP–12 b. *ApJ*, 714, L222–L227 {**132, 141, 176**}

Fossey SJ, Waldmann IP, Kipping DM, 2009, Detection of a transit by the planetary companion of HD 80606. *MNRAS*, 396, L16–L20 {**109**}

Foster RS, Fischer J, 1996, Search for protoplanetary and debris disks around millisecond pulsars. *ApJ*, 460, 902–905 {**77**}

Fragner MM, Nelson RP, 2009, Giant planet formation in stellar clusters: the effects of stellar fly-bys. *A&A*, 505, 873–889 {**109**}

Frakes LA, Francis JE, Syktus JI, 1992, *Climate Modes of the Phanerozoic*. Cambridge Monographs on Physics {**307**}

Franck S, von Bloh W, Bounama C, et al., 2001, Limits of photosynthesis in extrasolar planetary systems for Earth-like planets. *Advances in Space Research*, 28, 695–700 {**289**}

Frandsen S, Douglas NG, Butcher HR, 1993, An astronomical seismometer. *A&A*, 279, 310–321 {**25**}

Frank J, King A, Raine DJ, 2002, *Accretion Power in Astrophysics*. Cambridge University Press, Third Edition {220}

Frank MR, Fei Y, Hu J, 2004, Constraining the equation of state of fluid H_2O to 80 GPa using the melting curve, bulk modulus, and thermal expansivity of ice VII. *Geochim. Cosmochim. Acta*, 68, 2781–2790 {263}

Freed M, Close LM, McCarthy DW, 2003, MEDI: an instrument for direct-detection of massive extrasolar planets. *SPIE Conf Ser*, volume 4839, 1132–1141 {158}

Freedman RS, Marley MS, Lodders K, 2008, Line and mean opacities for ultracool dwarfs and extrasolar planets. *ApJS*, 174, 504–513 {264, 265, 272}

Fregeau JM, Chatterjee S, Rasio FA, 2006, Dynamical interactions of planetary systems in dense stellar environments. *ApJ*, 640, 1086–1098 {78, 109}

Freistetter F, Krivov AV, Löhne T, 2007, Planets of β Pic revisited. *A&A*, 466, 389–393 {224}

French M, Mattsson TR, Nettelmann N, et al., 2009, Equation of state and phase diagram of water at ultrahigh pressures as in planetary interiors. *Phys. Rev. B*, 79(5), 054107 {263}

Fressin F, Guillot T, Bouchy F, et al., 2005, Antarctica search for transiting extrasolar planets. *EAS Publications Series*, volume 14, 309–312 {110}

Fressin F, Guillot T, Morello V, et al., 2007a, Interpreting and predicting the yield of transit surveys: giant planets in the OGLE fields. *A&A*, 475, 729–746 {106}

Fressin F, Guillot T, Nesta L, 2009, Interpreting the yield of transit surveys: are there groups in the known transiting planets population? *A&A*, 504, 605–615 {146, 147}

Fressin F, Guillot T, Schmider FX, et al., 2007b, ASTEP: towards a large photometric survey for exoplanets at Dome C. *EAS Publications Series*, volume 25, 225–232 {110}

Fridlund CVM, 2000, Darwin: the infrared space interferometer. *Darwin and Astronomy: the Infrared Space Interferometer*, volume 451 of *ESA Special Publication*, 11–18 {165}

—, 2004, The Darwin mission. *Advances in Space Research*, 34, 613–617 {165}

Fridlund CVM, Eiroa C, Henning T, et al., 2010a, A roadmap for the detection and characterisation of other Earths. *Astrobiology*, 10, 113–119 {283}

Fridlund CVM, Hébrard G, Alonso R, et al., 2010b, Transiting exoplanets from the CoRoT space mission. IX. CoRoT-6 b: a transiting hot Jupiter planet in an 8.9 d orbit around a low-metallicity star. *A&A*, 512, A14 {326}

Fridlund CVM, Henning T, Lacoste H, 2003, *Towards other Earths: DARWIN/TPF and the search for extrasolar terrestrial planets*, volume 539 of *ESA Special Publication*. ESA {165}

Fried DL, 1965, Statistics of a geometric representation of wavefront distortion. *Journal of the Optical Society of America*, 55, 1427–1431 {151}

—, 1966, Optical resolution through a randomly inhomogeneous medium for very long and very short exposures. *Journal of the Optical Society of America*, 56, 1372–1379 {151}

Friedrich S, Zinnecker H, Brandner W, et al., 2005, A NICMOS direct imaging search for giant planets around the single white dwarfs in the Hyades. *14th European Workshop on White Dwarfs*, volume 334 of *ASP Conf Ser*, 431–434 {173}

Friel E, Cayrel de Strobel G, Chmielewski Y, et al., 1993, In search of real solar twins. III. *A&A*, 274, 825–837 {199}

Frink S, Mitchell DS, Quirrenbach A, et al., 2002, Discovery of a substellar companion to the K2 III giant ι Dra. *ApJ*, 576, 478–484 {32, 70}

Froeschlé C, 1984, The Lyapunov characteristic exponents and applications. *Journal de Mécanique Théorique et Appliquée Supplement*, 101–132 {45}

Froeschlé C, Lega E, Gonczi R, 1997, Fast Lyapunov indicators: application to asteroidal motion. *Celestial Mechanics and Dynamical Astronomy*, 67, 41–62 {46}

Fromang S, Papaloizou J, 2006, Dust settling in local simulations of turbulent protoplanetary disks. *A&A*, 452, 751–762 {225}

Fuhrmann K, 1998, Nearby stars of the Galactic disk and halo. *A&A*, 338, 161–183 {190}

—, 2004, Nearby stars of the Galactic disk and halo. III. *Astronomische Nachrichten*, 325, 3–80 {193}

Fuhrmann K, Pfeiffer MJ, Bernkopf J, 1997, Solar-type stars with planetary companions: 51 Peg and 47 UMa. *A&A*, 326, 1081–1089 {187, 188}

—, 1998, F- and G-type stars with planetary companions: υ And, ρ^1 Cnc, τ Boo, 16 Cyg and ρ CrB. *A&A*, 336, 942–952 {185, 187, 188}

Fukagawa M, Itoh Y, Tamura M, et al., 2009, H-band image of a planetary companion around HR 8799 in 2002. *ApJ*, 696, L1–L5 {223}

Fukui A, Abe F, Bond IA, et al., 2009, Transiting exoplanets search for MOA-I data. *IAU Symposium*, volume 253, 366–369 {106}

Funk B, Pilat-Lohinger E, Dvorak R, et al., 2004, Resonances in multiple planetary systems. *Celestial Mechanics and Dynamical Astronomy*, 90, 43–50 {44}

Fusco T, Blanc A, Nicolle M, et al., 2006a, Sky coverage estimation for multiconjugate adaptive optics systems: strategies and results. *MNRAS*, 370, 174–184 {152}

Fusco T, Petit C, Rousset G, et al., 2006b, Design of the extreme adaptive optics system for SPHERE, the planet finder instrument of the VLT. *SPIE Conf Ser*, volume 6272, 17 {159}

Fusco T, Verinaud C, Rousset G, et al., 2006c, Extreme adaptive optics for extrasolar planet detection with ELTs: application to OWL. *The Scientific Requirements for Extremely Large Telescopes*, volume 232 of *IAU Symposium*, 376–380 {159}

Gahm GF, Grenman T, Fredriksson S, et al., 2007, Globulettes as seeds of brown dwarfs and free-floating planetary-mass objects. *AJ*, 133, 1795–1809 {215}

Gaidos E, Williams DM, 2004, Seasonality on terrestrial extrasolar planets: inferring obliquity and surface conditions from infrared light curves. *New Astronomy*, 10, 67–77 {131}

Gail H, 2002, Radial mixing in protoplanetary accretion disks. III. Carbon dust oxidation and abundance of hydrocarbons in comets. *A&A*, 390, 253–265 {257, 258}

Gail H, Hoppe P, 2010, The origins of protoplanetary dust and the formation of accretion disks. *Protoplanetary Dust: Astrophysical and Cosmochemical Perspectives*, 27–65, Cambridge University Press {217, 219, 225}

Gail HP, 2004, Radial mixing in protoplanetary accretion disks. IV. Metamorphosis of the silicate dust complex. *A&A*, 413, 571–591 {257, 258}

Galeev AI, Bikmaev IF, Musaev FA, et al., 2004, Chemical composition of 15 photometric analogues of the Sun. *Astronomy Reports*, 48, 492–510 {286}

Galicher R, Baudoz P, Rousset G, 2008, Wavefront error correction and Earth-like planet detection by a self-coherent camera in space. *A&A*, 488, L9–L12 {157}

Galland F, Lagrange AM, Udry S, et al., 2005a, Extrasolar planets and brown dwarfs around A-F type stars. I. Performances of radial velocity measurements, first analyses of variations. *A&A*, 443, 337–345 {32, 34}

—, 2005b, Extrasolar planets and brown dwarfs around A-F type stars. II. A planet found with ELODIE around the F6V star HD 33564. *A&A*, 444, L21–L24 {32}

—, 2006a, Extrasolar planets and brown dwarfs around A-F type stars. III. β Pic: looking for planets, finding pulsations. *A&A*, 447, 355–359 {224}

—, 2006b, Extrasolar planets and brown dwarfs around A-F type stars. IV. A candidate brown dwarf around the A9V pulsating star HD 180777. *A&A*, 452, 709–714 {32}

Gammie CF, 1996, Layered accretion in T Tauri disks. *ApJ*, 457, 355–362 {221, 222}

—, 2001, Nonlinear outcome of gravitational instability in cooling, gaseous disks. *ApJ*, 553, 174–183 {236}

Gandolfi D, Hébrard G, Alonso R, et al., 2010, Transiting exoplanets from the CoRoT space mission. XIV. CoRoT-11 b: a transiting massive hot-Jupiter in a prograde orbit around a rapidly rotating F-type star. *A&A*, 524, A55 {129, 249, 326}

Gänsicke BT, Marsh TR, Southworth J, et al., 2006, A gaseous metal disk around a white dwarf. *Science*, 314, 1908–1910 {173}

Garaud P, Lin DNC, 2004, On the evolution and stability of a protoplanetary disk dust layer. *ApJ*, 608, 1050–1075 {226}

—, 2007, The effect of internal dissipation and surface irradiation on the structure of disks and the location of the snow line around Sun-like stars. *ApJ*, 654, 606–624 {260, 299}

Garcia Lopez RJ, Perez de Taoro MR, 1998, Beryllium abundances in parent stars of extrasolar planets: 16 Cyg A and B and 55 Cnc. *A&A*, 334, 599–605 {201}

References

Garcia-Melendo E, McCullough PR, 2009, Photometric detection of a transit of HD 80606 b. *ApJ*, 698, 558–561 {**109**}

García-Sánchez J, Preston RA, Jones DL, et al., 1999, Stellar encounters with the Oort Cloud based on Hipparcos data. *AJ*, 117, 1042–1055 {**307**}

García-Sánchez J, Weissman PR, Preston RA, et al., 2001, Stellar encounters with the solar system. *A&A*, 379, 634–659 {**307**}

Garrison R, 2000, Classification of stellar spectra. *Encyclopedia of Astronomy and Astrophysics* {**211**}

Gary DE, Linsky JL, 1981, First detection of nonflare microwave emissions from the coronae of single late-type dwarf stars. *ApJ*, 250, 284–292 {**173**}

Gatewood G, 1987, The multichannel astrometric photometer and atmospheric limitations in the measurement of relative positions. *AJ*, 94, 213–224 {**62, 64**}

—, 1996, Lalande 21185. *Bulletin of the American Astronomical Society*, volume 28, 885 {**64**}

Gatewood G, Eichhorn H, 1973, An unsuccessful search for a planetary companion of Barnard's star. *AJ*, 78, 769–776 {**64**}

Gatewood G, Han I, Black DC, 2001, A combined Hipparcos and multichannel astrometric photometer study of the proposed planetary system of ρ CrB. *ApJ*, 548, L61–L63 {**70**}

Gatewood G, Stein J, de Jonge JK, et al., 1992, Multichannel astrometric photometer and photographic astrometric studies in the regions of Lalande 21185, BD 56 2966, and HR 4784. *AJ*, 104, 1237–1247 {**64**}

Gaudi BS, 1998, Distinguishing between binary-source and planetary microlensing perturbations. *ApJ*, 506, 533–539 {**88**}

—, 2002, Interpreting the M22 spike events. *ApJ*, 566, 452–462 {**96**}

—, 2005, On the size distribution of close-in extrasolar giant planets. *ApJ*, 628, L73–L76 {**145**}

—, 2008, Microlensing searches for planets: results and future prospects. *ASP Conf Ser*, volume 398, 479–487 {**84**}

Gaudi BS, Albrow MD, An J, et al., 2002, Microlensing constraints on the frequency of Jupiter-mass companions: analysis of 5 years of planet photometry. *ApJ*, 566, 463–499 {**96**}

Gaudi BS, Bennett DP, Udalski A, et al., 2008, Discovery of a Jupiter/Saturn analogue with gravitational microlensing. *Science*, 319, 927–930 {**2, 92, 93, 98, 99, 101**}

Gaudi BS, Chang H, Han C, 2003, Probing structures of distant extrasolar planets with microlensing. *ApJ*, 586, 527–539 {**95**}

Gaudi BS, Gould A, 1997, Planet parameters in microlensing events. *ApJ*, 486, 85–99 {**88, 94**}

Gaudi BS, Han C, 2004, The many possible interpretations of microlensing event OGLE–2002–BLG–055. *ApJ*, 611, 528–536 {**96**}

Gaudi BS, Naber RM, Sackett PD, 1998, Microlensing by multiple planets in high-magnification events. *ApJ*, 502, L33–L37 {**88**}

Gaudi BS, Sackett PD, 2000, Detection efficiencies of microlensing data sets to stellar and planetary companions. *ApJ*, 528, 56–73 {**88, 90, 96**}

Gaudi BS, Winn JN, 2007, Prospects for the characterisation and confirmation of transiting exoplanets via the Rossiter–McLaughlin effect. *ApJ*, 655, 550–563 {**128**}

Gautier D, Hersant F, Mousis O, et al., 2001a, Enrichments in volatiles in Jupiter: a new interpretation of the Galileo measurements. *ApJ*, 550, L227–L230 {**271, 294**}

—, 2001b, Erratum: Enrichments in volatiles in Jupiter: a new interpretation of the Galileo measurements. *ApJ*, 559, L183–L183 {**271**}

Gautier TN, Borucki WJ, Caldwell DA, et al., 2007, The Kepler follow-up observation programme. *Transiting Extrasolar Planets Workshop*, volume 366 of *ASP Conf Ser*, 219–224 {**111**}

Gay J, Rabbia Y, 1996, An interferometric method for coronography. *Academie des Science Paris Comptes Rendus Serie B Sciences Physiques*, 322, 265–271 {**153**}

Gayon J, Bois E, 2008a, Are retrograde resonances possible in multi-planet systems? *A&A*, 482, 665–672 {**47, 54, 55**}

—, 2008b, Retrograde resonances in compact multi-planetary systems: a feasible stabilising mechanism. *IAU Symposium*, volume 249 of *IAU Symposium*, 511–516 {**54**}

Gayon J, Bois E, Scholl H, 2009, Dynamics of planets in retrograde mean motion resonance. *Celestial Mechanics and Dynamical Astronomy*, 103, 267–279 {**54**}

Gayon J, Marzari F, Scholl H, 2008, Stable chaos in the 55 Cnc exoplanetary system? *MNRAS*, 389, L1–L3 {**47, 49**}

Gayon-Markt J, Bois E, 2009, On fitting planetary systems in counter-revolving configurations. *MNRAS*, 399, L137–L140 {**55**}

Ge J, 2002, Fixed delay interferometry for Doppler extrasolar planet detection. *ApJ*, 571, L165–L168, erratum: 2003, ApJ, 593, L147 {**25, 26**}

—, 2007, An all-sky extrasolar planet survey with multiple object, dispersed fixed-delay interferometers. *Revista Mexicana de Astronomia y Astrofisica Conference Series*, volume 28, 31–37 {**27**}

Ge J, Angel JRP, Jacobsen B, et al., 2002a, An optical ultrahigh-resolution cross-dispersed echelle spectrograph with adaptive optics. *PASP*, 114, 879–891 {**17**}

Ge J, Eisenstein D, 2009, MARVELS: revealing the formation and dynamical evolution of giant planet systems. *Astro2010: The Astronomy and Astrophysics Decadal Survey*, volume 2010 of *Astronomy*, 86 {**27**}

Ge J, Erskine DJ, Rushford M, 2002b, An externally dispersed interferometer for sensitive Doppler extrasolar planet searches. *PASP*, 114, 1016–1028 {**25, 26**}

Ge J, Lee B, Mahadevan S, et al., 2009, The SDSS-III Multi-object APO Radial-Velocity Exoplanet Large-area Survey (MARVELS) and its early results. *American Astronomical Society Meeting Abstracts*, volume 213, 336.02 {**24, 27, 29, 32, 33**}

Ge J, van Eyken J, Mahadevan S, et al., 2006, The first extrasolar planet discovered with a new-generation high-throughput Doppler instrument. *ApJ*, 648, 683–695 {**27**}

Ge J, van Eyken JC, Mahadevan S, et al., 2007, An all sky extrasolar planet survey with new generation multiple object Doppler instruments at Sloan telescope. *Revista Mexicana de Astronomia y Astrofisica Conference Series*, volume 29, 30–36 {**27**}

Geballe TR, Knapp GR, Leggett SK, et al., 2002, Toward spectral classification of L and T dwarfs: infrared and optical spectroscopy and analysis. *ApJ*, 564, 466–481 {**210**}

Gehrels N, 2010, The Joint Dark Energy Mission (JDEM) Omega. *ArXiv e-prints* {**102**}

Gehrels T (ed.), 1976, *Jupiter: Studies of the Interior, Atmosphere, Magnetosphere and Satellites* {**293**}

Gehrels T, Matthews MS, 1984, *Saturn*. University of Arizona Press {**293**}

Gelino C, Kirkpatrick JD, Burgasser AJ, 2010, Photometry, spectroscopy, and astrometry of M, L, and T dwarfs. *www.dwarfarchives.org* {**210**}

Genda H, Abe Y, 2003, Survival of a proto-atmosphere through the stage of giant impacts: the mechanical aspects. *Icarus*, 164, 149–162 {**305**}

—, 2005, Enhanced atmospheric loss on protoplanets at the giant impact phase in the presence of oceans. *Nature*, 433, 842–844 {**305**}

George SJ, Stevens IR, 2007, Giant metrewave radio telescope low-frequency observations of extrasolar planetary systems. *MNRAS*, 382, 455–460 {**176, 177**}

Gezari DY, Nisenson P, Papaliolios CD, et al., 2003, ExPO: a discovery-class apodised square aperture exo-planet imaging space telescope concept. *SPIE Conf Ser*, volume 4860, 302–310 {**166**}

Ghezzi L, Cunha K, Smith VV, et al., 2009, Measurements of the isotopic ratio ^{6}Li/^{7}Li in stars with planets. *ApJ*, 698, 451–460 {**201**}

Ghil M, Le Treut H, 1981, A climate model with cryodynamics and geodynamics. *J. Geophys. Res.*, 86, 5262–5270 {**301**}

Ghosh A, Weidenschilling SJ, Amelin Y, et al., 2004, Planet formation and early solar system heating: recent advancements. *AGU Fall Meeting Abstracts*, C5 {**297**}

Gibson NP, Pollacco D, Simpson EK, et al., 2008, Updated parameters for the transiting exoplanet WASP–3 b using RISE, a new fast camera for the Liverpool Telescope. *A&A*, 492, 603–607 {**135**}

—, 2009, A transit timing analysis of nine RISE light curves of the exoplanet system TrES–3. *ApJ*, 700, 1078–1085 {**135**}

Gibson NP, Pollacco DL, Barros S, et al., 2010, A transit timing analysis of seven RISE light curves of the exoplanet system HAT–P–3. *MNRAS*, 401, 1917–1923 {**133, 135**}

Gies DR, Helsel JW, 2005, Ice age epochs and the Sun's path through the Galaxy. *ApJ*, 626, 844–848 {**307**}

Gil-Merino R, Lewis GF, 2005, Interpreting microlensing signal in QSO 2237+0305: stars or planets? *A&A*, 437, L15–L18 {**96**}

Gilles L, Ellerbroek B, Véran J, 2006, Laser guide star multi-conjugate adaptive optics performance of the Thirty Meter Telescope with elongated beacons and matched filtering. *SPIE Conf Ser*, volume 6272, 99 {**152**}

Gillet S, Riaud P, Lardière O, et al., 2003, Imaging capabilities of hypertelescopes with a pair of micro-lens arrays. *A&A*, 400, 393–396 {**168**}

Gilli G, Israelian G, Ecuvillon A, et al., 2006, Abundances of refractory elements in the atmospheres of stars with extrasolar planets. *A&A*, 449, 723–736 {**189, 195, 198**}

Gilliland RL, Brown TM, 1988, Time-resolved CCD photometry of an ensemble of stars. *PASP*, 100, 754–765 {**104**}

Gilliland RL, Brown TM, Guhathakurta P, et al., 2000, A lack of planets in 47 Tuc from an HST search. *ApJ*, 545, L47–L51 {**109**}

Gillon M, Anderson DR, Triaud AHMJ, et al., 2009, Discovery and characterisation of WASP-6 b, an inflated sub-Jupiter mass planet transiting a solar-type star. *A&A*, 501, 785–792 {**129, 327**}

Gillon M, Courbin F, Magain P, et al., 2005, On the potential of extrasolar planet transit surveys. *A&A*, 442, 731–744 {**104**}

Gillon M, Hatzes A, Csizmadia S, et al., 2010, Transiting exoplanets from the CoRoT space mission. XII. CoRoT-12 b: a short-period low-density planet transiting a solar analogue star. *A&A*, 520, A97 {**326**}

Gillon M, Pont F, Demory BO, et al., 2007, Detection of transits of the nearby hot Neptune GJ 436 b. *A&A*, 472, L13–L16 {**144, 326**}

Gilmour JD, Middleton CA, 2009, Anthropic selection of a solar system with a high ^{26}Al/^{27}Al ratio: implications and a possible mechanism. *Icarus*, 201, 821–823 {**197**}

Giménez A, 2000, uvby photometry of stars with planets. *A&A*, 356, 213–217 {**188**}

—, 2006a, Equations for the analysis of the light curves of extrasolar planetary transits. *A&A*, 450, 1231–1237, erratum: 2007 A&A, 474, 1049 {**119**}

—, 2006b, Equations for the analysis of the Rossiter–McLaughlin effect in extrasolar planetary transits. *ApJ*, 650, 408–413 {**128**}

Giménez A, Diaz-Cordovés J, 1993, Improving the light curve synthesis program EBOP: variable position of the periastron and second-order limb darkening. *IAU Commission on Close Binary Stars*, 21, 125–129 {**119**}

Girardi L, Bressan A, Bertelli G, et al., 2000, Evolutionary tracks and isochrones for low- and intermediate-mass stars. *A&AS*, 141, 371–383 {**185, 310**}

Give'on A, Kasdin NJ, Vanderbei RJ, et al., 2005, Amplitude and phase correction for high-contrast imaging using Fourier decomposition. *SPIE Conf Ser*, volume 5905, 368–378 {**157**}

Gizis JE, Kirkpatrick JD, Burgasser AJ, et al., 2001, Substellar companions to main-sequence stars: no brown dwarf desert at wide separations. *ApJ*, 551, L163–L166 {**210, 213**}

Gizis JE, Reid IN, Hawley SL, 2002, The Palomar/MSU nearby star spectroscopic survey. III. Chromospheric activity, M dwarf ages, and the local star formation history. *AJ*, 123, 3356–3369 {**213**}

Gizon L, Solanki SK, 2003, Determining the inclination of the rotation axis of a Sun-like star. *ApJ*, 589, 1009–1019 {**128, 203**}

Gladman B, 1993, Dynamics of systems of two close planets. *Icarus*, 106, 247–263 {**43, 76**}

Gladman B, Duncan M, 1990, On the fates of minor bodies in the outer solar system. *AJ*, 100, 1680–1693 {**44, 300**}

Glasby GP, 2006, Abiogenic origin of hydrocarbons: an historical overview. *Resource Geology*, 56(1), 85–98 {**279**}

Gliese W, 1957, Katalog der Sterne näher also 20 Parsek für 1950.0. *Astron. Rechen-Institut, Heidelberg*, 8, 1–89 {**182**}

—, 1969, Catalogue of Nearby Stars (CNS2). *Veroeffentlichungen des Astronomischen Rechen-Instituts Heidelberg*, 22, 1–117 {**182**}

—, 1982, Detectable perturbations in the proper motions of the nearest stars caused by Jupiter-like companions? *The Scientific Aspects of the Hipparcos Space Astrometry Mission*, volume 177 of *ESA Special Publication*, 193–194 {**69**}

Gliese W, Jahreiß H, 1991, Preliminary Version of the Third Catalogue of Nearby Stars. Technical report, The Astronomical Data Center CD-ROM: Selected Astronomical Catalogs, NASA/Astronomical Data Center {**182**}

Glindemann A, Algomedo J, Amestica R, et al., 2003, The VLTI: a status report. *SPIE Conf Ser*, volume 4838, 89–100 {**162**}

Gnevishev MN, Ohl AI, 1948, About the 22-year cycle of solar activity. *Astronomical Journal*, 25(18-20) {**67**}

Godon P, Livio M, 1999a, On the nonlinear hydrodynamic stability of thin Keplerian disks. *ApJ*, 521, 319–327 {**221**}

—, 1999b, Vortices in protoplanetary disks. *ApJ*, 523, 350–356 {**221**}

Gogarten JP, 1998, Origin and early evolution of life: deciphering the molecular record. *Origins*, volume 148 of *ASP Conf Ser*, 435–448 {**282**}

Goicoechea JR, Swinyard B, 2010, Exoplanetary systems with SAFARI: a far infrared imaging spectrometer for SPICA. *ASP Conf Ser*, volume 430, 448–449 {**114**}

Goicoechea JR, Swinyard B, Tinetti G, et al., 2008, Using SPICA space telescope to characterise exoplanets. *ArXiv e-prints* {**114**}

Gold T, 1979, Terrestrial sources of carbon and earthquake outgassing. *J. Petroleum Geology*, 1(3), 3–19 {**279**}

—, 1985, The origin of natural gas and petroleum, and the prognosis for future supplies. *Ann. Rev. Energy*, 10, 53–77 {**279**}

—, 1993, The origin of methane in the crust of the Earth. *US Geol. Surv. Prof. Paper*, 1570, 57–70 {**279**}

Gold T, Soter S, 1980, The deep-Earth gas hypothesis. *Scientific American*, 242(6), 130–137 {**279**}

Goldberg D, Mazeh T, Latham DW, 2003, On the mass-ratio distribution of spectroscopic binaries. *ApJ*, 591, 397–405 {**207**}

Goldberg DE, 1989, *Genetic Algorithms in Search, Optimisation and Machine Learning*. Addison-Wesley {**14**}

Goldin A, Makarov VV, 2006, Unconstrained astrometric orbits for Hipparcos stars with stochastic solutions. *ApJS*, 166, 341–350 {**70**}

Goldman B, Cushing MC, Marley MS, et al., 2008, CLOUDS search for variability in brown dwarf atmospheres. Infrared spectroscopic time series of L/T transition brown dwarfs. *A&A*, 487, 277–292 {**211**}

Goldman B, Marsat S, Henning T, et al., 2010, A new benchmark T8-9 brown dwarf and a couple of new mid-T dwarfs from the UKIDSS DR5+ LAS. *MNRAS*, 405, 1140–1152 {**211**}

Goldreich P, 1963, On the eccentricity of satellite orbits in the solar system. *MNRAS*, 126, 257–268 {**245**}

—, 1965, An explanation of the frequent occurrence of commensurable mean motions in the solar system. *MNRAS*, 130, 159–181 {**41**}

Goldreich P, Lithwick Y, Sari R, 2004a, Final stages of planet formation. *ApJ*, 614, 497–507 {**229**}

—, 2004b, Planet formation by coagulation: a focus on Uranus and Neptune. *ARA&A*, 42, 549–601 {**228**}

Goldreich P, Nicholson PD, 1977, Turbulent viscosity and Jupiter's tidal Q. *Icarus*, 30, 301–304 {**246, 247**}

Goldreich P, Peale S, 1966, Spin-orbit coupling in the solar system. *AJ*, 71, 425–437 {**246**}

Goldreich P, Sari R, 2003, Eccentricity evolution for planets in gaseous disks. *ApJ*, 585, 1024–1037 {**242**}

Goldreich P, Soter S, 1966, Q in the solar system. *Icarus*, 5, 375–389 {**245–247**}

Goldreich P, Tremaine S, 1979, The excitation of density waves at the Lindblad and corotation resonances by an external potential. *ApJ*, 233, 857–871 {**237, 238**}

—, 1980, Disk-satellite interactions. *ApJ*, 241, 425–441 {**192, 237, 238, 240, 242**}

Goldreich P, Ward WR, 1973, The formation of planetesimals. *ApJ*, 183, 1051–1062 {**225, 226**}

Goldsmith D, 1988, Who will speak for Earth? *IAU Colloq. 99: Bioastronomy – The Next Steps*, 425–428, IAU {**291**}

Golimowski DA, Ardila DR, Krist JE, et al., 2006, Hubble Space Telescope ACS multiband coronagraphic imaging of the debris disk around β Pic. *AJ*, 131, 3109–3130 {**224**}

Golimowski DA, Clampin M, Durrance ST, et al., 1992, High-resolution ground-based coronagraphy using image-motion compensation. *Appl. Opt.*, 31, 4405–4416 {**151, 152**}

Gomes RS, Levison HF, Tsiganis K, et al., 2005, Origin of the cataclysmic Late Heavy Bombardment period of the terrestrial planets. *Nature*, 435, 466–469 {**304**}

Gomes RS, Matese JJ, Lissauer JJ, 2006, A distant planetary-mass solar companion may have produced distant detached objects. *Icarus*, 184, 589–601 {**301**}

Gomes RS, Morbidelli A, Levison HF, 2004, Planetary migration in a planetesimal disk: why did Neptune stop at 30 AU? *Icarus*, 170, 492–507 {**243, 304**}

Goncharov AV, Dainty JC, Esposito S, et al., 2005, Laboratory MCAO testbed for developing wavefront sensing concepts. *Optics Express*, 13, 5580–5590 {**152**}

Goncharov AV, Owner-Petersen M, Puryayev DT, 2002, Intrinsic apodisation effect in a compact two-mirror system with a spherical primary mirror. *Optical Engineering*, 41, 3111–3118 {**155**}

Gondoin P, den Hartog R, Fridlund M, et al., 2008, GENIE: a Ground-

References

Based European Nulling Instrument at ESO VLTI. *The Power of Optical/IR Interferometry: Recent Scientific Results and Second Generation*, 445–458 {**163, 166**}

Gong X, Wang L, Cui X, et al., 2010, Dome A site testing and future plans. *EAS Publications Series*, volume 40 of *EAS Publications Series*, 65–72 {**161**}

Gonzalez G, 1997, The stellar metallicity-giant planet connection. *MNRAS*, 285, 403–412 {**37, 188, 241**}

—, 1998, Spectroscopic analyses of the parent stars of extrasolar planetary system candidates. *A&A*, 334, 221–238 {**188, 192, 199**}

—, 1999a, Are stars with planets anomalous? *MNRAS*, 308, 447–458 {**286**}

—, 1999b, Is the Sun anomalous? *Astronomy and Geophysics*, 40, 25–29 {**285**}

—, 2003, Colloquium: Stars, planets, and metals. *Reviews of Modern Physics*, 75, 101–120 {**191**}

—, 2006a, The chemical compositions of stars with planets: a review. *PASP*, 118, 1494–1505 {**185, 186, 191**}

—, 2006b, Condensation temperature trends among stars with planets. *MNRAS*, 367, L37–L41 {**193, 195, 196**}

—, 2008, Parent stars of extrasolar planets. IX. Lithium abundances. *MNRAS*, 386, 928–934 {**199, 200**}

Gonzalez G, Brownlee D, Ward P, 2001a, The Galactic habitable zone: Galactic chemical evolution. *Icarus*, 152, 185–200 {**196**}

Gonzalez G, Laws C, 2000, Parent stars of extrasolar planets. V. HD 75289. *AJ*, 119, 390–396 {**188, 196, 199**}

—, 2007, Parent stars of extrasolar planets. VIII. Chemical abundances for 18 elements in 31 stars. *MNRAS*, 378, 1141–1152 {**195, 198, 199**}

Gonzalez G, Laws C, Tyagi S, et al., 2001b, Parent stars of extrasolar planets. VI. Abundance analyses of 20 new systems. *AJ*, 121, 432–452 {**188, 192, 195, 196, 198**}

Gonzalez G, Vanture AD, 1998, Parent stars of extrasolar planets. III. ρ^1 Cnc revisited. *A&A*, 339, L29–L32 {**188**}

Gonzalez G, Wallerstein G, Saar SH, 1999, Parent stars of extrasolar planets. IV. 14 Her, HD 187123, and HD 210277. *ApJ*, 511, L111–L114 {**188**}

Goodman J, Lackner C, 2009, Dynamical tides in rotating planets and stars. *ApJ*, 696, 2054–2067 {**246**}

Goodman J, Pindor B, 2000, Secular instability and planetesimal formation in the dust layer. *Icarus*, 148, 537–549 {**226**}

Goodman J, Rafikov RR, 2001, Planetary torques as the viscosity of protoplanetary disks. *ApJ*, 552, 793–802 {**200**}

Goodwin SP, Whitworth AP, 2007, Brown dwarf formation by binary disruption. *A&A*, 466, 943–948 {**214**}

Gorkavyi N, Heap S, Ozernoy L, et al., 2004, Indicator of exoplanet(s) in the circumstellar disk around β Pic. *Planetary Systems in the Universe*, volume 202 of *IAU Symposium*, 331–334 {**224**}

Gorlova N, Rieke GH, Muzerolle J, et al., 2006, Spitzer 24 μm survey of debris disks in the Pleiades. *ApJ*, 649, 1028–1042 {**224**}

Gott JR, 1981, Are heavy halos made of low mass stars: a gravitational lens test. *ApJ*, 243, 140–146 {**86**}

Gouda N, Kobayashi Y, Yamada Y, et al., 2008, Infrared space astrometry project JASMINE. *IAU Symposium*, volume 248 of *IAU Symposium*, 248–251 {**73**}

Gough DO, Thompson MJ, 1990, The effect of rotation and a buried magnetic field on stellar oscillations. *MNRAS*, 242, 25–55 {**202, 203**}

Gough DO, Toomre J, 1991, Seismic observations of the solar interior. *ARA&A*, 29, 627–685 {**201**}

Gould A, 1992, Extending the MACHO search to about 10^6 solar masses. *ApJ*, 392, 442–451 {**93**}

—, 1996, Microlensing and the stellar mass function. *PASP*, 108, 465–476 {**86**}

—, 1997, Extreme microlensing toward the Galactic bulge. *ApJ*, 480, 188–195 {**94**}

—, 2005, Microlensing search for planets. *New Astronomy Reviews*, 49, 424–429 {**84**}

—, 2008, Hexadecapole approximation in planetary microlensing. *ApJ*, 681, 1593–1598 {**90**}

—, 2009, Recent developments in gravitational microlensing. *ASP Conf Ser*, volume 403, 86–109 {**92**}

Gould A, An JH, 2002, Resolving microlens blends using image subtraction. *ApJ*, 565, 1381–1385 {**92**}

Gould A, Dong S, Gaudi BS, et al., 2010, Frequency of solar-like systems and of ice and gas giants beyond the snow line from high-magnification microlensing events in 2005–2008. *ApJ*, 720, 1073–1089 {**100**}

Gould A, Dorsher S, Gaudi BS, et al., 2006a, Frequency of hot Jupiters and very hot Jupiters from the OGLE–III transit surveys toward the Galactic bulge and Carina. *Acta Astronomica*, 56, 1–50 {**106**}

Gould A, Ford EB, Fischer DA, 2003a, Early-type stars: most favorable targets for astrometrically detectable planets in the habitable zone. *ApJ*, 591, L155–L158 {**62**}

Gould A, Gaudi BS, Bennett DP, 2007, Ground-based microlensing surveys. *ArXiv e-prints* {**102**}

Gould A, Gaudi BS, Han C, 2003b, Resolving the microlens mass degeneracy for Earth-mass planets. *ApJ*, 591, L53–L56 {**93**}

Gould A, Kilic M, 2008, Finding planets around white dwarf remnants of massive stars. *ApJ*, 673, L75–L78 {**172**}

Gould A, Loeb A, 1992, Discovering planetary systems through gravitational microlenses. *ApJ*, 396, 104–114 {**88–90, 97**}

Gould A, Pepper J, DePoy DL, 2003c, Sensitivity of transit searches to habitable-zone planets. *ApJ*, 594, 533–537 {**115**}

Gould A, Udalski A, An D, et al., 2006b, Microlens OGLE–2005–BLG–169 implies that cool Neptune-like planets are common. *ApJ*, 644, L37–L40 {**90, 98, 99**}

Gould A, Udalski A, Monard B, et al., 2009, The extreme microlensing event OGLE–2007–BLG–224: terrestrial parallax observation of a thick-disk brown dwarf. *ApJ*, 698, L147–L151 {**94**}

Goullioud R, Catanzarite JH, Dekens FG, et al., 2008, Overview of the SIM PlanetQuest Light mission concept. *SPIE Conf Ser*, volume 7013, 151 {**72**}

Gounelle M, Meibom A, 2008, The origin of short-lived radionuclides and the astrophysical environment of solar system formation. *ApJ*, 680, 781–792 {**293**}

Goździewski K, 2002, Stability of the 47 UMa planetary system. *A&A*, 393, 997–1013 {**285**}

—, 2003a, A dynamical analysis of the HD 37124 planetary system. *A&A*, 398, 315–325 {**47**}

—, 2003b, Stability of the HD 12661 planetary system. *A&A*, 398, 1151–1161 {**47, 52**}

Goździewski K, Bois E, Maciejewski AJ, et al., 2001, Global dynamics of planetary systems with the MEGNO criterion. *A&A*, 378, 569–586 {**46, 47**}

Goździewski K, Breiter S, Borczyk W, 2008a, The long-term stability of extrasolar system HD 37124: numerical study of resonance effects. *MNRAS*, 383, 989–999 {**47**}

Goździewski K, Konacki M, 2004, Dynamical properties of the multi-planet system around HD 169830. *ApJ*, 610, 1093–1106 {**43, 47, 55**}

—, 2006, Trojan pairs in the HD 128311 and HD 82943 planetary systems? *ApJ*, 647, 573–586 {**50, 54, 55**}

Goździewski K, Konacki M, Maciejewski AJ, 2003, Where is the second planet in the HD 160691 planetary system? *ApJ*, 594, 1019–1032 {**47, 48**}

—, 2005a, Orbital solutions to the HD 160691 (μ Ara) Doppler signal. *ApJ*, 622, 1136–1148 {**47, 48**}

—, 2006, Orbital configurations and dynamical stability of multiplanet systems around Sun-like stars HD 202206, 14 Her, HD 37124, and HD 108874. *ApJ*, 645, 688–703 {**47, 52**}

Goździewski K, Konacki M, Wolszczan A, 2005b, Long-term stability and dynamical environment of the PSR B1257+12 planetary system. *ApJ*, 619, 1084–1097 {**76, 77**}

Goździewski K, Maciejewski AJ, 2001, Dynamical analysis of the orbital parameters of the HD 82943 planetary system. *ApJ*, 563, L81–L85 {**52**}

—, 2003, The Janus head of the HD 12661 planetary system. *ApJ*, 586, L153–L156 {**14, 47, 55**}

Goździewski K, Maciejewski AJ, Migaszewski C, 2007, On the extrasolar multiplanet system around HD 160691. *ApJ*, 657, 546–558 {**48**}

Goździewski K, Migaszewski C, 2009, Is the HR 8799 extrasolar system destined for planetary scattering? *MNRAS*, 397, L16–L20 {**223**}

Goździewski K, Migaszewski C, Konacki M, 2008b, A dynamical analysis of the 14 Her planetary system. *MNRAS*, 385, 957–966 {**52**}

Grady CA, Proffitt CR, Malumuth E, et al., 2003, Coronagraphic imaging with HST–STIS. *PASP*, 115, 1036–1049 {**163**}

Grady CA, Sitko ML, Bjorkman KS, et al., 1997, The star-grazing extrasolar comets in the HD 100546 system. *ApJ*, 483, 449–456 {**224**}

Grady CA, Woodgate B, Heap SR, et al., 2005, Resolving the inner cavity of the HD 100546 disk: a candidate young planetary system? *ApJ*, 620, 470–480 {**224**}

Graff DS, Gaudi BS, 2000, Direct detection of large close-in planets around the source stars of caustic-crossing microlensing events. *ApJ*, 538, L133–L136 {**88, 95**}

Graham JB, Dudley R, Aguilar N, et al., 1995, Implications of the late Paleozoic oxygen pulse for physiology and evolution. *Nature*, 375, 117–120 {**307**}

Graham JR, Matthews K, Neugebauer G, et al., 1990, The infrared excess of G29–38: brown dwarf or dust? *ApJ*, 357, 216–223 {**210**}

Grasset O, Schneider J, Sotin C, 2009, A study of the accuracy of mass-radius relationships for silicate-rich and ice-rich planets up to 100 Earth masses. *ApJ*, 693, 722–733 {**260, 267, 268**}

Gratadour D, Rouan D, Boccaletti A, et al., 2005, Four quadrant phase mask K-band coronagraphy of NGC 1068 with NAOS-CONICA at VLT. *A&A*, 429, 433–437 {**155**}

Gratton RG, Bonanno G, Claudi RU, et al., 2001, Non-interacting main-sequence binaries with different chemical compositions: evidence of infall of rocky material? *A&A*, 377, 123–131 {**192**}

Gratton RG, Carretta E, Claudi RU, et al., 2003, The SARG planet search: hunting for planets around stars in wide binaries. *Scientific Frontiers in Research on Extrasolar Planets*, volume 294 of *ASP Conf Ser*, 47–50 {**33, 58**}

—, 2004, The SARG exoplanet search. *Memorie della Societa Astronomica Italiana*, 75, 97–102 {**24**}

Gray DF, 1982, Observations of spectral line asymmetries and convective velocities in F, G, and K stars. *ApJ*, 255, 200–209 {**23**}

—, 1983, On the constancy of spectral-line bisectors. *PASP*, 95, 252–255 {**23**}

—, 1989, The morphology of reversed spectral-line bisectors. *PASP*, 101, 832–838 {**23**}

—, 1997, Absence of a planetary signature in the spectra of the star 51 Peg. *Nature*, 385, 795–796 {**28**}

—, 1998, A planetary companion for 51 Peg implied by absence of pulsations in the stellar spectra. *Nature*, 391, 153–154 {**28**}

—, 1999, Stellar rotation and precise radial velocities. *IAU Colloq. 170: Precise Stellar Radial Velocities*, volume 185 of *ASP Conf Ser*, 243–254 {**18**}

Gray DF, Brown KIT, 2006, Precise spectroscopic radial velocity measurements using telluric lines. *PASP*, 118, 399–404 {**19**}

Gray DF, Carney BW, Yong D, 2008, Asymmetries in the spectral lines of evolved halo stars. *AJ*, 135, 2033–2037 {**23**}

Gray DF, Hatzes AP, 1997, Non-radial oscillation in the solar-temperature star 51 Peg. *ApJ*, 490, 412–424 {**28**}

Gray DF, Nagar P, 1985, The rotational discontinuity shown by luminosity class IV stars. *ApJ*, 298, 756–760 {**32**}

Gray R, 2000, Effective temperature scale and bolometric corrections. *Encyclopedia of Astronomy and Astrophysics* {**184, 185**}

Gray RO, Corbally CJ, Garrison RF, et al., 2006, Contributions to the nearby stars (NStars) project: spectroscopy of stars earlier than M0 within 40 pc: the southern sample. *AJ*, 132, 161–170 {**286**}

Graziani F, Black DC, 1981, Orbital stability constraints on the nature of planetary systems. *ApJ*, 251, 337–341 {**56**}

Grbic A, Eleftheriades GV, 2004, Overcoming the diffraction limit with a planar left-handed transmission-line lens. *Physical Review Letters*, 92(11), 117403–7406 {**169**}

Greaves JS, 2004, Dense gas disks around T Tauri stars. *MNRAS*, 351, L99–L104 {**224**}

Greaves JS, Fischer DA, Wyatt MC, 2006, Metallicity, debris disks and planets. *MNRAS*, 366, 283–286 {**191**}

Greaves JS, Holland WS, Jayawardhana R, et al., 2004a, A search for debris disks around stars with giant planets. *MNRAS*, 348, 1097–1104 {**224**}

Greaves JS, Holland WS, Moriarty-Schieven G, et al., 1998, A dust ring around ϵ Eri: analogue to the young solar system. *ApJ*, 506, L133–L137 {**223**}

Greaves JS, Holland WS, Wyatt MC, et al., 2005, Structure in the ϵ Eri debris disk. *ApJ*, 619, L187–L190 {**223**}

Greaves JS, Wyatt MC, Holland WS, et al., 2004b, The debris disk around τ Cet: a massive analogue to the Kuiper belt. *MNRAS*, 351, L54–L58 {**224**}

Green RM, 1985, *Spherical Astronomy*. Cambridge University Press {**68**}

Greenaway AH, Spaan FHP, Mourai V, 2005, Pupil replication for exoplanet imaging. *ApJ*, 618, L165–L165 {**156**}

Greenberg R, 1982, Orbital evolution of the Galilean satellites. *Satellites of Jupiter*, 65–92, University of Arizona Press {**247**}

—, 1989, Time-varying orbits and tidal heating of the Galilean satellites. *NASA Special Publication*, 494, 100–115 {**247**}

—, 2005, *Europa – the Ocean Moon: Search for an Alien Biosphere*. Springer–Praxis {**250**}

—, 2009, Frequency dependence of tidal q. *ApJ*, 698, L42–L45 {**246**}

Greenberg R, Brahic A, 1984, *Planetary rings*. University of Arizona Press {**293**}

Greenstein JL, 1988, The companion of the white dwarf G29–38 as a brown dwarf. *AJ*, 95, 1494–1504 {**210**}

Greenzweig Y, Lissauer JJ, 1990, Accretion rates of protoplanets. *Icarus*, 87, 40–77 {**227**}

Gregory PC, 2005, A Bayesian analysis of extrasolar planet data for HD 73526. *ApJ*, 631, 1198–1214 {**14**}

—, 2007a, A Bayesian Kepler periodogram detects a second planet in HD 208487. *MNRAS*, 374, 1321–1333 {**14**}

—, 2007b, A Bayesian periodogram finds evidence for three planets in HD 11964. *MNRAS*, 381, 1607–1616 {**14**}

Gregoryanz E, Goncharov AF, Matsuishi K, et al., 2003, Raman spectroscopy of hot dense hydrogen. *Physical Review Letters*, 90(17), 175701 {**262**}

Grenfell JL, Stracke B, von Paris P, et al., 2007, The response of atmospheric chemistry on earthlike planets around F, G and K stars to small variations in orbital distance. *Planet. Space Sci.*, 55, 661–671 {**288**}

Grether D, Lineweaver CH, 2006, How dry is the brown dwarf desert? Quantifying the relative number of planets, brown dwarfs, and stellar companions around nearby Sun-like stars. *ApJ*, 640, 1051–1062 {**241**}

Grevesse N, Noels A, 1993, Cosmic abundances of the elements. *Origin and Evolution of the Elements* (eds. Prantzos N, Vangioni-Flam E, Casse M), 14–25, Cambridge University Press {**295, 309, 310**}

Grevesse N, Noels A, Sauval AJ, 1996, Standard Abundances. *Cosmic Abundances*, volume 99 of *ASP Conf Ser*, 117–126 {**295**}

Grevesse N, Sauval AJ, 1998, Standard solar composition. *Space Sci. Rev.*, 85, 161–174 {**295**}

—, 2002, The composition of the solar photosphere. *Advances in Space Research*, 30, 3–11 {**295**}

Grießmeier JM, Motschmann U, Mann G, et al., 2005, The influence of stellar wind conditions on the detectability of planetary radio emissions. *A&A*, 437, 717–726 {**176**}

Grießmeier JM, Preusse S, Khodachenko M, et al., 2007a, Exoplanetary radio emission under different stellar wind conditions. *Planet. Space Sci.*, 55, 618–630 {**176**}

Grießmeier JM, Stadelmann A, Penz T, et al., 2004, The effect of tidal locking on the magnetospheric and atmospheric evolution of Hot Jupiters. *A&A*, 425, 753–762 {**176**}

Grießmeier JM, Zarka P, Spreeuw H, 2007b, Predicting low-frequency radio fluxes of known extrasolar planets. *A&A*, 475, 359–368 {**176**}

Griest K, Safizadeh N, 1998, The use of high-magnification microlensing events in discovering extrasolar planets. *ApJ*, 500, 37–50 {**88, 89**}

Grieve RAF, Pesonen LJ, 1996, Terrestrial impact craters: their spatial and temporal distribution and impacting bodies. *Earth Moon and Planets*, 72, 357–376 {**229**}

Griffin REM, David M, Verschueren W, 2000, Accuracy of radial-velocity measurements for early-type stars. II. Investigations of spectrum mismatch from high-resolution observations. *A&AS*, 147, 299–321 {**32**}

Griffin RF, 1967, A photoelectric radial-velocity spectrometer. *ApJ*, 148, 465–476 {**17**}

—, 1973, On the possibility of determining stellar radial velocities to 0.01 km s^{-1}. *MNRAS*, 162, 243–253 {**19**}

Griffin RF, Griffin REM, 1973, Accurate wavelengths of stellar and telluric absorption lines near 700 nm. *MNRAS*, 162, 255–260 {**19**}

Grillmair CJ, Burrows A, Charbonneau D, et al., 2008, Strong water absorption in the dayside emission spectrum of the planet HD 189733 b. *Nature*, 456, 767–769 {**140, 142**}

Grillmair CJ, Charbonneau D, Burrows A, et al., 2007, A Spitzer spectrum of the exoplanet HD 189733 b. *ApJ*, 658, L115–L118 {**2, 140**}

Grossman AS, Graboske HC, 1973, Evolution of low-mass stars. V. Minimum mass for the deuterium main sequence. *ApJ*, 180, 195–198 {**209**}

References

Grossman L, 1972, Condensation in the primitive solar nebula. *Geochim. Cosmochim. Acta*, 36, 597–619 {**257, 297**}

Grossman L, Larimer JW, 1974, Early chemical history of the solar system. *Reviews of Geophysics and Space Physics*, 12, 71–101 {**257**}

Gu PG, Bodenheimer PH, Lin DNC, 2004, The internal structural adjustment due to tidal heating of short-period inflated giant planets. *ApJ*, 608, 1076–1094 {**145**}

Güdel M, 2002, Stellar radio astronomy: probing stellar atmospheres from protostars to giants. *ARA&A*, 40, 217–261 {**173**}

Guenther DB, Demarque P, 1997, Seismic tests of the Sun's interior structure, composition, and age, and implications for solar neutrinos. *ApJ*, 484, 937–959 {**296**}

Guenther EW, Hartmann M, Esposito M, et al., 2009, A substellar component orbiting the F-star 30 Arietis B. *A&A*, 507, 1659–1665 {**249**}

Guenther EW, Wuchterl G, 2003, Companions of old brown dwarfs, and very low mass stars. *A&A*, 401, 677–683 {**33**}

Guildner LA, Johnson DP, Jones FE, 1976, Vapour pressure of water at its triple point: highly accurate value. *Science*, 191, 1261–1263 {**262**}

Guillot T, 1999, A comparison of the interiors of Jupiter and Saturn. *Planet. Space Sci.*, 47, 1183–1200 {**234, 264, 294**}

—, 2005, The interiors of giant planets: models and outstanding questions. *Annual Review of Earth and Planetary Sciences*, 33, 493–530 {**228, 260–264, 266, 270, 294, 296**}

Guillot T, Burrows A, Hubbard WB, et al., 1996, Giant planets at small orbital distances. *ApJ*, 459, L35–L38 {**145, 260, 280, 281**}

Guillot T, Chabrier G, Morel P, et al., 1994a, Nonadiabatic models of Jupiter and Saturn. *Icarus*, 112, 354–367 {**264**}

Guillot T, Gautier D, Chabrier G, et al., 1994b, Are the giant planets fully convective? *Icarus*, 112, 337–353 {**264**}

Guillot T, Gautier D, Hubbard WB, 1997, New constraints on the composition of Jupiter from Galileo measurements and interior models. *Icarus*, 130, 534–539 {**234, 294**}

Guillot T, Hueso R, 2006, The composition of Jupiter: sign of a (relatively) late formation in a chemically evolved protosolar disc. *MNRAS*, 367, L47–L51 {**271**}

Guillot T, Santos NC, Pont F, et al., 2006, A correlation between the heavy element content of transiting extrasolar planets and the metallicity of their parent stars. *A&A*, 453, L21–L24 {**146, 189, 190**}

Guillot T, Showman AP, 2002, Evolution of 51 Peg-like planets. *A&A*, 385, 156–165 {**145, 260, 265**}

Guillot T, Stevenson DJ, Hubbard WB, et al., 2004, The interior of Jupiter. *Jupiter. The Planet, Satellites and Magnetosphere*, 35–57, Cambridge University Press {**174, 234, 264, 294**}

Gullbring E, Hartmann L, Briceno C, et al., 1998, Disk accretion rates for T Tauri stars. *ApJ*, 492, 323–341 {**221**}

Gurfil P, Kasdin J, Arrell R, et al., 2002, Infrared space observatories: how to mitigate zodiacal dust interference. *ApJ*, 567, 1250–1261 {**165**}

Gurnett DA, Kurth WS, Hospodarsky GB, et al., 2002, Control of Jupiter's radio emission and aurorae by the solar wind. *Nature*, 415, 985–987 {**174**}

Gusev A, Kitiashvili I, 2006, Transition from a direct rotation to the reverse rotation of exoplanets by action of the basic perturbations. *European Planetary Science Congress 2006*, 260 {**129**}

Guyon O, 2003, Phase-induced amplitude apodisation of telescope pupils for extrasolar terrestrial planet imaging. *A&A*, 404, 379–387 {**153**}

—, 2004, Imaging faint sources within a speckle halo with synchronous interferometric speckle subtraction. *ApJ*, 615, 562–572 {**157**}

—, 2005, Limits of adaptive optics for high-contrast imaging. *ApJ*, 629, 592–614 {**151, 157**}

—, 2007, A theoretical look at coronagraph design and performance for direct imaging of exoplanets. *Comptes Rendus Physique*, 8, 323–332 {**153**}

—, 2010, High sensitivity wavefront sensing with a nonlinear curvature wavefront sensor. *PASP*, 122, 49–62 {**151**}

Guyon O, Angel JRP, Backman D, et al., 2008, Pupil mapping Exoplanet Coronagraphic Observer (PECO). *SPIE Conf Ser*, volume 7010, 59 {**166**}

Guyon O, Angel JRP, Bowers C, et al., 2006a, Telescope to Observe Planetary Systems (TOPS): a high throughput 1.2-m visible telescope with a small inner working angle. *SPIE Conf Ser*, volume 6265, 52 {**166**}

—, 2007, Direct imaging of nearby exoplanets with a small size space telescope: Telescope to Observe Planetary System (TOPS). *In the Spirit of Bernard Lyot: The Direct Detection of Planets and Circumstellar Disks in the 21st Century*, 37 {**166**}

Guyon O, Matsuo T, Angel R, 2009, Coronagraphic low-order wave-front sensor: principle and application to a phase-induced amplitude coronagraph. *ApJ*, 693, 75–84 {**151**}

Guyon O, Pluzhnik EA, Galicher R, et al., 2005, Exoplanet imaging with a phase-induced amplitude apodisation coronagraph. I. Principle. *ApJ*, 622, 744–758 {**153, 154**}

Guyon O, Pluzhnik EA, Kuchner MJ, et al., 2006b, Theoretical limits on extrasolar terrestrial planet detection with coronagraphs. *ApJS*, 167, 81–99 {**153–155**}

Guyon O, Shao M, 2006, The pupil-swapping coronagraph. *PASP*, 118, 860–865 {**153**}

Guzzo M, Lega E, Froeschlé C, 2002, On the numerical detection of the effective stability of chaotic motions in quasi-integrable systems. *Physica D Nonlinear Phenomena*, 163, 1–25 {**46**}

Haberl F, Turolla R, de Vries CP, et al., 2006, Evidence for precession of the isolated neutron star RX J0720.4–3125. *A&A*, 451, L17–L21 {**79**}

Hadjidemetriou JD, Psychoyos D, Voyatzis G, 2009, The 1/1 resonance in extrasolar planetary systems. *Celestial Mechanics and Dynamical Astronomy*, 104, 23–38 {**54**}

Haghighipour N, 2004, On the dynamical stability of γ Cep, an S-type binary planetary system. *The Search for Other Worlds*, volume 713 of *American Institute of Physics Conference Series*, 269–272 {**57, 60**}

—, 2006, Dynamical stability and habitability of the γ Cep binary-planetary system. *ApJ*, 644, 543–550 {**57, 60, 285**}

—, 2008, Formation, dynamical evolution, and habitability of planets in binary star systems. *Exoplanets: Detection, Formation, Properties, Habitability*, 223–257, Springer {**56, 57**}

Haghighipour N, Boss AP, 2003, On gas drag-induced rapid migration of solids in a nonuniform solar nebula. *ApJ*, 598, 1301–1311 {**226**}

Haghighipour N, Raymond SN, 2007, Habitable planet formation in binary planetary systems. *ApJ*, 666, 436–446 {**58**}

Haghighipour N, Vogt SS, Butler RP, et al., 2010, The Lick–Carnegie exoplanet survey: a Saturn-mass planet in the habitable zone of the nearby M4V star HIP 57050. *ApJ*, 715, 271–276 {**103, 287**}

Hahn JM, Malhotra R, 1999, Orbital evolution of planets embedded in a planetesimal disk. *AJ*, 117, 3041–3053 {**295**}

—, 2005, Neptune's migration into a stirred-up Kuiper belt: a detailed comparison of simulations to observations. *AJ*, 130, 2392–2414 {**303**}

Hainaut OR, Rahoui F, Gilmozzi R, 2007, Down to Earths, with OWL. *Exploring the Cosmic Frontier: Astrophysical Instruments for the 21st Century*, 253–256, Springer-Verlag {**159**}

Haisch KE, Lada EA, Lada CJ, 2001, Disk frequencies and lifetimes in young clusters. *ApJ*, 553, L153–L156 {**239**}

Hajian AR, Behr BB, Cenko AT, et al., 2007, Initial results from the USNO dispersed Fourier Transform Spectrograph. *ApJ*, 661, 616–633 {**27**}

Halbwachs JL, Arenou F, Mayor M, et al., 2000, Exploring the brown dwarf desert with Hipparcos. *A&A*, 355, 581–594 {**35**}

Halbwachs JL, Mayor M, Udry S, et al., 2003, Multiplicity among solar-type stars. III. Statistical properties of the F7-K binaries with periods up to 10 years. *A&A*, 397, 159–175 {**59, 207**}

Hale A, 1994, Orbital coplanarity in solar-type binary systems: implications for planetary system formation and detection. *AJ*, 107, 306–332 {**114**}

Hale A, Doyle LR, 1994, The photometric method of extrasolar planet detection revisited. *Ap&SS*, 212, 335–348 {**104**}

Hale GE, 1895, On a new method of mapping the solar corona without an eclipse. *ApJ*, 1, 318–334 {**152**}

Hales SEG, Baldwin JE, Warner PJ, 1993, The 6C survey of radio sources. VI. *MNRAS*, 263, 25–30 {**176**}

Hales SEG, Waldram EM, Rees N, et al., 1995, A revised machine-readable source list for the Rees 38-MHz survey. *MNRAS*, 274, 447–451 {**176**}

Halliday AN, 2000, Terrestrial accretion rates and the origin of the Moon. *Earth and Planetary Science Letters*, 176, 17–30 {**302**}

—, 2004, Mixing, volatile loss and compositional change during impact-driven accretion of the Earth. *Nature*, 427, 505–509 {**297**}

—, 2008, A young Moon-forming giant impact at 70–110 million years accompanied by late-stage mixing, core formation and degassing of the Earth. *Royal Society of London Philosophical Transactions Series A*, 366, 4163–4181 {**302**}

Hamdani S, Arnold L, Foellmi C, et al., 2006, Biomarkers in disk-averaged near-ultraviolet to near-infrared Earth spectra using Earthshine observations. *A&A*, 460, 617–624 {**289**}

Hamilton DP, Burns JA, 1992, Orbital stability zones about asteroids. II. The destabilising effects of eccentric orbits and of solar radiation. *Icarus*, 96, 43–64 {**43**}

Hamilton DP, Ward WR, 2004, Tilting Saturn. II. Numerical model. *AJ*, 128, 2510–2517 {**301**}

Han C, 2002, Astrometric method for breaking the photometric degeneracy between binary-source and planetary microlensing perturbations. *ApJ*, 564, 1015–1018 {**94**}

—, 2005a, Analysis of microlensing light curves induced by multiple-planet systems. *ApJ*, 629, 1102–1109 {**88**}

—, 2005b, On the feasibility of characterising lens stars in future space-based microlensing surveys. *ApJ*, 633, 414–417 {**102**}

—, 2006a, Properties of planetary caustics in gravitational microlensing. *ApJ*, 638, 1080–1085 {**88**}

—, 2006b, Secure identification of free-floating planets. *ApJ*, 644, 1232–1236 {**96**}

—, 2007a, Criteria in the selection of target events for planetary microlensing follow-up observations. *ApJ*, 661, 1202–1207 {**97**}

—, 2007b, Expansion of planet detection methods in next-generation microlensing surveys. *ApJ*, 670, 1361–1366 {**89, 90, 102**}

—, 2008a, Microlensing detections of moons of exoplanets. *ApJ*, 684, 684–690 {**95**}

—, 2008b, Microlensing search for planets with two simultaneously rising suns. *ApJ*, 676, L53–L56 {**95**}

—, 2009, Characterisation of microlensing planets with moderately wide separations. *ApJ*, 700, 945–948 {**102**}

Han C, Chang HY, An JH, et al., 2001a, Properties of microlensing light curve anomalies induced by multiple planets. *MNRAS*, 328, 986–992 {**88**}

Han C, Chang K, 1999, The applicability of the astrometric method for determining the physical parameters of gravitational microlenses. *MNRAS*, 304, 845–850 {**94**}

—, 2003, Signs of planetary microlensing signals. *ApJ*, 597, 1070–1075 {**94**}

Han C, Chung SJ, Kim D, et al., 2004, Gravitational microlensing: a tool for detecting and characterising free-floating planets. *ApJ*, 604, 372–378 {**96**}

Han C, Gaudi BS, 2008, A characteristic planetary feature in double-peaked, high-magnification microlensing events. *ApJ*, 689, 53–58 {**88**}

Han C, Gaudi BS, An JH, et al., 2005, Microlensing detection and characterisation of wide-separation planets. *ApJ*, 618, 962–972 {**88**}

Han C, Han W, 2002, On the feasibility of detecting satellites of extrasolar planets via microlensing. *ApJ*, 580, 490–493 {**95**}

Han C, Kang YW, 2003, Probing the spatial distribution of extrasolar planets with gravitational microlensing. *ApJ*, 596, 1320–1326 {**88**}

Han C, Kim YG, 2001, Comparison of the two follow-up observation strategies for gravitational microlensing planet searches. *ApJ*, 546, 975–979 {**97**}

Han C, Lee C, 2002, Properties of planet-induced deviations in the astrometric microlensing centroid shift trajectory. *MNRAS*, 329, 163–174 {**94**}

Han I, 1989, The accuracy of differential astrometry limited by the atmospheric turbulence. *AJ*, 97, 607–610 {**62**}

Han I, Black DC, Gatewood G, 2001b, Preliminary astrometric masses for proposed extrasolar planetary companions. *ApJ*, 548, L57–L60 {**70**}

Han I, Lee BC, Kim KM, et al., 2010, Detection of a planetary companion around the giant star γ^1 Leo. *A&A*, 509, A24 {**32, 33**}

Han Z, Podsiadlowski P, Maxted PFL, et al., 2002, The origin of subdwarf B stars. I. The formation channels. *MNRAS*, 336, 449–466 {**81**}

Hanna DS, Ball J, Covault CE, et al., 2009, OSETI with STACEE: a search for nanosecond optical transients from nearby stars. *Astrobiology*, 9, 345–357 {**291**}

Hansen BMS, 2004, The astrophysics of cool white dwarfs. *Phys. Rep.*, 399, 1–70 {**79**}

—, 2008, On the absorption and redistribution of energy in irradiated planets. *ApJS*, 179, 484–508 {**138**}

—, 2009, Formation of the terrestrial planets from a narrow annulus. *ApJ*, 703, 1131–1140 {**235**}

Hansen BMS, Barman T, 2007, Two classes of hot Jupiters. *ApJ*, 671, 861–871, erratum: 2008, ApJ, 682, 7006 {**145–147, 241**}

Hansen BMS, Kulkarni S, Wiktorowicz S, 2006, A Spitzer search for infrared excesses around massive young white dwarfs. *AJ*, 131, 1106–1118 {**173**}

Harakawa H, Sato B, Fischer DA, et al., 2010, Detection of a low-eccentricity and super-massive planet to the subgiant HD 38801. *ApJ*, 715, 550–553 {**32**}

Hardorp J, 1978, The Sun among the stars. I. A search for solar spectral analogues. *A&A*, 63, 383–390 {**286**}

Hardy JH, 1982, Active optics – don't build a telescope without it. *SPIE Conf Ser*, volume 332, 252–259 {**151**}

Hardy JW, 1998, *Adaptive Optics for Astronomical Telescopes*. Oxford University Press {**151**}

Harrington J, Hansen BM, Luszcz SH, et al., 2006, The phase-dependent infrared brightness of the extrasolar planet υ And b. *Science*, 314, 623–626 {**2, 125, 138, 142**}

Harrington J, Luszcz S, Seager S, et al., 2007, The hottest planet. *Nature*, 447, 691–693 {**142**}

Harrington RS, 1977, Planetary orbits in binary stars. *AJ*, 82, 753–756 {**56, 114**}

Harrington RS, Dahn CC, 1980, Summary of US Naval Observatory parallaxes. *AJ*, 85, 454–465 {**62**}

Harris AW, Ward WR, 1982, Dynamical constraints on the formation and evolution of planetary bodies. *Annual Review of Earth and Planetary Sciences*, 10, 61–108 {**301**}

Hart MH, 1978, The evolution of the atmosphere of the Earth. *Icarus*, 33, 23–39 {**283, 284**}

—, 1979, Habitable zones about main sequence stars. *Icarus*, 37, 351–357 {**283**}

Hartman JD, Bakos GÁ, Kipping DM, et al., 2011a, HAT-P-26 b: a low-density Neptune-mass planet transiting a K star. *ApJ*, 728, 138–141 {**106, 326**}

Hartman JD, Bakos GÁ, Sato B, et al., 2011b, HAT-P-18 b and HAT-P-19 b: two low-density Saturn-mass planets transiting metal-rich K stars. *ApJ*, 726, 52–56 {**326**}

Hartman JD, Bakos GÁ, Torres G, et al., 2009, HAT-P-12 b: a low-density sub-Saturn mass planet transiting a metal-poor K dwarf. *ApJ*, 706, 785–796 {**326**}

Hartman JD, Gaudi BS, Holman MJ, et al., 2008a, Deep MMT transit survey of the open cluster M37. I. Observations and cluster parameters. *ApJ*, 675, 1233–1253 {**109**}

—, 2008b, Deep MMT transit survey of the open cluster M37. II. Variable stars. *ApJ*, 675, 1254–1277 {**109**}

Hartmann L, 2002, On disk braking of T Tauri rotation. *ApJ*, 566, L29–L32 {**200**}

Hartmann L, Calvet N, Gullbring E, et al., 1998, Accretion and the evolution of T Tauri disks. *ApJ*, 495, 385–400 {**221**}

Hartmann WK, Davis DR, 1975, Satellite-sized planetesimals and lunar origin. *Icarus*, 24, 504–514 {**302**}

Hartmann WK, Ryder G, Dones L, et al., 2000, The time-dependent intense bombardment of the primordial Earth–Moon system. *Origin of the Earth and Moon*, 493–512 {**304**}

Harwit M, 2003, Photon orbital angular momentum in astrophysics. *ApJ*, 597, 1266–1270 {**155**}

Hatzes AP, 1996, Simulations of stellar radial velocity and spectral line bisector variations. I. Nonradial pulsations. *PASP*, 108, 839–843 {**23**}

—, 2002, Starspots and exoplanets. *Astronomische Nachrichten*, 323, 392–394 {**18, 65**}

Hatzes AP, Cochran WD, 1993, Long-period radial velocity variations in three K giants. *ApJ*, 413, 339–348 {**2, 28**}

Hatzes AP, Cochran WD, Bakker EJ, 1998a, Further evidence for the planet around 51 Peg. *Nature*, 391, 154–155 {**28**}

—, 1998b, The lack of spectral variability in 51 Peg: confirmation of the planet hypothesis. *ApJ*, 508, 380–386 {**28**}

Hatzes AP, Cochran WD, Endl M, et al., 2003a, A planetary companion to γ Cep a. *ApJ*, 599, 1383–1394 {**28, 58, 60**}

—, 2006, Confirmation of the planet hypothesis for the long-period radial velocity variations of β Gem. *A&A*, 457, 335–341 {**28**}

Hatzes AP, Cochran WD, Johns-Krull CM, 1997, Testing the planet hypothesis: a search for variability in the spectral-line shapes of 51 Peg. *ApJ*, 478, 374–380 {**28**}

Hatzes AP, Cochran WD, McArthur B, et al., 2000, Evidence for a long-period planet orbiting ϵ Eri. *ApJ*, 544, L145–L148 {**223**}

Hatzes AP, Dvorak R, Wuchterl G, et al., 2010, An investigation into the radial velocity variations of CoRoT-7. *A&A*, 520, A93 {**111**}

Hatzes AP, Guenther E, Kürster M, et al., 2003b, The planet search programme of the Thüringer Landessternwarte Tautenburg. *Earths: Darwin/TPF and the Search for Extrasolar Terrestrial Planets*, volume 539 of *ESA Special Publication*, 441–445 {**24**}

Hatzes AP, Guenther EW, Endl M, et al., 2005, A giant planet around the massive giant star HD 13189. *A&A*, 437, 743–751 {**32**}

Hatzes AP, Kürster M, Cochran WD, et al., 1996, The European Southern observatory planetary search programme: preliminary results. *J. Geophys. Res.*, 101, 9285–9290 {**24**}

Hatzes AP, Zechmeister M, 2008, Stellar oscillations in planet-hosting giant stars. *Journal of Physics Conference Series*, 118(1), 012016 {**204**}

Hauser HM, Marcy GW, 1999, The orbit of 16 Cyg AB. *PASP*, 111, 321–334 {**60, 70**}

Hawarden TG, Dravins D, Gilmore GF, et al., 2003, Critical science for the largest telescopes: science drivers for a 100 m ground-based optical-IR telescope. *SPIE Conf Ser*, volume 4840, 299–308 {**157**}

Hawley JF, Balbus SA, 1991, A powerful local shear instability in weakly magnetised disks. II. Nonlinear evolution. *ApJ*, 376, 223–233 {**221**}

Hawley SL, Reid IN, 2003, An outsiders view of extrasolar planets. *The Future of Cool-Star Astrophysics: 12th Cambridge Workshop on Cool Stars, Stellar Systems, and the Sun*, volume 12, 128–140 {**187**}

Hayashi C, 1981, Structure of the solar nebula, growth and decay of magnetic fields and effects of magnetic and turbulent viscosities on the nebula. *Progress of Theoretical Physics Supplement*, 70, 35–53 {**220**}

Hayashi C, Nakazawa K, Mizuno H, 1979, Earth's melting due to the blanketing effect of the primordial dense atmosphere. *Earth and Planetary Science Letters*, 43, 22–28 {**278**}

Hayes W, Tremaine S, 1998, Fitting selected random planetary systems to Titius–Bode laws. *Icarus*, 135, 549–557 {**43, 48, 229**}

Hays JD, Imbrie J, Shackleton NJ, 1976, Variations in the Earth's orbit: pacemaker of the ice ages. *Science*, 194, 1121–1132 {**301**}

Haywood M, 2008, A peculiarity of metal-poor stars with planets? *A&A*, 482, 673–676 {**190, 193**}

—, 2009, On the correlation between metallicity and the presence of giant planets. *ApJ*, 698, L1–L5 {**193, 194**}

Heacox WD, 1986, On the application of optical-fiber image scramblers to astronomical spectroscopy. *AJ*, 92, 219–229 {**20**}

—, 1996, Statistical characteristics of extrasolar planetary transits. *J. Geophys. Res.*, 101, 14815–14822 {**104**}

Hearnshaw JB, 1974, Carbon and iron abundances for twenty F and G type stars. *A&A*, 36, 191–199 {**188**}

Hebb L, Collier-Cameron A, Loeillet B, et al., 2009, WASP-12 b: the hottest transiting extrasolar planet yet discovered. *ApJ*, 693, 1920–1928 {**250, 327**}

Hebb L, Collier-Cameron A, Triaud AHMJ, et al., 2010, WASP-19 b: the shortest period transiting exoplanet yet discovered. *ApJ*, 708, 224–231 {**108, 327**}

Hébrard G, Bouchy F, Pont F, et al., 2008, Misaligned spin-orbit in the XO-3 planetary system? *A&A*, 488, 763–770 {**129**}

Hébrard G, Lecavelier des Etangs A, 2006, A posteriori detection of the planetary transit of HD 189733 b in the Hipparcos photometry. *A&A*, 445, 341–346 {**113, 141**}

Hébrard G, Lecavelier des Etangs A, Vidal-Madjar A, et al., 2004, Evaporation rate of hot Jupiters and formation of chthonian planets. *Extrasolar Planets: Today and Tomorrow*, volume 321 of *ASP Conf Ser*, 203–204 {**281**}

Hébrard G, Robichon N, Pont F, et al., 2006, Search for transiting planets in the Hipparcos database. *Tenth Anniversary of 51 Peg-b: Status of and prospects for hot Jupiter studies*, 193–195 {**113**}

Heintz WD, 1978a, *Double Stars*, volume 15 of *Geophysics and Astrophysics Monographs*. Reidel, Dordrecht {**9, 68**}

—, 1978b, Reexamination of suspected unresolved binaries. *ApJ*, 220, 931–934 {**64**}

Heinze AN, Hinz P, Sivanandam S, et al., 2006, High contrast L′ band adaptive optics imaging to detect extrasolar planets. *SPIE Conf Ser*, volume 6272, 121 {**158**}

Heinze AN, Hinz PM, Kenworthy M, et al., 2008, Deep L′- and M-band imaging for planets around Vega and ϵ Eri. *ApJ*, 688, 583–596 {**223, 224**}

Heiter U, Luck RE, 2003, Abundance analysis of planetary host stars. I. Differential iron abundances. *AJ*, 126, 2015–2036 {**189**}

Hekker S, Reffert S, Quirrenbach A, et al., 2006, Precise radial velocities of giant stars. I. Stable stars. *A&A*, 454, 943–949 {**21, 32, 33**}

Hekker S, Snellen IAG, Aerts C, et al., 2008, Precise radial velocities of giant stars. IV. A correlation between surface gravity and radial velocity variation and a statistical investigation of companion properties. *A&A*, 480, 215–222 {**32, 33**}

Helled R, Schubert G, 2008, Core formation in giant gaseous protoplanets. *Icarus*, 198, 156–162 {**234**}

Heller CH, 1993, Encounters with protostellar disks. I. Disk tilt and the nonzero solar obliquity. *ApJ*, 408, 337–346 {**301**}

Heller R, Mislis D, Antoniadis J, 2009, Transit detections of extrasolar planets around main-sequence stars. I. Sky maps for hot Jupiters. *A&A*, 508, 1509–1516 {**104**}

Hellier C, Anderson DR, Collier Cameron A, et al., 2009a, An orbital period of 0.94 days for the hot-Jupiter planet WASP-18 b. *Nature*, 460, 1098–1100 {**250, 327**}

—, 2010, WASP-29 b: a Saturn-sized transiting exoplanet. *ApJ*, 723, L60–L63 {**108, 327**}

Hellier C, Anderson DR, Gillon M, et al., 2009b, WASP-7: a bright transiting-exoplanet system in the southern hemisphere. *ApJ*, 690, L89–L91 {**327**}

Helmi A, White SDM, de Zeeuw PT, et al., 1999, Debris streams in the solar neighbourhood as relicts from the formation of the Milky Way. *Nature*, 402, 53–55 {**34**}

Heng K, Keeton CR, 2009, Planetesimal disk microlensing. *ApJ*, 707, 621–631 {**96**}

Hennessy GS, Lane BF, Veillette D, et al., 2010, Achieving milli-arcsecond residual astrometric error for the JMAPS Mission. *SPIE Conf Ser*, volume 7731, 149 {**73**}

Henning T, Dullemond CP, Wolf S, et al., 2006, Dust coagulation in protoplanetary disks. *Planet Formation*, 112–128, Cambridge University Press {**226**}

Hénon M, Guyot M, 1970, Stability of periodic orbits in the restricted problem. *Periodic Orbits Stability and Resonances*, 349–374 {**57**}

Hénon M, Heiles C, 1964, The applicability of the third integral of motion: some numerical experiments. *AJ*, 69, 73–79 {**45**}

Henry GW, Donahue RA, Baliunas SL, 2002, A false planet around HD 192263. *ApJ*, 577, L111–L114 {**65**}

Henry GW, Marcy G, Butler RP, et al., 1999, HD 209458. *IAU Circ.*, 7307, 1 {**2, 103, 139, 140**}

Henry GW, Marcy GW, Butler RP, et al., 2000, A transiting 51 Peg-like planet. *ApJ*, 529, L41–L44 {**103, 113, 139, 140**}

Henry TJ, 2009, RECONS census 2010. *Bulletin of the American Astronomical Society*, volume 41, 203 {**182**}

Henry TJ, Jao W, Subasavage JP, et al., 2006, The solar neighborhood. XVII. Parallax results from the CTIOPI 0.9-m programme: 20 new members of the RECONS 10 pc sample. *AJ*, 132, 2360–2371 {**182**}

Henry TJ, Soderblom DR, Donahue RA, et al., 1996, A survey of Ca II H and K chromospheric emission in southern solar-type stars. *AJ*, 111, 439–465 {**187**}

Heppenheimer TA, 1978, On the formation of planets in binary star systems. *A&A*, 65, 421–426 {**58, 114, 283**}

Herbst W, Hamilton CM, Leduc K, et al., 2008, Reflected light from sand grains in the terrestrial zone of a protoplanetary disk. *Nature*, 452, 194–197 {**226**}

Hernández J, Hartmann L, Megeath T, et al., 2007, A Spitzer space telescope study of disks in the young σ Ori cluster. *ApJ*, 662, 1067–1081 {**222**}

Herrero E, Morales JC, Ribas I, et al., 2011, WASP-33: the first δ Scuti exoplanet host star. *A&A*, 526, L10–L13 {**108, 134**}

Hersant F, Gautier D, Lunine JI, 2004, Enrichment in volatiles in the giant planets of the Solar System. *Planet. Space Sci.*, 52, 623–641 {**294**}

Hershey JL, Lippincott SL, 1982, A study of the intensive 40-year Sproul plate series on Lalande 21185 and BD +5 1668. *AJ*, 87, 840–844 {**64**}

Hess S, Mottez F, Zarka P, et al., 2008, Generation of the Jovian radio decametric arcs from the Io Flux Tube. *Journal of Geophysical Research (Space Physics)*, 113(12), 3209–3218 {**174**}

Hester JJ, Desch SJ, 2005, Understanding our origins: star formation in H II region environments. *Chondrites and the Protoplanetary Disk*, volume 341 of *ASP Conf Ser*, 107–129 {**293**}

Hester JJ, Desch SJ, Healy KR, et al., 2004, The cradle of the solar system. *Science*, 304, 1116–1117 {293}

Hewitt JN, Turner EL, Schneider DP, et al., 1988, Unusual radio source MG1131+0456: a possible Einstein ring. *Nature*, 333, 537–540 {84}

Heyl JS, 2007, Orbital evolution with white-dwarf kicks. *MNRAS*, 382, 915–920 {79}

Heyl JS, Gladman BJ, 2007, Using long-term transit timing to detect terrestrial planets. *MNRAS*, 377, 1511–1519 {135}

Heyrovsky D, Loeb A, 1997, Microlensing of an elliptical source by a point mass. *ApJ*, 490, 38–50 {95}

Hidas MG, Ashley MCB, Webb JK, et al., 2005, The University of New South Wales extrasolar planet search: methods and first results from a field centred on NGC 6633. *MNRAS*, 360, 703–717 {109}

Hidas MG, Hawkins E, Walker Z, et al., 2008, Las Cumbres Observatory Global Telescope: a homogeneous telescope network. *Astronomische Nachrichten*, 329, 269–270 {97}

Hidas MG, Tsapras Y, Mislis D, et al., 2010, An ingress and a complete transit of HD 80606 b. *MNRAS*, 406, 1146–1151 {109}

Higgins CA, Carr TD, Reyes F, et al., 1997, A redefinition of Jupiter's rotation period. *J. Geophys. Res.*, 102, 22033–22042 {174}

Hilditch RW, 2001, *An Introduction to Close Binary Stars*. Cambridge University Press {9}

Hill GW, 1878, Researches in the lunar theory. *Am. J. Math.*, 1(129-147) {43}

Hill JM, 2010, The Large Binocular Telescope. *Appl. Opt.*, 49, 115–122 {**150, 152, 162**}

Hillenbrand LA, Carpenter JM, 2000, Constraints on the stellar/substellar mass function in the inner Orion nebula cluster. *ApJ*, 540, 236–254 {**215**}

Hilliard RL, Shepherd GG, 1966, Wide-angle Michelson interferometer for measuring Doppler line widths. *Journal of the Optical Society of America*, 56, 362–369 {**25, 26**}

Hines DC, Backman DE, Bouwman J, et al., 2006, The formation and evolution of planetary systems: Discovery of an unusual debris system associated with HD 12039. *ApJ*, 638, 1070–1079 {224}

Hinkle KH, Joyce RR, Hedden A, et al., 2001, Wavelength calibration of near-infrared spectra. *PASP*, 113, 548–566 {20}

Hinse TC, Michelsen R, Jørgensen UG, et al., 2008, Dynamics and stability of telluric planets within the habitable zone of extrasolar planetary systems. Numerical simulations of test particles within the HD 4208 and HD 70642 systems. *A&A*, 488, 1133–1147 {285}

Hinz PM, Angel JRP, Hoffmann WF, et al., 1998, Imaging circumstellar environments with a nulling interferometer. *Nature*, 395, 251–253 {163}

Hinz PM, Angel JRP, Woolf NJ, et al., 2000, BLINC: a testbed for nulling interferometry in the thermal infrared. *SPIE Conf Ser*, volume 4006, 349–353 {163}

Hinz PM, Heinze AN, Sivanandam S, et al., 2006, Thermal infrared constraint to a planetary companion of Vega with the MMT adaptive optics system. *ApJ*, 653, 1486–1492 {224}

Hinz PM, Hoffmann WF, Hora JL, 2001, Constraints on disk sizes around young intermediate-mass stars: nulling interferometric observations of Herbig Ae objects. *ApJ*, 561, L131–L134 {163}

Hinz PM, Solheid E, Durney O, et al., 2008, NIC: LBTI's nulling and imaging camera. *SPIE Conf Ser*, volume 7013, 100 {163}

Hobbs GB, Bailes M, Bhat NDR, et al., 2009, Gravitational-wave detection using pulsars: status of the Parkes pulsar timing array project. *Publications of the Astronomical Society of Australia*, 26, 103–109 {79}

Hobbs PV, 1974, *Ice Physics*. Clarendon Press, Oxford {262}

Hodgkin ST, Irwin JM, Aigrain S, et al., 2006, Monitor: transiting planets and brown dwarfs in star forming regions and young open clusters. *Astronomische Nachrichten*, 327, 9–13 {109}

Hoffman PF, Schrag DP, 2002, The snowball Earth hypothesis: testing the limits of global change. *Terra Nova*, 14, 129–155 {306}

Høg E, Bässgen G, Bastian U, et al., 1997, The Tycho catalogue. *A&A*, 323, L57–L60 {**69, 181**}

Høg E, Fabricius C, Makarov VV, et al., 2000, The Tycho 2 catalogue of the 2.5 million brightest stars. *A&A*, 355, L27–L30 {**69, 181**}

Høg E, Novikov ID, Polnarev AG, 1995, MACHO photometry and astrometry. *A&A*, 294, 287–294 {94}

Holder J, 2005, Optical SETI with imaging Cherenkov telescopes. *International Cosmic Ray Conference*, volume 5, 387 {**291**}

Holland HD, 2002, Volcanic gases, black smokers, and the great oxidation event. *Geochim. Cosmochim. Acta*, 66, 3811–3826 {306}

—, 2006, The oxygenation of the atmosphere and oceans. *Phil. Trans. R. Soc. B*, 361, 903–915 {306}

—, 2009, Why the atmosphere became oxygenated: a proposal. *Geochim. Cosmochim. Acta*, 73, 5241–5255 {306}

Holland WS, Greaves JS, Dent WRF, et al., 2003, Submillimeter observations of an asymmetric dust disk around Fomalhaut. *ApJ*, 582, 1141–1146 {223}

Hollenbach D, Johnstone D, Lizano S, et al., 1994, Photoevaporation of disks around massive stars and application to ultracompact H II regions. *ApJ*, 428, 654–669 {222}

Holman MJ, 2010, Transits in multiple planet systems. *EAS Publications Series*, volume 42, 39–54 {135}

Holman MJ, Fabrycky DC, Ragozzine D, et al., 2010, Kepler-9: a system of multiple planets transiting a Sun-like star, confirmed by timing variations. *Science*, 330, 51–54 {**2, 112, 122, 135, 326**}

Holman MJ, Murray NW, 2005, The use of transit timing to detect terrestrial-mass extrasolar planets. *Science*, 307, 1288–1291 {**133, 135**}

Holman MJ, Touma J, Tremaine S, 1997, Chaotic variations in the eccentricity of the planet orbiting 16 Cyg B. *Nature*, 386, 254–256 {**47, 60, 237**}

Holman MJ, Wiegert PA, 1999, Long-term stability of planets in binary systems. *AJ*, 117, 621–628 {**57, 81**}

Holman MJ, Winn JN, Latham DW, et al., 2007, The Transit Light Curve project. VI. Three transits of the exoplanet TrES–2. *ApJ*, 664, 1185–1189 {123}

Holman MJ, Wisdom J, 1993, Dynamical stability in the outer solar system and the delivery of short period comets. *AJ*, 105, 1987–1999 {**44, 300**}

Holmberg E, 1938, Invisible companions of parallax stars revealed by means of modern trigonometric parallax observations. *Medd. Lund Astron.Obs. Ser. II*, 92 {64}

Holmberg J, Flynn C, 2000, The local density of matter mapped by Hipparcos. *MNRAS*, 313, 209–216 {310}

—, 2004, The local surface density of disk matter mapped by Hipparcos. *MNRAS*, 352, 440–446 {221}

Holz DE, Wald RM, 1996, Photon statistics limits for Earth-based parallax measurements of MACHO events. *ApJ*, 471, 64–67 {94}

Honda M, Kataza H, Okamoto YK, et al., 2004, Crystalline silicate feature of the Vega-like star HD 145263. *ApJ*, 610, L49–L52 {224}

Hood B, Collier Cameron A, Kane SR, et al., 2005, A dearth of planetary transits in the direction of NGC 6940. *MNRAS*, 360, 791–800 {109}

Horne JH, Baliunas SL, 1986, A prescription for period analysis of unevenly sampled time series. *ApJ*, 302, 757–763 {13}

Horne K, 2001, Planetary transit searches: hot Jupiters galore. *Techniques for the Detection of Planets and Life beyond the Solar System*, 5 {104}

Horne K, Snodgrass C, Tsapras Y, 2009, A metric and optimisation scheme for microlens planet searches. *MNRAS*, 396, 2087–2102 {**89, 97**}

Horowitz P, Coldwell CM, Howard AB, et al., 2001, Targeted and all-sky search for nanosecond optical pulses at Harvard–Smithsonian. *SPIE Conf Ser*, volume 4273, 119–127 {291}

Hosseinbor AP, Edgar RG, Quillen AC, et al., 2007, The formation of an eccentric gap in a gas disk by a planet in an eccentric orbit. *MNRAS*, 378, 966–972 {240}

Hough JH, Lucas PW, Bailey JA, et al., 2003, A high sensitivity polarimeter for the direct detection and characterisation of extrasolar planets. *SPIE Conf Ser*, volume 4843, 517–523 {126}

—, 2006, Detecting the polarisation signatures of extrasolar planets. *SPIE Conf Ser*, volume 6269 {126}

Houghton JT, 2002, *The Physics of Atmospheres*. Cambridge University Press, Third Edition {272}

Hourigan K, Ward WR, 1984, Radial migration of preplanetary material: implications for the accretion time scale problem. *Icarus*, 60, 29–39 {238}

Howard A, Horowitz P, Mead C, et al., 2007, Initial results from Harvard all-sky optical SETI. *Acta Astronautica*, 61, 78–87 {291}

Howard AW, Bakos GÁ, Hartman J, et al., 2010, HAT–P–17 b,c: a transiting, eccentric, hot Saturn and a long-period, cold Jupiter. *ArXiv e-prints* {**106, 326**}

Howard AW, Horowitz P, Wilkinson DT, et al., 2004, Search for nanosecond optical pulses from nearby solar-type stars. *ApJ*, 613, 1270–1284

{291}

Howard AW, Johnson JA, Marcy GW, et al., 2009, The NASA–UC Eta–Earth programme. I. A super-Earth orbiting HD 7924. *ApJ*, 696, 75–83 {33}

—, 2011a, The NASA–UC Eta–Earth programme. II. A planet orbiting HD 156668 with a minimum mass of four Earth Masses. *ApJ*, 726, 73–77 {**33, 35**}

—, 2011b, The NASA–UC Eta–Earth programme. III. A super-Earth orbiting HD 97658 and a Neptune-mass planet orbiting GJ 785. *ApJ*, 730, 10–16 {33}

Howell SB, VanOutryve C, Tonry JL, et al., 2005, A search for variable stars and planetary occultations in NGC 2301. II. Variability. *PASP*, 117, 1187–1203 {**109**}

Howland JL, 2000, *The Surprising Archaea*. Oxford University Press {290}

Hrudková M, Skillen I, Benn CR, et al., 2010, Tight constraints on the existence of additional planets around HD 189733. *MNRAS*, 403, 2111–2119 {135}

Huang C, Zhao G, Zhang HW, et al., 2005, Chemical abundances of 22 extrasolar planet host stars. *MNRAS*, 363, 71–78 {**195, 196, 198**}

Huang S, 1959, The problem of life in the Universe and the mode of star formation. *PASP*, 71, 421–424 {**283**}

—, 1960, The sizes of habitable planets. *PASP*, 72, 489–493 {**283, 285**}

Hubbard WB, 1968, Thermal structure of Jupiter. *ApJ*, 152, 745–754 {**262**}

—, 1974, Tides in the giant planets. *Icarus*, 23, 42–50 {**246**}

—, 1977, The Jovian surface condition and cooling rate. *Icarus*, 30, 305–310 {**265**}

—, 1984, *Planetary Interiors*. Van Nostrand Reinhold Co., New York {**261**}

Hubbard WB, Burrows A, Lunine JI, 2002, Theory of giant planets. *ARA&A*, 40, 103–136 {**175, 264**}

Hubbard WB, DeWitt HE, 1985, Statistical mechanics of light elements at high pressure. VII. A perturbative free energy for arbitrary mixtures of H and He. *ApJ*, 290, 388–393 {**262**}

Hubbard WB, Hattori MF, Burrows A, et al., 2007, Effects of mass loss for highly-irradiated giant planets. *Icarus*, 187, 358–364 {**281**}

Hubbard WB, Macfarlane JJ, 1980, Structure and evolution of Uranus and Neptune. *J. Geophys. Res.*, 85, 225–234 {**261**}

Hubbard WB, Podolak M, Stevenson DJ, 1995, The interior of Neptune. *Neptune and Triton*, 109–138 {**269**}

Hubeny I, 1988, A computer program for calculating non-LTE model stellar atmospheres. *Computer Physics Communications*, 52, 103–132 {272}

Hubeny I, Burrows A, Sudarsky D, 2003, A possible bifurcation in atmospheres of strongly irradiated stars and planets. *ApJ*, 594, 1011–1018 {**142, 272, 277**}

Hubeny I, Lanz T, 1995, Non-LTE line-blanketed model atmospheres of hot stars. I. Hybrid complete linearisation/accelerated lambda iteration method. *ApJ*, 439, 875–904 {**272**}

Hubickyj O, 2006, The core accretion-gas capture model for gas-giant planet formation. *Planet Formation*, 163–178, Cambridge University Press {**232**}

—, 2010, Core accretion model. *Formation and Evolution of Exoplanets*, 101–122, Wiley {**232**}

Hubickyj O, Bodenheimer P, Lissauer JJ, 2004, Evolution of gas giant planets using the core accretion model. *Revista Mexicana de Astronomia y Astrofisica Conference Series*, volume 22, 83–86 {**36**}

—, 2005, Accretion of the gaseous envelope of Jupiter around a 5–10 Earth-mass core. *Icarus*, 179, 415–431 {**234, 274**}

Hubin N, Arsenault R, Conzelmann R, et al., 2005, Ground layer adaptive optics. *Comptes Rendus Physique*, 6, 1099–1109 {**152**}

Hubin N, Noethe L, 1993, Active optics, adaptive optics, and laser guide stars. *Science*, 262, 1390–1394 {**151**}

Huebner WF, Merts AL, Magee NH, et al., 1977, *Los Alamos Sci. Rep.* LA-6760-M {**264**}

Huey RB, Ward PD, 2005, Hypoxia, global warming, and terrestrial late Permian extinctions. *Science*, 308, 398–401 {**307**}

Hügelmeyer SD, Dreizler S, Homeier D, et al., 2007, Investigation of transit-selected exoplanet candidates from the MACHO survey. *A&A*, 469, 1163–1168 {**106**}

Hui L, Seager S, 2002, Atmospheric lensing and oblateness effects during an extrasolar planetary transit. *ApJ*, 572, 540–555 {**131**}

Hummer DG, Mihalas D, 1988, The equation of state for stellar envelopes. I. An occupation probability formalism for the truncation of internal partition functions. *ApJ*, 331, 794–814 {**261**}

Hundertmark M, Hessman FV, Dreizler S, 2009, Detecting circumstellar disks around gravitational microlenses. *A&A*, 500, 929–934 {**96**}

Hunten DM, 2002, Exospheres and planetary escape. *Atmospheres in the Solar System: Comparative Aeronomy*, 191–202, American Geophysical Union {**280**}

Hunten DM, Watson AJ, 1982, Stability of Pluto's atmosphere. *Icarus*, 51, 665–667 {**281**}

Hunter TR, Ramsey LW, 1992, Scrambling properties of optical fibers and the performance of a double scrambler. *PASP*, 104, 1244–1251 {**20**}

Hut P, 1980, Stability of tidal equilibrium. *A&A*, 92, 167–170 {**245**}

—, 1981, Tidal evolution in close binary systems. *A&A*, 99, 126–140 {**245, 246**}

Hutchison R, Alexander CMO, Barber DJ, 1988, Chondrules: chemical, mineralogical and isotopic constraints on theories of their origin. *Royal Society of London Philosophical Transactions Series A*, 325, 445–458 {**298**}

Hutchison R, Graham AL, 1975, Significance of calcium-rich differentiates in chondritic meteorites. *Nature*, 255, 471 {**298**}

IAU, 2003, IAU Working Group on Extrasolar Planets: position statement on the defintion of a planet. www.dtm.ciw.edu/boss/definition.html {7}

—, 2006, Definition of a planet in the solar system. www.iau.org/static/resolutions/ {7}

Ibanoğlu C, Çakırlı Ö, Taş G, et al., 2004, High-speed photometry of the pre-cataclysmic binary HW Vir and its orbital period change. *A&A*, 414, 1043–1048 {**82**}

Ibgui L, Burrows A, 2009, Coupled evolution with tides of the radius and orbit of transiting giant planets: general results. *ApJ*, 700, 1921–1932 {145}

Ibgui L, Burrows A, Spiegel DS, 2010, Tidal heating models for the radii of the inflated transiting giant planets WASP–4 b, WASP–6 b, WASP–12 b, WASP–15 b, and TrES–4. *ApJ*, 713, 751–763 {**145**}

Ida S, Larwood J, Burkert A, 2000, Evidence for early stellar encounters in the orbital distribution of Edgeworth–Kuiper belt objects. *ApJ*, 528, 351–356 {**109**}

Ida S, Lin DNC, 2004a, Toward a deterministic model of planetary formation. I. A desert in the mass and semi-major axis distributions of extrasolar planets. *ApJ*, 604, 388–413 {**251**}

—, 2004b, Toward a deterministic model of planetary formation. II. The formation and retention of gas giant planets around stars with a range of metallicities. *ApJ*, 616, 567–572 {**36, 146, 191, 251**}

—, 2005a, Dependence of exoplanets on host star metallicity and mass. *Progress of Theoretical Physics Supplement*, 158, 68–85 {**191**}

—, 2005b, Toward a deterministic model of planetary formation. III. Mass distribution of short-period planets around stars of various masses. *ApJ*, 626, 1045–1060 {**36, 37, 251**}

—, 2008a, Toward a deterministic model of planetary formation. IV. Effects of type I migration. *ApJ*, 673, 487–501 {**251**}

—, 2008b, Toward a deterministic model of planetary formation. V. Accumulation near the ice line and super-Earths. *ApJ*, 685, 584–595 {**251**}

—, 2010, Toward a deterministic model of planetary formation. VI. Dynamical interaction and coagulation of multiple rocky embryos and super-Earth systems around solar-type stars. *ApJ*, 719, 810–830 {**251**}

Iglesias CA, Rogers FJ, 1996, Updated OPAL opacities. *ApJ*, 464, 943–953 {**202, 264**}

Ignace R, 2001, Spectral energy distribution signatures of Jovian planets around white dwarf stars. *PASP*, 113, 1227–1231 {**172**}

Ikoma M, Genda H, 2006, Constraints on the mass of a habitable planet with water of nebular origin. *ApJ*, 648, 696–706 {**278**}

Ikoma M, Guillot T, Genda H, et al., 2006, On the origin of HD 149026 b. *ApJ*, 650, 1150–1159 {**267**}

Ikoma M, Nakazawa K, Emori H, 2000, Formation of giant planets: dependences on core accretion rate and grain opacity. *ApJ*, 537, 1013–1025 {**234**}

Imbrie J, 1982, Astronomical theory of the Pleistocene ice ages: a brief historical review. *Icarus*, 50, 408–422 {**301**}

Imbrie J, Imbrie JZ, 1980, Modeling the climatic response to orbital variations. *Science*, 207, 943–953 {**301**}

Impey C (ed.), 2010, *Talking About Life: Conversations on Astrobiology*. Cambridge University Press {**282**}

Inaba S, Tanaka H, Nakazawa K, et al., 2001, High-accuracy statistical simulation of planetary accretion. II. Comparison with N-body simulation. *Icarus*, 149, 235–250 {228}

Ingersoll AP, Porco CC, 1978, Solar heating and internal heat flow on Jupiter. *Icarus*, 35, 27–43 {273}

Ingrosso G, Novati SC, de Paolis F, et al., 2009, Pixel lensing as a way to detect extrasolar planets in M31. *MNRAS*, 399, 219–228 {101}

Innanen KA, Zheng JQ, Mikkola S, et al., 1997, The Kozai mechanism and the stability of planetary orbits in binary star systems. *AJ*, 113, 1915–1919 {57, 58, 60}

Inoue AK, Honda M, Nakamoto T, et al., 2008, Observational possibility of the snow line on the surface of circumstellar disks with the scattered light. *PASJ*, 60, 557–563 {284}

Ioannou PJ, Kakouris A, 2001, Stochastic dynamics of Keplerian accretion disks. *ApJ*, 550, 931–943 {221}

Iorio L, 2006a, Are we far from testing general relativity with the transiting extrasolar planet HD 209458 b? *New Astronomy*, 11, 490–494 {133}

—, 2006b, Dynamical constraints on the quadrupole mass moment of the HD 209458 star. *ArXiv General Relativity and Quantum Cosmology e-prints* {134}

—, 2011, Classical and relativistic node precessional effects in WASP–33 b and perspectives for detecting them. *Ap&SS*, 331, 485–496 {108, 134}

Ip WH, Kopp A, Hu JH, 2004, On the star-magnetosphere interaction of close-in exoplanets. *ApJ*, 602, L53–L56 {176}

Irwin J, Charbonneau D, Berta ZK, et al., 2009, GJ 3236: a new bright, very low mass eclipsing binary system discovered by the MEARTH observatory. *ApJ*, 701, 1436–1449 {108}

Irwin PGJ, 2006, *Giant Planets of our Solar System: An Introduction*. Springer–Praxis {280}

Ishiguro M, Ueno M, 2003, Prospects for the exosolar planetary systems based on the zodiacal light observations. *Astronomical Herald*, 96, 206–209 {178}

Israelian G, 2008, Abundances in stars with planetary systems. *Extrasolar Planets*, 150–161 {198, 199}

Israelian G, Delgado Mena E, Santos NC, et al., 2009, Enhanced lithium depletion in Sun-like stars with orbiting planets. *Nature*, 462, 189–191 {199, 200}

Israelian G, Santos NC, Mayor M, et al., 2001, Evidence for planet engulfment by the star HD 82943. *Nature*, 411, 163–166 {178, 192, 193, 201}

—, 2003, New measurement of the ^{6}Li/^{7}Li isotopic ratio in the extrasolar planet host star HD 82943 and line blending in the Li 670.8 nm region. *A&A*, 405, 753–762 {192, 193, 201}

—, 2004, Lithium in stars with exoplanets. *A&A*, 414, 601–611 {189, 199, 200}

Itoh Y, Hayashi M, Tamura M, et al., 2005, A young brown dwarf companion to DH Tauri. *ApJ*, 620, 984–993 {168}

Ivanov PB, Papaloizou JCB, 2004a, On equilibrium tides in fully convective planets and stars. *MNRAS*, 353, 1161–1175 {246}

—, 2004b, On the tidal interaction of massive extrasolar planets on highly eccentric orbits. *MNRAS*, 347, 437–453 {246}

—, 2007, Dynamic tides in rotating objects: orbital circularisation of extrasolar planets for realistic planet models. *MNRAS*, 376, 682–704 {246}

Ivezic Z, Axelrod T, Brandt WN, et al., 2008, Large Synoptic Survey Telescope: from science drivers to reference design. *Serbian Astronomical Journal*, 176, 1–13 {211}

Izumiura H, 2005, An East-Asian extrasolar planet search network. *Journal of Korean Astronomical Society*, 38, 81–84 {24}

Jackson B, 2010, Tides and exoplanets. *Formation and Evolution of Exoplanets*, 243–266, Wiley {244}

Jackson B, Barnes R, Greenberg R, 2008a, Tidal heating of terrestrial extrasolar planets and implications for their habitability. *MNRAS*, 391, 237–245 {250}

—, 2009, Observational evidence for tidal destruction of exoplanets. *ApJ*, 698, 1357–1366 {241}

Jackson B, Greenberg R, Barnes R, 2008b, Tidal evolution of close-in extrasolar planets. *ApJ*, 678, 1396–1406 {245–248}

—, 2008c, Tidal heating of extrasolar planets. *ApJ*, 681, 1631–1638 {248, 249}

Jackson JM, Rathborne JM, Shah RY, et al., 2006, The Boston University-Five College Radio Astronomy Observatory Galactic ring survey. *ApJS*, 163, 145–159 {194}

Jahreiß H, Wielen R, 1997, The impact of Hipparcos on the Catalogue of Nearby Stars: the stellar luminosity function and local kinematics. *Hipparcos – Venice '97*, volume 402 of *ESA Special Publication*, 675–680 {182}

—, 2000, The census of nearby star binaries. *IAU Symposium*, volume 200, 129 {182}

Janczak J, Fukui A, Dong S, et al., 2010, Sub-Saturn planet MOA–2008–BLG–310Lb: likely to be in the Galactic bulge. *ApJ*, 711, 731–743 {99}

Janes K, 1996, Star clusters: optimal targets for a photometric planetary search programme. *J. Geophys. Res.*, 101, 14853–14860 {104}

Jang-Condell H, 2007, Constraints on the formation of the planet in HD 188753. *ApJ*, 654, 641–649 {60}

Jang-Condell H, Boss AP, 2007, Signatures of planet formation in gravitationally unstable disks. *ApJ*, 659, L169–L172 {237}

Jang-Condell H, Sasselov DD, 2005, Type I migration in a nonisothermal protoplanetary disk. *ApJ*, 619, 1123–1131 {239}

Janson M, 2007, Celestial exoplanet survey occulter: a concept for direct imaging of extrasolar Earth-like planets from the ground. *PASP*, 119, 214–227 {156}

Janson M, Bergfors C, Goto M, et al., 2010, Spatially resolved spectroscopy of the exoplanet HR 8799 c. *ApJ*, 710, L35–L38 {2, 171, 223}

Janson M, Brandner W, Henning T, et al., 2007, NACO–SDI direct imaging search for the exoplanet ϵ Eri b. *AJ*, 133, 2442–2456 {170, 223}

Janson M, Reffert S, Brandner W, et al., 2008, A comprehensive examination of the ϵ Eri system: verification of a 4 μm narrow-band high-contrast imaging approach for planet searches. *A&A*, 488, 771–780 {223}

Jardine M, Collier Cameron A, 2008, Radio emission from exoplanets: the role of the stellar coronal density and magnetic field strength. *A&A*, 490, 843–851 {176}

Jaritz GF, Endler S, Langmayr D, et al., 2005, Roche lobe effects on expanded upper atmospheres of short-periodic giant exoplanets. *A&A*, 439, 771–775 {281}

Jaroszynski M, Paczyński B, 2002, A possible planetary event OGLE–2002–BLG–055. *Acta Astronomica*, 52, 361–367 {96}

Javaraiah J, 2005, Sun's retrograde motion and violation of even-odd cycle rule in sunspot activity. *MNRAS*, 362, 1311–1318 {66}

Jayawardhana R, Holland WS, Kalas P, et al., 2002, New submillimeter limits on dust in the 55 Cnc planetary system. *ApJ*, 570, L93–L96 {224}

Jeans JH, 1902, The stability of a spherical nebula. *Royal Society of London Philosophical Transactions Series A*, 199, 1–53 {218}

—, 1917, The part played by rotation in cosmic evolution. *MNRAS*, 77, 186–199 {224}

—, 1943, Non-solar planetary systems. *Nature*, 152, 721 {64}

Jeffery CS, Bailey ME, Chambers JE, 1997, Fractionated accretion and the solar neutrino problem. *The Observatory*, 117, 224–228 {193}

Jeffreys H, 1929a, Collision and the origin of rotation in the solar system. *MNRAS*, 89, 636–641 {224}

—, 1929b, The early history of the solar system on the collision theory. *MNRAS*, 89, 731–739 {224}

—, 1961, The effect of tidal friction on eccentricity and inclination. *MNRAS*, 122, 339–343 {245}

Jeffries RD, Totten EJ, Harmer S, et al., 2002, Membership, metallicity and lithium abundances for solar-type stars in NGC 6633. *MNRAS*, 336, 1109–1128 {187}

Jenkins C, 2008, Optical vortex coronagraphs on ground-based telescopes. *MNRAS*, 384, 515–524 {155}

Jenkins JM, 2002, The impact of solar-like variability on the detectability of transiting terrestrial planets. *ApJ*, 575, 493–505 {115}

Jenkins JM, Borucki WJ, Koch DG, et al., 2010, Discovery and Rossiter–Mclaughlin effect of exoplanet Kepler–8 b. *ApJ*, 724, 1108–1119 {129, 326}

Jenkins JM, Caldwell DA, Borucki WJ, 2002, Some tests to establish confidence in planets discovered by transit photometry. *ApJ*, 564, 495–507 {113}

Jenkins JM, Doyle LR, 2003, Detecting reflected light from close-in extrasolar giant planets with the Kepler photometer. *ApJ*, 595, 429–445 {112, 125}

Jenkins JS, Jones HRA, Goździewski K, et al., 2009a, First results from the Calan–Hertfordshire extrasolar planet search: exoplanets and the discovery of an eccentric brown dwarf in the desert. *MNRAS*, 398, 911–917 {33}

Jenkins JS, Jones HRA, Tinney CG, et al., 2006, An activity catalogue of

References

southern stars. *MNRAS*, 372, 163–173 {21}

Jenkins JS, Ramsey LW, Jones HRA, et al., 2009b, Rotational velocities for M dwarfs. *ApJ*, 704, 975–988 {186}

Jetsu L, Pelt J, 2000, Spurious periods in the terrestrial impact crater record. *A&A*, 353, 409–418 {307}

Jewitt D, 1999, Kuiper belt objects. *Annual Review of Earth and Planetary Sciences*, 27, 287–312 {295}

Jewitt D, Luu J, 1993, Discovery of the candidate Kuiper belt object 1992 QB1. *Nature*, 362, 730–732 {295}

Jha S, Charbonneau D, Garnavich PM, et al., 2000, Multicolour observations of a planetary transit of HD 209458. *ApJ*, 540, L45–L48 {140}

Ji H, Burin M, Schartman E, et al., 2006, Hydrodynamic turbulence cannot transport angular momentum effectively in astrophysical disks. *Nature*, 444, 343–346 {221}

Ji J, Kinoshita H, Liu L, et al., 2003a, The apsidal antialignment of the HD 82943 system. *Celestial Mechanics and Dynamical Astronomy*, 87, 113–120 {52}

—, 2003b, Could the 55 Cnc planetary system really be in the 3:1 mean motion resonance? *ApJ*, 585, L139–L142 {49, 52}

—, 2007, The secular evolution and dynamical architecture of the Neptunian triplet planetary system HD 69830. *ApJ*, 657, 1092–1097 {36}

Ji J, Liu L, Kinoshita H, et al., 2003c, The librating companions in HD 37124, HD 12661, HD 82943, 47 UMa, and GJ 876: alignment or antialignment? *ApJ*, 591, L57–L60 {42}

—, 2005, Could the 47 UMa planetary system be a second solar system? Predicting the Earth-like planets. *ApJ*, 631, 1191–1197 {136}

—, 2008, Habitable zones for Earth-mass planets in multiple planetary systems. *IAU Symposium*, volume 249 of *IAU Symposium*, 499–502 {285}

Ji JH, Liu L, Zhou JL, et al., 2003d, The apsidal motion in multiple planetary systems. *Chinese Astronomy and Astrophysics*, 27, 127–132 {42}

Jiang IG, Ip WH, 2001, The planetary system of υ And. *A&A*, 367, 943–948 {47}

Jiang IG, Laughlin G, Lin DNC, 2004, On the formation of brown dwarfs. *AJ*, 127, 455–459 {215}

Jiang IG, Yeh LC, Chang YC, et al., 2007, On the mass-period distributions and correlations of extrasolar planets. *AJ*, 134, 2061–2066 {37}

Jin L, Arnett WD, Sui N, et al., 2008, An interpretation of the anomalously low mass of Mars. *ApJ*, 674, L105–L108 {298}

Joergens V, 2006, Radial velocity survey for planets and brown dwarf companions to very young brown dwarfs and very low-mass stars in Chamaeleon I with UVES at the VLT. *A&A*, 446, 1165–1176 {33}

—, 2008, Binary frequency of very young brown dwarfs at separations smaller than 3 AU. *A&A*, 492, 545–555 {24}

Joergens V, Quirrenbach A, 2004, Towards characterisation of exoplanetary spectra with the VLT interferometer. *Astronomische Nachrichten Supplement*, 325, 3–7 {162}

Johansen A, Brauer F, Dullemond C, et al., 2008, A coagulation-fragmentation model for the turbulent growth and destruction of preplanetesimals. *A&A*, 486, 597–611 {226, 227}

Johansen A, Henning T, Klahr H, 2006, Dust sedimentation and self-sustained Kelvin–Helmholtz turbulence in protoplanetary disk midplanes. *ApJ*, 643, 1219–1232 {226}

Johansen A, Klahr H, 2005, Dust diffusion in protoplanetary disks by magnetorotational turbulence. *ApJ*, 634, 1353–1371 {225}

Johansen A, Oishi JS, Mac Low M, et al., 2007, Rapid planetesimal formation in turbulent circumstellar disks. *Nature*, 448, 1022–1025 {226}

Johns M, 2008a, Progress on the GMT. *SPIE Conf Ser*, volume 7012, 45 {24}

—, 2008b, The Giant Magellan Telescope (GMT). *SPIE Conf Ser*, volume 6986, 3 {159}

Johns-Krull CM, McCullough PR, Burke CJ, et al., 2008, XO-3 b: a massive planet in an eccentric orbit transiting an F5V star. *ApJ*, 677, 657–670 {108, 327}

Johnson BM, Gammie CF, 2005, Vortices in thin, compressible, unmagnetised disks. *ApJ*, 635, 149–156 {221}

Johnson ET, Goodman J, Menou K, 2006a, Diffusive migration of low-mass protoplanets in turbulent disks. *ApJ*, 647, 1413–1425 {239}

Johnson HL, Morgan WW, 1953, Fundamental stellar photometry for standards of spectral type on the revised system of the Yerkes spectral atlas. *ApJ*, 117, 313–352 {211}

Johnson JA, 2008, Planets around massive subgiants. *ASP Conf Ser*, volume 398, 59–66 {32, 34}

—, 2009, International Year of Astronomy Invited Review on Exoplanets. *PASP*, 121, 309–315 {34}

Johnson JA, Apps K, 2009, On the metal richness of M dwarfs with planets. *ApJ*, 699, 933–937 {190}

Johnson JA, Bowler BP, Howard AW, et al., 2010a, A hot Jupiter orbiting the 1.7 solar mass subgiant HD 102956. *ApJ*, 721, L153–L157 {32}

Johnson JA, Butler RP, Marcy GW, et al., 2007a, A new planet around an M dwarf: revealing a correlation between exoplanets and stellar mass. *ApJ*, 670, 833–840 {32}

Johnson JA, Fischer DA, Marcy GW, et al., 2007b, Retired a stars and their companions: exoplanets orbiting three intermediate-mass subgiants. *ApJ*, 665, 785–793 {32}

Johnson JA, Howard AW, Bowler BP, et al., 2010b, Retired A stars and their companions. IV. Seven Jovian exoplanets from Keck Observatory. *PASP*, 122, 701–711 {32, 33}

Johnson JA, Howard AW, Marcy GW, et al., 2010c, The California planet survey. II. A Saturn-mass planet orbiting the M dwarf GJ 649. *PASP*, 122, 149–155 {32, 190}

Johnson JA, Marcy GW, Fischer DA, et al., 2006b, An eccentric hot Jupiter orbiting the subgiant HD 185269. *ApJ*, 652, 1724–1728 {32, 33}

—, 2006c, The N2K consortium. VI. Doppler shifts without templates and three new short-period planets. *ApJ*, 647, 600–611 {19}

Johnson JA, Payne M, Howard AW, et al., 2011, Retired A stars and their companions. VI. A pair of interacting exoplanet pairs around the subgiants 24 Sex and HD 200964. *AJ*, 141, 16–22 {32}

Johnson JA, Winn JN, Cabrera NE, 2009, Submillimag photometry of transiting exoplanets with an orthogonal transfer array. *Bulletin of the American Astronomical Society*, volume 41, 192 {133}

Johnson JA, Winn JN, Narita N, et al., 2008a, Measurement of the spin-orbit angle of exoplanet HAT–P–1 b. *ApJ*, 686, 649–657 {129}

Johnson RE, Carlson RW, Cooper JF, et al., 2004, Radiation effects on the surfaces of the Galilean satellites. *Jupiter. The Planet, Satellites and Magnetosphere*, 485–512, Cambridge University Press {250}

Johnson RE, Huggins PJ, 2006, Toroidal atmospheres around extrasolar planets. *PASP*, 118, 1136–1143 {130}

Johnson TV, 2005, Geology of the icy satellites. *Space Sci. Rev.*, 116, 401–420 {269}

Johnson TV, Castillo-Rogez JC, Matson DL, et al., 2008b, Constraints on outer solar system chronology. *Lunar and Planetary Institute Science Conference Abstracts*, volume 39, 2314 {304}

Johnston KJ, 2003, The FAME mission. *SPIE Conf Ser*, volume 4854, 303–310 {73}

Johnston KJ, Benson JA, Hutter DJ, et al., 2006a, The Navy Prototype Optical Interferometer: recent developments since 2004. *SPIE Conf Ser*, volume 6268, 6 {162}

Johnston KJ, Dorland B, Gaume R, et al., 2006b, The Origins Billions Star Survey (OBSS): Galactic explorer. *PASP*, 118, 1428–1442 {73}

Johnstone D, Hollenbach D, Bally J, 1998, Photoevaporation of disks and clumps by nearby massive stars: application to disk destruction in the Orion nebula. *ApJ*, 499, 758–776 {222}

Jones BW, Sleep PN, 2002, The stability of the orbits of Earth-mass planets in the habitable zone of 47 UMa. *A&A*, 393, 1015–1026 {285}

Jones BW, Sleep PN, Chambers JE, 2001, The stability of the orbits of terrestrial planets in the habitable zones of known exoplanetary systems. *A&A*, 366, 254–262 {285}

Jones BW, Sleep PN, Underwood DR, 2006a, Habitability of known exoplanetary systems based on measured stellar properties. *ApJ*, 649, 1010–1019 {241, 285}

Jones BW, Underwood DR, Sleep PN, 2005, Prospects for habitable Earths in known exoplanetary systems. *ApJ*, 622, 1091–1101 {285}

Jones CA, Thompson MJ, Tobias SM, 2010, The solar dynamo. *Space Science Reviews*, 152, 591–616 {66}

Jones DJ, Diddams SA, Ranka JK, et al., 2000a, Carrier-envelope phase control of femtosecond mode-locked lasers and direct optical frequency synthesis. *Science*, 288, 635–640 {20}

Jones HRA, Butler RP, Marcy GW, et al., 2002, Extrasolar planets around HD 196050, HD 216437 and HD 160691. *MNRAS*, 337, 1170–1178 {33}

Jones HRA, Butler RP, Tinney CG, et al., 2006b, High-eccentricity planets from the Anglo–Australian Planet Search. *MNRAS*, 369, 249–256 {58}

Jones HRA, Butler RP, Wright JT, et al., 2008a, A catalogue of nearby exoplanets. *Precision Spectroscopy in Astrophysics*, 205–206 {7, 29}

Jones HRA, Jenkins JS, Barnes JR, 2008b, Close-orbiting exoplanets: formation, migration mechanisms and properties. *Exoplanets: Detection, Formation, Properties, Habitability*, 153–175, Springer {**191**}

Jones HRA, Rayner J, Ramsey L, et al., 2008c, Precision radial velocity spectrograph. *SPIE Conf Ser*, volume 7014 {**24**}

Jones RH, Lee T, Connolly HC, et al., 2000b, Formation of chondrules and CAIs: theory versus observation. *Protostars and Planets IV*, 927–962 {**297**}

Jordán A, Bakos GÁ, 2008, Observability of the general relativistic precession of periastra in exoplanets. *ApJ*, 685, 543–552 {**133**}

Jørgensen BR, Lindegren L, 2005, Determination of stellar ages from isochrones: Bayesian estimation versus isochrone fitting. *A&A*, 436, 127–143 {**187**}

Jorgensen UG, 1991, Advanced stages in the evolution of the Sun. *A&A*, 246, 118–136 {**80**}

Jorissen A, Mayor M, Udry S, 2001, The distribution of exoplanet masses. *A&A*, 379, 992–998 {**13, 35**}

Jose PD, 1965, Sun's motion and sunspots. *AJ*, 70, 193–200 {**66**}

Joshi KJ, Rasio FA, 1997, Distant companions and planets around millisecond pulsars. *ApJ*, 479, 948–959 {**76, 78**}

Joshi M, 2003, Climate model studies of synchronously rotating planets. *Astrobiology*, 3, 415–427 {**285**}

Joshi M, Haberle RM, 1997, On the ability of synchronously rotating planets to support atmospheres. *IAU Colloq. 161: Astronomical and Biochemical Origins and the Search for Life in the Universe*, 351–357 {**285**}

Joshi M, Haberle RM, Reynolds RT, 1997, Simulations of the atmospheres of synchronously rotating terrestrial planets orbiting M dwarfs: conditions for atmospheric collapse and the implications for habitability. *Icarus*, 129, 450–465 {**285**}

Joshi YC, Pollacco D, Collier Cameron A, et al., 2009, WASP-14 b: 7.3M_J transiting planet in an eccentric orbit. *MNRAS*, 392, 1532–1538 {**129, 250, 327**}

Joung MKR, Mac Low M, Ebel DS, 2004, Chondrule formation and protoplanetary disk heating by current sheets in nonideal magnetohydrodynamic turbulence. *ApJ*, 606, 532–541 {**297**}

Juckett DA, 2000, Solar activity cycles, north/south asymmetries, and differential rotation associated with solar spin-orbit variations. *Sol. Phys.*, 191, 201–226 {**66, 67**}

—, 2003, Temporal variations of low-order spherical harmonic representations of sunspot group patterns: evidence for solar spin-orbit coupling. *A&A*, 399, 731–741 {**66**}

Jura M, 2003, A tidally disrupted asteroid around the white dwarf G29–38. *ApJ*, 584, L91–L94 {**173**}

—, 2004, An observational signature of evolved oceans on extrasolar terrestrial planets. *ApJ*, 605, L65–L68 {**278, 281**}

—, 2006, Carbon deficiency in externally polluted white dwarfs: evidence for accretion of asteroids. *ApJ*, 653, 613–620 {**173, 193**}

—, 2008, Pollution of single white dwarfs by accretion of many small asteroids. *AJ*, 135, 1785–1792 {**173**}

Jura M, Farihi J, Zuckerman B, 2007, Externally polluted white dwarfs with dust disks. *ApJ*, 663, 1285–1290 {**173**}

Jurić M, Tremaine S, 2008, Dynamical origin of extrasolar planet eccentricity distribution. *ApJ*, 686, 603–620 {**127, 243**}

Juselius K, Kaufmann R, 2009, Long-run relationships among temperature, CO_2, methane, ice and dust over the last 420 000 years: cointegration analysis of the Vostok ice core data. *IOP Conference Series: Earth and Environmental Science*, 6(7), 072033 {**307**}

Kaiser N, 2007, The Pan-STARRS survey telescope project. *Advanced Maui Optical and Space Surveillance Technologies Conference* {**211**}

Kalas P, 2010, Dusty debris disks: first light from exosolar planetary systems. *EAS Publications Series*, volume 41 of *EAS Publications Series*, 133–154 {**223**}

Kalas P, Deltorn JM, Larwood J, 2001, Stellar encounters with the β Pic planetesimal system. *ApJ*, 553, 410–420 {**223**}

Kalas P, Duchene G, Fitzgerald MP, et al., 2007, Discovery of an extended debris disk around the F2V star HD 15745. *ApJ*, 671, L161–L164 {**224**}

Kalas P, Graham JR, Chiang E, et al., 2008, Optical images of an exosolar planet 25 light-years from Earth. *Science*, 322, 1345–1348 {**2, 163, 168, 171, 172, 223, 243**}

Kalas P, Graham JR, Clampin M, 2005, A planetary system as the origin of structure in Fomalhaut's dust belt. *Nature*, 435, 1067–1070 {**2, 171, 223**}

Kalas P, Graham JR, Clampin MC, et al., 2006, First scattered light images of debris disks around HD 53143 and HD 139664. *ApJ*, 637, L57–L60 {**224**}

Kalas P, Larwood J, Smith BA, et al., 2000, Rings in the planetesimal disk of β Pic. *ApJ*, 530, L133–L137 {**223**}

Kalas P, Liu MC, Matthews BC, 2004, Discovery of a large dust disk around the nearby star AU Mic. *Science*, 303, 1990–1992 {**224**}

Kaltenegger L, Eiroa C, Stankov A, et al., 2006, Target star catalogue for Darwin: nearby habitable star systems. *Direct Imaging of Exoplanets: Science and Techniques*, 89–92 {**181**}

Kaltenegger L, Selsis F, 2007, Biomarkers set in context. *Extrasolar Planets. Formation, Detection and Dynamics*, 79–87, Wiley {**285**}

Kaltenegger L, Traub WA, Jucks KW, 2007, Spectral evolution of an Earth-like planet. *ApJ*, 658, 598–616 {**290**}

Kane SR, 2007, Detectability of exoplanetary transits from radial velocity surveys. *MNRAS*, 380, 1488–1496 {**103, 109**}

Kane SR, Clarkson WI, West RG, et al., 2008, SuperWASP-north extrasolar planet candidates between 06^h < RA < 16^h. *MNRAS*, 384, 1097–1108 {**107**}

Kane SR, Collier Cameron AC, Horne K, et al., 2004, Results from the wide-angle search for planets prototype (WASP0). I. Analysis of the Pegasus field. *MNRAS*, 353, 689–696 {**107**}

—, 2005a, Results from the wide angle search for planets prototype (WASP0). III. Planet hunting in the Draco field. *MNRAS*, 364, 1091–1103 {**107**}

Kane SR, Lister TA, Collier Cameron A, et al., 2005b, Results from the wide angle search for planets prototype (WASP0). II. Stellar variability in the Pegasus field. *MNRAS*, 362, 117–126 {**107**}

Kane SR, Mahadevan S, von Braun K, et al., 2009, Refining exoplanet ephemerides and transit observing strategies. *PASP*, 121, 1386–1394 {**103, 109**}

Kane SR, Schneider DP, Ge J, 2007, Simulations for multi-object spectrograph planet surveys. *MNRAS*, 377, 1610–1622 {**27**}

Kane SR, von Braun K, 2008, Constraining orbital parameters through planetary transit monitoring. *ApJ*, 689, 492–498 {**122**}

—, 2009, Exoplanetary transit constraints based upon secondary eclipse observations. *PASP*, 121, 1096–1103 {**127**}

Kane TJ, Gardner CS, 1993, Lidar observations of the meteoric deposition of mesospheric metals. *Science*, 259, 1297–1300 {**151**}

Karmann C, Beust H, Klinger J, 2003, The physico-chemical history of falling evaporating bodies around β Pic: the sublimation of refractory material. *A&A*, 409, 347–359 {**223**}

Karttunen H, 1987, *Fundamental Astronomy*. Springer–Verlag {**309**}

Kasdin NJ, Braems I, 2006, Linear and Bayesian planet detection algorithms for the Terrestrial Planet Finder. *ApJ*, 646, 1260–1274 {**158**}

Kasdin NJ, Vanderbei RJ, Spergel DN, et al., 2003, Extrasolar planet finding via optimal apodised-pupil and shaped-pupil coronagraphs. *ApJ*, 582, 1147–1161 {**153**}

Kashyap VL, Drake JJ, Saar SH, 2008, Extrasolar giant planets and X-ray activity. *ApJ*, 687, 1339–1354 {**175, 206**}

Kasper ME, Beuzit J, Verinaud C, et al., 2008, EPICS: the exoplanet imager for the E-ELT. *SPIE Conf Ser*, volume 7015, 46 {**161**}

Kasting JF, 1988, Runaway and moist greenhouse atmospheres and the evolution of Earth and Venus. *Icarus*, 74, 472–494 {**80, 283, 284**}

—, 1996, Planetary atmosphere evolution: do other habitable planets exist and can we detect them? *Ap&SS*, 241, 3–24 {**283, 284, 288**}

—, 2004, When methane made climate. *Scientific American*, 291, 78–85 {**305, 306**}

—, 2008, Habitable planets around the Sun and other stars. *Extrasolar Planets*, 217–244 {**283**}

Kasting JF, Catling D, 2003, Evolution of a habitable planet. *ARA&A*, 41, 429–463 {**284**}

Kasting JF, Donahue TM, 1980, The evolution of atmospheric ozone. *J. Geophys. Res.*, 85, 3255–3263 {**288**}

Kasting JF, Siefert JL, 2002, Life and the evolution of Earth's atmosphere. *Science*, 296, 1066–1068 {**306**}

Kasting JF, Whitmire DP, Reynolds RT, 1993, Habitable zones around main sequence stars. *Icarus*, 101, 108–128 {**61, 283–285**}

Kaufer A, Stahl O, Tubbesing S, et al., 1999, Commissioning FEROS, the new high-resolution spectrograph at La Silla. *The Messenger*, 95, 8–12 {**24**}

References

Käufl H, Ballester P, Biereichel P, et al., 2004, CRIRES: a high-resolution infrared spectrograph for ESO's VLT. *SPIE Conf Ser*, volume 5492, 1218–1227 {**24**}

Kaula WM, 1968, *An Introduction to Planetary Physics: The Terrestrial Planets*. Wiley {**245**}

Kaula WM, Head JW III, Merrill RB, et al., 1981, *Basaltic Volcanism on the Terrestrial Planets*. Pergamon Press {**280**}

Kawaler SD, 1988, Angular momentum loss in low-mass stars. *ApJ*, 333, 236–247 {**186**}

Kawamura K, Parrenin F, Lisiecki L, et al., 2007, Northern hemisphere forcing of climatic cycles in Antarctica over the past 360 000 years. *Nature*, 448, 912–916 {**307**}

Kayser R, Refsdal S, Stabell R, 1986, Astrophysical applications of gravitational microlensing. *A&A*, 166, 36–52 {**90**}

Keenan PC, 1985, The MK classification and its calibration. *Calibration of Fundamental Stellar Quantities*, volume 111 of *IAU Symposium*, 121–135 {**211**}

Keller CF, 2004, A thousand years of climate change. *Advances in Space Research*, 34, 315–322 {**301**}

Kemp JC, Henson GD, Steiner CT, et al., 1987, The optical polarisation of the Sun measured at a sensitivity of parts in ten million. *Nature*, 326, 270–273 {**126**}

Kendall TR, Delfosse X, Martín EL, et al., 2004, Discovery of very nearby ultracool dwarfs from DENIS. *A&A*, 416, L17–L20 {**210**}

Kendall TR, Jones HRA, Pinfield DJ, et al., 2007a, New nearby, bright southern ultracool dwarfs. *MNRAS*, 374, 445–454 {**211**}

Kendall TR, Tamura M, Tinney CG, et al., 2007b, Two T dwarfs from the UKIDSS early data release. *A&A*, 466, 1059–1064 {**211**}

KenKnight CE, 1977, Methods of detecting extrasolar planets. I. Imaging. *Icarus*, 30, 422–433 {**164**}

Kennedy GM, Kenyon SJ, 2008a, Planet formation around stars of various masses: hot super-Earths. *ApJ*, 682, 1264–1276 {**238**}

—, 2008b, Planet formation around stars of various masses: the snow line and the frequency of giant planets. *ApJ*, 673, 502–512 {**37, 260**}

Kennedy GM, Kenyon SJ, Bromley BC, 2006, Planet formation around low-mass stars: the moving snow line and super-Earths. *ApJ*, 650, L139–L142 {**260**}

—, 2007, Planet formation around M-dwarfs: the moving snow line and super-Earths. *Ap&SS*, 311, 9–13 {**260**}

Kennicutt RC, 2008, The Schmidt Law: is it universal and what are its implications? *Pathways Through an Eclectic Universe*, volume 390 of *ASP Conf Ser*, 149–160 {**194**}

Kenyon SJ, 2002, Planet formation in the outer solar system. *PASP*, 114, 265–283 {**295**}

Kenyon SJ, Bromley BC, 2002a, Collisional cascades in planetesimal disks. I. Stellar flybys. *AJ*, 123, 1757–1775 {**109**}

—, 2002b, Dusty rings: signposts of recent planet formation. *ApJ*, 577, L35–L38 {**222**}

—, 2004a, Collisional cascades in planetesimal disks. II. Embedded planets. *AJ*, 127, 513–530 {**224**}

—, 2004b, Detecting the dusty debris of terrestrial planet formation. *ApJ*, 602, L133–L136 {**224**}

—, 2004c, Stellar encounters as the origin of distant solar system objects in highly eccentric orbits. *Nature*, 432, 598–602 {**295**}

—, 2005, Prospects for detection of catastrophic collisions in debris disks. *AJ*, 130, 269–279 {**178**}

—, 2006, Terrestrial planet formation. I. The transition from oligarchic growth to chaotic growth. *AJ*, 131, 1837–1850 {**228, 229**}

—, 2008, Variations on debris disks: icy planet formation at 30–150 AU for $1 - 3 M_\odot$ main-sequence stars. *ApJS*, 179, 451–483 {**260**}

Kenyon SJ, Hartmann L, 1987, Spectral energy distributions of T Tauri stars: disk flaring and limits on accretion. *ApJ*, 323, 714–733 {**220**}

Kenyon SJ, Luu JX, 1998, Accretion in the early Kuiper belt. I. Coagulation and velocity evolution. *AJ*, 115, 2136–2160 {**228**}

Kenyon SL, Lawrence JS, Ashley MCB, et al., 2006, Atmospheric scintillation at Dome C, Antarctica: implications for photometry and astrometry. *PASP*, 118, 924–932 {**63**}

Kenyon SL, Storey JWV, 2006, A review of optical sky brightness and extinction at Dome C, Antarctica. *PASP*, 118, 489–502 {**63**}

Kepler SO, Costa JES, Castanheira BG, et al., 2005, Measuring the evolution of the most stable optical clock G117–B15A. *ApJ*, 634, 1311–1318 {**80**}

Kepler SO, Mukadam A, Winget DE, et al., 2000, Evolutionary timescale of the pulsating white dwarf G117–B15A: the most stable optical clock known. *ApJ*, 534, L185–L188 {**80**}

Kepler SO, Winget DE, Nather RE, et al., 1991, A detection of the evolutionary time scale of the DA white dwarf G117–B15A with the Whole Earth Telescope. *ApJ*, 378, L45–L48 {**80**}

Keränen P, Ouyed R, 2003, Planets orbiting quark nova compact remnants. *A&A*, 407, L51–L54 {**77**}

Khodachenko ML, Ribas I, Lammer H, et al., 2007, Coronal mass ejection (CME) activity of low mass M stars as an important factor for the habitability of terrestrial exoplanets. I. CME impact on expected magnetospheres of Earth-like exoplanets in close-in habitable zones. *Astrobiology*, 7, 167–184 {**285**}

Kholshevnikov KV, Kuznetsov ED, 2002, Selection effect in semi-major axes of orbits of extrasolar planets. *Solar System Research*, 36, 466–477 {**36**}

Khonina SN, Kotlyar VV, Shinkaryev MV, et al., 1992, The phase rotor filter. *Journal of Modern Optics*, 39, 1147–1154 {**155**}

Kiang NY, Segura A, Tinetti G, et al., 2007a, Spectral signatures of photosynthesis. II. Coevolution with other stars and the atmosphere on extrasolar worlds. *Astrobiology*, 7, 252–274 {**289**}

Kiang NY, Siefert J, Govindjee, et al., 2007b, Spectral signatures of photosynthesis. I. Review of Earth organisms. *Astrobiology*, 7, 222–251 {**289**}

Kibrick RI, Clarke DA, Deich WTS, et al., 2006, A comparison of exposure meter systems for three exoplanet-hunting spectrometers: Hamilton, HIRES and APF. *SPIE Conf Ser*, volume 6274 {**21, 24**}

Kieffer HH, Jakosky BM, Snyder CW, et al., 1992, *Mars*. University of Arizona Press {**293**}

Kilic M, Gould A, Koester D, 2009, Limits on unresolved planetary companions to white dwarf remnants of 14 intermediate-mass stars. *ApJ*, 705, 1219–1225 {**173**}

Kilic M, von Hippel T, Leggett SK, et al., 2005, Excess infrared radiation from the massive DAZ white dwarf GD 362: a debris disk? *ApJ*, 632, L115–L118 {**173**}

Kilkenny D, 2007, Pulsating hot subdwarfs: an observational review. *Communications in Asteroseismology*, 150, 234–240 {**81**}

Kilkenny D, van Wyk F, Marang F, 2003, The sdB eclipsing system HW Vir: a substellar companion? *The Observatory*, 123, 31–36 {**82**}

Kim KM, Mkrtichian DE, Lee BC, et al., 2006, Precise radial velocities with BOES: detection of low-amplitude pulsations in the K-giant α Arietis. *A&A*, 454, 839–844 {**24**}

King JR, Boesgaard AM, Schuler SC, 2005, Keck HIRES spectroscopy of four candidate solar twins. *AJ*, 130, 2318–2325 {**286**}

King JR, Deliyannis CP, Boesgaard AM, 1997a, The ^{9}Be abundances of α Cen A and B and the Sun: implications for stellar evolution and mixing. *ApJ*, 478, 778–786 {**201**}

King JR, Deliyannis CP, Hiltgen DD, et al., 1997b, Lithium abundances in the solar twins 16 Cyg A and B and the solar analogue α Cen A. *AJ*, 113, 1871–1883 {**199**}

King LJ, Jackson N, Blandford RD, et al., 1998, A complete infrared Einstein ring in the gravitational lens system B1938+666. *MNRAS*, 295, L41–L44 {**84**}

Kingsley SA, 2001, Optical SETI observatories in the new millennium: a review. *SPIE Conf Ser* (eds. Kingsley SA, Bhathal R), volume 4273, 72–92 {**291**}

Kinoshita H, Yoshida H, Nakai H, 1991, Symplectic integrators and their application to dynamical astronomy. *Celestial Mechanics and Dynamical Astronomy*, 50, 59–71 {**44**}

Kipping DM, 2008, Transiting planets: light-curve analysis for eccentric orbits. *MNRAS*, 389, 1383–1390 {**121, 140**}

—, 2009a, Transit timing effects due to an exomoon. *MNRAS*, 392, 181–189 {**135**}

—, 2009b, Transit timing effects due to an exomoon. II. *MNRAS*, 396, 1797–1804 {**135, 136**}

Kipping DM, Bakos GÁ, Hartman J, et al., 2010, HAT-P-24 b: an inflated hot Jupiter on a 3.36-day period transiting a hot, metal-poor star. *ApJ*, 725, 2017–2028 {**326**}

Kipping DM, Fossey SJ, Campanella G, 2009, On the detectability of habitable exomoons with Kepler-class photometry. *MNRAS*, 400, 398–405 {**135, 136**}

Kirkpatrick JD, 2005, New spectral types L and T. *ARA&A*, 43, 195–245 {**210, 212**}

Kirkpatrick JD, Cruz KL, Barman TS, et al., 2008, A sample of very young field L dwarfs and implications for the brown dwarf lithium test at early ages. *ApJ*, 689, 1295–1326 {**212**}

Kirkpatrick JD, Reid IN, Liebert J, et al., 1999, Dwarfs cooler than M: the definition of spectral type L using discoveries from the 2 Micron All-Sky Survey (2MASS). *ApJ*, 519, 802–833 {**210, 212**}

Kita NT, Huss GR, Tachibana S, et al., 2005, Constraints on the origin of chondrules and CAIs from short-lived and long-lived radionuclides. *Chondrites and the Protoplanetary Disk*, volume 341 of *ASP Conf Ser*, 558–587 {**297**}

Kite ES, Manga M, Gaidos E, 2008, Geodynamics and rate of volcanism on massive Earth-like planets. *AGU Fall Meeting Abstracts*, C1327 {**279**}

Kivelson MG, Warnecke J, Bennett L, et al., 1998, Ganymede's magnetosphere: magnetometer overview. *J. Geophys. Res.*, 103, 19963–19972 {**250**}

Kjeldsen H, Bedding TR, Butler RP, et al., 2005, Solar-like oscillations in α Cen B. *ApJ*, 635, 1281–1290 {**22, 202**}

Kjeldsen H, Bedding TR, Christensen-Dalsgaard J, 2009, Measurements of stellar properties through asteroseismology: a tool for planet transit studies. *IAU Symposium*, volume 253 of *IAU Symposium*, 309–317 {**203**}

Kjeldsen H, Frandsen S, 1992, High-precision time-resolved CCD photometry. *PASP*, 104, 413–434 {**104**}

Klahr H, Bodenheimer P, 2006, Formation of giant planets by concurrent accretion of solids and gas inside an anticyclonic vortex. *ApJ*, 639, 432–440 {**221**}

Klahr H, Rózyczka M, Dziourkevitch N, et al., 2006, Turbulence in protoplanetary accretion disks: driving mechanisms and role in planet formation. *Planet Formation*, 42–63, Cambridge University Press {**221**}

Klein B, Jura M, Koester D, et al., 2010, Chemical abundances in the externally polluted white dwarf GD 40: evidence of a rocky extrasolar minor planet. *ApJ*, 709, 950–962 {**173**}

Klein R, Apai D, Pascucci I, et al., 2003, First detection of millimeter dust emission from brown dwarf disks. *ApJ*, 593, L57–L60 {**220**}

Kleine T, Mezger K, Palme H, et al., 2004, The W isotope evolution of the bulk silicate Earth: constraints on the timing and mechanisms of core formation and accretion. *Earth and Planetary Science Letters*, 228, 109–123 {**297**}

—, 2005a, Early core formation in asteroids and late accretion of chondrite parent bodies: evidence from ^{182}Hf–^{182}W in CAIs, metal-rich chondrites, and iron meteorites. *Geochim. Cosmochim. Acta*, 69, 5805–5818 {**297**}

Kleine T, Münker C, Mezger K, et al., 2002, Rapid accretion and early core formation on asteroids and the terrestrial planets from Hf–W chronometry. *Nature*, 418, 952–955 {**297**}

Kleine T, Palme H, Mezger K, et al., 2005b, Hf–W chronometry of lunar metals and the age and early differentiation of the Moon. *Science*, 310, 1671–1674 {**297**}

Kleinman SJ, Nather RE, Winget DE, et al., 1994, Observational limits on companions to G29–38. *ApJ*, 436, 875–884 {**80**}

Kley W, 2000, On the migration of a system of protoplanets. *MNRAS*, 313, L47–L51 {**41**}

—, 2010, Planets in mean motion resonance. *Formation and Evolution of Exoplanets*, 203–222, Wiley {**40**}

Kley W, Bitsch B, Klahr H, 2009, Planet migration in three-dimensional radiative disks. *A&A*, 506, 971–987 {**234**}

Kley W, Crida A, 2008, Migration of protoplanets in radiative disks. *A&A*, 487, L9–L12 {**239**}

Kley W, Lee MH, Murray N, et al., 2005, Modeling the resonant planetary system GJ 876. *A&A*, 437, 727–742 {**36, 41, 49**}

Kley W, Peitz J, Bryden G, 2004, Evolution of planetary systems in resonance. *A&A*, 414, 735–747 {**14, 41, 42, 82, 242**}

Kley W, Sándor Z, 2007, The formation of resonant planetary systems. *Extrasolar Planets. Formation, Detection and Dynamics*, 99–115, Wiley {**242**}

Klimov YG, 1963, The deflection of light rays in the gravitational fields of galaxies. *Soviet Physics Doklady*, 8, 119 {**84**}

Klioner SA, 2003, A practical relativistic model for microarcsec astrometry in space. *AJ*, 125, 1580–1597 {**64, 309, 310**}

—, 2004, Physically adequate proper reference system of a test observer and relativistic description of the Gaia attitude. *Phys. Rev. D*, 69(12), 124001–124009 {**64**}

—, 2008, Relativistic scaling of astronomical quantities and the system of astronomical units. *A&A*, 478, 951–958 {**309**}

Knutson HA, Charbonneau D, Allen LE, et al., 2007a, A map of the day-night contrast of the extrasolar planet HD 189733 b. *Nature*, 447, 183–186 {**138–142**}

—, 2008, The 3.6-8.0 μm broad-band emission spectrum of HD 209458 b: evidence for an atmospheric temperature inversion. *ApJ*, 673, 526–531 {**140–142**}

Knutson HA, Charbonneau D, Burrows A, et al., 2009a, Detection of a temperature inversion in the broad-band infrared emission spectrum of TrES-4. *ApJ*, 691, 866–874 {**142**}

Knutson HA, Charbonneau D, Cowan NB, et al., 2009b, Multiwavelength constraints on the day-night circulation patterns of HD 189733 b. *ApJ*, 690, 822–836 {**142**}

Knutson HA, Charbonneau D, Noyes RW, et al., 2007b, Using stellar limb-darkening to refine the properties of HD 209458 b. *ApJ*, 655, 564–575 {**140**}

Kobayashi H, Ida S, 2001, The effects of a stellar encounter on a planetesimal disk. *Icarus*, 153, 416–429 {**109**}

Kobayashi Y, Gouda N, Yano T, et al., 2008, The current status of the Nano-JASMINE project. *IAU Symposium*, volume 248 of *IAU Symposium*, 270–271 {**73**}

Koch A, Woehl H, 1984, The use of molecular iodine absorption lines as wavelength references for solar Doppler shift measurements. *A&A*, 134, 134–138 {**19**}

Koch D, Borucki W, Basri G, et al., 2006, The Kepler mission: astrophysics and eclipsing binaries. *Ap&SS*, 304, 391–395 {**111**}

Koch D, Borucki W, Cullers K, et al., 1996, System design of a mission to detect Earth-sized planets in the inner orbits of solar-like stars. *J. Geophys. Res.*, 101, 9297–9302 {**111**}

Koch D, Borucki WJ, Basri G, et al., 2010a, Kepler mission design, realised photometric performance, and early science. *ApJ*, 713, L79–L86 {**111**}

Koch D, Borucki WJ, Rowe JF, et al., 2010b, Discovery of the transiting planet Kepler–5 b. *ApJ*, 713, L131–L135 {**326**}

Koechlin L, Serre D, Duchon P, 2005, High resolution imaging with Fresnel interferometric arrays: suitability for exoplanet detection. *A&A*, 443, 709–720 {**157**}

Koen C, Lombard F, 2002, Testing photometry of stars for planetary transits. *Stellar Structure and Habitable Planet Finding*, volume 485 of *ESA Special Publication*, 159–161 {**113**}

Kohler R, Leinert C, 1998, Multiplicity of T Tauri stars in Taurus after ROSAT. *A&A*, 331, 977–988 {**207**}

Kokkotas KD, Schafer G, 1995, Tidal and tidal-resonant effects in coalescing binaries. *MNRAS*, 275, 301–308 {**204**}

Kokubo E, Ida S, 1996, On runaway growth of planetesimals. *Icarus*, 123, 180–191 {**228**}

—, 1998, Oligarchic growth of protoplanets. *Icarus*, 131, 171–178 {**228**}

—, 2000, Formation of protoplanets from planetesimals in the solar nebula. *Icarus*, 143, 15–27 {**228, 232**}

—, 2002, Formation of protoplanet systems and diversity of planetary systems. *ApJ*, 581, 666–680 {**228**}

—, 2007, Formation of terrestrial planets from protoplanets. II. Statistics of planetary spin. *ApJ*, 671, 2082–2090 {**129, 300**}

Kokubo E, Ida S, Makino J, 2000, Evolution of a circumterrestrial disk and formation of a single moon. *Icarus*, 148, 419–436 {**302**}

Kokubo E, Kominami J, Ida S, 2006, Formation of terrestrial planets from protoplanets. I. Statistics of basic dynamical properties. *ApJ*, 642, 1131–1139 {**229**}

Kolesnikov A, Kutcherov VG, Goncharov AF, 2009, Methane-derived hydrocarbons produced under upper-mantle conditions. *Nature Geoscience*, 2, 566–570 {**279**}

Kominami J, Ida S, 2002, The effect of tidal interaction with a gas disk on formation of terrestrial planets. *Icarus*, 157, 43–56 {**242**}

Komitov B, Bonev B, 2001, Amplitude variations of the 11 year cycle and the current solar maximum 23. *ApJ*, 554, L119–L122 {**67**}

Konacki M, 2005a, An extrasolar giant planet in a close triple-star system. *Nature*, 436, 230–233 {**33, 60**}

—, 2005b, Precision radial velocities of double-lined spectroscopic binaries with an iodine absorption cell. *ApJ*, 626, 431–438 {**19, 33, 59**}

Konacki M, Lewandowski W, Wolszczan A, et al., 1999a, Are there planets around the pulsar PSR B0329+54? *ApJ*, 519, L81–L84 {**78**}

Konacki M, Maciejewski AJ, 1999, Frequency analysis of reflex velocities of stars with planets. *ApJ*, 518, 442–449 {14}

Konacki M, Maciejewski AJ, Wolszczan A, 1999b, Resonance in the PSR B1257+12 planetary system. *ApJ*, 513, 471–476 {76}

—, 2000, Improved timing formula for the PSR B1257+12 planetary system. *ApJ*, 544, 921–926 {76}

—, 2002, Frequency decomposition of astrometric signature of planetary systems. *ApJ*, 567, 566–578 {67}

Konacki M, Torres G, Jha S, et al., 2003a, An extrasolar planet that transits the disk of its parent star. *Nature*, 421, 507–509 {**2, 28, 106, 327**}

Konacki M, Torres G, Sasselov DD, et al., 2003b, High-resolution spectroscopic follow-up of OGLE planetary transit candidates in the Galactic bulge: two possible Jupiter-mass planets and two blends. *ApJ*, 597, 1076–1091 {106}

—, 2005, A transiting extrasolar giant planet around the star OGLE–TR–10. *ApJ*, 624, 372–377 {**106, 327**}

Konacki M, Wolszczan A, 2003, Masses and orbital inclinations of planets in the PSR B1257+12 system. *ApJ*, 591, L147–L150 {**76, 77**}

Königl A, 1991, Disk accretion onto magnetic T Tauri stars. *ApJ*, 370, L39–L43 {186}

Konigl A, Pudritz RE, 2000, Disk winds and the accretion-outflow connection. *Protostars and Planets IV*, 759–788 {219}

Konopliv AS, Yoder CF, Standish EM, et al., 2006, A global solution for the Mars static and seasonal gravity, Mars orientation, Phobos and Deimos masses, and Mars ephemeris. *Icarus*, 182, 23–50 {**18, 309**}

Kopal Z, 1977, Fourier analysis of the light curves of eclipsing variables. XI. *Ap&SS*, 50, 225–246 {119}

Koppenhoefer J, Afonso C, Saglia RP, et al., 2009, Investigating the potential of the Pan-Planets project using Monte Carlo simulations. *A&A*, 494, 707–717 {110}

Korechoff RP, Diner DJ, Tubbs EF, et al., 1994, Extrasolar planet detection. *Ap&SS*, 212, 369–383 {164}

Kornacki AS, Fegley B, 1986, The abundance and relative volatility of refractory trace elements in Allende Ca-Al-rich inclusions: implications for chemical and physical processes in the solar nebula. *Earth and Planetary Science Letters*, 79, 217–234 {257}

Kornet K, Bodenheimer P, Rózyczka M, et al., 2005, Formation of giant planets in disks with different metallicities. *A&A*, 430, 1133–1138 {191}

Kornet K, Wolf S, 2006, Radial distribution of planets: predictions based on the core-accretion gas-capture planet-formation model. *A&A*, 454, 989–995 {**234, 251**}

Kornet K, Wolf S, Rózyczka M, 2006, Formation of giant planets around stars with various masses. *A&A*, 458, 661–668 {234}

Korycansky DG, Pollack JB, 1993, Numerical calculations of the linear response of a gaseous disk to a protoplanet. *Icarus*, 102, 150–165 {238}

Koskinen TT, Aylward AD, Miller S, 2007a, A stability limit for the atmospheres of giant extrasolar planets. *Nature*, 450, 845–848 {138}

Koskinen TT, Aylward AD, Smith CGA, et al., 2007b, A thermospheric circulation model for extrasolar giant planets. *ApJ*, 661, 515–526 {138}

Kóspál Á, Ardila DR, Moór A, et al., 2009, On the relationship between debris disks and planets. *ApJ*, 700, L73–L77 {224}

Kotani T, Lacour S, Perrin G, et al., 2009, Pupil remapping for high contrast astronomy: results from an optical testbed. *Optics Express*, 17, 1925–1934 {156}

Kotliarov I, 2008, A structural law of planetary systems. *MNRAS*, 390, 1411–1412 {48}

—, 2009, An invalid evidence for the Titius–Bode law. *Philippine Journal of Astronomy, Volume 1, Issue 2, p.5-10*, 1(2), 020000–10 {48}

Kovács G, Bakos GÁ, Hartman JD, et al., 2010, HAT–P–15 b: a 10.9-day extrasolar planet transiting a solar-type star. *ApJ*, 724, 866–877 {326}

Kovács G, Bakos GÁ, Noyes RW, 2005, A trend filtering algorithm for wide-field variability surveys. *MNRAS*, 356, 557–567 {104}

Kovács G, Bakos GÁ, Torres G, et al., 2007, HAT–P–4 b: a metal-rich low-density transiting hot Jupiter. *ApJ*, 670, L41–L44 {326}

Kovács G, Zucker S, Mazeh T, 2002, A box-fitting algorithm in the search for periodic transits. *A&A*, 391, 369–377 {**105, 114**}

Kovtyukh VV, Soubiran C, Belik SI, et al., 2003, High precision effective temperatures for 181 F-K dwarfs from line-depth ratios. *A&A*, 411, 559–564 {185}

Kozai Y, 1962, Secular perturbations of asteroids with high inclination and eccentricity. *AJ*, 67, 591–598 {**57, 300**}

Kozhevatov IE, Kulikova EK, Cheragin NP, 1995, An integrating interference spectrometer. *Astronomy Letters*, 21, 418–422 {25}

Krasinsky GA, Brumberg VA, 2004, Secular increase of astronomical unit from analysis of the major planet motions, and its interpretation. *Celestial Mechanics and Dynamical Astronomy*, 90, 267–288 {309}

Kraus JD, 1966, *Radio Astronomy*. McGraw-Hill {174}

Krist JE, Balasubramanian K, Beichman CA, et al., 2009, The JWST–NIRCam coronagraph: mask design and fabrication. *SPIE Conf Ser*, volume 7440, 28 {163}

Krist JE, Beichman CA, Trauger JT, et al., 2007, Hunting planets and observing disks with the JWST–NIRCam coronagraph. *SPIE Conf Ser*, volume 6693, 16 {**155, 158, 163**}

Krivov AV, Müller S, Löhne T, et al., 2008, Collisional and thermal emission models of debris disks: toward planetesimal population properties. *ApJ*, 687, 608–622 {223}

Krivov AV, Queck M, Löhne T, et al., 2007, On the nature of clumps in debris disks. *A&A*, 462, 199–210 {223}

Krot AN, Amelin Y, Cassen P, et al., 2005a, Young chondrules in CB chondrites from a giant impact in the early solar system. *Nature*, 436, 989–992 {298}

Krot AN, Keil K, Scott ERD, et al., 2007, Classification of meteorites. *Treatise on Geochemistry* (eds. Holland HD, Turekian KK) {297}

Krot AN, Yurimoto H, Hutcheon ID, et al., 2005b, Chronology of the early solar system from chondrule-bearing calcium-aluminium-rich inclusions. *Nature*, 434, 998–1001 {298}

Kroupa P, Bouvier J, 2003, On the origin of brown dwarfs and free-floating planetary-mass objects. *MNRAS*, 346, 369–380 {215}

Krymolowski Y, Mazeh T, 1999, Studies of multiple stellar systems. II. Second-order averaged Hamiltonian to follow long-term orbital modulations of hierarchical triple systems. *MNRAS*, 304, 720–732 {47}

Kubala A, Black D, Szebehely V, 1993, Stability of outer planetary orbits around binary stars: a comparison of Hill's and Laplace's stability criteria. *Celestial Mechanics and Dynamical Astronomy*, 56, 51–68 {57}

Kubas D, Cassan A, Dominik M, et al., 2008, Limits on additional planetary companions to OGLE–2005–BLG–390L. *A&A*, 483, 317–324 {98}

Kuchner MJ, 2003, Volatile-rich Earth-mass planets in the habitable zone. *ApJ*, 596, L105–L108 {**267, 269, 280, 281**}

—, 2004, A minimum-mass extrasolar nebula. *ApJ*, 612, 1147–1151 {220}

Kuchner MJ, Crepp J, Ge J, 2005, Eighth-order image masks for terrestrial planet finding. *ApJ*, 628, 466–473 {153}

Kuchner MJ, Koresko CD, Brown ME, 1998, Keck speckle imaging of the white dwarf G29–38: no brown dwarf companion detected. *ApJ*, 508, L81–L83 {210}

Kuchner MJ, Lecar M, 2002, Halting planet migration in the evacuated centers of protoplanetary disks. *ApJ*, 574, L87–L89 {241}

Kuchner MJ, Seager S, 2005, Extrasolar carbon planets. *ArXiv Astrophysics e-prints* {267}

Kuchner MJ, Spergel DN, 2003, Notch-filter masks: practical image masks for planet-finding coronagraphs. *ApJ*, 594, 617–626 {153}

Kuchner MJ, Traub WA, 2002, A coronagraph with a band-limited mask for finding terrestrial planets. *ApJ*, 570, 900–908 {153}

Kudryavtsev NA, 1951, Petroleum economy (in Russian). *Neftianoye Khozyaistvo*, 9(17-29) {279}

Kuhn JR, Bush RI, Scherrer P, et al., 1998, The Sun's shape and brightness. *Nature*, 392, 155–157 {134}

Kuhn JR, Potter D, Parise B, 2001, Imaging polarimetric observations of a new circumstellar disk system. *ApJ*, 553, L189–L191 {158}

Kuiper GP, 1951, On the origin of the solar system. *Proceedings of the National Academy of Science*, 37, 1–14 {235}

Kumar SS, 1963, The structure of stars of very low mass. *ApJ*, 137, 1121–1123 {209}

Kürster M, Endl M, Rodler F, 2006, In search of terrestrial planets in the habitable zone of M dwarfs. *The Messenger*, 123, 21–24 {**18, 19**}

Kürster M, Endl M, Rouesnel F, et al., 2003, The low-level radial velocity variability in Barnard's star: secular acceleration, indications for convective redshift, and planet mass limits. *A&A*, 403, 1077–1087 {19}

Kürster M, Hatzes AP, Cochran WD, et al., 1999, The ESO precise radial-velocity survey for extrasolar planets: results from the first five years. *IAU Colloq. 170: Precise Stellar Radial Velocities*, volume 185 of *ASP Conf Ser*, 154–161 {24}

Kurucz RL, 1991, New opacity calculations. *NATO ASIC Proc. 341: Stellar Atmospheres – Beyond Classical Models*, 441–449 {264}

—, 1993, ATLAS9 stellar atmosphere programs and $2\,\mathrm{km\,s^{-1}}$ grid. *CD-ROM No. 13*. Cambridge, Mass, 13 {**185**}

Kusaka T, Nakano T, Hayashi C, 1970, Growth of solid particles in the primordial solar nebula. *Progress of Theoretical Physics*, 44, 1580–1595 {**225**}

Kutzbach JE, 1981, Monsoon climate of the early Holocene: climate experiment with the Earth's orbital parameters for 9000 years ago. *Science*, 214, 59–61 {**301**}

Kuzuhara M, Tamura M, Kudo T, et al., 2009, A substellar companion with a very wide separation from a binary T Tauri star. *American Institute of Physics Conference Series*, volume 1158, 253–254 {**168**}

Laakso T, Rantala J, Kaasalainen M, 2006, Gravitational scattering by giant planets. *A&A*, 456, 373–378 {**285**}

Labeyrie A, 1970, Attainment of diffraction-limited resolution in large telescopes by Fourier analysing speckle patterns in star images. *A&A*, 6, 85–87 {**157**}

—, 1995, Images of exoplanets obtainable from dark speckles in adaptive telescopes. *A&A*, 298, 544–548 {**157**}

—, 1996, Resolved imaging of extrasolar planets with future 10–100 km optical interferometric arrays. *A&AS*, 118, 517–524 {**168**}

—, 2002, Hypertelescopes and exo-Earth coronagraphy. *Earth-like Planets and Moons*, volume 514 of *ESA Special Publication*, 245–250 {**168**}

—, 2003, Detecting exo-Earths with hypertelescopes in space: the exo-Earth discoverer concept. *EAS Publications Series*, volume 8, 327–342 {**168**}

Labeyrie A, Le Coroller H, 2004, Extrasolar planet imaging. *SPIE Conf Ser*, volume 5491, 90–96 {**157, 168**}

Labeyrie A, Le Coroller H, Dejonghe J, et al., 2003, Hypertelescope imaging: from exoplanets to neutron stars. *SPIE Conf Ser*, volume 4852, 236–247 {**168**}

Lacour S, Thiébaut E, Perrin G, 2007, High dynamic range imaging with a single-mode pupil remapping system: a self-calibration algorithm for redundant interferometric arrays. *MNRAS*, 374, 832–846 {**156**}

Lada CJ, Lada EA, 2003, Embedded clusters in molecular clouds. *ARA&A*, 41, 57–115 {**293**}

Lafrenière D, Doyon R, Marois C, et al., 2007a, The Gemini deep planet survey. *ApJ*, 670, 1367–1390 {**36, 170**}

Lafrenière D, Jayawardhana R, van Kerkwijk MH, 2008, Direct imaging and spectroscopy of a planetary-mass candidate companion to a young solar analogue. *ApJ*, 689, L153–L156 {**152, 168**}

Lafrenière D, Marois C, Doyon R, et al., 2007b, A new algorithm for point-spread function subtraction in high-contrast imaging: a demonstration with angular differential imaging. *ApJ*, 660, 770–780 {**158**}

—, 2009, HST–NICMOS detection of HR 8799 b in 1998. *ApJ*, 694, L148–L152 {**223**}

Lagrange A, Desort M, Meunier N, 2010, Using the Sun to estimate Earth-like planets detection capabilities. I. Impact of cold spots. *A&A*, 512, A38 {**22**}

Lagrange A, Kasper M, Boccaletti A, et al., 2009a, Constraining the orbit of the possible companion to β Pic: new deep imaging observations. *A&A*, 506, 927–934 {**224**}

Lagrange AM, Beust H, Udry S, et al., 2006, New constrains on GJ 86B: VLT near infrared coronagraphic imaging survey of planetary hosts. *A&A*, 459, 955–963 {**170**}

Lagrange AM, Gratadour D, Chauvin G, et al., 2009b, A probable giant planet imaged in the β Pic disk: VLT–NACO deep L'-band imaging. *A&A*, 493, L21–L25 {**2, 168, 171, 223, 224**}

Lai D, Foucart F, Lin DNC, 2010, Evolution of spin direction of accreting magnetic protostars and spin-orbit misalignment in exoplanetary systems. *ArXiv e-prints* {**129**}

Laine RO, Lin DNC, Dong S, 2008, Interaction of close-in planets with the magnetosphere of their host stars. I. Diffusion, ohmic dissipation of time-dependent field, planetary inflation, and mass loss. *ApJ*, 685, 521–542 {**241**}

Lambeck K, 1977, Tidal dissipation in the oceans: astronomical, geophysical and oceanographic consequences. *Royal Society of London Philosophical Transactions Series A*, 287, 545–594 {**247**}

Lambert DL, Heath JE, Edvardsson B, 1991, Lithium abundances for 81 F dwarfs. *MNRAS*, 253, 610–618 {**199**}

Lammer H, Bredehöft JH, Coustenis A, et al., 2009a, What makes a planet habitable? *A&A Rev.*, 17, 181–249 {**283**}

Lammer H, Kasting JF, Chassefière E, et al., 2008, Atmospheric escape and evolution of terrestrial planets and satellites. *Space Science Reviews*, 139, 399–436 {**281**}

Lammer H, Lichtenegger HIM, Kulikov YN, et al., 2007, Coronal mass ejection (CME) activity of low mass M stars as an important factor for the habitability of terrestrial exoplanets. II. CME-induced ion pick up of Earth-like exoplanets in close-in habitable zones. *Astrobiology*, 7, 185–207 {**281**}

Lammer H, Odert P, Leitzinger M, et al., 2009b, Determining the mass loss limit for close-in exoplanets: what can we learn from transit observations? *A&A*, 506, 399–410 {**146**}

Lammer H, Selsis F, Ribas I, et al., 2003, Atmospheric loss of exoplanets resulting from stellar x-ray and extreme-ultraviolet heating. *ApJ*, 598, L121–L124 {**281**}

—, 2004, Hydrodynamic escape of exo-planetary atmospheres. *Stellar Structure and Habitable Planet Finding*, volume 538 of *ESA Special Publication*, 339–342 {**281**}

Landgraf M, Jehn R, 2001, Zodiacal infrared foreground prediction for space based infrared interferometer missions. *Ap&SS*, 278, 357–365 {**165**}

Landgraf M, Liou JC, Zook HA, et al., 2002, Origins of solar system dust beyond Jupiter. *AJ*, 123, 2857–2861 {**295**}

Landscheidt T, 1981, Swinging Sun, 79-year cycle, and climatic change. *J. Interdiscipl. Cycle Res.*, 12, 3–19 {**66**}

—, 1999, Extrema in sunspot cycle linked to the Sun's motion. *Sol. Phys.*, 189, 413–424 {**66**}

Lane BF, Colavita MM, Boden AF, et al., 2000, Palomar Testbed Interferometer: update. *SPIE Conf Ser*, volume 4006, 452–458 {**69**}

Lane BF, Muterspaugh MW, 2004, Differential astrometry of subarcsecond scale binaries at the Palomar Testbed Interferometer. *ApJ*, 601, 1129–1135 {**69**}

Lane BF, Muterspaugh MW, Shao M, 2006, Calibrating an interferometric null. *ApJ*, 648, 1276–1284 {**163**}

Lane WM, Cohen AS, Cotton WD, et al., 2008, The VLA Low-Frequency Sky Survey (VLSS). *Frontiers of Astrophysics: A Celebration of NRAO's 50th Anniversary*, volume 395 of *ASP Conf Ser*, 370 {**176**}

Lang KR, 1980, *Astrophysical Formulae: a Compendium for the Physicist and Astrophysicist*. Springer–Verlag {**16, 18**}

—, 1992, *Astrophysical Data I. Planets and Stars*. Springer-=Verlag {**311**}

Langford SV, Wyithe JSB, Turner EL, 2009, Photometric variability in Earthshine observations. *Astrobiology*, 9, 305–310 {**289**}

Langlois M, Burrows A, Hinz P, 2006, Ground-based direct detection of close-in extrasolar planets with nulling and high order adaptive optics. *A&A*, 445, 1143–1149 {**163**}

Langston GI, Orban CM, Bastian TS, 2002, A search for cyclotron emission from known extrasolar planets. *Bulletin of the American Astronomical Society*, volume 34, 1176 {**176**}

Langton J, Laughlin G, 2007, Observational consequences of hydrodynamic flows on hot Jupiters. *ApJ*, 657, L113–L116 {**138**}

Lanza AF, 2008, Hot Jupiters and stellar magnetic activity. *A&A*, 487, 1163–1170 {**175, 205**}

—, 2009, Stellar coronal magnetic fields and star-planet interaction. *A&A*, 505, 339–350 {**175**}

—, 2010, Hot Jupiters and the evolution of stellar angular momentum. *A&A*, 512, A77 {**249**}

Lanza AF, Bonomo AS, Moutou C, et al., 2010, Photospheric activity, rotation, and radial velocity variations of the planet-hosting star CoRoT–7. *A&A*, 520, A53 {**22, 111, 206**}

Lanza AF, De Martino C, Rodonò M, 2008, Astrometric effects of solar-like magnetic activity in late-type stars and their relevance for the detection of extrasolar planets. *New Astronomy*, 13, 77–84 {**65**}

Lanza AF, Rodonò M, Pagano I, et al., 2003, Modeling the rotational modulation of the Sun as a star. *A&A*, 403, 1135–1149 {**22**}

Laplace PS, 1796, *Exposition du Système du Monde*. Circle-Sociale, Paris {**224**}

Larimer JW, 1967, Chemical fractionations in meteorites. I. Condensation of the elements. *Geochim. Cosmochim. Acta*, 31, 1215–1238 {**257, 297**}

—, 1988, The cosmochemical classification of the elements. *Meteorites and the Early Solar System*, 375–389, University of Arizona Press {**195, 259**}

Larson RB, 1984, Gravitational torques and star formation. *MNRAS*, 206, 197–207 {**301**}

—, 1992, Towards understanding the stellar initial mass function. *MNRAS*, 256, 641–646 {**214**}

Larwood JD, Kalas PG, 2001, Close stellar encounters with planetesimal disks: the dynamics of asymmetry in the β Pic system. *MNRAS*, 323, 402–416 {**223**}

Laskar J, 1990, The chaotic motion of the solar system: a numerical estimate of the size of the chaotic zones. *Icarus*, 88, 266–291 {**46**}

—, 1992, Le mouvement chaotique du système solaire. *Journées Scientifiques du Service des Calculs et de Mécanique Céleste du Bureau des Longitudes*, 52–56 {**299**}

—, 1993, Frequency analysis for multi-dimensional systems: global dynamics and diffusion. *Physica D Nonlinear Phenomena*, 67, 257–281 {**46**}

—, 1994, Large-scale chaos in the solar system. *A&A*, 287, L9–L12 {**44, 299–301**}

—, 1996, Marginal stability and chaos in the solar system. *Dynamics, Ephemerides, and Astrometry of the Solar System*, volume 172 of *IAU Symposium*, 75–88 {**45, 299, 300**}

—, 2000, On the spacing of planetary systems. *Physical Review Letters*, 84, 3240–3243 {**229**}

—, 2003, Chaos in the solar system. *Annales Henri Poincaré*, 4, 693–705 {**129**}

Laskar J, Correia ACM, 2009, HD 60532, a planetary system in a 3:1 mean motion resonance. *A&A*, 496, L5–L8 {**50**}

Laskar J, Gastineau M, 2009, Existence of collisional trajectories of Mercury, Mars and Venus with the Earth. *Nature*, 459, 817–819 {**299**}

Laskar J, Joutel F, Robutel P, 1993, Stabilisation of the Earth's obliquity by the moon. *Nature*, 361, 615–617 {**286, 301**}

Laskar J, Robutel P, 1993, The chaotic obliquity of the planets. *Nature*, 361, 608–612 {**286, 300, 301**}

—, 2001, High order symplectic integrators for perturbed Hamiltonian systems. *Celestial Mechanics and Dynamical Astronomy*, 80, 39–62 {**44, 46, 299**}

Latham DW, 1997, Radial-velocity searches for low-mass companions orbiting solar-type stars. *Planets Beyond the Solar System and the Next Generation of Space Missions*, volume 119 of *ASP Conf Ser*, 19–27 {**29**}

—, 2008a, Characterisation of terrestrial planets identified by the Kepler mission. *Physica Scripta Volume T*, 130(1), 014034–37 {**111**}

—, 2008b, Kepler and follow-up science. *ASP Conf Ser*, volume 398, 461–466 {**21, 24, 25, 114**}

Latham DW, Bakos GÁ, Torres G, et al., 2009, Discovery of a transiting planet and eight eclipsing binaries in HATNet field G205. *ApJ*, 704, 1107–1119 {**106, 326**}

Latham DW, Borucki WJ, Koch DG, et al., 2010, Kepler-7 b: a transiting planet with unusually low density. *ApJ*, 713, L140–L144 {**326**}

Latham DW, Stefanik RP, Mazeh T, et al., 1989, The unseen companion of HD 114762: a probable brown dwarf. *Nature*, 339, 38–40 {**2, 28, 210**}

—, 1998, Low-mass companions found in a large radial-velocity survey. *Brown Dwarfs and Extrasolar Planets*, volume 134 of *ASP Conf Ser*, 178–187 {**29**}

Lattanzi MG, Spagna A, Sozzetti A, et al., 2000, Space-borne global astrometric surveys: the hunt for extrasolar planets. *MNRAS*, 317, 211–224 {**72**}

Laughlin G, 2000, Mining the metal-rich stars for planets. *ApJ*, 545, 1064–1073 {**113, 192**}

Laughlin G, Adams FC, 1997, Possible stellar metallicity enhancements from the accretion of planets. *ApJ*, 491, L51–L54 {**192, 241**}

—, 1998, The modification of planetary orbits in dense open clusters. *ApJ*, 508, L171–L174 {**237**}

—, 1999, Stability and chaos in the υ And planetary system. *ApJ*, 526, 881–889 {**47**}

—, 2000, The frozen Earth: binary scattering events and the fate of the solar system. *Icarus*, 145, 614–627 {**282**}

Laughlin G, Bodenheimer P, 1994, Nonaxisymmetric evolution in protostellar disks. *ApJ*, 436, 335–354 {**236**}

Laughlin G, Bodenheimer P, Adams FC, 2004a, The core accretion model predicts few Jovian-mass planets orbiting red dwarfs. *ApJ*, 612, L73–L76 {**37**}

Laughlin G, Butler RP, Fischer DA, et al., 2005a, The GJ 876 planetary system: a progress report. *ApJ*, 622, 1182–1190 {**49, 50**}

Laughlin G, Chambers J, Fischer D, 2002, A dynamical analysis of the 47 UMa planetary system. *ApJ*, 579, 455–467 {**52**}

Laughlin G, Chambers JE, 2001, Short-term dynamical interactions among extrasolar planets. *ApJ*, 551, L109–L113 {**14, 49, 79**}

—, 2002, Extrasolar Trojans: the viability and detectability of planets in the 1:1 resonance. *AJ*, 124, 592–600 {**53, 54, 136**}

Laughlin G, Deming D, Langton J, et al., 2009, Rapid heating of the atmosphere of an extrasolar planet. *Nature*, 457, 562–564 {**109**}

Laughlin G, Marcy GW, Vogt SS, et al., 2005b, On the eccentricity of HD 209458 b. *ApJ*, 629, L121–L124 {**140, 145**}

Laughlin G, Steinacker A, Adams FC, 2004b, Type I planetary migration with magnetohydrodynamic turbulence. *ApJ*, 608, 489–496 {**239, 241**}

Laughlin G, Wolf A, Vanmunster T, et al., 2005c, A comparison of observationally determined radii with theoretical radius predictions for short-period transiting extrasolar planets. *ApJ*, 621, 1072–1078 {**260**}

Launhardt R, 2009, Exoplanet search with astrometry. *ArXiv e-prints* {**69**}

Launhardt R, Henning T, Queloz D, et al., 2008, The ESPRI project: narrow-angle astrometry with VLTI–PRIMA. *IAU Symposium*, volume 248, 417–420 {**69, 94, 162**}

Laureijs RJ, Jourdain de Muizon M, Leech K, et al., 2002, A 25 μm search for Vega-like disks around main-sequence stars with ISO. *A&A*, 387, 285–293 {**222, 224**}

Lauretta DS, McSween HY, 2006, *Meteorites and the Early Solar System II*. University of Arizona Press {**293**}

Lauretta DS, Nagahara H, Alexander CMO, 2006, Petrology and origin of ferromagnesian silicate chondrules. *Meteorites and the Early Solar System II*, 431–459, University of Arizona Press {**297**}

Lawler SM, Beichman CA, Bryden G, et al., 2009, Explorations beyond the snow line: Spitzer–IRS spectra of debris disks around solar-type stars. *ApJ*, 705, 89–111 {**224**}

Lawrence A, Warren SJ, Almaini O, et al., 2007, The UKIRT Infrared Deep Sky Survey (UKIDSS). *MNRAS*, 379, 1599–1617 {**211**}

Lawrence JS, 2004a, Adaptive-optics performance of Antarctic telescopes. *Appl. Opt.*, 43, 1435–1449 {**162**}

—, 2004b, Infrared and submillimetre atmospheric characteristics of high Antarctic plateau sites. *PASP*, 116, 482–492 {**161**}

Lawrence JS, Ashley MCB, Burton MG, et al., 2006, Exoplanet detection from Dome C, Antarctica: opportunities and challenges. *IAU Colloq. 200: Direct Imaging of Exoplanets: Science and Techniques*, 297–300 {**162**}

Lawrence JS, Ashley MCB, Tokovinin A, et al., 2004, Exceptional astronomical seeing conditions above Dome C in Antarctica. *Nature*, 431, 278–281 {**63, 161**}

Laws C, Gonzalez G, 2001, A differential spectroscopic analysis of 16 Cyg A and B. *ApJ*, 553, 405–409 {**192**}

—, 2003, A reevaluation of the super-lithium-rich star in NGC 6633. *ApJ*, 595, 1148–1153 {**193**}

Laws C, Gonzalez G, Walker KM, et al., 2003, Parent stars of extrasolar planets. VII. New abundance analyses of 30 systems. *AJ*, 125, 2664–2677 {**188**}

Lawson PR, Ahmed A, Gappinger RO, et al., 2006, Terrestrial Planet Finder Interferometer technology status and plans. *SPIE Conf Ser*, volume 6268, 70 {**166**}

Lawson PR, Lay OP, Martin SR, et al., 2008, Terrestrial Planet Finder Interferometer: 2007–2008 progress and plans. *SPIE Conf Ser*, volume 7013, 80 {**166**}

Laymand M, Vauclair S, 2007, Asteroseismology of exoplanets host stars: the special case of ι Hor. *A&A*, 463, 657–662 {**203**}

Lazio TJW, 2008, The Square Kilometer Array. *The Evolution of Galaxies Through the Neutral Hydrogen Window*, volume 1035 of *American Institute of Physics Conference Series*, 303–309 {**177**}

Lazio TJW, Farrell WM, 2007, Magnetospheric emissions from the planet orbiting τ Boo: a multi-epoch search. *ApJ*, 668, 1182–1188 {**175, 177**}

Lazio TJW, Farrell WM, Dietrick J, et al., 2004a, The radiometric Bode's law and extrasolar planets. *ApJ*, 612, 511–518 {**176**}

Lazio TJW, Fischer J, 2004, Mid- and far-infrared infrared space observatory limits on dust disks around millisecond pulsars. *AJ*, 128, 842–845 {**77**}

Lazio TJW, Tarter JC, Wilner DJ, 2004b, The cradle of life. *New Astronomy Reviews*, 48, 985–991 {**177**}

Lazorenko PF, 2006, Astrometric precision of observations at VLT–FORS2. *A&A*, 449, 1271–1279 {**63**}

Lazorenko PF, Lazorenko GA, 2004, Filtration of atmospheric noise in narrow-field astrometry with very large telescopes. *A&A*, 427, 1127–

1143 {**62**}

Lazorenko PF, Mayor M, Dominik M, et al., 2007, High-precision astrometry on the VLT–FORS1 at time scales of few days. *A&A*, 471, 1057–1067 {**63**}

—, 2009, Precision multi-epoch astrometry with VLT cameras FORS1/2. *A&A*, 505, 903–918 {**63**}

Le Bouquin J, Absil O, Benisty M, et al., 2009, The spin-orbit alignment of the Fomalhaut planetary system probed by optical long baseline interferometry. *A&A*, 498, L41–L44 {**223**}

Le Coroller H, Dejonghe J, Arpesella C, et al., 2004, Tests with a Carlinatype hypertelescope prototype. I. Demonstration of star tracking and fringe acquisition with a balloon-suspended focal camera. *A&A*, 426, 721–728 {**168**}

Le Roux B, Ragazzoni R, 2005, Beating the Poisson limit by coupling an occulting mask to wavefront sensing. *MNRAS*, 359, L23–L26 {**151**}

Lean J, Rind D, 1998, Climate forcing by changing solar radiation. *Journal of Climate*, 11, 3069–3094 {**284**}

Lecar M, Franklin FA, Holman MJ, et al., 2001, Chaos in the solar system. *ARA&A*, 39, 581–631 {**45, 46**}

Lecar M, Podolak M, Sasselov D, et al., 2006, On the location of the snow line in a protoplanetary disk. *ApJ*, 640, 1115–1118 {**260**}

Lecar M, Sasselov DD, 2003, Dispersing the gaseous protoplanetary disk and halting type II migration. *ApJ*, 596, L99–L100 {**241**}

Lecavelier des Etangs A, 2007, A diagram to determine the evaporation status of extrasolar planets. *A&A*, 461, 1185–1193 {**144**}

Lecavelier des Etangs A, Pont F, Vidal-Madjar A, et al., 2008, Rayleigh scattering in the transit spectrum of HD 189733 b. *A&A*, 481, L83–L86 {**140**}

Lecavelier des Etangs A, Scholl H, Roques F, et al., 1996, Perturbations of a planet on the β Pic circumstellar dust disk. *Icarus*, 123, 168–179 {**224**}

Lecavelier des Etangs A, Sirothia SK, Gopal-Krishna, et al., 2009, GMRT radio observations of the transiting extrasolar planet HD 189733 b at 244 and 614 MHz. *A&A*, 500, L51–L54 {**177**}

Lecavelier des Etangs A, Vidal-Madjar A, 2009, Is β Pic b the transiting planet of November 1981? *A&A*, 497, 557–562 {**224**}

Lecavelier des Etangs A, Vidal-Madjar A, Burki G, et al., 1997, β Pic light variations. I. The planetary hypothesis. *A&A*, 328, 311–320 {**130, 224**}

Lecavelier des Etangs A, Vidal-Madjar A, Ferlet R, 1999, Photometric stellar variation due to extrasolar comets. *A&A*, 343, 916–922 {**130, 131**}

Lecavelier des Etangs A, Vidal-Madjar A, McConnell JC, et al., 2004, Atmospheric escape from hot Jupiters. *A&A*, 418, L1–L4 {**146, 281**}

Lee BL, Ge J, Fleming SW, et al., 2011, MARVELS-1 b: a short-period, brown dwarf desert candidate from the SDSS–III MARVELS planet search. *ApJ*, 728, 32–35 {**29**}

Lee DW, Lee CU, Park BG, et al., 2008, Microlensing detections of planets in binary stellar systems. *ApJ*, 672, 623–628 {**95**}

Lee JW, Kim S, Kim C, et al., 2009, The sdB+M eclipsing system HW Vir and its circumbinary planets. *AJ*, 137, 3181–3190 {**2, 79, 82, 133**}

Lee MH, 2000, On the validity of the coagulation equation and the nature of runaway growth. *Icarus*, 143, 74–86 {**228**}

—, 2004, Diversity and origin of 2:1 orbital resonances in extrasolar planetary systems. *ApJ*, 611, 517–527 {**41, 42**}

Lee MH, Butler RP, Fischer DA, et al., 2006, On the 2:1 orbital resonance in the HD 82943 planetary system. *ApJ*, 641, 1178–1187 {**52**}

Lee MH, Peale SJ, 2002, Dynamics and origin of the 2:1 orbital resonances of the GJ 876 planets. *ApJ*, 567, 596–609 {**14, 41–43, 49**}

—, 2003, Secular evolution of hierarchical planetary systems. *ApJ*, 592, 1201–1216 {**43, 52**}

Lee MH, Peale SJ, Pfahl E, et al., 2007, Evolution of the obliquities of the giant planets in encounters during migration. *Icarus*, 190, 103–109 {**301**}

Lee MH, Thommes EW, 2009, Planetary migration and eccentricity and inclination resonances in extrasolar planetary systems. *ApJ*, 702, 1662–1672 {**53**}

Léger A, 1999, Strategies for remote detection of life. *NATO ASIC Proc. 532: Planets Outside the Solar System: Theory and Observations*, 397–412 {**282**}

Léger A, Mariotti J, Ollivier M, et al., 1998, Search for extrasolar life through planetary spectroscopy. *Origins*, volume 148 of *ASP Conf Ser*, 458–471 {**165**}

Léger A, Mariotti JM, Mennesson B, et al., 1996, The Darwin project. *Ap&SS*, 241, 135–146 {**165**}

Léger A, Ollivier M, Altwegg K, et al., 1999, Is the presence of H_2O and O_3 in an exoplanet a reliable signature of a biological activity? *A&A*, 341, 304–311 {**288**}

Léger A, Pirre M, Marceau FJ, 1994a, How to evidence life on a distant planet. *Ap&SS*, 212, 327–333 {**288**}

—, 1994b, Relevance of oxygen and ozone detections in the search for primitive life in extrasolar planets. *Advances in Space Research*, 14, 117–122 {**288**}

Léger A, Rouan D, Schneider J, et al., 2009, Transiting exoplanets from the CoRoT space mission. VIII. CoRoT-7 b: the first super-Earth with measured radius. *A&A*, 506, 287–302 {**2, 138, 246–248, 250, 326**}

Léger A, Selsis F, Sotin C, et al., 2004, A new family of planets? Ocean planets. *Icarus*, 169, 499–504 {**267–270, 281**}

Leggett SK, Cushing MC, Saumon D, et al., 2009, The physical properties of four 600 K T dwarfs. *ApJ*, 695, 1517–1526 {**212**}

Leigh C, Collier Cameron A, Guillot T, 2003a, Prospects for spectroscopic reflected-light planet searches. *MNRAS*, 346, 890–896 {**124**}

Leigh C, Collier Cameron A, Horne K, et al., 2003b, A new upper limit on the reflected starlight from τ Boo b. *MNRAS*, 344, 1271–1282 {**125**}

Leigh C, Collier Cameron AC, Udry S, et al., 2003c, A search for starlight reflected from HD 75289 b. *MNRAS*, 346, L16–L20 {**125**}

Leigh D, Horowitz P, 1997, Millions and billions: the META and BETA searches at Harvard. *IAU Colloq. 161: Astronomical and Biochemical Origins and the Search for Life in the Universe*, 601–610 {**290**}

Leinert C, Zinnecker H, Weitzel N, et al., 1993, A systematic approach for young binaries in Taurus. *A&A*, 278, 129–149 {**207**}

Leinhardt ZM, Richardson DC, 2002, N-body simulations of planetesimal evolution: effect of varying impactor mass ratio. *Icarus*, 159, 306–313 {**228**}

Lenzen R, Brandl B, Brandner W, 2006, The science case for exoplanets and star formation using mid-IR instrumentation at the OWL telescope. *The Scientific Requirements for Extremely Large Telescopes*, volume 232 of *IAU Symposium*, 329–333 {**159**}

Lenzen R, Close L, Brandner W, et al., 2004, A novel simultaneous differential imager for the direct imaging of giant planets. *SPIE Conf Ser*, volume 5492, 970–977 {**158**}

Lenzuni P, Chernoff DF, Salpeter EE, 1991, Rosseland and Planck mean opacities of a zero-metallicity gas. *ApJS*, 76, 759–801 {**264**}

Lépine S, 2005, Nearby stars from the LSPM-north proper motion catalogue. I. Main-sequence dwarfs and giants within 33 pc of the Sun. *AJ*, 130, 1680–1692 {**183**}

Leroy J, 2000, *Polarisation of Light and Astronomical Observation*. Gordon and Breach, Amsterdam {**126**}

Lestrade J, 2008, Astrometry with ALMA: a giant step from 0.1 arcsec to 0.1 mas in the sub-millimeter. *IAU Symposium*, volume 248, 170–177 {**177**}

Lestrade J, Phillips RB, Jones DL, et al., 1996, Search for extrasolar planets around radio-emitting stars by very long baseline interferometry astrometry. *J. Geophys. Res.*, 101, 14837–14842 {**173**}

Lestrade J, Preston RA, Jones DL, et al., 1999, High-precision VLBI astrometry of radio-emitting stars. *A&A*, 344, 1014–1026 {**173**}

Lestrade JF, 2003, Future astrometry with ALMA to characterise extrasolar planet orbits. *Scientific Frontiers in Research on Extrasolar Planets*, volume 294 of *ASP Conf Ser*, 587–590 {**177**}

Levison HF, Agnor C, 2003, The role of giant planets in terrestrial planet formation. *AJ*, 125, 2692–2713 {**229**}

Levison HF, Duncan MJ, 1993, The gravitational sculpting of the Kuiper belt. *ApJ*, 406, L35–L38 {**44, 300**}

—, 1994, The long-term dynamical behaviour of short-period comets. *Icarus*, 108, 18–36 {**44**}

Levison HF, Lissauer JJ, Duncan MJ, 1998, Modeling the diversity of outer planetary systems. *AJ*, 116, 1998–2014 {**243**}

Levrard B, Correia ACM, Chabrier G, et al., 2007, Tidal dissipation within hot Jupiters: a new appraisal. *A&A*, 462, L5–L8 {**145**}

Levrard B, Winisdoerffer C, Chabrier G, 2009, Falling transiting extrasolar giant planets. *ApJ*, 692, L9–L13 {**245**}

Lewis GF, 2001, Gravitational microlensing of stars with transiting planets. *A&A*, 380, 292–299 {**96**}

Lewis GF, Ibata RA, 2000, Probing the atmospheres of planets orbiting microlensed stars via polarisation variability. *ApJ*, 539, L63–L66 {**95**}

Lewis JS, 1969, Observability of spectroscopically active compounds in the atmosphere of Jupiter. *Icarus*, 10, 393–409 {**276**}

Lewis JS, Prinn RG, 1984, *Planets and Their Atmospheres: Origin and Evolution*, volume 33. Orlando Academic Press {**305**}

Lewis KM, Sackett PD, Mardling RA, 2008, Possibility of detecting moons of pulsar planets through time-of-arrival analysis. *ApJ*, 685, L153–L156 {**78**}

Li C, Benedick AJ, Fendel P, et al., 2008a, A laser frequency comb that enables radial velocity measurements with a precision of $1\,\mathrm{cm\,s}^{-1}$. *Nature*, 452, 610–612 {**20, 21**}

Li SL, Lin DNC, Liu XW, 2008b, Extent of pollution in planet-bearing stars. *ApJ*, 685, 1210–1219 {**192**}

Liang MC, Seager S, Parkinson CD, et al., 2004, On the insignificance of photochemical hydrocarbon aerosols in the atmospheres of close-in extrasolar giant planets. *ApJ*, 605, L61–L64 {**277**}

Libert A, Tsiganis K, 2009a, Kozai resonance in extrasolar systems. *A&A*, 493, 677–686 {**53**}

—, 2009b, Trapping in high-order orbital resonances and inclination excitation in extrasolar systems. *MNRAS*, 400, 1373–1382 {**53**}

Libert AS, Henrard J, 2006, Secular apsidal configuration of non-resonant exoplanetary systems. *Icarus*, 183, 186–192 {**42**}

—, 2007, Analytical study of the proximity of exoplanetary systems to mean-motion resonances. *A&A*, 461, 759–763 {**52**}

Lichtenberg AJ, Lieberman MA, 1983, *Regular and Stochastic Motion*. Springer, New York {**45**}

Liebes S, 1964, Gravitational lenses. *Physical Review*, 133, 835–844 {**84**}

Liebig C, Wambsganss J, 2010, Detectability of extrasolar moons as gravitational microlenses. *A&A*, 520, A68 {**95, 96**}

Lin DNC, 1997, Planetary formation in protostellar disks. *IAU Colloq. 163: Accretion Phenomena and Related Outflows*, volume 121 of *ASP Conf Ser*, 321–330 {**241**}

Lin DNC, Bodenheimer P, Richardson DC, 1996, Orbital migration of the planetary companion of 51 Peg to its present location. *Nature*, 380, 606–607 {**127, 192, 238, 241**}

Lin DNC, Ida S, 1997, On the origin of massive eccentric planets. *ApJ*, 477, 781–784 {**243**}

Lin DNC, Papaloizou J, 1979, Tidal torques on accretion disks in binary systems with extreme mass ratios. *MNRAS*, 186, 799–812 {**229**}

—, 1980, On the structure and evolution of the primordial solar nebula. *MNRAS*, 191, 37–48 {**240**}

—, 1986a, On the tidal interaction between protoplanets and the primordial solar nebula. II. Self-consistent nonlinear interaction. *ApJ*, 307, 395–409 {**237**}

—, 1986b, On the tidal interaction between protoplanets and the protoplanetary disk. III. Orbital migration of protoplanets. *ApJ*, 309, 846–857 {**237**}

Lin DNC, Papaloizou JCB, 1993, On the tidal interaction between protostellar disks and companions. *Protostars and Planets III*, 749–835 {**238**}

Lin DNC, Papaloizou JCB, Terquem C, et al., 2000, Orbital evolution and planet-star tidal interaction. *Protostars and Planets IV*, 1111–1134 {**238, 241**}

Lin DNC, Pringle JE, 1990, The formation and initial evolution of protostellar disks. *ApJ*, 358, 515–524 {**236**}

Lindal GF, 1992, The atmosphere of Neptune: an analysis of radio occultation data acquired with Voyager 2. *AJ*, 103, 967–982 {**294**}

Lindal GF, Sweetnam DN, Eshleman VR, 1985, The atmosphere of Saturn: an analysis of the Voyager radio occultation measurements. *AJ*, 90, 1136–1146 {**294**}

Lindal GF, Wood GE, Levy GS, et al., 1981, The atmosphere of Jupiter: an analysis of the Voyager radio occultation measurements. *J. Geophys. Res.*, 86, 8721–8727 {**294**}

Lindegren L, 1978, Photoelectric astrometry: a comparison of methods for precise image location. *IAU Colloq. 48: Modern Astrometry*, 197–217 {**62**}

—, 1980, Atmospheric limitations of narrow-field optical astrometry. *A&A*, 89, 41–47 {**62, 63**}

—, 2009, Gaia: astrometric performance and current status of the project. *IAU Symposium 261. Relativity in Fundamental Astronomy: Dynamics, Reference Frames, and Data Analysis*, 261, 1601 {**72**}

Lindegren L, Dravins D, 2003, The fundamental definition of radial velocity. *A&A*, 401, 1185–1201 {**18, 309**}

Lindegren L, Mignard F, Söderhjelm S, et al., 1997, Double star data in the Hipparcos Catalogue. *A&A*, 323, L53–L56 {**69**}

Lineweaver CH, 2001, An estimate of the age distribution of terrestrial planets in the universe: quantifying metallicity as a selection effect. *Icarus*, 151, 307–313 {**197**}

Lineweaver CH, Fenner Y, Gibson BK, 2004, The Galactic habitable zone and the age distribution of complex life in the Milky Way. *Science*, 303, 59–62 {**286**}

Link F, 1936, Sur les conséquences photométriques de la déviation d'Einstein. *C R Acad Sci Paris*, 202, 917–919 {**84**}

Liou JC, Zook HA, Greaves JS, et al., 2002, Structure of the Edgeworth–Kuiper belt dust disk and implications for extrasolar planet(s) in ε Eri. *IAU Colloq. 181: Dust in the Solar System and Other Planetary Systems*, 225–228 {**223**}

Lippincott SL, 1960, The unseen companion of the fourth nearest star, Lalande 21185. *AJ*, 65, 349 {**64**}

Liseau R, Risacher C, Brandeker A, et al., 2008, q^1 Eri: a solar-type star with a planet and a dust belt. *A&A*, 480, L47–L50 {**224**}

Liske J, Pasquini L, Bonifacio P, et al., 2009, From Espresso to Codex. *Science with the VLT in the ELT Era*, 243–248 {**24**}

Lissauer JJ, 1987, Timescales for planetary accretion and the structure of the protoplanetary disk. *Icarus*, 69, 249–265 {**232, 233**}

—, 1993, Planet formation. *ARA&A*, 31, 129–174 {**210, 225, 228**}

—, 1995, On the diversity of plausible planetary systems. *Icarus*, 114, 217–236 {**37, 225**}

—, 1997, Formation, frequency and spacing of habitable planets. *IAU Colloq. 161: Astronomical and Biochemical Origins and the Search for Life in the Universe*, 289–297 {**285**}

—, 1999, Three planets for υ And. *Nature*, 398, 659–660 {**47**}

—, 2007, Planets formed in habitable zones of M dwarf stars probably are deficient in volatiles. *ApJ*, 660, L149–L152 {**285**}

Lissauer JJ, Fabrycky DC, Ford EB, et al., 2011, A closely packed system of low-mass, low-density planets transiting Kepler-11. *Nature*, 470, 53–58 {**122**}

Lissauer JJ, Kary DM, 1991, The origin of the systematic component of planetary rotation. I. Planet on a circular orbit. *Icarus*, 94, 126–159 {**301**}

Lissauer JJ, Quintana EV, Chambers JE, et al., 2004, Terrestrial planet formation in binary star systems. *Revista Mexicana de Astronomia y Astrofisica Conference Series*, volume 22, 99–103 {**58**}

Lissauer JJ, Rivera EJ, 2001, Stability analysis of the planetary system orbiting υ And. II. Simulations using new Lick observatory fits. *ApJ*, 554, 1141–1150 {**43, 47**}

Lissauer JJ, Safronov VS, 1991, The random component of planetary rotation. *Icarus*, 93, 288–297 {**301**}

Lister TA, Anderson DR, Gillon M, et al., 2009, WASP-16 b: a new Jupiter-like planet transiting a southern solar analogue. *ApJ*, 703, 752–756 {**327**}

Lister TA, West RG, Wilson DM, et al., 2007, SuperWASP-north extrasolar planet candidates between $17^\mathrm{h} < \mathrm{RA} < 18^\mathrm{h}$. *MNRAS*, 379, 647–662 {**107**}

Liu K, Yue YL, Xu RX, 2007, PSR B1828–11: a precession pulsar torqued by a quark planet? *MNRAS*, 381, L1–L5 {**78, 79**}

Liu MC, Fischer DA, Graham JR, et al., 2002, Crossing the brown dwarf desert using adaptive optics: a very close L dwarf companion to the nearby solar analogue HR 7672. *ApJ*, 571, 519–527 {**190**}

Liu MC, Matthews BC, Williams JP, et al., 2004, A submillimeter search of nearby young stars for cold dust: discovery of debris disks around two low-mass stars. *ApJ*, 608, 526–532 {**224**}

Liu WM, Hinz PM, Meyer MR, et al., 2003, A resolved circumstellar disk around the Herbig Ae star HD 100546 in the thermal infrared. *ApJ*, 598, L111–L114 {**224**}

Livengood TA, A'Hearn MF, Deming D, et al., 2008, EPOXI empirical test of optical characterisation of an Earth-like planet. *AAS/Division for Planetary Sciences Meeting Abstracts*, volume 40, 01.03 {**113**}

Livio M, Pringle JE, 2003, Metallicity, planetary formation and migration. *MNRAS*, 346, L42–L44 {**146, 191**}

Livio M, Pringle JE, Saffer RA, 1992, Planets around massive white dwarfs. *MNRAS*, 257, 15P–16P {**79**}

Livio M, Pringle JE, Wood K, 2005, Disks and planets around massive white dwarfs. *ApJ*, 632, L37–L39 {**172**}

Livio M, Soker N, 1984, Star-planet systems as possible progenitors of cataclysmic binaries. *MNRAS*, 208, 763–781 {**79, 179**}

Lloyd JP, 2006, Detection of exoplanets from the Antarctic plateau. *IAU Colloq. 200: Direct Imaging of Exoplanets: Science and Techniques*, 301–304 {**162**}

Lloyd JP, Oppenheimer BR, Graham JR, 2002, The potential of differential astrometric interferometry from the high Antarctic plateau. *Publications of the Astronomical Society of Australia*, 19, 318–322 {**63**}

Lloyd-Hart M, Angel JRP, Groesbeck TD, et al., 1998, First astronomical images sharpened with adaptive optics using a sodium laser guide star. *ApJ*, 493, 950–954 {**151**}

Lo Curto G, Mayor M, Benz W, et al., 2010, The HARPS search for southern extrasolar planets. XXII. Multiple planet systems from the HARPS volume-limited sample. *A&A*, 512, A48 {**35, 36**}

Lodders K, 1999, Alkali element chemistry in cool dwarf atmospheres. *ApJ*, 519, 793–801 {**276**}

—, 2003, Solar system abundances and condensation temperatures of the elements. *ApJ*, 591, 1220–1247 {**195, 257–260, 265, 272, 296**}

—, 2004a, Brown dwarfs: faint at heart, rich in chemistry. *Science*, 303(323-324) {**277**}

—, 2004b, Jupiter formed with more tar than ice. *ApJ*, 611, 587–597 {**271**}

—, 2010, Exoplanet chemistry. *Formation and Evolution of Exoplanets*, 157–186, Wiley {**256, 257, 271, 273–276**}

Lodders K, Fegley B, 1993, Lanthanide and actinide chemistry at high C/O ratios in the solar nebula. *Earth and Planetary Science Letters*, 117, 125–145 {**258**}

—, 2002, Atmospheric chemistry in giant planets, brown dwarfs, and low-mass dwarf stars. I. Carbon, nitrogen, and oxygen. *Icarus*, 155, 393–424 {**259, 265, 271, 275**}

Lodge OJ, 1919, Gravitation and light. *Nature*, 104, 354–355 {**84**}

Lodieu N, Dobbie PD, Deacon NR, et al., 2009, Two distant brown dwarfs in the UKIRT Infrared Deep Sky Survey Deep Extragalactic Survey Data Release 2. *MNRAS*, 395, 1631–1639 {**211**}

Lodieu N, Hambly NC, Jameson RF, et al., 2007, New brown dwarfs in Upper Scorpius using UKIDSS Galactic Cluster Survey science verification data. *MNRAS*, 374, 372–384 {**211**}

—, 2008, Near-infrared cross-dispersed spectroscopy of brown dwarf candidates in the Upper Scorpius association. *MNRAS*, 383, 1385–1396 {**215**}

Loeb A, 2005, A dynamical method for measuring the masses of stars with transiting planets. *ApJ*, 623, L45–L48 {**131**}

—, 2009, Long-term evolution in transit duration of extrasolar planets from magnetic activity in their parent stars. *New Astronomy*, 14, 363–364 {**131**}

Loeb A, Gaudi BS, 2003, Periodic flux variability of stars due to the reflex doppler effect induced by planetary companions. *ApJ*, 588, L117–L120 {**126, 132**}

Loeillet B, Bouchy F, Deleuil M, et al., 2008a, Doppler search for exoplanet candidates and binary stars in a CoRoT field using a multifiber spectrograph. II. Global analysis and first results. *A&A*, 479, 865–875 {**111**}

Loeillet B, Shporer A, Bouchy F, et al., 2008b, Refined parameters and spectroscopic transit of the super-massive planet HD 147506 b. *A&A*, 481, 529–533 {**129**}

Löhmer O, Wolszczan A, Wielebinski R, 2004, A search for cold dust around neutron stars. *A&A*, 425, 763–766 {**77**}

Looney LW, Tobin JJ, Fields BD, 2006, Radioactive probes of the supernova-contaminated solar nebula: evidence that the Sun was born in a cluster. *ApJ*, 652, 1755–1762 {**293**}

Lopez B, Petrov RG, Mennesson B, et al., 2000, A Darwin proposal for the observation of planets orbiting evolved stars. *Darwin and Astronomy: the Infrared Space Interferometer*, volume 451 of *ESA Special Publication*, 77–80 {**285**}

Lopez B, Schneider J, Danchi WC, 2005, Can life develop in the expanded habitable zones around red giant stars? *ApJ*, 627, 974–985 {**285**}

López-Morales M, Butler RP, Fischer DA, et al., 2008, Two Jupiter-mass planets orbiting HD 154672 and HD 205739. *AJ*, 136, 1901–1905 {**24, 29**}

Lord HC, 1965, Molecular equilibria and condensation in a solar nebula and cool stellar atmospheres. *Icarus*, 4, 279–288 {**257**}

Loredo TJ, 2004, Bayesian adaptive exploration. *Bayesian Inference and Maximum Entropy Methods in Science and Engineering*, volume 707 of *American Institute of Physics Conference Series*, 330–346 {**15**}

Lorimer DR, 2005, Binary and millisecond pulsars. *Living Reviews in Relativity*, 8, 7 {**77**}

Loubeyre P, Letoullec R, Hausermann D, et al., 1996, X-ray diffraction and equation of state of hydrogen at megabar pressures. *Nature*, 383, 702–704 {**262, 294**}

Love SG, Keil K, Scott ERD, 1995, Electrical discharge heating of chondrules in the solar nebula. *Icarus*, 115, 97–108 {**297**}

Lovelace RVE, Romanova MM, Barnard AW, 2008, Planet migration and disk destruction due to magneto-centrifugal stellar winds. *MNRAS*, 389, 1233–1239 {**241**}

Lovis C, Mayor M, 2007, Planets around evolved intermediate-mass stars. I. Two substellar companions in the open clusters NGC 2423 and NGC 4349. *A&A*, 472, 657–664 {**32–34, 37**}

Lovis C, Mayor M, Pepe F, et al., 2006, An extrasolar planetary system with three Neptune-mass planets. *Nature*, 441, 305–309 {**2, 36, 55, 252, 267**}

—, 2008, Pushing down the limits of the radial velocity technique. *Precision Spectroscopy in Astrophysics*, 181–184 {**25**}

Lovis C, Pepe F, 2007, A new list of thorium and argon spectral lines in the visible. *A&A*, 468, 1115–1121 {**20**}

Lovis C, Ségransan D, Mayor M, et al., 2011, The HARPS search for southern extrasolar planets. XXVIII. Up to seven planets orbiting HD 10180: probing the architecture of low-mass planetary systems. *A&A*, 528, A112 {**2**}

Lowrance PJ, Becklin EE, Schneider G, et al., 2005, An infrared coronagraphic survey for substellar companions. *AJ*, 130, 1845–1861 {**213**}

Lubow SH, D'Angelo G, 2006, Gas flow across gaps in protoplanetary disks. *ApJ*, 641, 526–533 {**237, 240**}

Lubow SH, Ogilvie GI, 1998, Three-dimensional waves generated at Lindblad resonances in thermally stratified disks. *ApJ*, 504, 983–995 {**238**}

Lubow SH, Seibert M, Artymowicz P, 1999, Disk accretion onto high-mass planets. *ApJ*, 526, 1001–1012 {**237, 240**}

Lucas PW, Hough JH, Bailey JA, et al., 2009, Planetpol polarimetry of the exoplanet systems 55 Cnc and τ Boo. *MNRAS*, 393, 229–244 {**126**}

Lucas PW, Roche PF, 2000, A population of very young brown dwarfs and free-floating planets in Orion. *MNRAS*, 314, 858–864 {**214, 215**}

Lucas PW, Roche PF, Allard F, et al., 2001, Infrared spectroscopy of substellar objects in Orion. *MNRAS*, 326, 695–721 {**215**}

Lucas PW, Roche PF, Tamura M, 2005, A deep survey of brown dwarfs in Orion with Gemini. *MNRAS*, 361, 211–232 {**215**}

Lucas PW, Weights DJ, Roche PF, et al., 2006, Spectroscopy of planetary mass brown dwarfs in Orion. *MNRAS*, 373, L60–L64 {**215**}

Luck RE, Heiter U, 2005, Stars within 15 pc: abundances for a northern sample. *AJ*, 129, 1063–1083 {**189**}

—, 2006, Dwarfs in the local region. *AJ*, 131, 3069–3092 {**183, 185, 188, 189, 199**}

Ludwig H, 2006, Hydrodynamical simulations of convection-related stellar micro-variability. I. Statistical relations for photometric and photocentric variability. *A&A*, 445, 661–671 {**65, 115**}

Lufkin G, Richardson DC, Mundy LG, 2006, Planetesimals in the presence of giant planet migration. *ApJ*, 653, 1464–1468 {**241**}

Luhman KL, 2008, Distinguishing giant planets and brown dwarfs. *ASP Conf Ser*, volume 398, 357–360 {**212**}

—, 2010, Brown dwarfs. *Formation and Evolution of Exoplanets*, 145–156, Wiley {**210**}

Luhman KL, Adame L, D'Alessio P, et al., 2005a, Discovery of a planetary-mass brown dwarf with a circumstellar disk. *ApJ*, 635, L93–L96 {**215**}

Luhman KL, Allen LE, Allen PR, et al., 2008, The disk population of the Chamaeleon I star-forming region. *ApJ*, 675, 1375–1406 {**215**}

Luhman KL, D'Alessio P, Calvet N, et al., 2005b, Spitzer identification of the least massive known brown dwarf with a circumstellar disk. *ApJ*, 620, L51–L54 {**214**}

Luhman KL, Jayawardhana R, 2002, An adaptive optics search for companions to stars with planets. *ApJ*, 566, 1132–1146 {**170**}

Luhman KL, Muench AA, 2008, New low-mass stars and brown dwarfs with disks in the Chamaeleon I star-forming region. *ApJ*, 684, 654–662 {**214**}

Luhman KL, Patten BM, Marengo M, et al., 2007, Discovery of two T dwarf companions with the Spitzer Space Telescope. *ApJ*, 654, 570–579 {**168**}

Luhman KL, Stauffer JR, Muench AA, et al., 2003, A census of the young cluster IC 348. *ApJ*, 593, 1093–1115 {**214**}

Luhman KL, Wilson JC, Brandner W, et al., 2006, Discovery of a young substellar companion in Chamaeleon. *ApJ*, 649, 894–899 {**168**}

Lumer E, Forestini M, Arnould M, 1990, Application of an extended mixing length model to the convective envelope of the Sun and its Li and Be content. *A&A*, 240, 515–519 {**198, 201**}

Lunine JI, 1993, The atmospheres of Uranus and Neptune. *ARA&A*, 31, 217–263 {**294**}

—, 1999a, *Earth: Evolution of a Habitable World*. Cambridge University Press {**284**}

—, 1999b, In search of planets and life around other stars. *Proceedings of the National Academy of Science*, 96, 5353–5355 {**284**}

—, 2006, Origin of water ice in the solar system. *Meteorites and the Early Solar System II*, 309–319, University of Arizona Press {**299**}

Lunine JI, Stevenson DJ, 1982, Formation of the Galilean satellites in a gaseous nebula. *Icarus*, 52, 14–39 {**269**}

Lutz R, Schuh S, Silvotti R, et al., 2009, The planet-hosting subdwarf B star V391 Peg is a hybrid pulsator. *A&A*, 496, 469–473 {**81**}

Luu JX, Jewitt DC, 2002, Kuiper belt objects: relics from the accretion disk of the Sun. *ARA&A*, 40, 63–101 {**178**}

Luyten WJ, 1979, *LHS Catalogue: A Catalogue of Stars with Proper Motions Exceeding 0.5 arcsec Annually*. University of Minnesota {**181**}

Lyapunov AM, 1892, *General Problem of the Stability of Motion*. Translated from the Russian, Ann. Math. Studies 17, 1949; Princeton University Press {**45**}

Lydon TJ, Sofia S, 1996, A measurement of the shape of the solar disk: the solar quadrupole moment, the solar octopole moment, and the advance of perihelion of the planet Mercury. *Physical Review Letters*, 76, 177–179 {**134**}

Lykawka PS, Horner J, Jones BW, et al., 2010, Formation and dynamical evolution of the Neptune Trojans: the influence of the initial solar system architecture. *MNRAS*, 404, 1272–1280 {**304**}

Lynch P, 2003, On the significance of the Titius–Bode law for the distribution of the planets. *MNRAS*, 341, 1174–1178 {**48**}

Lynden-Bell D, Pringle JE, 1974, The evolution of viscous disks and the origin of the nebular variables. *MNRAS*, 168, 603–637 {**221**}

Lynds R, Petrosian V, 1986, Giant luminous arcs in galaxy clusters. *Bulletin of the American Astronomical Society*, volume 18, 1014 {**84**}

Lyne AG, Bailes M, 1992, No planet orbiting PSR 1829–10. *Nature*, 355, 213–214 {**78**}

Lyne AG, Biggs JD, Brinklow A, et al., 1988, Discovery of a binary millisecond pulsar in the globular cluster M4. *Nature*, 332, 45–47 {**77**}

Lyo AR, Song I, Lawson WA, et al., 2006, A deep photometric survey of the η Chamaeleontis cluster down to the brown dwarf–planet boundary. *MNRAS*, 368, 1451–1455 {**215**}

Lyon RG, Gezari DY, Melnick GJ, et al., 2003, Extrasolar planetary imager (ESPI) for space-based Jovian planetary detection. *SPIE Conf Ser*, volume 4860, 45–53 {**166**}

Lyon SP, Johnson JD, 1992, SESAME: the Los Alamos National Laboratory equation of state data base. *LA–UR–92–3407*, t1web.lanl.gov/doc/SESAME_3Ddatabase_1992.html {**261**}

Lyot B, 1939, The study of the solar corona and prominences without eclipses. *MNRAS*, 99, 580–594 {**152**}

Lyttleton RA, 1961, An accretion hypothesis for the origin of the solar system. *MNRAS*, 122, 399–407 {**224**}

MacDonald GJF, 1964, Tidal friction. *Reviews of Geophysics and Space Physics*, 2, 467–541 {**245**}

Machalek P, McCullough PR, Burrows A, et al., 2009, Detection of thermal emission of XO–2 b: evidence for a weak temperature inversion. *ApJ*, 701, 514–520 {**142**}

Maciejewski G, Dimitrov D, Neuhäuser R, et al., 2010, Transit timing variation in exoplanet WASP-3 b. *MNRAS*, 407, 2625–2631 {**135**}

Maciel WJ, Costa RDD, 2009, Abundance gradients in the Galactic disk: space and time variations. *IAU Symposium*, volume 254, 38–43 {**193**}

Macintosh BA, 2007, Direct detection of extrasolar planets with the Thirty Meter Telescope. *In the Spirit of Bernard Lyot: The Direct Detection of Planets and Circumstellar Disks in the 21st Century*, 38 {**161**}

Macintosh BA, Becklin EE, Kaisler D, et al., 2003, Deep Keck adaptive optics searches for extrasolar planets in the dust of ϵ Eri and Vega. *ApJ*, 594, 538–544 {**223, 224**}

Macintosh BA, Graham JR, Palmer DW, et al., 2008, The Gemini Planet Imager: from science to design to construction. *SPIE Conf Ser*, volume 7015, 31 {**152, 159**}

Macintosh BA, Troy M, Doyon R, et al., 2006, Extreme adaptive optics for the Thirty Meter Telescope. *SPIE Conf Ser*, volume 6272, 20 {**161**}

Madhusudhan N, Winn JN, 2009, Empirical constraints on Trojan companions and orbital eccentricities in 25 transiting exoplanetary systems. *ApJ*, 693, 784–793 {**136**}

Mahadevan S, Ge J, 2009, The use of absorption cells as a wavelength reference for precision radial velocity measurements in the near-infrared. *ApJ*, 692, 1590–1596 {**20**}

Mahadevan S, Ge J, Fleming SW, et al., 2008a, An inexpensive field-widened monolithic Michelson interferometer for precision radial velocity measurements. *PASP*, 120, 1001–1015 {**27**}

Mahadevan S, van Eyken J, Ge J, et al., 2008b, Measuring stellar radial velocities with a dispersed fixed-delay interferometer. *ApJ*, 678, 1505–1510 {**25, 27**}

Majid W, Winterhalter D, Chandra I, et al., 2006, Search for radio emission from extrasolar planets: preliminary analysis of GMRT data. *European Planetary Science Congress 2006*, 266 {**177**}

Makarov VV, Beichman CA, Catanzarite JH, et al., 2009, Starspot jitter in photometry, astrometry, and radial velocity measurements. *ApJ*, 707, L73–L76 {**65**}

Malbet F, 1996, High angular resolution coronography for adaptive optics. *A&AS*, 115, 161–174 {**153**}

Malbet F, Yu JW, Shao M, 1995, High-dynamic-range imaging using a deformable mirror for space coronography. *PASP*, 107, 386–398 {**166**}

Malhotra R, 1993a, The origin of Pluto's peculiar orbit. *Nature*, 365, 819–821 {**295, 303**}

—, 1993b, Three-body effects in the PSR B1257+12 planetary system. *ApJ*, 407, 266–275 {**76**}

—, 1994, A mapping method for the gravitational few-body problem with dissipation. *Celestial Mechanics and Dynamical Astronomy*, 60, 373–385 {**44**}

—, 1995, The origin of Pluto's orbit: implications for the solar system beyond Neptune. *AJ*, 110, 420–429 {**243, 303**}

—, 2002, A dynamical mechanism for establishing apsidal resonance. *ApJ*, 575, L33–L36 {**42, 43**}

Malhotra R, Black D, Eck A, et al., 1992, Resonant orbital evolution in the putative planetary system of PSR B1257+12. *Nature*, 356, 583–585 {**76**}

Malhotra R, Minton DA, 2008, Prospects for the habitability of OGLE–2006–BLG–109L. *ApJ*, 683, L67–L70 {**99**}

Mallik SV, 1999, Lithium abundance and mass. *A&A*, 352, 495–507 {**198**}

Malmberg D, Davies MB, 2009, On the origin of eccentricities among extrasolar planets. *MNRAS*, 394, L26–L30 {**237**}

Malmberg D, Davies MB, Chambers JE, 2007a, The instability of planetary systems in binaries: how the Kozai mechanism leads to strong planet-planet interactions. *MNRAS*, 377, L1–L4 {**57**}

Malmberg D, de Angeli F, Davies MB, et al., 2007b, Close encounters in young stellar clusters: implications for planetary systems in the solar neighbourhood. *MNRAS*, 378, 1207–1216 {**109**}

Mal'Nev AG, Orlov VV, Petrova AV, 2006, The dynamical evolution of stellar-planetary systems. *Astronomy Reports*, 50, 405–410 {**241**}

Mamajek EE, Meyer MR, 2007, An improbable solution to the underluminosity of 2M1207b: a hot protoplanet collision afterglow. *ApJ*, 668, L175–L178 {**170, 178**}

Mandel K, Agol E, 2002, Analytic light curves for planetary transit searches. *ApJ*, 580, L171–L175 {**118, 119**}

Mandell AM, 2008, Expanding and improving the search for habitable worlds. *New Horizons in Astronomy*, volume 393 of *ASP Conf Ser*, 19–34 {**284**}

Mandell AM, Ge J, Murray N, 2004, A search for ^{6}Li in lithium-poor stars with planets. *AJ*, 127, 1147–1157 {**201**}

Mandell AM, Raymond SN, Sigurdsson S, 2007, Formation of Earth-like planets during and after giant planet migration. *ApJ*, 660, 823–844 {**241, 285**}

Mandell AM, Sigurdsson S, 2003, Survival of terrestrial planets in the presence of giant planet migration. *ApJ*, 599, L111–L114 {**241**}

Mandushev G, O'Donovan FT, Charbonneau D, et al., 2007, TrES-4: a transiting hot Jupiter of very low density. *ApJ*, 667, L195–L198 {**107, 327**}

Mann I, Köhler M, Kimura H, et al., 2006, Dust in the solar system and in extrasolar planetary systems. *A&A Rev.*, 13, 159–228 {**222**}

Mao H, Hemley RJ, 1994, Ultrahigh-pressure transitions in solid hydrogen. *Reviews of Modern Physics*, 66, 671–692 {262}

Mao S, Paczyński B, 1991, Gravitational microlensing by double stars and planetary systems. *ApJ*, 374, L37–L40 {84, 87, 88, 94}

Marboeuf U, Mousis O, Ehrenreich D, et al., 2008, Composition of ices in low-mass extrasolar planets. *ApJ*, 681, 1624–1630 {260}

Marchal C, Bozis G, 1982, Hill stability and distance curves for the general three-body problem. *Celestial Mechanics*, 26, 311–333 {43}

Marchi S, 2007, Extrasolar planet taxonomy: a new statistical approach. *ApJ*, 666, 475–485 {**34, 251**}

Marchi S, Ortolani S, Nagasawa M, et al., 2009, On the various origins of close-in extrasolar planets. *MNRAS*, 394, L93–L96 {130}

Marcus RA, Stewart ST, Sasselov D, et al., 2009, Collisional stripping and disruption of super-Earths. *ApJ*, 700, L118–L122 {238}

Marcy GW, 2009a, Extrasolar planets: water world larger than Earth. *Nature*, 462, 853–854 {270}

—, 2009b, Planet Hunter: a astrometric search of 65 nearby stars for Earth-mass planets. *Bulletin of the American Astronomical Society*, volume 41, 507 {73}

Marcy GW, Benitz KJ, 1989, A search for substellar companions to low-mass stars. *ApJ*, 344, 441–453 {28}

Marcy GW, Butler RP, 1992, Precision radial velocities with an iodine absorption cell. *PASP*, 104, 270–277 {**17, 19, 24**}

—, 1996, A planetary companion to 70 Vir. *ApJ*, 464, L147–L149 {**2, 28**}

—, 1998a, Detection of extrasolar giant planets. *ARA&A*, 36, 57–98 {33}

—, 1998b, Doppler detection of extrasolar planets. *Cool Stars, Stellar Systems, and the Sun*, volume 154 of *ASP Conf Ser*, 9–24 {29}

—, 2000, Planets orbiting other suns. *PASP*, 112, 137–140 {**29, 213**}

Marcy GW, Butler RP, Fischer D, et al., 2001a, A pair of resonant planets orbiting GJ 876. *ApJ*, 556, 296–301 {**2, 38, 49**}

—, 2005a, Observed properties of exoplanets: masses, orbits, and metallicities. *Progress of Theoretical Physics Supplement*, 158, 24–42 {**36, 252, 253**}

Marcy GW, Butler RP, Fischer DA, et al., 2002, A planet at 5 AU around 55 Cnc. *ApJ*, 581, 1375–1388 {**14, 48, 49, 52**}

Marcy GW, Butler RP, Vogt SS, et al., 1998, A planetary companion to a nearby M4 dwarf, GJ 876. *ApJ*, 505, L147–L149 {**33, 49**}

—, 1999, Two new candidate planets in eccentric orbits. *ApJ*, 520, 239–247 {237}

—, 2001b, Two substellar companions orbiting HD 168443. *ApJ*, 555, 418–425 {55}

—, 2005b, Five new extrasolar planets. *ApJ*, 619, 570–584 {58}

—, 2008, Exoplanet properties from Lick, Keck and AAT. *Physica Scripta Volume T*, 130(1), 014001 {**28, 35, 38**}

Marcy GW, Butler RP, Williams E, et al., 1997, The planet around 51 Peg. *ApJ*, 481, 926–935 {28}

Marcy GW, Fischer DA, Butler RP, et al., 2006, Properties of exoplanets: a Doppler study of 1330 stars. *Planet Formation*, 179–191, Cambridge University Press {34}

Marcy GW, Lindsay V, Bergengren J, et al., 1986, A dynamical search for sub-stellar objects. *Astrophysics of Brown Dwarfs*, 50–56 {**30, 33**}

Marcy GW, Moore D, 1989, The extremely low mass companion to GJ 623. *ApJ*, 341, 961–967 {28}

Mardling RA, 2007, Long-term tidal evolution of short-period planets with companions. *MNRAS*, 382, 1768–1790 {**145, 248**}

Mardling RA, Lin DNC, 2002, Calculating the tidal, spin, and dynamical evolution of extrasolar planetary systems. *ApJ*, 573, 829–844 {**245–248**}

—, 2004, On the survival of short-period terrestrial planets. *ApJ*, 614, 955–959 {246}

Marengo M, Stapelfeldt K, Werner MW, et al., 2009, Spitzer–IRAC limits to planetary companions of Fomalhaut and ε Eri. *ApJ*, 700, 1647–1657 {223}

Margulis L, Lovelock JE, 1974, Biological modulation of the Earth's atmosphere. *Icarus*, 21, 471–484 {288}

Mariotti JM, Leger A, Mennesson B, et al., 1997, Detection and characterisation of Earth-like planets. *IAU Colloq. 161: Astronomical and Biochemical Origins and the Search for Life in the Universe*, 299–311 {**165, 282**}

Marks RD, Vernin J, Azouit M, et al., 1999, Measurement of optical seeing on the high Antarctic plateau. *A&AS*, 134, 161–172 {63}

Markwardt CB, 2009, Non-linear least-squares fitting in IDL with MPFIT. *ASP Conf Ser*, volume 411, 251–254 {14}

Marley MS, 1998, Atmospheres of giant planets from Neptune to GJ 229B. *Brown Dwarfs and Extrasolar Planets*, volume 134 of *ASP Conf Ser*, 383–393 {277}

—, 2010, The atmospheres of extrasolar planets. *EAS Publications Series*, volume 41, 411–428 {278}

Marley MS, Fortney J, Seager S, et al., 2007a, Atmospheres of extrasolar giant planets. *Protostars and Planets V*, 733–747 {**265, 277**}

Marley MS, Fortney JJ, Hubickyj O, et al., 2007b, On the luminosity of young Jupiters. *ApJ*, 655, 541–549 {**265, 272, 274**}

Marley MS, Gelino C, Stephens D, et al., 1999, Reflected spectra and albedos of extrasolar giant planets. I. Clear and cloudy atmospheres. *ApJ*, 513, 879–893 {**265, 272**}

Marley MS, Hubbard WB, 1988, Thermodynamics of dense molecular hydrogen–helium mixtures at high pressure. *Icarus*, 73, 536–544 {262}

Marley MS, McKay CP, 1999, Thermal structure of Uranus' atmosphere. *Icarus*, 138, 268–286 {272}

Marley MS, Saumon D, Guillot T, et al., 1996, Atmospheric, evolutionary, and spectral models of the brown dwarf GJ 229 B. *Science*, 272, 1919–1921 {**210, 211, 272**}

Marley MS, Seager S, Saumon D, et al., 2002, Clouds and chemistry: ultracool dwarf atmospheric properties from optical and infrared colours. *ApJ*, 568, 335–342 {**272, 276**}

Marois C, Doyon R, Nadeau D, et al., 2005, TRIDENT: an infrared differential imaging camera optimised for the detection of methanated substellar companions. *PASP*, 117, 745–756 {158}

Marois C, Doyon R, Racine R, et al., 2000, Efficient speckle noise attenuation in faint companion imaging. *PASP*, 112, 91–96 {**157, 158**}

Marois C, Lafrenière D, Doyon R, et al., 2006, Angular differential imaging: a powerful high-contrast imaging technique. *ApJ*, 641, 556–564 {158}

Marois C, Lafrenière D, Macintosh B, et al., 2008a, Confidence level and sensitivity limits in high-contrast imaging. *ApJ*, 673, 647–656 {**157, 158**}

Marois C, Macintosh B, Barman T, et al., 2008b, Direct imaging of multiple planets orbiting the star HR 8799. *Science*, 322, 1348–1352 {**2, 156, 158, 168, 170–172, 204, 223**}

Marois C, Racine R, Doyon R, et al., 2004, Differential imaging with a multicolour detector assembly: a new exoplanet finder concept. *ApJ*, 615, L61–L64 {158}

Marsh KA, Dowell CD, Velusamy T, et al., 2006a, Images of the Vega dust ring at 350 and 450 μm: new clues to the trapping of multiple-sized dust particles in planetary resonances. *ApJ*, 646, L77–L80 {224}

Marsh KA, Kirkpatrick JD, Plavchan P, 2010, A young planetary-mass object in the ρ Oph cloud core. *ApJ*, 709, L158–L162 {215}

Marsh KA, Silverstone MD, Becklin EE, et al., 2002, Mid-infrared images of the debris disk around HD 141569. *ApJ*, 573, 425–430 {224}

Marsh KA, Velusamy T, Ware B, 2006b, Point process algorithm: a new Bayesian approach for planet signal extraction with the Terrestrial Planet Finder-interferometer. *AJ*, 132, 1789–1795 {158}

Martin EL, 1997, Quantitative spectroscopic criteria for the classification of pre-main sequence low-mass stars. *A&A*, 321, 492–496 {187}

Martín EL, Delfosse X, Guieu S, 2004, Spectroscopic identification of DENIS-selected brown dwarf candidates in the Upper Scorpius OB association. *AJ*, 127, 449–454 {210}

Martín EL, Osorio MRZ, 2003, Spectroscopic estimate of surface gravity for a planetary member in the σ Ori cluster. *ApJ*, 593, L113–L116 {215}

Martín EL, Zapatero Osorio MR, Barrado y Navascués D, et al., 2001, Keck NIRC observations of planetary-mass candidate members in the σ Ori open cluster. *ApJ*, 558, L117–L121 {215}

Martin H, Albarède F, Claeys P, et al., 2006a, From Suns to life: a chronological approach to the history of life on Earth. IV. Building of a habitable planet. *Earth Moon and Planets*, 98, 97–151 {284}

Martin H, Claeys P, Gargaud M, et al., 2006b, From Suns to life: a chronological approach to the history of life on Earth. VI. Environmental context. *Earth Moon and Planets*, 98, 205–245 {284}

Martin RG, Lubow SH, Pringle JE, et al., 2007, Planetary migration to large radii. *MNRAS*, 378, 1589–1600 {242}

Martinache F, 2004, PIZZA: a phase-induced zonal Zernike apodisation designed for stellar coronagraphy. *Journal of Optics A: Pure and Applied Optics*, 6, 809–814 {153}

Martinache F, Lardière O, 2006, Pupil densification: a panorama. *EAS*

References

Publications Series, volume 22, 367–377 {**168**}

Martinez P, Boccaletti A, Kasper M, et al., 2008, Comparison of coronagraphs for high-contrast imaging in the context of extremely large telescopes. *A&A*, 492, 289–300 {**155**}

Martínez Fiorenzano AF, Gratton RG, Desidera S, et al., 2005, Line bisectors and radial velocity jitter from SARG spectra. *A&A*, 442, 775–784 {**23**}

Martioli E, McArthur BE, Benedict GF, et al., 2010, The mass of the candidate exoplanet companion to HD 136118 from HST astrometry and high-precision radial velocities. *ApJ*, 708, 625–634 {**71**}

Mary DL, 2006, A statistical analysis of the detection limits of fast photometry. *A&A*, 452, 715–726 {**116**}

Marzari F, 2010, Planet–planet gravitational scattering. *Formation and Evolution of Exoplanets*, chapter 223-242, Wiley {**243**}

Marzari F, Barbieri M, 2007a, Planet dispersal in binary systems during transient multiple star phases. *A&A*, 472, 643–647 {**57**}

—, 2007b, Planets in binary systems: is the present configuration indicative of the formation process? *A&A*, 467, 347–351 {**57**}

Marzari F, Scholl H, 2000, Planetesimal accretion in binary star systems. *ApJ*, 543, 328–339 {**58**}

—, 2007, Dynamics of Jupiter Trojans during the 2:1 mean motion resonance crossing of Jupiter and Saturn. *MNRAS*, 380, 479–488 {**303**}

Marzari F, Scholl H, Tricarico P, 2006, A numerical study of the 2:1 planetary resonance. *A&A*, 453, 341–348 {**42**}

Marzari F, Thébault P, Scholl H, 2009, Planet formation in highly inclined binaries. *A&A*, 507, 505–511 {**58**}

Marzari F, Weidenschilling SJ, 2002, Eccentric extrasolar planets: the jumping Jupiter model. *Icarus*, 156, 570–579 {**127, 241, 243**}

Masciadri E, Lascaux F, Hagelin S, et al., 2010a, Optical turbulence above the internal Antarctic plateau. *EAS Publications Series*, volume 40, 55–64 {**162**}

Masciadri E, Mundt R, Henning T, et al., 2005, A search for hot massive extrasolar planets around nearby young stars with the adaptive optics system NACO. *ApJ*, 625, 1004–1018 {**170**}

Masciadri E, Raga A, 2004, Exoplanet recognition using a wavelet analysis technique. *ApJ*, 611, L137–L140 {**158**}

Masciadri E, Stoesz J, Hagelin S, et al., 2010b, Optical turbulence vertical distribution with standard and high resolution at Mt Graham. *MNRAS*, 404, 144–158 {**151**}

Massarotti A, 2008, Stellar rotation and planet ingestion in giants. *AJ*, 135, 2287–2290 {**186**}

Masset FS, D'Angelo G, Kley W, 2006, On the migration of protogiant solid cores. *ApJ*, 652, 730–745 {**239, 240**}

Masset FS, Kley W, 2006, Disk-planet interaction and migration. *Planet Formation*, 216–235, Cambridge University Press {**225**}

Masset FS, Ogilvie GI, 2004, On the saturation of corotation resonances: a numerical study. *ApJ*, 615, 1000–1010 {**242**}

Masset FS, Papaloizou JCB, 2003, Runaway migration and the formation of hot Jupiters. *ApJ*, 588, 494–508 {**240**}

Masset FS, Snellgrove M, 2001, Reversing Type II migration: resonance trapping of a lighter giant protoplanet. *MNRAS*, 320, L55–L59 {**242**}

Mathieu RD, 1994, Pre-main-sequence binary stars. *ARA&A*, 32, 465–530 {**56**}

Mathieu RD, Ghez AM, Jensen ELN, et al., 2000, Young binary stars and associated disks. *Protostars and Planets IV*, 703–709 {**56**}

Matsui T, Abe Y, 1986, Impact-induced atmospheres and oceans on Earth and Venus. *Nature*, 322, 526–528 {**269, 278**}

Matsumura S, Peale SJ, Rasio FA, 2010, Tidal evolution of close-in planets. *ArXiv e-prints* {**129**}

Matsumura S, Pudritz RE, 2003, The origin of Jovian planets in protostellar disks: the role of dead zones. *ApJ*, 598, 645–656 {**222**}

—, 2005, Dead zones and the origin of planetary masses. *ApJ*, 618, L137–L140 {**222, 241**}

—, 2006, Dead zones and extrasolar planetary properties. *MNRAS*, 365, 572–584 {**222**}

Matsumura S, Pudritz RE, Thommes EW, 2007, Saving planetary systems: dead zones and planetary migration. *ApJ*, 660, 1609–1623 {**222**}

Matsumura S, Takeda G, Rasio FA, 2008, On the origins of eccentric close-in planets. *ApJ*, 686, L29–L32 {**247, 250**}

Matsuo T, Shibai H, Ootsubo T, et al., 2007, Planetary formation scenarios revisited: core-accretion versus disk instability. *ApJ*, 662, 1282–1292 {**189, 236**}

Matsuyama I, Johnstone D, Murray N, 2003, Halting planet migration by photoevaporation from the central source. *ApJ*, 585, L143–L146 {**241**}

Matthews BC, Greaves JS, Holland WS, et al., 2007, An unbiased survey of 500 nearby stars for debris disks: a JCMT legacy programme. *PASP*, 119, 842–854 {**224**}

Matthews J, 2006, Asteroseismology with the MOST space mission. *IAU Joint Discussion*, volume 17 of *IAU Joint Discussion*, 21 {**202**}

Mattox JR, Halpern JP, Caraveo PA, 1998, Timing the Geminga pulsar with gamma-ray observations. *ApJ*, 493, 891–893 {**78**}

Matzner CD, Levin Y, 2005, Protostellar disks: formation, fragmentation, and the brown dwarf desert. *ApJ*, 628, 817–831 {**214, 241**}

Mawet D, Riaud P, Absil O, et al., 2005, Annular groove phase mask coronagraph. *ApJ*, 633, 1191–1200 {**153, 155**}

Mawet D, Serabyn E, Liewer K, et al., 2009a, Optical vectorial vortex coronagraphs using liquid crystal polymers: theory, manufacturing and laboratory demonstration. *Optics Express*, 17, 1902–1918 {**156**}

—, 2010, The vector vortex coronagraph: laboratory results and first light at Palomar Observatory. *ApJ*, 709, 53–57 {**156**}

Mawet D, Serabyn E, Stapelfeldt K, et al., 2009b, Imaging the debris disk of HD 32297 with a phase-mask coronagraph at high Strehl ratio. *ApJ*, 702, L47–L50 {**224**}

Maxted PFL, Anderson DR, Collier Cameron A, et al., 2010a, WASP-32 b: a transiting hot Jupiter planet orbiting a lithium-poor, solar-type star. *PASP*, 122, 1465–1470 {**327**}

Maxted PFL, Anderson DR, Gillon M, et al., 2010b, WASP-22 b: a transiting hot Jupiter planet in a hierarchical triple system. *AJ*, 140, 2007–2012 {**327**}

May JL, Jennetti T, 2004, Telescope resolution using negative refractive index materials. *SPIE Conf Ser*, volume 5166, 220–227 {**169**}

Mayer L, 2010, Formation via disk instability. *Formation and Evolution of Exoplanets*, 71–100, Wiley {**235**}

Mayer L, Lufkin G, Quinn T, et al., 2007, Fragmentation of gravitationally unstable gaseous protoplanetary disks with radiative transfer. *ApJ*, 661, L77–L80 {**236**}

Mayer L, Wadsley J, Quinn T, et al., 2005, Gravitational instability in binary protoplanetary disks: new constraints on giant planet formation. *MNRAS*, 363, 641–648 {**58, 59**}

Mayor M, 1980, Metal abundances of F and G dwarfs determined by the radial velocity scanner CORAVEL. *A&A*, 87, L1–L2 {**17**}

Mayor M, Bonfils X, Forveille T, et al., 2009a, The HARPS search for southern extrasolar planets. XVIII. An Earth-mass planet in the GJ 581 planetary system. *A&A*, 507, 487–494 {**2, 24, 30, 33, 35, 286**}

Mayor M, Pepe F, Queloz D, et al., 2003, Setting new standards with HARPS. *The Messenger*, 114, 20–24 {**20, 24**}

Mayor M, Queloz D, 1995, A Jupiter-mass companion to a solar-type star. *Nature*, 378, 355–359 {**2, 20, 24, 28**}

Mayor M, Udry S, Lovis C, et al., 2009b, The HARPS search for southern extrasolar planets. XIII. A planetary system with 3 super-Earths (4.2, 6.9, and 9.2 $M_\oplus$). *A&A*, 493, 639–644 {**30, 31, 35, 36, 55**}

Mayor M, Udry S, Naef D, et al., 2004, The CORALIE survey for southern extrasolar planets. XII. Orbital solutions for 16 extrasolar planets discovered with CORALIE. *A&A*, 415, 391–402 {**43, 52, 173**}

Mazeh T, 2008, Observational evidence for tidal interaction in close binary systems. *EAS Publications Series*, volume 29, 1–65 {**245**}

Mazeh T, Guterman P, Aigrain S, et al., 2009a, Removing systematics from the CoRoT light curves. I. Magnitude-dependent zero point. *A&A*, 506, 431–434 {**110**}

Mazeh T, Krymolowski Y, Rosenfeld G, 1997a, The high eccentricity of the planet orbiting 16 Cyg B. *ApJ*, 477, L103–L106 {**60, 237**}

Mazeh T, Mayor M, Latham DW, 1997b, Eccentricity versus mass for low-mass secondaries and planets. *ApJ*, 478, 367–373 {**238**}

Mazeh T, Naef D, Torres G, et al., 2000, The spectroscopic orbit of the planetary companion transiting HD 209458. *ApJ*, 532, L55–L58 {**113**}

Mazeh T, Tsodikovich Y, Segal Y, et al., 2009b, TRIMOR: three-dimensional correlation technique to analyse multi-order spectra of triple stellar systems: application to HD 188753. *MNRAS*, 399, 906–913 {**22**}

Mazeh T, Zucker S, 1994, TODCOR: a two-dimensional correlation technique to analyze stellar spectra in search of faint companions. *Ap&SS*, 212, 349–356 {**22**}

Mazeh T, Zucker S, dalla Torre A, et al., 1999, Analysis of the Hipparcos measurements of υ And: a mass estimate of its outermost known planetary companion. *ApJ*, 522, L149–L151 {**47, 70**}

Mazeh T, Zucker S, Pont F, 2005, An intriguing correlation between the masses and periods of the transiting planets. *MNRAS*, 356, 955–957 {**146**}

McArthur BE, Benedict GF, Barnes R, et al., 2010, New observational constraints on the v And system with data from the Hubble Space Telescope and Hobby–Eberly Telescope. *ApJ*, 715, 1203–1220 {**2, 53, 71, 72**}

McArthur BE, Endl M, Cochran WD, et al., 2004, Detection of a Neptune-mass planet in the ρ^1 Cnc system using the Hobby–Eberly telescope. *ApJ*, 614, L81–L84 {**48, 71**}

McBreen B, Hanlon L, 1999, Gamma-ray bursts and the origin of chondrules and planets. *A&A*, 351, 759–765 {**298**}

McBreen B, Winston E, McBreen S, et al., 2005, Gamma-ray bursts and other sources of giant lightning discharges in protoplanetary systems. *A&A*, 429, L41–L45 {**298**}

McCarthy C, Butler RP, Tinney CG, et al., 2004, Multiple companions to HD 154857 and HD 160691. *ApJ*, 617, 575–579 {**48**}

McCarthy C, Zuckerman B, 2004, The brown dwarf desert at 75–1200 AU. *AJ*, 127, 2871–2884 {**35, 170, 173, 213**}

McCarthy DW, Probst RG, Low FJ, 1985, Infrared detection of a close cool companion to Van Biesbroeck 8. *ApJ*, 290, L9–L13 {**210**}

McCrea WH, 1960, The origin of the solar system. *Royal Society of London Proceedings Series A*, 256, 245–266 {**224**}

—, 1988, Formation of the solar system: brief review and revised protoplanet theory. *The Physics of the Planets*, 421–439 {**224**}

McCullough PR, Burke CJ, Valenti JA, et al., 2008, XO–4 b: an extrasolar planet transiting an F5V star. *ArXiv e-prints* {**108, 327**}

McCullough PR, Stys JE, Valenti JA, et al., 2005, The XO project: searching for transiting extrasolar planet candidates. *PASP*, 117, 783–795 {**108**}

—, 2006, A transiting planet of a Sun-like star. *ApJ*, 648, 1228–1238 {**108, 327**}

McElroy MB, Prather MJ, 1981, Noble gases in the terrestrial planets. *Nature*, 293, 535–539 {**280**}

McEwen AS, Isbell NR, Edwards KE, et al., 1992, New Voyager 1 hot spot identifications and the heat flow of Io. *Bulletin of the American Astronomical Society*, volume 24, 935 {**250**}

McFadden L, Thomas PC, Carcich B, et al., 2007, Observations of Vesta with HST–WFPC2 in 2007. *Bulletin of the American Astronomical Society*, volume 38, 469 {**230**}

McGrath MA, Nelan E, Black DC, et al., 2002, An upper limit to the mass of the radial velocity companion to ρ^1 Cnc. *ApJ*, 564, L27–L30 {**70, 71**}

McIvor T, Jardine M, Holzwarth V, 2006, Extrasolar planets, stellar winds and chromospheric hotspots. *MNRAS*, 367, L1–L5 {**175, 205**}

McKay CP, 1998, Life in the planetary context. *Origins*, volume 148 of *ASP Conf Ser*, 449–455 {**282**}

McKay CP, Pollack JB, Courtin R, 1989, The thermal structure of Titan's atmosphere. *Icarus*, 80, 23–53 {**272**}

McKee CF, Ostriker EC, 2007, Theory of star formation. *ARA&A*, 45, 565–687 {**217, 218**}

McKee CF, Taylor JH, 2000, *Astronomy and Astrophysics in the New Millennium: Report of the Astronomy and Astrophysics Survey Committee*. National Academy Press, Washington DC {**166**}

McKenna J, Lyne AG, 1988, Timing measurements of the binary millisecond pulsar in the globular cluster M4. *Nature*, 336, 226–228 {**77**}

McLaughlin DB, 1924, Some results of a spectrographic study of the Algol system. *ApJ*, 60, 22–31 {**127**}

McMillan RS, Moore TL, Perry ML, et al., 1994, Long, accurate time series measurements of radial velocities of solar-type stars. *Ap&SS*, 212, 271–280 {**24**}

McMillan RS, Smith PH, Perry ML, et al., 1990, Long-term stability of a Fabry–Perot interferometer used for measurement of stellar Doppler shift. *SPIE Conf Ser*, volume 1235, 601–609 {**24, 28**}

McNeil D, Duncan M, Levison HF, 2005, Effects of Type I migration on terrestrial planet formation. *AJ*, 130, 2884–2899 {**239**}

McNeil DS, Nelson RP, 2009, New methods for large dynamic range problems in planetary formation. *MNRAS*, 392, 537–552 {**44**}

—, 2010, On the formation of hot Neptunes and super-Earths. *MNRAS*, 401, 1691–1708 {**234**}

McSween HY, Huss GR, 2010, *Cosmochemistry*. Cambridge University Press {**257**}

Meadows VS, 2008, Planetary environmental signatures for habitability and life. *Exoplanets: Detection, Formation, Properties, Habitability*, 259–284, Springer {**289**}

Mecheri R, Abdelatif T, Irbah A, et al., 2004, New values of gravitational moments J_2 and J_4 deduced from helioseismology. *Sol. Phys.*, 222, 191–197 {**134**}

Meeus J, Vitagliano A, 2004, Simultaneous transits. *Journal of the British Astronomical Association*, 114, 132–135 {**123**}

Melo C, Santos NC, Pont F, et al., 2006, On the age of stars harbouring transiting planets. *A&A*, 460, 251–256 {**187**}

Melosh HJ, Vickery AM, 1989, Impact erosion of the primordial atmosphere of Mars. *Nature*, 338, 487–489 {**280, 281**}

Mendeleev DI, 1877, L'origine du pétrole. *Revue Scientifique, 2e Ser.*, 8, 409–416 {**279**}

Mennesson B, Akeson R, Appleby E, et al., 2006, Long baseline nulling interferometry with the Keck telescopes: a progress report. *IAU Colloq. 200: Direct Imaging of Exoplanets: Science and Techniques*, 227–232 {**163**}

Mennesson B, Léger A, Ollivier M, 2005, Direct detection and characterisation of extrasolar planets: the Mariotti space interferometer. *Icarus*, 178, 570–588 {**165**}

Mennesson B, Mariotti JM, 1997, Array configurations for a space infrared nulling interferometer dedicated to the search for Earth-like extrasolar planets. *Icarus*, 128, 202–212 {**165**}

Mennesson B, Shao M, Levine BM, et al., 2003, Optical planet discoverer: how to turn a 1.5-m class space telescope into a powerful exoplanetary systems imager. *SPIE Conf Ser*, volume 4860, 32–44 {**153, 166**}

Menou K, Tabachnik S, 2003, Dynamical habitability of known extrasolar planetary systems. *ApJ*, 583, 473–488 {**44, 285**}

Meru F, Bate MR, 2010, Exploring the conditions required to form giant planets via gravitational instability in massive protoplanetary disks. *MNRAS*, 406, 2279–2288 {**236**}

Meschiari S, Laughlin G, 2008, The potential impact of groove modes on Type II planetary migration. *ApJ*, 679, L135–L138 {**240**}

Metchev SA, Hillenbrand LA, 2006, HD 203030B: an unusually cool young substellar companion near the L/T transition. *ApJ*, 651, 1166–1176 {**168**}

Metchev SA, Hillenbrand LA, White RJ, 2003, Adaptive optics observations of Vega: eight detected sources and upper limits to planetary-mass companions. *ApJ*, 582, 1102–1108 {**224**}

Metchev SA, Marois C, Zuckerman B, 2009, Pre-discovery 2007 image of the HR 8799 planetary system. *ApJ*, 705, L204–L207 {**223**}

Meunier N, Desort M, Lagrange A, 2010, Using the Sun to estimate Earth-like planets detection capabilities. II. Impact of plages. *A&A*, 512, A39 {**22**}

Meyer BS, Zinner E, 2006, Nucleosynthesis. *Meteorites and the Early Solar System II*, 69–108, University of Arizona Press {**298**}

Meyer MR, Hillenbrand LA, Backman DE, et al., 2004, The formation and evolution of planetary systems: first results from a Spitzer legacy science programme. *ApJS*, 154, 422–427 {**224**}

Michel E, Baglin A, Auvergne M, et al., 2006, The seismology programme of CoRoT. *ESA Special Publication*, volume 1306, 39–50 {**202**}

Michikoshi S, Inutsuka Si, Kokubo E, et al., 2007, N-body simulation of planetesimal formation through gravitational instability of a dust layer. *ApJ*, 657, 521–532 {**235**}

Michtchenko TA, Beaugé C, Ferraz-Mello S, 2008a, Dynamic portrait of the planetary 2:1 mean-motion resonance. I. Systems with a more massive outer planet. *MNRAS*, 387, 747–758 {**41**}

—, 2008b, Dynamic portrait of the planetary 2:1 mean-motion resonance. II. Systems with a more massive inner planet. *MNRAS*, 391, 215–227 {**41**}

Michtchenko TA, Ferraz-Mello S, 2001a, Modeling the 5:2 mean motion resonance in the Jupiter–Saturn planetary system. *Icarus*, 149, 357–374 {**52**}

—, 2001b, Resonant structure of the outer solar system in the neighbourhood of the planets. *AJ*, 122, 474–481 {**46**}

Michtchenko TA, Ferraz-Mello S, Beaugé C, 2006, Modeling the 3d secular planetary three-body problem: discussion on the outer v And planetary system. *Icarus*, 181, 555–571 {**47**}

Michtchenko TA, Malhotra R, 2004, Secular dynamics of the three-body problem: application to the v And planetary system. *Icarus*, 168, 237–248 {**47**}

Mieremet AL, Braat JJM, 2003, Deep nulling by means of multiple-beam recombination. *Appl. Opt.*, 42, 1867–1875 {**163**}

Mihalas D, 1978, *Stellar Atmospheres*. Freeman, Second Edition {**264**}

Mikkola S, Innanen K, Muinonen K, et al., 1994, A preliminary analysis of the orbit of the Mars Trojan asteroid (5261) Eureka. *Celestial Mechanics and Dynamical Astronomy*, 58, 53–64 {**52**}

Milani A, Nobili AM, Carpino M, 1989, Dynamics of Pluto. *Icarus*, 82, 200–217 {**299**}

Milankovitch M, 1941, Kanon der Erdbestrahlungen und seine Anwendung auf das Eiszeitenproblem. *Roy. Serbian Acad. Spec. Publ.*, 133, 1–633 {**301**}

—, 1969, Canon of Insolation and the Ice Age Problem. *Israel Program for Scientific Translations* {**301**}

Miller AA, Irwin J, Aigrain S, et al., 2008, The Monitor project: the search for transits in the open cluster NGC 2362. *MNRAS*, 387, 349–363 {**109**}

Miller N, Fortney JJ, Jackson B, 2009, Inflating and deflating hot Jupiters: coupled tidal and thermal evolution of known transiting planets. *ApJ*, 702, 1413–1427 {**145**}

Miller-Ricci E, Meyer MR, Seager S, et al., 2009a, On the emergent spectra of hot protoplanet collision afterglows. *ApJ*, 704, 770–780 {**178**}

Miller-Ricci E, Rowe JF, Sasselov D, et al., 2008a, MOST space-based photometry of the transiting exoplanet system HD 189733: precise timing measurements for transits across an active star. *ApJ*, 682, 593–601 {**113, 135**}

—, 2008b, MOST space-based photometry of the transiting exoplanet system HD 209458: transit timing to search for additional planets. *ApJ*, 682, 586–592 {**113, 135**}

Miller-Ricci E, Seager S, Sasselov D, 2009b, The atmospheric signatures of super-Earths: how to distinguish between hydrogen-rich and hydrogen-poor atmospheres. *ApJ*, 690, 1056–1067 {**137**}

Millman PM, 1961, Meteor news. *JRASC*, 55, 265–267 {**295**}

Millour F, Vannier M, Petrov RG, et al., 2006, Extrasolar planets with VLTI–AMBER: what can we expect from current performances? *IAU Colloq. 200: Direct Imaging of Exoplanets: Science and Techniques*, 291–296 {**162**}

Min M, 2010, Dust composition in protoplanetary disks. *Protoplanetary Dust: Astrophysical and Cosmochemical Perspectives*, 161–190, Cambridge University Press {**220**}

Minier V, Lineweaver C, 2006, A search for water masers toward extrasolar planets. *A&A*, 449, 805–808 {**284**}

Minniti D, Butler RP, López-Morales M, et al., 2009, Low-mass companions for five solar-type stars from the Magellan planet search programme. *ApJ*, 693, 1424–1430 {**24**}

Minton DA, Malhotra R, 2009, A record of planet migration in the main asteroid belt. *Nature*, 457, 1109–1111 {**304**}

Miralda-Escudé J, 1996, Microlensing events from measurements of the deflection. *ApJ*, 470, L113–L116 {**94**}

Miralda-Escudé J, 2002, Orbital perturbations of transiting planets: a possible method to measure stellar quadrupoles and to detect Earth-mass planets. *ApJ*, 564, 1019–1023 {**133, 135**}

Mislis D, Schmitt JHMM, 2009, Detection of orbital parameter changes in the TrES–2 exoplanet. *A&A*, 500, L45–L49 {**135**}

Misner CW, Thorne KS, Wheeler JA, 1973, *Gravitation*. W.H. Freeman and Co. {**18**}

Miyake N, Sumi T, Dong S, et al., 2011, A sub-Saturn mass planet, MOA–2009–BLG–319Lb. *ApJ*, 728, 120–124 {**99**}

Miyamoto M, Yoshii Y, 1995, Astrometry for determining the MACHO mass and trajectory. *AJ*, 110, 1427–1432 {**94**}

Miyoshi K, Takeuchi T, Tanaka H, et al., 1999, Gravitational Interaction between a protoplanet and a protoplanetary disk. I. Local three-dimensional simulations. *ApJ*, 516, 451–464 {**238, 239**}

Mizuno H, 1980, Formation of the giant planets. *Progress of Theoretical Physics*, 64, 544–557 {**232**}

Mizuno H, Nakazawa K, Hayashi C, 1978, Instability of a gaseous envelope surrounding a planetary core and formation of giant planets. *Progress of Theoretical Physics*, 60, 699–710 {**232**}

Mochejska BJ, Stanek KZ, Sasselov DD, et al., 2002, Planets in stellar clusters extensive search. I. Discovery of 47 low-amplitude variables in the metal-rich cluster NGC 6791 with millimagnitude image subtraction photometry. *AJ*, 123, 3460–3472 {**109**}

—, 2004, Planets in stellar clusters extensive search. II. Discovery of 57 variables in the cluster NGC 2158 with millimagnitude image subtraction photometry. *AJ*, 128, 312–322 {**109**}

—, 2005, Planets in stellar clusters extensive search. III. A search for transiting planets in the metal-rich open cluster NGC 6791. *AJ*, 129, 2856–2868 {**109**}

—, 2006, Planets in stellar clusters extensive search. IV. A detection of a possible transiting planet candidate in the open cluster NGC 2158. *AJ*, 131, 1090–1105 {**109**}

—, 2008, Planets in stellar clusters extensive search. V. Search for planets and identification of 18 new variable stars in the old open cluster NGC 188. *Acta Astronomica*, 58, 263–278 {**109**}

Moeckel N, Raymond SN, Armitage PJ, 2008, Extrasolar planet eccentricities from scattering in the presence of residual gas disks. *ApJ*, 688, 1361–1367 {**243**}

Mohanty S, Jayawardhana R, Huélamo N, et al., 2007, The planetary mass companion 2MASS 1207-3932B: temperature, mass, and evidence for an edge-on disk. *ApJ*, 657, 1064–1091 {**170, 178**}

Mojzsis SJ, Harrison TM, Pidgeon RT, 2001, Oxygen-isotope evidence from ancient zircons for liquid water at the Earth's surface 4300 Myr ago. *Nature*, 409, 178–181 {**299**}

Mollerach S, Roulet E, 2002, *Gravitational lensing and microlensing*. STScI {**84**}

Monet DG, Dahn CC, Vrba FJ, et al., 1992, US Naval Observatory CCD parallaxes of faint stars. I. Programme description and first results. *AJ*, 103, 638–665 {**62**}

Monin J, Clarke CJ, Prato L, et al., 2007, Disk evolution in young binaries: from observations to theory. *Protostars and Planets V*, 395–409 {**56**}

Monnier JD, 2003, Optical interferometry in astronomy. *Reports on Progress in Physics*, 66, 789–857 {**162**}

—, 2007, Phases in interferometry. *New Astronomy Reviews*, 51, 604–616 {**124**}

Monnier JD, Pedretti E, Thureau N, et al., 2006, Michigan Infrared Combiner (MIRC): commissioning results at the CHARA array. *SPIE Conf Ser*, volume 6268 {**124**}

Montañés-Rodriguez P, Pallé E, Goode PR, et al., 2005, Globally integrated measurements of the Earth's visible spectral albedo. *ApJ*, 629, 1175–1182 {**289**}

Montañés-Rodríguez P, Pallé E, Goode PR, et al., 2006, Vegetation signature in the observed globally integrated spectrum of Earth considering simultaneous cloud data: applications for extrasolar planets. *ApJ*, 651, 544–552 {**289**}

Montalbán J, Rebolo R, 2002, Planet accretion and the abundances of lithium isotopes. *A&A*, 386, 1039–1043 {**198–200**}

Montalto M, Piotto G, Desidera S, et al., 2007, A new search for planet transits in NGC 6791. *A&A*, 470, 1137–1156 {**109**}

Moorhead AV, Adams FC, 2005, Giant planet migration through the action of disk torques and planet planet scattering. *Icarus*, 178, 517–539 {**229, 243**}

—, 2008, Eccentricity evolution of giant planet orbits due to circumstellar disk torques. *Icarus*, 193, 475–484 {**242**}

Morais MHM, Correia ACM, 2008, Stellar wobble caused by a binary system: can it really be mistaken as an extrasolar planet? *A&A*, 491, 899–906 {**23**}

Moran SM, Kuchner MJ, Holman MJ, 2004, The dynamical influence of a planet at semi-major axis 3.4 AU on the dust around ϵ Eri. *ApJ*, 612, 1163–1170 {**223**}

Morbidelli A, Brasser R, Tsiganis K, et al., 2009, Constructing the secular architecture of the solar system. I. The giant planets. *A&A*, 507, 1041–1052 {**304**}

Morbidelli A, Chambers J, Lunine JI, et al., 2000, Source regions and time scales for the delivery of water to Earth. *Meteoritics and Planetary Science*, 35, 1309–1320 {**260, 298**}

Morbidelli A, Crida A, 2007, The dynamics of Jupiter and Saturn in the gaseous protoplanetary disk. *Icarus*, 191, 158–171 {**242**}

Morbidelli A, Crida A, Masset F, et al., 2008, Building giant-planet cores at a planet trap. *A&A*, 478, 929–937 {**239**}

Morbidelli A, Levison HF, 2004, Scenarios for the origin of the orbits of the trans-Neptunian objects 2000 CR_{105} and 2003 VB_{12} (Sedna). *AJ*, 128, 2564–2576 {**293**}

—, 2008, Late evolution of planetary systems. *Physica Scripta Volume T*, 130(1), 014028 {**243**}

Morbidelli A, Levison HF, Tsiganis K, et al., 2005, Chaotic capture of Jupiter's Trojan asteroids in the early solar system. *Nature*, 435, 462–

465 {136, 304}

Morbidelli A, Petit J, Gladman B, et al., 2001, A plausible cause of the Late Heavy Bombardment. *Meteoritics and Planetary Science*, 36, 371–380 {304}

Mordasini C, Alibert Y, Benz W, 2009a, Extrasolar planet population synthesis. I. Method, formation tracks, and mass-distance distribution. *A&A*, 501, 1139–1160 {251, 252}

Mordasini C, Alibert Y, Benz W, et al., 2008, Giant planet formation by core accretion. *ASP Conf Ser*, volume 398, 235–242 {36, 232}

—, 2009b, Extrasolar planet population synthesis. II. Statistical comparison with observations. *A&A*, 501, 1161–1184 {252, 253}

Morgan WW, Keenan PC, 1973, Spectral classification. *ARA&A*, 11, 29–50 {211}

Morgan WW, Keenan PC, Kellman E, 1943, *An Atlas of Stellar Spectra, with an Outline of Spectral Classification*. University of Chicago Press {211}

Morison I, 2006, SETI in the new millennium. *Astronomy and Geophysics*, 47(4), 040000–4 {290}

Moriwaki K, Nakagawa Y, 2004, A planetesimal accretion zone in a circumbinary disk. *ApJ*, 609, 1065–1070 {57, 81}

Moro-Martín A, Malhotra R, Carpenter JM, et al., 2007, The dust, planetesimals, and planets of HD 38529. *ApJ*, 668, 1165–1173 {55}

Moro-Martín A, Wolf S, Malhotra R, 2005, Signatures of planets in spatially unresolved debris disks. *ApJ*, 621, 1079–1097 {224}

Morrison D, 1982, *Satellites of Jupiter*. University of Arizona Press {293}

Mörth HT, Schlamminger L, 1979, Planetary motion, sunspots and climate. *Solar-Terrestrial Influences on Weather and Climate*, 193–207 {66}

Mosqueira I, Estrada PR, 2006, Jupiter's obliquity and a long-lived circumplanetary disk. *Icarus*, 180, 93–97 {301}

Moulton FR, 1905, On the evolution of the solar system. *ApJ*, 22, 165–180 {224}

Mourard D, Blazit A, Bonneau D, et al., 2006, Recent progress and future prospects of the GI2T interferometer. *SPIE Conf Ser*, volume 6268, 7 {162}

Moutou C, Coustenis A, Schneider J, et al., 2001, Search for spectroscopical signatures of transiting HD 209458 b exosphere. *A&A*, 371, 260–266 {140}

—, 2003, Searching for helium in the exosphere of HD 209458 b. *A&A*, 405, 341–348 {140}

Moutou C, Hébrard G, Bouchy F, et al., 2009, Photometric and spectroscopic detection of the primary transit of the 111-day period planet HD 80606 b. *A&A*, 498, L5–L8 {60, 109, 326}

Moutou C, Pont F, Barge P, et al., 2005, Comparative blind test of five planetary transit detection algorithms on realistic synthetic light curves. *A&A*, 437, 355–368 {105}

Moya A, Amado PJ, Barrado D, et al., 2010, Age determination of the HR 8799 planetary system using asteroseismology. *MNRAS*, 405, L81–L85 {171, 204, 223}

Mudryk LR, Wu Y, 2006, Resonance overlap is responsible for ejecting planets in binary systems. *ApJ*, 639, 423–431 {57}

Mugrauer M, Neuhäuser R, 2005, Gl86B: a white dwarf orbits an exoplanet host star. *MNRAS*, 361, L15–L19 {173}

Mugrauer M, Neuhäuser R, Mazeh T, 2007a, The multiplicity of exoplanet host stars. Spectroscopic confirmation of the companions GJ 3021B and HD 27442B, one new planet host triple-star system, and global statistics. *A&A*, 469, 755–770 {59, 173, 207}

Mugrauer M, Neuhäuser R, Mazeh T, et al., 2004, A low-mass stellar companion of the planet host star HD 75289. *A&A*, 425, 249–253 {207}

Mugrauer M, Neuhäuser R, Seifahrt A, et al., 2005, Four new wide binaries among exoplanet host stars. *A&A*, 440, 1051–1060 {207}

Mugrauer M, Seifahrt A, Neuhäuser R, 2007b, The multiplicity of planet host stars: new low-mass companions to planet host stars. *MNRAS*, 378, 1328–1334 {207}

Mulet-Marquis C, Baraffe I, Aigrain S, et al., 2009, Accuracy of stellar parameters of exoplanet-host stars determined from asteroseismology. *A&A*, 506, 153–158 {203}

Mullally F, Reach WT, Degennaro S, et al., 2009, Spitzer planet limits around the pulsating white dwarf GD 66. *ApJ*, 694, 327–331 {80, 81, 173}

Mullally F, von Hippel T, Winget DE, 2007, Spitzer white dwarf planet limits. *15th European Workshop on White Dwarfs*, volume 372 of *ASP Conf Ser*, 355–358 {173}

Mullally F, Winget DE, Degennaro S, et al., 2008, Limits on planets around pulsating white dwarf stars. *ApJ*, 676, 573–583 {2, 80}

Müller TG, Ábrahám P, Crovisier J, 2005, Comets, asteroids and zodiacal light as seen by ISO. *Space Science Reviews*, 119, 141–155 {178}

Murakami N, Uemura R, Baba N, et al., 2008, An eight-octant phase-mask coronagraph. *PASP*, 120, 1112–1118 {153}

Muraki Y, Sumi T, Abe F, et al., 1999, Search for Machos by the MOA collaboration. *Progress of Theoretical Physics Supplement*, 133, 233–246 {86}

Murphy MT, Udem T, Holzwarth R, et al., 2007, High-precision wavelength calibration of astronomical spectrographs with laser frequency combs. *MNRAS*, 380, 839–847 {20}

Murray CD, 1998, Chaotic motion in the solar system. *Encyclopedia of the Solar System*, Academic Press, Orlando {45}

Murray CD, Dermott SF, 2000, *Solar System Dynamics*. CUP {9, 40, 41, 43–45, 48, 134, 246, 300}

Murray N, Chaboyer B, 2002, Are stars with planets polluted? *ApJ*, 566, 442–451 {188, 191, 192}

Murray N, Chaboyer B, Arras P, et al., 2001, Stellar pollution in the solar neighbourhood. *ApJ*, 555, 801–815 {188}

Murray N, Hansen B, Holman M, et al., 1998, Migrating planets. *Science*, 279, 69–72 {127, 238}

Murray N, Holman M, 2001, The role of chaotic resonances in the solar system. *Nature*, 410, 773–779 {45}

Murray N, Paskowitz M, Holman M, 2002, Eccentricity evolution of migrating planets. *ApJ*, 565, 608–620 {242}

Murray N, Weingartner JC, Capobianco C, 2004, On the flux of extrasolar dust in Earth's atmosphere. *ApJ*, 600, 804–827 {295}

Murray-Clay RA, Chiang EI, 2005, A signature of planetary migration: the origin of asymmetric capture in the 2:1 resonance. *ApJ*, 619, 623–638 {303}

—, 2006, Brownian motion in planetary migration. *ApJ*, 651, 1194–1208 {243, 303}

Musielak ZE, Cuntz M, Marshall EA, et al., 2005, Stability of planetary orbits in binary systems. *A&A*, 434, 355–364 {57}

Muterspaugh MW, Lane BF, Kulkarni SR, et al., 2006, Limits to tertiary astrometric companions in binary systems. *ApJ*, 653, 1469–1479 {69}

—, 2010, The PHASES differential astrometry data archive. I. Measurements and description. *AJ*, 140, 1579–1622 {69}

Naef D, Latham DW, Mayor M, et al., 2001, HD 80606 b, a planet on an extremely elongated orbit. *A&A*, 375, L27–L30 {109}

Naef D, Mayor M, Beuzit JL, et al., 2004, The ELODIE survey for northern extrasolar planets. III. Three planetary candidates detected with ELODIE. *A&A*, 414, 351–359 {48}

Nagasawa M, Ida S, Bessho T, 2008, Formation of hot planets by a combination of planet scattering, tidal circularisation, and the Kozai mechanism. *ApJ*, 678, 498–508 {55, 127, 243}

Nagasawa M, Lin DNC, Thommes E, 2005, Dynamical shake-up of planetary systems. I. Embryo trapping and induced collisions by the sweeping secular resonance and embryo-disk tidal interaction. *ApJ*, 635, 578–598 {234, 302}

Nagovitsyn YA, Nagovitsyna EY, Makarova VV, 2009, The Gnevyshev–Ohl rule for physical parameters of the solar magnetic field: the 400-year interval. *Astronomy Letters*, 35, 564–571 {67}

Najita JR, Tiede GP, Carr JS, 2000, From stars to superplanets: the low-mass initial mass function in the young cluster IC 348. *ApJ*, 541, 977–1003 {215}

Nakajima T, Oppenheimer BR, Kulkarni SR, et al., 1995, Discovery of a cool brown dwarf. *Nature*, 378, 463–465 {152, 168, 170, 210, 212}

Nakamoto T, Kita NT, Tachiban S, 2005, Chondrule age distribution and rate of heating events for chondrule formation. *Antarctic Meteorite Research*, 18, 253–272 {297}

Narayan R, Cumming A, Lin DNC, 2005, Radial velocity detectability of low-mass extrasolar planets in close orbits. *ApJ*, 620, 1002–1009 {14, 15}

Narita N, Enya K, Sato B, et al., 2007, Measurement of the Rossiter–McLaughlin effect in the transiting exoplanetary system TrES–1. *PASJ*, 59, 763–770 {107, 129}

Narita N, Hirano T, Sato B, et al., 2009a, Improved measurement of the Rossiter–McLaughlin effect in the exoplanetary system HD 17156. *PASJ*, 61, 991–997 {129}

References

Narita N, Sato B, Hirano T, et al., 2009b, First evidence of a retrograde orbit of a transiting exoplanet HAT–P–7 b. *PASJ*, 61, L35–L40 {**129**}

—, 2010, Spin-orbit alignment of the TrES–4 transiting planetary system and possible additional radial-velocity variation. *PASJ*, 62, 653–660 {**129**}

Narita N, Suto Y, Winn JN, et al., 2005, Subaru HD S transmission spectroscopy of the transiting extrasolar planet HD 209458 b. *PASJ*, 57, 471–480, erratum: 2005, PASJ, 57, 705 {**140**}

NASA, 1976, US Standard Atmosphere. *modelweb.gsfc.nasa.gov/atmos/* {**278**}

Nauenberg M, 2002a, Determination of masses and other properties of extrasolar planetary systems with more than one planet. *ApJ*, 568, 369–376 {**14, 49**}

—, 2002b, Stability and eccentricity for two planets in a 1:1 resonance, and their possible occurrence in extrasolar planetary systems. *AJ*, 124, 2332–2338 {**54**}

Nellis WJ, 2000, Metallisation of fluid hydrogen at 140 GPa (1.4 Mbar): implications for Jupiter. *Planet. Space Sci.*, 48, 671–677 {**175**}

Nelson AF, 2000, Planet formation is unlikely in equal-mass binary systems with A~50 AU. *ApJ*, 537, L65–L68 {**58, 59**}

Nelson AF, Angel JRP, 1998, The range of masses and periods explored by radial velocity searches for planetary companions. *ApJ*, 500, 940–957 {**14, 29**}

Nelson RP, Papaloizou JCB, 2002, Possible commensurabilities among pairs of extrasolar planets. *MNRAS*, 333, L26–L30 {**41**}

—, 2003, The interaction of a giant planet with a disk with MHD turbulence. II. The interaction of the planet with the disk. *MNRAS*, 339, 993–1005 {**238**}

—, 2004, The interaction of giant planets with a disk with MHD turbulence. IV. Migration rates of embedded protoplanets. *MNRAS*, 350, 849–864 {**238, 241**}

Nelson RP, Papaloizou JCB, Masset F, et al., 2000, The migration and growth of protoplanets in protostellar disks. *MNRAS*, 318, 18–36 {**238**}

Nemchin AA, Pidgeon RT, Whitehouse MJ, 2006, Re-evaluation of the origin and evolution of > 4.2 Ga zircons from the Jack Hills metasedimentary rocks. *Earth and Planetary Science Letters*, 244, 218–233 {**299**}

Nero D, Bjorkman JE, 2009, Did Fomalhaut, HR 8799, and HL Tau form planets via the gravitational instability? Placing limits on the required disk masses. *ApJ*, 702, L163–L167 {**236**}

Neron de Surgy O, Laskar J, 1997, On the long-term evolution of the spin of the Earth. *A&A*, 318, 975–989 {**286, 301**}

Neslušan L, 2004, The significance of the Titius–Bode law and the peculiar location of the Earth's orbit. *MNRAS*, 351, 133–136 {**48**}

Ness NF, Acuna MH, Behannon KW, et al., 1986, Magnetic fields at Uranus. *Science*, 233, 85–89 {**174**}

Ness NF, Acuna MH, Burlaga LF, et al., 1989, Magnetic fields at Neptune. *Science*, 246, 1473–1478 {**174**}

Nesvorný D, 2009, Transit timing variations for eccentric and inclined exoplanets. *ApJ*, 701, 1116–1122 {**135**}

Nesvorný D, Alvarellos JLA, Dones L, et al., 2003, Orbital and collisional evolution of the irregular satellites. *AJ*, 126, 398–429 {**237**}

Nesvorný D, Bottke WF, Dones L, et al., 2002, The recent breakup of an asteroid in the main-belt region. *Nature*, 417, 720–771 {**228**}

Nesvorný D, Morbidelli A, 2008, Mass and orbit determination from transit timing variations of exoplanets. *ApJ*, 688, 636–646 {**135**}

Neugebauer G, Leighton RB, 1969, *Two-Micron Sky Survey: A Preliminary Catalogue*. NASA SP, Washington {**210**}

Neuhäuser R, Comeron F, 1998, ROSAT X-ray detection of a young brown dwarf in the Chamaeleon I dark cloud. *Science*, 282, 83–85 {**213**}

Neuhäuser R, Guenther EW, 2004, Infrared spectroscopy of a brown dwarf companion candidate near the young star GSC 08047-00232 in Horologium. *A&A*, 420, 647–653 {**213**}

Neuhäuser R, Guenther EW, Wuchterl G, et al., 2005, Evidence for a comoving sub-stellar companion of GQ Lup. *A&A*, 435, L13–L16 {**2, 168, 215**}

Neuhäuser R, Seifahrt A, Röll T, et al., 2007, Detectability of planets in wide binaries by ground-based relative astrometry with AO. *IAU Symposium*, volume 240 of *IAU Symposium*, 261–263 {**63**}

Newhall XX, Standish EM, Williams JG, 1983, DE 102: a numerically integrated ephemeris of the moon and planets spanning forty-four centuries. *A&A*, 125, 150–167 {**299**}

Newman WI, Haynes MP, Terzian Y, 1994, Redshift data and statistical inference. *ApJ*, 431, 147–155 {**48**}

Newman WI, Symbalisty EMD, Ahrens TJ, et al., 1999, Impact erosion of planetary atmospheres: some surprising results. *Icarus*, 138, 224–240 {**280**}

Nicholson PD, Hedman MM, Clark RN, et al., 2008, A close look at Saturn's rings with Cassini–VIMS. *Icarus*, 193, 182–212 {**226**}

Nidever DL, Marcy GW, Butler RP, et al., 2002, Radial velocities for 889 late-type stars. *ApJS*, 141, 503–522 {**18, 32, 33**}

Niedzielski A, Goździewski K, Wolszczan A, et al., 2009a, A planet in a 0.6 AU orbit around the K0 giant HD 102272. *ApJ*, 693, 276–280 {**32**}

Niedzielski A, Konacki M, Wolszczan A, et al., 2007, A planetary-mass companion to the K0 giant HD 17092. *ApJ*, 669, 1354–1358 {**32**}

Niedzielski A, Nowak G, Adamów M, et al., 2009b, Substellar-mass companions to the K-dwarf BD+14 4559 and the K-giants HD 240210 and BD+20 2457. *ApJ*, 707, 768–777 {**32**}

Niedzielski A, Wolszczan A, 2008, A HET search for planets around evolved stars. *IAU Symposium*, volume 249, 43–47 {**32, 33**}

Niemann HB, Atreya SK, Carignan GR, et al., 1998, The composition of the Jovian atmosphere as determined by the Galileo probe mass spectrometer. *J. Geophys. Res.*, 103, 22831–22846 {**271, 276**}

Nieto MM, 1972, *The Titius–Bode Law of Planetary Distances: Its History and Theory*. Pergamon Press, Oxford {**48**}

Nimmo F, Kleine T, 2007, How rapidly did Mars accrete? Uncertainties in the Hf-W timing of core formation. *Icarus*, 191, 497–504 {**297**}

Nisenson P, Papaliolios C, 2001, Detection of Earth-like planets using apodised telescopes. *ApJ*, 548, L201–L205 {**164**}

Nissen PE, Edvardsson B, 1992, Oxygen abundances in F and G dwarfs derived from the forbidden OI line at 630 nm. *A&A*, 261, 255–262 {**198**}

Nobili AM, 1988, Long term dynamics of the outer solar system: review of LONGSTOP project. *IAU Colloq. 96: The Few Body Problem*, volume 140 of *Astrophysics and Space Science Library*, 147–163 {**299**}

Noble M, Musielak ZE, Cuntz M, 2002, Orbital stability of terrestrial planets inside the habitable zones of extrasolar planetary systems. *ApJ*, 572, 1024–1030 {**285**}

Noguchi K, Aoki W, Kawanomoto S, et al., 2002, High Dispersion Spectrograph for the Subaru telescope. *PASJ*, 54, 855–864 {**24**}

Noll KS, Geballe TR, Marley MS, 1997a, Detection of abundant carbon monoxide in the brown dwarf GJ 229 B. *ApJ*, 489, L87–90 {**275**}

Noll KS, Roush TL, Cruikshank DP, et al., 1997b, Detection of ozone on Saturn's satellites Rhea and Dione. *Nature*, 388, 45–47 {**288**}

Nordström B, Andersen J, Holmberg J, et al., 2004a, The Geneva–Copenhagen survey of the solar neighbourhood. *Publications of the Astronomical Society of Australia*, 21, 129–133 {**189**}

Nordström B, Mayor M, Andersen J, et al., 2004b, The Geneva–Copenhagen survey of the solar neighbourhood: ages, metallicities, and kinematic properties of ~14 000 F and G dwarfs. *A&A*, 418, 989–1019 {**187–189**}

Nowak G, Niedzielski A, 2008, The PSU/TCfA search for planets around evolved stars: bisector analysis of activity of a sample of red giants. *ASP Conf Ser*, volume 398, 171–172 {**23, 32**}

Noyes RW, Bakos GÁ, Torres G, et al., 2008, HAT–P–6 b: a hot Jupiter transiting a bright F star. *ApJ*, 673, L79–L82 {**326**}

Noyes RW, Jha S, Korzennik S, et al., 1997, The AFOE programme of extrasolar planet research. *Planets Beyond the Solar System and the Next Generation of Space Missions*, volume 119 of *ASP Conf Ser*, 119–122 {**24**}

Nutzman P, Charbonneau D, 2008, Design considerations for a ground-based transit search for habitable planets orbiting M dwarfs. *PASP*, 120, 317–327 {**108, 115**}

Oakley PHH, Cash W, 2009, Construction of an Earth model: analysis of exoplanet light curves and mapping the next Earth with the new worlds observer. *ApJ*, 700, 1428–1439 {**131**}

Oasa Y, Tamura M, Sugitani K, 1999, A deep near-infrared survey of the Chamaeleon I dark cloud core. *ApJ*, 526, 336–343 {**215**}

O'Brien DP, Geissler P, Greenberg R, 2002, A melt-through model for chaos formation on Europa. *Icarus*, 156, 152–161 {**250**}

O'Brien DP, Morbidelli A, Levison HF, 2006, Terrestrial planet formation with strong dynamical friction. *Icarus*, 184, 39–58 {**229, 299, 304**}

O'Dell CR, Wen Z, Hu X, 1993, Discovery of new objects in the Orion nebula on HST images: shocks, compact sources, and protoplanetary disks. *ApJ*, 410, 696–700 {222}

O'Donovan FT, Charbonneau D, Bakos GÁ, et al., 2007, TrES–3: a nearby, massive, transiting hot Jupiter in a 31-hour orbit. *ApJ*, 663, L37–L40 {**107, 327**}

O'Donovan FT, Charbonneau D, Mandushev G, et al., 2006a, TrES-2: the first transiting planet in the Kepler field. *ApJ*, 651, L61–L64 {**107, 123, 327**}

O'Donovan FT, Charbonneau D, Torres G, et al., 2006b, Rejecting astrophysical false positives from the TrES transiting planet survey: the example of GSC 03885-00829. *ApJ*, 644, 1237–1245 {**107**}

Ofir A, 2008, An algorithm for photometric identification of transiting circumbinary planets. *MNRAS*, 387, 1597–1604 {**59, 82, 114**}

Ofir A, Deeg HJ, Lacy CHS, 2009, Searching for transiting circumbinary planets in CoRoT and ground-based data using CB-BLS. *A&A*, 506, 445–453 {**115**}

Ogihara M, Ida S, 2009, N-body simulations of planetary accretion around M dwarf stars. *ApJ*, 699, 824–838 {**229**}

Ogihara M, Ida S, Morbidelli A, 2007, Accretion of terrestrial planets from oligarchs in a turbulent disk. *Icarus*, 188, 522–534 {**234**}

Ogilvie GI, Lin DNC, 2004, Tidal dissipation in rotating giant planets. *ApJ*, 610, 477–509 {**246, 247**}

—, 2007, Tidal dissipation in rotating solar-type stars. *ApJ*, 661, 1180–1191 {**246**}

Ogilvie GI, Lubow SH, 2003, Saturation of the corotation resonance in a gaseous disk. *ApJ*, 587, 398–406 {**242**}

Ohishi N, Yoshizawa M, Nishikawa J, et al., 2008, Recent progress at the MIRA: development of fringe tracking system. *SPIE Conf Ser*, volume 7013, 4 {**162**}

Ohta Y, Taruya A, Suto Y, 2005, The Rossiter–McLaughlin effect and analytic radial velocity curves for transiting extrasolar planetary systems. *ApJ*, 622, 1118–1135 {**128**}

Oishi JS, Mac Low M, Menou K, 2007, Turbulent torques on protoplanets in a dead zone. *ApJ*, 670, 805–819 {**222**}

Okamoto YK, Kataza H, Honda M, et al., 2004, An early extrasolar planetary system revealed by planetesimal belts in β Pic. *Nature*, 431, 660–663 {**223**}

O'Keefe JD, Ahrens TJ, 1982, Cometary and meteorite swarm impact on planetary surfaces. *J. Geophys. Res.*, 87, 6668–6680 {**280**}

Ollivier M, 2007, Towards the spectroscopic analysis of Earth-like planets: the Darwin/TPF project. *Comptes Rendus Physique*, 8, 408–414 {**165**}

Ollivier M, Absil O, Allard F, et al., 2009, PEGASE, an infrared interferometer to study stellar environments and low mass companions around nearby stars. *Experimental Astronomy*, 23, 403–434 {**166**}

Ollivier M, Le Duigou JM, Mourard D, et al., 2006, PEGASE: a Darwin/TPF pathfinder. *IAU Colloq. 200: Direct Imaging of Exoplanets: Science and Techniques*, 241–246 {**166**}

O'Neill C, Lenardic A, 2007, Geological consequences of super-sized Earths. *Geophys. Res. Lett.*, 34, 19204 {**250**}

Oppenheimer BR, Brenner D, Hinkley S, et al., 2008, The solar system-scale disk around AB Aur. *ApJ*, 679, 1574–1581 {**224**}

Oppenheimer BR, Golimowski DA, Kulkarni SR, et al., 2001, A coronagraphic survey for companions of stars within 8 pc. *AJ*, 121, 2189–2211 {**170, 213**}

Oppenheimer BR, Hinkley S, 2009, High-contrast observations in optical and infrared astronomy. *ARA&A*, 47, 253–289 {**153**}

Oppenheimer BR, Kulkarni SR, Matthews K, et al., 1995, Infrared spectrum of the cool brown dwarf GJ 229 B. *Science*, 270, 1478–1479 {**210**}

Orosz JA, Hauschildt PH, 2000, The use of the NextGen model atmospheres for cool giants in a light curve synthesis code. *A&A*, 364, 265–281 {**119, 132**}

Ortiz JL, Moreno F, Molina A, et al., 2007, Possible patterns in the distribution of planetary formation regions. *MNRAS*, 379, 1222–1226 {**48**}

Osborn J, Wilson RW, Dhillon VS, et al., 2011, Conjugate-plane photometry: reducing scintillation in ground-based photometry. *MNRAS*, 411, 1223–1230 {**116**}

Oseledec VI, 1968, A multiplicative ergodic theorem: the Lyapunov characteristic numbers of dynamical systems. *Moscow Math. Soc.*, 19(197-231) {**45**}

Oskvarek JD, Perry EC, 1976, Temperature limits on the early Archaean ocean from oxygen isotope variations in the Isua supracrustal sequence, West Greenland. *Nature*, 259, 192–194 {**269**}

Oti JE, Canales VF, Cagigal MP, 2005a, Improvements on the optical differentiation wavefront sensor. *MNRAS*, 360, 1448–1454 {**151**}

—, 2005b, The optical differentiation coronagraph. *ApJ*, 630, 631–636 {**153**}

O'Toole SJ, Jones HRA, Tinney CG, et al., 2009a, The frequency of low-mass exoplanets. *ApJ*, 701, 1732–1741 {**33**}

O'Toole SJ, Tinney CG, Butler RP, et al., 2009b, A Neptune-mass planet orbiting the nearby G dwarf HD 16417. *ApJ*, 697, 1263–1268 {**22, 33**}

O'Toole SJ, Tinney CG, Jones HRA, 2008, The impact of stellar oscillations on Doppler velocity planet searches. *MNRAS*, 386, 516–520 {**21**}

Owen T, 1980, The search for early forms of life in other planetary systems: future possibilities afforded by spectroscopic techniques. *Strategies for the Search for Life in the Universe*, volume 83 of *Astrophysics and Space Science Library*, 177–183 {**283, 288**}

Owen T, Mahaffy P, Niemann HB, et al., 1999, A low-temperature origin for the planetesimals that formed Jupiter. *Nature*, 402, 269–270 {**271, 294**}

Ozima M, Miura YN, Podosek FA, 2004, Orphan radiogenic noble gases in lunar breccias: evidence for planet pollution of the Sun? *Icarus*, 170, 17–23 {**295**}

Paardekooper S, Mellema G, 2006, Halting type I planet migration in non-isothermal disks. *A&A*, 459, L17–L20 {**239, 240**}

Paardekooper S, Papaloizou JCB, 2009, On corotation torques, horseshoe drag and the possibility of sustained stalled or outward protoplanetary migration. *MNRAS*, 394, 2283–2296 {**234**}

Paardekooper SJ, 2007, Dust accretion onto high-mass planets. *A&A*, 462, 355–369 {**232**}

Paardekooper SJ, Thébault P, Mellema G, 2008, Planetesimal and gas dynamics in binaries. *MNRAS*, 386, 973–988 {**58**}

Pace G, Pasquini L, 2004, The age-activity-rotation relationship in solar-type stars. *A&A*, 426, 1021–1034 {**187**}

Paczyński B, 1986a, Gravitational microlensing at large optical depth. *ApJ*, 301, 503–516 {**83, 84**}

—, 1986b, Gravitational microlensing by the Galactic halo. *ApJ*, 304, 1–5 {**84, 86**}

—, 1991, Gravitational microlensing of the Galactic bulge stars. *ApJ*, 371, L63–L67 {**84**}

—, 1995, The masses of nearby dwarfs can be determined with gravitational microlensing. *Acta Astronomica*, 45, 345–348 {**94**}

—, 1996, Gravitational microlensing in the local group. *ARA&A*, 34, 419–460 {**84–86, 94**}

—, 1998, Gravitational microlensing with the Space Interferometry Mission. *ApJ*, 494, L23–26 {**94**}

Padoan P, Nordlund Å, 2002, The stellar initial mass function from turbulent fragmentation. *ApJ*, 576, 870–879 {**214**}

—, 2004, The 'mysterious' origin of brown dwarfs. *ApJ*, 617, 559–564 {**210, 214**}

Padoan P, Nordlund A, Jones BJT, 1997, The universality of the stellar initial mass function. *MNRAS*, 288, 145–152 {**214**}

Pahlevan K, Stevenson DJ, 2007, Equilibration in the aftermath of the lunar-forming giant impact. *Earth and Planetary Science Letters*, 262, 438–449 {**302**}

Pajdosz G, 1995, Non-evolutionary secular period increase in pulsating DA white dwarfs. *A&A*, 295, L17–L19 {**80**}

Pál A, 2008, Properties of analytic transit light-curve models. *MNRAS*, 390, 281–288 {**118**}

Pál A, Bakos GÁ, 2006, Astrometry in wide-field surveys. *PASP*, 118, 1474–1483 {**104**}

Pál A, Bakos GÁ, Torres G, et al., 2008, HAT-P-7 b: an extremely hot massive planet transiting a bright star in the Kepler field. *ApJ*, 680, 1450–1456 {**205, 326**}

—, 2010, Refined stellar, orbital and planetary parameters of the eccentric HAT-P-2 planetary system. *MNRAS*, 401, 2665–2674 {**144**}

Pál A, Kocsis B, 2008, Periastron precession measurements in transiting extrasolar planetary systems at the level of general relativity. *MNRAS*, 389, 191–198 {**133**}

Palacios DM, 2005, An optical vortex coronagraph. *SPIE Conf Ser*, volume 5905, 196–205 {**153**}

Palacios DM, Hunyadi SL, 2006, Low-order aberration sensitivity of an optical vortex coronagraph. *Optics Letters*, 31, 2981–2983 {**155**}

Pallé E, Ford EB, Seager S, et al., 2008a, Identifying the rotation rate and the presence of dynamic weather on extrasolar Earth-like planets

References

from photometric observations. *ApJ*, 676, 1319–1329 {**131**}

Pallé E, Montañs-Rodríguez P, Vazquez M, et al., 2008b, Cloudiness and apparent rotation rate of Earth-like planets. *ASP Conf Ser*, volume 398, 399–402 {**277**}

Pallé E, Osorio MRZ, Barrena R, et al., 2009, Earth's transmission spectrum from lunar eclipse observations. *Nature*, 459, 814–816 {**289**}

Palme H, Fegley B, 1990, High-temperature condensation of iron-rich olivine in the solar nebula. *Earth and Planetary Science Letters*, 101, 180–195 {**257**}

Palmer BA, Engleman R, 1983, *Atlas of the Thorium Spectrum*. Los Alamos National Laboratory {**20**}

Panov KP, 2009, The orbital distances law in planetary systems. *The Open Astronomy Journal*, 2, 90–94 {**48**}

Papaloizou JCB, 2003, Disk-planet interactions: migration and resonances in extrasolar planetary systems. *Celestial Mechanics and Dynamical Astronomy*, 87, 53–83 {**41, 242**}

—, 2005, Disk planet interactions and early evolution in young planetary systems. *Celestial Mechanics and Dynamical Astronomy*, 91, 33–57 {**240**}

Papaloizou JCB, Larwood JD, 2000, On the orbital evolution and growth of protoplanets embedded in a gaseous disk. *MNRAS*, 315, 823–833 {**238**}

Papaloizou JCB, Lin DNC, 1984, On the tidal interaction between protoplanets and the primordial solar nebula. I . Linear calculation of the role of angular momentum exchange. *ApJ*, 285, 818–834 {**240**}

Papaloizou JCB, Nelson RP, 2003, The interaction of a giant planet with a disk with MHD turbulence. I. The initial turbulent disk models. *MNRAS*, 339, 983–992 {**238**}

Papaloizou JCB, Nelson RP, Kley W, et al., 2007, Disk-planet interactions during planet formation. *Protostars and Planets V*, 655–668 {**241**}

Papaloizou JCB, Nelson RP, Masset F, 2001, Orbital eccentricity growth through disk-companion tidal interaction. *A&A*, 366, 263–275 {**242**}

Papaloizou JCB, Szuszkiewicz E, 2005, On the migration-induced resonances in a system of two planets with masses in the Earth mass range. *MNRAS*, 363, 153–176 {**41**}

Papaloizou JCB, Terquem C, 1999, Critical protoplanetary core masses in protoplanetary disks and the formation of short-period giant planets. *ApJ*, 521, 823–838 {**232, 234**}

—, 2001, Dynamical relaxation and massive extrasolar planets. *MNRAS*, 325, 221–230 {**127**}

—, 2006, Planet formation and migration. *Reports on Progress in Physics*, 69, 119–180 {**225, 238, 240**}

Papuc AM, Davies GF, 2008, The internal activity and thermal evolution of Earth-like planets. *Icarus*, 195, 447–458 {**279**}

Park BG, Jeon YB, Lee CU, et al., 2006, Microlensing sensitivity to Earth-mass planets in the habitable zone. *ApJ*, 643, 1233–1238 {**88**}

Parker RJ, Goodwin SP, 2009, The role of cluster evolution in disrupting planetary systems and disks: the Kozai mechanism. *MNRAS*, 397, 1041–1045 {**57**}

Parsons SG, Marsh TR, Copperwheat CM, et al., 2010a, Orbital period variations in eclipsing post-common-envelope binaries. *MNRAS*, 407, 2362–2382 {**82**}

—, 2010b, Precise mass and radius values for the white dwarf and low mass M dwarf in the pre-cataclysmic binary NN Ser. *MNRAS*, 402, 2591–2608 {**82**}

Pascucci I, Apai D, Hardegree-Ullman EE, et al., 2008, Medium-separation binaries do not affect the first steps of planet formation. *ApJ*, 673, 477–486 {**58**}

Pascucci I, Gorti U, Hollenbach D, et al., 2006, Formation and evolution of planetary systems: upper limits to the gas mass in disks around Sun-like stars. *ApJ*, 651, 1177–1193 {**242**}

Pascucci I, Tachibana S, 2010, The clearing of protoplanetary disks and of the protosolar nebula. *Protoplanetary Dust: Astrophysical and Cosmochemical Perspectives*, 263–298, Cambridge University Press {**222**}

Pasquini L, Avila G, Dekker H, et al., 2008a, CODEX: the high-resolution visual spectrograph for the E-ELT. *SPIE Conf Ser*, volume 7014, 51 {**24, 27**}

Pasquini L, Avila G, Delabre B, et al., 2008b, Codex. *Precision Spectroscopy in Astrophysics*, 249–253 {**27**}

Pasquini L, Döllinger MP, Weiss A, et al., 2007, Evolved stars suggest an external origin of the enhanced metallicity in planet-hosting stars. *A&A*, 473, 979–982 {**37, 189**}

Pasquini L, Liu Q, Pallavicini R, 1994, Lithium abundances of nearby solar-like stars. *A&A*, 287, 191–205 {**198**}

Pasquini L, Manescau A, Avila G, et al., 2009, ESPRESSO: a high resolution spectrograph for the combined coudé focus of the VLT. *Science with the VLT in the ELT Era*, 395–400 {**24**}

Patel SG, Vogt SS, Marcy GW, et al., 2007, Fourteen new companions from the Keck and Lick radial velocity survey including five brown dwarf candidates. *ApJ*, 665, 744–753 {**35**}

Patience J, White RJ, Ghez AM, et al., 2002, Stellar companions to stars with planets. *ApJ*, 581, 654–665 {**207**}

Patterson C, 1956, Age of meteorites and the Earth. *Geochim. Cosmochim. Acta*, 10, 230–237 {**298**}

Pätzold M, Carone L, Rauer H, 2004, Tidal interactions of close-in extrasolar planets: the OGLE cases. *A&A*, 427, 1075–1080 {**248**}

Pätzold M, Rauer H, 2002, Where are the massive close-in extrasolar planets? *ApJ*, 568, L117–L120 {**242**}

Paulson DB, Cochran WD, Hatzes AP, 2004a, Searching for planets in the Hyades. V. Limits on planet detection in the presence of stellar activity. *AJ*, 127, 3579–3586 {**34**}

Paulson DB, Saar SH, Cochran WD, et al., 2002, Searching for planets in the Hyades. II. Some implications of stellar magnetic activity. *AJ*, 124, 572–582 {**34**}

—, 2004b, Searching for planets in the Hyades. III. The quest for short-period planets. *AJ*, 127, 1644–1652 {**34, 65**}

Paulson DB, Sneden C, Cochran WD, 2003, Searching for planets in the Hyades. IV. Differential abundance analysis of Hyades dwarfs. *AJ*, 125, 3185–3195 {**34**}

Paulson DB, Yelda S, 2006, Differential radial velocities and stellar parameters of nearby young stars. *PASP*, 118, 706–715 {**191**}

Pavlov AA, Brown LL, Kasting JF, 2001, Ultraviolet shielding of NH_3 and O_2 by organic hazes in the Archean atmosphere. *J. Geophys. Res.*, 106, 23267–23288 {**306**}

Pavlov AA, Hurtgen MT, Kasting JF, et al., 2003, Methane-rich Proterozoic atmosphere? *Geology*, 31, 87–91 {**306**}

Pavlov AA, Kasting JF, Brown LL, et al., 2000, Greenhouse warming by CH_4 in the atmosphere of early Earth. *J. Geophys. Res.*, 105, 11981–11990 {**306**}

Payne MJ, Lodato G, 2007, The potential for Earth-mass planet formation around brown dwarfs. *MNRAS*, 381, 1597–1606 {**214**}

Payne MJ, Wyatt MC, Thébault P, 2009, Outward migration of terrestrial embryos in binary systems. *MNRAS*, 400, 1936–1944 {**58**}

Peacock A, Verhoeve P, Rando N, et al., 1996, Single optical photon detection with a superconducting tunnel junction. *Nature*, 381, 135–137 {**133**}

Peale SJ, 1976, Orbital resonances in the solar system. *ARA&A*, 14, 215–246 {**40, 41, 51, 300**}

—, 1977, Rotation histories of the natural satellites. *Planetary Satellites*, 87–112 {**246**}

—, 1994, On the detection of mutual perturbations as proof of planets around PSR B1257+12. *Ap&SS*, 212, 77–89 {**76**}

—, 1997, Expectations from a microlensing search for planets. *Icarus*, 127, 269–289 {**88**}

—, 2001, Probability of detecting a planetary companion during a microlensing event. *ApJ*, 552, 889–911 {**88**}

—, 2003, Comparison of a ground-based microlensing search for planets with a search from space. *AJ*, 126, 1595–1603 {**102**}

—, 2008, Obliquity tides in hot Jupiters. *ASP Conf Ser*, volume 398, 281–292 {**145**}

Peale SJ, Cassen P, 1978, Contribution of tidal dissipation to lunar thermal history. *Icarus*, 36, 245–269 {**282**}

Peale SJ, Cassen P, Reynolds RT, 1979, Melting of Io by tidal dissipation. *Science*, 203, 892–894 {**250, 282**}

Peale SJ, Gold T, 1965, Rotation of the planet Mercury. *Nature*, 206, 1240–1241 {**246, 300**}

Pearson TJ, Readhead ACS, 1984, Image formation by self-calibration in radio astronomy. *ARA&A*, 22, 97–130 {**124**}

Pedretti E, Labeyrie A, Arnold L, et al., 2000, First images on the sky from a hyper telescope. *A&AS*, 147, 285–290 {**168**}

Pendleton YJ, Black DC, 1983, Further studies on criteria for the onset of dynamical instability in general three-body systems. *AJ*, 88, 1415–1419 {**56**}

Pendry JB, 2000, Negative refraction makes a perfect lens. *Physical Re-*

view Letters, 85, 3966–3969 {169}

Penz T, Micela G, 2008, X-ray induced mass loss effects on exoplanets orbiting dM stars. *A&A*, 479, 579–584 {281}

Penz T, Micela G, Lammer H, 2008, Influence of the evolving stellar X-ray luminosity distribution on exoplanetary mass loss. *A&A*, 477, 309–314 {281}

Pepe F, Correia ACM, Mayor M, et al., 2007, The HARPS search for southern extrasolar planets. VIII. μ Ara, a system with four planets. *A&A*, 462, 769–776 {**14, 46–48, 52, 54, 203**}

Pepe F, Lovis C, 2008, From HARPS to CODEX: exploring the limits of Doppler measurements. *Physica Scripta Volume T*, 130(1), 014007 {**21, 24, 25, 27**}

Pepe F, Mayor M, Galland F, et al., 2002, The CORALIE survey for southern extrasolar planets. VII. Two short-period Saturnian companions to HD 108147 and HD 168746. *A&A*, 388, 632–638 {**17**}

Pepe F, Mayor M, Queloz D, et al., 2004, The HARPS search for southern extrasolar planets. I. HD 330075b: a new hot Jupiter. *A&A*, 423, 385–389 {**2**}

Pepliński A, Artymowicz P, Mellema G, 2008a, Numerical simulations of Type III planetary migration. I. Disk model and convergence tests. *MNRAS*, 386, 164–178 {**240**}

—, 2008b, Numerical simulations of Type III planetary migration. II. Inward migration of massive planets. *MNRAS*, 386, 179–198 {**240**}

—, 2008c, Numerical simulations of Type III planetary migration. III. Outward migration of massive planets. *MNRAS*, 387, 1063–1079 {**240**}

Pepper J, Gaudi BS, 2005, Searching for transiting planets in stellar systems. *ApJ*, 631, 581–596 {**109**}

—, 2006, Toward the detection of transiting hot Earths and hot Neptunes in open clusters. *Acta Astronomica*, 56, 183–197 {**109**}

Perrier C, Mariotti J, 1987, On the binary nature of Van Biesbroeck 8. *ApJ*, 312, L27–L30 {**210**}

Perrier C, Sivan JP, Naef D, et al., 2003, The ELODIE survey for northern extrasolar planets. I. Six new extrasolar planet candidates. *A&A*, 410, 1039–1049 {**58**}

Perrin G, Lacour S, Woillez J, et al., 2006, High dynamic range imaging by pupil single-mode filtering and remapping. *MNRAS*, 373, 747–751 {**156**}

Perrin MD, Sivaramakrishnan A, Makidon RB, et al., 2003, The structure of high Strehl ratio point-spread functions. *ApJ*, 596, 702–712 {**157**}

Perruchot S, Kohler D, Bouchy F, et al., 2008, The SOPHIE spectrograph: design and technical key-points for high throughput and high stability. *SPIE Conf Ser*, volume 7014, 17 {**24**}

Perryman MAC, 1997, *The Hipparcos and Tycho catalogues. Astrometric and photometric star catalogues derived from the ESA Hipparcos space astrometry mission*, volume SP-1200 of *ESA Special Publication*. European Space Agency {**69, 70, 181–183**}

—, 2000, Extrasolar planets. *Reports on Progress in Physics*, 63, 1209–1272 {**4**}

—, 2009, *Astronomical Applications of Astrometry: Ten Years of Exploitation of the Hipparcos Satellite Data*. Cambridge University Press {**69, 181, 186, 307, 309, 310**}

Perryman MAC, de Boer KS, Gilmore G, et al., 2001, Gaia: composition, formation and evolution of the Galaxy. *A&A*, 369, 339–363 {**65, 72**}

Perryman MAC, Favata F, Peacock A, et al., 1999, Optical STJ observations of the Crab pulsar. *A&A*, 346, L30–L32 {**133**}

Perryman MAC, Foden CL, Peacock A, 1993, Optical photon counting using superconducting tunnel junctions. *Nuclear Instruments and Methods in Physics Research A*, 325, 319–325 {**133**}

Perryman MAC, Hainaut O, Dravins D, et al., 2005, ESA–ESO Working Group on Extrasolar Planets. Technical report, ESA/ESO {**160**}

Perryman MAC, Lindegren L, Arenou F, et al., 1996, Hipparcos distances and mass limits for the planetary candidates: 47 UMa, 70 Vir, and 51 Peg. *A&A*, 310, L21–L24 {**70**}

Perryman MAC, Lindegren L, Kovalevsky J, et al., 1997, The Hipparcos Catalogue. *A&A*, 323, L49–L52 {**6, 69, 181**}

Perryman MAC, Schulze-Hartung T, 2011, The barycentric motion of exoplanet host stars: tests of solar spin-orbit coupling. *A&A*, 525, A65 {**66, 67**}

Peter D, Feldt M, Henning T, et al., 2010, PYRAMIR: exploring the on-sky performance of the world's first near-infrared pyramid wavefront sensor. *PASP*, 122, 63–70 {**151**}

Petit C, Fusco T, Charton J, et al., 2008, The SPHERE XAO system: design and performance. *SPIE Conf Ser*, volume 7015, 35 {**152, 159**}

Petit JR, Jouzel J, Raynaud D, et al., 1999, Climate and atmospheric history of the past 420 000 years from the Vostok ice core, Antarctica. *Nature*, 399, 429–436 {**307**}

Pettengill GH, Dyce RB, 1965, A radar determination of the rotation of the planet Mercury. *Nature*, 206, 1240–1240 {**246, 300**}

Pfahl E, 2005, Cluster origin of the triple star HD 188753 and its planet. *ApJ*, 635, L89–L92 {**60**}

Pfahl E, Arras P, Paxton B, 2008, Ellipsoidal oscillations induced by substellar companions: a prospect for the Kepler mission. *ApJ*, 679, 783–796 {**132**}

Pfahl E, Muterspaugh M, 2006, Impact of stellar dynamics on the frequency of giant planets in close binaries. *ApJ*, 652, 1694–1697 {**58**}

Phan-Bao N, Bessell MS, Martín EL, et al., 2008, Discovery of new nearby L and late-M dwarfs at low Galactic latitude from the DENIS data base. *MNRAS*, 383, 831–844 {**210**}

Phillips JA, Thorsett SE, 1994, Planets around pulsars: a review. *Ap&SS*, 212, 91–106 {**76, 77**}

Phinney ES, Hansen BMS, 1993, The pulsar planet production process. *Planets Around Pulsars*, volume 36 of *ASP Conf Ser*, 371–390 {**77**}

Pichardo B, Sparke LS, Aguilar LA, 2005, Circumstellar and circumbinary disks in eccentric stellar binaries. *MNRAS*, 359, 521–530 {**58**}

Pierens A, Nelson RP, 2008a, Constraints on resonant-trapping for two planets embedded in a protoplanetary disk. *A&A*, 482, 333–340 {**82, 242**}

—, 2008b, On the formation and migration of giant planets in circumbinary disks. *A&A*, 483, 633–642 {**57, 81**}

Piétu V, Dutrey A, Guilloteau S, 2007, Probing the structure of protoplanetary disks: a comparative study of DM Tau, LkCa 15, and MWC 480. *A&A*, 467, 163–178 {**220**}

Pilat-Lohinger E, Dvorak R, 2002, Stability of S-type orbits in binaries. *Celestial Mechanics and Dynamical Astronomy*, 82, 143–153 {**57**}

Pilat-Lohinger E, Dvorak R, Bois E, et al., 2004, Stable planetary motion in double stars. *Extrasolar Planets: Today and Tomorrow*, volume 321 of *ASP Conf Ser*, 410–418 {**57**}

Pilat-Lohinger E, Funk B, Dvorak R, 2003, Stability limits in double stars. A study of inclined planetary orbits. *A&A*, 400, 1085–1094 {**57**}

Pilat-Lohinger E, Robutel P, Süli Á, et al., 2008a, On the stability of Earth-like planets in multi-planet systems. *Celestial Mechanics and Dynamical Astronomy*, 102, 83–95 {**285**}

Pilat-Lohinger E, Süli Á, Robutel P, et al., 2008b, The influence of giant planets near a mean motion resonance on Earth-like planets in the habitable zone of Sun-like stars. *ApJ*, 681, 1639–1645 {**285**}

Pilcher CB, 2003, Biosignatures of early Earths. *Astrobiology*, 3, 471–486 {**306**}

Pilipp W, Hartquist TW, Morfill GE, 1992, Large electric fields in acoustic waves and the stimulation of lightning discharges. *ApJ*, 387, 364–371 {**297**}

Pilipp W, Hartquist TW, Morfill GE, et al., 1998, Chondrule formation by lightning in the protosolar nebula? *A&A*, 331, 121–146 {**297**}

Pinho LGF, Porto de Mello GF, 2003, Astrobiologically interesting stars in the solar neighborhood. *Bulletin of the Astronomical Society of Brazil*, 23, 128–128 {**286**}

Pinho LGF, Porto de Mello GF, de Medeiros JR, et al., 2003, The Sol project: the Sun in time. *Bulletin of the Astronomical Society of Brazil*, 23, 126–126 {**286**}

Pinotti R, Arany-Prado L, Lyra W, et al., 2005, A link between the semi-major axis of extrasolar gas giant planets and stellar metallicity. *MNRAS*, 364, 29–36 {**191**}

Pinsonneault MH, 1997, Mixing in stars. *ARA&A*, 35, 557–605 {**198, 296**}

Pinsonneault MH, Deliyannis CP, Demarque P, 1992, Evolutionary models of halo stars with rotation. II. Effects of metallicity on lithium depletion, and possible implications for the primordial lithium abundance. *ApJS*, 78, 179–203 {**198**}

Pinsonneault MH, DePoy DL, Coffee M, 2001, The mass of the convective zone in FGK main-sequence stars and the effect of accreted planetary material on apparent metallicity determinations. *ApJ*, 556, L59–L62 {**192**}

Pinsonneault MH, Kawaler SD, Sofia S, et al., 1989, Evolutionary models of the rotating Sun. *ApJ*, 338, 424–452 {**200**}

Pitjeva EV, 2005, Relativistic effects and solar oblateness from radar observations of planets and spacecraft. *Astronomy Letters*, 31, 340–349 {309}

Plavchan P, Jura M, Kirkpatrick JD, et al., 2008, Near-infrared variability in the 2MASS calibration fields: a search for planetary transit candidates. *ApJS*, 175, 191–228 {115}

Poddaný S, Brát L, Pejcha O, 2010, Exoplanet Transit Database: reduction and processing of the photometric data of exoplanet transits. *New Astronomy*, 15, 297–301 {105}

Podolak M, 2003, The contribution of small grains to the opacity of protoplanetary atmospheres. *Icarus*, 165, 428–437 {234}

Podolak M, Hubbard WB, Stevenson DJ, 1991, Model of Uranus' interior and magnetic field. *Uranus*, 29–61, University of Arizona Press {269}

Podolak M, Pollack JB, Reynolds RT, 1988, Interactions of planetesimals with protoplanetary atmospheres. *Icarus*, 73, 163–179 {232}

Podolak M, Weizman A, Marley M, 1995, Comparative models of Uranus and Neptune. *Planet. Space Sci.*, 43, 1517–1522 {266, 294}

Podsiadlowski P, 1993, Planet formation scenarios. *Planets Around Pulsars*, volume 36 of *ASP Conf Ser*, 149–165 {77}

Poindexter S, Afonso C, Bennett DP, et al., 2005, Systematic analysis of 22 microlensing parallax candidates. *ApJ*, 633, 914–930 {93, 210}

Pollacco D, Skillen I, Collier Cameron A, et al., 2006, The WASP project and the SuperWASP cameras. *PASP*, 118, 1407–1418 {107}

—, 2008, WASP-3 b: a strongly irradiated transiting gas-giant planet. *MNRAS*, 385, 1576–1584 {107, 327}

Pollack JB, 1984, Origin and history of the outer planets: theoretical models and observations–constraints. *ARA&A*, 22, 389–424 {232, 235}

Pollack JB, Black DC, 1979, Implications of the gas compositional measurements of Pioneer Venus for the origin of planetary atmospheres. *Science*, 205, 56–59 {280}

—, 1982, Noble gases in planetary atmospheres: implications for the origin and evolution of atmospheres. *Icarus*, 51, 169–198 {278, 280}

Pollack JB, Hubickyj O, Bodenheimer P, et al., 1996, Formation of the giant planets by concurrent accretion of solids and gas. *Icarus*, 124, 62–85 {146, 191, 226, 228, 232, 233}

Pollack JB, Podolak M, Bodenheimer P, et al., 1986, Planetesimal dissolution in the envelopes of the forming, giant planets. *Icarus*, 67, 409–443 {232, 233}

Pols OR, Tout CA, Eggleton PP, et al., 1995, Approximate input physics for stellar modeling. *MNRAS*, 274, 964–974 {261}

Pont F, 2008, Ground-based searches for transiting planets. *ASP Conf Ser*, volume 398, 87–92 {104}

—, 2009, Empirical evidence for tidal evolution in transiting planetary systems. *MNRAS*, 396, 1789–1796 {146, 248}

Pont F, Bouchy F, 2005, Exoplanet transit search at Dome C. *EAS Publications Series*, volume 14, 155–160 {110}

Pont F, Bouchy F, Melo C, et al., 2005, Doppler follow-up of OGLE planetary transit candidates in Carina. *A&A*, 438, 1123–1140 {106}

Pont F, Bouchy F, Queloz D, et al., 2004, The missing link: a 4-day period transiting exoplanet around OGLE–TR–111. *A&A*, 426, L15–L18 {106, 327}

Pont F, Endl M, Cochran WD, et al., 2010, The spin-orbit angle of the transiting hot Jupiter CoRoT–1 b. *MNRAS*, 402, L1–L5 {129}

Pont F, Gilliland RL, Moutou C, et al., 2007, HST time-series photometry of the planetary transit of HD 189733: no moon, no rings, starspots. *A&A*, 476, 1347–1355 {112, 130, 131, 140}

Pont F, Hébrard G, Irwin JM, et al., 2009, Spin-orbit misalignment in the HD 80606 planetary system. *A&A*, 502, 695–703 {129}

Pont F, Knutson H, Gilliland RL, et al., 2008a, Detection of atmospheric haze on an extrasolar planet: the 0.55–1.05 μm transmission spectrum of HD 189733 b with the HubbleSpaceTelescope. *MNRAS*, 385, 109–118 {140}

Pont F, Tamuz O, Udalski A, et al., 2008b, A transiting planet among 23 new near-threshold candidates from the OGLE survey: OGLE–TR–182. *A&A*, 487, 749–754 {105, 106, 327}

Pont F, Zucker S, Queloz D, 2006, The effect of red noise on planetary transit detection. *MNRAS*, 373, 231–242 {115}

Poppenhaeger K, Robrade J, Schmitt JHMM, 2010, Coronal properties of planet-bearing stars. *A&A*, 515, A98 {206}

Popper K, 1976, *Unended Quest, An Intellectual Autobiography*. Fontana/Collins {288}

Porco CC, Baker E, Barbara J, et al., 2005, Cassini imaging science: initial results on Saturn's rings and small satellites. *Science*, 307, 1226–1236 {226}

Portegies Zwart SF, 2009, The lost siblings of the Sun. *ApJ*, 696, L13–L16 {293}

Portegies Zwart SF, McMillan SLW, 2005, Planets in triple star systems: the case of HD 188753. *ApJ*, 633, L141–L144 {60}

Porto de Mello GF, da Silva L, 1997, HR 6060: the closest ever solar twin? *ApJ*, 482, L89–92 {286}

Porto de Mello GF, da Silva R, da Silva L, 2000, A survey of solar twin stars within 50 pc of the Sun. *Bioastronomy 99*, volume 213 of *ASP Conf Ser*, 73–79 {286}

Porto de Mello GF, del Peloso EF, Ghezzi L, 2006, Astrobiologically interesting stars within 10 pc of the Sun. *Astrobiology*, 6, 308–331 {286}

Pott J, Woillez J, Akeson RL, et al., 2009, Astrometry with the Keck Interferometer: the ASTRA project and its science. *New Astronomy Reviews*, 53, 363–372 {69}

Poulton CJ, Greaves JS, Collier Cameron A, 2006, Detecting a rotation in the ϵ Eri debris disk. *MNRAS*, 372, 53–59 {223}

Pourbaix D, 2001, The Hipparcos observations and the mass of substellar objects. *A&A*, 369, L22–L25 {70}

—, 2002, Precision and accuracy of the orbital parameters derived from 2D and 1D space observations of visual or astrometric binaries. *A&A*, 385, 686–692 {67}

Pourbaix D, Arenou F, 2001, Screening the Hipparcos-based astrometric orbits of sub-stellar objects. *A&A*, 372, 935–944 {70}

Pourbaix D, Jorissen A, 2000, Re-processing the Hipparcos Transit Data and Intermediate Astrometric Data of spectroscopic binaries. I. Ba, CH and Tc-poor S stars. *A&AS*, 145, 161–183 {68}

Poveda A, Lara P, 2008, The exo-planetary system of 55 Cnc and the Titius–Bode Law. *Revista Mexicana de Astronomia y Astrofisica*, 44, 243–246 {48}

Povich MS, Giampapa MS, Valenti JA, et al., 2001, Limits on line bisector variability for stars with extrasolar planets. *AJ*, 121, 1136–1146 {23}

Pradel N, 2009, Presentation of the Low Frequency Array (LOFAR). *SF2A-2009: Proceedings of the Annual meeting of the French Society of Astronomy and Astrophysics*, 19–20 {177}

Pradel N, Charlot P, Lestrade J, 2006, Astrometric accuracy of phase-referenced observations with the VLBA and EVN. *A&A*, 452, 1099–1106 {173}

Prantzos N, 2008, On the Galactic habitable zone. *Space Science Reviews*, 135, 313–322 {286}

Pratt MR, Alcock C, Allsman RA, et al., 1996, Real-time detection of gravitational microlensing. *Astrophysical Applications of Gravitational Lensing*, volume 173 of *IAU Symposium*, 221–226 {97}

Pravdo SH, Shaklan SB, 1996, Astrometric detection of extrasolar planets: results of a feasibility study with the Palomar 5m telescope. *ApJ*, 465, 264–277 {62, 68}

—, 2009a, Ten years of STEPS astrometry. *American Astronomical Society Meeting Abstracts*, volume 214, 306.07 {68}

—, 2009b, An ultracool star's candidate planet. *ApJ*, 700, 623–632 {68}

Pravdo SH, Shaklan SB, Lloyd J, et al., 2005, Discovering M-dwarf companions with STEPS. *Astrometry in the Age of the Next Generation of Large Telescopes*, volume 338 of *ASP Conf Ser*, 288–292 {68}

Prentice AJR, 1978a, Origin of the solar system. I. Gravitational contraction of the turbulent proto-Sun and the shedding of a concentric system of gaseous Laplacian rings. *Moon and Planets*, 19, 341–398 {225}

—, 1978b, Towards a modern Laplacian theory for the formation of the solar system. *Origin of the Solar System*, 111–161, Wiley {225}

Press WH, Teukolsky SA, Vetterling WT, et al., 2007, *Numerical Recipes: The Art of Scientific Computing*. Cambridge University Press, Third Edition {14, 44, 119}

Preusse S, Kopp A, Büchner J, et al., 2006, A magnetic communication scenario for hot Jupiters. *A&A*, 460, 317–322 {175, 205}

Pringle JE, 1981, Accretion disks in astrophysics. *ARA&A*, 19, 137–162 {220}

Proffitt CR, Sahu K, Livio M, et al., 2004, Limits on the optical brightness of the ϵ Eri dust ring. *ApJ*, 612, 481–495 {223}

Proskurowski G, Lilley MD, Seewald JS, et al., 2008, Abiogenic hydrocarbon production at Lost City Hydrothermal Field. *Science*, 319, 604–607 {279}

Protopapas P, Jimenez R, Alcock C, 2005, Fast identification of transits from light-curves. *MNRAS*, 362, 460–468 {105}

Provencal JL, 1997, White dwarfs and planetary systems. *Planets Beyond the Solar System and the Next Generation of Space Missions*, volume 119 of *ASP Conf Ser*, 123–126 {**80**}

Pruzan P, Chervin JC, Wolanin E, et al., 2003, Phase diagram of ice in the VII–VIII–X domain: vibrational and structural data for strongly compressed ice VIII. *Journal of Raman Spectroscopy*, 34, 591–610 {**263**}

Pshirkov MS, Baskaran D, Postnov KA, 2010, Observing gravitational wave bursts in pulsar timing measurements. *MNRAS*, 402, 417–423 {**79**}

Pudritz RE, Ouyed R, Fendt C, et al., 2007, Disk winds, jets, and outflows: theoretical and computational foundations. *Protostars and Planets V*, 277–294 {**219**}

Qian S, Dai Z, Liao W, et al., 2009, A substellar companion to the white dwarf-red dwarf eclipsing binary NN Ser. *ApJ*, 706, L96–L99 {**82**}

Qian S, Liao W, Zhu L, et al., 2010a, A giant planet in orbit around a magnetic-braking hibernating cataclysmic variable. *MNRAS*, 401, L34–L38 {**82**}

—, 2010b, Detection of a giant extrasolar planet orbiting the eclipsing polar DP Leo. *ApJ*, 708, L66–L68 {**79, 82**}

Qian Y, 2003, The origin of the heavy elements: recent progress in the understanding of the r-process. *Progress in Particle and Nuclear Physics*, 50, 153–199 {**197**}

Quanz SP, Goldman B, Henning T, et al., 2010, Search for very low-mass brown dwarfs and free-floating planetary-mass objects in Taurus. *ApJ*, 708, 770–784 {**215**}

Queloz D, 1995, Echelle spectroscopy with a CCD at low signal-to-noise ratio. *New Developments in Array Technology and Applications*, volume 167 of *IAU Symposium*, 221–28 {**17, 18**}

Queloz D, Anderson D, Collier Cameron A, et al., 2010, WASP-8 b: a retrograde transiting planet in a multiple system. *A&A*, 517, L1–L4 {**108, 129, 327**}

Queloz D, Bouchy F, Moutou C, et al., 2009, The CoRoT-7 planetary system: two orbiting super-Earths. *A&A*, 506, 303–319 {**111**}

Queloz D, Eggenberger A, Mayor M, et al., 2000a, Detection of a spectroscopic transit by the planet orbiting the star HD 209458. *A&A*, 359, L13–L17 {**2, 127–129**}

Queloz D, Henry GW, Sivan JP, et al., 2001, No planet for HD 166435. *A&A*, 379, 279–287 {**23, 33, 65**}

Queloz D, Mayor M, Weber L, et al., 2000b, The CORALIE survey for southern extrasolar planets. I. A planet orbiting the star GJ 86. *A&A*, 354, 99–102 {**24, 58, 192**}

Quillen AC, 2002, Using a Hipparcos-derived Hertzsprung-Russell diagram to limit the metallicity scatter of stars in the Hyades: are stars polluted? *AJ*, 124, 400–403 {**192**}

—, 2006a, Predictions for a planet just inside Fomalhaut's eccentric ring. *MNRAS*, 372, L14–L18 {**171, 223**}

—, 2006b, Reducing the probability of capture into resonance. *MNRAS*, 365, 1367–1382 {**42**}

—, 2010, Pinpointing planets in circumstellar disks. *Formation and Evolution of Exoplanets*, 27–48, Wiley {**223**}

Quillen AC, Faber P, 2006, Chaotic zone boundary for low free eccentricity particles near an eccentric planet. *MNRAS*, 373, 1245–1250 {**42**}

Quillen AC, Holman M, 2000, Production of star-grazing and star-impacting planetesimals via orbital migration of extrasolar planets. *AJ*, 119, 397–402 {**192**}

Quillen AC, Morbidelli A, Moore A, 2007, Planetary embryos and planetesimals residing in thin debris disks. *MNRAS*, 380, 1642–1648 {**224**}

Quillen AC, Thorndike S, 2002, Structure in the ϵ Eri dusty disk caused by mean motion resonances with a 0.3 eccentricity planet at periastron. *ApJ*, 578, L149–L152 {**223**}

Quinn SN, Bakos GÁ, Hartman J, et al., 2010, HAT–P–25 b: a hot-Jupiter transiting a moderately faint G star. *ArXiv e-prints* {**326**}

Quintana EV, Adams FC, Lissauer JJ, et al., 2007, Terrestrial planet formation around individual stars within binary star systems. *ApJ*, 660, 807–822 {**57, 81**}

Quintana EV, Lissauer JJ, 2006, Terrestrial planet formation surrounding close binary stars. *Icarus*, 185, 1–20 {**57, 81**}

Quintana EV, Lissauer JJ, Chambers JE, et al., 2002, Terrestrial planet formation in the α Cen System. *ApJ*, 576, 982–996 {**57, 58**}

Quirrenbach A, 2001, Optical Interferometry. *ARA&A*, 39, 353–401 {**162**}

—, 2005, Coronographic methods for the detection of terrestrial planets. *ArXiv Astrophysics e-prints* {**153, 165**}

Quist CF, 2001, Astrometric detection of sub-stellar companions with Gaia. *A&A*, 370, 672–679 {**72**}

Rabien S, Ageorges N, Angel R, et al., 2008, The laser guide star programme for the LBT. *SPIE Conf Ser*, volume 7015, 28 {**152**}

Rabl G, Dvorak R, 1988, Satellite-type planetary orbits in double stars: a numerical approach. *A&A*, 191, 385–391 {**57**}

Rabus M, Alonso R, Belmonte JA, et al., 2009a, A cool starspot or a second transiting planet in the TrES–1 system? *A&A*, 494, 391–397 {**107, 131**}

Rabus M, Alonso R, Deeg HJ, et al., 2009b, Transit timing variability in TrES–1. *IAU Symposium*, volume 253 of *IAU Symposium*, 432–435 {**135**}

Rabus M, Brown TM, Deeg HJ, et al., 2007, Update and recent results of the STARE instrument. *Transiting Extrasolar Planets Workshop*, volume 366 of *ASP Conf Ser*, 96–98 {**107**}

Rabus M, Deeg HJ, Alonso R, et al., 2009c, Transit timing analysis of the exoplanets TrES–1 and TrES–2. *A&A*, 508, 1011–1020 {**135**}

Racine R, Walker GAH, Nadeau D, et al., 1999, Speckle noise and the detection of faint companions. *PASP*, 111, 587–594 {**157, 158**}

Rafikov RR, 2003a, The growth of planetary embryos: orderly, runaway, or oligarchic? *AJ*, 125, 942–961 {**227–229**}

—, 2003b, Planetesimal disk evolution driven by embryo-planetesimal gravitational scattering. *AJ*, 125, 922–941 {**227**}

—, 2005, Can giant planets form by direct gravitational instability? *ApJ*, 621, L69–L72 {**236**}

—, 2007, Convective cooling and fragmentation of gravitationally unstable disks. *ApJ*, 662, 642–650 {**236**}

—, 2009a, Properties of gravitoturbulent accretion disks. *ApJ*, 704, 281–291 {**236**}

—, 2009b, Stellar proper motion and the timing of planetary transits. *ApJ*, 700, 965–970 {**133, 136**}

Ragazzoni R, 1996, Pupil plane wavefront sensing with an oscillating prism. *Journal of Modern Optics*, 43, 289–293 {**151**}

Raghavan D, Henry TJ, Mason BD, et al., 2006, Two suns in the sky: stellar multiplicity in exoplanet systems. *ApJ*, 646, 523–542 {**181, 207**}

Ragland S, Wizinowich P, Akeson R, et al., 2008, Recent progress at the Keck Interferometer: operations and science. *SPIE Conf Ser*, volume 7013, 10 {**69**}

Ragozzine D, Wolf AS, 2009, Probing the interiors of very hot Jupiters using transit light curves. *ApJ*, 698, 1778–1794 {**103, 133, 134, 143**}

Ramakrishna SA, 2005, Physics of negative refractive index materials. *Reports on Progress in Physics*, 68, 449–521 {**169**}

Ramírez I, Meléndez J, 2004, Cooler and bigger than previously thought? Planetary host stellar parameters from the infrared flux method. *ApJ*, 609, 417–422 {**185**}

—, 2005a, The effective temperature scale of FGK stars. I. Determination of temperatures and angular diameters with the infrared flux method. *ApJ*, 626, 446–464 {**185**}

—, 2005b, The effective temperature scale of FGK stars. II. T_{eff}–colour–[Fe/H] calibrations. *ApJ*, 626, 465–485 {**185**}

Ramsey LW, Barnes J, Redman SL, et al., 2008, A pathfinder instrument for precision radial velocities in the near-infrared. *PASP*, 120, 887–894 {**24**}

Rasio FA, Ford EB, 1996, Dynamical instabilities and the formation of extrasolar planetary systems. *Science*, 274, 954–956 {**192, 237, 243, 245**}

Rasio FA, Nicholson PD, Shapiro SL, et al., 1992, An observational test for the existence of a planetary system orbiting PSR B1257+12. *Nature*, 355, 325–326 {**76**}

Rasio FA, Tout CA, Lubow SH, et al., 1996, Tidal decay of close planetary orbits. *ApJ*, 470, 1187–1191 {**44, 238, 245, 246**}

Rasool SI, de Bergh C, 1970, The runaway greenhouse and the accumulation of CO_2 in the Venus atmosphere. *Nature*, 226, 1037–1039 {**283**}

Rattenbury NJ, Bond IA, Skuljan J, et al., 2002, Planetary microlensing at high magnification. *MNRAS*, 335, 159–169 {**88, 89**}

Rauch KP, Hamilton DP, 2002, The HNBody package for symplectic integration of nearly-Keplerian systems. *Bulletin of the American Astronomical Society*, volume 34, 938 {**44**}

Rauer H, Fruth T, Erikson A, 2008, Prospects of long-time-series observations from Dome C for transit search. *PASP*, 120, 852–859 {**110**}

Rauer H, Queloz D, Csizmadia S, et al., 2009, Transiting exoplanets from the CoRoT space mission. VII. The hot-Jupiter planet CoRoT–5 b. *A&A*, 506, 281–286 {**326**}

Raup D, Sepkoski JA, 1982, Mass extinctions in the marine fossil record. *Science*, 215(1501-1503) {**307**}

Rauscher E, Menou K, Cho JYK, et al., 2007a, Hot Jupiter variability in eclipse depth. *ApJ*, 662, L115–L118 {**131**}

Rauscher E, Menou K, Seager S, et al., 2007b, Toward eclipse mapping of hot Jupiters. *ApJ*, 664, 1199–1209 {**138**}

Raven JA, Cockell C, 2006, Influence on photosynthesis of starlight, moonlight, planetlight, and light pollution: reflections on photosynthetically active radiation in the universe. *Astrobiology*, 6, 668–675 {**289**}

Raymond J, Segrè D, 2006, The effect of oxygen on biochemical networks and the evolution of complex life. *Science*, 311, 1764–1767 {**306**}

Raymond SN, 2006, The search for other Earths: limits on the giant planet orbits that allow habitable terrestrial planets to form. *ApJ*, 643, L131–L134 {**241, 299**}

—, 2010, Formation of terrestrial planets. *Formation and Evolution of Exoplanets*, 123–144, Wiley {**225**}

Raymond SN, Armitage PJ, Gorelick N, 2009a, Planet–planet scattering in planetesimal disks. *ApJ*, 699, L88–L92 {**243, 244**}

—, 2010, Planet–planet scattering in planetesimal disks. II. Predictions for outer extrasolar planetary systems. *ApJ*, 711, 772–795 {**243, 244**}

Raymond SN, Barnes R, 2005, Predicting planets in known extrasolar planetary systems. II. Testing for Saturn mass planets. *ApJ*, 619, 549–557 {**44, 48**}

Raymond SN, Barnes R, Armitage PJ, et al., 2008a, Mean motion resonances from planet–planet scattering. *ApJ*, 687, L107–L110 {**243, 244**}

Raymond SN, Barnes R, Gorelick N, 2008b, A dynamical perspective on additional planets in 55 Cnc. *ApJ*, 689, 478–491 {**49**}

Raymond SN, Barnes R, Kaib NA, 2006a, Predicting planets in known extrasolar planetary systems. III. Forming terrestrial planets. *ApJ*, 644, 1223–1231 {**44**}

Raymond SN, Barnes R, Mandell AM, 2008c, Observable consequences of planet formation models in systems with close-in terrestrial planets. *MNRAS*, 384, 663–674 {**238**}

Raymond SN, Barnes R, Veras D, et al., 2009b, Planet–planet scattering leads to tightly packed planetary systems. *ApJ*, 696, L98–L101 {**243**}

Raymond SN, Mandell AM, Sigurdsson S, 2006b, Exotic earths: forming habitable worlds with giant planet migration. *Science*, 313, 1413–1416 {**285**}

Raymond SN, O'Brien DP, Morbidelli A, et al., 2009c, Building the terrestrial planets: constrained accretion in the inner solar system. *Icarus*, 203, 644–662 {**304**}

Raymond SN, Quinn T, Lunine JI, 2004, Making other Earths: dynamical simulations of terrestrial planet formation and water delivery. *Icarus*, 168, 1–17 {**298**}

—, 2005a, The formation and habitability of terrestrial planets in the presence of close-in giant planets. *Icarus*, 177, 256–263 {**241, 285, 298**}

—, 2005b, Terrestrial planet formation in disks with varying surface density profiles. *ApJ*, 632, 670–676 {**229**}

—, 2006c, High-resolution simulations of the final assembly of Earth-like planets. I. Terrestrial accretion and dynamics. *Icarus*, 183, 265–282 {**304**}

—, 2007a, High-resolution simulations of the final assembly of Earth-like planets.II. Water delivery and planetary habitability. *Astrobiology*, 7, 66–84 {**285, 298**}

Raymond SN, Scalo J, Meadows VS, 2007b, A decreased probability of habitable planet formation around low-mass stars. *ApJ*, 669, 606–614 {**250**}

Reach WT, Kuchner MJ, von Hippel T, et al., 2005, The dust cloud around the white dwarf G29–38. *ApJ*, 635, L161–L164 {**172**}

Reach WT, Lisse C, von Hippel T, et al., 2009, The dust cloud around the white dwarf G29–38. II. Spectrum from 5–40 μm and mid-infrared photometric variability. *ApJ*, 693, 697–712 {**173**}

Rebolo R, Martin EL, Basri G, et al., 1996, Brown dwarfs in the Pleiades cluster confirmed by the lithium test. *ApJ*, 469, L53–56 {**198**}

Rebolo R, Martin EL, Magazzu A, 1992, Spectroscopy of a brown dwarf candidate in the Alpha Persei open cluster. *ApJ*, 389, L83–L86 {**198**}

Rebolo R, Zapatero Osorio MR, Martín EL, 1995, Discovery of a brown dwarf in the Pleiades star cluster. *Nature*, 377, 129–131 {**210**}

Rebull LM, 2001, Rotation of young low-mass stars in the Orion Nebula cluster flanking fields. *AJ*, 121, 1676–1709 {**200**}

Rebull LM, Wolff SC, Strom SE, et al., 2002, The early angular momentum history of low-mass stars: evidence for a regulation mechanism. *AJ*, 124, 546–559 {**200**}

Reche R, Beust H, Augereau JC, et al., 2008, On the observability of resonant structures in planetesimal disks due to planetary migration. *A&A*, 480, 551–561 {**223**}

Reddy BE, Lambert DL, Allende Prieto C, 2006, Elemental abundance survey of the Galactic thick disk. *MNRAS*, 367, 1329–1366 {**190, 198**}

Reddy BE, Lambert DL, Laws C, et al., 2002, A search for ^{6}Li in stars with planets. *MNRAS*, 335, 1005–1016 {**201**}

Reddy BE, Tomkin J, Lambert DL, et al., 2003, The chemical compositions of Galactic disk F and G dwarfs. *MNRAS*, 340, 304–340 {**190, 197**}

Redfield S, Endl M, Cochran WD, et al., 2008, Sodium absorption from the exoplanetary atmosphere of HD 189733 b detected in the optical transmission spectrum. *ApJ*, 673, L87–L90 {**140**}

Reffert S, Launhardt R, Hekker S, et al., 2005, Choosing suitable target, reference and calibration stars for the PRIMA astrometric planet search. *Astrometry in the Age of the Next Generation of Large Telescopes*, volume 338 of *ASP Conf Ser*, 81–85 {**65**}

Reffert S, Quirrenbach A, 2006, Hipparcos astrometric orbits for two brown dwarf companions: HD 38529 and HD 168443. *A&A*, 449, 699–702 {**70**}

Refsdal S, 1964, The gravitational lens effect. *MNRAS*, 128, 295–306 {**84**}

—, 1966, On the possibility of determining the distances and masses of stars from the gravitational lens effect. *MNRAS*, 134, 315–319 {**93**}

Regenauer-Lieb K, Yuen DA, Branlund J, 2001, The initiation of subduction: criticality by addition of water? *Science*, 294, 578–581 {**250**}

Régulo C, Almenara JM, Alonso R, et al., 2007, TRUFAS, a wavelet-based algorithm for the rapid detection of planetary transits. *A&A*, 467, 1345–1352 {**105**}

Reichert J, Holzwarth R, Udem T, et al., 1999, Measuring the frequency of light with mode-locked lasers. *Optics Communications*, 172, 59–68 {**20**}

Reid IN, 1998, HIpparcos subdwarf parallaxes: metal-rich clusters and the thick disk. *AJ*, 115, 204–228 {**194**}

—, 2002, On the nature of stars with planets. *PASP*, 114, 306–329 {**187, 188**}

Reid IN, Cruz KL, Burgasser AJ, et al., 2008, L-dwarf binaries in the 20-pc sample. *AJ*, 135, 580–587 {**211**}

Reid IN, Gizis JE, Hawley SL, 2002, The Palomar/MSU nearby star spectroscopic survey. IV. The luminosity function in the solar neighbourhood and M dwarf kinematics. *AJ*, 124, 2721–2738 {**187**}

Reid IN, Kirkpatrick JD, Liebert J, et al., 1999, L dwarfs and the substellar mass function. *ApJ*, 521, 613–629 {**210, 214**}

Reid IN, Metchev SA, 2008, The brown dwarf–exoplanet connection. *Exoplanets: Detection, Formation, Properties, Habitability*, 115–152, Springer {**210, 212, 213**}

Reid IN, Turner EL, Turnbull MC, et al., 2007, Searching for Earth analogues around the nearest stars: the disk age-metallicity relation and the age distribution in the solar neighbourhood. *ApJ*, 665, 767–784 {**187, 190**}

Reid MJ, 2008, Micro-arcsecond astrometry with the VLBA. *IAU Symposium*, volume 248, 141–147 {**173**}

Reidemeister M, Krivov AV, Schmidt TOB, et al., 2009, A possible architecture of the planetary system HR 8799. *A&A*, 503, 247–258 {**223, 244**}

Reiger SH, 1963, Starlight scintillation and atmospheric turbulence. *AJ*, 68, 395–406 {**116**}

Rein H, Papaloizou JCB, Kley W, 2010, The dynamical origin of the multi-planetary system HD 45364. *A&A*, 510, A4 {**52**}

Reipurth B, Clarke C, 2001, The formation of brown dwarfs as ejected stellar embryos. *AJ*, 122, 432–439 {**215**}

Reipurth B, Jewitt D, Keil K, 2007, *Protostars and Planets V*. University of Arizona Press {**293**}

Renard S, Absil O, Berger JP, et al., 2008, Prospects for near-infrared characterisation of hot Jupiters with the VLTI Spectro–Imager (VSI). *SPIE Conf Ser*, volume 7013, 91 {**162**}

Retter A, Marom A, 2003, A model of an expanding giant that swallowed planets for the eruption of V838 Monocerotis. *MNRAS*, 345, L25–L28 {**179**}

Retter A, Zhang B, Siess L, et al., 2006, The planets capture model of V838 Monocerotis: conclusions for the penetration depth of the planet(s). *MNRAS*, 370, 1573–1580 {**178, 179**}

Reuyl D, Holmberg E, 1943, On the existence of a third component in the system 70 Oph. *ApJ*, 97, 41–45 {**64**}

Reynaud F, Delage L, 2007, Proposal for a temporal version of a hypertelescope. A&A, 465, 1093–1097 {168}

Reynolds AP, de Bruijne JHJ, Perryman MAC, et al., 2003, Temperature determination via STJ optical spectroscopy. A&A, 400, 1209–1217 {133}

Reynolds RT, Squyres SW, Colburn DS, et al., 1983, On the habitability of Europa. Icarus, 56, 246–254 {250}

Rhie SH, Becker AC, Bennett DP, et al., 1999, Observations of the binary microlens event MACHO–1998–SMC–1 by the microlensing planet search collaboration. ApJ, 522, 1037–1045 {97}

Rhie SH, Bennett DP, Becker AC, et al., 2000, On planetary companions to the MACHO–1998–BLG–35 microlens star. ApJ, 533, 378–391 {96}

Riaud P, Boccaletti A, Baudrand J, et al., 2003, The four-quadrant phase mask coronagraph. III. Laboratory performance. PASP, 115, 712–719 {155}

Riaud P, Boccaletti A, Gillet S, et al., 2002, Coronagraphic search for exoplanets with a hypertelescope. I. In the thermal infrared. A&A, 396, 345–352 {164, 168}

Riaud P, Boccaletti A, Rouan D, et al., 2001, The four-quadrant phase-mask coronagraph. II. Simulations. PASP, 113, 1145–1154 {153, 155}

Riaud P, Mawet D, Absil O, 2005, Limitation of the pupil replication technique in the presence of instrumental defects. ApJ, 628, L81–L84 {156}

Riaud P, Mawet D, Absil O, et al., 2006, Coronagraphic imaging of three weak-line T Tauri stars: evidence of planetary formation around PDS 70. A&A, 458, 317–325 {155}

Riaud P, Schneider J, 2007, Improving Earth-like planet detection with an ELT: the differential radial velocity experiment. A&A, 469, 355–361 {27}

Ribas I, Miralda-Escudé J, 2007, The eccentricity-mass distribution of exoplanets: signatures of different formation mechanisms? A&A, 464, 779–785 {60}

Ribas I, Solano E, Masana E, et al., 2003, Effective temperatures and radii of planet-hosting stars from infrared photometry. A&A, 411, L501–L504 {185}

Ricci D, Le Coroller H, Labeyrie A, 2009, Extreme coronagraphy with an adaptive hologram. Simulations of exoplanet imaging. A&A, 503, 301–308 {157}

Ricci L, Robberto M, Soderblom DR, 2008, The HST–ACS atlas of protoplanetary disks in the great Orion Nebula. AJ, 136, 2136–2151 {224}

Rice WKM, Armitage PJ, 2003, On the formation timescale and core masses of gas giant planets. ApJ, 598, L55–L58 {234}

—, 2005, Quantifying orbital migration from exoplanet statistics and host metallicities. ApJ, 630, 1107–1113 {192}

—, 2009, Time-dependent models of the structure and stability of self-gravitating protoplanetary disks. MNRAS, 396, 2228–2236 {236}

Rice WKM, Armitage PJ, Bate MR, et al., 2003a, Astrometric signatures of self-gravitating protoplanetary disks. MNRAS, 338, 227–232 {65}

—, 2003b, The effect of cooling on the global stability of self-gravitating protoplanetary disks. MNRAS, 339, 1025–1030 {236}

Rice WKM, Armitage PJ, Hogg DF, 2008, Why are there so few hot Jupiters? MNRAS, 384, 1242–1248 {241}

Rice WKM, Lodato G, Armitage PJ, 2005, Investigating fragmentation conditions in self-gravitating accretion disks. MNRAS, 364, L56–L60 {236}

Rice WKM, Lodato G, Pringle JE, et al., 2004, Accelerated planetesimal growth in self-gravitating protoplanetary disks. MNRAS, 355, 543–552 {226}

—, 2006, Planetesimal formation via fragmentation in self-gravitating protoplanetary disks. MNRAS, 372, L9–L13 {227}

Rice WKM, Mayo JH, Armitage PJ, 2010, The role of disk self-gravity in the formation of protostars and protostellar disks. MNRAS, 402, 1740–1749 {236}

Richard D, Zahn J, 1999, Turbulence in differentially rotating flows: what can be learned from the Couette–Taylor experiment. A&A, 347, 734–738 {221}

Richardson LJ, Deming D, Horning K, et al., 2007, A spectrum of an extrasolar planet. Nature, 445, 892–895 {139, 140}

Richardson LJ, Deming D, Seager S, 2003a, Infrared observations during the secondary eclipse of HD 209458 b. II. Strong limits on the infrared spectrum near 2.2 μm. ApJ, 597, 581–589 {139, 140}

Richardson LJ, Deming D, Wiedemann G, et al., 2003b, Infrared observations during the secondary eclipse of HD 209458 b. I. 3.6 μm occultation spectroscopy using the VLT. ApJ, 584, 1053–1062 {139, 140}

Richardson LJ, Harrington J, Seager S, et al., 2006, A Spitzer infrared radius for the transiting extrasolar planet HD 209458 b. ApJ, 649, 1043–1047 {121, 140}

Richichi A, 2003, Lunar occultations of stars with exoplanet candidates. A&A, 397, 1123–1127 {157}

Richling S, Hollenbach D, Yorke HW, 2006, Destruction of protoplanetary disks by photoevaporation. Planet Formation, 31–41, Cambridge University Press {222}

Rieke GH, Su KYL, Stansberry JA, et al., 2005, Decay of planetary debris disks. ApJ, 620, 1010–1026 {224}

Rigaut FJ, Ellerbroek BL, Flicker R, 2000, Principles, limitations, and performance of multiconjugate adaptive optics. SPIE Conf Ser, volume 4007, 1022–1031 {152}

Riley JMW, Waldram EM, Riley JM, 1999, The 7C survey of radio sources at 151 MHz: 33 regions in the range 7h < RA < 17h, 30 deg < Dec < 58 deg. MNRAS, 306, 31–34 {176}

Rivera EJ, Haghighipour N, 2007, On the stability of test particles in extrasolar multiple planet systems. MNRAS, 374, 599–613 {44}

Rivera EJ, Laughlin G, Butler RP, et al., 2010, The Lick–Carnegie exoplanet survey: a Uranus-mass fourth planet for GJ 876 in an extrasolar Laplace configuration. ApJ, 719, 890–899 {2, 32, 50, 51}

Rivera EJ, Lissauer JJ, 2000, Stability analysis of the planetary system orbiting υ And. ApJ, 530, 454–463 {47}

—, 2001, Dynamical models of the resonant pair of planets orbiting the star GJ 876. ApJ, 558, 392–402 {14, 49}

Rivera EJ, Lissauer JJ, Butler RP, et al., 2005, A 7.5 Earth-mass planet orbiting the nearby star, GJ 876. ApJ, 634, 625–640 {2, 32, 36, 49, 267}

Roberge A, Feldman PD, Weinberger AJ, et al., 2006a, Stabilisation of the disk around β Pic by extremely carbon-rich gas. Nature, 441, 724–726 {224}

Roberge A, Lecavelier des Etangs A, Vidal-Madjar A, et al., 2006b, Evidence for comet-like bodies around the 12 Myr old star β Pic. Astrophysics in the Far Ultraviolet: Five Years of Discovery with FUSE, volume 348 of ASP Conf Ser, 294–296 {223}

Robichon N, 2002, Detection of transits of extrasolar planets with Gaia. EAS Publications Series, volume 2, 215–221 {113}

Robichon N, Arenou F, 2000, HD 209458 planetary transits from Hipparcos photometry. A&A, 355, 295–298 {113, 139}

Robinson SE, Ammons SM, Kretke KA, et al., 2007, The N2K consortium. VII. Atmospheric parameters of 1907 metal-rich stars: finding planet-search targets. ApJS, 169, 430–438 {187, 188}

Robinson SE, Laughlin G, Bodenheimer P, et al., 2006a, Silicon and nickel enrichment in planet host stars: observations and implications for the core accretion theory of planet formation. ApJ, 643, 484–500 {195, 251}

Robinson SE, Strader J, Ammons SM, et al., 2006b, The N2K consortium. V. Identifying very metal-rich stars with low-resolution spectra: finding planet-search targets. ApJ, 637, 1102–1112 {29}

Robutel P, Bodossian J, 2009, The resonant structure of Jupiter's Trojan asteroids. II. What happens for different configurations of the planetary system. MNRAS, 399, 69–87 {304}

Roddier F, 1988, Curvature sensing and compensation: a new concept in adaptive optics. Appl. Opt., 27, 1223–1225 {151}

Roddier F, Northcott M, Graves JE, 1991, A simple low-order adaptive optics system for near-infrared applications. PASP, 103, 131–149 {151}

Roddier F, Roddier C, 1997, Stellar coronagraph with phase mask. PASP, 109, 815–820 {153, 155}

Rodigas TJ, Hinz PM, 2009, Which radial velocity exoplanets have undetected outer companions? ApJ, 702, 716–723 {15}

Rodler F, Kürster M, Henning T, 2008, HD 75289Ab revisited: searching for starlight reflected from a hot Jupiter. A&A, 485, 859–864 {125}

Rodríguez A, Gallardo T, 2005, The dynamics of the HD 12661 extrasolar planetary system. ApJ, 628, 1006–1013 {52}

Rogers FJ, Iglesias CA, 1992, Radiative atomic Rosseland mean opacity tables. ApJS, 79, 507–568 {264}

Rogers FJ, Swenson FJ, Iglesias CA, 1996, OPAL equation-of-state tables for astrophysical applications. ApJ, 456, 902–908 {202}

Rogers JC, Apai D, López-Morales M, et al., 2009, Ks-band detection of thermal emission and colour constraints to CoRoT–1 b: a low-albedo planet with inefficient atmospheric energy redistribution and a temperature inversion. *ApJ*, 707, 1707–1716 {**111, 143**}

Rohde RA, Muller RA, 2005, Cycles in fossil diversity. *Nature*, 434, 208–210 {**307**}

Röll T, Seifahrt A, Neuhäuser R, 2008, Micro-arcsecond relative astrometry by ground-based and single-aperture observations. *IAU Symposium*, volume 248, 48–51 {**63**}

Romanova MM, Lovelace RVE, 2006, The magnetospheric gap and the accumulation of giant planets close to a star. *ApJ*, 645, L73–L76 {**241**}

Rosenblatt F, 1971, A two-colour photometric method for detection of extrasolar planetary systems. *Icarus*, 14, 71–93 {**104, 118**}

Röser S, 1999, DIVA: beyond Hipparcos and towards Gaia. *Reviews in Modern Astronomy*, volume 12 of *Reviews in Modern Astronomy*, 97–106 {**73**}

Rossiter RA, 1924, On the detection of an effect of rotation during eclipse in the velocity of the brighter component of β Lyr, and on the constancy of velocity of this system. *ApJ*, 60, 15–21 {**127**}

Rosvick JM, Robb R, 2006, A photometric search for planets in the open cluster NGC 7086. *AJ*, 132, 2309–2317 {**109**}

Rothman LS, Gordon IE, Barbe A, et al., 2009, The HITRAN 2008 molecular spectroscopic database. *J. Quant. Spec. Radiat. Transf.*, 110, 533–572 {**264**}

Rouan D, Pelat D, 2008, The achromatic chessboard, a new concept of a phase shifter for nulling interferometry. II. Theory. *A&A*, 484, 581–589 {**153**}

Rouan D, Riaud P, Boccaletti A, et al., 2000, The four-quadrant phase-mask coronagraph. I. Principle. *PASP*, 112, 1479–1486 {**153, 155**}

Rousset G, Lacombe F, Puget P, et al., 2003, NAOS, the first adaptive optics system of the VLT: on-sky performance. *SPIE Conf Ser*, volume 4839, 140–149 {**151, 152**}

Routh EJ, 1875, On Laplace's three particles with a supplement on the stability of their motion. *Proc. London Math. Soc.*, 6(86-97) {**52**}

Rowe JF, Matthews JM, Seager S, et al., 2006, An upper limit on the albedo of HD 209458 b: direct imaging photometry with the MOST satellite. *ApJ*, 646, 1241–1251 {**113**}

—, 2008, The very low albedo of an extrasolar planet: MOST space-based photometry of HD 209458. *ApJ*, 689, 1345–1353 {**113, 125, 138, 140**}

Roy AE, 1978, *Orbital Motion*. Adam Hilger {**9**}

Roy AE, Walker IW, MacDonald AJ, et al., 1988, Project LONGSTOP. *Vistas in Astronomy*, 32, 95–116 {**299**}

Rubenstein EP, Schaefer BE, 2000, Are superflares on solar analogues caused by extrasolar planets? *ApJ*, 529, 1031–1033 {**176**}

Rubin AE, 1995, Petrologic evidence for collisional heating of chondritic asteroids. *Icarus*, 113, 156–167 {**298**}

—, 2000, Petrologic, geochemical and experimental constraints on models of chondrule formation. *Earth Science Reviews*, 50, 3–27 {**297**}

Ruden SP, 1993, The evolution of protoplanetary disks. *Planets Around Pulsars*, volume 36 of *ASP Conf Ser*, 197–215 {**77**}

—, 1999, The formation of planets. *NATO ASIC Proc. 540: The Origin of Stars and Planetary Systems*, 643–680 {**238**}

Ruden SP, Lin DNC, 1986, The global evolution of the primordial solar nebula. *ApJ*, 308, 883–901 {**221, 298**}

Rupprecht G, Pepe F, Mayor M, et al., 2004, The exoplanet hunter HARPS: performance and first results. *SPIE Conf Ser*, volume 5492, 148–159 {**24**}

Russell CT, 1993, Planetary magnetospheres. *Reports on Progress in Physics*, 56, 687–732 {**174**}

Russell HN, 1934, Molecules in the Sun and stars. *ApJ*, 79, 317–342 {**257**}

Rutledge RE, Basri G, Martín EL, et al., 2000, Chandra detection of an X-Ray flare from the brown dwarf LP 944–20. *ApJ*, 538, L141–L144 {**213**}

Ryabov VB, Zarka P, Ryabov BP, 2003, Search for exoplanetary radio bursts in decameter wave band: statistical enhancement of sensitivity under severe interference conditions. *AGU Fall Meeting Abstracts*, 1131 {**177**}

Ryan EV, Melosh HJ, 1998, Impact fragmentation: from the laboratory to asteroids. *Icarus*, 133, 1–24 {**228**}

Ryan SG, 2000, The host stars of extrasolar planets have normal lithium abundances. *MNRAS*, 316, L35–L39 {**199**}

Rybicki KR, 2006, On the energy flux reaching planets during the parent star's evolutionary track: the Earth-Sun system. *PASP*, 118, 1124–1135 {**277**}

Rybicki KR, Denis C, 2001, On the final destiny of the Earth and the solar system. *Icarus*, 151, 130–137 {**80**}

Rye R, Kuo PH, Holland HD, 1995, Atmospheric carbon dioxide concentrations before 2.2 billion years ago. *Nature*, 378, 603–605 {**306**}

Ryu D, Goodman J, 1992, Convective instability in differentially rotating disks. *ApJ*, 388, 438–450 {**221**}

Saar SH, 2009, The radial velocity effects of stellar surface phenomena. *American Inst. of Phys. Conf. Ser.*, volume 1094, 152–161 {**21**}

Saar SH, Butler RP, Marcy GW, 1998, Magnetic activity-related radial velocity variations in cool stars: first results from the Lick extrasolar planet survey. *ApJ*, 498, L153–L157 {**21, 22, 32**}

Saar SH, Cuntz M, 2001, A search for Ca II emission enhancement in stars resulting from nearby giant planets. *MNRAS*, 325, 55–59 {**21**}

Saar SH, Cuntz M, Kashyap VL, et al., 2008, First observation of planet-induced X-ray emission: the system HD 179949. *IAU Symposium*, volume 249, 79–81 {**175, 287**}

Saar SH, Cuntz M, Shkolnik E, 2004, Stellar activity enhancement by planets: theory and observations. *Stars as Suns: Activity, Evolution and Planets*, volume 219 of *IAU Symposium*, 355–366 {**205**}

Saar SH, Donahue RA, 1997, Activity-related radial velocity variation in cool stars. *ApJ*, 485, 319–327 {**18, 21, 22, 65**}

Saar SH, Hatzes A, Cochran W, et al., 2003, Stellar intrinsic radial velocity noise: causes and possible cures. *The Future of Cool-Star Astrophysics: 12th Cambridge Workshop on Cool Stars, Stellar Systems, and the Sun*, volume 12, 694–698 {**65**}

Saar SH, Seager S, 2003, Uses of linear polarisation as a probe of extrasolar planet atmospheres. *Scientific Frontiers in Research on Extrasolar Planets*, volume 294 of *ASP Conf Ser*, 529–534 {**126**}

Sackett PD, 2004, Results from microlensing searches for extrasolar planets. *Planetary Systems in the Universe*, volume 202 of *IAU Symposium*, 44–54 {**84**}

Sackett PD, Albrow MD, Beaulieu JP, et al., 2004, PLANET II: a microlensing and transit search for extrasolar planets. *Bioastronomy 2002: Life Among the Stars*, volume 213 of *IAU Symposium*, 35–40 {**97**}

Sackmann I, Boothroyd AI, Kraemer KE, 1993, Our Sun. III. Present and future. *ApJ*, 418, 457–468 {**79, 80**}

Sadakane K, Ohkubo M, Takeda Y, et al., 2002, Abundance analyses of 12 parent stars of extrasolar planets observed with the Subaru/HD S. *PASJ*, 54, 911–931 {**188, 195, 196, 198**}

Saffe C, Gómez M, 2004, A search for disks around exoplanet host stars. *A&A*, 423, 221–233 {**224**}

Saffe C, Gómez M, Chavero C, 2005, On the ages of exoplanet host stars. *A&A*, 443, 609–626 {**187**}

Safizadeh N, Dalal N, Griest K, 1999, Astrometric microlensing as a method of discovering and characterising extrasolar planets. *ApJ*, 522, 512–517 {**94, 95**}

Safronov VS, 1969, Evolution of the protoplanetary cloud and formation of the Earth and planets. *Nauka Press, Moscow; English translation: NASA-677, 1972* {**225, 228, 298**}

Safronov VS, 1972, *Evolution of the Protoplanetary Cloud and Formation of the Earth and Planets*. Israel Program for Scientific Translation {**147, 225–228**}

Sagan C, 1973, Ultraviolet selection pressure on the earliest organisms. *J. Theor. Biol.*, 39(195-200) {**284**}

Sagan C, Mullen G, 1972, Earth and Mars: evolution of atmospheres and surface temperatures. *Science*, 177, 52–56 {**305**}

Sagan C, Thompson WR, Carlson R, et al., 1993, A search for life on Earth from the Galileo spacecraft. *Nature*, 365, 715–721 {**289**}

Saha P, Tremaine S, 1992, Symplectic integrators for solar system dynamics. *AJ*, 104, 1633–1640 {**44**}

—, 1994, Long-term planetary integration with individual time steps. *AJ*, 108, 1962–1969 {**44**}

Sahu KC, Anderson J, King IR, 2002, A reexamination of the planetary lensing events in M22. *ApJ*, 565, L21–L24 {**96**}

Sahu KC, Casertano S, Bond HE, et al., 2006, Transiting extrasolar planetary candidates in the Galactic bulge. *Nature*, 443, 534–540 {**112, 327**}

Sahu KC, Casertano S, Livio M, et al., 2001, Gravitational microlensing by low-mass objects in the globular cluster M22. *Nature*, 411, 1022–1024 {**96**}

Salmeron R, Wardle M, 2005, Magnetorotational instability in protoplanetary disks. *MNRAS*, 361, 45–69 {**221**}

Salzmann CG, Radaelli PG, Mayer E, et al., 2009, Ice XV: a new thermodynamically stable phase of ice. *Physical Review Letters*, 103(10), 105701 {**263**}

Sánchez-Lavega A, 2004, The magnetic field in giant extrasolar planets. *ApJ*, 609, L87–L90 {**175**}

Sándor Z, Érdi B, Efthymiopoulos C, 2000, The phase space structure around L4 in the restricted three-body problem. *Celestial Mechanics and Dynamical Astronomy*, 78, 113–123 {**47**}

Sándor Z, Érdi B, Széll A, et al., 2004, The relative Lyapunov indicator: an efficient method of chaos detection. *Celestial Mechanics and Dynamical Astronomy*, 90, 127–138 {**47**}

Sándor Z, Kley W, 2006, On the evolution of the resonant planetary system HD 128311. *A&A*, 451, L31–L34 {**50, 51**}

—, 2010, Formation of the resonant system HD 60532. *A&A*, 517, A31 {**50, 52, 53**}

Sándor Z, Kley W, Klagyivik P, 2007a, Stability and formation of the resonant system HD 73526. *A&A*, 472, 981–992 {**52, 55**}

Sándor Z, Süli Á, Érdi B, et al., 2007b, A stability catalogue of the habitable zones in extrasolar planetary systems. *MNRAS*, 375, 1495–1502 {**47, 55, 56, 285**}

Sandquist EL, Dokter JJ, Lin DNC, et al., 2002, A critical examination of Li pollution and giant-planet consumption by a host star. *ApJ*, 572, 1012–1023 {**192, 201**}

Sandquist EL, Taam RE, Lin DNC, et al., 1998, Planet consumption and stellar metallicity enhancements. *ApJ*, 506, L65–L68 {**192**}

Santos NC, 2008, Extrasolar planets: detection methods and results. *New Astronomy Reviews*, 52, 154–166 {**34**}

Santos NC, Bouchy F, Mayor M, et al., 2004a, The HARPS search for southern extrasolar planets. II. A 14 Earth-masses exoplanet around μ Ara. *A&A*, 426, L19–L23 {**48**}

Santos NC, García López RJ, Israelian G, et al., 2002a, Beryllium abundances in stars hosting giant planets. *A&A*, 386, 1028–1038 {**201**}

Santos NC, Israelian G, García López RJ, et al., 2004b, Are beryllium abundances anomalous in stars with giant planets? *A&A*, 427, 1085–1096 {**201**}

Santos NC, Israelian G, Mayor M, 2000, Chemical analysis of 8 recently discovered extrasolar planet host stars. *A&A*, 363, 228–238 {**188, 196**}

—, 2001, The metal-rich nature of stars with planets. *A&A*, 373, 1019–1031 {**188, 196**}

—, 2004c, Spectroscopic [Fe/H] for 98 extrasolar planet-host stars. Exploring the probability of planet formation. *A&A*, 415, 1153–1166 {**185, 188, 189**}

Santos NC, Israelian G, Mayor M, et al., 2003, Statistical properties of exoplanets. II. Metallicity, orbital parameters, and space velocities. *A&A*, 398, 363–376 {**188, 189**}

—, 2005, Spectroscopic metallicities for planet-host stars: extending the samples. *A&A*, 437, 1127–1133 {**188, 189, 191**}

Santos NC, Israelian G, Randich S, et al., 2004d, Beryllium anomalies in solar-type field stars. *A&A*, 425, 1013–1027 {**201**}

Santos NC, Lovis C, Pace G, et al., 2009, Metallicities for 13 nearby open clusters from high-resolution spectroscopy of dwarf and giant stars. Stellar metallicity, stellar mass, and giant planets. *A&A*, 493, 309–316 {**190**}

Santos NC, Mayor M, Bouchy F, et al., 2007, The HARPS search for southern extrasolar planets. XII. A giant planet orbiting the metal-poor star HD 171028. *A&A*, 474, 647–651 {**33, 34**}

Santos NC, Mayor M, Naef D, et al., 2002b, The CORALIE survey for southern extrasolar planets. IX. A 1.3-day period brown dwarf disguised as a planet. *A&A*, 392, 215–229 {**22, 23**}

Santos NC, Udry S, Bouchy F, et al., 2008, ELODIE metallicity-biased search for transiting hot Jupiters. V. An intermediate-period jovian planet orbiting HD 45652. *A&A*, 487, 369–372 {**29**}

Sargent WLW, Schechter PL, Boksenberg A, et al., 1977, Velocity dispersions for 13 galaxies. *ApJ*, 212, 326–334 {**17**}

Sari R, Goldreich P, 2004, Planet-disk symbiosis. *ApJ*, 606, L77–L80 {**200, 242**}

Sartoretti P, Schneider J, 1999, On the detection of satellites of extrasolar planets with the method of transits. *A&AS*, 134, 553–560 {**104, 130, 135**}

Sasselov DD, 2003, The new transiting planet OGLE–TR–56 b: orbit and atmosphere. *ApJ*, 596, 1327–1331 {**106, 133, 248**}

Sato B, Ando H, Kambe E, et al., 2003, A planetary companion to the G-type giant star HD 104985. *ApJ*, 597, L157–L160 {**32**}

Sato B, Fischer DA, Henry GW, et al., 2005a, The N2K consortium. II. A transiting hot Saturn around HD 149026 with a large dense core. *ApJ*, 633, 465–473 {**109, 145, 266, 267, 326**}

Sato B, Fischer DA, Ida S, et al., 2009, A substellar companion in a 1.3-yr nearly circular orbit of HD 16760. *ApJ*, 703, 671–674 {**24**}

Sato B, Izumiura H, Toyota E, et al., 2007, A planetary companion to the Hyades giant ϵ Tau. *ApJ*, 661, 527–531 {**2, 33, 34**}

—, 2008a, Planetary companions around three intermediate-mass G and K giants: 18 Del, ξ Aql, and HD 81688. *PASJ*, 60, 539–550 {**24, 32, 33**}

Sato B, Kambe E, Takeda Y, et al., 2005b, Radial velocity variability of G-type giants: first three years of the Okayama planet search programme. *PASJ*, 57, 97–107 {**24, 32**}

Sato B, Omiya M, Liu Y, et al., 2010, Substellar companions to evolved intermediate-mass stars: HD 145457 and HD 180314. *PASJ*, 62, 1063–1069 {**32, 33**}

Sato B, Toyota E, Omiya M, et al., 2008b, Planetary companions to evolved intermediate-mass stars: 14 And, 81 Cet, 6 Lyn, and HD 167042. *PASJ*, 60, 1317–1326 {**32, 33**}

Sato M, Asada H, 2009, Effects of mutual transits by extrasolar planet-companion systems on light curves. *PASJ*, 61, L29–L34 {**130**}

Saumon D, Chabrier G, van Horn HM, 1995, An equation of state for low-mass stars and giant planets. *ApJS*, 99, 713–741 {**261, 262**}

Saumon D, Chabrier G, Wagner DJ, et al., 2000, Modeling pressure-ionisation of hydrogen in the context of astrophysics. *High Pressure Research*, 16, 331–343 {**262, 294**}

Saumon D, Hubbard WB, Burrows A, et al., 1996, A theory of extrasolar giant planets. *ApJ*, 460, 993–1018 {**209, 218, 260**}

Saumon D, Marley MS, 2008, The evolution of L and T dwarfs in colour–magnitude diagrams. *ApJ*, 689, 1327–1344 {**272**}

Saumon D, Marley MS, Cushing MC, et al., 2006, Ammonia as a tracer of chemical equilibrium in the T7.5 dwarf GJ 570 D. *ApJ*, 647, 552–557 {**272**}

Saumon D, Marley MS, Leggett SK, et al., 2007, Physical parameters of two very cool T dwarfs. *ApJ*, 656, 1136–1149 {**272, 275**}

Saxena SK, Eriksson G, 1983, Low- to medium-temperature phase equilibria in a gas of solar composition. *Earth and Planetary Science Letters*, 65, 7–16 {**257**}

Scafetta N, 2010, Empirical evidence for a celestial origin of the climate oscillations and its implications. *Journal of Atmospheric and Solar-Terrestrial Physics*, 72, 951–970 {**66**}

Scalo J, Kaltenegger L, Segura AG, et al., 2007, M stars as targets for terrestrial exoplanet searches and biosignature detection. *Astrobiology*, 7, 85–166 {**285**}

Scargle JD, 1982, Studies in astronomical time series analysis. II. Statistical aspects of spectral analysis of unevenly spaced data. *ApJ*, 263, 835–853 {**13**}

Schaefer BE, King JR, Deliyannis CP, 2000, Superflares on ordinary solar-type stars. *ApJ*, 529, 1026–1030 {**176**}

Schaefer L, Fegley B, 2009, Chemistry of silicate atmospheres of evaporating super-Earths. *ApJ*, 703, L113–L117 {**267**}

Scharf CA, 2006, The potential for tidally heated icy and temperate moons around exoplanets. *ApJ*, 648, 1196–1205 {**250**}

—, 2007, Exoplanet transit parallax. *ApJ*, 661, 1218–1221 {**136**}

—, 2008, Moons of exoplanets: habitats for life? *Exoplanets: Detection, Formation, Properties, Habitability*, 285–303, Springer {**237, 250, 251**}

Scharf CA, Menou K, 2009, Long-period exoplanets from dynamical relaxation. *ApJ*, 693, L113–L117 {**243**}

Scherer K, Fichtner H, Anderson JD, et al., 1997, A pulsar, the heliosphere, and Pioneer 10: probable mimicking of a planet of PSR B1257+12 by solar rotation. *Science*, 278, 1919–1923 {**76, 79**}

Schindler TL, Kasting JF, 2000, Synthetic spectra of simulated terrestrial atmospheres containing possible biomarker gases. *Icarus*, 145, 262–271 {**289**}

Schlaufman KC, 2010, Evidence of possible spin–orbit misalignment along the line of sight in transiting exoplanet systems. *ApJ*, 719, 602–611 {**130**}

Schlaufman KC, Lin DNC, Ida S, 2010, A population of very hot super-Earths in multiple-planet systems should be uncovered by Kepler. *ApJ*, 724, L53–L58 {**112**}

Schlesinger F, 1910, The Algol-variable δ Lib. *Publications of the Allegheny Observatory of the University of Pittsburgh*, 1, 123–134 {**127**}

—, 1916, The orbit of λ Tau. *Publications of the Allegheny Observatory of the University of Pittsburgh*, 3, 23–30 {**127**}

Schlichting HE, Sari R, 2007, The effect of semicollisional accretion on planetary spins. *ApJ*, 658, 593–597 {**300**}

Schloerb FP, Berger J, Carleton NP, et al., 2006, IOTA: recent science and technology. *SPIE Conf Ser*, volume 6268, 18 {**162**}

Schmid HM, Beuzit J, Feldt M, et al., 2006, Search and investigation of extrasolar planets with polarimetry. *IAU Colloq. 200: Direct Imaging of Exoplanets: Science and Techniques*, 165–170 {**289**}

Schmidt OY, 1944, Dok. Akad. Nauk. *USSR*, 45(6) {**224**}

Schmidt RM, Housen KR, 1987, Some recent advances in the scaling of impact and explosion cratering. *International Journal of Impact Engineering*, 5, 543–560 {**280**}

Schmidt TOB, Neuhäuser R, Seifahrt A, 2009, Homogeneous comparison of planet candidates imaged directly until 2008. *American Institute of Physics Conference Series*, volume 1158, 231–234 {**169**}

Schmidt TOB, Neuhäuser R, Seifahrt A, et al., 2008, Direct evidence of a sub-stellar companion around CT Cha. *A&A*, 491, 311–320 {**168**}

Schmitt J, 1997, *Étude et réalisation en laboratoire d'un accéléromètre astronomique absolu*. Ph.D. thesis, Thesis, University of Paris VI {**27**}

Schneider G, Becklin EE, Smith BA, et al., 2001, NICMOS coronagraphic observations of 55 Cnc. *AJ*, 121, 525–537 {**170**}

Schneider G, Silverstone MD, Hines DC, 2005, Discovery of a nearly edge-on disk around HD 32297. *ApJ*, 629, L117–L120 {**224**}

Schneider G, Silverstone MD, Hines DC, et al., 2006a, Discovery of an 86 AU radius debris ring around HD 181327. *ApJ*, 650, 414–431 {**224**}

Schneider G, Smith BA, Becklin EE, et al., 1999, NICMOS imaging of the HR 4796A circumstellar disk. *ApJ*, 513, L127–L130 {**223, 224**}

Schneider G, Weinberger AJ, Becklin EE, et al., 2009a, STIS imaging of the HR 4796A circumstellar debris ring. *AJ*, 137, 53–61 {**223**}

Schneider J, 1994, On the search for O_2 in extrasolar planets. *Ap&SS*, 212, 321–325 {**104, 282, 288**}

—, 1996, Photometric search for extrasolar planets. *Ap&SS*, 241, 35–42 {**104, 114**}

—, 1999, The study of extrasolar planets: methods of detection, first discoveries and future perspectives. *Academie des Science Paris Comptes Rendus Serie B Sciences Physiques*, 327, 621–634 {**130, 288**}

—, 2000, Extrasolar planets transits: detection and follow-up. *From Extrasolar Planets to Cosmology: The VLT Opening Symposium*, 499–501 {**127**}

—, 2002a, Biosignatures and exoplanet characterisation: visible versus thermal infrared imaging. *Exo-Astrobiology*, volume 518 of *ESA Special Publication*, 409–412 {**165**}

—, 2002b, Characterising extrasolar planets in reflected light and thermal emission. *SF2A-2002: Semaine de l'Astrophysique Française*, 597–602 {**165**}

—, 2003, Biosignatures and extrasolar planet characterisation: visible versus infrared. *Earths: Darwin/TPF and the Search for Extrasolar Terrestrial Planets*, volume 539 of *ESA Special Publication*, 205–213 {**165**}

—, 2005, Light-time effect and exoplanets. *The Light-Time Effect in Astrophysics: Causes and Cures of the O-C diagram*, volume 335 of *ASP Conf Ser*, 191–198 {**135**}

—, 2008a, Characterising super-Earths in reflected light. *EAS Publications Series*, volume 33, 71–74 {**287**}

—, 2008b, Exoplanets: which wavelengths? *SPIE Conf Ser*, volume 6986 {**287**}

—, 2010, The Extrasolar Planets Encyclopaedia. *www.exoplanet.eu* {**215**}

Schneider J, Auvergne M, Baglin A, et al., 1998, The COROT mission: from structure of stars to origin of planetary systems. *Origins*, volume 148 of *ASP Conf Ser*, 298–303 {**130**}

Schneider J, Boccaletti A, Mawet D, et al., 2009b, Super Earth explorer: a coronagraphic off-axis space telescope. *Experimental Astronomy*, 23, 357–377 {**166**}

Schneider J, Cabrera J, 2006, Can stellar wobble in triple systems mimic a planet? *A&A*, 445, 1159–1163 {**23, 78**}

Schneider J, Chevreton M, 1990, The photometric search for Earth-sized extrasolar planets by occultation in binary systems. *A&A*, 232, 251–257 {**104, 114**}

Schneider J, Riaud P, Tinetti G, et al., 2006b, SEE-COAST: the super-Earth explorer. *SF2A-2006: Semaine de l'Astrophysique Française*, 429–432 {**166**}

Schneider P, 1985, A new formulation of gravitational lens theory, time-delay, and Fermat's principle. *A&A*, 143, 413–420 {**95**}

Schneider P, Kochanek CS, Wambsganss J, 2006c, *Gravitational Lensing: Strong, Weak and Micro*. Springer {**84**}

Schneider P, Weiss A, 1986, The two-point-mass lens: detailed investigation of a special asymmetric gravitational lens. *A&A*, 164, 237–259 {**87**}

—, 1987, A gravitational lens origin for AGN-variability? Consequences of microlensing. *A&A*, 171, 49–65 {**90**}

Schneider S, 1979, Ice ages and orbital variations: some simple theory and modeling. *Quaternary Research*, 12, 188–203 {**301**}

Schoenlein RW, Peteanu LA, Mathies RA, et al., 1991, The first step in vision: femtosecond isomerisation of rhodopsin. *Science*, 254, 412–415 {**277**}

Scholz R, 2010, ULAS J141623.94+134836.3: a faint common proper motion companion of a nearby L dwarf. Serendipitous discovery of a cool brown dwarf in UKIDSS DR6. *A&A*, 510, L8–4 {**211**}

Schrag DP, Berner RA, Hoffman PF, et al., 2002, On the initiation of a snowball Earth. *Geochemistry, Geophysics, Geosystems*, 1 {**306**}

Schrijver CJ, Pols OR, 1993, Rotation, magnetic braking, and dynamos in cool giants and subgiants. *A&A*, 278, 51–67 {**32**}

Schröder KP, Connon Smith R, 2008, Distant future of the Sun and Earth revisited. *MNRAS*, 386, 155–163 {**80**}

Schultz AB, Jordan IJ, Kochte M, et al., 2003, UMBRAS: a matched occulter and telescope for imaging extrasolar planets. *SPIE Conf Ser*, volume 4860, 54–61 {**156, 166**}

Schulz R, 2002, Trans-Neptunian objects. *A&A Rev.*, 11, 1–31 {**295**}

Schuster A, 1911, The influence of planets on the formation of sunspots. *Royal Society of London Proceedings Series A*, 85, 309–323 {**66**}

Schütz O, Böhnhardt H, Pantin E, et al., 2004a, A search for circumstellar dust disks with ADONIS. *A&A*, 424, 613–618 {**224**}

Schütz O, Nielbock M, Wolf S, et al., 2004b, SIMBA's view of the ϵ Eri disk. *A&A*, 414, L9–L12 {**223**}

Schwartz RN, Townes CH, 1961, Interstellar and interplanetary communication by optical masers. *Nature*, 190, 205–208 {**291**}

Schwarz R, Dvorak R, Pilat Lohinger E, et al., 2007a, Trojan planets in HD 108874? *A&A*, 462, 1165–1170 {**54**}

Schwarz R, Dvorak R, Süli Á, et al., 2007b, Stability of fictitious Trojan planets in extrasolar systems. *Astronomische Nachrichten*, 328, 785–789 {**54**}

—, 2007c, Survey of the stability region of hypothetical habitable Trojan planets. *A&A*, 474, 1023–1029 {**54**}

Schwarz R, Pilat-Lohinger E, Dvorak R, et al., 2005, Trojans in habitable zones. *Astrobiology*, 5, 579–586 {**136**}

Schwarzenberg-Czerny A, Beaulieu JP, 2006, Efficient analysis in planet transit surveys. *MNRAS*, 365, 165–170 {**105**}

Schwarzschild M, 1958, *Structure and Evolution of the Stars*. Princeton University Press {**264**}

Scott AC, 2000, The Pre-Quaternary history of fire. *Palaeogeogr Palaeoclimatol Palaeoecol*, 164, 281–329 {**307**}

Scott ERD, 2007, Chondrites and the protoplanetary disk. *Annual Review of Earth and Planetary Sciences*, 35, 577–620 {**297**}

Scott HP, Hemley RJ, Mao H, et al., 2004, Generation of methane in the Earth's mantle: in situ high pressure-temperature measurements of carbonate reduction. *Proceedings of the National Academy of Science*, 101, 14023–14026 {**279**}

Seager S, 2010, *Exoplanet Atmospheres: Physical Processes*. Princeton University Press {**271**}

Seager S, Deming D, 2010, Exoplanet atmospheres. *ARA&A*, 48, 631–672 {**271**}

Seager S, Deming D, Valenti JA, 2009, Transiting exoplanets with JWST. *Astrophysics in the Next Decade, Astrophysics and Space Science Proceedings*, 123–130 {**114, 137**}

Seager S, Hui L, 2002, Constraining the rotation rate of transiting extrasolar planets by oblateness measurements. *ApJ*, 574, 1004–1010 {**131, 270**}

Seager S, Kuchner M, Hier-Majumder CA, et al., 2007, Mass–radius relationships for solid exoplanets. *ApJ*, 669, 1279–1297 {**260, 261, 267, 268**}

Seager S, Mallén-Ornelas G, 2003, A unique solution of planet and star parameters from an extrasolar planet transit light curve. *ApJ*, 585, 1038–1055 {**117–121**}

Seager S, Richardson LJ, Hansen BMS, et al., 2005a, On the dayside thermal emission of hot Jupiters. *ApJ*, 632, 1122–1131 {**138**}

Seager S, Sasselov DD, 1998, Extrasolar giant planets under strong stellar irradiation. *ApJ*, 502, L157–L161 {**124**}

—, 2000, Theoretical transmission spectra during extrasolar giant planet transits. *ApJ*, 537, 916–921 {**136**}

Seager S, Turner EL, Schafer J, et al., 2005b, Vegetation's red edge: a possible spectroscopic biosignature of extraterrestrial plants. *Astrobiology*, 5, 372–390 {**289**}

Seager S, Whitney BA, Sasselov DD, 2000, Photometric light curves and polarisation of close-in extrasolar giant planets. *ApJ*, 540, 504–520 {**124, 126**}

Seagroves S, Harker J, Laughlin G, et al., 2003, Detection of intermediate-period transiting planets with a network of small telescopes: Transitsearch.org. *PASP*, 115, 1355–1362 {**109, 122**}

Sears DW, 1978, Condensation and the composition of iron meteorites. *Earth and Planetary Science Letters*, 41, 128–138 {**257**}

Seaton MJ, Yan Y, Mihalas D, et al., 1994, Opacities for stellar envelopes. *MNRAS*, 266, 805–828 {**264**}

Segura A, Kasting JF, Meadows V, et al., 2005, Biosignatures from Earth-like planets around M dwarfs. *Astrobiology*, 5, 706–725 {**288**}

Segura A, Krelove K, Kasting JF, et al., 2003, Ozone concentrations and ultraviolet fluxes on Earth-like planets around other stars. *Astrobiology*, 3, 689–708 {**288**}

Segura A, Meadows VS, Kasting JF, et al., 2007, Abiotic formation of O_2 and O_3 in high-CO_2 terrestrial atmospheres. *A&A*, 472, 665–679 {**288**}

Seidelmann PK, 1992, *Explanatory Supplement to the Astronomical Almanac*. University Science Books, New York {**68, 310**}

Seifahrt A, Käufl HU, 2008, High precision radial velocity measurements in the infrared. A first assessment of the radial velocity stability of CRIRES. *A&A*, 491, 929–939 {**20**}

Sekiya M, 1998, Quasi-equilibrium density distributions of small dust aggregations in the solar nebula. *Icarus*, 133, 298–309 {**226**}

Sellwood JA, Binney JJ, 2002, Radial mixing in galactic disks. *MNRAS*, 336, 785–796 {**193**}

Selsis F, Despois D, Parisot JP, 2002, Signature of life on exoplanets: can Darwin produce false positive detections? *A&A*, 388, 985–1003 {**165**}

Selsis F, Kaltenegger L, Paillet J, 2008a, Terrestrial exoplanets: diversity, habitability and characterisation. *Physica Scripta Volume T*, 130(1), 014032 {**288, 289**}

Selsis F, Kasting JF, Levrard B, et al., 2007, Habitable planets around the star GJ 581? *A&A*, 476, 1373–1387 {**250, 284, 287**}

Selsis F, Paillet J, Allard F, 2008b, Biomarkers of extrasolar planets and their observability. *Extrasolar Planets*, 245–262 {**288**}

Semenov D, Chakraborty S, Thiemens M, 2010, Chemical and isotopic evolution of the solar nebula and protoplanetary disks. *Protoplanetary Dust: Astrophysical and Cosmochemical Perspectives*, 97–127, Cambridge University Press {**217, 220**}

Sengupta S, 2008, Cloudy atmosphere of the extrasolar planet HD 189733 b: a possible explanation of the detected B-band polarisation. *ApJ*, 683, L195–L198 {**126**}

Sengupta S, Maiti M, 2006, Polarisation of starlight by an unresolved and oblate extrasolar planet in an elliptical orbit. *ApJ*, 639, 1147–1152 {**126**}

Sephton MA, Court RW, 2010, Meteorite gases and planetary atmospheres. *Astronomy and Geophysics*, 51(5), 050000–5 {**278**}

Sepinsky JF, Willems B, Kalogera V, 2007, Equipotential surfaces and Lagrangian points in nonsynchronous, eccentric binary and planetary systems. *ApJ*, 660, 1624–1635 {**57**}

Sepkoski JA, 2002, *A Compendium of Fossil Marine Animal Genera*, volume 363. Paleontological Research Institute, Ithaca {**307**}

Serabyn E, 2009, High-contrast, narrow-field exoplanet imaging with a multi-aperture telescope phased-array coronagraph. *ApJ*, 697, 1334–1340 {**156**}

Serabyn E, Mawet D, Burruss R, 2010, An image of an exoplanet separated by two diffraction beamwidths from a star. *Nature*, 464, 1018–1020 {**156, 171**}

Sestito P, Randich S, 2005, Time scales of Li evolution: a homogeneous analysis of open clusters from zero-age main sequence to late-main sequence. *A&A*, 442, 615–627 {**187**}

Setiawan J, Hatzes AP, von der Lühe O, et al., 2003a, Evidence of a substellar companion around HD 47536. *A&A*, 398, L19–L23 {**32–34**}

Setiawan J, Henning T, Launhardt R, et al., 2008a, A young massive planet in a star-disk system. *Nature*, 451, 38–41 {**33, 34**}

Setiawan J, Klement RJ, Henning T, et al., 2010, A giant planet around a metal-poor star of extragalactic origin. *Science*, 330, 1642–1644 {**33, 34**}

Setiawan J, Pasquini L, da Silva L, et al., 2003b, Precise radial velocity measurements of G and K giants: first results. *A&A*, 397, 1151–1159 {**32, 33**}

—, 2004, Precise radial velocity measurements of G and K giants: multiple systems and variability trend along the red giant branch. *A&A*, 421, 241–254 {**32**}

Setiawan J, Rodmann J, da Silva L, et al., 2005, A substellar companion around the intermediate-mass giant star HD 11977. *A&A*, 437, L31–L34 {**70**}

Setiawan J, Weise P, Henning T, et al., 2007, Evidence for a planetary companion around a nearby young star. *ApJ*, 660, L145–L148 {**34**}

—, 2008b, Planets around active stars. *Precision Spectroscopy in Astrophysics*, 201–204 {**24, 33**}

Setiawan J, Weldrake D, Afonso C, et al., 2008c, MAESTRO–1 b: a transiting planet in a close binary? *ASP Conf Ser*, volume 398, 113 {**106**}

Seto N, 2008, Detecting planets around compact binaries with gravitational wave detectors in space. *ApJ*, 677, L55–L58 {**179**}

Shabanova TV, 1995, Evidence for a planet around the pulsar PSR B0329+54. *ApJ*, 453, 779–782 {**78**}

Shaklan SB, Green JJ, 2005, Low-order aberration sensitivity of eighth-order coronagraph masks. *ApJ*, 628, 474–477 {**153**}

Shakura NI, Sunyaev RA, 1973, Black holes in binary systems: observational appearance. *A&A*, 24, 337–355 {**221, 240**}

Shankland PD, Blank DL, Boboltz DA, et al., 2008, Further constraints on the presence of a debris disk in the multiplanet system GJ 876. *AJ*, 135, 2194–2198 {**224**}

Shankland PD, Rivera EJ, Laughlin G, et al., 2006, On the search for transits of the planets orbiting GJ 876. *ApJ*, 653, 700–707 {**109**}

Shao M, 1993, Orbiting stellar interferometer. *SPIE Conf Ser*, volume 1947, 89–90 {**72**}

Shao M, Colavita MM, 1992, Potential of long-baseline infrared interferometry for narrow-angle astrometry. *A&A*, 262, 353–358 {**63**}

Shao M, Colavita MM, Hines BE, et al., 1988, The Mark III stellar interferometer. *A&A*, 193, 357–371 {**162**}

Sharp CM, Burrows A, 2007, Atomic and molecular opacities for brown dwarf and giant planet atmospheres. *ApJS*, 168, 140–166 {**264, 272**}

Shaviv NJ, 2002, Cosmic ray diffusion from the Galactic spiral arms, iron meteorites, and a possible climatic connection. *Physical Review Letters*, 89(5), 051102 {**307**}

—, 2003, The spiral structure of the Milky Way, cosmic rays, and ice age epochs on Earth. *New Astronomy*, 8, 39–77 {**307**}

Shen Y, Stone JM, Gardiner TA, 2006, Three-dimensional compressible hydrodynamic simulations of vortices in disks. *ApJ*, 653, 513–524 {**221**}

Shen Y, Turner EL, 2008, On the eccentricity distribution of exoplanets from radial velocity surveys. *ApJ*, 685, 553–559 {**15**}

Shirley JH, 2006, Axial rotation, orbital revolution and solar spin-orbit coupling. *MNRAS*, 368, 280–282 {**67**}

Shizgal BD, Arkos GG, 1996, Nonthermal escape of the atmospheres of Venus, Earth, and Mars. *Reviews of Geophysics*, 34, 483–505 {**281**}

Shklovskii IS, 1970, Possible causes of the secular increase in pulsar periods. *Soviet Astronomy*, 13, 562–565 {**136**}

Shklovskii IS, Sagan C, 1966, *Intelligent Life in the Universe*. Holden-Day {**198**}

Shkolnik E, Bohlender DA, Walker GAH, et al., 2008, The on/off nature of star-planet interactions. *ApJ*, 676, 628–638 {**205, 206**}

Shkolnik E, Walker GAH, Bohlender DA, 2003, Evidence for planet-induced chromospheric activity on HD 179949. *ApJ*, 597, 1092–1096 {**175, 205**}

Shkolnik E, Walker GAH, Bohlender DA, et al., 2005, Hot Jupiters and hot spots: the short- and long-term chromospheric activity on stars with giant planets. *ApJ*, 622, 1075–1090 {**21, 175, 205, 206**}

Shoemaker EM, Shoemaker CS, Wolfe RF, 1989, Trojan asteroids: populations, dynamical structure and origin of the L4 and L5 swarms. *Asteroids II*, 487–523, University of Arizona Press {**52**}

Showman AP, Cooper CS, Fortney JJ, et al., 2008a, Atmospheric circulation of hot Jupiters: three-dimensional circulation models of HD 209458 b and HD 189733 b with simplified forcing. *ApJ*, 682, 559–576, erratum: 2008 ApJ, 685, 1324 {**138**}

Showman AP, Fortney JJ, Lian Y, et al., 2009, Atmospheric circulation of hot Jupiters: coupled radiative-dynamical general circulation model simulations of HD 189733 b and HD 209458 b. *ApJ*, 699, 564–584 {**138**}

Showman AP, Gierasch PJ, Lian Y, 2006, Deep zonal winds can result from shallow driving in a giant-planet atmosphere. *Icarus*, 182, 513–526 {**138**}

Showman AP, Guillot T, 2002, Atmospheric circulation and tides of 51 Peg b-like planets. *A&A*, 385, 166–180 {**137, 145**}

Showman AP, Menou K, Cho JYK, 2008b, Atmospheric circulation of hot Jupiters: a review of current understanding. *ASP Conf Ser*, volume 398, 419–442 {**138**}

Shporer A, Bakos GÁ, Bouchy F, et al., 2009, HAT–P–9 b: a low-density planet transiting a moderately faint F star. *ApJ*, 690, 1393–1400 {**326**}

Shu FH, 2001, The X-wind theory for the origin of chondritic meteorites. *Eleventh Annual V. M. Goldschmidt Conference*, 3700 {**297**}

Shu FH, Adams FC, Lizano S, 1987, Star formation in molecular clouds: observation and theory. *ARA&A*, 25, 23–81 {**217**}

Shu FH, Johnstone D, Hollenbach D, 1993, Photoevaporation of the solar nebula and the formation of the giant planets. *Icarus*, 106, 92–101 {**222**}

Shu FH, Najita JR, Shang H, et al., 2000, X-winds theory and observations. *Protostars and Planets IV*, 789–795 {**241, 297**}

Shu FH, Shang H, Glassgold AE, et al., 1997, X-rays and fluctuating X-winds from protostars. *Science*, 277, 1475–1479 {**297**}

Shu FH, Shang H, Lee T, 1996, Toward an astrophysical theory of chondrites. *Science*, 271, 1545–1552 {**297**}

Shuch HP, 2001, Optical SETI comes of age. *SPIE Conf Ser* (eds. Kingsley SA, Bhathal R), volume 4273, 128–135 {**291**}

Sicilia-Aguilar A, Hartmann LW, Briceño C, et al., 2004, Low-mass stars and accretion at the ages of planet formation in the Cepheus OB2 region. *AJ*, 128, 805–821 {**221**}

Siess L, Livio M, 1999a, The accretion of brown dwarfs and planets by giant stars. I. Asymptotic giant branch stars. *MNRAS*, 304, 925–937 {**178, 179**}

—, 1999b, The accretion of brown dwarfs and planets by giant stars. II. Solar-mass stars on the red giant branch. *MNRAS*, 308, 1133–1149 {**178, 179, 192, 200**}

Sigurdsson S, 1992, Planets in globular clusters? *ApJ*, 399, L95–L97 {**77**}

—, 1993, Genesis of a planet in Messier 4. *ApJ*, 415, L43–L46 {**77**}

—, 1995, Assessing the environmental impact on PSR B1620−26 in M4. *ApJ*, 452, 323–331 {**77**}

Sigurdsson S, Richer HB, Hansen BM, et al., 2003, A young white dwarf companion to pulsar PSR B1620−26: evidence for early planet formation. *Science*, 301, 193–196 {**77**}

Sigurdsson S, Stairs IH, Moody K, et al., 2008, Planets around pulsars in globular clusters. *ASP Conf Ser*, volume 398, 119–132 {**77**}

Silva AVR, 2003, Method for spot detection on solar-like stars. *ApJ*, 585, L147–L150 {**131**}

—, 2008, Estimating stellar rotation from starspot detection during planetary transits. *ApJ*, 683, L179–L182 {**131**}

Silva AVR, Cruz PC, 2006, Search for planetary candidates within the OGLE stars. *ApJ*, 642, 488–494 {**106**}

Silvotti R, Schuh S, Janulis R, et al., 2007, A giant planet orbiting the extreme horizontal branch star V391 Peg. *Nature*, 449, 189–191 {**2, 79, 81**}

Simkin SM, 1974, Measurements of velocity dispersions and Doppler shifts from digitised optical spectra. *A&A*, 31, 129–136 {**17**}

Simon A, Szatmáry K, Szabó GM, 2007, Determination of the size, mass, and density of exomoons from photometric transit timing variations. *A&A*, 470, 727–731 {**135**}

Simon M, Prato L, 1995, Disk dissipation in single and binary young star systems in Taurus. *ApJ*, 450, 824–829 {**222**}

Simon NR, 1982, A plea for reexamining heavy element opacities in stars. *ApJ*, 260, L87–L90 {**264**}

Simonelli DP, Pollack JB, McKay CP, et al., 1989, The carbon budget in the outer solar nebula. *Icarus*, 82, 1–35 {**232**}

Simpson EK, Baliunas SL, Henry GW, et al., 2010a, Rotation periods of exoplanet host stars. *MNRAS*, 408, 1666–1679 {**186**}

Simpson EK, Barros SCC, Brown DJA, et al., 2010b, Independent discovery and refined parameters of the transiting exoplanet HAT-P-14 b. *ArXiv e-prints* {**327**}

Simpson EK, Faedi F, Barros SCC, et al., 2011, WASP-37 b: a 1.8 Jupiter-mass exoplanet transiting a metal-poor star. *AJ*, 141, 8–12 {**327**}

Sing DK, Désert J, Lecavelier Des Etangs A, et al., 2009, Transit spectrophotometry of the exoplanet HD 189733 b. I. Searching for water but finding haze with HST NICMOS. *A&A*, 505, 891–899 {**140**}

Sing DK, López-Morales M, 2009, Ground-based secondary eclipse detection of the very-hot Jupiter OGLE–TR–56 b. *A&A*, 493, L31–L34 {**106, 143**}

Sion EM, Holberg JB, Oswalt TD, et al., 2009, The white dwarfs within 20 pc of the Sun: kinematics and statistics. *AJ*, 138, 1681–1689 {**79**}

Sivaramakrishnan A, Koresko CD, Makidon RB, et al., 2001, Ground-based coronagraphy with high-order adaptive optics. *ApJ*, 552, 397–408 {**153**}

Sivaramakrishnan A, Lloyd JP, Hodge PE, et al., 2002, Speckle decorrelation and dynamic range in speckle noise-limited imaging. *ApJ*, 581, L59–L62 {**157**}

Sivaramakrishnan A, Soummer R, Pueyo L, et al., 2008, Sensing phase aberrations behind Lyot coronagraphs. *ApJ*, 688, 701–708 {**157**}

Sivaramakrishnan A, Yaitskova N, 2005, Lyot coronagraphy on giant segmented-mirror telescopes. *ApJ*, 626, L65–L68 {**153**}

Skillen I, Pollacco D, Collier Cameron A, et al., 2009, The $0.5 M_J$ transiting exoplanet WASP-13 b. *A&A*, 502, 391–394 {**327**}

Skrutskie MF, Cutri RM, Stiening R, et al., 2006, The Two Micron All Sky Survey (2MASS). *AJ*, 131, 1163–1183 {**210**}

Sleep NH, 2010, The Hadean–Achaean environment. *Cold Spring Harbor Prespectives in Biology*, 2, a002527 {**305**}

Sleep NH, Zahnle K, 2001, Carbon dioxide cycling and implications for climate on ancient Earth. *J. Geophys. Res.*, 106, 1373–1400 {**305**}

Smalley B, Anderson DR, Collier Cameron A, et al., 2010a, WASP-26 b: a 1-Jupiter-mass planet around an early G-type star. *A&A*, 520, A56 {**327**}

—, 2010b, WASP-34 b: a near-grazing transiting sub-Jupiter-mass exoplanet in a hierarchical triple system. *ArXiv e-prints* {**327**}

Smith AMS, Collier Cameron A, Christian DJ, et al., 2006, The impact of correlated noise on SuperWASP detection rates for transiting extrasolar planets. *MNRAS*, 373, 1151–1158 {**104**}

Smith AMS, Collier Cameron A, Greaves J, et al., 2009a, Radio cyclotron emission from extrasolar planets. *IAU Symposium*, volume 253, 456–458 {**177**}

Smith AMS, Collier Cameron AC, Greaves J, et al., 2009b, Secondary radio eclipse of the transiting planet HD 189733 b: an upper limit at 307–347 MHz. *MNRAS*, 395, 335–341 {**177**}

Smith AW, Lissauer JJ, 2009, Orbital stability of systems of closely-spaced planets. *Icarus*, 201, 381–394 {**44, 55, 285**}

Smith BA, Terrile RJ, 1984, A circumstellar disk around β Pic. *Science*, 226, 1421–1424 {**223**}

Smith KW, Bonnell IA, 2001, Free-floating planets in stellar clusters? *MNRAS*, 322, L1–L4 {**215**}

Smith MC, Mao S, Paczyński B, 2003, Acceleration and parallax effects in gravitational microlensing. *MNRAS*, 339, 925–936 {**93**}

Smith MC, Mao S, Woźniak P, 2002, Parallax microlensing events in the OGLE II data base toward the Galactic bulge. *MNRAS*, 332, 962–970 {**93**}

Smith R, Wyatt MC, Haniff CA, 2009c, Resolving the hot dust around HD 69830 and η Corvi with MIDI and VISIR. *A&A*, 503, 265–279 {**224**}

Smith VV, Cunha K, Lazzaro D, 2001, The abundance distribution in the exolanet host star HD 19994. *AJ*, 121, 3207–3218 {**188, 195**}

Smoluchowski MV, 1916, Drei Vortrage uber Diffusion, Brownsche Bewegung und Koagulation von Kolloidteilchen. *Zeitschrift fur Physik*, 17, 557–585 {**228**}

Snellen IAG, 2005, High-precision K-band photometry of the secondary eclipse of HD 209458. *MNRAS*, 363, 211–215 {**115, 140**}

Snellen IAG, Albrecht S, de Mooij EJW, et al., 2008, Ground-based detection of sodium in the transmission spectrum of exoplanet HD 209458 b. *A&A*, 487, 357–362 {**140**}

Snellen IAG, de Mooij EJW, Albrecht S, 2009a, The changing phases of extrasolar planet CoRoT–1 b. *Nature*, 459, 543–545 {**2, 111, 124, 137, 138, 143**}

Snellen IAG, Koppenhoefer J, van der Burg RFJ, et al., 2009b, OGLE2-TR-L9b: an exoplanet transiting a rapidly rotating F3 star. *A&A*, 497, 545–550 {**106, 249, 327**}

Snellen IAG, van der Burg RFJ, de Hoon MDJ, et al., 2007, A search for transiting extrasolar planet candidates in the OGLE–II microlens database of the galactic plane. *A&A*, 476, 1357–1363 {**106**}

Snellgrove MD, Papaloizou JCB, Nelson RP, 2001, On disk driven inward migration of resonantly coupled planets with application to the system around GJ 876. *A&A*, 374, 1092–1099 {**41, 49, 242**}

Snodgrass C, Horne K, Tsapras Y, 2004, The abundance of Galactic planets from OGLE–III 2002 microlensing data. *MNRAS*, 351, 967–975 {**96**}

Sobolev VV, 1975, *Light Scattering in Planetary Atmospheres*. Pergamon Press {**124**}

Söderhjelm S, Robichon N, Arenou F, 1999, HD 209458. *IAU Circ.*, 7323, 3–4 {**113**}

Sohl F, Spohn T, Breuer D, et al., 2002, Implications from Galileo observations on the interior structure and chemistry of the Galilean satellites. *Icarus*, 157, 104–119 {**271**}

Soker N, 1994, The expected morphology of the solar system planetary nebula. *PASP*, 106, 59–62 {**79, 80**}

—, 1996, What planetary nebulae can tell us about planetary systems. *ApJ*, 460, L53–L56 {**79**}

—, 1998, Can planets influence the horizontal branch morphology? *AJ*, 116, 1308–1313 {**81, 179**}

—, 1999, Detecting planets in planetary nebulae. *MNRAS*, 306, 806–808 {**79**}

Soker N, Hershenhorn A, 2007, Expected planets in globular clusters. *MNRAS*, 381, 334–340 {**110**}

Soker N, Rappaport S, Fregeau J, 2001, Collisions of free-floating planets with evolved stars in globular clusters. *ApJ*, 563, L87–L90 {**243**}

Solano E, von Braun K, Velasco A, et al., 2009, The LAEX and NASA portals for CoRoT public data. *A&A*, 506, 455–463 {**111**}

Solomon S, Qin D, Manning M, et al., 2007, *Climate Change 2007: The Physical Science Basis*. Cambridge University Press {**307**}

Song I, Schneider G, Zuckerman B, et al., 2006, HST–NICMOS imaging of the planetary-mass companion to the young brown dwarf 2MASSW J1207334–393254. *ApJ*, 652, 724–729 {**163, 170, 171**}

Song I, Zuckerman B, Weinberger AJ, et al., 2005, Extreme collisions between planetesimals as the origin of warm dust around a Sun-like star. *Nature*, 436, 363–365 {**178**}

Soper P, Franklin F, Lecar M, 1990, On the original distribution of the asteroids. III. Orbits between Jupiter and Saturn. *Icarus*, 87, 265–284 {**46**}

Soriano M, Vauclair S, 2010, New seismic analysis of the exoplanet-host star μ Ara. *A&A*, 513, A49, 1–8 {**203**}

Soriano M, Vauclair S, Vauclair G, et al., 2007, The CoRoT primary target HD 52265: models and seismic tests. *A&A*, 471, 885–892 {**204**}

Soter S, 2006, What is a planet? *AJ*, 132, 2513–2519 {**7**}

Sotin C, Grasset O, Mocquet A, 2007, Mass radius curve for extrasolar Earth-like planets and ocean planets. *Icarus*, 191, 337–351 {**250, 260, 261, 267, 268, 270**}

Soubiran C, Bienaymé O, Siebert A, 2003, Vertical distribution of Galactic disk stars. I. Kinematics and metallicity. *A&A*, 398, 141–151 {**194**}

Soucail G, Fort B, Mellier Y, et al., 1987, A blue ring-like structure, in the center of the A370 cluster of galaxies. *A&A*, 172, L14–L16 {**84**}

Soummer R, Aime C, Falloon PE, 2003a, Stellar coronagraphy with prolate apodised circular apertures. *A&A*, 397, 1161–1172 {**153**}

Soummer R, Aime C, Ferrari A, et al., 2006, Apodised pupil Lyot coronagraphs: concepts and application to the Gemini Planet Imager. *IAU Colloq. 200: Direct Imaging of Exoplanets: Science and Techniques*, 367–372 {**159**}

Soummer R, Dohlen K, Aime C, 2003b, Achromatic dual-zone phase mask stellar coronagraph. *A&A*, 403, 369–381 {**153**}

Soummer R, Ferrari A, Aime C, et al., 2007, Speckle noise and dynamic range in coronagraphic images. *ApJ*, 669, 642–656 {**157**}

Sousa SG, Santos NC, Israelian G, et al., 2006, Spectroscopic parameters for a sample of metal-rich solar-type stars. *A&A*, 458, 873–880 {**189**}

Sousa SG, Santos NC, Mayor M, et al., 2008, Spectroscopic parameters for 451 stars in the HARPS GTO planet search programme. Stellar [Fe/H] and the frequency of exo-Neptunes. *A&A*, 487, 373–381 {**25, 185, 188, 190, 194**}

Southworth J, 2008, Homogeneous studies of transiting extrasolar planets. I. Light-curve analyses. *MNRAS*, 386, 1644–1666 {**123, 146**}

—, 2009, Homogeneous studies of transiting extrasolar planets. II. Physical properties. *MNRAS*, 394, 272–294 {**146, 147**}

Southworth J, Hinse TC, Burgdorf MJ, et al., 2009a, High-precision photometry by telescope defocusing. II. The transiting planetary system WASP-4. *MNRAS*, 399, 287–294 {**115**}

Southworth J, Hinse TC, Jørgensen UG, et al., 2009b, High-precision photometry by telescope defocusing. I. The transiting planetary system WASP-5. *MNRAS*, 396, 1023–1031 {**115**}

Southworth J, Mancini L, Novati SC, et al., 2010, High-precision photometry by telescope defocusing. III. The transiting planetary system WASP-2. *MNRAS*, 408, 1680–1688 {**115**}

Southworth J, Maxted PFL, Smalley B, 2004, Eclipsing binaries in open clusters. II. V453 Cyg in NGC 6871. *MNRAS*, 351, 1277–1289 {**119**}

Southworth J, Wheatley PJ, Sams G, 2007, A method for the direct determination of the surface gravities of transiting extrasolar planets. *MNRAS*, 379, L11–L15 {**119, 123, 140, 146**}

Sozzetti A, 2004, On the possible correlation between the orbital periods of extrasolar planets and the metallicity of the host stars. *MNRAS*, 354, 1194–1200 {**191**}

—, 2005, Astrometric methods and instrumentation to identify and characterise extrasolar planets: a review. *PASP*, 117, 1021–1048 {**65, 67**}

Sozzetti A, Casertano S, Brown RA, et al., 2002, Narrow-angle astrometry with the Space Interferometry Mission: the search for extrasolar planets. I. Detection and characterisation of single planets. *PASP*, 114, 1173–1196 {**73**}

—, 2003, Narrow-angle astrometry with the Space Interferometry Mission: the search for extrasolar planets. II. Detection and characterisation of planetary systems. *PASP*, 115, 1072–1104 {**73**}

Sozzetti A, Casertano S, Lattanzi MG, et al., 2001, Detection and measurement of planetary systems with Gaia. *A&A*, 373, L21–L24 {**72**}

Sozzetti A, Desidera S, 2010, Hipparcos preliminary astrometric masses for the two close-in companions to HD 131664 and HD 43848. A brown dwarf and a low-mass star. *A&A*, 509, A103–114 {**70**}

Sozzetti A, Torres G, Charbonneau D, et al., 2007, Improving stellar and planetary parameters of transiting planet systems: the case of TrES–2. *ApJ*, 664, 1190–1198 {**123**}

Sozzetti A, Torres G, Latham DW, et al., 2006, A Keck HIRES Doppler search for planets orbiting metal-poor dwarfs. II. Testing giant planet formation and migration scenarios. *ApJ*, 649, 428–435 {**33, 34, 146**}

—, 2009, A Keck HIRES Doppler search for planets orbiting metal-poor dwarfs. II. On the frequency of giant planets in the metal-poor regime. *ApJ*, 697, 544–556 {**33, 34**}

Spaan FHP, Greenaway AH, 2007, Analysis of pupil replication. *ApJ*, 658, 1380–1385 {**156**}

Sparks WB, Ford HC, 2002, Imaging spectroscopy for extrasolar planet detection. *ApJ*, 578, 543–564 {**158**}

Sparks WB, Hough J, Germer TA, et al., 2009, Detection of circular polarisation in light scattered from photosynthetic microbes. *Proceedings of the National Academy of Science*, 106, 7816–7821 {**126**}

Sperber KR, Fairbridge RW, Shirley JH, 1990, Sun's inertial motion and luminosity. *Sol. Phys.*, 127, 379–392 {**66**}

Spiegel DS, Haiman Z, Gaudi BS, 2007, On constraining a transiting exoplanet's rotation rate with its transit spectrum. *ApJ*, 669, 1324–1335 {**136**}

Spiegel DS, Menou K, Scharf CA, 2008, Habitable climates. *ApJ*, 681, 1609–1623 {**284**}

—, 2009a, Habitable climates: the influence of obliquity. *ApJ*, 691, 596–610 {**284**}

Spiegel DS, Silverio K, Burrows A, 2009b, Can TiO explain thermal inversions in the upper atmospheres of irradiated giant planets? *ApJ*, 699, 1487–1500 {**277**}

Spiegel DS, Zamojski M, Gersch A, et al., 2005, Can we probe the atmospheric composition of an extrasolar planet from its reflection spectrum in a high-magnification microlensing event? *ApJ*, 628, 478–486 {**95**}

Spite F, Spite M, 1982a, Abundance of lithium in unevolved halo stars and old disk stars: interpretation and consequences. *A&A*, 115, 357–366 {**198**}

Spite M, Spite F, 1982b, Lithium abundance at the formation of the Galaxy. *Nature*, 297, 483–485 {**198**}

Spohn T, Schubert G, 2003, Oceans in the icy Galilean satellites of Jupiter? *Icarus*, 161, 456–467 {**271**}

Spronck JFP, Pereira SF, 2009, The effect of the longitudinal polarisation component in multi-axial nulling interferometry for exoplanet detection. *A&A*, 498, 931–947 {**154**}

Stacey FD, Davis PM, 2008, *Physics of the Earth*. Cambridge University Press, Fourth Edition {**281, 282**}

Stahl SM, Sandler DG, 1995, Optimisation and performance of adaptive optics for imaging extrasolar planets. *ApJ*, 454, L153–L156 {**151**}

Stairs IH, Lyne AG, Shemar SL, 2000, Evidence for free precession in a pulsar. *Nature*, 406, 484–486 {**79**}

Stam DM, 2004, Polarisation spectra of extrasolar giant planets. *Extrasolar Planets: Today and Tomorrow*, volume 321 of *ASP Conf Ser*, 195–196 {**126**}

—, 2008, Spectropolarimetric signatures of Earth-like extrasolar planets. *A&A*, 482, 989–1007 {**126**}

Stam DM, de Rooij WA, Cornet G, et al., 2006, Integrating polarised light over a planetary disk applied to starlight reflected by extrasolar planets. *A&A*, 452, 669–683 {**126, 289**}

Stam DM, Hovenier JW, 2005, Errors in calculated planetary phase functions and albedos due to neglecting polarisation. *A&A*, 444, 275–286 {**126**}

Stam DM, Hovenier JW, Waters LBFM, 2004, Using polarimetry to detect and characterise Jupiter-like extrasolar planets. *A&A*, 428, 663–672 {**126**}

Stamatellos D, Hubber DA, Whitworth AP, 2007, Brown dwarf formation by gravitational fragmentation of massive, extended protostellar discs. *MNRAS*, 382, L30–L34 {**214**}

Stamatellos D, Whitworth AP, 2008, Can giant planets form by gravitational fragmentation of disks? *A&A*, 480, 879–887 {**236**}

Stankov A, Martin D, Schulz R, et al., 2007, High temporal resolution transit observations with ESA's cryogenic camera. *Transiting Extrapolar Planets Workshop*, volume 366 of *ASP Conf Ser*, 268–270 {**133**}

Stapelfeldt KR, Holmes EK, Chen C, et al., 2004, First look at the Fomalhaut debris disk with the Spitzer space telescope. *ApJS*, 154, 458–462 {**171, 223**}

Stark CC, Kuchner MJ, 2008, The detectability of exo-Earths and super-Earths via resonant signatures in exozodiacal clouds. *ApJ*, 686, 637–648 {**223**}

Stassun KG, Mathieu RD, Mazeh T, et al., 1999, The rotation period distribution of pre-main-sequence stars in and around the Orion Nebula. *AJ*, 117, 2941–2979 {**200**}

Stassun KG, Mathieu RD, Valenti JA, 2006, Discovery of two young brown dwarfs in an eclipsing binary system. *Nature*, 440, 311–314 {**213**}

Steele IA, Bates SD, Gibson N, et al., 2008, RISE: a fast-readout imager for exoplanet transit timing. *SPIE Conf Ser*, volume 7014, 217 {**133**}

Steffen JH, Agol E, 2005, An analysis of the transit times of TrES–1 b. *MNRAS*, 364, L96–L100 {**107, 135**}

Steinmetz T, Wilken T, Araujo-Hauck C, et al., 2008, Laser frequency combs for astronomical observations. *Science*, 321, 1335–1337 {**20**}

Stephens A, Boesgaard AM, King JR, et al., 1997, Beryllium in lithium-deficient F and G stars. *ApJ*, 491, 339–358 {**198**}

Stepinski TF, Black DC, 2000, Statistics of low-mass companions to stars: implications for their origin. *A&A*, 356, 903–912 {**37**}

Stepinski TF, Malhotra R, Black DC, 2000, The v And system: models and stability. *ApJ*, 545, 1044–1057 {**14, 47**}

Stern SA, 1994, The detectability of extrasolar terrestrial and giant planets during their luminous final accretion. *AJ*, 108, 2312–2317 {**177**}

Stern SA, Colwell JE, 1997a, Accretion in the Edgeworth–Kuiper belt: forming 100–1000 km radius bodies at 30 AU and beyond. *AJ*, 114, 841–848 {**228**}

—, 1997b, Collisional erosion in the primordial Edgeworth–Kuiper belt and the generation of the 30–50 AU Kuiper gap. *ApJ*, 490, 879–882 {**228**}

Stern SA, Tholen DJ, 1997, *Pluto and Charon*. University of Arizona Press {**293**}

Stevens IR, 2005, Magnetospheric radio emission from extrasolar giant planets: the role of the host stars. *MNRAS*, 356, 1053–1063 {**176**}

Stevenson DJ, 1975, Thermodynamics and phase separation of dense fully ionised hydrogen-helium fluid mixtures. *Phys. Rev. B*, 12, 3999–4007 {**262**}

—, 1982, Formation of the giant planets. *Planet. Space Sci.*, 30, 755–764 {**232, 235, 260, 281**}

—, 1985, Cosmochemistry and structure of the giant planets and their satellites. *Icarus*, 62, 4–15 {**271**}

—, 1991, The search for brown dwarfs. *ARA&A*, 29, 163–193 {**209**}

—, 1999, Life-sustaining planets in interstellar space? *Nature*, 400, 32–33 {**281–283**}

—, 2003, Planetary magnetic fields. *Earth and Planetary Science Letters*, 208, 1–11 {**174**}

Stevenson DJ, Harris AW, Lunine JI, 1986, Origins of satellites. *IAU Colloq. 77: Some Background about Satellites*, 39–88 {**237**}

Stevenson DJ, Salpeter EE, 1977a, The dynamics and helium distribution in hydrogen–helium fluid planets. *ApJS*, 35, 239–261 {**271**}

—, 1977b, The phase diagram and transport properties for hydrogen–helium fluid planets. *ApJS*, 35, 221–237 {**262**}

Stewart GR, Wetherill GW, 1988, Evolution of planetesimal velocities. *Icarus*, 74, 542–553 {**227**}

Stokes GG, 1851, On the effect of the internal friction of fluids on the motion of pendulums. *Transactions of the Cambridge Philosophical Society*, 9, 8–106 {**226**}

Stone JM, Balbus SA, 1996, Angular momentum transport in accretion disks via convection. *ApJ*, 464, 364–372 {**221**}

Stone JM, Hawley JF, Gammie CF, et al., 1996, Three-dimensional magnetohydrodynamical simulations of vertically stratified accretion disks. *ApJ*, 463, 656–673 {**221**}

Stone RPS, Wright SA, Drake F, et al., 2005, Lick Observatory optical SETI: targeted search and new directions. *Astrobiology*, 5, 604–611 {**291**}

Storey JWV, 2009, Astronomy and astrophysics from Antarctica. *Assoc Asia Pacific Phys Soc Bull*, volume 19, 4–10 {**161**}

Storey JWV, Ashley MCB, Lawrence JS, et al., 2003, Dome C—the best astronomical site in the world? *Memorie della Societa Astronomica Italiana Supplement*, 2, 13–18 {**161**}

Stothers RB, 1998, Galactic disk dark matter, terrestrial impact cratering and the law of large numbers. *MNRAS*, 300, 1098–1104 {**229, 295, 307**}

Strand KA, 1943, 61 Cyg as a triple system. *PASP*, 55, 29–32 {**64**}

Strassmeier KG, Andersen MI, Granzer T, et al., 2007, The International Concordia Explorer Telescope (ICE-T): an ultimate transit-search experiment for Dome C. *Transiting Extrasolar Planets Workshop*, volume 366 of *ASP Conf Ser*, 332–336 {**110**}

Street RA, Christian DJ, Clarkson WI, et al., 2007, SuperWASP-north extrasolar planet candidates between $18^h < RA < 21^h$. *MNRAS*, 379, 816–832 {**107**}

Street RA, Horne K, Lister TA, et al., 2003, Searching for planetary transits in the field of open cluster NGC 6819. I. *MNRAS*, 340, 1287–1297 {**109**}

Street RA, Simpson E, Barros SCC, et al., 2010, WASP–24 b: a new transiting close-in hot Jupiter orbiting a late F-star. *ApJ*, 720, 337–343 {**327**}

Strobel DF, 2002, Aeronomic systems on planets, moons, and comets. *Atmospheres in the Solar System: Comparative Aeronomy*, 7–22, American Geophysical Union {**280**}

—, 2005, Photochemistry in outer solar system atmospheres. *Space Sci. Rev.*, 116, 155–170 {**277**}

Strom RG, Malhotra R, Ito T, et al., 2005, The origin of planetary impactors in the inner solar system. *Science*, 309, 1847–1850 {**304**}

Struve O, 1952, Proposal for a project of high-precision stellar radial velocity work. *The Observatory*, 72, 199–200 {**36, 64, 104**}

Stumpff P, 1980, Two self-consistent FORTRAN subroutines for the computation of the Earth's motion. *A&AS*, 41, 1–8 {**18**}

Su KYL, Rieke GH, Misselt KA, et al., 2005, The Vega debris disk: a surprise from Spitzer. *ApJ*, 628, 487–500 {**224**}

Su KYL, Rieke GH, Stapelfeldt KR, et al., 2008, The exceptionally large debris disk around γ Oph. *ApJ*, 679, L125–L129 {**224**}

—, 2009, The debris disk around HR 8799. *ApJ*, 705, 314–327 {**223**}

Sudarsky D, Burrows A, Hubeny I, 2003, Theoretical spectra and atmospheres of extrasolar giant planets. *ApJ*, 588, 1121–1148 {**272**}

Sudarsky D, Burrows A, Hubeny I, et al., 2005, Phase functions and light curves of wide-separation extrasolar giant planets. *ApJ*, 627, 520–533 {**272**}

Sudarsky D, Burrows A, Pinto P, 2000, Albedo and reflection spectra of extrasolar giant planets. *ApJ*, 538, 885–903 {**265, 272, 278**}

Sumi T, Bennett DP, Bond IA, et al., 2010, A cold Neptune-mass planet OGLE–2007–BLG–368Lb: cold Neptunes are common. *ApJ*, 710, 1641–1653 {**99**}

Supulver KD, Bridges FG, Tiscareno S, et al., 1997, The sticking properties of water frost produced under various ambient conditions. *Icarus*, 129, 539–554 {**226**}

Sussman GJ, Wisdom J, 1988, Numerical evidence that the motion of Pluto is chaotic. *Science*, 241, 433–437 {**45, 299**}

—, 1992, Chaotic evolution of the solar system. *Science*, 257, 56–62 {**44, 45, 299**}

Svensmark H, 2006a, Cosmic rays and the biosphere over 4 billion years. *Astronomische Nachrichten*, 327, 871–875 {**307**}

—, 2006b, Imprint of Galactic dynamics on Earth's climate. *Astronomische Nachrichten*, 327, 866–870 {**307**}

—, 2007, Cosmoclimatology: a new theory emerges. *Astronomy and Geophysics*, 48(1), 010000–1 {**307**}

Svensson F, Ludwig H, 2005, Hydrodynamical simulations of convection-related stellar micro-variability. *13th Cambridge Workshop on Cool Stars, Stellar Systems and the Sun*, volume 560 of *ESA Special Publication*, 979–984 {**65, 115**}

Swain MR, Bouwman J, Akeson RL, et al., 2008a, The mid-infrared spectrum of the transiting exoplanet HD 209458 b. *ApJ*, 674, 482–497 {**140**}

Swain MR, Coude du Foresto V, Fossat E, et al., 2003, The Antarctic planet interferometer and the potential for interferometric observations of extrasolar planets from Dome C. *Memorie della Societa Astronomica Italiana Supplement*, 2, 207–211 {**162**}

Swain MR, Deming D, Vasisht G, et al., 2009a, THESIS: the Terrestrial and Habitable-zone Exoplanet Spectroscopy Infrared Spacecraft. *Astro2010: The Astronomy and Astrophysics Decadal Survey*, volume 2010 of *Astronomy*, 61–79 {**114**}

Swain MR, Tinetti G, Vasisht G, et al., 2009b, Water, methane, and carbon dioxide present in the dayside spectrum of the exoplanet HD 209458 b. *ApJ*, 704, 1616–1621 {**139–141**}

Swain MR, Vasisht G, Henning T, et al., 2010, THESIS: the terrestrial habitable-zone exoplanet spectroscopy infrared spacecraft. *SPIE Conf Ser*, volume 7731 {**114**}

Swain MR, Vasisht G, Tinetti G, 2008b, The presence of methane in the atmosphere of an extrasolar planet. *Nature*, 452, 329–331 {**140–142**}

Swain MR, Vasisht G, Tinetti G, et al., 2009c, Molecular signatures in the near-infrared dayside spectrum of HD 189733 b. *ApJ*, 690, L114–L117 {**2**}

Swartzlander GA, 2001, Peering into darkness with a vortex spatial filter. *Optics Letters*, 26, 497–499 {**155**}

—, 2006, Achromatic optical vortex lens. *Optics Letters*, 31, 2042–2044 {**155, 156**}

Swartzlander GA, Ford EL, Abdul-Malik RS, et al., 2008, Astronomical demonstration of an optical vortex coronagraph. *Optics Express*, 16, 10200–10207 {**156**}

Syer D, Clarke CJ, 1995, Satellites in disks: regulating the accretion luminosity. *MNRAS*, 277, 758–766 {**240**}

Szabó GM, Kiss LL, Benkő JM, et al., 2010, A multi-site campaign to detect the transit of the second planet in HAT–P–13. *A&A*, 523, A84 {**106**}

Szabó GM, Szatmáry K, Divéki Z, et al., 2006, Possibility of a photometric detection of exomoons. *A&A*, 450, 395–398 {**135**}

Szebehely V, 1980, Stability of planetary orbits in binary systems. *Celestial Mechanics*, 22, 7–12 {**56**}

—, 1984, Review of concepts of stability. *Celestial Mechanics*, 34, 49–64 {**56**}

Szebehely V, McKenzie R, 1981, Stability of outer planetary systems. *Celestial Mechanics*, 23, 3–7 {**57**}

Szebehely V, Zare K, 1977, Stability of classical triplets and of their hierarchy. *A&A*, 58, 145–152 {**43**}

Szenkovits F, Makó Z, 2008, About the Hill stability of extrasolar planets in stellar binary systems. *Celestial Mechanics and Dynamical Astronomy*, 101, 273–287 {**57**}

Szentgyorgyi AH, Furész G, 2007, Precision radial velocities for the Kepler era. *Revista Mexicana de Astronomia y Astrofisica Conference Series*, volume 28, 129–133 {**27**}

Tajika E, 2008, Snowball planets as a possible type of water-rich terrestrial planet in extrasolar planetary systems. *ApJ*, 680, L53–L56 {**284**}

Takeda G, Ford EB, Sills A, et al., 2007a, Structure and evolution of nearby stars with planets. II. Physical properties of 1000 cool stars from the SPOCS catalog. *ApJS*, 168, 297–318 {**43**}

Takeda G, Kita R, Rasio FA, 2008a, Planetary systems in binaries. I. Dynamical classification. *ApJ*, 683, 1063–1075 {**57**}

Takeda G, Rasio FA, 2005, High orbital eccentricities of extrasolar planets induced by the Kozai mechanism. *ApJ*, 627, 1001–1010 {**58, 60**}

—, 2006, Eccentricities of planets in binary systems. *Ap&SS*, 304, 239–242 {**57, 60**}

Takeda Y, Honda S, 2005, Photospheric CNO abundances of solar-type stars. *PASJ*, 57, 65–82 {**188, 189, 196**}

Takeda Y, Kawanomoto S, 2005, Lithium abundances of F-, G-, and K-type stars: profile-fitting analysis of the Li I 670.8 nm doublet. *PASJ*, 57, 45–63 {**199**}

Takeda Y, Kawanomoto S, Honda S, et al., 2007b, Behaviour of Li abundances in solar analogue stars. Evidence for line-width dependence. *A&A*, 468, 663–677 {**199**}

Takeda Y, Sato B, Kambe E, et al., 2001, Photospheric abundances of volatile and refractory elements in planet-harbouring stars. *PASJ*, 53, 1211–1221 {**195, 196**}

—, 2005, High-dispersion spectra collection of nearby F–K stars at Okayama Astrophysical Observatory: a basis for spectroscopic abundance standards. *PASJ*, 57, 13–25 {**189**}

Takeda Y, Sato B, Murata D, 2008b, Stellar parameters and elemental abundances of late-G giants. *PASJ*, 60, 781–802 {**190, 194**}

Takeuchi T, Clarke CJ, Lin DNC, 2005a, The differential lifetimes of protostellar gas and dust disks. *ApJ*, 627, 286–292 {**226**}

Takeuchi T, Lin DNC, 2002, Radial flow of dust particles in accretion disks. *ApJ*, 581, 1344–1355 {**226**}

Takeuchi T, Miyama SM, 1998, Wave excitation in isothermal disks by external gravity. *PASJ*, 50, 141–148 {**238**}

Takeuchi T, Miyama SM, Lin DNC, 1996, Gap formation in protoplanetary disks. *ApJ*, 460, 832–847 {**240**}

Takeuchi T, Velusamy T, Lin DNC, 2005b, Apparent stellar wobble by a planet in a circumstellar disk: limitations on planet detection by astrometry. *ApJ*, 618, 987–1000 {**65**}

Talon S, 2008, Transport processes in stars: diffusion, rotation, magnetic fields and internal waves. *EAS Publications Series*, volume 32, 81–130 {**296**}

Tamburini F, Ortolani S, Bianchini A, 2002, Polarisation statistics of extrasolar systems. *A&A*, 394, 675–678 {**126**}

Tamburini F, Umbriaco G, Anzolin G, et al., 2006, FrogEye, the quantum coronagraphic mask: the photon orbital angular momentum and its applications to astronomy. *Memorie della Societa Astronomica Italiana Supplement*, 9, 484 {**155**}

Tamura M, 2009, Subaru Strategic Exploration of Exoplanets and Disks with HiCIAO/AO188. *American Institute of Physics Conference Series*, volume 1158, 11–16 {**159**}

Tamura M, Hodapp K, Takami H, et al., 2006, Concept and science of HiCIAO: high contrast instrument for the Subaru next generation adaptive optics. *SPIE Conf Ser*, volume 6269, 28 {**158, 159**}

Tamura M, Itoh Y, Oasa Y, et al., 1998, Isolated and companion young brown dwarfs in the Taurus and Chamaeleon molecular clouds. *Science*, 282, 1095–1097 {**215**}

Tamura M, Suto H, Itoh Y, et al., 2000, Coronagraph imager with adaptive optics (CIAO): description and first results. *SPIE Conf Ser*, volume 4008, 1153–1161 {**153**}

Tamuz O, Mazeh T, Zucker S, 2005, Correcting systematic effects in a large set of photometric light curves. *MNRAS*, 356, 1466–1470 {**104**}

Tamuz O, Ségransan D, Udry S, et al., 2008, The CORALIE survey for southern extrasolar planets. XV. Discovery of two eccentric planets orbiting HD 4113 and HD 156846. *A&A*, 480, L33–L36 {**12, 24, 29–31, 33, 60, 181**}

Tanaka H, Ida S, 1999, Growth of a migrating protoplanet. *Icarus*, 139, 350–366 {**234**}

Tanaka H, Takeuchi T, Ward WR, 2002, Three-dimensional interaction between a planet and an isothermal gaseous disk. I. Corotation and Lindblad torques and planet migration. *ApJ*, 565, 1257–1274 {**238, 239**}

Tanaka H, Ward WR, 2004, Three-dimensional interaction between a planet and an isothermal gaseous disk. II. Eccentricity waves and bending waves. *ApJ*, 602, 388–395 {**242**}

Tanigawa T, Ikoma M, 2007, A systematic study of the final masses of gas giant planets. *ApJ*, 667, 557–570 {**232**}

Tarter JC, 1976, Brown dwarfs, Lilliputian stars, giant planets and missing mass problems. *Bulletin of the American Astronomical Society*, volume 8, 517 {**209**}

—, 2001, The Search for Extraterrestrial Intelligence (SETI). *araa*, 39, 511–548 {**290**}

—, 2004, Astrobiology and SETI. *New Astronomy Reviews*, 48, 1543–1549 {**177**}

—, 2006, The history of SETI at the Hat Creek Radio Observatory. *Revealing the Molecular Universe: One Antenna is Never Enough*, volume 356 of *ASP Conf Ser*, 117–125 {**291**}

Tarter JC, Backus PR, Mancinelli RL, et al., 2007, A reappraisal of the habitability of planets around M dwarf stars. *Astrobiology*, 7, 30–65 {**115,

References

250, 285}

Tavrov AV, Kobayashi Y, Tanaka Y, et al., 2005, Common-path achromatic interferometer-coronagraph: nulling of polychromatic light. *Optics Letters*, 30, 2224–2226 {153}

Taylor SR, 2001, *Solar System Evolution: A New Perspective*. Cambridge University Press {195, 259}

Telesco CM, Fisher RS, Wyatt MC, et al., 2005, Mid-infrared images of β Pic and the possible role of planetesimal collisions in the central disk. *Nature*, 433, 133–136 {178, 223}

ten Brummelaar TA, McAlister HA, Ridgway ST, et al., 2003, An update of the CHARA array. *SPIE Conf Ser*, volume 4838, 69–78 {162}

Tera F, Papanastassiou DA, Wasserburg GJ, 1974, Isotopic evidence for a terminal lunar cataclysm. *Earth and Planetary Science Letters*, 22, 1–21 {304}

Terquem C, 2003, Stopping inward planetary migration by a toroidal magnetic field. *MNRAS*, 341, 1157–1173 {239, 241}

Terquem C, Papaloizou JCB, 2002, Dynamical relaxation and the orbits of low-mass extrasolar planets. *MNRAS*, 332, L39–L43 {127}

—, 2007, Migration and the formation of systems of hot super-Earths and Neptunes. *ApJ*, 654, 1110–1120 {239}

—, 2008, Forming hot super–Earths. *ASP Conf Ser*, volume 398, 265–273 {267}

Terquem C, Papaloizou JCB, Nelson RP, 2000, Disks, extrasolar planets and migration. *Space Science Reviews*, 92, 323–340 {238}

Terquem C, Papaloizou JCB, Nelson RP, et al., 1998, On the tidal interaction of a solar-type star with an orbiting companion: excitation of g-mode oscillation and orbital evolution. *ApJ*, 502, 788–801 {205, 237}

Terrile RJ, Ftaclas C, 1997, Direct detection of extrasolar planetary systems from balloon borne telescopes. *IAU Colloq. 161: Astronomical and Biochemical Origins and the Search for Life in the Universe*, 359–366 {164}

Thalmann C, Carson J, Janson M, et al., 2009a, Discovery of the coldest imaged companion of a Sun-like star. *ApJ*, 707, L123–L127 {168}

Thalmann C, Goto M, Carson J, et al., 2009b, SPOTS: search for planets of two stars: a direct imaging survey for exoplanets in binary systems. *American Institute of Physics Conference Series*, volume 1158, 271–272 {207}

Thalmann C, Schmid HM, Boccaletti A, et al., 2008, SPHERE ZIMPOL: overview and performance simulation. *SPIE Conf Ser*, volume 7014, 112 {126}

Thébault P, Augereau JC, Beust H, 2003, Dust production from collisions in extrasolar planetary systems. The inner β Pic disk. *A&A*, 408, 775–788 {223}

Thébault P, Beust H, 2001, Falling evaporating bodies in the β Pic system: resonance refilling and long-term duration of the phenomenon. *A&A*, 376, 621–640 {223}

Thébault P, Marzari F, Scholl H, 2006, Relative velocities among accreting planetesimals in binary systems: The circumprimary case. *Icarus*, 183, 193–206 {58, 59}

—, 2008, Planet formation in α Cen a revisited: not so accretion friendly after all. *MNRAS*, 388, 1528–1536 {58}

Thébault P, Marzari F, Scholl H, et al., 2004, Planetary formation in the γ Cep system. *A&A*, 427, 1097–1104 {58, 59}

Thiele TN, 1883, Neue Methode zur Berechung von Doppelsternbahnen. *Astronomische Nachrichten*, 104, 245–253 {68}

Thielemann F, 2002, Nucleosynthesis. *Encyclopedia of Astronomy and Astrophysics* {197}

Thomas PC, Parker JW, McFadden LA, et al., 2005, Differentiation of the asteroid Ceres as revealed by its shape. *Nature*, 437, 224–226 {230}

Thommes EW, 2007, Terrestrial planet formation. *Planetary Systems and the Origins of Life*, 41–61, Cambridge University Press {225}

Thommes EW, Bryden G, Wu Y, et al., 2008a, From mean motion resonances to scattered planets: producing the solar system, eccentric exoplanets, and late heavy bombardments. *ApJ*, 675, 1538–1548 {243, 304}

Thommes EW, Duncan MJ, 2006, The accretion of giant-planet cores. *Planet Formation*, 129–146, Cambridge University Press {232}

Thommes EW, Duncan MJ, Levison HF, 1999, The formation of Uranus and Neptune in the Jupiter–Saturn region of the solar system. *Nature*, 402, 635–638 {303}

—, 2003, Oligarchic growth of giant planets. *Icarus*, 161, 431–455 {228}

Thommes EW, Lissauer JJ, 2003, Resonant inclination excitation of migrating giant planets. *ApJ*, 597, 566–580 {39, 41, 53, 127}

—, 2005, Planet migration. *Astrophysics of Life*, 41–53 {225}

Thommes EW, Matsumura S, Rasio FA, 2008b, Gas disks to gas giants: simulating the birth of planetary systems. *Science*, 321, 814–817 {243, 251}

Thommes EW, Murray N, 2006, Giant planet accretion and migration: surviving the Type I regime. *ApJ*, 644, 1214–1222 {234}

Thommes EW, Nagasawa M, Lin DNC, 2008c, Dynamical shake-up of planetary systems. II. N-body simulations of solar system terrestrial planet formation induced by secular resonance sweeping. *ApJ*, 676, 728–739 {234, 299, 302}

Thommes EW, Nilsson L, Murray N, 2007, Overcoming migration during giant planet formation. *ApJ*, 656, L25–L28 {234}

Thompson SL, 1990, ANEOS: analytic equations of state for shock physics codes. *Sandia National Laboratory Doc. SAND89–2951*, www.fas.org/sgp/othergov/doe/lanl/lib-www/sand/892951.pdf {261}

Thorsett SE, Arzoumanian Z, Camilo F, et al., 1999, The triple pulsar system PSR B1620–26 in M4. *ApJ*, 523, 763–770 {77}

Thorsett SE, Arzoumanian Z, Taylor JH, 1993, PSR B1620–26: a binary radio pulsar with a planetary companion? *ApJ*, 412, L33–L36 {77, 79}

Thronson HA, 1997, Our cosmic origins: NASA's Origins theme and the search for Earth-like planets. *Planets Beyond the Solar System and the Next Generation of Space Missions*, volume 119 of *ASP Conf Ser*, 3–7 {166}

Throop HB, Bally J, 2005, Can photoevaporation trigger planetesimal formation? *ApJ*, 623, L149–L152 {226}

Tian F, 2009, Thermal escape from super Earth atmospheres in the habitable zones of M stars. *ApJ*, 703, 905–909 {281}

Tian F, Toon OB, 2005, Hydrodynamic escape of nitrogen from Pluto. *Geophys. Res. Lett.*, 32, 18201–204 {281}

Tian F, Toon OB, Pavlov AA, et al., 2005, Transonic hydrodynamic escape of hydrogen from extrasolar planetary atmospheres. *ApJ*, 621, 1049–1060 {281}

Tinetti G, Liang MC, Vidal-Madjar A, et al., 2007a, Infrared transmission spectra for extrasolar giant planets. *ApJ*, 654, L99–L102 {137}

Tinetti G, Meadows VS, Crisp D, et al., 2006a, Detectability of planetary characteristics in disk-averaged spectra. I. The Earth model. *Astrobiology*, 6, 34–47 {289}

—, 2006b, Detectability of planetary characteristics in disk-averaged spectra. II. Synthetic spectra and light-curves of Earth. *Astrobiology*, 6, 881–900 {289}

Tinetti G, Rashby S, Yung YL, 2006c, Detectability of red-edge-shifted vegetation on terrestrial planets orbiting M stars. *ApJ*, 644, L129–L132 {289}

Tinetti G, Vidal-Madjar A, Liang MC, et al., 2007b, Water vapour in the atmosphere of a transiting extrasolar planet. *Nature*, 448, 169–171 {2, 140}

Tingley B, 2003, A rigorous comparison of different planet detection algorithms. *A&A*, 403, 329–337 {105}

—, 2004, Using colour photometry to separate transiting exoplanets from false positives. *A&A*, 425, 1125–1131 {105}

Tingley B, Endl M, Gazzano J, et al., 2011, Transiting exoplanets from the CoRoT space mission: XIII. CoRoT–14 b: an unusually dense very hot Jupiter. *ArXiv e-prints* {326}

Tingley B, Sackett PD, 2005, A photometric diagnostic to aid in the identification of transiting extrasolar planets. *ApJ*, 627, 1011–1018 {104, 121}

Tinney CG, 1999, Brown dwarfs: the stars that failed. *Nature*, 397, 37–40 {209, 210}

Tinney CG, Butler RP, Marcy GW, et al., 2001, First results from the Anglo–Australian planet search: a brown dwarf candidate and a 51 Peg-like planet. *ApJ*, 551, 507–511 {24, 70}

—, 2003, Four new planets orbiting metal-enriched stars. *ApJ*, 587, 423–428 {33, 52}

—, 2006, The 2:1 resonant exoplanetary system orbiting HD 73526. *ApJ*, 647, 594–599 {52, 55}

Tinney CG, McCarthy C, Jones HRA, et al., 2002, Echelle spectroscopy of Ca II HK activity in Southern Hemisphere planet search targets. *MNRAS*, 332, 759–763 {21}

Tipler FJ, 2003, Intelligent life in cosmology. *International Journal of Astrobiology*, 2, 141–148 {290}

Tiscareno MS, Thomas PC, Burns JA, 2009, The rotation of Janus and Epimetheus. *Icarus*, 204, 254–261 {**52**}

Tisserand P, Le Guillou L, Afonso C, et al., 2007, Limits on the Macho content of the Galactic halo from the EROS-2 Survey of the Magellanic Clouds. *A&A*, 469, 387–404 {**86**}

Todorov K, Deming D, Harrington J, et al., 2010a, Spitzer–IRAC secondary eclipse photometry of the transiting extrasolar planet HAT–P–1 b. *ApJ*, 708, 498–504 {**143**}

Todorov K, Luhman KL, McLeod KK, 2010b, Discovery of a planetary-mass companion to a brown dwarf in Taurus. *ApJ*, 714, L84–L88 {**168, 214**}

Tokovinin AA, 1992, The frequency of low-mass companions to K and M stars in the solar neighbourhood. *A&A*, 256, 121–132 {**28**}

Tokunaga AT, Bond T, Elias J, et al., 2006, Design tradeoffs for a high spectral resolution mid-infrared echelle spectrograph on the Thirty-Meter Telescope. *SPIE Conf Ser*, volume 6269 {**24**}

Toner CG, Gray DF, 1988, The starpatch on the G8 dwarf ξ Boo A. *ApJ*, 334, 1008–1020 {**23**}

Tonks WB, Melosh HJ, 1992, Core formation by giant impacts. *Icarus*, 100, 326–346 {**229, 302**}

Tonry J, Davis M, 1979, A survey of galaxy redshifts. I Data reduction techniques. *AJ*, 84, 1511–1525 {**17**}

Tonry JL, Howell SB, Everett ME, et al., 2005, A search for variable stars and planetary occultations in NGC 2301. I. Techniques. *PASP*, 117, 281–289 {**109**}

Toomre A, 1964, On the gravitational stability of a disk of stars. *ApJ*, 139, 1217–1238 {**218, 221**}

Toon OB, McKay CP, Ackerman TP, et al., 1989, Rapid calculation of radiative heating rates and photodissociation rates in inhomogeneous multiple scattering atmospheres. *J. Geophys. Res.*, 94, 16287–16301 {**272**}

Torres G, 2007, The planet host star γ Cep: physical properties, the binary orbit, and the mass of the substellar companion. *ApJ*, 654, 1095–1109 {**70**}

Torres G, Andersen J, Giménez A, 2010a, Accurate masses and radii of normal stars: modern results and applications. *A&A Rev.*, 18, 67–126 {**61**}

Torres G, Bakos GÁ, Hartman J, et al., 2010b, HAT–P–14 b: a 2.2 Jupiter-mass exoplanet transiting a bright F star. *ApJ*, 715, 458–467 {**326**}

Torres G, Bakos GÁ, Kovács G, et al., 2007, HAT–P–3 b: a heavy-element-rich planet transiting a K dwarf star. *ApJ*, 666, L121–L124 {**146, 326**}

Torres G, Winn JN, Holman MJ, 2008, Improved parameters for extrasolar transiting planets. *ApJ*, 677, 1324–1342 {**12, 111, 120, 123, 145–147**}

Touboul M, Kleine T, Bourdon B, et al., 2007, Late formation and prolonged differentiation of the Moon inferred from W isotopes in lunar metals. *Nature*, 450, 1206–1209 {**297, 302**}

—, 2009, Tungsten isotopes in ferroan anorthosites: implications for the age of the Moon and lifetime of its magma ocean. *Icarus*, 199, 245–249 {**297**}

Townes CH, 1997, Optical and infrared SETI. *IAU Colloq. 161: Astronomical and Biochemical Origins and the Search for Life in the Universe*, 585–594 {**290**}

Townes CH, Wishnow EH, 2008, Interferometry at mid-infrared wavelengths: the ISI system. *SPIE Conf Ser*, volume 7013, 12 {**162**}

Toyota E, Itoh Y, Ishiguma S, et al., 2009, Radial velocity search for extrasolar planets in visual binary systems. *PASJ*, 61, 19–28 {**33, 58**}

Tracy AJ, Hankla AK, Lopez CA, et al., 2004, High-power solid-state sodium beacon laser guide star for the Gemini North Observatory. *SPIE Conf Ser*, volume 5490, 998–1009 {**152**}

Trafton LM, Hunten DM, Zahnle KJ, et al., 1997, Escape processes at Pluto and Charon. *Pluto and Charon*, 475–522, University of Arizona Press {**281**}

Traub WA, Beichman C, Boden AF, et al., 2010, Detectability of Earth-like planets in multi-planet systems: preliminary report. *EAS Publications Series*, volume 42, 191–199 {**67, 73**}

Traub WA, Chen P, Kern B, et al., 2008, Planetscope: an exoplanet coronagraph on a balloon platform. *SPIE Conf Ser*, volume 7010, 110 {**156, 162**}

Traub WA, Shaklan S, Lawson P, 2007, Prospects for Terrestrial Planet Finder: TPF-C, TPF-I, and TPF-O. *In the Spirit of Bernard Lyot: The Direct Detection of Planets and Circumstellar Disks in the 21st Century*, 36 {**166**}

Traub WA, Vanderbei RJ, 2003, Two-mirror apodisation for high-contrast imaging. *ApJ*, 599, 695–701 {**155**}

Trauger J, 2007, Eclipse: a case study for direct coronagraphic imaging of planetary systems from space. *In the Spirit of Bernard Lyot: The Direct Detection of Planets and Circumstellar Disks in the 21st Century* {**166**}

Trauger J, Stapelfeldt K, Traub W, et al., 2008, ACCESS: a NASA mission concept study of an Actively Corrected Coronagraph for Exoplanet System Studies. *SPIE Conf Ser*, volume 7010, 69 {**166**}

Trauger J, Traub WA, 2007, A laboratory demonstration of the capability to image an Earth-like extrasolar planet. *Nature*, 446, 771–773 {**157**}

Travouillon T, Jolissaint L, Ashley MCB, et al., 2009, Overcoming the boundary layer turbulence at Dome C: ground-layer adaptive optics versus tower. *PASP*, 121, 668–679 {**162**}

Tremaine S, 1991, On the origin of the obliquities of the outer planets. *Icarus*, 89, 85–92 {**300, 301**}

Tremaine S, Dones L, 1993, On the statistical distribution of massive impactors. *Icarus*, 106, 335–341 {**280**}

Triaud AHMJ, Collier Cameron A, Queloz D, et al., 2010, Spin-orbit angle measurements for six southern transiting planets: new insights into the dynamical origins of hot Jupiters. *ArXiv e-prints* {**129**}

Triaud AHMJ, Queloz D, Bouchy F, et al., 2009, The Rossiter–McLaughlin effect of CoRoT–3 b and HD 189733 b. *A&A*, 506, 377–384 {**129**}

Triaud AHMJ, Queloz D, Hellier C, et al., 2011, WASP–23 b: a transiting hot Jupiter around a K dwarf and its Rossiter-McLaughlin effect. *ArXiv e-prints* {**327**}

Trieloff M, Palme H, 2006, The origin of solids in the early Solar System. *Planet Formation*, 64–89, Cambridge University Press {**226**}

Trilling DE, 2000, Tidal constraints on the masses of extrasolar planets. *ApJ*, 537, L61–L64 {**247**}

Trilling DE, Benz W, Guillot T, et al., 1998, Orbital evolution and migration of giant planets: modeling extrasolar planets. *ApJ*, 500, 428–439 {**238, 241, 281**}

Trilling DE, Brown RH, 1998, A circumstellar dust disk around a star with a known planetary companion. *Nature*, 395, 775–777 {**170, 224**}

Trilling DE, Lunine JI, Benz W, 2002, Orbital migration and the frequency of giant planet formation. *A&A*, 394, 241–251 {**237, 241, 281**}

Trilling DE, Stansberry JA, Stapelfeldt KR, et al., 2007, Debris disks in main-sequence binary systems. *ApJ*, 658, 1289–1311 {**57, 81**}

Tripathi A, Winn JN, Johnson JA, et al., 2010, A prograde, low-inclination orbit for the very hot Jupiter WASP–3 b. *ApJ*, 715, 421–428 {**129**}

Troyer J, Moses JI, Fegley B, et al., 2007, Disequilibrium chemistry on GJ 229 B. *Bulletin of the American Astronomical Society*, volume 38, 450 {**277**}

Tsapras Y, Horne K, Kane S, et al., 2003, Microlensing limits on numbers and orbits of extrasolar planets from the 1998-2000 OGLE events. *MNRAS*, 343, 1131–1144 {**96**}

Tsapras Y, Street R, Horne K, et al., 2009, RoboNet-II: follow-up observations of microlensing events with a robotic network of telescopes. *Astronomische Nachrichten*, 330, 4–11 {**97**}

Tsiganis K, Gomes R, Morbidelli A, et al., 2005, Origin of the orbital architecture of the giant planets of the solar system. *Nature*, 435, 459–461 {**243, 303**}

Tsokolov SA, 2009, Why is the definition of life so elusive? Epistemological considerations. *Astrobiology*, 9, 401–412 {**288**}

Tsujimoto T, Nomoto K, Yoshii Y, et al., 1995a, Relative frequencies of Type Ia and Type II supernovae in the chemical evolution of the Galaxy, LMC and SMC. *MNRAS*, 277, 945–958 {**198**}

Tsujimoto T, Yoshii Y, Nomoto K, et al., 1995b, Abundance gradients in the star-forming viscous disk and chemical properties of the bulge. *A&A*, 302, 704–712 {**199**}

Tsukamoto Y, Makino J, 2007, Formation of protoplanets from massive planetesimals in binary systems. *ApJ*, 669, 1316–1323 {**58**}

Tucker CJ, Fung IY, Keeling CD, et al., 1986, Relationship between atmospheric CO_2 variations and a satellite-derived vegetation index. *Nature*, 319, 195–199 {**289**}

Tull RG, 1998, High-resolution fiber-coupled spectrograph of the Hobby–Eberly Telescope. *SPIE Conf Ser*, volume 3355, 387–398 {**24**}

Turcotte DL, Schubert G, 2002, *Geodynamics*. Cambridge University Press, Second Edition {**230, 282**}

Turnbull MC, Tarter JC, 2003a, Target selection for SETI. I. A catalog of nearby habitable stellar systems. *ApJS*, 145, 181–198 {**291**}

—, 2003b, Target selection for SETI. II. Tycho-2 dwarfs, old open clusters,

References

and the nearest 100 stars. *ApJS*, 149, 423–436 {291}

Turnbull MC, Traub WA, Jucks KW, et al., 2006, Spectrum of a habitable world: Earthshine in the near-infrared. *ApJ*, 644, 551–559 {289}

Turner NJ, Sano T, Dziourkevitch N, 2007, Turbulent mixing and the dead zone in protostellar disks. *ApJ*, 659, 729–737 {**222, 225**}

Turner NJ, Willacy K, Bryden G, et al., 2006, Turbulent mixing in the outer solar nebula. *ApJ*, 639, 1218–1226 {**225**}

Tyler GL, Sweetnam DN, Anderson JD, et al., 1989, Voyager radio science observations of Neptune and Triton. *Science*, 246, 1466–1473 {**294**}

Udalski A, 2003, The Optical Gravitational Lensing Experiment: real time data analysis systems in the OGLE–III survey. *Acta Astronomica*, 53, 291–305 {97}

—, 2007, Transit campaigns of the OGLE–III survey. *Transiting Extrasolar Planets Workshop*, volume 366 of *ASP Conf Ser*, 51–57 {**106**}

Udalski A, Jaroszyński M, Paczyński B, et al., 2005, A Jovian-mass planet in microlensing event OGLE–2005–BLG–071. *ApJ*, 628, L109–L112 {**2, 97–100**}

Udalski A, Paczynski B, Zebrun K, et al., 2002a, The Optical Gravitational Lensing Experiment: search for planetary and low-luminosity object transits in the Galactic disk: results of 2001 campaign. *Acta Astronomica*, 52, 1–37 {**106**}

Udalski A, Pont F, Naef D, et al., 2008, OGLE–TR–211: a new transiting inflated hot Jupiter from the OGLE survey and ESO LP666 spectroscopic follow-up programme. *A&A*, 482, 299–304 {**106, 327**}

Udalski A, Szewczyk O, Zebrun K, et al., 2002b, The Optical Gravitational Lensing Experiment: planetary and low-luminosity object transits in the Carina fields of the Galactic disk. *Acta Astronomica*, 52, 317–359 {**106**}

Udalski A, Szymanski M, Kaluzny J, et al., 1993, The Optical Gravitational Lensing Experiment: discovery of the first candidate microlensing event in the direction of the Galactic Bulge. *Acta Astronomica*, 43, 289–294 {86}

Udalski A, Szymanski MK, Kubiak M, et al., 2004, The optical gravitational lensing experiment: planetary and low-luminosity object transits in the fields of Galactic disk; results of the 2003 OGLE observing campaigns. *Acta Astronomica*, 54, 313–345 {**106**}

Udem T, Holzwarth R, Hänsch TW, 2002, Optical frequency metrology. *Nature*, 416, 233–237 {20}

Udry S, Bonfils X, Delfosse X, et al., 2007, The HARPS search for southern extrasolar planets. XI. Super-Earths (5 and 8 M⊕) in a 3-planet system. *A&A*, 469, L43–L47 {**2, 30, 35, 36, 55, 286, 287**}

Udry S, Mayor M, 2008, Exoplanets: the golden age of radial velocities. *ASP Conf Ser*, volume 398, 13–26 {**22, 35, 36**}

Udry S, Mayor M, Benz W, et al., 2006, The HARPS search for southern extrasolar planets. V. A 14 Earth-mass planet orbiting HD 4308. *A&A*, 447, 361–367 {**37, 190**}

Udry S, Mayor M, Clausen JV, et al., 2003a, The CORALIE survey for southern extrasolar planets. X. A Hot Jupiter orbiting HD 73256. *A&A*, 407, 679–684 {12}

Udry S, Mayor M, Maurice E, et al., 1999a, 20 years of CORAVEL monitoring of radial-velocity standard stars. *IAU Colloq. 170: Precise Stellar Radial Velocities*, volume 185 of *ASP Conf Ser*, 383–389 {18}

Udry S, Mayor M, Naef D, et al., 2000, The CORALIE survey for southern extrasolar planets. II. The short-period planetary companions to HD 75289 and HD 130322. *A&A*, 356, 590–598 {**35, 58**}

—, 2002, The CORALIE survey for southern extrasolar planets. VIII. The very low-mass companions of HD 141937, HD 162020, HD 168443 and HD 202206: brown dwarfs or superplanets? *A&A*, 390, 267–279 {**37, 52**}

Udry S, Mayor M, Queloz D, 1999b, Towards a new set of high-precision radial-velocity standard stars. *IAU Colloq. 170: Precise Stellar Radial Velocities*, volume 185 of *ASP Conf Ser*, 367–377 {18}

Udry S, Mayor M, Santos NC, 2003b, Statistical properties of exoplanets. I. The period distribution: Constraints for the migration scenario. *A&A*, 407, 369–376 {37}

Udry S, Santos NC, 2007, Statistical properties of exoplanets. *ARA&A*, 45, 397–439 {**34, 193**}

Ueno Y, Yamada K, Yoshida N, et al., 2006, Evidence from fluid inclusions for microbial methanogenesis in the early Archaean era. *Nature*, 440, 516–519 {**305**}

Umebayashi T, 1983, The densities of charged particles in very dense interstellar clouds. *Progress of Theoretical Physics*, 69, 480–502 {**221**}

Umebayashi T, Nakano T, 1981, Fluxes of energetic particles and the ionisation rate in very dense interstellar clouds. *PASJ*, 33, 617–635 {**221**}

Underwood DR, Jones BW, Sleep PN, 2003, The evolution of habitable zones during stellar lifetimes and its implications on the search for extraterrestrial life. *International Journal of Astrobiology*, 2, 289–299 {**283**}

Unwin SC, Shao M, Tanner AM, et al., 2008, Taking the measure of the Universe: precision astrometry with SIM PlanetQuest. *PASP*, 120, 38–88 {**72, 73, 94**}

Urakawa S, Yamada T, Suto Y, et al., 2006, An extrasolar planet transit search with Subaru Suprime-cam. *PASJ*, 58, 869–881 {**110**}

Urey HC, 1952, On the early chemical history of the Earth and the origin of life. *Proceedings of the National Academy of Science*, 38, 351–363 {**305**}

—, 1955, The cosmic abundances of potassium, uranium, and thorium and the heat balances of the Earth, the Moon, and Mars. *Proceedings of the National Academy of Science*, 41, 127–144 {**196, 197**}

Ushikubo T, Kita NT, Cavosie AJ, et al., 2008, Lithium in Jack Hills zircons: evidence for extensive weathering of Earth's earliest crust. *Earth and Planetary Science Letters*, 272, 666–676 {**299**}

Vakili F, Aristidi E, Schmider FX, et al., 2005, KEOPS: towards exo-Earths from Dome C of Antarctica. *EAS Publications Series*, volume 14, 211–217 {**162**}

Valat B, Schmider FX, Lopez B, et al., 2006, Study of the scientific potential of a three 40cm telescopes interferometer at Dome C. *SPIE Conf Ser*, volume 6268, 138 {**162**}

Valencia D, O'Connell RJ, Sasselov D, 2006, Internal structure of massive terrestrial planets. *Icarus*, 181, 545–554 {**260, 263, 267, 268**}

Valencia D, O'Connell RJ, Sasselov DD, 2007a, Inevitability of plate tectonics on super-Earths. *ApJ*, 670, L45–L48 {**249, 250**}

Valencia D, Sasselov DD, O'Connell RJ, 2007b, Detailed models of super-Earths: how well can we infer bulk properties? *ApJ*, 665, 1413–1420 {**260, 263, 267–269**}

—, 2007c, Radius and structure models of the first super-Earth planet. *ApJ*, 656, 545–551 {**249, 267, 268**}

Valenti JA, Butler RP, Marcy GW, 1995, Determining spectrometer instrumental profiles using FTS reference spectra. *PASP*, 107, 966–976 {19}

Valenti JA, Fischer D, Marcy GW, et al., 2009, Two exoplanets discovered at Keck observatory. *ApJ*, 702, 989–997 {15}

Valenti JA, Fischer DA, 2005, Spectroscopic properties of cool stars. I. 1040 F, G, and K dwarfs from Keck, Lick, and AAT planet search programs. *ApJS*, 159, 141–166 {**29, 33, 37, 185, 187–189, 192, 195**}

van Altena WF, Lee JT, Hoffleit ED, 1995, *The General Catalogue of Trigonometric Stellar Parallaxes*. Yale University Observatory, Fourth Edition {**182**}

van Belle GT, 2008, Closure phase signatures of planet transit events. *PASP*, 120, 617–624 {**123, 124**}

van de Kamp P, 1963, Astrometric study of Barnard's star from plates taken with the 24-inch Sproul refractor. *AJ*, 68, 515–521 {64}

—, 1981, *Stellar Paths: Photographic Astrometry with Long-Focus Instruments*. D. Reidel, Dordrecht (Astrophysics and Space Science Library. Volume 85) {**67, 68**}

—, 1982, The planetary system of Barnard's star. *Vistas in Astronomy*, 26, 141–157 {64}

van Eyken JC, Ge J, Mahadevan S, et al., 2004, First planet confirmation with a dispersed fixed-delay interferometer. *ApJ*, 600, L79–L82 {**25, 27**}

van Eyken JC, Ge J, Wan X, et al., 2007, New results from the multi-object Keck exoplanet tracker. *Revista Mexicana de Astronomia y Astrofisica Conference Series*, volume 29, 151–151 {**24, 27**}

van Leeuwen F, 2007, *Hipparcos, the new reduction of the raw data*, volume 350. Astrophysics and Space Science Library {**69, 181**}

van Leeuwen F, Evans DW, Grenon M, et al., 1997, The Hipparcos mission: photometric data. *A&A*, 323, L61–L64 {**184**}

van Straten W, Bailes M, Britton M, et al., 2001, A test of general relativity from the three-dimensional orbital geometry of a binary pulsar. *Nature*, 412, 158–160 {75}

Vance S, Harnmeijer J, Kimura J, et al., 2007, Hydrothermal systems in small ocean planets. *Astrobiology*, 7, 987–1005 {**250**}

Vanderbei RJ, Cady E, Kasdin NJ, 2007, Optimal occulter design for finding extrasolar planets. *ApJ*, 665, 794–798 {**156**}

Vanderbei RJ, Kasdin NJ, Spergel DN, 2004, Checkerboard-mask coronagraphs for high-contrast imaging. *ApJ*, 615, 555–561 {153}

Vanderbei RJ, Spergel DN, Kasdin NJ, 2003a, Circularly symmetric apodisation via star-shaped masks. *ApJ*, 599, 686–694 {153}

—, 2003b, Spiderweb masks for high-contrast imaging. *ApJ*, 590, 593–603 {153}

Vanderbei RJ, Traub WA, 2005, Pupil mapping in two dimensions for high-contrast imaging. *ApJ*, 626, 1079–1090 {155}

Vannier M, Petrov RG, Lopez B, et al., 2006, Colour-differential interferometry for the observation of extrasolar planets. *MNRAS*, 367, 825–837 {162}

Varadi F, Ghil M, Kaula WM, 1999, Jupiter, Saturn, and the edge of chaos. *Icarus*, 139, 286–294 {52}

Varadi F, Runnegar B, Ghil M, 2003, Successive refinements in long-term integrations of planetary orbits. *ApJ*, 592, 620–630 {299}

Varnière P, Blackman EG, Frank A, et al., 2006, Planets rapidly create holes in young circumstellar disks. *ApJ*, 640, 1110–1114 {224}

Vasisht G, Crossfield IJ, Dumont PJ, et al., 2006, Post-coronagraph wavefront sensing for the TMT Planet Formation Imager. *SPIE Conf Ser*, volume 6272, 161 {157, 161}

Vauclair S, 2010, What do stars tell us about planets? Asteroseismology of exoplanet-host stars. *EAS Publications Series*, volume 41, 77–84 {202}

Vauclair S, Laymand M, Bouchy F, et al., 2008, The exoplanet-host star ι Hor: an evaporated member of the primordial Hyades cluster. *A&A*, 482, L5–L8 {204}

Vázquez M, Pallé E, Montañés Rodríguez P, 2010, *The Earth as a Distant Planet*. Springer {289}

Veeder GJ, Matson DL, Johnson TV, et al., 1994, Io's heat flow from infrared radiometry: 1983-1993. *J. Geophys. Res.*, 99, 17095–17162 {282}

Veras D, Armitage PJ, 2004a, The dynamics of two massive planets on inclined orbits. *Icarus*, 172, 349–371 {43}

—, 2004b, Outward migration of extrasolar planets to large orbital radii. *MNRAS*, 347, 613–624 {242}

—, 2005, The influence of massive planet scattering on nascent terrestrial planets. *ApJ*, 620, L111–L114 {243}

—, 2006, Predictions for the correlation between giant and terrestrial extrasolar planets in dynamically evolved systems. *ApJ*, 645, 1509–1515 {241}

Veras D, Crepp JR, Ford EB, 2009, Formation, survival, and detectability of planets beyond 100 AU. *ApJ*, 696, 1600–1611 {242, 243}

Veras D, Ford EB, 2009, Secular evolution of HD 12661: a system caught at an unlikely time. *ApJ*, 690, L1–L4 {52}

Vérinaud C, Hubin N, Kasper M, et al., 2006, The EPICS project: exoplanets detection with OWL. *IAU Colloq. 200: Direct Imaging of Exoplanets: Science and Techniques*, 507–512 {159, 161}

Vérinaud C, Le Louarn M, Korkiakoski V, et al., 2005, Adaptive optics for high-contrast imaging: pyramid sensor versus spatially filtered Shack–Hartmann sensor. *MNRAS*, 357, L26–L30 {151}

Vernet-Viard E, Arcidiacono C, Bagnara P, et al., 2005, Layer-oriented wavefront sensor for a multiconjugate adaptive optics demonstrator. *Optical Engineering*, 44(9), 096601 {152}

Verrier PE, Evans NW, 2006, Planets and asteroids in the γ Cep system. *MNRAS*, 368, 1599–1608 {57, 58, 60}

—, 2007, Planetary stability zones in hierarchical triple star systems. *MNRAS*, 382, 1432–1446 {57}

—, 2008, HD 98800: a most unusual debris disk. *MNRAS*, 390, 1377–1387 {224}

—, 2009, High-inclination planets and asteroids in multistellar systems. *MNRAS*, 394, 1721–1726 {57}

Veselago VG, 1968, The electrodynamics of substances with simultaneously negative values of ϵ and μ. *Soviet Physics Uspekhi*, 10, 509–514 {168}

Vidal-Madjar A, Arnold L, Ehrenreich D, et al., 2010, The Earth as an extrasolar transiting planet: Earth's atmospheric composition and thickness revealed by Lunar eclipse observations. *A&A*, 523, A57 {289}

Vidal-Madjar A, Désert JM, Lecavelier des Etangs A, et al., 2004, Detection of oxygen and carbon in the hydrodynamically escaping atmosphere of the extrasolar planet HD 209458 b. *ApJ*, 604, L69–L72 {2, 139, 140, 281}

Vidal-Madjar A, Lecavelier des Etangs A, Désert JM, et al., 2003, An extended upper atmosphere around the extrasolar planet HD 209458 b. *Nature*, 422, 143–146 {139, 140, 281}

Vidotto AA, Jardine M, Helling C, 2010, Early ultraviolet ingress in WASP-12 b: measuring planetary magnetic fields. *ApJ*, 722, L168–L172 {132, 176}

Vigan A, Langlois M, Moutou C, et al., 2007, Characterising extrasolar planets with long slit spectroscopy. *In the Spirit of Bernard Lyot: The Direct Detection of Planets and Circumstellar Disks in the 21st Century* {158, 159}

—, 2008, Exoplanet characterisation with long slit spectroscopy. *A&A*, 489, 1345–1354 {158, 159}

Vilas F, Chapman CR, Matthews MS, 1988, *Mercury*. University of Arizona Press {293}

Villaver E, Livio M, 2007, Can planets survive stellar evolution? *ApJ*, 661, 1192–1201 {79}

Vinet P, Rose JH, Ferrante J, et al., 1989, Universal features of the equation of state of solids. *Journal of Physics Condensed Matter*, 1, 1941–1963 {267}

Viotti RF, Badiali M, Boattini A, et al., 2003, Wide-field observations at Dome Concordia. *Memorie della Societa Astronomica Italiana Supplement*, 2, 177–180 {110}

Visscher C, Fegley B, 2005, Chemical constraints on the water and total oxygen abundances in the deep atmosphere of Saturn. *ApJ*, 623, 1221–1227 {271}

Visscher C, Lodders K, Fegley B, 2006, Atmospheric chemistry in giant planets, brown dwarfs, and low-mass dwarf stars. II. Sulphur and phosphorus. *ApJ*, 648, 1181–1195 {259}

—, 2010, Atmospheric chemistry in giant planets, brown dwarfs, and low-mass dwarf stars. III. Iron, magnesium, and silicon. *ApJ*, 716, 1060–1075 {259}

Vogt N, 2006, FH Leo, the first dwarf nova member of a multiple star system? *A&A*, 452, 985–986 {179}

Vogt SS, 1987, The Lick Observatory Hamilton echelle spectrometer. *PASP*, 99, 1214–1228 {17}

Vogt SS, Allen SL, Bigelow BC, et al., 1994, HIRES: the high-resolution échelle spectrometer on the Keck 10-m telescope. *SPIE Conf Ser*, volume 2198, 362–375 {17, 24, 25}

Vogt SS, Butler RP, Marcy GW, et al., 2002, Ten low-mass companions from the Keck precision velocity survey. *ApJ*, 568, 352–362 {70, 181}

—, 2005, Five new multicomponent planetary systems. *ApJ*, 632, 638–658 {47, 50, 51, 55}

Vogt SS, Butler RP, Rivera EJ, et al., 2010a, The Lick–Carnegie exoplanet survey: a 3.1 Earth-mass planet in the habitable zone of the nearby M3V star GJ 581. *ApJ*, 723, 954–965 {36, 55, 286, 287}

Vogt SS, Marcy GW, Butler RP, et al., 2000, Six new planets from the Keck precision velocity survey. *ApJ*, 536, 902–914 {2, 24, 25, 33, 47}

Vogt SS, Wittenmyer RA, Butler RP, et al., 2010b, A super-Earth and two Neptunes orbiting the nearby Sun-like star 61 Vir. *ApJ*, 708, 1366–1375 {30, 31, 35}

Voigt H, 1956, Drei-Strom-Modell der Sonnenphotosphäre und Asymmetrie der Linien des infraroten Sauerstoff-Tripletts. *Zeitschrift fur Astrophysik*, 40, 157–190 {23}

von Bloh W, Bounama C, Cuntz M, et al., 2007a, The habitability of super-Earths in GJ 581. *A&A*, 476, 1365–1371 {287}

von Bloh W, Bounama C, Franck S, 2003, Cambrian explosion triggered by geosphere-biosphere feedbacks. *Geophys. Res. Lett.*, 30(18), 180000–1 {307}

—, 2007b, Dynamic habitability for Earth-like planets in 86 extrasolar planetary systems. *Planet. Space Sci.*, 55, 651–660 {285}

von Braun K, Abajian M, Ali B, et al., 2009a, The NStED exoplanet transit survey service. *IAU Symposium*, volume 253 of *IAU Symposium*, 478–481 {105}

von Braun K, Kane SR, Ciardi DR, 2009b, Observational window functions in planet transit surveys. *ApJ*, 702, 779–790 {104}

von Braun K, Lee BL, Seager S, et al., 2005, Searching for planetary transits in Galactic open clusters: EXPLORE/OC. *PASP*, 117, 141–159 {109}

von Braun K, van Belle G, 2008, Directly determined radii and effective surface temperatures of exoplanet host stars. *Bulletin of the American Astronomical Society*, volume 40, 242 {185}

von Hoerner S, 1961, The search for signals from other civilisations. *Science*, 134, 1839–1843 {290}

—, 1973, Astronomical aspects of interstellar communication. *Acta Astronautica*, 18, 421–430 {290}

von Zeipel H, 1924, The radiative equilibrium of a rotating system of gaseous masses. *MNRAS*, 84, 665–683 {**131**}

Vorobyov EI, Basu S, 2010, The burst mode of accretion and disk fragmentation in the early embedded stages of star formation. *ApJ*, 719, 1896–1911 {**236**}

Voyatzis G, 2008, Chaos, order, and periodic orbits in 3:1 resonant planetary dynamics. *ApJ*, 675, 802–816 {**43**}

Voyatzis G, Hadjidemetriou JD, 2005, Symmetric and asymmetric librations in planetary and satellite systems at the 2:1 resonance. *Celestial Mechanics and Dynamical Astronomy*, 93, 263–294 {**42**}

—, 2006, Symmetric and asymmetric 3:1 resonant periodic orbits with an application to the 55 Cnc extrasolar system. *Celestial Mechanics and Dynamical Astronomy*, 95, 259–271 {**49**}

Wadhwa M, Amelin Y, Davis AM, et al., 2007, From dust to planetesimals: implications for the solar protoplanetary disk from short-lived radionuclides. *Protostars and Planets V*, 835–848 {**297**}

Wagner W, Pruß A, 2002, The IAPWS formulation 1995 for the thermodynamic properties of ordinary water substance for general and scientific use. *Journal of Physical and Chemical Reference Data*, 31, 387–535 {**262**}

Wahhaj Z, Koerner DW, Backman DE, et al., 2005, Radial distribution of dust grains around HR 4796A. *ApJ*, 618, 385–396 {**224**}

Wahhaj Z, Koerner DW, Ressler ME, et al., 2003, The inner rings of β Pic. *ApJ*, 584, L27–L31 {**223**}

Wai CM, Wasson JT, 1977, Nebular condensation of moderately volatile elements and their abundances in ordinary chondrites. *Earth and Planetary Science Letters*, 36, 1–13 {**257**}

—, 1979, Nebular condensation of Ga, Ge and Sb and the chemical classification of iron meteorites. *Nature*, 282, 790–793 {**257**}

Walker GAH, 2008, The first high-precision radial velocity search for extrasolar planets. *ArXiv e-prints* {**28**}

Walker GAH, Bohlender DA, Walker AR, et al., 1992, γ Cep: rotation or planetary companion? *ApJ*, 396, L91–L94 {**28**}

Walker GAH, Buchholz V, Fahlman GG, et al., 1973, Hα observations of Algol on 2 September 1972. *AJ*, 78, 681–683 {**19**}

Walker GAH, Croll B, Matthews JM, et al., 2008, MOST detects variability on τ Boo a possibly induced by its planetary companion. *A&A*, 482, 691–697 {**113, 205**}

Walker GAH, Matthews J, Kuschnig R, et al., 2003a, The MOST asteroseismology mission: ultraprecise photometry from space. *PASP*, 115, 1023–1035 {**113**}

Walker GAH, Shkolnik E, Bohlender DA, et al., 2003b, The radial velocity precision of fiber-fed spectrographs. *PASP*, 115, 700–705 {**20**}

Walker GAH, Walker AR, Irwin AW, et al., 1995, A search for Jupiter-mass companions to nearby stars. *Icarus*, 116, 359–375 {**24**}

Walker JCG, Hays PB, Kasting JF, 1981, A negative feedback mechanism for the long-term stabilisation of the earth's surface temperature. *J. Geophys. Res.*, 86, 9776–9782 {**250**}

Walker MA, 1995, Microlensed image motions. *ApJ*, 453, 37–39 {**94**}

Wallace K, Hardy G, Serabyn E, 2000, Deep and stable interferometric nulling of broad-band light with implications for observing planets around nearby stars. *Nature*, 406, 700–702 {**163**}

Walsh D, Carswell RF, Weymann RJ, 1979, 0957+561 AB: twin quasistellar objects or gravitational lens. *Nature*, 279, 381–384 {**84**}

Wambsganss J, 1997, Discovering Galactic planets by gravitational microlensing: magnification patterns and light curves. *MNRAS*, 284, 172–188 {**88, 90**}

—, 2004, Microlensing surveys in search of extrasolar planets. *Extrasolar Planets: Today and Tomorrow*, volume 321 of *ASP Conf Ser*, 47–65 {**84**}

—, 2006, Gravitational Microlensing. *Gravitational Lensing: Strong, Weak and Micro, Saas-Fee Advanced Courses, Volume 33, p. 453*, Springer {**84**}

Wan X, Ge J, Guo P, et al., 2006, A fiber feed system for a multiple object Doppler instrument at Sloan Telescope. *SPIE Conf Ser*, volume 6269, 88–97 {**27**}

Wang JY, Markey JK, 1978, Modal compensation of atmospheric turbulence phase distortion. *Journal of the Optical Society of America*, 68, 78–87 {**151**}

Wang S, Zhao G, Zhou J, 2009, Dynamics and eccentricity formation of planets in OGLE–2006–BLG–109L system. *ApJ*, 706, 772–784 {**99**}

Wang Z, Chakrabarty D, Kaplan DL, 2006, A debris disk around an isolated young neutron star. *Nature*, 440, 772–775 {**77**}

Wang Z, Gurnett DA, Fischer G, et al., 2010, Cassini observations of narrowband radio emissions in Saturn's magnetosphere. *Journal of Geophysical Research (Space Physics)*, 115, 6213–6218 {**174**}

Ward WR, 1973, Large-scale variations in the obliquity of Mars. *Science*, 181, 260–262 {**300, 301**}

—, 1974, Climatic variations on Mars. I. Astronomical theory of insolation. *J. Geophys. Res.*, 79, 3375–3386 {**286**}

—, 1986, Density waves in the solar nebula: differential Lindblad torque. *Icarus*, 67, 164–180 {**238**}

—, 1989, On the rapid formation of giant planet cores. *ApJ*, 345, L99–L102 {**234**}

—, 1997a, Protoplanet migration by nebula tides. *Icarus*, 126, 261–281 {**127, 238, 240**}

—, 1997b, Survival of planetary systems. *ApJ*, 482, L211–L214 {**238**}

—, 1998, On planet formation and migration. *Origins*, volume 148 of *ASP Conf Ser*, 338–346 {**238**}

Ward WR, Hamilton DP, 2004, Tilting Saturn. I. Analytic model. *AJ*, 128, 2501–2509 {**301**}

Ward WR, Hourigan K, 1989, Orbital migration of protoplanets: the inertial limit. *ApJ*, 347, 490–495 {**237**}

Ward WR, Reid MJ, 1973, Solar tidal friction and satellite loss. *MNRAS*, 164, 21–32 {**237**}

Warren SJ, Mortlock DJ, Leggett SK, et al., 2007, A very cool brown dwarf in UKIDSS DR1. *MNRAS*, 381, 1400–1412 {**211**}

Warwick JW, Evans DR, Peltzer GR, et al., 1989, Voyager planetary radio astronomy at Neptune. *Science*, 246, 1498–1501 {**294**}

Warwick JW, Evans DR, Romig JH, et al., 1986, Voyager 2 radio observations of Uranus. *Science*, 233, 102–106 {**294**}

Watanabe M, Takami H, Takato N, et al., 2004, Design of the Subaru laser guide star adaptive optics module. *SPIE Conf Ser*, volume 5490, 1096–1104 {**152**}

Watson AJ, Donahue TM, Walker JCG, 1981, The dynamics of a rapidly escaping atmosphere: applications to the evolution of Earth and Venus. *Icarus*, 48, 150–166 {**281**}

Watson CA, Littlefair SP, Cameron AC, et al., 2010, Estimating the masses of extrasolar planets. *MNRAS*, 408, 1606–1622 {**186**}

Weber EJ, Davis L Jr, 1967, The angular momentum of the solar wind. *ApJ*, 148, 217–227 {**249**}

Weertman J, 1976, Milankovitch solar radiation variations and ice age ice sheet sizes. *Nature*, 261, 17–20 {**301**}

Weidenschilling SJ, 1975, Mass loss from the region of Mars and the asteroid belt. *Icarus*, 26, 361–366 {**298**}

—, 1977a, Aerodynamics of solid bodies in the solar nebula. *MNRAS*, 180, 57–70 {**226**}

—, 1977b, The distribution of mass in the planetary system and solar nebula. *Ap&SS*, 51, 153–158 {**220, 298**}

—, 1980, Dust to planetesimals: settling and coagulation in the solar nebula. *Icarus*, 44, 172–189 {**226**}

—, 2005, Formation of the cores of the outer planets. *Space Science Reviews*, 116, 53–66 {**232**}

Weidenschilling SJ, Donn BD, Meakin P, 1989, The physics of planetesimal formation. *The Formation and Evolution of Planetary Systems*, 131–146, Cambridge University Press {**225**}

Weidenschilling SJ, Marzari F, 1996, Gravitational scattering as a possible origin for giant planets at small stellar distances. *Nature*, 384, 619–621 {**237, 243**}

Weidenschilling SJ, Spaute D, Davis DR, et al., 1997, Accretional evolution of a planetesimal swarm. *Icarus*, 128, 429–455 {**228**}

Weigelt G, Kraus S, Driebe T, et al., 2007, Near-infrared interferometry of η Car with spectral resolutions of 1500 and 12 000 using VLTI–AMBER. *A&A*, 464, 87–106 {**124**}

Weights DJ, Lucas PW, Roche PF, et al., 2009, Infrared spectroscopy and analysis of brown dwarf and planetary mass objects in the Orion nebula cluster. *MNRAS*, 392, 817–846 {**215**}

Weinberger AJ, Becklin EE, Schneider G, et al., 1999, The circumstellar disk of HD 141569 imaged with NICMOS. *ApJ*, 525, L53–L56 {**222–224**}

Weinberger AJ, Becklin EE, Zuckerman B, 2003, The first spatially resolved mid-infrared spectroscopy of β Pic. *ApJ*, 584, L33–L37 {**223**}

Weir ST, Mitchell AC, Nellis WJ, 1996, Metallisation of fluid molecular hydrogen at 140 GPa. *Physical Review Letters*, 76, 1860–1863 {**262, 294**}

Weiss NO, Thompson MJ, 2009, The solar dynamo. *Space Science Reviews*, 144, 53–66 {**66**}

Weldrake DTF, 2008, Searching for planetary transits in star clusters. *ASP Conf Ser*, volume 398, 133–136 {**109**}

Weldrake DTF, Bayliss DDR, Sackett PD, et al., 2008a, Lupus-TR-3b: a low-mass transiting hot Jupiter in the Galactic plane? *ApJ*, 675, L37–L40 {**108, 327**}

Weldrake DTF, Sackett PD, 2005, A method for the detection of planetary transits in large time series data sets. *ApJ*, 620, 1033–1042 {**105**}

Weldrake DTF, Sackett PD, Bridges TJ, 2008b, The frequency of large-radius hot and very hot Jupiters in ω Cen. *ApJ*, 674, 1117–1129 {**109**}

Weldrake DTF, Sackett PD, Bridges TJ, et al., 2005, An absence of hot Jupiter planets in 47 Tuc: results of a wide-field transit search. *ApJ*, 620, 1043–1051 {**109**}

Welsh WF, Orosz JA, Seager S, et al., 2010, The discovery of ellipsoidal variations in the Kepler light curve of HAT–P–7. *ApJ*, 713, L145–L149 {**106, 119, 132**}

Werthimer D, Anderson D, Bowyer CS, et al., 2001, Berkeley radio and optical SETI programmes. *SPIE Conf Ser*, volume 4273, 104–109 {**291**}

Wesson PS, 1984, Protostars and the origin of the angular momentum of the solar system. *Earth Moon and Planets*, 30, 275–280 {**301**}

West AA, Hawley SL, Bochanski JJ, et al., 2008, Constraining the age-activity relation for cool stars: the Sloan Digital Sky Survey data release 5 low-mass star spectroscopic sample. *AJ*, 135, 785–795 {**281**}

West RG, Anderson DR, Gillon M, et al., 2009a, The low density transiting exoplanet WASP–15 b. *AJ*, 137, 4834–4836 {**327**}

West RG, Collier Cameron A, Hebb L, et al., 2009b, The sub-Jupiter mass transiting exoplanet WASP–11 b. *A&A*, 502, 395–400 {**327**}

Wetherill GW, 1980, Formation of the terrestrial planets. *ARA&A*, 18, 77–113 {**280**}

—, 1990, Formation of the Earth. *Annual Review of Earth and Planetary Sciences*, 18, 205–256 {**225**}

—, 1994, Possible consequences of absence of Jupiters in planetary systems. *Ap&SS*, 212, 23–32 {**285**}

—, 1996, The formation and habitability of extrasolar planets. *Icarus*, 119, 219–238 {**225, 285**}

Wetherill GW, Stewart GR, 1989, Accumulation of a swarm of small planetesimals. *Icarus*, 77, 330–357 {**227**}

—, 1993, Formation of planetary embryos: effects of fragmentation, low relative velocity, and independent variation of eccentricity and inclination. *Icarus*, 106, 190–209 {**228**}

Whipple FL, 1966, Chondrules: suggestion concerning the origin. *Science*, 153, 54–56 {**297**}

White RJ, Ghez AM, 2001, Observational constraints on the formation and evolution of binary stars. *ApJ*, 556, 265–295 {**56**}

Whitehouse SC, Bate MR, 2006, The thermodynamics of collapsing molecular cloud cores using smoothed particle hydrodynamics with radiative transfer. *MNRAS*, 367, 32–38 {**214**}

Whitmire DP, Matese JJ, Criswell L, et al., 1998, Habitable planet formation in binary star systems. *Icarus*, 132, 196–203 {**58**}

Whitworth AP, Bate MR, Nordlund Å, et al., 2007, The formation of brown dwarfs: theory. *Protostars and Planets V*, 459–476 {**214**}

Whitworth AP, Stamatellos D, 2006, The minimum mass for star formation, and the origin of binary brown dwarfs. *A&A*, 458, 817–829 {**214**}

Whitworth AP, Zinnecker H, 2004, The formation of free-floating brown dwarves and planetary-mass objects by photo-erosion of prestellar cores. *A&A*, 427, 299–306 {**215**}

Wiedemann G, Deming D, Bjoraker G, 2001, A sensitive search for methane in the infrared spectrum of τ Boo. *ApJ*, 546, 1068–1074 {**125**}

Wiegert PA, Holman MJ, 1997, The stability of planets in the α Cen system. *AJ*, 113, 1445–1450 {**57**}

Wiegert PA, Innanen KA, Mikkola S, 1997, An asteroidal companion to the Earth. *Nature*, 387, 685–686 {**52**}

Wielen R, Fuchs B, Dettbarn C, 1996, On the birth-place of the Sun and the places of formation of other nearby stars. *A&A*, 314, 438–447 {**193, 194**}

Wigley TML, Brimblecombe P, 1981, Carbon dioxide, ammonia and the origin of life. *Nature*, 291, 213–215 {**306**}

Wigner E, Huntington HB, 1935, On the possibility of a metallic modification of hydrogen. *J. Chem. Phys.*, 3, 764–770 {**262, 294**}

Wiktorowicz SJ, 2009, Non-detection of polarised, scattered light from the HD 189733 b hot Jupiter. *ApJ*, 696, 1116–1124 {**126**}

Wiktorowicz SJ, Ingersoll AP, 2007, Liquid water oceans in ice giants. *Icarus*, 186, 436–447 {**269**}

Wilde SA, Valley JW, Peck WH, et al., 2001, Evidence from detrital zircons for the existence of continental crust and oceans on the Earth 4.4 Gyr ago. *Nature*, 409, 175–178 {**297, 299**}

Wildman RA, Hickey LJ, Dickinson MB, et al., 2004, Burning of forest materials under late Paleozoic high atmospheric oxygen levels. *Geology*, 32, 457–460 {**307**}

Wildt R, 1933, Kondensation in Sternatmosphären. *ZAp*, 6, 345–354 {**257**}

Will CM, 1993, *Theory and Experiment in Gravitational Physics*. Cambridge University Press, Second Edition {**64, 84, 133**}

—, 2003, Propagation speed of gravity and the relativistic time delay. *ApJ*, 590, 683–690 {**309**}

Willems B, Kolb U, Justham S, 2006, Eclipsing binaries in extrasolar planet transit surveys: the case of SuperWASP. *MNRAS*, 367, 1103–1112 {**105**}

Willems B, Van Hoolst T, Smeyers P, et al., 1997, On the possibility of a tidally excited low-frequency g-mode in 51 Peg. *A&A*, 326, L37–L40 {**28**}

Willes AJ, Wu K, 2004, Electron-cyclotron maser emission from white dwarf pairs and white dwarf planetary systems. *MNRAS*, 348, 285–296 {**176**}

—, 2005, Radio emissions from terrestrial planets around white dwarfs. *A&A*, 432, 1091–1100 {**176**}

Williams DM, Gaidos E, 2008, Detecting the glint of starlight on the oceans of distant planets. *Icarus*, 195, 927–937 {**131**}

Williams DM, Kasting JF, 1997, Habitable planets with high obliquities. *Icarus*, 129, 254–267 {**286**}

Williams DM, Kasting JF, Wade RA, 1997, Habitable moons around extrasolar giant planets. *Nature*, 385, 234–236 {**250, 283**}

Williams DM, Pollard D, 2002, Earth-like worlds on eccentric orbits: excursions beyond the habitable zone. *International Journal of Astrobiology*, 1, 61–69 {**284**}

Williams GE, 1993, History of the Earth's obliquity. *Earth Science Reviews*, 34, 1–45 {**301**}

Williams IP, 1997, The trans-Neptunian region. *Reports on Progress in Physics*, 60, 1–22 {**295**}

Williams JP, Andrews SM, Wilner DJ, 2005, The masses of the Orion proplyds from submillimeter dust emission. *ApJ*, 634, 495–500 {**220**}

Williams PKG, Charbonneau D, Cooper CS, et al., 2006, Resolving the surfaces of extrasolar planets with secondary eclipse light curves. *ApJ*, 649, 1020–1027 {**138**}

Williams S, Woolfson MM, 1983, Planetary spin and satellite formation. *MNRAS*, 204, 853–863 {**225**}

Wilner DJ, Holman MJ, Kuchner MJ, et al., 2002, Structure in the dusty debris around Vega. *ApJ*, 569, L115–L119 {**224**}

Wilson DM, Gillon M, Hellier C, et al., 2008a, WASP-4 b: a 12th magnitude transiting hot Jupiter in the southern hemisphere. *ApJ*, 675, L113–L116 {**327**}

Wilson IRG, Carter BD, Waite IA, 2008b, Does a spin-orbit coupling between the Sun and the Jovian planets govern the solar cycle? *Publ. Astron. Soc. Australia*, 25, 85–93 {**67**}

Wilson RE, 1990, Accuracy and efficiency in the binary star reflection effect. *ApJ*, 356, 613–622 {**132**}

Wilson RN, 1991, Active optics and the New Technology Telescope (NTT): the key to improved optical quality at lower cost in large astronomical telescopes. *Contemporary Physics*, 32, 157–172 {**150**}

Winglee RM, Dulk GA, Bastian TS, 1986, A search for cyclotron maser radiation from substellar and planet-like companions of nearby stars. *ApJ*, 309, L59–L62 {**176**}

Winn JN, 2008, Precise photometry and spectroscopy of transits. *ASP Conf Ser*, volume 398, 101–108 {**127**}

—, 2009, Measuring accurate transit parameters. *IAU Symposium*, volume 253 of *IAU Symposium*, 99–109 {**117**}

Winn JN, Fabrycky D, Albrecht S, et al., 2010a, Hot stars with hot Jupiters have high obliquities. *ApJ*, 718, L145–L149 {**129, 130**}

Winn JN, Henry GW, Torres G, et al., 2008a, Five new transits of the super-Neptune HD 149026 b. *ApJ*, 675, 1531–1537 {**127**}

Winn JN, Holman MJ, 2005, Obliquity tides on hot Jupiters. *ApJ*, 628, L159–L162 {**145, 246**}

Winn JN, Holman MJ, Henry GW, et al., 2007, The transit light curve project. V. System parameters and stellar rotation period of HD 189733. *AJ*, 133, 1828–1835 {**119, 128, 140**}

Winn JN, Howard AW, Johnson JA, et al., 2011, Orbital orientations of exoplanets: HAT–P–4 b is prograde and HAT–P–14 b is retrograde. *AJ*, 141, 63–67 {**106, 129, 130**}

Winn JN, Johnson JA, Albrecht S, et al., 2009, HAT–P–7 b: a retrograde or polar orbit, and a third body. *ApJ*, 703, L99–L103 {**2, 129**}

Winn JN, Johnson JA, Howard AW, et al., 2010b, The HAT–P–13 exoplanetary system: evidence for spin-orbit alignment and a third companion. *ApJ*, 718, 575–582 {**129**}

—, 2010c, The oblique orbit of the super-Neptune HAT–P–11 b. *ApJ*, 723, L223–L227 {**129, 131**}

Winn JN, Johnson JA, Marcy GW, et al., 2006, Measurement of the spin-orbit alignment in the exoplanetary system HD 189733. *ApJ*, 653, L69–L72 {**128, 129**}

Winn JN, Johnson JA, Narita N, et al., 2008b, The prograde orbit of exoplanet TrES–2 b. *ApJ*, 682, 1283–1288 {**129**}

Winn JN, Noyes RW, Holman MJ, et al., 2005, Measurement of spin-orbit alignment in an extrasolar planetary system. *ApJ*, 631, 1215–1226 {**121, 127, 128**}

Winn JN, Suto Y, Turner EL, et al., 2004, A search for Hα absorption in the exosphere of the transiting extrasolar planet HD 209458 b. *PASJ*, 56, 655–662 {**140**}

Wisdom J, 1983, Chaotic behaviour and the origin of the 3:1 Kirkwood gap. *Icarus*, 56, 51–74 {**45**}

—, 1987a, Chaotic behaviour in the solar system. *Royal Society of London Proceedings Series A*, 413, 109–129 {**45**}

—, 1987b, Chaotic dynamics in the solar system. *Icarus*, 72, 241–275 {**45**}

—, 2004, Spin-orbit secondary resonance dynamics of Enceladus. *AJ*, 128, 484–491 {**250**}

Wisdom J, Holman M, 1991, Symplectic maps for the N-body problem. *AJ*, 102, 1528–1538 {**44, 299**}

Wisdom J, Peale SJ, Mignard F, 1984, The chaotic rotation of Hyperion. *Icarus*, 58, 137–152 {**45**}

Witt HJ, 1990, Investigation of high amplification events in light curves of gravitationally lensed quasars. *A&A*, 236, 311–322 {**89**}

Wittenmyer RA, Endl M, Cochran WD, 2007, Long-period objects in the extrasolar planetary systems 47 UMa and 14 Her. *ApJ*, 654, 625–632 {**36, 52**}

Wittenmyer RA, Endl M, Cochran WD, et al., 2006, Detection limits from the McDonald Observatory planet search programme. *AJ*, 132, 177–188 {**36**}

—, 2009, A search for multi-planet systems using the Hobby–Eberly telescope. *ApJS*, 182, 97–119 {**50**}

Wittenmyer RA, Welsh WF, Orosz JA, et al., 2005, System parameters of the transiting extrasolar planet HD 209458 b. *ApJ*, 632, 1157–1167 {**132, 140**}

Wizinowich P, Acton DS, Shelton C, et al., 2000, First light adaptive optics images from the Keck II telescope: a new era of high angular resolution imagery. *PASP*, 112, 315–319 {**151**}

Wizinowich P, Akeson R, Colavita M, et al., 2006a, Recent progress at the Keck Interferometer. *SPIE Conf Ser*, volume 6268, 21 {**162**}

Wizinowich P, Le Mignant D, Bouchez AH, et al., 2006b, The W.M. Keck Observatory laser guide star adaptive optics system: overview. *PASP*, 118, 297–309 {**152**}

Woillez J, Akeson R, Colavita M, et al., 2010, ASTRA: astrometry and phase-referencing astronomy on the Keck interferometer. *SPIE Conf Ser*, volume 7734 {**69**}

Wolf AS, Laughlin G, Henry GW, et al., 2007, A determination of the spin-orbit alignment of the anomalously dense planet orbiting HD 149026. *ApJ*, 667, 549–556 {**129**}

Wolf R, 1859, Extract of a letter to Mr. Carrington. *MNRAS*, 19, 85–86 {**66**}

Wolf S, 2008, Detecting protoplanets with ALMA. *Ap&SS*, 313, 109–112 {**177**}

Wolf S, D'Angelo G, 2005, On the observability of giant protoplanets in circumstellar disks. *ApJ*, 619, 1114–1122 {**177, 224, 240**}

Wolf S, Gueth F, Henning T, et al., 2002, Detecting planets in protoplanetary disks: a prospective study. *ApJ*, 566, L97–L99 {**224, 240**}

Wolff EW, Barbante C, Becagli S, et al., 2010, Changes in environment over the last 800 000 years from chemical analysis of the EPICA Dome C ice core. *Quaternary Science Reviews*, 29, 285–295 {**307**}

Wolff SC, Strom SE, Hillenbrand LA, 2004, The angular momentum evolution of $0.1 - 10 M_\odot$ stars from the birth line to the main sequence. *ApJ*, 601, 979–999 {**200**}

Wolfram S, 2010, The Wolfram demonstrations project. *http://demonstrations.wolfram.com/SolarSystemMandalas/* {**66**}

Wolk SJ, Walter FM, 1996, A search for protoplanetary disks around naked T Tauri stars. *AJ*, 111, 2066–2076 {**222**}

Wolstencroft RD, Raven JA, 2002, Photosynthesis: likelihood of occurrence and possibility of detection on Earth-like planets. *Icarus*, 157, 535–548 {**289**}

Wolszczan A, 1994a, Confirmation of Earth-mass planets orbiting the millisecond pulsar PSR B1257+12. *Science*, 264, 538–542 {**76, 77, 79**}

—, 1994b, Toward planets around neutron stars. *Ap&SS*, 212, 67–75 {**76**}

—, 1997, The pulsar planets update. *Planets Beyond the Solar System and the Next Generation of Space Missions*, volume 119 of *ASP Conf Ser*, 135–138 {**75, 76**}

—, 2008, Planets around the pulsar PSR B1257+12. *ASP Conf Ser*, volume 398, 3–12 {**76, 77**}

Wolszczan A, Frail DA, 1992, A planetary system around the millisecond pulsar PSR B1257+12. *Nature*, 355, 145–147 {**2, 75, 79**}

Wood BE, Müller H, Zank GP, et al., 2002a, Measured mass-loss rates of solar-like stars as a function of age and activity. *ApJ*, 574, 412–425 {**176**}

Wood K, Lada CJ, Bjorkman JE, et al., 2002b, Infrared signatures of protoplanetary disk evolution. *ApJ*, 567, 1183–1191 {**224**}

Wood RM, Wood KD, 1965, Solar motion and sunspot comparison. *Nature*, 208, 129–131 {**66**}

Wooden D, Desch S, Harker D, et al., 2007, Comet grains and implications for heating and radial mixing in the protoplanetary disk. *Protostars and Planets V*, 815–833 {**297**}

Woodward CE, Shull JM, Thronson HA (eds.), 1998, *Origins*, volume 148 of *ASP Conf Ser* {**282**}

Woolf NJ, 2001, *Very large optics for the study of extrasolar terrestrial planets: Life Finder*. NASA Institute for Advanced Concepts {**167**}

Woolf NJ, Angel JRP, 1997, Planet Finder options. I. New linear nulling array configurations. *Planets Beyond the Solar System and the Next Generation of Space Missions*, volume 119 of *ASP Conf Ser*, 285–293 {**164, 288**}

Woolf NJ, Smith PS, Traub WA, et al., 2002, The spectrum of Earthshine: a pale blue dot observed from the ground. *ApJ*, 574, 430–433 {**289**}

Woolfson MM, 1964, A capture theory of the origin of the solar system. *Royal Society of London Proceedings Series A*, 282, 485–507 {**225**}

—, 1993, The solar system: its origin and evolution. *QJRAS*, 34, 1–20 {**225**}

—, 2000a, The origin and evolution of the solar system. *Astronomy and Geophysics*, 41(1), 12–19 {**225**}

—, 2000b, *The Origin and Evolution of the Solar System*. Institute of Physics Publishing {**225**}

—, 2003, Commentary on an ailing theory. *Space Sci. Rev.*, 107, 651–663 {**225**}

—, 2004, The stability of evolving planetary orbits in an embedded cluster. *MNRAS*, 348, 1150–1156 {**77**}

—, 2007, *The Formation of the Solar System: Theories Old and New*. Imperial College Press {**225**}

Woolley RvdR, 1970, Catalogue of stars within twenty-five parsecs of the Sun. *Royal Observatory Annals*, 5 {**182**}

Wozniak P, Paczyński B, 1997, Microlensing of blended stellar images. *ApJ*, 487, 55–60 {**92**}

Wright EL, 2008, WISE the Wide-field Infrared Survey Explorer. *EAS Publications Series*, volume 33, 57–62 {**211**}

—, 2010a, WISE, the Wide-field Infrared Survey Explorer: status and early results. *American Astronomical Society Meeting Abstracts*, volume 216, 104.01 {**211**}

Wright JT, 2005, Radial velocity jitter in stars from the California and Carnegie planet search at Keck Observatory. *PASP*, 117, 657–664 {**21, 22, 32**}

—, 2010b, A survey of multiple planet systems. *EAS Publications Series*, volume 42 of *EAS Publications Series*, 3–17 {**38, 43**}

Wright JT, Fischer DA, Ford EB, et al., 2009a, A third giant planet orbiting HIP 14810. *ApJ*, 699, L97–L101 {**38**}

Wright JT, Howard AW, 2009, Efficient fitting of multiplanet Keplerian models to radial velocity and astrometry data. *ApJS*, 182, 205–215 {**13, 14, 67, 68**}

Wright JT, Marcy GW, Butler RP, et al., 2004, Chromospheric Ca II emission in nearby F, G, K, and M stars. *ApJS*, 152, 261–295 {**35, 187**}

Wright JT, Marcy GW, Fischer DA, et al., 2007, Four new exoplanets and hints of additional substellar companions to exoplanet host stars.

ApJ, 657, 533–545 {**16, 37, 55**}

Wright JT, Upadhyay S, Marcy GW, et al., 2009b, Ten new and updated multiplanet systems and a survey of exoplanetary systems. *ApJ*, 693, 1084–1099 {**39, 43, 45, 55**}

Wu Y, Murray N, 2003, Planet migration and binary companions: the case of HD 80606 b. *ApJ*, 589, 605–614 {**60**}

Wu Y, Murray NW, Ramsahai JM, 2007, Hot Jupiters in binary star systems. *ApJ*, 670, 820–825 {**127, 129**}

Wuchterl G, 1990, Hydrodynamics of giant planet formation. I. Overviewing the kappa-mechanism. *A&A*, 238, 83–94 {**232**}

—, 1991a, Hydrodynamics of giant planet formation. II. Model equations and critical mass. *Icarus*, 91, 39–64 {**232**}

—, 1991b, Hydrodynamics of giant planet formation. III. Jupiter's nucleated instability. *Icarus*, 91, 53–64 {**232**}

Wuchterl G, Tscharnuter WM, 2003, From clouds to stars: protostellar collapse and the evolution to the pre-main sequence I. Equations and evolution in the Hertzsprung–Russell diagram. *A&A*, 398, 1081–1090 {**218**}

Wurm G, 2007, Light-induced disassembly of dusty bodies in inner protoplanetary disks: implications for the formation of planets. *MNRAS*, 380, 683–690 {**227**}

Wurm G, Blum J, 2006, Experiments on planetesimal formation. *Planet Formation*, 90–111, Cambridge University Press {**226**}

Wyatt MC, 2003, Resonant trapping of planetesimals by planet migration: debris disk clumps and Vega's similarity to the solar system. *ApJ*, 598, 1321–1340 {**224, 242**}

Wyatt MC, Clarke CJ, Greaves JS, 2007, Origin of the metallicity dependence of exoplanet host stars in the protoplanetary disk mass distribution. *MNRAS*, 380, 1737–1743 {**191**}

Wyatt MC, Dent WRF, 2002, Collisional processes in extrasolar planetesimal disks: dust clumps in Fomalhaut's debris disk. *MNRAS*, 334, 589–607 {**223**}

Wyatt SP, Whipple FL, 1950, The Poynting–Robertson effect on meteor orbits. *ApJ*, 111, 134–141 {**178**}

Xie J, Zhou J, 2009, Planetesimal accretion in binary systems: role of the companion's orbital inclination. *ApJ*, 698, 2066–2074 {**58**}

Xie JW, Zhou JL, 2008, Planetesimal accretion in binary systems: the effects of gas dissipation. *ApJ*, 686, 570–579 {**58**}

Yang W, Kostinski AB, 2004, One-sided achromatic phase apodisation for imaging of extrasolar planets. *ApJ*, 605, 892–901 {**153**}

Yantis WF, Sullivan WT III, Erickson WC, 1977, A search for extrasolar Jovian planets by radio techniques. *Bulletin of the American Astronomical Society*, volume 9, 453 {**176**}

Yee JC, Gaudi BS, 2008, Characterising long-period transiting planets observed by Kepler. *ApJ*, 688, 616–627 {**111**}

Yee JC, Udalski A, Sumi T, et al., 2009, Extreme magnification microlensing event OGLE–2008–BLG–279: strong limits on planetary companions to the lens star. *ApJ*, 703, 2082–2090 {**94**}

Yeghikyan A, Fahr H, 2004a, Effects induced by the passage of the Sun through dense molecular clouds. I. Flow outside of the compressed heliosphere. *A&A*, 415, 763–770 {**307**}

—, 2004b, Terrestrial atmospheric effects induced by counterstreaming dense interstellar cloud material. *A&A*, 425, 1113–1118 {**307**}

Yelle RV, 2004, Aeronomy of extrasolar giant planets at small orbital distances. *Icarus*, 170, 167–179 {**281**}

Yelle RV, Lammer H, Ip WH, 2008, Aeronomy of extrasolar giant planets. *Space Science Reviews*, 139, 437–451 {**281**}

Yin Q, Jacobsen SB, Yamashita K, et al., 2002, A short timescale for terrestrial planet formation from Hf–W chronometry of meteorites. *Nature*, 418, 949–952 {**297**}

Yock P, 2006, Detecting Earth-like extrasolar planets from Antarctica by gravitational microlensing. *Acta Astronomica Sinica*, 47, 410–417 {**102**}

Yoder CF, 1979, How tidal heating in Io drives the Galilean orbital resonance locks. *Nature*, 279, 767–770 {**250**}

—, 1995, *Astrometric and Geodetic Properties of Earth and the Solar System*. American Geophysical Union, editor T. J. Ahrens {**311**}

Yoder CF, Peale SJ, 1981, The tides of Io. *Icarus*, 47, 1–35 {**247, 250**}

Yoo J, DePoy DL, Gal-Yam A, et al., 2004a, Constraints on planetary companions in the magnification A = 256 microlensing event OGLE–2003–BLG–423. *ApJ*, 616, 1204–1214 {**96**}

—, 2004b, OGLE–2003–BLG–262: finite-source effects from a point-mass lens. *ApJ*, 603, 139–151 {**91, 97**}

Yoshizawa M, Nishikawa J, Ohishi N, et al., 2006, MIRA status report: recent progress of MIRA-I.2 and future plans. *SPIE Conf Ser*, volume 6268, 8 {**162**}

You XP, Hobbs GB, Coles WA, et al., 2007, An improved solar wind electron density model for pulsar timing. *ApJ*, 671, 907–911 {**79**}

Youdin AN, 2010, From grains to planetesimals. *EAS Publications Series*, volume 41, 187–207 {**225, 226**}

Youdin AN, Chiang EI, 2004, Particle pileups and planetesimal formation. *ApJ*, 601, 1109–1119 {**226**}

Young AT, 1967, Photometric error analysis. VI. Confirmation of Reiger's theory of scintillation. *AJ*, 72, 747–753 {**116**}

Yu Q, Tremaine S, 2001, Resonant capture by inward-migrating planets. *AJ*, 121, 1736–1740 {**127**}

Yuan X, Cui X, Gong X, et al., 2010, Progress of Antarctic Schmidt telescopes (AST3) for Dome A. *SPIE Conf Ser*, volume 7733, 57 {**110**}

Yung YL, Demore WB, 1999, *Photochemistry of Planetary Atmospheres*. Oxford University Press {**277**}

Zacharias N, Dorland B, 2006, The concept of a stare-mode astrometric space mission. *PASP*, 118, 1419–1427 {**73**}

Zahn J, 1977, Tidal friction in close binary stars. *A&A*, 57, 383–394 {**245, 246**}

Zahnle KJ, 2008, Atmospheric chemistry: her dark materials. *Nature*, 454, 41–42 {**278**}

Zahnle KJ, Arndt N, Cockell C, et al., 2007, Emergence of a habitable planet. *Space Sci. Rev.*, 129, 35–78 {**284, 305**}

Zahnle KJ, Kasting JF, Pollack JB, 1988, Evolution of a steam atmosphere during Earth's accretion. *Icarus*, 74, 62–97 {**269**}

Zahnle KJ, Marley MS, Freedman RS, et al., 2009, Atmospheric sulphur photochemistry on hot Jupiters. *ApJ*, 701, L20–L24 {**142**}

Zahnle KJ, Pollack JB, Kasting JF, 1990, Mass fractionation of noble gases in diffusion-limited hydrodynamic hydrogen escape. *Icarus*, 84, 503–527 {**281**}

Zahnle KJ, Schaefer L, Fegley B, 2010, Earth's earliest atmosphere. *Cold Spring Harbor Prespectives in Biology*, 2, a004895 {**278, 305**}

Zapatero Osorio MR, Béjar VJS, Martín EL, et al., 2000, Discovery of young, isolated planetary mass objects in the σ Ori star cluster. *Science*, 290, 103–107 {**215**}

—, 2002, A methane, isolated, planetary-mass object in Orion. *ApJ*, 578, 536–542 {**2, 215**}

Zapatero Osorio MR, Caballero JA, Béjar VJS, et al., 2007, Disks of planetary-mass objects in σ Ori. *A&A*, 472, L9–L12 {**214**}

Zapatero Osorio MR, Martín EL, del Burgo C, et al., 2009, Infrared radial velocities of vB 10. *A&A*, 505, L5–L8 {**69**}

Zapolsky HS, Salpeter EE, 1969, The mass–radius relation for cold spheres of low mass. *ApJ*, 158, 809–813 {**260**}

Zaqarashvili TV, 1997, On a possible generation mechanism for the solar cycle. *ApJ*, 487, 930–935 {**67**}

Zaqarashvili TV, Javakhishvili G, Belvedere G, 2002, On a mechanism for enhancing magnetic activity in tidally interacting binaries. *ApJ*, 579, 810–816 {**205**}

Zarka P, 1992, The auroral radio emissions from planetary magnetospheres: what do we know, what don't we know, what do we learn from them? *Advances in Space Research*, 12, 99–115 {**174, 175**}

—, 2004, Non-thermal radio emissions from extrasolar planets. *Extrasolar Planets: Today and Tomorrow*, volume 321 of *ASP Conf Ser*, 160–169 {**174, 177**}

—, 2007, Plasma interactions of exoplanets with their parent star and associated radio emissions. *Planet. Space Sci.*, 55, 598–617 {**176**}

Zarka P, Cecconi B, Kurth WS, 2004, Jupiter's low-frequency radio spectrum from Cassini–Radio and Plasma Wave Science absolute flux density measurements. *Journal of Geophysical Research (Space Physics)*, 109(18), 9–26 {**174**}

Zarka P, Halbwachs JL, 2006, Plasma interactions of exoplanets with their parent stars and associated radio emissions. *Formation Planétaire et Exoplanètes, Ecole Thématique du CNRS, Goutelas*, 28, 190–241 {**176**}

Zarka P, Kurth WS, 2005, Radio wave emission from the outer planets before Cassini. *Space Sci. Rev.*, 116, 371–397 {**174**}

Zarka P, Treumann RA, Ryabov BP, et al., 2001, Magnetically-driven planetary radio emissions and application to extrasolar planets. *Ap&SS*, 277, 293–300 {**174–176**}

Zebker HA, Marouf EA, Tyler GL, 1985, Saturn's rings: particle size distributions for thin layer model. *Icarus*, 64, 531–548 {**226**}

Zechmeister M, Kürster M, 2009, The generalised Lomb–Scargle periodogram: a new formalism for the floating-mean and Keplerian periodograms. *A&A*, 496, 577–584 {**13**}

Zechmeister M, Kürster M, Endl M, 2009, The M dwarf planet search programme at the ESO VLT–UVES. A search for terrestrial planets in the habitable zone of M dwarfs. *A&A*, 505, 859–871 {**33**}

Zhai C, Yu J, Shao M, et al., 2008, Picometer accuracy white light fringe modeling for SIM PlanetQuest spectral calibration development unit. *SPIE Conf Ser*, volume 7013, 157 {**72**}

Zhang B, Sigurdsson S, 2003, Electromagnetic signals from planetary collisions. *ApJ*, 596, L95–L98 {**177**}

Zhang XD, Zhou JL, 2006, The formation and stability of the configuration of the planetary system HD 12661. *Chinese Astronomy and Astrophysics*, 30, 420–430 {**52**}

Zhao G, Chen YQ, Qiu HM, et al., 2002, Chemical abundances of 15 extrasolar planet host stars. *AJ*, 124, 2224–2232 {**188**}

Zhao M, Monnier JD, ten Brummelaar T, et al., 2008, Exoplanet studies with CHARA–MIRC. *SPIE Conf Ser*, volume 7013 {**124, 125**}

Zharkov VN, Gudkova TV, 1991, Models of giant planets with a variable ratio of ice to rock. *Annales Geophysicae*, 9, 357–366 {**294**}

Zharkov VN, Trubitsyn VP, 1971, The figure of planets with a uniform or two-component density distribution. *Soviet Ast.*, 14, 1012–1018 {**270, 294**}

—, 1974, Determination of the equation of state of the molecular envelopes of Jupiter and Saturn from their gravitational moments. *Icarus*, 21, 152–156 {**270**}

—, 1978, *Physics of Planetary Interiors*. Pachart, Tucson {**270**}

Zhou JL, Lin DNC, 2007, Planetesimal accretion onto growing proto-gas giant planets. *ApJ*, 666, 447–465 {**232**}

Zhou JL, Lin DNC, Sun YS, 2007, Post-oligarchic evolution of protoplanetary embryos and the stability of planetary systems. *ApJ*, 666, 423–435 {**228**}

Zhou JL, Sun YS, 2003, Occurrence and stability of apsidal resonance in multiple planetary systems. *ApJ*, 598, 1290–1300 {**42**}

Ziglin SL, 1975, Secular evolution of the orbit of a planet in a binary-star system. *Soviet Astronomy Letters*, 1, 194–195 {**57**}

Zinnecker H, 1982, Prediction of the protostellar mass spectrum in the Orion near-infrared cluster. *Annals of the New York Academy of Sciences*, 395, 226–235 {**217**}

Zollinger R, Armstrong JC, 2009, Additional planets in the habitable zone of GJ 581? *A&A*, 497, 583–587 {**287**}

Zsom A, Ormel CW, Güttler C, et al., 2010, The outcome of protoplanetary dust growth: pebbles, boulders, or planetesimals? II. Introducing the bouncing barrier. *A&A*, 513, A57 {**226**}

Zucker S, Mazeh T, 2000, Analysis of the Hipparcos measurements of HD 10697: a mass determination of a brown dwarf secondary. *ApJ*, 531, L67–L69 {**70**}

—, 2001, Analysis of the Hipparcos observations of the extrasolar planets and the brown dwarf candidates. *ApJ*, 562, 549–557 {**70**}

—, 2002, On the mass-period correlation of the extrasolar planets. *ApJ*, 568, L113–L116 {**37, 59, 146**}

Zucker S, Mazeh T, Alexander T, 2007, Beaming binaries: a new observational category of photometric binary stars. *ApJ*, 670, 1326–1330 {**126**}

Zucker S, Mazeh T, Santos NC, et al., 2003, Multi-order TODCOR: application to observations taken with the CORALIE echelle spectrograph. I. The system HD 41004. *A&A*, 404, 775–781 {**22**}

—, 2004, Multi-order TODCOR: application to observations taken with the CORALIE echelle spectrograph. II. A planet in the system HD 41004. *A&A*, 426, 695–698 {**22**}

Zuckerman B, 2001, Dusty circumstellar disks. *ARA&A*, 39, 549–580 {**178**}

Zuckerman B, Becklin EE, 1987a, A search for brown dwarfs and late M dwarfs in the Hyades and the Pleiades. *ApJ*, 319, L99–L102 {**209**}

—, 1987b, Excess infrared radiation from a white dwarf: an orbiting brown dwarf? *Nature*, 330, 138–140 {**210**}

Zuckerman B, Song I, 2004, Dusty debris disks as signposts of planets: implications for Spitzer space telescope. *ApJ*, 603, 738–743 {**224**}

Zwicky F, 1937a, Nebulae as gravitational lenses. *Physical Review*, 51, 290–290 {**84**}

—, 1937b, On the probability of detecting nebulae which act as gravitational lenses. *Physical Review*, 51, 679–679 {**84**}

Index

albedos, 124, 278
ALMA, astrometry/imaging, 177
Antarctic
    astrometry, 63
    imaging, 161–162
    transits, 110
apocentre/apoapsis, 9
apsis/apsides, 9
argument of pericentre, 10
asteroseismology, *see* host stars
astrometry, 61–73
    aberration, 64
    accuracy from the ground, 62–63
    accuracy limits, 65
    Antarctic, 63
    astrometric signature, 61
    barycentric motion, 66
    early investigations, 64
    effects due to disk instabilities, 65
    future observations from space, 71–73
    ground-based searches, 68–69
    instruments/programmes
        ALMA, 177
        AMEX, 73
        ASTRA, 69
        CAPS, 69
        CHARA, 123–125, 162, 184, 185
        DIVA, 73
        ESPRI, 69
        FAME, 73
        Gaia, 72
        Hipparcos, 69–70
        HST–FGS, *see* HST–FGS
        JASMINE, 73
        JMAPS, 73
        LIDA, 73
        Nano-JASMINE, 73
        OBSS, 73
        OSIRIS, 73
        PHASES, 69
        SIM PlanetQuest/SIM Lite, 72–73
        STEPS, 68
        VLT–PRIMA, 69
    interferometry, 63
    limits due to surface structure, 65
    microarcsec accuracy, 64–65
    multiple planets, 66–68
        linearisation, 68
        modeling, 67–68
        relative orbit inclination, 67
    perspective acceleration, 64
    secular change in parallax, 64
    source motion, 64
    space-based searches, 69–73
    Thiele–Innes constants, 68
atmospheres, *see* interiors and atmospheres

BD Catalogue, 6
binary and multiple stars, 206–207
    planets around, 55–60
        circumbinary, 56
        circumstellar, 56
        dynamical stability, 56
        examples, 60
        formation, 58
        P-type orbits, 57
        properties, 59
        S-type orbits, 56–57
bisector analysis, 23
bolometric corrections, 184
brown dwarf desert, 35, 213, 241
brown dwarfs, 209–215
    census, 210
    classification, 212
    definition, 7
    detection, 209–211
    disks and planets, 213
    formation, 214–215
    free-floating objects of planetary mass, 215
    fusion, 209
    luminosity and age, 211
    nomenclature, 210
    occurrence as binary companions, 213
    recognition of, 212
    spectral features, 212
    surveys, 211
    X-ray and radio emission, 213

Catalogue of Nearby Stars, 6, 182
chaotic orbits, 44–47
    Hyperion, 45
    indicators of chaos, 46–47
        fast Lyapunov indicators, 46
        MEGNO, 46
        relative Lyapunov indicators, 47
    Kirkwood gap, 45
    Lyapunov exponent, 45
    Lyapunov time, 45
    Pluto, 45
collisional signatures
    accretion onto the central star, 178
        FH Leo, 179
        V838 Mon, 178

collisional debris, 178
proto-planet collisions, 177
comets, *see* solar system, minor bodies
CoRoT, *see* transits
cross-correlation spectroscopy, 17

detection methods, 4
Doppler shift, 16
Doppler variability, 126
dwarf planet, definition, 7

Earth, *see* solar system
eccentric anomaly, 10
eccentricity, 9
    distribution, 37
eclipsing binaries, 81–82
    CM Dra, 82, 114
    DP Leo, 82
    NN Ser, 82
    HW Vir, 81
    QS Vir, 82
emission spectroscopy, 137–143

formation and evolution, 217–253
    debris disks, 222–224
    disk dispersal, 222
    disk formation, 218–219
    disk viscosity and turbulence, 221
    giant planet formation, 231–237
        core accretion, 231–235
        disk instability, 235–237
        dynamical shake-up, 234
        effects of migration, 234
    Goldschmidt element classification, 231, 232
        atmophiles, 231
        chalcophiles, 231
        lithophiles, 231
        siderophiles, 231
    layered accretion disks, 221
    magnetorotational instability, 221
    minimum mass solar nebula, 220, 233–235, 242
    molecular clouds, 217
    orbital migration, 237–244
        and existence of terrestrial planets, 241
        and the brown dwarf desert, 240
        evidence for, 237
        gas disk migration, 238–243
        halting migration, 241–242
        hot Jupiters, 237
        hot super-Earths, 238
        planet–planet scattering, 243–244
        planetesimal disk migration, 243
        resonant migration, 242–243
        scattering into resonances, 244
        type I migration, 238–239
        type II migration, 239–240
        type III migration, 240
    overview, 217
    population synthesis, 251–253
    protoplanetary disks, 220
    protostars and protostellar collapse, 217–218
    protostellar, protoplanetary, and debris disks, 218
    satellite formation, 237
    star formation, 217–218
    terrestrial planet formation, 224–229
        chemical differentiation, 230
        coagulation equation, 228
        embryos, 226
        Epstein drag, 226
        feeding zone, 228
        Goldreich–Ward mechanism, 226
        gravitational focusing, 227
        isolation mass, 228
        meter-size barrier, 227
        oligarchic growth, 228
        orderly growth, 227
        planetesimal growth, 226
        post-oligarchic growth, 228
        runaway growth, 227
        size and shape of bodies, 230
        solar nebula theory, 224
        stages of growth, 225–229
        Stokes drag, 226
        thermophoresis, 227
    tidal effects, 244–251
        application to known exoplanets, 247
        constraints on planet mass and radius, 246
        constraints on tidal dissipation parameters, 247
        orbital evolution, 245–246
        planet shape, 248
        specific dissipation function, 245
        spin–orbit resonance, 246
        spin-up of host stars, 248–249
        tidal heating, 249–251
        tidal heating and habitability, 250–251
        tidal locking, 246
    young stellar objects, 219–220
free-floating planets, 95–96, 101, 215

Gaia, 61, 63, 64, 67, 72, 94, 95, 113, 178
Galactic coordinates, 183
genetic algorithm minimisation, 14, 55
Giant Magellan Telescope, GMT, 27, 159, 161
giant planet formation, *see* formation and evolution
Gliese Catalogue, 6

habitability, 282–291
    anthropic principle, 286
    astrobiology roadmaps (NASA/ESA), 283
    characterising Earth-like atmospheres, 289
    continuously habitable zone, 283–284
    dynamical stability, 55
    Earth's spectrum over geological history, 289–290
    exoplanets in the habitable zone
        GJ 581, 286
        HIP 5750, 287
    frequency of habitable planets, 287
    Galactic habitable zone, 286
    habitable zone, 283
    importance of the Moon, 285
    orbital dependence, 284–285
    planet radius, 285
    plant life and photosynthesis, 289
    searches for water, 284
    SETI, 290–291
        Catalogue of Nearby Habitable Systems, 291
        optical surveys, 291
        radio/microwave surveys, 290–291

    solar analogues, 286
    solar twins, 286
    spectroscopic indicators of life, 287–290
    stellar spectral type, 285
    system stability and architecture, 285
    ultraviolet radiation zone, 284
HD Catalogue, 6
Hertzsprung–Russell diagram, 187
hierarchical stability, 43
Hill radius, 43
Hill stability, 43
Hipparcos, 6, 22, 32, 33, 47, 49, 61, 69–70, 95, 112, 168, 178, 181, 184, 186, 187, 203, 286, 291
    bolometric corrections, 184
    Galactic coordinates, 183
    Hertzsprung–Russell diagram, 187
    nearby star census, 181–182
    photometry, 113, 139, 141, 179, 184, 185
    space velocities, 183
host stars, 181–207
    ages, 186, 187
    alpha elements, 198
    angular diameters, 184
    asteroseismology, 201–205
        excitation of g-modes, 204
        masses of giant host stars, 204
        masses of imaged companions, 204
        radii of transiting planets, 204
        tests of self-enrichment, 203
    beryllium abundance, 201
    bolometric magnitudes, 184
    effective temperature, 184
    element abundances, 185, 188–201
    Hertzsprung–Russell diagram, 187
    lithium abundance, 199–201
    magnetic and chromospheric activity, 205–206
    masses and radii, 186
    metallicity dependence, 188–201
        biases, 191
        different Galactic origins, 193
        primordial occurrence, 191
        self-enrichment, 192
    r- and s-process elements, 196, 198
    refractory and volatile elements, 194–198
    rotation velocities, 186
    X-ray activity, 206
hot Jupiter
    definition, 103
hot subdwarfs, 81
    V391 Peg, 2, 81
HST, 112, 130, 131, 137, 139, 140, 153, 167, 170, 213, 220, 222, 230, 231
    ACS, 2, 112, 131, 140, 163, 168, 171, 172
    COS, 141
    FGS, 2, 39, 48, 53, 68, 70–71, 73
    NICMOS, 140, 142, 163, 170–172, 223
    STIS, 2, 139, 140, 163, 223
    WFPC2, 168
Hyperion, 45, 231

imaging, 149–179
    active optics, 150
    adaptive optics, 150–152
    Antarctic, 161–162
    coronagraphy, 152–157
        four-quadrant phase mask, 155
        free-flying occulters, 156–157
        lunar occultation, 157
        optical vortex coronagraph, 155–156
    Fomalhaut, 171, 172
    ground-based interferometry, 162
    HR 8799, 171, 172
    instruments/programmes
        ALMA, 177
        Darwin, 163–167, 181, 287, 288, 290
        European Extremely Large Telescope, 159–161
        extremely large telescopes, 159–161
        Gemini Planet Imager, 159
        Giant Magellan Telescope, 159–161
        Subaru–AO188–HiCIAO, 159
        Thirty Metre Telescope, 159–161
        TPF, 153, 158, 163–167, 288, 290
        VLT–NACO, 2, 36, 99, 153, 155, 158, 168, 170, 171, 173, 223
        VLT–SPHERE, 159
        other, 158
    integral field spectroscopy, 158
    laser guide stars, 151
    mm/sub-mm wavelengths (ALMA), 177
    nulling interferometry, 163
    β Pic, 171
    planet compilation, 168
    radio wavelengths, 173–177
    resolved imaging
        hypertelescopes, 167
        Life Finder, 167
        Planet Imager, 167
        superlenses, 168
    searches around exoplanet host stars, 170
    searches around nearby stars, 170
    searches around white dwarfs, 172–173
    searches in debris disk systems, 171
    space interferometry (Darwin, TPF), 163–167
    space-based (HST, Spitzer, JWST), 162–163
    speckle calibration, 157
    speckle noise, 157–158
    speckle suppression, 157
    star–planet brightness ratio, 149
    star–planet separation, 149
    stratospheric observations, 162
interiors and atmospheres, 255–291
    atmospheres of gas giants, 271–278
        albedos, 124, 278
        atmospheric circulation, 273–274
        carbon chemistry, 275
        carbon/oxygen ratio, 275
        chemical composition, 274
        condensate clouds, 276
        core accretion start models, 274
        dependence on initial conditions, 274
        effects of external heating, 273
        hot start models, 274
        models, 271
        oxygen chemistry, 275
        photochemistry, 277–278
        photolysis, 277–278
        pressure–temperature relations, 272
        rainout, 276

stratospheres, 276–277
    temperature inversions, 276–277
  atmospheres of terrestrial planets, 278–282
    capture of nebular gases, 278
    chthonian planets, 281
    ejected planets, 281–282
    erosion, 279–281
    formation, 278–279
    hydrodynamic escape, 281
    Jeans escape, 280–281
    outgassing through accretion, 278–279
    outgassing through tectonic activity, 279
    photolytic dissociation, 281
    planetesimal erosion, 279–280
    stripping of hot Jupiters and Neptunes, 281
  chemical composition and condensation, 257–260
  composition of the solar system planets, 257
  diagnostics from rotation, 270–271
    Galilean satellites, 270
    solar system giants, 270
  equations of state, 261
  gas, rock, and ice, 255–257
  giant planet interiors, 260–271
    dependence on composition, 265–267
    effects of stellar irradiation, 265
    H/He gas giants, 266
    model predictions, 265–267
  ocean planets, 269–270
    atmospheres, 269
    candidates, 270
    hypothetical existence, 269
    ocean depth, 270
  opacities, 264–265
  phase diagram for hydrogen, 261–262
  phase diagram for water, 262–263
  refractory and volatile elements, 259
  snow line, 259–260
  structural models, 263–264
  super-Earths, 267–269
    compositional extremes, 267
    equation of state, 267
    mass–radius relation, 268–269
    ternary diagrams, 267–268

JWST, 114, 150, 153, 156, 162, 163, 167
  MIRI, 114, 155, 163
  NIRCam, 114, 155, 158, 163
  TFI, 163

Kepler, *see* transits
Kepler's laws, 11
kinematic properties, 183
Kozai migration, 60, 127, 128
Kozai resonance, *see* resonance

Laplace resonance, 2, 3, 32, 50, 51, 244, 299, 300
laser frequency combs, 20
Levenberg–Marquardt minimisation, 13, 14, 67, 119
line of apsides, 9
Lomb–Scargle period algorithm, 13
longitude of ascending node, 10
longitude of pericentre, 10
Love number, 134
Lyapunov exponent/indicators, 45–47

mandalas, 66
Markov Chain Monte Carlo minimisation, 14, 67, 119
mass function, 12
mean anomaly, 10
mean longitude, 10
mean motion, 10
metallicity of host star, 37
microlensing, 2, 83–102
  astrometric, 94–95
  binary lens, 87
  blending, 92
  caustics and critical curves, 87–90
  classification of event types, 90
  early surveys, 86
  Einstein radius, 85
  finite source size effects, 90
  formulation, 84–87
  free-floating planets, 95–96, 101
  future developments, 102
  high-magnification events, 88–90
  historical background, 84
  individual events
    MOA–2007–BLG–192, 98, 99
    MOA–2007–BLG–400, 98, 99
    MOA–2008–BLG–310, 98, 99
    MOA–2009–BLG–319, 98, 99
    OGLE–2003–BLG–235, 2, 98, 100
    OGLE–2003–BLG–390, 2, 98
    OGLE–2005–BLG–71, 2, 98, 100
    OGLE–2005–BLG–169, 98
    OGLE–2006–BLG–109, 2, 98, 101
    OGLE–2007–BLG–368, 98, 99
  instruments/programmes
    Euclid, 102
    Galactic Exoplanet Survey Telescope, GEST, 102
    JDEM–Omega, 102
    MicroFUN, 96–99
    Microlensing Planet Finder, MPF, 102
    MiNDSTEp, 96–99
    MOA, 96–99
    OGLE, 96–99
    PLANET/RoboNet, 96–99
    WFIRST, 102
  light bending, 84
  light-curve modeling, 90
  limb darkening, 91
  limitations and strengths, 100–102
  magnification, 85–87
  magnification maps, 88
  microlens parallax effects, 92–94
  naming conventions, 98
  observations, 96–99
  orbital motion, 91–92
  planet compilation, 98
  planet orbiting the source star, 95
  ray shooting, 89
  satellite orbiting a planet, 95
  single lens, 87
  star spots, 91
  statistical results, 100
  terrestrial parallax, 94
migration, *see* formation and evolution
minimum mass solar nebula, *see* formation and evolution
multiple planet systems, 37–55

Index

coplanarity, 39
dynamical modeling, 39
resonances, *see* resonance
statistics, 39
theories of formation, 38

nearby stars, 181–183
nomenclature
    exoplanet names, 6
    microlensing events, 98
    star names, 6
notation, 313–316
numerical quantities, 309–311

open clusters
    Hyades, 34
    NGC 2423, 34
    NGC 4349, 34
    radial velocity searches, 34
orbit determination
    multiple planet fitting, 13
    single planet fitting, 13
orbit inclination, 10
orbits, 9–14

packed planetary systems, 44
pericentre/periapsis, 9
planet, definition, 7
    difficulties of formulating, 7
    IAU 2003 recommendation, 7
    IAU 2006 resolution, 7
planets/systems
    2M J1207, 2, 163, 168, 170, 171, 178, 214, 215
    υ And, 2, 14, 38, 39, 42, 46, 47, 52, 53, 70–72, 124, 125, 142, 175–177, 185, 187, 188, 205, 206
    μ Ara, 14, 38, 46, 47, 52, 54, 66, 71, 203, 204
    Barnard's star, 18, 19, 64, 94
    τ Boo, 113, 125, 126, 138, 174–177, 185, 187, 188, 205–206, 246–248, 276
    γ Cep, 2, 28, 57, 58, 60, 70, 71
    55 Cnc, 2, 3, 13, 14, 30, 31, 37, 38, 44, 47–49, 52, 70, 71, 126, 170, 185, 187, 188, 201, 224, 281
    CoRoT–1, 2, 111, 124, 129, 135, 137, 138, 143, 248
    CoRoT–2, 111, 129, 143, 146, 248
    CoRoT–3, 111, 129, 143–145
    CoRoT–4, 135
    CoRoT–7, 2, 22, 35, 111, 143, 206, 247, 248, 250, 281
    CoRoT–9, 143
    CoRoT–10, 143
    CoRoT–11, 129, 249
    CoRoT–14, 143
    CoRoT–17, 111, 143
    16 Cyg, 14, 60, 70, 185, 187, 192, 199, 201, 286
    61 Cyg, 64
    CM Dra, 82, 114
    ϵ Eri, 2, 71, 170, 177, 222, 223
    Fomalhaut, 2, 163, 168, 171, 172, 222, 223, 236, 242, 243
    G117–B15A, 80
    G29–38, 80
    GD 66, 80
    GJ 229 B, 168, 170, 210, 212, 272, 273, 277
    GJ 436, 113, 123, 127, 131, 135, 136, 142–144, 249, 281
    GJ 581, 2, 3, 30, 35, 36, 38, 55, 286–287
    GJ 876, 2, 14, 32, 36, 38, 41–43, 49–50, 67, 71, 177, 249, 267, 268, 299
    GJ 1214, 2, 35, 108, 115, 143, 270
    HAT–P–1, 2, 129, 135, 143, 248, 249
    HAT–P–2, 106, 121, 129, 143, 144, 147
    HAT–P–3, 135, 143, 146
    HAT–P–4, 113, 129
    HAT–P–7, 2, 106, 112, 113, 119, 129, 132, 142, 143, 204, 205
    HAT–P–10, 104
    HAT–P–11, 112, 129, 131, 204
    HAT–P–13, 2, 106, 129, 134
    HAT–P–14, 104, 106, 129
    HAT–P–23, 106
    HAT–P–26, 106
    HD 4113, 12, 30, 31
    HD 10180, 2, 3, 38
    HD 12661, 14, 46, 52, 53, 55
    HD 17156, 2, 109, 122, 142, 143
    HD 37124, 42, 47, 52, 66
    HD 40307, 30, 31, 35, 36, 55
    HD 41004, 22, 23, 248
    HD 45364, 52
    HD 52609, 2, 28
    HD 60532, 42, 50–53
    HD 69830, 2, 36, 55, 224, 234, 267
    HD 73256, 12, 248
    HD 74156, 44, 47, 53, 67
    HD 80606, 109, 122, 143
    HD 82943, 43, 52, 178, 193, 201
    HD 108874, 52, 53, 55
    HD 114762, 2, 28, 210
    HD 128311, 50, 51, 54, 55, 71, 177
    HD 141569, 222–224
    HD 142022, 12, 13, 17
    HD 149026, 109, 113, 123, 136, 142, 143, 145, 189, 266, 272
    HD 168443, 55, 67, 71
    HD 169830, 47, 53, 55
    HD 188753, 22, 60
    HD 189733, 2, 109, 112, 113, 123–126, 128–131, 133, 135, 138–142, 158, 177, 248, 272
    HD 202206, 52, 55, 59, 71
    HD 209458, 2, 103, 104, 107, 112, 113, 118, 121, 123, 125, 127, 129–131, 133–136, 138–143, 145, 158, 189, 248, 249, 262, 272, 273, 277, 281
    14 Her, 52, 188
    HR 4796A, 223, 224
    HR 8799, 2, 156, 158, 168, 170–172, 204, 223, 236, 242, 244
    Kepler–1/3, 112
    Kepler–4, 2
    Kepler–6, 112
    Kepler–8, 129
    Kepler–9, 2, 112, 122, 135
    Lalande 21185, 64
    DP Leo, 82
    Lupus–TR–3, 108
    MOA microlensing events, *see* microlensing
    OGLE microlensing events, *see* microlensing
    OGLE–TR–10, 106, 131, 189
    OGLE–TR–56, 2, 28, 106, 143, 241, 247
    OGLE–TR–111, 106, 135
    OGLE–TR–113, 106, 135
    OGLE–TR–122, 248
    OGLE–TR–123, 248
    OGLE–TR–132, 106

OGLE–TR–182, 105, 106
OGLE–TR–211, 106
OGLE2–TR–L9, 106, 143, 249
70 Oph, 64
51 Peg, 2, 20, 27, 28, 70, 138, 187, 188, 280, 286
V391 Peg, 2, 81
β Pic, 2, 130, 168, 171, 178, 222, 223
PSR B1257+12, 2, 28, 75–77
PSR B1620–26, 77–78
NN Ser, 82
SWEEPS–4/11, 112
TrES–1, 2, 107, 112, 125, 129, 131, 133, 135, 138, 142, 272, 273
TrES–2, 112, 113, 115, 123, 129, 135, 204
TrES–3, 113, 133, 135, 273
TrES–4, 115, 129, 142, 143, 145, 277
4 UMa, 2
47 UMa, 28, 38, 52, 70, 187, 276, 286
61 Vir, 30, 31, 35
70 Vir, 2, 28, 70, 174, 177
HW Vir, 81
QS Vir, 82
WASP–1, 2
WASP–2, 2, 115, 129, 143
WASP–3, 107, 108, 113, 129, 135
WASP–4, 129, 248
WASP–5, 129, 248
WASP–6, 129
WASP–8, 107, 129
WASP–9, 107
WASP–10, 133
WASP–11, 104
WASP–12, 132, 134, 141, 143, 176, 241, 250
WASP–14, 129, 250
WASP–15, 129
WASP–17, 108, 129, 143
WASP–18, 129, 143, 250
WASP–19, 108, 143
WASP–21, 108, 181
WASP–27, 104
WASP–29, 108
WASP–33, 108, 130, 134, 143, 181, 249
WASP–37, 181
XO–1, 135
XO–2, 113, 142
XO–3, 129, 143, 248
polarisation, 126
populations synthesis, 251–253
pulsars, 75–79
    planet detection limits, 78
    planet formation, 77–78
    planetary satellites, 78
    PSR B1257+12, 75–77
    PSR B1620–26, 77–78
    unconfirmed planets, 78
pulsating stars, 79–81

radial velocity, 9–60
    accuracy limits, 21
    adaptive scheduling, 15
    binary companions, 22
    bisector analysis, 23
    bisector curvature, 23
    bisector inverse slope, 23

    bisector velocity span, 23
    convective line shift, 18
    cross-correlation spectroscopy, 17
    discovery by instrument, 29
    discovery status, 28
    dispersed Fourier transform spectrometer, 27
    Earth motion, 18
    eccentricity distribution, 37
    échelle spectroscopy, 23–25
    example orbits, 12
    exposure metering, 21
    externally dispersed interferometry, 25–27
    first detections, 28
    Fourier transform spectroscopy, 25
    frequency of massive planets, 35
    fringing spectrometer, 25
    gravitational redshift, 18
    holographic heterodyne spectroscopy, 25
    instruments/programmes, 23–27
        AAT, 19, 21, 22, 24, 29, 31, 33, 187, 189, 193, 286, 291
        ASEPS, 27
        CODEX, 24, 27
        CORALIE, 13, 22, 24, 29–33, 49, 59, 107, 129, 181, 183, 188
        ELODIE, 2, 17, 20, 21, 24, 29, 32–34, 48, 59, 108, 129
        HARPS, 2, 17, 20, 22–24, 27, 29, 31–34, 36, 129, 181, 190, 199, 200, 286
        HARPS–north, 25
        Hectochelle, 27
        Keck–HIRES, 2, 17–19, 21, 24, 25, 29, 30, 32–34, 45, 47, 49, 70, 129, 189, 193, 286, 291
        Lick, 2, 17–19, 21, 22, 24, 28–30, 32, 33, 46, 152, 187, 189, 193, 286, 291
        MARVELS, 27, 29, 32
        SHARPS, 24, 27
        TEDI, 27, 30
        other, 24, 29, 33
    jitter due to solar cycle, 22
    low-mass planets, 35
    mass distribution, 35
    mass of host star, 37
    mass–period relation, 37
    measurement from accelerations, 27
    metallicity of host star, 37
    Michelson interferometer, 25
    multiple planet systems, 37–55
    on-line compilations, 28
    orbital period distributions, 36
    orbits, 9–14
    photon noise, 22
    planet compilation, 317–323
    selection effects, 14
    semi-amplitude, 12
    spectral line bisector, 23
    stellar noise/jitter, 21, 22
    stellar space motion, 18
    super-Earth systems, 55
    surveys
        binary and multiple stars, 34, 58
        early-type dwarfs, 32
        GK dwarfs, 29
        GK giants, 32
        infrared, 30
        metal-poor stars, 34

Index

    M dwarfs, 30
    open clusters, 34
    subgiants, 32
    young stars, 34
  systemic velocity, 13
  wavelength calibration, 18–21
    fibre-optic scrambling, 20
    hydrogen fluoride, 19
    infrared, 20
    iodine, 19
    laser frequency combs, 20
    telluric lines, 19
    thorium–argon, 20
  zero point, 18
radio emission
  astrometry, 173
  Blackett's law, 176
  exoplanet flux limits, 176
  exoplanet flux predictions, 176
  exoplanet magnetic fields
    evidence, 175
    super-flares, 176
    theory, 175
  future radio surveys (LOFAR, SKA), 177
  imaging, 173–177
  magnetic white dwarfs, 176
  M dwarfs, 173
  radio Bode's law, 174
  RS CVn binaries, 173
  solar system decametric emission, 174
reflected light, 124–125
resonance
  1:1 eccentric resonance, 54
  1:1 resonance, 52, 53
  2:1 resonance, 40
  3:1 resonance, 40
  4:1, 5:1, 3:2, 5:2, etc., 52
  apsidal alignment, 42
  apsidal corotation, 42
  apsidal libration, 42
  apsidal motion, 42–43
  capture into, 41
  deep resonance, 40
  disturbing function, 39
  exact resonance, 40
  horseshoe orbits, 52
  inclination resonance, 52
  Kozai, 40, 53, 55, 57, 58, 60, 128, 130, 243, 300
  Lagrange equilibrium points $L_1 - L_5$, 52
  Laplace, 2, 3, 32, 50, 51, 244, 299, 300
  mean motion, 40
  physics of, 41
  proximity to, 52
  resonance order, 40
  resonant argument, 40
  resonant theory, 40
  retrograde, 47, 54–55
  separatrix, 41
  tadpole orbits, 52
  Trojans, 52–54, 56, 59, 136, 300, 303, 304
retrograde orbits, 128–129
Rossiter–McLaughlin effect, *see* transits

Safronov number, 147

secondary eclipse, *see* transits
snow line, *see* interiors and atmospheres
SOHO, 22, 52, 115, 134
solar nebula, *see* solar system
solar nebula theory, *see* formation and evolution
solar system, 293–307
  age and early chronology, 296–297
  birth in star cluster, 293
  decametric emission, 174
  dynamical shake-up, 234
  Earth
    atmosphere from ice-core records, 307
    early atmosphere, 305
    evolution of CO_2, CH_4, O_2, 305–306
    growth and patterns in biodiversity, 307
    mass extinctions, 307
    Milankovitch cycles, 301
    origin of atmosphere, 305
    origin of Moon, 302
    origin of water, 298–299, 304
    Phanerozoic evolution, 306
    snowball Earth, 306
  giant planets, 293–295
    Jupiter, 293
    masses and radii, 293
    Neptune, 294
    noble gas enrichment, 294
    Saturn, 293
    Uranus, 294
  Mars, anomalous mass of, 298
  minor bodies
    collisional debris, 295
    comets, 295
    Edgewoth–Kuiper belt, 295
    meteorite taxonomy, 297
    Oort cloud, 295
    planetesimals and protoplanets, 295
  Nice model, 303–305
  numerical integration of orbits, 299–300
  obliquites, 300–302
  orbital stability and chaos, 299–300
  orbits of terrestrial planets, 302
  planet properties, 309–311
  planet survival at late age, 80
  planetary satellites, 302
  planetesimal migration, 302–305
    capture of irregular satellites, 304
    Kuiper belt resonances, 303
    lunar heavy bombardment, 304
    migration of Neptune, 304
    orbit of Pluto, 303
    water on Earth, 304
  resonances, 299, 300
  solar barycentric motion, 66–67
  solar nebula, 220–222, 226, 233, 255–260, 268, 271, 272, 280, 294, 295, 297, 298
  transient heating events, 297–298
space velocities, 183
spectral line bisector, 23
Spitzer Space Telescope, 2, 55, 93, 94, 103, 107, 109, 112, 121, 125, 131, 137, 139–142, 163, 167, 168, 172, 220, 272
  IRAC, 81, 112
  IRS, 112
  MIPS, 77, 112, 125, 224

stability
  chaos, *see* chaotic orbits
  dynamical classification, 44
  dynamical packing, 44
  hierarchical, 43
  Hill criterion, 43
  Hill radius, 43
  Lagrange, 43
star spots, 18, 23, 65, 91, 111, 115, 131
sub-brown dwarf, definition, 7
symplectic integrators, 44
systemic velocity, 13

terrestrial planet formation, *see* formation and evolution
tidal effects, *see* formation and evolution
timing, 75–82
  eclipsing binaries, 81–82
  planet compilation, 78
  pulsars, 75–79
Titius–Bode law, 48
transits, 103–147
  anomalous radii of hot Jupiters, 145
  Antarctic, 110
  apsidal precession, 133–134
  astrophysical limits, 115
  atmospheric limits, 115–116
  atmospheric transparency, 116
  bow shocks, 132
  candidate identification/confirmation, 105
  circular orbits, 119–121
  circumbinary systems, 114
  conjugate-plane photometry, 116
  Doppler variability, 126
  early studies, 104
  eccentric orbits, 121
  eclipsing binaries, 114
  effects of other planets, 135
  effects of parallax and space motion, 136
  effects of satellites, 135
  ellipsoidal variations, 132
  emission spectroscopy, 137–143
  equilibrium temperatures, 138
  future searches from ground, 110
  geometric formulation, 118
  giant stars, 115
  higher-order photometric effects, 130–132
  higher-order spectroscopic effects, 136
  higher-order timing effects, 132–136
  instrumentation for transit time determinations, 133
  instruments/programmes
    CoRoT, 2, 7, 110–111, 115, 146, 202–204
    EPOXI–EPOCh, 113, 231
    HAT/HATNet, 105–106
    Hipparcos, 113
    HST, 112
    JWST, 114
    Kepler, 2, 7, 111–112, 115, 125, 126, 132, 133, 136, 202
    MACHO, 106
    MEarth, 108, 115
    MOA, 106
    MOST, 113
    OGLE, 106
    PLATO, 114
    SPICA, 114
    Spitzer, 112
    STJ, 133
    SWEEPS, 112
    TESS, 114
    THESIS, 114
    TrES, 106
    ULTRACAM, 82, 108, 133
    WASP/SuperWASP, 107–108
    WFI/Lupus, 108
    XO, 108
  interferometric observations, 123–124
  light curves, 117–123
  limb darkening, 118
  line-profile tomography, 130
  long-period planets, 109
  mass versus period, 146
  mass–radius relation, 143–146
  methods from eclipsing binaries, 119
  multiple transiting planets, 122
  M dwarfs, 103, 115
  nodal precession, 134
  observables, 117
  planet compilation, 325–327
  planet surface gravity, 123
  planetary oblateness and rotation, 131
  planetary satellites, 130
  pM and pL classes, 142
  polarisation, 126
  properties of transiting planets, 143
  radial velocity discoveries, 108
  reflected light, 124–125
  retrograde orbits, 128–129
  rings and comets, 130
  Rossiter–McLaughlin effect, 2, 107, 127–130, 241
  Safronov number, 147
  scintillation noise, 116
  searches from ground, 105–110
  searches from space, 110–114
  searches in globular clusters, 109–110
  searches in open clusters, 109
  secondary eclipse, 103, 126
  simultaneous transits, 122
  star spots, 131
  stellar density, 123
  stellar rotation, 131
  surface gravity versus period, 146
  tidal locking and asymmetric heating, 137–138
  timing precision, 133
  transit duration variations, 132–136
  transit time variations, 132–136
  transmission spectroscopy, 137–143
  Trojan planets, 136
  white dwarfs, 103
transmission spectroscopy, 137–143
Trojans, *see* resonance
true anomaly, 10
true longitude, 10

white dwarfs, 79–81
  G117–B15A, 80
  G29–38, 80
  GD 66, 80
  pulsating, 79–81
  transits, 103